VUV and Soft X-Ray Photoionization

PHYSICS OF ATOMS AND MOLECULES

Series Editors

P. G. Burke, *The Queen's University of Belfast, Northern Ireland*
H. Kleinpoppen, *Atomic Physics Laboratory, University of Stirling, Scotland*

Editorial Advisory Board

R. B. Bernstein *(New York, U.S.A.)*
J. C. Cohen-Tannoudji *(Paris, France)*
R. W. Crompton *(Canberra, Australia)*
Y. N. Demkov *(St. Petersburg, Russia)*
C. J. Joachain *(Brussels, Belgium)*

W. E. Lamb, Jr. *(Tucson, U.S.A.)*
P. -O. Löwdin *(Gainesville, U.S.A.)*
H. O. Lutz *(Bielefeld, Germany)*
M. C. Standage *(Brisbane, Australia)*
K. Takayanagi *(Tokyo, Japan)*

Recent volumes in this series:

ATOMIC PHOTOEFFECT
M. Ya. Amusia

ATOMIC SPECTRA AND COLLISIONS IN EXTERNAL FIELDS
Edited by K. T. Taylor, M. H. Nayfeh, and C. W. Clark

ATOMS AND LIGHT: INTERACTIONS
John N. Dodd

ELECTRON COLLISIONS WITH MOLECULES, CLUSTERS, AND SURFACES
Edited by H. Ehrhardt and L. A. Morgan

ELECTRON–MOLECULE SCATTERING AND PHOTOIONIZATION
Edited by P. G. Burke and J. B. West

THE HANLE EFFECT AND LEVEL-CROSSING SPECTROSCOPY
Edited by Giovanni Moruzzi and Franco Strumia

INTRODUCTION TO THE THEORY OF LASER–ATOM INTERACTIONS, Second Edition
Marvin H. Mittleman

INTRODUCTION TO THE THEORY OF X-RAY AND ELECTRONIC SPECTRA OF FREE ATOMS
Romas Karazija

MOLECULAR PROCESSES IN SPACE
Edited by Tsutomu Watanabe, Isao Shimamura, Mikio Shimizu, and Yukikazu Itikawa

POLARIZATION BREMSSTRAHLUNG
Edited by V. N. Tsytovich and I. M. Ojringel

POLARIZED ELECTRON/POLARIZED PHOTON PHYSICS
Edited by H. Kleinpoppen and W. R. Newell

THEORY OF ELECTRON–ATOM COLLISIONS, Part 1: Potential Scattering
Philip G. Burke and Charles J. Joachain

VUV AND SOFT X-RAY PHOTOIONIZATION
Edited by Uwe Becker and David A. Shirley

A Chronological Listing of Volumes in this series appears at the back of this volume.

A Continuation Order Plan is available for this series. A continuation order will bring delivery of each new volume immediately upon publication. Volumes are billed only upon actual shipment. For further information please contact the publisher.

VUV and Soft X-Ray Photoionization

Edited by

Uwe Becker
Fritz-Haber Institute of the Max Planck Society
Berlin, Germany

and

David A. Shirley
Pennsylvania State University
University Park, Pennsylvania

Plenum Press • New York and London

Library of Congress Cataloging-in-Publication Data

VUV and soft X-ray photoionization / edited by Uwe Becker and David A.
  Shirley.
      p.   cm. -- (Physics of atoms and molecules)
    Includes bibliographical references and index.
    ISBN 0-306-45038-0 (alk. paper)
    1. Photoionization. 2. Grenz rays. 3. Far ultraviolet radiation.
  I. Becker, U. (Uwe)   II. Shirley, David Allen. III. Series.
  QC702.7.P48V88 1996
  535.01'4--dc20   96-5250
                                                                 CIP

ISBN 0-306-45038-0

©1996 Plenum Press, New York
A Division of Plenum Publishing Corporation
233 Spring Street, New York, N.Y. 10013

10 9 8 7 6 5 4 3 2 1

All rights reserved

No part of this book may be reproduced, stored in a retrieval system, or transmitted in any
form or by any means, electronic, mechanical, photocopying, microfilming, recording, or
otherwise, without written permission from the Publisher

Printed in the United States of America

CONTRIBUTORS

H. AKSELA • *Department of Physics, University of Oulu, 90570 Oulu 10, Finland*

S. AKSELA • *Department of Physics, University of Oulu, 90570 Oulu 10, Finland*

M. YA. AMUSIA • *A. F. Ioffe Physical-Technical Institute, 194021 St. Petersburg, Russia*

H. BAUMGÄRTEL • *Institut für Physikalische und Theoretische Chemie, Freie Universität Berlin, 14159 Berlin, Germany*

U. BECKER • *Fritz-Haber-Institut der Max-Planck-Gesellschaft, 14195 Berlin, Germany*

J. BERKOWITZ • *Chemistry Division, Argonne National Laboratory, Argonne, Illinois 60439*

MATTHEW BRAUNSTEIN • *Institute for Defense Analyses, Alexandria, Virginia 22311-1772*

C. D. CALDWELL • *Department of Physics, University of Central Florida, Orlando, Florida 32816-2385*

N. A. CHEREPKOV • *State Academy of Aerospace Instrumentation, 190000 St. Petersburg, Russia*

J.-P. CONNERADE • *The Blackett Laboratory, Imperial College, London SW7 2BZ, United Kingdom*

J. H. D. ELAND • *Physical and Theoretical Chemistry Laboratory, Oxford University, Oxford OX1 3QZ, United Kingdom*

T. HAYAISHI • *Institute of Applied Physics, University of Tsukuba, Tsukuba, Ibaraki 305, Japan*

U. HEINZMANN • *Fakultät für Physik, Universität Bielefeld, 33615 Bielefeld, Germany*

J. HORMES • *Physikalisches Institut der Universität Bonn, 53115 Bonn, Germany*

N. KABACHNIK • *Institute of Nuclear Physics, Moscow State University, 119899 Moscow, Russia*

G. C. KING • *Department of Physics and Astronomy, Schuster Laboratory, The University of Manchester, Manchester M13 9PL, United Kingdom*

M. O. KRAUSE • *Oak Ridge National Laboratory, Oak Ridge, Tennessee 37831-6201*

M. KUTZNER • *Department of Physics, Andrews University, Berrien Springs, Michigan 49104*

MU-TAO LEE • *Departmento de Quimica, Universidade Federal de São Paulo, São Carlos 13560, Brazil*

STEVEN T. MANSON • *Department of Physics and Astronomy, Georgia State University, Atlanta, Georgia 30303-3083*

VINCENT MCKOY • *Arthur Amos Noyes Laboratory of Chemical Physics, California Institute of Technology, Pasadena, California 91125*

PAUL MORIN • *CEA/DSM/DRECAM, Service des Photons, Atomes et Molécules, Centre d'Études de Saclay. 91191 Gif sur Yvette Cedex, France, and Laboratoire pour l'Utilisation du Rayonnement Electromagnétique, Laboratoire mixte, CNRS, CEA et MESR, Centre Universitaire, 91405 Orsay Cedex, France*

IRÈNE NENNER • *CEA/DSM/DRECAM, Service des Photons, Atomes et Molécules, Centre d'Études de Saclay, 91191 Gif sur Yvette Cedex, France, and Laboratoire pour l'Utilisation du Rayonnement Electromagnétique, Laboratoire mixte, CNRS, CEA et MESR, Centre Universitaire, 91405 Orsay Cedex, France*

E. RÜHL • *Institut für Physikalische und Theoretische Chemie, Freie Universität Berlin, 14159 Berlin, Germany. Current address: Institut für Physik, Johannes-Gutenberg–Universität Mainz 55099 Mainz, Germany*

K.-H. SCHARTNER • *I. Physikalisches Institut der Justus-Liebig-Universität, 35392 Giessen, Germany*

JOCHEN SCHIRMER • *Physikalisch-Chemisches Institut, Universität Heidelberg, 69120 Heidelberg, Germany*

V. SCHMIDT • *Fakultät für Physik, Albert-Ludwigs-Universität Freiburg, 79104 Freiburg, Germany*

G. SCHÖNHENSE • *Institut für Physik, Johannes-Gutenberg-Universität Mainz, 55099 Mainz, Germany*

D. A. SHIRLEY • *Departments of Chemistry and Physics, Pennsylvania State University, University Park, Pennsylvania 16802*

ANTHONY F. STARACE • *Department of Physics and Astronomy, The University of Nebraska, 116 Brace Laboratory, Lincoln, Nebraska 68588-0111*

J. B. WEST • *Daresbury Laboratory, SERC, Daresbury, Warrington WA4 4AD, United Kingdom*

F. J. WUILLEUMIER • *Laboratoire de Spectroscopie Atomique et Ionique, Unité de l'Université Paris Sud Associée au CNRS, 91405 Orsay Cedex, France*

P. ZIMMERMANN • *Institut für Strahlungs- und Kernphysik, Technische Universität Berlin, 10623 Berlin, Germany*

PREFACE

VUV- and soft x-ray photoionization studies have undergone dramatic advances in recent years. New experimental tools, in particular new synchrotron radiation light sources, have greatly enhanced not only the number of experiments in this field but also the scope of problems that are accessible. On the other hand, similarly fast developments in computer technology and computational techniques made complementary theoretical studies on photoionization phenomena possible. A fruitful interplay between theory and experiment began, in which these partners intermittently moved ahead of each other. Theoretical predictions were followed by experimental confirmation or disproof and vice versa. One main theme showing up in most of the questions raised by photoionization studies is the problem of electron correlation in its various forms. This problem deals with the fact that the motion of electrons in matter is, on one hand, well described in many aspects by a so-called single particle model. In this approach each electron is only affected by the mean field generated by all other electrons, neglecting all individual interactions. However, specific electron–electron interactions may play a dominant role in certain processes, including binding-energy spectra and partial cross-section behavior. The broader field of photoionization had diversified so much that the task of covering all aspects of this rapidly growing field by one author seemed to be nearly impossible. Rather, it made more sense to combine contributions from different authors, each of them among the leading figures in a certain subfield. This would give the opportunity to combine several viewpoints and aspects within the field of photoionization in a comprehensive overview. Nevertheless, even this approach is still somewhat fragmentary because the selection of topics in the present volume is necessarily somewhat arbitrary, reflecting the personal views of the editors as to where the most important and exciting developments in this field are taking place. On the other hand, since the early days of photoionization studies, the photoionization process has found its way into fundamental research and application in physics, chemistry, materials science, and many other fields! Because of this, and the dramatic advances made, particularly in VUV and soft x-ray regions, knowledge of the photoionization process in atoms and molecules in these energy regions has become a necessity in research and development. Moreover, the widespread nature of this information hampers easy access to these data. Recollection and comprehensive presentation seemed to be propitious at the present time, just as a new generation of even more advanced light sources comes into operation, multiplying the available body of information on photoionization data. In this sense, such a volume would also serve as a practical reference work for extended and more detailed studies, including completely new experimental and theoretical approaches. It is the intention of this volume particularly to help younger scientists and students to explore this fast-growing field by making it easier to find the relevant data in one comprehensive overview, supplementing the collection of earlier reviews mentioned in the different chapters of this book. This is not meant in the sense of being a complete critical data compilation for atomic and molecular photoionization, but it may help to call attention to the necessity of such a compilation and may become at some time a first step toward such a project. The editors are grateful to the contributors for their painstaking effort to present thorough and carefully elaborated reviews. They are also indebted to the patience of the authors with frustrating delays that often seem to be unavoidable in the production of a work of this kind, which combines so many different contributions. In this context, they consider it as a matter of special concern to remind the reader that one of the major contributors, Hugh Kelly, died during work on his chapter on fundamental aspects of photoionization phenomena in the light of many-body perturbation theory. His work was taken over by his former student and co-worker Mikey Kutzner, who completed a chapter particularly dedicated to the memory of the pioneering role of Hugh Kelly in the field of photoionization. One of the editors (U.B.) would like to thank his long-time co-worker Burkhard Langer and his technical assistant Evelyn Prohn for their untiring help in preparing many of the figures, particularly those in Chapter 3.

Uwe Becker
Fritz-Haber-Institut der Max-Planck-Gesellschaft
Berlin-Dahlem

David A. Shirley
Pennylvania State University
University Park

Contents

CHAPTER 1
THEORY OF PHOTOIONIZATION: VUV AND SOFT X-RAY FREQUENCY REGION
M. Ya. Amusia

1. Introduction 1
   1.1. Main Physical Conceptions Important for Understanding the Photoionization Process 2
   1.2. Theoretical Approaches and the Mathematical Apparatus 3
   1.3. Similarities and Specifics in Applying Theoretical Ideas to Different Objects 3
2. Main Achievements of Photoionization Studies 4
3. Theoretical Description of the Photoionization Cross Section:
   One-Electron Hartree–Fock (HF) Approach 4
4. Random Phase Approximation with Exchange (RPAE)—A Method to Include
   Many-Electron Correlations 7
5. RPAE: Some Results of Calculations 9
6. New Objects of RPAE Studies 12
7. Beyond the RPAE Frame: Inclusion of Rearrangement Effects 16
8. Explicit Inclusion of Two-Electron–Two-Vacancy Excitations 19
9. Direct Knock-Out 21
10. Two-Electron Excitations and Ionization 24
11. Formation of Multiply Charged Ions 27
12. Near- and Below-Threshold Processes 31
13. Inclusion of Higher-Order Corrections 36
14. Future Objects of High-Accuracy Calculations 38
15. Forthcoming Development of the Photoionization Theory 40
16. Some Concluding Remarks 43
    References 44

CHAPTER 2
MANY-ELECTRON EFFECTS IN PHOTOIONIZATION
M. Kutzner

1. Introduction to Many-Body Theory 47
2. Review of the Theory 48
3. One-Electron Photoionization 55
   3.1. Argon 55
   3.2. Chlorine 57
   3.3. Giant Resonances 60
   3.4. Shape Resonances in Photoionization of the $4d^{10}$ Subshell 66
4. Photoionization with Excitation and Double-Electron Resonances 69
5. Double Photoionization 73
6. Conclusion 76
   References 77

CHAPTER 3

FUNDAMENTAL ASPECTS OF ATOMIC PHOTOIONIZATION WITH HIGH-BRIGHTNESS LIGHT SOURCES

Anthony F. Starace and Steven T. Manson

1. Introduction 81
2. Brief Review of Theory 81
3. Examples of the Essential Physics Underlying Complex Photoionization Phenomena 83
   3.1. Doubly Excited State Spectra of H^- and He 83
   3.2. The Xe $5s$-Subshell Photoelectron Angular Distribution 85
   3.3. Final Ionic State Branching Ratios 89
   3.4. Fluorescence Spectroscopy 93
   3.5. Photoionization of Inner Shells of Excited States 94
   3.6. Photoelectron Angular Distributions from Parity-Unfavored Transitions 97
   3.7. Double Photoionization of Helium 98
   3.8. Multichannel Aspects of Photoionization 99
4. Concluding Remarks 101
   References 102

CHAPTER 4

CORE RELAXATION EFFECTS IN MOLECULAR PHOTOIONIZATION

Jochen Schirmer, Matthew Braunstein, Mu-Tao Lee, and Vincent McKoy

1. Introduction 105
2. Photoionization Cross Sections in the Relaxed-Core Hartree–Fock Approximation 106
   2.1. Single-Particle Representation of the N-Electron Transition Moment 106
   2.2. Core–Valence Separation 108
   2.3. Generalization to Open-Shell Systems and Equivalent Core Holes 110
3. Generation of Relaxed Continuum Orbitals 112
   3.1. Static-Exchange Potentials for K-Shell Photoelectrons 112
   3.2. Direct Method for Determining Molecular Continuum Orbitals 113
4. Analysis of the RCHF Model 114
   4.1. Relaxation Correction to the Ionic Core Energy 114
   4.2. Term Values and Resonance Positions 117
   4.3. Transition Amplitudes 120
5. Applications 121
   5.1. $C1s$ Photoionization in CO 121
   5.2. $O1s$ Photoionization in CO 124
   5.3. $C1s$ Photoionization in H_2CO 125
   5.4. $C1s$ Photoionization in Ethylene 127
6. General Discussion and Conclusions 130
   References 132

CHAPTER 5

PARTIAL CROSS SECTIONS AND ANGULAR DISTRIBUTIONS

U. Becker and D. A. Shirley

1. Introduction 135
2. Definitions and Experimental Methods 137
3. Helium 140
4. Other Rare Gases 144
   4.1. Neon 144
   4.2. Argon 146

    4.3. Krypton 150
    4.4. Xenon 153
5. Transition Elements 156
    5.1. Cadmium 157
    5.2. Mercury 158
    5.3. Manganese 160
    5.4. Europium 162
6. CO: A Showcase Molecule for Molecular Photoionization 165
    6.1. Valence Electrons 166
    6.2. Core Ionization 168
7. Fullerenes 171
   References 173

CHAPTER 6

HIGH-RESOLUTION ELECTRON SPECTROMETRY OF ATOMS

M. O. Krause and C. D. Caldwell

1. Introduction 181
2. Some Experimental Details 182
3. Photoelectron Spectra of Atoms 191
4. Correlation Phenomena in Atoms 196
    4.1. The Rare Gases 197
    4.2. Open-Shell Atoms 201
5. Electron Spectrometry of Radiationless Processes 202
    5.1. Autoionization Resonances 203
    5.2. The Auger Process 208
    5.3. Decay of Excited States 209
6. Higher-Order Effects in Photoelectron and Auger Spectra 212
7. Prospects 216
   References 217

CHAPTER 7

VALENCE IONIZATION PROCESSES IN THE VUV REGION

J. Berkowitz, E. Rühl, and H. Baumgärtel

1. Introduction 221
2. Detailed Characterization (Photoabsorption, Photoionization, and Photoelectron
   Spectroscopy) of Small Molecules 222
    2.1. ClO_2 222
    2.2. Ammonia and Some Fluoroamines 227
    2.3. Chlorofluoroethylenes 232
3. Photoionization Studies of Transient Species 236
    3.1. Open-Shell Atoms 238
    3.2. Group V and Group VI Hydrides 240
    3.3. Group V Fluorides 243
    3.4. Group IV Hydrides 243
    3.5. CH_2OH/CH_3O and CH_2SH/CH_3S 247
4. Photoionization and Coincidence Studies of Some Large Molecules 251
    4.1. Benzene Isomers 251
    4.2. Naphthalene, Azulene 251
    4.3. Larger PAHs 255
    4.4. Buckminsterfullerene 255

CHAPTER 8

PHOTOION-PAIR FORMATION

J. Berkowitz

1. Brief Summary and Overview of Ion-Pair Processes 263
2. Recent Studies 272
   2.1. Diatomic Molecules 272
   2.2. Triatomic Molecules 276
   2.3. Polyatomic Molecules 280
   2.4. Ion-Pair Formation at Higher Energies 286
3. Conclusions 287
   References 288

CHAPTER 9

ELECTRONIC AND NUCLEAR RELAXATION OF CORE-EXCITED MOLECULES

Irène Nenner and Paul Morin

1. Introduction 291
2. Photoabsorption 295
   2.1. "One-Electron" Transitions 297
   2.2. "Two-Electron" Transitions 301
3. Electronic Relaxation 304
   3.1. The Core Level Photoelectron Spectrum: A General Description 304
   3.2. Shape Resonances 307
   3.3. The Auger Process 308
   3.4. The Resonant Auger Process 311
   3.5. Double Auger Effects, Cascades, and Multiionization 315
4. Electron–Nuclear Motion Interaction 318
   4.1. Vibrational Effects 319
   4.2. Direct Dissociation and Competition with Electronic Relaxation 323
5. Fragmentation of Core-Excited Molecules 327
   5.1. Experimental Techniques 328
   5.2. Dissociation of Simple Polyatomic Molecules 332
   5.3. Dissociation of Complex Molecules 340
6. Future Trends 344
   References 345

CHAPTER 10

RESONANCES AND NEAR-THRESHOLD PROCESSES

G. C. King and K.-H. Schartner

1. Introduction 355
2. Threshold Photoelectron Spectroscopy 356
   2.1. Principles and Experimental Methods 356
   2.2. Threshold Photoelectron Studies in Atoms 359
3. Constant Ion State Spectroscopy 368
   3.1. Principles and Experimental Methods 368
   3.2. Constant Ion State Studies in Atoms 370

Previous entries:

5. Conclusions 256
   References 257

4. Photon-Induced Fluorescence Spectroscopy 377
   4.1. Principles 377
   4.2. Experimental Procedure 381
   4.3. Photoionization of the Outer s Electrons of the Rare Gases near Threshold 383
   4.4. Examples of Autoionization Resonances in Satellite Production,
        Comparison with Calculations 393
5. Summary and Future Prospects 397
   References 397

CHAPTER 11

RESONANT AND NONRESONANT AUGER RECOMBINATION

H. Aksela, S. Aksela, and N. Kabachnik

1. Introduction 401
2. Instrumentation and Methods 403
   2.1. Conventional Auger Spectroscopy 403
   2.2. Angular Resolving Spectrometers 404
   2.3. Synchrotron Radiation Excited Auger and Resonant Auger Spectroscopies 404
   2.4. Metal Vapor Ovens 405
3. Basic Ideas of Theoretical Treatment 406
   3.1. Auger Transition Energies 406
   3.2. Transition Probabilities 407
   3.3. Configuration Interaction, Channel Interaction, and Relaxation 408
   3.4. Molecular Auger Transitions 409
4. Normal Auger Electron Spectra: Experiment versus Theory 409
   4.1. Auger Transition Energies of Free Atoms 409
   4.2. Spectra of Rare Gases and Other Closed-Shell Atoms 409
   4.3. Spectra of Open-Shell Atoms 412
   4.4. Total Auger Decay Width, Group Rate, and Multiplet Structure 413
   4.5. Coster–Kronig Transitions 416
   4.6. Molecular Auger Spectra 416
   4.7. Near-Threshold Phenomena 417
5. Satellite Transitions 418
   5.1. Satellites Produced during Auger Decay 418
   5.2. Satellites Caused by the Initial-State Effects 418
   5.3. Cascades 418
6. Resonant Auger Recombination 420
   6.1. Decay Spectra of Core-Excited Atoms 420
   6.2. Shake Modified Resonant Auger Decay 421
   6.3. Auger Cascades and Their Role in Producing Multiply Charged Ions 421
   6.4. Decay of Two-Electron Excitations 423
   6.5. Decay Channels of Core-Excited Molecules 423
7. Angular Distribution of Auger Electrons 423
   7.1. Alignment of Atoms in Photoionization and Photoexcitation 424
   7.2. Anisotropy of Normal Auger Transitions 425
   7.3. Anisotropy of Resonant Auger Decay 428
8. Photoelectron–Auger Electron Angular Correlation 429
   8.1. Theoretical Background of Coincidence Measurements 429
   8.2. Experimental Investigation 430
9. Spin Polarization of Auger Electrons 432
   9.1. Dynamic Polarization in Photoionization 432
   9.2. Polarization Transfer in Photoionization by Circularly Polarized Light 433

9.3. On the Possibility of a "Complete" Experiment in Auger Recombination 434
References 435

CHAPTER 12

LASER-BASED UV AND VUV SPECTROSCOPY OF DOUBLY EXCITED ATOMS

J.-P. Connerade

1. Introduction 441
2. Laser Sources for VUV Spectroscopy 442
3. Detection 445
4. Wavelength Calibration 447
5. Faraday Rotation 447
6. Studies of Doubly Excited Series 450
   6.1. Perturbations in the Principal Series of Barium 450
   6.2. Double Excitations in Calcium 452
   6.3. The Doubly Excited Spectrum of Strontium 455
   6.4. The Doubly Excited Spectrum of Barium 457
   6.5. Energy Degeneracies between Single and Double Excitation Spectra 458
7. Inner-Shell Excitation 459
8. Collisions and External Fields 460
9. The Use of Collisions to Extend the Wavelength Range 461
10. From Order to Chaos in Many-Electron Spectra 461
11. Conclusion 462
    References 462

CHAPTER 13

ION YIELD SPECTROSCOPY WITH SOFT X RAYS

T. Hayaishi and P. Zimmermann

1. Introduction 465
2. Experimental Techniques 465
   2.1. Time-of-Flight Techniques 465
   2.2. Atomic Beam Technique 467
3. Results 469
   3.1. Rare Gases 469
   3.2. Alkali and Alkali-Earth Metals 479
   3.3. $3d$ Transition Metals (Iron Group) 482
   3.4. $4f$ Elements (Rare Earths) 487
   References 492

CHAPTER 14

COINCIDENCE MEASUREMENTS ON IONS AND ELECTRONS

J. H. D. Eland and V. Schmidt

1. Introduction 495
2. Experimental Methods 498
3. Illustrative Examples 505
   3.1. Photoionization of Atoms 505
   3.2. Photoionization of Molecules 511
4. Conclusions and Outlook 518
   References 518

CHAPTER 15

SPIN POLARIZATION IN PHOTOIONIZATION

U. Heinzmann and N. A. Cherepkov

1. Introduction 521
2. General Description of Photoelectron Spin Polarization 522
   2.1. Theoretical Description 522
   2.2. Experimental Examples of the Angular Dependence of Spin Polarization 525
   2.3. The Possibility of Complete Experiments 529
   2.4. Other Types of Complete Experiments 530
3. Techniques and Particular Results 533
   3.1. Experimental Techniques 533
   3.2. Theoretical Calculations 535
   3.3. Comparison between Theory and Experiment for Rare Gas Atoms 539
   3.4. Experimental Results for Other Atoms and Comparison with Theory 540
4. Photoionization of Molecules 544
   4.1. Differences between Atomic and Molecular Cases 544
   4.2. Influence of Molecular Vibration and Rotation onto Polarization of Photoelectrons . . 547
5. Photoionization and Auger Decay of Oriented Systems 551
   5.1. Polarized Atoms 551
   5.2. Oriented Molecules 553
   5.3. Polarized Auger Electrons 554
6. Conclusion 555
   References 556

CHAPTER 16

PHOTOIONIZATION OF EXCITED AND IONIZED SYSTEMS

F. J. Wuilleumier and J. B. West

1. Introduction 561
2. Experimental Techniques 562
   2.1. Spark and Plasma Sources (DLPP) 562
   2.2. Resonant Laser-Driven Ionization (RLDI) 565
   2.3. Ion Beam Experiments 566
   2.4. Combined Laser–Synchrotron Experiments 568
3. Experimental Results 572
   3.1. Excited Atoms 572
   3.2. Atomic Ions 579
4. Developments in Theoretical Interpretation 585
   4.1. Inadequacy of the One-Electron Approximation 585
   4.2. Recent Theoretical Developments for Excited Atoms 587
   4.3. Calculations for Ions 589
5. Future Work 597
   5.1. Extension of the Present Experiments 597
   5.2. New Storage Rings and New ECR/EBIT/EBIS Ion Sources 597
   References 600

CHAPTER 17

PHOTOIONIZATION OF ORIENTED SYSTEMS AND CIRCULAR DICHROISM

G. Schönhense and J. Hormes

1. Introduction 607

2. What Is Circular Dichroism? 608
   2.1. Empirical Description of CD 608
   2.2. Basis for a Quantum-Mechanical Description of CD 609
   2.3. Importance of Optically Active Molecules in Biology and Pharmacy 611
3. How to Measure CD Spectra? 611
   3.1. Conventional Methods 611
   3.2. Measurements with Synchrotron Radiation 613
   3.3. Present Experimental Limits 614
4. What Can Be Learned from CD Spectra? 615
   4.1. Models for the Origin of Optical Activity 616
   4.2. The Dynamic Polarization Mechanism and the Octant Rule 617
   4.3. The Exciton Coupling Model 618
5. Some Typical Applications of CD Measurements 620
   5.1. Applications in Chemistry 621
   5.2. Applications in Biology 622
6. Circular Dichroism in the Angular Distribution of Photoelectrons (CDAD) 624
   6.1. Physical Nature of CDAD in an Electric Dipole Transition 624
   6.2. A Quantitative Example 626
   6.3. Experimental Technique 628
   6.4. What Can Be Learned from CDAD Measurements? 629
7. Case Studies of CDAD for Oriented Molecules and Comparison with *ab initio* Calculations 630
   7.1. Experiments on Gas-Phase Molecules with Laser-Induced Alignment 630
   7.2. Oriented CO and NO on Surfaces 633
   7.3. Adsorbed Benzene and Formate as Model Systems of Totally Oriented Species . . 636
   7.4. Atomic Oxygen on a Copper Surface 640
   7.5. CDAD in X-Ray Photoemission from Core Levels 641
8. Future Developments 645
   8.1. Circularly Polarized Radiation from Insertion Devices 646
   8.2. Related Techniques: Circular Intensity Differential Scattering (CIDS)
        and CDAD Microscopy 646
   8.3. Measurements in the X-Ray Region 647
   References 648

AUTHOR INDEX 653

SUBJECT INDEX 661

CHAPTER 1

THEORY OF PHOTOIONIZATION

VUV AND SOFT X-RAY FREQUENCY REGION

M. YA. AMUSIA

1. INTRODUCTION

In this book photoionization is considered as a process of interaction between a low-intensity electromagnetic field and atomic or multiatomic targets, which results in removing or exciting their electrons. Although studied for about a century, this process still remains a subject of intense experimental and theoretical investigation and will definitely keep its importance at least in the foreseeable future.

The data obtained in photoionization studies are of great value for other domains of science, for example astro- and plasma physics, chemistry, biology, as well as for technological applications.

However, these data have their own purely scientific interest. Indeed, atoms with many electrons as well as multiatomic targets are systems in which the interaction between constituents—electrons and nuclei—is known with a very high accuracy. Well investigated and comparatively weak is the interaction between the electrons and low-intensity electromagnetic field. Therefore, everything in photoionization which cannot be understood qualitatively or even quantitatively in the framework of the simplest hydrogenlike approach for isolated atoms (noninteracting electrons in the field of a nucleus) is a consequence of the many-electron nature of atoms or multiatomic targets. Thus, one of the most important goals of atomic physics is to study atoms as many-body systems.

Multiatomic formations, such as molecules, clusters, and solid bodies, are in essence much more complicated objects, in which the relative motion of constituent nuclei is of importance.

It appeared that the multielectron nature of an atom or multiatomic formation plays a considerable or even a decisive role in determining to a large extent the entire mechanism of the photoionization process. The more the delicate features of this process became the objects of investigation, the more important the multielectron nature of the target proved to be.

Although the interelectron interaction itself is not too strong, its action after multiplying by the number of actively participating electrons becomes very important, at least as important as the action of the nuclear Coulomb field.

Typically, the interelectron interaction energy is in the range from 10 eV for outer shells to hundreds of electron volts for intermediate shells in heavy atoms. The reaction of atoms and multiatomic formations to the action of photons in this frequency region, i.e., in the region of vacuum ultraviolet (VUV) and soft x rays, is most sensitive to the multielectron nature of the targets.

Many-electron atoms deserve special attention as an object of photoionization studies. They are much simpler than other multiatomic formations because of the absence of rather complicated internuclear motion. On the other hand, they are the bricks, the fundamental or "elementary particles"

M. YA. AMUSIA • A. F. Ioffe Physical-Technical Institute, 194021 St. Petersburg, Russia.

VUV and Soft X-Ray Photoionization. Edited by Uwe Becker and David A. Shirley. Plenum Press, New York, 1996.

of macroscopical bodies. It is their properties that are reflected in a number of processes in multiatomic formations, namely, those connected with ionization of intermediate and inner shells or transitions between them.

A lot of experimental data on photoionization and their theoretical interpretation have already been collected. These data unambiguously demonstrate that not only the simplest hydrogenlike picture, but also much more sophisticated approaches, which consider electrons as moving independently in a common rather complex nonlocal single particle field, are insufficient. It appeared that the interelectron interaction leading to dependent, or correlated motion of atomic electrons must be taken into account. The development in this direction leads to creation of a number of theoretical approaches, which permit one to describe quantitatively the photoionization data. In turn, it leads to considerable development of experimental technique, making ever more sophisticated studies possible.

1.1. Main Physical Conceptions Important for Understanding the Photoionization Process

1.1.1. Self-Consistent Field. This a common static field acting on each electron and created by the nuclei and all of the electrons of the system under consideration. The most well known and precise is the Hartree–Fock (HF) approximation for this field.[1]

It is implied that the simplest (one-electron) approach to photoionization considers this process as removing or exciting one of the electrons which are all under the action of the same field in initial and final states of the system. Part of the interelectron interaction is included in the self-consistent field making it different from the pure nuclear Coulomb field. But the remaining part of this interelectron interaction leads to an alteration of the common self-consistent field while even a single electron changes its state. Therefore, this field is in fact different for initial and final states and taking this into account improves considerably the description of this process, leading, however, beyond the one-electron approximation.

While in the hydrogenic approximation the photoionization cross section of each electron shell or even subshell is represented by a rapidly decreasing monotonic function, in the HF approximation the maxima very often proved to be shifted to higher energies from ionization thresholds and new maxima (and minima, called Cooper minima), not connected with the thresholds, appear.

1.1.2. Polarization of the Atomic Electron Cloud. This is the deformation of the electron system of the target, as well as the redistribution of its nuclei due to action of the incoming photon. Large maxima in the response function, i.e., in the target polarizabilities, represent collective oscillations of the target electrons. For a homogeneous electron gas this is the so-called plasma oscillation. The damping of these oscillations is due either to inhomogeneity of the electron distribution or to the interaction with "two-electron–two-vacancy" and even more complicated target excitations. Note that the collective oscillations are a coherent superposition of the "one-electron–one-vacancy" excitations. The coupling between them determines the oscillation frequency. The simplest description of electron collective excitations is achieved in the frame of the random phase approximation (RPA)[2] or the random phase approximation with exchange (RPAE).[3]

1.1.3. Polarization Potential. This contribution is, in addition to the self-consistent HF potential, acting on the photoelectron and other target electrons in their ground states. This potential appears due to deformation of the electron distribution either by the excited electron or by an electron in its ground state. Any of the target electrons in its motion polarizes the distribution of all of the others, which alters the total field acting on the considered electron. The polarization potential alters the ionization potential of the target, shifting it off the HF value. It also affects the outgoing or excited electrons, modifying the photoionization cross section.

1.1.4. Rearrangement or Relaxation. This is the modification of the field acting on the outgoing electron due to its elimination of the target system (or, in other words, due to creation of a vacancy). After elimination of an electron from an intermediate or inner shell, the states of all others are altered. It changes the interaction between the outgoing electron and the vacancy, leading to screening of the bare interelectron interaction. Rearrangement or relaxation manifests itself also in modification of the potential acting on all other target electrons. The alteration of their states leads not only to temporal (or virtual), but also to real excitations, thus resulting in multiple excitation or ionization of the target electron system.

The simplest way to describe these effects theoretically is to use the generalized random phase approximation with exchange (GRPAE), namely, two versions of this method: GRPAE and GRPAE II.[3]

1.1.5. *Interchannel Interaction.* This is that part of interelectron interaction which leads to autoionization, decay of vacancies, and creation of satellites. It is responsible for formation of states, which are more complex than a simple one-electron excitation. These states proved to be easily excited in the processes of the photon interaction with atoms or multiatomic formations.

Interchannel interaction leads to strong modification of the partial photoionization cross section, i.e., that which corresponds to elimination of electrons from a given subshell.

This interaction manifests itself in excitation of satellites to main states, i.e., to vacancies in electron shells. A satellite includes together with its mother-vacancy also a discrete excitation of at least one other electron.

1.2. Theoretical Approaches and the Mathematical Apparatus

Almost all theoretical approaches developed to investigate photoionization use HF[1] as its starting approximation: this is the basis of the perturbative interelectron interaction approaches, such as many-body perturbation theory (MBPT)[4] and RPAE[3] as well as multiconfiguration Hartree–Fock (MCHF).[4] As an approach, going beyond HF even at its first step, the local density approximation (LDA)[5] was suggested. The interaction of atoms and multiatomic formations with an alternating electromagnetic field, including photoionization processes, can be considered using the time-dependent local density approximation (TDLDA).[6] This approach is a generalization of LDA which permits one to take into account the action of a time-dependent external field on the constituents of the target system just as the time-dependent Hartree–Fock (TDHF)[7] generalizes the HF approach for the case when all electrons of the atom or multiatomic formation are under the action of the time-dependent external field.

A number of results in the photoionization domain are obtained using the R-matrix approach.[8]

All theoretical methods mentioned above, except HF, permit one to account for the interelectron interaction out of the self-consistent one-electron field frame. They take into account the deviations from the picture of independent electron motion, i.e., electron correlations. However, they differ by the degree to which they are able to describe more or less sophisticated correlation effects.

In the discussion of photoionization in this chapter, the many-body diagrammatic technique[3] is used. This technique employs special pictures or diagrams. The terms used are: "vacuum," for the ground state of the closed shell target system; "electron," for an electron at a discrete excited level or in the continuum; "vacancy," for an absence of an electron at an initially occupied level; and "effective" external field and "effective" interelectron interaction. The word "effective" emphasizes the role of virtual electron excitations in modification of the external field and the interelectron interaction.

While studying many-electron effects in systems with a comparatively small number of electrons, one must keep in mind the difficulty in drawing a definite border between one-electron and multielectron effect.[3] It appeared, for instance, that some versions of HF are able to include a considerable fraction of electron correlations. Therefore, the role of electron correlations depends on the approximation used in describing the independent electron motion.

1.3. Similarities and Specifics in Applying Theoretical Ideas to Different Objects

While the general ideas, aims, and methods in studying photoionization of atoms and multiatomic formations are the same, there are important differences in applying them to concrete systems.

The similarities come from the same aim of investigation—to disclose the electronic structure and to disclose the role played by electron correlation; the same main observed features—big powerful maxima or giant resonances in the photoionization cross sections, autoionization resonances, and satellite spectra; qualitatively the same picture of the process—the cross sections of multiatomic formations are very close to that of the constituent atoms.

The specific features in the consideration of atoms and multiatomic formations are due to different inhomogeneity of their electron distribution, different shell structure and its role, different degree of collectivization in electron motion which is greatest for homogeneous multielectron systems without shell structure, and the different speed of the decay of collective excitations.

2. MAIN ACHIEVEMENTS OF PHOTOIONIZATION STUDIES

Very important results in the investigation of photoionization of atoms and multiatomic formations were obtained which improved considerably the understanding of their electron structure:

1. It was demonstrated that giant resonances in photoionization cross sections of all systems discussed in this chapter are of multielectron nature. It proved to be that these resonances cannot be described in the frame of any pure one-electron model and represent results of coherent action of all electrons in a considered shell.
2. It was demonstrated that multielectron shells can strongly and even qualitatively modify the photoionization cross section of a few-electron shell, leading to new continuous spectrum resonances. These resonances cannot be reproduced by any one-electron model and are a direct consequence of interchannel interaction, leading to interference phenomena in the photoionization process.
3. It was shown that together with the main line, satellites or shadow lines of a pure multielectron nature exist in the photoelectron spectrum. Their comparatively large strength is a direct consequence and manifestation of the important role of interelectron interaction, which cannot be accounted for by any choice of a one-electron potential.
4. It was demonstrated that the interaction of two or even more continuous spectra grounded on different levels can lead to new resonances. Because of this origin, these maxima and corresponding minima are of an interfering multielectron nature.
5. Strong autoionization resonances were observed coming from the interaction of a discrete "two-electron–two-vacancy" excitation from one subshell with a "one-electron–one-vacancy" continuum of another. Therefore, it is possible that almost everywhere except in the close vicinity of the first ionization potential, the photoionization cross section has a fine structure of more or less narrow resonances of autoionization-like nature.
6. It was demonstrated that the Auger line is shifted to higher energies when an inner vacancy is created near its threshold. This shift increases with the decrease of photoelectron energy and is a manifestation of instantaneous variation of the field acting on the photoelectron due to the decay of initially created vacancy.
7. A strong similarity was observed for the photoionization cross sections in different phase states of the same composition of atoms, e.g., gases, mixtures, and solid bodies, for intermediate and inner shells.

3. THEORETICAL DESCRIPTION OF THE PHOTOIONIZATION CROSS SECTION: ONE-ELECTRON HARTREE–FOCK (HF) APPROACH

The photoionization process is determined by several quantities which are functions of the photon frequency ω.[*] For atomic photoionization, which will be the main subject of this chapter, these are the total $\sigma(\omega)$ and partial $\sigma_i(\omega)$ photoionization cross sections, angular anisotropy parameters $\beta_i(\omega)$, index i denoting the quantum numbers of the ionized subshell.

Obviously, one has

---

[*]Atomic units will be used throughout this chapter: $e = m_e = \hbar = 1$, when e and m_e are the electron charge and mass.

$$\sigma(\omega) = \sum_i \sigma_i(\omega) \qquad (1)$$

In the dipole approximation for the incoming photon which is valid in the VUV and the soft x-ray region, the angular distribution is given by

$$\frac{d\sigma_i(\omega,\theta)}{d\Omega} = \frac{\sigma_i(\omega)}{4\pi}[1 + \beta_i(\omega)P_2(\cos\theta)] \qquad (2)$$

Here Ω is the solid angle, θ is the angle between the incoming photon and the emitted photoelectron momenta, and $P_2(\cos\theta)$ is the second-order Legendre polynomial. The photoelectron energy ε is connected with the photon frequency ω by the relation

$$\varepsilon = \omega - I_i \qquad (3)$$

where I_i is the ionization potential of the ith subshell.

Expression (2) is valid for unpolarized photons only. Other cases, namely linear and circular polarizations, are considered in Ref. 3.

Photoelectrons are characterized not only by their velocity, but also by the spin direction or spin polarization. This is entirely determined by three projections of the spin vector. Each of these projections is described by a frequency-dependent function $\gamma_i(\omega)$, $\zeta_i(\omega)$, and $A_i(\omega)$.[9]

The partial photoionization cross section is expressed via a photoionization amplitude, which is a matrix element of the operation $\hat{d}$ describing the interaction of an electron with the electromagnetic field. In the dipole approximation, the latter can be presented in one of the following forms:

"length" $\qquad \vec{e}\,\vec{r}_j \qquad (4a)$

"velocity" $\qquad \frac{1}{c}\vec{e}\,\vec{\nabla}_j \qquad (4b)$

"acceleration" I $\qquad \frac{Z}{\omega}\frac{\vec{e}\,\vec{r}_j}{r_j^3} \qquad (4c)$

"acceleration" II $\qquad \frac{1}{\omega}\left[\frac{Z(\vec{e}\,\vec{r}_j)}{r_j^3} - \vec{e}\,\vec{\nabla}W(r_j)\right] \qquad (4d)$

Here $\vec{e}$ is the photon's polarization vector, $\vec{r}_j$ and $\vec{\nabla}_j$ are the coordinate and momentum operators of the jth electron, c is the speed of light, Z is the nuclear charge, and $W(r_j)$ is the local self-consistent field acting on the jth electron.

In one-electron approximation the dipole matrix element of photoionization amplitude $d_{i\varepsilon} \equiv (i|\hat{d}|\varepsilon)$, where $(i|$ and $|\varepsilon)$ are the initial- and final-state one-electron wave functions, permits one to calculate the partial photoionization cross section $\sigma_i(\omega)$ using the following formula[3]:

$$\sigma_i(\omega) = \frac{4\pi^2}{\omega c}|(i|d|\varepsilon)|^2 \qquad (5)$$

Usually the atomic self-consistent field is considered to be spherically symmetric, which is correct only for a closed-shell atom. In this case the state of each electron is characterized by its principal quantum number n or energy ε, angular momentum ℓ, its projection m, and in the case of not too strong spin-orbit interaction, by the spin projection s also. Having in mind a rather general description of the photoionization process, the LS coupling scheme can be used which assumes that there is almost no interaction between the angular momentum and spin so that the total angular momentum L of an atom and its total spin S are both good quantum numbers. Presenting then the one-electron wave function $\varphi_{n(\varepsilon)}(\vec{r})$ as a product of the radial $R_{n(\varepsilon)\ell}(r)$, angular $Y_{\ell m}(\theta,\varphi)$, and spin χs parts, one can perform analytically the integration over angle and summation over spin of the matrix element of the photoionization amplitude. If the angular momentum of the ionizing electron is ℓ_i, the photoelectron's angular momentum is either $\ell_i + 1$ or $\ell_i - 1$. The expression for the partial cross section $\sigma_{n\ell}(\omega)$ is given by

$$\sigma_{n\ell} = 8\pi^2 \frac{N_{n\ell}}{3\omega c(2\ell+1)} [\,|(n\ell|\hat{d}|\varepsilon,\ell+1)|^2 + |(n\ell|\hat{d}|\varepsilon,\ell-1)|^2\,] \tag{6}$$

where $(n\ell|\hat{d}|\varepsilon,\ell \pm 1)$ are pure radial matrix elements.

The angular anisotropy parameter $\beta_{n\ell}(\omega)$ is given by

$$\beta_{n\ell}(\omega) = [(2\ell+1)(\,|(n\ell|\hat{d}|\varepsilon,\ell+1)|^2 + |(n\ell|\hat{d}|\varepsilon,\ell-1)|^2)]^{-1} \times \{(\ell+2)|(n\ell|\hat{d}|\varepsilon,\ell+1)|^2 \tag{7}$$

$$+ (\ell-1)|(n\ell|d|\varepsilon,\ell-1)|^2 + 6\sqrt{\ell(\ell+1)}\operatorname{Re}((n\ell|d|\varepsilon,\ell+1)(\varepsilon,\ell-1|d|n\ell)\exp[i(\delta_{\ell+1} - \delta_{\ell-1})])\}$$

where $\delta_{\ell+1}, \delta_{\ell-1}$ are the elastic scattering phase shifts of the photoelectron with angular momentum $\ell_i + 1$ and $\ell_i - 1$ moving in the field of the residual ion.

The formulas for spin polarization parameters $\gamma_i(\omega)$, $\xi_i(\omega)$, and $A_i(\omega)$ which include the same radial matrix elements and phase differences, but in combinations other than in (6) and (7), are given by rather complex expressions and can be found in Ref. 9 or 3.

The self-consistent field in the HF approximation is something much more complicated than a simple radial dependent potential. Indeed, the HF equations are given[1] by a system of N expressions, where N is the number of electrons:

$$\left(-\frac{\Delta}{2} - \frac{Z}{r} - \varepsilon_i\right)\varphi_i(r) + \sum_{j=1}^{N} \int d\vec{r}\,' \frac{1}{|\vec{r}-\vec{r}\,'|} \varphi_j^*(\vec{r}\,')(\phi_j(\vec{r}\,')\varphi_i(\vec{r}) - \phi_i(\vec{r}\,')\varphi_j(\vec{r})) = 0 \tag{8}$$

In HF the total electron density $\rho(r)$ is given by $\sum_{j=1}^{N} |\varphi_j(r)|^2$. The term in (8) which includes $\rho(r')$ is called the Hartree potential and it is local, while the other one containing $\sum_{j=1}^{N} \varphi_j^*(r')\varphi_j(r)$ is the Fock term, which is a nonlocal interaction. This feature of the Fock term is important and must be taken into account. It essentially affects the choice of the operator describing the photon–electron interaction (4). These are equivalent only if as initial and final states the exact wave function of the total atomic Hamiltonian $\hat{\mathcal{H}}$ is used. Different approximations can destroy this equivalence. If the self-consistent potential is local, (4a) and (4b) are equivalent but different from (4c) or (4d). The presence of nonlocality, as in HF, destroys this equivalence and the problem arises as to which of the forms (4) is preferential. In general, for any potential—local or nonlocal—the most fundamental is the length form (4a), which follows directly from the expression $\vec{j}\vec{A}$ for the photon–electron interaction energy, where $\vec{A}$ is the vector potential of the electromagnetic field and $\vec{j}$ is the electron current, proportional to its velocity $\vec{v} \equiv \dot{\vec{r}} = i[\hat{\mathcal{H}}\vec{r}]$, but not to the momentum. For nonlocal potentials, velocity and momentum are not directly proportional. However, the HF potential is an approximate one and its nonlocality is not a fundamental feature of the atomic Hamiltonian $\hat{\mathcal{H}}$. This permits one to be more pragmatic while choosing the best form of (4), preferring that which leads to a better agreement with experimental data. There is also another possibility—to go beyond the HF frame and to choose, if possible, an approximation for which all forms (4) are equivalent. This feature of an approximate approach is essential evidence of its internal consistency.

In the HF approximation the total cross section is reproduced reasonably, with a large deviation from the simple hydrogenic picture, with maxima shifted from the ionization thresholds and extra maxima and minima far from the thresholds. Quantitatively, the deviation from experiment even for total cross section can be as large as 2–3 times. Usually the difference for calculations in the "length" and "velocity" forms can also reach several times, the "length" form being better near threshold, while the "velocity" form is better far from threshold. Photoionization of subshells, whose contributions to the total cross section are small, failed to be reproduced in HF even qualitatively. The HF approach is definitely insufficient and unable to describe multielectron and even two-electron excitations and/or ionization of atoms by single photons. To describe these effects, other methods, such as RPAE, GRPAE, and MBPT (see Section 1.2), were developed.

4. RANDOM PHASE APPROXIMATION WITH EXCHANGE (RPAE)—A METHOD TO INCLUDE MANY-ELECTRON CORRELATIONS

Important and nontrivial features of the photoionization process are connected with correlational or collective effects in the electron motion. RPAE forms a very convenient basis to study these effects in multiparticle systems, such as an electron gas in metals, atoms, and solid bodies. This approach takes into account the direct elimination or excitation of an electron after absorbing a photon and an indirect action: elimination of the considered electron by the alternative part of the atomic self-consistent field appearing due to the virtual or real excitation of any other electron after absorbing a photon.[3] The total photoionization amplitude $D(\omega)$ in RPAE is determined by the integral equation:

$$(\varepsilon|D(\omega)|i) = (\varepsilon|\hat{d}|i) + \sum_{\substack{\varepsilon'>F \\ i'\leq F}} [(\varepsilon'|D(\omega)|i')(\omega - \varepsilon' - I_{i'} - i\delta)^{-1}(\varepsilon'i|u|i'\varepsilon)$$

$$+ (i'|D(\omega)|\varepsilon')(\omega + \varepsilon' + I_{i'})^{-1}(i'i|u|\varepsilon'\varepsilon)] \qquad (9)$$

Here u stands for a combination of the direct and exchange Coulomb interelectron interaction $|\vec{r}-\vec{r}'|^{-1}$ matrix elements:

$$(\varepsilon'i|u|i'\varepsilon) = \int d\vec{r}\,d\vec{r}\,'\varphi_{\varepsilon'}^*(\vec{r})\varphi_i(\vec{r}')|\vec{r}-\vec{r}'|^{-1}(\varphi_{i'}^*(r)\varphi_\varepsilon(r') - \varphi_i^*(r')\varphi_\varepsilon(r)) \qquad (10)$$

One-electron wave functions $\varphi_{\varepsilon(i)}(r)$ are calculated in the HF approximation, by solving equation (8). Summation and integration in (9) are performed over initially occupied $i' \leq F$ as well as vacant and continuous spectrum $\varepsilon' > F$ atomic levels.

In the language of many-body diagrams, which originates from the Feynman technique, equation (9) can be presented in the following form:

(a)   (b)   (c_1)   (d_1)
      (c_2)   (d_2)   (11)

In (11) a dashed line denotes a photon, a wavy line represents the Coulomb interelectron interaction, and a line with an arrow to the right (left) stands for an electron (vacancy). The first term on the right of (11) represents the direct photon action, leading to creation of an electron ε and a vacancy i while the other terms (11c) and (11d) describe the formation of intermediate states $\varepsilon'i'$ with their subsequent "decay" into εi electron–vacancy state.

The photoionization cross section in RPAE, as well as other characteristics of this process can be calculated by substituting in (5), (6), and (7) the matrix elements of $\hat{d}$ by $(i|D(\omega)|\varepsilon)$. It is essential to have in mind that all forms of the dipole operator (4) are equivalent in RPAE, and the dipole golden sum rule is fulfilled which serves as a check of the accuracy of numerical calculations performed. It is seen from (9) and (11) that RPAE takes into account the polarization of the electron cloud by the incoming photon, due to virtual $\varepsilon'i'$ excitations which transform the real photon field into an effective one, their ratio being equal to $D(\omega)/d$.

The electron shell structure of the ionized object leads to the amplification of a contribution of a given level or a group of levels, e.g., those belonging to the same shell. For a finite system it is difficult to mark a strict border between one-electron and correlational multielectron effects. It is necessary to have in mind that some fraction of terms in (11) may be taken into account by a proper choice of the self-consistent field, in which the photoelectron is moving. Indeed, let us construct a photoionization amplitude, which would include only those RPAE terms which as intermediate state have only "one-electron–one-vacancy" excitations [(11d) are neglected] and vacancies with the same energy, i.e., with the same principal quantum number. This amplitude is determined by an integral equation:

$$\text{(12)}$$

The bracket with $L=1$ and $S=0$ emphasize that the total momentum and spin of the electron–vacancy state is equal to that of a dipole photon. In (12) only terms with ionization potentials $I_{i'}$ equal to I_i are included. It was demonstrated[3] that the matrix element $(\varepsilon|\tilde{D}(\omega)|i)$ can be transformed into $(\tilde{\varepsilon}|d|i)$, where the outgoing electron wave function $\tilde{\varphi}_\varepsilon(\vec{r})$ is calculated in the so-called term-dependent HF approximation, where the "term-dependency" means that only the total angular momentum and spin of the electron–vacancy pair are conserved, being good quantum numbers, but not separately $\ell m s$ for the electron and the vacancy.

A semiphenomenological choice of the self-consistent field can effectively include many more terms which in the diagrammatical language belong to multielectron correlations.

The degree of collectivization in electron motion depends on the number of electrons in a given or neighboring subshell and on the relative role of the nuclear charge for that particular shell. The smaller the role of nuclear charge, the larger is the degree of electron collectivization.

It is seen from (12) that while in the intermediate state i' summations are performed over all angular momentum and spin projections, it includes the contribution of all $2(2\ell_i+1)=2(2\ell_{i'}+1)$ electrons of the ionizing subshell.[*] Therefore, term-dependent HF is in essence a many-body approximation, which takes into account the coherent action of at least all electrons of the ionizing subshell.

Due to the energy separation of atomic shells and even subshells, it is natural to distinguish intra- and intershell correlations. The former arise from the interaction of electrons of the ionizing shell only,

---

*It is assumed throughout this chapter that targets have closed subshells only.

and the latter from the interaction with neighbors, among which the two nearest in energy are usually most important.

Only intrashell correlations are included if the summation over i' in (9) is restricted to those vacancy states which have the same ionization potential as that of vacancy $i - I_i$. These correlations are the most important and usually sufficient to be taken into account for multielectron shells, whose photoionization cross section is large. Small cross sections are affected strongly by intershell correlations and are accounted for by the summation over i', including two subshells nearest to the ionizing one.

In numerical calculations, in order to improve the accuracy of RPAE results it is convenient to start from the term-dependent HF as an initial approximation. It means that all diagrams of type (12) are included from the beginning, which leads to modification of (9) by introducing in front of the interelectron potential (10) a projection operator P, which eliminates from the sum over i' the states i' with $I_{i'} = I_i$.

Most straightforward and simple application of RPAE and other many-body approaches is to closed-shell atoms or ions. However, with some modification RPAE can be applied to open-shell systems also.

A large group of such objects are semifilled shell atoms. Due to Hund's rule, all electrons in their half-filled shells have the same spin projection—either "up" or "down." Then due to exchange between them and other atomic electrons, all subshells are split, each into two—"up" and "down"—levels. Therefore, a semifilled shell object can be considered as having only completely filled levels, the number of which is almost doubled, with "up" and "down" electrons in it. Then RPAE can be easily applied to these objects simply by using equations (9) generalized to two-component systems.

At first glance, as is seen from (9) and (11), RPAE describes only one-electron photoionization. However, double vacancy state formation after Auger decay of the initially created vacancy or due to inelastic scattering of the photoelectron with the outer shell is also described by RPAE.

5. RPAE: SOME RESULTS OF CALCULATIONS

RPAE has been applied to many atoms, some molecules, and electron gases in metals. For atoms it leads in general to a quite reasonable agreement, within 15–20%, between theory and experiment for the total and partial cross section of outer and intermediate electron shells. The difference between RPAE and HF is therefore rather large, even if the HF term-dependent version is used.

RPAE corrections affect and considerably alter not only the cross sections, but also other characteristics of the photoionization process: the angular anisotropy parameter and the spin polarization of the photoelectron. Let us concentrate, however, on the cross sections.

Very large and impressive is the role of RPAE correlations in outer or intermediate multielectron shells, such as nd^{10} and nf^{14}. For these orbitals, the photoionization cross sections display larger powerful maxima—so-called giant resonances—with the oscillator strength accumulating almost completely the number of electrons in the ionizing subshells. The maxima are located well above the corresponding ionization thresholds. Such structures in the cross section which are described satisfactorily in RPAE, but not HF, should be called collective resonances, thus emphasizing their many-electron origin. Good examples of such resonances are the photoionization cross sections of $4d^{10}$ subshells in Xe, Cs, and Ba, which is schematically presented in Figure 1.

Sometimes these resonances are erroneously called shape resonances, because to some extent they can be reproduced within a single particle model where they arise due to the effect of the centrifugal barrier of the effective potential on the outgoing electron. However, as mentioned above, many of these approximations such as time-dependent HF are by no means pure single-electron ones. None of the real single-electron approaches, e.g., the Hartree–Slater method (HS), describing the photoelectron as moving in the field of a residual ion with a vacancy whose state is determined by n_i, ℓ_i as well as by m_i and s_i, can reproduce these maxima quantitatively reasonably accurately.

The collective giant resonances exist not only in atoms, but also in multiatomic formations. A recent impressive example is the fullerene C_{60}, where an RPA calculation[10] demonstrated the existence of a resonance at ≈ 23 eV with a tremendous oscillator strength of about 180, while the total number

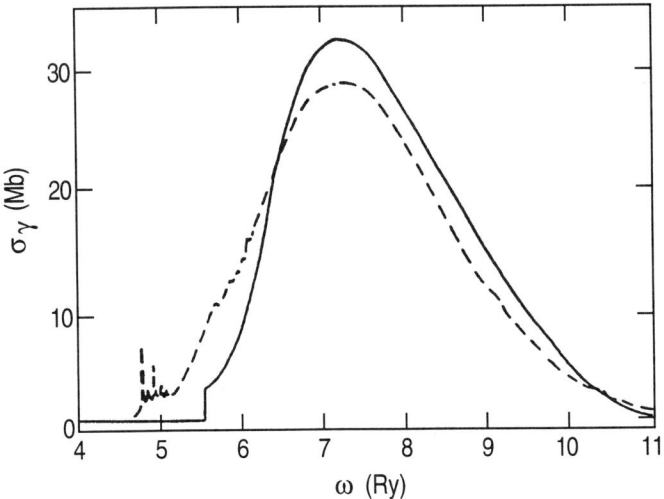

FIGURE 1. Giant resonance in photoionization of $4d^{10}$ subshell in Xe. Total photoabsorption cross section: solid line—RPAE, calculations; dashed line—experiment.

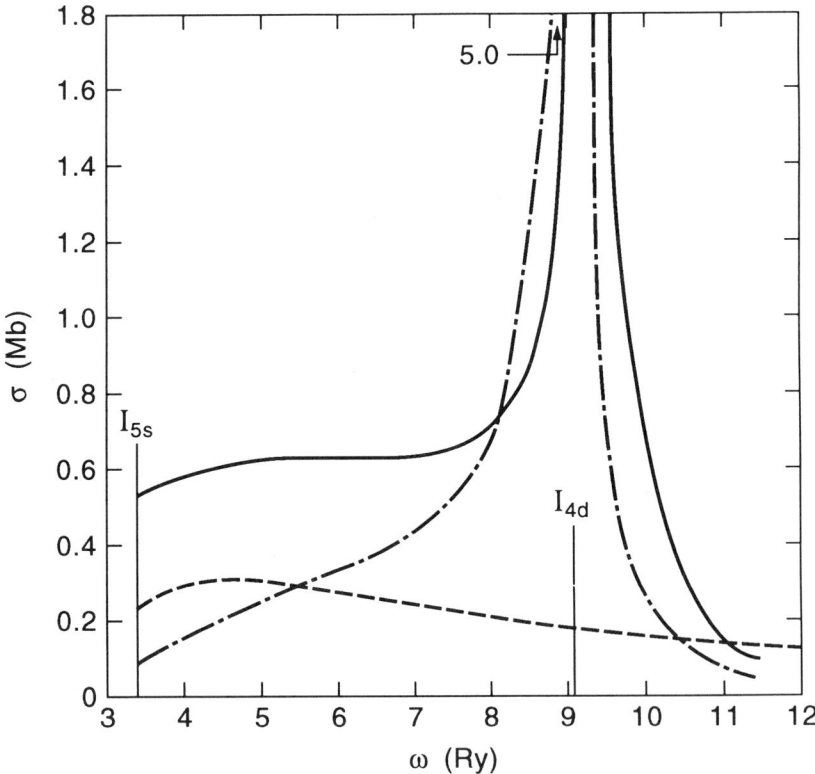

FIGURE 2. Photoionization cross section of $5s^2$ electrons in La. Calculations: solid line—RPAE, with intershell interaction taken into account; dashed line—intrashell correlations; dashed-dotted line—with account of only the $4d^{10}$ subshell.

Theory of Photoionization

of common electrons is 240, 4 coming from each of the participating carbon atoms. The results of these calculations are in qualitative agreement with experiment.[11]

Strong collective resonances are exhibited also by metallic clusters.[12] The theory of photoabsorption for them and the fullerenes could be considerably improved by consistent inclusion of the exchange effects among common electrons as well as with the core electrons of the ions. For clusters, a first step would be a model with the positive charge smeared out homogeneously (except for the surface region) over its volume, with HF and RPAE applied then to electrons moving in such a smeared nuclear-type field.

Not only do RPAE intrashell correlations lead to giant resonances in continuous spectrum, but they can also affect considerably the discrete excitations of the same shell. This influence can be very important, particularly if the resonance and excitation energies are close.

Very important can be the intershell correlations which take into account the interaction of electrons belonging to different subshells. Usually it is quite sufficient to include the two nearest to the considered subshells. This interaction in many cases modifies the partial photoionization cross section even qualitatively. After including the intershell correlations already in RPAE the affected and in some cases affecting subshell cross sections acquire additional maxima and minima. These are of multielectron nature being a manifestation of strong interference between different pathways for photoionization—direct elimination of i-electron and indirect, via i' states. A good illustration of the strength of intershell correlations is the $5s^2$ subshell of the La atom, which is strongly affected by neighboring $5p^6$ and $4d^{10}$ subshells.[13] The results of RPAE calculations are given in Figure 2. It is seen that the $5s^2$ subshell cross section acquires a minimum, which is entirely of a many-electron nature

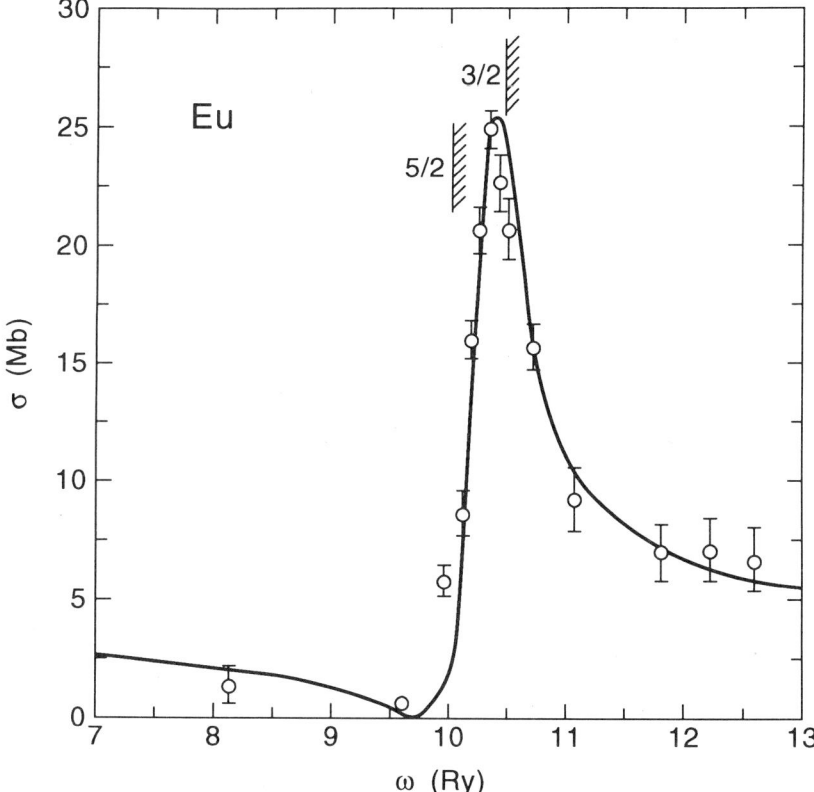

FIGURE 3. Photoionization cross section of a semifilled $4f^7$ shell in the Eu atom.[15] Solid line—RPAE, calculations; circles—experiment.

resulting from the action of six $5p$ and ten $4d$ electrons on the two $5s$. Sometimes these minima are erroneously called Cooper minima, the latter being by definition of a one-electron nature. Indeed, Cooper minima appear as a result of the sign variation of the one-electron matrix element $(i|d|\varepsilon)$ due to the nodal structure of the corresponding wave function as seen in several rare gases. On the contrary, the minimum in Figure 2 appears due to the interference of the two or even three pathways of the $5s^2$ electron photoionization suggesting interference minimum as a more appropriate name in this case.

Similar effects were found to be important in photoionization of molecules whose outer electron structure resembles that of valent and subvalent subshells of noble gases, e.g., SiH_4.[14] The outer electron level is similar to that of Ar and the RPAE calculations demonstrate the strong influence of "np^6" electrons which completely modifies the "ns^2" photoionization cross section.

RPAE describes mutual interaction of several continuous spectra, corresponding to different subshells and also the effect of discrete excitations on discrete and continuous spectrum of other shells. A rather interesting example is the Eu atom, in which extremely strong $4d \rightarrow 4f$ discrete transitions (to the semifilled $4f^7$ subshell) acquire a large width mainly due to the interaction with the $4f \rightarrow \varepsilon g, d$ continuous spectrum transitions, illustrated in Figure 3.[15] Note that the maximum in the $4f^7$-photoionization cross section is similar to that observed in the $4d^{10}$ of Xe and in this sense can also be considered as a giant collective resonance. But their nature is considerably different—in $4d^{10}$ the resonance is due to interaction of ten $4d$ electrons and is decaying by emitting comparatively slow electrons, with an energy equal to $\omega - \ell_{4d}$. On the contrary, the $4f^7$ subshell maximum is due to strong action of $4d$-electron discrete transition on $4f$ electrons. This resonance is decaying by emitting comparatively fast electrons, with energy $(\omega - \ell_{4f})$ which is much larger than $(\omega - \ell_{4d})$.

6. NEW OBJECTS OF RPAE STUDIES

Although most suitable for closed-shell systems, RPAE was applied recently to a number of other objects, starting from positive and negative closed-shell ions and including atoms and ions in excited states.

Let us begin with a positive ion A^{n+}, which forms a closed-shell system. The application of RPAE in this case is straightforward, but the results can be considerably different from those of a neutral atom with the same electronic shells. The reason is that the increase of the field acting on the photoelectron increases considerably the strength and the role of discrete transitions. It also opens levels which are occupied in atoms for transitions after the photon absorption. The photoionization maximum is shifted to the higher-energy side and the role of correlations remains as important, as in neighboring atoms, at least for A^+ and A^{2+}. The RPAE cross section for $K^{+(3)}$ together with the results of measurements[16] are given in Figure 4, demonstrating quite reasonable agreement. The ion K^+ has the outer $3p^6$ subshell just as Ar and the cross section is much closer to the threshold than in Ar. The calculations for K^+ were recently considerably improved in Ref. 17.

The role of RPAE correlations is large and very important in negative ions A^- which form a closed-shell system. Good examples are the halogen negative ions, such as Cl^- (similar to Ar) and I^- (similar to Xe).[18] The correlational effects are stronger than in corresponding noble gases, because the nuclear field is weaker. Qualitatively, however, the situation is similar to that in noble neighbors, demonstrating strong intrashell correlations in outer ($3p^6$ in Cl^- and $5p^6$ in I^-) and intermediate ($4d^{10}$ in I^-) subshells. Rather strong are the intershell correlations in subvalent shells, e.g., $3s^2$ in Cl^- and $5s^2$ in I^-. For A^- the cross sections, both partial and total, are equal to zero at the first ionization threshold. However, within a narrow frequency region of about several tenths of an electron volt they increase very fast and reach values close to those of corresponding noble gas atoms.[18] As an example of an A^- cross section, consider I^-, its $5s^2$ subshell, which is depicted in Figure 5,[18] illustrating the important role of intershell correlations.

As mentioned at the end of the previous section, atoms with semifilled subshells can be treated in RPAE almost as simply as closed-shell atoms. Recently, some activity was connected with studies of photoionization of such atoms as Mn and Cr. The most important effect found is the interaction of a strong discrete transition to the vacant states in the semifilled shell $3d^5 - 3p \rightarrow 3d$ with the excitations

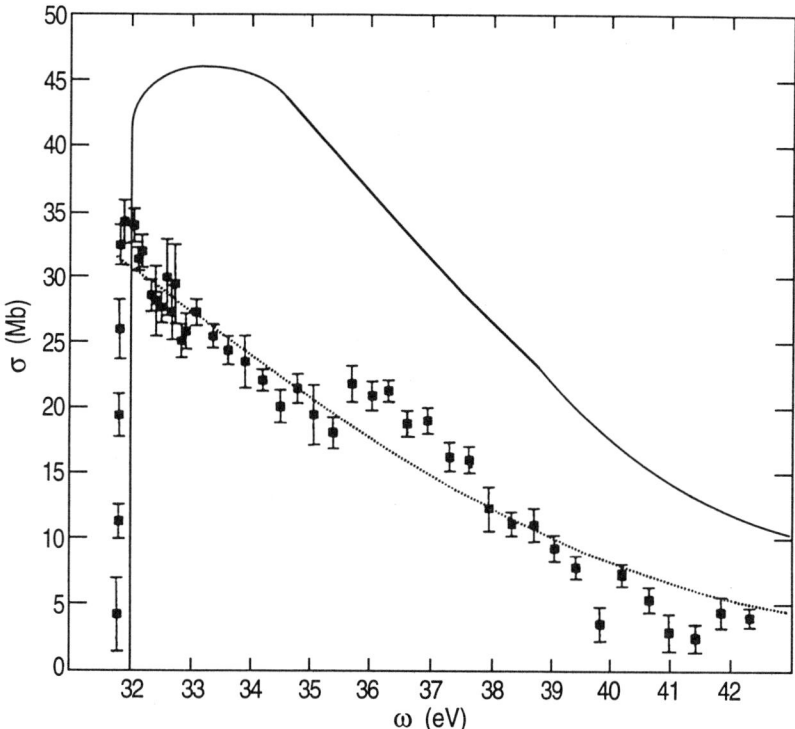

FIGURE 4. Photoionization cross section of $3p^6$ electrons in the K^+ ion. Solid line—RPAE; dotted line—from Ref. 17, calculations; squares—see Ref. 17, experimental data.

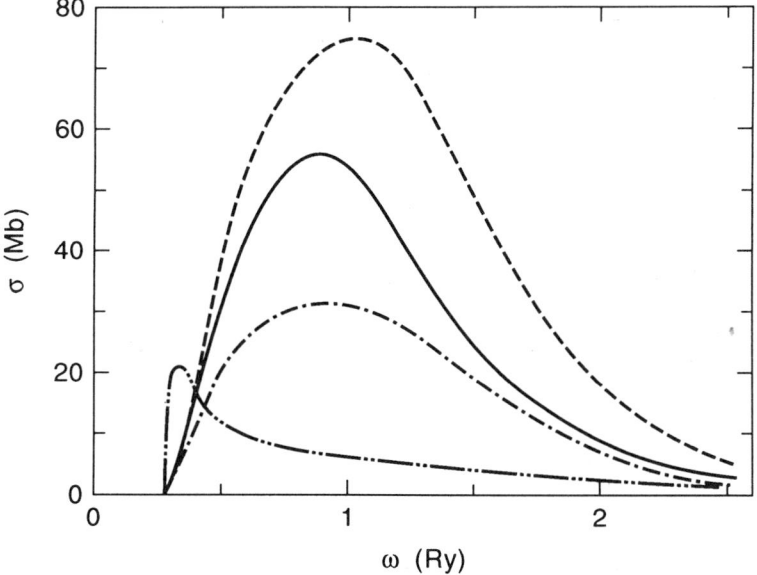

FIGURE 5. Photodetachment cross section of outer $5p^6$ electrons in I^-. Calculations: solid line—RPAE, $5p \to \varepsilon d$ transition; dashed and dashed-dotted line—HF,r and v forms for $5p \to \varepsilon d$ transition; dashed-double-dotted line—RPAE, $5p \to \varepsilon s$ transition.

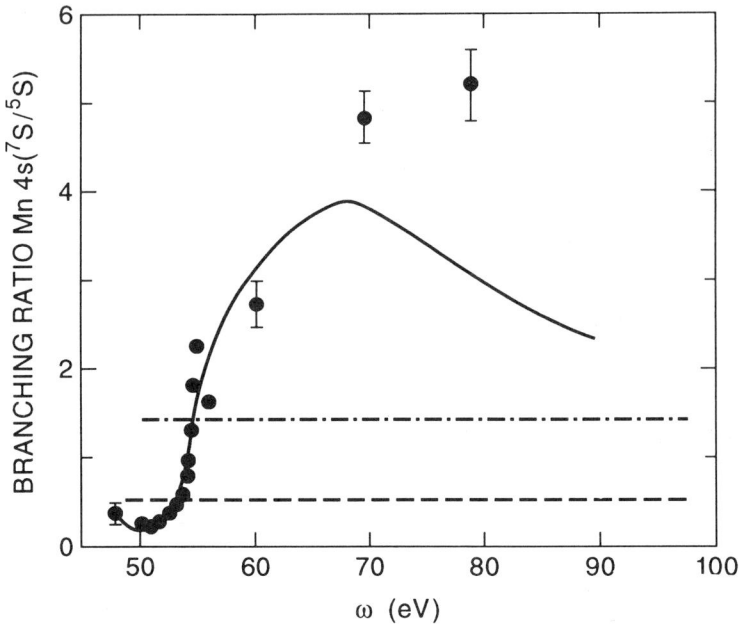

FIGURE 6. The ratio of photoionization cross sections of outer "up" and "down" 4s electrons in the Mn atom.[19] Solid line—spin-polarized RPAE; dashed line—spin-polarized HF; dashed-dotted line—the statistical ratio, calculations; dots—experimental data.

of 3d electrons. It appeared, however, that there are also more complicated effects. For example, the photoionization cross section of 4s "up" and "down" electrons (see Section 4) in Mn is strongly affected by both the discrete transition $3p \rightarrow 3d$ and the continuous spectrum excitations $3d \rightarrow \varepsilon f, p$. The role of the latter proved to be very important, leading to big difference in 4s-"up" and 4s-"down" photoionization cross section, which is illustrated in Figure 6.[19] It is seen that 4s is strongly affected by the virtual excitation or polarization of the inner $3p^6$ and $3d^5$ electrons.

The convenience in applying RPAE to semifilled shell atoms opens the possibility to study a large group of atoms in which positive or negative ions in their ground state have a semifilled shell structure. Examples are the single charged ions of the VIA group of periodic table elements, such as O^+ or Se^+. Adding electrons to the group IVA elements, negative ions with semifilled shells are formed. After including RPAE correlations, the cross sections acquire in some cases more than one extra maximum (and minimum). A very good example is the Si^- atom, the $3p^3$ subshell cross section of which is depicted in Figure 7.[20] As mentioned in the previous section, the intershell or interlevel correlations are particularly strong for semi-closed-shell atoms and ions, leading to fast variations of the Si^- cross section in Figure 7. Of course, Si^- is not an exceptional example and the calculations in the RPAE frame can reveal many interesting variations of the cross section caused by the electron correlations in the new domain of semi-closed-shell atoms and ions.

Atoms or ions with one electron on an s level can be treated as semi-filled-shell systems and thus considered in the RPAE frame. This includes the alkali atoms and single charged ions of group II, such as Ca^+. The photoionization cross section has a prominent autoionization structure and RPAE intra- and intershell correlations proved to be very important. Recently this was illustrated by calculations of the 4s, $3p^6$, and $3s^2$ subshells in Ca^+.[21]

The RPAE intershell correlations strongly affect the photoionization cross section of excited outer atomic electrons. The photoionization of the latter can be described within the HF frame. Their cross section, however, is modified by nearly all inner electron shells, mainly in the vicinity of their thresholds. As an example, Figure 8 presents the photoionization cross section of the outer Cs excited

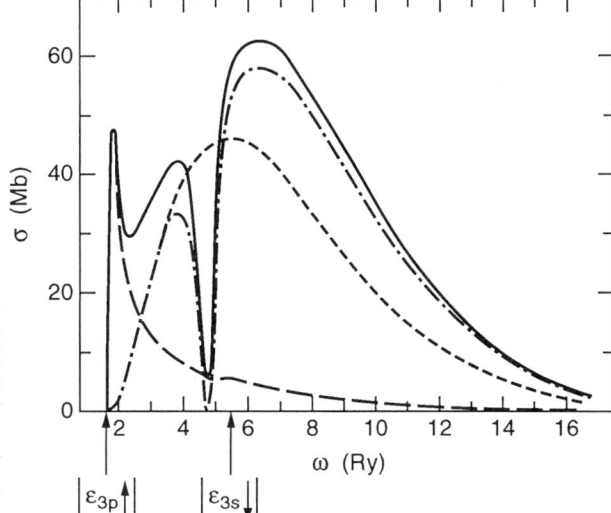

FIGURE 7. Photodetachment cross section in Si⁻. Calculation[18,20] for the ground state of Si⁻, $[Ne]3s^2 3p^3\uparrow$: solid line—total $3p\uparrow$ cross section; dashed line—$3p\uparrow \to \varepsilon d \uparrow$ channel with only $3p^3\uparrow$ electron correlations included; dashed-dotted and long-dashed lines —$3p\uparrow \to \varepsilon d \uparrow$ and $3p\uparrow \to \varepsilon s \uparrow$ respectively, with all RPAE correlations included.

$8s$ electron, modified due to interaction with the core $5p^6$ electrons.[22] It is seen that the $5p^6$ action is very strong.

Some excited states are quasistable and therefore can be studied by photoionization. Interesting examples are the triplet states of He and the elements of the second group in the periodic table $ns, n(s+1)\,^3S$. The exchange interaction between the outer s electrons is, in this case, quite different from that in the ns^2 ground state, which inevitably leads to prominent deviations in corresponding RPAE photoionization cross sections.

An unusual object, which, however, can be studied in the RPAE frame, is a highly excited atomic state, all electrons of which have the same spin projection. The lifetime of such a state for light and medium atoms can be as long as several microseconds, opening, in principle, the possibility of photoionization studies.

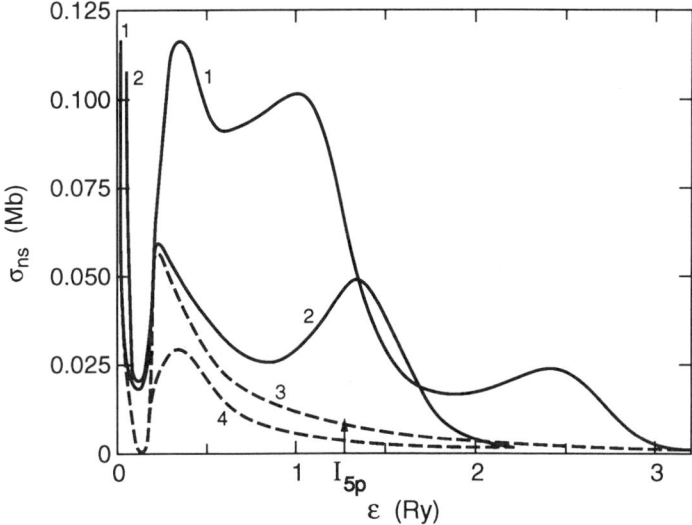

FIGURE 8. Photoionization cross section of the outer electron, excited to levels $7s$ and $8s$ in the Cs atom. Calculations: solid line—RPAE; dashed line—HF; 1, 3—$7s$ level; 2, 4—$8s$ level.

7. BEYOND THE RPAE FRAME: INCLUSION OF REARRANGEMENT EFFECTS

From a purely physical point of view, it is evident that the ionized or virtually excited electron [ε' and ε in (9) and (11)] is moving not in the frozen core field of an atom with a vacancy i (or i'). In fact, as soon as the vacancy is created it also affects the other electron states, leading to more or less slow rearrangement, which is often called static rearrangement. Two major effects determine this process:

1. The screening of the electron–vacancy interaction due to virtual excitations of other atomic electrons. As a result, the photoelectron in its real ε or virtual ε' states does not feel a pure Coulomb field of the vacancies i or i'', but a screened one.
2. The shift of the ionization potential due to polarization of electron shells by inner (or intermediate) vacancies.

The neglect of these effects in the RPAE frame leads to noticeable deviations from experimental data, particularly for frequencies ω close to the ionization threshold of inner and, in some cases, intermediate subshells. For inner shells the static rearrangement is even stronger than the RPAE effect. Diagrammatically, static rearrangement can be illustrated by following two figures. The first is

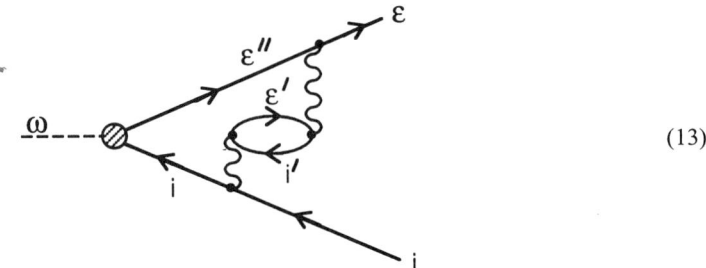

(13)

which exemplifies screening of the electron $\varepsilon(\varepsilon')$–vacancy i interaction due to polarization of the electron cloud, or in other words, its virtual excitation by creating "electron ε'–vacancy i'''" states. Here the dashed circle stands for the RPAE photoionization amplitude given by (11). The second figure is

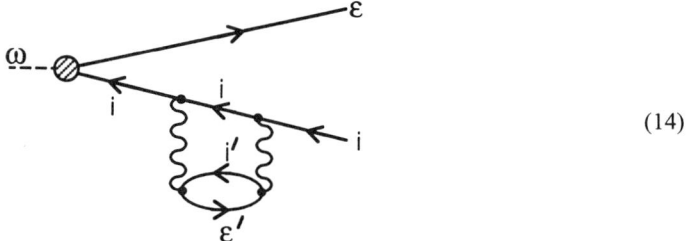

(14)

which describes one of the contributions to the shift of the ionization potential I_i due to virtual excitations of other atomic electrons or, in other words, due to the formation of "electron ε'–vacancy i'''" virtually excited states.

It is essential to have in mind that static rearrangement corresponds only to parts of these diagrams even in the order considered of the perturbation in the interelectron interaction theory: instead of a summation over all intermediate vacancy states, only the term with the vacancy i is taken into account. By including in the RPAE scheme the corrections of the type discussed, the generalized RPAE (GRPAE) is formed, in which the photoionization amplitude $\tilde{D}(\omega)$ is given by[3]:

$$(\underline{\varepsilon}|\tilde{D}(\omega)|\underline{i}) = (\underline{\varepsilon}|d|\underline{i}) + \sum_{\varepsilon > F, i' \leq F} [(\underline{\varepsilon}|\tilde{D}(\omega)|\underline{i'})(\omega - \varepsilon' - I_{i'})^{-1}(\varepsilon'\underline{i}|u|\underline{i'}\varepsilon)$$

$$+ (\underline{i'}|\tilde{D}(\omega)|\underline{\varepsilon'})(\omega + \varepsilon' + I_{i'})^{-1}(\underline{i'i}|u|\varepsilon'\varepsilon)] \quad (15)$$

Here the sign ~ under an electron or vacancy wave function emphasizes that these states are calculated in the HF field of an ion with a vacancy, namely, a state $(\tilde{\varepsilon}, \tilde{i})$ in the presence of the vacancy i and $(\tilde{\varepsilon}, \tilde{i}')$ in the presence of the vacancy i'. The ionization potentials I_i in GRPAE are determined as differences between the self-consistent HF energy of an ion with the vacancy i and the HF energy of a neutral atom. This quantity is smaller than the frozen core ionization potential I_i HF which is equal to the absolute value of the HF energy ε_i of the vacancy i.

The difference between $(\varepsilon|\tilde{D}(\omega)|\tilde{i})$ and $(\varepsilon|D(\omega)|i)$ is particularly large in the near-threshold region of intermediate and inner shells, where the effect of relaxation and vacancy screening by decreasing the strength of the field acting on the outgoing photoelectron shifts the photoionization maximum from the threshold to higher energies. With the growth of the frequency ω above the threshold, the rearrangement effects decrease. GRPAE effects are important not only for atoms, but also for multiatomic formations, because their outer-shell polarizabilities, which determine the screening of the photoelectron–vacancy interaction and the ionization potential alteration, are larger than in most of the atoms.

To illustrate the strong effect of static rearrangement successfully taken into account in the GRPAE frame, the cross section of La $4d^{10}$ subshell photoionization is given in Figure 9.[13]

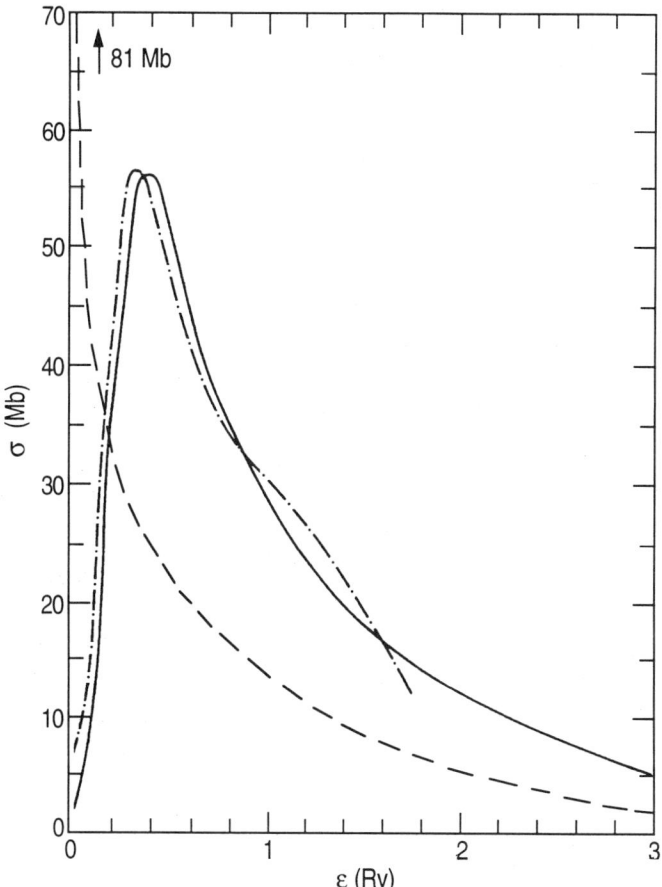

FIGURE 9. Static rearrangement: photoionization of the $4d^{10}$ electrons in the La atom.[13] Solid line—GRPAE; dashed line—RPAE, calculations; dashed-dotted line—experiment.

GRPAE strongly affects the near-threshold region of intermediate shells not only of atoms, but also of negative ions.

Another kind of rearrangement effect takes place when a deep vacancy is created which decays via an Auger process. In this case the outgoing electron, if sufficiently slow, can find itself instantly in a field not of a single, but of a double or even multiple vacancies. This field is of course stronger than that of a single vacancy, so this dynamic rearrangement leads to an increase of the cross section at threshold, as compared to the RPAE case. Diagrammatically, this process is exemplified by the following figure:

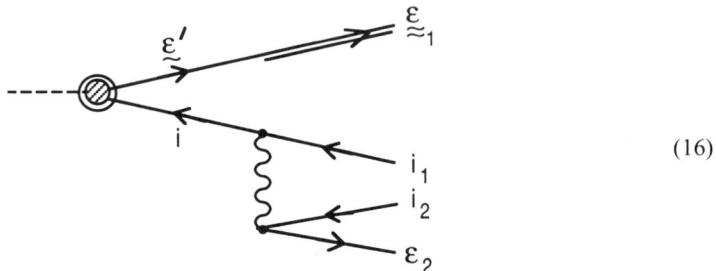

(16)

Here the doubled line starting from the moment when the inner vacancy i decays via an Auger process, denotes that the photoelectron moves in a double (i_1 and i_2) instead of a single i-vacancy field. This leads to an extra factor in the diagram (16)—an overlap integral $(\varepsilon' | \varepsilon_1)$ of wave functions of electrons, moving in the statically rearranged fields of a single vacancy i (denoted by ~) and a double vacancy $i_1 i_2$ (denoted by ≈). The doubled circle denotes the GRPAE photoionization amplitude. The simplest way to account for dynamic rearrangement is to calculate the photoionization amplitude as $(i | \tilde{D}(\omega) | \varepsilon_1)$, determining ε_1 via energy conservation low $\varepsilon_1 = \omega - I_i$.

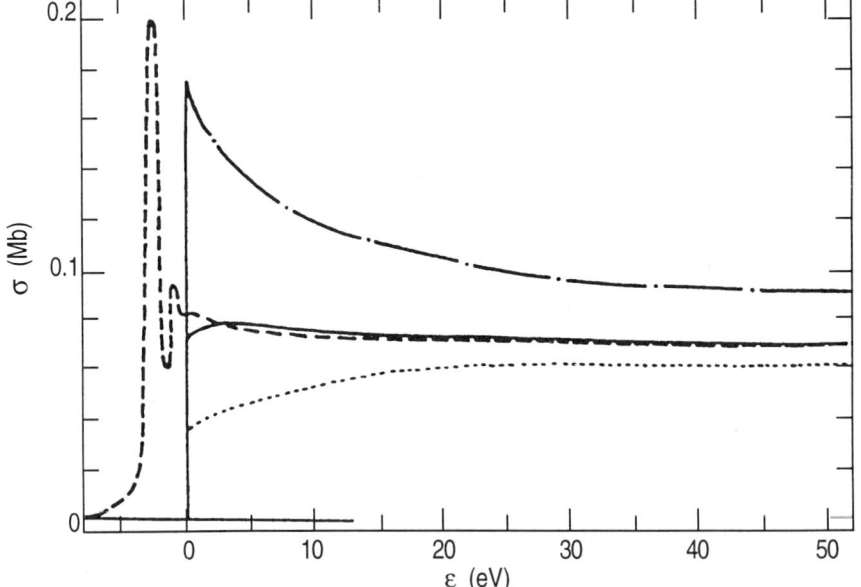

FIGURE 10. Dynamic rearrangement: photoionization cross section of $1s^2$ electrons in Ar.[23] Solid line—GRPAE II; dotted line—GRPAE; dashed-dotted line—HF, calculations; dashed line—experiment.

If the wave functions of the final-state electron ε_1 and of the intermediate RPAE states are calculated taking into account the possibility of Auger vacancy decay, the GRPAE II is formed and the corresponding amplitude is determined by (15), with the electron (vacancy) functions calculated in the self-consistent HF field of an ion having two vacancies $i_1 i_2$ instead of one vacancy i.

The role of dynamic rearrangement is illustrated in Figure 10 by the case of the Ar $1s$ shell.[23] Neither HF nor RPAE is sufficient in this case, GRPAE overestimates considerably the role of the vacancy field screening while GRPAE II (or an almost equivalent approach developed in Ref. 24 and applied to $1s$-shell in Kr) leads to good agreement with experiment. Quite unexpected is the large role which correlation plays even in the inner-shell region.

8. EXPLICIT INCLUSION OF TWO-ELECTRON–TWO-VACANCY EXCITATIONS

The corrections taken into account by GRPAE I, II are due to inclusion, as is seen from (13), (14), and (16), of "two-electron–two-vacancy" excitations. The precise inclusion of these effects is now impossible because it requires taking into account the interaction of two electrons moving in the field of an atom with vacancies, i.e., to find a solution of at least the three-body Schrödinger equation, which must then be incorporated into a rather complicated scheme, including multielectron correlations. The three-body problem itself is not yet completely solved for such a simple system as He. On the other hand, it seems that the main features of the photoionization cross section are reasonably well described in the RPAE frame and even GRPAE correlations are essential only close to the ionization threshold. Of course, there are exceptions: the photoionization cross section of La above the $4d$ threshold is entirely determined by static rearrangement. Without it, instead of a broad maximum above threshold a small monotonically decreasing cross section would be obtained. So, even the description of total photoionization cross sections require the development of a theory which would include GRPAE effects not semiphenomenologically, as was described in previous sections, but by a direct summation of the corresponding perturbation theory diagrams.

When considering characteristics more delicate than the total cross section of the photoionization process, the role of virtual excitations, other than those included in RPAE, becomes evident. The simplest way to demonstrate this is to investigate small partial cross sections and their characteristics, such as the angular anisotropy parameter β. A good example here is the $5s$ subshell of Xe, where the cross section can be understood only if "two-electron–two-vacancy" excitations are taken into account. In describing this small (about 1–4% of the total value) cross section, several effects proved to be very important. The RPAE modifies the one-electron cross section of the $5s^2$ subshell in Xe just as it does with $5s^2$ in La (see Figure 2). While near threshold the agreement with experiment proved to be quite good, the second maximum which is mainly due to the $4d^{10}$ subshell action is too large by a factor of 2.5. In order to achieve better agreement corrections out of the RPAE frame, namely those which are due to interaction with "two-electron–two-vacancy" excitations should be taken into account. Specifically, the double electron excitations of the $5p^6$ subshell are most important, and were accounted for, leading to an alteration of the direct $5s$- and $5p$-electron photoionization amplitude and to a modification of the intershell interaction of the $5p^6$ ($4d^{10}$) and $5s^2$ subshells, as well as to mixing of the pure $5s$ vacancy in the final state with more complicated configurations of the outer subshell.

Another case of obvious importance of "two-electron–two-vacancy" virtual excitations is the formation and photoionization of those negative ions whose bound states appear due to the polarization of the atom by an extra electron. Examples are Ca^- as well as other negative ions of the second group elements. To describe their photoionization, the following corrections must be taken into account:

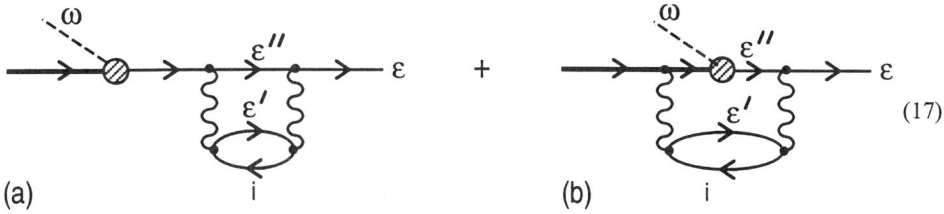

(a) (b) (17)

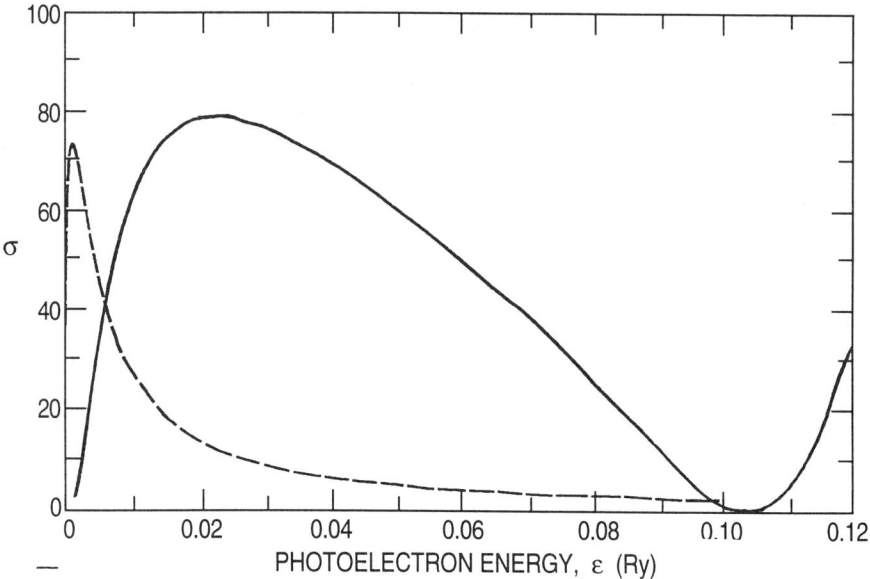

FIGURE 11. Photodetachment of Ca⁻ outer electron. Calculations[26]: solid line:—$4p^2P \to \varepsilon^2 S$; dashed line—$4p^2P \to \varepsilon^2 D$ transitions.

Here again the dashed circle stands for the RPAE amplitude, given by (9). The diagrams (17) describe the effect of atomic polarization, i.e., of virtual excitations of the $\varepsilon' i$ states on the extra electron photoionization process. The heavy line in (17) denotes that this is an outer electron bound to the atom only when its polarization by this outer electron is taken into account. It appeared that the binding of Ca⁻-type negative ions with an extra electron in an empty np shell can be achieved by accounting for the polarization of the atom in the second order of the perturbation in the interelectron interaction theory, namely by including the effects of $\varepsilon' i$ excitations of the outer ns^2 subshell.[25] It is essential that the correlation corrections presented in (17) for Ca⁻ ions are very large and therefore must be taken into account nonperturbatively by summing an infinite sequence of diagrams, which is achieved by solving an integral equation, called in quantum electrodynamics the Dyson equation. It is presented diagrammatically in the following way:

$$\varepsilon \longrightarrow = \varepsilon \longrightarrow + \varepsilon \longrightarrow \overset{\varepsilon''}{\longrightarrow} \overset{\varepsilon'''}{\longrightarrow} \quad (18)$$

In the analytical form (18) is similar to an ordinary Schrödinger equation of an electron moving in an external field. However, the "external" field in (18) created due to the polarization of the target by the particle ε is energy ε dependent and for $\varepsilon > I_i$ also has an imaginary part. This makes solving equation (18) rather complicated. To simplify this, the dependence of the polarization field on the energy ε must be neglected.

The electron affinities were calculated for all negative ions formed by the elements of the second group of the periodic table.[25] The photoionization of the Ca⁻ outer electron was also considered,[26] leading to a cross section dominated by two collective resonances, each determined by virtual excitations of $4s^2$ electrons due to their interaction with either an s or d photoelectron wave.* The

---

*The ionizing electron is at the level $n = 4$, $\ell = 1$, so the photoelectron's angular momentum is either $\ell = 0$ or $\ell = 2$.

frequency dependence of the outer electron is quite interesting, as demonstrated in Figure 11. However, in calculations, the contribution of (17b) was neglected. It appeared also that the simple approximation employed considerably overestimates the Ca⁻ electron affinity, the modern experimental value being 0.018 eV,[27] while the result of calculations is about 0.058 eV.[25] So, it is necessary to modify the polarization potential, which causes formation of A⁻ and acts on the photoelectron, as well as to take into account other corrections, described by diagrams similar to (17b). As to the overestimation of the polarization potential strength, it could be corrected at least to some extent by including the virtual excitations of inner subshells, mainly the $3p^6$ in the case of Ca⁻.

9. DIRECT KNOCK-OUT

The methods described in the previous sections consider one-electron photoionization including automatically only those multielectron transitions which result from the Auger decay of the initially created vacancy. It appeared, however, that the probability is rather high for a photoelectron on its way out of the ionizing atom to hit other atomic electrons, ionizing or exciting them. A large difference, up to 30–40%, was observed between the total photoabsorption cross section $\sigma_{tot}(\omega)$ and the sum of all one-electron partial contributions $\sigma_i(\omega)$, obtained from electron spectroscopy data, of the $4d^{10}$-subshell threshold region in Xe and Ba.[28,29] The difference at $\omega > I_{4d}$ between $\sigma_{tot}(\omega)$ and the sum of partial contributions $\sigma_{5p}(\omega)$, $\sigma_{5s}(\omega)$, $\sigma_{4d}(\omega)$ [and $\sigma_{6s}(\omega)$ for Ba], among which the largest is $\sigma_{4d}(\omega)$, is increasing fast after the inelastic scattering channel for the $4d$ photoelectron opens, i.e., at photoelectron energies higher than the excitation and ionization thresholds of Xe⁺ or Ba⁺ ions.

The simplest way to describe this process is to combine GRPAE and perturbation theory. In the lowest order in the interaction between the photoelectron and all other electrons, the inelastic collision is described by a diagram:

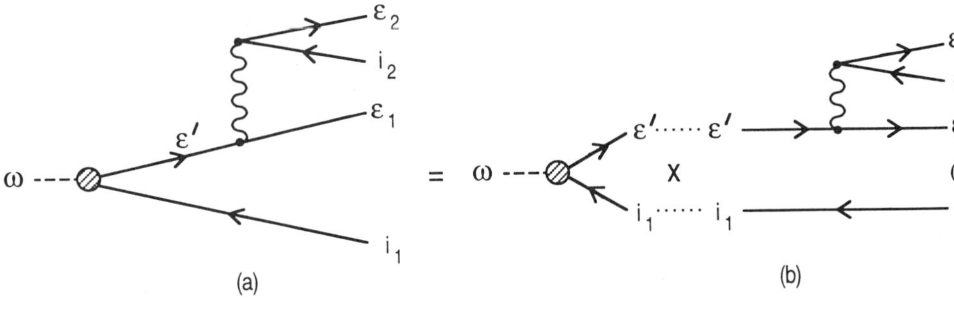

(19)

which is presented also as a product of two processes, the first being the one-electron ionization and the second the inelastic scattering of the photoelectron, created in the first step. This forms a good approach for estimating the direct knock-out amplitude and the corresponding cross section as a product of the photoionization and inelastic collision amplitudes and cross sections.

This process leads to corrections of the one-electron elimination amplitude, which comes from the next order terms in perturbation theory. The respective diagram appears as follows:

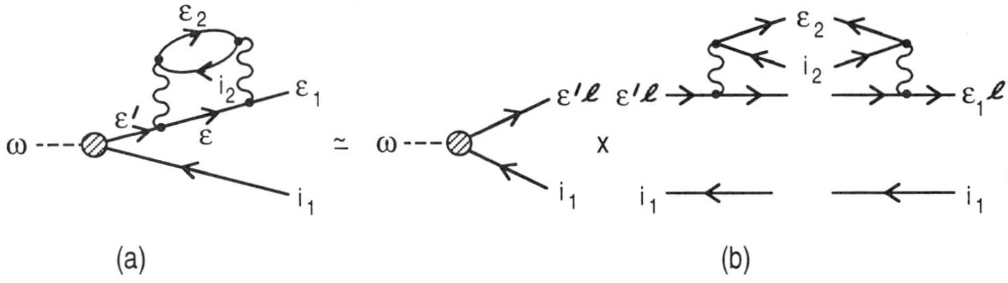

(20)

It describes the alteration of the i-electron elimination amplitude due to outer electrons polarized by the photoelectron. Iterating the polarization part of (20) an expression for the i-electron ionization cross section is derived:

$$\sigma_i^+(\omega) \approx \sum_{\ell'=\ell\pm 1} \sigma_{i\ell'}^{GRPAE}(\omega) \times \exp[-2\mathrm{Im}\,\delta_{\ell'}(\varepsilon)] \approx \sum_{\ell'=\ell\pm 1} \sigma_{i\ell'}^{GRPAE}(\omega)[1 - 2\mathrm{Im}\,\delta_{\ell'}(\varepsilon)] \quad (21)$$

where $\sigma_{i\ell'}^{GRPAE}(\omega)$ denotes the contribution to the GRPAE photoionization cross section of the i subshell with angular momentum ℓ. The last equality in (21) is valid for small $\mathrm{Im}\,\delta_{\ell'}(\varepsilon)$, where $\delta_{\ell'}(\varepsilon)$ is the elastic scattering phase shift of the photoelectron's partial wave ℓ' with the kinetic energy ε. The cross section for double electron ionization and ionization with excitation due to the photoelectron inelastic scattering is given by

$$\sigma_i^{2+}(\omega) + \sigma_i^{+*}(\omega) = \sum_{\ell'=\ell\pm 1} \sigma_{i\ell'}^{GRPAE}(\omega)\{1 - \exp[-2\mathrm{Im}\,\delta_{\ell'}(\varepsilon)]\} \approx \sum_{\ell'=\ell\pm 1} 2\sigma_{i\ell'}^{GRPAE}(\omega) \times \mathrm{Im}\,\delta_{\ell'}(\varepsilon) \quad (22)$$

Expressions (21) and (22) can, of course, be used for not only the phase shifts determined by the scattering equation (18). They are valid even for precise phase shifts which would mean taking into account in the polarization potential, i.e., in the second term on the right-hand side of (18), much more complicated diagrams, including those of RPAE type.

The direct knock-out is accompanied by some other interesting processes:

1. If the photoelectron energy is close to the electron ionization or excitation thresholds in the cross section, a Wigner–Baz singularity appears modified due to the fact that the outgoing electron is moving in a Couloub field instead of a short-range field.
2. The secondary photon emission, both direct and indirect, i.e., via other electron virtual excitations, can take place, as described by diagrams (23a) and (23b), respectively:

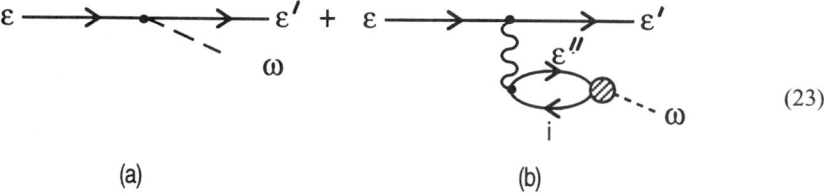

(23)

(a)   (b)

In (23) the first term is the ordinary bremsstrahlung of the photoelectron in the field of the ion. The second term presents a more complicated and more interesting process of so-called atomic bremsstrahlung, the amplitude of which is proportional to the interelectron interaction and the dynamic polarizability of the residual ion.

3. The so-called post-collision interaction (see Section 12) due to inelastic photoelectron scattering on other atomic electrons can take place when its energy and the autoionization resonance energies are close. This will modify the energy distribution of the photoelectrons. Such a process is represented by the diagram

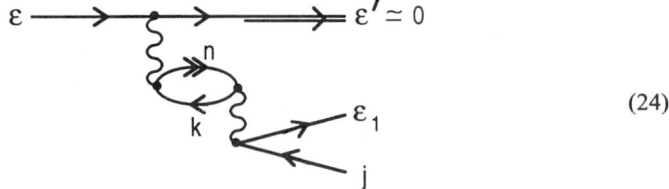

(24)

where kn is the resonance state with energy equal to $\omega_{kn} = \varepsilon_k - \varepsilon_n$ and j denotes the outer vacancy, which is created in the process of autoionization decay. The doubled arrow denotes the excited discrete level, while the doubled line emphasizes the instant change of the field acting upon the outgoing photoelectron which, after inelastic scattering, loses almost all of its energy. This alteration of the field leads to an increase of the ionization decay energy ε_1.

4. Multistep processes can become important, which is exemplified by the Mn atom, where the photoelectron from an inner shell can excite the most intensive discrete level $3p \to 3d$ decaying subsequently into semifilled shell excitation $3d \to \varepsilon f, \varepsilon p$, and only after that to other channels:

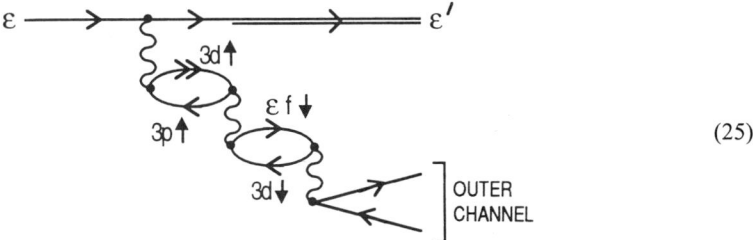

(25)

The direct knock-out affects not only the partial cross sections, but also the angular distributions and spin orientation of the outgoing electron. This process must be important in isolated atoms and in multiatomic formations, particularly in cases where the target system has many loosely bound electrons. This process is of interest for negative ions, especially if there are resonances in the corresponding electron–atom scattering amplitude, which may considerably enhance or suppress the direct knock-out probability.

Until now concrete calculations were performed for single electron photoionization of Xe and Ba atoms, in the vicinity of their $4d^{10}$ thresholds.[30] The amplitudes (20) without their factorization, which lead to (21) and (22), were used. The results along with the experimental data[28,29] are presented in Figures 12 and 13 for Xe and Ba, respectively. Quite good agreement is achieved. Calculations of σ^+ for these atoms, which include effects of rearrangement but neglect the direct knock-out, were performed in Ref. 31, leaving a noticeable difference between theory and experiment.

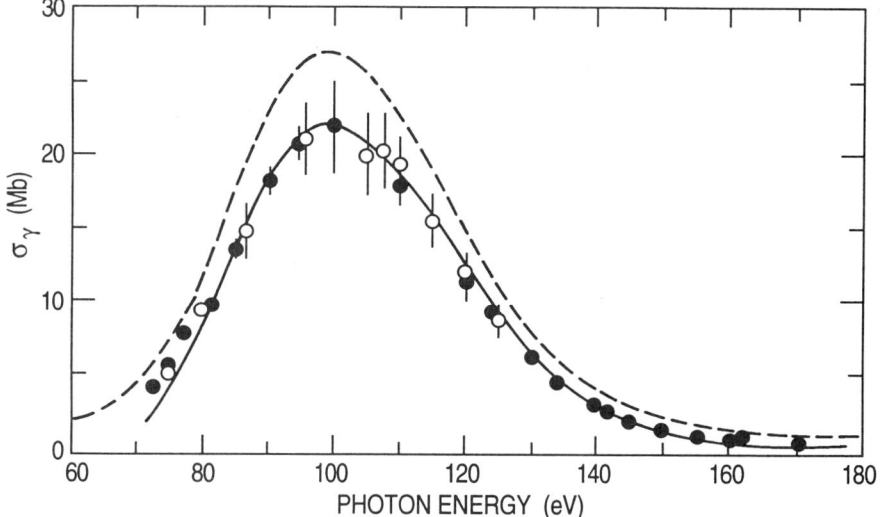

FIGURE 12. Single-electron photoionization cross section of the $4d^{10}$ subshell in Xe. Dashed line—RPAE; solid line—GRPAE with direct knock-out[30]; open circles—Ref. 28; black circles—Ref. 29, experiment.

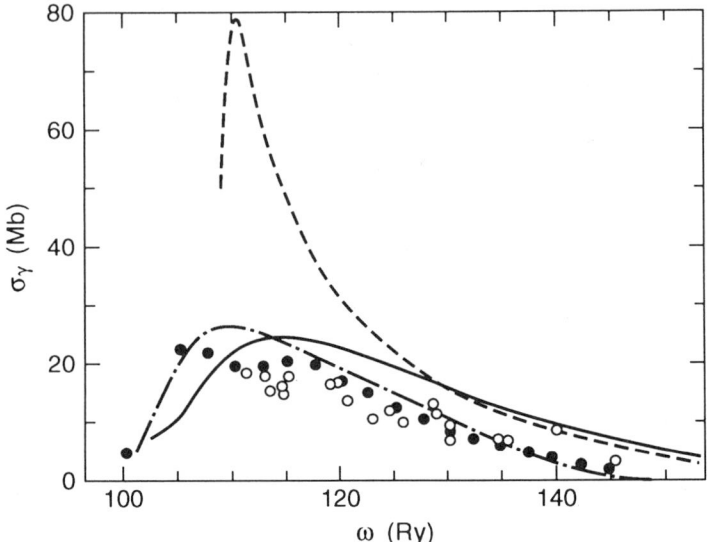

FIGURE 13. Single-electron photoionization cross section of the $4d^{10}$ subshell in Ba. Dashed line—RPAE; dashed-dotted line—GRPAE with direct knock-out; solid line—GRPAE only, calculations; circles—see Ref. 32, experiment.

10. TWO-ELECTRON EXCITATIONS AND IONIZATION

In this section we will discuss the photoionization process, in which one electron is leaving the atom or multiatomic formation, while the other remains at an excited level. As a result, the target is left with a vacancy and an extra excitation. Such state is called a satellite (to the vacancy itself) or a shadow. To clarify their difference, let us consider the process of their creation well above the ionization threshold. In this energy region, the interaction between the photoelectron and the rest of the atom can be neglected and satellites are defined as those states which appear as a consequence of instant creation of the vacancy acting on the electron shells. This mechanism is called "shakeup." "Shadows" are created via quasi-Auger decay, i.e., a transition in which the vacancy does not have enough energy for a real Auger decay. The simplest diagram of a "shakeup" is (26a) while that of quasi-Auger is (26b):

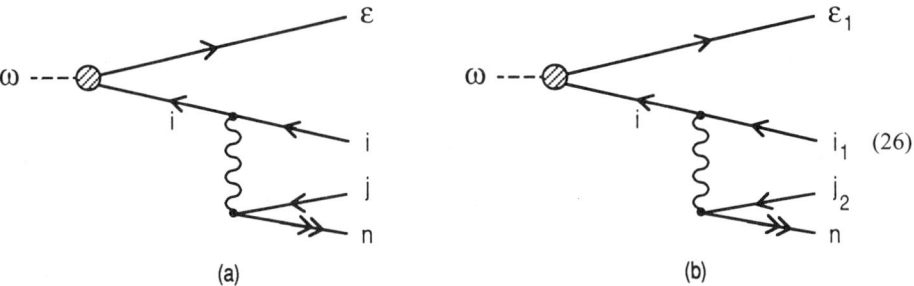

(26)

In both cases the final state is reached via the i-vacancy formation stage. Although (26) presents the lowest-order perturbative description of these processes (formation of the vacancy i is described in RPAE or GRPAE), the probability of exciting satellite (shadow) states, with n being a discrete level, is often large. So in principle they should be accounted for nonperturbatively.

At least far above threshold the cross section of a photoionization process which leaves the ion in a satellite state is directly connected to the cross section of the vacancy formation. Quite a few examples of such cross sections were calculated. A crude estimate can be given by the following expressions:

$$\sigma_{i,nj}(\omega) \approx \sigma_i(\omega - \omega_{nj}) \frac{|(in|u|ij)|^2}{\omega_{nj}^2} \tag{27}$$

for the "shakeup" process and

$$\sigma_{i,ni_1 i_2}(\omega) \approx \sigma_i(\omega - \Delta_{i,ni_1 n_2}) \frac{|(in|u|i_1 i_2)|^2}{\Delta_{i,ni_1 i_2}^2} \tag{28}$$

for the quasi-Auger. In (27) the energy ω_{nj} is equal to $\varepsilon_n - \varepsilon_i$, while $\Delta_{i,ni_1 i_2}$ denotes the energy deficiency for an Auger decay of vacancy i into a state $ni_1 i_2$ to be real: $\Delta_{i,ni_1 i_2} = -\varepsilon_i - \varepsilon_n + \varepsilon_{i_1} + \varepsilon_{i_2} > 0$. It is seen from (27) and (28) that the perturbative approach is applicable if the factors standing after the cross section namely, $|(in|u|ij)|^2/\omega_{nj}^2$ and $|(in|u|i_1 i_2)|^2/\Delta_{i,ni_1 i_2}^2$, are much smaller than 1. However, (27) and (28) are more general than the amplitudes (26) used in their derivation. They can be corrected by substituting the lowest-order perturbative expression for the "shakeup" and Auger-decay probabilities, as well as energy deficiency $\Delta_{i,ni_1 i_2}$ by their precise values.

If some of the corrections (26) to the photoionization amplitude are large, they must be included nonperturbatively, by iterating the insertion into the vacancy line:

$$\text{(diagrammatic equation)} \tag{29}$$

This summation in (29) can be achieved by solving an integral equation, similar to (18) with the straight electron line substituted by that of a vacancy.

Concrete calculations for the ionization process leaving the atom with a satellite state were performed in few cases. The most interesting among them is the group of states "$\widetilde{3s^{-1}}$" = $3p^{-2} 3d(4d,5d,\ldots)^2 S$ being shadows of the $3s^{-1}$ main state. Because the direct formation of this shadow is most improbable, the process goes via the excitation of the main state. In the near-threshold region the $3s^{-1}$ formation is completely determined by the intershell interaction with $3p^6$ electrons, by their resonating action on $3s^2$, and are quite well described in RPAE by equation (9).[3] The same holds for "$\widetilde{3s^{-1}}$" levels.[3,33,34] For high ω the ratio $\sigma_{shad}(\omega)/\sigma_{main}(\omega)$ becomes almost independent of ω.[35] While in Refs. 33,34 the numerical calculations were performed, a simple analytical formula, which describes an "integrated" cross section of formation of all "shadows" $\sigma_{\tilde{i}}$ in a photoionization process, was derived in Ref. 3:

$$\sigma_{\tilde{i}}(\varepsilon + I_{\tilde{i}})/\sigma_i(\varepsilon + I_i) = (1 - Z_i)|(\varepsilon|D(\varepsilon + I_{\tilde{i}})|i)/(\varepsilon|D(\varepsilon + I_i)|i)|^2 \tag{30}$$

where Z_i is the relative intensity of the shadows at $\omega \to \infty$. The notation $\sim$ in (30) marks the shadow states. In order to derive this expression, the reasonable assumption was made that the direct formation of a shadow state, without having the main line created at an intermediate step, as in (26), is impossible.

At low ω different states and kinds of satellites can be excited. The probability that an outgoing photoelectron will collide with another electron exciting it to a discrete level is rather high if the photoelectron is not too fast. Thus, a satellite $ni_2 i_1$ to the vacancy i_1 can be formed. The simplest picture of this process is represented by a diagram:

$$\text{(diagram)} \tag{31}$$

It describes either a two-step process, namely, creation of an "electron ε'–vacancy i_1" pair with a subsequent photoelectron inelastic collision leading to $(\varepsilon_1\ell,ni_2)$ state, or a one-step process, in which the intermediate state $\varepsilon'\ell'$ is virtual. Higher-order corrections to the diagram (31) may be important, specifically if the excited i_2 electron is from a multielectron shell.

The mechanism (31) is a specific case of that given by (19). The satellite or shadow states, excited via the mechanism represented by diagram (31), can be distinguished from that of (26) only via the frequency dependence of their intensity: the contributions coming from (31) lead to relative intensities "satellite-to-main line" decreasing with the growth of ω. Indeed, the probability of any inelastic electron–atom or electron–ion collision as a function of impact energy E is increasing from the threshold of the inelastic channel I_{ex}, reaches a maximum at $E \approx 3I_{ex}$, and then starts to decrease as E^{-1}.

In the inelastic collision of the electron ε', the angular momentum and spin projection can change, thus leading to a modification of the angular distribution and polarization of the outgoing electron ε. Such a mechanism as (31), contrary to (26a), creates not just satellites, which are monopole excitations. While not proceeding via an i-vacancy state, the process presented in (31) leads to another kind of satellite, which one could call a direct collisional or simply a collisional satellite. Its total momentum can be different from that of the vacancy i.

What was said can be illustrated by the example of Ca. Consider $i_1 = 3s$, $i_2 = 4s$, and the first excitation level $n = 4p$. Then the outgoing electron can have angular momentum either $\ell = 2$ or $\ell = 0$. This leads to an energy-dependent angular anisotropy parameter given by (7), while without excitation of the state $4s \rightarrow 4p$ the outgoing electron would be described by only one partial wave $\ell = 1$ and the parameter $\beta_{3s}(\omega)$ would be frequency independent and equal to 2. The cross section of formation of this collisional satellite is determined by the intershell interaction between $3s^2$ and $3p^6$ electrons. Another satellite can be formed in the same atom by ionization of the $3p$ and excitation of the $4s$ electrons to the $n = 4p$ level. The angular momentum of the intermediate state ℓ' is either 0 or 2 and the photoelectron leaves the atom with momentum $\ell = 3$ or $\ell = 1$. Therefore, the anisotropy parameter for the satellite is determined by the elastic scattering phase shift $(\delta_3 - \delta_1)$ instead of $(\delta_2 - \delta_0)$ for the main line.

The probability of excitation of a satellite ni_2i_1, described by the amplitude (31), can be strongly enhanced if the process goes via a "doorway" state, i.e., via a powerful discrete excitation $n''i_1$ instead of that of a continuous spectrum $\varepsilon'i$. Note that in (31) the dashed circle includes only RPAE (GRPAE) states, i.e., "one-electron–one-vacancy" excitations.

Together with the outgoing photoelectron the satellite or shadow forms a "two-electron–two-vacancy" state. Due to interelectron interaction such states also exist in the ground-state wave function of the target. The mechanisms given by (26) and (31) describe formation of the "two-electron–two-vacancy" state after a photon is absorbed by the target. It can be formed also when a photon is absorbed by a virtual state of this kind which exists before a photon absorption takes place. This is illustrated by the diagram:

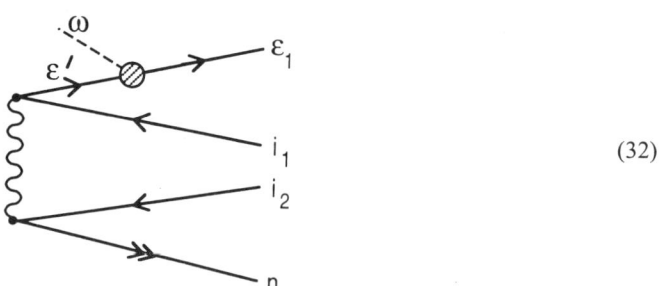

(32)

The probabilities of excitation of satellites and shadows by these mechanisms depend very critically on the photon frequency, having no direct connection with the partial photoionization cross section $\sigma_{i_1}(\omega)$.

11. FORMATION OF MULTIPLY CHARGED IONS

The role of interelectron interaction and the electron structure of the atom or multiatomic formation is transparently demonstrated in creation of multiply charged ions. The simplest way to form these ions is to proceed step by step via Auger processes. Of course, each state created via an allowed Auger decay is dominant. But it is almost self-evident that not all possible degrees of ionization can be reached in such a way. On the contrary, there are photon frequency regions and corresponding ionization degrees which can be explained theoretically only if other mechanisms of multiple ionization, namely, shakeoff, quasi-Auger decay, and direct knock-out by the photoelectron, are taken into account.

The multiply charged ion formation was recently investigated for a number of atoms in very broad frequency regions.[36,37] The main features of corresponding cross sections as functions of frequency resemble those of the total photoionization cross section $\sigma_{tot}(\omega)$, having large maxima of $\sigma_{tot}(\omega)$ reflected in almost all cross sections $\sigma^{n+}(\omega)$ of the A^{n+} ion formation. This demonstrates that the dominant mechanism of A^{n+} creation is that via an inner vacancy with its subsequent Auger decay and shakeoff processes. A noticeable similarity was observed between $\sigma^{n+}(\omega)$ and corresponding partial photoionization cross section $\sigma_i(\omega)$.[38] The reasons for this similarity are not completely understood. But in noble gases the RPAE partial photoionization cross section of ns electrons $\sigma_{ns}(\omega)$ calculated with HF thresholds almost coincides above the double electron threshold with the double electron photoionization cross section σ^{2+} of the outer np^6 electrons. It is unclear whether such a coincidence exists for other systems.

Some qualitative features of the A^{n+} formation process are well understood. An example is the transition from dominating A^{2+} formation via the Auger effect to A^+ via the autoionization on the way from Xe to Eu in the $4d$-shell frequency region. While in Xe the photoionization of the $4d$ electron dominates, in Eu the discrete transition $4d \rightarrow 4d$ with subsequent autoionization of one of the seven $4d$ electrons is most important.

Of great interest for understanding the electronic structure of atoms and multiatomic formations is the non-Auger doubly charged ion formation. The few existing calculations describe this process in the lowest order of perturbation theory in the interelectron interaction.[39] The corresponding four lowest-order diagrams include two that are due to the interelectron interaction in the final state and are similar to (19a) and (26) with discrete excitations n substituted by a continuous state with energy ε_2. The other two are the contributions of the initial state interaction, represented by (32), with the continuum state ε_2 instead of a discrete level n, and other similar diagrams an example of which is given by

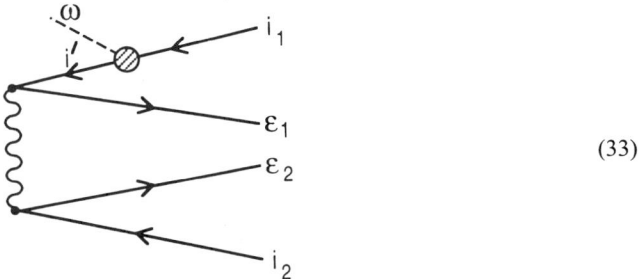

(33)

Contrary to (32), here the photon interacts with a vacancy i' instead of an electron ε'. However, these four diagrams are in many cases insufficient. One example is the near double ionization threshold region, where the interaction of the two outgoing electrons must be taken into account nonperturbatively, as is described by an infinite sequence of diagrams:

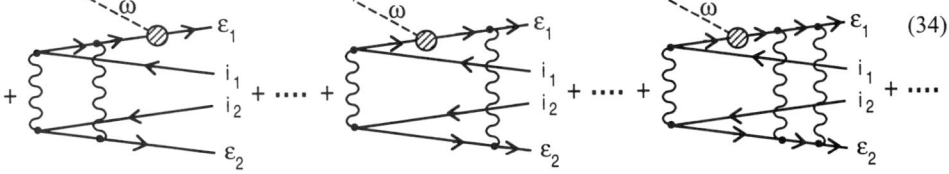

(34)

The summation of (34) corresponds to a solution of a three-body problem, namely, the description of two interacting electrons moving in the self-consistent field of an atom with two vacancies—i_1 and i_2.

The dynamic screening of the interelectron Coulomb interaction due to a virtual excitation of other atomic electrons is of importance, especially when the transferred electron energy [$\varepsilon_2 - \varepsilon_{i_2}$ in (33)] is close to one of the outer or intermediate multielectron shell thresholds. In the RPAE frame this screening is achieved by substituting the Coulomb interaction with the effective one $\Gamma(\omega)$, which is determined by an integral equation, diagrammatically represented in the following way:

$$\text{(35)}$$

Here ω is the energy transferred via the effective interaction $\Gamma(\omega)$, and the "electron ε'–vacancy i'" pair represents the excitations of all atomic electrons.

The "doorway" states, i.e., the resonances in the "one-electron–one-vacancy" channel, lead to an enhancement of the double electron photoionization probability. However, there are not enough theoretical or experimental data on double ionization of complex multielectron atoms, where the "doorway" can have resonances and virtual excitations of the core are also important.

The angular and energy distributions of both outgoing electrons are particularly interesting, being strongly affected by the interelectron interaction. Near an inner-shell ionization threshold the doubly charged ion production is determined to a large extent by postcollision interaction effects, which will be discussed at length in the next section.

Although of not perfect quality, calculations for double electron ionization, mainly for the He atom, exist in the literature. There are no calculations at all for A^{n+} production for $n \geq 3$.

Recently, it was demonstrated experimentally[40] that in a number of cases a large contribution to A^{2+} comes from decays of discrete states in complicated processes, an example of which for Ar is given by the following diagram:

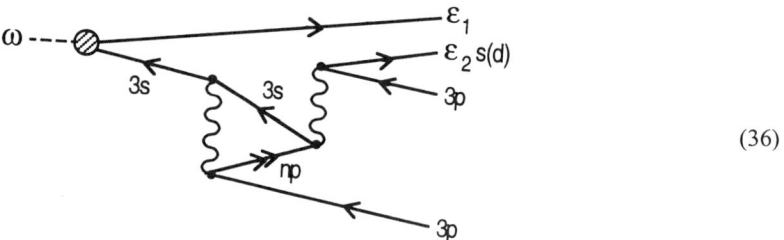

$$\text{(36)}$$

Here, the double-electron photoionization proceeds in two successive steps which can be presented at least by a second-order diagram only, the first step being the formation of a satellite $np3s3p$, while the second is an autoionization decay of a discrete excitation $np3s$ into the continuous spectrum of the $\varepsilon_2 s(d)$ states. The satellite is a resonance, strongly modifying the photoelectron energy distribution in the edge part of the spectrum, i.e., where the ε_2 electron is slow. For Ar about 30–40% of A^{2+} are produced via resonance decay of satellites.[40]

While in He the ratio $\eta \equiv \sigma^{2+}(\omega)/\sigma^+(\omega)$ is no more than several percent, it is much larger for more complex atoms, reaching about unity near the $4d^{10}$-subshell ionization threshold in Xe. There the cross section $\sigma^{2+}(\omega)$ is very strongly affected by virtual excitations of this subshell.

The ratio η can be more than one when $\sigma^+(\omega)$ is very small, for example near Cooper minima. η is large for a system with one loosely bound electron, such as negative ions, highly excited atoms, and similar states in multiatomic formations. Being loosely bound, this electron is easily excited or eliminated via the shakeoff mechanism when another electron is ionized and a vacancy is formed, which is diagrammatically represented by (26a), with n substituted by a continuous spectrum one-electron state ε_2. The smaller the outer electron binding energy, the higher is its probability of being eliminated from the target. Indeed, using (26a) one has for the ratio $(d\sigma_i^{2+}(\omega)/d\varepsilon_2)/\sigma_i^+(\omega)$ at $\varepsilon_2 = 0$:

$$\frac{d\sigma_i^{2+}(\omega)}{d\varepsilon_2}\bigg|_{\varepsilon_2=0} /\sigma_i^+(\omega) \approx |(n\ell|r^{-1}|\varepsilon_2 0)/(\varepsilon_2 - \varepsilon_{n\ell})|^2\bigg|_{\varepsilon_2=0} \tag{37}$$

Here $n\ell$ are the quantum numbers of the excited electron which is shaken off after the vacancy i is created. For high excitations ($n \gg 1$), (37) may be estimated using pure hydrogenic energies $\varepsilon_{n\ell}$ and wave functions $\varphi_{n\ell}(r)$ and $\varphi_{\varepsilon_2\ell}(r)$, demonstrating an unlimited increase with n growth. It means that the double-electron ionization can become dominant over the single-electron one.

Very interesting and comparatively simple for theoretical interpretation are the $\sigma^+(\omega)$ cross sections. All maxima and minima in them are due to intershell interaction. For instance, $\sigma^+(\omega)$ in Xe has a maximum just above the $4d^{10}$-subshell ionization threshold which is entirely due to the very strong action of the giant resonance on the outer $5p^6$ and $5s^2$ electrons.

It is rather difficult to calculate $\sigma^{n+}(\omega)$ for $n \geq 3$ as a function of ω in a very broad region: it requires consideration of different Auger-decay chains, numerous shakeoff, direct knock-out, and ground-state interaction mechanisms. However, recently the mean charge of the Ar ions produced in photoionization $N(\omega)$ was analyzed and it was demonstrated[37] that this quantity as a function of ω is increasing while ω is approaching the 1s shell ionization potential I_{1s}. It appeared that in a rather broad region of ω, about 30–40 eV below the 1s threshold, the mean charge of the ions produced increases from about 3.1 to 3.8. This increase leaves no doubt that it is connected with the virtual excitation of the inner 1s shell. In the lowest order in interelectron interaction there are only two diagrams, which include virtual formation of the 1s vacancy:

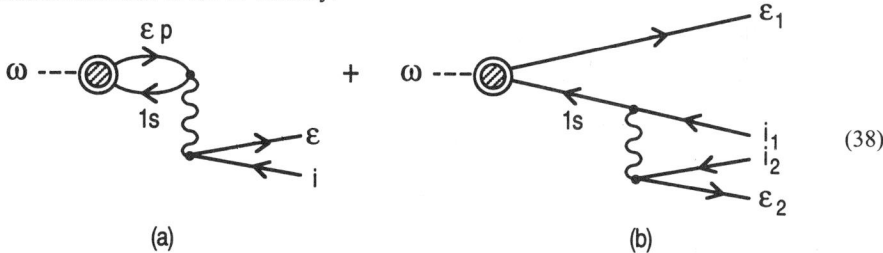

(38)

However, the mechanism (38a) leads at most to 2s vacancy formation which after Coster–Kronig and Auger decay leads to Ar^{3+} and can only increase the Ar^{3+} yield, but cannot lead to $N(\omega) > 3$. It means that the contribution (38b) becomes decisively important.[40] This mechanism leads to production of $2s^{-2}$, $2p^{-1} 2s^{-1}$ states and therefore to Ar^{4+}–Ar^{6+} ions with reasonably high probability. Diagram (38b) includes virtual (before its threshold) creation of the 1s vacancy with its subsequent decay into $i_1 i_2 \varepsilon_1$ state and can be called Auger decay of an intermediate state. The contribution of (38b) is easy to calculate and it has a rather simple form assuming that the dependences of the photoionization amplitude $(1s|D(\omega)|\varepsilon_1 p)$ on energy ε_1 and the Coulomb matrix element $(1s\varepsilon_2|u|i_1 i_2)$ on ε_2 are very weak. As a result, one has[41]

$$\sigma_{``1s"}(\omega) = \sigma_{1s}(I_{1s})\left[\frac{1}{2} + \frac{1}{\pi}\arctan\frac{\omega - I_{1s}}{\Gamma_{1s}/2}\right] \tag{39}$$

where $\sigma_{1s}(I_{1s})$ is the 1s-shell photoionization cross section at threshold. Well below threshold, at $(I_{1s} - \omega) \gg \Gamma_{1s}/2$, equation (39) reduces to

$$\sigma_{"1s"}(\omega) \cong \frac{\sigma_{1s}(I_{1s})}{2\pi} \frac{\Gamma_{1s}}{I_{1s} - \omega} \tag{40}$$

Equations (39) and (40) are valid not only for 1s but for any subshell or shell in which the photoionization cross section slowly decreases above its threshold. To obtain these more general expressions, one should substitute index 1s by i, which denotes some other shells also.

The mean ion charge is determined by the relation

$$N(\omega) \cong \frac{\left(\sum_i n_i \sigma_i(\omega)\right)}{\left(\sum_i \sigma_i(\omega)\right)} \tag{41}$$

where n_i is the ion charge formed after i-subshell ionization. Then a formula can be derived for the mean ion charge variation $\Delta N(\xi)$ in the vicinity of a threshold:

$$\Delta N(\xi) \equiv N(\xi) - N_- = \Delta N \frac{\beta}{\xi + \beta} \tag{42}$$

Here $\xi = 2(I_i - \omega)/\Gamma_i$ and $\beta = \sigma_i(I_i)/\pi\sigma_{i+1}(I_i)$, $\sigma_i(\omega)$ is the photoionization cross section of the i shell, while the $i+1$ is the outer shell as compared to the ith one. In (42) N_- is the average ion charge well below the ith threshold, $\Delta N = N_+ - N_-$, N_+ being the ion charge far above threshold. It is implied in (42) that ϵ varies from $\xi = \infty$ to $\xi = 0$. The results of calculations using (42) and the experimental data are presented in Figure 14, demonstrating satisfactory agreement. The following experimental parameters were used: $I_{1s} = 3203.5$ eV, $N_- = 3.07$, $\Gamma_{1s} = 0.69$ eV, $\sigma_{1s}(I_{1s})/\sigma_{2s}(I_{1s}) = 14$.[(42)]

A relation similar to (42) can be derived for the cross section $\sigma^{n+}(\omega)$:

$$\sigma^{n+}(\omega) \approx \frac{1}{2\pi} \frac{\Gamma_i}{I_i - \omega} \sigma_i^{n+}(I_i), \quad \omega < I_i \tag{43}$$

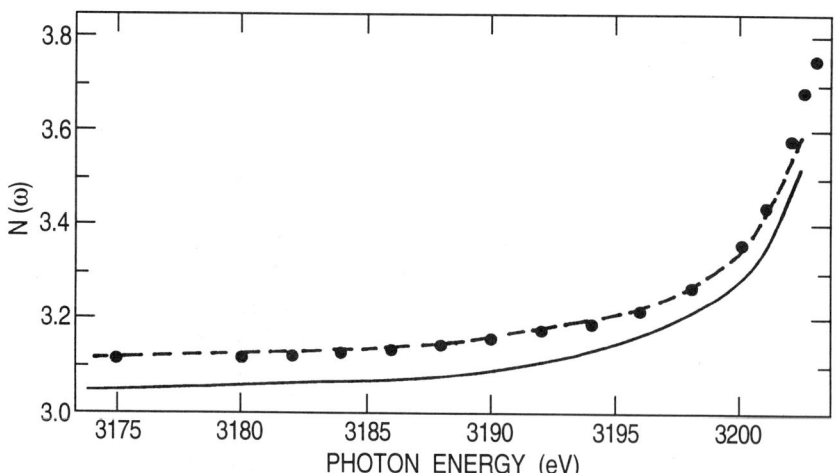

FIGURE 14. The average charge N as a function of photon frequency in Ar. Solid points—experiment, from Ref. 37. Calculations: solid line—with experimental parameters for the 1s ionization potential $I_{1s} = 3203.5$ eV, Auger width $\Gamma_{1s} = 0.69$ eV, and the cross-section ratio $\sigma_{1s}(I_{1s})/\sigma_{2s}(I_{1s}) = 14$;[(42)] dashed line—the same, with $N(\infty) = 3.07$.

Theory of Photoionization

Here $\sigma^{n+}(\omega)$ is the total cross section of A^{n+} formation and $\sigma_i^{n+}(I_i)$ is the threshold value of the A^{n+} production at $\omega = I_i$. Note that deriving (42) and (43), the series of excited states, the so-called Rydberg series, was neglected. Its manifestation will be discussed in the next section.

From the parameters used in describing $N(\omega)$ and $\sigma^{n+}(\omega)$, we see the kind of information that can be derived from ion yield studies: different partial cross sections and Auger widths, as well as the mean ion charges far from the subshell and shell thresholds.

12. NEAR- AND BELOW-THRESHOLD PROCESSES

If the photoionization process takes place close to its threshold, the corresponding cross section is strongly modified by the mutual Coulomb repulsion of the outgoing particles. If sufficiently slow, they interact for quite a long time. Therefore, the repulsion is able to force them to move in opposite directions. They must also move with equal velocities, otherwise the faster will overcome the slower leaving it in a stronger field of the residual vacancies. This slower electron will then be recaptured by the ion without participating in the multiply charged ion formation. So, to leave the atom the electrons must have equal velocities and total zero momentum. For two-electron ionization this classical picture was developed long ago by Wannier,[43] its more modern and to some extent quantum-mechanical versions being presented in Ref. 44. The requirement of equal velocities or that of equal separation from the ion's center while leaving the atom decreases the final-state phase space, leading to a smaller cross section[43]:

$$\sigma^{2+}(\omega) = \alpha(\omega - I_2)^{1.052} \quad (44)$$

where I_2 is the two-electron ionization potential. The classical picture given above corresponds to summation of an infinite series (34). The extra 0.052 comes only from accounting for strong mutual electron repulsion.

It is not completely clear, however, what the range of validity of equation (44) is. But it can be distinguished from the usual Wigner threshold law $\sigma^{2+}(\omega) \sim (\omega - I_2)$ which neglects the long-range interaction between outgoing electrons only very close to the threshold, mainly at tenths or even hundredths of an electron volt above threshold.

For atoms more complex than He, the double ionization is affected by the multielectron structure even in the threshold region. For multiatomic formations, where there are loosely bound electrons and the system is easily polarizable, the long-range Coulomb repulsion between the outgoing particles and the attraction between them and the ion can be screened and the cross section of the ionization of two or more particles is modified.

In the one-electron photoionization cross sections of multiatomic formations, this screening can eliminate the "jumps" at thresholds, which are typical for isolated atoms. For slow photoelectrons, the inelastic process, such as photon emission (23a) or low-energy electron–vacancy excitation, (19a) can considerably and even qualitatively modify the near-threshold behavior of the cross section. Instead of having a "jump," it starts to increase fast in accord with the Wigner threshold law, reaching the one-electron cross section values only at some photon energy above threshold. A width Γ_p of a photoelectron state can be introduced, which is acquired due to any inelastic processes and determines the frequency region over which the "jump" in the cross section is smeared out. If in photoionization a vacancy is created which decays radiatively or via Auger process, its width Γ_v determines the frequency region where the cross section smoothly increases from zero to its one-electron value. Usually $\Gamma_v \gg \Gamma_p$, but for the outer subshells in atoms or outer electrons in multiatomic formations, $\Gamma_v = 0$ and the near-threshold behavior of the one-electron cross section is determined by Γ_p, which is very small.

If the doubly charged ions A^{2+} are formed near the inner or intermediate shell threshold, the spectrum of emitted electrons is considerably modified due to Auger vacancy decay, which leads to an increase of the force acting on the slow photoelectron. This effect is called postcollision interaction (PCI) and its first quantum-mechanical description was given in Ref. 45. Diagrammatically, PCI is represented by (16). The instant change of the field acting on the slow electron modifies the spectrum of both outgoing particles. The theoretical investigation of PCI is essentially simplified if the Auger electron is fast enough, so one can neglect its interaction with the other electron, which is slow. The energy loss of the slow-electron $\Delta\varepsilon_i$ is transferred to the Auger electron, whose energy is shifted to the higher side of the spectrum by the magnitude $\Delta\varepsilon_i$, which can be estimated as

$$\Delta\varepsilon_i \approx \Gamma_i^{(A)}/v_1 \tag{45}$$

v_1 being the slow photoelectron's speed. In (45) $\Gamma_i^{(A)}$ is the Auger width of the vacancy i. The upper value of $\Delta\varepsilon_i$ can be estimated by the order of magnitude as $\Delta\varepsilon_i^{max} \approx (\Gamma_i^{(A)})^{1/2}$ which is obtained from (45), substituting v_1 by $v_{1_{min}} = (2\varepsilon_{1_{min}})^{1/2} \sim (\Gamma_i^{(A)})^{1/2}$.

It is easy to take into account the interaction of two outgoing electrons ε_1 and ε_2 with velocities $\vec{v}_1$ and $\vec{v}_2$ semiclassically, having in mind that the distance between them after the period of decay $(\Gamma_i^{(A)})^{-1}$ is $|\vec{v}_2 - \vec{v}_1|/\Gamma_i^{(A)}$. This modifies the expression (45) leading to

$$\Delta\tilde{\varepsilon}_i \approx \Gamma_i^{(A)} \left(\frac{1}{v_1} - \frac{1}{|\vec{v}_2 - \vec{v}_1|} \right) \tag{46}$$

It is seen from (46), according to what was demonstrated in Ref. 46, that the Auger line energy shift is dependent on the angle between $\vec{v}_2$ and $\vec{v}_1$ and reduces to (45) only at very high velocities $v_2 \gg v_1$.

Particularly large is $\Delta\varepsilon_i$ for rapidly decaying vacancies, although the relative value $\Delta\varepsilon_i/\Gamma_i$ decreases with Γ_i growth. Giant are the PCI effects for the vacancies or autoionization resonances having extraordinarily large widths, as for example the $4p$ vacancy in Xe or $4d \rightarrow 4f$ resonance in Eu. In the first case the width is as large as 20 eV, while in the second it is about 10 eV. For such values, the Auger line shifts are considerable even for comparatively rather high photoelectron energies.

If a vacancy is created in an inner shell, its Auger decay proceeds in several steps, which is illustrated by the following diagram:

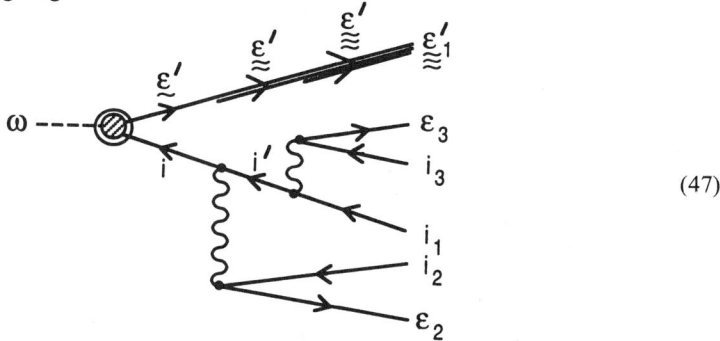

(47)

Here the photoelectron ε', which was at the beginning in the field of the vacancy i, after the first Auger decay finds itself in the field of the vacancies i_1 and i_2 and then after the second decay in the field of $i_1 i_2 i_3$. It is very complicated to describe this process quantum mechanically, but it is comparatively easy to derive a classical formula for the total energy loss of the photoelectron in PCI with a two-step Auger decay:

$$\Delta\varepsilon_i \approx -\frac{1}{v_1} \left(\Gamma_i + \sum_j \frac{\Gamma_{2j}}{\Gamma_2} \frac{\Gamma_i \Gamma_{2j}}{\Gamma_i + \Gamma_{2j}} \right) \tag{48}$$

Here $\Gamma_2 = \Sigma_j \Gamma_{2j}$, and Γ_2 is the total second-step Auger-decay width, while j denotes each partial channel.

If the energy loss of the slow electron $(-\Delta\varepsilon_i)$, given by (45), (46), or (48), is larger than its kinetic energy ε_i, it will be captured to a discrete level, the entire process leading to ionization with excitation instead of a pure multiple ionization. However, a considerable fraction of these highly excited states in complex atoms and multiatomic formations will not decay via photon emission, but via autoionization, i.e., by emitting electrons. A concrete example of such a process is given in the following diagram, which presents a result of a PCI near the 2p-subshell threshold in Ar:

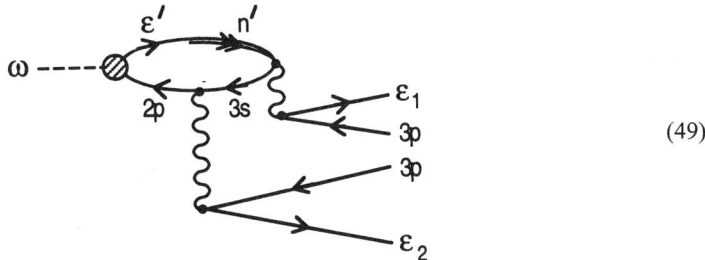
(49)

In this process a discrete state n' and a subvalent vacancy $3s$ decay by creating a valent vacancy $3p$ and a continuous spectrum electron ε_1. There is a possibility that discrete excitations formed due to PCI are able to decay by emitting an electron while the total spin of two vacancies changes from $S = 0$ to $S = 1$.[40]

For frequencies below the inner-shell threshold as a first step of the process an electron is excited to one of the discrete levels, which is followed by the vacancy decay or something more complicated. The excited electron can be either a passive spectator or an active participant of the subsequent process. Very often it is an active participant which is illustrated by the decay of the $2p^{-1}4d$ state of Ar. It was demonstrated that the presence of a $4d$ excited electron suppresses the $2p$-vacancy Auger decay.[47] This process is represented by the diagram:

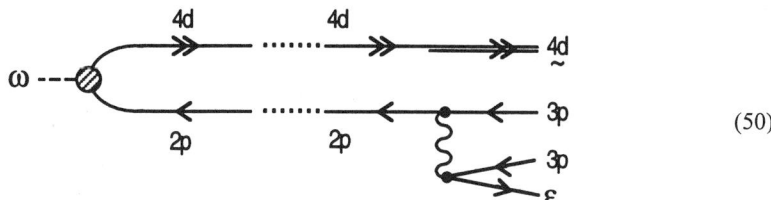
(50)

At first glance it is a surprise that a highly excited electron can affect the inner vacancy decay. However, the $4d$ electron, being loosely bound, can easily change its state due to the alteration of the field acting on this electron after Auger decay. Indeed, it appeared that the overlap integral $(4d|\widetilde{4d})$ is nearly zero and most probable is the decay of the $2p$ vacancy accompanied by the $4d$ electron transitions to higher excited levels—$5d$, $6d$, and some others. Thus, the total decay probability is not altered, while the intensities of its particular channels are redistributed.

Due to PCI the slow electron loses its energy, becoming slower or even being recaptured into the high discrete levels. On the contrary, in a process similar to (50), the initially excited electron prefers to acquire energy and go to higher levels.

Not only the Auger electron energy, but also its angular momentum and spin can be affected by PCI. Because the momentum and spin of the initially created vacancy i in (16) is in general different from the corresponding total momentum and spin values of the two vacancies i_1 and i_2 created in the Auger decay. Therefore the slow photoelectron finds itself instantly in a field with another radial dependence, angular momentum, and spin. This can affect the angular distribution and spin orientation of the slow electron. If the velocity of the fast electron is not too large, its interaction with the slow one is important, leading to energy redistribution between two outgoing particles which is demonstrated by (46). This same interaction makes it possible to exchange angular momentum and spin between the outgoing particles.

While PCI due to emission of an electron was and is studied extensively both experimentally and theoretically, at least some attention is due a similar process with a photon emitted. This is exemplified by the following diagram:

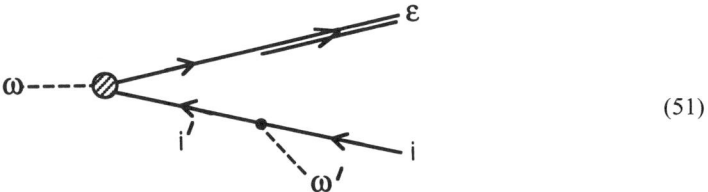

(51)

The fields of vacancies i and i' have the same radial asymptotics. Therefore, the modification of the field due to radiative decay which alters the vacancies' energy and angular momentum leads to much smaller energy shifts than (45) or (46). A rough estimate for this energy shift $\Delta\varepsilon_{i'}^{(\gamma)}$ is given by

$$\Delta\varepsilon_{i'}^{(\gamma)} = U_i^{HF}(v/\Gamma_{i'}^{\gamma}) - U_{i'}^{HF}(v/\Gamma_{i'}^{\gamma}) \tag{52}$$

where $U_{i(i')}^{HF}(r)$ is the Hartree–Fock potential with vacancies $i(i')$, respectively. The difference in (52) is taken at the distance $r = v/\Gamma_{i'}^{\gamma}$, with $v = \sqrt{2\varepsilon}$ and $(\Gamma_{i'}^{\gamma})^{-1}$ being the vacancy i' radiative decay lifetime.

Although the energy difference is small, the effect of this decay on the probability of the process (51) is large. This is most transparent for subthreshold excitation instead of ionization, represented by (51). The exchange of angular momentum after a subthreshold excitation can open forbidden channels, making possible the forbidden radiative decay. Such an example is represented by the diagram:

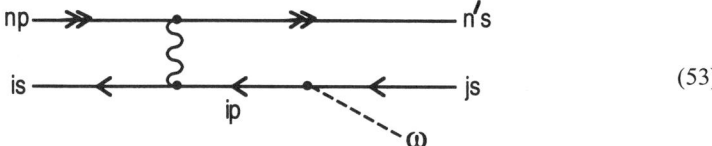

(53)

This process is suppressed, as compared to the direct decay $ip \to js + \gamma$ by a factor $[(np, ip|v|n's, is)/(\varepsilon_{is} - \varepsilon_{ip})]^2 \ll 1$.

The presence of an excited electron can strongly affect the probability of a radiative decay by considerably altering the contribution of the correlational term. The $4p^{-1} \to 4d^{-1} + \gamma$ decay can serve as an example. It is almost completely forbidden in an ion, where the direct transition amplitude is compensated by the indirect one, which describes a decay via virtually excited states with two $4d$ vacancies. The situation in the presence of an excited electron can be considerably different. The corresponding diagrams are the following:

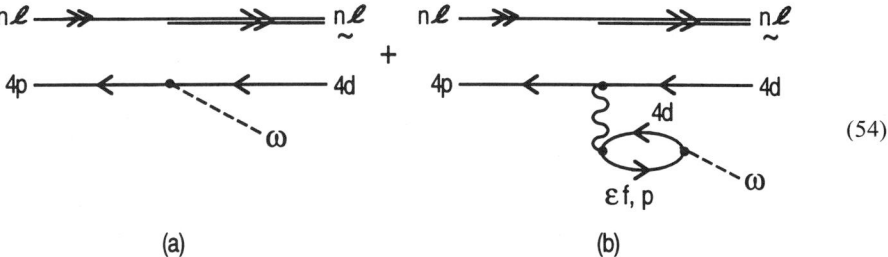

(54)

(a)   (b)

The first term (54a) describes the direct, while the second term (54b) presents the correlation decay amplitude. The presence of an excited electron $n\ell$ strongly affects the contribution of (54b) by modifying the transition $4p^{-1}, n\ell \to 4d^{-1}, n\ell$ and the virtual intermediate state $\varepsilon f(p), 4d^{-1}4pn\ell$ energies as compared to their values in an ion.

The below-threshold excitation leads to the formation of states with several vacancies and excited electrons, their decay enriching the Auger spectrum with new lines, which yield information on the photoionization mechanism.

The presence of excited electrons can open new decay channels for states which are more complex than (53). As an example consider the radiative decay of a "two-vacancy–one-electron" state with two 4s vacancies and one 5p excited electron into a "two 4p vacancies–one 4d excited electron" state in Kr.[48] The simplest corresponding diagram is as follows:

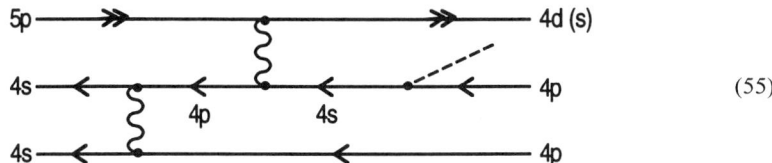
(55)

The initial state in (55) is formed in an Auger decay of the 3d vacancy of the "5p electron–3d vacancy state" excited by an absorbed photon in Kr.[48] Without the participation of the excited electron leading to the alteration of its energy and angular momentum, the radiative decay (55) is almost forbidden.

While for inner-shell vacancies the excited electron is a "spectator," for intermediate shells the competition is very strong between independent vacancy Auger decay, i.e., the "spectator" picture and that of "participant," with mainly the autoionization decay of the initially created state. Diagrammatically, the "spectator" Auger decay is illustrated by (56a), while the "participant" decay of an intermediate shell vacancy is given by (56b), i.e., by the very simple pictures:

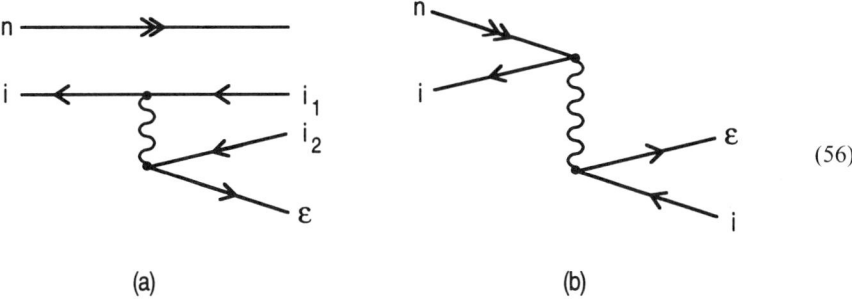
(56)

(a) (b)

In higher orders these processes can combine, leading to complicated and competitive paths of the decay.

The "two-vacancy–one-excited-electron" state can be formed not in the first, but in the second step of the ionization process: after a real or a quasi-Auger decay of the vacancy [see (56a)]. Of interest is the autoionization of these states which can proceed in a manner similar to the scattering process or via a one-vacancy bottleneck. These possibilities are illustrated by the following lowest-order diagrams:

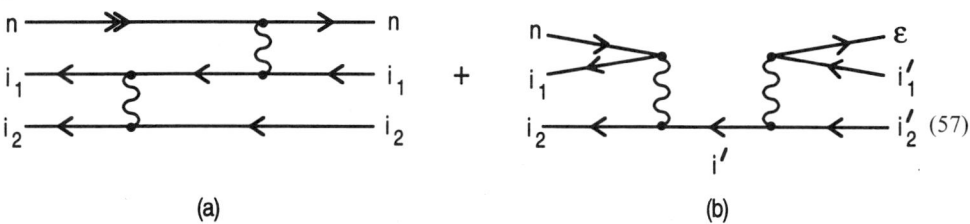
(57)

(a) (b)

Diagram (57a) presents the inelastic scattering of an excited electron on a vacancy. The energy is taken not only from the vacancy i_1 but also i_2, the latter being able to transfer energy due to intervacancy interaction. Diagram (57b) describes the connection of two complicated configurations via a one-vacancy state.

The theory of all correlational processes in which either two or more electrons are excited and/or ionized or multivacancy states decay is in its infancy: only shake-type and lowest-order perturbative approaches have been developed. The latter are definitely more general and open the possibilities of the consistent improvement without the danger of double counting the same corrections.

13. INCLUSION OF HIGHER-ORDER CORRECTIONS

In the previous sections two approaches were combined very often: the interaction of the incoming photon with the target electron was described in RPAE or GRPAE while a number of secondary processes were treated in the lowest order of perturbation theory. An approach, which applies many-body perturbation theory (MBPT) to the photoionization of atoms, was initiated long ago[4] and is developing now.[49] A number of successful calculations were performed using this approach. It forms a convenient first step for considering almost any physical process. Usually, it is easy to draw a diagram, having in mind a particular physical process. Then special correspondence rules exist,[3] which permit one to write down an analytical expression which is an equivalent to a given diagram depicting this process. Of course, to draw is much easier than to calculate, so obtaining numerical results even for simple diagrams is usually very complicated and becomes more and more difficult when considering not a few electron, but real multielectron atoms. Simplifications are inevitable, mainly accounting for only the nearest to the considered subshells, including only a rather limited part of the angular momentum values transferred via interelectron interaction and so on.

The summation of all diagrams is equivalent to a precise quantum-mechanical calculation which means solution of the Schrödinger equation for an atom. Because this is impossible for all but hydrogenlike atoms, it is reasonable to start with the lowest nonvanishing order for a particular process and consider the simplest corrections to it. Whether this is sufficient or not depends on the relative importance of the interelectron interaction in the atoms or multiatomic formations, i.e., on the ratio of interelectron interaction matrix element to the virtual excitation energy. If this parameter is small, then the lowest-order perturbative calculation is sufficient. If not, higher-order corrections must be taken into account.

In general, the interelectron interaction in atoms and multiatomic formations is not weak. But a large fraction of it is usually taken into account by considering electrons as moving not in a Coulomb field of the nuclei but in a self-consistent HF field, thus taking into account a large part of the interelectron interaction. However, what remains is still strong enough to lead to considerable RPAE-type corrections. Most important among them are the intrashell effects, and particularly those which are described by (12) and can be (and in fact are) taken into account by a suitable choice of the one-electron approximation. What remains is still not always small. However, it proved to be that quite often while in a given order each correction is large, they are almost compensating each other and their sum is quite small. Therefore, usually all diagrams in a given order must be taken into account. Practically, corrections not higher than that of the third order were included, because the number of different diagrams N_n in a given order n increases very fast, being connected by the relation $N_n \sim n!$.

Some sequences of corrections, even infinite, are easy to take into account by using experimental instead of calculated energy values. Let us start with ionization potentials I_i. If a vacancy state is represented only by a line directed to the left, as I_i the HF value $|\varepsilon_i|$ of the vacancy is taken. If any insertions into the vacancy line are included, not only those presented by (29), the sequence of corresponding diagrams forms a geometrical progression:

$$\overset{i}{\longleftarrow} + \overset{i}{\longleftarrow}\boxed{\Sigma}\overset{i}{\longleftarrow} + \overset{i}{\longleftarrow}\boxed{\Sigma}\overset{i}{\longleftarrow}\boxed{\Sigma}\overset{i}{\longleftarrow} + \cdots \quad (58)$$

Analytically, this sequence is represented by the following series:

$$\frac{1}{E+\varepsilon_i} + \frac{1}{(E+\varepsilon_i)}\Sigma_{ii}(E)\frac{1}{(E+\varepsilon_i)} + \ldots = \frac{1}{E+\varepsilon_i - \Sigma_{ii}(E)} \quad (59)$$

The quantity $\Sigma_{ii}(E)$ is the diagonal matrix element of an operator called in many-body theory the "self-energy part"[3] of the one-particle Green's function. The main effect of it is to shift the vacancy energy to a new value:

$$E = -\varepsilon_i - \Sigma_{ii}(E) \approx I_i^{HF} - \Sigma_{ii}(I_i^{HF}) \quad (60)$$

In principle, the summation (59) not only shifts the energy, at which the sum in (59) has a pole from the I_i^{HF} value, but also leads to a residue Z_i which is less than one. However, neglecting the Z_i

Theory of Photoionization

correction, all of the sequence (59) is taken into account by a shift of the ionization potential to its experimental value.

Now let us consider another case, an excitation of an electron to a discrete level n from the initial state i. The excitation energy ω_{ni} in the lowest order of perturbation theory is $\omega_{ni} = \varepsilon_n - \varepsilon_i = \varepsilon_n + I_i^{HF}$. Substituting this quantity by its experimental value includes an infinite sequence of many-body correlational corrections:

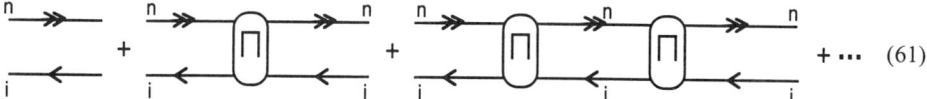

Similar to (58), the sequence (61) is summed as a geometrical series, leading to a new expression for the excitation energy

$$\tilde{\omega}_{ni} \cong \varepsilon_n - \varepsilon_i + \Pi_{ni}(\varepsilon_n - \varepsilon_i) \tag{62}$$

If corrections to vacancy (and electron) energies given by (59) are taken into account, the excitation energy is modified again, leading to

$$\tilde{\omega}_{ni} \cong \varepsilon_n - \varepsilon_i + \Sigma_{nn}(\varepsilon_n) - \Sigma_{ii}(\varepsilon_i) + \Pi_{ni}(\varepsilon_n + \varepsilon_i) \tag{63}$$

For precise values of Σ and Π the quantity $\tilde{\omega}_{ni}$ must be equal to its experimental value. Thus, substituting ω_{ni} by the corresponding experimental values, an essential fraction of terms, which belong to many-electron correlational corrections, are taken into account.

In the same way the corrections to two-vacancy energies may be taken into account. Instead of being a sum of ε_{i_1} and ε_{i_2} it is modified, at first by including the diagonal term of pure Coulomb intervacancy interaction, and then by including diagonal terms of higher-order contributions. Again, by using the experimental two-vacancy state energy instead of a sum of pure HF values an important group of correlational corrections are included.

A separate problem which is to some extent solvable in MBPT is the calculation of the corresponding energy shifts, namely, the operators Σ and Π. Each of them can be derived in the lowest nonvanishing order of MBPT, introducing the corrections to the energies which were discussed above in this section. The framework of MBPT is in principle very flexible and a number of higher-order corrections may be introduced semiphenomenologically, improving considerably the results of the calculations.

One of the most difficult problems in MBPT calculations (and in any other method of describing electronic structure of atoms and multiatomic formations) is to take into account the interaction of two continuous spectrum electrons in intermediate and final states. In both cases [see (18) and (34)] this interaction is quite strong and must be accounted for nonperturbatively, by solving the Schrödinger equation describing the motion of the two electrons in the self-consistent field of an ion. This is a complicated three-body problem, which is far from being solved for even heliumlike atoms. For intermediate states like $\varepsilon'\varepsilon''i$ in (18) this problem can be considerably simplified by using an approach similar to that discussed above. Indeed, by calculating contributions of two subsequent diagrams from a total series presented by the following pictures

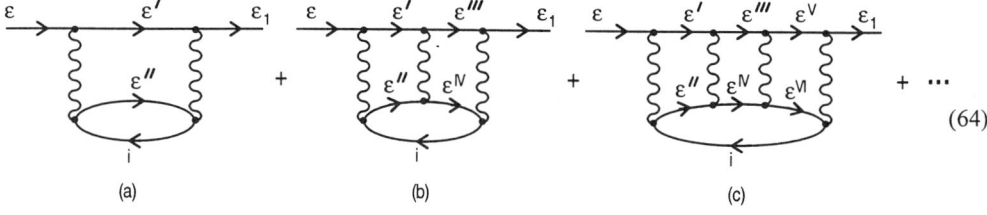

the ratio of contributions (64b) and (64a) or (64c) and (64b) ξ can be found. Then, assuming that (64) is a geometrical series, the sum of (64) can be estimated as the contribution of (64a), multiplied by

$(1 - \xi)^{-1}$. Of course, this approximation is crude and is far from being a real calculation of the contribution of (64). It is essential to have in mind, however, that it is not many-body theory that creates the difficulties, such as a solution of the three-body equation. On the contrary, they are inherited in the problem of the electronic structure of complex atoms and many-body theory makes these difficulties more transparent.

It is desirable to have a gauge-invariant theory of photoionization, for which all operators describing the interaction between the electromagnetic field and electrons (4) are equivalent. Otherwise, a recipe must be given on which form is preferred. It was mentioned above that RPAE is gauge-invariant. Practical calculations demonstrated that for GRPAE at least "length" and "velocity" forms are equivalent. If the perturbation theory uses the pure Couloub field of the atomic nuclei as its one-particle potential, the gauge invariance is preserved at each order of perturbation theory. If as one-electron the HF wave functions are used the photon–electron interaction in each diagram must be taken in RPAE. So, all diagrams of a given order out of the RPAE frame must be taken into account and in each of them the photon–electron interaction must be included in RPAE, i.e., determined by the $\hat{D}(\omega)$ operator from (9) or (11).

Of course, if experimental energies are used or some other higher-order corrections are taken into account nonconsistently, the gauge invariance will be violated. There is no general method which can help to restore it in such cases.

14. FUTURE OBJECTS OF HIGH-ACCURACY CALCULATIONS

The theoretical understanding of the photoionization of isolated atoms has reached a very high level. A number of new features of the process are discussed, including those in which two or more atomic electrons participate either explicitly, as in multiple ionization, or implicitly, as in the one-electron ionization channel modified and strongly perturbed by the many-electron virtual excitations or ionizations. The most transparent manifestation of the latter is the resonance structure of the one-electron continuum, which is acquired as a result of its strong interaction with discrete excitations of a "two-electron–two-vacancy" nature. Improvement in experimental resolution will give evidence of the presence of even more complicated excitations in the cross section resonance structure.

The role of many-electron effects proved to be very important, and the quality of calculations, particularly those performed using HF, RPAE, and MBPT methods, has reached perhaps the highest level in physics. Still a lot can be done in the theoretical description of isolated atoms and ions, including the investigation of their excited states both low- and high-lying (see Section 6). But it is also attractive and timely to apply these same methods to investigating the photoionization of multiatomic formations or atoms in an external field in order to achieve the same accuracy, which is already achieved for isolated atoms. Of particular interest are the atomlike systems, such as clusters of metals and fullerenes. The complicated structure of multiatomic formations is determined by the large number of both the electrons and the nuclei. The latter can move relatively to each other and are forming a number of centers of attraction for electrons. The average field, acting on electrons, loses its spherical symmetry, which makes even the one-electron problem three-dimensional and, therefore, very complex.

Systems in which all of the nuclei and the electrons are tightly bound to them, forming something similar to a large "smeared" nucleus, in the field of which the outer electrons are moving, deserve special attention. The magnitude of this field is smaller, while the interelectron interaction is larger than for isolated atoms and, therefore, the many-electron correlational effects are more profound and important. This feature of the quasiatomic formations or quasiatoms makes them attractive objects for photoionization studies. These objects are small and inhomogeneous enough to permit a direct photon absorption by an electron and have such large multielectron subshells that the correlational effects are very strong.

The theory of photoionization of such objects must start with construction of an HF self-consistent field, which is determined by (8), with the nuclear potential substituted by a field created by a smeared positive charge, occupying a given volume of space. An independent and important problem is the determination of this volume for a given number of constituent atoms in the multiatomic system. After the HF wave functions are found, the RPAE calculations similar to those for isolated atoms can be performed,[50] leading to photoionization cross sections with giant resonance structure in them. In order to estimate their power, one should have in mind that the total oscillator strength of resonance is a large fraction of the total number of common electrons N. The number N, which for alkali metal clusters is equal to the number of atoms in it, can be in the range from several units to one or two thousand. Note that the largest oscillator strength of an atomic giant resonance is smaller than ten. For fullerenes C_{60} the number N is as large as 240.

The large oscillator strength of the giant resonance leads inevitably to strong intershell correlations, which considerably alters the partial photoionization cross sections and the multiply charged ion yields. Other characteristics of the photoionization process are also strongly affected by the giant resonance. Not only RPAE but also GRPAE effects deserve attention. These, as well as secondary processes such as direct knock-out, can be studied before corresponding experimental measurements are performed. It is unclear now whether the open-shell metallic clusters or fullerenes will obey something like Hund's rules having all electrons in a given shell with maximal possible spin. If a Hund-like rule is valid, the RPAE formalism, as well as other many-body approaches, might be easily generalized to systems with two kinds of electrons, namely, "up" and "down" electrons, thus increasing the number of targets, in which photoionization can be studied quantitatively.

This approach, which neglects the possible nuclear motion and the alteration of the interaction between valence electrons due to atomic core polarization, may require serious corrections. Thus, an important step in the theory is to take into account in the calculations starting from the HF level instead of direct Coulomb interelectron interaction the screened one, modified due to inclusion of virtual core excitations. This is a difficult problem which can require a lot of effort.

If the HF equation could be solved without assumption that the self-consistent field is spherically symmetric, so that the radial equations are not one-, but two- or even three-dimensional, a number of very complicated systems such as molecules and many-electron atoms in strong external fields could be considered theoretically with high accuracy. The electron correlation effects are in such systems at least not weaker than in isolated atoms and in this sense molecules formed from alkalies and halogens, such as KCl, CsI, and so on, are of particular interest. In the first approximation, these have outer multielectron shells formed from two np^6 valent atomic subshells. The correlational effects are very strong, as is known[3] for these shells in isolated atoms, so they should also be strong in the above-mentioned molecules.

Application of a strong external electrical field deforms even an initially spherically symmetric atom. Such a field increases the mean distance between atomic electrons and the nucleus, thus decreasing its role, while not modifying the direct interelectron interaction. This means an increased role of electron correlation. Under the action of a strong electrical field an atom is distorted, stretched, and its polarizability becomes larger because with an increase of the external field strength, less and less energy is required for atomic ionization. Growth of polarizability increases the screening in interelectron interaction, which leads in turn to modification of the role of correlations. As a result, the correlations start to be dependent on the external field, which is an interesting subject for investigation. This problem can be studied by considering the photoionization of atoms in a low-frequency and comparatively high-intensity field.

To study the modification of correlations under the external field action, such intensity of this field must be considered, at which the ionization under the direct action of this field still has a small probability. If, however, the external (laser) field is so strong that it is able to eliminate electrons from the atom, these are forming groups far from the nucleus and the problem of ionization under the photon action becomes a problem of eliminating not a single particle, but the entire group.

If the target system, either a multiatomic formation or an atom in an intensive field, is soft enough after absorption of a photon, it can be so strongly deformed that RPAE or its modifications such as GRPAE will be insufficient. In this case as a step, next to HF the time-dependent Hartree–Fock (TDHF) approximation[7] should be applied. This approach assumes that the wave function of the system at each moment is a Slater determinant of the one-electron wave function, just as in HF, but with a nontrivial time-dependence. TDHF is reduced to RPAE if not only the photon field itself is weak but also the response of the target to its action is small enough.

It seems possible that atoms can have highly excited states with all electrons located very symmetrically and far from the nuclei. Such a state would be long lived and, therefore, could be an object of photoionization studies. Being weakly affected by the nucleus, it will be strongly perturbed by the photon field, thus requiring for its description TDHF instead of RPAE.

The investigation of the photoionization of the objects mentioned in this section is a challenge for both theory and experiment.

15. FORTHCOMING DEVELOPMENT OF THE PHOTOIONIZATION THEORY

While the general agreement between theory and experiment regarding photoionization cross sections and other characteristics of this process for isolated atoms is reasonably good, some important limitations of the methods described in the previous sections were observed. They require considerable improvement of the photoionization theory for isolated atoms. Although the RPAE is an approximation to the precise theory, it is, on the other hand, self-consistent while preserving gauge invariance, having correct sum rules,[3] and giving reasonable results in describing photoionization. The generalization of it has been until now either semiphenomenological, as in the case of GRPAE, or by applying MBPT after taking into account the RPAE effects. The semiphenomenological introduction of the corrections, which relies largely on experimental data, is an undesirable feature of a theory. As to the combination of RPAE and perturbation theory in its lowest orders, it is far from being universal, because the interelectron interaction is in general too strong. If, however, it were so strong that all corrections of any order would be large and equally important, there would be no chance of constructing a microscopical theory allowing high-accuracy *ab initio* calculations of atomic properties, the photoionization cross sections among them. In such a case, only phenomenology, similar to that developed for homogeneous Fermi liquids,[51] could be used. Such phenomenology has a number of similarities to RPAE, differing from it by the choice of the interelectron interaction. The latter is pure Coulombic in RPAE and is much more complicated, in general altered by the virtual excitation of atomic electrons. This interaction is nonlocal and energy dependent. Diagrammatically, the photoionization amplitude is represented [similarly to (11)] by

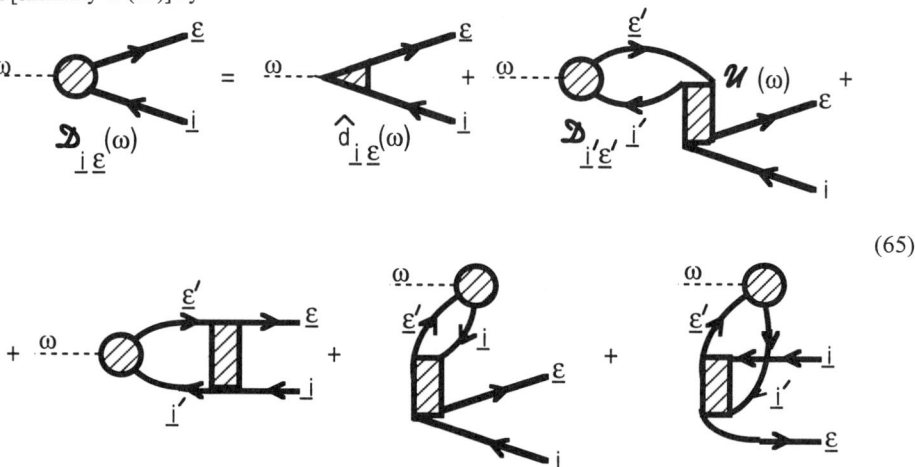

(65)

Here the heavy lines stand for one-electron states which are determined by an equation, almost identical to the HF one, in which the direct Coulomb interelectron interaction is substituted by the dashed rectangle:

(66)

THEORY OF PHOTOIONIZATION

Similar are the equations for the wave functions of the vacancies. In fact, (66) describes the so-called quasiparticles, which are coherent sums of pure one-particle and more complicated states, for instance, "two-particle–one-vacancy" excitations and so on.

The dashed triangle in (65) stands for the effective operator $\hat{d}(\omega)$ describing photon–electron (vacancy) interaction, while the dashed rectangle presents the effective interelectron interaction $U(\omega)$. Both are modified by inclusion of "two-electron–two-vacancy" as well as more complicated excitations. There is no closed form equation for these quantities. But using them an analytical equation can be presented for the photoionization amplitude:

$$(\varepsilon|\mathcal{D}(\omega)|i) = (\varepsilon|\hat{d}(\omega)|i) + \sum_{\varepsilon>F, i\leq F} [(\varepsilon|\mathcal{D}(\omega)|\underline{i}')(\omega - \varepsilon' - \underline{I}_{i'})^{-1}(\underline{\varepsilon}'\underline{i}|U(\omega)|\underline{i}'\underline{\varepsilon})$$

$$+ (\underline{i}'|\mathcal{D}(\omega)|\underline{\varepsilon}')(\omega + \varepsilon' + \underline{I}_{i'})^{-1}(\underline{i}'\underline{i}|U(\omega)|\underline{\varepsilon}'\underline{\varepsilon})] \tag{67}$$

Here the dash under $i(i')$ or $\varepsilon(\varepsilon')$ denotes that these states are determined by equation (66). As ionization potentials $\underline{I}_i$ the precise values must be used. Equation (67) is similar to (15). To solve it, operators $U(\omega)$ and $\hat{d}(\omega)$ must be found.

Quite good results already achieved when using RPAE suggest an approximate but very promising approach: as a next step toward the precise theory (67) to use the next step of expansion in powers of RPAE effective interaction (35). It means that in all exchange terms of the HF and RPAE equations (8) and (9) the exchange part of the interelectron potential u must be substituted by $\Gamma(\omega')$ from (35). This corresponds to accounting for the following diagrams:

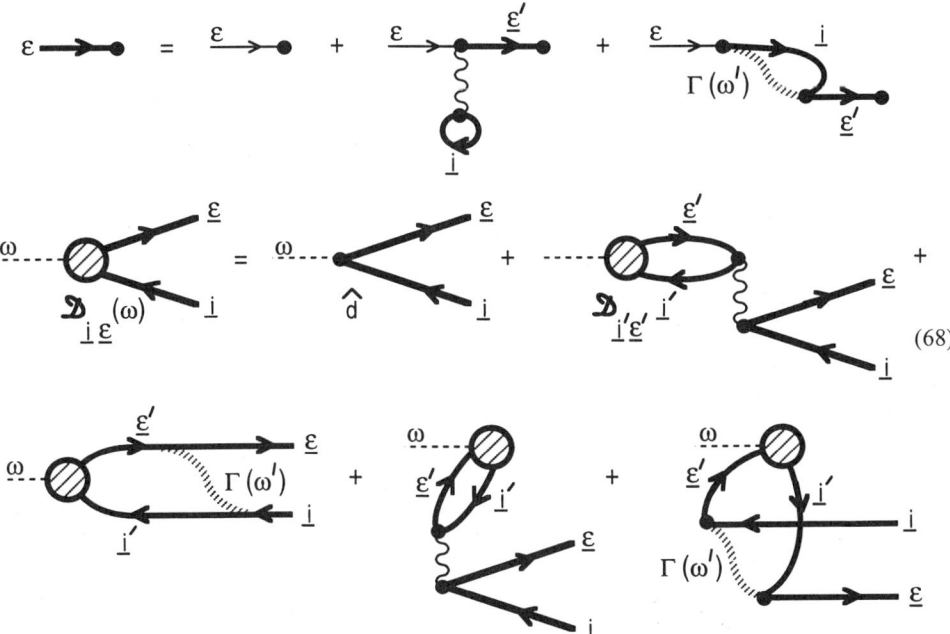

(68)

Here the effective interaction $\Gamma(\omega')$ is determined by (35), so that (68) and (35) form a closed system of equations. Expanding $\Gamma(\omega)$ in powers of the Coulomb interaction u one sees that (68) includes the "two-electron–two-vacancy" excitations, discussed in Section 8. However, this approximation is not sufficient for considering these excitations alone, because they do not take into account the electron–electron and vacancy–vacancy direct interaction in the intermediate and final states. Hopefully, the approximation (68) allows one to reach very high accuracy in cross section calculations, thus helping to achieve an important goal for future theoretical research in photoionization: to have a calculation accuracy of about 1%. The ability to achieve this goal in the above-mentioned way is supported by the high accuracy of the results of *ab initio* calculations of energy levels and transition probabilities, which were performed in connection with the investigation of parity nonconservation in atoms.[52] These calculations support the idea of a fast convergence of the expansion in powers of the effective interaction $\Gamma(\omega')$, given by (35).

Qualitatively, the success of RPAE in describing atomic properties became understood only recently, when it was demonstrated that the range of validity of this approximation includes not only high but also medium and even low densities of an electron gas.[53] It is the inclusion of exchange that permits the application of RPAE in a very broad electron gas density region, while RPA is valid only for high densities.

The solution of (68) is a very complicated and challenging problem even without including nonperturbatively electron–electron and vacancy–vacancy interaction. The latter would require the solution of a three-body problem with long-range Coulomb interaction, i.e., solving the Schrödinger equation describing the motion of two electrons in the self-consistent field of an ion as an intermediate step to any further calculations.

In principle, if the exact expression for the ground-state energy as a function of density of an atom or any other multiparticle system is known, expressions for the effective interaction can be derived. This forms the foundation of the density functional method (DFM)[54] which is basic to the time-dependent local density approximation (TDLDA).[6] In such an approximation, as well as in the Fermi liquid theory, an important relation is valid, connecting $\hat{d}(\omega)$ and d: $\hat{d}(\omega) = d$.[55] In TDLDA the interaction $U(\omega)$ is energy-independent and derived using very simple interpolational formulas for the electron gas exchange and correlational energies as a function of electron density. The interaction U is presented as a sum of two terms: a pure Coulomb part without exchange and zero range exchange and correlational correction. Thus, $U(\omega)$ is parameterized as

$$U(\omega) \equiv U(\vec{r}_1, \vec{r}_2, \omega) \approx |\vec{r}_1 - \vec{r}_2|^{-1} + f(\rho(r_1))\delta(\vec{r}_1 - \vec{r}_2) \tag{69}$$

The function $f(\rho(r))$ is expressed via the exchange and correlational energy as its second variational derivative in density. The self-consistent potential corresponding to (69) is local, energy-dependent, and has a zero-range part. It fails, however, to reproduce the correct asymptotic behavior, $\sim -1/r^4$ at $r \to \infty$ of the self-consistent one-particle.

The new expression for the ground-state energy, obtained in the framework of a recently developed approach to the theory of many-particle systems,[53] allowed for the derivation of another formula for the interaction U[56]:

$$U(r_1, r_2) = |\vec{r}_1 - \vec{r}_2|^{-1}\left\{1 + \frac{1}{2\chi^2}\left[1 - \frac{p_F^2(\vec{R})}{\pi}\frac{d^2}{d\rho^2}[\rho(\vec{R})\varepsilon_c(\rho(\vec{R}))]\right] \times [\cos 2\chi - \frac{\sin 2\chi}{\chi} + \frac{\sin^2\chi}{\chi^2}]\right\} \tag{70}$$

where $\chi = |\vec{r}_1 P_F(\vec{r}_1) - \vec{r}_2 P_F(\vec{r}_2)|$, $\vec{R} = (\vec{r}_1 + \vec{r}_2)/2$, and the Fermi momentum $P_F(\vec{R})$ is connected with the electron density $\rho(\vec{R})$ by the relation $P_F(\vec{R}) = [3\pi^2\rho(\vec{R})]^{1/3}$, $\varepsilon_c(\rho)$ is the correlational energy as a function of the density ρ. All details necessary for calculation of the one-electron potential, correlational energy, linear response function, and finally the photoionization cross section are given in Ref. 56.

The expression (70) presents a finite-range effective interaction. The corresponding self-consistent potential has a correct asymptotic behavior, but is local. This fact considerably simplifies the numerical procedure of solving the equations (66) and (67), but introduces some inaccuracy because the real self-consistent field is both energy-dependent and nonlocal. By using (70) as an effective interaction in (66) and (67) an important step toward solution of the precise system of equations, describing photoionization, can be made.

Entirely, there are two very important directions in future development of photoionization theory: going beyond RPAE along the lines suggested by equation (69) and including the three-particle effects in "two-electron–two-vacancy" excitations. The success in these directions will formulate a new basis for forthcoming experimental and theoretical studies in this domain.

16. SOME CONCLUDING REMARKS

The previous sections presented the theory of the photoionization of multielectron atoms. Main consideration was given to physical ideas and only some illustrative data were presented. In order to continue the development of the photoionization theory of multielectron systems the efforts of theorists themselves are definitely not sufficient: due to the very complicated nature of these objects, each step forward in a theoretical description must be accompanied by an experimental response and vice versa. Therefore, it seems appropriate to conclude this chapter by mentioning some directions and domains of experimental research, the results of which are of great importance for understanding the electronic structure of atoms and multiatomic formations.

1. Systematic studies of partial cross sections of different atoms, comparing the results obtained by electron spectroscopy, by photoinduced fluorescence spectroscopy, and by measuring the multiply charged ion yield. The aim is to study the role of intra- and intershell as well as other types of correlations, and of such processes as shakeup and shakeoff.
2. Systematic investigation of outer-shell ionization near the intermediate- and inner-shell thresholds using either the electron spectroscopy technique or by measuring the singly charged ion yield. Close to an inner threshold postcollision interaction is important, while far away only intershell correlations are essential.
3. Investigation of satellite levels due to ground-state electron correlations, real or virtual decay of vacancies or due to direct excitation of the residual ion by the photoelectron. Photoionization with formation of these states is a direct manifestation of the role played by electron correlation.
4. Measurement of the angular distribution and polarization of photoelectrons emitted from inner s, mainly $1s$ shells. The aim is to study the role of nondipole components of the photon and the angular momentum exchange between photoelectron and the core after the Auger decay of the inner vacancy. The nondipole part of the angular distribution comes from the nondipole component of the photon and the angular core–photoelectron momentum exchange.
5. Photoionization near inner- or intermediate-shell thresholds. The aim is to investigate postcollision interaction, mainly for those shells and frequency regions where the velocities of the two outgoing electrons, termed "fast" and "slow," are about the same. Until now, mainly the fast electron energy distribution was measured, but not the angular correlations. It is essential to find the cross section and angular distribution variation due to radiative vacancy decay.
6. Photoionization with the production of electrons whose energy is close to the outer-shell ionization or excitation potential. The aim is to study the cross section variation due to opening of some other channels, which are known for systems with short-range interaction (Wigner–Baz singularity) but was almost not considered for the case of the coulombic interaction.
7. The investigation of double-electron photoionization for almost all atoms, starting from the lightest and proceeding to medium and heavy. The aim is to study different mechanisms of interelectron interaction leading to double-electron photoionization including initial- or final-state correlation. The energy and angular distribution of electrons as a function of incoming photon energy give information on interaction in both initial and final states starting from the threshold region where the Wannier regime dominates, to short-range correlations for high photon frequencies. Double-electron photoionization gives information on the role of different kinds of virtual excitations, i.e., on discrete and continuous spectra of other than the ionized atomic electrons.
8. Study of photoionization of excited (including highly excited) atoms with one electron out of the core. The aim is not only to find the deviations from simple hydrogenlike behavior of the photoionization cross section, but also to disclose the influence of the core on this cross section. It is of interest to learn how the probability of inner-shell ionization is modified by

the excited electron and how the latter affects the double-ionization cross section of the excited atom.

9. Investigation of negative ion photodetachment. The aim is to find specific features of this process in the threshold region of photodetachment and near inner-shell thresholds. Of particular interest are Si^-, Ca^-, and Ba^-, where quite impressive variations of the cross sections near threshold are predicted. Of interest also is the threshold law of photodetachment, which is strongly affected by the polarization of the atom by the detached electron. Inner electron elimination deserves to be investigated because it may be affected by the presence of the extra electron in negative ions.
10. Investigation of positive ion photoionization. The aim is to study the role of electron correlations in ions, which seems to be very important. With the growth of the degree of ionization the contribution of correlations in general decreases, but not linearly. In singly and doubly charged ions they may be even stronger than in the corresponding atoms.
11. Creation and then photoionization of exotic atomic states, such as those with all electrons having the same spin projection, planetary-type and hollow structures. All of these states are so unusual that their photoionization characteristics will have little in common with what is already known for ordinary atoms.
12. Studying ion formation in coincidence with secondary photons, which originate due to internal bremsstrahlung. The aim is to achieve x-ray graphy of an atom. The most transparent example is the He^{2+} formation in He ionization. Such an investigation permits one to study the interaction of two outgoing electrons after He ionization in the case when the pair's total momentum is zero, while in double-electron photoionization itself the total momentum is equal to one.
13. Investigation of elastic and inelastic photon scattering by atoms. The aim is to measure the frequency dependence of the atomic polarizability, which is strongly affected by electron correlation. Inelastic scattering permits one to study all mechanisms of photoemission and the role of electron correlations in these processes.
14. Consideration of photoionization of multiatomic formations well above the inner- or intermediate-shell thresholds of the constituent atoms. The aim is to study the extra multiatomic potential by a procedure similar to Fourier transformation but on the basis of accurate numerical wave functions for isolated constituent atoms.

Similar processes are of interest not only for isolated atoms in their ground and excited states, positive and negative ions, but also for multiatomic formations, such as molecules, clusters and fullerenes and atoms in an external (laser) electromagnetic field.

ACKNOWLEDGMENT. This chapter was written while the author was a Fellow of Argonne National Laboratory, to which—particularly to the Physics Division and its Atomic Physics Group—he is indebted for their hospitality. This work was supported by the U.S. Department of Energy, Office of Basic Energy Sciences, under contract W-31-109-ENG-38.

REFERENCES

1. D. HARTREE, *The Calculation of Atomic Structures* (Wiley, New York, 1957).
2. D. PINES AND P. NOZIERES, *The Theory of Quantum Liquids* (Benjamin, New York, 1966).
3. M. YA. AMUSIA, *Atomic Photoeffect* (Plenum Press, New York, 1990).
4. H. P. KELLY, in *Atomic Physics* **8**, edited by I. Lindgren, et al., (Plenum Press, New York, 1983), p. 305.
5. A. R. WILLIAMS AND U. BARTH, in *Theory of Inhomogeneous Electron Gas*, edited by S. Lundqvist and N. H. March (Plenum Press, New York, 1983).
6. A. ZANGWILL AND P. SOVEN, *Phys. Rev. A* **21**, 1561 (1980).
7. D. J. THOULESS, *The Quantum Mechanics of Many-body Systems* (Academic Press, New York, 1961).
8. P. G. BURKE, in *Electronic and Atomic Collisions*, edited by G. Watel (North-Holland, Amsterdam, 1978), p. 201.

9. N. A. Cherepkov, *Adv. At. Mol. Phys.* **19**, 395 (1983).
10. G. F. Bertsch, A. Bulgac, D. Tománek, and Yang Wang, *Phys. Rev. Lett.* **67**, 2690 (1991).
11. I. V. Hertel, H. Steger, J. de Vries, B. Weisser, C. Menzel, B. Kamke, and W. Kamke, *Phys. Rev. Lett.* **68**, 784 (1992).
12. C. Bréchignac, P. Cahuzac, F. Carlier, and J. Leygnier, *Phys. Rev. Lett.* **63**, 1368 (1989).
13. M. Ya. Amusia, L. V. Chernysheva, V. K. Ivanov, and V. A. Kupchenko, *Z. Phys. D* **14**, 215 (1989).
14. S. V. Lavrentjev, M. E. Vasiljeva, I. D. Petrov, and V. L. Sukhorukov, *Opt. Spectrosc. (USSR)* **69**(2), 186 (1990).
15. M. Ya. Amusia, V. K. Ivanov, and V. A. Kupchenko, *Z. Phys. D* **14**, 219 (1989).
16. B. Peart and I. C. Lyon, *J. Phys. B* **20**, L673 (1987).
17. J. Tulkki, *Phys. Rev. A* **48**, 2048 (1993).
18. M. Ya. Amusia, G. F. Gribakin, V. K. Ivanov, and L. V. Chernysheva, *J. Phys. B* **23**, 385 (1990).
19. M. Ya. Amusia, V. K. Dolmatov, and M. M. Mansurov, *J. Phys. B* **23**, L491 (1990).
20. G. F. Gribakin, A. A. Gribakina, B. V. Gul'tsev, and V. K. Ivanov, *J. Phys. B* **25**, 1767 (1992).
21. V. K. Ivanov and J. B. West, *J. Phys. B* **26**, 2099 (1993).
22. M. Ya. Amusia and N. B. Avdonina, *Z. Phys. D* **14**, 191 (1980).
23. M. Ya. Amusia, V. K. Ivanov, and V. A. Kupchenko, *J. Phys. B* **14**, L667 (1981).
24. S. J. Schaphorst, A. F. Kodre, J. Ruscheinski, B. Crasemann, T. Åberg, J. Tulkki, M. H. Chen, Y. Azuma, and G. S. Brown, *Phys. Rev. A* **47**, 1953 (1993).
25. G. F. Gribakin, B. V. Gul'tsev, V. K. Ivanov, and M. Yu. Kushiev, *J. Phys. B* **23**, 4505 (1990).
26. C. Froese-Fischer and J. E. Hansen, *Phys. Rev. A* **44**, 1559 (1991).
27. C. W. Walter and J. R. Peterson, *Phys. Rev. Lett.* **68**, 2281 (1992).
28. B. Kämmerling, H. Kossmann, and V. Schmidt, *J. Phys. B* **22**, 841 (1989).
29. U. Becker, D. Szostak, H. G. Kerkhoff, M. Kupsch, B. Langer, R. Wehlitz, A. Yagishita, and T. Hayishi, *Phys. Rev. A* **39**, 3902 (1989).
30. M. Ya. Amusia, L. V. Chernysheva, and K. L. Tsemekhman, *J. Phys. B* **23**, 393 (1990).
31. V. Radojevic', M. Kutzner, and H. P. Kelly, *Phys. Rev. A* **40**, 727 (1989).
32. J. M. Bizau, D. Cubaynes, P. Gérard, and F. J. Wuilleumier, *Phys. Rev. A* **40**, 3002 (1989).
33. W. Wijesundera and H. P. Kelly, *Phys. Rev. A* **39**, 634 (1989).
34. V. L. Sukhorukov, B. M. Lagutin, H. Schmoranzer, I. D. Petrov, and K.-H. Schartner, *Phys. Lett. A* **169**, 445 (1992).
35. U. Becker and D. Shirley, *Phys. Scr.* **T31**, 56 (1990).
36. T. Nagata, M. Yoshino, T. Hayaishi, Y. Itikawa, Y. Itoh, T. Koizumi, T. Matsuo, Y. Sato, E. Shigemasa, Y. Takizawa, and A. Yagishita, *Phys. Scr.* **41**, 47 (1990).
37. J. Doppelfeld, N. Anders, B. Esser, F. von Busch, H. Scherer, and S. Zinz, *J. Phys. B* **26**, 445 (1993).
38. P. Zimmermann, *Comments At. Mol. Phys.* **23**, 45 (1989).
39. M. Ya. Amusia, *J. Phys. IV* **3**, Colloq. 6, Suppl. JP II, N11, 91–106 (1993).
40. U. Becker and R. Wehlitz, *Phys. Scr.* **T41**, 127 (1992).
41. M. Ya. Amusia, *Phys. Lett. A* **183**, 201 (1993).
42. R. D. Deslattes, R. E. LaVilla, P. L. Cowan, and A. Henins, *Phys. Rev. A* **27**, 923 (1983).
43. G. H. Wannier, *Phys. Rev.* **90**, 817 (1953).
44. A. K. Kazansky and V. N. Ostrovsky, *J. Phys. B* **25**, 2121 (1992).
45. M. Ya. Amusia, M. Yu. Kuchiev, and S. A. Sheinerman, in *Coherence and Correlations in Atomic Collisions*, edited by H. Kleinpoppen and J. F. Williams (Plenum Press, New York, 1980), p. 297.
46. M. Yu. Kuchiev and S. A. Sheinerman, *Sov. Phys. Usp.* **32**, 569 (1989).
47. M. Meyer, E. von Raven, B. Sonntag, and J. E. Hansen, *Phys. Rev. A* **43**, 177 (1991).
48. A. Ehresmann, V. A. Kilini, L. V. Chernysheva, H. Schmoranzer, M. Ya. Amusia, and K.-H. Schartner, *J. Phys. B* **26**, L97 (1993).
49. M. Kutzner, this volume.
50. C. Guet and W. R. Johnson, *Phys. Rev. B* **45**, 11283 (1992).
51. L. D. Landau, *Zh. Eksp. Teor. Fiz.* **30**, 1058 (1956); **32**, 59 (1957) (in Russian).
52. S. A. Blundell, J. Saperstein, and W. Johnson, *Phys. Rev. D* **45**, 1602 (1992).
53. V. R. Shaginyan, *Solid State Commun.* **55**, 9 (1985).
54. R. M. Dreizler and E. K. U. Gross, *Density Functional Theory* (Springer-Verlag, Berlin, 1990).
55. A. B. Migdal, *Theory of Finite Fermi Systems and Applications to Atomic Nuclei* (Interscience Publishers, New York, 1967).
56. M. Ya. Amusia and V. R. Shaginyan, *J. Phys. II* **3**, 449 (1993).

CHAPTER 2

MANY-ELECTRON EFFECTS IN PHOTOIONIZATION

M. KUTZNER

1. INTRODUCTION TO MANY-BODY THEORY

Photoionization is the process by which a photon of electromagnetic radiation is absorbed by a quantum system followed by the ejection of an electron. Such a transition from a well-defined initial state of the atom/molecule to various continuum states provides an excellent probe of atomic/molecular structure as well as a test for various theoretical models. Although satisfactory results for photoionization parameters such as cross section, $\sigma(\omega)$, and photoelectron angular distribution asymmetry parameter, $\beta(\omega)$, are frequently obtained within the one-electron approximation, these parameters are often complicated by the presence of electron correlation. Photoionization phenomena which are outside the single-particle model include:

1. Interchannel coupling (where two distinct continuum channels interact)
2. Autoionization (resonances appearing in the photoionization cross section because of the degeneracy of bound-excited states with continuum photoionization channels)
3. Photoionization with excitation (one electron is ejected from the atom with the remaining ionic core left in an excited state)
4. Ground-state correlation (correlations occurring because of pair excitations in the ground state of the atom before the interaction with the radiation field)
5. Multiple photoionization (one photon ejects two or more photoelectrons)

Several methods have been developed to successfully include electron correlations in photoionization. The excellent review by Starace[1] should be consulted for a detailed comparison of these methods. The random-phase approximation with exchange (RPAE)[2,3] and the relativistic random-phase approximation (RRPA)[4] provide excellent descriptions of many correlation effects (including intra- and interchannel coupling) in photoionization for closed-shell atomic systems. A related model is the time-dependent local density approximation (TDLDA)[5] which solves the RPA equations, with orbitals obtained from a local density approximation (LDA) and a different one-electron potential which is zero in the asymptotic region rather than coulombic. The R-matrix method[6] has been very effective and widely applied to open- and closed-shell atoms as well as molecules. Configuration interaction (CI)[7] and the multiconfiguration Hartree–Fock method (MCHF)[8] have also been used to improve both the initial states and the final ionic states. In certain cases, the hyperspherical coordinate approach has given a good description of photoionization with excitation and double photoionization.[9]

As was the case for the RPAE, TDLDA, and R-matrix theories, the many-body perturbation theory (MBPT) was developed initially to attack problems in nuclear physics.[10,11] The first application of MBPT to atomic systems was a calculation of the correlation energy of beryllium by Kelly in 1963.[12] Over the subsequent years MBPT has been applied to a wide range of photoionization problems including open- and closed-shell atoms, photoionization with excitation, and double pho-

---

M. KUTZNER • Department of Physics, Andrews University, Berrien Springs, Michigan 49104.

VUV and Soft X-Ray Photoionization. Edited by Uwe Becker and David A. Shirley. Plenum Press, New York, 1996.

toionization. Much of the success of the MBPT relates to its ability to assess the importance of various many-body effects by starting with a one-electron approximation and evaluating subsequent corrections. Some of the other approaches to photoionization (i.e., the RPAE and RRPA) can be described within the paradigm of MBPT.[13]

The purpose of this chapter is to review the applications of MBPT to the photoionization of atoms. We will begin with a review of the theoretical methods of MBPT focusing in particular on photoionization. Then, we will examine specific applications of the theory to single-electron photoionization, photoionization with excitation, and double photoionization.

2. REVIEW OF THE THEORY

In this section the diagrammatic expansion of MBPT as applied to atomic photoionization is discussed. The choice of single-particle states used in the perturbation expansion is considered and many of the low-order diagrams are presented.

The photoionization cross section, $\sigma(\omega)$, is a parameter defined by the relation

$$\sigma(\omega) = P(\omega)/I \tag{1}$$

where $P(\omega)$ is the ionization probability per atom per unit time and I is the incident photon flux of radiation in photons per unit area per unit time. On absorbing a photon the system makes a transition from its ground state Ψ_0 to an excited state Ψ_k. The photon energy ω and the momentum of the continuum state of the ejected photoelectron k are related through the Einstein equation

$$\omega + E_0 = k^2/2 + E_{\text{ion}} \tag{2}$$

which describes the balance of energy where E_0 and E_{ion} are the total energies of the initial and the ionized atoms, respectively. Hartree's atomic units ($\hbar = m = e = 1$) are employed throughout this chapter.

The transition from Ψ_0 to Ψ_k is the result of a perturbation interaction

$$H_{\text{int}} = \mathbf{E} \cdot \mathbf{r} = \mathbf{E} \cdot \Sigma_i \mathbf{r}_i \tag{3}$$

where $\mathbf{r}_i$ is the position of the ith atomic electron and the dipole approximation has been assumed. Application of Fermi's second Golden Rule from time-dependent perturbation theory gives

$$P = 2\pi |\langle \Psi_k | H_{\text{int}} | \Psi_0 \rangle|^2 \rho(E_k) \tag{4}$$

The expression for the density of final states, $\rho(E_k)$, depends on the normalization scheme used for the continuum states. Assuming a k-scale normalization,[14] continuum orbitals vary asymptotically as

$$R(k,l,r) = 1/r \cos(X) \tag{5}$$

with $X = kr + q/k \ln(2kr) - (l+1)\pi/2 + \sigma_l$, and σ_l is the total phase shift. In this case, the formula for the photoionization cross section becomes

$$\sigma_L(\omega) = 8\pi\omega/kc \, |\langle \Psi_k | Z | \Psi_0 \rangle|^2 \tag{6}$$

This is the so-called "length" form of the cross section. The many-body operator $Z = \Sigma_i z_i$ assumes incident radiation polarized in the $\mathbf{z}$ direction.

The "velocity" form may be derived from the "length" form by using the following commutation relation:

$$[\mathbf{r}, H] = i\mathbf{p} \tag{7}$$

with H being the exact Hamiltonian for the system. This yields the dipole matrix element relation

$$\langle\Psi_k|Z|\Psi_0\rangle = (E_0 - E_k)^{-1} \langle\Psi_k|\partial/\partial Z|\Psi_0\rangle \tag{8}$$

involving the velocity operator, $\partial/\partial Z = \Sigma_i \partial/\partial z_i$, where Ψ_k and Ψ_0 are exact eigenstates of H with eigenvalues E_k and E_0, respectively. The velocity form for the cross section then appears as

$$\sigma_V(\omega) = 8\pi/\omega kc \, |\langle\Psi_k|\partial/\partial Z|\Psi_0\rangle|^2 \tag{9}$$

Agreement between the length and velocity forms of the cross section is considered an indicator of the degree to which correlation has been included. However, length and velocity agreement is a necessary but not sufficient condition for correlation. For example, length and velocity cross sections always agree when a local potential is used for calculation of excited states.

In MBPT, the absorption cross section is obtained from the imaginary part of the frequency-dependent electric dipole polarizability, $\alpha(\omega)$, given by[15]

$$\alpha(\omega) = -\sum_{k'} |\langle\Psi_{k'}|\sum_{i=1}^{N} z_i|\Psi_0\rangle|^2 \left[\frac{1}{E_0 - E_{k'} - \omega} + \frac{1}{E_0 - E_{k'} + \omega}\right] \tag{10}$$

for an electric field $F\mathbf{z}\cos\omega t$ polarized in the $\hat{\mathbf{z}}$ direction.

In equation (10) Ψ_0 and $\Psi_{k'}$ are the exact initial and final many-particle states, respectively. Of course it is impossible in practice to obtain exact solutions for the $(N+1)$-body problem of N electrons plus the nucleus, hence the need for various approximation methods. In general, the difference between the exact energies of the final state and the ground state, $E_0 - E_{k'}$ will be a negative quantity causing the denominator of the second term in equation (10) to vanish for certain values of ω and k'. Where this occurs the denominator is treated by shifting the pole off the real axis by adding a small imaginary contribution $i\eta$ to the denominator and taking the limit as $\eta \to 0$.[16] We then have

$$\lim_{\eta \to 0}(D + i\eta)^{-1} = PD^{-1} - i\pi\delta(D) \tag{11}$$

where P represents a principal value integration. The imaginary part of $\alpha(\omega)$ obtained from equation (11) leads to photoabsorption in the following way. The imaginary part of $\alpha(\omega)$ which results from combining equations (10) and (11) is

$$\text{Im}\,\alpha(\omega) = \sum_{k'} |\langle\Psi_{k'}|Z|\Psi_0\rangle|^2 \times \pi\delta(E_0 - E_{k'} + \omega) \tag{12}$$

For a continuum of states, again using the k-scale normalization, the summation in equation (12) becomes an integral

$$\sum_{k'} \to \frac{2}{\pi}\int dk' \tag{13}$$

Using the identity

$$\int g(k')\delta([f(k') - a])dk' = \frac{g(k')}{|df/dk'|} \Bigg|_{\substack{k'=k \\ f(k)=a}} \tag{14}$$

and the relation $E_{k'} = k'^2/2$, the imaginary part of the frequency-dependent polarizability becomes

$$\text{Im}\,\alpha(\omega) = \frac{2}{k}|\langle\Psi_k|Z|\Psi_0\rangle|^2 \tag{15}$$

Comparison of equations (6) and (15) shows that the cross section is proportional to Im $\alpha(\omega)$ with the relationship being

$$\sigma(\omega) = 4\pi\omega/c \ \text{Im} \ \alpha(\omega) \tag{16}$$

At this point the perturbation expansion for $\alpha(\omega)$ developed by Kelly[12] and based on the expansion for the energy may be used to derive an expansion for the many-particle matrix element $\langle \Psi_k | Z | \Psi_0 \rangle$. There are terms for the expansion of $\alpha(\omega)$ representing normalization corrections which are not included in the perturbation expansion for $\langle \Psi_k | Z | \Psi_0 \rangle$. This leads to a normalization factor appearing in equations (6) and (9). This normalization factor can be determined by evaluating normalization diagrams[17]; their effect is to decrease the photoionization cross section by a small amount (usually on the order of 2%).

To develop the expansion it is assumed that an N-electron atom having nuclear charge Z may be described by the nonrelativistic Hamiltonian

$$H = \sum_{i=1}^{N} T_i + \sum_{i<j} 1/r_{ij} \tag{17}$$

where $T_i = -\nabla_i^2/2 - Z/r_i$ is the sum of the kinetic energy operators for the ith particle and all one-body potentials acting on the ith particle; the Coulomb repulsion between electrons is accounted for in the term involving $1/r_{ij}$. The presence of this last term prevents exact solutions to the problem and the success of an approximate method depends on its effectiveness in treating this last term.

To simplify the problem initially, we may approximate the potential for the ith particle from the $N-1$ other particles by a single-particle potential $V_i = V(r_i)$ (frequently the V^{N-1} Hartree–Fock potential) and let $H = H_0 + H'$, where

$$H_0 = \sum_{i=1}^{N} (T_i + V_i) \tag{18}$$

and the correlation Hamiltonian

$$H' = \sum_{i<j} 1/r_{ij} - \sum_i V_i \tag{19}$$

is considered the perturbation to be used in the expansion.

The perturbation calculations are carried out using a "complete set" of unperturbed wave functions Φ_n satisfying the approximate eigenvalue equation

$$H_0 \Phi_0 = E_0 \Phi_0 \tag{20}$$

where the Φ_0 represent linear combinations of Slater determinants constructed of single-particle states ϕ_n which satisfy $(T+V)\phi_n = \varepsilon_n \phi_n$. The single-particle states ϕ_n which are occupied in Φ_0 are referred to as unexcited states and all others are called excited states. The unoccupied, unexcited states are called holes and occupied, excited states are called particles. In practice, it is sufficient to calculate a finite number of bound-excited states (approximately 10) and a finite number of continuum states (approximately 40) since the continuum is discretized. The number of orbital angular momentum values is also limited (to approximately 4).

It is customary to choose V to be the Hartree–Fock potential for the orbitals ϕ_n in Φ_0. For excited states the appropriate potential can be written as[18]

$$V = R_{HF} + (1-P)\Omega(1-P) \tag{21}$$

where R_{HF} is the Hartree–Fock potential of the ground state, Ω is an arbitrary Hermitian operator, and

$$P = \sum_{n_{occ}} |n\rangle \langle n| \qquad (22)$$

is a projection operator with n_{occ} running over the occupied ground-state orbitals. It is seen that $V|n\rangle = R_{HF}|n\rangle$ when $|n\rangle$ is a ground-state orbital. The operator Ω may be chosen to ensure that excited states are calculated in the LS-coupled V^{N-1} Hartree–Fock potential appropriate to the particular photoionization channel. However, all orbitals are actually being calculated in the same general, nonlocal potential.

Following the development of Brueckner and Goldstone,[10,11] the perturbation expansion for the many-body ground state is given by

$$\Psi_0 = \sum_{\substack{m=0 \\ \text{linked}}}^{\infty} \left\{ \frac{1}{E_0 - H_0} H' \right\}^m \Phi_0 \qquad (23)$$

with only "linked" terms included in the sum. The expression for Ψ_n is similar. Individual terms in the series of equation (23) may be represented by Goldstone diagrams where an electron in an excited state is represented by a line directed upward and a hole state is represented by a line directed downward. An unlinked part of a diagram is defined as any part of a diagram that is completely disconnected from the rest and has no external lines attached. A linked diagram is defined as one containing no unlinked parts.

The matrix elements may be calculated by evaluating the series of all open diagrams that contain one dipole interaction and any number of interactions with the electron-correlation perturbation of equation (19). The lowest-order diagrams in the dipole expansion for a transition in which orbital p is excited to k are shown in Figure 1. In these diagrams, time increases from bottom to top, although many workers use a notation in which time increases from left to right (see Amusia[13]). Dashed lines terminated by a dot are dipole interactions with the operator z (or $\partial/\partial z$ in the velocity case) and other dashed lines are interactions with the correlation perturbation, H' of equation (19). Exchange diagrams, though not all shown, are also included. The lowest-order term in the dipole expansion shown in Figure 1a is $\langle k|z|p\rangle$ in the length formalism and $1/\omega \langle k|\partial/\partial z|p\rangle$ in the velocity formalism, where k and p are one-electron states calculated in the potential $V(r)$. The diagrams are read from bottom to top corresponding to right to left in the matrix elements. The diagram of Figure 1b is first order in the correlation interaction and accounts for ground-state correlations (GSC); these occur prior to interactions with the radiation field. Closed loops in these diagrams imply sums over the (virtual) excited electron states. The expression represented by Figure 1b is

$$\sum_{k'} \frac{\langle q|z|k'\rangle \langle kk'|v|pq\rangle}{\varepsilon_p + \varepsilon_q - \varepsilon_k - \varepsilon_{k'}'} \qquad (24)$$

where the sum over k' is limited to excited states. The exchange diagram corresponding to Figure 1b must be included but is not shown in the figure.

First-order final-state correlations (FSC) are represented by the diagram of Figure 1c and the corresponding exchange diagram of Figure 1d. Figure 1c is given by

$$\sum_{k'} \frac{\langle kq|v|pk'\rangle \langle k'|z|q\rangle}{\varepsilon_q - \varepsilon_{k'}' + \omega + i\eta} \qquad (25)$$

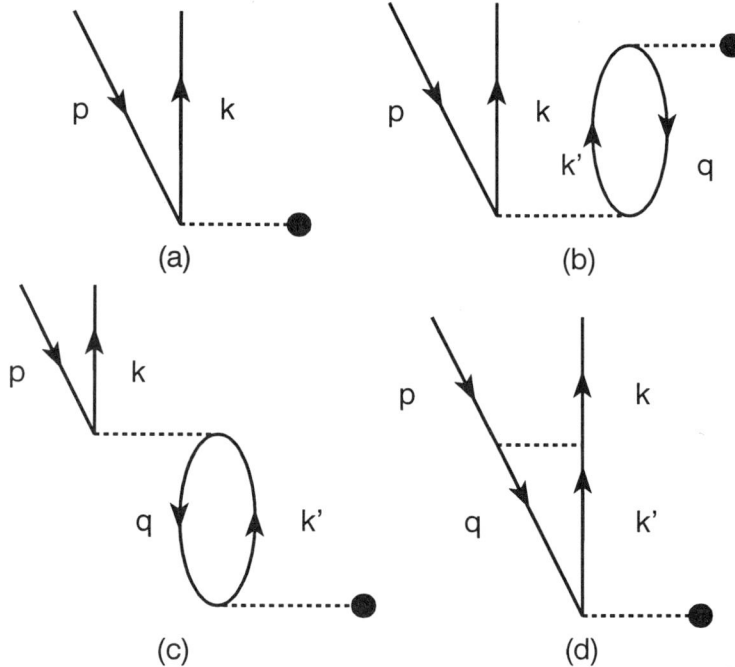

FIGURE 1. Diagrams contributing to the matrix element $\langle \Psi_k | Z | \Psi_0 \rangle$, which describes the transition $p \to k$. Dashed lines ending with a solid dot indicate a dipole matrix element (length or velocity). The other dashed lines represent Coulomb interactions. Diagrams are to be read from bottom to top corresponding to increasing time. (a) Zeroth-order diagram representing $\langle k | z | p \rangle$. (b) First-order ground-state correlation diagram. There is also a corresponding exchange diagram. (c) First-order final-state correlation diagram. (d) Exchange diagram corresponding to (c).

The term $i\eta$ has been added to the denominator of equation (25) to deal with the cases when $\varepsilon_q - \varepsilon_{k'} + \omega = 0$. The matrix elements then become complex according to equation (11). Interchannel coupling can be described in lowest order by the diagrams of Figure 1c and d with the imaginary part accounting for the exchange of oscillator strength from one photoionization channel to another. For the case when k' is a bound-excited state n', the diagrams have simple poles corresponding to the excitation $q \to n'$. Thus, the diagram of Figure 1c is the lowest-order description of an autoionizing resonance. An examination of higher-order diagrams[15] shows that certain higher-order diagrams containing the same pole repeated many times may be summed to infinite order to shift the denominator $\varepsilon_q - \varepsilon_{k'} + \omega + i\eta$ to

$$\varepsilon_q - \varepsilon_{k'} + \omega + \Delta(\omega) + i\Gamma/2 \qquad (26)$$

with $\Delta(\omega)$ being the shift in the energy position of the resonance and Γ being the linewidth of the resonance.

When the LS-coupled Hartree–Fock potential is used to calculate the excited orbitals, the diagram of Figure 1c and d vanish when p and q refer to the same subshell and k and k' have the same orbital angular momentum.[19] These interactions will be canceled by interactions with the potential in higher orders as well.

Some of the most important diagrams of second order in the perturbation H' are shown in Figure 2. The diagrams of Figure 2a and b and their exchange diagrams are part of a class of diagrams known as "bubble" diagrams which at any vertex have only a single particle line and a single hole line. Summing this class of diagrams to all orders is equivalent to the RPAE method.[13] The remaining

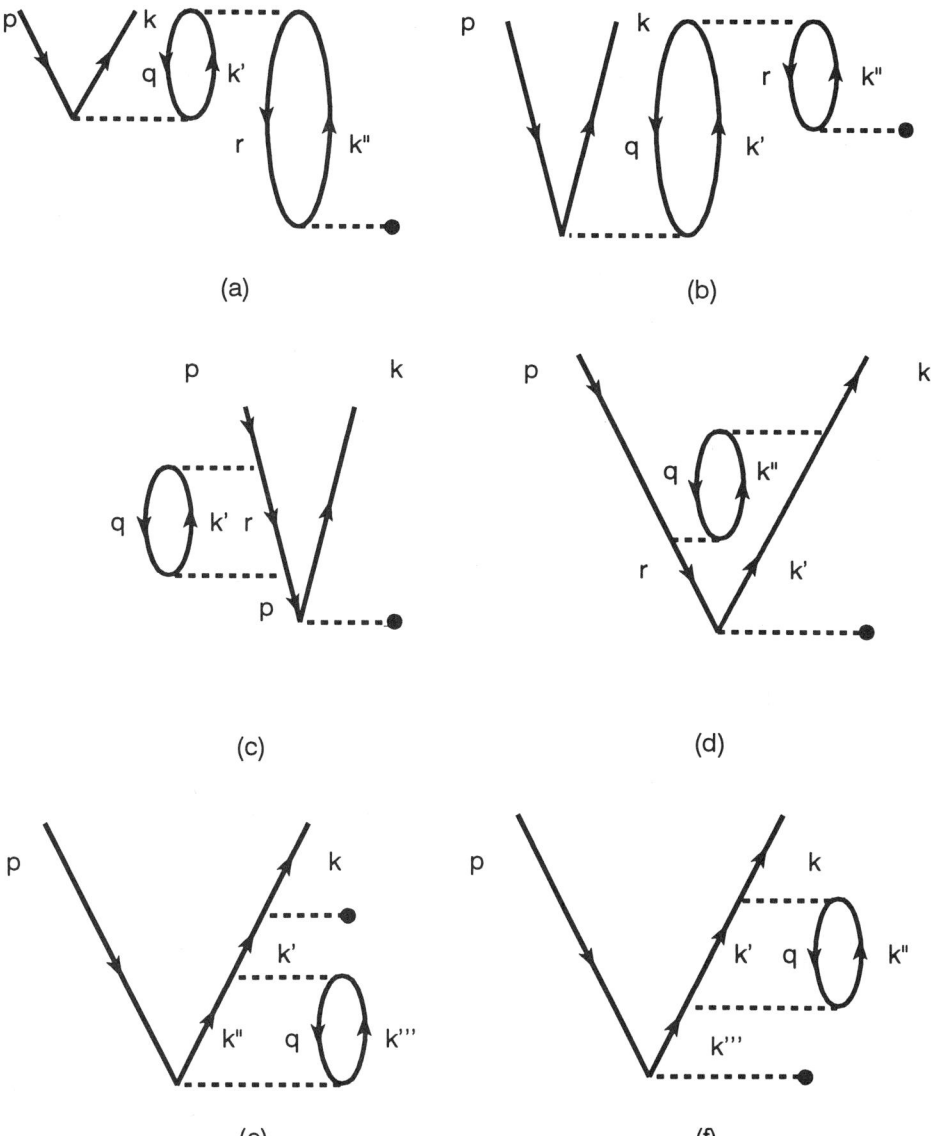

FIGURE 2. Second-order diagrams contributing to the transition $p \to k$. (a,b) Diagrams included to all orders in the RPAE. (c) Energy correction diagram on the hole line. (d) Relaxation diagram. The time ordering where the interaction on the particle line precedes the interaction on the hole line should also be included. (e) Polarization in the ground state. This diagrams contributes to the "Brueckner orbital" p. (f) Polarization diagram for the final-state orbital or inelastic scattering of photoelectron. All diagrams have exchange counterparts.

diagrams of Figure 2 are not included within the RPAE framework since at some of the vertices a hole state is scattered into a hole state or a particle state is scattered into a particle state. The diagram of Figure 2c corresponds to a shift in the single-particle energy when $r = p$ and will bring the hole energy ε_p into agreement with experiment when taken to all orders.

Relaxation effects may also be included by calculating the appropriate MBPT diagrams. The diagram shown in Figure 2d with $p = r$ and the similar diagram with the order of interaction with the hole line and particle line reversed represent relaxation. The orbital q in Figure 2d is perturbed as a result of the presence of the hole r and then interacts with the outgoing photoelectron. Diagrams shown in Figure 2e and f represent polarization in the initial state and final state, respectively. In polarization, the outgoing photoelectron excites the core and interacts with the excitation. The inclusion of diagrams typified by Figure 2e is equivalent to the use of Brueckner orbitals in the ground state. Polarization diagrams account for the effects of doubly excited channels on the main-line channel. In Figure 2f, the single-excitation channel ($p \rightarrow k'''$) interacts with the doubly excited channel ($p\, q \rightarrow k'\, k''$) which in turn couples with the single-excitation channel ($p \rightarrow k$). The interaction is with satellite channels whenever either k' or k'' is a bound-excited state and with double-photoionization channels when k' and k'' are both continuum states. Note that when both denominators in Figure 2f produce an imaginary contribution, the product is negative and the polarization diagram is of opposite sign to the lowest-order diagram (Figure 1a). It is possible for absorption flux to be distributed from the main-line channel to doubly excited channels in this way.

A technique frequently used to couple various photoionization channels is the coupled equations method[20] sometimes referred to as a K-matrix evaluation.[1] In this approach many of the final-state correlation diagrams are effectively summed to infinite order. The coupled integral equations may be written explicitly as

$$\underline{D}_p(k) - \sum_{s=q,r} \int \frac{V_{ps}(k',k)\underline{D}_s(k')}{\varepsilon_s - \varepsilon'_k + \omega}\, k' = D_p(k) \qquad (27)$$

where $\underline{D}_p$ (D_p) is the correlated (uncorrelated) dipole matrix element for channel p, $V_{p,s}(k',k)$ is the perturbation interaction matrix element between channels p and s, ε_s is the energy of an electron in state s, $\varepsilon_{k'}$ is the energy of an electron in state k', and ω is the photon energy. There is a sum and integral over excited bound and continuum states denoted by k'. By discretizing the continuum to a limited number of values of momentum k, equation (27) is converted into a matrix equation which is solved for the correlated dipole. The coupled equations are illustrated as diagrams in Figure 3, where the double line represents a correlated dipole matrix element. The coupled equations method is diagrammatically equivalent to the Tamm–Dancoff approximation including exchange contributions (RPAE with time-backward diagram contributions neglected).[13]

Besides the cross section, $\sigma(\omega)$, another frequently measured photoionization quantity is the angular distribution asymmetry parameter $\beta(\omega)$ defined by the relation

$$\frac{d\sigma}{d\Omega} = \frac{\sigma}{4\pi}[1 + \beta P_2(\cos\theta)] \qquad (28)$$

where θ is the angle between the polarization direction of the incident radiation and the momentum vector **k** of the ejected photoelectron. For closed shells in the LS-coupling approximation, β can be expressed as[1,19]

$$\beta = \frac{\ell(\ell-1)|R_{\ell-1}|^2 + (\ell+1)(\ell+2)|R_{\ell+1}|^2 - 6\ell(\ell+1)Re(\chi)}{(2\ell+1)[\ell|R_{\ell-1}|^2 + (\ell+1)|R_{\ell+1}|^2]} \qquad (29)$$

where

$$\chi = R^*_{\ell-1}R_{\ell+1}e^{i(\delta_{\ell+1}-\delta_{\ell-1})} \qquad (30)$$

In equations (29) and (30), $R_{\ell\pm1}$ represent the radial parts of the dipole matrix elements which are in general complex as a result of the final-state correlations. The phase shifts $\delta_{\ell\pm1}$ in equation (30) represent the sum of the Coulomb and non-Coulomb parts. In the case of open-shell atoms the expression for β becomes more complicated but may still be written in closed form as demonstrated by Salomonson and Boyle[21] using angular momentum graphical techniques.

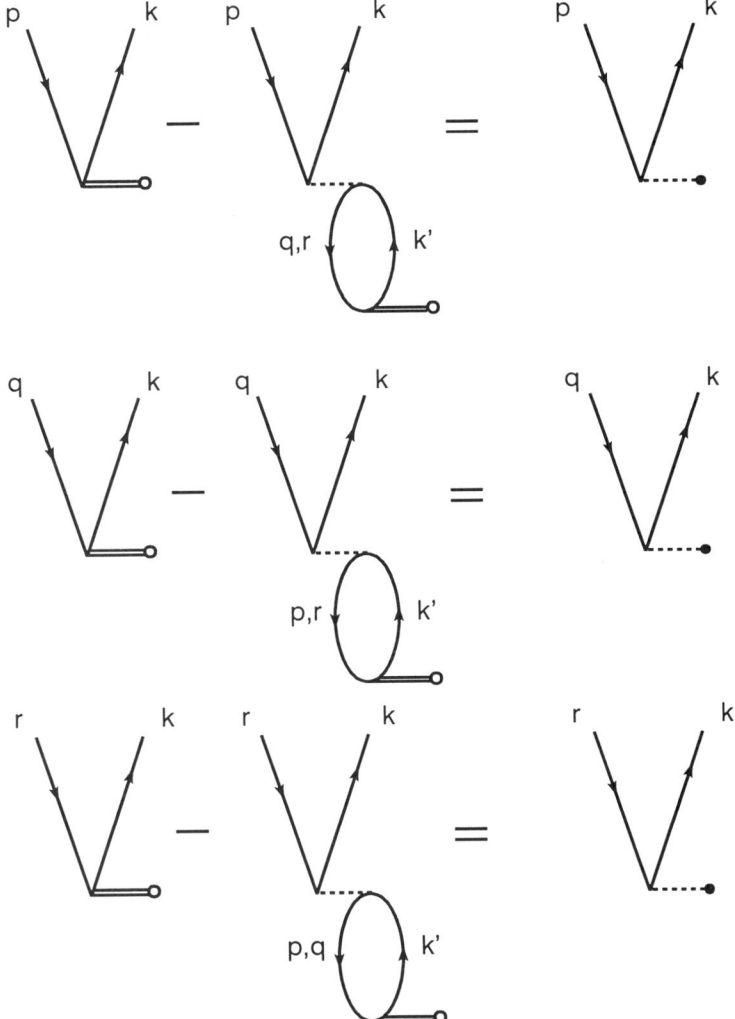

FIGURE 3. Symbolic representation of the coupled equations for three-channel coupling. The double bar line represents the correlated dipole matrix element.

3. ONE-ELECTRON PHOTOIONIZATION

In this section specific examples are given to illustrate the application of MBPT to calculations in which one photon is absorbed and one electron is ejected. Effects of many of the diagrams will be illustrated. Photoionization with excitation will be reserved for the next section.

3.1. Argon

It is not always necessary to carry out an MBPT calculation to high order. Consider the fairly early calculation of the $3p^6$ subshell of atomic argon by Kelly and Simons[22] shown in Figure 4. The curves labeled HFL and HFV are the lowest-order (Hartree–Fock) length and velocity curves. Although neither the length nor velocity results are in good agreement with experiment, the geometric mean of length and velocity is in fair agreement. This is the result of an argument given by Hansen[23] that the

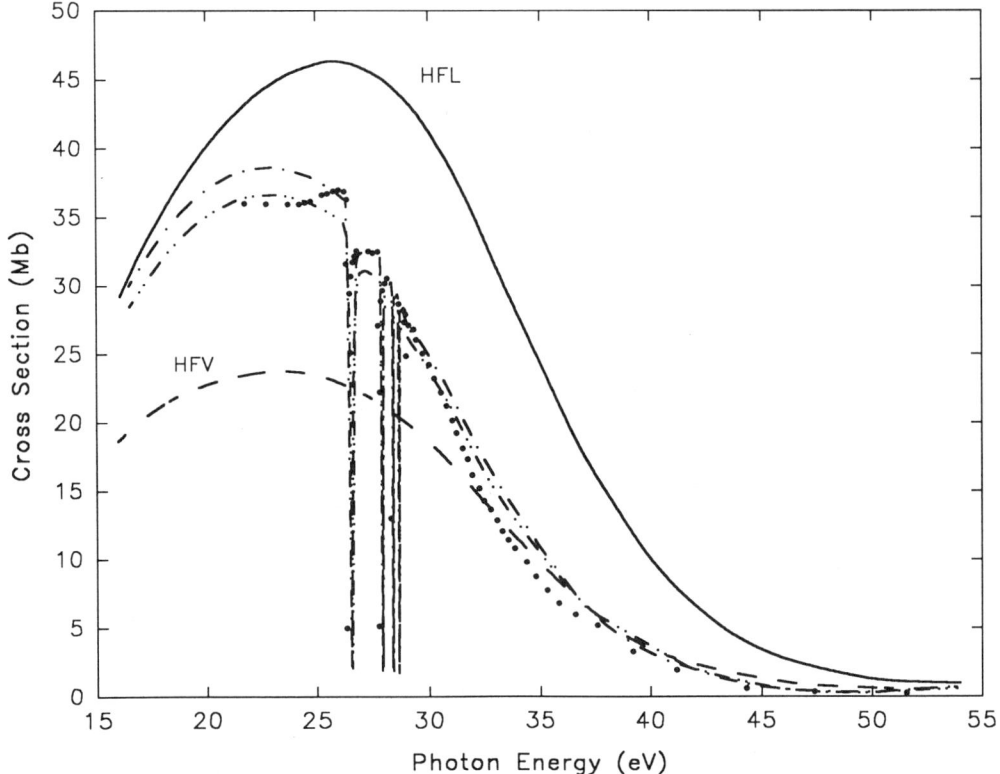

FIGURE 4. Photoionization cross section for argon (from Ref. 22). The solid and dashed lines represent the Hartree–Fock length and velocity cross sections, respectively. The dot–dashed line and the double-dot–dashed line represent length and velocity calculations including higher-order terms. The solid dots represent experimental data from Ref. 24. Only the first few $3s \rightarrow np$ resonances preceding the $3s$ threshold at 29.24 eV are shown.

geometric mean of the length and velocity dipoles is less sensitive to correlations of the type shown diagrammatically in Figure 1b than either the length or velocity dipoles individually. Including the ground-state correlation diagrams typified by Figure 1b brought the length and velocity cross sections into much closer agreement with one another.

Final-state correlations were included by evaluating the basic resonance diagram of Figure 5a. The autoionizing window resonances preceding the $3s$ threshold at 29.24 eV are generated when $np = 4p, 5p$, etc. The diagram of Figure 5b was also included with the imaginary part taken for the denominator of the $3p \rightarrow k'$ excitation. The diagrams of Figure 5a and b are the first two members of a geometric series with ratios given by Figure 5c divided by an energy denominator. This series can be summed to all orders resulting in a $3s \rightarrow np$ denominator shift from $\varepsilon_{3s} - \varepsilon_{np} + \omega$ to $\varepsilon_{3s} - \varepsilon_{np} + \Delta + i\Gamma_n/2$. It is also necessary for argon to consider correlation modifications of the dipole matrix element of the basic resonance diagram of Figure 5a as shown in Figure 5d. These are important in this case since the dipole matrix elements $\langle kd|z|3p\rangle$ are larger than $\langle np|z|3s\rangle$. At the resonance, the diagram of Figure 5d with the bottom denominator treated according to $-i\pi\delta(\varepsilon_{3p} - \varepsilon_{k'} + \omega)$ cancels the lowest-order dipole of Figure 1a. This cancelation leads to the "windowlike" resonances shown in Figure 4. The correlated curves of Figure 4 also include the second-order RPAE-type diagrams of Figure 2a and b.

The agreement with the experimental results for the total cross section of Madden et al.[24] is very good even for this low-order calculation. It should be noted that some second-order effects such as core relaxation and polarization have not been included and that photoionization-with-excitation

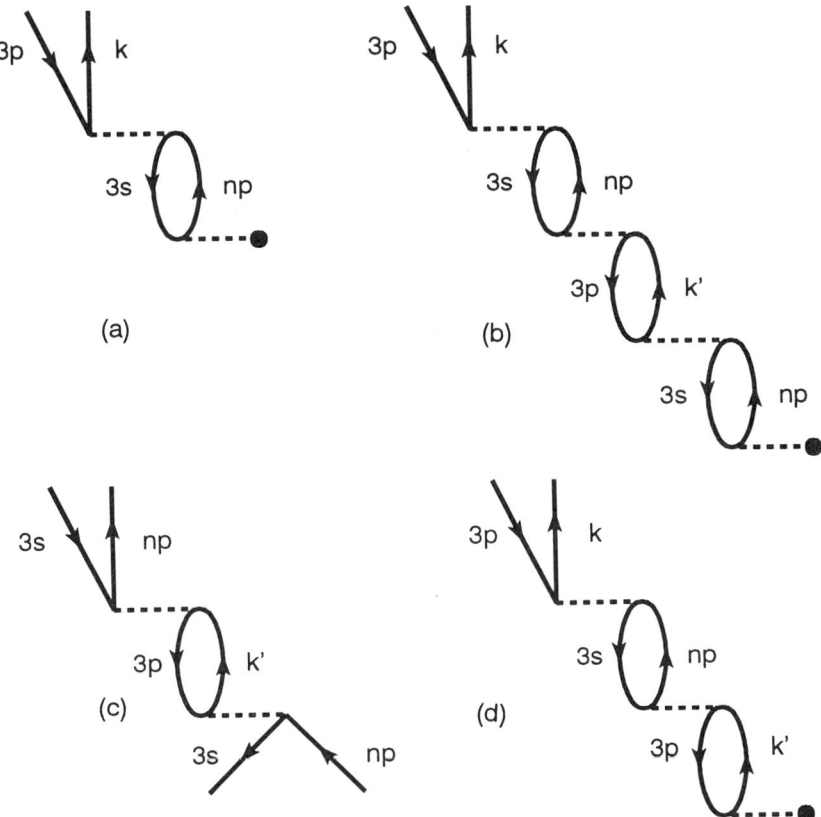

FIGURE 5. Resonance diagrams for argon calculation of Ref. 22. (a) Basic resonance diagram. (b) Next diagram in geometric series which results in shifted denominator of diagram (a). (c) Segment by which (b) differs from (a). (d) Diagram modifying dipole matrix element $\langle np|z|3s \rangle$.

channels have been neglected in this calculation. Calculations of argon that include photoionization with excitation and double ionization will be discussed later.

3.2. Chlorine

Open-shell systems present particular challenges to theorists. Calculations become more complex because of the existence of a larger number of strongly interacting channels. Although neutral chlorine, with ground-state configuration $3s^2 3p^5(^2P)$, is only one electron short of being a closed-shell system, there are nine possible photoionization channels for $3p$ electrons in LS coupling as compared with only two channels for argon. When a $3p$ electron is photoionized ($3p \to kd$ or $3p \to ks$), three core couplings (^1S, ^1D, ^3P) combine with three final couplings (^2S, ^2P, ^2D).

Brown et al.[20] found large effects related to interchannel coupling as represented by the diagram of Figure 1c in which p and q refer to different ionic cores. For example, $p \to k$ might refer to $3p^5(^2P) \to 3p^4(^3P)kd(^2D)$ with $q \to k'$ referring to $3p^4(^1D)k'd(^2D)$ or $3p^4(^1S)k'd(^2D)$. Diagonal contributions are eliminated by choosing the potential as mentioned above. The ionic state with the lowest ionization energy is the ^3P ionic core. Thus, resonances are found in the ^3P cross section leading up to the ^1D and ^1S ionization edges. These resonances also are accounted for by the final-state correlation diagrams of Figure 1c. Slow convergence of the time-forward diagrams led Brown et al.[20] to develop the coupled integral equations approach, discussed above and shown diagrammatically in Figure 3, which calculated certain final-state correlation diagrams to all orders. Note that only channels with the

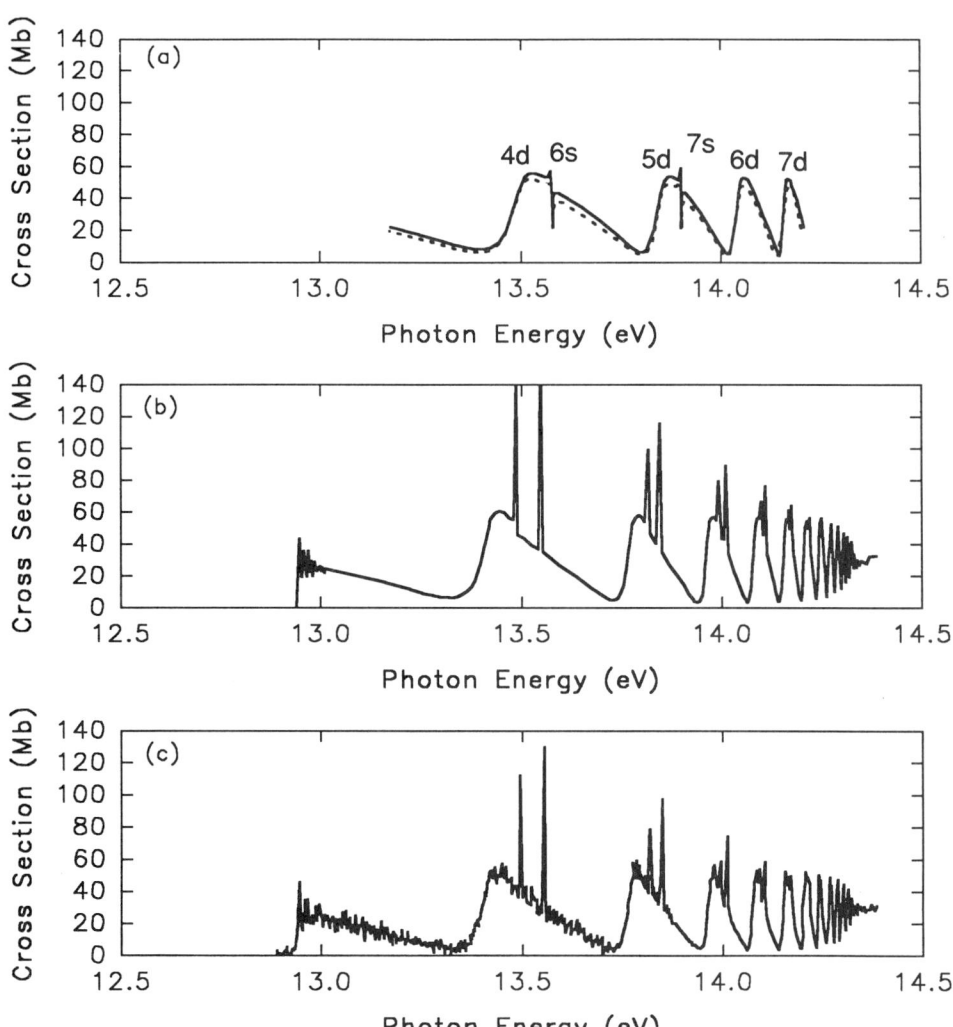

FIGURE 6. Photoionization cross section of the ^2P ground state of Cl between the ^3P and ^1S thresholds. (a) Solid and dotted lines represent calculated length and velocity cross section, respectively, from Ref. 20. (b) Eigenchannel R-matrix length calculation from Ref. 25. (c) Experimental cross section from Ref. 26.

same total L and S may be coupled in this way. The coupled equations calculation would be identical to solving the K-matrix or performing a close-coupling calculation on the continuum orbitals except that Brown et al.[20] have also included ground-state correlations to first order which tend to bring the length and velocity cross sections closer together.

The sum of the ^1D nd,ns resonances in all of the ^2P, ^2D channels is shown in Figure 6. The resonances were obtained via the coupled-equations method. The $(^1D)nd(^2D)$ and $(^1D)nd(^2P)$ series of resonances are predicted to be broad and add to look like a single broad series. A narrow series for $(^1D)ns(^2D)$ series was also predicted. Experiment shows the single broad resonance series as predicted but the narrow series is spin-orbit split into two series. Very recently, Robicheaux and Greene[25] have published impressive results for the Cl resonances using the eigenchannel R-matrix approach and the multichannel quantum-defect theory (MQDT) to extend the wave functions to distances larger than the R-matrix box radius. These recent results illustrate the importance of spin-orbit mixing. The recent R-matrix results are shown in Figure 6 along with the earlier predictions of Brown et al.[20] and the experiment.[26]

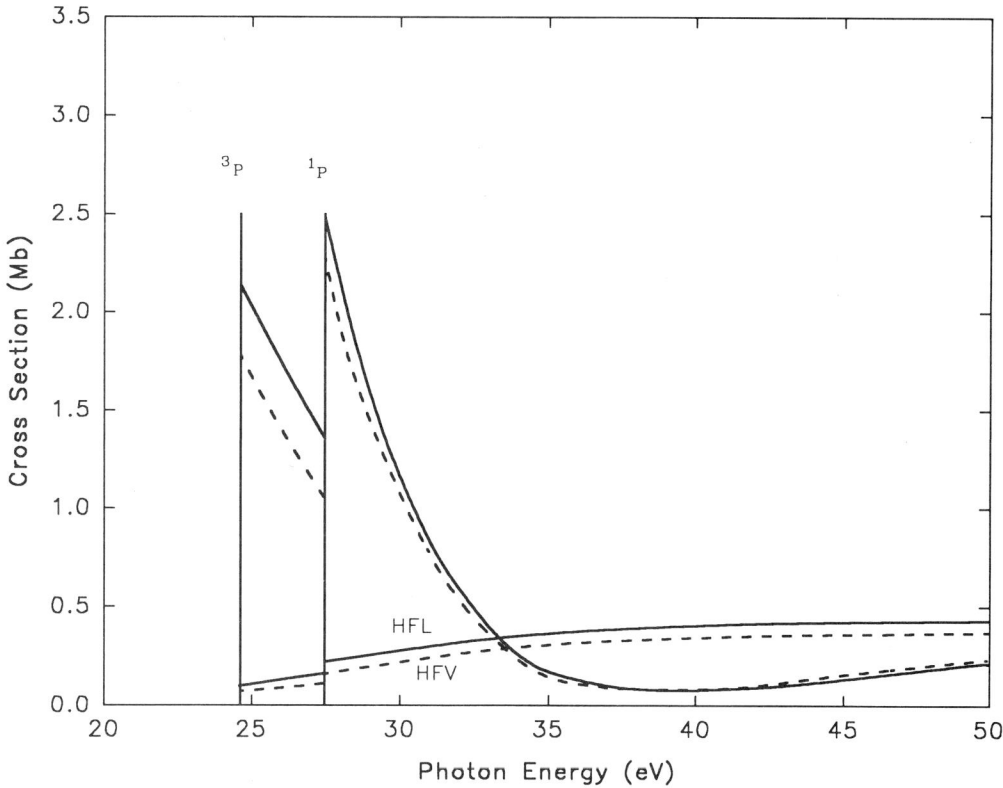

FIGURE 7. Correlated photoionization cross section for the sum of the $3s \to kp$ transitions from Ref. 20. Solid line is length calculation, dashed is velocity. Lowest-order values are labeled HFL and HFV and are shown for comparison. Resonance structure is not included in this figure.

The $3s$ photoionization cross section shown in Figure 7 demonstrates the large amount of correlation occurring in this system. Interchannel coupling with the $3p^5$ shell causes a large increase in the $3s$ cross section near threshold, an effect also to be noted for argon.[27] A type of Cooper minimum has been introduced into the $3s$ cross section because of correlations.

The total photoionization cross sections as calculated by a variety of techniques[28–31] are shown in Figure 8. The experimental measurements by Samson et al.[32] are also shown for comparison. There is general agreement among the theories that there is important coupling between final-state channels with the same L and S, e.g., $3p^4(^3P)kd(^2D)$, $3p^4(^1D)kd(^2D)$, $3p^4(^1S)ks(^2D)$, and $3p^4(^1S)kd(^2D)$. Such coupling has the effect of both reducing the cross section and broadening the peak cross section.

It is of interest to note that the sophisticated calculations lead to cross sections that have features more similar to the argon cross section than to the chlorine Hartree–Fock cross section. This led to the search for a single-particle potential that would reproduce the effects of coupling between the various multiplet channels. Qian et al.[33] and Boyle et al.[34] were able to find such an "effective potential" by considering the ionic multiplets as degenerate in energy. The form of the "effective potential" derived by Qian et al.[33] is for given M_L

$$V_{LS_{\text{ave}}}(M_L) = \frac{\sum_F V_F |\langle F|Z|G\rangle|^2}{\sum_F |\langle F|Z|G\rangle|^2} \tag{31}$$

where $|G\rangle$ represents the LS-coupled ground state, $|F\rangle$ represents the final channel of interest, $|I\rangle$ may be any channel of the same ℓ, and V_F is given by

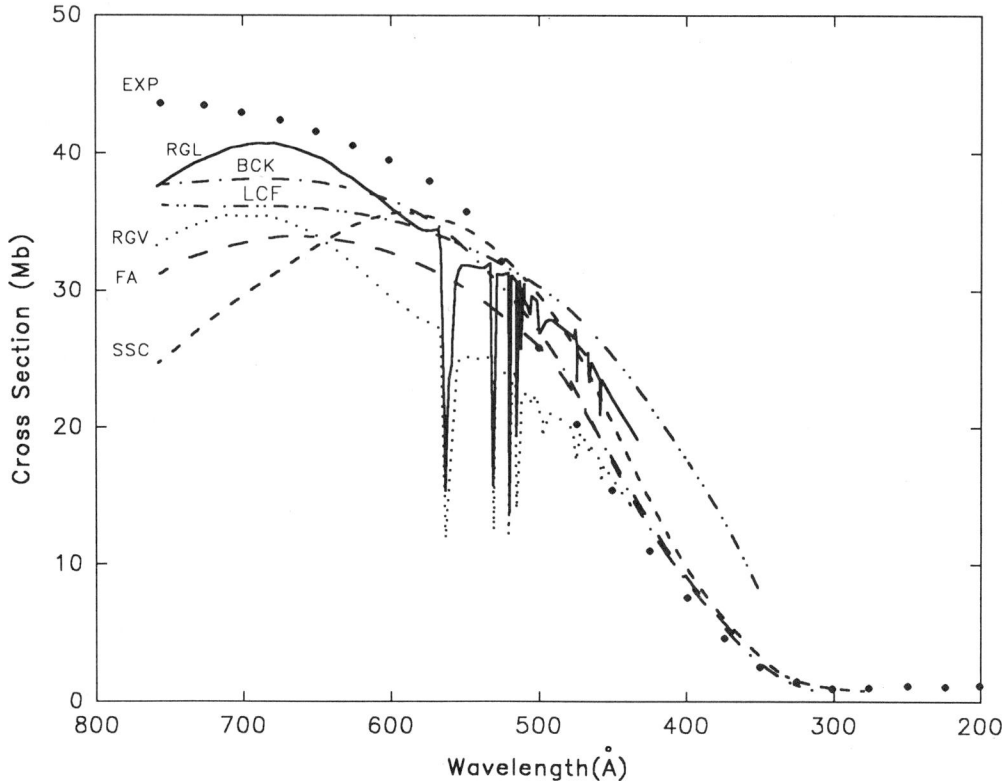

FIGURE 8. Comparison of the absolute photoionization cross sections of atomic chlorine with theoretical results. Experiment (EXP) by Samson et al.[32] Theoretical: BCK is MBPT result from Ref. 20, RGL and RGV are length and velocity eigenchannel R-matrix calculations from Ref. 31, LCF is R-matrix calculation from Ref. 30, FA and SSC are open-shell RPAE from Refs. 28 and 29, respectively.

$$V_F = \frac{\sum_I \langle F|v|I\rangle \langle I|Z|G\rangle}{\langle F|Z|G\rangle} \tag{32}$$

This potential for specific M_L is then averaged over the possible M_L's to give

$$V_{LS_{ave}} = \frac{1}{(2L+1)} \sum_{M_L} V_{LS_{ave}}(M_L) \tag{33}$$

In Figure 9 a comparison is made between the geometric means of the Hartree–Fock, the coupled equations result, and the effective single-particle potential methods. The effective potential has in an approximate way included the effects of interchannel interaction using a minimal amount of computer resources. In fact, because this method greatly reduces the number of different continuum states based on different ionic multiplets for open-shell atoms, the effective potential is even more convenient than the Hartree–Fock method.

3.3. Giant Resonances

An interesting phenomenon observed in open-shell atoms occurs when an electron makes a resonant transition into an open shell having the same principal quantum number. For example, in

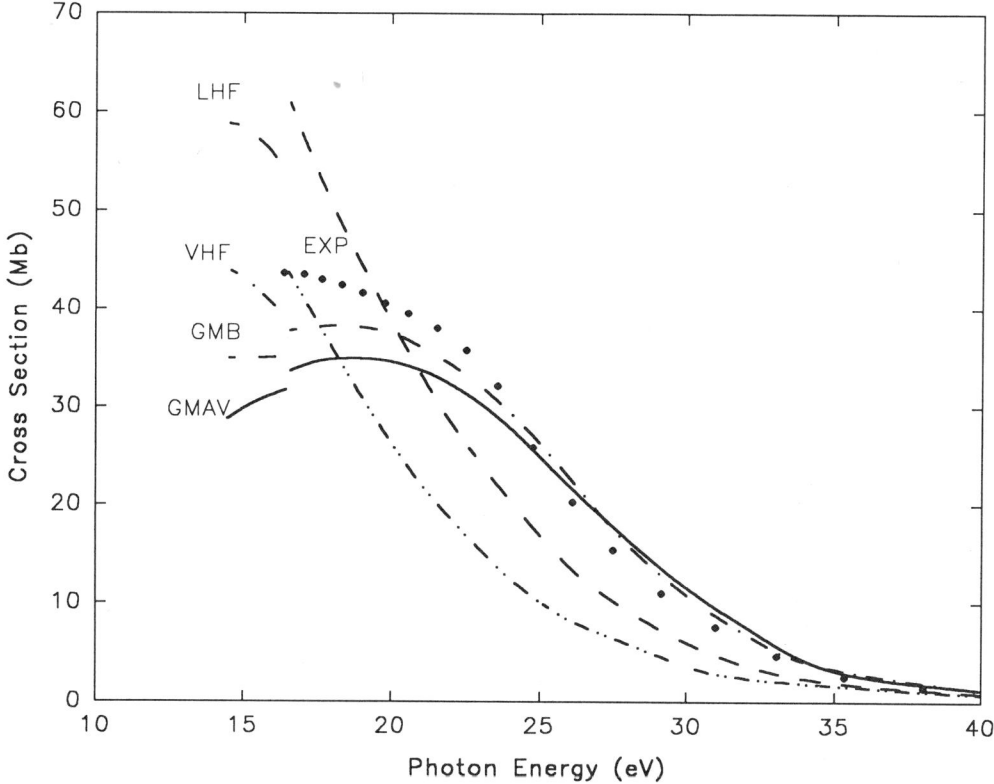

FIGURE 9. Photoionization cross section of $3p$ subshell of chlorine. Experiment (EXP) by Samson et al.[32] LHF and VHF are length and velocity Hartree–Fock calculations, respectively. GMB is the geometric mean of correlated length and velocity calculations from Ref. 20. GMAV is the geometric mean of low-order calculations using the effective potential from Ref. 33.

transition metal atoms with ground-state configurations $3p^63d^n4s^2$ or $3p^63d^n4s$ transitions of $3p$ electrons into the open $3d$ shell give rise to resonances in the $3d$ and $4s$ cross sections. The lowest-order MBPT diagrams giving rise to these resonances are shown in Figure 1c with $p = 3d$ or $4s$ and $q = 3p$, $k' = 3d$.

Atomic manganese with ground-state configuration $3p^63d^54s^2(^6S)$ is particularly simple in LS coupling since the $3d$ subshell is half filled with spin-aligned electrons. Garvin et al.[35] calculated the photoionization cross section in the resonance region using both the coupled equations approach in LS coupling and the interacting resonance method which included spin-orbit splitting as a perturbative effect.

The geometric mean of length and velocity results for the two methods employed by Garvin et al.[35] are shown in Figure 10. The upper dot–dashed curve is the $3d$ coupled equations result which included RPAE diagrams of Figure 2a and b. It is in excellent agreement with the spin-polarized RPAE calculation of Amusia et al.[36] The lower curve is from the interacting resonance method.[35]

Recent experimental measurements are from Cooper et al.[37] and were normalized to the MBPT at 65 eV. With this normalization the interacting resonance calculations are in good agreement with experiment near the $3p \rightarrow 3d$ peak.

The $3p \rightarrow 3d$ excitation energies were calculated by first obtaining the first-order ΔSCF (difference between the self-consistent Hartree–Fock energies for the $3p^53d^6$ excited state and the ground state) value of 53.4 eV. Corrections to the ΔSCF value were computed by evaluating second-order energy correlation diagrams for transitions into the $3p$ hole and estimates of pair

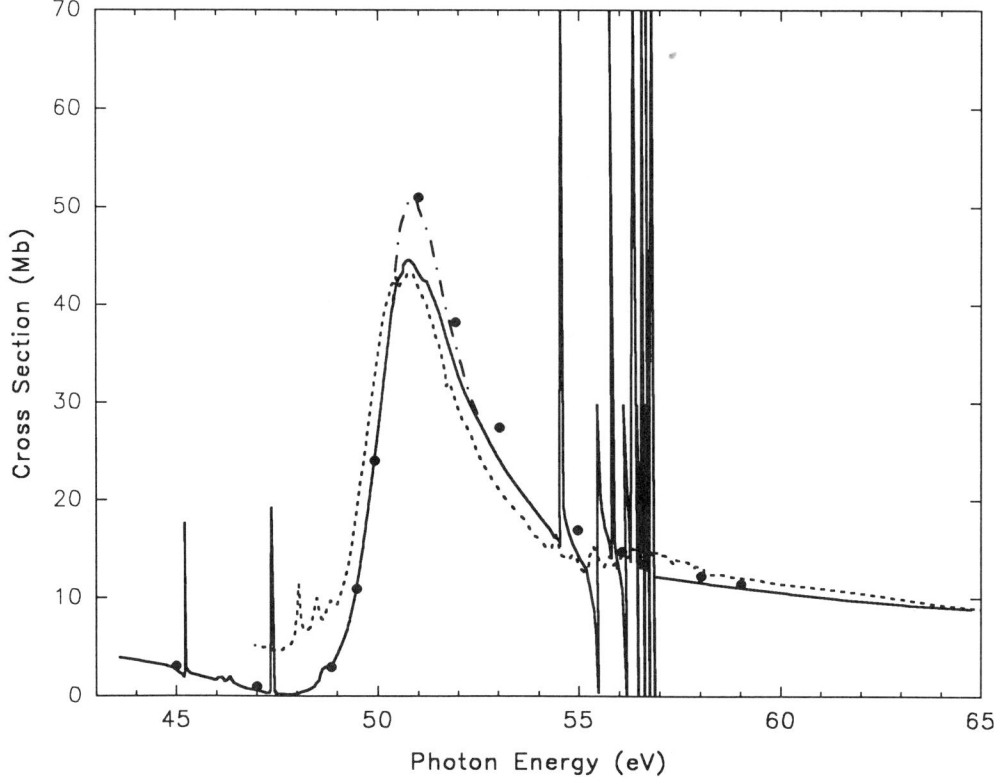

FIGURE 10. Resonance structure in $3d$ subshell cross section of atomic manganese. Experimental data are from Ref. 37 and have been normalized to the MBPT calculations at 65 eV. The solid line is by Garvin et al.[35] and uses the interacting resonance theory including spin-orbit effects. The upper dot–dashed line is also by Garvin et al.[35] but uses the coupled equations method.[20] Solid dots represent the RPAE calculation of Amusia et al.[36]

correlations among $3p$ and $3d$ electrons. The corrected energy value of 50.4 eV is in reasonable agreement with experiment.

Additional resonances related to $3p^53d^5(^7P)nd$ and ns bound-excited states preceding the $3p^53d^5(^7P)$ edge at 57 eV were predicted by the MBPT calculations but were not detected by experiments. The calculated resonances are narrow and tall because of the long lifetimes of the excited states. Spin prevents $3d$ electrons in the excited state from filling the $3p$ hole. A recent calculation by Chang et al.[38] has demonstrated that including the decay of $4s$ electrons has the effect of broadening the series. Resonances leading to the $3p^53d^5(^5P)$ edge at 74.3 eV were effectively eliminated in the calculations of Garvin et al.[35] by $3d$ electrons decaying into the $3p$ hole.

The lanthanides with ground-state structure $4d^{10}4f^n5s^25p^66s^2$ also have giant resonance structure with $4d$ electrons excited into the open $4f$ shell. Atomic europium with a half-filled $4f$ shell has a particularly simple structure in LS coupling. Pan et al.[39] employed MBPT utilizing the coupled equations method to obtain partial cross sections for $4d$, $4f$, and $5p$ cross sections in the vicinity of the $4d \rightarrow 4f$ resonance. One of the interesting features found by Pan et al.[39] is that in LS coupling the bound-excited state $4d^94f^8(^8P)$ at 141 eV lies above the $4d^94f^7(^9D)$ threshold at 139.6 eV. Thus, a resonance occurs just above threshold in the $4d^94f^7(^9D)\varepsilon f$ and εp cross sections.

Results of the MBPT calculation of Pan et al.[39] are shown in Figure 11 along with recent photoemission measurements of Becker et al.,[40] Meyer et al.,[41] and Richter et al.[42] Other theoretical calculations compared in the figure are those of Amusia et al.[43,44] and relativistic time-dependent local density approximation (RTDLDA) calculations by Zangwill.[45] The $4f$ photoemission measurement

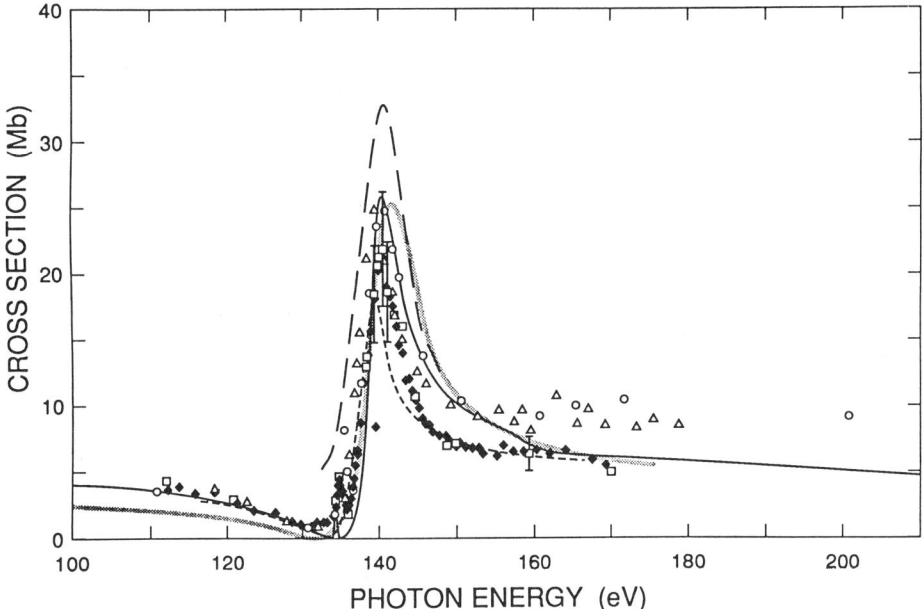

FIGURE 11. Partial 4f photoionization cross section in the vicinity of the $4d \to 4f$ resonance from Ref. 39. Theoretical curves: Solid line is MBPT calculation, Ref. 39, long dashed line is $4f \to kg$ cross section of Ref. 43, short dashed line is relativistic time-dependent local density approximation calculation of Ref. 45, and the thick gray line is the spin-polarized RPAE calculation of Ref. 44. Experimental: Circles are from Ref. 40, triangles are from Ref. 41, squares and diamonds are from Ref. 42.

by Becker et al.[40] has been normalized to the MBPT calculation at 111 eV; the experimental results by Meyer et al.[41] are normalized to those of Becker et al.[40] at the peak of the resonance; the measurements of Richter et al.[42] are normalized to the MBPT between 160 and 170 eV. The calculations compare reasonably well with the experiments except at higher energies where two of the experimental results indicate more absorption in the 4f channels than the theory predicts. In Figure 12, the europium 4d and 5p partial cross sections are compared with experiment,[40–42] with the spin-polarized RPAE calculation by Amusia et al.,[44] and with the RTDLDA calculations by Zangwill[45] and Zangwill and Doolen.[46] The MBPT calculations of angular asymmetry parameters β_{4f}, β_{4d}, and β_{5p} were also performed by Pan et al.[39] with the average of β_{5p} for $5p^{-1}(^9P)$ and $5p^{-1}(^7P)$ ionic states weighted by the cross sections as follows:

$$\beta = \frac{\sum_i \sigma_i \beta_i}{\sum_i \sigma_i} \qquad (34)$$

The β parameters for the 4f and 5p subshells are shown in Figures 13 and 14. The Hartree–Fock calculation is also given for comparison. The importance of correlation in the calculation of β is obvious in this case.

An example of the application of MBPT to a more complex open-shell system is the recent calculation of photoionization of atomic tungsten by Boyle et al.[34] With ground-state configuration $5p^6 4f^{14} 5d^4 6s^2 (^5D_0)$ strong resonances related to transitions from the $5p^6$ level into the $5d^4$ level are expected. Photoionization measurements of tungsten using laser ablation have recently been reported by Costello et al.[47] and experimental results are also available for metallic tungsten.[48]

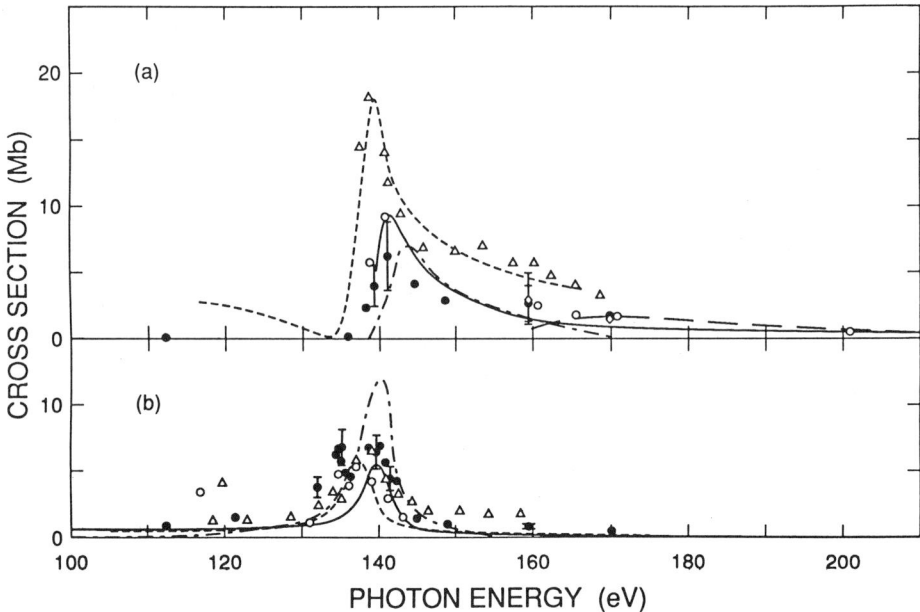

FIGURE 12. Partial photoionization cross sections for the $4d$ and $5p$ subshells near the $4d \rightarrow 4f$ resonance from Ref. 39. (a) $4d$ cross section and (b) $5p$ cross section. Theoretical: Solid line is MBPT calculation leaving the ion in $4d^{-1}(^9D)$ ionic state, long dashed line MBPT calculation leaving the ion in $4d^{-1}(^7D)$ ionic state from Ref. 39. Short dashed curve is results of relativistic time-dependent local density approximation, Refs. 45 and 46. Dot–dashed line is spin-polarized RPAE calculation of Ref. 44. Experimental: Circles are experimental results of Ref. 40, triangles are from Ref. 41, and solid circles and diamonds are from Ref. 42.

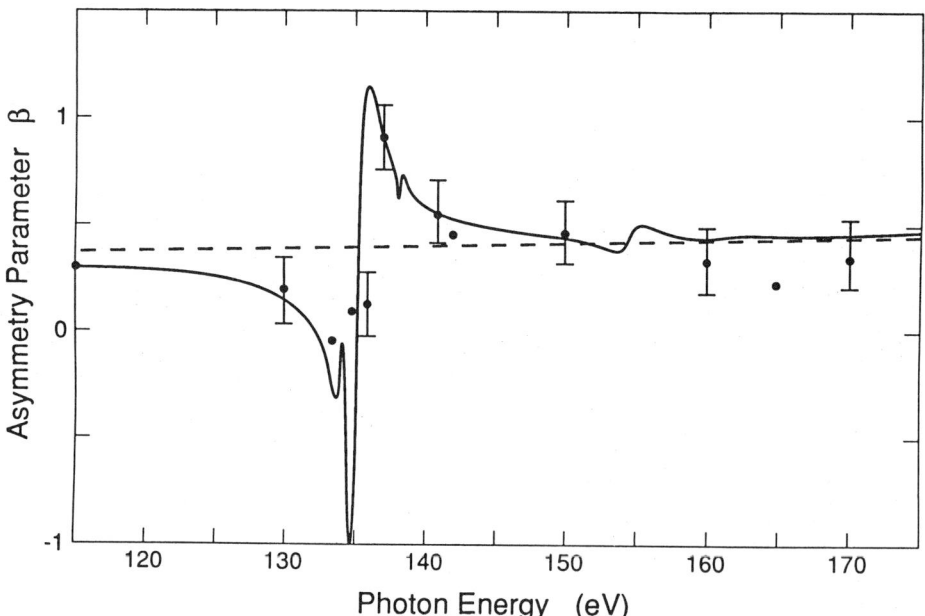

FIGURE 13. Asymmetry parameter β_{4f} near the $4d \rightarrow 4f$ resonance from Ref. 39. Experimental data points (solid dots) are from Ref. 51. Solid curve represents MBPT results including correlation effects from Ref. 39. Dashed line represents Hartree–Fock results, Ref. 39.

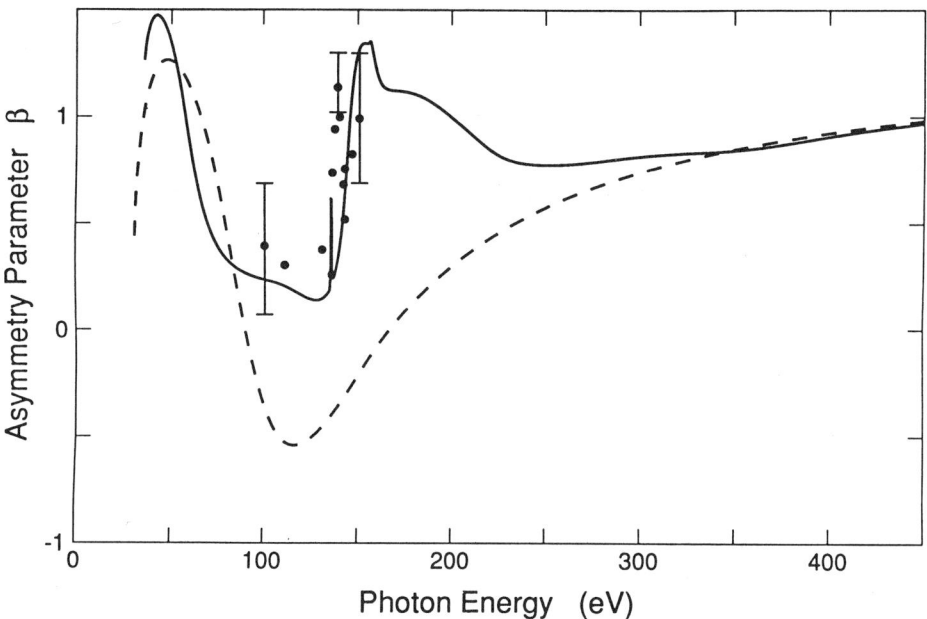

FIGURE 14. Asymmetry parameter β_{5p} from Ref. 39. Average of $5p^{-1}(^9P)$ and $5p^{-1}(^7P)$ β parameters have been taken according to the prescription in equation (34). Curves and symbols are defined as in Figure 13.

Three different calculations were made each having a different representation for the ground state. The different approximations were

1. A single configuration ground state $5d^4(^5D_{J=0})$
2. A ground state diagonalized with respect to the spin-orbit Hamiltonian using the eigenstates $5d^4(^5D, {}_2^3P, {}_4^3P, {}_0^1S, {}_4^1S)$ all coupled to $J=0$ where the subscripts indicate seniority number
3. A sum over nearly degenerate ground states $5d^4(^5D_{J=0, 1, 2, 3, 4})$ weighting each partial cross section by $(2J+1)$.

Each of these three approximations gave very different results. The $5d \to 5d$ and $4f \to 5d$ resonances were calculated using the "generalized resonance" approach developed by Garvin[49] but applied for the first time by Boyle et al.[34] in the tungsten calculation. The generalized resonance method includes the real parts of resonance diagrams to first order in the correlation and the imaginary part to all orders. The coupled equations method was impractical for tungsten because of the 78 resonance states in *LSJ* coupling and 43 possible continuum decay channels. The resulting matrix to be inverted for each ω value by the coupled equations is intractably large. The generalized resonance scheme requires the inversion of only a 43 × 43 matrix and is therefore much more efficient.

The tungsten calculation that reproduces the laser-ablation experiment best is the second one mentioned above with spin-orbit effects included in the ground state. The result of this calculation is shown in Figure 15 where it is compared with experiment.[47] Above approximately 43 eV, the theory and experiment compare favorably; however, the large structure at approximately 37 eV has not been reproduced by the theory. Boyle[21] has suggested that including the ground state $3d^54s$ may resolve the discrepancy. The first type of calculation where spin-orbit effects are omitted from the ground state yields a deeper dip at approximately 42 eV but does not agree well with experiment at higher energies. The third method of treatment for the ground state yields a broader, reduced peak in the resonances near 47 eV and appears to agree better with the experiment for tungsten metal[48] than with the laser-ablated results of Costello et al.[47]

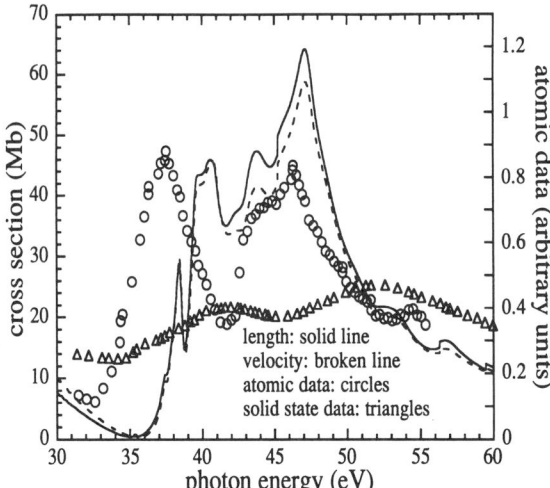

FIGURE 15. Total photoabsorption cross section for tungsten from Ref. 34. Solid curve is MBPT length form and dashed curve is velocity form for the intermediate coupling initial state $5d^4(^5D, {}^3_2P, {}^3_4P, {}^1_0S, {}^1_4S)_{J=0}$. Open circles represent the atomic data of Ref. 47. Triangles represent the solid state data of Ref. 48.

3.4. Shape Resonances in Photoionization of the $4d^{10}$ Subshell

The inner-shell $4d$ photoionization spectra of many elements are characterized by a broad, delayed absorption peak. These strong absorption peaks have been referred to as giant resonances and have been the topic of much recent discussion. Three closed-shell systems with an inner $4d$ subshell—cadmium, xenon, and barium—have been the subjects of many theoretical studies. Much of the interest in these systems is related to the sensitivity of the calculations to correlation effects. The MBPT can be useful in determining the role of such effects.

The ground-state configuration of neutral cadmium is $4d^{10}5s^2(^1S)$. Photoelectrons are emitted from the $4d^{10}$ subshell as either kf or kp orbitals. Herman–Skillman calculations which use a local potential to approximate the exchange interaction do not agree well with photoabsorption experiments. Carter and Kelly[50] obtained much better agreement with experiment using the Hartree–Fock approximation. They also used low-order MBPT including ground-state correlation diagrams (see Figure 1b) and RPAE diagrams (see Figure 2a and b) to obtain better length and velocity agreement. Another interesting correlation effect, namely, relaxation, was included in an approximate way and was found to more than cancel the effect of the RPAE diagrams.

Relaxation occurs when the ionic core readjusts to the hole left by the outgoing photoelectron. Near the photoionization threshold, the slow photoelectrons experience a change of potential as the core orbitals relax to become ionic orbitals. At sufficiently small photoelectron energies, it is reasonable to assume that relaxation is complete and to approximate the relaxation process near the photoionization threshold by calculating final-state orbitals in the potential of a relaxed ionic core.

The calculations of Carter and Kelly[50] are shown in Figure 16 along with the experimental measurements of the partial $4d$ cross section of Becker.[51] The solid curve is the geometric mean of the first-order MBPT calculation for the $4d \rightarrow kf$ plus $4d \rightarrow kp$ cross sections. The broken curve is the result including relaxation.

The importance of relaxation in the photoionization of $4d$ electrons in atomic xenon $(4d^{10}5s^25p^6)$ has been investigated by a number of workers. Experiments have been performed measuring total absorption[52,53] and photoelectron spectroscopy has been used to isolate the amount of absorption resulting from single photoionization of $4d$ electrons.[54,55] The measurements have found that a substantial fraction of the total absorption is related to satellite channels (photoionization with

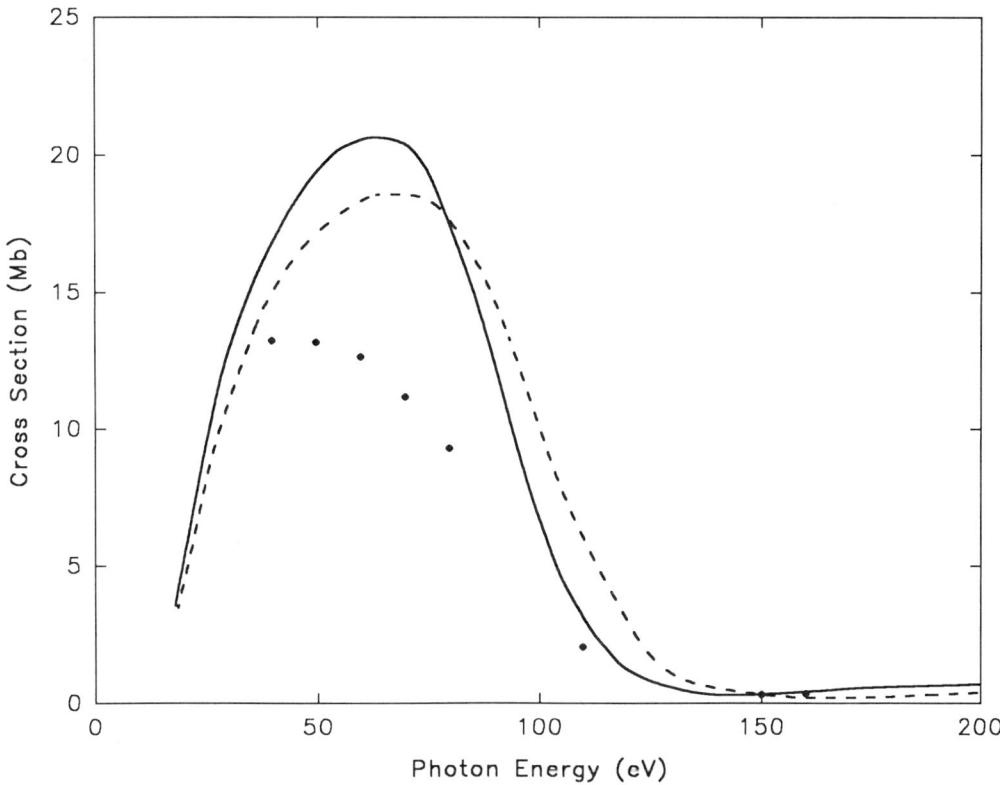

FIGURE 16. Cadmium photoionization cross section in region of 4d excitation. Experimental points from Becker.[51] Full curve is geometric mean of length and velocity results from the first-order MBPT calculation from Ref. 50. Broken curve includes relaxation effects and is also from Ref. 50.

excitation) and double photoionization. It is interesting to determine how well the theoretical calculations can reproduce the photoemission measurements.

One approximate method for accounting for the reduction in the 4d main-line absorption flux related to satellite channels is to use the "sudden" or "shake-off" approximation where the probability amplitude for doubly excited channels is calculated from the overlap integral between relaxed and unrelaxed atomic orbitals. This method has been discussed extensively by Åberg[56] and is only applicable at high photoelectron energies. Low-order MBPT calculations were carried out[57] for xenon with excited states calculated in the potential of a relaxed ion and evaluating a determinant of the relevant matrices of overlap integrals. This method somewhat improves the theoretical partial 4d cross section. The results shown in Figure 17 match the peak in the 4d cross section at approximately 100 eV reasonably well but are below the experimental points near threshold and above the experimental beyond the peak.

While the effects of relaxation may be approximated by the choice of a relaxed potential, the energy dependence of the relaxation process is ignored by this technique. An alternative approach is to evaluate the second-order MBPT diagrams which correspond to relaxation. One time ordering of the relaxation diagrams was shown in Figure 2d with $p = r$. When these diagrams and higher-order iterations of the same diagrams were included in the MBPT calculations,[57] the resulting 4d cross section for xenon was found to be similar to those obtained via the relaxed orbital method (see Figure 17). Both methods are larger than experiment beyond the peak suggesting that polarization diagrams (Figure 2f) representing loss of flux to channels with multiple excitations also need to be included.

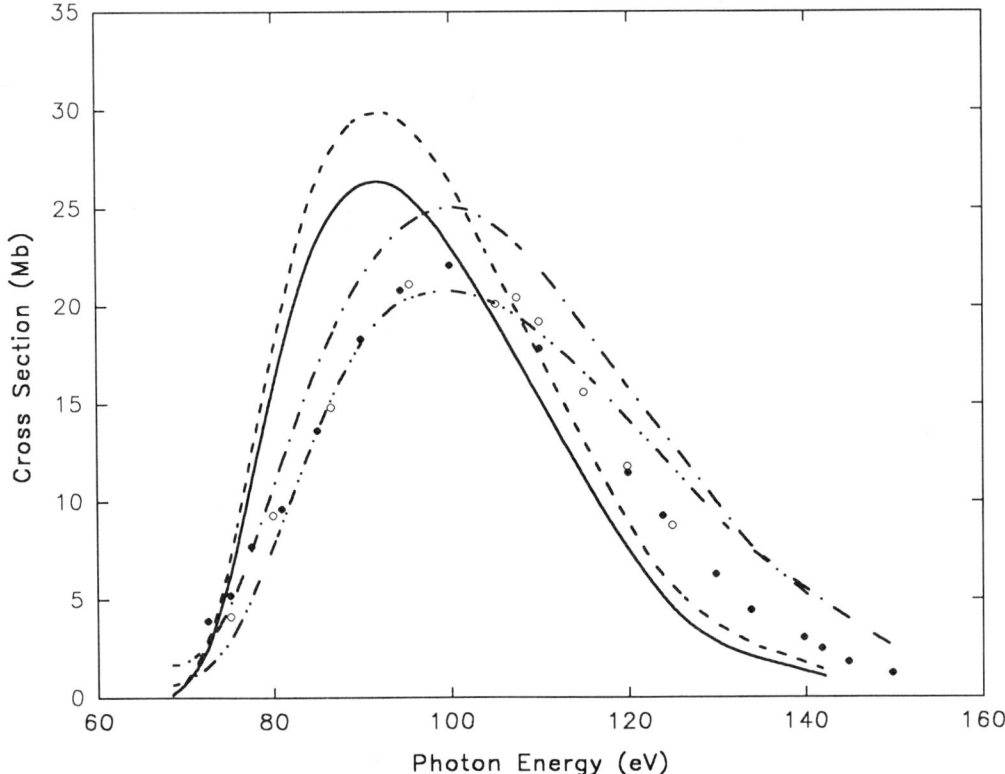

FIGURE 17. Photoionization cross sections of atomic xenon as a function of photon energy. Solid dots are measured $4d$ partial cross section data from Ref. 54. Open circles are from Ref. 55. The full curve and the dashed curve are MBPT length and velocity calculations (Ref. 57) of the $4d$ partial cross section using orbitals calculated in the frozen core potential; no relaxation effects are included in these calculations. The dot–dashed line is a length MBPT calculation (Ref. 57) including relaxation diagrams. The double-dot–dashed line is a length MBPT calculation (Ref. 57) using relaxed orbitals and including overlap factors.

Few cases of giant resonances have been studied as carefully by experimentalists and theorists as the shape resonance observed in barium ($Z = 56$) above the $4d$ ionization threshold. The first data available for barium were relative measurements of the total absorption cross section. These total absorption measurements were reproduced quite well by calculations which include relaxation such as the RPAE suitably modified to include relaxation effects.[58,59] Low-order MBPT calculations using the relaxed orbital technique[60] and more sophisticated MBPT calculations including among other things relaxation diagrams[61] appropriate to the $4d$ hole were also in agreement with the total absorption cross section. The TDLDA[62] also was able to reproduce the experimental curve of total absorption even though it does not explicitly include relaxation effects. Calculations (other than the TDLDA) which do not include relaxation yield a peak in the cross section that is far too narrow and too close to the $4d$ threshold. The $4d$ cross section of barium is very sensitive to the inclusion of higher-order effects such as relaxation, polarization, RPAE diagrams, and even relativistic effects; this makes it a showcase for a variety of many-body calculations.

Recently, photoelectron spectroscopy has been used to partition the total cross section into partial cross sections.[63,64] The photoemission studies found that the satellite emission is nearly equal to the $4d$ emission for some energies. At the peak of the cross section the measured total absorption is 38.8 Mb and the partial $4d$ cross section is only 21.7 Mb. This presented a new challenge to theorists because the techniques mentioned above all gave results comparable with the total absorption cross section rather than the $4d$ partial cross section.

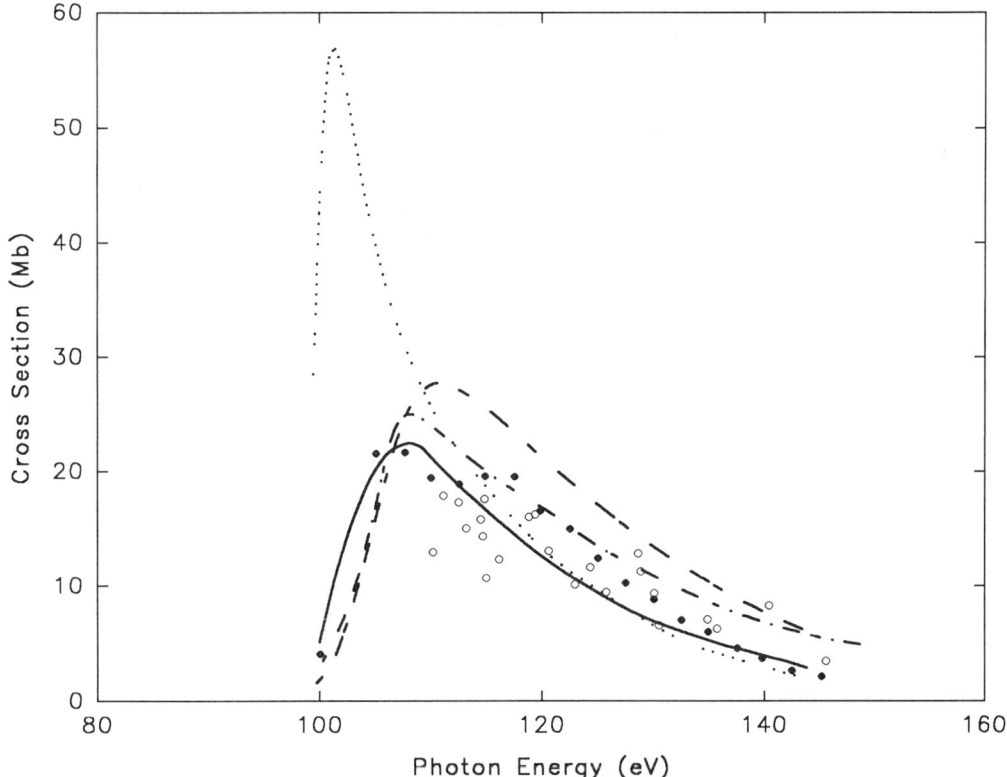

FIGURE 18. Barium $4d$ photoionization cross sections. Solid dots are experimental data from Ref. 63. Open circles are experimental data from Ref. 64. Dotted line represents length correlated MBPT calculation not including the effects of relaxation from Ref. 61. The dashed line represents an MBPT length calculation (Ref. 61) using relaxed orbitals in the potential and overlap factors are included. The dot–dashed line represents the RPAE calculation with relaxed orbitals and polarization of the final-state orbital included (Ref. 65). The solid line represents an MBPT length calculation including correlation, relaxation and polarization diagrams (Ref. 61).

The inclusion of overlap effects in the relaxed orbital method reduces the cross section by approximately 22%. The overlap effects may be interpreted roughly as loss of flux to satellite channels and double photoionization channels. This adjustment led to much better agreement with the measured partial $4d$ cross section.[61]

When the polarization diagram of Figure 2f and the relaxation diagram of Figure 2d were included along with higher-order iterations of this process, the theoretical cross section thus obtained[61] was in good agreement with the experimental results. Similar results were also obtained recently by Amusia et al.[65] who incorporated this diagram and higher-order iterations into the particle lines of the RPAE equations. These results are shown in Figure 18.

4. PHOTOIONIZATION WITH EXCITATION AND DOUBLE-ELECTRON RESONANCES

The photoionization process where one electron is ejected and another is promoted to an excited state depends critically on electron correlation. The photoelectron spectrum will show discrete photoemission lines near the line for direct single-electron photoionization but smaller and with lower kinetic energy. They are often referred to as satellite lines and the photoionization-with-excitation channel partial cross sections are often called satellite cross sections.

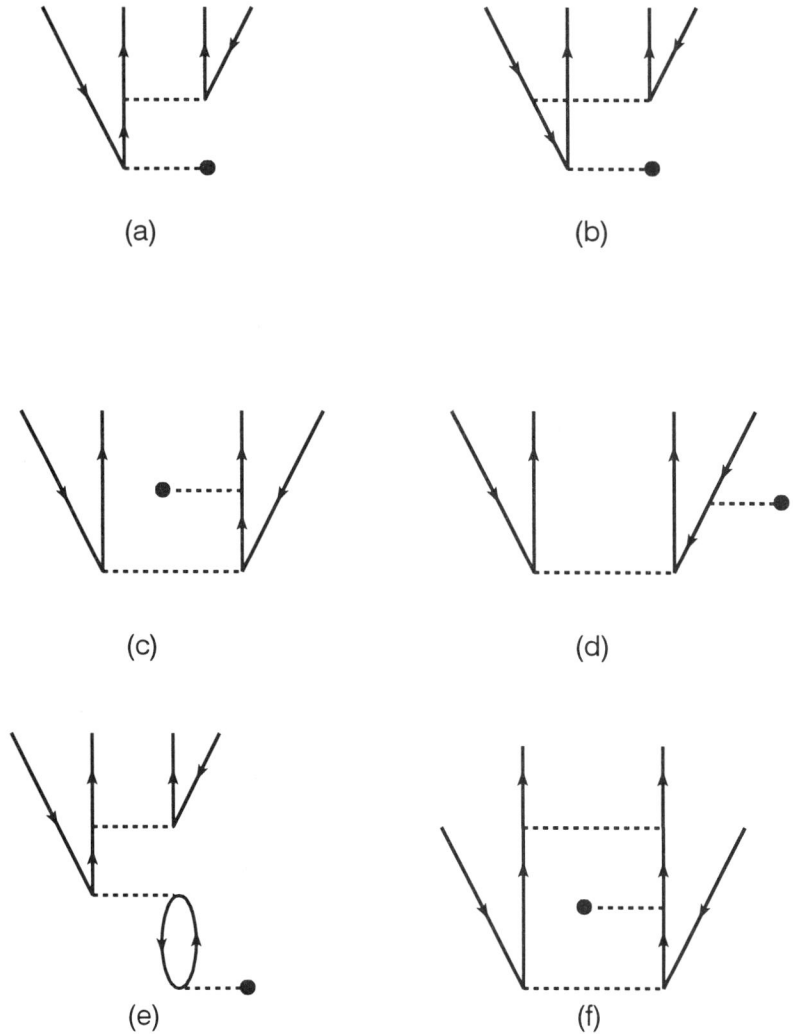

FIGURE 19. Low-order diagrams contributing to photoionization with excitation.

Some of the low-order MBPT diagrams corresponding to photoionization with excitation are shown in Figure 19. The diagrams represented by Figure 19a and b have the correlation in the final state and the diagrams represented by Figure 19c and d have the correlation in the initial state. The diagrams of Figure 19e and f represent coupling between final-state channels. Common to all of the diagrams is that the final state involves two hole states and two particle states. For photoionization with excitation one of the particle states is bound and the other is a continuum state. Diagrams representing double photoionization are the same but with two continuum states for the upwardly directed particle lines.

The related process of double-electron resonances occurs when a doubly excited state with two bound states decays into the direct single-photoionization channel. This second-order effect is represented by the diagrams of Figure 2d and f with k' and k'' both bound-excited states. These doubly excited states often give rise to series of resonances leading up to the threshold of a photoionization-with-excitation channel.

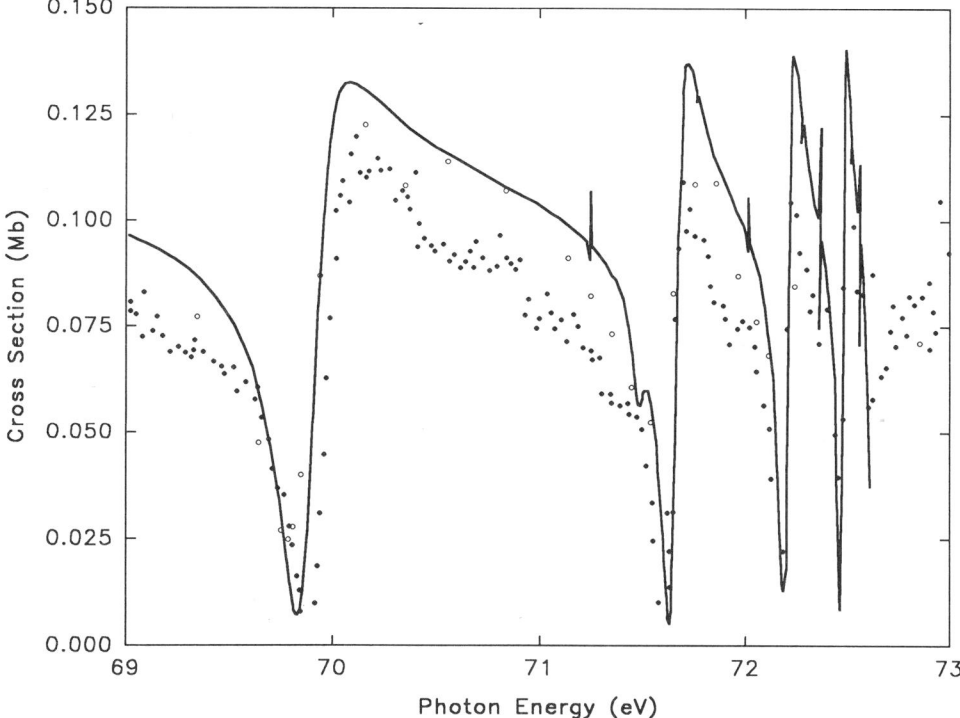

FIGURE 20. The total $n = 2$ cross section for helium in the region leading to the $n = 3$ threshold at 73.0 eV. The full curve is the MBPT length calculation of Ref. 67. Solid dots are experimental measurements of Ref. 69 and open circles are experimental measurements of Ref. 70.

Double-electron resonance structure is especially dominant above the valence thresholds of helium-like systems. Helium itself is the simplest system that exhibits electron correlation, yet it has a very complex photoionization cross section containing many resonances related to double excitations. Salomonson et al.[66,67] used MBPT to describe the complicated structure seen in the cross section to leave He^+ in the $n = 2$ level. In the calculations of Salomonson et al.[66,67] ground-state correlations were carried out to all orders using the pair equation method[68] and final-state correlations were performed using the coupled equations method.[20] A total of nine interacting final-state channels based on exact He^+ states with $n = 1$, 2, and 3 were included in the coupled equations. The total $n = 2$ cross section for helium in the region leading up to the $n = 3$ threshold at 73.0 eV is shown in Figure 20 and compared with the experiments of Woodruff and Samson[69] as well as Lindle et al.[70] Because of the final state mixing performed by the coupled equations, one cannot definitely say which excited state corresponds to a particular resonance; however, the first resonance is described as being mostly due to the $3s3p$ configuration followed by $3p3d$, $3s4p$, etc. Recent calculations by Burkov et al.[71] are also in good agreement with experiment.

Salomonson et al.[66,67] and Burkov et al.[71] have also carried out calculations for the total $n = 2$ angular distribution asymmetry parameter $\beta(\omega)$ as prescribed in equation (29). These results and the experimental $\beta(\omega)$ are shown in Figure 21. The agreement with experiment is reasonable once the instrumental broadening has been deconvolved from the experiment.

Double-electron resonances in alkaline earth metal atoms have also been studied by a variety of theoretical methods. Altun[72] and Altun and co-workers[73] have applied MBPT to the study of Mg and Ca above the $3s^2$ and $4s^2$ valence thresholds, respectively. Frye and Kelly[74] have carried out similar calculations above the $5s^2$ threshold for Sr. Relaxation and polarization diagrams as shown in Figure 2d and f with k'' and k' both bound-excited states are used in MBPT to calculate the double-electron

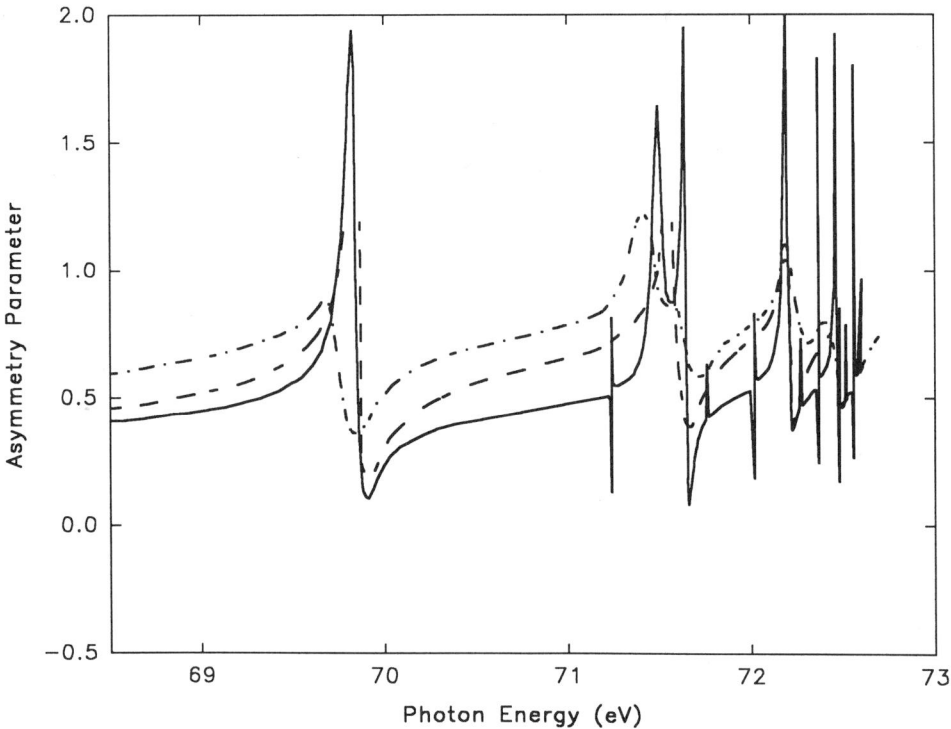

FIGURE 21. The asymmetry parameter β for the total $n = 2$ cross section of helium. Solid line is length MBPT calculation of Ref. 67. Dot–dashed line is a calculation by Burkov et al.[71] The dashed curve is a fit to experimental data (Ref. 70) with monochromator broadening deconvolved.

resonances. Analogous calculations have been carried out using the eigenchannel R-matrix method[75] combined with MQDT for Sr, Ca, and Mg. The MBPT and R-matrix[75] results for Mg are compared with experiment[76,77] in Figure 22. The theoretical description of double-electron resonances, especially to open-shell systems, continues to be a challenging problem.

A further example of the application of MBPT to photoionization with excitation is a calculation of argon by Wijesundera and Kelly.[78] The RPAE approach had already been applied to single-electron photoionization of $3s$ and $3p$ electrons from argon. It was found that coupling between the $3s$ and $3p$ channels radically modified the $3s$ channel cross section[79] and gave good agreement with experiment.[80–82] This is similar to the case of chlorine mentioned above. There has also been considerable interest in the satellite spectrum following ionization in the $3s$ shell. The satellite lines belonging to the $3s^23p^4md(^2S)$ series are particularly strong and have been measured by photoelectron spectroscopy[82–89] (γ,e) and electron momentum spectroscopy[90] ($e,2e$) methods. "Shake-up" satellite lines, $3s^23p^4(^3P)4p(^2P)$, $3s^23p^4(^1D)4p(^2P)$, and $3s^23p^4(^1S)4p(^2P)$, from the main-line channels with ionic state $3s^23p^5(^2P)$ have also been observed.[82,84,86,91]

In the MBPT calculation, Wijesundera and Kelly[78] coupled the three direct photoionization channels from the $3s$ and $3p$ subshells with the photoionization-with-excitation channels mentioned above in a coupled equations calculation to obtain individual cross sections for main-line and satellite channels. To account for the strong interaction between the $3s3p^6(^2S)$ ionic configuration and the $3s^23p^4(^1D)md(^2S)$ ionic states a CI calculation was performed using ten md orbitals ($m = 3$–12). The mixing coefficients obtained from the CI were used to calculate appropriate potentials for bound and continuum states of $3s3p^6(^2S)kp$ and $3s^23p^4md(^2S)kp$ channels. The results for the $3s$ cross section, shown along with experiment[80–82] in Figure 23, are similar to the RPAE[79] and R-matrix results except

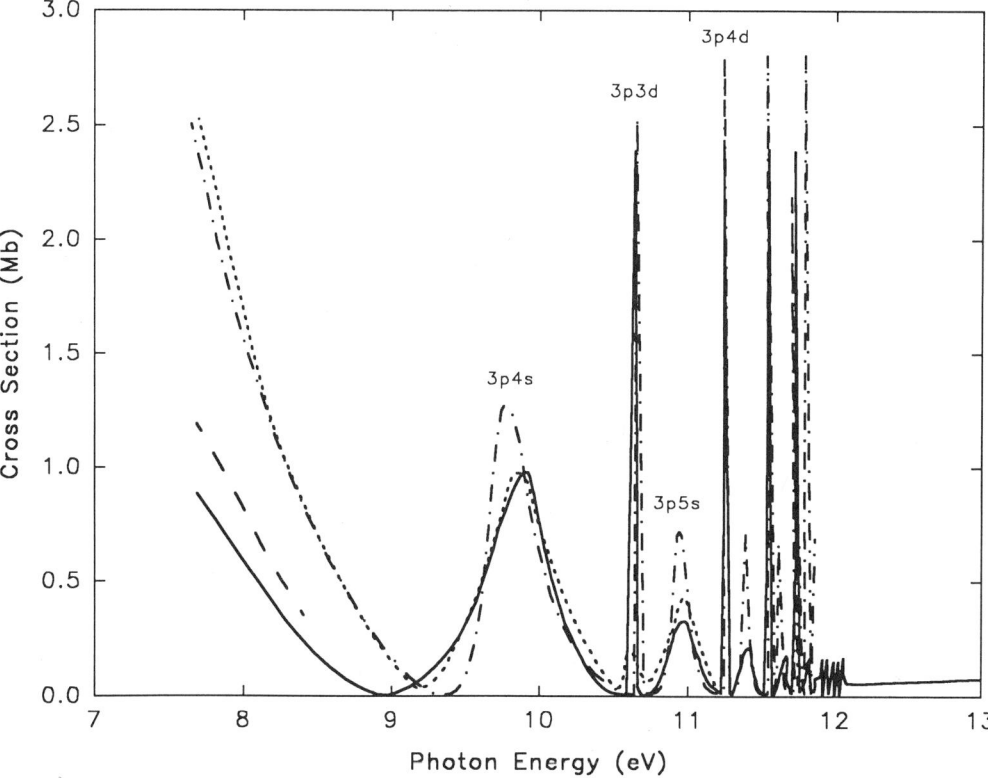

FIGURE 22. Photoionization cross section for the $3s^2 \to 3skp(^1P)$ transition involving the double-electron resonances as a function of photon energy. The solid curve is the MBPT velocity calculation of Ref. 72. The dot–dashed line is by Aymar and Greene.[75] Experimental measurements are represented by the long dashed line (Ref. 77) and the short dashed line (Ref. 76).

that for the first time double-electron resonances are seen below the thresholds for photoionization-with-excitation channels. In Figure 24 the $3s^23p^4(^1D)md(^2S)kp$ satellite channel cross sections are shown and compared with experiment.[88,89] Here the experimental values were obtained from intensities relative to the $3s$ cross section scaled to the calculated $3s$ cross section of Figure 23. The overall shape of the satellite cross sections seem to follow that of the main-line channel $3s3p^6(^2S)$ which is driving the satellites. The success of this calculation suggests that this method of using CI to mix the ionic cores and MBPT to couple the continuum channels could be useful for the calculation of satellite photoionization in other systems.

5. DOUBLE PHOTOIONIZATION

The process of double photoionization, like photoionization-with-excitation, depends critically on electron correlation both in the ground and final states. At energies well above threshold, the "sudden" or "shake-off" approximation applies. However, near threshold where the photoelectrons interact strongly a more detailed many-body treatment is needed.

In the MBPT one treats double photoionization[92] by developing a perturbation expansion for the imaginary part of the frequency-dependent dipole polarizability $\alpha(\omega)$ and using the relation

$$\sigma^{2+}(\omega) = (4\pi\omega/c)\mathrm{Im}\alpha(\omega) \qquad (35)$$

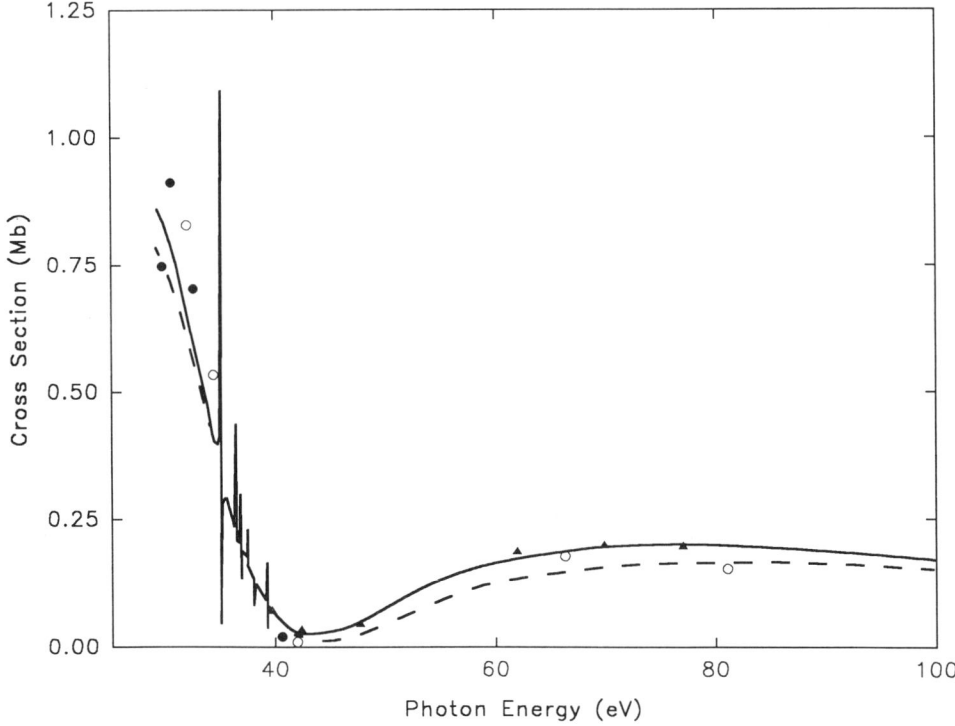

FIGURE 23. Partial photoionization cross section calculated in dipole length (solid line) and velocity (dashed line) gauges for $3s^23p^6 \rightarrow 3s3p^6kp$ from Ref. 78. Experiment: solid dots from Ref. 79, open circles from Ref. 80, and solid triangles from Ref. 81.

With the excited-state, single-particle continuum orbitals normalized as in equation (5) one obtains the cross section formula[92]

$$\sigma^{2+}(\omega) = \frac{16\omega}{c} \int_0^{k_{max}} dk \frac{|Z(pq \rightarrow k'k)|^2}{k'} \tag{36}$$

where

$$k' = [2(\varepsilon_p + \varepsilon_q - \tfrac{1}{2}k^2 + \omega)]^{1/2} \tag{37}$$

and

$$k_{max} = [2(\varepsilon_p + \varepsilon_q + \omega)]^{1/2} \tag{38}$$

The dipole matrix elements $Z(pq \rightarrow k'k)$ are between the correlated ground state and correlated final states with excitation of an electron pair from ground-state orbitals p and q to excited-state orbitals k' and k. Individual terms in the perturbation expansion for $Z(pq \rightarrow k'k)$ involve single-particle states and are represented in low order by the diagrams shown in Figure 19. Note that there is no diagram of zeroth order in the interaction Hamiltonian.

The helium atom is the simplest system in which one-photon two-electron photoionization can occur and several experiments have been performed to determine either the total cross section $\sigma^{2+}(\omega)$ or the ratio σ^{2+}/σ^+. An early MBPT calculation performed by Carter and Kelly[93] provided an accurate description of double photoionization in helium. In that calculation the number of final-state

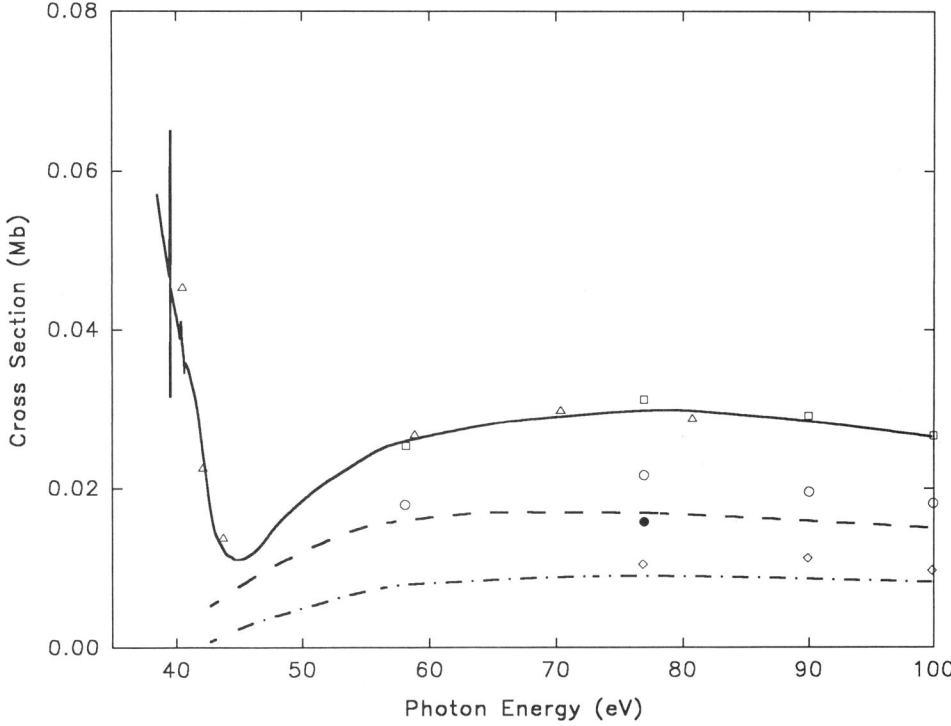

FIGURE 24. Photoionization cross sections of $3s^2 3p^4(^1D)md(^2S)$ satellites. Solid line, dashed line, and dot–dashed line are MBPT length calculations of $3d(^2S)$, $4d(^2S)$, and $5d(^2S)$ satellites, respectively, from Ref. 78. Experiment: Open triangles and open squares are $3d(^2S)$ cross section from Ref. 83. Open circles are $4d(^2S)$ cross section from Ref. 83. Open diamonds are $5d(^2S)$ cross section from Ref. 83. Solid dot is $4d(^2S)$ cross section from Ref. 84.

correlation diagrams was substantially reduced by calculating one set of continuum orbitals (for electron k, say) in a V^{N-1} potential and the remaining continuum orbitals (for electron k', say) in a V^{N-2}. This procedure eliminates to a large degree electron-screening effects which would otherwise be difficult to evaluate. Both ground- and final-state correlations were found to be important. Final-state correlations dominated the $kskp$ channels, but the $kpkd$ cross section displayed a delicate cancelation between ground- and final-state correlations. Contributions from additional partial-wave channels of the form $k_l k_{l+1}$ were not included. More recently, Ishihara et al.[94] performed a low-order MBPT calculation using orbitals calculated in a V^{N-1} potential which agrees with a recent experiment[95] at higher energies. Ishihara et al.[94] also find a large cancelation between ground- and final-state correlation diagrams (only length results are given, however). The MBPT results[93] are compared with experiment[96] in Figure 25.

Double photoionization studies have also been carried out for the argon atom.[97] This is more complicated since it is a $(3p)^2$ pair which is ejected from the system and the residual ion may be left in the states $3p^4(^3P, ^1D, \text{ or } ^1S)$ and an infinite number of partial-wave channels ($kpkd$, $kdkf$, $kskp$, $kskf$, etc.) are possible. At higher energies it is also possible to eject a $3s3p$ pair.

The results of MBPT calculations by Carter and Kelly[97] are shown in Figure 26 along with the experimental measurements.[98–102] These calculations were carried out with excited orbitals calculated in a V^{N-1} potential. The calculated results near threshold are too small compared with experiment; this is related to the V^{N-1} choice for the excited orbitals. It would be of interest to have a calculation involving a combination of orbitals computed in both V^{N-1} and V^{N-2} potentials as in the case of helium. Beyond the peak in the double photoionization cross section the agreement between theory and

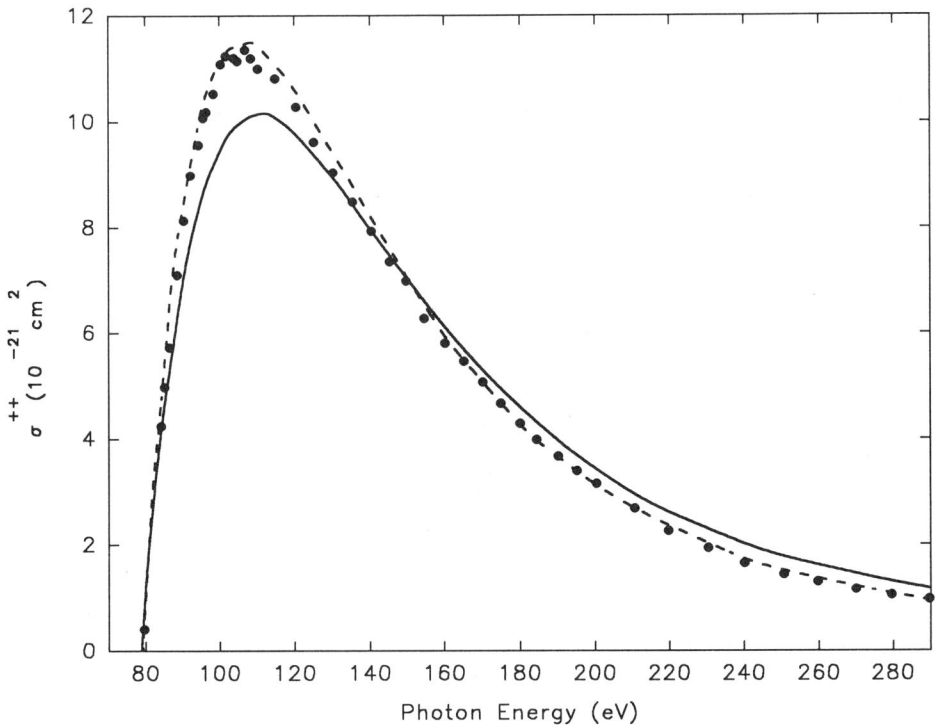

FIGURE 25. Theoretical calculations of the total double photoionization cross section for helium from Ref. 93. Curves for the correlated length (solid line) and velocity (dashed line) cross sections include contributions from both $kskp$ and $kpkd$ channels. Experimental data are from Bizau et al.[96]

experiment is quite good until the threshold for single photoionization from the $2p^6$ subshell is approached. In the region just below the $2p$ threshold, the measured double photoionization cross section undergoes a large increase. A possible cause for increase is the coupling between the direct $2p$ photoionization channel and the double photoionization channels. Recent calculations by Pan and Kelly[103] (also shown in Figure 26) which include diagrams of Figure 19a–c and higher-order iterations to account for the interchannel coupling do not show an effect as large as the measured increase seen in the experiment. This situation deserves further study; double photoionization experiments and calculations on other atomic systems near inner-shell single-electron photoionization channels would be of interest.

Studies using MBPT have also been carried out for double photoionization of a small number of other atoms as well. Calculations have been carried out for neon,[104] beryllium,[105] and the open-shell system carbon.[106]

6. CONCLUSION

Over the past two decades there have been many applications of MBPT to photoionization problems. Techniques, such as that of the coupled equations, have been developed to sum certain classes of diagrams to infinite order and methods have been developed to include some many-body effects in the single-particle potentials used to generate excited states. MBPT has been used to compute photoionization cross sections of complex open-shell atoms, estimate relaxation and polarization effects in photoionization, calculate photoionization-with-excitation cross sections and double-elec-

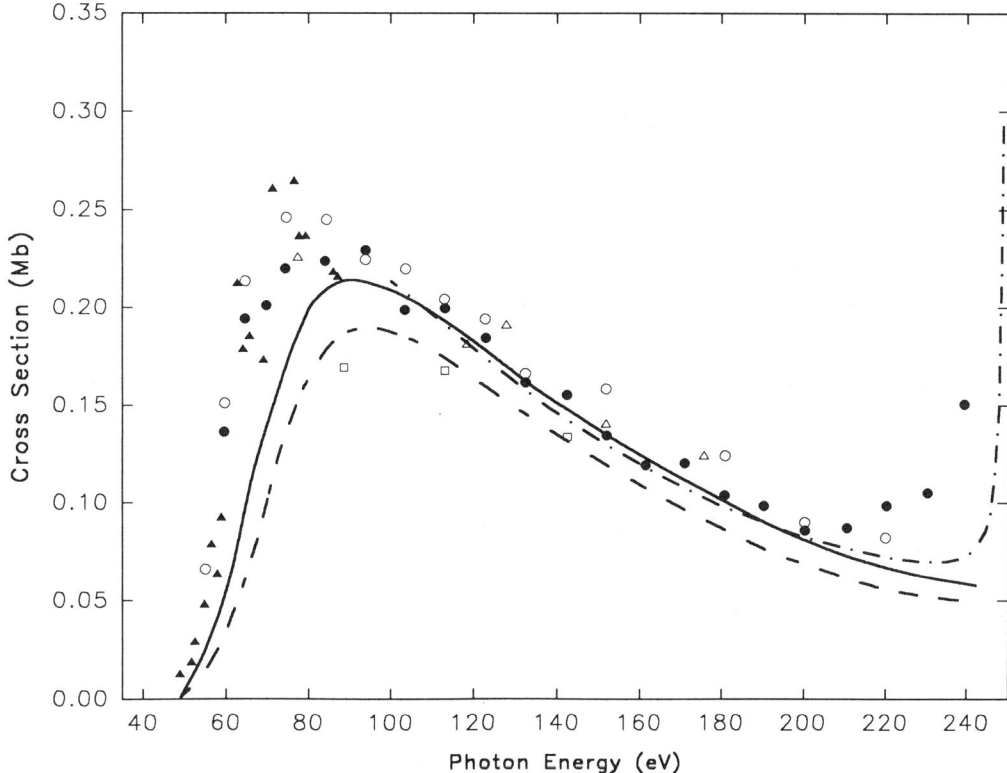

FIGURE 26. Double photoionization cross section σ^{++} argon. Solid and dashed lines are length and velocity cross sections calculated by Carter and Kelly.[97] Dot–dashed line is calculation by Pan and Kelly[103] which includes coupling with $2p^5$ single-excitation channel. Experiment: Solid dots from Ref. 98, open triangles from Ref. 99, open squares from Ref. 100, solid triangles from Ref. 101, and open circles from Ref. 102.

tron resonances. Double photoionization in closed- and open-shell systems has also been studied using MBPT.

Much remains to be carefully studied with these techniques. More and more complicated systems should be considered as computational resources expand. Many open-shell atoms have yet to be investigated. Photoionization of molecules, clusters, and the solid state deserves theoretical attention. Triple photoionization is a challenging area yet to be studied and multiphoton ionization should also be tackled with the insightful methods begun by the late Dr. Hugh P. Kelly.

REFERENCES

1. See first chapter of this book or A. F. STARACE, in *Corpuscles and Radiation in Matter I*, Vol. 31 of *Handbüch der Physik*, edited by W. Mehlhorn (Springer-Verlag, Berlin, 1982), p. 1.
2. P. L. ALTICK AND A. E. GLASSGOLD, *Phys. Rev. A* **133**, 632 (1964).
3. M. YA. AMUSIA, V. K. IVANOV, N. A. CHEREPKOV, AND L. V. CHERNYSHEVA, *Zh. Eksp. Teor. Fiz.* **66**, 1537 (1974) [*Sov. Phys. JETP* **39**, 752 (1975)]; M. YA. AMUSIA, N. A. CHEREPKOV, D. ŽIVANOVIĆ, AND V. RADOJEVIĆ, *Phys. Rev. A* **13**, 1466 (1976).
4. W. R. JOHNSON AND C. D. LIN, *Phys. Rev. A* **20**, 964 (1979); W. R. JOHNSON, C. D. LIN, K. T. CHENG, AND C. M. LEE, *Phys. Scr.* **21**, 409 (1980).
5. A. ZANGWILL AND P. SOVEN, *Phys. Rev. Lett.* **45**, 204 (1980).
6. P. G. BURKE AND K. T. TAYLOR, *J. Phys. B* **8**, 2620 (1975).
7. H. P. SAHA, *Phys. Rev. A* **39**, 628 (1989).

8. J. R. Swanson and L. Armstrong, Jr., *Phys. Rev. A* **15**, 661 (1977); **16**, 1117 (1977).
9. C. H. Greene, *Phys. Rev. A* **23**, 661 (1981); U. Fano and A. R. P. Rau, *Atomic Collisions and Spectra* (Academic Press, Orlando, 1986).
10. K. A. Brueckner, *Phys. Rev.* **97**, 1353 (1955); **100**, 36 (1955).
11. J. Goldstone, *Proc. R. Soc. London Ser. A* **239**, 267 (1957).
12. H. P. Kelly, *Phys. Rev.* **131**, 684 (1963).
13. M. Ya. Amusia, *Atomic Photoeffect* (Plenum Press, New York, 1990).
14. U. Fano and A. R. P. Rau, *Atomic Collisions and Spectra* (Academic Press, Orlando, 1986), p. 66.
15. U. Fano and J. W. Cooper, *Rev. Mod. Phys.* **40**, 441 (1968).
16. B. Lippman and J. Schwinger, *Phys. Rev.* **79**, 469 (1950).
17. H. P. Kelly, *Adv. Chem. Phys.* **XIV**, 129 (1969).
18. S. Huzinaga and C. Arnau, *Phys. Rev. A* **1**, 1285 (1970); H. J. Silverstone and M. L. Yin, *J. Chem. Phys.* **49**, 2026 (1968).
19. M. Ya. Amusia and N. A. Cherepkov, *Case Studies in Atomic Physics* **5**, 47 (1975).
20. E. R. Brown, S. L. Carter, and H. P. Kelly, *Phys. Rev. A* **21**, 1237 (1980).
21. J. Boyle, Ph.D. Thesis, University of Virginia (1992).
22. H. P. Kelly and R. L. Simons, *Phys. Rev. Lett.* **30**, 529 (1973).
23. A. E. Hansen, *Mol. Phys.* **13**, 425 (1967).
24. R. P. Madden, D. L. Ederer, and K. Codling, *Phys. Rev.* **177**, 136 (1969).
25. F. Robicheaux and C. H. Greene, *Phys. Rev. A* **46**, 3821 (1992).
26. B. Rušić and J. Berkowitz, *Phys. Rev. Lett.* **50**, 675 (1983).
27. W. Wijesundera and H. P. Kelly, *Phys. Rev. A* **39**, 634 (1989).
28. W. R. Fielder and L. Armstrong, Jr., *Phys. Rev. A* **28**, 218 (1983).
29. S. Shahabi, A. F. Starace, and T. N. Chang, *Phys. Rev. A* **30**, 1819 (1984).
30. M. Lamoureux and F. Combet-Farnoux, *J. Phys. (Paris)* **40**, 545 (1979).
31. F. Robicheaux and C. H. Greene, *Phys. Rev. A* **47**, 1066 (1993).
32. J. A. R. Samson, Y. Shefer, and G. C. Angel, *Phys. Rev. Lett.* **56**, 2020 (1986).
33. Z.-D. Qian, S. L. Carter, and H. P. Kelly, *Phys. Rev. A* **33**, 1751 (1986).
34. J. Boyle, Z. Altun, and H. P. Kelly, *Phys. Rev. A* **47**, 4811 (1993).
35. L. J. Garvin, E. R. Brown, S. L. Carter, and H. P. Kelly, *J. Phys. B* **16**, L269 (1983).
36. M. Ya. Amusia, V. K. Ivanov, and L. V. Chernysheva, *J. Phys. B* **14**, L19 (1981).
37. J. W. Cooper, C. W. Clark, C. R. Cromer, T. B. Lucatorto, B. F. Sonntag, E. T. Kennedy, and J. T. Costello, private communication.
38. J. Chang, Z. Liu, and H. P. Kelly, unpublished.
39. C. Pan, S. L. Carter, and H. P. Kelly, *Phys. Rev. A* **43**, 1290 (1991).
40. U. Becker, H. G. Kerkhoff, D. W. Lindle, P. H. Kobrin, T. A. Ferrett, P. A. Heimann, C. M. Truesdale, and D. A. Shirley, *Phys. Rev. A* **34**, 2858 (1986).
41. M. Meyer, T. Prescher, E. von Raven, M. Richter, E. Schmidt, B. Sonntag, and H. E. Wetzel, in *Giant Resonances in Atoms, Molecules, and Solids*, edited by J. P. Connerade, J. M. Esteva, and R. C. Karnatak (Plenum Press, New York, 1987); *Z. Phys. D* **2**, 347 (1986).
42. M. Richter, M. Meyer, M. Pahler, T. Prescher, E. von Raven, B. Sonntag, and H. E. Wetzel, *Phys. Rev. A* **40**, 7007 (1989).
43. M. Ya. Amusia, S. I. Sheftel, and L. V. Chernysheva, *Zh. Tekh. Fiz.* **51**, 2441 (1981) [*Sov. Phys. Tech. Phys.* **26**, 1444 (1981)].
44. M. Ya. Amusia, V. K. Ivanov, and V. A. Kupchenko, *Z. Phys. D* **14**, 219 (1989).
45. A. Zangwill, *J. Phys. C* **20**, L627 (1987).
46. A. Zangwill and G. Doolen, unpublished, as presented in Ref. 42.
47. J. T. Costello, E. T. Kennedy, B. F. Sonntag, and C. L. Cromer, *J. Phys. B* **24**, 5063 (1991).
48. R. Haensel, K. Radler, and B. Sonntag, *Solid State Commun.* **7**, 1495 (1969).
49. L. J. Garvin, Ph.D. Thesis, University of Virginia (unpublished) (1983).
50. S. L. Carter and H. P. Kelly, *J. Phys. B* **11**, 2467 (1978).
51. U. Becker, private communication.
52. R. Haensel, G. Keitel, P. Schreiber, and C. Kunz, *Phys. Rev.* **188**, 1375 (1969).
53. R. Rabe, K. Radler, and H. W. Wolff, in *VUV Radiation Physics*, edited by E. E. Koch *et al.* (Vieweg-Pergamon, Berlin, 1974), p. 247.
54. U. Becker, D. Szostak, H. G. Kerkhoff, M. Kupsch, B. Langer, R. Wehlitz, A. Yagishita, and T. Hayaishi, *Phys. Rev. A* **39**, 3902 (1989).
55. B. Kämmerling, H. Kossmann, and V. Schmidt, *J. Phys. B* **22**, 841 (1989).

56. T. ÅBERG, *Ann. Acad. Sci. Fenn. Ser. A* VI, **308**, 1 (1969).
57. Z. ALTUN, M. KUTZNER, AND H. P. KELLY, *Phys. Rev. A* **37**, 4671 (1988).
58. G. WENDIN, *Phys. Lett.* **51A**, 291 (1975).
59. M. YA. AMUSIA, V. K. IVANOV, AND L. V. CHERNYSHEVA, *Phys. Lett.* **59A**, 191 (1976).
60. H. P. KELLY, S. L. CARTER, AND B. E. NORUM, *Phys. Rev. A* **25**, 2052 (1982).
61. M. KUTZNER, Z. ALTUN, AND H. P. KELLY, *Phys. Rev. A* **41**, 3612 (1990).
62. A. ZANGWILL AND P. SOVEN, *Phys. Rev. Lett.* **45**, 204 (1980).
63. U. BECKER, in *Giant Resonances in Atoms, Molecules, and Solids*, Vol. 151 of *NATO Advanced Study Institute, Series B: Physics*, edited by J. P. Connerade, J. M. Esteva, and R. C. Karnatak (Plenum Press, New York, 1987), p. 473.
64. J. M. BIZAU, D. CUBAYNES, P. GÉRARD, AND F. J. WUILLEUMIER, *Phys. Rev. A* **40**, 3002 (1989).
65. M. YA. AMUSIA, L. V. CHERNYSHEVA, G. F. GRIBAKIN, AND K. L. TSEMEKHMAN, *J. Phys. B* **23**, 393 (1990).
66. S. SALOMONSON, S. L. CARTER, AND H. P. KELLY, *J. Phys. B* **18**, L149 (1985).
67. S. SALOMONSON, S. L. CARTER, AND H. P. KELLY, *Phys. Rev. A* **39**, 5111 (1989).
68. A.-M. MÅRTENSSON, *J. Phys. B* **12**, 3995 (1979).
69. P. R. WOODRUFF AND J. A. R. SAMSON, *Phys. Rev. A* **25**, 848 (1982).
70. D. W. LINDLE, T. A. FERRETT, U. BECKER, P. H. KOBRIN, C. M. TRUESDALE, H. G. KERKHOFF, AND D. A. SHIRLEY, *Phys. Rev. A* **31**, 714 (1985).
71. S. M. BURKOV, N. A. LETYAEV, S. I. STRAKHOVA, AND T. M. ZAJAC, *J. Phys. B* **21**, 1195 (1988).
72. Z. ALTUN, *Phys. Rev. A* **40**, 4968 (1989).
73. Z. ALTUN, S. L. CARTER, AND H. P. KELLY, *Phys. Rev. A* **27**, 1943 (1983).
74. D. FRYE AND H. P. KELLY, *J. Phys. B* **20**, L677 (1987); *Phys. Rev. A* **36**, 5143 (1987).
75. C. H. GREENE AND M. AYMAR, *Phys. Rev. A* **44**, 6271 (1991), method based on C. H. GREENE AND M. AYMAR, *Phys. Rev. A* **44**, 1773 (1991).
76. J. M. PRESES, C. E. BURKHARDT, W. P. GARVER, AND J. J. LEVENTHAL, *Phys. Rev. A* **29**, 985 (1984).
77. R. W. DITCHBURN AND G. B. MARR, *Proc. Phys. Soc. London Sect. A* **66**, 655 (1953).
78. W. WIJESUNDERA AND H. P. KELLY, *Phys Rev. A* **39**, 634 (1989).
79. M. Y. AMUSIA, N. A. CHEREPKOV, AND L. V. CHERNYSHEVA, *Phys. Lett. A* **40**, 15 (1972).
80. J. A. R. SAMSON AND J. L. GARDNER, *Phys. Rev. Lett.* **33**, 671 (1974).
81. R. G. HOULGATE, J. B. WEST, K. CODLING, AND G. V. MARR, *J. Phys. B* **7**, L470 (1974).
82. M. Y. ADAM, P. MORIN, AND G. WENDIN, *Phys. Rev. A* **31**, 1426 (1985).
83. D. P. SPEARS, H. J. FISCHBECK, AND T. A. CARLSON, *Phys. Rev. A* **9**, 1603 (1974).
84. M. Y. ADAM, F. WUILLEUMIER, S. KRUMMACHER, V. SCHMIDT, AND W. MEHLHORN, *J. Phys. B* **11**, L413 (1978).
85. V. SCHMIDT, *Z. Phys. D* **2**, 275 (1986).
86. H. KOSSMANN, B. KRÄSSIG, V. SCHMIDT, AND J. E. HANSEN, *Phys. Rev. Lett.* **58**, 1620 (1987).
87. S. SVENSSON, K. HELENELUND, AND U. GELIUS, *Phys. Rev. Lett.* **58**, 1624 (1987).
88. J. A. R. SAMSON, Y. CHUNG, AND E. LEE, *Phys. Lett. A* **127**, 171 (1988).
89. U. BECKER, B. LANGER, H. G. KERKHOFF, M. KUPSCH, D. SZOSTAK, R. WEHLITZ, P. A. HEIMAN, S. H. LIU, D. W. LINDLE, T. A. FERRETT, AND D. A. SHIRLEY, *Phys. Rev. Lett.* **60**, 1490 (1988).
90. I. E. MCCARTHY AND E. WEIGOLD, *Phys. Rep.* **27C**, 275 (1976); K. T. Leung AND C. E. BRION, *Chem. Phys.* **82**, 87 (1983); I. E. MCCARTHY AND E. WEIGOLD, *Phys. Rev. A* **31**, 160 (1985); J. P. D. COOK, I. E. MCCARTHY, J. MITROY, AND E. WEIGOLD, *Phys. Rev. A* **33**, 211 (1986).
91. W. T. SILFVAST, D. Y. AL-SALAMEH, AND O. R. WOOD II, *Phys. Rev. A* **34**, 5164 (1986).
92. H. P. KELLY, *At. Phys.* **8**, 305 (1983).
93. S. L. CARTER AND H. P. KELLY, *Phys. Rev. A* **24**, 170 (1981).
94. T. ISHIHARA, K. HINO, AND J. H. MCGUIRE, *Phys. Rev. A* **44**, R6980 (1991).
95. J. C. LEVIN, D. W. LINDLE, N. KELLER, R. D. MILLER, Y. AZUMA, N. BERRAH MANSOUR, H. G. BERRY, AND I. A. SELLIN, *Phys. Rev. Lett.* **67**, 968 (1991).
96. J. M. BIZAU, F. WUILLEUMIER, D. EDERER, P. DHEZ, S. KRUMMACHER, AND V. SCHMIDT, unpublished.
97. S. L. CARTER AND H. P. KELLY, *Phys. Rev. A* **16**, 1525 (1977).
98. D. M. P. HOLLAND, K. CODLING, J. B. WEST, AND G. V. MARR, *J. Phys. B* **12**, 2465 (1979).
99. V. SCHMIDT, N. SANDNER, H. KUNTZEMÜLLER, P. DHEZ, F. WUILLEUMIER, AND E. KÄLLNE, *Phys. Rev. A* **13**, 1748 (1976).
100. T. A. CARLSON, *Phys. Rev.* **156**, 142 (1967).
101. J. A. R. SAMSON AND G. N. HADDAD, *Phys. Rev. Lett.* **33**, 875 (1974).
102. G. R. WIGHT AND M. J. VAN DER WIEL, *J. Phys. B* **9**, 1319 (1976).
103. C. PAN AND H. P. KELLY, *Phys. Rev. A* **39**, 6232 (1989).

104. T. N. Chang and R. T. Poe, *Phys. Rev. A* **12**, 1432 (1975); S. L. Carter and H. P. Kelly, *Phys. Rev. A* **16**, 1525 (1977).
105. P. J. Winkler, *J. Phys. B* **10**, L693 (1977).
106. S. L. Carter and H. P. Kelly, *J. Phys. B* **9**, 1887 (1976).

CHAPTER 3

FUNDAMENTAL ASPECTS OF ATOMIC PHOTOIONIZATION WITH HIGH-BRIGHTNESS LIGHT SOURCES

ANTHONY F. STARACE AND STEVEN T. MANSON

1. INTRODUCTION

The construction of numerous new synchrotron light sources worldwide is providing the means to study atomic and molecular photoionization processes at a level of detail that is unparalleled in even the recent past. Correspondingly, in theory, the availability of supercomputers and, perhaps more important, relatively inexpensive computer workstations, has permitted theoretical calculations to tackle more complex processes than ever before. In short, these technical developments are creating an unprecedented amount of data to be analyzed and understood.

In this volume, many of the experimental and theoretical advances of recent years will be described. We take the point of view in this introductory paper that while agreement of theory and experiment is always desirable, the main goal of researchers in this field should be to advance our understanding of the key physics governing photoionization processes in as simple a way as possible. As there can be no general prescription for achieving such a goal, we therefore present an eclectic set of examples of recent advances in experimental techniques or theoretical analysis which have achieved such simplicity in the face of complexity. For clarity, however, we review briefly beforehand some essentials of the theory of photoionization.

2. BRIEF REVIEW OF THEORY

In the dipole approximation, which is excellent for low-energy photoionization, the cross section for a photoionizing transition induced by an unpolarized beam of photons of energy $h\nu$ from an initial state $|i\rangle$ to a final state $|f\rangle$ given by[1]

$$\sigma_{if} = \frac{4\pi^2 \alpha\, a_0^2}{3 g_i}(h\nu) |M_{if}|^2 \qquad (1)$$

where α is the fine structure constant, a_0 is the Bohr radius, and g_i is the statistical weight of the initial state. The absolute square of the matrix element is given by[2]

ANTHONY F. STARACE • Department of Physics and Astronomy, The University of Nebraska, 116 Brace Laboratory, Lincoln, Nebraska 68588-0111.   STEVEN T. MANSON • Department of Physics and Astronomy, Georgia State University, Atlanta, Georgia 30303-3083.

VUV and Soft X-Ray Photoionization. Edited by Uwe Becker and David A. Shirley. Plenum Press, New York, 1996.

$$|M_{if}|^2 = \sum_{i,f} | \langle f| \sum_j \vec{r}_j |i\rangle|^2 \qquad (2)$$

where $\vec{r}_j$ is the position coordinate of the jth electron, the sums are over the degenerate initial and final magnetic substates, and the wave functions are normalized such that

$$\langle i|i\rangle = 1, \quad \langle f|f'\rangle = \delta(\varepsilon - \varepsilon') \qquad (3)$$

where ε is the photoelectron energy. Details and examples for a number of cases are given in Refs. 3–7.

The photoelectron angular distribution resulting from photoionization of state $|i\rangle$ by linearly polarized photons leaving the ion in state $|j\rangle$ is given by[8–10]

$$\frac{d\sigma_{ij}}{d\Omega} = \frac{\sigma_{ij}}{4\pi}[1 + \beta_{ij}P_2(\cos\theta)] \qquad (4)$$

where σ_{ij} is the total photoionization cross section for producing state $|j\rangle$ of the ion, θ is the angle between the photon's polarization vector and the photoelectron's momentum direction, $P_2(x) = (3x^2 - 1)/2$, and β_{ij} is the asymmetry parameter. There are various equivalent ways of obtaining β_{ij}[11,12]; the angular momentum transfer formulation[11,12] is presented here since the essential determinants of the angular distribution emerge most clearly in this formulation.

Consider the ejection of a photoelectron from an unpolarized atom A,

$$A(J_0\pi_0) + (j_\gamma = 1, \pi_\gamma = -1) \rightarrow (J_c\pi_c) + e^-(lsj, \pi_e = (-1)^l) \qquad (5)$$

where π denotes the parity quantum number of the various constituents of the process. The cross section can be partitioned into incoherent contributions characterized by alternative values of the angular momentum transfer,[13]

$$\vec{j}_t = \vec{j}_\gamma - \vec{l} = \vec{J}_c + \vec{s} - \vec{J}_0 . \qquad (6)$$

The possible values of j_t are determined from equation (6), subject to the constraints of angular momentum and parity conservation. The allowed values of j_t are characterized by their parity according to $\pi_0\pi_c = \pm (-1)^{j_t}$, where values of j_t for which the plus sign is required are called "parity-favored" and those requiring the minus sign are called "parity-unfavored." The key point here is that for parity-unfavored values of j_t, $\beta(j_t) = -1$, independent of energy; for parity-favored values of j_t, $\beta(j_t)$ is in general energy-dependent and varies between the limits of -1 and $+2$. The asymmetry parameter for the entire transition is then given by the following weighted average of the various $\beta(j_t)$ parameters:

$$\beta_{ij} = \left(\sum_{j_t}^{\text{fav}} \sigma(j_t)_{\text{fav}} \beta(j_t)_{\text{fav}} - \sum_{j_t}^{\text{unf}} \sigma(j_t)_{\text{unf}} \right) \Big/ \sigma_{ij} \qquad (7)$$

$$\sigma_{ij} = \sum_{j_t} \sigma(j_t) \qquad (8)$$

Detailed expressions for $\sigma(j_t)$ and $\beta(j_t)_{\text{fav}}$ are given elsewhere.[6,11–13]

Of course, in order to actually compute cross sections and angular distributions, wave functions for initial and final states are required. Discussion of these wave functions is beyond the scope of this chapter, but numerous reviews[3–7] along with several chapters of the present volume deal with this subject extensively.

3. EXAMPLES OF THE ESSENTIAL PHYSICS UNDERLYING COMPLEX PHOTOIONIZATION PHENOMENA

In this main section we discuss a number of instances in which rather complex photoionization phenomena have been interpreted simply by novel theoretical or experimental means. We hasten to add that these examples are not intended to be complete. Many of them have involved one or the other of the authors of this chapter.

3.1. Doubly Excited State Spectra of H^- and He

The helium atom and the negative ion of the hydrogen atom are fundamental three-body Coulomb systems. Their photoionization and photodetachment spectra have attracted much experimental and theoretical interest for decades. Only recently, however, through the use of lasers and of high-brightness synchrotron light, has it been possible to obtain experimental spectra on their doubly excited states converging to thresholds higher than $n = 2$. Thus, Harris et al.[14] observed doubly excited spectra of H^- converging to the $n = 4-8$ thresholds of H^+. Similarly, Domke et al.[15] recently reported the high-resolution photoionization study of the doubly excited He states below the $n = 2-7$ thresholds of He^+. Both of these experimental spectra are very rich, having a wealth of detail. Furthermore, in addition to providing much more detail on doubly excited state spectra than earlier measurements focused only on the $n = 2$ threshold,[16] so too did these measurements permit theorists to devise a more general interpretation of doubly excited state spectra than provided in earlier work.[17,18]

The pioneering measurements of Madden and Codling[16] on resonances converging to the He^+ ($n = 2$) threshold were surprising in two respects. First, an independent electron model picture leads one to expect three $^1P^o$ Rydberg series of resonances: $He2snp(^1P^o)$, $He2pns(^1P^o)$, and $He2pnd(^1P^o)$. However, experiment observed the presence of only two series. Second, of the two series that were observed, one was very prominent and the other was barely observable. Cooper et al.[17] interpreted the two observed series as different linear combinations of the first two independent particle model series listed above, i.e., $He(2snp \pm 2pns)\ ^1P^o$. They postulated that the "+" series had much more intensity than the "−" series as a result of cancellations in the dipole amplitude to the "−" states. Their interpretation, however, did not explain the absence of the $He2pnd(^1P^o)$ series from the experimental observations.

It was Macek's use of an adiabatic hyperspherical representation for these two-electron states that finally explained (at least qualitatively) all observed characteristics of these spectra.[18] In hyperspherical coordinates the six independent electron coordinates $r_1, r_2, \mathbf{r}_1$, and $\mathbf{r}_2$ are replaced by $R, \alpha, \hat{r}_1$, and $\hat{r}_2$, where the hyperradius R defined by

$$R \equiv (r_1^2 + r_2^2)^{1/2} \qquad (9)$$

measures the "size" of a two-electron state, and the radial angle α, defined by

$$\alpha \equiv \arctan(r_2/r_1) \qquad (10)$$

measures the radial correlation of the two electrons. In particular, $\alpha = 0$ or $\pi/2$ corresponds to independent particle motion in which one electron is near the nucleus and the other is very far away; $\alpha = \pi/4$ on the other hand corresponds to a doubly excited state in which the two electrons are at equal distances from the nucleus and hence are comparably excited. Macek[18] proposed an adiabatic approximation in which the Schrödinger equation for a two-electron system is expressed in hyperspherical coordinates and the angular equation in the angles $\hat{r}_1, \hat{r}_2$, and α is solved at fixed radius R. The eigenvalues of this angular equation represent radial potentials $U_\mu(R)$ that describe the radial motion, where μ is a channel index.

The essential point of Macek's work[18] is that the radial potentials $U_\mu(R)$ explain at a glance the key features of the observed He spectra near the He^+ ($n = 2$) threshold. Macek found that there are three $^1P^o$ channels μ, designated $\mu = {}^1P^o+$, $\mu = {}^1P^o-$, and $\mu = {}^1P^o(d)$. These three channels have potentials $U_\mu(R)$ with very different centrifugal potential barriers when $R \rightarrow 0$. The intensities of the expected three series are determined by these radial potential barriers because they control the penetration of the excited channel radial wave function in the small R region where the ground-state radial wave function is localized. The smaller the penetration, the less likely is significant overlap with the ground state, and hence the smaller will be the magnitudes of the dipole transition amplitudes which govern the observed intensities. Furthermore, the two potentials having the weakest centrifugal potential barriers, i.e., the "+" and "−" adiabatic hyperspherical potentials, support states that correspond to the He($2snp \pm 2pns$) $^1P^o$ states postulated by Cooper et al.,[17] thus confirming their interpretation.

The recent experimental work on H^- and on He doubly excited spectra near $n > 2$ ionization thresholds requires a generalization of this theoretical picture, however. The reason is that in general there are more than one "+" type series converging to higher thresholds. While the selection of those "+" type potentials $U_\mu(R)$ having the weakest centrifugal potential barriers turns out to be the key to this generalization, the various centrifugal barriers are so much closer in energy at these higher thresholds that another explanation of the intensities of the observed spectra is called for. Consider first the case of H^-.

Sadeghpour and Greene[19] calculated the adiabatic hyperspherical potential curves for very highly excited states of H^- converging to $H(n \leq 12)$. Keeping only the lowest "+" states converging to each threshold $H(n)$, they were able to interpret the doubly excited resonance structures converging to

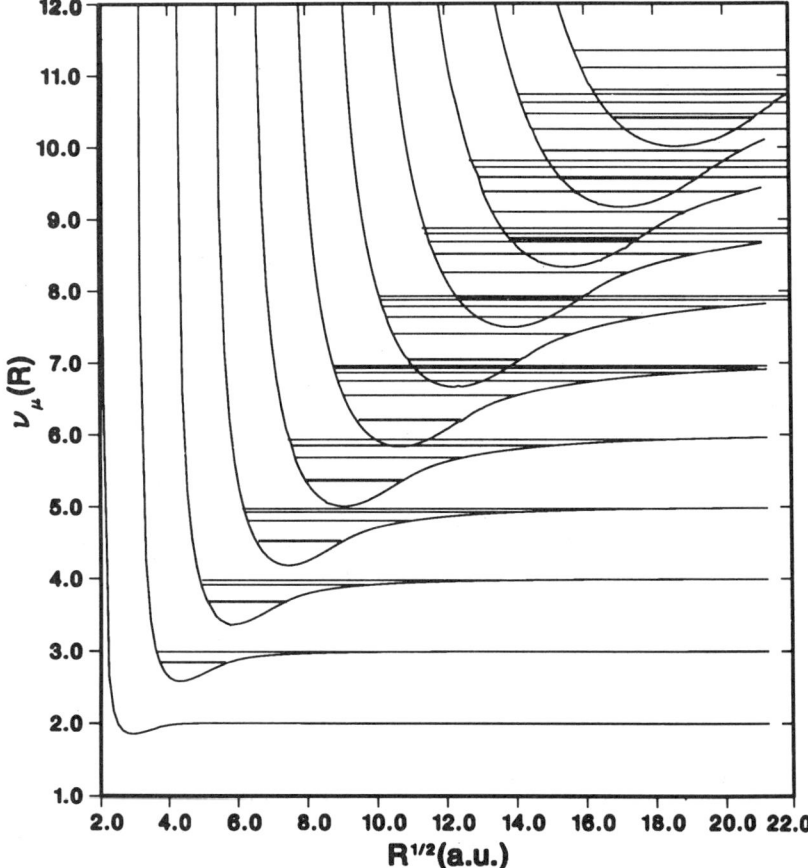

FIGURE 1. Adiabatic hyperspherical potentials for the lowest $^1P^o+$ channels of H^- plotted as effective quantum numbers $\nu_\mu(R) \equiv [-2U_\mu(R)]^{-1/2}$ versus $R^{1/2}$. Doubly excited state level positions supported within each potential are indicated by horizontal lines. (From Ref. 19.)

the $n = 4$–8 thresholds that were observed in the photodetachment measurements of Harris et al.[14] Sadeghpour and Greene[19] interpreted these observed resonances as the doubly excited states supported by the lowest "+" adiabatic hyperspherical radial potentials, as shown in Figure 1.

The interpretation given by Sadeghpour and Greene for the observed photodetachment spectra[14] implies that all of the other allowed levels supported by the many other $^1P^o$ adiabatic hyperspherical potentials converging to each H(n) threshold are not populated in the photodetachment process. Sadeghpour and Greene justified their interpretation by noting that the states corresponding to the lowest + channels converging to each H(n) threshold have no nodes in the angle θ_{12} between the two electrons. For example, Figure 2 shows the adiabatic hyperspherical two-electron density[20] as a contour plot in θ_{12} and α for the two lowest + channels converging to the H($n = 6$) threshold. One can see clearly that the density plot for the lowest + channel in Figure 2a has no θ_{12} nodes, whereas that for the next higher + channel in Figure 2b has a node along $\theta_{12} \approx 0.75 \pi$. Sadeghpour and Greene therefore postulated the propensity rule that in photodetachment of the H⁻ ground state, only doubly excited states having no θ_{12} nodes are populated with significant intensity.

Sadeghpour and Greene used the bending vibrational quantum number v to quantify the number of nodes in θ_{12}. They postulated that, in general, photoexcitation processes for the ground state of two-electron systems obey the rule, $\Delta v = 0$, reasoning that nonadiabatic couplings of transitions from the lowest + channel to higher + channels with $v > 0$ are negligible because of the different nodal structures. Note that Rost et al.[21] have pointed out that these nodal structures can be alternatively described in the separable spheroidal coordinates of the molecular orbital picture of two-electron systems.[22,23]

Very recently, Sadeghpour et al.[24] have carried out eigenchannel R-matrix calculations of the photodetachment cross sections for H⁻ with excitation of the $n = 2$, 3, and 4 levels of H. These calculations give quantitative confirmation of the propensity rules postulated in Ref. 19 on the basis of the adiabatic hyperspherical model.

The interpretation of the recent He doubly excited state spectrum is similar. Domke et al.[15] reported a high-resolution photoionization study of the doubly excited He states below the $n = 2$–7 thresholds of He⁺. Sadeghpour[25] has shown that the adiabatic hyperspherical representation for highly excited states of He gives a picture similar to that for H⁻. Namely, the energy levels of doubly excited states calculated in the lowest + adiabatic hyperspherical potentials agree very well with the positions of the experimentally observed[15] resonances for the lowest n levels. Furthermore, the density plots for He display the same kinds of nodal structures, leading to the same conclusion that $\Delta v = 0$ is a good propensity rule. For higher n levels, beginning at about $n = 6$, overlapping of Rydberg levels corresponding to different n manifolds requires, however, explicit treatment of nonadiabatic coupling terms.

Thus, the newly obtained high-resolution data on doubly excited state spectra of H⁻ and He are permitting theory to develop new understanding of the dynamics of these fundamental three-body Coulomb systems.

3.2. The Xe 5s-Subshell Photoelectron Angular Distribution

A particularly striking example of the effect of relativistic interactions is provided by the photoelectron angular distribution for the 5s subshell in xenon. Since xenon is a closed-shell atom, its ground state is spherically symmetric. Hence, the only symmetry axis in electric dipole transitions from the ground state is provided by the polarization vector of the incident light. If one assumes only single configurations to describe initial and final states, then photoionization of the 5s subshell may be represented as the following process:

$$\text{Xe } 5s^2 5p^6(^1S_0) + \gamma \rightarrow \text{Xe}^+ 5s 5p^6(^2S_{1/2})\varepsilon p \ ^{2S+1}P_{J=1} \tag{11}$$

In the absence of relativistic interactions the spin S of the final state in equation (11) is conserved in this transition and hence has the value $S = 0$. In this case there is only a single final state channel allowed and hence the asymmetry parameter β becomes a constant, independent of the photon

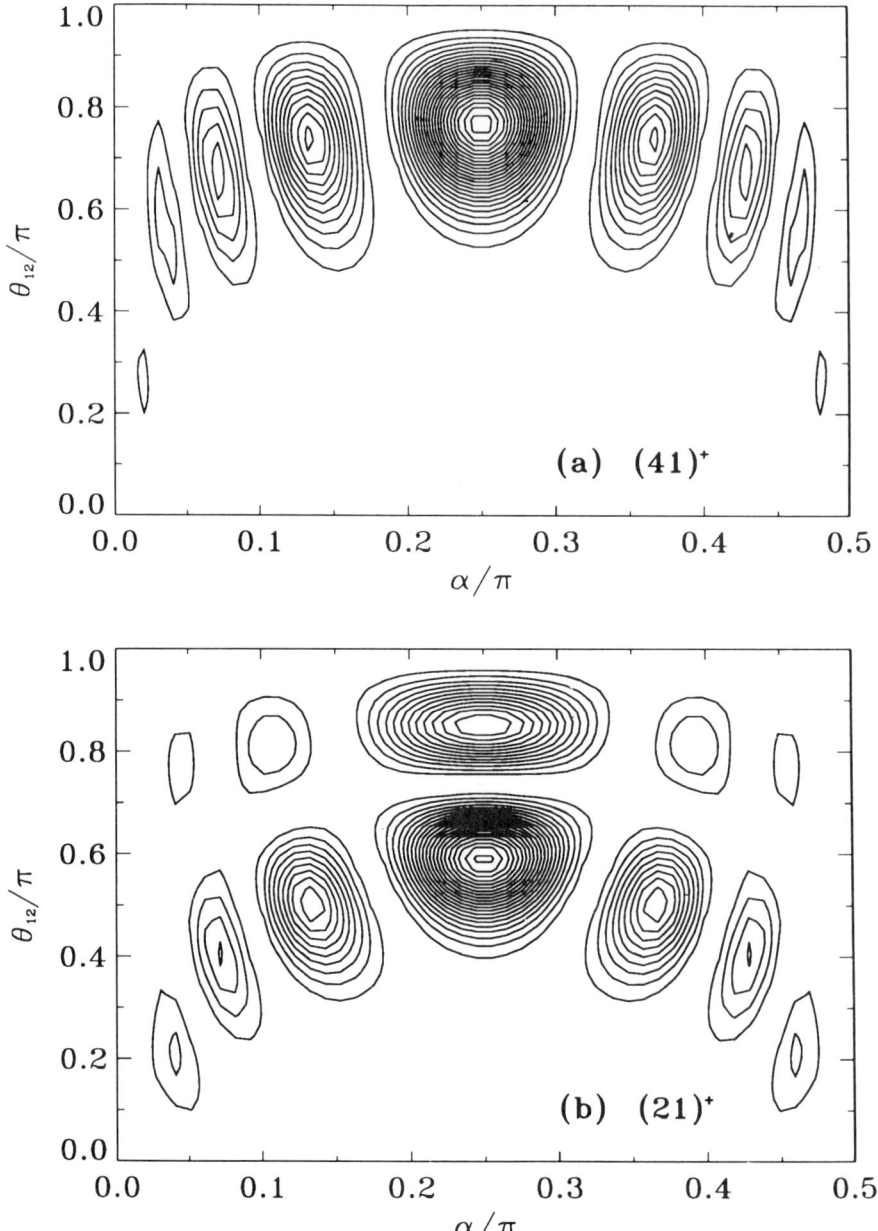

FIGURE 2. The adiabatic hyperspherical two-electron density function shown as a contour plot versus α and θ_{12}, displaying the nodal patterns for the two lowest + channels in the $n = 6$ manifold at $R = 80$ a.u. (a) and (b) correspond, respectively, to $(KT)^A = (41)^+$ and $(21)^+$ channels, i.e., $v^A = 0^+$ and 1^+. (From Ref. 19.)

energy.[11] The value of this constant is $\beta = 2$, which implies that the photoelectron has a $\cos^2\theta$ angular distribution about the direction of linear polarization in the case of linearly polarized light.[11]

The first experimental measurement reported by Dehmer and Dill[26] found, however, that $\beta = 1.4$ at the photon wavelength 584 Å. It was noted that relativistic interactions were likely responsible for this drop in the value of β below the nonrelativistically expected value, $\beta = 2$. In terms of equation

(11), one may note that such relativistic interactions permit the $S = 1$ (triplet) channel to have a nonzero amplitude. Within the electric dipole approximation the photoelectron angular distribution asymmetry parameter is then given by

$$\beta = (2 - r)/(1 + r) \qquad (12)$$

where $r \equiv \sigma(^3P)/\sigma(^1P)$. Here $\sigma(^1P)$ and $\sigma(^3P)$ are the photoionization cross sections for the ^1P and ^3P transitions. In the absence of relativistic interactions $\sigma(^3P)$ is zero and equation (4) shows that β is equal to 2. In practice, $\sigma(^3P) \ll \sigma(^1P)$ so that β is close to 2 except in the region of the well-known, near-threshold minimum in $\sigma(^1P)$. [Note that $\sigma(^1P)$ never becomes exactly zero because of interchannel interactions with photoelectrons from the 5p and 4d subshells, among others.]

The Dehmer and Dill result[26] stimulated much theoretical work to account for these relativistic interactions. The Dirac–Fock (DF) result of Ong and Manson[27] as well as the two relativistic random-phase approximation (RRPA) results of Johnson and Cheng,[28] which included coupling between the 5s- and 5p-subshell channels and between the 4d-, 5s-, and 5p-subshell channels, respectively, all passed close to the single experimentally measured point.[26] These three calculations made very different predictions, however, for the location of the minimum in β, which is apparently very sensitive to the electron correlations included in the calculation. Only the most detailed calculation, the RRPA (4d + 5s + 5p) one, was in agreement with the next measured experimental points at lower energy of White et al.[29] The K-matrix results of Huang and Starace,[30] which treated the effect of final-state spin-orbit interactions in the Breit–Pauli approximation within a basis of HF nonrelativistic wave functions, did not give nearly so large a drop in β as did the other, purely relativistic calculations.[27,28] This result appeared to indicate the importance of having a fully relativistic treatment. The state of affairs as of 1980 is summarized in Figure 3.

The excellent agreement of experiment with the most detailed theoretical predictions (see Figure 3) seemed to provide further evidence that the essential physics of atomic photoionization, at least for rare gas atoms, was well understood. However, this confidence was undermined in 1983 by two new experimental measurements which explored the energy region of the minimum in the Xe 5s-subshell cross section.[31,32] These measurements found a significantly higher value for β in the energy region of the minimum, as well as a somewhat lower value for β at higher energies, than predicted by the most detailed (RRPA) calculation.[28]

Wendin and Starace[33] postulated a reason for these unexpected discrepancies between theory and experiment, not only for the β parameter but also for the partial cross sections both near threshold and at higher energies. They noted that there is a usually weak interaction, besides the spin-orbit interaction, which has typically been ignored when describing photoionization processes theoretically, but which may have measurable effects when the dominant photoionization transition amplitude is small: final-state ionic configuration interaction. Specifically, the excited $5s^2 5p^4(^1D)5d(^2S)$ configuration in Xe$^+$ is very strongly mixed[34–37] with the usual ionic configuration $5s^1 5p^6(^2S)$. In a two-level theoretical treatment of process (11) the ionic Hamiltonian would be diagonalized to obtain two new eigenstates, each represented as a linear combination of the configurations $5s5p^6(^2S)$ and $5s^2 5p^4(^1D)5d(^2S)$. The eigenstate with the lower energy would be a better representation for the ionic state than the single configuration $5s5p^6(^2S)$. One effect of such ionic configuration mixing in process (11) would be on the kinetic energy and wave function of the continuum electron, which would see both a lower ionization threshold and a less attractive ionic ground state. Another effect would be the modification of the 5s–5p intershell interaction, which produces the drop in value of the β parameter. To the extent that the ratio $\sigma(^3P)/\sigma(^1P)$ decreases in the neighborhood of the minimum in $\sigma(^1P)$ as a result of these two effects, this configuration mixing, they suggested, might explain the discrepancy between theoretical and experimental values for β.

It was Tulkki[38] in 1989 who showed, in a tour-de-force multichannel, multiconfiguration Dirac–Fock calculation, that the $5p^4 5d(^2S)$ excited ionic configuration was essential to describe the Xe 5s-subshell photoelectron cross section and angular distribution. He first carried out a configuration interaction calculation for the Xe$^+ 5s^{-1}(J = \frac{1}{2})$ configuration with the five jj-coupled configurations Xe$^+ 5p^{-2} 5d(J = \frac{1}{2})$. Then final-state interactions were treated among the 23 channels corresponding to

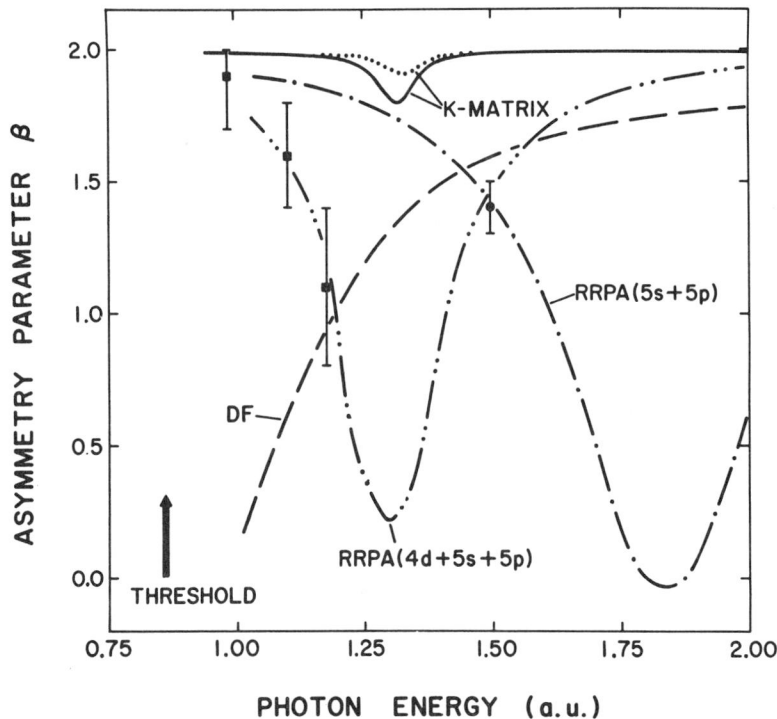

FIGURE 3. Photoelectron angular distribution asymmetry parameter β for the 5s subshell of xenon. *DF*: Dirac–Fock results of Ong and Manson.[27] *RRPA*: relativistic random-phase approximation results of Johnson and Cheng[28] including interchannel correlations between the 5s- and 5p-subshell channels (dash–dot line) and between the 4d-, 5s-, and 5p-subshell channels (dash–double dot line). *K-matrix*: results of Huang and Starace[30] including final-state spin-orbit interactions and coupling between the 5s-, and 5p-subshell channels in dipole length (dotted line) and dipole velocity (solid line) approximation. *Solid circle*: Experimental result of Dehmer and Dill.[26] *Solid squares*: Experimental results of White et al.[29] (From Ref. 11.)

these six ionic states as well as those corresponding to 4d and 5p hole states. For comparison, calculations were also carried out which included only single-excitation channels corresponding to final 4d, 5s, and 5p hole states (13 channels) and only 5s and 5p hole states (7 channels).

Tulkki's results[38] are given in Figure 4, which shows that his 23-channel calculation including final-state ionic configuration interaction is in excellent agreement with the new experimental results for β[31,32] as well as with experimental results for the Xe 5s-subshell cross section.[32,39–41] This contrasts with the results of his 13- and 7-channel "reference" calculations, which do not include final-state ionic configuration interaction effects. In Tulkki's words,

> The effect of interchannel interaction becomes decisive at low photon energies where the cross section and especially the β parameter are increased, when the double-excitation channels [i.e., those produced by final state ionic configuration interaction] are included....[We conclude] that the behavior of the Xe 5s cross section in the threshold region is largely determined by the interaction between the 5s single-hole and $5p^45d$ double-hole ionization channels.

In this way, Tulkki's large-scale calculation demonstrated conclusively that the new physics required to describe process (11) was well understood. Curiously, these new theoretical results[38] and the new experimental results[31,32] lie significantly higher than the original measurement[26] at 584 Å which stimulated all of the succeeding intensive theoretical and experimental efforts.

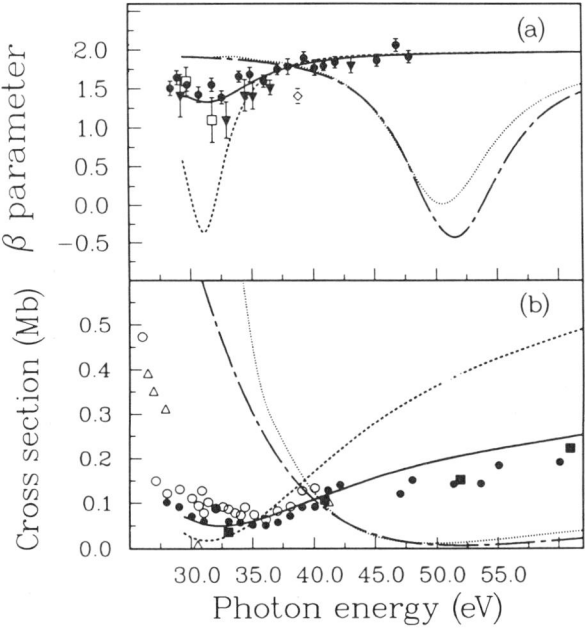

FIGURE 4. Theoretical and experimental cross section and photoelectron asymmetry parameter for Xe 5s photoionization. *Theoretical curves of Tulkki*[38]: Solid line, 23-channel calculation; dashed line, 13-channel calculation; long–short-dashed line, 7-channel calculation; dotted line, 7-channel calculation excluding relaxation. *Experimental data*: Solid dots, Ref. 32; open circles, Ref. 39; solid squares, Ref. 40; open squares, Ref. 29; solid inverted triangles, Ref. 31; open triangles, Ref. 41; diamond, Ref. 26. (From Ref. 38.)

3.3. Final Ionic State Branching Ratios

As we have just seen in the case of Xe 5s-subshell photoionization, configuration interaction effects are very important for describing inner-shell photoionization processes quantitatively. Nevertheless, often a single-configuration point of view allows one to provide a simple interpretation of experimental results. We give here two examples.

3.3.1. Resonant Photoionization of the Be 1s Subshell. In examining the decay of photon-produced Be $1s2s^2np(^1P)$ for $n = 2,3$, Caldwell et al.[42] found that the predominant decay mode is to the final excited ionic state $Be^+ 1s^2np$, rather than to the ground ionic state, $Be^+ 1s^22s$. For $n = 2$, the 2p ionic state accounts for 95% of the total cross section, with $Be^+ 1s^23p$ the other main contributor. For $n = 3$, 3p is dominant, again with a very small fraction of 2s. The experimental data are shown in Figure 5.

One may understand the large observed ratio of 2p:2s production for the $n = 2$ case using an analogue of the compound nucleus model for resonant nuclear scattering,[43] i.e., assuming the decay of the Be $1s2s^22p(^1P)$ state is independent of its formation. The relative intensities of the $Be^+ 1s^22p(^2P)$ and $Be^+ 1s^22s(^2S)$ final ionic states are then proportional to the squares of the following Coulomb matrix elements:

$$V_{2p} \equiv \langle 1s2s^22p(^1P) | \sum_{i>j} \frac{1}{r_{ij}} | 1s^22p(^2P)\varepsilon s(^1P)\rangle = -\int dr_1 \int dr_2 P_{1s}(r_1) P_{\varepsilon s}(r_2) r_>^{-1} [P_{2s}(r_1)P_{2s}(r_2)] \quad (13)$$

$$V_{2s} \equiv \langle 1s2s^22p(^1P) | \sum_{i>j} \frac{1}{r_{ij}} | 1s^22s(^2S)\varepsilon' p(^1P)\rangle$$

$$= \int dr_1 \int dr_2 P_{1s}(r_1) P_{\varepsilon'p}(r_2) r_>^{-1} [P_{2s}(r_1)P_{2p}(r_2) - (2r_</3r_>)P_{2s}(r_2)P_{2p}(r_1)] \quad (14)$$

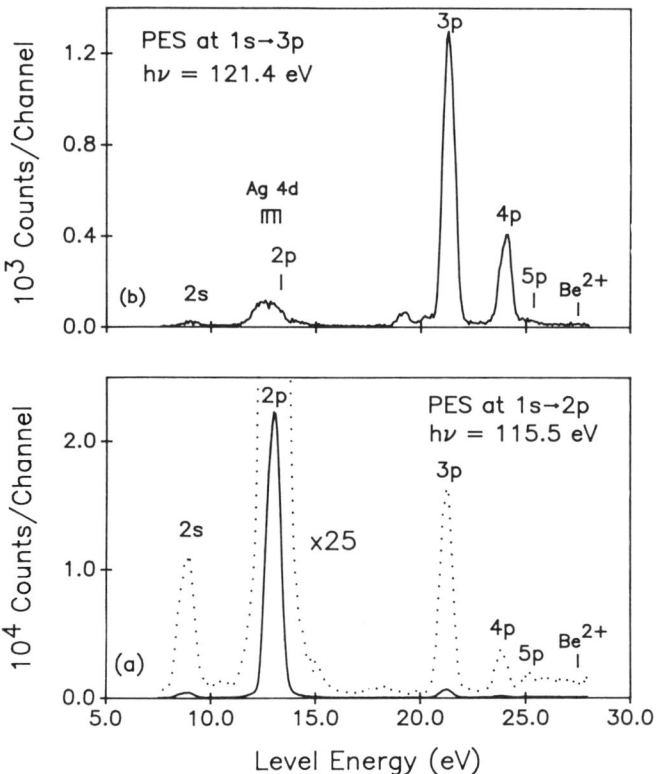

FIGURE 5. Electron spectra resulting from the decay of the Be $1s2s^2np(^1P)$ excited states. (a) $n = 2$, $h\nu = 115.5$; (b) $n = 3$, $h\nu = 121.4$ eV. The $3p$ spectrum shows some contamination of the Be^+ $1s^22p$ peak resulting from the presence of silver in the oven. Note the predominance of the Be^+ $1s^2np$, $n = 2,3$, final states of the ion, and the very small amount of Be^+ $1s^22s$. (From Ref. 42.)

In equations (13) and (14), both initial and final states are described by a single configuration. The right-hand side represents the matrix element in terms of radial Slater integrals over the one-electron radial wave functions $P_{nl}(r)$; ε and ε' are the continuum kinetic energies of the photoionized electron for each of the two final ionic states. Finally, $r_> \equiv \max(r_1,r_2)$ and $r_< \equiv \min(r_1,r_2)$.

One sees clearly from (13) and (14) that where the Coulomb repulsion between the jumping electrons is largest, i.e., $r_1 = r_2$, V_{2p} has its maximum value [with the factor in brackets $\propto P_{2s}^2(r)$] while V_{2s} has extensive cancellation [with the factor in brackets $\propto P_{2s}(r)P_{2p}(r)/3$]. Because of this factor $\frac{1}{3}$ in V_{2s} for $r_1 = r_2$, one expects the squares of (13) and (14) to differ by an order of magnitude.

This expectation is easily confirmed numerically using a basis of bound Hartree–Fock (HF) wave functions calculated for the Be $1s2s^22p(^1P)$ resonant state as well as HF continuum orbitals εs and εp calculated in the field of the appropriate relaxed Be^+ state. Effects of nonorthogonal overlap integrals should also be included.[(44)] The $1s$, $2s$, and $2p$ orbitals are shown in Figure 6. Clearly the region $r_1 = r_2 = 1.8$ is where P_{2s} and P_{2p} have their maximum values. Use of these wave functions leads to a ratio of 24 for the Be^+ $2p$ to $2s$ final ionic states, as compared with the experimental branching ratio of 56 ± 3.

Matrix elements similar to those in (13) and (14) apply for the resonant state Be $1s2s^23p(^1P)$. V_{3p} has the same analysis as V_{2p}. However, the analysis of V_{2s} requires more care since the first two antinodes of P_{3p}, shown in Figure 6, are both relevant. Such an analysis leads one to expect again that the square of V_{2s} is at least an order of magnitude smaller than the square of V_{3p}. The HF result reveals it is two orders of magnitude smaller, giving a $3p:2s$ ratio of 250. This compares with the experimental result of 78 ± 3.

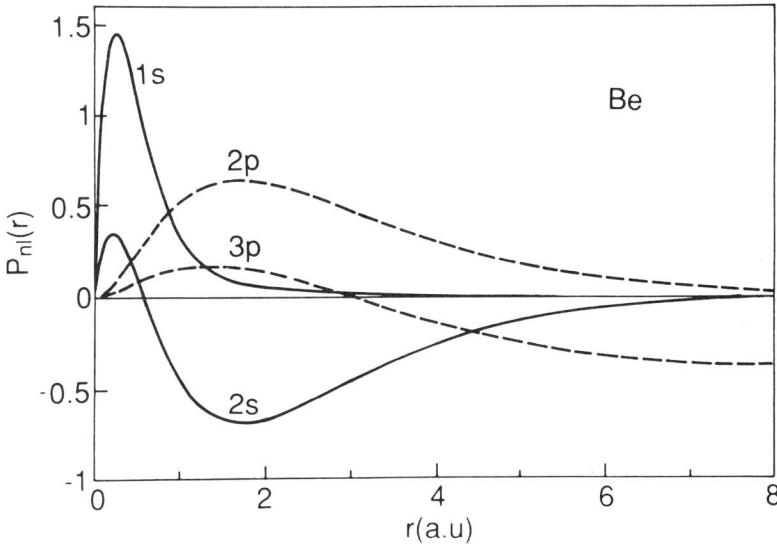

FIGURE 6. Radial orbital wave functions for $1s$, $2s$, and $2p$ in Be calculated in HF approximation for the state Be $1s2s^22p(^1P)$. The $3p$ orbital shown is calculated for the state Be $1s2s^23p(^1P)$. (From Ref. 42.)

Much more detailed calculations which include treatment of initial state and final ionic state configuration interactions as well as final state interchannel coupling effects are needed to obtain quantitative agreement with experiment.[45] This effort is necessary because the Be$^+$ $1s^22s$ ionic state cross section is very small, making its quantitative prediction subject to many competing influences. Clearly, however, the simple single-configuration picture presented above gives a qualitative, order-of-magnitude understanding of why the Be$^+$ $1s^22s$ ionic state cross section is so small. Namely, the transition matrix element to the $2s$ state is subject to extensive cancellations, as revealed in (14), whereas the transition matrix elements to the np states are not, as revealed in (13).

3.3.2. Relative Cross Sections for Fine-Structure Transitions. High-resolution photoelectron spectroscopy experiments[46] can provide the partial cross section for a transition from a particular fine-structure level of an initial atomic state to a particular fine-structure level of the residual ion. Such partial cross sections offer more severe tests of theoretical approximation methods than do the cross sections measured without regard to individual fine-structure levels. However, it is useful to distinguish geometric from dynamic effects particularly since the geometric effects alone can result in intensity distributions that are quite different from those expected for a statistically averaged atomic state.

Theoretical efforts to disentangle geometrical effects from dynamical effects in atomic photo-processes have a long history. The initial stimuli appear to have been the measurements of Lineberger and co-workers on fine-structure transitions in photodetachment of negative ions in the early 1970s.[47,48] Rau and Fano[49,50] were able to provide interpretations of the observed[47,48] branching ratios for photodetachment of negative ions using a primarily geometrical analysis. The key point of their analysis is that in the region of space where photoprocesses occur, near the origin, the transition amplitudes are unaffected by the generally small spin-orbit and other relativistic interactions that lead to fine structure. Those weak interactions affect primarily the kinetic energy of the electron at asymptotic distances. Hence, the relative magnitudes for particular fine-structure transitions may be calculated analytically using only the geometric (i.e., angular momentum) properties of the transitions. The process of photodetachment is much simpler than that of photoionization since near the threshold for the former process only a single photoelectron orbital angular momentum dominates. In particular, the photodetachment processes considered by Rau and Fano[49,50] involved primarily s-wave photoelectrons, which simplifies the theoretical analysis significantly. While the general case of photoionization was sketched in the review of Rau,[50] it was not analyzed in detail. However, the work of Fano and Rau on photodetachment stimulated several other researchers to examine the general case of atomic photoionization.[51–55] All of these works make the additional assumption from the start that the dynamical transition amplitudes are independent of the orbital and spin angular momentum quantum numbers relevant to the transition under consideration except for the orbital angular momentum of the photoelectron. In the case that LS-coupled single configurations are used to represent the initial and final states, all of these authors obtain the same results. While the assumption of LS-independent transition amplitudes is not generally a good one, it permits theory to predict the relative cross sections for photoinduced fine-structure transitions completely analytically and hence provides experimenters with benchmarks against which to compare.

Recently, Pan and Starace[56] have presented a more detailed formulation. Within a single-configuration representation of the initial and final states, they have retained the LS-coupling dependence of the dynamical (electric-dipole) transition amplitudes. In special cases (in particular, for the case that the LS-dependence of the radial dipole amplitudes and photoelectron phase shifts is ignored), their results for the partial cross sections for particular fine-structure transitions are shown to reduce to those obtained by others for photodetachment[49,50] and photoionization.[51–55]

The utility of their more general analysis is highlighted by a detailed examination of photoionization of inner s subshells in open-shell atoms, i.e., the process

$$A n_0 s^2 n \ell^N (L_0 S_0 J_0) + \gamma \to A^+ n_0 s n \ell^N (L_c S_c J_c) + e^- \qquad (15)$$

When the transition amplitudes are assumed to be LS-independent, then any transition not obeying the triangle inequality,

$$|J_0 - \tfrac{1}{2}| \le J_c \le J_0 + \tfrac{1}{2} \qquad (16)$$

may be shown to have a cross section equal to zero.[56] This selection rule indicates that the photoionization of an s electron removes only an angular momentum of $\tfrac{1}{2}$ from the residual core. However, the more general treatment of Pan and Starace,[56] in which the transition amplitudes are assumed to retain their LS-dependence, indicates that no such restriction on the allowed transitions applies: transitions violating the triangle relation in (16) (but satisfying the usual angular momentum and parity conservation laws) are allowed. There is both experimental and theoretical evidence, however, that fine-structure transitions which do not satisfy (16) are nevertheless quasiforbidden, indicating therefore that there are hitherto unsuspected cancellation effects between different transition amplitudes contributing to these particular fine-structure transitions.

Consider the case of atomic chlorine,

$$\text{Cl } 3s^2 3p^5 (^2 P_{J_0}) + \gamma \to \text{Cl } 3s 3p^5 (^3 P_{J_c}) + e^- \qquad (17)$$

for which Pan and Starace[56] presented LS-dependent HF results for the relative partial cross sections and compared these to the purely geometric predictions one may make by assuming that the transition amplitudes are LS-independent. This case was chosen because there exist experimental data[57] for this process and also because Robicheaux and Greene have recently carried out multiconfiguration, eigenchannel R-matrix calculations for photoionization of both the $3p$ subshell[58] and $3s$ subshell[59] of atomic chlorine. Reference 59 does not contain plots of the fine-structure partial cross sections; however, Robicheaux's results are presented in Ref. 56.

The relative partial cross sections for process (17) for the case $J_0 = \tfrac{3}{2}$ are presented in Figure 7. The purely geometrical predictions, which are obtained by assuming LS-independent dynamical amplitudes, are indicated by the horizontal solid lines. These predictions agree surprisingly well with those obtained using the more general equations of Pan and Starace,[56] in which the LS-dependent amplitudes and phase shifts are calculated in HF approximation. These single-configuration predictions are shown in Figure 7 to agree very well also with the multiconfiguration, eigenchannel R-matrix

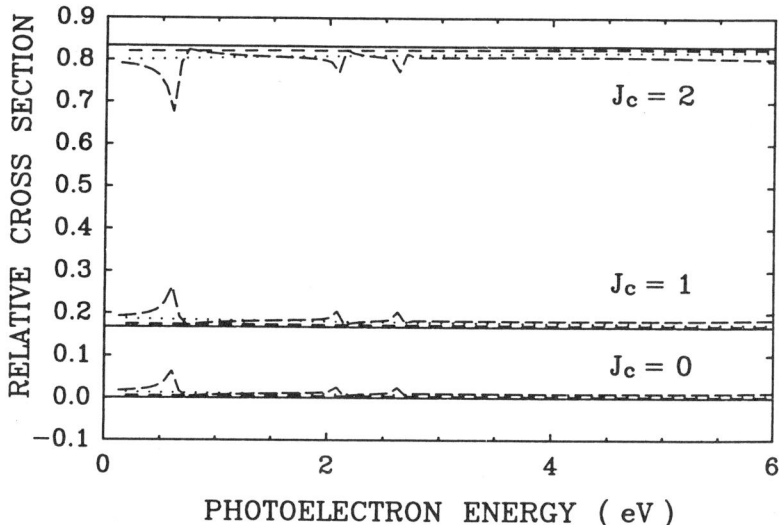

FIGURE 7. Relative cross sections for the transitions $Cl3s^23p^5(^2P_{3/2}) + \gamma \rightarrow Cl^+ 3s3p^5(^3P_{J_c}) + e^-$ in three levels of approximation. *Solid lines*: purely geometrical predictions assuming *LS*-independent transition amplitudes.[56] *Short-dashed (dotted) lines*: *LS*-dependent HF predictions in dipole length (velocity) approximation of Pan and Starace.[56] *Long-dashed lines*: multiconfiguration, eigenchannel *R*-matrix results of Robicheaux and Greene.[59] (From Ref. 56.)

results of Robicheaux and Greene,[59] except of course in the vicinity of autoionizing resonances. This agreement is all the more remarkable since they find that the $^3P^o$ state of the ion only contains 71% $3s3p^5$ and 29% other configurations.[59] Furthermore, a recent experiment[57] was unable to discern any signal for the quasiforbidden transition $J_0 = \frac{3}{2} \rightarrow J_c = 0$ shown in Figure 7.

In conclusion, based on comparisons with more general theoretical results[56,59] as well as with experiment,[57] it appears that inner *s*-subshell transitions of the form of (15) which do not obey the triangle inequality in (16) are quasiforbidden. Thus, simple geometrical arguments based on a single-configuration, *LS*-independent analysis appear to give a reliable indicator of otherwise unsuspected cancellation among the amplitudes contributing to these processes. This is true not only in the case of the more general single-configuration treatment of Pan and Starace,[56] but also for a multiconfiguration treatment.[59]

3.4. Fluorescence Spectroscopy

High-brightness light sources enable one to study relatively weak processes, such as satellite structures attendant on the main photoionization line. One method of studying the often complex satellite spectrum is fluorescence spectroscopy.[60] It provides complementary information to photoelectron spectroscopy. Its advantages are its ability to study satellite structures near threshold, and its ability to distinguish satellite states having similar energies (since they fluoresce with different wavelengths). Also, it obviates the need for coincidence techniques. A disadvantage is the necessity for cascade corrections.

Fluorescence spectroscopy has had in recent years a number of successes in elucidating photoionization spectra. One of the earliest of such successes was the first clear evidence of doubly excited states converging to the He^+ ($n = 3$) threshold by Woodruff and Samson.[61] This experiment measured the He^+ $2p$ fluorescence decay at 304 Å and saw evidence in this spectrum (as a function of incident photon energy) of helium doubly excited states autoionizing to the He^+ ($n = 2$) state. Also, the He^+ $2s$ population was observed by quenching with an electric field.

Another such success was the finding by Schartner et al.[60,62] that doubly excited resonance structures are prominent in the region of the threshold for the Ar 3s-subshell photoionization cross section. These findings underlined the need for theoretical calculations to treat such satellite structures in inner-shell photoionization cross section calculations. They also indicated the need for high-brightness synchrotron light sources to enable experiment to measure the line shapes for these resonances.

Very recent work on Ne photoionization by Samson et al.[63] provides a further example of the complementary information gained by fluorescence spectroscopy. Consider the neon photoexcitation process,

$$\text{Ne } 2p^6(^1S_0) + \gamma \rightarrow \text{Ne}^{**} 2p^4(^3P)3pnl \ (^{2S+1}P) \tag{18}$$

that is followed by the autoionization process

$$\text{Ne}^{**} 2p^4(^3P)3pnl \ (^{2S+1}P) \rightarrow (\text{Ne}^+)^* 2p^4(^3P)3p(^4P) + e^- \tag{19}$$

By angular momentum selection rules, process (19) is only possible for the triplet doubly excited state (i.e., $S = 1$). This state is produced in process (18) only through final-state spin-orbit interactions; in LS-coupling, the triplet state is forbidden by electric dipole selection rules.

In a photoelectron spectroscopy experiment,[64] it is difficult to distinguish the $(\text{Ne}^+)^* 2p^4(^3P)3p(^4P)$ satellite from the $(\text{Ne}^+)^* 2p^4(^1D)3s(^2D)$ satellite since the two states differ in energy by only a few millielectron volts. Fluorescence spectroscopy, however, permits one to distinguish the ^4P satellite by its fluorescence in the wavelength range 3665–3778 Å, i.e.,

$$(\text{Ne}^+)^* 2p^4(^3P)3p(^4P_J) \rightarrow (\text{Ne}^+)^* 2p^4(^3P)3s(^4P_{J'}) + \gamma' \tag{20}$$

where the wavelength range arises from transitions between the various allowed J and J' levels. In contrast, the $(\text{Ne}^+)^* 2p^4(^1D)3s(^2D)$ satellite fluoresces in the neighborhood of 406 Å.

A comparison of the photoelectron[64] and fluorescence[63] spectra is shown in Figure 8. The bottom panel shows the fluorescence spectrum[63] [see process (20)] as a function of the incident photon energy [see process (18)]. The top panel shows the corresponding photoelectron spectroscopy (PES) spectrum,[64] which cannot distinguish between the $3p(^4P)$ and $3s(^2D)$ satellites. It is plotted also versus incident photon energy. Notice that the fluorescence spectrum is obtained at the very threshold for producing the $3p(^4P)$ satellite, whereas the PES spectrum starts about 0.35 eV above threshold.

Clearly, fluorescence spectroscopy is a valuable complementary tool for studying detailed features of photoionization spectra.

3.5. Photoionization of Inner Shells of Excited States

Inner-shell photoionization differs from outer-shell processes in that the relaxation of the ion with an inner-shell vacancy is considerably more complex, even aside from the Auger or x-ray processes which fill the vacancy. From a physical point of view, removal of an outer-shell electron strongly affects the other outer-shell electrons which it partially screens. But, if the outer-shell electron is thought of as roughly a shell of charge at the outer-shell radius, Gauss's law of electrostatics tells us that this electron exerts no force inside the shell, i.e., on inner-shell electrons. Thus, the primary effect of the removal of an outer-shell electron on inner shells is a constant change in the potential energy and little else.

Removing an inner-shell electron, on the other hand, changes the screening (and, thus, the potential) of all electrons in that shell and in all shells outside it, thereby leading to much more significant relaxation effects. The effect is likely to be even more pronounced for an excited electron since more of the electron density is outside the inner shell and removal of an inner-shell electron changes the screening by close to a full electron charge; the more highly excited the electron, the greater the change in screening.

As an example, $2p$ inner-shell photoionization has been investigated experimentally[65] for both ground-state sodium atoms ($1s^2 2s^2 2p^6 3s$) and excited-state sodium atoms ($1s^2 2s^2 2p^6 3p$). As a practical

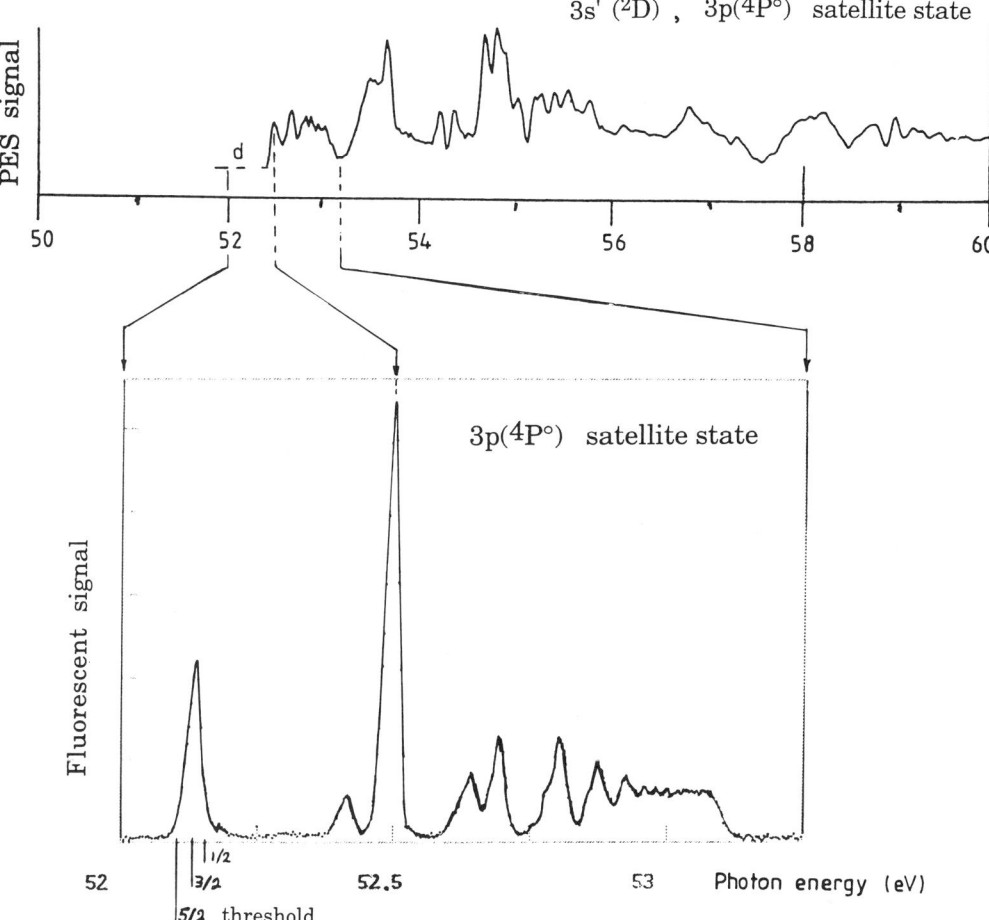

FIGURE 8. Comparison of photoelectron spectrum resulting from autoionization of the doubly excited states Ne**$2p^4(^3P)3pns,d$ (top figure, from Ref. 64) with the fluorescence spectrum of the (Ne$^+$)* $2p^4(^3P)3p(^4P^o)$ satellite state (bottom figure, from Ref. 63). Both figures are plotted versus initial photon energy [see process (18)]. For further details, see text description.

matter, relaxation shows up as satellite transitions, i.e., photoionization processes where one electron is ionized and another is excited. The intensity of the satellites for 2p ionization from the ground state, relative to the main transition in which all electrons other than the photoelectron remain spectators, is shown in Figure 9. As a function of energy, the ratio remains relatively constant, just below 20%. Also shown in Figure 9 is the corresponding ratio for the 3p excited state of Na; it is also independent of energy, but about 40%. Thus, the excitation of the 3s electron to 3p in the initial state of Na roughly doubles the cross section for satellite transitions.

These results can be understood in terms of changes in screening, as discussed above. The cross section for the main transition for 2p photoionization in the presence of an outer nl spectator electron is proportional to $|\langle nl_i | nl_f \rangle|^2$, where the subscripts i and f refer to the initial and final states, respectively. This overlap factor, which differs from unity because of relaxation, is about 0.85 for 3s and 0.7 for 3p, leading to the results shown in Figure 9. Since 3s is such a penetrating orbital, removal of a 2p electron does not have as large an effect as for the 3p which has a larger radius and which is much less penetrating. Thus, relaxation is more significant for the 3p initial state, leading to the enhanced satellite intensity, in agreement with theoretical predictions.[66]

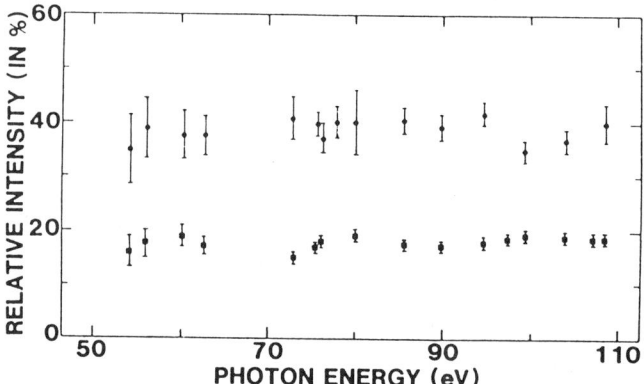

FIGURE 9. Variation, as a function of photon energy, of the relative intensity of shakeup satellites for sodium atoms in the ground state (■; $2p^54s$ final ionic states) and in the $3p$ excited state (●; $2p^54p$ final ionic states), respectively. (From Ref. 65.)

Clearly, by these arguments, this relaxation effect will be even more pronounced for more highly excited states. A recent calculation[67] investigated inner-shell photoionization for excited states of lithium. The results for $1s^23p$ are given in Figure 10, which shows the cross sections for leaving the Li^+ ion in the $1s2p$, $1s3p$, and $1s4p$ $^{3,1}P$ states. The outstanding feature of these results is that the dominant cross section is that of the $1s^23p(^2P) \rightarrow 1s4p(^3P) + e^-$ channel, which results from an ionization plus satellite excitation transition. The next largest cross section is for the $1s^23p(^2P) \rightarrow 1s4p(^1P) + e^-$ transition, which also involves ionization plus excitation. In fact, away from threshold, the $4p$ "satellites" represent 76% of the total cross section, whereas the $3p$ "main" transitions account for only 13%. As in the case of sodium, the explanation is the relaxation of the $3p$ orbital of the initial state, as reflected in the overlap factors.

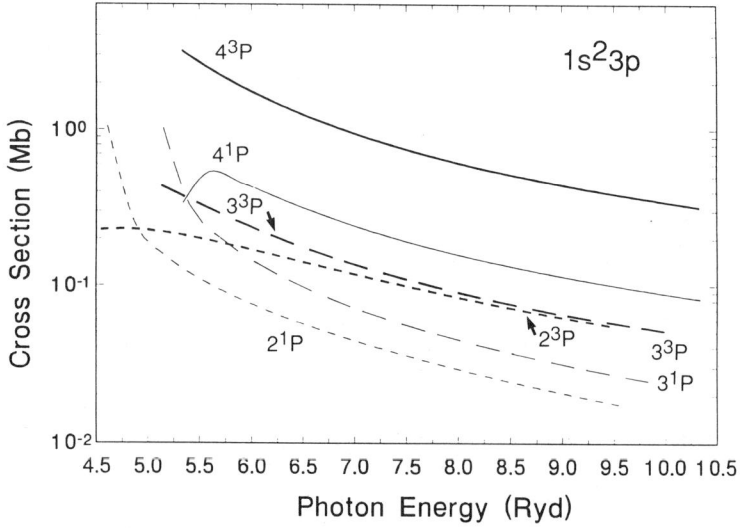

FIGURE 10. Photoionization cross sections calculated for $1s$ ionization from the Li $1s^23p$ excited state to various n^3P and n^1P states of the Li^+ ion plotted versus photon energy. The curves are the dipole-length results, which are in good agreement with the dipole velocity results (not shown). (From Ref. 67.)

That the overlaps favor $3p \rightarrow 4p$ so strongly is related to the fact that a reasonably highly excited state of lithium is virtually completely screened by the $1s^2$ core; the excited electron "sees" a charge of 1. Similarly, in the Li$^+$ ion, the excited electron "sees" a charge of 2, so that the $3p_f$ orbital of the final state is considerably more compact than the $3p_i$ orbital of the initial state. Thus, the principal overlap of the $3p_i$ wave function is with the $4p_f$ orbital which occupies the same region of space owing to the increased charge it "sees." One may easily verify this from even the simple Bohr atomic model. For lower excited states, or ground states, the effect is not nearly so dramatic, as seen for sodium, because the change in effective charge is far smaller than the factor of 2 change that occurs for highly excited states.

While this phenomenon has not yet been seen experimentally for highly excited states, the work on sodium[65,68,69] and more recent work on potassium,[70] are certainly indicative of it. In addition, it has been observed experimentally in multiphoton ionization of excited barium[71] as well as in autoionization of core-excited argon.[72] In any case, with the emergence of the technology to create and ionize such excited states[73,74] one has a new "laboratory" in which to study multielectron transitions.

3.6. Photoelectron Angular Distributions from Parity-Unfavored Transitions

Photoelectron angular distributions give in general information about amplitudes and phases of the dipole matrix elements involved in the photoionizing transitions. The parameter β [equation (7)], which describes the angular distribution, is normally energy-dependent since the amplitudes and phases of the matrix elements are always energy-dependent. However, the energy-dependence of β arises from the interference of the transition amplitudes for alternative photoelectron emission channels, analogous to the interference underlying a double slit experiment. In cases where there is only a single photoelectron emission channel, no interference is possible and β must be independent of energy. Such a situation occurs in the photoionization of the ground state of the hydrogen atom, or any alkali atom, if relativistic effects are ignored. In these cases, there is only an $s \rightarrow p$ transition and β = 2. This corresponds to a $\cos^2\theta$ distribution, which attains its maximum in the direction of the photon's electric vector. This result is consistent with the intuitive picture of photoionization as a response to the impulse of the photon's electric field. In addition, in certain cases, β may have the constant value 0.2.[11]

Most curious, however, is the prediction that under certain conditions, β can take the energy-independent value of −1, giving rise to a $\sin^2\theta$ distribution. This distribution attains its maximum at 90° to the photon's electric vector, an entirely counterintuitive result. In the language of the angular momentum transfer analysis of photoelectron angular distributions, this is called a "parity-unfavored" transition, discussed in Section 2, where it was seen that such a transition is characterized by

$$\pi_0 \pi_c = -(-1)^{j_t} \tag{21}$$

where j_t is the angular momentum transfer and π_0 and π_c are the parities of the initial atomic and the final ionic states, respectively. In many photoionization processes, the possible values of j_t include one or more which are parity-unfavored.[11] But transitions which are purely parity-unfavored were thought to be relatively rare. For example, the transition

$$p^3(^2D) \rightarrow p^2(^1S)\varepsilon d(^2D) \tag{22}$$

in atomic nitrogen or phosphorus is such a process.[75] Within the context of LS coupling, equation (6) for j_t becomes

$$\vec{j}_t = \vec{L}_c - \vec{L}_0 = \vec{j}_\gamma - \vec{l} \tag{23}$$

with $\vec{L}_0$, $\vec{L}_c$, and $\vec{l}$ the orbital angular momenta of the initial atomic state, the final ionic core, and the photoelectron, respectively, while $\vec{j}_\gamma$ is the photon angular momentum (having a magnitude of unity). Thus, for process (22), in which $L_c = 0$ and $L_0 = 2$, equation (23) shows that $j_t = 2$ is the only possibility.

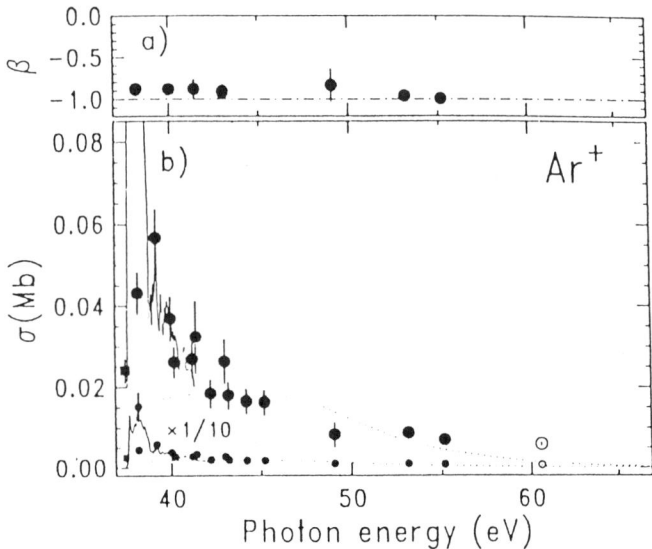

FIGURE 11. (a) Angular distribution parameter β and (b) partial cross section σ of the $3p^4(^1D)3d(^2P)$ parity-unfavored satellite transition in argon (E_B = 37.4 eV). The dotted curve represents the $3p$ partial cross section scaled to the satellite intensity at equal kinetic energy. (From Ref. 77.)

Furthermore, $\pi_0 = -1$ and $\pi_c = 1$ so that this is a parity-unfavored transition. Thus, theory predicts that β = −1. However, experiment has not yet confirmed this result.

Previously, purely parity-unfavored transitions have been investigated experimentally only for autoionizing resonances.[76] Recently, however, it has been pointed out that a whole class of states exist which exhibit this phenomenon,[77,78] namely, a subset of satellite transitions where one electron is ionized and another is excited. As an example, consider photoionization of the ground state of argon in which one $3p$ electron is ionized and another is promoted to a $3d$ state,

$$3p^6(^1S) \rightarrow 3p^4(^1D)3d(^2P)\varepsilon p(^1P) \tag{24}$$

In this case $L_0 = 0$ and $L_c = 1$ so that, by equation (23), the only possible value for j_t is $j_t = 1$. Then, since $\pi_0 = \pi_c = 1$, according to (21) this is a purely parity-unfavored transition. The experimental cross section and β parameter are shown in Figure 11, in which it is seen that, although the cross section is highly dependent on the photon energy, β is constant at −1. From this discussion, it is clear that there exist a whole class of such satellite transitions for all of the noble gases, as well as for the rest of the periodic system.

3.7. Double Photoionization of Helium

The double photoionization process is particularly interesting since the final state, an ion plus two outgoing electrons, is an example of the three-body continuum Coulomb problem. This is not merely one of the oldest unsolved problems in atomic physics, it is one of the oldest unsolved problems in physics! From a theoretical point of view, the difficulty is that, although the boundary conditions are now known,[79] realistic solutions to the Schrödinger equation which conform to these boundary conditions have not been obtained. In addition, in single photoionization, the energy of the photoelectron generated by a photon of a given energy is fixed. In double photoionization, on the other hand, since two electrons are ejected, only the sum of the two energies is fixed. Thus, the total double photoionization cross section is a sum over all of the possible energy sharings between the two electrons. This, of course, makes experimental photoelectron spectroscopy of the double photoionization process considerably more difficult than for the single photoionization process.

Despite the inherent extra complexity of the double photoionization process, it can be shown that at high photon energies the behavior of the double photoionization cross section for He is proportional to $E^{-7/2}$, where E is the total energy of the two photoelectrons.[80] The same asymptotic behavior is predicted for single photoionization. These asymptotic behaviors depend only on the character of the initial state and the fact that the photoionization process is an electric dipole transition.[80] Thus, since both cross sections approach the same asymptotic form at high energy, the ratio of double to single photoionization must also approach a constant limit. But knowing that the ratio approaches a constant limit does not tell us at what energy the asymptotic ratio is reached, or the value of the asymptotic ratio.

Experimental measurements of the ratio of double to single ionization were limited to fairly low energies until recently when it was measured in the range of 2–12 keV.[81,82] The ratio was found to be $1.5 \pm 0.2\%$, and essentially constant over this entire range. This indicates that a constant ratio has been reached. Also, this ratio is in good agreement with recent calculations as well as with a number of earlier ones.[80] The results disagree, however, with a recent prediction, based on an analogy with electron impact ionization, that the ratio should be decreasing in this region, despite remarkable success of that model at low energy.[83a] The difference may indicate the greater importance of shake-off effects as compared to electron scattering effects at high photon energies.[83b]

These high-energy experiments are only possible with high-brightness sources since the cross sections become so small, particularly for double ionization. Furthermore, measurement of the angular and energy correlations between the two photoelectrons produced in double photoionization requires a coincidence measurement, which in turn necessitates a still higher photon intensity. But such detailed measurements are precisely what is needed to shed further light on this fundamental three-body Coulomb problem.

In fact, both experimental and theoretical investigations are continuing at a rapid pace. Among the recent works is a measurement of the double to single ionization ratio with a fairly fine energy grid over a broad energy range[84] which shows good agreement in the intermediate energy region with the recent theoretical work of Pan and Kelly;[85] this indicates that some higher order perturbations, which are effectively included in Pan and Kelly's work,[85] but not included in recent calculations of Hino et al.,[86] are important. It has also been pointed out that, with increasing energy, the atomic Compton effect becomes an increasingly important contributor to both the single and the double ionization processes, and that it dominates above the 10 keV region.[87] A recent calculation which includes the Compton contribution[88] is in excellent agreement with experiment at the higher energies, but there is some debate as to the high energy Compton ratio. Further, the angular distribution of photoelectrons resulting from double photoionization has been measured[89,90] and good agreement is found with recent calculations.[91] Finally, at the low energies, a very recent experiment[92] using cold ion recoil momentum spectroscopy (COLTRIMS) has found a ratio substantially lower (~25%) than previous experiments. This new result also makes the photon ratio consistent with the charged particle impact He^{++}/He^{+} ratio.[92]

3.8. Multichannel Aspects of Photoionization

Photoionization is often thought of as a single electron process owing to the fact that the transition operator responsible for low-energy photoionizing transitions, the electric dipole operator, is a single particle operator. However, although it is quite true that, in many cases, the single electron aspects of photoionization are most important, it is equally true that in many cases multielectron aspects dominate.[7] These multielectron aspects arise from the electron–electron interaction and are an example of the general problem of electron correlations.

From a theoretical point of view, electron correlation can be included explicitly, by including the r_{ij} dependence in the wave function, or implicitly, via expansion in a complete basis set. When the initial or final ionic discrete states are expanded, multiconfiguration or configuration interaction (CI) wave functions are formed. Multiconfiguration effects can cause a serious breakdown in the single electron picture, as double photoionization (see Section 3.7) amply demonstrates.[80] The total final continuum wave function can also be expanded, and this is referred to as a multichannel expansion.

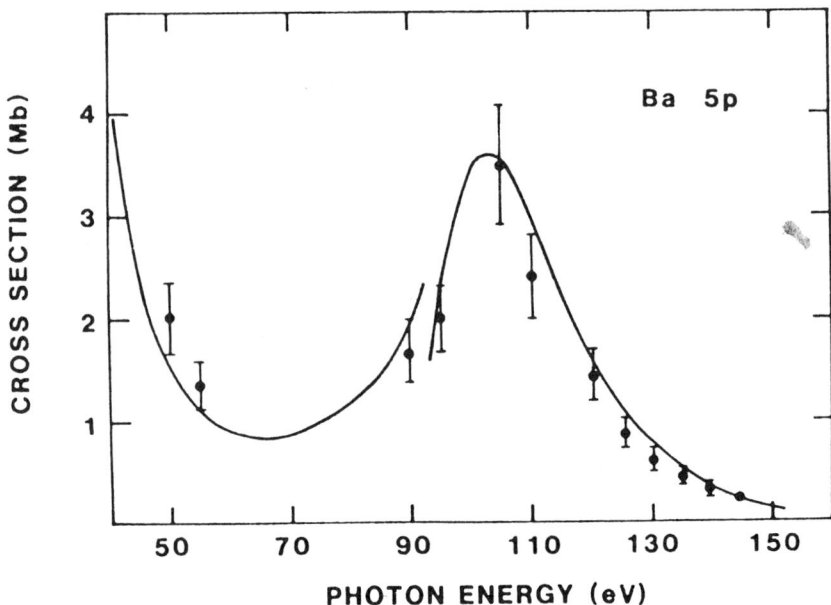

FIGURE 12. Experimental and theoretical data for the Ba 5p photoionization cross section. (From Ref. 93.)

The Coulomb matrix elements between these channel wave functions are known generally as interchannel coupling matrix elements.

The effects of interchannel interactions can be exhibited using simple perturbation theory. Let $|f_0\rangle$ and $|f_1\rangle$ denote the final-state wave functions of two different channels of a photoionization process resulting from the initial state $|i\rangle$. If the Hamiltonian of the system is H, and the photoionization transition operator is T, then the matrix element for the $|i\rangle \rightarrow |f_0\rangle$ photoionization process, including the effects of interchannel interaction, is given essentially by

$$\langle i|T|f_0\rangle + \sum_{f_1} \frac{\langle i|T|f_1\rangle \langle f_1|H|f_0\rangle}{\Delta E} \qquad (25)$$

The first term in (25) is the direct electric dipole transition to the final state $|f_0\rangle$. The second term indicates schematically the indirect transition amplitude resulting from an electric dipole transition to the intermediate states $|f_1\rangle$ followed by an interchannel interaction to the final state $|f_0\rangle$. ΔE denotes the energy difference between the final state and the intermediate state and Σ_f denotes a sum over degenerate quantum numbers in the intermediate state $|f_1\rangle$ as well as an integration over intermediate state energy. The question then looms as to when the second term in the above equation has a significant influence relative to the first term. The mixing coefficient $\langle f_0|H|f_1\rangle/\Delta E$ will be most significant in general when ΔE is small; thus, the second term is most important when channels f_0 and f_1 are degenerate, i.e., they both are accessible at the given photon energy. Then if the basic single-particle dipole matrix element for channel 1, $\langle i|T|f_1\rangle$, is significantly larger than $\langle i|T|f_0\rangle$, the second term in the above equation can be quite important.

As an example, consider the case of Ba 5p photoionization,[93] shown in Figure 12. The cross section is dropping from threshold (which is actually below the lowest energy shown in this curve), but then there is a huge hump extending over some 60 eV and centered about $h\nu \cong 100$ eV. This hump is seen to be the major feature of the cross section over a broad energy region, and is the result of interchannel coupling of the 5p channels with the 4d channels.[93] If it were not for this coupling, the 5p photoionization cross section would be a few tenths of a megabarn, as can be inferred from Figure

12. The $4d$-subshell threshold is around 100 eV and its cross section has a maximum near threshold of about 20 Mb,[94] which indicates that $\langle i|T|f_1\rangle$ is about an order of magnitude larger than $\langle i|T|f_0\rangle$ at this energy since the $4d$ cross section is about two orders of magnitude larger. Thus, the Ba $5p$-subshell cross section, in this energy range, is dominated by its interchannel coupling with the $4d$ channels. From a theoretical viewpoint, this occurs because the true wave function of the final continuum state is a mixture of the $5p \to \varepsilon l$ channels and the $4d \to \varepsilon'l'$ channels, i.e.,

$$\langle i|T|5p^{-1}\varepsilon l\rangle + \int d\varepsilon' \sum_{l'} \langle i|T|4d^{-1}\varepsilon'l'\rangle \frac{\langle 4d^{-1}\varepsilon'l'|T|5p^{-1}\varepsilon l\rangle}{\Delta E} \qquad (26)$$

where $4d^{-1}$ and $5d^{-1}$ represent vacancies in the respective subshells. The second term in (26) dominates in the region where the $4d$ cross section is large.

This result is of importance for a number of reasons. First, such phenomena are by no means limited to this particular case, but occur throughout the periodic table, for both ground and excited states,[94] for almost all subshell cross sections in the vicinity of a new threshold. Second, measurement of the cross section, and the effect of interchannel coupling as exemplified by Figure 12, gives an experimental characterization of the strength of multichannel/interchannel effects. While they affect photoionization cross sections in other situations as well, in the type of experiment reported in Figure 12 multichannel interactions show a clear fingerprint. Finally, the analysis discussed in this section give us a clear *a priori* guide as to where multichannel effects are likely to be important: in situations where a weak channel is degenerate with a strong one.

4. CONCLUDING REMARKS

In this introductory paper on atomic photoionization with high-brightness light sources we have presented a large number of examples of phenomena whose study would be impossible without such sources. Not surprising, perhaps, is the fact that nearly all of these examples concern processes involving either s subshells or satellite states (or both) as such processes have small cross sections. Nevertheless, much is gained by the study of these weak processes precisely because they are weak and therefore are susceptible to influences that have little effect on stronger transition processes.

Among the examples we have presented are many that exhibit our stated goal of attempting to extract simplicity from the complexity of detail often found in present-day measurements. Thus, measurements of doubly excited state spectra of H⁻ and He have led to a significant amount of theoretical activity on three-body Coulomb states that has already resulted in new propensity rules for photoionization to these states (Section 3.1). Key features of satellite states have been shown to be amenable to interpretation based on single-configuration theoretical models (Sections 3.3.1, 3.5, and 3.6) and to measurement by means of fluorescence spectroscopy (Section 3.4). Also, satellite states have been demonstrated to have major influences on weak inner-shell photoionization channels near threshold (Section 3.2), as do strong photoionization channels at energies near their threshold (Section 3.8). Both fine-structure branching ratios (Section 3.3.2) and photoelectron angular distribution asymmetry parameters for satellite states (Section 3.6) have been shown to be interpretable using primarily angular momentum analyses. Finally, double photoionization processes are becoming much better understood, particularly in the asymptotic region (Section 3.7).

All of these examples show that photoionization studies of atoms with high-brightness light sources represent an ideal laboratory for uncovering fundamental new physics. This new physics includes further advances in descriptions of three-body Coulomb systems. It also includes the development of propensity rules and selection rules for processes which illustrate the effects of relativistic interactions and of electron correlations on complex many-body systems.

ACKNOWLEDGMENTS. A.F.S. gratefully acknowledges numerous discussions with James A. R. Samson, particularly concerning fluorescence spectroscopy. The work of A.F.S. reported herein was supported in part by the National Science Foundation under Grant Nos. PHY-8026055, PHY-8908605, and

PHY-9108002. The work of S.T.M. reported herein was supported in part by the National Science Foundation under Grant No. PHY-9107539 and in part by the U.S. Army Research Office under Grant No. DAAL03-89-K-0098.

REFERENCES

1. D. R. BATES, *Mon. Not Ry. Astron. Soc.* **106**, 432 (1946).
2. H. A. BETHE AND E. E. SALPETER, *Quantum Mechanics of One- and Two-Electron Atoms* (Springer, Berlin, 1957), pp. 247–323.
3. U. FANO AND J. COOPER, *Rev. Mod. Phys.* **40**, 441 (1968).
4. S. T. MANSON, *Adv. Electron. Electron Phys.* **41**, 73 (1976).
5. S. T. MANSON AND D. DILL, in *Electron Spectroscopy*, Vol. 2, edited by C. R. Brundle and A. D. Baker (Academic Press, New York, 1978), pp. 157–195.
6. A. F. STARACE, in *Handbuch der Physik*, Vol. 31, edited by W. Mehlhorn (Springer, Berlin, 1982), pp. 1–121.
7. M. Y. AMUSIA, *Atomic Photoeffect* (Plenum Press, New York, 1990).
8. C. N. YANG, *Phys. Rev.* **74**, 764 (1948).
9. H. A. BETHE, in *Handbuch der Physik*, Vol. 24/1, edited by H. Geiger and K. Scheel (Springer, Berlin, 1933), p. 482.
10. J. COOPER AND R. N. ZARE, *J. Chem. Phys.* **48**, 942 (1968).
11. S. T. MANSON AND A. F. STARACE, *Rev. Mod. Phys.* **54**, 389 (1982), and references therein.
12. D. DILL, *Phys. Rev. A* **7**, 1976 (1973), and references therein.
13. U. FANO AND D. DILL, *Phys. Rev. A* **6**, 185 (1972).
14. P. G. HARRIS, H. C. BRYANT, A. H. MOHAGHEGHI, R. A. REEDER, H. SHARIFIAN, C. Y. TANG, H. TOOTOONCHI, J. B. DONAHUE, C. R. QUICK, D. C. RISLOVE, W. W. SMITH, AND J. E. STEWART, *Phys. Rev. Lett.* **65**, 309 (1990).
15. M. DOMKE, C. XUE, A. PUSCHMANN, T. MANDEL, E. HUDSON, D. A. SHIRLEY, G. KAINDL, C. H. GREENE, H. R. SADEGHPOUR, AND H. PETERSEN, *Phys. Rev. Lett.* **66**, 1306 (1991).
16. R. P. MADDEN AND K. CODLING, *Astrophys. J.* **141**, 364 (1965).
17. J. W. COOPER, U. FANO, AND F. PRATS, *Phys. Rev. Lett.* **10**, 518 (1963).
18. J. MACEK, *J. Phys. B* **1**, 831 (1968).
19. H. R. SADEGHPOUR AND C. H. GREENE, *Phys. Rev. Lett.* **65**, 313 (1990).
20. X.-H. LIU, Z. CHEN, AND C. D. LIN, *Phys. Rev. A* **44**, 5468 (1991).
21. J. M. ROST, J. S. BRIGGS, AND J. M. FEAGIN, *Phys Rev. Lett.* **66**, 1642 (1991).
22. J. M. FEAGIN AND J. S. BRIGGS, *Phys. Rev. A* **37**, 4599 (1988); J. M. FEAGIN, in *Fundamental Processes of Atomic Dynamics*, edited by J. S. Briggs, H. Kleinpoppen, and H. O. Lutz (Plenum Press, New York, 1988), pp. 275–300.
23. J. M. ROST AND J. S. BRIGGS, *J. Phys. B* **23**, L339 (1990).
24. H. R. SADEGHPOUR, C. H. GREENE, AND M. CAVAGNERO, *Phys. Rev. A* **45**, 1587 (1992).
25. H. R SADEGHPOUR, *Phys. Rev. A* **43**, 5821 (1991).
26. J. L. DEHMER AND D. DILL, *Phys. Rev. Lett.* **37**, 1049 (1976).
27. W. ONG AND S. T. MANSON, *J. Phys. B* **11**, L65 (1978); *Phys. Rev. A* **19**, 688 (1979).
28. W. R. JOHNSON AND K. T. CHENG, *Phys. Rev. Lett.* **40**, 1167 (1978); *Phys. Rev. A* **20**, 978 (1979).
29. M. G. WHITE, S. H. SOUTHWORTH, P. KOBRIN, E. D. POLIAKOFF, R. A. ROSENBERG, AND D. A. SHIRLEY, *Phys. Rev. Lett.* **43**, 1661 (1979).
30. K.-N. HUANG AND A. F. STARACE, *Phys. Rev. A* **21**, 697 (1980).
31. H. DERENBACH AND V. SCHMIDT, *J. Phys. B* **16**, L337 (1983).
32. A. FAHLMAN, T. A. CARLSON, AND M. O. KRAUSE, *Phys. Rev. Lett.* **50**, 1114 (1983).
33. G. WENDIN AND A. F. STARACE, *Phys. Rev. A* **28**, 3143 (1983).
34. See, e.g., G. WENDIN, *Struct. Bonding (Berlin)* **45**, 1 (1981), and references therein.
35. G. WENDIN, *Phys. Scr.* **16**, 296 (1977); see also G. WENDIN AND M. OHNO, *Phys. Scr.* **14**, 148 (1976) for the analogous case of a 4p hole state.
36. J. E. HANSEN, *J. Opt. Soc. Am.* **67**, 754 (1977); J. E. HANSEN AND W. PERSSON, *Phys. Rev. A* **18**, 1459 (1978).
37. M. Y. ADAM, F. WUILLEUMIER, N. SANDNER, V. SCHMIDT, AND G. WENDIN, *J. Phys. (Paris)* **39**, 129 (1978).
38. J. TULKKI, *Phys. Rev. Lett.* **62**, 2817 (1989).
39. T. GUSTAFSSON, *Chem. Phys. Lett.* **51**, 383 (1977).
40. M. Y. ADAM, Thesis, L'Université de Paris-Sud, Centre d'Orsay, France (1978).
41. J. A. R. SAMSON AND J. L. GARDNER, *Phys. Rev. Lett.* **33**, 671 (1974).

42. C. D. CALDWELL, M. G. FLEMMING, M. O. KRAUSE, P. VAN DER MEULEN, C. PAN, AND A. F. STARACE, *Phys. Rev. A* **41**, 542 (1990).
43. J. M. BLATT AND V. F. WEISSKOPF, *Theoretical Nuclear Physics* (Wiley, New York, 1952), Chap. VIII, Sect. 3.
44. G. HOWAT, T. ÅBERG, AND O. GOSCINSKI, *J. Phys. B* **11**, 1575 (1978).
45. D. PETRINI (unpublished); see also *J. Phys. B* **14**, 3839 (1981) for analogous calculations for B^+.
46. See, e.g., J.-Z. WU, S. B. WHITFIELD, C. D. CALDWELL, M. O. KRAUSE, P. VAN DER MEULEN, AND A. FAHLMAN, *Phys. Rev. A* **42**, 1350 (1990); C. D. CALDWELL AND M. O. KRAUSE, *J. Phys. B* **23**, 2233 (1990); M. G. FLEMMING, J.-Z. WU, C. D. CALDWELL, AND M. O. KRAUSE, *Phys. Rev. A* **44**, 1733 (1991); P. VAN DER MEULEN, M. O. KRAUSE, AND C. A. DE LANGE, *J. Phys. B* **25**, 97 (1992).
47. W. C. LINEBERGER AND B. W. WOODWARD, *Phys. Rev. Lett.* **25**, 424 (1970).
48. H. HOTOP, T. A. PATTERSON, AND W. C. LINEBERGER, *Phys. Rev. A* **8**, 762 (1973).
49. A. R. P. RAU AND U. FANO, *Phys. Rev. A* **4**, 1751 (1971).
50. A. R. P. RAU, in *Electron and Photon Interactions with Atoms*, edited by H. Kleinpoppen and M. R. C. McDowell (Plenum Press, New York, 1976), pp. 141–148.
51. P. A. COX, *Struct. Bonding* **24**, 59 (1975).
52. P. C. ENGELKING AND W. C. LINEBERGER, *Phys. Rev. A* **19**, 149 (1979).
53. J. BERKOWITZ AND G. L. GOODMAN, *J. Chem. Phys.* **71**, 1754 (1979).
54. J. SCHIRMER, L. S. CEDERBAUM, AND J. KIESSLING, *Phys. Rev. A* **22**, 2696 (1980).
55. G. L. GOODMAN AND J. BERKOWITZ, *J. Chem. Phys.* **94**, 321 (1991).
56. C. PAN AND A. F. STARACE, *Phys. Rev. A* **47**, 295 (1993).
57. M. O. KRAUSE, C. D. CALDWELL, S. B. WHIFFLELD, C. A. DE LANGE, AND P. VAN DER MEULEN (unpublished).
58. F. ROBICHEAUX AND C. H. GREENE, *Phys. Rev. A* **46**, 3821 (1992).
59. F. ROBICHEAUX AND C. H. GREENE (unpublished); F. ROBICHEAUX (private communication).
60. See, e.g., the recent review by K. H. SCHARTNER, in *The Physics of Electronic and Atomic Collisions: XVI International Conference, New York, NY 1989*, edited by A. Dalgarno, R. S. Freund, P. M. Koch, M. S. Lubell, and T. B. Lucatorto (AIP Conference Proceedings No. 205, New York, 1990), pp. 215–223.
61. P. R. WOODRUFF AND J. A. R. SAMSON, *Phys. Rev. Lett.* **45**, 110 (1980); *Phys. Rev. A* **25**, 848 (1982).
62. K.-H. SCHARTNER, B. MÖBUS, P. LENZ, H. SCHMORANZER, AND M. WILDBERGER, *Phys. Rev. Lett.* **61**, 2744 (1988).
63. J. A. R. SAMSON, Y. CHUNG, AND E. M. LEE, *Phys. Rev. A* **45**, 259 (1992); J. A. R. SAMSON (private communication).
64. A. A. WILLS, A. A. CAFOLLA, A. SVENSSON, AND J. COMER, *J. Phys. B* **23**, 2013 (1990).
65. D. CUBAYNES, J. M. BIZAU, F. J. WUILLEUMIER, B. CARRÉ, AND F. GOUNAND, *Phys. Rev. Lett.* **63**, 2460 (1989).
66. B. I. CRAIG AND F. P. LARKINS, *J. Phys. B* **18**, 3569 (1985); **18**, 3713 (1985).
67. Z. FELFLI AND S. T. MANSON, *Phys. Rev. Lett.* **68**, 1687 (1992).
68. M. RICHTER, J. M. BIZAU, D. CUBAYNES, T. MENZEL, F. J. WUILLEUMIER, AND B. CARRÉ, *Europhys. Lett.* **12**, 35 (1990).
69. D. CUBAYNES, J. M. BIZAU, M. RICHTER, AND F. J. WUILLEUMIER, in *VUV10. Scientific Program and Abstracts* (Maison de la Chemie, Paris, 1992), p. Mo71.
70. D. CUBAYNES, J. M. BIZAU, T. J. MORGAN, B. CARRÉ, AND F. J. WUILLEUMIER, in *VUV10. Scientific Program and Abstracts* (Maison de la Chemie, Paris, 1992), p. Tu79.
71. H. STAPELFELDT, D. G. PAPAIOANNOU, L. D. NOORDAM, AND T. F. GALLAGHER, *Phys. Rev. Lett.* **67**, 3223 (1991).
72. M. MEYER, E. V. RAVEN, B. SONNTAG, AND J. E. HANSEN, *Phys. Rev. A* **43**, 177 (1991).
73. F. J. WUILLEUMIER, D. L. EDERER, AND J. L. PICQUÉ, *Adv. At. Mol. Phys.* **23**, 197 (1988).
74. F. J. WUILLEUMIER, D. CUBAYNES, AND J. M. BIZAU, in *Atomic and Molecular Physics*, edited by C. Cisneros, T. J. Morgan, and I. Alvarez (World Scientific, Singapore, 1991), pp. 474–495.
75. E. S. CHANG AND K. T. TAYLOR, *J. Phys. B* **11**, L507 (1978).
76. T. A. CARLSON, D. R. MULLINS, C. E. BEALL, B. W. YATES, J. W. TAYLOR, D. W. LINDLE, B. P. PULLEN, AND F. A. GRIMM, *Phys. Rev. Lett.* **60**, 1382 (1988).
77. B. LANGER, J. VIEFHAUS, O. HEMMERS, A. MENZEL, R. WEHLITZ, AND U. BECKER, in *ICAP XIII. Book of Abstracts* (XIII ICAP, Munich, 1992), p. A41.
78. B. LANGER, *Anregungsenergieabhängigkeit von Photoelektronensatelliten Unter Spezieller Berücksichtigung des Schwellbereiches*, Ph.D. thesis, Tech. Univ. Berlin (1992).
79. M. BRAUNER, J. S. BRIGGS, AND H. KLAR, *J. Phys. B* **22**, 2265 (1989), and references therein.
80. A. DALGARNO AND H. R. SADEGHPOUR, *Phys. Rev. A* **46**, R3591 (1992), and references therein.

81. J. C. Levin, D. W. Lindle, N. Keller, R. D. Miller, Y. Azuma, N. Berrah Mansour, H. G. Berry, and I. A. Sellin, *Phys. Rev. Lett.* **67**, 968 (1991); J. C. Levin, I. A. Sellin, B. M. Johnson, D. W. Lindle, R. D. Miller, N. Berrah, Y. Azuma, H. G. Berry, and D.-H. Lee, *Phys. Rev. A* **47**, R16 (1993).
82. D. Cubaynes, J. M. Bizau, M. Richter, and F. J. Wuilleumier, in *VUV10. Scientific Program and Abstracts* (Maison de la Chemie, Paris, 1992), p. Mo80.
83. (a) J. A. R. Samson, *Phys. Rev. Lett.* **65**, 2861 (1990); R. J. Bartlett, P. J. Walsh, Z. X. He, Y. Chung, E.-M. Lee, and J. A. R. Samson, *Phys. Rev. A* **46**, 5574 (1992); J. A. R. Samson, R. J. Bartlett, and Z. X. He, *Phys. Rev.* **46**, 7277 (1992). (b) J. A. R. Samson (private communication).
84. N. Berrah, F. Heiser, R. Wehlitz, J. Levin, S. B. Whitfield, J. Viefhaus, I. A. Sellin, and U. Becker, *J. Phys. IV Colloq.* **3**(C6), 197 (1993).
85. C. Pan and H. P. Kelly, *J. Phys. B* **28**, 5001 (1995).
86. K. Hino, *Phys. Rev. A* **47**, 4845 (1993); K. Hino, T. Ishihara, F. Shimizu, N. Toshima, and J. H. McGuire, *Phys. Rev. A* **48**, 1271 (1993).
87. J. A. R. Samson, C. H. Greene, and R. J. Bartlett, *Phys. Rev. Lett.* **71**, 201 (1993).
88. L. R. Andersson and J. Burgdörfer, *Phys. Rev. Lett.* **71**, 50 (1993).
89. R. I. Hall, A. G. McConkey, L. Avaldi, K. Ellis, M. A. MacDonald, G. Dawber, and G. C. King, *J. Phys. B* **25**, 1195 (1992).
90. R. Wehlitz, O. Hemmers, B. Langer, A. Menzel, and U. Becker, in *XVIII ICPEAC, Abstracts of Papers*, Vol. 1, edited by T. Andersen, B. Fastrup, F. Folkmann, and H. Knudsen (ICPEAC, Aarhus, Denmark, 1993), p. 7.
91. F. Maulbetsch and J. S. Briggs, *J. Phys. B* **26**, 1679 (1993); **26**, L647 (1993).
92. R. Dörner, C. L. Cocke, J. Feagin, T. Vogt, V. Mergel, H. Khemliche, S. Kravis, J. Ullrich, M. Unversagt, L. Spielberger, M. Damrau, O. Jagutzki, I. Ali, K. Ullman, M. Jung, E. P. Kanter, B. Sonntag, M. H. Prior, E. Rotenberg, J. Denliger, T. Warwick, S. T. Manson, and H. Schmidt-Böcking, *Phys. Rev. Lett.* (submitted).
93. J. M. Bizau, D. Cubaynes, P. Gérard, and F. J. Wuilleumier, *Phys. Rev. A* **40**, 3002 (1989).
94. J. M. Bizau, D. Cubaynes, P. Gérard, F. J. Wuilleumier, J. L. Picqué, D. L. Ederer, B. Carré, and G. Wendin, *Phys. Rev. Lett.* **57**, 306 (1986).

CHAPTER 4

CORE RELAXATION EFFECTS IN MOLECULAR PHOTOIONIZATION

JOCHEN SCHIRMER, MATTHEW BRAUNSTEIN,
MU-TAO LEE, AND VINCENT MCKOY

1. INTRODUCTION

Ionization of K-shell or, more generally, of deep inner-shell electrons in atoms and molecules is accompanied by a considerable rearrangement of the valence (outer-shell) electrons in response to the reduced shielding of the nuclear attraction.[1] This adjustment of the valence electrons, referred to as electronic relaxation, leads to a significant energy lowering of the final ionic state relative to a state where the valence electron distribution of the initial state is maintained ("frozen"). The magnitude of this relaxation energy scales with the number of valence electrons.[2,3] In the case of the K-shell ionization of second-row atoms ($Z = 3$–10), for example, the relaxation energies (in eV) are approximately given by $E^R(Z) = 3.1(Z - 2.2)$. In a molecular environment the corresponding relaxation energies are typically 2–3 eV larger than the values for the free atom. Relaxation not only plays a role in the ionic core but also affects the motion of the outgoing photoelectron. The relaxation of the valence electrons, being essentially a contraction of the valence charge distribution, quite effectively screens the inner-shell hole potential experienced by the photoelectron. This means that the potential of the relaxed ionic core is less attractive than its unrelaxed (frozen) counterpart. As a consequence, resonances in the photoionization cross section will appear at higher energy for a relaxed core than for an unrelaxed (frozen) core. Concomitantly with the shift to higher energy, the resonance peaks will be lowered and broadened as a result of relaxation.

Although relaxation effects are expected to play an essential role in molecular inner-shell photoionization, most of the previous theoretical studies were based on an unrelaxed representation of the ionic core potential.[4–11] A first theoretical description of molecular K-shell photoionization cross sections was given by Dehmer and Dill[4] in the case of N_2 and CO using the multiple scattering (MS) method. This work focused primarily on establishing and clarifying the occurrence of shape resonances observed in the experimental K-shell excitation spectra of these and other molecules.[12] The Stieltjes–Tchebycheff moment theory (STMT) was used with some success in K-shell photoionization studies of several molecules by Rescigno and Langhoff[5] and by Padial et al.[6,7] In this method, the photoionization cross section is extracted from a discrete pseudospectrum. However, as noted by Daasch et al.[8] in their study of the $O1s$ ionization in CO_2, the cross section obtained by the STMT procedure can be sensitive to the order of the moment theory used in the analysis. Also a direct

---

JOCHEN SCHIRMER • Physikalisch-Chemisches Institut, Universität Heidelberg, 69120 Heidelberg, Germany. MATTHEW BRAUNSTEIN • Institute for Defense Analyses, Alexandria, Virginia 22311-1772. MU-TAO LEE • Departmento de Quimica, Universidade Federal de São Paulo, São Carlos 13560, Brazil. VINCENT MCKOY • Arthur Amos Noyes Laboratory of Chemical Physics, California Institute of Technology, Pasadena, California 91125.

VUV and Soft X-Ray Photoionization. Edited by Uwe Becker and David A. Shirley. Plenum Press, New York, 1996.

method[13] based on the Schwinger variational principle and single-center expansion techniques was used to obtain K-shell photoelectron continua and cross sections in molecules.[9,10]

A common feature of the approaches used in these studies is that the photoelectron continuum is determined by a single-particle (Hartree–Fock) equation with a suitably chosen potential for the electron–ion interaction. As a particularly simple choice, the frozen-core Hartree–Fock (FCHF) approximation was adopted in several of the above calculations. In this model, which is also widely used in valence photoionization, the valence charge distribution of the Hartree–Fock (HF) molecular ground state is assumed for the final ionic state. A similar one-particle potential corresponding to the frozen valence charge distribution was used in the MS calculations. However, because of the local treatment of the exchange interaction and the "muffin-tin" form of the potential, some deviations from the FCHF potential are to be expected.

An obvious way to take relaxation effects into account is to express the ionic state and the associated potential in terms of relaxed orbitals generated by a separate HF calculation for the ionic state. In such a relaxed-core Hartree–Fock (RCHF) approach, however, specific difficulties arise in the calculation of the cross sections because of the use of mutually nonorthogonal frozen and relaxed orbital sets in the initial and final states, respectively. One serious complication, for example, is the lack of strict orthogonality of the initial and final N-electron states which may lead to a dependence of the cross section on the origin of the molecular coordinate frame. These complexities may explain the previous preference for the simpler FCHF model. It should be noted that the relaxed-core approximation has also been criticized on more fundamental grounds.[15,17,18] For example, one may well ask whether the assumption of a static fully relaxed ion-core potential could provide an adequate description of electronic relaxation accompanying the ejection of an inner-shell electron, an obviously dynamic process.

For atoms where more sophisticated theoretical methods are available,[14] the relaxation of the ionic core has been considered in a number of studies of inner-shell photoionization.[15–22] Unequivocally these investigations show a substantial effect of relaxation on the cross sections. The inevitable nonorthogonality problem arising from the mixed frozen and relaxed orbital representations of the initial and final state, respectively, seems to be a smaller obstacle in the atomic case. The orthogonality, for example, of the initial and final state is usually assured simply on the basis of symmetry. Unfortunately, the details of the numerical procedures used in many of these calculations which would be needed to reproduce such calculations and to disentangle relaxation and other higher-order effects in the cross sections are often not given.

In this chapter we report on molecular K-shell ionization studies in the framework of the RCHF model. Studies of this kind were first performed by Lynch and McKoy[23] for the $N1s$ ionization in N_2 and were later continued by the present authors in Ref. 24. In the latter paper, henceforth referred to as I, a new efficient method was developed for evaluating the N-electron transition moment and eliminating the unphysical nonorthogonality contributions to the cross sections. This new formulation, which will be briefly reviewed here, greatly facilitates the applicability of the RCHF approach and requires a numerical effort of the same order of that of the familiar FCHF model. Three examples including those of our previous studies of CO, the nonlinear H_2CO molecule, and C_2H_4 as a species with equivalent core-levels, will be presented here allowing us to give a representative picture of the relaxation effects expected in molecular inner-shell ionization. Also, as in I we shall analyze the RCHF results from a more basic point of view using perturbation theory. This will allow us to obtain a better understanding of the achievements as well as of the shortcomings of the RCHF model.

2. PHOTOIONIZATION CROSS SECTIONS IN THE RELAXED-CORE HARTREE–FOCK APPROXIMATION

2.1. Single-Particle Representation of the N-Electron Transition Moment

In photoionization a photon of energy ω is absorbed by the molecule initially in its ground state $|\Psi_0\rangle$, and an ion in the state $|\Psi_n^{N-1}\rangle$ and a photoelectron with kinetic energy $k^2/2$ are generated. The corresponding partial cross section is given by the familiar golden rule expression (in atomic units):

$$\sigma_n(\omega) = \frac{4\pi^2}{3c} \omega \sum_\mu |\langle \Psi^{(-)}_{k,n} | \hat{D}^{(\mu)} | \Psi_0 \rangle|^2 \tag{1}$$

Here, $|\Psi^{(-)}_{k,n}\rangle$ represents the final N-electron state describing asymptotically the ion in the state $|\Psi^{N-1}_n\rangle$ and a photoelectron of kinetic energy $k^2/2$ in the one-electron scattering state $|\psi^{(-)}_k\rangle$ (with incoming-wave boundary conditions). $\hat{D}^{(\mu)}$, $\mu = x,y,z$ denotes the Cartesian (or spherical) components of the N-electron dipole operator

$$\hat{D} = \sum_{i=1}^{N} \mathbf{r}_i \tag{2}$$

In the case of degeneracy in the final or initial state, equation (1) has to be generalized accordingly. Energy conservation requires

$$k^2/2 = \omega - E^{N-1}_n + E^N_0 \tag{3}$$

where E^N_0 and E^{N-1}_n are the energies of the initial ground state and the ionic state, respectively. The form of equation (1) assumes the final states to be energy normalized according to

$$\langle \Psi^{(-)}_{k,n} | \Psi^{(-)}_{k',n} \rangle = \delta(k^2/2 - k'^2/2) \tag{4}$$

In this chapter we shall not consider vibrational and rotational degrees of freedom; a fixed nuclear geometry will be assumed for all electronic states.

In the following we will evaluate the N-electron transition moment

$$A^{(\mu)}_{k,n} = \langle \Psi^{(-)}_{k,n} | \hat{D}^{(\mu)} | \Psi_0 \rangle \tag{5}$$

introduced in (1) assuming the following distinct single-particle or Hartree–Fock (HF) representations for the initial and final state:

1. The ground-state HF representation

$$|\phi_0\rangle = |\varphi_1 \varphi_2 \ldots \varphi_N| \tag{6}$$

    approximates the initial state $|\Psi_0\rangle$. The orbitals $|\varphi_i\rangle$ are referred to as "frozen" orbitals.
2. The final excited state $|\Psi^{(-)}_{k,n}\rangle$ is represented in terms of "relaxed" orbitals $|\psi_r\rangle$ generated by a suitably chosen Hartree–Fock calculation for the ionic core and a consistent scattering orbital $|\psi^{(-)}_k\rangle$. Using creation (annihilation) operators $c^\dagger_r$ (c_r) of second quantization associated with the relaxed orbitals, a singly excited final state can be written as

$$|\Psi_{k,h}\rangle = c^\dagger_k c_h |\bar{\phi}_0\rangle, \quad h \leq N \tag{7a}$$

where

$$|\bar{\phi}_0\rangle = |\psi_1 \psi_2 \ldots \psi_N| \tag{7b}$$

denotes a "relaxed" N-particle state defined analogously to (6) as the Slater determinant of the N lowest relaxed orbitals $|\psi_r\rangle$. The ionic core corresponding to the singly excited state of (7a) is, of course, the relaxed single-hole state

$$|\bar{\phi}_h\rangle = c_h |\bar{\phi}_0\rangle \tag{8}$$

We note that final states of proper spin and spatial symmetry can be generated as linear combinations of the "primitive" product states of (7a).

Now the transition moment can readily be evaluated further using the biorthogonalization method described in I. The result is

$$A_{k,h} = \sum_{r=1}^{N} \{\langle \psi_k^{(-)} | \hat{d} | \varphi_r \rangle (\mathbf{S}^{-1})_{rh} - \langle \psi_k^{(-)} | \psi_r \rangle (\mathbf{S}^{-1}\mathbf{dS}^{-1} - \mathrm{Tr}(\mathbf{dS}^{-1})\mathbf{S}^{-1})_{rh}\} \langle \bar{\phi}_0 | \phi_0 \rangle \tag{9a}$$

where $\mathbf{S}$ and $\mathbf{d}$ denote $N \times N$ matrices of "relaxed–frozen" overlap and dipole integrals, respectively,

$$S_{ij} = \langle \psi_i | \varphi_j \rangle, \quad i,j \leq N \tag{9b}$$

$$d_{ij} = \langle \psi_i | \hat{d} | \varphi_j \rangle, \quad i,j \leq N \tag{9c}$$

For convenience of notation, the superscript μ denoting the transition (dipole) operator component has been dropped. We note that because of the restriction to occupied orbitals ($i,j \leq N$) $\mathbf{S}$ and $\mathbf{d}$ are small matrices which can easily be handled as required by (9a); the continuum orbital $|\psi_k^{(-)}\rangle$ enters this expression only via the bound-free dipole and overlap integrals $\langle \psi_k^{(-)} | \hat{d} | \varphi_r \rangle$ and $\langle \psi_k^{(-)} | \varphi_r \rangle$, respectively, $r \leq N$. The overlap $\langle \bar{\phi}_0 | \phi_0 \rangle$ of the relaxed and frozen Slater determinants on the right-hand side of (9a) is given by

$$\langle \bar{\phi}_0 | \phi_0 \rangle = \det(\mathbf{S}) \tag{10}$$

The expression (9) for the N-electron transition moment simplifies to the familiar FCHF result

$$A_{k,h}^{\mathrm{FC}} = \langle \varphi_k^{(-)} | \hat{d} | \varphi_h \rangle, \quad h \leq N \tag{11}$$

if the frozen orbitals are also used to represent the final state. In the more general mixed relaxed–frozen representation assumed here, complexities arise because of the nonorthogonality between the initial and final states. This can be seen by writing the overlap of the initial and final states more explicitly (following the recipe given in I) as

$$\langle \Psi_{k,h} | \phi_0 \rangle = \sum_{r=1}^{N} \langle \psi_k^{(-)} | \varphi_r \rangle (\mathbf{S}^{-1})_{rh} \langle \bar{\phi}_0 | \phi_0 \rangle \tag{12}$$

It is readily seen that, in general, this overlap does not vanish. As a consequence, the transition moment of (9a) introduces artificial contributions depending on the choice of the origin of the molecular coordinate frame. As will be discussed below, this nonorthogonality problem can be circumvented quite naturally in applications to K-shell (or more general core-level) photoionization.

2.2. Core–Valence Separation

In the following we will specialize our considerations to the case of ionization out of a K- or inner-shell (core) level henceforth designated by c. Characteristically, the core and valence levels are separated by a large energy gap ΔE_{cv}. Moreover, any coupling matrix elements V_{cv} (of the N-electron Hamiltonian) between states with different core level occupations are usually of rather small magnitude. Consequently, the overlap between initial and final states (12) which is of the order

$$\langle \Psi_{kc} | \phi_0 \rangle \sim V_{cv}/\Delta E_{cv}$$

can be neglected to a good approximation. This is the essence of the so-called core–valence separation (CVS) approximation which may be introduced formally by neglecting certain types of Coulomb integrals.[25] On the orbital level, the consequences of the CVS approximation are as follows. First, the frozen and the relaxed core orbitals become identical (possibly up to a phase factor),

$$\langle \psi_c | \varphi_c \rangle = 1 \tag{13a}$$

and, second, the orthogonality relations

$$\langle \psi_r | \varphi_c \rangle = 0, \quad r \neq c \tag{13b}$$

$$\langle \psi_k^{(-)} | \varphi_c \rangle = 0 \tag{13c}$$

hold. The relations (13) make explicit that the overlap between initial and final states (12) vanishes. Moreover, a considerable simplification results for the transition moment (9a), which now becomes

$$A_{kc} = [\langle \psi_k^{(-)} | \hat{d} | \varphi_c \rangle - \sum_{r,r' \leq N} \langle \psi_k^{(-)} | \varphi_r \rangle (\mathbf{S}^{-1})_{rr'} \langle \psi_{r'} | \hat{d} | \varphi_c \rangle] \langle \bar{\phi}_0 | \phi_0 \rangle \tag{14}$$

The leading term on the right-hand side of (14) is the familiar one-electron dipole integral for the transition of interest. The additional term arising in (14) introduces relaxed–frozen overlap integrals $\langle \psi_k^{(-)} | \varphi_r \rangle$. The effect of this latter term, referred to as nonorthogonality or conjugate part, can be seen more clearly by writing (14) in the form

$$A_{kc} = \langle \psi_k^{(-)} | (1 - \hat{P}) \hat{d} | \varphi_c \rangle \langle \bar{\phi}_0 | \phi_0 \rangle \tag{15a}$$

where

$$\hat{P} = \sum_{r,r' \leq N} | \varphi_r \rangle (\mathbf{S}^{-1})_{rr'} \langle \psi_{r'} | \tag{15b}$$

is a (non-Hermitian) projection operator. Since

$$\langle \varphi_i | \hat{P} = \langle \varphi_i | + \sum_r (\mathbf{S}^{-1})_{ir} \langle \bar{\psi}_r | \tag{16a}$$

where

$$| \bar{\psi}_r \rangle = \sum_{j > N} | \varphi_j \rangle \langle \varphi_j | \psi_r \rangle \tag{16b}$$

is the projection of the relaxed state $| \psi_r \rangle$ onto the space spanned by the unoccupied (virtual) frozen orbitals, we see that the action of $1 - \hat{P}$ in (15a) is essentially to project out all occupied frozen–core orbital admixtures in $| \psi_k^{(-)} \rangle$. This consideration also shows that the summation on the right-hand side of (14) should include terms with $r,r' = c$ which vanish in the strict CVS approximation. Usually, however, the CVS approximation is not enforced in the generation of the relaxed orbitals so that the relations (13a–c) may not be strictly fulfilled. In such a case the extended summation eliminates any admixtures of $| \varphi_c \rangle$ in $| \psi_k^{(-)} \rangle$ and thus precludes the unphysical (origin-dependent) contributions from entering the transition moment. The expression (14) with the summation including $r,r' = c$ will be referred to as the RCHF transition moment (for single excitations). It should be recalled that this expression applies to the "primitive" excitation (7a) and is written in terms of spin-orbitals. The spin-free formulation of the transition moment corresponding to a singlet (or generally a symmetry-adapted) excitation is obvious (see I for details).

We note that the overall factor $\langle \bar{\phi}_0 | \phi_0 \rangle$ on the right-hand side of (14) can be identified as the ΔSCF approximation for the spectroscopic factor $x_c = \langle \Psi_c^{N-1} | c_c | \Psi_0 \rangle$. This quantity is encountered in core-level electron spectroscopy where $|x_c|^2$ measures the relative intensity of the single-hole main line with respect to the total (c^{-1}) ionization intensity including shake-up and shake-off satellite contributions.

2.3. Generalization to Open-Shell Systems and Equivalent Core Holes

In the derivation of the RCHF transition moment (14) a nondegenerate closed-shell ground state was tacitly assumed. In the open-shell case more complicated moments of the type

$$A_{k,nij} = \langle \phi_{k,ni} | \hat{D} | \phi_{0j} \rangle \tag{17}$$

occur. Here $|\phi_{k,ni}\rangle$ and $|\phi_{0j}\rangle$ are "primitive" states (single Slater determinants) used in the construction (linear combination) of the symmetry-adapted initial and final states, e.g.,

$$|\phi_{k,n\nu}\rangle = \sum_i a_{\nu i} |\phi_{k,ni}\rangle \tag{18a}$$

$$|\phi_{0\lambda}\rangle = \sum_j b_{\lambda j} |\phi_{0j}\rangle \tag{18b}$$

The primitive transition moments $A_{k,nij}$ of (17) can be evaluated in the RCHF approximation in a similar way as described above for the closed-shell case. Explicit results depend on the particular example under consideration and shall not be worked out here. We note that the partial cross section for a photoionization experiment where the initial and final state degeneracies λ,ν are described by a statistical mixture with equal weights is given by

$$\sigma_n(\omega) = \frac{4\pi^2}{3c} \omega \sum_{\nu,\lambda,\mu} |\langle \phi_{k,n\nu} | \hat{D}^{(\mu)} | \phi_{0\lambda} \rangle|^2 \tag{19}$$

Often one may make use of the fact that the trace with respect to the states $|\phi_{0\lambda}\rangle$ or $|\phi_{k,n\nu}\rangle$ can be replaced by a trace with respect to the primitive states $|\phi_{0j}\rangle$ or $|\phi_{k,ni}\rangle$, respectively.

Some further comments are required for the case of molecules with two or more equivalent core levels such as N_2, C_2H_4, BF_3, C_6H_6. To be specific, let us consider the example of N_2 or C_2H_4 with two equivalent atomic $1s$ orbitals which we designate by s_1 and s_2. With symmetric or antisymmetric linear combinations of the localized orbitals,

$$\sigma_g = \frac{1}{\sqrt{2}} (s_1 + s_2) \tag{20a}$$

$$\sigma_u = \frac{1}{\sqrt{2}} (s_1 - s_2) \tag{20b}$$

one can form molecular orbitals of g and u symmetry (with respect to inversion) with nearly degenerate orbital energies (typically, the energy difference is of the order of 0.1 eV). At the many-electron level, this core-orbital degeneracy is reflected by the fact that states with a core-level vacancy occur in nearly degenerate (g,u) pairs. In analogy to (20), one may transform the symmetry-adapted N-electron states $|\Psi_g\rangle$, $|\Psi_u\rangle$ into "localized" states $|\Psi_1\rangle$ and $|\Psi_2\rangle$.

The dipole selection rules for a transition from a (closed-shell) ground state of g symmetry lead only to u final states. Thus, in the particular case of $1s$ to continuum excitations two distinct types of final u states contribute to the partial cross section. In a single-particle approximation these states may be written as

$$|\phi_{k_u, \sigma_g}\rangle = c_{k_u}^\dagger |\phi_g^{N-1}\rangle \tag{21a}$$

$$|\phi_{k_g, \sigma_u}\rangle = c^\dagger_{k_g} |\phi_u^{N-1}\rangle \tag{21b}$$

where symmetry-adapted states $|\phi_g^{N-1}\rangle$, $|\phi_u^{N-1}\rangle$ and $|k_g\rangle, |k_u\rangle$ have been chosen to represent the ionic core and the continuum orbitals, respectively; one may note that only even (odd) l-waves occur in a single-center partial wave expansion of k_g (k_u). The integral cross section is given by

$$\sigma_{1s} \sim |A_{k_u, \sigma_g}|^2 + |A_{k_g, \sigma_u}|^2 \tag{22}$$

where the transition moments

$$A_{k_u, \sigma_g} = \langle \phi_{k_u, \sigma_g} | \hat{D} | \phi_0 \rangle \tag{23a}$$

$$A_{k_g, \sigma_u} = \langle \phi_{k_g, \sigma_u} | \hat{D} | \phi_0 \rangle \tag{23b}$$

are defined with respect to the symmetry-adapted final states (21a,b). These symmetry-adapted transition moments can be evaluated straightforwardly in the RCHF model as described in Sections 2.1 and 2.2.

As an alternative approach one may formulate the RCHF model in terms of localized orbitals. It is well known (see Section 4.1) that the localized HF representation for the ion

$$|\phi_{s_1}^{N-1}\rangle = c_{s_1} |\bar{\phi}_0^{(1)}\rangle \tag{24}$$

is a much better approximation to the exact ionic state than the delocalized form $|\phi_{g(u)}^{N-1}\rangle$, the latter accounting for only half of the total relaxation energy. In (24), $|\bar{\phi}_0^{(1)}\rangle$ denotes the formal N-electron Slater determinant of localized orbitals. As will be discussed in Section 4.2, the choice of the orbital representation is less crucial in the calculation of the cross section since the essential RCHF correction, i.e., the relaxation-induced screening of the potential, is equally well described in the localized and delocalized approach. We now briefly sketch the localized formulation. The relation

$$|\phi_{g(u)}^{N-1}\rangle = \frac{1}{\sqrt{2}} (c_{s_1} |\bar{\phi}_0^{(1)}\rangle \stackrel{+}{(-)} c_{s_2} |\bar{\phi}_0^{(2)}\rangle) \tag{25}$$

defining the associated symmetry-adapted ionic states allows one to replace the symmetry-adapted transition moments (23a,b) by localized ones

$$A_{k_u, \sigma_g} = \sqrt{2} \, A_{k_u, s_1} \tag{26a}$$

$$A_{k_g, \sigma_u} = \sqrt{2} \, A_{k_g, s_1} \tag{26b}$$

Since the individual l-contributions add independently to the integral cross section we find the expected result

$$\sigma_{1s} \sim 2 \cdot |A_{k, s_1}|^2 \tag{27}$$

where the (left) localized RCHF transition moment

$$A_{k, s_1} = \langle \bar{\phi}_0^{(1)} | c^\dagger_{s_1} c_k \hat{D} | \phi_0 \rangle \tag{28}$$

now contains both the even and odd l-contributions to the continuum orbital. A_{k, s_1} may be evaluated further as described in Sections 2.1 and 2.2. The factor of two in (27) indicates that both the (identical) contributions of the left and right hole s_1 and s_2, respectively, have to be taken into account.

In a similar way the case of three or more equivalent core levels may be analyzed. In every case the result will be an obvious generalization of (27, 28).

3. GENERATION OF RELAXED CONTINUUM ORBITALS

While the occupied relaxed orbitals $|\psi_r\rangle$, $r \leq N$ required in the RCHF representation of the excited N-electron state can be obtained by performing an HF calculation for the ionic single-hole state $|\phi_c^{N-1}\rangle$ [see equation (8)], it is far more demanding to generate the appropriate molecular continuum orbitals $|\psi_k^{(-)}\rangle$. In the following sections we shall briefly address the pertinent single-particle scattering equation and the methods used for its solution.

3.1. Static-Exchange Potentials for K-Shell Photoelectrons

An obvious approach to the motion of the photoelectron is to assume that these orbitals are solutions of a one-electron Schrödinger equation, containing the potential of the relaxed ionic charge distribution (relaxed V^{N-1} potential[26]). For a singlet excitation this spin-free equation is

$$(\hat{t} + \sum_{i \neq c}(2\bar{J}_i - \bar{K}_i)n_i + \bar{J}_c + \bar{K}_c - \frac{k^2}{2})|\psi_k^{(-)}\rangle = 0 \tag{29}$$

Here the operator $\hat{t}$ contains the kinetic energy operator and the Coulomb attraction by the nuclei, and $\bar{J}_i$ and $\bar{K}_i$ denote the Coulomb and exchange operators, respectively, associated with the relaxed ionic orbitals $|\psi_i\rangle$. In (29) all subscripts refer to spatial quantum numbers and the familiar (ground state) HF occupation numbers n_i are used to restrict the summation to occupied orbitals ($n_i = 1$). Equation (29) is not yet complete, but has to be augmented by the conditions

$$\langle\psi_k^{(-)}|\psi_r\rangle = 0, \quad n_r = 1 \tag{30}$$

imposing orthogonality between $|\psi_k^{(-)}\rangle$ and all occupied relaxed orbitals.

Single-particle scattering equations such as (29) and (30) can be deduced from the variational expression[27]

$$\langle\delta\Psi_{k,h}|\hat{H} - E|\Psi_{k,h}\rangle = 0 \tag{31}$$

where, as in (7a,b), $|\Psi_{k,h}\rangle$ denotes a representation of the N-electron final state expressed in terms of target-state (ionic) orbitals that are kept fixed and a continuum orbital $|\psi_k^{(-)}\rangle$ which is to be determined; here the variation $|\delta\Psi_{k,h}\rangle$ indicates only a variation $|\delta\psi_k^{(-)}\rangle$ of the continuum orbital. For the FCHF representation of the final state $|\Psi_{k,h}\rangle$, the derivation of these equations using the variational principle (31) has been demonstrated by Lucchese et al.[13] A generalization of this procedure to the RCHF case was considered in I. In the latter case complications arise because the variational principle does not exclude admixtures of the occupied orbital $|\psi_h\rangle$ in the continuum orbital $|\psi_k^{(-)}\rangle$. By decomposing $|\psi_k^{(-)}\rangle$ according to

$$|\psi_k^{(-)}\rangle = |\tilde{\psi}\rangle + \langle\psi_h|\psi_k^{(-)}\rangle|\psi_h\rangle \tag{32}$$

where $\langle\psi_h|\tilde{\psi}\rangle = 0$, the variational equation may be rewritten in the form of a set of relatively complicated coupled equations for the orthogonal part $|\tilde{\psi}_k\rangle$ and its complement $\langle\psi_h|\psi_k^{(-)}\rangle|\psi_h\rangle$. Here the explicit expression

$$\langle\psi_h|\psi_k^{(-)}\rangle = (\varepsilon_h - \tfrac{1}{2}k^2)^{-1}\langle\psi_h|\hat{t} + \sum_{i \neq h}(2\bar{J}_i - \bar{K}_i)n_i|\tilde{\psi}_k\rangle \tag{33}$$

relates the bound-free overlap to the orthogonal part $|\tilde{\psi}_k\rangle$. Fortunately, these complexities disappear for K-shell ionization on the basis of the CVS approximation. If $|\psi_h\rangle$ is a K- or inner-shell orbital, the absolute value of the denominator in (33) is of the order of the core–valence energy gap, while the matrix elements in the numerator are of the core–valence coupling type V_{cv}. Hence, only a small error is introduced by setting $\langle\psi_h|\psi_k^{(-)}\rangle = 0$ and, thus, we may safely adopt (29) and (30) to calculate K-shell photoionization continua.

The analysis of the RCHF approximation in Section 4 will show that the relaxed equations (29), (30) correctly account for the relaxation effects and, in particular, for the screening interaction $\Delta E^S(k)$ experienced by the continuum orbital. On the other hand, as in any static approximation, the response of the target to the photoelectron leading to attractive polarization forces is not considered.

3.2. Direct Method for Determining Molecular Continuum Orbitals

The methods used in this study to obtain the photoelectron continuum wave functions as solutions of (29), (30) have been amply described elsewhere[13,28] and here we just briefly sketch the basic features. The continuum function is written in a partial wave expansion

$$\Psi_k^{(-)}(r) = \left(\frac{2}{\pi}\right)^{1/2} \sum_{l,m} \frac{i^l}{k} \psi_{klm}^{(-)}(r) Y_{lm}^*(k) \tag{34}$$

Each partial wave $\psi_{klm}(r)$ satisfies the Lippmann–Schwinger (LS) equation

$$\psi_{klm}^{(-)} = \phi_{klm}^c + G_c^{(-)}(v + 1/r)\psi_{klm} \tag{35}$$

where $G_c^{(-)}$ is the Coulomb Green's function with incoming-wave boundary conditions and v is the relaxed V^{N-1} potential for the molecular ion. Orthogonalization between the continuum and the occupied orbitals is taken into account by means of the Philips–Kleinman pseudopotential as described in Ref. 13. The Schwinger variational method is used to solve the LS equations (35). By assuming a separable approximation to the potential

$$V = v + 1/r \tag{36}$$

of the form

$$V \cong V^s = \sum_{i,j} \langle \mathbf{r}|V|\alpha_i\rangle(\mathbf{V}^{-1})_{ij}\langle\alpha_j|V|\mathbf{r}'\rangle \tag{37}$$

the solutions of the LS equations (35) with the separable potential of (37) can be written as

$$\tilde{\psi}_{klm}^{(-)}(r) = \varphi_{klm}^c(\mathbf{r}) + \sum \langle\mathbf{r}|G_c^{(-)}V|\alpha_i\rangle(\mathbf{D}^{-1})_{ij}\langle\alpha_j|V|\varphi_{klm}^c\rangle \tag{38a}$$

Here, $\mathbf{V}$ and $\mathbf{D}$ denote matrices with the matrix elements

$$V_{ij} = \langle\alpha_i|V|\alpha_j\rangle \tag{38b}$$

and

$$D_{ij} = \langle\alpha_i|V - VG_c^{(-)}V|\alpha_j\rangle \tag{38c}$$

The set of functions $|\alpha_i\rangle$, referred to as the initial scattering basis set, consists of suitably chosen discrete functions. Expansion of $\tilde{\psi}_{klm}^{(-)}(\mathbf{r})$ in partial waves

$$\tilde{\psi}_{klm}^{(-)} = \sum_{l',m'} g_{lm,l'm'}^{(0)}(r) Y_{l'm'}(\hat{a}) \tag{38d}$$

and substitution of this expansion into (38a) leads to a set of coupled equations for the radial functions $g^{(0)}_{lm,l'm'}(r)$. These radial functions can be readily obtained using single-center expansions of all quantities, e.g., $\hat{V}_c |\alpha_i\rangle$, and $G_c(\mathbf{r},\mathbf{r}')$, appearing in (38a) and simple quadratures to evaluate the radial integrals associated with the matrix elements of (38a). The overlap and dipole integrals required in the RCHF expression (14) are also evaluated using similar single-center expansion techniques.

These $\tilde{\psi}^{(-)}_{klm}$ are not the true solutions of (35) but just approximations to these solutions. How good these solutions $\tilde{\psi}^{(-)}_{klm}$ are depends on the expansion basis $|\alpha_i\rangle$. It is clearly desirable to have a systematic procedure for improving these $\tilde{\psi}^{(-)}_{klm}$ and hence obtaining the true $\tilde{\psi}^{(-)}_{klm}$. We have developed and implemented such a procedure.[13] It begins by viewing the approximate solution $\tilde{\psi}^{(-)}_{klm}$ in numerical form as a new basis function which is now added to the initial discrete scattering basis $|\alpha\rangle$ to form a new and larger basis. This expanded basis is then used in (37) and (38a) to obtain an improved approximation to $\psi^{(-)}_{klm}$. Criteria for assessing the convergence of this iterative Schwinger variational procedure have been established and extensively applied. This procedure which generally converges in a few steps has been used to obtain the photoelectron wave function and matrix elements needed in these studies.

4. ANALYSIS OF THE RCHF MODEL

Before discussing the results of applications of the RCHF approach it may be useful to investigate its potential in a more general way. For this purpose we shall use perturbation theory which allows us to compare the approximate results with the exact quantities through some low, however nontrivial, order in the electronic repulsion. Following the discussion given in I, we consider a singlet core excitation $|\Psi_{vc}\rangle$ obtained from the unperturbed singly excited state

$$|\phi_{vc}\rangle = \frac{1}{\sqrt{2}} (a^\dagger_{v\alpha} a_{c\alpha} + a^\dagger_{v\beta} a_{c\beta}) |\phi_0\rangle \qquad (39)$$

where $a^\dagger(a)$ denote frozen orbital operators. As above, c stands for a K-shell or inner-shell level, while v refers to a general virtual orbital which may represent likewise an unoccupied valence orbital above or below threshold, a Rydberg orbital, or a continuum orbital. In Sections 4.1–4.3 we shall separately consider the ionization potential IP_c of the corresponding ionic core state $|\Psi_c^{N-1}\rangle$, the energy E_{vc} of the excited state relative to the ionic energy E_c^{N-1}, and the transition moment A_{vc}, respectively. Our notation is as follows. Except when explicitly stated otherwise, Latin letters are used in this section to denote spatial orbitals, while spin quantum numbers m_s are specified by Greek letters. As usual, the values $m_s = \pm 1/2$ are denoted by α,β. An overbar is used to distinguish relaxed from frozen quantities. Coulomb integrals and orbital energies are denoted by V_{ijkl} and ε_i, respectively; we also use the notations $J_{ij} = V_{ijij}$ and $K_{ij} = V_{ijji}$. Throughout we shall take advantage of the simplifications introduced by the CVS approximation of Section 2.

4.1. Relaxation Correction to the Ionic Core Energy

The occupied frozen orbitals $|r\rangle$ are eigenfunctions of the HF operator

$$\hat{f}_0 = \hat{t} + \sum_l (2\hat{J}_l - \hat{K}_l) n_l \qquad (40)$$

where $\hat{J}_i$ and $\hat{K}_i$ are the frozen orbital Coulomb and exchange operators. The occupied relaxed orbitals $|\bar{r}\rangle$, on the other hand, are determined by an HF calculation for the $(N-1)$-particle state $|\bar{\phi}_c^{N-1}\rangle = c_c |\bar{\phi}_0\rangle$, which in this case is not equivalent to diagonalizing a single-particle operator. However, if one is not interested in the subtleties arising from the core–valence exchange contribution, one may assume that both the core and valence orbitals are eigenfunctions of the same operator

$$\hat{f} = \hat{t} + \sum_{\substack{l \\ l \ne c}} (2\bar{J}_l - \bar{K}_l)n_l + \hat{J}_c - \hat{K}_c \qquad (41)$$

As a starting point for our perturbation treatment we write the relaxed operator in the form

$$\hat{f} = \hat{f}_0 - \hat{J}_c + \hat{\Delta\rho}_c \qquad (41a)$$

where

$$\hat{\Delta\rho}_c = \sum_{\substack{l \\ l \ne c}} (2\bar{J}_l - \bar{K}_l - 2\hat{J}_l + \hat{K}_l)n_l \qquad (41b)$$

will be referred to as a screening operator. We identify $\hat{f}_0$ and $\hat{f}_1 = -\hat{J}_c + \hat{\Delta\rho}_c$ with the unperturbed and interaction part, respectively. An unusual feature here is that the interaction part $\hat{f}_1$ contains a contribution ($\hat{\Delta\rho}_c$) which itself has a perturbation expansion beginning in second order.[29] In the expansion of the relaxed orbitals through first order

$$|\bar{r}\rangle = |r\rangle - \sum_{\substack{q \\ q \ne r}} \frac{V_{qcrc}}{\varepsilon_r - \varepsilon_q} |q\rangle + O(2) \qquad (42)$$

only the first-order part $-\hat{J}_c$ of $\hat{f}_1$ comes into play. For the orbital energy, the expansion through second order given by

$$\bar{\varepsilon}_r = \varepsilon_r - J_{rc} + \sum_q \frac{V_{rqc}^2}{\varepsilon_r - \varepsilon_q} + \Delta E^S(r) + O(3) \qquad (43a)$$

introduces the (second-order) screening energy for the orbital r,

$$\Delta E^S(r) = \langle r | \hat{\Delta\rho}_c | r \rangle \qquad (43b)$$

describing the change in the Coulomb (and exchange) interaction between the (frozen) orbital $|r\rangle$ and the valence charge distribution induced by the relaxation. Using the first-order expansions (42) for the relaxed orbitals, the second-order screening energy can be written explicitly as

$$\langle r | \hat{\Delta\rho}_c | r \rangle^{(2)} = 2 \sum_{\substack{l,q \\ l \ne c}} (2V_{rqrl} - V_{rqlr}) \frac{V_{qclc}}{\varepsilon_q - \varepsilon_l} n_l \bar{n}_q \qquad (44)$$

Here, for notational convenience, real Coulomb integrals are assumed. Note that the summation over q is restricted to unoccupied orbitals as indicated by the inverse occupation number $\bar{n} = 1 - n_q$.

A special comment applies to the case of the core orbital c. It is readily seen that all but the zeroth-order term in the perturbation expansion for $|\bar{c}\rangle$ vanishes if the CVS approximation is assumed. Thus, within the validity of the CVS we may suppose $|c\rangle = |\bar{c}\rangle$, which was already exploited in Section 2.2. In particular, there is no need to distinguish between $\bar{J}_c$ and $\hat{J}_c$ in (41). Of course, the frozen and relaxed core orbital energies differ according to

$$\bar{\varepsilon}_c = \varepsilon_c - J_{cc} + \Delta E^S(c) + O(3) \qquad (45)$$

where

$$\Delta E^S(c) = \langle c | \hat{\Delta \rho}_c | c \rangle = 2 \sum_{\substack{l,q \\ l \neq c}} (2V_{cqcl} - V_{cqlc}) \frac{V_{lcqc}}{\varepsilon_q - \varepsilon_l} n_l \bar{n}_q + O(3) \tag{46}$$

is the screening energy for the core orbital.

Let us now briefly review the ionization energy $IP_c = E_c^{N-1} - E_0$. In the ΔSCF method this quantity is determined as the difference

$$IP_c^{\Delta SCF} = \langle \bar{\phi}_c^{N-1} | \hat{H} | \bar{\phi}_c^{N-1} \rangle - \langle \phi_0 | \hat{H} | \phi_0 \rangle \tag{47}$$

of the relaxed ionic and frozen ground state HF energies. Equation (47) may be written in the form

$$IP_c^{\Delta SCF} = -\varepsilon_c - \Delta E_c^R \tag{48}$$

where

$$\Delta E_c^R = \langle \phi_c^{N-1} | \hat{H} | \phi_c^{N-1} \rangle - \langle \bar{\phi}_c^{N-1} | \hat{H} | \bar{\phi}_c^{N-1} \rangle \tag{49}$$

may be viewed as a definition of the relaxation energy (excluding for the moment the possibility of two or more equivalent core levels). The right-hand side of (49) can be straightforwardly evaluated through second order using the first-order expansions (42) for the relaxed orbitals. The final result is simply

$$\Delta E_c^R = 1/2 \, \Delta E^S(c) + O(3) \tag{50}$$

This result states that through second order the relaxation energy is given by one-half of the screening energy $\Delta E^S(c)$ for the core orbital.[30] More explicitly, this is the outcome of two opposing effects. The single-particle contributions in the relaxed ionic energy are lowered with respect to the frozen expression by $-3/2\Delta E^S(c)$, whereas the Coulomb (and exchange) interaction between the remaining core orbital and the relaxed valence charge distribution increases by $\Delta E^S(c)$. As is well known, (50) provides the justification for the transition operator method[31,32] in which the ionization potential is calculated as the orbital energy of a half-electron hole state.

The perturbation expansion of the exact ionization potential through second order has the following form:

$$IP_c = -\varepsilon_c + U_c^{(2)}(2h-1p) + U_c^{(2)}(3h-2p) - U_0^{(2)}(2p-2h) + O(3) \tag{51}$$

where the second and third terms represent the second-order contributions arising from the interaction of the single-hole state $|\phi_c^{N-1}\rangle$ with two-hole–one-particle ($2h-1p$) states and $3h-2p$ states; $U_0^{(2)}(2p-2h)$ denotes the second-order contribution to the ground-state energy. In the CVS approximation the latter term is canceled exactly by the $3h-2p$ contributions to the ionic energy. Moreover, the CVS approximation leads to an obvious restriction on the $2h-1p$ states contributing to $U_c^{(2)}(2h-1p)$, namely, to states of the form $(q)^1(c')^{-1}(l)^{-1}$ with exactly one hole in the orbital c or, possibly, in a level c', equivalent to c. The explicit result for the expansion of IP_c through second order is given by

$$IP_c = -\varepsilon_c - 2 \sum_{c',q,l} \frac{V_{cqc'l}^2}{\varepsilon_q - \varepsilon_l} \bar{n}_q n_l + O(3) \tag{52}$$

where, for simplicity, the unessential core–valence exchange integrals V_{cqlc} have been omitted. In the case of a single core level this result is (up to core–valence exchange contribution) identical with the ΔSCF result according to (46), (48–50).

The approximate equivalence of the ΔSCF result and the exact second-order ionization potential is no longer guaranteed in the case of two or more equivalent core levels. As is well known, the ΔSCF approach here may differ significantly for different core orbital representations such as symmetry-adapted and localized orbitals, respectively. On the other hand, the full second-order ionization potential in the CVS approximation according to (52) is, of course, independent of the chosen orbital sets, and, thus, may serve as a more general definition of the relaxation energy than (48) and (49). As has been shown previously, only the localized version of the ΔSCF relaxation energy agrees (up to core–valence exchange contributions) with the exact result.[33,34]

4.2. Term Values and Resonance Positions

The single-particle energy ε_v of a virtual orbital v generated using either the frozen or relaxed core potential according to (29) is directly related to the term value

$$T_v = E_{vc} - E_c^{N-1} \tag{53}$$

that is, the excitation energy relative to the ionization threshold. In the following we will analyze the FCHF and RCHF results for this quantity through second order of perturbation theory and compare them with the exact second-order expression. The discussion of term values directly applies also to the position of resonances in the cross section, as resonances may be viewed as originating from a valence excitation interacting with a background continuum.

In the FCHF model the virtual orbital $|v^{FC}\rangle$ is an eigenfunction of the frozen-core operator

$$\hat{f}^{FC} = \hat{f}_0 - \hat{J}_c + 2\hat{K}_c \tag{54}$$

where $\hat{f}_0$ is the ground state HF operator [equation (40)]. In addition, $|v^{FC}\rangle$ satisfies the orthogonality constraints

$$\langle v^{FC} | r \rangle = 0, \quad n_r = 1 \tag{55}$$

The resulting perturbation expansion for the term value T_v^{FC} given by the corresponding orbital energy ε_v^{FC} is, through second order,

$$T_v^{FC} = \varepsilon_v^{FC} = \varepsilon_v - J_{vc} + 2K_{vc} + \sum_{\substack{q \\ q \neq v}} \frac{(V_{vcqc} - 2V_{vccq})^2}{\varepsilon_v - \varepsilon_q} \bar{n}_q + O(3) \tag{56}$$

Note that the summation on the right-hand side is restricted to unoccupied orbitals ($\bar{n}_q = 1$) in accordance with the orthogonality constraints (55). The first-order Coulomb contribution $-J_{vc}$ accounts for the lowering of the electrostatic repulsion on removal of a core electron. The second-order term reflects the adjustment (relaxation) of the virtual orbital $|v\rangle$ to the core level vacancy.

The relaxed core operator [equation (29)] used in the RCHF model differs from $\hat{f}^{FC}$ by the screening operator $\hat{\Delta}\rho_c$ of 41)

$$\hat{f}^{RC} = \hat{f}_0 - \hat{J}_c + 2\hat{K}_c + \hat{\Delta}\rho_c \tag{57}$$

Apart from the core exchange operator, $\hat{f}^{RC}$ is identical with the relaxed operator $\hat{f}$ of (41) assumed for the occupied relaxed orbitals. Neglecting core–valence exchange integrals both the virtual and occupied orbitals are eigenfunctions of the same operator $\hat{f}$. Then the first-order expansion of $|\bar{v}\rangle$ is as given by (42); in particular, the orthogonality constraints (30) augmenting the relaxed core operator (57) are satisfied through first order. Similar to (43), the second-order expansion of the relaxed term value $T_v^{RC}(=\bar{\varepsilon}_v)$ may be written as

$$T_v^{RC} = \bar{\varepsilon}_v = \varepsilon_v - J_{vc} + 2K_{vc} + \sum_{r \neq v} \frac{V_{vcrc}^2}{\varepsilon_v - \varepsilon_r} + \Delta E^S(v) + O(3) \tag{58a}$$

where

$$\Delta E^S(v) = \langle v | \hat{\Delta \rho}_c v \rangle \tag{58b}$$

is the screening energy for the orbital v. The explicit second-order form of $\Delta E^S(v)$ is as specified by (44). Note that the first-order contributions on the right-hand side of (58a) correspond to the full core operator (57). Comparing (56) and (58) we see that T_v^{RC} and T_v^{FC} differ by two second-order contributions

$$T_v^{RC} = T_v^{FC} + \sum_r \frac{V_{vcrc}^2}{\varepsilon_v - \varepsilon_r} n_r + \Delta E^S(v) + O(3) \tag{59}$$

As will be seen below, the first contribution accounts for the difference of the relaxation energy in the excited and the ionic states. The physical origin of the second contribution has already been discussed: $\Delta E^S(v)$ describes the change in the electrostatic repulsion between the orbital $|v\rangle$ and the occupied valence orbitals upon relaxation of the latter. Both contributions are positive, that is, they enhance the term values and resonance positions of the FCHF model.

Now we will compare the FCHF and RCHF approximations for T_v with the exact result through second order. T_v may also be obtained [see (53)] from

$$T_v = \Delta E_{vc} - IP_c \tag{60}$$

as the difference of the excitation energy $\Delta E_{vc} = E_{vc} - E_0$ and the ionization potential IP_c. The perturbation expansion for the latter quantity was already considered in Section 4.1. Following Ref. 35 the excitation energy can be written through second order in the form

$$\Delta E_{vc} = \varepsilon_v - \varepsilon_c - J_{vc} + 2K_{vc} + U_{vc}^{(2)}(p-h) + U_{vc}^{(2)}(2p-2h) + R_{vc}^{(2)} + O(3) \tag{61}$$

Here $U_{vc}^{(2)}(p-h)$ and $U_{vc}^{(2)}(2p-2h)$ represent the second-order contributions arising from the interaction of the excitation $|\phi_{vc}\rangle$ with other $p-h$ states and $2p-2h$ states, respectively; the last second-order term

$$R_{vc}^{(2)} = U_{vc}^{(2)}(3p-3h) - U_0^{(2)}(2p-2h) \tag{62}$$

results from a partial compensation of the $3p-3h$ contributions in the excited-state energy and the second-order ground-state correlation energy. Combining the expansions (51) and (61) yields for the term value

$$T_v = \varepsilon_v - J_{vc} + 2K_{vc} + U_{vc}^{(2)}(p-h) + U_{vc}^{(2)}(2p-2h) - U_c^{(2)}(2h-1p) + R_{vc}^{(2)} + O(3) \tag{63}$$

Explicit expressions, however in spin-orbital form, of the second-order contributions appearing in (61) can be found in Ref. 35. Their more complicated spin-free formulation appropriate for a singlet excitation (39) was given in I.

As is readily seen, the first four terms of the exact expansion (63) including $U_{vc}^{(2)}(p-h)$ are identical to those of the FCHF model given by (56). Here and in the following the CVS approximation is assumed. The first deviation arises in the $2p-2h$ term $U_{vc}^{(2)}(2p-2h)$. According to the CVS approximation we need consider here only excitations of the form $(q)^1(p)^1(c')^{-1}(r)^{-1}$, that is, excitations with exactly one hole in the core level c' where the prime indicates the possibility of several equivalent core levels. It is useful to further analyze excitations of the form $(q)^1(v)^1(c')^{-1}(r)^{-1}$. Accordingly, $U_{vc}(2p-2h)$ may be split into a main part, $U'_{vc}(2p-2h)$ and a remainder $U''_{vc}(2p-2h)$, the latter part being associated with the excitations where $q,p \neq v$. Neglecting all exchange integrals which are not essential for the present analysis, the main part can be put into the following transparent form:

$$U'_{vc}(2p-2h) = -2 \sum_{c',q,r} \frac{(1 - 1/2\delta_{v,q})}{\varepsilon_q - \varepsilon_r} (V_{c'rcq} - \delta_{cc'}V_{vrvq})^2 n_r \bar{n}_q \qquad (64)$$

Apart from the neglect of exchange integrals of the form $V_{c'rqc}$ and V_{vrqv}, (64) is more general than the corresponding equation (75) in I since it applies also to the case of equivalent core levels. Evaluating the square of the bracket on the right-hand side of (64) leads to the decomposition

$$U'_{vc}(2p-2h) = R + S + P \qquad (65)$$

where the parts R, S, and P are associated with the quadratic term $V^2_{c'rcq}$, the mixed term $\delta_{cc'}V_{c'rcq}V_{vrvq}$, and the quadratic term $\delta_{cc'}V^2_{vrvq}$, respectively. Each of these parts has a distinct physical meaning.

The first part

$$R = -2 \sum_{c',q,r} \frac{(1 - 1/2\delta_{q,v})}{\varepsilon_q - \varepsilon_r} V^2_{c'rcq} n_r \bar{n}_q \qquad (66)$$

can be regarded as the relaxation energy for the excited state. It is compensated largely by the part $U_c^{(2)}(2h-1p)$, that is the relaxation energy of the ion, the remainder being

$$R - U_c^{(2)}(2h-1p) = \sum_{c',r} \frac{V^2_{c'rcv}}{\varepsilon_v - \varepsilon_r} n_r \qquad (67)$$

If there is only a single core level ($c' = c$), this part is fully reproduced in the term value T_v^{RC} of the RCHF model, namely, by the second term on the right-hand side of (59). In the case of equivalent core levels, the latter term agrees with the exact relaxation energy change (67) only for the localized core-orbital representation. The mixed part

$$S = +4 \sum_{q,r} \frac{(1 - 1/2\delta_{v,q})}{\varepsilon_q - \varepsilon_r} V_{crcq} V_{vrvq} n_r \bar{n}_q \qquad (68)$$

can obviously be identified with the (relaxation-induced) screening energy $\Delta E^S(v) = \langle v | \hat{\Delta} p_c | v \rangle$ for the orbital v which constitutes the principal ingredient of the RCHF result (58). It should be noted that S and $\Delta E^S(v)$ also agree in the case of equivalent core holes irrespective of the orbital representation chosen. The last part of $U'_{vc}(2p-2h)$,

$$P = -2 \sum_{q,r} \frac{(1 - 1/2\delta_{v,q})}{\varepsilon_q - \varepsilon_r} V^2_{vrvq} n_r \bar{n}_q \qquad (69)$$

resembles R except that v now plays the role of c. This part arises from the adjustment of the valence electrons to the presence of an electron in the virtual orbital v, which is referred to as the polarization of the ionic core by the excited electron. In a similar way as in Section 4.1 one may establish the relation $P = 1/2 \langle v | \hat{\Delta} p_v | v \rangle + O(3)$, where $\hat{\Delta} p_v$ is the polarization induced screening operator analogous to [41b]. The energy associated with the polarization is negative, thus shifting the term value or the resonance position to lower energy. Obviously, the polarization correction is not accounted for by the present FCHF and RCHF models. We note that the interpretation given here for P allows also for another interpretation of S. Indeed, up to a sign, S may as well be seen as a screening energy $\Delta E^P(c)$ for the core orbital c arising from the polarization of the valence charge distribution by the excited electron in the orbital v, that is $S = -\langle c | \hat{\Delta} p_v | c \rangle$.

The remaining second-order contributions to T_v, that is, $U''_{cv}(2p-2h)$ and $R_{vc}^{(2)}$, represent genuine many-body effects. The former term arises from the admixture of double excitations

$(p)^1(q)^1(c)^{-1}(l)^{-1}$ where $p,q \neq v$ in the final excited state; the latter reflects the difference of correlation energies in the excited and the ground states. Both contributions, being absent in the single-particle models, are expected to be only of secondary importance.

In conclusion, we have seen that the second-order contributions to the RCHF result for T_v have been retrieved from the exact expression. This means that both the relaxation induced screening effect and the relaxation energy change are correctly described by the RCHF model. Moreover, our analysis has shown that, other than the relaxation effect, the screening effect can be likewise treated in a localized or a symmetry-adapted single-particle representation in the case of several equivalent core levels. The polarization effect, on the other hand, is not accounted for by the RCHF (and the FCHF) model which will result in term values and resonance positions that are too high. The neglect of polarization may prove to be a particular shortcoming of the RCHF scheme, since the second-order energy shifts R, S, and P are not independent of each other. Using the Schwartz inequality for a suitably defined scalar product one may readily derive from (66), (68), and (69) an inequality relating the absolute values of the (second-order) relaxation, screening, and polarization contributions as follows:

$$|P||R| \geq S^2/4 \tag{70}$$

Here the case of a single core level has been assumed. The inequality (70) establishes a lower bound to the error introduced by neglecting the polarization contribution. According to (70) a large screening shift will be accompanied by a large polarization correction. The joint effect of $R + S + P$ always results in a lowering of the excited-state energy, as can be seen from (64).

4.3. Transition Amplitudes

In the perturbation theoretical analysis of the transition moment $A_{vc} = \langle \Psi_{vc} | \hat{D} | \Psi_0 \rangle$ we will confine ourselves to first order. For the simple FCHF model the corresponding expansion reads

$$A_{vc}^{FC} = \sqrt{2}(\langle v|\hat{d}|c \rangle + \langle v^{(1)}|\hat{d}|c \rangle) + O(2) \tag{71}$$

where the first-order contribution $|v^{(1)}\rangle$ to the virtual orbital (of the frozen V^{N-1} potential) is

$$|v^{(1)}\rangle = \sum_{\substack{q \\ q \neq v}} \frac{-V_{vcqc} + 2V_{vccq}}{\varepsilon_v - \varepsilon_q} \bar{n}_q |q\rangle \tag{72}$$

The factor $\sqrt{2}$ on the right-hand side of (71) arises from the spin-free formulation for a singlet transition. The first-order expansion for the more complicated RCHF expression (14) can be written in the form

$$A_{vc}^{RC} = \sqrt{2}\left\{ \langle v|\hat{d}|c \rangle + \langle \bar{v}^{(1)}|\hat{d}|c \rangle - \sum_r \langle \bar{v}^{(1)}|r\rangle\langle r|\hat{d}|c \rangle n_r \right\} + O(2) \tag{73}$$

where $|\bar{v}^{(1)}\rangle$ denotes the first-order contribution to the relaxed virtual orbital as specified (up to core–valence exchange integrals) by (42). Since

$$|\bar{v}^{(1)}\rangle - \sum_r |r\rangle\langle r|\bar{v}^{(1)}\rangle = |v^{(1)}\rangle \tag{74}$$

we see that the overlap part of the RCHF amplitude cancels the occupied frozen-core orbital admixtures in $|\bar{v}^{(1)}\rangle$ and the RCHF and FCHF transition moments are identical through first order.

The exact transition moment

$$A_{vc} = \sqrt{2}(\langle v|\hat{d}|c \rangle + \langle \Psi_{vc}^{(1)}|\hat{D}|\phi_0^N \rangle + \langle \phi_{vc}|\hat{D}|\Psi_0^{(1)} \rangle) + O(2) \tag{75}$$

contains two first-order contributions arising from the first-order corrections $|\Psi_{cv}^{(1)}\rangle$ and $|\phi_0^{(1)}\rangle$ to the excited and ground states, respectively. It is readily seen that

$$\langle \Psi_0^{(1)} | \hat{D} | \phi_0 \rangle = \langle v^{(1)} | \hat{a} | c \rangle \tag{76}$$

while the ground-state correlation contribution $\langle \phi_{vc} | \hat{D} | \Psi_0^{(1)} \rangle$ vanishes in the CVS approximation. Thus, we may conclude that for K-shell ionization both the FCHF and RCHF transition moment are consistent through first order with the exact result. The preceding considerations also show that relaxation and polarization contributions to the transition moment arise only in second and higher order. An analysis of A_{vc} is, however, rather tedious and we shall not discuss them here.

5. APPLICATIONS

In this section we will discuss the performance of the RCHF method in the calculation of K-shell photoionization cross sections. As examples we shall consider the molecules CO, H_2CO, and C_2H_4. In the case of CO an ample description of the computational procedures has been given in I, to which the reader is referred for any details. In I the spin-free working equations for the cross section and the asymmetry parameter β are also given. The studies of H_2CO and C_2H_4 are new and have not been described elsewhere. Largely, these calculations are similar to those for CO, and it may suffice here to specify briefly the SCF calculations performed for the neutral molecules and the cationic $1s$-hole states. A new feature of the calculations for H_2CO and C_2H_4 is that, whereas for linear systems m must always be equal to m' in (38d), i.e., the projection of angular momentum along the intermolecular axis is a good quantum number, in a nonlinear polyatomic molecule m' need not be equal to m. The numerical procedures required here have been described at some detail elsewhere.[36]

For formaldehyde we used Dunning's[37] ($10s6p$) basis contracted to $[6s4p]$ on the carbon and the oxygen. This basis set was supplemented by each two d functions with exponents 0.75 and 0.25 on C, and 0.85 and 0.225 on O, respectively. For ethylene the $[6s4p]$ contraction of Dunning was augmented by two s functions with exponents 0.00125, 0.0045 and by two p functions with exponents 0.00125 and 0.0035 at the center of the molecule. Dunning's ($4s$) basis contracted to $[3s]$ was used for the hydrogen atoms. The resulting SCF ground-state energies are −113.8972 and −78.0466 a.u. for H_2CO and C_2H_4, respectively. The following energies were obtained for the C$1s$-hole states: −103.0488 a.u. (H_2CO), −67.0372 a.u. ($C_2H_4/1a_g^{-1}$), −67.0344 a.u. ($C_2H_4/1b_u^{-1}$).

5.1. C1s Photoionization in CO

The K-shell ionization in the prototypical diatomic molecule CO has been investigated extensively using electron energy-loss spectroscopy (EELS)[38–40] and photoabsorption spectroscopy with synchrotron radiation.[41–43] While these techniques provide the total (photo)absorption cross sections, more recently the $1s$ photoelectron intensities as a function of photon energy have also been measured.[44,45] Several electron volts above threshold, the experimental spectra show a broad peak, which has been widely interpreted as the σ^* shape resonance. From the molecular orbital point of view, this resonance is associated with the excitation of a K-shell electron into the antibonding σ^* orbital formed by the linear combination of the $2p_z$ atomic orbitals (oriented along the molecular axis). As is well known, a similar excitation to the virtual π^* orbital gives rise to a prominent absorption peak below the $1s$ ionization threshold.

The FCHF and RCHF result for the C$1s$ photoionization cross section in CO are shown in Figure 1, together with experimental data of Kay et al.[40] and of Schmidbauer et al.[45] The two calculated cross sections are seen to differ considerably with respect to the position, height, and width of the σ^* shape resonance. The relatively narrow and large "frozen" peak centered at about 5 eV above the C$1s$ ionization threshold at 296.2 eV turns into a considerably lower and broader resonance feature, shifted by 6.2 eV toward higher energy when valence electron relaxation is taken into account. As discussed in Section 4, these differences between the frozen and relaxed cross sections are mainly the result of screening of the $1s$ hole potential by the relaxed valence charge distribution which makes the potential experienced by the photoelectron less attractive than in the case of a frozen ionic core. As a further, but less important, relaxation effect, the nonuniform relaxation energies in the ionic $1s$-hole state and the neutral $1s$-excited state contribute to the enhancement of the resonance energy. For the bound

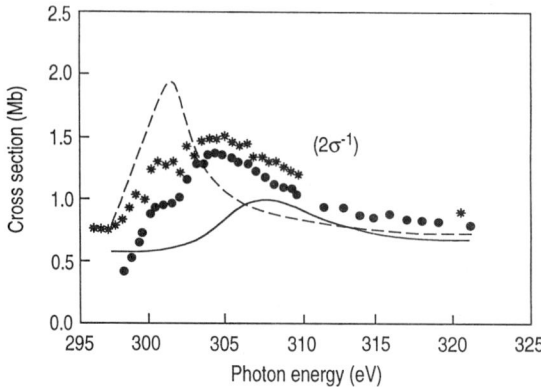

FIGURE 1. Photoionization cross section leading to the $2\sigma^{-1}(C1s^{-1})$ state of CO^+. ---, present results (length) in the FCHF approximation; ——, present results (length) in the RCHF approximation; *, experimental results of Ref. 40; ●, experimental results of Ref. 45.

$C1s$–π^* excitation below threshold this contribution can be estimated using equation (59) to be of the order of 0.5–1.0 eV. An even smaller shift will be expected for the position of the σ^* shape resonance.

It should be noted that the different magnitudes of the frozen and relaxed cross sections result to some extent from the overall spectroscopic factor

$$|x_{1s}|^2 = |\langle \bar{\phi}_0^N | \phi_0^N \rangle|^2$$

appearing in the RCHF description [see equation (14)]. The deviation of this quantity from 1 is a measure of the relative spectral strength diverted from the single-hole state to shake-up and shake-off satellites. The present RCHF calculations give a value of $|x_{1s}|^2 = 0.79$ which is somewhat larger than the value of 0.68 of a previous many-body Green's functions study.[46] In the FCHF model the spectroscopic factor is strictly 1, which means that shake-up and shake-off states have vanishing intensity in this approximation.

As seen in Figure 1, the σ^* resonance is at about 9 eV above threshold in the measured spectrum, 4 eV higher and 2.2 eV lower than the frozen-core and the relaxed-core result, respectively. For the FCHF calculation the discrepancy between theory and experiment clearly reflects the absence of relaxation. The overshooting of the relaxed-core result, on the other hand, seems to be primarily caused by the neglect of core polarization by the excited electron. As was discussed in Section 4.2, the inclusion of the attractive polarization potential would lower the position and increase the height of the resonance peak. To what extent dynamic effects, which are not included in the present static description of the relaxation, also play a role is an open question (see Section 6).

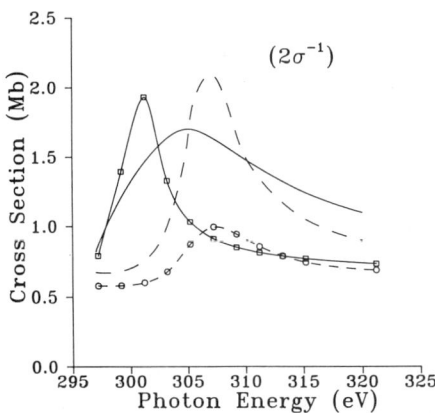

FIGURE 2. Photoionization cross section leading to the $2\sigma^{-1}(C1s^{-1})$ state of CO^+. □, present results (length) in the FCHF approximation; ○, present results (length) in the RCHF approximation; ——, calculated FCHF results of Ref. 6 using the STMT method; ---, calculated results of Ref. 47 using the MS method.

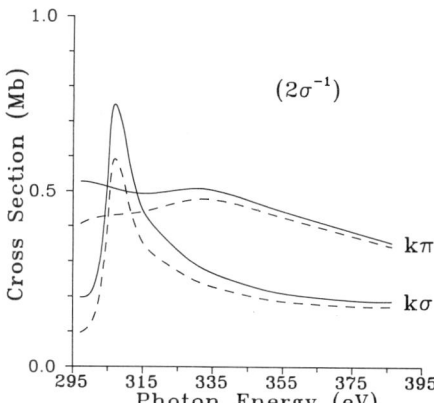

FIGURE 3. Calculated partial channel cross section for photoionization from the $2\sigma(C1s)$ orbital of CO in the RCHF approximation. ———, direct plus conjugate cross section as per equation (14); – – –, only direct cross section using the first term of equation (14). Note that these cross sections omit the overall spectroscopic factor.

In Figure 2 we compare our calculated frozen- and relaxed-core photoionization cross sections with previous theoretical results obtained by the Stieltjes–Tchebycheff moment theory[6] (STMT) and the multiple scattering (MS) method.[47] All calculations assume a frozen valence charge distribution and hence should provide a consistent set of cross sections. Nevertheless, substantial discrepancies are seen: The STMT cross section shows a very broad resonance with a maximum 9 eV above threshold in good agreement with the experimental data of Kay et al.[40]; in the MS calculation, the shape of the resonance peak resembles that of our FCHF curve, but the position of the maximum is shifted to even higher energy, almost coinciding with our relaxed-core result. According to the discussion in Section 4.2, it seems unrealistic to expect the σ^* resonance at or even above the experimental positions without taking the screening of the $1s$-hole potential resulting from relaxation into account. Similar discrepancies between different FCHF calculations have been noticed in the case of $N1s$ ionization in N_2.[5,23]

The decomposition of the integral $C1s$ cross section into the $k\sigma$ and $k\pi$ contributions shown in Figure 3 shows clearly that, as expected, only the $k\sigma$ channel is resonant. Figure 3 also gives plots of the "direct" $k\sigma$ and $k\pi$ cross sections obtained by omitting the "nonorthogonality" (overlap) part of the amplitude (14). In both subchannels we see a substantial enhancement of the cross sections at low photon energies as a result of inclusion of these overlap contributions. At threshold the full $k\sigma$ cross section is about twice as large as the direct contribution, while at the resonance maximum the nonorthogonality enhancement is still about 30%. For the $k\pi$ channel, the absolute enhancement at threshold is even larger than in the $k\sigma$ channel, the relative increase being 30%. Since the bound–free overlap integrals vanish more rapidly than the corresponding dipole integrals, the direct curves approach the full curves at larger energies.

Figure 4 shows the photoelectron asymmetry parameter β calculated in the frozen- and relaxed-core models, together with recent experimental data of Schmidbauer et al.[45] Both calculated curves show a rapid change of the β parameter in the vicinity of the σ^* resonance energy, and a distinct minimum at lower energy. Besides the shift in the position of the turning point, the relaxed curve assumes a somewhat higher β value at the minimum and a lower β value at higher energy than the frozen one. Except for a horizontal shift of 2–3 eV, the relaxed curve is in rather good agreement with the experimental data of Schmidbauer et al.[45] The $C1s$ photoelectron asymmetry parameters for CO have been calculated previously by Dill et al.[48] and by Grimm[49] using the MS model. Both MS β-curves exhibit the typical resonance structure seen also in the present results. There are differences in the minimum values and the resonance positions. The minimum β value in both MS calculations is larger than 0.2, while the present RCHF curve drops to –0.2. In Grimm's calculation, the resonance position derived from the turning point of the β-curve is in excellent agreement with the experimental value. The latter calculation was based on a "transition state" potential in which one-half of an electron

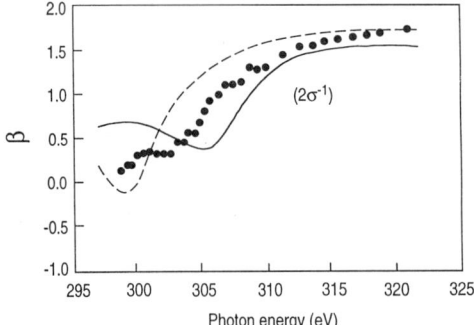

FIGURE 4. Photoelectron asymmetry parameters for photoionization leading to the $2\sigma^{-1}(C1s^{-1})$ state of CO^+. ---, present results (length) in the FCHF approximation; ——, present results (length) in the RCHF approximation; ●, experimental results of Ref. 45.

was removed from the K-shell. The resulting "half-relaxed" valence charge distribution leads to a correspondingly reduced screening of the $1s$-hole potential. In spite of the usually better agreement with experiment, the physical justification for this procedure is questionable, since the essential shortcoming of the RCHF model is not "overscreening" but neglect of target polarization.

5.2. O1s Photoionization in CO

In Figure 5 the $O1s$ photoionization cross sections calculated both in the frozen- and relaxed-core approximations are shown, along with experimental results of Barrus et al.[50] (photoabsorption) and Truesdale et al.,[44] and previous theoretical results[6] obtained using the FCHF model and the STMT method. The relaxation effect in the $O1s$ ionization is considerably larger than in the $C1s$ case discussed above, as reflected by the relaxation energies of 13 and 21 eV of the $C1s$- and $O1s$-hole states, respectively. Correspondingly, we expect a larger relaxation-induced screening effect for the $O1s$-hole potential. Figure 5 shows indeed a quite dramatic difference between the frozen and the relaxed curve. The peak height drops from 1.7 Mb in the frozen-core calculation to one-third of this value on inclusion of relaxation. The peak maximum, being only 2.5 eV above threshold in the frozen-core cross section, is shifted by 10.5 eV to a value of 13 eV above threshold in the relaxed curve. Thus, the shift is about 4 eV larger than in the $C1s$ cross section. Moreover, the relaxed $O1s$ resonance peak is substantially

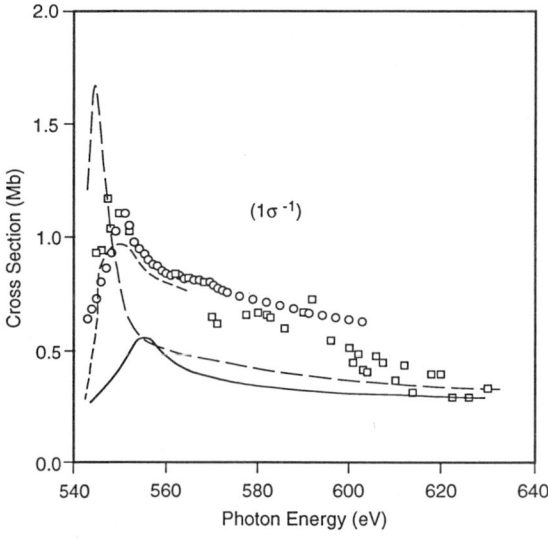

FIGURE 5. Photoionization cross section leading to the $1\sigma^{-1}(O1s^{-1})$ state of CO^+. ---, present results (length) in the FCHF approximation; ——, present results (length) in the RCHF approximation; -----, calculated FCHF results of Ref. 6 using the STMT method; ○, experimental results (photoabsorption) of Ref. 50; □, experimental results of Ref. 44.

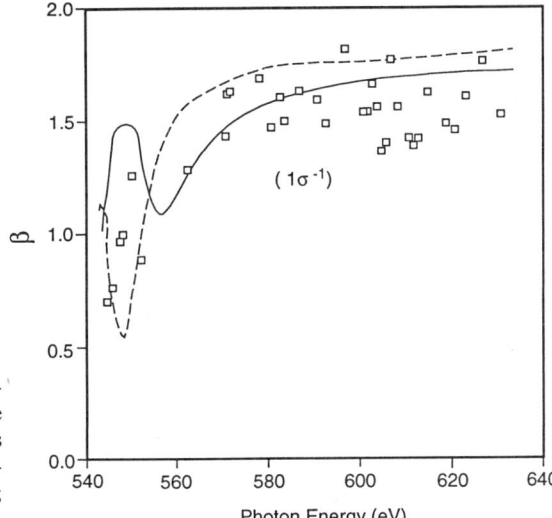

FIGURE 6. Photoelectron asymmetry parameters for photoionization leading to the $1\sigma^{-1}(O1s^{-1})$ state of CO^+. – – –, present results (length) in the FCHF approximation; ———, present results (length) in the RCHF approximation; □, experimental results of Ref. 44.

broader than its $C1s$ counterpart. The overall spectroscopic factor $|x_{1\sigma}|^2$ [see equation (14)] contained in the relaxed-core cross section was found to be 0.74, which may be compared with a previous Green's functions result[46] of 0.61.

The σ^* resonance position of about 7.5 eV in the experimental $O1s$ spectrum, that is, 1.5 eV lower than in the $C1s$ spectrum, lies almost in the middle of the frozen- and relaxed-core results of 2.5 and 13 eV, respectively. This results in an overshooting of the relaxed-core resonance position by more than 5 eV. While we may expect that inclusion of polarization in the RCHF potential will lead to a better agreement with experiment, this discrepancy seems to be too large to be accounted for only by polarization. In view of the large width of the resonance it might be that dynamic effects play a role here. The $O1s$ cross section calculated by Padial et al.[6] in the FCHF approximation using the STMT method is again at variance with our frozen-core calculation.

The frozen and relaxed photoelectron asymmetry parameters for $O1s$ ionization are shown in Figure 6. The relaxed curve exhibits an interesting oscillation below the shape resonance which is absent in the frozen curve. A similar oscillatory behavior was predicted by the MS calculations of Dill et al.[48] and Grimm.[49] The experimental data of Truesdale et al.[44] are somewhat scarce and scattered, making the comparison with the theoretical curves inconclusive.

5.3. $C1s$ Photoionization in H_2CO

In its $^1A_1$ ground state, the formaldehyde molecule (C_{2v} symmetry group) has the electron configuration $(O1s)^2(C1s)^2(3a_1)^2(4a_1)^2(1b_2)^2(5a_1)^2(1b_1)^2(2b_2)^2$ assuming the molecule in the yz-plane with the C–O bond along the z-axis. Here $1b_1$ is the out-of-plane π orbital, and $2b_2$ is the oxygen lone-pair orbital (n). Minimal basis set considerations yield four valence-type orbitals in the virtual space which may lead to resonances in the $C1s$ excitation spectrum: a π^* orbital in b_1 symmetry, a C–H antibonding orbital of b_2 symmetry, and two orbitals of a_1 symmetry which may be characterized further as σ^* (C–H) and σ^* (C–O) orbitals. The experimental spectrum of Hitchcock and Brion[51] puts the $C1s$–π^* excitation below the ionization threshold. Also much of the b_2 and a_1(C–H) virtual valence character seems to be contained in the discrete excitation below threshold.[52] Thus, only the a_1(C–O) orbital remains as a possible candidate for a shape resonance in the $C1s$ cross section.

Let us first consider Figure 7 which shows separately the relaxed-core results for the ka_1, kb_1, and kb_2 partial subchannels contributing to the full $C1s$ cross section. We observe a resonance in both the ka_1 and kb_2 channel, while the kb_1 channel is nonresonant. The kb_2 resonance predicted slightly above the $C1s$ threshold (294.5 eV) is absent in the frozen-core result, because here the more attractive ionic potential shifts the corresponding transition below threshold. The same effect must be

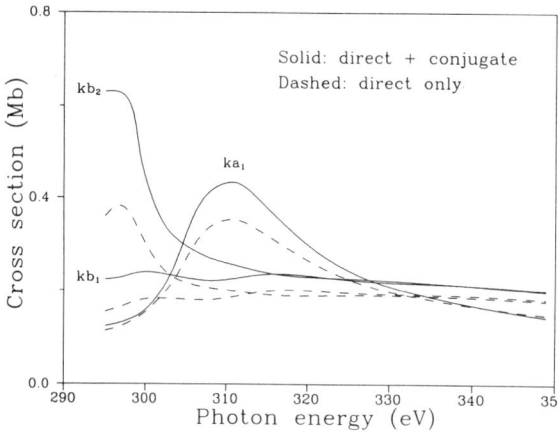

FIGURE 7. Partial channel photoionization cross sections leading to the $2a^{-1}(C1s^{-1})$ state of H_2CO^+ in the RCHF approximation. ———, direct plus conjugate cross section as per equation (14); — — —, only direct cross section using the first term of equation (14).

expected if the attractive polarization interaction is taken into account. This means that the resonance feature in the kb_2 cross section is probably an artifact of the RCHF model. With a more appropriate potential, this resonance would turn into a discrete excitation below threshold.

In the ka_1 channel we find a very broad resonance peak with a maximum at 311 eV, that is, 16.5 eV above threshold. We associate this structure with the expected σ^*(C–O) resonance. The ka_1 resonance feature changes dramatically in the frozen- and relaxed-core treatment (see Figure 8a). The resonance positions differ by about 14 eV, that is, more than twice the shift encountered in the case of $C1s$ ionization in CO. The height of the maximum drops from 1.2 Mb in the frozen-core description to 0.4 Mb in the relaxed-core calculation, while the width increases from about 5 eV to 15 eV. The reason for this huge relaxation effect seems to be a very effective intramolecular screening mechanism. A Mulliken population analysis of the ground state shows that both the π and n orbital are centered on the oxygen atom. This localization is particularly strong for the oxygen lone-pair orbital n. The creation of a hole in the $C1s$ orbital causes a substantial intramolecular charge transfer from the O atom to the C atom which in turn results in very efficient screening of the $C1s$ hole potential. The same effect was found to be responsible for the very low shake-up energies of the π–π^* and n–π^* satellites in the $C1s$ photoelectron spectrum.[53]

In Figure 8a we compare the frozen- and relaxed-core results for the full $C1s$ photoionization cross section with recent experimental data of Kilcoyne et al.[54] These data have been scaled to our RCHF results at high energy. The spectroscopic factor used in the RCHF cross section is $|x_{1s}|^2 = 0.83$, which may be compared to a value of 0.65 resulting from Green's function calculations.[53] As discussed above, the relaxed curve shows a spurious kb_2 resonance near threshold. The second peak in the relaxed curve as well as the prominent frozen-core peak are associated with the σ^*(C–O) shape resonance. The experimental spectra[51,54] show a corresponding peak at 301 eV. This means that the resonance position predicted by the relaxed-core model is 10 eV above the experimental value. The more recent data of Kilcoyne et al.[54] seem to be compatible with the view that at 301 eV autoionizing (double excitation) structure is superimposed on a broad shape resonance with a maximum at about 304 eV. Even under this more favorable assumption, the discrepancy between the calculated and observed σ^* resonance position would amount to 7 eV which appears too large to be explained by the neglect of polarization. As discussed in Section 4.2, however, a large relaxation screening, as in the present case, also indicates the presence of a large polarization effect. Using the inequality (70) with the values $S = |R| = 14$ eV deduced from the calculated relaxation shift of the resonance position and the relaxation energy of the ionic $C1s$-hole state,[53] respectively, one concludes that the polarization shift must exceed 3.5 eV. For comparison, similar estimates in the case of $C1s$ and $O1s$ ionization in CO give lower bounds for the polarization shift of 0.7 and 1.3 eV, respectively. These considerations show that a substantial polarization correction must be expected in the $C1s$ cross

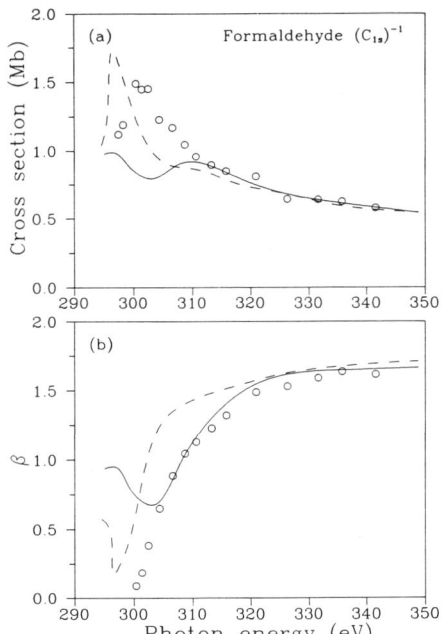

FIGURE 8. (a) Photoionization cross section leading to the $2a_1^{-1}(C1s^{-1})$ state of H_2CO^+. ---, present results (length) in the FCHF approximation; ——, present results (length) in the RCHF approximation, ○, experimental results of Ref. 54. (b) Photoelectron asymmetry parameters for the $(C1s^{-1})$ state of H_2CO^+. Symbols as in (a).

section in H_2CO. Of course, this may not be the only source of discrepancies between the RCHF result and experiment.

Figure 8b shows frozen- and relaxed-core results for the photoelectron asymmetry parameter β, together with experimental data of Kilcoyne et al.[54] For energies above 310 eV the relaxed curve agrees quite well with experiment. At lower energy, however, the theoretical curve fails to follow the data points to the small β values. Apart from a horizontal shift, the frozen-core result gives a better description here.

5.4. C1s Photoionization in Ethylene

A specific feature in the K-shell ionization of ethylene is the presence of two equivalent core levels. As discussed in Section 2.3, one may opt in such a case between a localized and a delocalized RCHF treatment, differing in the representation of the K-shell orbitals. In both ways the main relaxation effect, that is, the screening of the ionic core, is described correctly (see Section 4.2). Here we have used the delocalized approach formulated in terms of the symmetry-adapted core orbitals.

We begin with an overview of the possible resonances in the $C1s$ photoionization cross section. In the ground state ethylene is planar (D_{2h}) and has the following electron configuration:

$$^1A_g(1a_g)^2(1b_{1u})^2(2a_g)^2(2b_{1u})^2(1b_{2u})^2(3a_g)^2(1b_{3g})^2(1b_{3u})^2$$

Here we take the molecule to lie in the yz-plane with the C–C bond along the z-axis. The $1a_g$ and $1b_{1u}$ orbitals represent the symmetric and antisymmetric linear combinations of the two $C1s$ atomic orbitals with almost degenerate orbital energies (0.08 eV splitting). $1b_{3u}$ is the occupied π orbital formed essentially from the symmetric linear combination of the (out-of-plane) $C2p_x$ atomic orbitals. Its antisymmetric counterpart is the antibonding π^* orbital of b_{2g} symmetry. As is well known from earlier experimental[55] and theoretical[56] studies, the $C1S$-π^* excitation gives rise to a prominent band in the absorption spectrum below threshold. The four in-plane $C2p$ orbitals ($2p_y$, $2p_z$) and the four hydrogen $1s$ orbitals may be combined to form each two molecular orbitals with symmetry a_g, b_{1u}, b_{2u}, and b_{3g}, respectively. Three of these orbitals, namely, $1b_{2u}(2p_y + 2p_y)$, $3a_g(2p_z - 2p_z)$, and $1b_{3g}(2p_y - 2p_y)$, are occupied orbitals. Thus, there remain five valence-type (σ^*) orbitals in the virtual spectrum which may

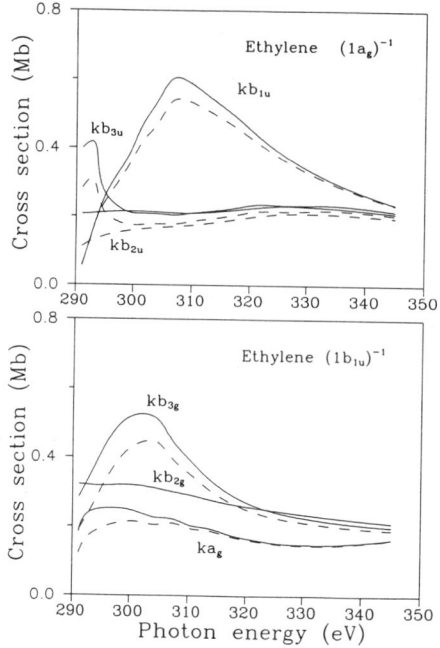

FIGURE 9. Partial channel photoionization cross sections leading to the $1a_g^{-1}$ state (upper panel) and to the $1b_{1u}^{-1}$ state (lower panel) of $C_2H_4^+$ in the RCHF approximation. Symbols as in Figure 7.

give rise to resonances in the C1s excitation spectrum. More specifically, there is one σ^* orbital in each of the symmetry species a_g, b_{2u}, and b_{3g}, respectively, and two σ^* orbitals of b_{1u} symmetry. The two energetically lowest orbitals $\sigma^*(b_{2g})$ and $\sigma^*(a_g)$ are characterized as C–C bonding and C–H antibonding, whereas the $\sigma^*(b_{3g})$ and the two $\sigma^*(b_{1u})$ orbitals are antibonding along both the C–C and the C–H bonds. The antibonding character of the two $\sigma^*(b_{1u})$ is, however, differently proportioned. In the energetically lower orbital the C–H antibonding character is particularly pronounced, whereas the energetically higher orbital is dominantly C–C antibonding. In the (minimal basis) FCHF approximation the term values of $\sigma^*(b_{2u})$, $\sigma^*(a_g)$, and $\sigma^*(b_{1u}\text{C–H})$ are negative, suggesting that the corresponding transitions will lead to discrete excitations below threshold.

The partial channel cross sections calculated in the relaxed-core approximation are presented in Figures 9a and b. Under dipole selection rules the gerade (g) continua ka_g, kb_{2g}, kb_{3g} (Figure 9b) and the ungerade (u) continua kb_{1u}, kb_{2u}, kb_{3u} (Figure 9b) are associated with the $1b_{1u}^{-1}$ and $1a_g^{-1}$ core holes, respectively. Let us first consider the $u \rightarrow g$ transitions. As expected, both the kb_{2g} and the ka_g partial channels are nonresonant. In the kb_{3g} cross section, we see a distinct resonance peak centered 11 eV above threshold (290.8 eV$^{(57)}$). Obviously, this feature is to be assigned to the C1s–$\sigma^*(b_{3g})$ transition predicted by the minimal basis MO model. Without relaxation, this resonance is found at 8 eV lower energy, close to threshold. The relatively large screening energy shift implies a correspondingly large polarization correction. The actual kb_{3g} resonance position will be expected somewhere between the frozen- and relaxed-core results. Weak features have been observed in the EELS spectra$^{(55,58)}$ at 1.8 and 5.0 eV, respectively, which were earlier attributed to double excitations. Possibly, the feature at higher energy is more adequately described as a kb_{3g} shape resonance.

Let us now turn to the three kb_u cross sections shown in Figure 9a. The most conspicuous feature is seen in the kb_{1u} channel where a $\sigma^*(\text{C–C})$ resonance is expected. Here the cross section rises steeply from a very low value at threshold to a maximum of 0.6 Mb at 307 eV (or 16 eV above threshold) and then steadily declines over the next 40 eV to one-third of its maximal value. The width of this almost triangular-shaped structure is in the range of 30 eV. An analysis of the partial waves contributing to the kb_{1u} cross section shows that the $l = 1$ component is dominant at lower energies, while $l = 3$ becomes the leading partial wave toward higher energy. The corresponding frozen-core result (see the integral FCHF curve in Figure 10) exhibits a more usual behavior. Here a strong resonance feature is centered

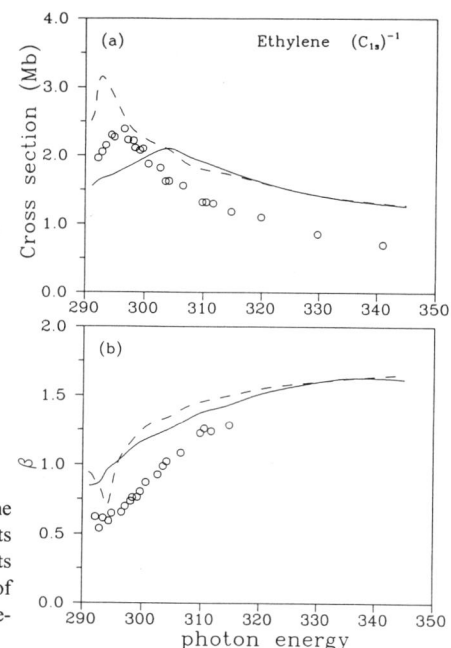

FIGURE 10. (a) Photoionization cross section leading to the $1a_g^{-1}$ plus $1b_{1u}^{-1}$, $(C1s^{-1})$ states of $C_2H_4^+$. ——, present results (length) in the FCHF approximation; ———, present results (length) in the RCHF approximation; ○, experimental results of Ref. 54. (b) Photoelectron asymmetry parameters for the corresponding photoionization channels of (a). Symbols as in (a).

3 eV above threshold. This feature shows a steep rise above threshold followed by a wide decaying slope with a full width at half maximum (FWHM) of about 17 eV. Recently, the $1a_g$–kb_{1u} channel cross section of C_2H_4 has been calculated by Farren et al.[59] using the STMT method. Their curve shows a single relatively broad σ^*(C–C) resonance peak with a maximum of 0.8 Mb at 7 eV above threshold. Qualitatively, these results agree with our frozen-core picture, but differ in detail. The resonance positions differ by about 4 eV, and the STMT curve has a distinctly larger value (0.6 Mb) at threshold.

Finally, referring to the kb_{2u} partial cross section in Figure 9b, a weak and narrow resonance feature is seen close to threshold. This resonance will turn into a discrete excitation below threshold for a slightly more attractive potential, e.g., by taking the polarization effects into account. The corresponding frozen-core cross section is clearly nonresonant here.

The integral $C1s$ photoionization cross section calculated in both the frozen- and the relaxed-core approximation is shown in Figure 10a, along with recent experimental data of Kilcoyne et al.[54] In the relaxed-core calculation, the kb_{2g} resonance (at 302 eV) and the kb_{1u} peak (at 307 eV) are merged together into a broad feature exhibiting a shallow maximum at 305 eV. The weak spurious resonance in the kb_{2u} partial channel is reflected by the small shoulder near threshold. In the frozen-core cross section, the joint resonance feature is distinctly more pronounced and shifted close to threshold. As in the previous cases, we may expect that the exact cross section will behave between the frozen and relaxed descriptions given here. In Figure 10a we also compare the theoretical results with recent experimental data for partial $C1s$ photoionization cross sections obtained by Kilcoyne et al. These data have been scaled to represent 74% of the total absorption cross section at 315 eV, as recorded by Hitchcock and Brion[55] and McLaren et al.[58] using EELS. An earlier measurement of the partial $C1s$ ionization cross section has been reported by Piancastelli et al.[60] A broad resonance enhancement is seen in all experimental cross sections, most pronounced in the partial $C1s$ data. However, the reported maximum positions differ considerably, ranging from 5 eV (Kilcoyne et al.) to 9 eV (Piancastelli et al.) and 10 eV (EELS) above threshold. These values lie within the bounds given by the FCHF and RCHF results. The discrepancy between the experimental cross section and the RCHF result may be partly due to the somewhat too large spectroscopic factor ($|x_{1s}| = 0.93$) obtained in the delocalized HF calculations. A value of $|x_{1s}|^2 = 0.77$ is predicted by Green's function calculations similar to those of [46].

The photoelectron asymmetry parameter β as a function of photon energy is depicted in Figure 10b. This quantity is calculated according to $\beta = (\sigma_g \beta_g + \sigma_u \beta_u)/(\sigma_g + \sigma_u)$ from the partial parameters β_g and β_u associated with the $1a_g$ and $1b_{1u}$ ionization, respectively. A typical resonance behavior is seen only in the frozen-core β-curve. The relaxed-core result, on the other hand, is monotonically growing with increasing photon energy. The experimental data points of Kilcoyne et al.,[54] inserted in Figure 10b, seem to agree better with the relaxed than the frozen theoretical curve.

6. GENERAL DISCUSSION AND CONCLUSIONS

The RCHF model introduces several approximations of different degrees of relevance into the calculation of the photoionization cross sections. To better understand the observed discrepancies between theory and experiment we will now briefly reiterate the underlying assumptions.

First, the RCHF method is based on a single-particle representation of the initial and the final N-electron state. This means that no correlation effects are considered. In particular, a single-channel approximation is assumed for the final excited state, which precludes, for example, inclusion of any effects related to interchannel coupling. As a consequence, autoionization or Auger decay of the 1s-hole states is beyond the scope of the RCHF scheme. Likewise, double excitations (representing Feshbach resonances) in the K-shell photosbsorption spectra cannot be accounted for. Although the neglect of correlation prevents one from studying some important processes, the impact of this approximation on the single-channel features, in particular, on the shape resonances, will be rather modest.

Of greater importance are the approximations made within the single-channel description. Here we may ask to what extent the chosen static electron–ion potential is appropriate, and, more generally, whether a static potential is appropriate at all. In Section 4.2, the limitations of the relaxed-core potential have been analyzed from a static point of view, by comparing the exact and RCHF term values for discrete core-level excitations through second-order of perturbation theory. This showed that the relaxation effects, consisting of the residual relaxation shift and the screening of the core hole, are correctly described. What is not accounted for is the polarization of the ionic core by the excited electron giving rise to an attractive force. The neglect of polarization broadens the resonance features in the RCHF cross section and shifts their positions toward higher energy. As emphasized in Section 4.2, the "relaxation" screening (in second order) can be viewed as arising from an interplay of relaxation and polarization contributions. The polarization energy shift was shown to be bounded from below by a bound which itself increases with the square of the screening energy. Thus, a substantial polarization correction has to be expected in cases of large screening energy.

Is a static approximation appropriate at all? Actually, both the valence electron relaxation and polarization in the photoionization process are dynamic effects, and the dynamic aspects are suppressed in any static potential model. Within the framework of the single-channel approximation, a dynamic description may be achieved by using a properly chosen energy-dependent (optical) potential for the electron–ion interaction. Optical potentials have been used successfully in electron–molecule scattering[61–64] providing an adequate description of the (dynamic) target polarization. In the field of photoionization, optical potentials may, in principle, be derived as specially devised approximations for the polarization propagator. Presently, however, no practical optical potential method seems to be available.

The error introduced by neglecting the dynamic aspects of relaxation and polarization will be small for a narrow, that is, long-lived, resonance. Obviously, here the static discrete state picture assumed in Section 4.2 is applicable to a good approximation. However, the shape resonances encountered in the core-level ionization spectra are certainly not of this type, and we must expect substantial discrepancies between the static description and the exact results. The present calculations showed that the resonance positions obtained in the relaxed static model were distinctly larger than the experimental values. We expect that the dynamically screened electron–ion potential will be more attractive than the fully relaxed static potential. A corresponding optical potential including polarization should lead to satisfactory agreement between theory and experiment.

In spite of the shortcomings of the static models, valuable qualitative information can be gained from the RCHF and the FCHF calculations which may be viewed as limiting cases bracketing the true result. As the examples in Section 5 have shown, the unrelaxed ionic core potential of the FCHF method is too attractive, whereas the relaxed core potential is, in general, not attractive enough. With respect to resonance positions, we observed shifts ranging from 6 eV in CO($C1s$) to 14 eV in H$_2$CO($C1s$). In some cases, low-lying resonances in the RCHF cross sections appeared as discrete excitations below threshold in the FCHF description. Also the shapes of the resonances differed distinctly, being much smaller and broader in the relaxed- than in the frozen-core description. It should be remembered that to some extent the reduction of the RCHF cross section is related to the spectroscopic factor x_{1s}^2 diverting spectral strength from the $1s$ single-hole channel to shake-up satellites.

For the purpose of assigning and characterizing shape resonances, the minimal basis set MO model proved very useful. For the examples studied in Section 5, its predictions were confirmed by our calculations. In the case of H$_2$CO we found that only one σ^* orbital (a_1(C–O)) gives rise to a resonance, while the valence character of two further σ^* orbitals appears in discrete excitations below threshold. Similarly, all five σ^* orbitals in C$_2$H$_4$ give rise to two resonances and three discrete below-threshold excitations. It should be mentioned that the RCHF model can easily be extended to calculate term values and oscillator strengths of discrete core-level excitation.[65]

The effects of nuclear motion have not been considered in the present study. All calculations were performed at the equilibrium geometry for the molecular ground state leading to so-called vertical cross sections. One may go beyond this static nuclear picture by calculating the N-electron transition moments and cross sections as a function of the nuclear coordinates. The transition moment functions, for example, allow one to determine (within the Born–Oppenheimer approximation) the vibrational structure of the final ionic state at a given photon energy. As is known from valence electron ionization, substantial deviations from the simple Franck–Condon picture have to be expected in the vicinity of resonances. A further important aspect of such calculations is to consider the variation of resonance positions with the change of selected bond lengths or angles. The specific antibonding character of the associated σ^* orbital will eventually be reflected by a distinct variation pattern, supporting the assignment and characterization of the resonance under consideration. An interesting dynamical situation may occur if a resonance approaches threshold and becomes a discrete excitation as a function of some nuclear displacements.

According to the present formulation of the RCHF model, the N-electron transition moment is written as the sum of a direct part containing the bound–free dipole integral and a conjugate (or nonorthogonality) part combining bound–free overlap and bound–bound dipole integrals. The latter part, arising from the use of different orbital sets in the initial and final state, modifies the bound–free dipole integral of the direct part, essentially by projecting out any occupied frozen orbital admixtures from the relaxed continuum orbital. Throughout, the conjugate part leads to an enhancement of the cross sections, which could become quite substantial at low photoelectron energy and in resonance regions. Toward higher energy, the direct and the full cross section converge rapidly, as the bound-free overlap integrals vanish.

In the present calculations the length form of the transition moment has been used throughout. While also the velocity form easily could have been used, we expect only small differences between these formulations. As is well known, the length and velocity form are equivalent in the random phase approximation (RPA) if a complete one-particle basis is used. For core-level excitations, the RPA reduces to the single-channel Tamm–Dancoff approximation (TDA) which is equivalent with the FCHF scheme. Here we assume the validity of the CVS approximation described in Section 2.2. As shown in Section 4.3, both the FCHF and RCHF transition moments are consistent through first order of perturbation theory. Differences appear only in second and higher order, which should not affect very much the agreement between the length and velocity results.

In conclusion, we hope to have demonstrated that the RCHF model is a simple and useful means to obtain at least a qualitative description of molecular core-level ionization. Though in a static and thus somewhat exaggerated way, the present applications have made apparent the paramount importance of valence electron relaxation here. Obviously, however, the RCHF approximation must be improved before quantitatively satisfactory results can also be expected. What is needed here is to go beyond the static picture and to take the polarization effect into account. These extensions may still be introduced within the framework of the single-channel approximation, but will require the use of appropriate optical potentials.

ACKNOWLEDGMENTS. Work at the California Institute of Technology was supported by the National Science Foundation. One of the authors (J.S.) acknowledges funding by the Deutsche Forschungsgemeinschaft.

REFERENCES

1. For example, see H. SIEGBAHN AND L. KARLSSON, in *Handbuch der Physik*, Vol. 31, edited by W. Mehlhorn (Springer, Berlin, 1982).
2. L. C. SNYDER, *J. Chem. Phys.* **55**, 95 (1971).
3. U. GELIUS, *Phys. Scr.* **9**, 133 (1974).
4. J. L. DEHMER AND D. DILL, *Phys. Rev. Lett.* **35**, 213 (1975); *J. Chem. Phys.* **65**, 5327 (1976).
5. T. N. RESCIGNO AND P. W. LANGHOFF, *Chem. Phys. Lett.* **51**, 65 (1977).
6. N. PADIAL, G. CSANAK, B. V. McKOY, AND P. W. LANGHOFF, *J. Chem. Phys.* **69**, 2992 (1978).
7. N. PADIAL, G. CSANAK, B. V. McKOY, AND P. W. LANGHOFF, *Phys. Rev. A* **23**, 218 (1981).
8. W. R. DAASCH, E. R. DAVIDSON, AND A. U. HAZI, *J. Chem. Phys.* **76**, 6031 (1982).
9. R. R. LUCCHESE AND V. McKOY, *Phys. Rev. A* **26**, 1406 (1982).
10. D. LYNCH, M.-T. LEE, R. R. LUCCHESE, AND V. McKOY, *J. Chem. Phys.* **80**, 1907 (1984).
11. J. A. STEPHENS, D. DILL, AND J. L. DEHMER, *J. Chem. Phys.* **84**, 3638 (1986).
12. For example, see J. L. DEHMER, A. C. PARR, AND S. H. SOUTHWORTH, in *Handbook on Synchrotron Radiation*, Vol. 2, edited by G. V. Marr (North-Holland, Amsterdam, 1987).
13. R. R. LUCCHESE, K. TAKATSUKA, AND V. McKOY, *Phys. Rep.* **131**, 147 (1986).
14. A. F. STARACE, in *Handbuch der Physik*, Vol. 31, edited by W. Mehlhorn (Springer, Berlin, 1982).
15. M. Y. AMUSIA, V. K. IVANOV, S. A. SHEINERMAN, S. I. SHEFTEL', AND A. F. IOFFE, *Zh. Eksp. Teor. Fiz.* **78**, 910 (1980) [*Sov. Phys. JETP* **51**, 458 (1980)].
16. M. Y. AMUSIA, *Adv. At. Mol. Phys.* **17**, 1 (1981).
17. H. P. KELLY, S. L. CARTER, AND B. E. NORUM, *Phys. Rev. A* **25**, 2052 (1982).
18. Z. ALTUN, M. KUTZNER, AND H. P. KELLY, *Phys. Rev. A* **37**, 4671 (1988).
19. F. P. LARKINS, P. D. ADENEY, AND K. G. DYALL, *J. Electron Spectrosc. Relat. Phenom.* **22**, 141 (1981).
20. G. B. ARMEN, B. I. CRAIG, F. P. LARKINS, AND J. A. RICHARDS, *J. Electron Spectrosc. Relat. Phenom.* **51**, 183 (1990).
21. J. TULKKI AND T. ÅBERG, *J. Phys. B* **18**, L489 (1985).
22. H. P. SAHA, *Phys. Rev. A* **42**, 6507 (1990).
23. D. L. LYNCH AND V. McKOY, *Phys. Rev. A* **30**, 1561 (1984).
24. J. SCHIRMER, M. BRAUNSTEIN, AND V. McKOY, *Phys. Rev. A* **41**, 283 (1990).
25. L. S. CEDERBAUM, W. DOMCKE, AND J. SCHIRMER, *Phys. Rev. A* **22**, 206 (1980).
26. H. P. KELLY, *Phys. Rev.* **136**, B896 (1964).
27. For example, see B. H. BRANSDEN, *Atomic Collision Theory* (Benjamin, New York, 1970).
28. R. R. LUCCHESE AND V. McKOY, *Phys. Rev. A* **21**, 112 (1980).
29. For example, see B. T. PICKUP AND O. GOSCINSKI, *Mol. Phys.* **26**, 1013 (1973).
30. L. HEDIN AND A. JOHANSSON, *J. Phys. B* **2**, 1336 (1969).
31. J. C. SLATER, *Adv. Quantum Chem.* **6**, 1 (1972).
32. O. GOSCINSKI, B. T. PICKUP, AND G. PURVIS, *Chem. Phys. Lett.* **22**, 167 (1973).
33. A. DENIS, J. LANGLET, AND J. P. MALRIEU, *Theor. Chim. Acta* **38**, 49 (1975).
34. L. S. CEDERBAUM AND W. DOMCKE, *J. Chem. Phys.* **66**, 5084 (1977).
35. J. SCHIRMER, *Phys. Rev. A* **26**, 2395 (1982).
36. M. BRAUNSTEIN, V. McKOY, L. E. MACHADO, L. M. BRESCANSIN, AND M. A. P. LIMA, *J. Chem. Phys.* **89**, 2998 (1988).
37. T. H. DUNNING, JR., *J. Chem. Phys.* **55**, 716 (1971).
38. G. R. WIGHT, C. E. BRION, AND M. J. VAN DER WIEL, *J. Electron Spectrosc. Relat. Phenom.* **1**, 457 (1973).
39. A. P. HITCHCOCK AND C. E. BRION, *J. Electron Spectrosc. Relat. Phenom.* **18**, 1 (1980).
40. R. B. KAY, Ph.E. VAN DER LEEUW, AND M. J. VAN DER WIEL, *J. Phys. B* **10**, 2513 (1977).
41. T. K. SHAM, B. X. YANG, J. KIRZ, AND J. S. TSE, *Phys. Rev. A* **40**, 652 (1989).

42. M. Domke, C. Xue, A. Puschmann, T. Mandel, E. Hudson, D. A. Shirley, and G. Kaindl, *Chem. Phys. Lett.* **173**, 122 (1990).
43. Y. Ma, C. T. Chen, G. Meigs, K. Randall, and F. Sette, *Phys. Rev. A* **44**, 1848 (1991).
44. C. M. Truesdale, D. W. Lindle, P. H. Kobrin, U. Becker, H. G. Kerkhoff, P. A. Heimann, T. A. Ferrett, and D. A. Shirley, *J. Chem. Phys.* **80**, 2319 (1984).
45. M. Schmidbauer, A. L. D. Kilcoyne, H.-M. Köppe, J. Feldhaus, and A. M. Bradshaw, *Chem. Phys. Lett.* **199**, 119 (1992).
46. G. Angonoa, O. Walter, and J. Schirmer, *J. Chem. Phys.* **87**, 6789 (1987).
47. J. L. Dehmer and D. Dill, Argonne National Laboratory Report No. ANL-77-65 (unpublished), p. 65.
48. D. Dill, S. Wallace, J. Siegel, and J. L. Dehmer, *Phys. Rev. Lett.* **41**, 1230 (1978); **42**, 411 (1979).
49. F. A. Grimm, *Chem. Phys.* **53**, 71 (1980).
50. D. M. Barrus, R. L. Blake, A. J. Burek, K. C. Chambers, and A. L. Pregenzer, *Phys. Rev. A* **20**, 1045 (1979).
51. A. P. Hitchcock and C. E. Brion, *J. Electron Spectros. Relat. Phenom.* **19**, 231 (1980).
52. J. Schirmer, A. Barth, and F. Tarantelli, *Chem. Phys.* **122**, 9 (1988).
53. G. Angonoa and J. Schirmer, *J. Mol. Struct.* (*Theochem*) **202**, 203 (1989).
54. A. L. D. Kilcoyne, M. Schmidbauer, A. Koch, K. J. Randall, and J. Feldhaus, *J. Chem. Phys.* **98**, 6735 (1993).
55. A. P. Hitchcock and C. E. Brion, *J. Electron Spectrosc. Relat. Phenom.* **10**, 317 (1977).
56. A. Barth, R. J. Buenker, S. D. Peyerimhoff, and W. Butscher, *Chem. Phys.* **46**, 149 (1980).
57. W. L. Jolly, K. D. Bomben, and C. J. Eyermann, *At. Data Nucl. Data Tables* **31**, 433 (1984).
58. R. McLaren, S. A. C. Clark, I. Ishii, and A. P. Hitchcock, *Phys. Rev. A* **36**, 1683 (1987).
59. R. E. Farren, J. A. Sheehy, and P. W. Langhoff, *Chem. Phys. Lett.* **177**, 307 (1991).
60. M. N. Piancastelli, T. A. Ferrett, D. W. Lindle, L. J. Medhurst, P. A. Heimann, S. H. Liu, and D. A. Shirley, *J. Chem. Phys.* **90**, 3004 (1989).
61. A. Klonover and U. Kaldor, *J. Phys. B* **11**, 1623 (1978); **12**, 323 (1979).
62. B. I. Schneider and L. A. Collins, *J. Phys. B* **15**, L335 (1982); *Phys. Rev. A* **27**, 2847 (1983).
63. M. Berman, O. Walter, and L. S. Cederbaum, *Phys. Rev. Lett.* **50**, 1979 (1983).
64. H.-D. Meyer, *Phys. Rev. A* **40**, 5605 (1989).
65. A. Schmitt and J. Schirmer, *Chem. Phys.* **164**, 1 (1992).

CHAPTER 5

PARTIAL CROSS SECTIONS AND ANGULAR DISTRIBUTIONS

U. BECKER AND D. A. SHIRLEY

1. INTRODUCTION

Photoionization cross sections are among those fundamental quantities of nature which provide direct insight into the orbital or, more generally, electronic structure of atoms, molecules, and solids. Therefore, since the discovery of the photoeffect by H. Hertz in 1887,[1] the determination of cross section data was one of the primary goals of photoionization studies. Starting from simple photoyield measurements, the determination of photoionization cross sections has become more and more sophisticated. A first step toward a more highly differentiated analysis was the energy dispersion of the emitted electrons via electron spectrometers [photoelectron spectroscopy (PES)] in order to determine partial photoionization cross sections rather than the sum of all emitted electrons in the form of the total photoionization cross section. Early PES studies concentrated solely on the line structure of the photoelectron spectra, determining line positions and deriving electron binding energies which reflect the orbital structure of the atomic or molecular system under study.[2,3] In fact, electron spectroscopy for chemical analysis (ESCA) became a major tool in chemical and materials sciences. However, beyond these purely spectroscopic objectives, quantitative analysis of the line intensities attracted increasing interest over time. These studies were limited at the beginning to a very select number of photon energies, corresponding to the atomic transitions excited in VUV discharge lamps and x-ray sources. Nevertheless, the pioneering measurements of partial photoionization cross sections were made using these discrete photon sources; this is particularly true for most of the early high-resolution data. On the other hand, the restriction to this limited number of photon energies was a severe handicap in the exploration of the photon-energy-dependent partial cross section behavior. It was at this time when theory made several advances in the calculation of partial cross section behavior, predicting characteristic features such as delayed onsets, shape resonances, and Cooper minima.[4,5] In addition to these "one-electron" features, "many-electron" effects were predicted to show up in the partial cross sections.[6,7]

To catch up with these developments, experimentalists had to employ continuously tunable light sources; the method of choice was monochromatized synchrotron radiation.[8] Even the very early experiments showed the enormous potential of this excitation source. Starting with the measurement of total cross sections via absorption measurement,[9,10] experimentalists moved rapidly to partial cross section measurements, and their dependence on the photon energy, sometimes known as electron spectroscopy using synchrotron radiation (ESSR).[11] This was the situation more than 17 years ago when J. Berkowitz wrote his remarkable monograph *Photoabsorption, Photoionization, and Photo-*

U. BECKER • Fritz-Haber-Institut der Max-Planck-Gesellschaft, 14195 Berlin, Germany.   D. A. SHIRLEY • Departments of Chemistry and Physics, Pennsylvania State University, University Park, Pennsylvania 16802.

VUV and Soft X-Ray Photoionization. Edited by Uwe Becker and David A. Shirley. Plenum Press, New York, 1996.

electron Spectroscopy[12] showing the first results of ESSR, along with the older discrete-photon-energy data. In the present chapter we shall use this classic presentation as a point of reference, and show how new developments have increased our knowledge of partial photoionization cross sections.

A further step toward a detailed understanding of the photoionization process and the structural constituents of the ionized target is provided by angle-resolved measurements studying the angular distribution of photoelectrons, so-called angle-resolved photoelectron spectroscopy (ARPES). At the time of Berkowitz's comprehensive monograph, very few data were available on angular distributions, mostly still from discrete photon sources. The present review will document the dramatic improvements in this area during the last 25 years. In the meantime, a step even further toward a complete description of the photoionization process has been achieved, in the form of spin-resolved detection of photoelectrons; however, this will be the subject of a separate chapter.[13] In the present chapter we shall neglect spin and take the partial cross section to be the intensity of all energetically degenerate components of a photoline. But there is still an extremely wide field of experimental effort in different areas. In order to systematize the progress during the last 25 years, and to look for possible future directions in this field, Figure 1 shows the development in photoionization studies, including partial cross sections and angular distributions, along three dimensions: systems, transitions, and energies. This characterization reflects, in a sense, the definition of an energy-dependent partial cross section $\sigma_{nl}(\varepsilon)$. Whereas 25 years ago most studies focused on simple systems such as rare-gas atoms, and strong transitions such as photoelectron main lines, and these in a very limited energy range, particularly in the vacuum-ultraviolet (VUV) region, today a whole variety of systems, including open-shell atoms, clusters, and ions, is under study. In addition, not only main lines, but also satellite transitions, multiple-electron ejection, and ionization of excited states have come under study. Ordinary main-line transitions, and also some of the more complicated processes, are now being studied over extended energy ranges throughout the VUV and soft x-ray regions and extending toward or into the hard x-ray regime, accompanied by dramatic improvements in resolution. This tremendous progress is illustrated schematically in Figure 1. It would be impossible to give even a summary description of these developments within the limits of such a short chapter. Indeed, such a task today would probably even be beyond the scope of a monograph, because of the sheer explosion of experimental data and corresponding theoretical calculations, particularly during the last 10 years. It will therefore be the aim of this chapter to illustrate this progress by means of selected showcase examples, putting special emphasis on photoionization behavior over extended energy ranges. Complex phenomena involving threshold behavior and autoionizing resonances,[14] open-shell systems,[15] larger molecules,[16] and studies of excited and ionized states[17] will be covered in separate chapters. Given this restriction, the number of systems remaining to be discussed is significantly reduced; again rare gases are the most

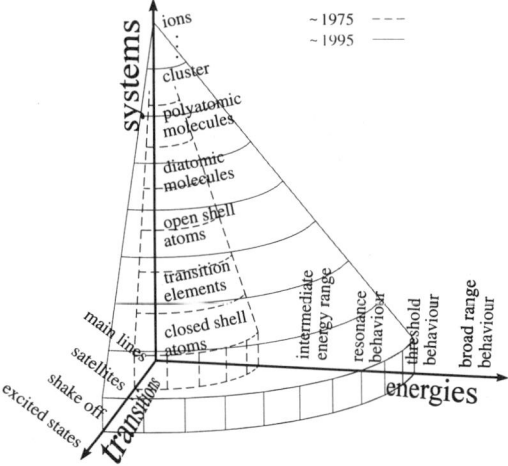

FIGURE 1. Present status (1995) of research in the field of photoionization cross sections and angular distribution compared to 1975, shown schematically. Progress evolved along three main directions: photon energies, target systems, and transitions under study.

thoroughly studied species, followed by some transition elements, particularly those with half-filled subshells. Among molecules the situation is even more limited, as very few molecules, including CO, N_2, SF_6, and CH_3I, have been studied over extended energy ranges. Although this selection represents the subjective view of the authors, it still provides a good basis for illustrating general trends, as well as specific features in the behavior of partial photoionization cross sections. A related survey, on the partial photoionization cross sections in solids, was very recently given by Lindau in *Synchrotron Radiation Research: Advances in Surface and Interface Sciences*.[18] The present chapter is subdivided into five more sections covering general definitions and experimental methods, helium as a showcase example of a two-electron system, the other rare gases Ne, Ar, Kr, and Xe, followed by transition and heavier elements. The chapter will close with a discussion of a prototype molecule, for which we have chosen CO, and a short overview of a new class of materials, the so-called fullerenes.

2. DEFINITIONS AND EXPERIMENTAL METHODS

The basis of all photoionization cross section data in the VUV and soft x-ray region is the total photoionization cross section, which gives the total probability of ionization of a system by electromagnetic radiation of a given energy. The literature on total photoionization cross sections is quite extensive, and the subject has just recently been critically revisited in a series of measurements performed by the alternative method of electron-energy-loss ("virtual-photon") spectroscopy.[19–23] Although the total cross section by itself is of interest, it is not subshell-sensitive because it is the sum of intensities from several subshells at the same photon energy. A more specific, and informative, quantity for subshell-dependent ionization behavior is the *partial* cross section, given by

$$\sigma_{if}(h\nu) = \frac{4\pi^2 \alpha a_0^2}{3} h\nu |M_{if}|^2 \qquad (1)$$

where α is the fine-structure constant, a_0 the Bohr radius, and $h\nu$ the photon energy in Rydberg units (1 Ry = 13.6 eV). The matrix element M_{if} describes, in the dipole approximation, the dipole transition between the initial state i and the final state f:

$$|M_{if}|^2 = |\langle f | \sum_\mu r_\mu | i \rangle|^2 \qquad (2)$$

with $\sum_\mu r_\mu$ being the dipole operator of the *n*-electron system.

This description of the partial cross section allows a separation of all ionization processes which give rise to energetically nondegenerate photoelectron lines, as independent partial cross sections with individual cross-sectional behavior. This behavior reflects not only the orbital structure of the particular subshell, but also the various interactions between different subshells, including excitation to unoccupied energy levels during the photoionization process. Although this is a highly differential measurement, requiring good experimental resolution, the description of the process is by no means complete if one considers the number of still-unresolved components in the outgoing photoelectron wave. Whereas in the discrete part of the photoabsorption spectrum the different 1 ± 1 components are in principle separated in energy, this is not true in the photoionization case. These partial-wave components in the continuum part of the absorption spectrum are energetically degenerate but show still different photon-energy-dependent behavior. To unravel this behavior is one of the objectives of partial cross section measurements. Angle-resolved studies are a welcome first step toward this goal. Because they are relatively easy to perform, angle-resolved photoemission measurements over wide energy ranges are feasible, determining the so-called angular distribution *asymmetry parameter* β, a characteristic quantity in the differential partial cross section:

$$\frac{d\sigma_{if}(h\nu)}{d\Omega}(\theta) = \frac{\sigma_{if}(h\nu)}{4\pi} [1 + \beta_{if}(h\nu) P_2(\cos\theta)] \qquad (3)$$

with θ being the angle between the electric vector of the photon beam and the direction of the outgoing electron, and $P_2(\cos\theta)$ the Legendre polynomial of second degree. At the so-called "magic angle" θ = 54.7° (assuming 100% linear polarization), this polynomial becomes zero and therefore the differential partial cross section becomes proportional to the integral partial cross section. Measurement at any other angle yields information about β if σ is known. The asymmetry parameter depends on the partial-wave matrix elements and the phase shift between the two partial waves of the outgoing electron. In a simplified model, assuming LS coupling of the system, it is given by

$$\beta = \frac{l(l-1)R_{l-1}^2 + (l+1)(l+2)R_{l+1}^2 - 6(l+1)R_{l-1}R_{l+1}\cos(\delta_{l+1} - \delta_{l-1})}{(2l+1)[lR_{l-1}^2 + (l+1)R_{l+1}^2]} \quad (4)$$

where $R_{l \pm 1}$ are radial integrals and $\delta_{l \pm 1}$ are phase shifts of the respective partial waves. This formulation, known as the Cooper–Zare model,[24] can be extended to a more general coupling-independent formulation, the angular-momentum transfer (AMT) theory, described by quite similar equations.[25,26] The most important result of the AMT formulation of the photoionization cross section is the fact that σ is essentially determined by the radial properties of the orbitals and wave functions involved in the photoionization process whereas the β parameter reflects basically the angular momenta involved. This statement will be illustrated by examples in the following sections.

Differential partial cross section measurements to date have been based largely on angle-resolved photoelectron spectroscopy, but in some cases other methods, such as ion yield measurement[27] and fluorescence detection,[14,28] have also been employed. For studying photon-energy-dependent behavior, synchrotron radiation has become the most commonly used excitation source. The main class of electron spectrometers used for this purpose are electrostatic analyzers, such as hemispherical or cylindrical-mirror analyzers.[29,30] These highly developed and widely used systems have been described in depth in the literature, and they are effective with conventional laboratory sources, which have no intrinsic time structure. In this chapter we are emphasizing the new areas of photoionization involving variable photon energy. We shall therefore concentrate on a less widely used electron spectrometry technique which is particularly well suited to synchrotron-radiation-based research: the time-of-flight method.[31–33] This method makes effective use of one characteristic feature of synchrotron radiation: its excellent time structure. Light pulses of a few hundred picoseconds' duration are emitted at time intervals ranging from nano- to microseconds, depending on the circumference of the electron storage ring and its operational mode. Operating the ring in a single- or few-bunch mode makes it possible to ionize the target during the very short period of time of the light pulse and to record the arrival times of the emitted electrons at a fixed detector, thereby energy-analyzing the electrons according to their velocities. A so-called "time-of-flight" (TOF) electron spectrum represents a kinetic energy spectrum on a nonlinear, but known, energy scale. The advantage of the method is threefold: first, it allows simultaneous sampling of the whole energy spectrum, thereby increasing the sensitivity while reducing the effects of dark counts and time-dependent photon-flux fluctuations; second, the method is very well suited to measure low-kinetic-energy electrons, a capability essential for all threshold and near-threshold studies; and finally, high-energy resolution can be achieved at high kinetic energies by applying a retarding voltage. Based on its simple design concept the TOF method is intrinsically angle-resolved, and the use of two or more spectrometers allows measurements of σ and β simultaneously. Figure 2 shows schematically the experimental setup for angle-resolved TOF-PES using synchrotron radiation excitation. The measurements require extensive control and/or calibration of several parameters, including the degree of polarization of the incoming radiation. In order to do this with sufficiently high accuracy a rotationally mounted chamber affords a great advantage, because it allows the choice of any direction of the spectrometer with respect to the electric vector of the photon beam. Nevertheless, this is only one possible experimental arrangement which has been extensively used in partial cross section measurements, among many others. With proper energy and transmission calibrations of the TOF detectors, the resulting data are still *relative* partial cross sections on an arbitrary intensity scale. To bring them to an absolute scale, absolute total cross section values are necessary. Alternatively, absolute partial cross section measurements may be performed for selected main lines using electron-impact excitation for calibration purposes, to determine weaker lines via branching

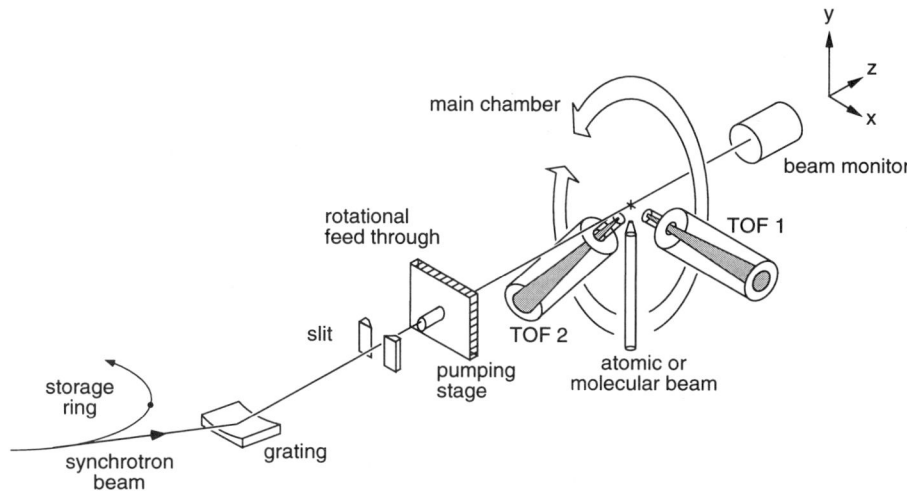

FIGURE 2. Schematic diagram of an angle-resolved photoemission experiment using synchrotron radiation as the excitation source and time-of-flight as the detection method.

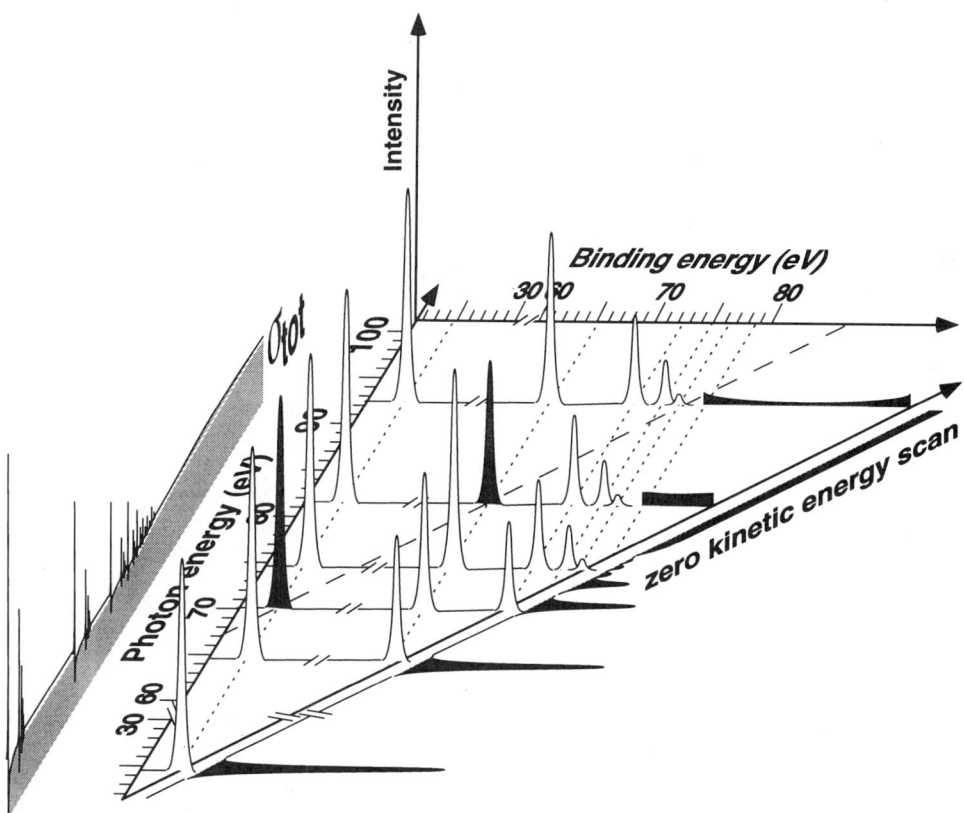

FIGURE 3. Evolution of photoelectron spectra from threshold to higher energies, shown schematically.

ratios obtained from regular electron spectroscopy. Measurements of the asymmetry parameter β are independent of such calibration procedures because they only require *relative* measurements of intensities at different observation angles. The quality and accuracy of measured partial cross sections and asymmetry parameters depends critically on the quality of these calibrations.

In order to visualize the information obtained from PES along certain observation directions, Figure 3 shows a schematic three-dimensional representation of a set of complete photoelectron spectra. Integration of the whole electron intensity in each spectrum gives a total yield spectrum with an intensity variation analogous to an absorption spectrum. Registration of all electron intensities right at threshold, which means at "zero kinetic energy," yields a zero-kinetic-energy (ZEKE) spectrum,[34,35] a special kind of constant-final-state (CFS) spectrum performed at zero kinetic energy. Tracing the intensities of one specific photoelectron line along the photon-energy axis gives photon-energy-dependent partial cross sections, with the kinetic energy of the photoelectron line varying along with the photon energy. A special case is given by the electrons emitted in a simultaneous double-ionization process, the so-called shakeoff electrons. Because these electrons are emitted in a continuous energy distribution rather than in discrete lines, determination of their partial cross sections requires integration over this energy distribution. Two-dimensional registration of shakeoff energy distributions along photon and kinetic energy allows further insight into the double-ionization process, particularly if angular distributions are additionally studied as well. In the following sections measurements along all of these directions will be presented, with main emphasis on the photon-energy-dependent behavior of σ and β over extended energy ranges.

3. HELIUM

Photoionization of helium is the showcase example for atomic photoionization processes[36] because it exhibits all of the possible processes, including those based purely on electron–electron correlation in a form which permits separation of each of them experimentally, thereby allowing an unambiguous theoretical interpretation. For some processes, such as shakeoff, helium is the only case where they show up in a form accessible to a rigid analysis and interpretation, establishing the basis for corresponding studies in other elements.

The total photoionization cross section of He was determined by Marr and West[37] and more recently by Samson *et al.*[38] There are small differences between the two data sets, but in general the overall agreement is quite good. Therefore, for purposes of considering the partial cross sections, the choice between these two data sets is not really of critical importance. We have adopted the more actual cross sections of Samson, in order to make the data more easily comparable to future results. Figure 4a is a photoelectron spectrum of He taken at $h\nu = 89.5$ eV showing the $1s$ main line and satellite lines from $n = 2$ to $n = 7$, which converge to the double-ionization threshold.[39] These satellites correspond to final ionic states in which the remaining electron stays in an orbital with principal quantum number n, whereas above the shakeoff threshold to which these satellites converge the second electron also becomes ejected into the continuum. Both processes are entirely related to electron–electron correlation, and are subjects of intense theoretical and experimental studies. A series of complete He photoelectron spectra taken between 84 and 120 eV[40] is plotted in a three-dimensional representation, with shading of the continuous electron distribution related to double photoionization in Figure 4b. Figure 5 shows the partial cross sections of all of these processes along with the total photoionization cross section, on a double logarithmic scale because the corresponding intensities vary over five orders of magnitude. The dashed lines represent the data situation as shown in Berkowitz's earlier review.[12] There is not much change relative to Berkowitz's representation on this scale with respect to the $1s$ main line compared to the data which we have adopted. Some changes are seen for the first He$^+$ ($n = 2$) satellite and for the He shakeoff; however, both partial cross sections were already relatively well known (within 25%) 25 years ago. The main progress for these processes appeared at higher photon energies and more generally for all satellites with $n > 2$ where no data were available until very recently. In particular, the progress in zero-kinetic-energy spectroscopy yielded threshold intensities up to very high n values[41,42] while high-resolution/high-flux monochromators made it possible to measure

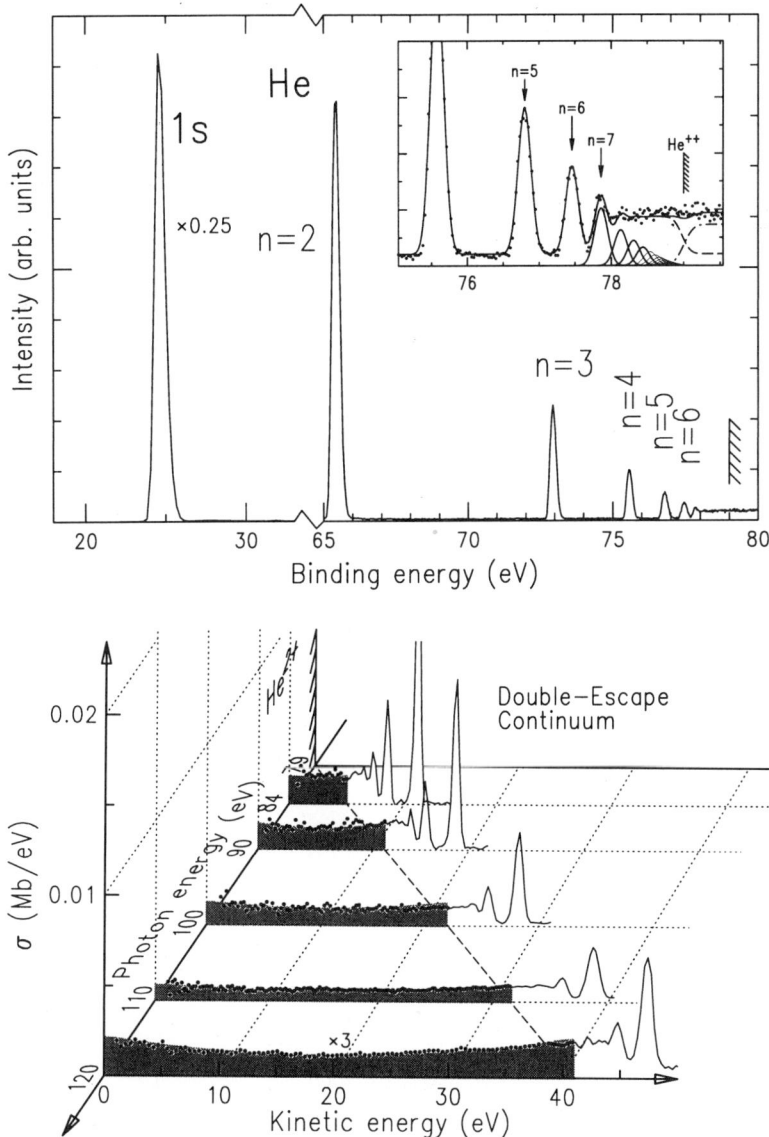

FIGURE 4. (a) High-resolution photoelectron spectrum of He taken $h\nu$ = 89.5 eV,[39] showing the $1s$ main line along with a series of satellite lines (n = 1–7). The inset shows an enlargement of the threshold region. (b) Three-dimensional representation of complete He photoelectron spectra with shading of the continuous electron distribution resulting from double photoionization.

satellite intensities still resolved up to several hundreds of electron volts.[39] The ratio of double-to-single ionization He^{2+}/He^+, a direct measurement of the shakeoff intensity, was even determined up to several kiloelectron volts.[43,44] This tremendous progress allowed an extrapolation of the relative intensities of the high-n satellite lines via an n^{-3} quantum defect progression so as to compare them with the difference between single and double ionization relative to the total photoionization cross section. This comparison shows that the relative partial cross sections in He are now known down to a fractional intensity of 10^{-3}. Theoretical predictions which describe the complete situation in the presently known photoionization of He are highly desirable.

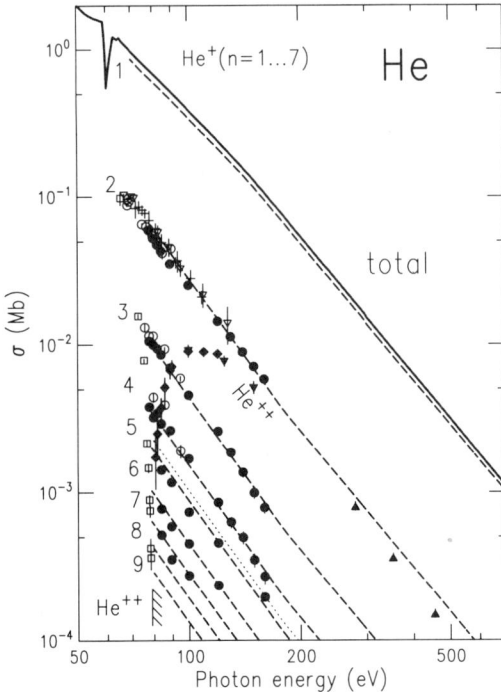

FIGURE 5. Partial cross sections for the different photoionization channels of He in a double logarithmic representation. The dashed lines are extrapolations of the measured data by exponential power laws, the dotted line is an analogue, quantum-defect-based extrapolation of the satellite intensity σ_n for all n > 8. The shaded line represents the adopted double ionization cross section, the solid line the total cross section of Ref. 45 and the dashed dotted line is the 1s partial cross section derived from these data. Experimental data points are taken from the following references: ●, Ref. 46 and unpublished results from the authors; ▼, Ref. 47; ○, Ref. 48; □, Ref. 41; ▽, Ref. 49; +, Ref. 50; ◆, Ref. 40; ▲, Ref. 51.

One area of strong theoretical activity concentrated on the angular distributions of satellite and shakeoff electrons. Interest has focused on the He^+ ($n = 2$) satellite, the threshold behavior of satellites with higher principal quantum number n, and the angular distributions of selected shakeoff electrons such as those with low kinetic energies. Figure 6 shows the angular distribution asymmetry parameters β of the helium satellites $n = 2-6$, from near threshold to several hundred electron volts,[39] showing how the β values converge to an asymptotic value of 2. Theoretical curves to compare with these results exist only for $n = 2$.[55,56] Both available calculations agree quite well with the experimental data. The theoretical curves of Jacobs and Burke[56] for 2s and 2p satellite photoionization were taken, with a variable s to p ratio depending on the principal quantum number n, in order to obtain β values for the higher n satellites. The results are shown as dashed curves. These semiempirical β values agree reasonably well with the experimental values, particularly at higher photon energies, showing that the influence of higher angular momentum components ($l > 0$) is quite small at these energies. The photon energy dependence of the satellites is similar for the different principal quantum numbers n but with a clear tendency toward reduced fractional intensity of s-like orbitals for states with higher n, as the semiempirical analysis shows. Furthermore, at lower photon energies the satellites with higher n tend to have β values somewhat lower than expected from our simplified model, reflecting the influence of higher angular momenta near threshold.

Right at threshold Greene[57,58] predicted on the basis of very general symmetry considerations that the angular distribution of the high-n satellites should approach $\beta = -1$. The tendency to more negative values is already clearly seen in Figure 6, although the data very close to threshold are not shown because they do not strongly deviate from the points at or slightly above 3 eV. In order to test the $\beta = -1$ prediction the low-energy part of the shakeoff distribution is much better suited than the satellites at very high n because the resolution becomes less crucial, although the intensity of the shakeoff is quite small at very low energies. In the shakeoff case two asymptotic limits exist, one at threshold where both electrons have nearly zero kinetic energy, the $\beta = -1$ case, and the second the sudden limit at high energies, where the fast electron approaches $\beta = 2$ because it takes the whole angular momentum of the ionizing photon, while the slow or "shaken" electron becomes isotropically emitted.[59] Between these two extreme cases there would be a smooth transition if electron–electron

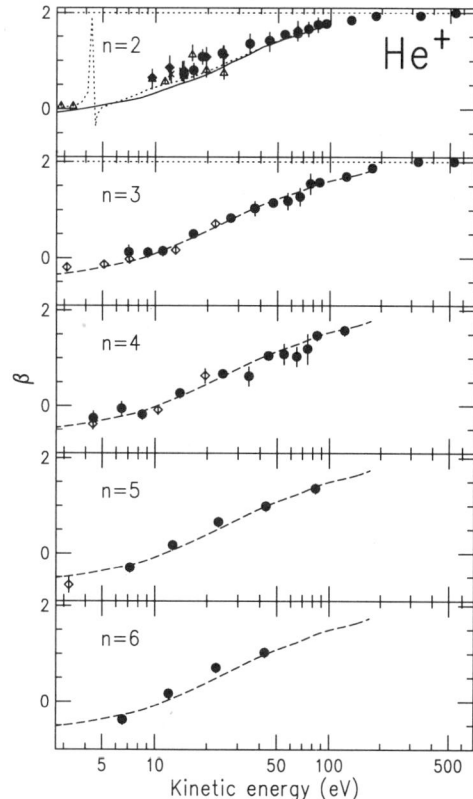

FIGURE 6. Angular distribution parameter β for the He (n = 2–6) satellites on an extended logarithmic energy scale. The experimental data are from the following references: ●, Ref. 37; ×, Ref. 52; △, Ref. 48; ♦, Ref. 53; ◇, Ref. 54. The dotted curve represents a calculation of Ref. 55 and the solid curve, Ref. 56. The broken curves were semiempirically derived on the basis of Ref. 56.

correlations could be neglected; however, because this is not the case, strong modulations resulting from the Coulomb repulsion forces are theoretically expected. Figure 7 shows the results of a calculation which takes electron correlations fully into account,[59,60] along with measurements at two slightly different positions of the kinetic energy distribution of the shakeoff electrons, at 0.25 and 1.24

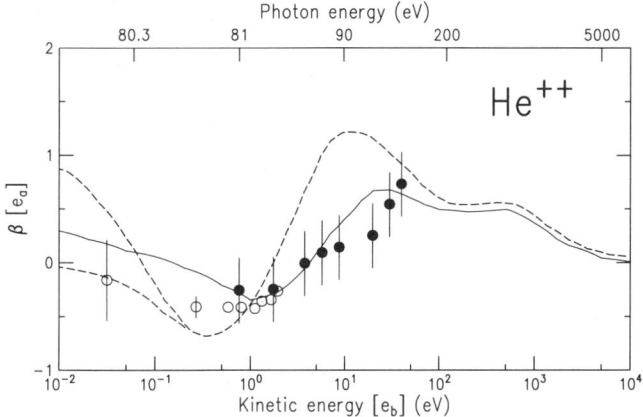

FIGURE 7. Angular distribution of shakeoff electrons with low kinetic energy. The experimental data (● and ○) are from Refs. 42 and 40, whereas the theoretical curves (dashed and solid lines) are from Refs. 60 and 59.

eV.[40,61] The experimental data confirm quite nicely the corresponding theoretical curves: only the high-energy behavior is still awaiting experimental confirmation. A further step in differentiation with respect to double electron ejection is the coincidence detection of the second electron with respect to the direction of the first electron. Measurements of such coincidence angular distribution are just beginning.[62,63]

Summarizing, one can say that the experimental situation concerning partial cross sections and angular distributions of the photoionization of He is at present close to a nearly complete picture, awaiting an adequate theoretical description, which is rapidly developing.

4. OTHER RARE GASES

A quite comprehensive description of the photoionization of most of the rare gases has been given very recently by Schmidt.[64] As mentioned in the introduction, we will therefore concentrate more closely on the general trends in photoionization over extended energy ranges. Whereas inner-shell photoionization is characterized by edgelike structures related to the long-range coulombic forces, outer-shell photoionization is, even in an independent particle description, much more controlled by the values of the principal quantum number n and the angular momenta l involved in the photoionization process. This gives rise to the well-known phenomena of Cooper minima and shape resonances.[4,5] The two quantum numbers n and l determine the maximum possible number of Cooper minima, $n-l-1$, whereas the exact position of these minima is sensitively dependent on the effect of electron–electron correlation. This sensitivity is so pronounced that in many cases, such as Ar $3s$, the Cooper minimum becomes shifted below threshold if electron–electron correlations are neglected, suggesting that one consider the existing minimum in this case as a correlational minimum rather than a "single particle" Cooper minimum.[45,65,66]

Shape resonances behave similarly; their pronounced appearance is determined by the angular momentum of the ionized orbital due to the centrifugal barrier seen by the outgoing electron in an effective potential, whereas the qualitative heights and positions of these resonances are again subject to correlation or, in more general terms, collective effects.[45,65] The shape resonances evolve along l starting from a plateau-like resonance for the p ionization over the classical shape resonance in d ionization to the oscillatory behavior in f photoionization. The principal quantum number n determines in this case the position of the ionization threshold, which may vary from the onset to the maximum as seen for Kr $3d$, Xe $4d$, and Hg $5d$ photoionization, for example.

Another point of strong theoretical interest is the coupling between different photoionization channels, the so-called interchannel coupling. In this case the cross-sectional behavior of one photoionization channel affects the behavior of another channel; in practice, weaker photoionization channels are forced to mimic the behavior of a stronger photoionization channel when this channel undergoes large intensity variations. The classic example of this kind of correlation effect is the modulation of the $5s$ and $5p$ partial cross sections in the shape resonance and Cooper minimum region of the Xe $4d$ partial cross sections.[6,7,45] Similar behavior is also observed for satellite channels which are influenced by the behavior of their main lines. All of these effects are reflected in both the partial cross section and angular distribution behavior, which behave approximately in a reciprocal way.

4.1. Neon

To begin, we consider the partial cross sections of neon, which has the ground-state electron configuration $(1s)^2(2s)^2(2p)^6$. Figure 8 shows a valence and an inner-shell photoelectron spectrum of Ne. We see, besides the $2p$ valence, $2s$ inner-valence and $1s$ core-level main lines, a whole variety of satellite lines associated with each of these main lines, as well as several shakeoff continua. This strikingly shows the increase in complexity compared to He and confirms why He became such a unique showcase example for correlation effects. Nevertheless, although we are far from having a complete understanding of the situation in Ne, both theoretically and experimentally, many aspects are very well understood, particularly for the main line behavior and in many cases also for satellites and shakeoff continua. Figure 9 shows the partial cross sections of the two valence main lines of neon,

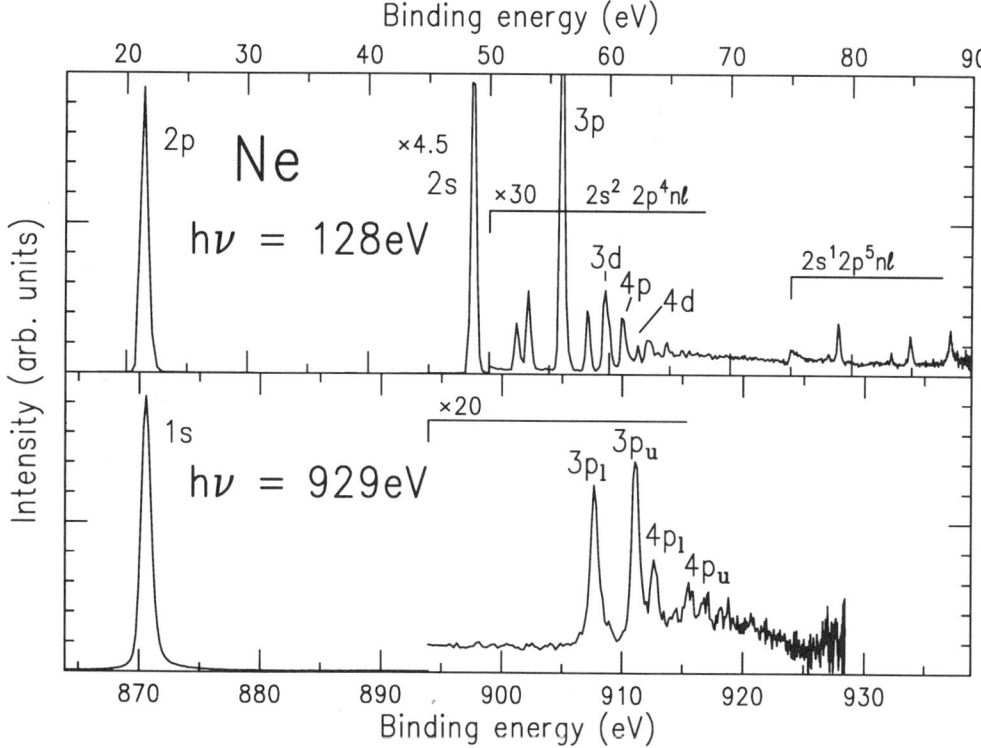

FIGURE 8. Photoelectron spectrum of the Ne valence shell taken at $h\nu = 128$ eV and the Ne 1s shell at $h\nu = 929$ eV. The spectra show the corresponding main lines along with a whole variety of satellite lines, some of them showing clearly asymmetric line shapes resulting from coupling with the underlying double ionization continuum. These data (unpublished) were obtained at BW3 of HASYLAB.

2p and 2s. The 2p cross sections are from Wuilleumier and Krause,[67] whereas the 2s data are supplemented by new measurements which, however, vary only slightly from the older data. The β values of Ne 2p in Figure 9 are also well known and quite similar to the corresponding values for satellite emission in helium.

The main progress in neon partial cross sections appeared in the study of satellite transitions. The progress here is twofold. First, high-resolution measurements revealed an enormous number of satellite lines over a wide binding energy range; second, the most prominent transitions could be followed over an extended energy range,[81-86] uncovering many unexpected new phenomena in satellite behavior. However, a simple quantity such as the total sum of all satellite transition intensities as a kind of "semiintegral" satellite partial cross section is still poorly known, particularly with respect to the photon-energy dependence. The same statement is in part true for the shakeoff probability. These problems are related because the distinction between satellite and shakeoff contributions to a photoelectron spectrum is by no means trivial. The problem became even more complicated after the discovery of valence Auger decay as an important channel for valence double ionization,[87] making ion-yield measurements questionable as a method for determining shakeoff intensities. Instead of showing such integral data, we will concentrate on the σ and β behavior of one prominent satellite in the neon valence and K-shell spectrum. Figure 10 shows the partial cross sections of the 3p shakeup satellites associated with 2p and 1s photoionization over a wide kinetic-energy range. Both are examples of shakeup behavior, but in the valence shell this behavior is less common than for core levels. A complete theoretical analysis of the satellite spectrum and its photon-energy-dependent behavior is still needed.

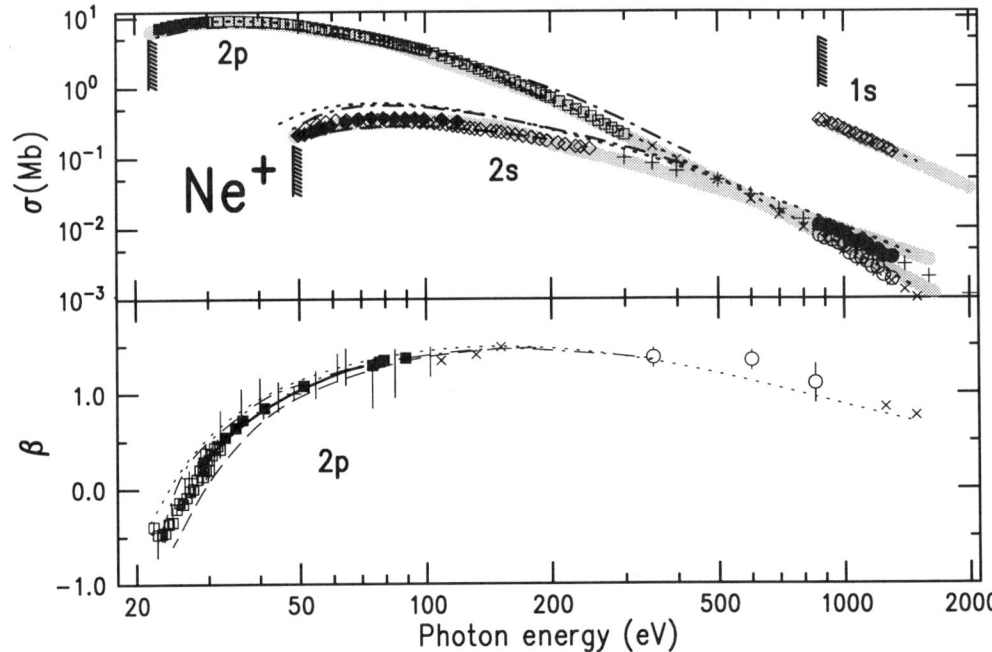

FIGURE 9. Partial cross section σ and angular distribution asymmetry parameter β for the 2p, 2s, and 1s main lines of neon. The dotted and the dashed dotted lines represent HS and HF calculations, respectively,[68,69] whereas the solid lines show the results of the RPAE[70,71] and RRPA[72] method. The experimental data are from Refs. 73 and 38, 67, 74–80. The shadowed areas represent the data compilation of Ref. 12.

4.2. Argon

Argon, with the electronic ground-state configuration $(1s)^2(2s)^2(2p)^6(3s)^2(3p)^6$, is the first element with a pronounced variation in several photoionization channels because of the appearance of a Cooper minimum. Furthermore, three subshells have to be considered in order to give a complete picture of its photoionization. Figure 11 shows a photoelectron spectrum of two of these three subshells: the valence shell spectrum and the 2p spectrum. Both spectra show, besides their main lines, pronounced satellite structure in the valence shell related in particular to the collapse of the 3d wave function in the final ionic state. Regarding the main lines, there has been considerable progress in the

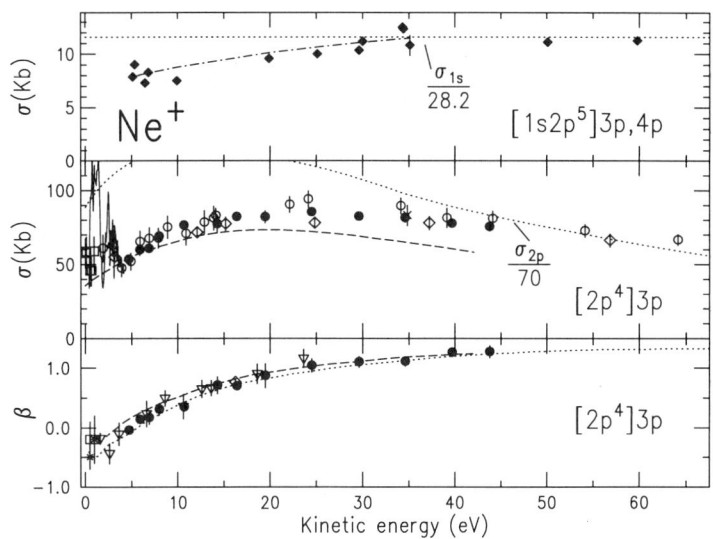

FIGURE 10. σ and β of the strongest valence and K-shell satellites [2p⁴] 3p and [1s 2p⁵] 3,4p compared to their corresponding main lines (shown as dotted lines). The long dashed curve represents a Greens function calculation of Ref. 88; the dashed dotted curve is a semiempirical curve according to Ref. 89. The experimental data points are from the following references: ◆(σ), ●(β), Ref. 90; ●(σ), ▽(β), Ref. 91; ○, Ref. 92; □, Ref. 41; ×, Ref. 82; ◇, Ref. 83. Satellite [2p⁴]3p is a superposition of one dominant satellite state with two smaller basically unresolved components; the low-energy β value of one of these minor components is shown by an asterisk.[41] The resonance structure shown by solid lines is from Ref. 93.

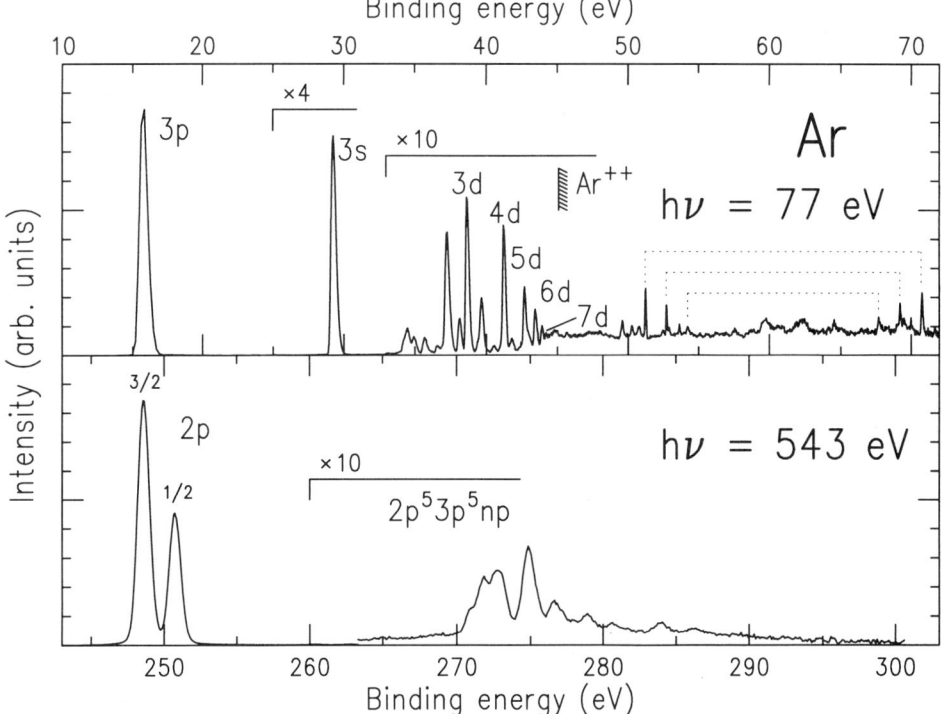

FIGURE 11. Photoelectron spectrum of the Ar valence shell taken at $h\nu = 77$ eV[98] and the Ar 2p shell at $h\nu = 543$ eV (unpublished data). The spectra are characterized by a complex satellite structure dominated by a Rydberg series of $nd(^2S)$ satellites resulting from the d wave-function collapse in the final ionic state.

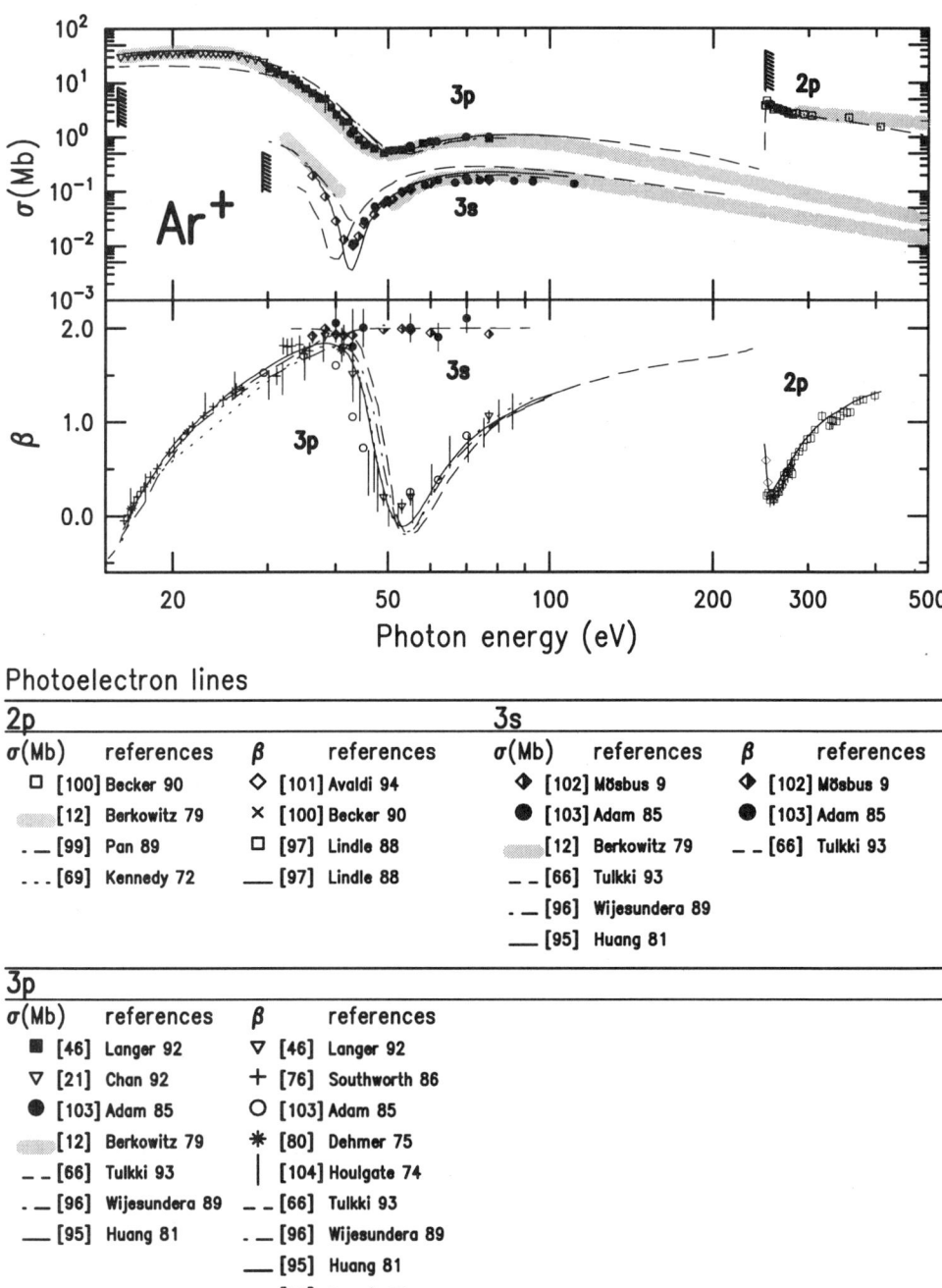

FIGURE 12. Partial cross section σ and angular distribution asymmetry parameter β for the 3p, 3s, and 2p subshells of argon. The dotted σ line shows an HF calculation of Ref. 69, the dashed dotted lines for σ and β MBPT calculations of refs. 99 and 96. The dashed lines represent MMCDF calculation of Ref. 66, whereas the dotted and solid lines show RPAE (β) and RRPA calculations of Refs. 70 and 95, respectively. The experimental data points are from Refs. 12, 21, 46, 76, 80, 97, 100–104 as shown in the figure.

determination of partial cross sections and β values during the last 10 years, in particular with respect to Cooper minima and near-threshold behavior. Figure 12 shows the partial cross section and β values of Ar 3p, 3s, and 2p ionization. The absolute scales of these data were derived using different total cross section measurements.[38,94] These experimental data are now quite well described by different theoretical methods, taking care of electron–electron correlation in one way or another.[66,70,95–97] A still-unsolved problem is the influence of resonances on the partial cross section, but this problem will be covered in more detail in a separate chapter on resonances and near-threshold processes.[14]

The very rich satellite spectrum of the argon valence shell exhibits many different types of satellite behavior,[81,86,93,105,106] of which we will present two examples, one representing the type of satellite originating from the s-d mixing in the final ionic state and another one representing a nice example for a type of satellite predicted by angular momentum transfer theory for more than 20 years, but never observed in nonresonant photoelectron spectra until very recently, the so-called parity-unfavored transitions. In a parity-unfavored transition the angular momenta of the electron and photon couple in such a way that no angular momentum component of the photon is transferred to the electron, which means that the transfer momentum j_t between the atom and the ion becomes equal to the electron momentum l. In this case there is no ambiguity in the l-composition of the outgoing electrons and β becomes –1, because the momentum projection along the photon quantization axis is zero. Figures 13 and 14 show σ and β of the two satellites. The intensively studied $3d(^2S)$ satellite exhibits its origin as a $3s$-based satellite by two characteristic features, first a pronounced Cooper minimum in its partial cross section, and second a photon-energy-independent β value of 2. Different theoretical methods, such as many-body perturbation theory (MBPT)[96] and configuration interaction (CI) calculations,[107] were able to describe the cross section behavior of this satellite quite well. The parity-unfavored $3d(^2P)$ satellite has a much lower overall intensity but is still clearly seen over a wide nonresonant

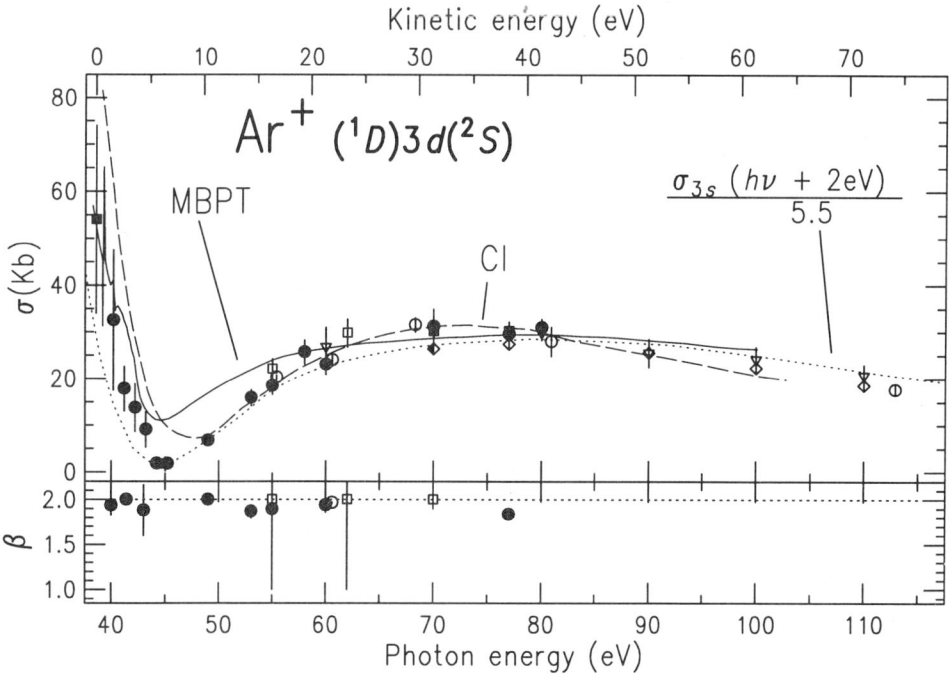

FIGURE 13. σ and β of the first nd-satellite of Ar coupled to a ^2S final ionic state $(^1D)3d(^2S)$. The solid curve represents an MBPT calculation of Ref. 96, whereas the dashed curve shows the result of a CI calculation of Ref. 107. The dotted curve gives the $3s$ partial cross section scaled by a factor 1/5.5 and shifted by 2 eV to higher photon energies. The experimental data points are from the following references: □, Ref. 103; ◆, Ref. 108; ◇, Ref. 109; ●, Ref. 110; ▽, Ref. 111; ■, Ref. 112; ○, Ref. 83.

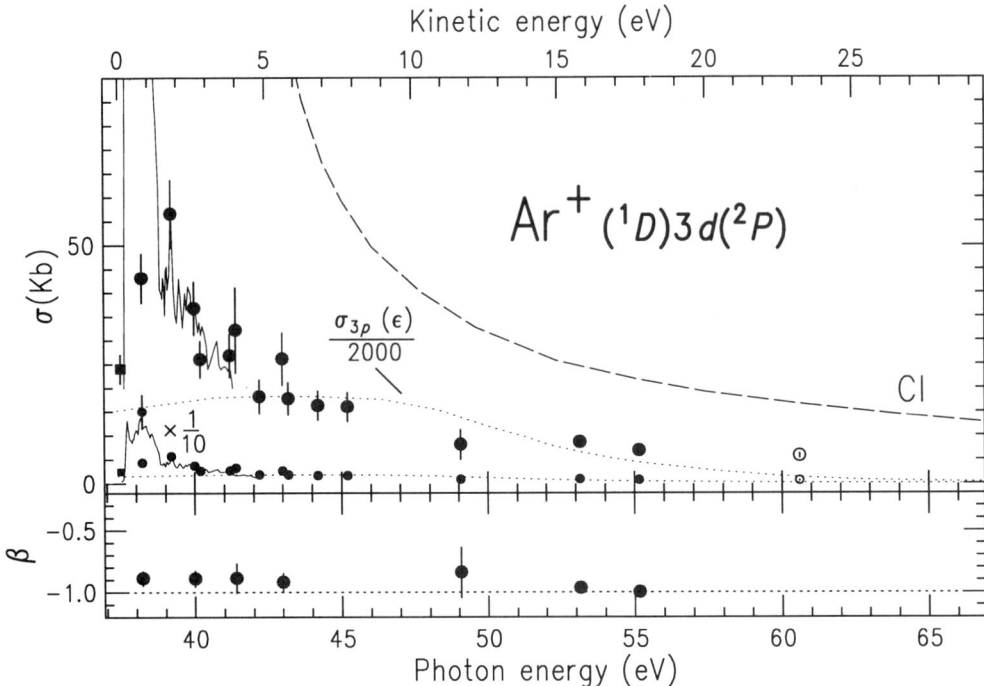

FIGURE 14. σ and β of the parity-unfavored 3d-satellite coupled to a ^2P field ionic state (^1D)3d(^2P). The short dashed curve shows a CI calculation of Ref. 107. The dotted curve gives the 3p partial cross section scaled by a factor 1/2000 and shifted by the binding energy difference of 21.59 eV; the solid line represents a partial cross section curve from Ref. 93 showing strong resonance enhancement near threshold. The other data points are from the following references: ●, Ref. 98; ■, Ref. 112; ○, Ref. 83.

energy region. The β values are, as theoretically predicted, always equal to −1. The good agreement with a scaled σ_{3p} curve indicates that this satellite is indeed of 3p origin. Theoretical calculations show, in this case, less reasonable agreement compared to the 3d(^2S) satellite, which is not so surprising in view of the weak intensity. Several other parity-unfavored satellites were observed in the argon valence satellite spectrum at different binding energies.[98]

The K-shell photoionization of argon has been little studied as far as partial cross sections are concerned. Most studies were performed using ion-yield measurements.[113–115] However, a few data do exist.[106,116,117] Figure 15 shows the partial cross sections of the 1s main line along with the strongest 4p satellite. A more detailed analysis of the satellite cross section shows that this satellite results from shakeup, as one would expect theoretically. Further, many-electron transitions such as 2p-based satellites and shakeoff transitions have not been widely studied quantitatively,[81] particularly concerning their photon-energy-dependent behavior. This will be the subject of future measurements at the synchrotron radiation light sources of the third generation, which are just now coming into operation.

4.3. Krypton

Krypton has in its ground state the electron configuration $(1s)^2(2s)^2(2p)^6(3s)^2(3p)^6(3d)^{10}(4s)^2(4p)^6$ and is the first rare gas element which shows all of the phenomena which may be expected in principle in a one-electron description. This is because it is the first element with a filled subshell with angular momentum $l = 2$, giving rise to the occurrence of a shape resonance in the partial cross section. Figure 16 shows a valence photoelectron spectrum of Kr exhibiting, besides the valence main line, a very clear Rydberg series of nd correlation satellites.[81,105,121] Figure 17 shows the partial cross sections

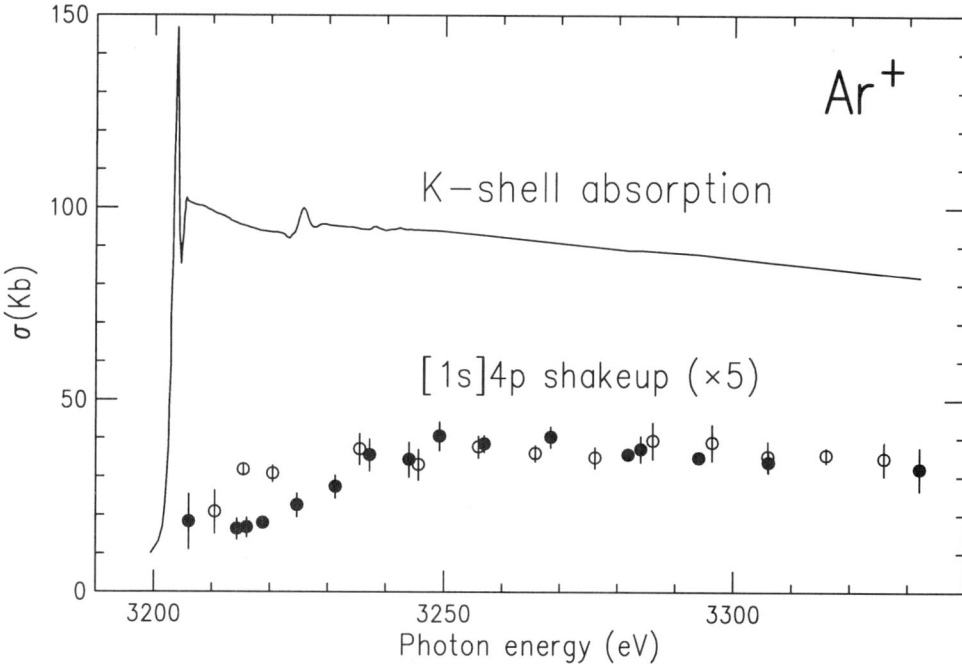

FIGURE 15. Partial cross section of the Ar 1s main line together with its strongest satellite $[1s^1\,3p^5]4p$. The solid line shows the absorption measurement of Ref. 118, whereas the satellite data are from Ref. 119 (●) and 117 (○).

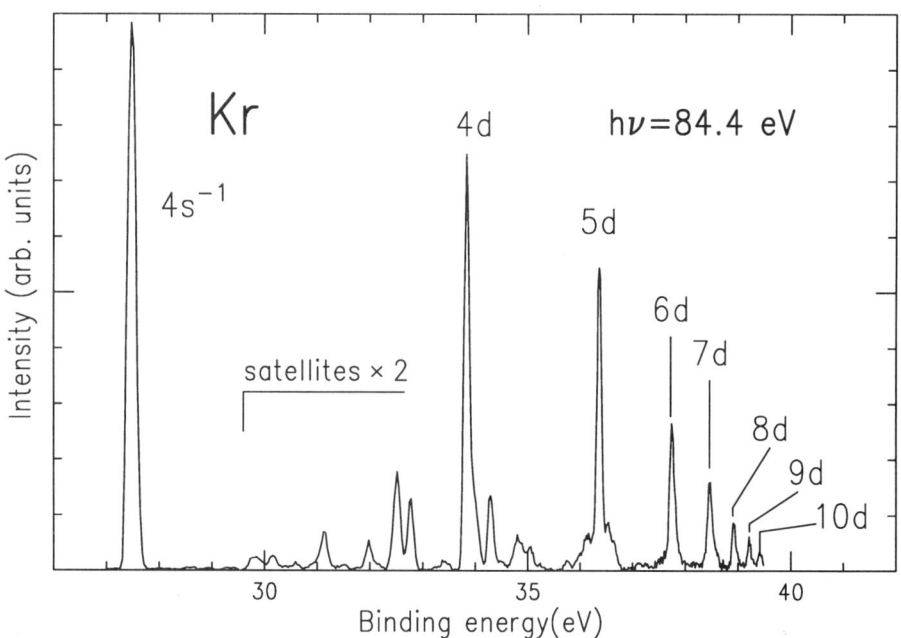

FIGURE 16. Photoelectron spectrum of the valence shell of Kr with the 4s main line and their associated satellite lines.[120] In order to distinguish more clearly between the most prominent satellite series nd, and the 4s main line, the latter was labeled by $4s^{-1}$.

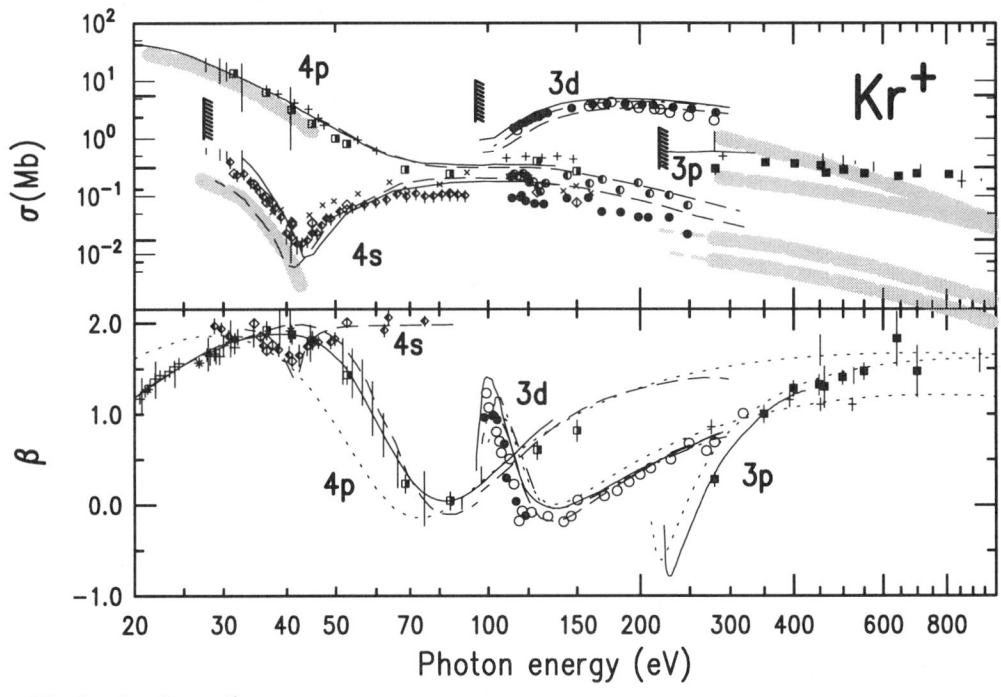

FIGURE 17. Partial cross section σ and angular distribution asymmetry parameter β for the 4p, 4s, and 3d subshells of krypton. The dotted and dashed dotted lines represent HS[68] and HF[69] calculations, respectively. The short dashed lines show an MMCDF calculation of Ref. 125, whereas the solid lines represent RRPA calculations of Refs. 95 and 126 as in the case of argon. The experimental data points are from Refs. 12, 76, 80, 120, 123, 124, 127–133 as shown in the figure.

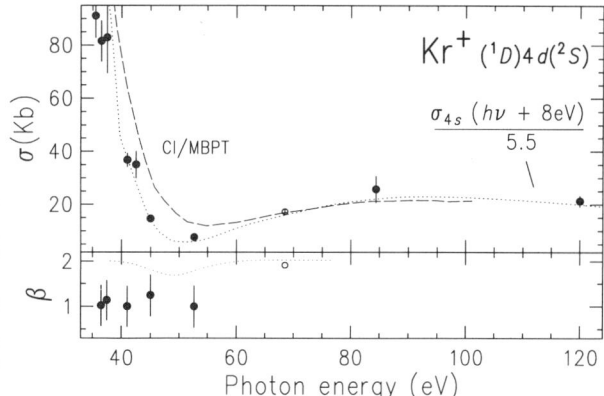

FIGURE 18. σ and β of the Kr (^1D) $5d(^2S)$ satellite state. Solid circles are from Ref. 120, open circles from Ref. 83. The dashed curve represents a CI/MBPT curve of Ref. 134. The dotted lines show the behavior of the 4s photoline shifted by 8 eV and scaled to the satellite intensity at higher kinetic energies.

of the 4p, 4s, and 3d main lines along with their β values. The accuracy of the data presented here has improved more than in the case of Ar, because Kr was much less intensively studied than Ar. The number of independent measurements of the Kr 3d photoline is still very small; further measurements, particularly for σ at higher energies, are absolutely necessary. The situation is even worse for the other inner-shell photoionization cross sections; only some β values for Kr 3p have been measured recently.[122] The discrete x-ray source measurements of Krause[123,124] are, for these subshells, still the only available partial cross section data. More progress was again made in the study of many-electron processes, particularly the correlation satellites. Figure 18 shows σ and β of the prominent $4d(^2S)$ satellite, in comparison to the corresponding but rescaled data of the 4s main line. The observed behavior is similar to the behavior of the Ar $3d(^2S)$ satellite shown in Figure 13.

The double ionization rate in Kr has been the subject of controversy because of the unusual oscillation in the partial cross section below the 3d threshold.[135] More recent results still show this structure, but it is much less pronounced.[136] In the light of what was said about double ionization in neon, it seems most likely that the oscillating structure results from a two-step process rather than from simultaneous shakeoff. In this way it would reflect the intensity variation of a satellite with a binding energy above the double ionization threshold. Such a satellite could decrease from threshold or/and exhibit a Cooper minimum, which would result in an oscillating structure of its partial cross section. Valence Auger decay[137] of this satellite would transfer this satellite-specific behavior to the double ionization rate measured by ion-yield spectroscopy. Although this is the most probable explanation of this behavior, it still has to be tested experimentally.

4.4. Xenon

Xenon, with a $(1s)^2(2s)^2(2p)^6(3s)^2(3d)^{10}(4s)^2(4p)^6(4d)^{10}(5s)^2(5p)^6$ ground-state electron configuration, has become the showcase for inner-shell phenomena in a very similar way as has helium for correlations in a two-electron system. The reason is not only that all "single-particle" effects show up even more pronounced in Xe than in Kr, but also because many-electron effects become dominant features of the partial cross section behavior here. The different subshells can no longer be treated as even approximately independent. Instead, coupling of the different photoionization channels becomes absolutely necessary in order to describe the cross section behavior correctly. It was the strong influence of the 4d partial cross section on the 5p and 5s photoionization cross section, particularly in the shape resonance, which brought the breakthrough for the random phase approximation (RPA) in its various forms. Following the pioneering work of Amusia[6,7,45] in this field, the calculations have finally covered extended energy ranges, showing that not only is the 4d shape resonance seen in the 5p and 5s cross sections, but so also is the subsequent Cooper minimum.[138] Theoretical calculations matched the experimental data almost perfectly for xenon. Fine-tuning of both experimental and theoretical

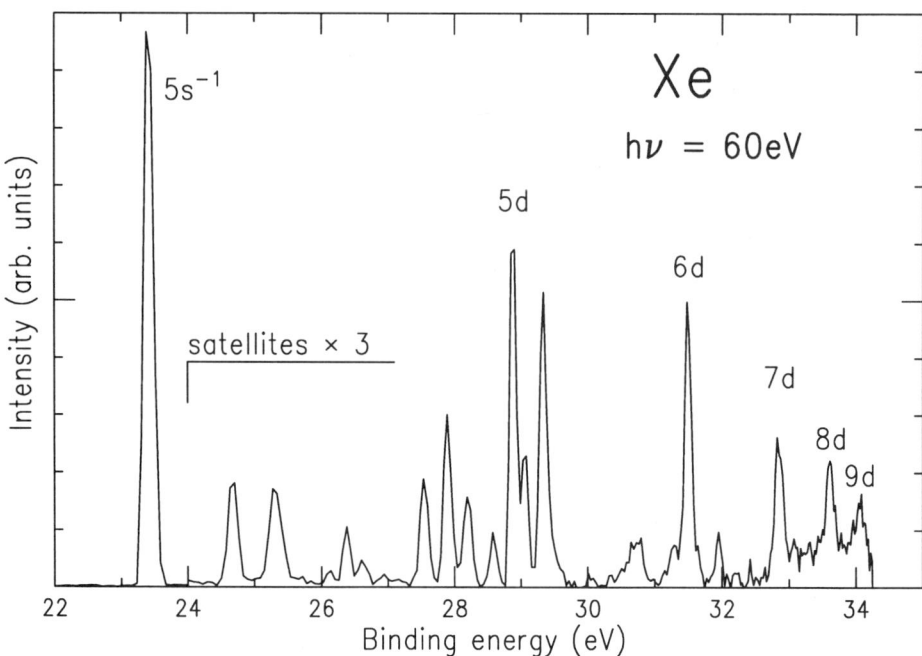

FIGURE 19. Photoelectron spectrum of the valence shell of Xe showing the 4s main line ($4s^{-1}$) along with a whole variety of satellite lines, among them the nd states analogue to Ar and Kr.[145]

results have reached a point of high accuracy in the case of the β values of the Xe 5s line in the Cooper minimum (see, e.g., Refs. 139, 140) and the partial cross section of the 4d in the shape resonance (see, e.g., Refs. 100, 141). Both cases were milestones in our understanding of atomic photoionization. Figure 19 shows photoelectron spectra of both the valence shell and the 4d shell. Similarly to Kr, we see a large variety of satellite transition intensity, dominated again by an nd series.[81,105,121] Satellites are also seen associated with the 4d photoline.[81,142] Figure 20 shows the partial cross sections and β values of Xe 5p, 5s, 4d, 4p, 4s, and 3d on a double logarithmic scale. The figure shows how strongly the different partial cross sections vary in this energy regime. For the sake of clarity, only one set of theoretical curves is shown, the RRPA calculation of Kutzner et al.,[138] because they are the only ones which cover the complete energy range. Only for the case of the controversial 5s photoionization are the results of an alternative approach, the MMCDF method,[140] are shown. The experimental data represent for the same reason a selection of measurements performed over wider energy ranges but showing good agreement with other less extended measurements, particularly in the shape resonance.[143,144] Considering satellite cross section behavior, we show in Figure 21 again the most prominent nd satellite in xenon: the $(^1D)5d(^2S)$ satellite. This satellite is the only one for which theoretical calculations have been performed: the agreement obtained with experiment is very promising.[140] The satellite follows the behavior of its main line, Xe 5s, in nearly all details, including the enhancement resulting from coupling with the 4d shell. This behavior is seen for many of the xenon valence satellites; however, the agreement between main and satellite line seen in the β values of the 5s satellites is not observed for the 5p satellites. In this case, further experimental work, particularly with better energy resolution, as well as more sophisticated theoretical work, seems to be necessary to understand the details of these many-electron processes in xenon.

In summary, one can say that the σ and β values of not only the main lines but also of selected satellites in the rare gas atoms are theoretically quite well understood and experimentally documented with good overall accuracy, in most cases. Further work on many-electron transitions is needed.

PARTIAL CROSS SECTIONS AND ANGULAR DISTRIBUTIONS 155

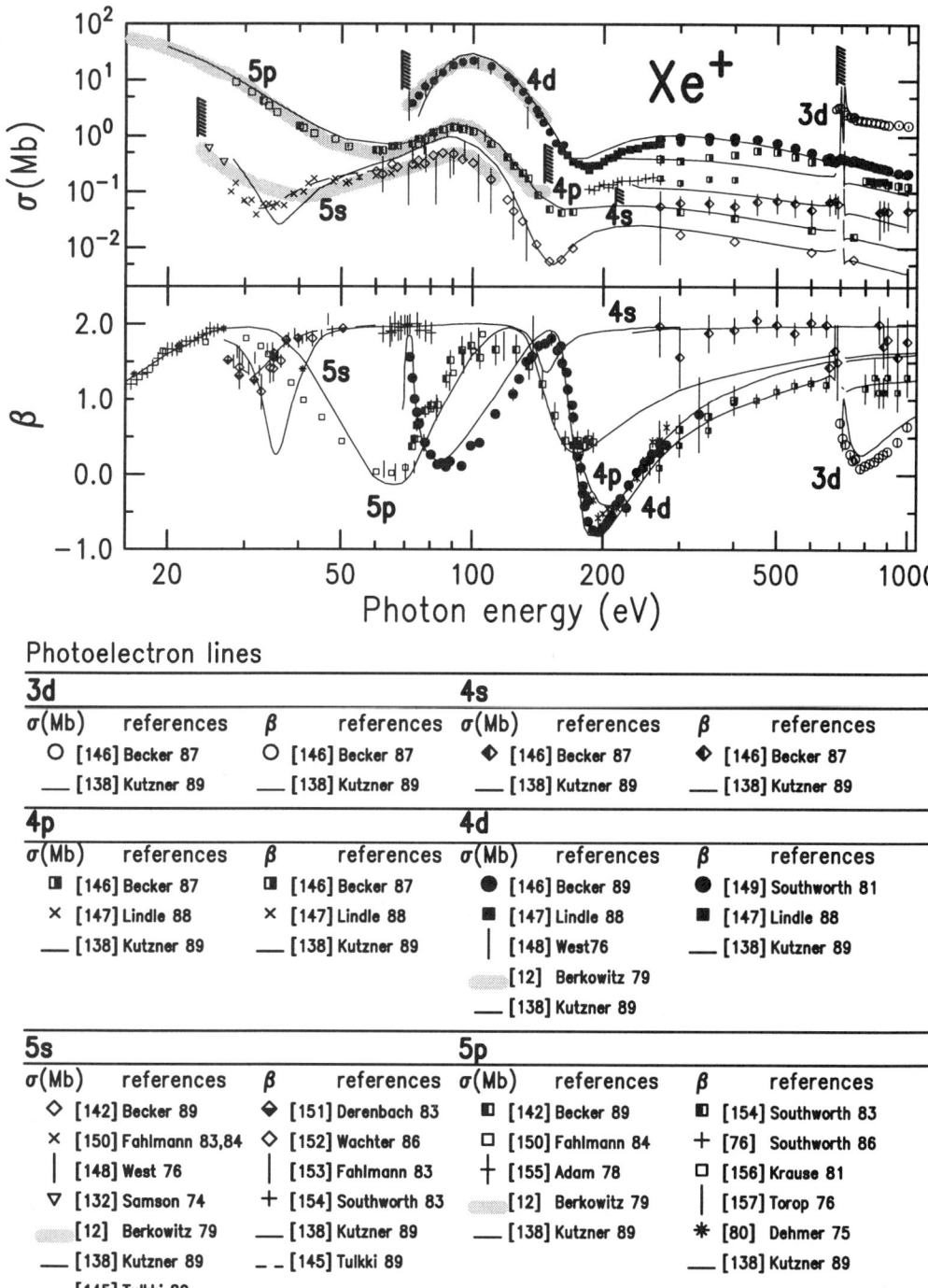

FIGURE 20. Partial cross section and angular distribution parameter β for the 5p, 5s, 4d, 4p, 4s and 3d satellites of xenon. The short dashed line shows an MMCDF calculation of Ref. 140 for Xe 5s, whereas the solid lines for all other photolines represent RRPA calculations of Ref. 138. The experimental data points are from Refs. 12, 76, 80, 132, 142, 146–157 according to the figure.

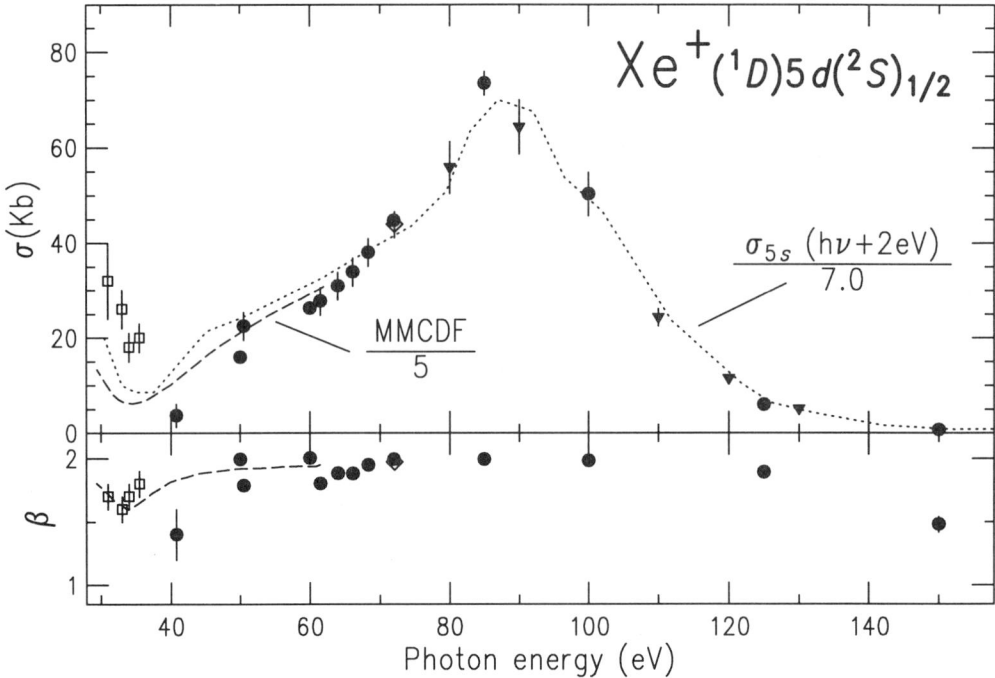

FIGURE 21. σ and β of the Xe $(^1D)5d(^2S)$ satellite line. The short dashed lines show MMCDF curves of Ref. 140 of which the σ curve is scaled by a factor of 3. The dotted line indicates the behavior of the $5s$ main line scaled by a factor of 5.5 and shifted by 4 eV. The experimental data points are from the following references: ●, Ref. 145; ◇, Ref. 83; ▼, Ref. 111, □, Ref. 150.

5. TRANSITION ELEMENTS

If one starts to study the photoionization of elements other than the rare gases, one will very soon realize that these elements fall practically into two groups, metal atoms and molecules, as the natural form of their appearance. This explains why these elements were so much less intensively studied than the rare gases. The production of an atomic beam of an adequate density to perform angle-resolved photoelectron spectrometry was by no means trivial; therefore, the number of experiments performed to date is still very limited. The electron spectrometry of metal vapors was reviewed quite recently by Sonntag and Zimmermann,[158] and many results obtained by ion spectrometry are treated within this volume in a separate chapter on "Ion Yield Spectroscopy with Soft X Rays."[27] Among the elements forming molecules, free atoms had to be produced via radio frequency or discharge dissociation. Therefore, such open-shell atoms have only recently become the subject of photoelectron spectrometry studies. They are addressed in the chapter on "High-Resolution Electron Spectrometry of Atoms."[15] Here we will concentrate again only on studies which were performed over extended energy ranges. For these purposes we chose as examples for transition metals the two elements with half-filled subshells, Mn and Eu, and with filled subshells, Cd and Hg. There may be other elements, such as Ba,[159] Yb,[160] or some other alkali earth elements,[161] which are similarly well studied, but for the purpose of this review, to compare principal types of subshell photoionization behavior, the examples selected are, in our opinion, best suited. These examples will be presented according to the specific features which show up in their cross section behavior according to certain systematic trends, rather than simply against atomic numbers. In this way we will start with Cd and proceed through Hg to the two transition metals Mn and Eu.

5.1. Cadmium

Cadmium, with the electron configuration [Kr] $(4d)^{10}(5s)^2$, was a favorite candidate for photoionization studies since the early days when discharge lamps were still employed.[162–164] However, a complete set of partial cross sections and angular distribution parameters over wide energy ranges, covering all of the single-particle and interchannel coupling effects, became available only very recently.[165] Therefore, Cd represents a case of continuous study over more than two decades, as in the case of the rare gases. The questions concerning Cd are very closely related to Xe, the corresponding rare gas. As in Xe, the 4d photoionization channel is subject to pronounced oscillations in σ and β. Also the influence of interchannel coupling is exhibited in the partial cross sections of the 5s and 4d subshells. All of these phenomena were extensively studied theoretically, applying the RRPA method.[166,167] The agreement obtained with the experimental data is quite reasonable. The 4d satellite structure consists of more clearly separated satellite lines than in Xe, where this structure is very compact and difficult to resolve. Also distinguishing Cd from Xe is the occurrence in Cd of an even broader "4p" peak in the spectrum.[168] The shape of this peak is attributed to the strong interaction with the 4d fluctuations, the so-called "super Coster–Kronig" fluctuations $4p^{-1} \Leftrightarrow 4d^{-2}mf,\varepsilon f$. This process gives rise to the broad peak structure, because the 4p intensity is spread over a wide continuum of energies also covered by the 4d double-hole satellite structure.[169] The existence of strong satellites in Cd 4d photoionization is associated with excitations into the 4f orbital resulting from the approximately collapsed character of this orbital in the final ionic state. Figure 22 shows a photoelectron spectrum of Cd with all of these lines taken at hv = 110 eV. Figure 23 shows the 4d partial cross section along with the corresponding angular distribution parameter. Also shown are theoretical curves obtained by the RRPA method.[166] Both RRPA calculations, made without[166] and with relaxation,[167] are in good agreement with the experimental data, although the absolute scales still differ by a factor

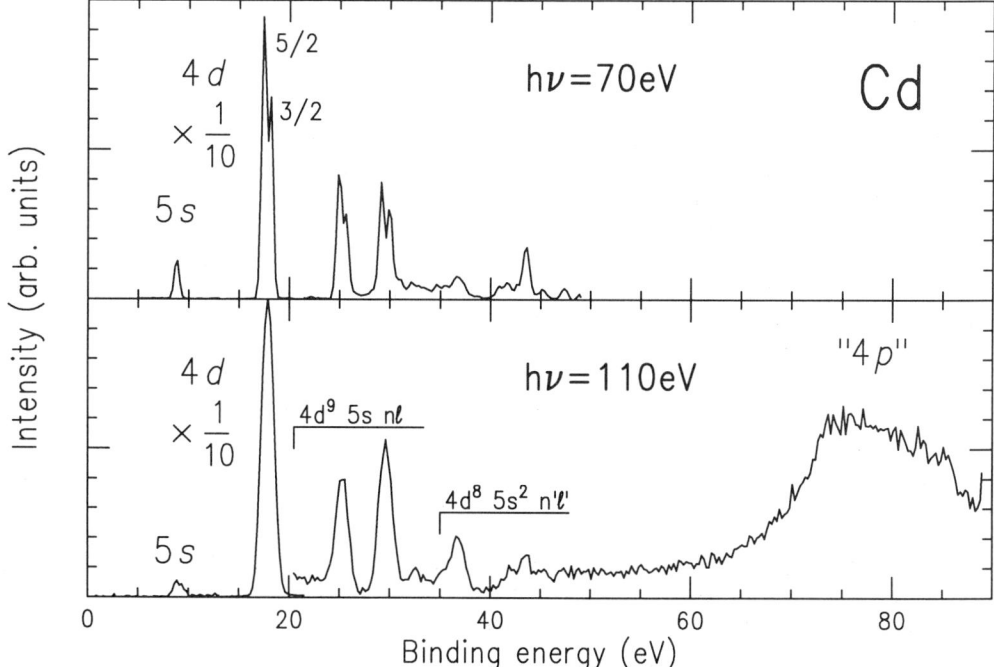

FIGURE 22. Photoelectron spectrum of Cd taken at photon energies of 70 and 110 eV.[165] The spectrum in the upper part with its higher resolution shows clearly the spin orbit splitting of both main and satellite transitions, whereas the lower spectrum with its higher photon energy shows in addition the broad structure of the "4p" photoemission line.

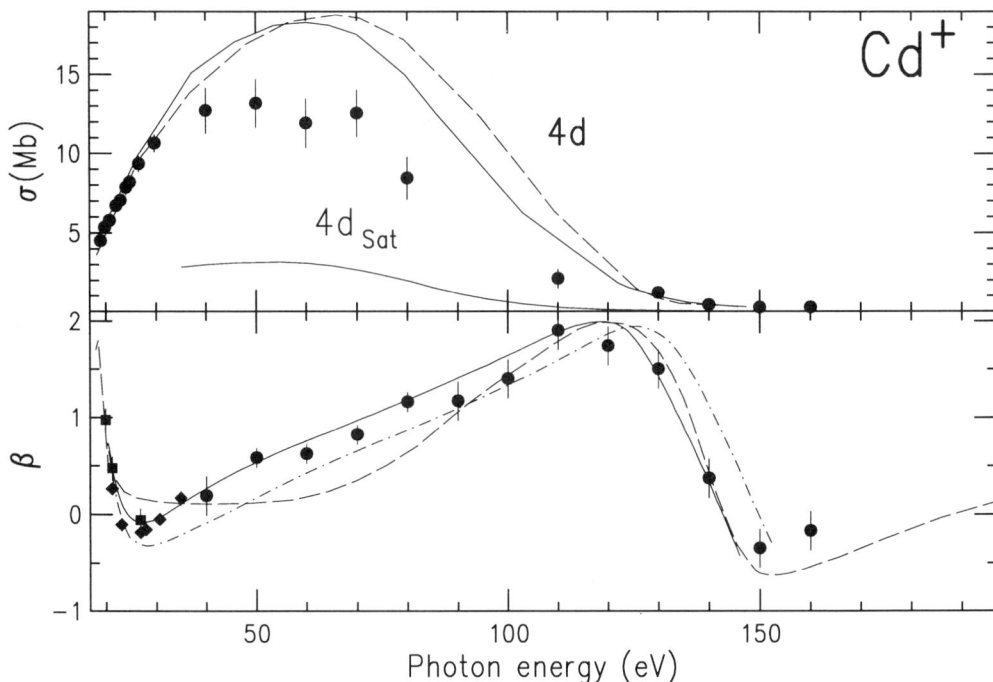

FIGURE 23. Partial cross section σ and β parameter for photoionization of the Cd 4d subshell. The dashed dotted curve shows the results of a DF calculation,[170] the long dashed lines show MBPT results of Ref. 171, and the solid lines represent σ and β values obtained by the RRPA method.[166] The dotted line indicates the integral intensity of the 4d satellite lines. The experimental data are from the following references: ♦, Ref. 164; ■, Ref. 172; ●, Ref. 165.

of 1.3–1.5 if compared with respect to the absorption measurements. The calculations including relaxation give some integral information on the strengths of the many-electron processes, which is in approximate agreement with the more recent measurements. The calculations predict a second Cooper minimum in the 5s partial cross sections, which should give rise to a small dip in the $\beta_{5s} = 2$ curve at the minimum position. There are indications in the experimental data supporting this prediction. The oscillation in the β_{4d} values are quite similar to Xe, the three theoretical curves deviating from each other only for the higher energies because relaxation tends to lower the total intensity at the maximum of the shape resonance but causes an increase in intensity at higher energies. To confirm these predictions, however, improved measurements still have to be performed.

5.2. Mercury

If we go down a row in the periodic table we come to mercury, an element so heavy that its corresponding rare gas element radon is unstable, and an easy comparison with rare-gas behavior cannot be done. Hg with its ground-state electron configuration [Xe] $(4f)^{14}(5d)^{10}(6s)^2$ is the only widely studied element with a filled 4f subshell, making this element particularly interesting for a critical comparison with theoretical calculations.[167,173–178] Figure 24 shows a photoelectron spectrum of mercury taken at $hv = 120$ eV. The 4f, 5p, and 5d photolines are, aside from an Auger peak containing a whole group of Auger lines, the prominent features in this spectrum. The most interesting partial cross sections are Hg 5d and 4f, because they represent the only examples of nonresonant f-shell photoionization besides Yb[160] and of d-shell photoionization with a principal quantum number $n = 5$. Figure 25 shows σ and β values of these two photolines in the energy range where both experience their strongest variations. Random-phase (RRPA)[166,177] and Dirac–Slater (DS)[178] calculations are

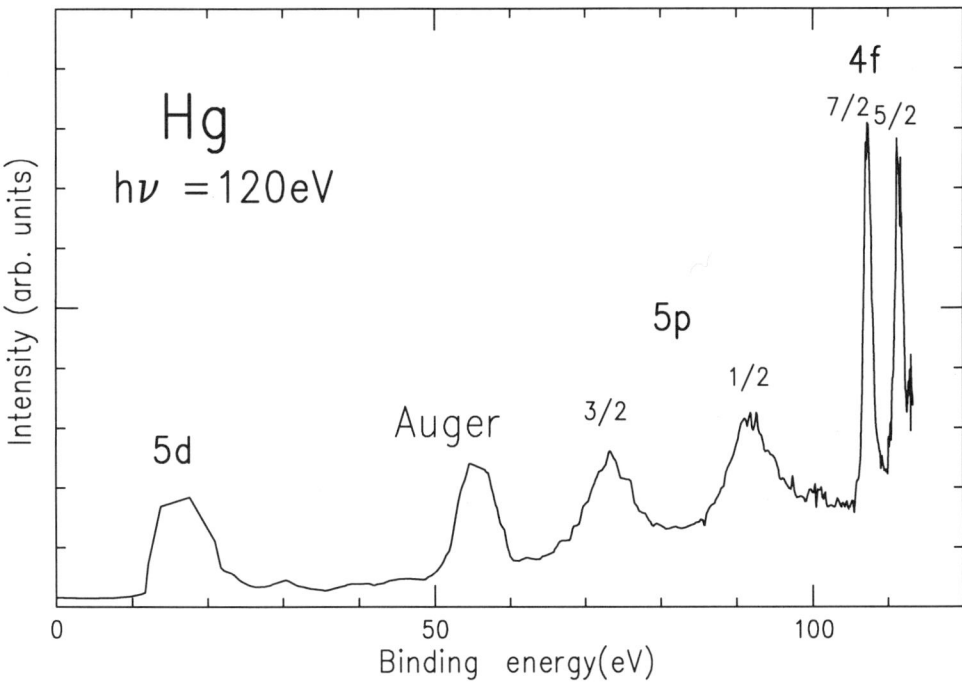

FIGURE 24. Photoelectron spectrum of Hg showing the 4f, 5p, and 5d photolines along with structure related to Auger transitions.[179]

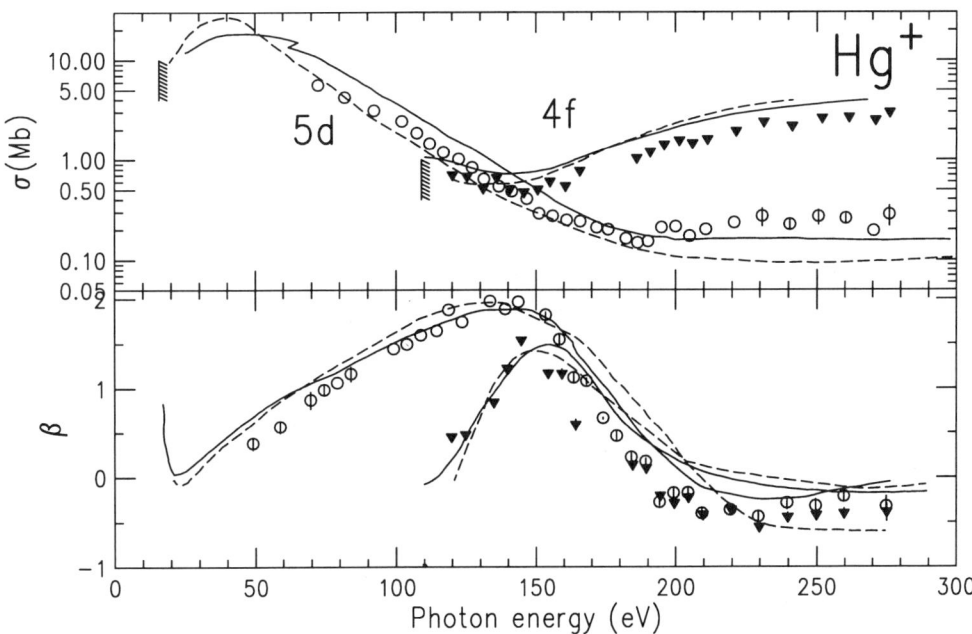

FIGURE 25. σ and β of the Hg 5d and 4f subshells. The dashed curves show DS calculations of Ref. 178 whereas the solid curves represent the results using the RRPA method.[166,177] The experimental data points are from Ref. 179 with open circles for 5d and filled triangles for 4f subshells.

shown for comparison. The similarity between the two curves shows quite obviously that single-particle effects are the dominant features in the partial cross section and angular distribution behavior. Comparing this behavior with that of the corresponding photolines, 3d and 4d in Kr, Xe, and Cd, one sees the following: the 4f-based shape resonance is less pronounced than the 4d shape resonance; on the other hand, the onset is moved to lower energies, starting even before the first minimum, because a small shape resonance occurs in the d-wave channel just above threshold.[173] This supports the trend toward lower onset energies relative to the shape resonance maximum with increasing l. The opposite is seen with respect to the principal quantum number n: here the onset moves toward higher energies relative to the maximum as n increases. This is nicely demonstrated by the difference in the cross section behaviors of Kr 3d, Xe 4d, and Hg 5d. Qualitatively, the binding energy of the d electron decreases rapidly with n in this sequence and the overlap of the outgoing εf wave with the less-localized nd orbitals becomes more and more favorable with increasing n, thereby shifting the onset of the partial cross section up along the shape resonance. These very general trends are clearly demonstrated both experimentally and theoretically for d-shell photoionization, and to a lesser extent also for p-shell photoionization. Most of the trends, which can still be explained in a single-particle picture, persist, at least qualitatively, even in the presence of strong many-electron interactions, provided that filled subshells are considered.

5.3. Manganese

The situation changes dramatically if unfilled subshells are taken into account. In this case, strong resonances based on Coster–Kronig and super-Coster–Kronig transitions start to dominate the cross section behavior near the thresholds of those inner-shell orbitals from which excitations into the unfilled orbital become possible. The 3d and 4f transition-metal atoms are the most prominent examples for such behavior. We will demonstrate the characteristic behavior of the inner-shell photoionization of these elements by two examples, both with a half-filled subshell. These are Mn and

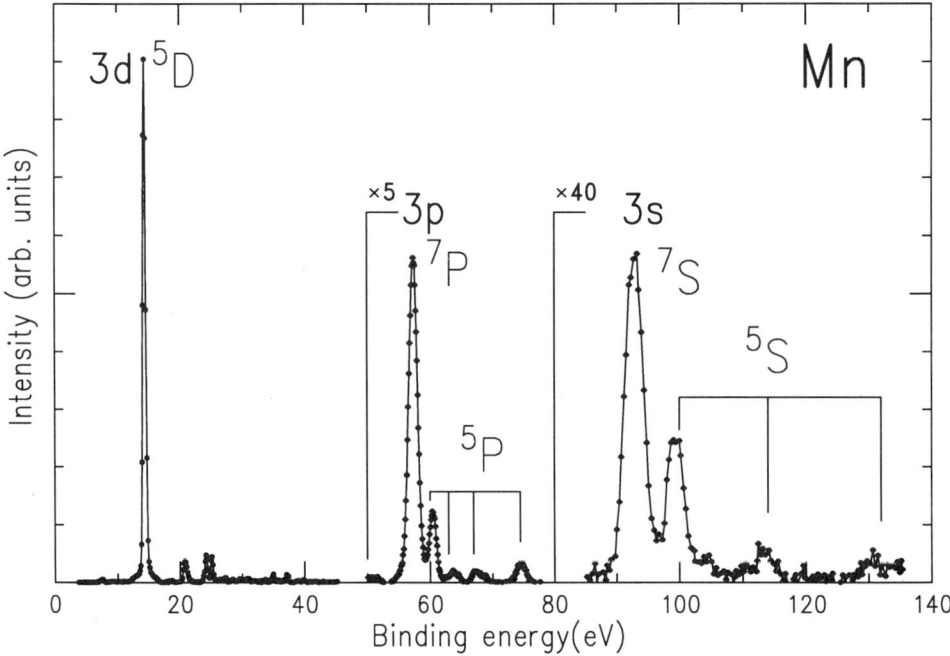

FIGURE 26. Photoelectron spectrum of Mn showing the 3d, 3p, and 3s photolines with their associated satellites.[180]

Eu, with the ground-state electronic configurations [Ar] $(3d)^5(4s)^2$ and [Xe] $(4f)^7(6s)^2$, respectively. These two elements have been the subjects of various experimental and theoretical investigations because they represent the simplest cases of atoms with unfilled subshell which are easily produced in the vapor phase. The photoionization of these two elements was studied over an extended energy range, including covering the cross section behavior outside the resonance regime, to which most studies of the other transition elements have been restricted.[183,188–194] Figure 26 shows a composite photoelectron spectrum of Mn taken at different photon energies in order to illustrate the structure of the different subshell photopeaks.[180] The unfilled 3d subshell allows excitation of the 3p electron into the 3d shell with subsequent decay of this core-excited state via the emission of another 3d electron, the so-called super-Coster–Kronig (sCK) decay.[195] The occurrence of these strong transitions which are, in a single-particle description, related to a phenomenon known as orbital collapse, are, however, so strongly governed by many-electron effects that a theoretical description within the single-particle model makes no sense. Indeed, the calculation of these resonances has become the domain of theories which implicitly take care of all of the many-electron effects involved, such as the random-phase approximation with exchange (RPAE)[182] and many-body perturbation theory (MBPT).[181,196] The calculation of the partial cross section behavior of Mn and Eu within the $3p \rightarrow 3d$ and $4d \rightarrow 4f$ resonances was one of the highlights of these approaches. Because of the many-electron or collective nature of these resonances they also became known as "giant resonances," in analogy to similar phenomena observed in nuclear physics. Figure 27 shows the partial cross sections and angular distribution parameters of Mn 3d along with a theoretical curve obtained by MBPT.[181] The $3p \rightarrow 3d$ resonance is seen as a broad and mostly unstructured feature in the 3d partial cross section in both the experimental data and the theoretical curves. However, there is still the question whether this resonance really consists of one broad resonance or perhaps of many, either unresolved or overlapping within their natural linewidths. This question may never be answered unambiguously by

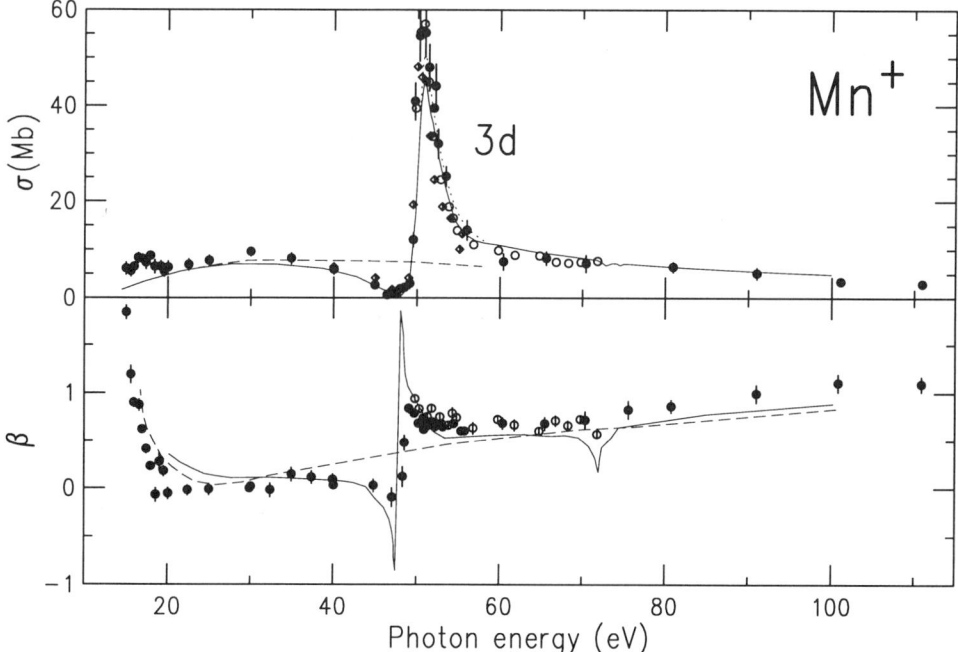

FIGURE 27. σ and β of the Mn 3d photoline. The dashed curves represent HF calculations,[181] whereas the long solid curves show MBPT calculations[181] and the dotted curve RPAE results.[182] The experimental results are from the following references: ◊, Ref. 183; ●, Ref. 184; ○, Ref. 185.

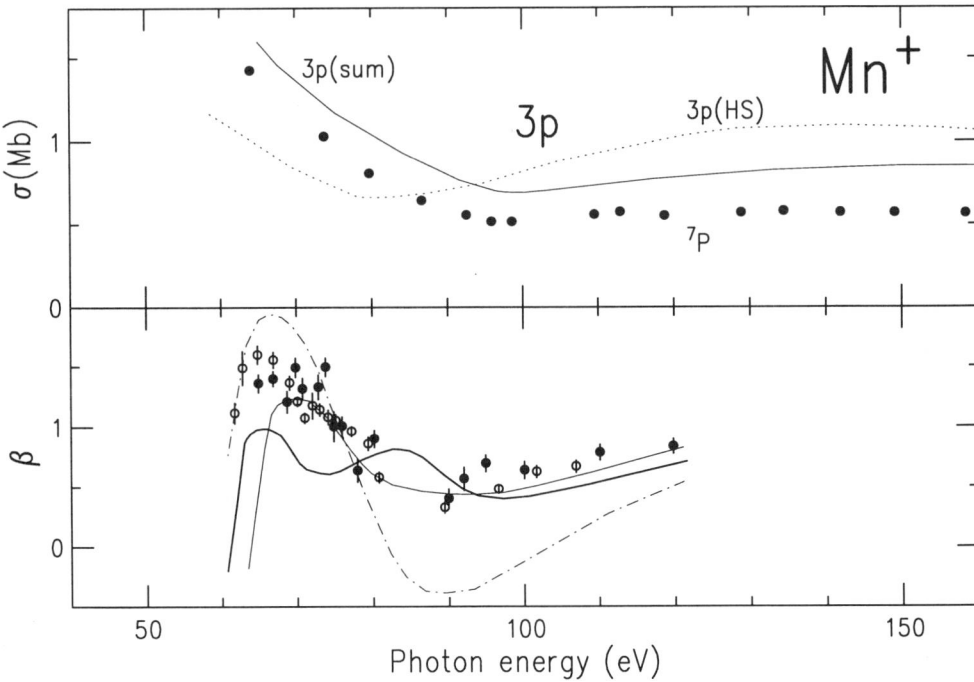

FIGURE 28. σ and β of the Mn 3p photoline. The partial cross sections are shown for the ^7P and the strongest components of the ^5P states and compared to the sum of the multiplet (solid line) and a 3p HS calculation[68] (dotted line). The β values are given for the $3p(^7P)$ photoelectrons only. The dashed dotted β curve shows an HF calculation,[186] whereas the two solid curves represent two different RPAE calculations, the thinner one including coupling with the 3p and 3d shell, the thicker one additional coupling with the 3s subshell.[186] The experimental data are from Ref. 187 (○) and Ref. 180 (●).

experiment alone, if only the 3d partial cross section is studied. However, there are other ways to look somewhat closer into the details of this resonance. It is well known that the enhancement of satellites via resonances is not uniformly distributed among all of the involved resonances; rather it is very selective, or in other words certain satellites may be enhanced on certain resonances while others are quite unaffected. This varies from resonance to resonance, depending on the configuration and coupling in the excited state. Therefore, one way to look for hidden structure in the Mn $3p \to 3d$ resonance may be to look for the satellite behavior within this resonance, because the Mn photoelectron spectrum exhibits a whole variety of satellite lines. And in fact, a very recent measurement of this kind indeed reveals very distinct structure within the satellite resonance profiles.[197] Figure 32 of Chapter 6 shows, as an example, the 3d satellites of Mn in the $3p \to 3d$ resonance region. The structure is completely different from what is known from the 3d partial cross section behavior. A theoretical interpretation of this observed satellite behavior in giant resonances is still needed. The behavior of the nonresonant photoionization of the 3p subshell of Mn is shown in Figure 28. In this case, because of the half-filled 3d subshell, a spin-polarized model has to be employed as zero-approximation, but again electron correlations have to be explicitly taken into account if one wants to describe the 3d subshell behavior also quantitatively well.[186]

5.4. Europium

The rare-earth, or lanthanide, elements, characterized by the filling of the 4f shell, are of particular interest in several respects. All of the phenomena seen in the 3d elements are also seen in the 4f elements; however, the phenomenon of orbital collapse now becomes the dominant process because

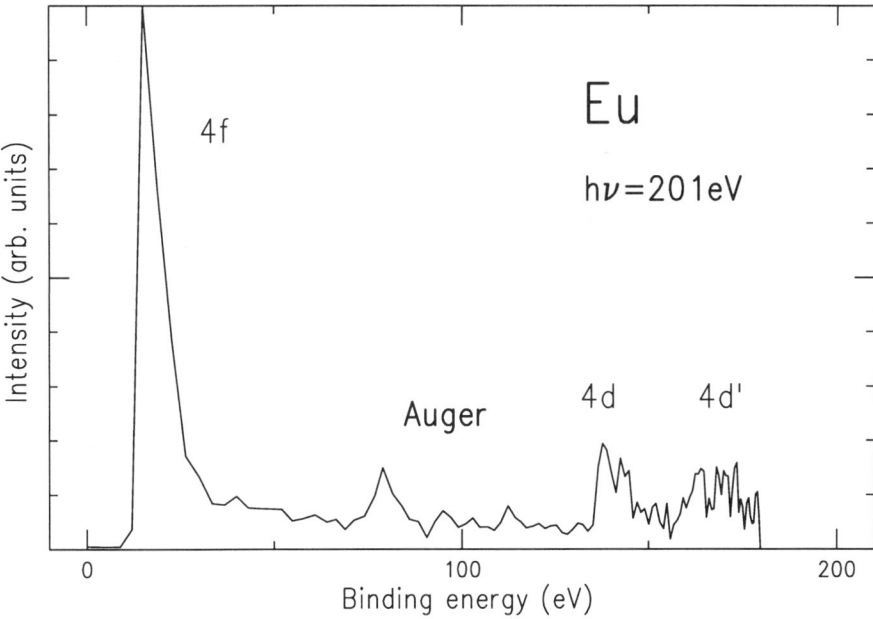

FIGURE 29. Photoelectron spectrum of Eu showing the 4f and 4d photolines along with their associated satellites and subsequent Auger transitions of the 4d hole states.[198]

of the larger angular momenta involved in the excitation and decay process. As a result, there is not only resonance enhancement in the sCK-4f channel,[199] in analogy to the 3d channel in the 3d elements, but also resonance enhancement in the 4d excitation channel itself,[200] a phenomenon not seen in the $3p \to 3d$ giant resonances. This 4d enhancement may be, roughly speaking, considered as a kind of shape resonance enhancement, as seen in Xe, where the 4f orbital is still in the outer valley of the potential curve governed by the delicate balance between coulombic and centrifugal forces. In a way, one may consider the fractional intensities between 4d and 4f resonant enhancement as a measure of the gradual collapse among the rare-earth elements. In practice, the situation is much more complicated, involving term dependencies which reflect the strong influence of electron correlations in terms of an extended one-electron model.[201,202] Nevertheless, the unique behavior of many rare-earth metal compounds in the solid state, such as mixed-valence and Kondo effects, are based on the localization properties of the 4f electrons in these materials. Photoionization studies of free rare-earth atoms provide detailed insight into the critical interplay between correlation and localization in 4f electron systems. Figure 29 shows a photoelectron spectrum of Eu, a 4f element with a half-filled 4f subshell.[198] The spectrum shows the 4f and 5p photolines along with Auger lines stemming from decay of the 4d hole states just above their first threshold. Concerning the $4d \to 4f$ resonance enhancement of Eu, the situation is even more complicated than in Mn because it is unclear in which of the 4d channels this enhancement actually occurs. It may occur in all fine-structure components, supporting the idea of an "autoionizing resonance in its own channel,"[203] or there may indeed only be excitations in the closed channel below threshold which autoionizes into the open channels of the lower-binding-energy 4d components. Both cases would give rise to the emission of 4d-based NOO Auger electrons. Because all experiments performed so far near threshold measured only the 4d Auger intensity, a decision between these two interpretations cannot be made. In fact, even the multiplet structure of the 4d hole states is still somewhat ambiguous because there were some indications for the existence of two more widely separated groups of 4d lines, perhaps the theoretically predicted ^7D and ^9D LS-coupled final states.[204] However, there were also indications that at least the first group consists of only two strong components, which may be interpreted as $d_{3/2}$ and $d_{5/2}$ hole components, which may also be written

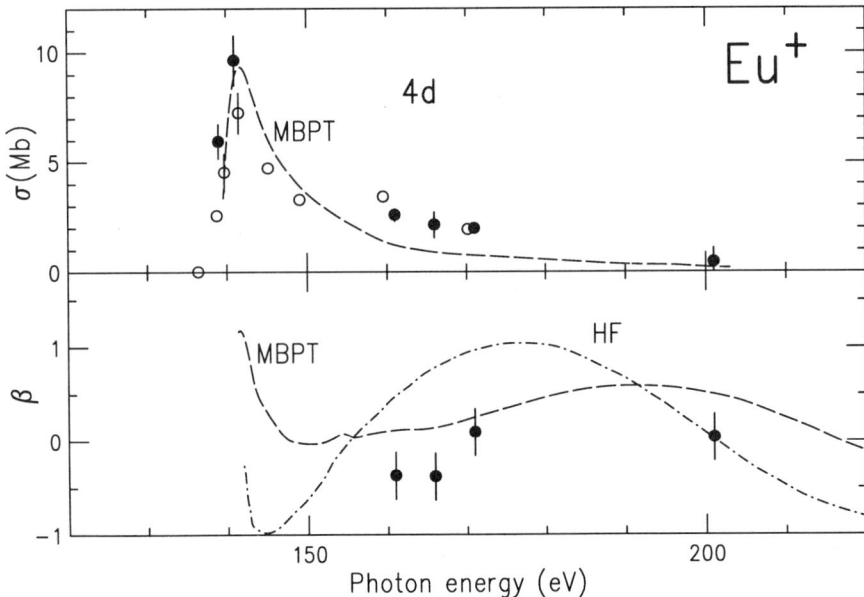

FIGURE 30. σ and β of the Eu 4f photoline. The dotted curves show HF calculations, whereas MBPT results are given by long dashed curves.[204] The experimental σ results are from Ref. 205 (●), Ref. 206 (○), and the experimental β values from Ref. 198 (●).

as $d_\downarrow^{-1}$ and $d_\uparrow^{-1}$ in terms of a spin-polarized model. Independent of these questions we are able to give partial cross sections for the first two 4d components, at approximately 137 and 142 eV, and to compare them with corresponding theoretical calculations, in this case with the intensities of the $4d^{-1}(^9D)$ photolines.

Figure 30 shows the partial cross section and angular distribution parameters of these lines, along with an MBPT calculation.[204] The experimental data are all relative partial cross sections which were brought to an absolute scale by scaling the total yield to the calculated total intensities. The good agreement in the case of the partial cross section shows that the MBPT calculations have succeeded in giving each photoionization channel its correct strength, a result no other method provides so far in a comparable way. There are only very few experimental data available to compare with the theoretical β curve; considering the error bars these points are in reasonable agreement with the MBPT calculation. However, more experimental data are urgently needed to characterize the overall β behavior. The situation is somewhat more favorable concerning the 4f partial cross sections and β values. This is because the $4f(^7F)$ photoline is clearly defined and its intensity can be unambiguously followed over a wide energy range covering resonant and nonresonant regions. Figure 31 shows these experimental data along with theoretical curves, in particular the MBPT curves. The comparison with the Hartree–Fock (HF) β curve shows nicely that the resonant β behavior is completely governed by electron correlation, in fact by the interchannel interaction between the $4f \to \varepsilon g$ channel and the $4d \to 4f$ resonant transition which changes the dipole matrix element of the continuum channel by a phase factor of approximately −1. This is different from the 4d β behavior where the oscillation near the threshold is not entirely related to correlation effects although it becomes appreciably affected by them. This supports our interpretation of the 4d enhancement as being the result of above-threshold shape-resonance-like enhancement. In the case of the 4f σ and β results there is quite good agreement between theory and experiment, demonstrating once more the unique capability of MBPT in treating such complex photoionization phenomena as seen in Eu and the other rare-earth elements. However, in contrast to all of our other selected examples, in this case even the main-line behavior has still to be studied with higher resolution, and satellite behavior has only barely been touched because of the

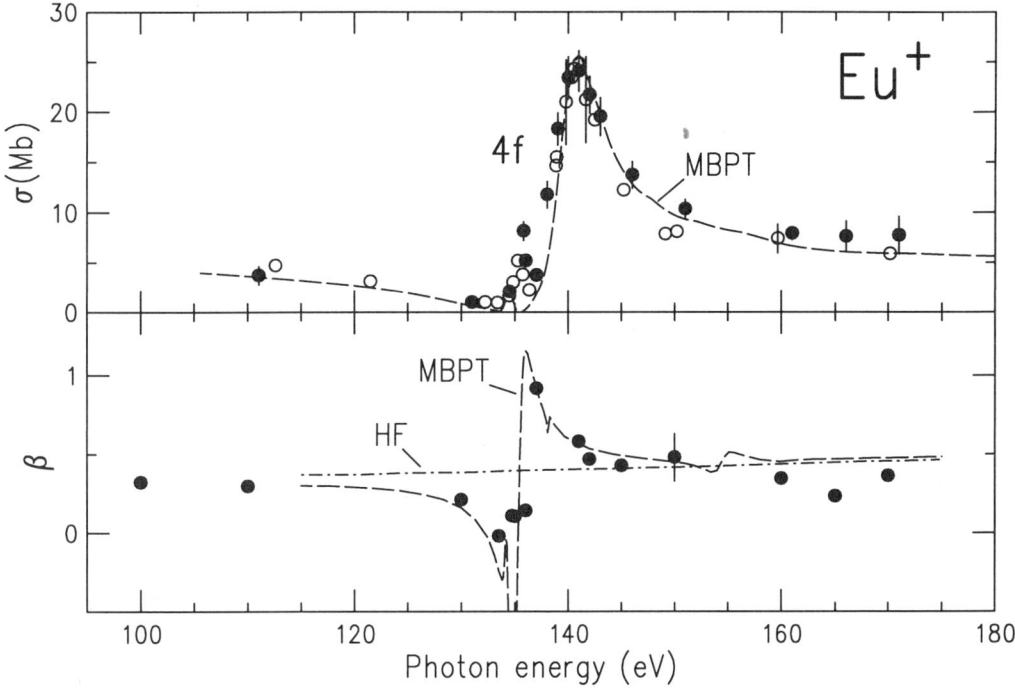

FIGURE 31. σ and β of the Eu 4d photoline. As in the case of Eu 4f the dotted curves represent HF calculations and the long dashed curves the MBPT results.[204] References for the experimental results are the same as in Figure 30.

experimental difficulties in performing such angle-resolved photoelectron experiments with metal vapors. However, the basic guidelines are drawn and further studies employing new-generation light sources will greatly extend the results discussed here, removing the remaining uncertainties.

6. CO: A SHOWCASE MOLECULE FOR MOLECULAR PHOTOIONIZATION

The task of summarizing molecular photoionization on an introductory level would require not simply a chapter but rather a separate volume dedicated to the subject (see, e.g., Refs. 207 and 208). Some aspects of current interest in this large field are presented in the chapters on "Valence Ionization Processes in the VUV Region," "Electronic and Nuclear Relaxation of Core-Excited Molecules," and "Photoion-Pair Formation." These experimental contributions are complemented by a theoretical chapter on a particular aspect of growing interest in molecular photoionization with soft x rays: core relaxation effects. Therefore, we will present in this chapter only one showcase example for molecular photoionization: carbon monoxide. The reason for choosing CO is the same as in the case of atoms, the availability of photoionization data over extended energy ranges, in this case covering all subshells which may be written in terms of molecular orbitals as $(1\sigma)^2(2\sigma)^2(3\sigma)^2(4\sigma)^2(1\pi)^4(5\sigma)^2$ from the inner- to the outermost. CO is isoelectronic to Ne and therefore characterized similarly by three general photoionization regimes: outer-valence, inner-valence, and core-level photoionization. The outer-valence behavior is the one most widely studied and consequently best understood, both theoretically and experimentally. The inner-valence part is the most complicated because its binding energy is so close to a whole variety of many-electron transitions including excitations to Rydberg orbitals, making this region a natural example for the breakdown of the orbital picture in molecular photoionization.[209] Interest in core-level photoionization is growing rapidly now because of dramatic progress in the

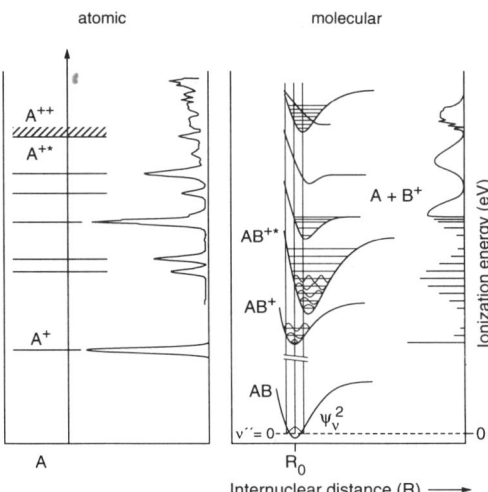

FIGURE 32. Structural representation of characteristic features which show up in molecular photoionization compared to a typical photoelectron spectrum obtained in atomic photoionization represented by the case of valence photoionization of neon.

resolving power of photon sources in the corresponding photon energy regime.[210–212] Many-electron effects in this field are just starting to become the subject of detailed studies.

There are three important points in which molecular photoionization differs from atomic photoionization: (1) the possibility of dissociation associated with photoionization or photoexcitation, (2) the influence of vibrational excitations on the ionization behavior, and (3) the existence of shape resonances in s-shell photoionization, particularly in the core-level regime. There are many other differences between photoionization in atoms and in molecules, but in our context of following general trends we will concentrate on these three, the first two of them shown schematically in Figure 32.

6.1. Valence Electrons

The upper-panel of Figure 33 shows a valence photoelectron spectrum of CO, and the lower panel a $C1s$ spectrum. The valence spectrum is a combined spectrum taken at different photon energies in order to show the vibrational structure of the outer valence lines and their first double-hole satellite transition in more detail. The other double-hole one-electron satellites, particularly $(n\lambda^{-1},n'\lambda'^{-1})2\pi$ and the 3σ to peak by itself, were recorded at a higher photon energy with comparable resolution;[213] however, the vibrational structure of the satellites requires even higher resolution. The situation is different in core-level photoionization. Here only one very recent measurement succeeded in resolving the vibrational structure of the $C1s$ photoline:[215] for satellites no vibrationally resolved data, except ZEKE measurements,[216] exist. Therefore, we show a spectrum taken at a fixed photon energy of $h\nu$ = 330 eV with sufficient resolution to resolve most of the satellite structure, but no vibrational fine structure. All three ionization regimes show quite different behaviors but also have much in common and interact with each other at distinct energies, such as threshold and resonance positions. The partial cross sections and β parameters of the valence electrons between 20 and 500 eV are shown in Figure 34. Two features are of interest in this figure: the occurrence of shape resonances in both σ and β and the influence of the π^* resonance on the σ valence photoionization. The existence of shape resonances in the partial cross sections of small molecules such as CO has been known for a long time and was studied in quite some detail as, e.g., their dependence on the vibrational excitations in the molecular ion after photoionization.[217–220] These shape resonances may be explained in a scattering model by

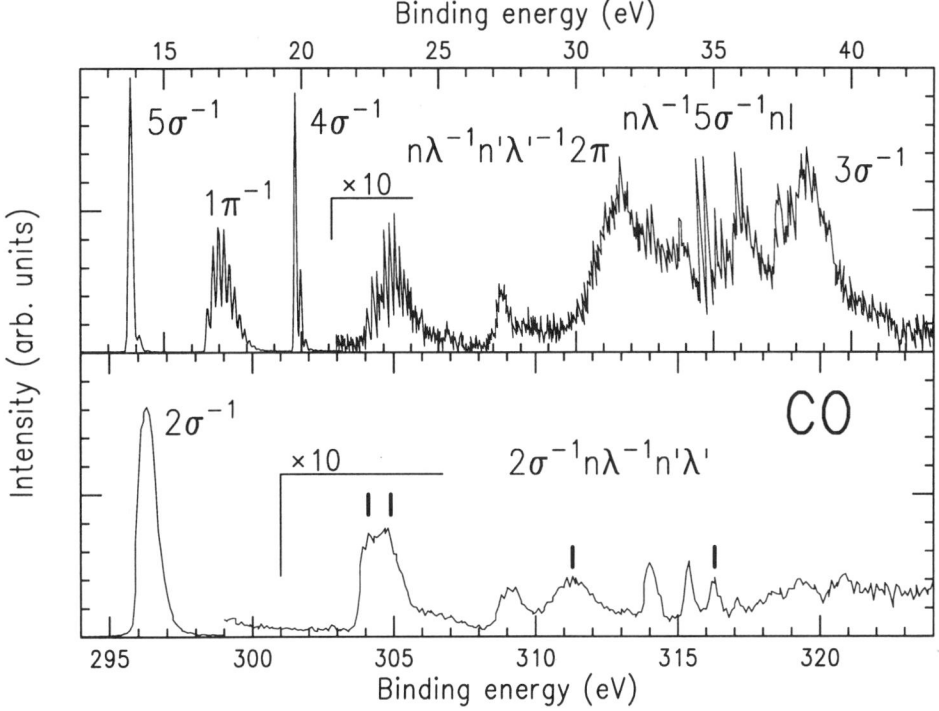

FIGURE 33. Combined high-resolution photoelectron spectrum of the CO valence main and satellite lines[213] (upper panel). The CO $C1s$ satellite spectrum in the lower panel is an example of a molecular inner-shell photoelectron spectrum.[214]

the emitted electron being scattered on its way out by the atoms of the molecule, thereby picking up angular momentum which in turn produces a centrifugal barrier seen by this electron, so that it becomes trapped within this barrier for some time.[221] This trapping time affects the width of the shape resonance, whereas the energy position is, in the first approximation, related to the distance between the two atoms, among other parameters. This aspect of molecular photoionization gave rise to the supposition about a semiempirical relationship between the position of molecular shape resonances and bond lengths, particularly in the core-level regime.[222,223] Attempts to prove this assumption by theoretical *ab initio* calculations with respect to its general validity, however, are still at the beginning.[224]

The partial cross sections of the valence electrons become dramatically enhanced at the position of the $C1s \to \pi^*$ resonance, demonstrating the strong overlap of these delocalized orbitals with the strongly localized $C1s$ core hole. This strong valence enhancement in core-excited CO is one of the most prominent examples of a behavior known in resonant photoelectron spectroscopy as participator transitions. Because the phases of the partial waves of these outgoing electrons become severely affected by the interaction with the resonance and the molecule also becomes strongly aligned as a result of the excitation, the β values of the valence photolines also become severely affected, and their β's drop by up to one β unit.

A particularly noteworthy point is the occurrence of valence β oscillations for the $n\sigma$ photolines within the σ^* resonance above the carbon K-edge, although no corresponding change in the partial cross section is seen. This effect seems to be restricted to the $n\sigma$ valence photolines, because only the 1π photoline is practically unaffected. All experimental studies on molecular core-level shape resonances so far have shown that a $1s$ shape resonance has no effect on the valence photoionization, at least concerning the partial cross sections.[231] This means that there are no significant participator

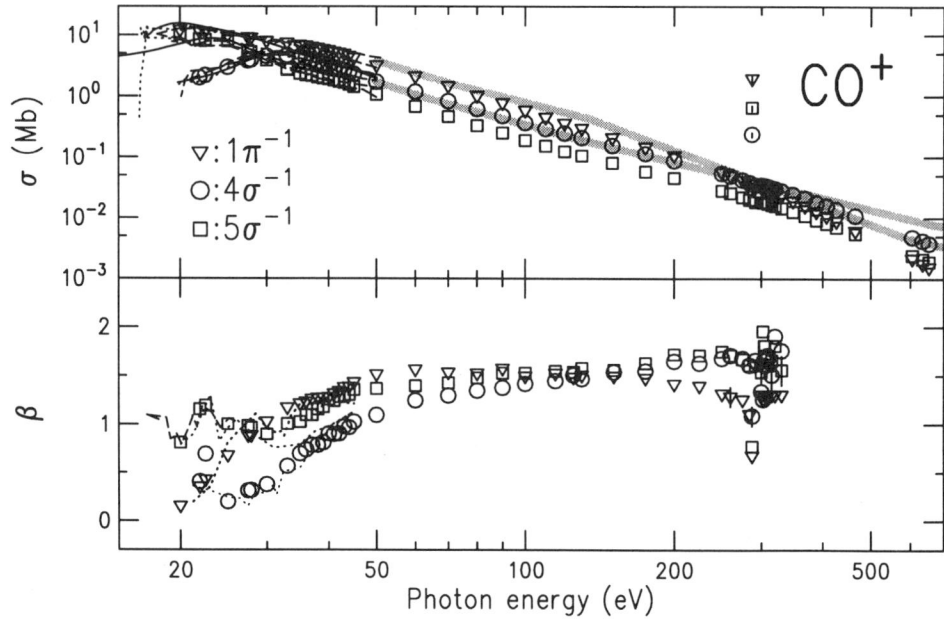

FIGURE 34. Photon energy dependences of the cross section σ and the angular distribution symmetry parameter β of the outer valence lines in CO. The shadowed curves represent the data compilation by Berkowitz. More recent σ data are given by: □: $5\sigma^{-1}$; ▽: $1\pi^{-1}$; ○: $4\sigma^{-1}$.[225] Solid line: theory, Ref. 226; dotted line: Ref. 227; (short dashed line): Ref. 228; (long dashed line): Ref. 229. Thin dotted line: $4\sigma^{-1}$; dotted line: $5\sigma^{-1}$ and thick dotted line: $1\pi^{-1}$, Ref. 230; (dashed line): $5\sigma^1$, Ref. 218.

Auger transitions in the shape resonance. On the other hand, the change in β shows that there seems to be an interaction which may result either from a phase change or, to a lesser extent, the alignment induced by the resonance. The overall σ and β behavior of the valence-level photoionization is now much better documented. However, a comparison with earlier results indicates that the σ curvature has changed only slightly. Real progress has been made in studying β behavior, for which very few data points existed until very recently. The progress is even more pronounced for the inner-valence lines, which means the 3σ and the adjacent satellite lines, as is seen in Figure 35. The satellite lines are dramatically enhanced on the π^* resonance because their final states are exactly the ones populated in the spectator decay of the resonance. Correspondingly dramatic is the change seen in β of these lines of slightly more than 2 β units resulting from the strong alignment of the molecule in the resonance.[232,233] On the other hand, these lines seem to be very little affected in the σ^* shape resonance, even including β. But one should keep in mind that all of these lines are superpositions of several unresolved states and therefore resonance effects may be present but averaged out. Near the 3σ peak double ionization starts to appear. It has been shown by different investigations that the first double-ionization threshold is associated with dissociative double ionization to $C^+ + O^+$, opening up before the metastable doubly charged final ionic state CO^{2+} $^1\Pi$ and/or $^1\Sigma^+$ is energetically reached.[234] Again, as seen for atoms, ions of both, two-step and one-step processes, contribute to the double-ionization yield.[235] This partial molecular photoionization yield, when compared to atoms, is twice as large, reaching 30% in the sudden limit. This may be an indication for the increased role of electron correlations in the inner-valence regime of molecules.

6.2 Core Ionization

We will close this section with a consideration of the $C1s$ and $O1s$ photoionization of CO. The first detailed variable-energy studies of the core levels of small molecules began about 10 years

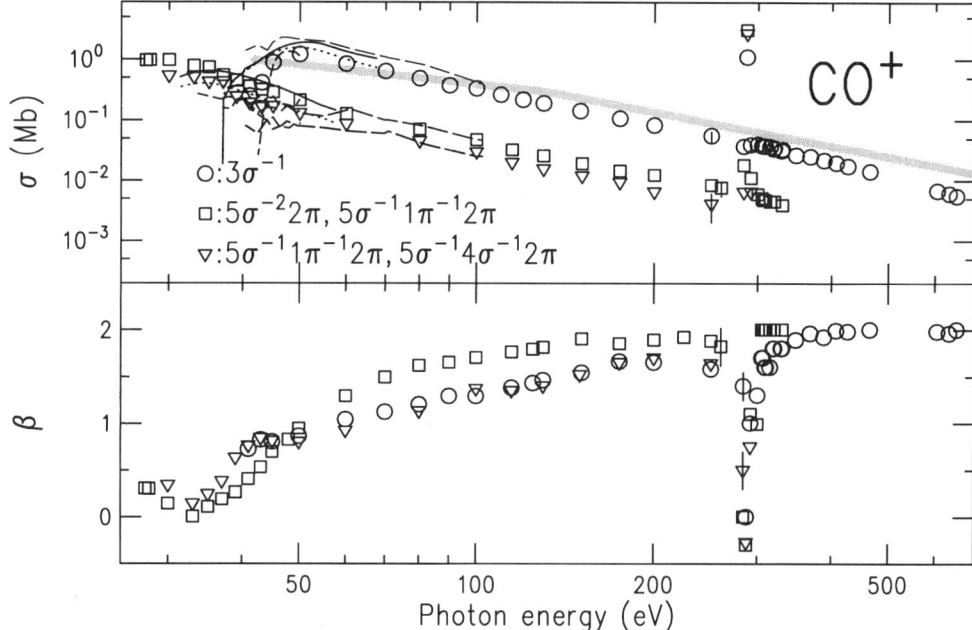

FIGURE 35. Photon energy dependences of the cross section σ and the angular distribution asymmetry parameter β of the inner valence lines in CO. The shadowed curves represent the data reported by Berkowitz. More recent results for σ and β are given by: (○) $3\sigma^{-1}$; (□) $5\sigma^{-2}\,2\pi + 5\sigma^{-1}\,1\pi^{-1}\,2\pi$ satellite lines[225]; (▽) $5\sigma^{-1}\,1\pi^{-1}\,2\pi + 5\sigma^{-1}\,4\sigma^{-1}\,2\pi$ satellite lines. Selection of previous experimental data sets of other authors for σ; (short dashed line): $3\sigma^{-1}$, Ref. 228; (thin long dashed line): $3\sigma^{-1}$; (long dashed line): $5\pi^{-2}\,2\pi + 5\pi^{-1}\,1\pi^{-1}\,2\pi$ satellite line; (thick long dashed line): $5\pi^{-1}\,1\pi^{-1}\,2\pi + 5\pi^{-1}\,4\pi^{-1}\,2\pi$ satellite line, Ref. 236. Theoretical results of Ref. 237 are given by (solid line): $3\sigma^{-1}$ main line and $5\pi^{-2}\,2\pi + 5\pi^{-1}\,1\pi^{-1}\,2\pi$ satellite lines (POLCI); (dotted line): $3\sigma^{-1}$ main line and $5\pi^{-2}\,2\pi + 5\pi^{-1}\,1\pi^{-1}\,2\pi$ satellite line (SECI).

ago.[238–240] Since that time, both the partial cross-sectional and angular-distribution behavior of the 1s main lines have been studied intensively, giving a quite clear picture of how σ and β vary from threshold up to the kiloelectron volt regime. There are still some uncertainties regarding the near-threshold behavior of the β parameter, so that it is not quite clear which theoretical approach really describes the experimental data best, but this is for a very limited, although quite important regime.

The situation changes when we come to satellite behavior. The K-shell satellites in CO are predominantly associated with an additional π* excitation of one of the valence electrons. Some few lines belong to excitation into higher-lying nσ levels. Most of the satellite lines are shakeup transitions, and some are assumed to be of the conjugate shakeup type.[241] For some time there was only one set of experimental data showing how differently various satellites may behave, some of them losing fractional intensity while others gained, when approaching threshold.[242] However, these measurements were done with somewhat limited resolution. Since the operation of high-resolution soft x-ray monochromators behind undulator beam lines began, a new era in inner-shell physics has started. At least three groups began to study highly resolved satellite spectra, with particular interest in their dependence on the photon energy.[215,243,244] The results show that the principal behavior of the different types of satellites are indeed reproduced, but there are many more satellites contributing to this behavior than thought before. Figure 36 shows the integrated satellite intensity underneath the peaks $S_0 + S'_0$ and $S_1 + S'_1$, along with the two $C1s$ and $O1s$ main lines. The β behavior of the satellites is similar to that of the main line, but the decrease in β toward threshold is less steep than for the main line; it reaches its minimum more gradually over an energy range of approximately 100 eV. There is still not very good agreement with theory in this respect but this may be related to resolution, i.e.,

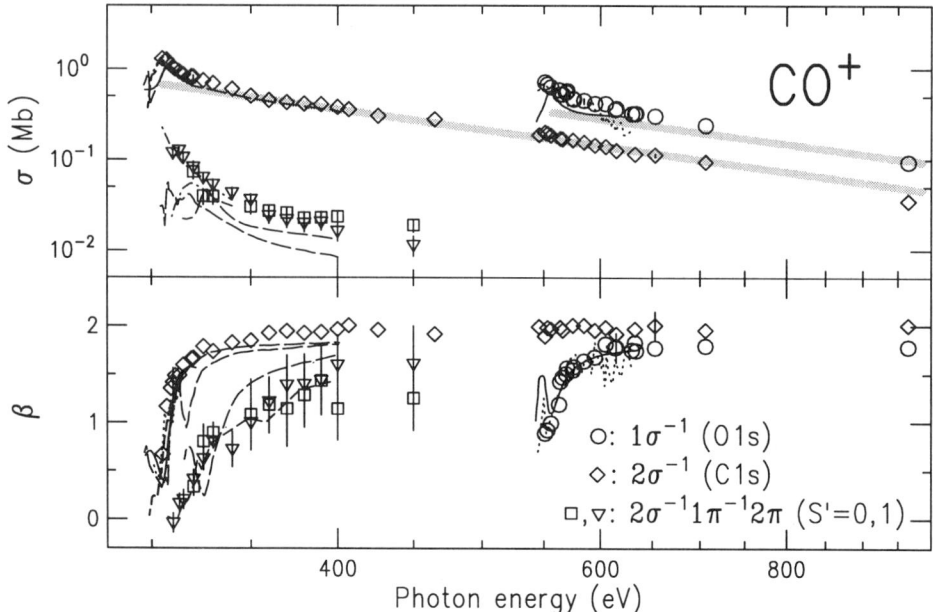

FIGURE 36. Photon energy dependences of the cross section σ and the angular distribution asymmetry parameter β of the inner shells in CO: (○) $1\sigma^{-1}$; (◇) $2\sigma^{-1}$; (▽) $2\sigma^{-1} 1\pi^{-1} 2\pi(S'=1)$; (□) $2\sigma^{-1} 1\pi^{-1} 2\pi\ (S'=0)$.[225] Further experimental and theoretical data sets are shown by solid lines: Ref. 246; long dashed lines: Ref. 247; (dotted lines): Ref. 238; (short dashed lines): Ref. 248; (dot-dashed lines): Ref. 243.

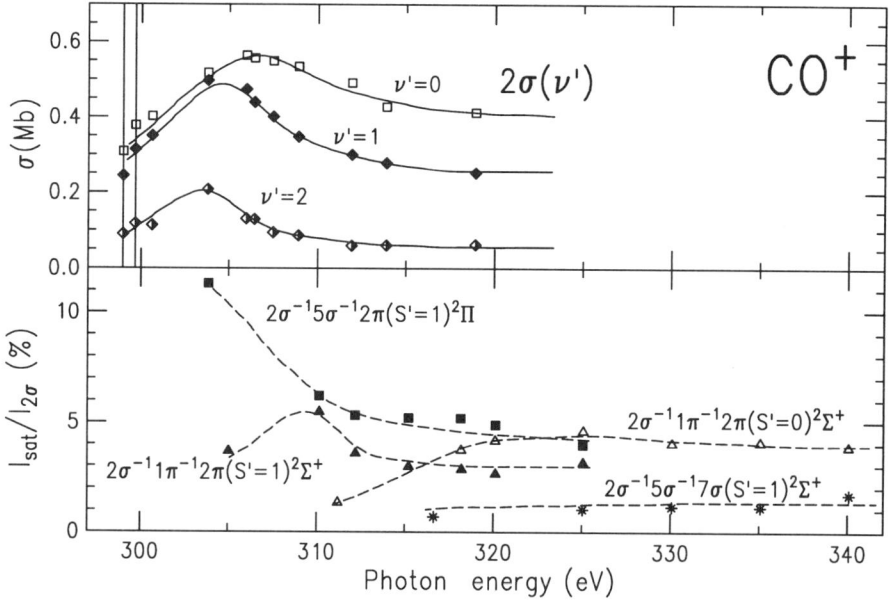

FIGURE 37. Partial cross section behavior of the vibrationally resolved components (v' = 0–2) of the 2σ main line in the shape resonance region.[215] The solid curves are drawn to guide the eye in order to visualize the change in the position of the shape resonance maximum more clearly. The lower panel shows the branching ratio behavior of some selected 2σ satellites, which are marked by bars in Figure 33. The dashed lines are again to guide the eye and underline the differences in the characteristic behavior of each satellite.[243]

comparing transitions which are not really equivalent. This assumption is supported if one compares the theoretical curves with the more recent high-resolution data, as shown in Figure 37. The total intensities here fit much better, although the details of the cross section behavior are still not very well reproduced. Nevertheless, the partial cross sections of these satellites corroborate, in a sense, the classification concept of intrinsic and dynamic correlations contributing to the production of these satellites.[90,245]

The vibrationally resolved partial cross sections of the $C1s$ main line demonstrate, on the other hand, how nicely the concept of changing vibrational structure in the presence of a shape resonance also works in the core-level regime.[212] This is the kind of study in which molecular photoionization, particularly concerning the inner shells, will develop in the near future when high-resolution, high-flux soft x-ray beam lines become more common. This will open new insights into the complex interplay between localized and nonlocalized behavior of electron correlation in small molecules.

7. FULLERENES

We will close this chapter with a short overview of a whole class of very large molecules which have just recently been discovered, the so-called fullerenes.[249,250] Fullerenes are highly symmetric carbon clusters which form stable species both as solids and in the gas phase. The well-known C_{60} species was the first carbon cluster and is the most thoroughly studied. Many others, including metal-doped species and compounds, followed. From the very beginning of this new field in molecular and material sciences people started to study among other properties, the photoionization behavior of these new materials (see, e.g., Refs. 251–253). One of the most surprising findings was the extremely close similarity between gas-phase and solid-state behavior.[254] The reason for this unusual similarity may be twofold: First, the large size and symmetric structure of this molecule gives rise to solid-state properties of some aspects of its electronic structure, e.g., the existence of a broad plasmon excitation.[255,256] On the other hand, the results of many investigations performed on thin films of fullerenes

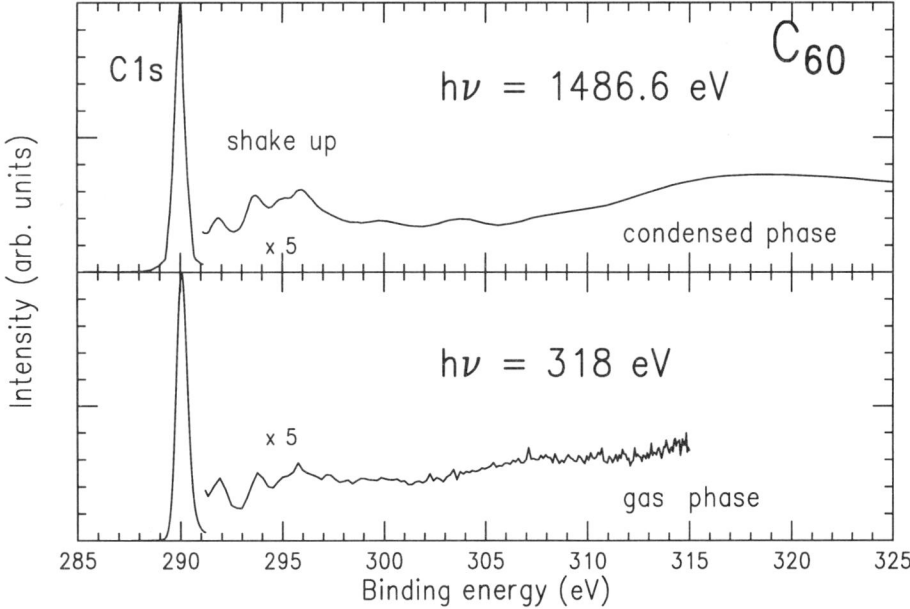

FIGURE 38. $C(1s)$ photoelectron spectrum of C_{60} taken from a condensed-phase target (upper panel)[251] and gas-phase target (lower panel).[259] One clearly sees the shakeup structure and the onset of a broad plasmonlike peak at higher binding energies in both spectra.

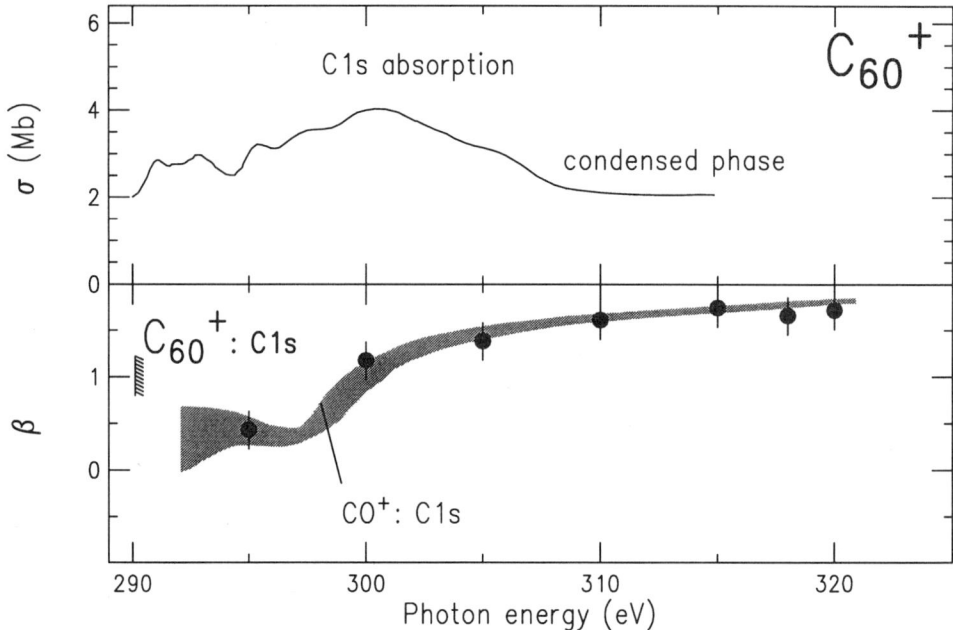

FIGURE 39. Total cross section σ and angular distribution parameter β of the $C(1s)$ photoline in the region of the $C\ 1s$ shape resonance. The upper panel shows a condensed-phase absorption spectrum[260] whereas the β values are taken from a gas-phase experiment.[259] The shaded region represents the corresponding $C1s$ β values of CO.

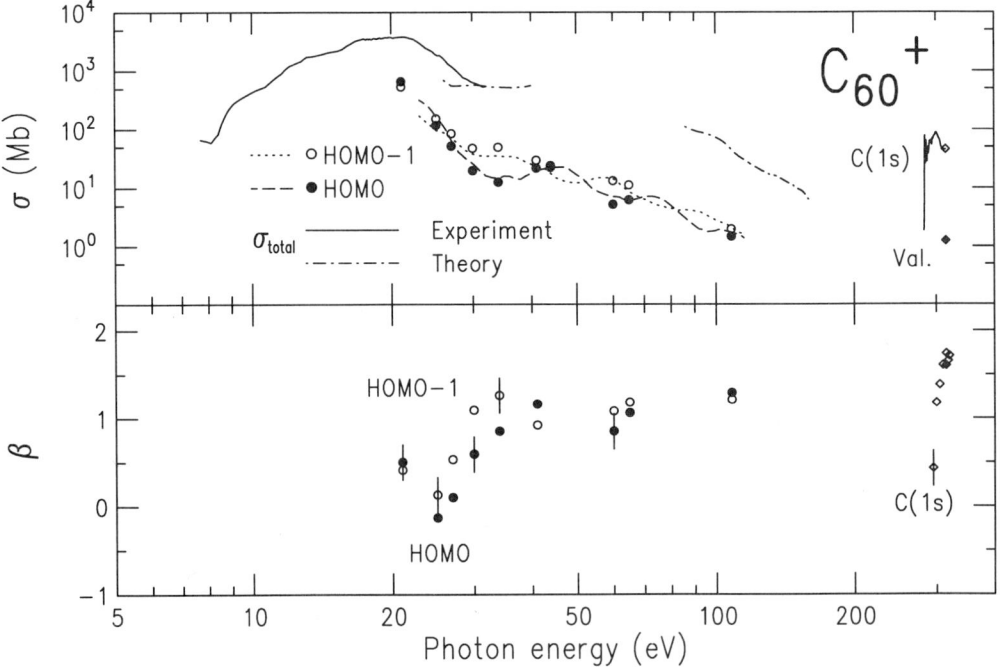

FIGURE 40. Total and partial cross section σ (upper panel) and angular distribution anisotropy parameter β (lower panel) of valence and core ionization of C_{60}. The solid lines represent two sets of measured total cross sections[256,260] related to each other by the calculation of Wendin and Wästberg[261] (dotted-dashed line). The absolute scale is derived from thin-film measurements.[262,263] First gas-phase data[259] are compared with condensed-phase measurements of Benning et al.[264] The dashed and the dotted line represent the partial cross sections of the HOMO and the HOMO-1 levels, respectively.

just reflect partly-free-molecule behavior, because the large fullerenes are in this case, in a strict sense, not in a crystal lattice, and sometimes show a more independent behavior, allowing for rotations and change of places. Therefore, both types of measurements, solid state and gas phase, should resemble each other. This was clearly demonstrated by the observation of a giant plasmon excitation[256] in the gas phase and the registration of very narrow resonances below the carbon K-edge in the solid state.[257,258] Regarding partial cross sections and angular distributions, the field is still in its infancy and systematic trends have hardly been considered as yet. However, some results, particularly in the core-level regime, were obtained very recently and compared to the corresponding behavior of other carbon-containing molecules such as CO. Figure 38 shows $C1s$ photoelectron spectra of C_{60}. These, in many respects, still very exploratory data show the following tendency: besides the difference in the electronic structure of the unoccupied orbitals or bands the basic behavior of $C1s$ in C_{60} and CO seems to be very similar. In both cases strong bound–bound transition resonances dominate the absorption cross section below threshold, whereas the above-threshold behavior is in both cases likewise dominated by the occurrence of a molecular shape resonance. The similarity with respect to the latter behavior is shown not only by the partial cross section, but also in the angular distribution. Figure 39 shows the partial cross section and angular distribution asymmetry parameter of the $C1s$ main line of C_{60}, compared to the corresponding data for the β of CO.[259] The similarity between the two molecules is striking. The satellite structure seen in Figure 38 shows σ and β behavior pointing to shakeup as their origin. Figure 40 shows σ and β of the most prominent lines in the photoelectron spectra for both valence and core ionization of CO between 21 and 318 eV exhibiting pronounced oscillations of the valence partial photoionization cross sections for both the molecule[259] and the solid.[263,264] But these studies are indeed at their very beginning. Comparison with other fullerenes promises further insight into the structure of these new molecular species.

ACKNOWLEDGMENTS. We are grateful to the many colleagues and collaborators for helpful discussions and providing us with their latest results. This work was sponsored by the Bundesminister für Bildung, Wissenscheft, Forschung and Technologie under (BMFB) and the Deutsche Forshungs—gemeinsheft.

REFERENCES

1. H. HERTZ, *Wied. Ann. d. Phys. v. Chem.* **31**, 983 (1887).
2. K. SIEGBAHN, C. NORDLING, A. FAHLMAN, R. NORDBERG, K. HAMRIN, J. HEDMAN, G. JOHANSSON, T. BERGMARK, S.-E. KARLSSON, I. LINDGREN, AND B. LINDBERG, *ESCA—Atomic, Molecular and Solid State Structure Studied by Means of Electron Spectroscopy* (Almquist & Wiksells, Uppsala, Sweden, 1967).
3. K. SIEGBAHN, C. NORDLING, G. JOHANSSON, J. HEDMAN, P. F. HEDÉN, K. HAMRIN, U. GELIUS, T. BERGMARK, L. O. WERME, R. MANNE, AND Y. BAER, *ESCA—Applied to Free Molecules* (North-Holland, Amsterdam, 1969).
4. J. W. COOPER, *Phys. Rev.* **128**, 681 (1962).
5. J. W. COOPER, *Phys. Rev. Lett.* **13**, 762 (1964).
6. M. Y. AMUSIA, V. K. IVANOV, AND L. V. CHERNYSHEVA, *Phys. Lett.* **59A**, 191 (1976).
7. M. Y. AMUSIA AND V. K. IVANOV, *Phys. Lett.* **59A**, 194 (1976).
8. C. KUNZ, in *Synchrotron Radiation*, edited by C. Kunz (Springer-Verlag, Berlin, 1979).
9. G. V. MARR, *Photoionization Processes in Gases* (Academic Press, New York, 1967).
10. J. A. R. SAMSON, *Methods of Experimental Physics*, Vol. 7A, edited by B. Bederson and W. L. Fite (Academic Press, New York, 1968).
11. M. O. KRAUSE, in *Synchrotron Radiation Research*, edited by H. Winnick and S. Doniach (Plenum Press, New York, 1980), p. 101.
12. J. BERKOWITZ, *Photoabsorption, Photoionization, and Photoelectron Spectroscopy* (Academic Press, New York, 1978).
13. U. HEINZMANN AND N. CHEREPKOV, in *VUV- and Soft X-ray Photoionization*, edited by U. Becker and D. A. Shirley (Plenum Press, New York, 1996).
14. G. C. KING AND K.-H. SCHARTNER, in *VUV- and Soft X-ray Photoionization*, edited by U. Becker and D. A. Shirley (Plenum Press, New York, 1996).

15. M. O. KRAUSE AND C. D. CALDWELL, in *VUV- and Soft X-ray Photoionization*, edited by U. Becker and D. A. Shirley (Plenum Press, New York, 1996).
16. J. BERKOWITZ, E. RÜHL, AND H. BAUMGÄRTEL, in *VUV- and Soft X-ray Photoionization*, edited by U. Becker and D. A Shirley (Plenum Press, New York, 1996).
17. F. J. WUILLEUMIER AND J. B. WEST, in *VUV- and Soft X-ray Photoionization*, edited by U. Becker and D. A. Shirley (Plenum Press, New York, 1996).
18. I. LINDAU, *Synchrotron Radiation Research: Advances in Surface and Interface Sciences* Vol. 2, edited by R. Z. Bachrach (Plenum Press, New York, 1992), p. 38.
19. W. F. CHAN, G. COOPER, AND C. E. BRION, *Phys. Rev. A* **44**, 186 (1991).
20. W. F. CHAN, G. COOPER, X. GUO, AND C. E. BRION, *Phys. Rev. A* **45**, 1420 (1992).
21. W. F. CHAN, G. COOPER, X. GUO, G. R. BURTON, AND C. E. BRION, *Phys. Rev. A* **46**, 149 (1992).
22. W. F. CHAN, G. COOPER, R. N. S. SODHI, AND C. E. BRION, *Chem. Phys.* **170**, 81 (1993).
23. W. F. CHAN, G. COOPER, AND C. E. BRION, *Chem. Phys.* **170**, 99 (1993).
24. J. COOPER AND R. N. ZARE, *J. Chem. Phys.* **48**, 942 (1968).
25. D. DILL AND U. FANO, *Phys. Rev. Lett.* **29**, 1203 (1972).
26. U. FANO AND D. DILL, *Phys. Rev. A* **6**, 185 (1972).
27. T. HAYAISHI AND P. ZIMMERMANN, in *VUV- and Soft X-ray Photoionization*, edited by U. Becker and D. A. Shirley (Plenum Press, New York, 1996).
28. J. A. R. SAMSON, in *Many-body Theory of Atomic Structure and Photoionization*, edited by T. N. Chang (World Scientific, Singapore, 1993).
29. J. A. R. SAMSON, in *Corpuscles and Radiation in Matter I, Encyclopedia of Physics* XXXI, edited by W. Mehlhorn (Springer-Verlag, Berlin, 1982), p. 123.
30. H. SIEGBAHN AND L. KARLSSON, in *Corpuscles and Radiation in Matter I, Encyclopedia of Physics* XXXI, edited by W. Mehlhorn (Springer-Verlag, Berlin, 1982), p. 215.
31. R. Z. BACHRACH, F. C. BROWN, AND S. B. M. HAGSTRÖM, *J. Vac. Sci. Technol.* **12**, 309 (1975).
32. B. TSAI, T. BAER, AND M. L. HOROVITZ, *Rev. Sci. Instrum.* **45**, 494 (1974).
33. M. G. WHITE, R. A. ROSENBERG, G. GABOR, E. D. POLIAKOFF, G. THORNTON, S. H. SOUTHWORTH, AND D. A. SHIRLEY, *Rev. Sci. Instrum.* **50**, 1268 (1979).
34. W. B. PEATMAN, T. B. BORNE, AND E. W. SCHLAG, *Chem. Phys. Lett.* **3**, 492 (1969).
35. P. M. GUYON, T. BAER, L. F. A. FERREIRA, I. NENNER, A. TABCHÉ-FOUHAILLÉ, R. BOTTER, AND T. R. GOVERS, *J. Phys. B* **11**, L141 (1978).
36. F. J. WUILLEUMIER, *Ann. Phys. Fr.* **4**, 231 (1982).
37. G. V. MARR AND J. B. WEST, *At. Data Nucl. Data Tables* **18**, 497 (1976).
38. J. A. R. SAMSON, Z. X. HE, L. YIN, AND G. N. HADDAD, *J. Phys B* **27**, 887 (1994).
39. R. WEHLITZ, B. LANGER, N. BERRAH, S. B. WHITFIELD, J. VIEFHAUS, AND U. BECKER, *J. Phys. B* **26**, L783 (1993).
40. R. WEHLITZ, F. HEISER, O. HEMMERS, B. LANGER, A. MENZEL, AND U. BECKER, *Phys. Rev. Lett.* **67**, 3764 (1991).
41. P. A. HEIMANN, U. BECKER, H. G. KERKHOFF, B. LANGER, D. SZOSTAK, R. WEHLITZ, D. W. LINDLE, T. A. FERRETT, AND D. A. SHIRLEY, *Phys. Rev. A* **34**, 3782 (1986).
42. R. I. HALL, L. AVALDI, G. DAWBER, M. ZUBEK, K. ELLIS, AND G. C. KING, *J. Phys. B* **24**, 115 (1991).
43. J. C. LEVIN, I. A. SELLIN, B. M. JOHNSON, D. W. LINDLE, R. D. MILLER, N. BERRAH, Y. AZUMA, H. G. BERRY, AND D.-H. LEE, *Phys. Rev. A* **47**, R16 (1993).
44. L. SPEILBERGER, O. JAGUTZKI, R. DÖRNER, J. ULLRICH, U. MEYER, V. MERGEL, M. UNVERZAGT, M. DAMRAU, T. VOGT, I. ALI, KH. KHAYYAT, D. BAHR, H. G. SCHMIDT, R. FRAHM, AND H. SCHMIDT-BÖCKING, *Phys. Rev. Lett.* **74**, 4615 (1995).
45. M. Y. AMUSIA, *Atomic Photoeffect* (Plenum Press, New York, 1990).
46. B. LANGER, in *Studies of Vacuum Ultraviolet and X-ray Processes* Vol. 2, edited by U. Becker (AMS Press, New York, 1992).
47. V. SCHMIDT, N. SANDNER, H. KUNTZEMÜLLER, P. DHEZ, F. WUILLEUMIER, AND E. KÄLLNE, *Phys. Rev. A* **13**, 1748 (1976).
48. D. W. LINDLE, T. A. FERRETT, U. BECKER, P. H. KOBRIN, C. M. TRUESDALE, H. G. KERKHOFF, AND D. A. SHIRLEY, *Phys. Rev. A* **31**, 714 (1985).
49. P. R. WOODRUFF AND J. A. R. SAMSON, *Phys. Rev. Lett.* **45**, 110 (1980).
50. F. WUILLEUMIER, M. Y. ADAM, N. SANDNER, AND V. SCHMIDT, *J. Phys. (Paris) Lett.* **41**, L373 (1980).
51. N. BERRAH, F. HEISER, R. WEHLITZ, J. LEVIN, S. B. WHITFIELD, J. VIEFHAUS, I. A. SELLIN, AND U. BECKER, *J. Phys. IV3, Colloq.* **C6**, 197 (1993).
52. V. SCHMIDT, H. DERENBACH, AND R. MALUTZKI, *J. Phys. B* **15**, L523 (1982).

53. J.-M. Bizau, F. Wuilleumier, P. Dhez, D. L. Ederer, T. N. Chang, S. Krummacher, and V. Schmidt, *Phys. Rev. Lett.* **48**, 588 (1982).
54. D. W. Lindle, P. A. Heimann, T. A. Ferrett, and D. A. Shirley, *Phys. Rev. A* **35**, 1128 (1987).
55. S. Salomonson, S. L. Carter, and H. P. Kelly, *Phys. Rev. A* **39**, 5111 (1989).
56. V. L. Jacobs and P. G. Burke, *J. Phys. B* **5**, L67 (1972).
57. C. H. Greene, *Phys. Rev. Lett.* **44**, 869 (1980).
58. C. H. Greene, *J. Phys. B* **20**, L357 (1987).
59. F. Maulbetsch and J. S. Briggs, *J. Phys. B* **26**, 1679 (1993).
60. F. Maulbetsch and J. S. Briggs, *Phys. Rev. Lett.* **68**, 2004 (1992).
61. R. I. Hall, A. G. McConkey, L. Avaldi, K. Ellis, M. A. MacDonald, G. Dawber, and G. C. King, *J. Phys. B* **25**, 1195 (1992).
62. O. Schwarzkopf, B. Krässig, J. Elmiger, and V. Schmidt, *Phys. Rev. Lett.* **70**, 3008 (1993).
63. A. Huetz, P. Lablanquie, L. Anric, P. Selles, and J. Mazeau, *J. Phys. B* **27**, L13 (1994).
64. V. Schmidt, *Rep. Prog. Phys.* **55**, 1483 (1992).
65. M. Y. Amusia, in *Advances in Atomic and Molecular Physics*, Vol. 17, edited by D. Bates (Academic Press, New York, 1981), p. 1.
66. J. Tulkki, *Phys. Rev. A* **48**, 2048 (1993).
67. F. Wuilleumier and M. O. Krause, *J. Electron Spectrosc. Relat. Phenom.* **15**, 15 (1979).
68. J.-J. Yeh, *Atomic Calculation of Photoionization Cross Sections and Asymmetry Parameters* (Gordon & Breach, New York, 1993).
69. D. J. Kennedy and S. T. Manson, *Phys. Rev. A* **5**, 227 (1972).
70. M. Y. Amusia, N. A. Cherepkov, and L. V. Chernysheva, *Phys. Lett.* **40A**, 15 (1972).
71. M. Y. Amusia, V. K. Ivanov, N. A. Cherepkov, and L. V. Chernysheva, *Phys. Lett.* **40A**, 361 (1972).
72. W. R. Johnson and K. T. Cheng, *Phys. Rev. A* **20**, 978 (1979).
73. N. Saito and I. H. Suzuki, *Int. J. Mass Spectrom. Ion Processes* **115**, 157 (1992).
74. K.-H. Schartner, B. Möbus, G. Mentzel, A. Ehresmann, F. Vollweiler, and H. Schmoranzer, *Phys. Lett. A* **169**, 393 (1990).
75. F. Wuilleumier and M. O. Krause, *Phys. Rev. A* **10**, 242 (1974).
76. S. H. Southworth, A. C. Parr, J. E. Hardis, J. L. Dehmer, and D. M. P. Holland, *Nucl. Instrum. Methods Phys. Res.* **A246**, 782 (1986).
77. U. Becker and R. Hentges, unpublished results, (1994).
78. H. Derenbach, R. Malutzki, and V. Schmidt, *Nucl. Instrum. Methods* **208**, 845 (1983).
79. K. Codling, R. G. Houlgate, J. B. West, and P. R. Woodruff, *J. Phys. B* **9**, L83 (1976).
80. J. L. Dehmer, W. A. Chupka, J. Berkowitz, and W. T. Jivery, *Phys. Rev. A* **12**, 1966 (1975).
81. S. Svensson, B. Eriksson, N. Mårtensson, G. Wendin, and U. Gelius, *J. Electron Spectrosc. Relat. Phenom.* **47**, 327 (1988).
82. A. D. O. Bawagan, B. J. Olsson, K. H. Tan, J. M. Chen, and G. M. Bancroft, *Chem. Phys. Lett.* **179**, 344 (1991).
83. M. O. Krause, S. B. Whitfield, C. D. Caldwell, J.-Z. Wu, P. van der Meulen, C. A. de Lange, and R. W. C. Hansen, *J. Electron Spectrosc. Relat. Phenom.* **58**, 79 (1992).
84. A. A. Wills, A. A. Cafolla, A. Svensson, and J. Comer, *J. Phys. B* **23**, 2013 (1990).
85. R. I. Hall, G. Dawber, K. Ellis, M. Zubek, L. Avaldi, and G. C. King, *J. Phys. B* **24**, 4133 (1991).
86. F. Heiser, U. Hergenhahn, J. Viefhaus, K. Wieliczek, and U. Becker, *J. Electron Spectrosc. Relat. Phenom.* **60**, 337 (1992).
87. U. Becker, R. Wehlitz, O. Hemmers, B. Langer, and A. Menzel, *Phys. Rev. Lett.* **63**, 1054 (1989).
88. T. M. Luke, J. D. Talman, H. Aksela, and M. Leväsalmi, *Phys. Rev. A* **41**, 1350 (1990).
89. T. D. Thomas, *Phys. Rev. Lett.* **52**, 417 (1984).
90. U. Becker and D. A. Shirley, *Phys. Scr.* **T31**, 56 (1990).
91. P. A. Heimann, C. M. Truesdale, H. G. Kerkhoff, D. W. Lindle, T. A. Ferrett, C. C. Bahr, W. D. Brewer, U. Becker, and D. A. Shirley, *Phys. Rev. A* **31**, 2260 (1985).
92. U. Becker, R. Hölzel, H. G. Kerkhoff, B. Langer, D. Szostak, and R. Wehlitz, *Phys. Rev. Lett.* **56**, 1120 (1986).
93. A. A. Wills, A. A. Cafolla, F. J. Currell, J. Comer, A. Svensson, and M. A. MacDonald, *J. Phys. B* **22**, 3217 (1989).
94. J. A. R. Samson, L. Lyn, G. N. Haddad, and G. C. Angel, *J. Phys. IV* **1**, C1 (1991).
95. K.-N. Huang, W. R. Johnson, and K. T. Cheng, *At. Data Nucl. Data Tables* **26**, 33 (1981).
96. W. Wijesundera and H. P. Kelly, *Phys. Rev. A* **39**, 634 (1989).

97. D. W. LINDLE, L. J. MEDHURST, T. A. FERRETT, P. A. HEIMANN, M. N. PIANCASTELLI, S. H. LIU, AND D. A. SHIRLEY, *Phys. Rev. A* **38**, 2371 (1988).
98. B. LANGER, J. VIEFHAUS, O. HEMMERS, A. MENZEL, R. WEHLITZ, AND U. BECKER, *Phys. Rev. A* **51**, R882 (1995).
99. C. PAN AND H. P. KELLY, *Phys. Rev. A* **39**, 6232 (1989).
100. U. BECKER. in *The Physics of Electronic and Atomic Collisions, AIP Conf. Proc.* 205, edited by A. Dalgarno, R. S. Freund, P. M. Koch, M. S. Lubell, and T. B. Lucatorto (AIP, New York, 1990), p. 162.
101. L. AVALDI, G. DAWBER, R. CAMILLONI, G. C. KING, M. ROPER, M. R. F. SIGGEL, G. STEFANI, AND M. ZITNIK, *J. Phys. B* **27**, 3953 (1994).
102. B. MÖBUS, B. MAGEL, K.-H. SCHARTNER, B. LANGER, U. BECKER, M. WILDBERGER, AND H. SCHMORANZER, *Phys. Rev. A* **47**, 3888 (1993).
103. M. Y. ADAM, P. MORIN, AND G. WENDIN, *Phys. Rev. A* **31**, 1426 (1985).
104. R. G. HOULGATE, J. B. WEST, K. CODLING, AND G. V. MARR, *J. Phys. B* **7**, L470 (1974).
105. M. O. KRAUSE, S. B. WHITFIELD, C. D. CALDWELL, J.-Z. WU, P. VAN DER MEULEN, C. A. DE LANGE, AND R. W. C. HANSEN, *J. Electron Spectrosc. Relat. Phenom.* **58**, 79 (1992).
106. F. HEISER, S. B. WHITFIELD, J. VIEFHAUS, U. BECKER, P. A. HEIMANN, AND D. A. SHIRLEY, *J. Phys. B* **27**, 19 (1994).
107. V. L. SUKHORUKOV, B. M. LAGUTIN, H. SCHMORANZER, I. D. PETROV, AND K.-H. SCHARTNER, *Phys. Lett. A* **169**, 445 (1992).
108. V. SCHMIDT, *Z. Phys. D* 275 (1986).
109. H. KOSSMANN, B. KRÄSSIG, V. SCHMIDT, AND J. E. HANSEN, *Phys. Rev. Lett.* **58**, 1620 (1987).
110. U. BECKER, B. LANGER, H. G. KERKHOFF, M. KUPSCH, D. SZOSTAK, R. WEHLITZ, P. A. HEIMANN, S. H. LIU, D. W. LINDLE, T. A. FERRETT, AND D. A. SHIRLEY, *Phys. Rev. Lett.* **60**, 1490 (1988).
111. C. E. BRION, A. O. BAWAGAN, AND K. H. TAN, *Can. J. Chem.* **66**, 1877 (1988).
112. L. J. MEDHURST, A. S. V. WITTENAU, R. D. VAN ZEE, S. ZHANG, S. H. LIU, D. A. SHIRLEY, AND D. W. LINDLE, *J. Electron Spectrosc. Relat. Phenom.* **52**, 671 (1990).
113. J. C. LEVIN, C. BIEDERMANN, N. KELLER, L. LILJEBY, C.-S. O, R. T. SHORT, I. A. SELLIN, AND D. W. LINDLE, *Phys. Rev. Lett.* **65**, 988 (1990).
114. K. UEDA, E. SHIGEMASA, Y. SATO, A. YAGISHITA, M. UKAI, H. MAEZAWA, T. HAYAISHI, AND T. SASAKI, *J. Phys. B* **24**, 605 (1991).
115. J. DOPPELFELD, N. ANDERS, B. ESSER, F. V. BUSCH, H. SCHERER, AND S. ZINZ, *J. Phys. B* **26**, 445 (1993).
116. P. H. KOBRIN, S. SOUTHWORTH, C. M. TRUESDALE, D. W. LINDLE, U. BECKER, AND D. A. SHIRLEY, *Phys. Rev. A* **29**, 194 (1984).
117. G. B. ARMEN, T. ÅBERG, K. R. KARIM, J. C. LEVIN, B. CRASEMANN, G. S. BROWN, M. H. CHEN, AND G. E. ICE, *Phys. Rev. Lett.* **54**, 182 (1985).
118. M. DEUTSCH, N. MASKIL, AND W. DRUBE, *Phys. Rev. A* **46**, 3963 (1992).
119. F. HEISER, S. B. WHITFIELD, J. VIEFHAUS, U. BECKER, P. A. HEIMANN, AND D. A. SHIRLEY, *J. Phys. B* **27**, 19 (1994).
120. N. BERRAH, B. LANGER, J. VIEFHAUS, S. B. WHITFIELD, F. HEISER, AND U. BECKER, in *Abstracts of Contributed Papers of the XVIII ICPEAC, Abstracts of Contributed Papers*, Vol. I, edited by T. Andersen, B. Fastrup, F. Folkmann and H. Knudsen (Aarhus, 1993), p. 17.
121. C. E. BRION, A. O. BAWAGAN, AND K. H. TAN, *Can. J. Chem.* **66**, 1877 (1988).
122. D. W. LINDLE, P. A. HEIMANN, T. A. FERRETT, P. H. KOBRIN, C. M. TRUESDALE, U. BECKER, M. G. KERKHOFF, AND D. A. SHIRLEY, *Phys. Rev. A* **33**, 319 (1986).
123. M. O. KRAUSE AND T. A. CARLSON, *Phys. Rev.* **149**, 52 (1966).
124. M. O. KRAUSE, *Phys. Rev.* **177**, 151 (1969).
125. J. TULKKI, S. AKSELA, H. AKSELA, E. SHIGEMASA, A. YAGISHITA, AND Y. FURUSAWA, *Phys. Rev. A* **45**, 4640 (1992).
126. N. SHANTI, P. C. DESHMUKH, AND S. T. MANSON, *Phys. Rev. A* **37**, 4720 (1988).
127. D. W. LINDLE, P. A. HEIMANN, T. A. FERRETT, P. H. KOBRIN, C. M. TRUESDALE, U. BECKER, H. G. KERKHOFF, AND D. A. SHIRLEY, *Phys. Rev. A* **33**, 319 (1986).
128. T. A. CARLSON, M. O. KRAUSE, F. A. GRIMM, P. R. KELLER, AND J. W. TAYLOR, *Chem. Phys. Lett.* **87**, 552 (1982).
129. A. EHRESMANN, F. VOLLWEILER, H. SCHMORANZER, V. L. SUKHORUKOV, B. M. LAGUTIN, I. D. PETROV, G. MENTZEL, AND K.-H. SCHARTNER, *J. Phys. B* **27**, 1489 (1994).
130. S. AKSELA, H. AKSELA, M. LEVASALMI, K. H. TAN, AND G. M. BANCROFT, *Phys. Rev. A* **36**, 3449 (1987).
131. H. DERENBACH AND V. SCHMIDT, *J. Phys. B* **17**, 83 (1984).
132. J. A. R. SAMSON AND J. L. GARDNER, *Phys. Rev. Lett.* **33**, 671 (1974).
133. D. L. MILLER, J. D. DOW, R. G. HOULGATE, G. V. MARR, AND J. B. WEST, *J. Phys. B* **10**, 3205 (1977).

134. H. Schmoranzer, A. Ehresmann, F. Vollweiler, V. L. Sukhorukov, B. M. Lagutin, I. D. Petrov, K. H. Schartner, and B. Möbus, *J. Phys. B* **26**, 2795 (1993).
135. J. A. R. Samson and G. N. Haddad, *Phys. Rev. Lett.* **33**, 875 (1974).
136. N. Saito and I. H. Suzuki, *Phys. Scr.* **49**, 80 (1994).
137. G. B. Armen and F. P. Larkins, *J. Phys. B* **24**, 741 (1991).
138. M. Kutzner, V. Radojevic, and H. P. Kelly, *Phys. Rev. A* **40**, 5052 (1989).
139. M. O. Krause, *Phys. Scr.* **T17**, 146 (1987).
140. J. Tulkki, *Phys. Rev. Lett.* **62**, 2817 (1989).
141. Z. Altun, M. Kutzner, and H. P. Kelly, *Phys. Rev. A* **37**, 4671 (1988).
142. U. Becker, D. Szostak, H. G. Kerkhoff, M. Kupsch, B. Langer, R. Wehlitz, A. Yagishita, and T. Hayaishi, *Phys. Rev. A* **39**, 3902 (1989).
143. M. Y. Adam, Thesis, Université de Paris-Sud, unpublished (1978).
144. B. Kämmerling, H. Kossmann, and V. Schmidt, *J. Phys. B* **22**, 841 (1989).
145. S. B. Whitfield, B. Langer, J. Viefhaus, R. Wehlitz, N. Berrah, W. Mahler, and U. Becker, *J. Phys. B* **27**, L359 (1994).
146. U. Becker, H. G. Kerkhoff, M. Kupsch, B. Langer, D. Szostak, and R. Wehlitz, *J. Phys. (Paris) Colloq.* **48**, C9 (1987).
147. D. W. Lindle, T. A. Ferrett, P. A. Heimann, and D. A. Shirley, *Phys. Rev. A* **37**, 3808 (1988).
148. J. B. West, P. R. Woodruff, K. Codling, and R. G. Houlgate, *J. Phys. B* **9**, 407 (1976).
149. S. H. Southworth, P. H. Kobrin, C. M. Truesdale, D. Lindle, S. Owaki, and D. A. Shirley, *Phys. Rev. A* **24**, 2257 (1981).
150. A. Fahlman, M. O. Krause, T. A. Carlson, and A. Svensson, *Phys. Rev. A* **30**, 812 (1984).
151. H. Derenbach and V. Schmidt, *J. Phys. B* **16**, L337 (1983).
152. A. Wachter, R. Malutzki, and V. Schmidt, in *8th International Conference on Vacuum Ultraviolet Radiation Physics, Book of Abstracts*, Vol. I, edited by P. O. Nielsen (Lund, 1986), p. 33.
153. A. Fahlman, T. A. Carlson, and M. O. Krause, *Phys. Rev. Lett.* **50**, 1114 (1983).
154. S. Southworth, U. Becker, C. M. Truesdale, P. H. Kobrin, D. W. Lindle, S. Owaki, and D. A. Shirley, *Phys. Rev. A* **28**, 261 (1983).
155. M. Y. Adam, F. Wuilleumier, N. Sandner, S. Krummacher, V. Schmidt, and W. Mehlhorn, *Jpn. J. Appl. Phys. Suppl. 17-2* **17**, 170 (1978).
156. M. O. Krause, T. A. Carlson, and P. R. Woodruff, *Phys. Rev. A* **24**, 1374 (1981).
157. L. Torop, J. Morton, and J. B. West, *J. Phys. B* **9**, 2035 (1976).
158. B. Sonntag and P. Zimmermann, *Rep. Prog. Phys.* **55**, 911 (1992).
159. J.-M. Bizau, D. Cubaynes, P. Gérard, and F. J. Wuilleumier, *Phys. Rev. A* **40**, 3002 (1989).
160. W. A. Svensson, M. O. Krause, T. A. Carlson, V. Radojevic, and W. R. Johnson, *Phys. Rev.* **33**, 1024 (1986).
161. J.-M. Bizau, P. Gérard, F. J. Wuilleumier, and G. Wendin, *Phys. Rev. A* **36**, 1220 (1987).
162. G. V. Marr and J. M. Austin, *Proc. R. Soc. London Ser. A* **310**, 137 (1969).
163. R. B. Cairns, H. Harrison, and R. I. Schoen, *Chem. Phys.* **51**, 5440 (1969).
164. G. Schönhense, *J. Phys. B* **14**, L187 (1981).
165. V. von Garnier, Diploma Thesis, Technische Universität Berlin, unpublished (1997).
166. W. R. Johnson, V. Radojevic, P. Deshmukh, and K. T. Cheng, *Phys. Rev. A* **25**, 337 (1982).
167. M. Kutzner, C. Tidwell, S. E. Vance, and V. Radojevic, *Phys. Rev. A* **49**, 300 (1994).
168. S. P. Kowalczyk, L. Ley, R. L. Martin, F. R. McFeely, and D. A. Shirley, *Faraday Discuss. Chem. Soc.* **60**, 7 (1975).
169. G. Wendin, *Breakdown of the One-Electron Pictures in Photoelectron Spectra* (Springer-Verlag, Berlin, 1981).
170. C. E. Theodosiou, A. F. Starace, B. R. Tambe, and S. T. Manson, *Phys. Rev. A* **24**, 301 (1981).
171. S. L. Carter and H. P. Kelly, *J. Phys. B* **11**, 2467 (1978).
172. P. H. Kobrin, U. Becker, S. Southworth, C. M. Truesdale, D. W. Lindle, and D. A. Shirley, *Phys. Rev.* **26**, 842 (1982).
173. J. S. Shyu and S. T. Manson, *Phys. Rev. A* **11**, 166 (1975).
174. F. Keller and F. C. Farnoux, *J. Phys. B* **12**, 2821 (1979).
175. T. E. H. Walker and J. T. Waber, *J. Phys. B* **7**, 674 (1974).
176. F. Keller and F. C. Farnoux, *J. Phys. B* **15**, 2657 (1982).
177. W. R. Johnson and V. Radojevic, *Phys. Lett.* **92A**, 75 (1982).
178. B. R. Tambe and S. T. Manson, *Phys. Rev. A* **30**, 256 (1984).

179. P. H. KOBRIN, P. A. HEIMANN, H. G. KERKHOFF, D. W. LINDLE, C. M. TRUESDALE, T. A. FERRETT, U. BECKER, AND D. A. SHIRLEY, *Phys. Rev. A* **27**, 3031 (1983).
180. J. JIMÉNEZ-MIER, M. O. KRAUSE, P. GERARD, B. HERMSMEIER, AND C. S. FADLEY, *Phys. Rev. A* **40**, 3712 (1989).
181. L. J. GARVIN, E. R. BROWN, S. L. CARTER, AND H. P. KELLY, *J. Phys. B* **16**, L269 (1983).
182. M. Y. AMUSIA, V. K. IVANOV, AND L. V. CHERNYSHEVA, *J. Phys. B* **14**, L19 (1981).
183. R. BRUHN, E. SCHMIDT, M. SCHRÖDER, AND B. SONNTAG, *Phys. Lett.* **90A**, 41 (1982).
184. M. O. KRAUSE, T. A. CARLSON, AND A. FAHLMAN, *Phys. Rev. A* **30**, 1316 (1984).
185. P. H. KOBRIN, U. BECKER, C. M. TRUESDALE, D. W. LINDLE, H. G. KERKHOFF, AND D. A. SHIRLEY, *J. Electron Spectrosc. Relat. Phenom.* **34**, 129 (1984).
186. M. Y. AMUSIA, V. K. DOLMATOV, AND V. K. IVANOV, *J. Phys. B* **16**, L753 (1983).
187. R. MALUTZKI, M. S. BANNA, W. BRAUN, AND V. SCHMIDT, *J. Phys. B* **18**, 1735 (1985).
188. E. SCHMIDT, H. SCHRÖDER, B. SONNTAG, H. VOSS, AND H. E. WETZEL, *J. Phys. B* **16**, 2961 (1983).
189. E. SCHMIDT, H. SCHRÖDER, B. SONNTAG, H. VOSS, AND H. E. WETZEL, *J. Phys. B* **18**, 79 (1985).
190. M. MEYER, T. PRESCHER, E. V. RAVEN, M. RICHTER, E. SCHMIDT, B. SONNTAG, AND H. E. WETZEL, *J. Phys. D* **2**, 347 (1986).
191. M. MEYER, T. PRESCHER, E. V. RAVEN, M. RICHTER, E. SCHMIDT, B. SONNTAG, AND H. E. WETZEL, in *Giant Resonances in Atoms, Molecules and Solids, NATO ASI Series* B 151 edited by J. P. Connerade, J. M. Esteva, and R. C. Karnatak (Plenum Press, New York, 1987), p. 251.
192. U. BECKER, in *Giant Resonances in Atoms, Molecules and Solids, NATO ASI Series* B 151, edited by J. P. Connerade, J. M. Esteva, and R. C. Karnatak (Plenum Press, New York, 1987), p. 473.
193. M. RICHTER, J. M. BIZAU, D. CUBAYNES, T. MENZEL, F. J. WUILLEUMIER, AND B. CARRÉ, *Europhys. Lett.* **12**, 35 (1990).
194. M. RICHTER, M. MEYER, M. PAHLER, T. PRESCHER, E. V. RAVEN, B. SONNTAG, AND H. E. WETZEL, *Phys. Rev. A* **39**, 5666 (1989).
195. L. C. DAVIS AND L. A. FELDKAMP, *Phys. Rev. A* **17**, 2012 (1978).
196. L. J. GARVIN, E. R. BROWN, S. L. CARTER, AND H. P. KELLY, *J. Phys. B* **16**, L643 (1983).
197. S. B. WHITFIELD, M. O. KRAUSE, P. V. D. MEULEN, AND C. D. CALDWELL, *Phys. Rev. A* **50**, 1269 (1994).
198. H. G. KERKHOFF, Thesis, Technische Universität Berlin, unpublished (1991).
199. M. Y. AMUSIA, S. I. SHEFTEL, AND L. V. CHERNYSHEVA, *Zh. Tekh. Fiz.* **51**, 2441 (1981).
200. M. W. D. MANSFIELD AND J. P. CONNERADE, *Proc. R. Soc. London Ser. A* **352**, 125 (1976).
201. K. T. CHENG AND C. F. FISCHER, *Phys. Rev. A* **28**, 2811 (1983).
202. K. T. CHENG AND W. R. JOHNSON, *Phys. Rev. A* **28**, 2820 (1983).
203. J. L. DEHMER, A. F. STARACE, U. FANO, J. SUGAR, AND J. W. COOPER, *Phys. Rev. Lett.* **26**, 1521 (1971).
204. C. PAN, S. L. STEVEN, AND H. P. KELLY, *Phys. Rev. A* **43**, 1290 (1991).
205. U. BECKER, H. G. KERKHOFF, D. W. LINDLE, P. H. KOBRIN, T. A. FERRETT, P. A. HEIMANN, C. M. TRUESDALE, AND D. A. SHIRLEY, *Phys. Rev. A* **34**, 2858 (1986).
206. M. RICHTER, M. MEYER, M. PAHLER, T. PRESCHER, E. V. RAVEN, B. SONNTAG, AND H.-E. WETZEL, *Phys. Rev. A* **40**, 7007 (1989).
207. J. L. DEHMER, A. C. PARR, AND S. H. SOUTHWORTH, in *Handbook on Synchrotron Radiation*, Vol. 2, edited by G. V. Marr (North-Holland, Amsterdam, 1987), p. 241.
208. I. NENNER AND J. A. BESWICK, in *Handbook on Synchrotron Radiation*, Vol. 2, edited by G. V. Marr (North-Holland, Amsterdam. 1987), p. 355.
209. L. S. CEDERBAUM, W. DOMCKE, J. SCHIRMER, AND W. V. NIESSEN, *Adv. Chem. Phys.* **65**, 115 (1986).
210. C. T. CHEN AND F. SETTE, *Phys. Scr.* **T31**, 119 (1990).
211. M. DOMKE, C. XUE, A. PUSCHMANN, T. MANDEL, E. HUDSON, D. A. SHIRLEY, AND G. KAINDL, *Chem. Phys. Lett.* **173**, 122 (1990).
212. K. J. RANDALL, J. FELDHAUS, W. ERLEBACH, A. M. BRADSHAW, W. EBERHARDT, Z. XU, Y. MA, AND P. D. JOHNSON, *Rev. Sci. Instrum.* **63**, 1367 (1992).
213. P. BALTZER, M. LUNDQVIST, B. WANNBERG, L. KARLSSON, M. LARSSON, M. A. HAYES, J. B. WEST, M. R. F. SIGGEL, A. C. PARR, AND J. L. DEHMER, *J. Phys. B* **27**, 4915 (1994).
214. O. HEMMERS, S. B. WHITFIELD, N. BERRAH, B. LANGER, R. WEHLITZ, AND U. BECKER, *J. Phys. B* **28**, L693 (1995).
215. K. J. RANDALL, A. L. D. KILCOYNE, H. M. KÖPPE, J. FELDHAUS, A. M. BRADSHAW, J.-E. RUBENSSON, W. EBERHARDT, Z. XU, P. D. JOHNSON, AND Y. MA, *Phys. Rev. Lett.* **71**, 1156 (1993).
216. L. J. MEDHURST, P. A. HEIMANN, M. R. F. SIGGEL, D. A. SHIRLEY, C. T. CHEN, Y. MA, S. MODESTI, AND F. SETTE, *Chem. Phys. Lett.* **193**, 493 (1992).
217. R. STOCKBAUER, B. E. COLE, D. L. EDERER, J. B. WEST, A. C. PARR, AND J. L. DEHMER, *Phys. Rev. Lett.* **43**, 757 (1979).

218. B. E. Cole, D. L. Ederer, R. Stockbauer, K. Codling, A. C. Parr, J. B. West, E. D. Poliakoff, and J. L. Dehmer, *J. Chem. Phys.* **72**, 6308 (1980).
219. S. Kakar, H. C. Choi, and E. D. Poliakoff, *Chem. Phys. Lett.* **190**, 489 (1992).
220. M. R. F. Siggel, J. B. West, M. A. Hayes, A. C. Parr, J. L. Dehmer, and I. Iga, *J. Chem. Phys.* **99**, 1556 (1993).
221. J. L. Dehmer and D. Dill, *Phys. Rev. Lett.* **35**, 213 (1975).
222. F. Sette, J. Stöhr, and A. P. Hitchcock, *J. Chem. Phys.* **81**, 4906 (1984).
223. A. P. Hitchcock and J. Stöhr, *J. Chem. Phys.* **87**, 3253 (1987).
224. T. J. Gil, C. L. Winstead, J. A. Sheehy, R. E. Farren, and P. W. Langhoff, *Phys. Scr.* **T31**, 179 (1990).
225. O. Hemmers, in *Studies of Vacuum Ultraviolet and X-ray Processes*, Vol. 3, edited by U. Becker (AMS Press, New York, 1993).
226. N. Padial, G. Csanak, B. V. McKoy, and P. W. Langhoff, *J. Chem. Phys.* **69**, 2992 (1978).
227. J. A. R. Samson and J. L. Gardner, *J. Electron Spectrosc. Relat. Phenom.* **8**, 35 (1976).
228. E. W. Plummer, T. Gustafsson, W. Gudat, and D. E. Eastman, *Phys. Rev. A* **15**, 2339 (1977).
229. A. Hamnett, W. Stoll, and C. E. Brion, *J. Electron Spectrosc. Relat. Phenom.* **8**, 367 (1976).
230. G. V. Marr, J. M. Morton, R. M. Holmes, and D. G. McCoy, *J. Phys. B* **12**, 43 (1979).
231. H. Kanamori, S. Iwata, A. Mikuni, and T. Sasaki, *J. Phys. B* **17**, 3887 (1984).
232. U. Becker, R. Holzel, H. G. Kerkhoff, B. Langer, D. Szostak, and R. Wehlitz, *Phys. Rev. Lett.* **56**, 1455 (1986).
233. O. Hemmers, F. Heiser, J. Eiben, R. Wehlitz, and U. Becker, *Phys. Rev. Lett.* **71**, 987 (1993).
234. P. Lablanquie, J. Delwiche, M.-J. Hubin-Franskin, I. Nenner, P. Morin, K. Ito, J. H. Eland, J.-M. Robbe, G. Gandara, J. Fournier, and P. G. Fournier, *Phys. Rev. A* **40**, 5673 (1989).
235. U. Becker, O. Hemmers, B. Langer, A. Menzel, R. Wehlitz, and W. B. Peatman, *Phys. Rev. A* **45**, R1295 (1992).
236. S. Krummacher, V. Schmidt, F. Wuilleumier, J. M. Bizau, and D. Ederer, *J. Phys. B* **16**, 1733 (1983).
237. P. W. Langhoff, S. R. Langhoff, T. N. Rescigno, J. Schirmer, L. S. Cederbaum, W. Domcke, and W. v. Niessen, *Chem. Phys.* **58**, 71 (1981).
238. C. M. Truesdale, S. H. Southworth, P. H. Kobrin, U. Becker, D. W. Lindle, H. G. Kerkhoff, and D. A. Shirley, *Phys. Rev. Lett.* **50**, 1265 (1983).
239. C. M. Truesdale, D. W. Lindle, P. H. Kobrin, U. E. Becker, H. G. Kerkhoff, P. A. Heimann, T. A. Ferrett, and D. A. Shirley, *J. Chem. Phys.* **80**, 2319 (1984).
240. D. W. Lindle, C. M. Truesdale, P. H. Kobrin, T. A. Ferrett, P. A. Heimann, U. Becker, H. G. Kerkhoff, and D. A. Shirley, *J. Chem. Phys.* **81**, 5375 (1984).
241. L. Ungier and T. D. Thomas, *Phys. Rev. Lett.* **53**, 435 (1984).
242. A. Reimer, J. Schirmer, J. Feldhaus, A. M. Bradshaw, U. Becker, H. G. Kerkhoff, B. Langer, D. Szostak, and R. Wehlitz, *Phys. Rev. Lett.* **57**, 1707 (1986).
243. T. Reich, P. A. Heimann, B. L. Petersen, E. Hudson, Z. Hussain, and D. A. Shirley, *Phys. Rev. A* **49**, 4570 (1994).
244. O. Hemmers, N. Berrah, B. Langer, R. Wehlitz, S. B. Whitfield, and U. Becker, *J. Phys. B* (to be published).
245. S. J. Desjardins, A. D. O. Bawagan, K. H. Tan, Y. Wang, and E. R. Davidson, *Chem. Phys. Lett.* **227**, 5 (1994).
246. J. Schirmer, M. Braunstein, and V. McKoy, *Phys. Rev. A* **41**, 283 (1990).
247. G. Bandarage and R. R. Lucchese, *Phys. Rev. A* **47**, 1989 (1993).
248. M. Schmidbauer, Ph.D. Thesis, Technische Universität Berlin, unpublished (1992).
249. H. W. Kroto, J. R. Heath, S. C. O'Brien, R. F. Curl, and R. E. Smalley, *Nature* **318**, 162 (1985).
250. W. Krätschmer, L. D. Lamb, K. Fostiropoulos, and D. R. Huffman, *Nature* **347**, 354 (1990).
251. J. H. Weaver, J. L. Martins, T. Komeda, Y. Chen, T. R. Ohno, G. H. Kroll, N. Troullier, R. E. Haufler, and R. E. Smalley, *Phys. Rev. Lett.* **66**, 1741 (1991).
252. J. H. Weaver, *J. Phys. Chem. Solids* **53**, 1433 (1992).
253. D. L. Lichtenberger, K. W. Nebesny, C. D. Ray, D. R. Huffman, and L. D. Lamb, *Chem. Phys. Lett.* **176**, 203 (1991).
254. S. Krummacher, M. Biermann, M. Neeb, A. Liebsch, and W. Eberhardt, *Phys. Rev. B* **48**, 8424 (1993).
255. G. F. Bertsch, A. Bulgac, D. Tomanek, and Y. Wang, *Phys. Rev. Lett.* **67**, 2690 (1991).
256. I. V. Hertel, H. Steger, J. de Vries, B. Weisser, C. Menzel, B. Kamke, and W. Kamke, *Phys. Rev. Lett.* **68**, 784 (1992).
257. L. J. Terminello, D. K. Shuh, F. J. Himpsel, D. A. Lapiano-Smith, J. Stöhr, D. S. Bethune, and G. Meijer, *Chem. Phys. Lett.* **182**, 491 (1991).

258. S. L. Molodtsov, A. Gutierrez, M. Domke, and G. Kaindl, *Europhys. Lett.* **19**, 369 (1992).
259. T. Liebsch, O. Plotzke, F. Heiser, U. Hergenhahn, O. Hemmers, R. Wehlitz, J. Viefhaus, B. Langer, S. B. Whitfield, and U. Becker, *Phys. Rev. A* **52**, 457 (1995).
260. H. Werner, T. Schedel-Niedrig, M. Wohlers, D. Herein, B. Herzog, R. Schlögl, M. Keil, A. M. Bradshaw, and J. Kirschner, *J. Chem. Soc. Faraday Trans.* **90**, 403 (1994).
261. G. Wendin and B. Wästberg, *Phys. Rev. B* **48**, 14764 (1993).
262. A. Ding, in *Electron collisions with molecules, clusters, and surfaces,* edited by H. Ehrhardt and L. Morgan (Plenum, New York, 1994).
263. S. Ren, Y. Wang, A. Rao, E. McRae, J. Holden, T. Hager, K. Wang, W. Lee, H. Ni, J. Selegue, and P. Ecklund, *Appl. Phys. Lett.* **59**, 2678 (1991).
264. P. Benning, D. Poirer, N. Troullier, J. Martins, J. Weaver, R. Haufler, L. Chibante, and R. Smalley, *Phys. Rev. B* **44**, 1962 (1991).
265. J. Wu, Z. Shen, D. Dessau, R. Cao, D. Marshall, P. Pianetta, I. Lindau, X. Yang, J. Terry, D. King, B. Wells, D. Elloway, H. Wendt, C. Brown, H. Hunziker, and M. de Vries, *Physica* **C197**, 251 (1992).

CHAPTER 6

HIGH-RESOLUTION ELECTRON SPECTROMETRY OF ATOMS

M. O. KRAUSE AND C. D. CALDWELL

1. INTRODUCTION

Electron spectrometry has been practiced under various conditions since it became a viable method in the 1960s for the study of the electronic structure and electron dynamics of free atoms. This technique, as an experimental tool, follows the dynamics of the electrons in the atom as they respond to an initial excitation, and the spectral resolution available depends on the conditions of the experiment. For example, in Auger electron spectrometry (AES), the determining instrumental factor dictating the overall resolution is the resolution of the electron energy analyzer for any mode of excitation, whether by photons, electrons, ions, or internal nuclear conversion. In electron energy-loss spectrometry (EELS), or in (e,2e) electron spectrometry, which can be used to simulate photoionization, the resolution of the electron beam source is a second instrumental factor to be considered. Similarly, in photoelectron spectrometry (PES), the resolution, or bandpass, of the photon source is one of the two basic instrumental factors which contribute to the overall resolution. In this chapter we discuss primarily electron spectrometry as it is practiced using synchrotron radiation as an excitation source (ESSR). Within this particular area, we will focus on the experimental results which have become available by taking advantage of the steady improvements which have been made over the years both in the electron spectrometer and in the photon source, with the most important advances having occurred more recently in the performance of the photon monochromators coupled to electron storage rings.

Of course, high resolution is a relative term, often invoked after instrumental and operational advances have occurred, but ultimately justified only in regard to the natural properties of the matter under scrutiny. In atoms, these are primarily the natural widths of the atomic levels and the fine-structure splittings of levels. With the tools which have become available the resolution of electron spectrometry of atoms is approaching the perimeter defined by the natural width in the case of shallow core levels and the spacing of fine-structure levels in the outer regions of the electron cortege of the atom. This applies to spectrometry carried out with fixed line sources, monochromatized synchrotron radiation, electron excitation, and, at the lowest energies, with laser radiation. In fact, a very specialized laser-based technique, zero-kinetic-energy electron spectrometry (ZEKE), already meets the natural limits in many instances by providing a resolution of 50 μ eV below 20 eV, as contrasted to the 5 meV to several tenths of one electron volt achievable by other means between 20 and 2000 eV photon energy.

At this point it is tempting to illustrate the progress that has occurred over the first 15 years of ESSR studies with the aid of a photoelectron spectrum which elucidates the photoeffect in the outer subshells of the argon atom. As seen in Figure 1, an early spectrum produced at a synchrotron radiation source[1] gave clear evidence of satellite structure due to electron correlation—an important topic in modern atomic and molecular science—but only a recent study[2] with advanced instrumentation was

---

M. O. KRAUSE • Oak Ridge National Laboratory, Oak Ridge, Tennessee 37831-6201.   C. D. CALDWELL • Department of Physics, University of Central Florida, Orlando, Florida 32816-2385.

VUV and Soft X-Ray Photoionization. Edited by Uwe Becker and David A. Shirley. Plenum Press, New York, 1996.

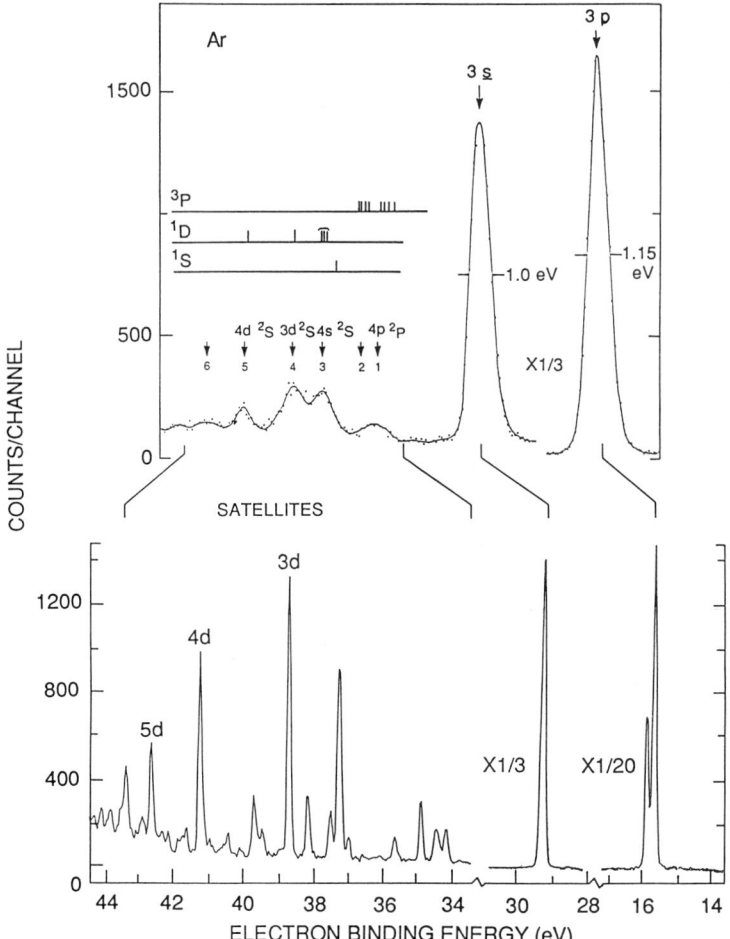

FIGURE 1. The argon photoelectron spectrum obtained with the use of synchrotron radiation, for the first time in 1978 at low resolution,[1] 1.0 eV, and again in 1992 with high resolution,[2] 0.1 eV (FWHM).

capable of definitively resolving the major components of the structure. Along with the greater differentiation afforded by the higher resolution has come a deeper understanding of the photoionization and electron correlation processes in many atomic and molecular systems.

Many treatises of electron spectrometry of atoms can be found in books and reviews. We name the early works by Siegbahn et al.,[3] Carlson,[4] Sevier,[5] and Krause[6] and the more recent works by Siegbahn and Karlsson,[7] Krause,[8] Samson,[9] Sonntag and Zimmermann,[10] and Schmidt.[11] An overview of the theory of photoionization has been given by Starace,[12] and advances in instrumentation associated with synchrotron radiation are periodically reported in proceedings of relevant conferences.[13] We also point to various chapters in this volume relating to other aspects of photoionization and electron spectrometry.

2. SOME EXPERIMENTAL DETAILS

Before beginning a discussion of high resolution it is appropriate to outline a few considerations which relate to the role which resolution plays in an experiment in electron spectrometry. We will restrict ourselves to experiments in the gas phase in which a dilute gas, which may be in a cell or in a beam, is intersected with photons from a suitably chosen monochromatic source, producing an electron (or electrons) of specified energy corresponding to an ion (or ions) in a particular final state (or states). The focus of measurements in electron spectrometry is to follow the process through the detection of the electrons.

The heart of the measurement is the photoelectron spectrum (PES) in which the distribution in energy of electrons ejected by photons of fixed energy is accumulated. The overall resolution of the experiment is determined by contributions from the photon source, the electron analyzer, physical broadening effects such as thermal motion, and the natural width of the feature being measured. The first two are completely experimental in nature and lend themselves to improvement as technology improves. The third can be overcome to a great extent by experimental techniques such as the use of an atomic beam rather than randomly moving particles in a gas target. However, the natural width represents a fundamental limitation on the additional information about the physics of the system which can be gained through improvements in the experimental resolution.

The line contour in electron spectrometry is of the general form:

$$V(x,\Gamma_v) = \int_{-\infty}^{+\infty} L(\xi,\Gamma_L)\, G(x-\xi,\Gamma_G)\, d\xi \tag{1}$$

where the contribution $L(\xi,\Gamma_L)$ is that related to the natural line profile with width Γ_L, and the function $G(x,\Gamma_G)$ represents the instrumental contour with width Γ_G. $G(x,\Gamma_G)$ consists of contributions to the width from all of the instrumental sources and will, in general, derive from an overlap of different Gaussian profiles, but in principle may have other functional shapes. In a typical photoelectron spectrum $G(x,\Gamma_G)$ will generally be dominated by the contributions from the monochromator and the electron energy analyzer. However, as instrumentation is being improved, the thermal or Doppler broadening gains in importance. So too do the effects of electronic noise.

Generally, Gaussian profiles $g_i(x)$, with i referring to the given parameter, are excellent representations of the real functional dependencies of all of these types of experimental broadening. When this is the case, the function $G(x,\Gamma_G)$ in equation (1) is the convolution of the individual Gaussian functions:

$$G(x,\Gamma_G) = \int\int\int\int g_P(x-\xi_1,\Gamma_P)\, g_E(\xi_1-\xi_2,\Gamma_E)\, g_T(\xi_2-\xi_3,\Gamma_T)\, g_N(\xi_3-\xi_4,\Gamma_N)$$
$$\cdot d\xi_1\, d\xi_2\, d\xi_3\, d\xi_4 \tag{2}$$

where $g_P(\xi)$ represents the contribution from the photon source, $g_E(\xi)$ the contribution from the electron energy analyzer, $g_T(\xi)$ the thermal contribution, and $g_N(\xi)$ the contribution from the electronic noise.

We note that the width coming from the thermal motion of the particles is given by

$$\Gamma_T = 0.723 \left(\frac{ET}{M}\right)^{1/2} \quad \text{(meV)} \tag{3}$$

where E is the kinetic energy of the photoelectron in eV, T the temperature in K, and M the atomic mass of the particle. For the case in which the Gaussian representations of the instrumental functions are valid, with all contributions considered, the overall instrumental width is given by

$$\Gamma_G^2 = \Gamma_P^2 + \Gamma_E^2 + \Gamma_T^2 + \Gamma_N^2 \tag{4}$$

where subscripts are as in Equation 2. As already pointed out, the functions may have other than Gaussian contours, including asymmetric profiles. If this is the case, then the width is not as simply attributable to the different sources. However, it is always possible, and often convenient, to determine experimentally the overall contour $G(x,\Gamma_G)$ by recording the photoline from an atomic level with negligible width, whether the photoline contour is of the simple form of equation (1) or of a more complicated form. The monochromator function $g_P(\xi)$ can be determined individually by direct observation using a suitable detection technique, such as the observation of the light transmitted by the monochromator in zero order or a scan over a very narrow autoionization resonance.

In practice the natural widths Γ_L in equation (1) are determined by the application of the equation to a spectrum using a convolution technique which permits the determination of the contribution of the instrumental factor to the line profile. The natural line profiles $L(\xi,\Gamma_L)$ are Lorentzian, except in certain special cases. (See Section 6.) Thus, the overall line shape given by the convolution in equation (1) is a Voigt profile. A particularly useful representation of the Voigt profile which is especially appropriate for fitting complicated, overlapping spectra is due to Pearson.[14] The so-called Pearson-7 function has the form

$$P_7(\varepsilon) = A\left[1 + \frac{(\varepsilon - \varepsilon_0)^2}{B^2 C}\right]^{-C} \qquad (5)$$

In this equation A is the peak height, ε_0 the peak position, B the nominal half-width-half-maximum of the peak, and C relates to the shape of the peak. For $C = 1$ the Pearson-7 function is identically Lorentzian, and in the limit $C \to \infty$ the function is identically Gaussian. However, for practical purposes, the function becomes essentially Gaussian for values of C greater than 20. A useful modification of the Pearson-7 function which can be employed for fitting asymmetric line shapes such as those which result from postcollision interaction (PCI) has also been derived.[15]

Although the calculation of the Voigt profile for a line shape cannot be performed analytically, it is possible to determine an analytical form of equation (1) for the case of a triangular instrumental profile and a Fano line shape.[16] In practice, the difference in widths as determined with either the use of a triangular profile or a Gaussian profile is negligible. This property is especially convenient for the analysis of autoionization profiles when the function $L(\xi,\Gamma_L)$ represents the more general form of a Fano or Shore profile. Using the analytic form of equation (1) as the line shape for a fitting algorithm, it is possible to determine the Fano parameters directly from the fit without having to proceed through a convolution routine. This both increases the accuracy of the fit and shortens the time required for the analysis by about an order of magnitude.

The application of equation (1) becomes simpler for experiments in photoionization which utilize other detection techniques, i.e., total absorption, ion yield measurements, total fluorescence yield measurements, and total or partial electron yield measurements. For these techniques the only instrumental contribution to the experiment is that related to the bandpass of the photon source and, at very high resolution, that arising from the Doppler effect. As these techniques sample the resonance channel when resonance excitation occurs, they can reveal the Fano-type line shapes which often characterize resonances provided the photon bandpass is adequately narrow.

Perhaps the most significant characteristic of emission studies is the potential for differentiation, which allows the determination of branching ratios into the various final states and the angular distributions of decay electrons. Such detail provides a more sensitive test of theoretical treatments of electron dynamics than is afforded by absorption alone. For a specific final state channel, the decay process is best followed through a partial electron yield experiment, or constant ionic state (CIS) scan. This mode of operation is particularly valuable because the bandwidth or resolution of the electron analyzer only makes a contribution insofar as the *isolation* of particular final ionic states is concerned. The electron analyzer will isolate all of those ionic levels which are separated by more than the analyzer resolution; decays into two or more ionic states which lie within the analyzer width cannot be distinguished. However, once the selection of final states to be studied has been made, the resolution of the CIS scan for these states is determined solely by the bandpass of the monochromator and the level width, exactly as it would be if other particles such as photoions were used for detection. When it is possible to isolate the final ionic levels, as, for example, in the case of the spin-orbit components of the rare gases or the multiplets in halogen atoms, then a CIS analysis forms an excellent means of measuring the characteristic parameters of autoionizing resonances[17,18] within the bandpass limits of the monochromator.

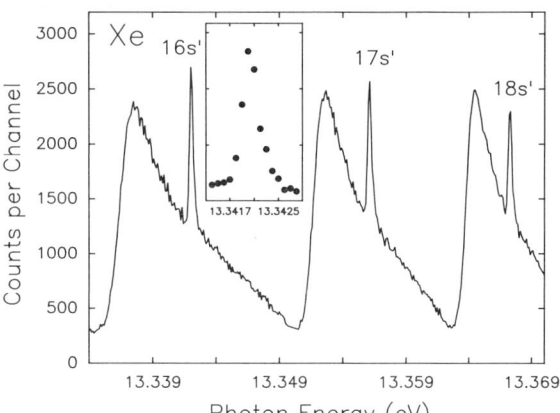

FIGURE 2. The higher Xe $ns',(n-2)d'$ autoionization resonances observed in emission by a constant-ionic-state scan using the Xe $5p_{3/2}$ photoline. The widths of the ns' features reflect the monochromator bandpass of 290 µeV. (From Ref. 19.)

In an interesting variation of the CIS method, a broad-band or white-radiation photon source is employed and an electron spectrometer serves as the dispersive element that determines the resolution. Since in general a photon monochromator is capable of a higher resolution than an electron spectrometer, dispersion by the monochromator is the preferred approach except near thresholds for a photoelectron of low kinetic energy. The two alternative modes of the CIS method have an analogue in traditional photoabsorption setups in which the radiation is dispersed either *before* the photon-target interaction by a monochromator or *after* the interaction by a spectrometer.

Let us consider as an example of the factors discussed in the preceding paragraphs the profile and bandpass of the photon beam emerging from the 4-m normal-incidence monochromator (NIM) at the Synchrotron Radiation Center (SRC) in Wisconsin. A narrow autoionization resonance, such as the $16s'$ state of the Rydberg series converging toward the $^2P_{1/2}$ ionization limit of xenon, is most suitable for determining the ultimate bandpass of the NIM by observing the intensity variation of the Xe $5p_{3/2}$ photoelectron signal across the resonance in a CIS experiment. The resulting CIS spectrum[19] is shown in Figure 2. As the natural width of the resonances is very small, about 60 µeV for the $16s'$ state,[20,21] the bandpass can, to a good approximation, be read off directly from the full width at half maximum of the peak. If a more accurate result is desired, the bandpass has to be calculated from equation (1) with $\Gamma_L = 60$ µeV for $L(\xi,\Gamma_L)$ as input and Γ_V in $V(x,\Gamma_V)$ being the experimentally determined FWHM of the peak. However, the use of a Lorentzian function is not exact. As will be shown later, all of the resonance features in this series have Fano-type line shapes and $L(\xi,\Gamma_L)$ has to be suitably substituted. Doing this, a bandpass of $\Delta E = 290$ µeV, corresponding to $\Delta\lambda = 20$ mÅ or a resolving power of $\lambda/\Delta\lambda = E/\Delta E = 44,000$ for slit settings of 20 µm,[19] was found for the spectrum

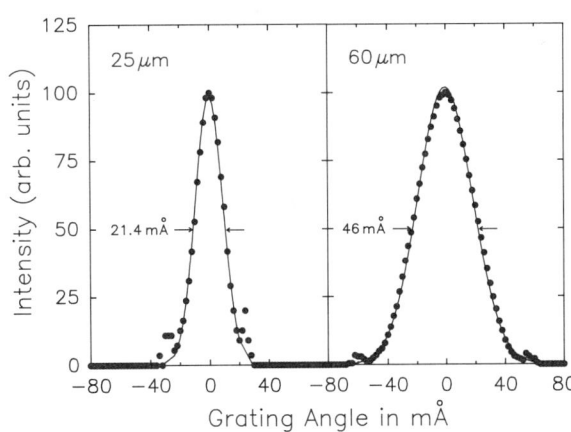

FIGURE 3. Photon beam profile of the 4-m NIM instrument at the Aladdin Storage Ring recorded with a photodiode for zero-order light at two narrow slit settings. Solid lines represent a fit to the data points with a Gaussian contour.

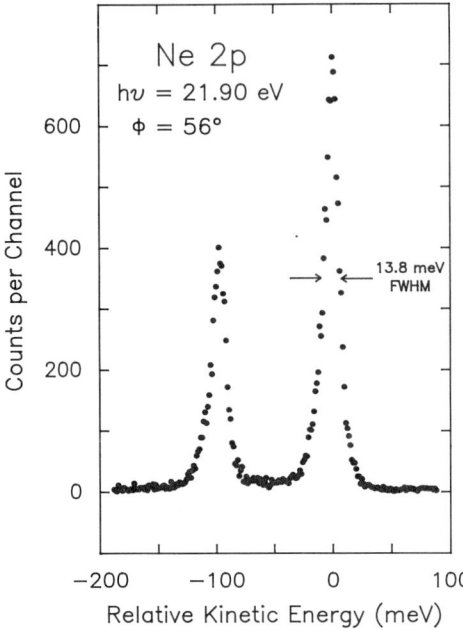

FIGURE 4. Ne $2p$ photoelectron spectrum measured at high resolution and low electron kinetic energy. The Ne $2p_{3/2}$ level energy is 21.56 eV, and the photon source was the 4-m NIM at SRC. (From Ref. 19.)

illustrated in Figure 2. This result can be corroborated in the case of the NIM by observing the signal from a photodiode while scanning the zero-order light across the monochromator exit slit. As seen in Figure 3, nearly the same value for the bandpass at equivalent slit settings is observed as in the CIS experiment. In addition, the contours are seen to be substantially Gaussian for these small apertures, justifying the use of a Gaussian function for $G(x,\Gamma_G)$ in equation (1).

The photoelectron spectrum of Ne $2p$, which is shown in Figure 4, exemplifies the determination of the width Γ_G. Here the natural width, Γ_L, is entirely negligible. Again the experiment[19] refers to the 4-m NIM at SRC and the same electron spectrometer as used for the CIS scan presented in Figure 2. With slit settings of 200 µm for the monochromator and an analyzer resolution of $\Delta E/E = 0.01$, the PES resolution, as given by the FWHM of the photoline, was measured to be 13.8 meV. This resolution is composed of the following contributions which could be determined separately: a photon bandpass of 5.2 meV, an analyzer resolution of 5.4 meV, a composition of ripple and high-frequency noise from the storage ring and other electronic sources amounting to 9 meV, and the thermal broadening of 1.6 meV. Assuming Gaussian distributions for these components a total width of 12.3 meV is calculated. This assumption, while being quite good, is not entirely accurate, as a comparison of the observed and calculated widths indicates, and as the slightly asymmetric peak profile implies. As discussed by Krause et al.,[19] a high precision in the definition of both the photon and the electron beams by the respective slit arrangements and, most importantly, a careful preparation of all surfaces seen by the electrons are of crucial importance to ensure a symmetric, virtually Gaussian profile for the photoline. Indeed, Baltzer et al.[22] realized these necessary conditions for the electron spectrometer, including a low noise level, and achieved an excellent line contour with a FWHM of 5 meV for the Ne $2p$ photolines excited by He Iβ photons ($h\nu = 23.09$ eV). Their PES result is shown in Figure 5.

As mentioned earlier, the natural width of the ionic level constitutes a fundamental limit on the extent to which improving instrumental resolving power can bring to light new information about the underlying physics and thus provides a working definition of the term *high resolution*. We can obtain a rough estimate of the requirements which we place on resolving power in order for the experiment to be understood as being appropriate to high-resolution electron spectrometry in terms of the energies and widths involved. We consider cases appropriate to two different energy regions. For outer levels we choose the characteristic widths to be those corresponding to the autoionization levels lying between the fine-structure components of the rare-gas ions; for inner shells we take the natural widths

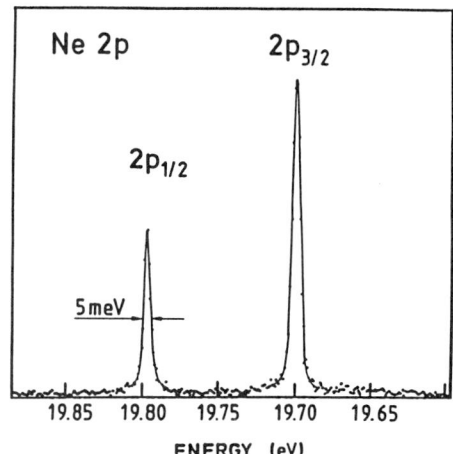

FIGURE 5. High-resolution Ne 2p photoelectron spectrum obtained in a laboratory setting with the He Iβ line ($h\nu = 23.09$ eV) from an ECR source. Note that thermal broadening is a major contributor to the width. (From Ref. 22.)

of the atomic levels as being representative. The corresponding energies are either those of the excitation selected or the inner-shell ionization (atomic level) energy. In Table I we give a listing of some examples of energies and widths for particular systems of interest, and from these we calculate the corresponding resolving powers required to match the natural width. In order for an experiment to follow the behavior of an excitation or decay in greater detail, it will be necessary that the instrumental resolving powers exceed the values listed in Table I by a factor of 3–5. Then the influence of the instrumental width on the line shape can be ignored in a first approximation or can be reliably assessed through a convolution procedure. The energies and widths given in Table I should be taken as being merely representative; they are not intended to serve as a reference table for calibration.

TABLE I
Typical Resolving Powers Required to Study Atomic Levels[a]

Principal shell	Element/subshell	E_B (eV)[b]	Γ (meV)[c]	Resolving power[a] E_B/Γ
K	Ne	871	270	3,200
	Ar	3203	680	4,700
	O	544	130	4,200
	O (O$_2$)	542	180	3,000
	N (N$_2$)	410	120	3,400
L	Ar 2p	248.4	127	2,000
	Kr 2p	1679	1170	1,400
	Xe 2p	4786	3130	1,500
M	Kr 3d	94.4	90	1,000
N	Xe 4d	68.5	110	700
Outer	He 2s3p$^+$	63.7	10.0	6,400
	Ne 14s'	21.6	0.022	981,000
	Ar 13s'	15.8	0.054	293,000
	Xe 9s'	12.9	0.852	15,000
	Xe 16s'	13.3	0.07	190,000

[a]Resolving power $E/\Delta E$ is given in terms of the level energy E_B and the level width Γ; values are rounded to nearest hundred.
[b]When spin-orbit components are involved, average values are used.
[c]Width values are taken from various sources and are approximate.

Although the greatest portion of the work which we will discuss in this chapter uses monochromatized synchrotron radiation as the photon source, high-energy laser sources will be essential for studies of the outer levels in which the resolution demands are highest. For example, the widths of the $1s^2 2s^2 2p^5(^2P_{1/2})ns', nd'$ neon resonances which decay into the $Ne^+(^2P_{3/2})$ ionic channel can at present only be measured through the combination of a laser source with a molecular beam.[21,23] Developments which extend laser sources deeper and deeper into the ultraviolet should continue to improve possibilities for very-high-resolution studies of the valence levels.[24]

For inner-shell processes where the energies are higher, synchrotron radiation should continue to remain the photon source par excellence. For the region of very low energy, 10–40 eV, which overlaps with laser sources at the lower end, monochromators of the normal incidence type are especially useful in emission experiments requiring high resolution. This has been demonstrated in several studies.[17,19,25] For example, in a CIS scan over the neon Beutler–Fano resonances a resolution of 730 μeV was achieved using a normal-incidence monochromator.[25]

Much improvement in resolution at higher energy should be realized with the construction of photon sources having unprecedented high brightness based on storage rings which are equipped with insertion devices. At the same time, notable advances have already taken place through the development of new and improved monochromators for high-energy measurements. Of these, the Dragon monochromator at NSLS,[26] the Spherical Grating Monochromators (SGM) currently in use at SSRL,[27] the 10-m Grazing Incidence Monochromator (GIM) at the Photon Factory,[28] and the SX-700 developed at BESSY[29] are noteworthy. Benchmark resolving powers of about 10,000 have been reported for these instruments in the photon energy range from 60 to 600 eV. This translates into 30 meV at the carbon edge and 40 meV at the nitrogen edge. In fact, the vibrational structure of the K level in N_2 has become the standard for assessing the performance of a monochromator. This reference, while problematic for an absolute determination of the resolution, because the natural width of the K level in N_2 is not accurately known, can nevertheless serve well as a comparative criterion. Recently, it was shown by a study of the elusive $(2p,nd)^1 P^o$ resonances associated with double excitation in helium[30] that the resolving power can be pushed to 16,000 for at least one of these advanced monochromators. While none of the new high-energy instruments have been tested in emission measurements, the SGM construction will form the basis of the monochromators at the new Advanced Light Source at Berkeley, where studies in emission will constitute a major component of the gas-phase activity.

Although lacking the tunability of the synchrotron, line radiation from a discharge in rare gases still forms a most useful source for studies in the vacuum ultraviolet. With a helium source based on the ECR (electron cyclotron resonance) principle, a photoelectron resolution of 50 meV could be achieved in a study[31] of the $5s$ correlation satellites in xenon at $h\nu = 40.8$ eV, (He IIα), while a resolution of about 5 meV was realized in a study of the inner valence levels of the O_2 molecule.[22] At the lowest photon energies, laser-based techniques such as ZEKE and resonance-enhanced multiphoton ionization (REMPI) are especially powerful for delineating the electronic structure of atoms and molecules at very high resolution (see Chapter 10).

Resolution is intimately intertwined with intensity. It is a truism that one must be traded for the other. A useful photon intensity on a sample at the optimum resolution capability of a monochromator requires a high brightness of the photon source and a high brightness in the interaction region, with a good focusing of the photon beam onto the sample. A bright photon beam focused on a small spot makes it possible, in turn, to optimize the performance of conventional electron energy analyzers by employing a narrow entrance slit system with or without entrance focusing elements. The inevitable reduction of signal strength by the small solid angle accepted by a high-resolution analyzer can be effectively obviated by the use of a multichannel, or position sensitive, electron detector at the analyzer exit. In recent years, great efforts have been placed on improving the various factors essential for high-resolution electron spectrometry. The gain that comes from the use of a multichannel detector is illustrated by Figure 6, which displays the photoelectron spectrum of atomic vanadium ionized by 21.2-eV He Iα photons.[32] The improved signal strength is evident at once by comparing the single-channel detector and multichannel detector spectra. At the same time, a better resolution and a

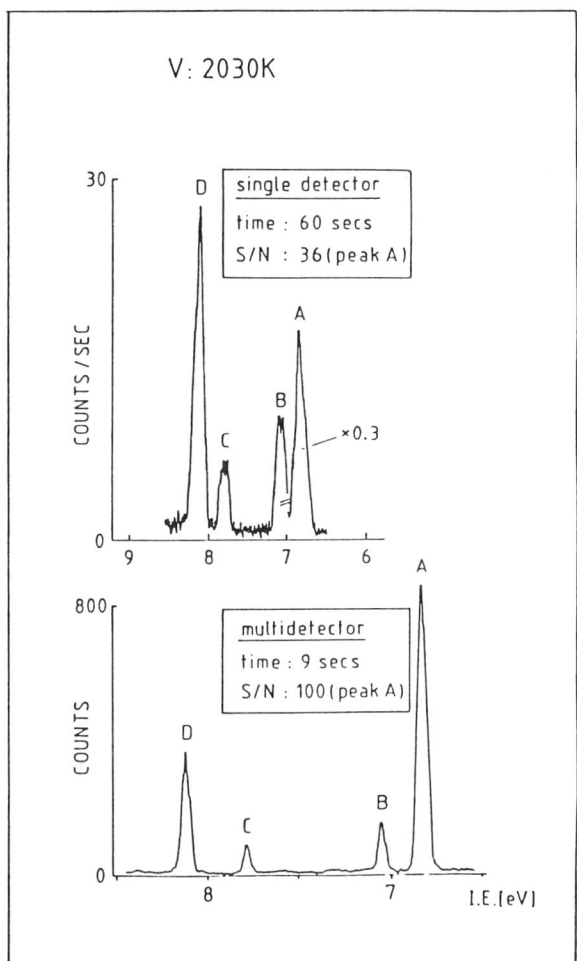

FIGURE 6. Photoelectron spectrum of atomic vanadium recorded with He Iα ($h\nu = 21.2$ eV) photons using either a single-channel or a multichannel detector. Note the superior multichannel detector spectrum. (From Ref. 32.)

greater signal-to-noise ratio could be realized by applying a multichannel detector alongside other improvements.

Multichannel detectors have relied on microchannel plates, preferably of high image quality, for efficient and uniform electron detection, but the handling of the electron bunch, following the multiplication avalanche, had remained problematic. Most recently, however, an integrated system has been developed which is especially suited for electron spectrometry with the use of a tunable photon source.[33] Because the charge-coupled solid-state device is integrated with the associated amplifiers and closely coupled with the microchannel electron multiplier, the resulting compact and ultrahigh vacuum compatible detector requires little space in the vacuum chamber, and the resulting signal pulses have a sufficient amplitude to be processed readily by a computer. Two-dimensional spectra are measured by recording a series of photoelectron spectra at fixed photon energies that are incremented in small steps.[34] The schematic diagrams of Figure 7 demonstrate the readout capabilities of such a system of data acquisition. As can be seen, the previously mentioned PES and CIS modes of operation with a single detector are now incorporated in a single two-dimensional array. In addition, a third complementary mode of operation, the constant kinetic energy (CKE) mode, forms part of the data handling process. For further illustration, a real data plot is presented in Figure 8 with the number of counts displayed cartographically in terms of altitude contour lines. As implied in the diagrams of Figure 7, conventional PES displays can be extracted by making horizontal cuts through the contour

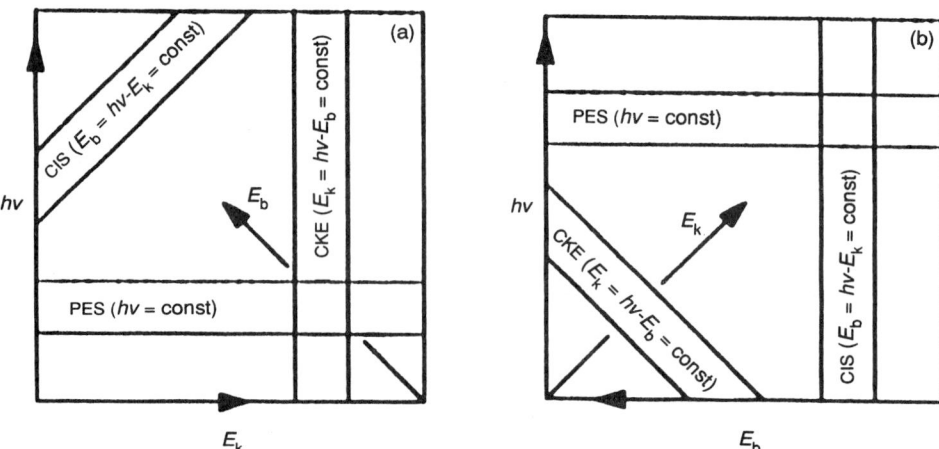

FIGURE 7. Schematic display of the various modes of operation used in electron spectrometry in an $h\nu$ versus E_{kin} or an $h\nu$ versus E_B frame. These modes are: PES, photoelectron spectrum; CIS, constant ionic state or constant initial state; and CKE, constant kinetic energy, which is also known as CFS, constant final state. Either frame is applicable to the data array that can be produced with a multichannel (or position sensitive) detector. (From Ref. 34.)

map of Figure 8 at the desired photon energies $h\nu$. The result of such a procedure is presented in Figure 9, where the Kr $3d_{5/2}$ and Kr $3d_{3/2}$ photolines, skewed by postcollision interaction, are displayed. From the experimental point of view these spectra must be regarded as high-resolution data since the instrumental factors, photon bandpass and analyzer resolution, contribute only moderately compared to the natural factors of the process, the width of the Kr $3d$ levels and the asymmetric line broadening (and shift) by the interaction between Auger and photoelectron.

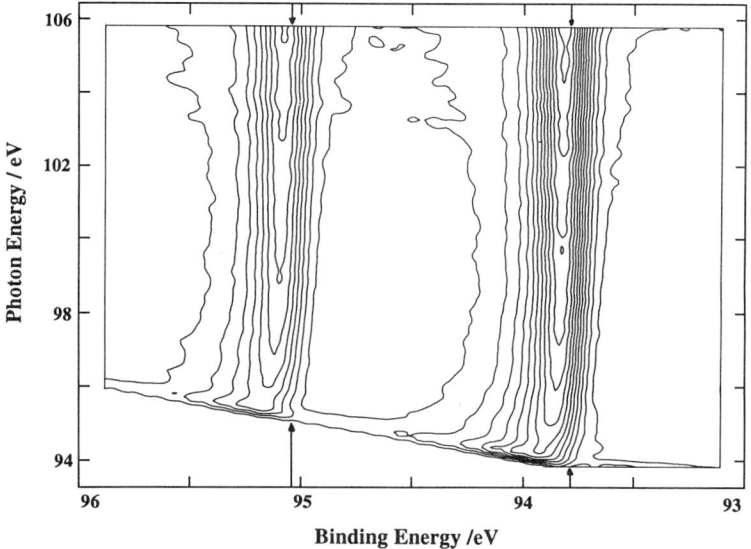

FIGURE 8. Contour map of data array (landscape) obtained with a multichannel detector according to Ref. 34. Arrows at 95.04 eV and 93.80 eV indicate the Kr $3d_{5/2}$ and Kr $3d_{3/2}$ level energies; the narrow ridges at slightly higher binding energies denote the maximum number of counts.

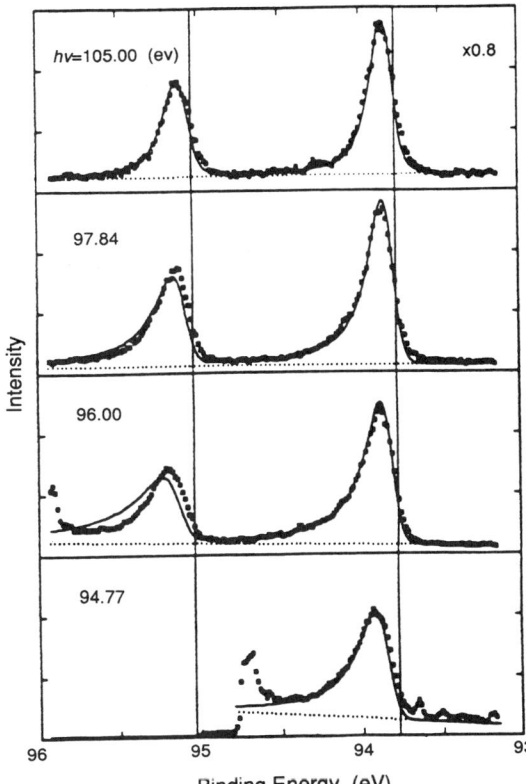

FIGURE 9. Horizontal cuts through the landscape shown in Figure 8 at the photon energies indicated. These PES cuts show Kr 3d photolines modified by postcollision interactions. A bandpass of 90 meV and an analyzer resolution of 30 meV were used in the experiment, denoted by points. (From Ref. 34.)

3. PHOTOELECTRON SPECTRA OF ATOMS

Photoelectron spectrometry reveals the electronic structure by virtue of the photoeffect. In the simplest, and usually most probable, photon–atom interaction, a *single* electron is ejected from one of the atomic subshells. This, in turn, gives direct information on various subshell properties, with the energy being the most basic property and spin polarization being the most detailed. The question arises whether the PES resolution is sufficient to differentiate between the various subshells, or more generally, the various atomic energy levels. For an overview of the requirements Table II presents the resolving powers needed to separate the spin-orbit components or the multiplet components in several closed-shell and open-shell atoms. The values apply to photon energies that are of comparable magnitude to the level energies. In contrast to Table I, where the level energies[35,36] are compared with the natural widths,[8,27,37] the comparison of the level energies with the spin-orbit or multiplet splittings in Table II indicates that a modest PES resolving power, less than 300, will suffice to resolve the components for all levels in the rare gases and for many core levels in other elements. However, if photon energies far above the level energy are to be used, as required for the measurement of cross sections over an extended energy range, then greater resolving powers are needed. For example, the spectrum of Figure 1 shows that even for a PES resolution corresponding to $E/\Delta E = 600$, the Ar 3p doublet is not completely separated at a photoelectron energy of 50 eV. A similar situation would prevail in the case of the Ne 2p doublet depicted in Figure 4 if the photon energy were raised above 35 eV under the conditions of that experiment.[19] As these examples illustrate, resolution may quickly become an issue even for the simplest process, that of single-electron photoionization of a rare-gas atom, or other atoms, having an $^1S_0$ ground state. This places a greater demand on the resolving power of monochromators at higher energy and on the flux requirements for the photon source. If a large flux

TABLE II
Resolving Power $E/\Delta E$ Needed for Separation of Spin-Orbit and Multiplet Components of Selected Atomic Subshells[a]

Element/subshell	Energy, E_B (eV)	Splitting, ΔE^b (eV)	Resolving power $E/\Delta E$
Ne 2p	21	0.1	210
Ar 3p	16	0.18	90
Ar 2p	250	2.2	110
Kr 3d	95	1.2	80
Kr 3p	220	8	30
Kr 2p	1700	52	30
Ag 4d	12	0.1	120
Mn 3p	60	3	20
Mn 2p	645	11	60

[a] All values are rounded off and apply to photon energies comparable to the subshell energies, namely, $h\nu \approx 2E_B$.
[b] In the case of a multiplet, the smallest component interval is given.

is available, then the contribution of the electron energy analyzer to the overall resolution can be reduced by retardation of the electrons.

Photoelectron spectra of open-shell atoms can become complex because of the multiplet states that are reached on ionization from a particular subshell. High resolution may be needed to separate the LS term components and is certainly mandatory for separating the fine-structure components. From an experimental point of view, it must be realized that in many instances an ensemble of open-shell atoms cannot be prepared with all atoms solely in the ground state but will contain an admixture of atoms in excited states. Generally, the excited states are reached by thermal population and, as the separation from the ground state is on the order of kT, they can rarely be resolved in a PES recording. We will discuss a few examples in the following, namely, vanadium, nickel, and manganese atoms produced by high-temperature vaporization of the metal, oxygen and chlorine atoms produced by dissociation of O_2 and Cl_2 in a microwave discharge, and, finally, bromine and carbon atoms produced by laser dissociation of suitable molecules.

The vanadium spectrum presented in Figure 6 shows four of the six photolines associated with ionization in the 4s and 3d subshells by He Iα (21.22 eV) photons. Peaks B and C correspond to the $3d^34s^1(^5F)$ and $3d^34s^1(^3F)$ ionic states reached by 4s electron removal from the $3d^34s^2(^4F)$ ground state. Peaks A and D arise nominally from 3d ionization of the atom in its $3d^44s^1(^6D)$ excited state and correspond to the $3d^34s^1(^5F)$ and $3d^34s^1$ (^5P) ionic states. Two additional peaks, which appear at higher level energies outside the range plotted in Figure 6, are the result of 3d photoionization from the $3d^34s^2(^4F)$ ground state and correspond to the $3d^24s^2(^3F)$ and $3d^24s^2(^3P)$ ionic states. Two major observations can be made about this photoelectron spectrum,[32] which is based on an instrumental resolution of 40 meV and a negligibly small bandpass of the He Iα photon source. First, the photolines are broad, 80 to 100 meV. A major cause for this is the strong population of the $J = 5/2, 7/2, 9/2$ components of the ^4F state at the vaporization temperature of 2000 K. The $^4F_{3/2 \text{ to } 9/2}$ manifold spans 68.6 meV, and the J components of the various final states cover ranges from 40 to 70 meV. To isolate the transitions connecting to the $^4F_{3/2}$ ground state, or any of the excited states, a resolution of better than 10 meV would be required. Ultimately, such a resolution is mandatory for a rigorous comparison of the experimental data with theoretical results, which must be based on the relevant population of the fine-structure states. Second, peaks A and D give evidence of the $3d^44s^1(^6D)$ excited state being admixed to the ^4F state in the neutral atom. Such configuration mixing is common and often pronounced, leading to the appearance of extra peaks in the PES. It is a type of correlation that complicates a spectrum. However, in vanadium the different LS components are spaced far enough apart to be separable.

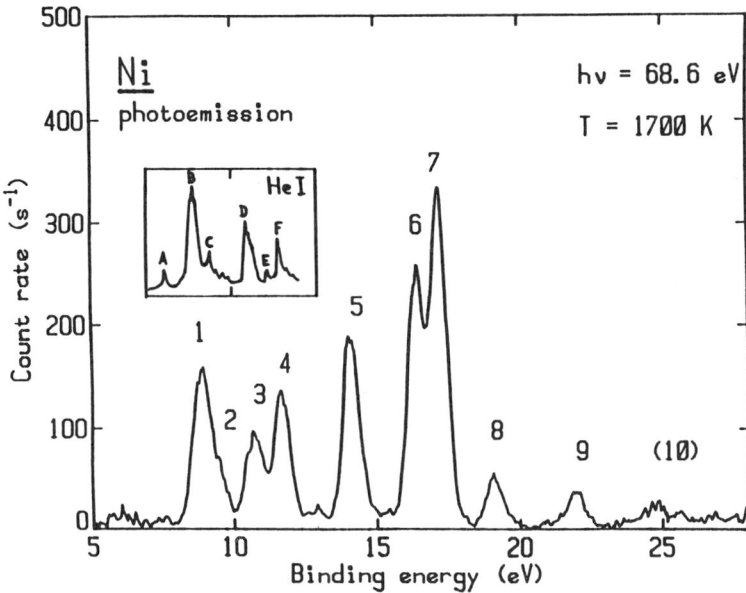

FIGURE 10. Photoelectron spectrum of atomic nickel recorded at the magic angle in the $3p \to 3d$ resonance.[38,39] The inset shows the He Iα spectrum recorded at an angle of 90°[40]; region B to F corresponds to the range of peaks 1 to 4.

In nickel the PES of the outer levels displays many lines, as seen in Figure 10. Because the $3d^8 4s^2(^3F)$ and $3d^9 4s^1(^3D)$ manifolds have nearly the same energy, with their respective J components overlapping, it becomes mute to speak of $3d$ or $4s$ electron ionization except perhaps in a few instances of clear separation. Instead, lines 1 to 4 of Figure 10 should be assigned to the $3d^8(L)4s(L')$ states and lines 5 to 10 to the $3d^7 4s^2(L'')$ states of the Ni$^+$ ion[38,39] without making specific reference to ionization from a specific subshell. The Ni spectrum was excited by 68.6-eV photons, and the spectral distribution reflects the influence of the strong $3p \to 3d$ resonance transition at this energy (see also Section 5.3). The doubtless respectable PES resolution of this study carried out at a synchrotron radiation source allowed the delineation of the final states according to the LS multiplets. A higher-resolution spectrum with a He Iα light source accentuates the separation of lines 1 to 4, as seen in the inset of Figure 10. The authors of this latter study[40] succeeded in bringing forth the J components buried in line B (or line 1) using a resolution of 65 meV. In that spectrum, Figure 6a of Ref. 40, the $^4F_{9/2,7/2,5/2,3/2}$ components of the $3d^8(^3F)4s^1$ final ionic state can be discerned, but the contributions from the $^3F_4$ and $^3D_3$ initial atomic states still merge together.

Among the $3d$ transition elements, manganese assumes a special place because of its half-filled $3d$ subshell, giving it a $3d^7 4s^2(^6S_{5/2})$ ground state. In the single-particle picture, one would expect ionization in the $3p$ and $3s$ subshells to each produce a doublet, ^7P, ^5P and ^7S, ^5S, respectively. However, photoelectron spectra do not show simply two distinct pairs of peaks whose widths are determined by the strengths of the radiationless transitions filling the $3p$ and $3s$ vacancies, but at least three more peaks at higher binding energies.[41] These additional peaks have been explained by configuration mixing, or "internal" excitation, of the type $3p^2 \leftrightarrow 3s3d$.[42] So far, the $3p$ and $3s$ PES of Mn have not been completely interpreted by theory, nor has the experimental resolution been high enough to reveal the fine structure of each observed state within the limits set by the natural widths of the levels.

The complexity of the $3d$ element spectra described above is not encountered in Cu nor in the other elements of this group, Ag and Au. For these elements, with an $(n-1)d^{10}ns(^2S_{1/2})$ ground state, ionization from the filled d subshell produces four photolines, which reflect the interaction with the single outer $4s$ electron. From the spacing of the components of the quartet recorded by He Iα excitation

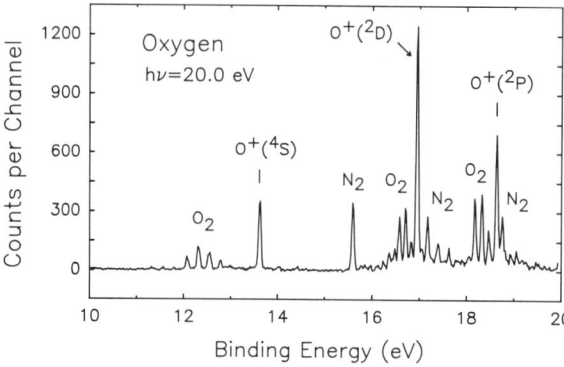

FIGURE 11. Photoelectron spectrum of atomic oxygen recorded at θ = 56° and 55-meV resolution.[45] Atoms originate from a microwave discharge in O_2 with N_2 added, and photons come from a 4-m normal-incidence monochromator. The spectrum contains 256 points.

with about 30-meV resolution,[43] one might view the lines as a $^3D_3$, $^3D_2$, $^3D_1$ group and a $^1D_2$ line in Cu and as two doublets, $5d_{5/2}$ and $5d_{3/2}$, split into two components, in Au. So far, no high-resolution spectra from synchrotron radiation-based work have been reported; however, an early study[44] did allow the four components to be distinguished in the case of Ag at $h\nu \leq 25$ eV.

Halogen and chalcogen atoms are special in the sense that they exhibit a clean multiplet structure because there are no close-lying excited states suitable for strong configuration mixing with the nominal ground state, as encountered in the $3d$ elements. However, conditions in the atomic source, which uses either a discharge or laser excitation for dissociation of a molecule, are conducive to the population of higher J states of the ground-state LSJ configuration. Ignoring the J splitting for the moment, we find that the three multiplet states created by ionization in the outer p subshell and the two multiplet states created by ionization in the s subshell are well separated and hence, it may seem, are accessible to studies with moderate resolution. In reality, however, the presence of bands from the precursor molecule and from other molecular species formed in the atomic source makes a high-resolution measurement desirable or even critical. The spectrum of the oxygen atom shown in Figure 11 serves to illustrate this point. While the PES resolution of 55 meV achieved in this experiment[45] is not a necessity for isolating the $[2p](^4S)$ photoline (where the square brackets denote a hole state), it is a great aid for isolating the $[2p](^2D)$ and $[2p](^2P)$ lines from nearby molecular features. A similar observation pertains to the spectrum of atomic chlorine presented in Figure 12.[45] More significantly, high resolution is crucial in the case of the chlorine atom for a study of the $^3P_{2,1,0}$ ionic state at the J level. In such a high-resolution study the $^3P_{2,1,0}$ fine-structure states could be resolved for $3p$ ionization and also for $3s$ ionization.[46] All arise from the $3s^23p^5(^2P_{3/2})$ ground state of Cl on photoionization. The populations, it was found, are determined by geometric factors based on Racah angular momentum coupling with the assumption of the single-particle approximation and the neglect of LS-dependence in the dynamical matrix elements.[47,48] For open-shell atoms, these populations differ from the simple $(2J + 1)$ rule that applies to closed-shell atoms.

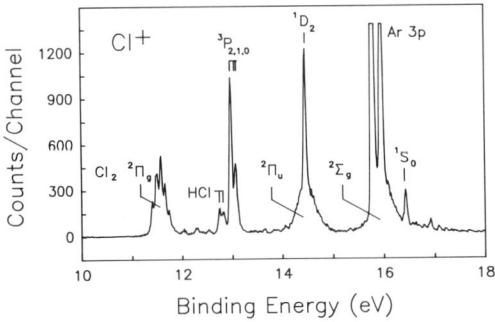

FIGURE 12. Photoelectron spectrum of atomic chlorine recorded at θ = 56° and $h\nu$ = 30 eV using a 6-m TGM.[45] PES resolution is 57 meV. Atoms effuse from a microwave discharge in Cl_2. The spectrum contains 512 channels.

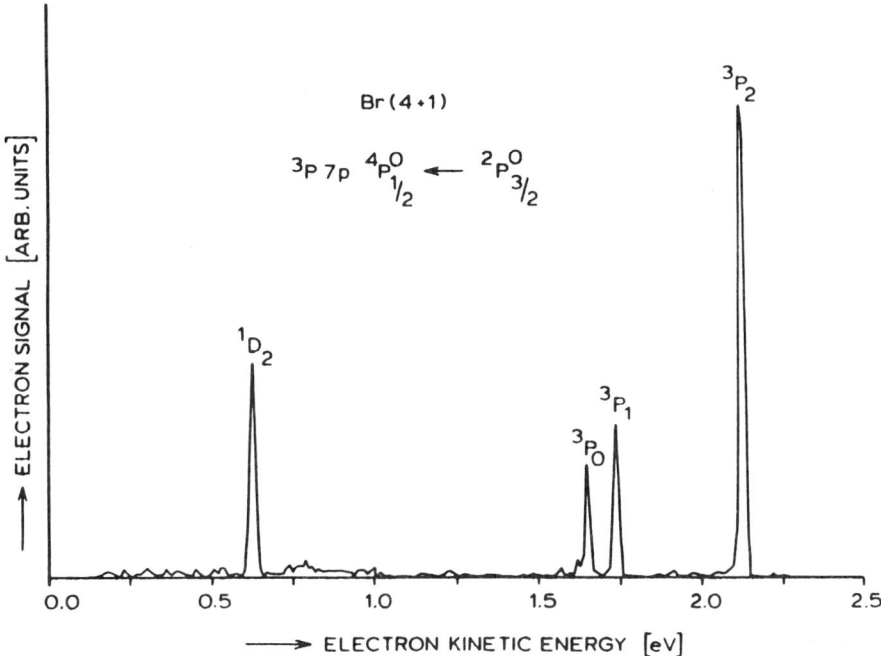

FIGURE 13. Photoelectron spectrum of atomic bromine obtained from the ground state via the intermediate state indicated in the plot by resonance multiphoton (4 + 1) ionization using a 22,490 cm^{-1} laser. PES resolution is about 15 meV. (From Ref. 49.)

The electronic structure of an atom can be delineated in great detail by REMPI and the use of a high-resolution electron spectrometer. An example of this technique is given in Figure 13 where the spectrum of the bromine atom is shown. Starting from the ground state $4s^24p^5(^2P^o_{3/2})$ the Br$^+(^3P_J,^1D_2)$ ionic states are reached via the intermediate Br$^*(^3P)7p(^4P^o_{1/2})$ state using (4 + 1) photons.[49] We note that the Br$^+(^1S_0)$ state is too energetic to be accessed in that study. Because of the excellent resolution of 15 meV, which is governed by the electron spectrometer, the three $^3P_{2,1,0}$ components are completely separated. While such experiments are usually aimed at the study of intermediate states, the ionic spectrum resulting from photoionization of the ground state can be obtained by an appropriate choice of photon energies and excitation steps. Because ionization proceeds via a specifically selected state, an extremely clean spectrum can be obtained.

In REMPI the same laser often serves the dual purpose of producing the atoms by dissociation of a suitable molecule and ionizing the atom. In the case of bromine the precursor molecule Br$_2$ can be used, and in the case of atomic carbon the CCl$_4$ molecule lends itself as a precursor.[50] The PES of atomic carbon resulting from (2 + 1) photoionization at two different energies are shown in Figure 14. Ionization from the ground state $2s^22p^2(^3P_0)$ proceeds via the $^3D_2$ intermediate resonance state, and ionization from the excited $3p(^1D_2)$ state (1.26 eV above $^3P_0$) proceeds via the $3p(^1S_0)$ intermediate state. The latter chain is interesting as actual ionization comes from an $^1S_0$ state which cannot have any alignment because of a nonstatistical distribution in magnetic sublevels. A measurement of the angular distribution of the photoelectrons demonstrated that none of the even Legendre polynomials beyond $P_2(\cos\theta)$ enter into the distribution for $^1S_0$ ionization. The measurement for $^1S_0$ ionization also revealed that the single-electron picture is inadequate to describe the photoprocess and that the coupling of the $3p$ electron with the core cannot be ignored.[50]

The examples discussed in this section emphasize the need for adequate resolution to unravel the multiplet structure in open-shell atoms and to delineate dynamic parameters, e.g., photoionization

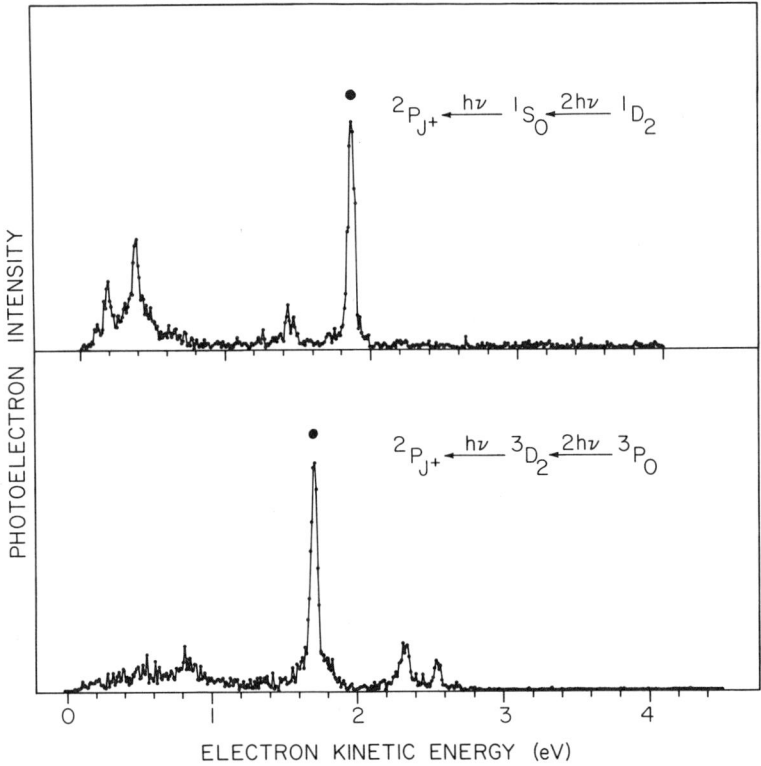

FIGURE 14. Photoelectron spectra of atomic carbon for ionization proceeding from the $^3P_0$ ground state via the $^3D_2$ intermediate state and from the $^1D_2$ excited state via $^1S_0$ by a three-photon process. The resulting $C^+(^2P)$ peak is the most intense line in each spectrum. (From Ref. 50.)

cross sections, at the *LS* term level and ultimately at the *J* fine-structure level. In the following, requirements for the detailed study of satellite structures will be stressed.

4. CORRELATION PHENOMENA IN ATOMS

In addition to the main photolines discussed in the preceding section as being representative of single-electron emission, photoelectron spectra often exhibit a host of satellite lines. These are the result of the emission of one electron and the simultaneous excitation of another electron. These satellites, referred to generically as shakeup satellites or correlation satellites, result as a direct consequence of electron correlation. Although evidence for such satellites was presented many years ago under low-resolution conditions,[1,51,52] high-resolution spectrometry is needed to be able to more fully understand the phenomenon of electron correlation, especially when electrons in outer shells are involved. As discussed in Chapters 1 to 4 and elsewhere,[2,8,11,53,54] the energy dependence of satellite production is closely related to the dynamics of electron correlation. To date, the different mechanisms that underlie electron correlation have been identified on the basis of experiments using a PES resolution of 100 to 300 meV for outer-shell ionization[11,53,54] and a resolution of 2 to 5 times the natural level width for inner-shell ionization.[3,55,56] However, for detailed scrutiny of complex satellite structures, truly high-resolution spectra are required. Most significantly, a superior resolution must be achieved not just at a few conveniently chosen photon energies, but at energies spanning an extended range. In the following, showcase electron spectra are presented to illustrate electron correlation as it pertains to various subshells in both closed-shell and open-shell atoms under vastly different excitation conditions.

4.1. The Rare Gases

Not surprisingly, direct manifestation of electron correlation was first seen in the rare gases, and most theoretical approaches have concentrated on these closed-shell atoms with an $^1S_0$ configuration as the ground state. Because of the ease of handling rare-gas atoms both experimentally and theoretically, they have continued to be a mainstay for studying many-electron effects. Partially for historical reasons—the neon 1s spectrum yielded the first evidence for the appearance of photoelectron satellites[51]—and partially because of the good quality achieved for this core-level spectrum, the Ne 1s PES obtained with monochromatized Al $K\alpha$ x rays is presented in Figure 15. This spectrum,[56] which omits the Ne 1s (Al $K\alpha$) main line of the single-electron process, shows the shakeup satellites corresponding to the excitation of a 2p electron to the np, n = 3 to 6, levels concomitant with the ionization of the 1s electron. Because the excited electron can have its spin parallel or antiparallel to the remaining 1s electron, two states are created, designated as upper (U) and lower (L) states. Also seen in the spectrum is a so-called conjugate shakeup transition in which the excited electron takes up the angular momentum, 2p → 3s, and the 1s electron will then retain its symmetry, 1s → εs, in the overall dipole photoionization process. The spectrum was recorded at an angle of 90° with respect to the direction of the unpolarized photon beam and at a PES resolution of about 0.5 eV. Because this resolution is within a factor of two of the natural limit given by the width of the 1s level, the spectrum yields virtually all of the principally accessible spectral information. The same degree of completeness is not readily achieved for outer-shell ionization because of the usually negligible level widths (see Section 6 for exceptions) and the large number of states that can be populated.

The Ne 2s,2p satellite spectrum is shown in Figures 16 to 18 under three greatly different excitation conditions: ionization at threshold, somewhat above threshold, and far above threshold. A comparison between the spectra demonstrates that the relative satellite intensities vary dramatically with the excitation energy. This can be readily verified by focusing on four peaks at four selected

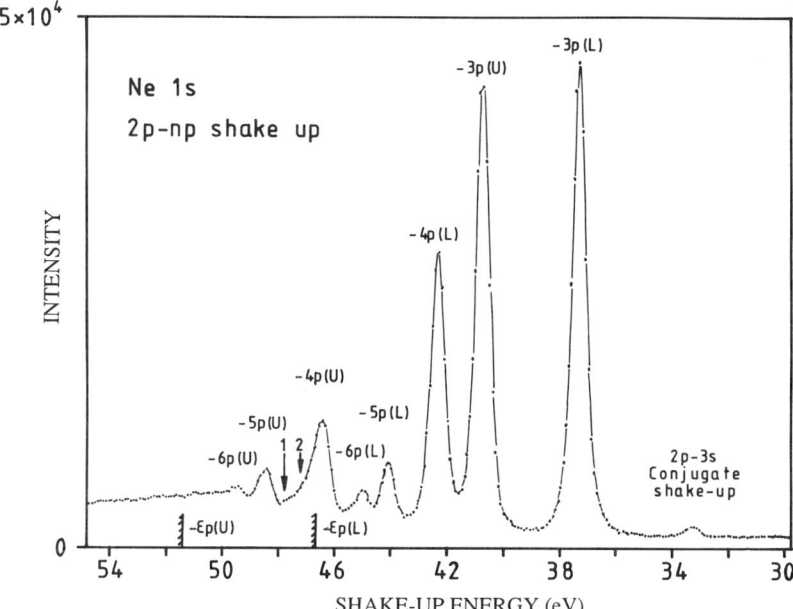

FIGURE 15. Photoelectron spectrum of correlation or shakeup satellites associated with photoionization in the 1s shell of neon; excitation by 1.5-keV unpolarized x rays. Resolution is about 0.5 eV and angle of observation is 90° with respect to the direction of photon propagation. (From Ref. 56.)

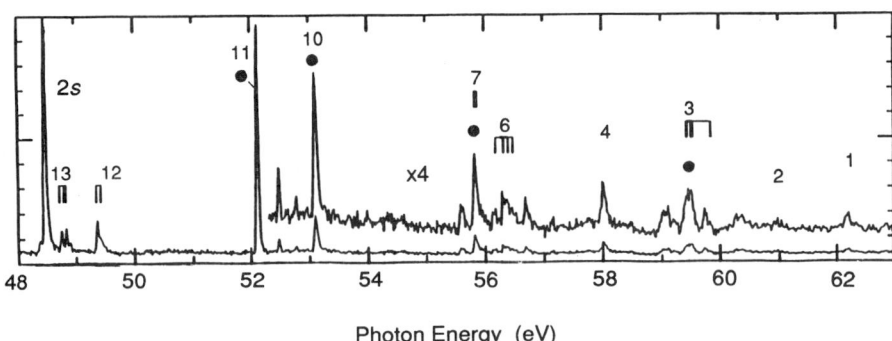

FIGURE 16. Threshold photoelectron spectrum of neon, obtained at a resolution of 48 meV. (From Ref. 57.)

binding energies, marked by solid dots. For example, at 1.5 keV,[56] satellites of ^2S symmetry are emphasized, e.g., peaks 2 and 5 in Figure 18, indicating a strong correlation with ionization from the $2s$ subshell, while in the zero-energy electron spectrum, when the photon energy is tuned to the threshold energies of satellite production,[57] transitions to low-lying, high-multiplicity states are intense. In this context, it should be stressed that satellite intensities may be strongly influenced by two-electron *excitation* processes in the threshold region.[11,57–59] The spectrum observed at $h\nu = 71$ eV reveals a spectral distribution intermediate to the threshold and high-energy spectra.[2] It is this type of spectrum that, if measured as a function of photon energy, can be profitably used to classify the various correlation satellites according to origin and mechanism within a given framework of theory.[8,11,53,54,58,60] We note that a measurement of the photoelectron angular distribution, as indicated in Figure 17, is an important complement to the cross section measurement. It can corroborate the conclusions drawn from the energy dependence as to which type of correlation is predominant. For example, both the intensities or cross sections and the angular distributions, or β parameters, for satellites of ^2P symmetry in neon should exhibit a similar energy dependence as those for $2p$ photoionization. This has, in fact, been found to be by and large true.[58,60] As a final remark on the Ne satellite spectra of Figures 16–18, attention is drawn to the differences arising from a purely instrumental effect, the PES resolution. At 50 meV of the threshold spectrum many of the $(LS)nl(L'S')$ states, and sometimes even the J components, can be distinguished, but at 0.5 eV of the Al $K\alpha$-excited spectrum, only estimates, though usually valid ones, can be made as to the identity of the major contributors.

In Ar, Kr, and Xe, a major driving force for satellite production in outer-shell photoionization is the strong final state interaction between the $ns^1np^6(^2S)$ and $ns^2np^4(^1D)n'd(^2S)$, $n' \geq n$, ionic configurations. By its nature, this interaction should have, at most, a weak dependence on photon energy. This is, in fact, predicted by theory,[61,62] and the calculated relative strengths of the $n'd(^2S)$ satellites in argon are seen to be in good agreement with experiment.[2,11] However, in the case of the $4d(^2S)$ line at 41.21 eV good agreement could only be achieved after a close-lying double line, corresponding to $(^3P)5d(^2D,^2P)$, could be removed. At a resolution of 60 meV, this was difficult to accomplish, although the different β values for the states aided in the procedure.[2] At 35 meV obtained in a threshold scan,[57] a decomposition is facilitated for these peaks 40 and 50 meV apart, albeit with the recognition that threshold intensities are more difficult to interpret. Many of the satellites can also be studied in fluorescence with a high resolution determined by the fluorescence spectrometer only, but with the disadvantage of the presence of a doublet which is split by 167 meV in the case of argon.[63]

Photoelectron spectrometry in the gas phase is not restricted to the soft x-ray regime, as illustrated by the Kr $1s$ spectrum excited by 15.23-keV photons from the 27-period undulator at the PEP-SSRL storage ring.[64] A double-crystal monochromator selected a photon band of less than 7 eV FWHM, which compares favorably with the Kr $1s$ level width of 2.75 eV.[37] The PES width of about 8 eV in this experiment allowed the separation of the satellite structure from the $1s$ photoline as depicted in

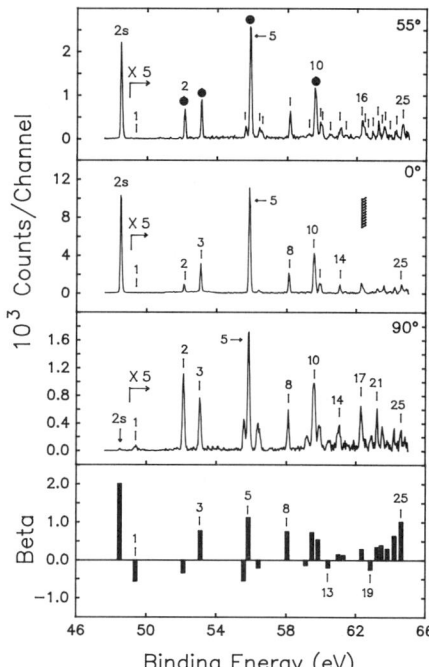

FIGURE 17. Photoelectron spectra of Ne $2s,2p$ satellites at $h\nu = 71$ eV obtained with a resolution of 91 meV at different angles. The 55° spectrum gives the relative cross sections; the β parameter is derived from the 0° and 90° spectra. (From Ref. 2.)

Figure 19. This remarkable spectrum provides a direct comparison with the predictions of the shake model. The experiment, carried out at an excitation energy at which the sudden approximation (shake model) should be valid, was seen to agree well with the prediction based on a single-channel, single-configuration Dirac–Fock calculation.[64] Perhaps even more important was the observation that

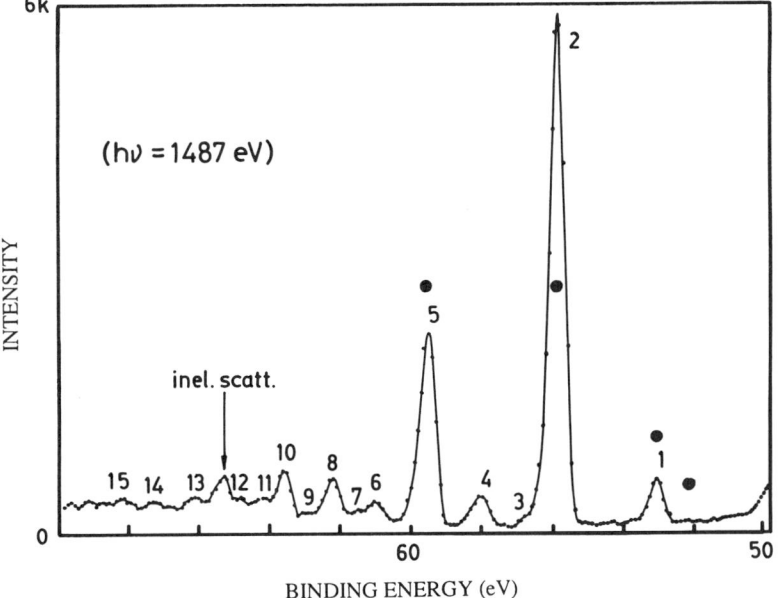

FIGURE 18. Neon $2s,2p$ correlation satellites observed at 90° and photoionization by 1.5-keV unpolarized x rays. The rise in intensity near 50 eV is related to the tail of the strong $2s$ photoline. The resolution is 0.5 eV. (From Ref. 56.)

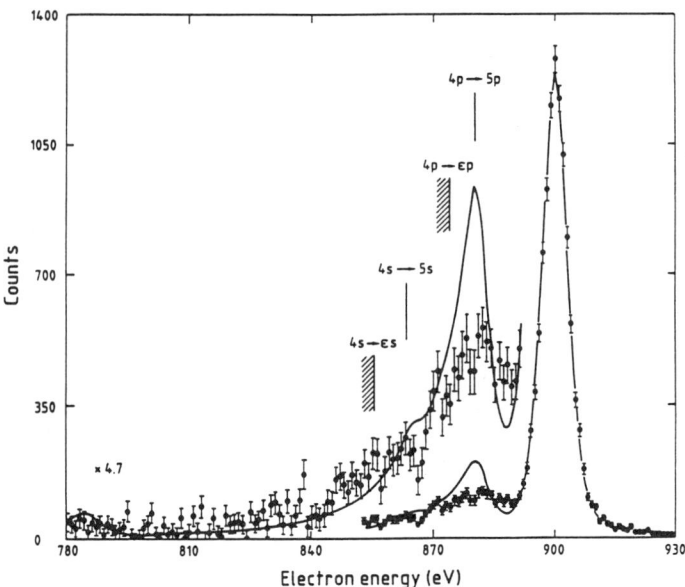

FIGURE 19. Krypton 1s photoelectron spectrum obtained at $h\nu = 15.2$ keV. Positions and thresholds for principal shake processes are indicated. The solid curve represents the theoretical spectrum convoluted with the experimental resolution of about 8 eV. (From Ref. 64.)

the Kr 1s conversion electron spectrum arising from the 32-keV isomeric transition in ^{83m}Kr and recorded with a toroidal-field magnetic spectrometer was found to be identical with the photoelectron spectrum within the experimental uncertainties. Thus, the experiments provide an underpinning to the theoretical prediction that the shake process in these cases should not depend on the differences in the excitation mode.

In addition to photoelectron spectra, Auger electron spectra lend themselves to the determination of partial photoionization cross sections for one- and two-electron processes in the initial ionization event.[65] In favorable cases, such as inner-shell photoionization in neon and argon, the Auger satellite lines can be grouped into shakeup and shakeoff categories with the latter signifying the simultaneous *ionization* of two electrons. In argon this opportunity for grouping was utilized to measure the energy dependence of shakeup satellites on the one hand and shakeoff satellites on the other. Figure 20 shows the relevant part of the spectrum, which was excited by 5.2-keV photons emerging from a double-crystal Bragg monochromator coupled to an eight-pole wiggler.[66] In addition to the ^1D diagram Auger line arising from the initial photoionization of a single K electron, two groups of lines appear in the spectrum, the first group arising from shakeup of M electrons, with a contribution from the ^1S diagram Auger line, and the second group arising from shakeoff of an M electron. The groups are clearly distinguished from one another and the ^1D line, although the resolution of the electron energy analyzer approached the natural width of the argon K-$L_{2,3}L_{2,3}$ Auger line only within a factor of two. Since the study was directed at determining the threshold behavior of both the shakeup and shakeoff events, the small photoelectron energies in the threshold region of photoionization would cause a broadening of the Auger lines in excess of the natural linewidths by the postcollision effect and, hence, would obviate any benefit from higher resolution in this experiment. By comparing the intensities of the shakeup and shakeoff groups with the intensity of the ^1D diagram line, the probabilities for multiple photoexcitation processes could be derived and, because of the high resolution of the experiment, combined with an adequate photon flux, the dependences of both the shakeup and the shakeoff probabilities could be established as a function of photon energy. A good accord with theoretical predictions was found, namely, for shakeoff a vanishing probability at threshold and a moderately steep rise into a plateau,

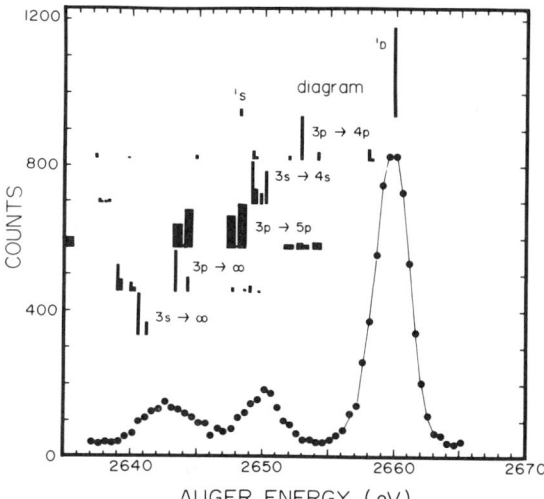

FIGURE 20. Argon K-$L_{2,3}L_{2,3}$ Auger spectrum excited by 5.2-keV photons, showing the ^1D diagram line and two satellite groups resulting from shakeup and shakeoff during the initial K-shell photoionizing event. The positions of the shake transitions are calculated. (From Ref. 66.)

and for shakeup a finite threshold value and a mild rise to a virtually constant probability at higher energies.

4.2. Open-Shell Atoms

Studies of electron correlation by wide-range photoelectron spectra are rare for open-shell systems. Work has concentrated up to now on the multiplet structure. However, as configuration mixing can transfer strength between multiplet components, resulting in nongeometric branching ratios, the distinction between multiplet and correlation effects can become fuzzy. In the case of inner-shell

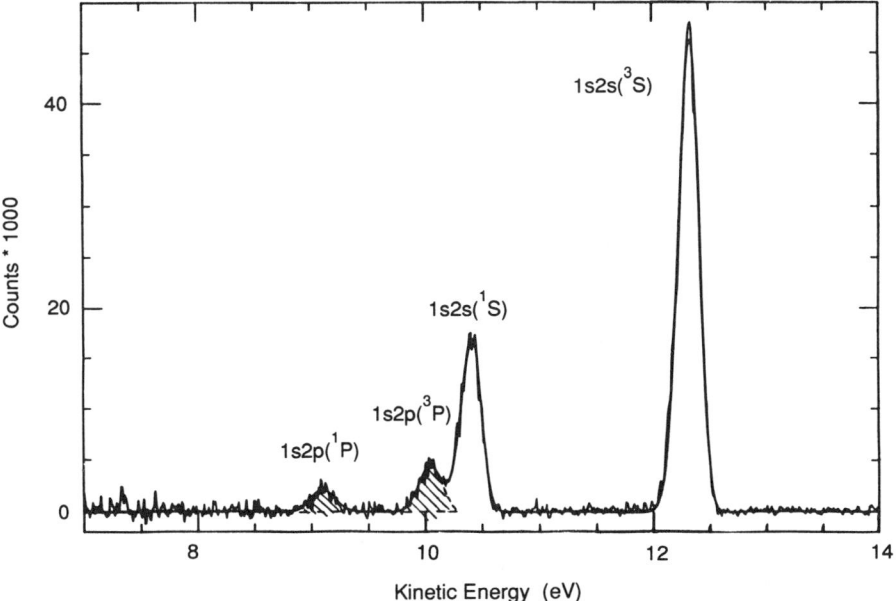

FIGURE 21. The lithium $1s$ photoelectron spectrum recorded at $hv = 77$ eV and a PES resolution of 0.2 eV. A TGM coupled to an undulator served as the photon source. (From Ref. 68.)

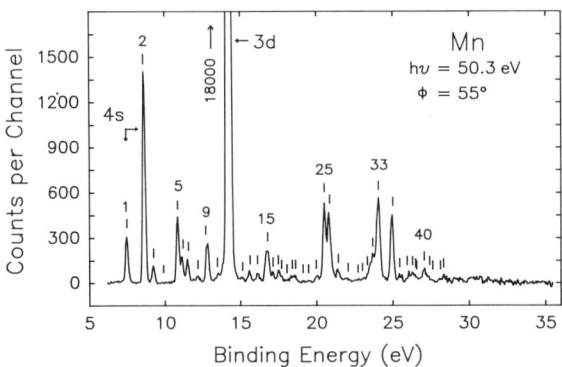

FIGURE 22. Photoelectron spectrum of atomic manganese excited in the $3p \to 3d$ resonance region. (From Ref. 70.)

ionization in manganese discussed in Section 3, configuration mixing was recognized as being responsible for the disintegration of the $[3s](^5S)$ and $[3p](^5P)$ multiplet states. Thus, a situation exists in Mn[41] that is akin to the configuration mixing described for Ar.

The most basic open-shell system is the Li atom. Following an early experiment,[67] a photoelectron spectrum of sufficient resolution to clearly demonstrate the effect of electron correlation has become available.[68] An ordinary doublet arises from the coupling of the $2s$ electron with the remaining $1s$ electron following $1s$ photoionization, namely, $1s^2 2s(^2S) \to 1s2s(^1S$ or $^3S)\varepsilon p(^2P)$. However, if the ionized electron goes into an εs continuum, a satellite doublet will arise according to $1s^2 2s(^2S) \to 1s2p(^1P$ or $^3P)\varepsilon s(^2P)$. The latter transitions correspond to the familiar conjugate shakeup process mentioned in Section 4.1 in context with Ne $1s$ photoionization. The experimental evidence is presented in Figure 21. Branching ratios and β parameters for all transitions were measured between photon energies of 74 and 150 eV,[68] and the results were found to be in satisfactory agreement with the predictions of a relaxed-orbital Hartree–Fock calculation.[69] It should be noted that no transitions to $n = 3$ were reported.

The photoelectron spectrum of Mn taken in the region of the broad $3p \to 3d$ resonance near 50 eV abounds with satellites, as seen in Figure 22. These satellites persist at all energies measured[41] but vary greatly in their intensities from threshold to about 70 eV because of the presence of many resonance states in this region.[70] The influence of resonances on satellite intensities will be taken up in Section 5; at this point, attention is drawn to the separation of many lines at the rather moderate resolution of 180 meV. Nevertheless, the multitude of states that can be reached on $4s$ or $3d$ ionization of Mn with its half-filled $3d$ subshell will require further substantial improvements for detailed study.

Evidence for electron correlation in atomic oxygen on photoionization in the $2p$ and $2s$ subshells has been obtained in a recent experiment.[71] The initial data, taken with a moderate resolution at different photon energies between $h\nu = 100$ and 500 eV, indicate a strong need for high resolution, comparable to that of the spectra in Figure 17, for a detailed scrutiny of the correlation processes in this open-shell atom.

5. ELECTRON SPECTROMETRY OF RADIATIONLESS PROCESSES

On excitation the response of the atom is usually by radiationless rearrangement unless the excitation involves the outermost electron. As a result the emitted electron presents itself as a natural probe of excitation and decay processes alike. Traditionally, we distinguish autoionization processes and Auger processes and, between the two extremes, resonance Auger processes. However, as particles with definite energies are created in all cases, the unifying picture of a resonance scattering process can be applied throughout.[72] For convenience, we will discuss the high-resolution results presented in this section in the usual terminology.

The presence of resonance structure in the absorption spectra above the first ionization limit of atoms has been known since the time of Beutler.[73] Excitation involves an electron in the penultimate fine-structure level of the outermost shell of atoms, at low photon energies, and there is usually a strong coupling of the indirect channel via the resonance state with the direct channel associated with the ionization of the least bound electron. The resulting interference between the alternate channels, displaying characteristic shapes and amplitudes, has become known as an autoionization or Beutler–Fano resonance.[12,73,74] While in the case of autoionization the resonance state couples to a single-ionization continuum, in the case of the Auger process, which is associated with the ionization of a core electron, the resonance state couples to a double-ionization continuum. Because the probability for direct transitions to a two-electron continuum is about two orders of magnitude less than the probability for single inner-shell ionization leading to the Auger process, as demonstrated in the case of neon,[75] interference between the alternate channels will have a negligible effect. As a result, Auger lines have typically a Lorentzian line shape and, overall, the Auger effect can be treated as a two-step process: core ionization and subsequent Auger decay. In the intermediate case of *excitation* of a core electron, rather than *ionization*, the excited state couples with single-ionization continua and it depends on the details of excitation and decay whether the process should be likened to an autoionization or to an Auger process. Experimentally, the line shape offers a good criterion as to which approximation might be applied most profitably. Generally, coupling to the direct channel of ionization is small for excitation from a deep core level, especially in the case of closed-shell atoms, and the line shape is essentially Lorentzian, as it is in the case of the Auger process. The similarity is expressed in the description of this event as a resonance Auger process. If excitation involves a shallow core level, coupling to the direct channels of ionization can be substantial, especially in certain open-shell atoms, and the line profile will resemble the familiar shape of an autoionization resonance.

In the following, a number of high-resolution recordings will be selected in the areas of autoionization, Auger processes, and resonance Auger processes. Other aspects of the phenomena are also discussed in various chapters of this book.

5.1. Autoionization Resonances

The specific autoionization resonances known as the Beutler–Fano resonances occur between the spin-orbit doublet components of the outer np subshells of the rare gases. Those Rydberg states converging toward the $^2P_{1/2}$ ionization limit and lying above the ionization energy of the $^2P_{3/2}$ ionic state are subject to radiationless decay into the $^2P_{3/2}$ continuum. Since only one continuum for ionization from the outermost level is available, several methods are at our disposal for the study of these resonances. Indeed, photoabsorption and ion spectrometry have served well to characterize the Beutler–Fano resonances. However, only electron emission experiments can give complete information on the process by a measurement of the photoelectron angular distribution and the photoelectron spin polarization in addition to a measurement of the integrated intensity or cross section. This has been demonstrated convincingly, for example in the case of xenon, by both theory[76] and experiment[77] (see also Chapter 15). An overall account is given in Ref. 11. For illustration of one aspect we present in Figure 23 the photoelectron angular distribution parameter, β, as obtained in a high-resolution measurement of the $8s'$ resonance in xenon. The sharp ns' resonances, which project above the broad $(n-2)d'$ resonances, as seen in Figure 2, exhibit an equally sharp dip in the β parameter. Figure 23 shows that two electron emission experiments of different types, one using monochromatized synchrotron radiation,[78] and the other a laser[24] for excitation, give virtually identical results. In both instances, the resolution, 0.33 meV for the synchrotron and 0.26 meV for the laser photon sources, is much less than the natural width, $\Gamma \approx 2$ meV, for the $8s'$ resonance. Thus, the natural properties can be traced with only minor instrumental disturbance. Together with the data on the other pertinent properties,[11] experiments of this type have provided us with one of the most rigorous tests of the theoretical models.

Some of the narrowest Beutler–Fano resonances are encountered in neon. Even with the best monochromators in existence at synchrotron radiation sources, photoelectron spectrometry of the Ne resonances will fall far short in matching the resonance widths, which are on the order of magnitude of 10 μeV at most.[21,23] However, the observation of the photoelectron in a study using a monochromator bandpass of 19 mÅ, equivalent to 0.73 meV at 21.6 eV photon energy, has proven valuable in

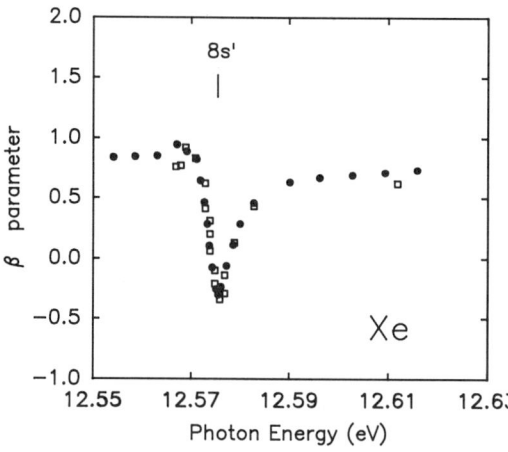

FIGURE 23. Photoelectron angular distribution parameter β across the $8s'$ Beutler–Fano resonance in xenon. Solid dots are from synchrotron radiation data,[78] with a bandpass of 26 mÅ; open squares are from laser data,[24] with a bandpass of 20 mÅ.

not only resolving the first closely spaced ns' and nd' autoionizing Rydberg states but also in measuring the β parameters for these resonances.[25] The resulting spectra are presented in Figure 24. Analogous to the other rare gases, experiment and theory[79] are seen to be in satisfactory, though not complete, accord concerning the various characteristics of the autoionization resonances.

Helium is a special case; for autoionization to occur both of the $1s$ electrons need to be excited. The highest resolution available above a photon energy of 60 eV has been applied to this prototype atom for the study of electron correlation—but only in photoabsorption or in the equivalent ion-yield spectrometry.[30,80] Actually, for the $(sp,2n\pm)^1P^°$ series, e.g., $(2s2p)^1P^°$, no gain would come from detecting the continuum electron because the β parameter remains equal to two through the resonance, as confirmed in a recent experiment with an accuracy of 0.05%.[2] However, for the higher series, which can decay into He^+ ($n\geq 2$), the β parameter for the electrons corresponding to the production of

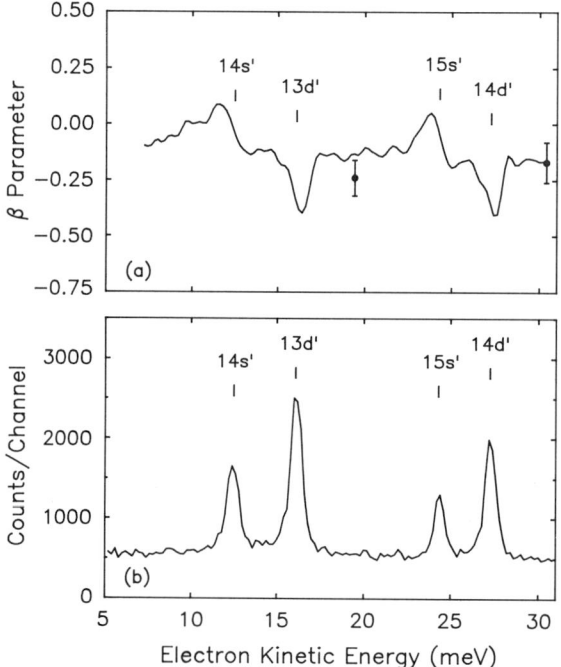

FIGURE 24. CIS scans across the first autoionization resonances between the $2p$ doublet of neon for (a) the β parameter and (b) the relative cross section obtained with a bandpass of 19 mÅ. (From Ref. 25.)

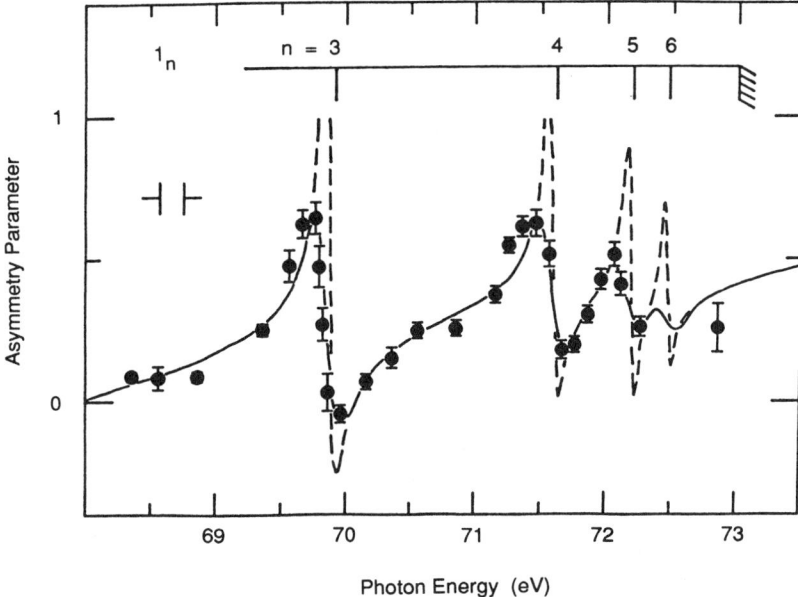

FIGURE 25. The β parameter of the He$^+$ ($n = 2$) satellite in the resonance region below the $n = 3$ threshold. The solid line represents a Fano-type fit to the data, and the dashed line follows from the removal of the monochromator bandpass of 0.43 Å. (From Ref. 81.)

the excited states will serve as one of the criteria against which our understanding of the correlation expressed in autoionization can be gauged. As seen in Figure 25, the β parameter is subject to excursions, as is the cross section, across the resonance series converging toward the $n=3$ threshold because of the presence of more than one continuum. The initial measurement with a moderate resolution of 170 meV clearly indicates this behavior through the relatively broad first member of the autoionizing series.[81] A critical comparison with theory,[82] as well as a β determination through the $(sp,3n^-)$ member at 71.3 eV, will have to await an experiment with improved resolution.

In autoionization events limited to a single continuum channel (in LS terminology), the multi-faceted information inherent in an electron emission study must be weighed for its value against the higher-resolution single-parameter datum that might be obtained in an absorption or total particle yield measurement. This holds true in the cases just discussed. However, in excitations subject to autoionization into two or more continuum channels, electron spectrometry becomes a necessity for characterization of the resonance. Again using the xenon atom for illustration, excitation of a 5s electron, or, alternatively, two 5p electrons, is recognized to lead to electron emission with the ion left either in the $^2P_{3/2}$ or $^2P_{1/2}$ state, the same states created on direct photoionization of either a $5p_{3/2}$ or $5p_{1/2}$ electron. Both the partial cross sections and β parameters for the two exit channels have been measured with the aid of the photoelectron using CIS scans and a resolution of 0.10 to 0.17 Å at a synchrotron radiation source.[17] The results for the cross sections are presented in Figure 26. The resonance profiles exhibited in the partial cross sections are similar for the 5s → 6p excitation, but may differ for various classes of double excitations. A measure of the similarities or dissimilarities in the two partial channels is given by the cross section ratios depicted in panel (a). Although energies, amplitudes, widths, profile parameters, and β parameters are now known for these resonances, no interpretation has been brought forth by theory thus far, in large part because of the difficulty in properly treating the two-electron excitation states.

Much complexity marks autoionization in open-shell atoms, as has been shown for atomic Cl[18,83] and O[45,84] by way of electron spectrometry. The multiplet structure of these atoms gives rise to a number of Rydberg series which can autoionize into many continuum channels. For example,

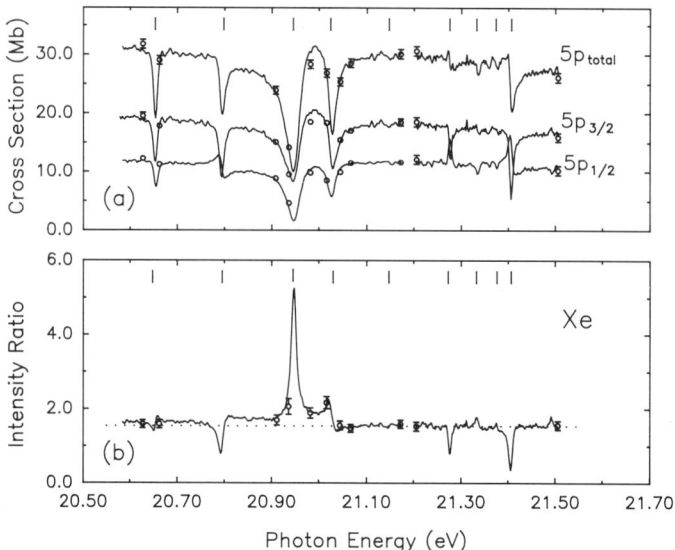

FIGURE 26. Photoionization in the vicinity of the Xe $5s5p^66p$ autoionization resonance. In panel (a) are shown the partial cross sections for production of the two final states as well as the sum. Differences in the exit channels are accentuated in the ratio of the $5p_{3/2}$ and $5p_{1/2}$ spectra (panel b). (From Ref. 17.)

excitation of a $3s$ electron in Cl leads to three Rydberg series terminating on the $^3P^o_{2,1,0}$ and $^1P^o_1$ inner-shell thresholds of the Cl$^+$ ion. Each of the members can autoionize into the three continua associated with the $[3p]$ $^3P^e$, $^1D^e_2$, and $^1S^e_0$ ionic states with the possibility of subdividing the $^3P^e$ channel into the $J = 2,1,0$ subchannels at a resolution sufficient to distinguish the $[3p]$ $^3P^e_2$, $^3P^e_1$, and $^3P^e_0$ photolines. Between the $^3P^o_{2,1,0}$ and $^1P^o_1$ thresholds, autoionization into the $[3s](^3P^o)$ channels becomes possible as well. An excerpt of an extensive study[18,83] is shown in Figure 27, where a CIS scan of the partial cross section over the threefold $^3P^o$ series is presented for the dominant Cl$^+$($^3P^e$) continuum exit channel. The bandpass of 0.16Å, or 7 meV, is smaller than the natural widths for the $n = 4$ states and is adequate for resolving the features dominating the $n = 5$ region. The $3s \rightarrow np$ excitations display window-type resonances in all exit channels. This behavior is analogous to that observed in argon for the excitation of the $3s$ electron with the exception that the window resonances in Ar lack fine-structure splitting. Spectra of the detailed nature evident in Figure 27 are ideal for challenging theory. In this instance this challenge was taken up in short order, and the results of an eigenchannel R-matrix procedure were found to be in astonishingly good agreement with experiment in all regards for these complex autoionization processes occurring in a halogen atom.[85]

Following the original measurement of the autoionization profiles for the $2s \rightarrow np$ Rydberg series in the N atom decaying into the $[2p](^3P)$ continuum of the N$^+$ ion,[86] a recent high-resolution scan of the $3p$ and $4p$ resonances confirmed the earlier results with improved accuracy.[87] In addition, an accompanying multiconfiguration Hartree–Fock calculation was seen to agree reasonably well with the experiment in both the cross section and the β parameter results.

A final discussion is devoted to the oxygen atom, one of the most abundant and important elements in the universe. As for the halogens, several resonance states occur for each np excitation out of the $2s$ level, and these states autoionize generally into the three continua associated with the $[2p]$ ^4S, ^2D, and ^2P ionic states. The first comprehensive study[84] clearly revealed the major features of the $3p$ manifold, consisting of the $^3D^o$, $^3S^o$, and $^3P^o$ terms, but lacked the resolution necessary for delineating the finer detail. As demonstrated in Figure 28, which represents the $[2p](^2D)$ partial cross section across the $3p$ manifold, the profiles and, especially, the widths differ greatly for the three terms. A resolution of 0.07 Å, or 3.9 meV, also reveals evidence, most clearly in the $^3S^o$ term, for the presence of fine structure related to the excited $^3P^e_1$ state of the neutral O atom. Being only 20 meV above the

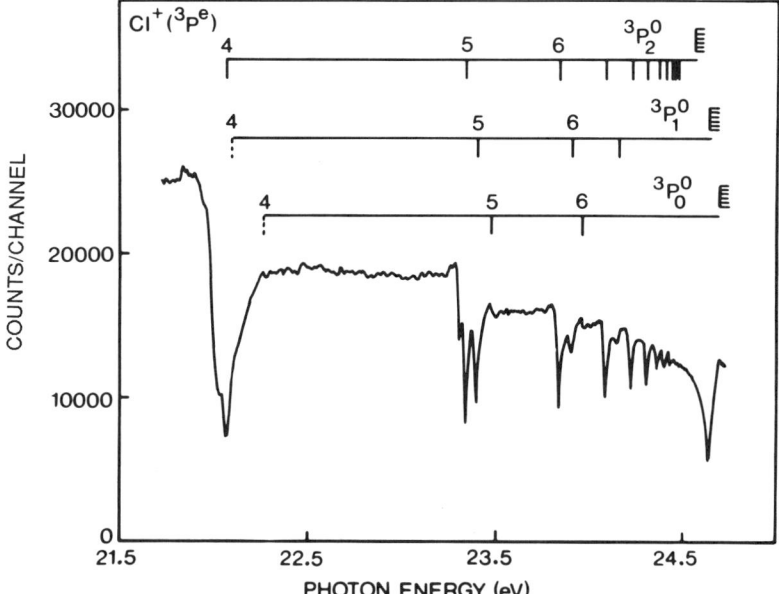

FIGURE 27. Relative partial cross section of the $Cl^+[3p](^3P^e)$ ionic state over the Rydberg series converging to the $^3P_{2,1,0}$ thresholds. CIS scan in 5-meV steps using a bandpass of 0.16 Å. (From Ref. 83.)

$^3P_2^e$ ground state, this state can be effectively populated under the conditions of the microwave discharge tube serving as the atomic source. In a low-resolution spectrum no clues can be extracted from an autoionization profile for the presence of excited states; instead the resulting tailing of the profile can be mistaken for lifetime broadening. The lower panel of Figure 28 shows the result when the high-resolution spectrum of the upper panel is convoluted with the bandpass of 0.71 Å, 38 meV, used in the earlier study. The convoluted spectrum reproduces the spectrum of Ref. 84 well, and a

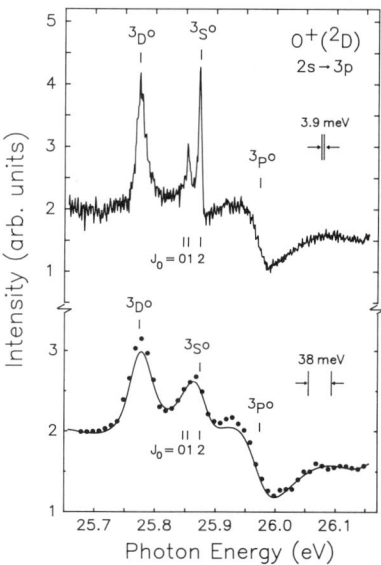

FIGURE 28. CIS scan of the $O^+[2p](^2D)$ ionic state over the $2s \rightarrow 3p$ ($^3S^o$, $^3P^o$, $^3D^o$) resonances in the oxygen atom with a bandpass of 0.07 Å (upper panel).[45] The result of a convolution with a bandpass of 0.71 Å, as used in an earlier experiment,[84] indicated by solid dots, is shown in the lower panel. Note the distinct "satellite" peak in the $^3S^o$ resonance giving evidence for the thermally excited $^3P_1^e$ state of the atom.

comparison between the two spectra serves to stress the importance of high resolution in order to be able to discern essential detail and obtain unambiguously defined parameter values.

In concluding this section, attention is drawn once again to the fact that the spectral resolution of the autoionization features depends on the bandpass of the photons (resolution of the monochromator) and the natural widths of the resonance states. The CIS scan as discussed in Section 2 represents the most appropriate mode of operation for the study of the resonances by way of the emitted electron. Since a resonance often interacts with several continua, great benefits accrue from the use of a multichannel detector system, or, at a minimum, two separate analyzers, with regard to the accuracy and acquisition time for the simultaneously recorded partial CIS spectra.

5.2. The Auger Process

Auger electron spectra of free atoms have been recorded with high resolution for many years.[3,4,65,88] Various means have been used to create the initial inner-shell vacancy, but electron bombardment has been the most common. The best resolution $\Delta E/E$ achieved so far is about 0.03% for the most frequently applied electrostatic electron energy analyzers. With this resolution, corresponding to a resolving power of 3300, the instrumental width approaches, or is even smaller than, the natural widths of the Auger lines, as can be estimated from the entries in Table I. While an account of Auger spectrometry is given in Chapter 11, the K-LL spectra of the rare gases neon, argon, and krypton are cited here because these are the spectra that place the greatest demand on the resolution capability of a spectrometer.

A resolution of $\Delta E/E = 0.03\%$ has been reported for the Ne K-LL spectrum.[88] For this case the instrumental broadening is equal to the linewidths and allows for a detailed analysis of the satellite Auger lines and the normal, or diagram, Auger lines. It should be remembered that the normal Auger lines result from single ionization in an inner shell, and the satellite Auger lines arise as a consequence of two-electron correlation processes. Because in neon the final states correspond to holes in the outermost shell with no possibility for further radiationless transitions, all K-LL lines have the same width. This no longer holds true in argon; here the linewidths vary for the different Auger transitions,[37] namely, $\Gamma(K$-$L_{2,3}L_{2,3}) = 0.93$ eV, $\Gamma(K$-$L_1L_{2,3}) = 2.4$ eV, and $\Gamma(K$-$L_1L_1) = 3.9$ eV. With an effective resolution $\Delta E/E = 0.02\%$ for the magnetic spectrometer used in the study of Ar,[89] the instrumental width of 0.5 eV is less than the smallest K-LL Auger width; a spectrometer width of 0.6 eV applicable to the K-$L_{2,3}M$ transitions is still narrower than the 0.8 eV for the Auger width. Hence, the Ar K Auger spectrum measured with $\Delta E/E = 0.02\%$ can reveal maximum information on both the Auger spectrum and the initial ionization process chosen.

Although the high energies of the K-LL transitions in krypton exceed the scope of this book, this Auger spectrum is included in the presentation because it further highlights the rare gas K-Auger series and because it was measured under rather unique conditions.[90] The spectrum arises as a consequence of the internal conversion in the K shell of a 32-keV E3 electromagnetic transition in the isomeric decay of ^{83m}Kr. The initial ionization event is then similar to the E1 photoionization process depicted in Figure 19. The remarkable achievement of measuring the K Auger spectrum of a heavier element in the gas phase, rather than in the solid phase as usual, allows for a full utilization of the instrumental resolution, precluding deterioration of resolution and tailing of peaks resulting from collisions of the Auger electrons while escaping a solid sample. The Kr Auger electrons were dispersed in a toroidal-field magnetic spectrometer ("β-spectrometer"), a type of electron spectrometer[5,91] almost forgotten in recent times, but well suited for the analysis of the energetic 11-keV electrons of the Kr K-LL Auger spectrum. The spectrum is displayed in Figure 29. The nine Auger lines predicted in intermediate coupling are clearly exposed, and the satellite structure related to shakeup and shakeoff from the other shells, especially the N shell, can be assessed from the spectrum, which was obtained with a resolution of $\Delta E = 5.5$ eV. As in the case of argon, the spectrometer width is comparable to the width for the K-$L_{2,3}L_{2,3}$ line, $\Gamma = 5.2$ eV, and considerably less than the widths for the other lines, $\Gamma(K$-$L_1L_{2,3}) = 8.3$ eV and $\Gamma(K$-$L_1L_1) = 11.3$ eV.

It should be noted that spectra of high resolution, such as presented in this section, are essential to rigorously test the theory of the Auger effect over wide Z ranges and to gain further insights into correlation processes such as those embodied in shakeup and shakeoff. Concerning the latter,

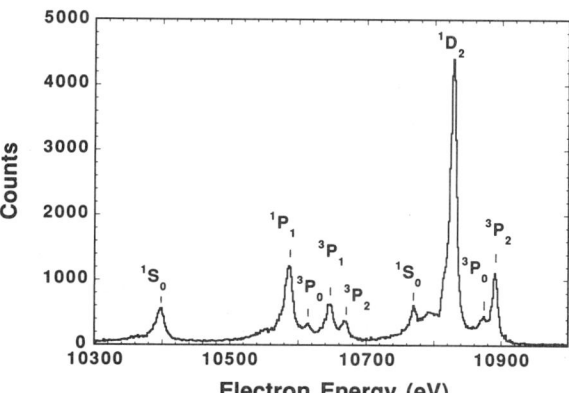

FIGURE 29. The krypton K-LL Auger spectrum resulting from the nuclear decay of gaseous ^{83m}Kr.[90] The widths of the nine normal Auger lines are essentially natural. Shakeup and shakeoff processes give rise to structure and tailing on the low-energy sides of the peaks.

high-resolution Auger spectrometry with a tunable photon source is especially advantageous, as demonstrated in the case of argon.[66] However, this potential remains to be fully exploited in future experiments.

5.3. Decay of Excited States

The radiationless decay of excited states typically leads to singly ionized species associated with one or more of the energetically accessible continua. Which of the electrons of the excited neutral atom will preferentially participate in the decay will depend on the particulars of the excitation. If an inner core electron is excited in a closed-shell atom, this electron is less likely to participate than the electrons that have remained passive during the absorption. Such a case bears the greatest similarity to an Auger decay with the excited electron remaining a spectator. If the excited electron originates from a shallow core level, it is unlikely to remain passive; it may either participate in the decay to the continuum or it may change its orbital in a shakeup or shakedown process while another electron makes the transition to the continuum. For example, in two excitations which have been studied extensively, $3d \rightarrow np(n \geq 5)$ in krypton and $4d \rightarrow np(n \geq 6)$ in xenon, the decay patterns can be reproduced well by the interpretation through resonance Auger decay accompanied by shake processes.[92,93] However, these two systems are rather unique in that the smallest value of n which can be reached corresponds to an empty *principal* shell, where the spectator approach is most likely to be valid. This is confirmed by the analysis of the decay of the high-lying $2p \rightarrow ns,nd$ excitations in magnesium.[94,95]

A frequent occurrence is one in which the excitation takes place to an empty subshell within a principal shell having otherwise filled subshells. This type of excitation is common, for example, in the alkaline earth elements, such as $4p \rightarrow 4d$ in strontium. This is also notably the case for the $2p \rightarrow 3d$ excitation in argon, where the simplest form of the resonance Auger description breaks down as a result of the collapse of the $3d$ orbital.[96,97] The strong configuration mixing related to the partially filled shells of these excited states can give rise to complex absorption spectra and even more complex electron distributions in the decay, requiring a full configuration-interaction description to explain the process.

Finally, another form of excitation is one in which the electron is promoted to a partially filled subshell. This situation applies to open-shell atoms, particularly the $np \rightarrow n'd$ excitations in the transition metals[10,39] and $nd \rightarrow n'f$ in the rare earths,[10,98,99] and it applies of course to the halogens and chalcogens, examples being $4d \rightarrow 5p$ in iodine,[100,101] $3d \rightarrow 4p$ in bromine,[101] and $1s \rightarrow 2p$ in oxygen.[71] In excitations of this type the excited electron is always intimately involved in the decay.

High-resolution decay patterns have been measured for the $3d$ resonances in krypton[93,102] and the $4d$ resonances in xenon.[102] Part of the decay spectrum for the $4d_{3/2} \rightarrow 6p$ and $4d_{5/2} \rightarrow 6p$ resonance transitions in xenon[102] is shown in Figure 30. The spectrum was recorded with a PES resolution of 90 meV using undulator radiation from the Aladdin storage ring as a source, with a monochromator bandpass of 50 meV. The diamonds at the top of the figure show the locations of the different

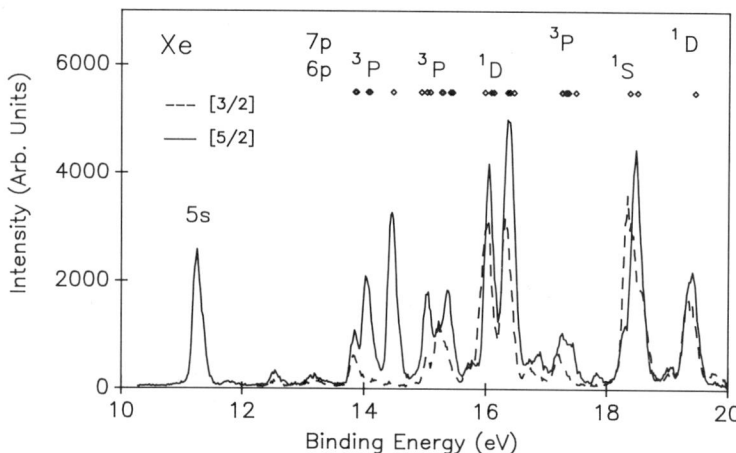

FIGURE 30. Decay patterns of the $[4d_{5/2}]6p$ (solid line) and $[4d_{3/2}]6p$ (dashed line) excitations in xenon. At the top are given the locations of the Xe$^+$ $5p^4(^{2S+1}L)6p,7p$ final ionic states to which the excitation states decay. The 5s photoline is visible on the left. $E(^5P_{3/2}) = 12.13$ eV should be added to the abscissa scale. (From Ref. 102.)

$5p^4(^{2S+1}L)6p_j$ fine-structure components of the Xe$^+$ final state into which each resonance decays. Clearly resolution plays a key role in the analysis of these spectra as the different j-states of the initial excitation decay very differently into the final j-states, and high resolution is required to separate these decay modes. As an example of the type of information contained in these patterns, notice that both spin components have intensity in the ^1S feature for both excitations. For the case of the $4d_{5/2} \rightarrow 6p$ excitation this implies that a spin flip has occurred in the course of the decay, indicating that the excited electron does not remain entirely a spectator but has some interaction with the core during the decay process.

In addition to the enhancement of the details of the decay spectrum over that of previous measurements,[93,103] resolution is also essential to the determination of the angular distributions of the decay electrons. This was illustrated in krypton[102] for a decay level consisting of two features having roughly equal and opposite β values, +0.43 and −0.42. In the lower-resolution work this gave a β value of −0.16 for the unresolved structure.[103]

The propensity for electron decay in the rare gases krypton and xenon to high-lying Rydberg orbitals has important consequences when the decay occurs to levels which lie between multiplets of the doubly charged ion. One outcome of such decay can be the production of large quantities of doubly charged ions through a two-step process which involves an initial resonance Auger decay into a bound state of the singly charged ion.[104] This observation has been confirmed by coincidence measurements[104,105] and, indirectly, by decay of the $4d \rightarrow np$ excitations in xenon.[15] The process has been revisited using the multichannel detector described in Section 2. Measurements of the low-energy electrons produced by the secondary decay were performed with an electron energy resolution of 30 meV.[34] In that work energies could be assigned to the original excitations, thus confirming the validity of the two-step approach.

As one moves from the rather simple resonance Auger spectra exemplified by the case of the rare gases, resolution becomes more critical, especially as the excitation may become equally as complex as the decay, placing greater demands on the monochromator resolution for isolation of the excitations. This is illustrated by way of the $4p \rightarrow 4d$ excitation in strontium. Actually, this designation is somewhat of a misnomer, as even the ground state of strontium requires a number of configurations for a proper description. Nonetheless, as illustrated in the absorption spectrum,[106] it is possible to locate a strong feature which is predominantly characterized by this designation.

Shown in Figure 31 are two selections from an in-depth analysis of the decay of the 4p excitations in strontium.[107] Using the 4-m NIM at Aladdin, it was possible to isolate more than 20 single features

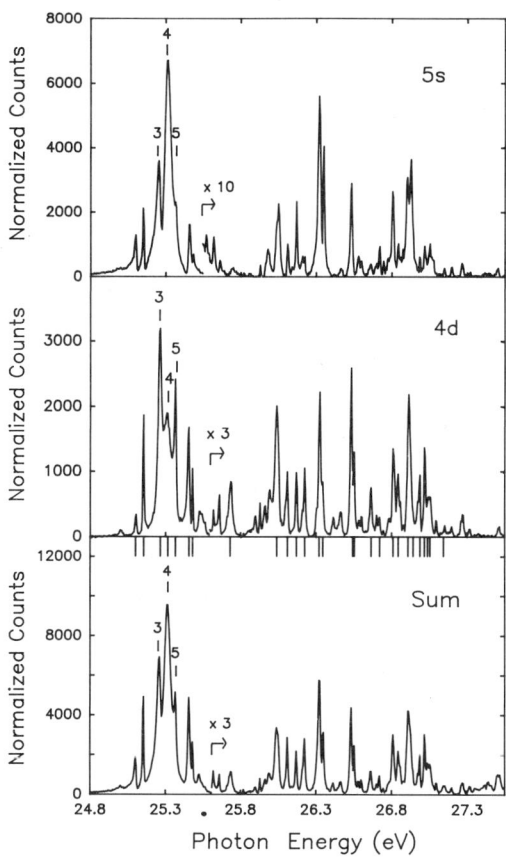

FIGURE 31. Partial 4p absorption spectrum of strontium as measured through decay into the lowest-lying Sr^+ ionic states. Decays into the Sr^+ 5s and 4d states are shown explicitly; the sum represents all measured partial contributions. At the top of the summed spectrum are shown the positions of the resonances as determined in an absorption spectrum. (From Ref. 107.)

in the absorption spectrum and determine the relative decay into the ten lowest-lying states of the ion. In the top panel are given the resonance excitations which produce the 5s state of the singly charged ion and in the center panel those which produce the 4d state of the singly charged ion. Both spectra are scans taken in the CIS mode and therefore the resolution is governed by the bandpass of the monochromator, namely, 0.18 Å or 6 meV at the photon energy used. In the bottom panel is shown the sum spectrum which contains the decays into the strongest final states. Each of the major peaks in the spectra can be assigned to a single excitation in the absorption spectrum.[106] The principal $4p \rightarrow 4d$ excitation corresponds to peak #4 in this spectrum. Notice that it is adjacent to neighboring narrow features 50 meV to either side. However, the high resolution achieved in this study makes it possible to separate the decay of each of these three resonances and separately determine the contributions for decay into the two predominant channels corresponding to final states $Sr^+(5s)$ and $Sr^+(4d)$. The intensity ratios for the various possible final states will obviously depend on the degree of differentiation of the various components making up the manifold. For example, the high-resolution spectra of Figure 31 yield ratios of $\sigma(4d):\sigma(5s) = 60:30$, 20:70, and 57:31 for features #3, #4, and #5, respectively. These values are to be contrasted with a value of 34:54 for the structure comprising peaks #3–#5 in a spectrum recorded with a bandpass of 60 meV.[108]

As an illustration of the decay behavior of excitations to partially filled subshells we select high-resolution results for the decay of the $3p \rightarrow 3d$ excitation in manganese.[70] As demonstrated in the PES recording of Figure 22, this resonance decays into a multitude of ionic states involving a variety of two-electron transitions. The $3p \rightarrow 3d$ excitation comprises many multiplet and fine-structure states, each with its own decay pattern. Correlating the photolines of Figure 22 with the various excitation states should provide insight into the structure and dynamics of the excitation states. In

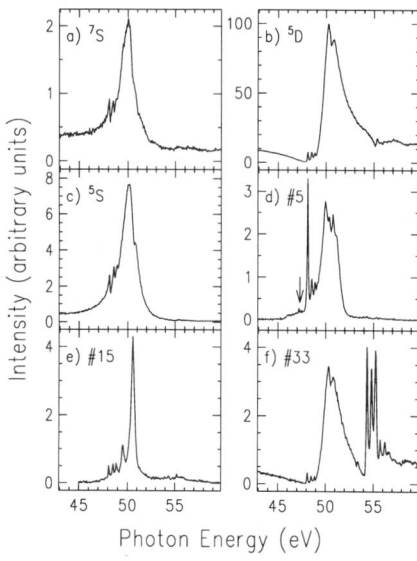

FIGURE 32. Partial absorption spectra of manganese in the vicinity of the $3p \rightarrow 3d$ resonance as measured through decay into selected Mn$^+$ ionic states: (a) Mn$^+$[4s](^7S); (b) Mn$^+$[3d](^5D); (c) Mn$^+$[4s](^5S); (d) satellite #5; (e) satellite #15; and (f) satellite #33 in Figure 22. (From Ref. 70.)

Figure 32 CIS scans over the region of $3p \rightarrow 3d$ excitation are depicted for several of the Mn$^+$ states shown in Figure 22. In the various panels are given the branching ratios as a function of energy across the resonance for decay into selected Mn$^+$ ionic states: panel (a), Mn$^+$[4s](^7S); panel (b), Mn$^+$[3d](^5D); panel (c), Mn$^+$[4s](^5S); panel (d), satellite #5; panel (e), satellite #15; and panel (f), satellite #33 in Figure 22. Evidently, there is considerable difference between the different patterns, and the appearance of certain features is closely linked with a specific excitation even if the final states have the same angular momentum, e.g., ^7S and ^5S. Thus, monochromator resolution becomes a key factor in the analysis of the behavior of these resonance transitions. The information which emerges from the analysis is that the broad feature in the manganese absorption, which is typified by the prominent decay shown in panel (b), is most likely produced by the overlap of a number of excited states, each with its own decay mode. This is best exemplified by the decay which gives rise to satellite #15 in which the broad feature disappears and is replaced by a single strong and one (or more) weak narrow lines. Similarly, the structure on the top of the broad feature shown in panel (d) indicates the presence of fine structure. Thus, through the use of high-resolution electron spectrometry, what appears as a broad feature in absorption becomes separable into a large number of individual features. The case of the $3p \rightarrow 3d$ resonance in Mn serves also as an example for the need of high resolution for both the monochromator and the electron spectrometer for a detailed scrutiny to be possible.

Perhaps a word is appropriate about the role which resolution can play in the analysis of broad features of a spectrum. Typically, one would not consider this to be significant, and it will not be as long as the predominant feature in the spectrum is the one broad feature. However, if this feature is interspersed with sharper features which ride atop it, as it seems to be the case in manganese, high resolution becomes essential for determining the background appropriate to the sharper features and to the determination of the correct shape of the broad feature. Conversely, a poor resolution applied to a closely spaced series of sharp lines, for example a vibrational progression in a molecule, may artificially enhance a weak underlying broad feature, or may simulate a nonexistent broad feature.

6. HIGHER-ORDER EFFECTS IN PHOTOELECTRON AND AUGER SPECTRA

The category of higher-order processes may include all those events in which inner-shell excitation and deexcitation are no longer separable, or radiationless and radiative deexcitations are not independent, or discrete shakeup states can interact with shakeoff continua. Some manifestations of higher-order phenomena have been alluded to in several places in the preceding sections; in the following a few specific examples will be highlighted.

The postcollision interaction (PCI) is an indicator of the limits of the two-step model of inner-shell ionization followed by radiationless deexcitation. A sufficiently slow continuum electron produced by photon or particle impact can interact with a faster Auger electron, and, thereby, can exchange energy. As a result, the "initial" electron is retarded and the Auger electron is accelerated. In the case of photoionization, neither the photoelectron nor the Auger electron are in their correct energy positions, and their line shapes are skewed because the interaction extends over a range of mutual distances as the electrons move away from the ionic core. For a long time, the effect was delineated only by studying the Auger electron, but recently it was also observed in a photoelectron spectrum, that of Kr $3d$.[34] As shown in Figure 9, PCI manifests itself by a shift of the photoelectron toward lower kinetic energy and a skewing of the peak toward the low-kinetic-energy side. Both features are the more pronounced the slower the photoelectron. The PCI effect on the corresponding Auger lines, e.g., the Kr $3d$-$4p4p(^1S_0)$ line or in the case of Xe the $4d$-$5p5p(^1S_0)$ line, is in the opposite direction toward greater energies. Relevant spectra can be viewed elsewhere, e.g., Refs. 65, 109, 110 and Chapter 11. Although the width of the Xe $4d$ level, or that of an associated Auger line, is rather broad as it is for Kr $3d$, a high resolution of the electron spectrometer is needed to clearly demonstrate the line distortion by PCI and, most importantly, to delimit the range of the effect. In accord with theoretical considerations,[111,112] experiment could thus demonstrate[109] that PCI ceases when the energy of the photoelectron exceeds that of the Auger electron.

In contrast to photoionization, the PCI effect will never vanish for ionization by electrons, because there exists a high probability for producing a slow continuum electron even at a great energy of the incident electron.[113] The residual effect on the Auger line shape can be seen in Figure 33, which presents the Ar $L_{2,3}$-$M_{2,3}M_{2,3}$ Auger spectrum excited by 2-keV electrons.[114] Because of the small instrumental width of 60 meV, which is about half the natural width, the small high-energy tail indicative of PCI becomes evident on all Auger lines of this spectrum. According to theoretical considerations,[113] the observed linewidth should exceed the decay width of the initial state under these excitation conditions by up to 10% even at asymptotically high impact energies. This broadening, as

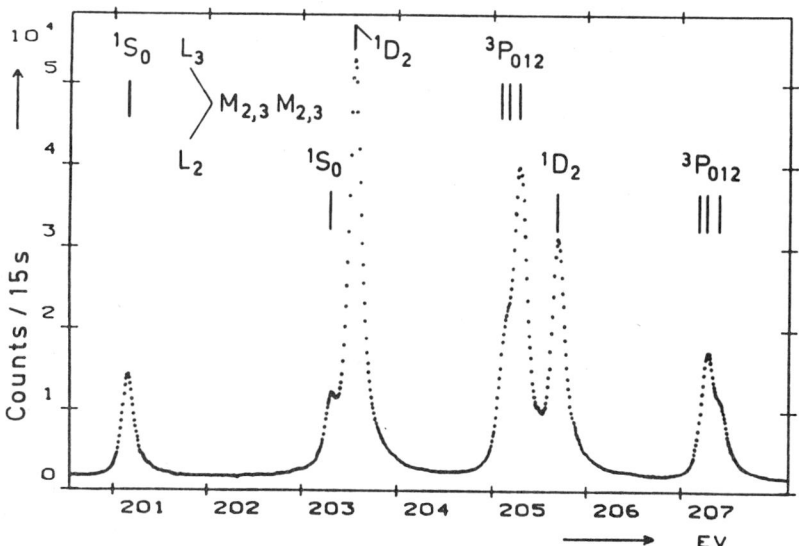

FIGURE 33. The argon $L_{2,3}$-$M_{2,3}M_{2,3}$ Auger spectrum excited by 2-keV electrons and recorded with a resolution of $\Delta E/E = 3 \times 10^{-4}$. Note the tails on the high-energy sides of the peaks resulting from postcollision interaction (PCI). (From Ref. 114.)

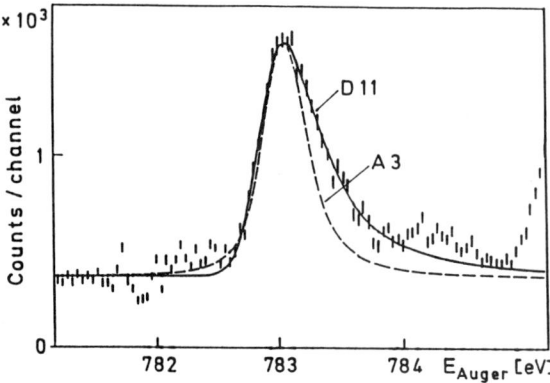

FIGURE 34. The effect of the postcollision interaction (PCI) on a diagram Auger line (A3) and a satellite Auger line (D11) in the electron ionization of neon. (From Ref. 88.)

well as an asymmetrical line distortion and line shift, can be avoided in photoionization by selecting a photon energy that produces a photoelectron of greater energy than that of the Auger electron.

If under electron impact the PCI affects the shape and position of a diagram Auger line, as it does, it will affect even more the shape and position of a satellite Auger line because in this case the probability of having *two* slow electrons near the ion is rather high. The resulting skewing of the satellite lines can be seen well in the Ne K-LL spectrum,[88,115,116] and the increased skewing over that of the diagram line has been demonstrated clearly for selected Ne satellite lines.[88,116] As an example, Figure 34 compares the profile of the $[1s2p](^3P) \rightarrow [2p^3](^2P)$ satellite lines, designated as D11, with the profile of the $[1s] \rightarrow [2s2p](^3P)$, or K-$L_1L_{2,3}$, diagram line, designated as A3. The figure shows distinctly the enhanced broadening and skewing of the satellite line over that of the diagram line. However, a delineation of the different line shifts for A3 and D11 would require additional experimentation. If photons are used for the initial ionization, PCI effects on the diagram Auger lines can be avoided by

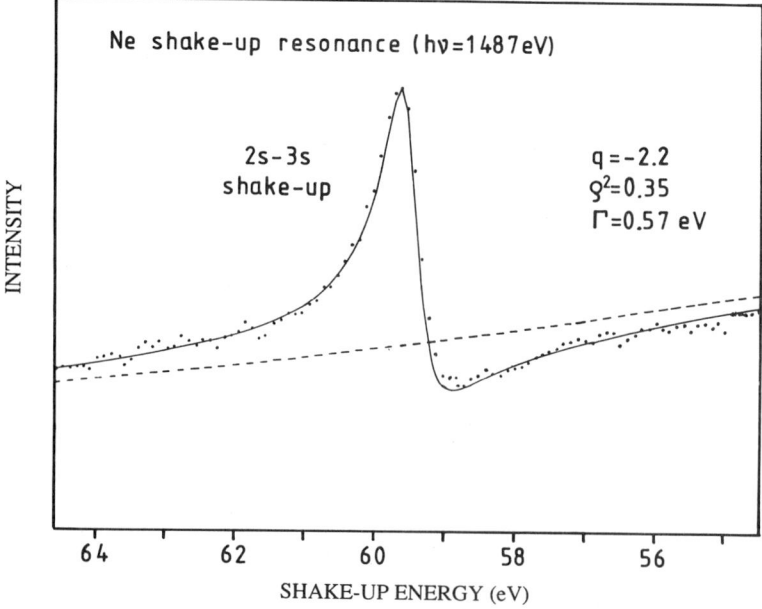

FIGURE 35. The neon $1s2s(^3S)2p^63s(^2S)$ core electron shakeup line, showing a resonance profile via the interaction with the underlying shakeoff continuum. (From Ref. 56.)

the proper choice of the photon energy, but the effects on the satellite lines will persist at any energy because the shakeoff electron plays a similar role as does the secondary electron in electron ionization.

Autoionization, described in Section 5.1, has an interesting analogue in the radiationless decay of shakeup states (again using the term *shakeup* generically). In the former case an excited state of the neutral atom can interact with continua belonging to the singly ionized atom, while in the latter case an excited state of a singly ionized atom can interact with continua belonging to the doubly ionized atom. The higher-order effect of a shakeup state interacting with a shakeoff continuum (continua) was demonstrated for photoionization of neon by 1.5-keV photons.[56] As shown in Figure 35, the $2s \to 3s$ shakeup satellite associated with $1s$ photoionization displays a typical Fano profile as a consequence of the interaction of the Ne^+ $1s2s2p^63s,\varepsilon p$ channel with the underlying Ne^{2+} $1s2s^22p^5\varepsilon\ell,\varepsilon p$ continuum. The width of the resonance, $\Gamma = 0.57$ eV, is about equal to the PES resolution and about twice as large as the Ne $1s$ width. The subsequent radiationless decay of the resonance state by an Auger transition involving the $1s$ vacancy gives rise to Auger electron satellites, as observed previously.[65,88,115] Because of the resonance pathway, the distinction between Auger satellites related to shakeup and those related to shakeoff becomes less sharp. The resonance is also a contributor to a higher ionic charge state than expected from a noninteracting shakeup state.

The shakeup resonances just discussed for inner-shell photoionization in neon are not unique to that particular system, but can occur whenever the shakeup state lies within at least one shakeoff continuum. This is the case for photoionization in the outer shell of neon, where the $2s2p^5(^{1,3}P)n\ell$ satellites display resonance behavior. In addition to the shakeup satellites, discrete Auger lines have been observed at low kinetic energies and attributed to the "subsequent" Auger decay of $2s2p^5(^1P)$ np states into $2s2p^5(^3P)\varepsilon\ell$ and $2s2p^4(L)\varepsilon\ell$ continua.[117] In analogy to the shakeup resonances concomitant with 1s ionization in neon, the process can be considered as autoionization, and on the basis of the discrete lines observed it could be considered an Auger process[118,119]; clearly, such differentiation should be viewed with caution. Energetically, radiationless decay of $2s2p^5n\ell$ states is allowed, and the consequences of this deexcitation are evident in both the widths and profiles of the shakeup satellite lines[56,119] and in the occurrence of discrete Auger lines.[117]

Figure 36 shows a photoelectron spectrum of several $2s2p^5(^{1,3}P)n\ell$ satellites taken with a moderately high PES resolution.[119] In the absence of radiationless processes, these lines would be intrinsically symmetric and extremely narrow. However, radiationless events broaden the $2s2p^5(^3P)3p$ ^2S ionic state, line 18 in Figure 36, to $\Gamma = 0.11$ eV, and $2s2p^5(^1P)3p$ ^2S, line 22, to $\Gamma = 0.41$ eV and, at the same time, create Fano-type line shapes. Line 22 is broader than line 18 because it can also undergo a multiplet-changing transition, $^1P \to ^3P$. This type of transition appears unusual from the standpoint of the Auger effect, yet rather commonplace from the standpoint of autoionization. Although the PES resolution of 0.20 eV applied in the study of the neon resonances limited the accuracy of the analysis, it proved sufficient to elucidate the phenomenon and make a comprehensive comparison with theoretical predictions.

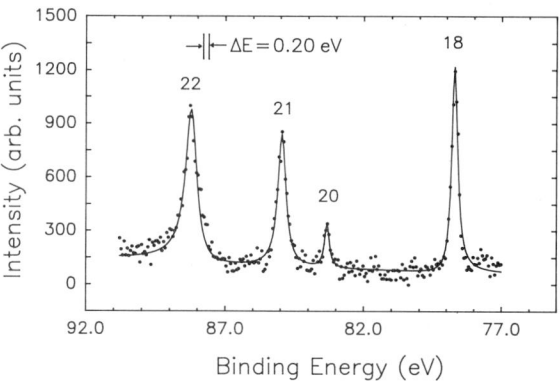

FIGURE 36. The neon $2s2p^5(^{1,3}P)n\ell^2L$ valence shell satellites displaying different widths and resonance profiles resulting from different radiationless deexcitations. Line 18 is the $(^3P)3p$ ^2S shakeup line, and line 22 the $(^1P)3p$ ^2S line which undergoes especially rapid decay. (From Ref. 119.)

7. PROSPECTS

Doubtless, much of the future electron spectrometry work on atoms will be centered at the new generation of synchrotron radiation sources equipped with insertion devices. Undulators are very bright light sources, and the high spatial concentration of the photons can be converted into high resolution without a serious trade-off in intensity. First, a photon beam of small dimensions can be dispersed and focused with smaller figure errors because smaller areas of the optical elements, gratings and mirrors, are needed. Resolving powers of 10,000 to 20,000 should be standard, and values of 50,000 and more possible, even for emission work at energies in the regions of 10 and 100 eV. Second, the small focal spot obtainable with a bright photon source will allow for the efficient utilization of narrow physical or electron-optical apertures needed for high resolution of the electron energy analyzer. Augmenting narrow entrance apertures by a multichannel detector with small pixel sizes will be essential to maintain signal strength even at a resolution approaching $\Delta E/E \approx 1 \times 10^{-4}$. With these prospects for improved resolution, the criteria laid out in Table I can be met by and large for both the CIS and PES modes of electron spectrometry.

For electron spectrometry with the use of synchrotron radiation, the advances will be substantial. Just as the resolution has improved by a factor of 10 over the first 15 years, another factor of 10 might be expected over the next 5 to 10 years. As an example, the argon satellite spectrum shown in Figure 1 was originally observed with a PES resolution of 1 eV and recently with a resolution of 100 meV. The prospects for a recording with 10-meV resolution are excellent. Similar prospects apply to other spectra of lightly bound electrons, especially the least studied open-shell atomic spectra. In these cases, the ensuing capability of elucidating the electronic structure and dynamics at the fine-structure level presents a particularly challenging opportunity. Electron spin analysis, in addition to energy, cross section, and electron angular distribution measurements, can also be carried out with higher resolution than previously possible, giving virtually complete information on the photoionization process. However, the limitations of the spin detector still place bounds on the resolution achievable for reasons of intensity.

A projected PES resolution of 10 meV for future work may appear conservative in view of several recent experiments that have achieved such a high resolution, or even higher. However, the same value of 10 meV is optimistic in view of the thermal broadening that occurs at higher electron kinetic energies in a randomly moving gaseous target. The use of a suitably directed atomic beam can alleviate this problem in first order, though at the expense of signal strength because of the reduced target density.

In the case of scanning electron emission spectrometry, as embodied by the CIS technique, the advances in resolution will be coming from an improved bandpass of the monochromator. An important corollary to this is the greater photon intensity once the monochromator is coupled to an undulator. Emission-type measurements should become feasible with resolving powers hitherto reserved for absorption-type experiments. At photon energies from 10 to 100 eV, a CIS resolution of less than a few millielectron volts, and generally in the region of 100 µeV, is then available for characterizing autoionization resonances and certain threshold phenomena. As an example, the determination of the photoelectron angular distribution parameter for such basic processes as the two-electron excitations to higher autoionizing resonance states of helium will then come within our reach.

In Auger electron spectrometry the advances will not be as dramatic as for photoelectron spectrometry because the improvements are largely restricted to the electron energy analyzer. However, the prospects of being able to excite Auger spectra at specific energies with a very narrow bandpass will be most valuable for gaining a better understanding of the resonance Auger transitions and the limits of the two-step model of excitation and deexcitation. A photon bandpass equal or better than the natural width of an atomic core level can then elicit optimal information about the physical phenomena.

Again consulting Table I and comparing the resolving powers needed to study inner shells with the resolving powers projected for both monochromators and electron spectrometers, it becomes apparent that photoelectron spectrometry of core levels can be carried out in the future over a wide range of photon energies with few instrumental limitations. This is, indeed, a bright prospect and especially important for the study of phenomena in the outer regions of the atom and around inner-shell thresholds.

ACKNOWLEDGMENTS. Research sponsored in part by the Division of Chemical Sciences, Office of Basic Energy Sciences, U.S. Department of Energy under Contract DE-AC05-84OR21400 with Martin Marietta Energy Systems, Inc., and in part by the National Science Foundation under Grant PHY-9207634.

REFERENCES

1. M. Y. ADAM, F. WUILLEUMIER, S. KRUMMACHER, V. SCHMIDT, AND W. MEHLHORN, *J. Phys. B* **11**, L413 (1978).
2. M. O. KRAUSE, S. B. WHITFIELD, C. D. CALDWELL, J.-Z. WU, P. VAN DER MEULEN, C. A. DE LANGE, AND R. W. C. HANSEN, *J. Electron Spectrosc. Relat. Phenom.* **58**, 79 (1992).
3. K. SIEGBAHN, C. NORDLING, G. JOHANSSON, J. HEDMAN, R. F. HEDEN, K. HAMRIN, U. GELIUS, T. BERGMARK, L. O. WERME, R. MANNE, AND Y. BAER, *ESCA Applied to Free Molecules* (North-Holland, Amsterdam, 1969).
4. T. A. CARLSON, *Photoelectron and Auger Spectroscopy* (Plenum Press, New York, 1975).
5. K. D. SEVIER, *Low Energy Electron Spectrometry* (Wiley–Interscience, New York, 1972).
6. M. O. KRAUSE, in *Atomic Inner-Shell Processes*, edited by B. Crasemann (Academic Press, New York, 1975), Vol. II, pp. 33–81.
7. H. SIEGBAHN AND L. KARLSSON, in *Corpuscles and Radiation in Matter*, edited by W. Mehlhorn, *Handbuch der Physik*, Vol. 31 (Springer-Verlag, Berlin, 1982), p. 215.
8. M. O. KRAUSE, in *Synchrotron Radiation Research*, edited by H. Winick and S. Doniach (Plenum Press, New York, 1980), Chapter 5.
9. J. A. R. SAMSON, in *Corpuscles and Radiation in Matter*, edited by W. Mehlhorn, *Handbuch der Physik*, Vol. 31 (Springer-Verlag, Berlin, 1982), p. 123.
10. B. SONNTAG AND P. ZIMMERMANN, *Rep. Prog. Phys.* **55**, 911 (1992).
11. V. SCHMIDT, *Rep. Prog. Phys.* **55**, 1483 (1992).
12. A. F. STARACE, in *Corpuscles and Radiation in Matter*, edited by W. Mehlhorn, *Handbuch der Physik*, Vol. 31 (Springer-Verlag, Berlin, 1982), p. 1.
13. For example, Proceedings of Conferences on Synchrotron Radiation Instrumentation published periodically in *Rev. Sci. Instrum.* and *Nucl. Instrum. Methods A*.
14. K. PEARSON, *Biometrika* **16**, 172 (1924).
15. S. B. WHITFIELD, C. D. CALDWELL, D. X. HUANG, AND M. O. KRAUSE, *J. Phys. B* **25**, 4755 (1992).
16. J. JIMÉNEZ-MIER, *J. Quant. Spectrosc. Radiat. Transfer* **51**, 741 (1994).
17. M. G. FLEMMING, J.-Z. WU, C. D. CALDWELL, AND M. O. KRAUSE, *Phys. Rev. A* **44**, 1733 (1991).
18. P. VAN DER MEULEN, M. O. KRAUSE, C. D. CALDWELL, S. B. WHITFIELD, AND C. A. DE LANGE, *Phys. Rev. A* **46**, 2468 (1992).
19. M. O. KRAUSE, C. D. CALDWELL, S. B. WHITFIELD, S. J. SCHAPHORST, AND Y. AZUMA, *Synchr. Radiat. News* **5**, 25 (1992).
20. K. MAEDA, K. UEDA, T. NAMIOKA, AND K. ITO, *Phys. Rev. A* **45**, 527 (1992).
21. D. KLAR, K. HARTH, J. GANZ, T. KRAFT, M.-W. RUF, H. HOTOP, V. TSEMEKHMAN, K. TSEMEKHMAN, AND M. Y. AMUSIA, *Z. Phys. D* **23**, 101 (1992).
22. P. BALTZER, B. WANNBERG, L. KARLSSON, M. CARLSSON-GÖTHE, AND M. LARSSON, *Phys. Rev. A* **45**, 4374 (1992).
23. J. GANZ, M. RAAB, H. HOTOP, AND J. GEIGER, *Phys. Rev. Lett.* **53**, 1547 (1984).
24. R. G. TONKYN AND M. G. WHITE, *Rev. Sci. Instrum.* **60**, 1245 (1989).
25. C. D. CALDWELL AND M. O. KRAUSE, *J. Phys. B* **23**, 2233 (1990).
26. C. T. CHEN AND F. SETTE, *Rev. Sci. Instrum.* **60**, 1616 (1989).
27. P. A. HEIMANN, F. SENF, W. MCKINNEY, M. HOWELLS, R. D. VAN ZEE, L. J. MEDHURST, T. LAURITZEN, J. CHIN, J. MENEGHETTI, W. GATH, H. HOGREFE, AND D. A. SHIRLEY, *Phys. Scr.* **T31**, 127 (1990).
28. A. YAGISHITA, S. MASUI, T. TOYOSHIMA, H. MAEZAWA, AND E. SHIGEMASA, *Rev. Sci. Instrum.* **63**, 1351 (1992).
29. H. PETERSEN, *Nucl. Instrum. Methods A* **246**, 260 (1986).
30. M. DOMKE, G. REMMERS, AND G. KAINDL, *Phys. Rev. Lett.* **69**, 1171 (1992).
31. M. CARLSSON-GÖTHE, P. BALTZER, AND B. WANNBERG, *J. Phys. B* **24**, 2477 (1991).
32. J. M. DYKE, B. W. J. GRAVENOR, M. P. HASTINGS, G. D. JOSLAND, AND A. MORRIS, *J. Electron Spectrosc. Relat. Phenom.* **35**, 65 (1985).
33. J. V. HATFIELD, S. A. BURKE, J. COMER, F. J. CURRELL, J. GOLDFINCH, T. A. YORK, AND P. J. HICKS, *Rev. Sci. Instrum.* **63**, 235 (1992).
34. D. ČUBRIĆ, A. A. WILLS, J. COMER, AND M. A. MACDONALD, *J. Phys. B* **25**, 5069 (1992).
35. J. A. BEARDEN AND A. F. BURR, *Rev. Mod. Phys.* **39**, 125 (1967).

36. C. E. MOORE, *Atomic Energy Levels*, Natl. Bur. Stand. (U.S.) Circ. No. 467 (U.S. Government Printing Office, Washington, DC, 1949, 1952, 1958); and revisions for selected elements, published by various authors in *J. Phys. Chem. Ref. Data*.
37. M. O. KRAUSE AND J. H. OLIVER, *J. Phys. Chem. Ref. Data* **8**, 329 (1979).
38. E. SCHMIDT, Dissertation Universität Hamburg (1985).
39. M. MEYER, T. PRESCHER, E. VON RAVEN, M. RICHTER, E. SCHMIDT, B. SONNTAG, AND H.-E. WETZEL, *Z. Phys. D* **2**, 347 (1986).
40. J. M. DYKE, B. W. J. GRAVENOR, R. A. LEWIS, AND A. MORRIS, *J. Phys. B* **15**, 4523 (1982).
41. J. JIMÉNEZ-MIER, M. O. KRAUSE, P. GERARD, B. HERMSMEIER, AND C. S. FADLEY, *Phys. Rev. A* **40**, 3712 (1989).
42. P. S. BAGUS, A. J. FREEMAN, AND B. SASAKI, *Phys. Rev. Lett.* **30**, 850 (1973).
43. J. M. DYKE, N. K. FAYAD, A. MORRIS, AND I. R. TRICKLE, *J. Phys. B* **12**, 2985 (1979).
44. M. O. KRAUSE, P. R. WOODRUFF, AND T. A. CARLSON, *J. Phys. B* **14**, L673 (1981).
45. Unpublished data of the authors.
46. M. O. KRAUSE, C. D. CALDWELL, S. B. WHITFIELD, C. A. DE LANGE, AND P. VAN DER MEULEN, *Phys. Rev. A* **47**, 3015 (1993).
47. C. PAN AND A. F. STARACE, *Phys. Rev. A* **47**, 295 (1993).
48. J. SCHIRMER, L. S. CEDERBAUM, AND J. KIESSLING, *Phys. Rev. A* **22**, 2696 (1980).
49. B. G. KOENDERS, K. E. DRABE, AND C. A. DE LANGE, *Chem. Phys. Lett.* **138**, 1 (1987).
50. S. T. PRATT, J. L. DEHMER, AND P. M. DEHMER, *J. Chem. Phys.* **82**, 676 (1985).
51. M. O. KRAUSE, T. A. CARLSON, AND R. D. DISMUKES, *Phys. Rev.* **170**, 37 (1968).
52. T. A. CARLSON, *Phys. Rev.* **156**, 142 (1967).
53. H. KOSSMANN, B. KRÄSSIG, V. SCHMIDT, AND J. E. HANSEN, *Phys. Rev. Lett.* **58**, 1620 (1987).
54. C. E. BRION, A. O. BAWAGAN, AND K. H. TAN, *Can. J. Chem.* **66**, 1877 (1988).
55. T. A. CARLSON, M. O. KRAUSE, AND W. E. MODDEMAN, *J. Phys. (Paris)* **32**, C4-76 (1971).
56. S. SVENSSON, B. ERIKSSON, N. MÅRTENSSON, G. WENDIN, AND U. GELIUS, *J. Electron Spectrosc. Relat. Phenom.* **47**, 327 (1988).
57. F. HEISER, U. HERGENHAHN, J. VIEFHAUS, K. WIELICZEK, AND U. BECKER, *J. Electron Spectrosc. Relat. Phenom.* **60**, 337 (1992).
58. U. BECKER AND D. A. SHIRLEY, *Phys. Scr.* **T31**, 56 (1990).
59. A. A. WILLS, A. A. CAFOLLA, F. J. CURRELL, J. COMER, A. SVENSSON, AND M. A. MACDONALD, *J. Phys. B* **22**, 3217 (1989).
60. P. A. HEIMANN, U. BECKER, H. G. KERKHOFF, B. LANGER, D. SZOSTAK, R. WEHLITZ, D. W. LINDLE, F. A. FERRETT, AND D. A. SHIRLEY, *Phys. Rev. A* **34**, 3782 (1986).
61. H. SMID AND J. E. HANSEN, *Phys. Rev. Lett.* **52**, 2138 (1984).
62. W. WIJESUNDERA AND H. P. KELLY, *Phys. Rev. A* **39**, 634 (1989).
63. J. A. R. SAMSON, Y. CHUNG, AND E. LEE, *Phys. Lett.* **127**, 171 (1988).
64. D. L. WARK, R. BARTLETT, T. J. BOWLES, R. G. H. ROBERTSON, D. S. SIVIA, W. TRELA, J. F. WILKERSON, G. S. BROWN, B. CRASEMANN, S. L. SORENSEN, S. J. SCHAPHORST, D. A. KNAPP, J. HENDERSON, J. TULKKI, AND T. ÅBERG, *Phys. Rev. Lett.* **67**, 2291 (1991).
65. W. MEHLHORN, in *Atomic Inner-Shell Physics*, edited by B. Crasemann (Plenum Press, New York, 1985), Chapter 4.
66. G. B. ARMEN, T. ÅBERG, K. R. KARIM, J. C. LEVIN, B. CRASEMANN, G. S. BROWN, M. H. CHEN, AND G. E. ICE, *Phys. Rev. Lett.* **54**, 182 (1985).
67. P. GERARD, Dissertation Université Paris-Sud (1984).
68. B. LANGER, J. VIEFHAUS, O. HEMMERS, A. MENZEL, R. WEHLITZ, AND U. BECKER, *Phys. Rev. A* **43**, 1652 (1991).
69. G. B. ARMEN AND F. P. LARKINS, *J. Phys. B* **24**, 2675 (1991).
70. S. B. WHITFIELD, M. O. KRAUSE, P. VAN DER MEULEN, AND C. D. CALDWELL, *Phys. Rev. A* **50**, 1269 (1994).
71. C. D. CALDWELL, S. J. SCHAPHORST, M. O. KRAUSE, AND J. JIMÉNEZ-MIER, *J. Electron Spectrosc. Relat. Phenom.* **67**, 243 (1994).
72. T. ÅBERG AND G. HOWAT, in *Corpuscles and Radiation in Matter*, edited by W. Mehlhorn, *Handbuch der Physik*, Vol. 31 (Springer-Verlag, Berlin, 1982), p. 469.
73. H. BEUTLER, *Z. Phys.* **93**, 177 (1935).
74. U. FANO, *Nuovo Cimento* **12**, 154 (1935); *Phys. Rev.* **124**, 1866 (1961).
75. F. WUILLEUMIER AND M. O. KRAUSE, *Phys. Rev. A* **10**, 242 (1974).
76. W. R. JOHNSON, K. T. CHENG, K.-N. HUANG, AND M. LE DOURNEUF, *Phys. Rev. A* **22**, 989 (1980).
77. U. HEINZMANN, F. SCHÄFERS, K. THIMM, A. WOLCKE, AND J. KESSLER, *J. Phys. B* **12**, L679 (1979).
78. J.-Z. WU, S. B. WHITFIELD, C. D. CALDWELL, M. O. KRAUSE, P. VAN DER MEULEN, AND A. FAHLMAN, *Phys. Rev. A* **42**, 1350 (1990).

79. V. Radojević and J. D. Talman, *J. Phys. B* **23**, 2241 (1990).
80. M. Domke, C. Xue, A. Puschmann, T. Mandel, E. Hudson, D. A. Shirley, G. Kaindl, C. H. Greene, H. R. Sadeghpour, and H. Petersen, *Phys. Rev. Lett.* **66**, 1306 (1991).
81. D. W. Lindle, T. A. Ferrett, U. Becker, P. H. Kobrin, C. M. Truesdale, H. G. Kerkhoff, and D. A. Shirley, *Phys. Rev. A* **31**, 714 (1985).
82. S. Salomonson, S. L. Carter, and H. P. Kelly, *Phys. Rev. A* **39**, 5111 (1989).
83. P. van der Meulen, M. O. Krause, C. D. Caldwell, S. B. Whitfield, and C. A. de Lange, *J. Phys. B* **24**, L573 (1991).
84. P. van der Meulen, M. O. Krause, and C. A. de Lange, *Phys. Rev. A* **43**, 5997 (1991).
85. F. Robicheaux and C. H. Greene, *Phys. Rev. A* **47**, 1066 (1993).
86. P. M. Dehmer, J. Berkowitz, and W. A. Chupka, *J. Chem. Phys.* **60**, 2676 (1974).
87. S. J. Schaphorst, S. B. Whitfield, H. P. Saha, and C. D. Caldwell, *Phys. Rev. A* **47**, 3007 (1993).
88. A. Albiez, M. Thoma, W. Weber, and W. Mehlhorn, *Z. Phys. D* **16**, 97 (1990).
89. L. Asplund, P. Kelfve, B. Blomster, H. Siegbahn, and K. Siegbahn, *Phys. Scr.* **16**, 268 (1977).
90. D. J. Decman and W. Stoeffl, in *15th International Conference on X-Ray and Inner-Shell Processes*, Book of Abstracts, p. G04 (unpublished July, 1990); and private communication.
91. K. Siegbahn, in *Alpha, Beta and Gamma-Ray Spectroscopy*, edited by K. Siegbahn (North-Holland, Amsterdam, 1965), Vol. I, p. 79.
92. T. Åberg, *Phys. Scr.* **21**, 495 (1980).
93. H. Aksela, G. M. Bancroft, and B. Olsson, *Phys. Rev. A* **46**, 1345 (1992).
94. S. B. Whitfield, C. D. Caldwell, and M. O. Krause, *Phys. Rev. A* **43**, 2338 (1991).
95. S. B. Whitfield, J. Tulkki, and T. Åberg, *Phys. Rev. A* **44**, R6983 (1991).
96. H. Aksela, S. Aksela, H. Pulkkinen, G. M. Bancroft, and K. H. Tan, *Phys. Rev. A* **37**, 1798 (1988).
97. M. Meyer, E. von Raven, B. Sonntag, and J. E. Hansen, *Phys. Rev. A* **43**, 177 (1991).
98. M. Richter, M. Meyer, M. Pahler, T. Prescher, E. von Raven, B. Sonntag, and H.-E. Wetzel, *Phys. Rev. A* **40**, 7007 (1989).
99. U. Becker, H. G. Kerkhoff, D. W. Lindle, P. H. Kobrin, T. A. Ferrett, P. A. Heimann, C. M. Truesdale, and D. A. Shirley, *Phys. Rev. A* **34**, 2858 (1986).
100. L. Nahon, L. Duffy, P. Morin, F. Combet-Farnoux, J. Tremblay, and M. Larzilliere, *Phys. Rev. A* **41**, 4879 (1990).
101. L. Nahon and P. Morin, *Phys. Rev. A* **45**, 2887 (1992).
102. C. D. Caldwell, in *X-Ray and Inner-Shell Processes*, edited by T. A. Carlson, M. O. Krause, and S. T. Manson, American Institute of Physics Proceeding No. 215, p. 685 (1990).
103. T. A. Carlson, D. R. Mullins, C. E. Beall, B. W. Yates, J. W. Taylor, D. W. Lindle, and F. A. Grimm, *Phys. Rev. A* **39**, 1170 (1989).
104. T. Hayaishi, A. Yagishita, E. Murakami, E. Shigemasa, Y. Morioka, and T. Sasaki, *J. Phys. B* **23**, 1633 (1990).
105. E. von Raven, M. Meyer, M. Pahler, and B. Sonntag, *J. Electron Spectrosc. Relat. Phenom.* **52**, 677 (1990).
106. M. W. D. Mansfield and G. H. Newsom, *Proc. R. Soc. London Ser. A* **337**, 431 (1981).
107. J. Jiménez-Mier, C. D. Caldwell, M. G. Flemming, S. B. Whitfield, and P. van der Meulen, *Phys. Rev. A* **48**, 442 (1993).
108. A. Yagishita, S. Aksela, T. Prescher, M. Meyer, M. Richter, E. von Raven, and B. Sonntag, *J. Phys. B* **21**, 945 (1988).
109. M. Borst and V. Schmidt, *Phys. Rev. A* **33**, 4456 (1986).
110. V. Schmidt, S. Krummacher, F. Wuilleumier, and P. Dhez, *Phys. Rev. A* **24**, 1803 (1981).
111. A. Russek and W. Mehlhorn, *J. Phys. B* **19**, 911 (1986).
112. G. B. Armen, J. Tulkki, T. Åberg, and B. Crasemann, *Phys. Rev. A* **36**, 5606 (1987).
113. W. Sandner and M. Völkel, *Phys. Rev. Lett.* **62**, 885 (1989).
114. R. Huster, as quoted in Ref. 65.
115. M. O. Krause, T. A. Carlson, and W. E. Moddeman, *J. Phys. (Paris)* **32**, C4-139 (1971).
116. K. Starke, D. Gräf, and W. Hink, *J. Phys. B* **21**, 4217 (1988).
117. U. Becker, R. Wehlitz, O. Hemmers, B. Langer, and A. Menzel, *Phys. Rev. Lett.* **63**, 1054 (1989).
118. G. B. Armen and F. P. Larkins, *J. Phys. B* **24**, 741 (1991).
119. M. Pahler, C. D. Caldwell, S. J. Schaphorst, and M. O. Krause, *J. Phys. B* **26**, 1617 (1993).

CHAPTER 7

Valence Ionization Processes in the VUV Region

J. Berkowitz, E. Rühl, and H. Baumgärtel

1. INTRODUCTION

The vacuum-ultraviolet (VUV) region has a long-wavelength limit (~2000 Å) where air (particularly oxygen) begins to absorb radiation, but its lower limit is a bit fuzzy. Terms such as extreme ultraviolet or soft x-ray radiation are sometimes used to describe shorter-wavelength regions. We shall here employ a more pragmatic, though still cloudy limit, namely the domain of normal incidence VUV spectrometers (~500 Å) which is roughly the domain of laboratory continuum light sources (~600 Å) and of He I photoelectron spectroscopy (584 Å). Photoionization experiments in this region began in earnest about 30 years ago. These experiments progressed from measurements of absolute cross sections to photoionization mass spectrometry and photoelectron spectroscopy, and later to photoion–photoelectron coincidence and photoelectron angular distribution measurements.[1–4] They have been supplemented by fluorescence studies and even coincidence measurements involving fluorescence and ionization.[5,6] Various types of partial cross sections can be extracted from these measurements. At the present time, a large body of data has accumulated, primarily on atoms and molecules which are stable, and can be readily introduced as gases into a photoionization apparatus. Investigations of the more difficult species, which are chemically reactive, unstable, explosive, or involatile, have been performed primarily in the recent decade. Some of these molecules or radicals have gained recent interest because of their occurrence in the atmosphere or interstellar media. It is the purpose of this chapter to focus on these studies.

The techniques employed are rather standard. In the present chapter, they involve a continuum source of VUV radiation (laboratory discharge or synchrotron), a VUV monochromator, and some type of mass spectrometer. Some photoelectron spectroscopic experiments as well as threshold photoelectron–photoion coincidence (T-PEPICO) studies are also discussed. Recent studies have begun to utilize VUV lasers, which can now achieve $\lambda \geq 700$ Å,[7] but they are beyond the scope of the present review.

The feature of the present chapter that distinguishes it somewhat from other reviews of VUV photoionization concerns sample preparation. Since this differs from one species to another, and perhaps from one laboratory to another, it is difficult to offer a generic prescription. Rather, each section of this chapter describes methods appropriate to that particular study. In the first section, a selected class of small molecules will be discussed. These may be chemically reactive or explosive, but they purport to be a systematic study of an entire class, such as NH_nF_{3-n} (where $n = 0–3$), or the genre called

---

J. Berkowitz • Chemistry Division, Argonne National Laboratory, Argonne, Illinois 60439.   E. Rühl and H. Baumgärtel • Institut für Physikalische und Theoretische Chemie, Freie Universität Berlin, 14159 Berlin, Germany; *Current address of E. Rühl*: Institut für Physik, Johannes-Gutenberg-Universität Mainz 55099 Mainz, Germany.

VUV and Soft X-Ray Photoionization. Edited by Uwe Becker and David A. Shirley. Plenum Press, New York, 1996.

chlorofluoroethylenes. The second section also discusses small molecules, but in this case they have a fleeting existence because of their reactivity. They must be prepared *in situ*, and are referred to as transient species. The last section concerns itself with large molecules, beginning with benzene and its isomers, continuing on to naphthalene and other polycyclic aromatic hydrocarbons and culminating in buckminsterfullerene, C_{60}.

2. DETAILED CHARACTERIZATION (PHOTOABSORPTION, PHOTOIONIZATION, AND PHOTOELECTRON SPECTROSCOPY) OF SMALL MOLECULES

The spectroscopy and ionic fragmentation of small stable molecules in the VUV regime have been studied for many years. Mostly stable molecules have been investigated in the past.[1] However, complete series of chemically related molecules are often not available since some species may be chemically reactive or unstable. Therefore, chemically reactive and explosive molecules as well as stereoisomers of stable molecules (e.g., *cis–trans* isomers), and free radicals have been investigated more recently. In contrast to transient species which are discussed in Section 3 of this chapter the purity and concentration can be as high as for stable molecules.

The goal of the studies in Section 2 is to obtain a better understanding of chemical trends in VUV spectroscopy and ionic fragmentation if substituents are exchanged or rearranged. The species that will be discussed in this section are chlorine dioxide, fluoroamines, and some chlorofluoroethylenes.

2.1. ClO_2

Two isomers of ClO_2 are known to exist. The chemically more stable isomer is the symmetric isomer OClO [ΔH_f^{298} (OClO) = 24.49 kcal/mol; ΔH_f^{298} (ClOO) = 21.3 kcal/mol].[8] The asymmetric isomer, ClOO, is thermodynamically more stable but because of the weak Cl–O bond dissociation energy (4.76 ± 0.49 kcal/mol)[9] it is unstable with respect to immediate dissociation into molecular oxygen and atomic chlorine.[10] Attempts to isolate ClOO in the gas phase have failed thus far, so that our knowledge about the asymmetric isomer comes exclusively from the condensed phase. Early experiments have been reported, where matrix-isolated OClO was bleached by irradiation with 360-nm light.[11] The product was identified by infrared spectroscopy to be ClOO. OClO is reactive at high temperature, where molecular chlorine and oxygen are formed in an explosion.[10] Recent interest in spectroscopic and photochemical properties of chlorine dioxide in the UV regime is motivated by its occurrence in the polar stratosphere.[12] Both isomers are formed in the polar stratosphere by the reaction of ClO with BrO[13,14]:

$$ClO + BrO \rightarrow ClOO + Br$$

$$ClO + BrO \rightarrow OClO + Br$$

$$ClO + BrO \rightarrow BrCl + O_2$$

The branching ratio between these product channels favors the first two reactions.

In the laboratory, however, OClO is prepared conveniently by the reaction of molecular chlorine with sodium chlorite.[15] This quantitative reaction makes it unnecessary to store or purify the yellow gas.

Numerous studies have concentrated on the photoreactivity of OClO in the UV excitation regime since this property is of importance for stratospheric photochemistry.[12,13,16–23]

The electronic configuration of OClO can be described as follows:

$$\text{core}...(7a_1)^2 (2b_1)^2 (4b_2)^2 (5b_2)^2 (1a_2)^2 (8a_1)^2 (3b_1)^1 \quad ^2B_1$$

The lowest electronic transition is associated with the transition $\tilde{A}(^2A_2) \leftarrow \tilde{X}(^2B_1)$ which gives rise to a broad vibrational progression between 270 and 476 nm.[12] Higher excited electronic states in the VUV are identified as Rydberg transitions. The lowest Rydberg state of OClO is found at 6.78 eV with

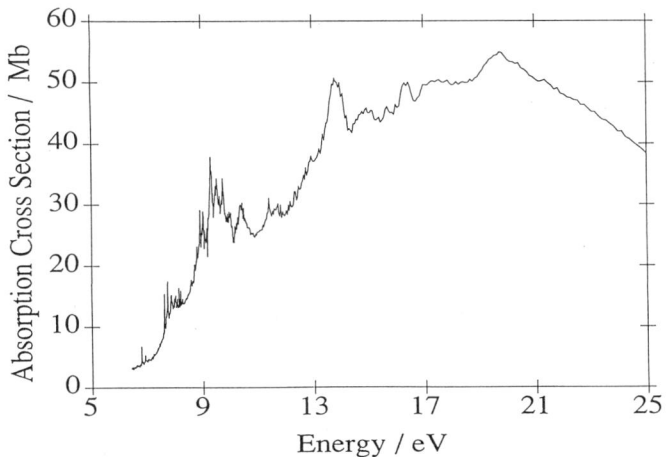

FIGURE 1. Vacuum-UV absorption spectrum of chlorine dioxide.

a short vibrational progression $\tilde{C}(^2A_1) \leftarrow \tilde{X}(^2B_1)$.[24] This progression has been assigned in earlier VUV absorption studies to the first member of an s-type Rydberg series. The following progression starts at 7.618 eV.[25]

Figure 1 shows the absorption cross section of OClO in the VUV.[26] Sharp features on an increasing continuum dominate the spectrum below 12 eV. These features are assigned to Rydberg series and their vibrational progressions converging to the first and higher ionization potentials. Above 12 eV broad features dominate the absorption cross section of OClO. Its maximum is found to be 55 Mb at 20 eV. This value is close to that of SO_2, which has a similar electronic structure as OClO, where the absorption cross section attains its maximum (56 Mb) at 18.5 eV.[27]

The convergence limits of Rydberg series are ionic states which may have singlet and triplet character.[28] The lowest ionization potential corresponds to the ejection from the singly occupied $3b_1$ orbital. Figure 2 shows the He I photoelectron spectrum (PES) of OClO. Assignments are given in Table I.

TABLE I
Ionization Potentials (IP) of OClO

Cation state	$IP_{adiabatic}$ (eV)	$IP_{vertical}$ (eV)
$^1A_1$	10.34	10.47
$^3B_1$	12.40	12.59
$^1B_1$	12.44	12.63
$^3B_2$	12.87	12.99
$^3A_2$	13.33	13.33
$^1A_2$	13.50	13.59
$^1B_2$	15.25	15.44
$^3A_1$	—	17.30
$^1B_1$	—	17.58
$^3A_2$	—	17.87
$^1A_2$	—	18.06
$^3A_1$	—	19.06
$^1A_1$	—	20.50

(a)

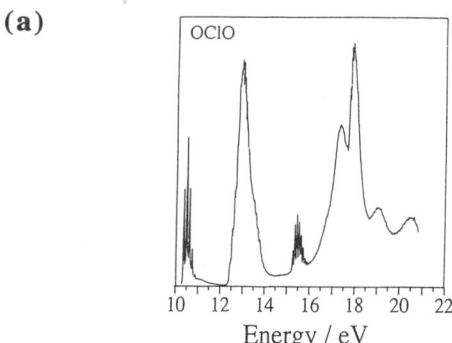

(b)

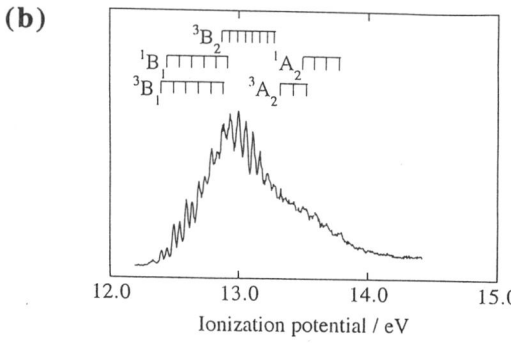

FIGURE 2. He I photoelectron spectrum of chlorine dioxide. (a) Overview spectrum; (b) detailed spectrum of the second photoelectron band.

Its shape is similar to PES published earlier,[28] but there are significant differences in the regime above 16 eV. The vibrational progression of the lowest PE band shows that the symmetrical stretch (v_1) is increased from 962.8 cm^{-1} in the neutral[29] to 1020 cm^{-1} in the cation. The bend (v_2) is also increased from 455.4 cm^{-1} to 520 cm^{-1}. This shows the slightly antibonding character of the highest occupied orbital ($3b_1$) in OClO. The following band systems in the PES (see Figure 2b) are the result of ionization from fully occupied orbitals leaving the cation with two half-filled MOs. SCF calculations indicate that the $8a_1$, $1a_2$, and $5b_2$ orbitals are almost of the same energy.[26] The corresponding electronic configurations with the singly occupied $3b_1$ orbital are related to singlet or triplet states which may differ considerably in energy. The splitting of singlet and triplet states depends on the overlap of the charge density of both half-filled MOs which determines the amount of self-repulsion. Therefore, singlet–triplet splitting can be estimated as twice the value of the exchange integral of both MOs. Consequently, a larger splitting is expected from orbitals with the same symmetry. Since the HOMO is of π-symmetry a large singlet–triplet splitting is expected for electronic cation states with other singly occupied orbitals of π-symmetry. This is the case for the $^3B_2$ (12.87 eV) and $^1B_2$ (15.25 eV) cation states which are split in energy by 2.38 eV. Other cation states, such as $^{1,3}A_2$ or $^{1,3}B_1$, consequently show a much smaller singlet–triplet splitting. These states are formed by ionization from the $5b_2$ and $8a_1$ orbital, respectively. The lowest portion of the second PE band is assigned to $^{1,3}B_1$ states of OClO$^+$ (at 12.400 and 12.440 eV). Both $^{1,3}A_2$ states at 13.33 and 13.50 eV also show a small energy shift. This assignment is also supported by a comparison with the PES of SO$_2$.[30,31] The $^{1,3}A_2$ states are in a regime where the vibrational fine structure is considerably broadened. This could be evidence for fast ionic fragmentation. The threshold of ClO$^+$ formation is consistently found at 13.40 ± 0.04 eV, as will be discussed below.

The threshold energy for OClO$^+$ formation is at 10.33 ± 0.02 eV (see Figure 3). The threshold regime of the OClO$^+$ photoion yield shows evidence for the dominant direct photoionization process: A staircase structure is observed as expected for the direct photoionization process (see Figure 3b), where the spacings are identical to those of the v_1 mode in the first PE band. The cation yield remains

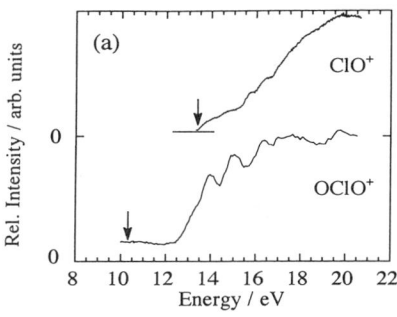

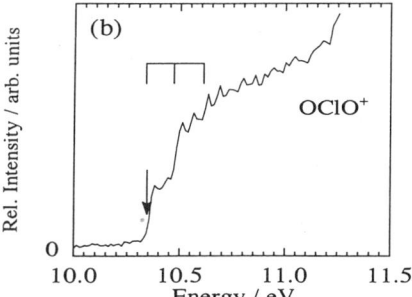

FIGURE 3. (a) Photoion yield curves of OClO$^+$ and ClO$^+$. The arrows indicate the onset energies. (b) Threshold region of the OClO$^+$ photoion yield. The spectrum is recorded with a lithium fluoride cutoff filter. The progression of vibrations corresponds to n_1 of OClO$^+$. The arrow indicates the ionization threshold.

weak until excited cation states are reached above 12.5 eV. Therefore, the threshold regime can only be investigated with a LiF cutoff filter, where second-order light is avoided. Further broad maxima in the parent cation yield spectrum are observed in the absorption cross section as well, and contribute via autoionization to the OClO$^+$ yield. It can be assumed that both fragments are produced at threshold in their electronic ground states: ClO$^+$ ($^3\Sigma^-$) + O (^3P). The following thermochemical considerations are corrected to $T = 0$ K. The fragmentation threshold of OClO is corrected by thermal contributions of vibration and rotation to 0 K: AP(ClO$^+$) = 13.45 ± 0.04 eV (310.1 ± 1 kcal/mol). The heat of formation of ClO is obtained from the atomic values [$\Delta H_f^{0\,K}$ (O) and $\Delta H_f^{0\,K}$ (Cl)][8] and the bond dissociation energy ($D_0^0 = 2.7505$ eV).[32] This gives $\Delta H_f^{0\,K}$ (ClO) = 24.21 kcal/mol. The heat of formation of the cation is calculated with the adiabatic ionization potential (10.87 eV[33]): $\Delta H_f^{0\,K}$ (ClO$^+$) = 274.8 kcal/mol. This value is lower than the currently accepted reference value of 277 kcal/mol.[34] The heat of formation of neutral OClO can be calculated according to

$$\Delta H_f^{0\,K}(\text{OClO}) = \Delta H_f^{0\,K}(\text{ClO}^+) + \Delta H_f^{0\,K}(\text{O}) - \text{AP}(\text{ClO}^+)$$

We obtain $\Delta H_f^{0\,K}$ (OClO) = 23.7 ± 1 kcal/mol and finally $\Delta H_f^{298\,K}$ (OClO) = 23.1 ± 1 kcal/mol by the use of the corresponding reference data.[8,34] This value is in excellent agreement with the reference values: $\Delta H_f^{298\,K}$ (OClO) = 23 ± 2 kcal/mol,[34] and $\Delta H_f^{298\,K}$ (OClO) = 24.49 kcal/mol.[8]

The thermochemical thresholds for other possible fragmentations are:

$$\text{OClO} \rightarrow \text{Cl} + \text{O}_2^+: 12.197 \text{ eV}$$

$$\text{OClO} \rightarrow \text{Cl}^+ + \text{O}_2: 13.093 \text{ eV}$$

These channels were observed in a very recent experiment, where a highly purified sample was attached directly to the gas inlet of the mass spectrometer.[35] At the same time as the vapor was leaking into the spectrometer it was simultaneously pumped into a liquid-nitrogen cold trap backed by a rotary pump, to remove traces of O$_2$. A weak onset of both fragmentations is observed at 13.3 eV. The low probability of these channels is very likely related to the fact that a rearrangement is necessary, since OClO and OClO$^+$ have angles near 120°. A union of both O atoms would require a transition state very different from the cation geometry formed by direct ionization. Considerably decreased bond angles of 92.7–94.2° have been found for electronically excited neutral OClO ($^2B_2$).[21,36] Recent results from *ab initio* calculations on OClO$^+$ suggest that the $^{1,3}A_2$ cation states show a decreased angle of 98°, which may promote isomerization of the cation.[37]

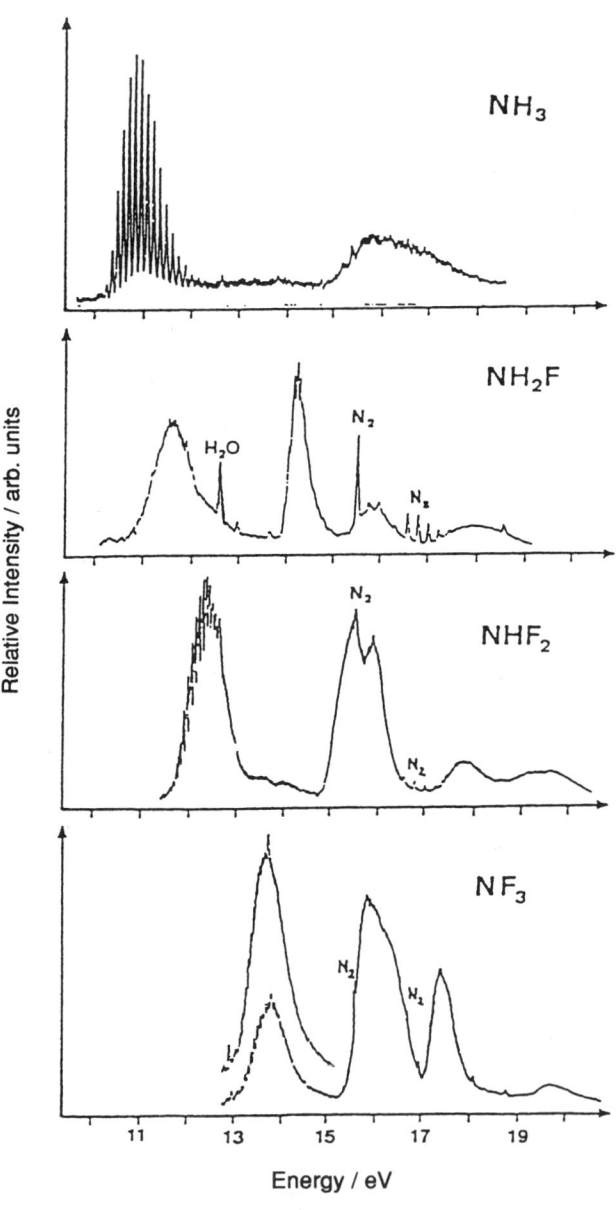

FIGURE 4. He I photoelectron spectra of ammonia and fluoroamines.

2.2. Ammonia and Some Fluoroamines

Both ammonia (NH_3) and trifluoroamine (NF_3) are thermally stable molecules. Therefore, several photoionization studies of these molecules have been performed in the past.[38–42] The other fluoroamines NH_2F and NHF_2 are extremely reactive and can be handled only at low temperature or at high dilution in the gas phase. Difluoroamine detonates even at low temperature by exothermic formation of N_2 + HF.[43,44] NH_2F has been prepared recently.[45] Photoionization of the entire series of fluoroamines has been reported by Baumgärtel et al.[46]

It is useful to estimate the total numbers of ionic states that can be accessed with He-I radiation (21.22 eV) by using a "rule of thumb" for fluoroamines $NH_{3-n}F_n$: $(3+3-n+5n)/2 = 2n+3$.[47] From this one expects 3 He-I ionizations for ammonia, and 5, 7, and 9 ionizations for NH_2F, NHF_2, and NF_3, respectively. The overview PES of the fluoroamines are shown in Figure 4. The lowest ionic states correspond to ionizations from the nitrogen lone pair (10–14 eV), whereas the higher ionic states are assigned to fluorine lone pairs, as well as two ionizations from σ_{NH} and σ_{NF} bonds. All ionization energies shift to higher energy with increasing number of fluorine substituents because of the higher effective nuclear charge of fluorine compared to hydrogen. This is consistent with the well-known perfluoro effect.[48,49]

The assignment of the radical cation states is done by the use of MNDO calculations and by comparison with iso(valence)electronic molecules (see Table II and Figure 5). The MNDO eigenvalues ε_J^{MNDO} are adapted according to the following regression to the vertical ionization energies IE_n^v:

$$IE_n^v = 0.578 + 0.937\,(\varepsilon_J^{MNDO})$$

This gives an acceptable correlation coefficient ($corr^2$) = 0.986, as well as reasonable results for the iso(valence)electronic fluoromethanes, where the "united atom" approach CH ↔ N is used.[46] This

TABLE II
Vertical Ionization Potentials (IP_{vert}) of Fluoroamines in Comparison with Results from MNDO Calculations

Molecule	IP_{vert} (eV)	$-\varepsilon_J^{MNDO}$ (eV)	Orbital
NH_3	10.82	11.2	$2a_1$
	15.8	16.7	$1e$
	16.5		
NH_2F	11.62	11.7	$5a'$
	14.25	14.4	$2a''$
	15.9	16.7	$4a'$
	18.1	19.1	$1a''$
		19.3	$3a'$
NHF_2	12.38	12.6	$6a'$
	15.6	15.7	$5a'$
		16.1	$4a''$
	16.0	16.5	$3a''$
	18.0	18.4	$4a'$
	19.3	20.6	$2a''$
	19.8	20.9	$3a'$
NF_3	13.83	13.9	$4a_1$
	15.9	16.8	$1a_2$
	16.3	17.0	$4e$
	17.4	17.4	$3e$
	19.8	20.0	$3a_1$
	21.1	21.9	$2e$

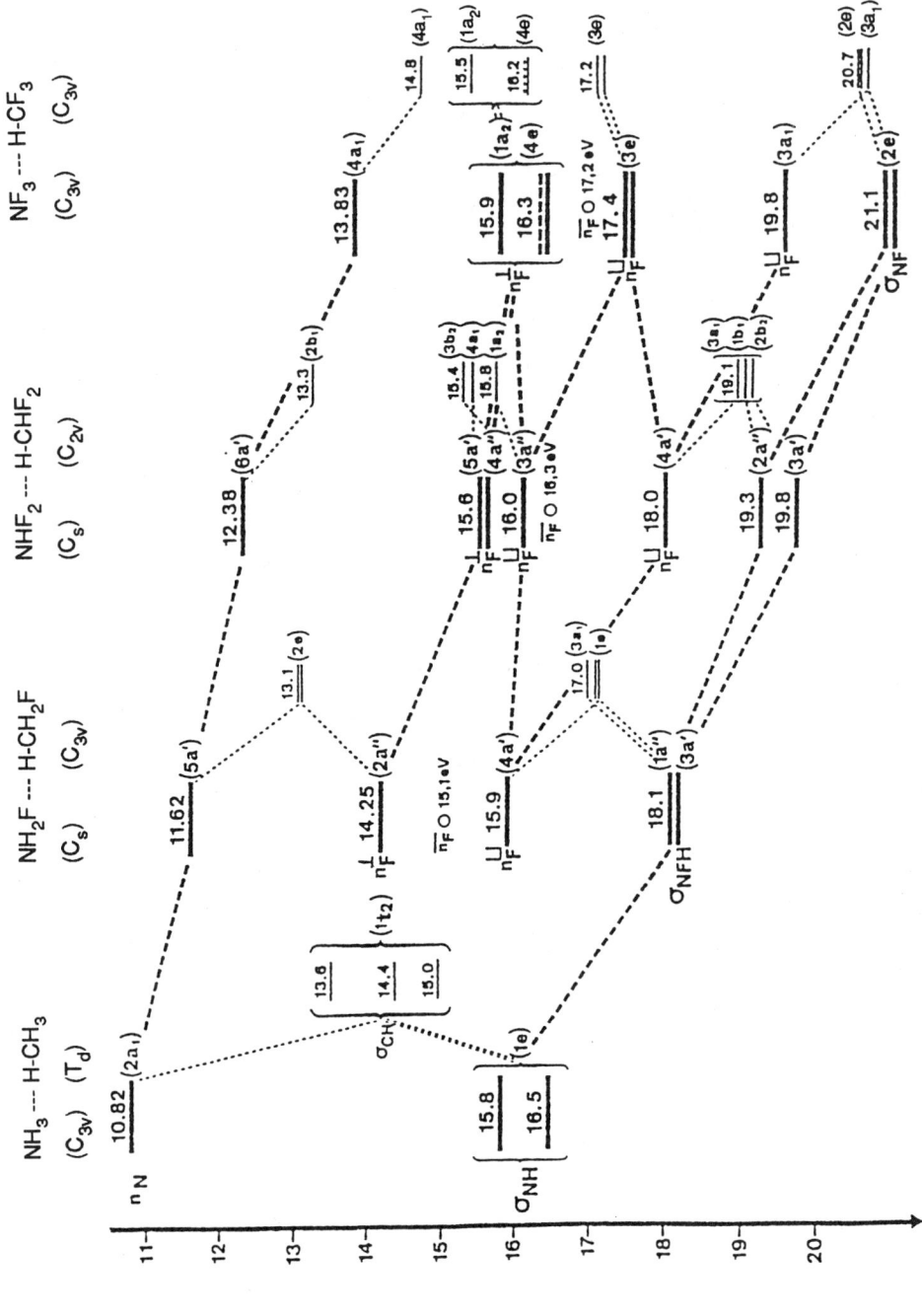

FIGURE 5. Correlation diagram of fluoroamines and fluoromethanes.

is necessary since it is known that the assignment of PES via Koopmans correlation $IE_n^v = -(\varepsilon_i^{SCF})$ often breaks down for small molecules containing several atoms of high effective nuclear charge such as F. This is because electron correlation and relaxation of the individual cation states no longer cancel.[50]

Table II shows the assignments for the PE bands of all fluoroamines as well as the results from MNDO calculations. A correlation diagram (Figure 5) visualizes the changes in orbital energies as well as the results from the CH ↔ N approach. Therefore, the PES of ammonia is compared to that of methane. The symmetry changes from T_d to C_{3v}. The triply degenerate Jahn–Teller split cation states σ_{CH} ($1t_2$) transform into the σ_{NH} ($1e$) states of NH_3^+.

In NH_2F both n_F lone-pair ionizations at 14.25 and 15.9 eV ($2a''$ and $4a'$) are inserted between the former n_N ($2a_1$) and σ_{NH} ($1e$) PE bands of NH_3. From the perturbation arguments ($CH_3F ↔ NH_2F$) the symmetry change from C_{3v} to C_s leads to $e \rightarrow a', a''$. Therefore, the n_N ($5a''$) and n_F ($2a''$) originate from the σ_{CH}($2e$) state of CH_3F^+. The removal of a proton from the eventual n_N lone-pair region leads to destabilization of the n_N ($5a'$) component and the n_F ($2a''$) is stabilized by the increase in effective nuclear charge from CH → N. A similar splitting is observed for the more strongly bound fluorine lone pairs ($1e$ in CH_3F and $4a'$ and $1a''$ in NH_2F). The signs ⊥ and ∥ shown in Figure 5 denote perpendicular or parallel orientation of the fluorine lone-pair lobes relative to the n_N ($5a'$) lobe direction.

Four fluorine lone-pair ionizations are expected for NHF_2, where unresolved and congested bands are found in the PES (see Figure 4). Two ionizations are observed at 15.6 eV, and two others are found at 16.0 and 18.0 eV, respectively. The effect of CH ↔ N perturbation shows that the n_F($4a'$) orbital is lower in energy than in CH_3F, whereas the other fluorine lone pairs ($4a''$ and $3a''$) show almost the same ionization energies as the equivalent orbitals in CH_2F_2.

The PES of NF_3 exhibits four PE bands (see Figure 4), where the second band contains probably three and the third two ionizations. This is deduced from MNDO calculations and by comparing the PES to that of the isovalent CHF_3.

Vibrational fine structure is observed in the first PE band for all fluoroamines including ammonia. In addition, the 14.25-eV ($2a''$) ionization of NH_2F also shows a vibrational progression. However, the vibrational spacings of these progressions differ considerably. This is rationalized by different geometries of the radical cations. Ammonia becomes planar in its cation ground state (D_{3h} symmetry), whereas NF_3^+ remains C_{3v} pyramidal. The cation ground states of NH_2F and NHF_2 are more or less planar according to SCF and MNDO calculations. Pyramidal molecules possess a double minimum potential in the inversion coordinate. With decreasing barrier height relative to the vibrational energy, the fine structure passes from clear separation at high barriers via a "chaotic" sequence of levels. Table III gives estimated values from MNDO calculations for inversion barriers of both neutral and ionic fluoroamines. The values for the neutral ground states are in agreement with results from ab initio SCF calculations,[51] as well as the experimental value for ammonia.[52]

The vibrational frequency of NF_3^+ is 560 cm^{-1} in its n_N ionization. For NHF_2^+ this frequency is 580 cm^{-1}. This has been extensively discussed earlier[53] and may be related either to a reduced NH bending mode (ν_2 = 1307 cm^{-1}) or to an increased NF_2 bending (ν_4 = 500 cm^{-1}). NH_2F^+ shows a complex vibrational structure with a more regular progression (ν^+ = 300 cm^{-1}) beyond the vertical ionization maximum. This would fit the expectation for a barrier penetration occurring at about 11.6 eV. However, MNDO calculations suggest a nearly planar structure of NH_2F^+, so that inversion doubling cannot be a plausible explanation for the complex vibrational structure.

The first PE band of NHF_2 shows a disappearance of vibrational structure above ν = 16 at 12.63 eV. However, the threshold of ionic fragmentation is found somewhat higher at 13.03 eV, excluding

TABLE III
Inversion Barriers from MNDO Calculations in kcal/mol of Neutral and Ionic Fluoroamines

	NH_3	NH_2F	NHF_2	NF_3
Neutral	11	18	29	120
Cation	0	0	1	49

TABLE IV

Mass Spectral Intensities at 20.65 eV Photon Energy and Appearance Potentials (AP) of Fluoroamines

Molecule	Cation	Rel. intensity (%)	AP (eV)
NF_3	NF_3^+	67	13.06
	NF_2^+	16	14.0
	NF^+	17	17.5
NHF_2	NHF_2^+	45	11.60
	NF_2^+	0.5	15.1
	NHF^+	43	14.2
	NF^+	10.5	13.03
	NH^+	1	
NH_2F	NH_2F^+	51.5	10.83
	NHF^+	1	15.2
	NF^+	1.5	14.85
	NH_2^+	44	14.10
	NH^+	2	14.50

any predissociation (see below). This difference in energy may be rationalized in terms of an ionic van der Waals complex (HF···NF$^+$) for which a binding energy of 9 kcal/mol is estimated and which would be analogous to carbene complexes like (H$_2$C···HF$^+$) formed by ionization of fluoromethane.[54] The absence of vibrational fine structure beyond the maximum of the first PE band can also be related to a barrier penetration, as discussed above for NH$_2$F$^+$.

Table IV shows the mass spectral intensities of the fluoroamine cations. Figure 6 displays the photoion yield curves of these cations and fragments.

The appearance potentials of NF$_3$ and its fragments agree with earlier work.[41,42] The onset of the parent cation occurs at the first ionization potential of the PES. The cation yield increases continuously to 14 eV, where the onset of cation fragmentation (NF$_2^+$ formation) is observed. A second onset of the fragment NF$_2^+$ at 15.5 eV is related to electronic predissociation of the first excited cation

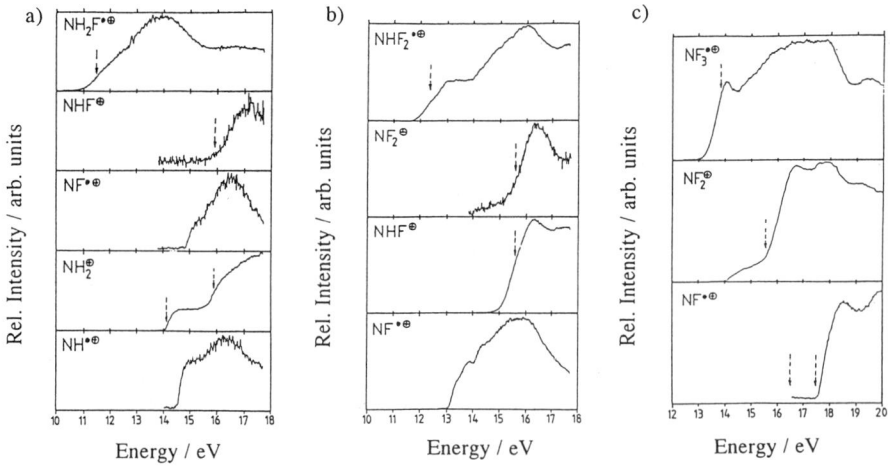

FIGURE 6. Photoion yield curves of fluoroamines. The dashed arrows indicate the energy positions of bands in the photoelectron spectra.

state of NF_3^+. The threshold of NF^+ formation is found at 17.5 eV. With the experimental material it cannot be distinguished between a simultaneous or sequential fragmentation mechanism that leads to NF^+.

Similar to the case of NF_3^+ a distinct onset of the parent cation yield NHF_2^+ is observed at the first ionization threshold. The parent cation yield remains constant above the threshold of NF^+ formation at 13 eV. A second onset of the NHF_2^+ yield is observed at 14 eV. This energy is located in the middle of the Franck–Condon gap where direct ionization does not occur, as can be seen from the PES (Figure 4). Resonant processes such as excitation into Rydberg orbitals ($3a'' \rightarrow 3p$) followed by autoionization may account for this onset.

Fragmentation of NH_2F^+ is observed above 14.1 eV, where the NH_2^+ fragment is formed. This fragment ion yield curve shows clearly two distinct onsets at 14.1 and 15.6 eV, respectively. This fragmentation is in competition with NH^+ formation (threshold at 14.5 eV), which results from predissociation of the second ionic state. At 15.6 eV the third ionization of NH_2F^+ gives another contribution via direct dissociation into the NH_2^+ channel.

Appearance potentials (AP) are used to derive thermochemical properties of fragment ions according to the approximation[46]:

$$AP = IP + D = \Delta H_f (\text{products}) - \Delta H_f (\text{neutral molecule})$$

where IP is the adiabatic ionization potential of the neutral molecule, D is the dissociation energy of the cation, and ΔH_f is the heat of formation at $T = 298$ K. Excess energies which may shift photofragmentation thresholds to higher energies are not considered in the equation. They can be estimated by comparison of the results from different fragmentation reactions leading to identical products. The physical nature of excess energies has been discussed extensively, as well as inherent problems with threshold determination from fragmentation processes in polyatomic molecules. It turns out that in the case of fluoroamines no significant differences in heats of formation occur for those cations whose heats of formation are known. Heats of formation of unknown species such as NHF^+ can be derived from two different routes:

$$NHF_2^+ \rightarrow NHF^+ + F \quad (AP = 14.2 \text{ eV})$$

$$NH_2F^+ \rightarrow NHF^+ + H \quad (AP = 15.2 \text{ eV})$$

From both reactions one obtains consistently $\Delta H_f(NHF^+) = 292 \pm 3$ kcal/mol, indicating that excess energies are negligible. This value is reasonable if it is compared to the heats of formation of NH_2^+ and NF_2^+, which are $\Delta H_f(NH_2^+) = 303.5$ kcal/mol and $\Delta H_f(NF_2^+) = 279.7$ kcal/mol, respectively.

Besides heats of formation of fragments one can also derive bond dissociation energies for both neutral and charged molecules. Table V shows these bond dissociation energies. The reference value for trifluoroamine[55] is in agreement with the value obtained in this work.

TABLE V
Bond Dissociation Energies of Neutral and Ionic Fluoroamines

Bond	D (neutral) (kcal/mol)	D (cation) (kcal/mol)
$F_2N–F$	61^a	27
FHN–F	60	75
$H_2N–F$	69	75
F_2NH	78	78
HFN–H	83	101
$H_2N–H$	109	129

^{a}58.2 kcal/mol is reported in Ref. 55.

The N–H bonds are stronger than the N–F bonds, which reflects predominantly the thermodynamic stability of H versus F as well as the decreasing stability of the neutral radicals: $NF_2 > NHF > NH_2$, as well as decreasing stability of the parent molecules, $NF_3 > NHF_2 > NH_2F > NH_3$.

2.3. Chlorofluoroethylenes

The spectroscopic and photochemical properties of some haloethylenes have been investigated in the past.[48,56,57] The entire series of all 27 chlorofluoroethylenes has been studied more recently by photoionization mass spectrometry and photoelectron spectroscopy.[58] This includes all possible combinations of hydrogen, fluorine, and chlorine substitutions including their stereoisomers. Some compounds are commercially available, whereas c-, t-$C_2H_2F_2$; C_2H_2ClF; 1,1-, c-, t-C_2HClF_2; 1,1-, c-, t-C_2HCl_2F; and 1,1-, c-, t-$C_2Cl_2F_2$ have to be prepared in the laboratory.

The emphasis of this work has been to obtain an understanding of fragmentation pathways of haloethylenes, as well as their thermochemical properties, when a complete data set of ionization and fragmentation thresholds is available.[58]

The cationic fragmentation of haloethylenes can be divided into nine different reactions:

$$C_2X_4 + h\nu \rightarrow C_2X_4^+ + e^- \quad (1)$$

$$\rightarrow C_2X_3^+ + X + e^- \quad (2)$$

$$\rightarrow X^+ + C_2X_3^+ + e^- \quad (3)$$

$$\rightarrow C_2X_2^+ + 2X\,(X_2) + e^- \quad (4)$$

$$\rightarrow X_2^+ + C_2X_2 + e^- \quad (5)$$

$$\rightarrow C_2X^+ + 3X\,(X + X_2) + e^- \quad (6)$$

$$\rightarrow CX_2^+ + CX_2 + e^- \quad (7)$$

$$\rightarrow CX_3^+ + CX + e^- \quad (8)$$

$$\rightarrow CX^+ + CX_3 + e^- \quad (9)$$

The reactions (3) and (5) are only observed for Cl^+, Cl_2^+, and HCl^+. The formation of CH^+ is observed above 21 eV excitation energy. These fragmentations are weak in intensity since their formation is not favored because the other moiety has typically a lower ionization potential, and carries therefore preferentially the positive charge, which is known as "Stevenson's rule."[59]

The fragmentations [reactions (2)–(9)] can be grouped together as

1. Formation of one substituent [reactions (2) and (3)]
2. Losses of more than one substituent [reactions (4)–(6)]
3. Formation of carbene cations [reaction (7)]
4. Fragmentation of rearranged cations (ethylidene cations) [reactions (8) and (9)]

The relative yield for the individual fragmentation routes depends strongly on the molecular system.

The VUV absorption spectra of some haloethylenes have been reported earlier in the energy regime below the first ionization potential.[48,60] The spectra show in this energy regime several Rydberg series converging to the first IP, as in the case of the above-mentioned chlorine dioxide. Broad absorption bands govern the regime from the first ionization threshold to 25 eV, which is typically the limit of the gratings used in normal incidence monochromators (see Figure 7). These features were tentatively assigned to transitions into Rydberg states with low principal quantum number. Evidence for this assignment comes from term value arguments and the similarity in shape of the absorption features to the corresponding photoelectron bands, which are the convergence limits of these Rydberg series. A firm assignment of these features has not been given. In particular, no other valence transitions than the well-known $\pi \rightarrow \pi^*$ valence transitions (centered at ~ 7 eV) have been assigned. It is therefore possible that these broad features may also have some contributions of valence transitions, such as $\sigma \rightarrow \sigma^*$. There are a few exceptions where Rydberg series converging to the second ionization potential are found. These are predominantly related to chlorine lone pair excitations (e.g., C_2H_3Cl, C_2H_2FCl, $C_2H_2Cl_2$, $C_2F_2Cl_2$, $1,1-C_2HClF_2$), where their lowest members have been assigned to so-called D-bands.[48]

We will discuss some results on ionic fragmentation of two dichlorofluoroethylene isomers in the following:

Photoion yield curves of cis- and trans-1-chloro-1,2-difluoroethylene have been measured.[58,61] The photoionization efficiency (PIE) curves of the parent and some intense fragments (losses of one substituent) are shown in Figure 8. The PIE curves of the parent cations differ considerably in shape, which reflects different autoionization processes, as observed in the VUV absorption cross section (see Figure 7). In contrast, the shapes of the corresponding fragments are often more similar since autoionization is much less pronounced than in the parent cation.

The appearance potentials and the relative intensities of the major fragmentation routes are compiled in Table VI. These thresholds differ slightly from each other in most fragmentation pathways.

TABLE VI
Relative Intensities (at 20.65 eV) and Appearance Potentials (AP) of Fragments from cis- and trans-1-Chloro-1,2-difluoroethylene

Cation	c-$C_2HClF_2^+$ relative intensity (%)	AP (eV)	t-$C_2HClF_2^+$ relative intensity (%)	AP (eV)
$C_2HClF_2^+$	100	9.85 ± 0.01	100	9.85 ± 0.01
$C_2ClF_2^+$	1.6	15.16 ± 0.08	0.9	15.48 ± 0.1
C_2HClF^+	19.6	14.81 ± 0.04	17.9	14.86 ± 0.05
$C_2HF_2^+$	20.5	14.16 ± 0.05	14.9	14.20 ± 0.04
C_2ClF^+	4.9	13.19 ± 0.02	2.6	13.27 ± 0.03
$C_2F_2^+$	7.3	13.37 ± 0.03	3.1	13.44 ± 0.04
C_2HCl^+	0.6	18.24 ± 0.1	0.8	18.60 ± 0.08
C_2HF^+	18.7	17.88 ± 0.08	11.8	17.65 ± 0.03
C_2Cl^+	0.4	17.74 ± 0.1	0.4	18.07 ± 0.1
C_2F^+	1.6	17.43 ± 0.05	1.0	17.53 ± 0.03
HCl^+	0.1	16.49 ± 0.1	0.1	16.50 ± 0.1
$CClF^+$	2.5	16.18 ± 0.08	1.8	16.10 ± 0.1
$CHCl^+$	15.5	13.73 ± 0.04	9.2	13.75 ± 0.05
CHF^+	5.7	15.90 ± 0.1	3.1	15.92 ± 0.1
$CClF_2^+$	0.4	13.41 ± 0.05	0.3	13.51 ± 0.04
$CHClF^+$	57.3	13.28 ± 0.02	43.8	13.38 ± 0.02
CHF_2^+	4.1	13.96 ± 0.02	1.8	14.04 ± 0.03
CCl^+	6.6	14.26 ± 0.05	4.0	14.38 ± 0.1
CF^+	25.4	14.42 ± 0.1	21.9	14.62 ± 0.1

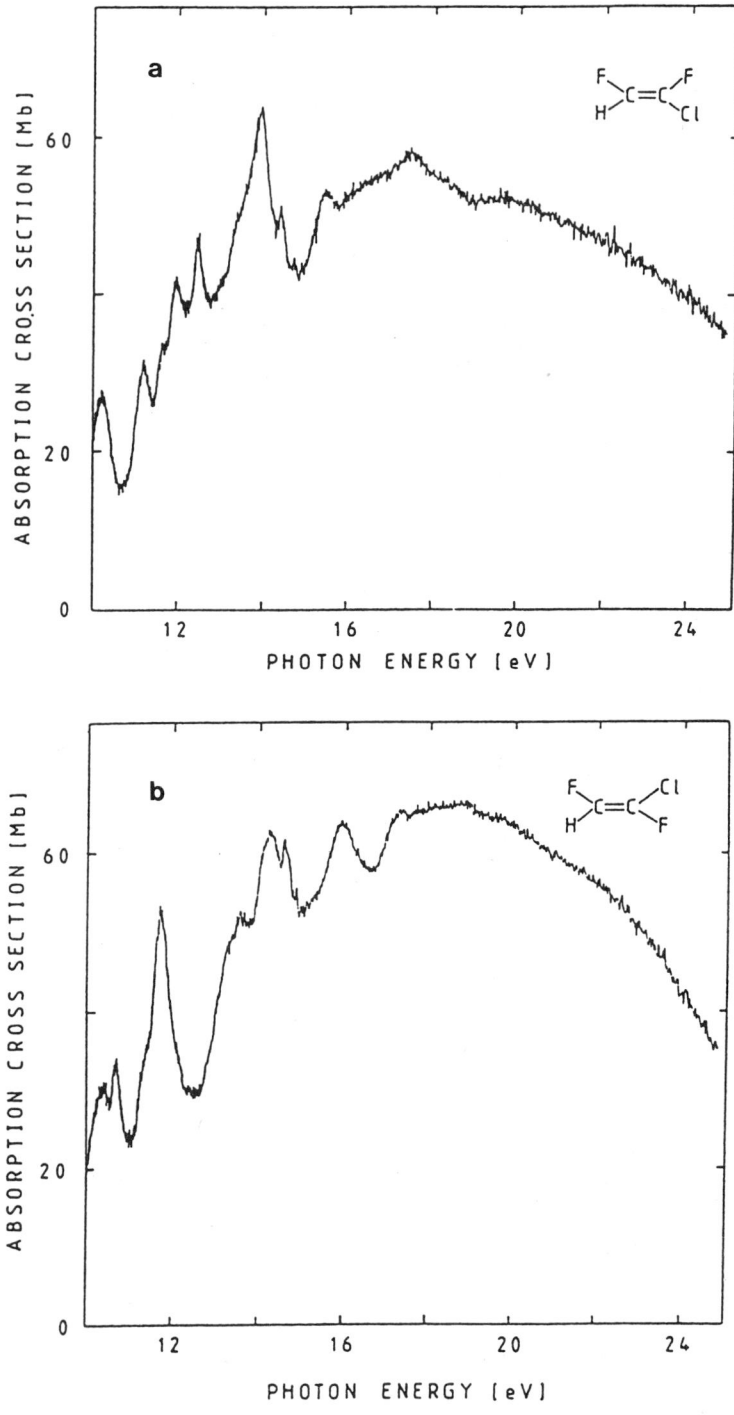

FIGURE 7. Vacuum-UV absorption spectra of *cis*- and *trans*-1-chloro-1,2-difluoroethylene.

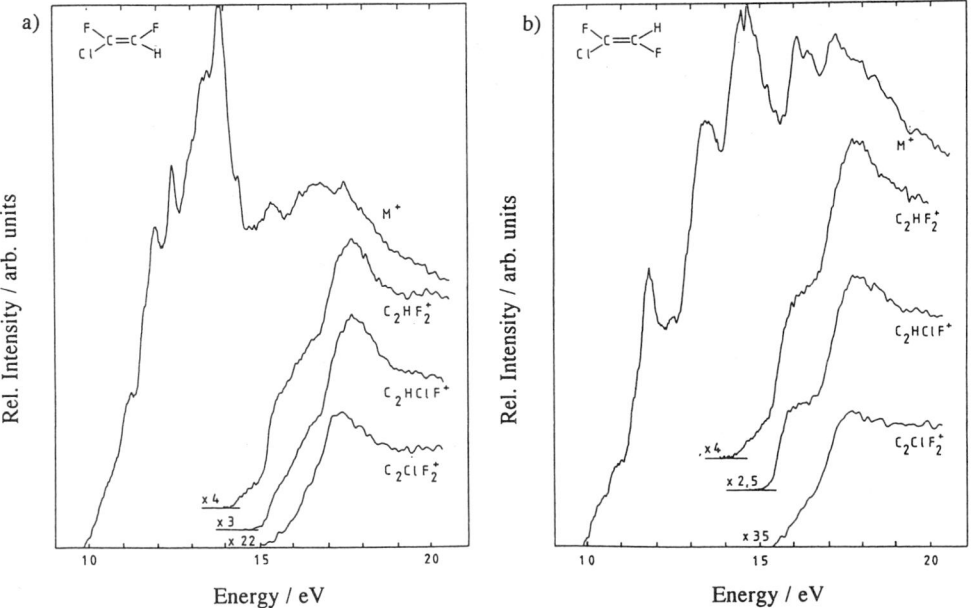

FIGURE 8. Photoion yield curves of *cis*- and *trans*-1-chloro-1,2-difluoroethylene and fragment ion yield curves from losses of one substituent.

It can be assumed that a common precursor or common reaction path with the same transition state may be the reason for this finding.

This behavior is investigated by semiempirical calculations on all stereoisomers of difluoroethylene.[62] Several reaction paths such as the fluorine and HF losses, as well as CF^+ and CH_2F^+ formation were studied. It turns out that all isomers have access to the same product channels since the barriers for isomerization of the ionic stereoisomers are relatively low. Therefore, the appearance potentials of these fragments are only different by the heat of formation of the neutrals. The fluorine loss from ground-state cations is expected to proceed via a deformed geminal isomer with a small reverse activation barrier. The HF loss leads via 1,2-elimination to a fluoroacetylene cation. The 1,1-elimination yields in a first step a fluorovinylidene cation and in a second step without an activation barrier to the fluoroacetylene cation. Both pathways have high activation barriers for the reverse reaction. Rearrangement processes lead to CF^+ and CH_2F^+. The key isomer is a fluoromethylcarbenium structure, which is found to be accessible from all isomers. This species is quite stable and decays via C–C bond rupture into CH_2F^+ with no reverse activation barrier. These detailed calculations have shown that a common precursor cation of octahedral shape which may also explain all ionic fragmentations[63] need not be postulated.

The difference in appearance potentials of major fragmentation pathways can be used to estimate the difference in heats of formation of both isomers. It turns out that the *cis* isomer shows systematically lower fragmentation thresholds than the *trans* isomer (see Table VI). The above-mentioned isomerization into an ethylidene cation seems to have no reverse activation barrier. This is consistent with calculations on difluoroethylene cations.[62] The difference in appearance potential is 100 meV, which corresponds to the difference in heat of formation. By using thermodynamical reference values for these fragmentation routes one obtains the heats of formation for both isomers:

$$c\text{-}C_2HClF_2: -79 \text{ kcal/mol}$$

$$t\text{-}C_2HClF_2: -81 \text{ kcal/mol}$$

TABLE VII
Heats of Formation (ΔH_f) of $C_2X_3^+$ Species in kcal/mol at 298° K

Cation	ΔH_f (cation)	Reference value[a]
$C_2H_3^+$	266 ± 1	265.9[b]
$C_2H_2F^+$	229 ± 1	227
$C_2HF_2^+$	222 ± 2	
$C_2F_3^+$	175 ± 3	189
$C_2H_2Cl^+$	250 ± 2	
$C_2HCl_2^+$	259 ± 3	
$C_2Cl_3^+$	261 ± 2	
C_2HClF^+	227 ± 3	
$C_2ClF_2^+$	208 ± 3	
$C_2Cl_2F^+$	228 ± 4	

[a] From Ref. 34.
[b] This value agrees with the most recent reference value: $\Delta H_{f0}^0(C_2H_3^+) = 265_{-2}^{+1}$ kcal/mol reported by M. Hawley and M. A. Smith, *J. Am. Chem. Soc.* **111**, 8293 (1989).

These values are in general agreement with increment approaches, which have been developed to predict heats of formation of unknown species. However, in some approaches no distinctions are made between the stereoisomers.

C_2HClF_2: −79.81 kcal/mol (1,1-, c-, t-C_2HClF_2)[64]

C_2HClF_2: −80.86 kcal/mol (1,1-C_2HClF_2); −76.65 kcal/mol (c-, t-C_2HClF_2)[65]

C_2HClF_2: −78.13 kcal/mol (1,1-C_2HClF_2); −84.73 kcal/mol (c-, t-C_2HClF_2)[66]

It is therefore advantageous to use fragmentation thresholds to estimate the difference in heats of formation of stereoisomers. The absolute values are obtained by using the proper thermochemical data for the products.[8,34,67]

Appearance potential spectroscopy has also been applied for the various fragmentation routes.[58,61,68] This extensive work cannot be discussed here in great detail. The advantage of this approach is the availability of the entire, self-consistent data set, which enabled the determination of thermochemical data such as heats of formation, and bond dissociation energies, by using many different reactions leading to identical products. Excess energies can be determined easily. The data were used for calculating increments, which verified the quality of the experimental data and allowed the determination of unknown heats of formation.[58]

Table VII shows as an example a compilation of heats of formation of all $C_2X_3^+$ species with X = H, Cl, or F. In this series of related molecules only a few reference values are reported in the literature. The data show that the heats of formation drop with increasing number of fluorine substituents. It is interesting that species with the same number of fluorine centers such as $C_2FX_2^+$ (X = H or Cl) show approximately the same heat of formation. In some cases it may be possible that the heats of formation are still too high, which may be caused by fragmentation thresholds that are affected by excess energies. One typical example is $\Delta H_f(C_2HF_2^+)$, where 222 ± 2 kcal/mol is derived from appearance potential measurements of most chlorofluoroethylenes. A comparison with $C_2ClF_2^+$ ($\Delta H_f = 208$ kcal/mol) shows that $\Delta H_f(C_2HF_2^+)$ should be found in the same energy regime.

3. PHOTOIONIZATION STUDIES OF TRANSIENT SPECIES

In the application of PIMS to the study of transient species, some modifications are required to the conventional photoionization apparatus, and one must recognize certain limitations. Usually, the conversion of the precursor to the desired transient species is not complete. Consequently, dissociative ionization of the precursor could lead to the same mass of photoion as direct formation of a parent ion from the transient species. This circumstance often limits the energy range for photoionization of the transient species to several electron volts above its threshold.

Typically, the number density of the target is approximately two or more orders of magnitude smaller than that of a permanent gas, resulting in lower counting rates. This may require compromising the photon energy resolution. Free radicals are usually chemically reactive on surfaces, and sometimes with other gaseous species such as their precursors. Unlike stable species, which can be introduced into a relatively tight (enclosed) ionization chamber with apertures for incoming and outgoing photon beam and exiting photoions and photoelectrons, free radical studies are best conducted with a *molecular beam* of free radicals crossed by the photon beam. The interaction should occur in a region of well-defined electrical potential, so that the photoions and photoelectrons can be efficiently extracted and focused. This can be achieved by making large openings in a conventional enclosed chamber, so that the beam of free radicals passes through the chamber and enters the throat of a pumping system.

Various methods have been developed for generating transient species *in situ* (see Table VIII). Pyrolysis of a suitable precursor is perhaps the simplest, but it is limited by the available precursors, and the resultant species may have a rather high Boltzmann temperature, unless subsequent cooling (e.g., supersonic expansion) is used. For example, N_2F_4 and P_2F_4, which are gases, can be pyrolyzed to yield NF_2 and PF_2.[55] To generate BiF_2 and BiF, one can heat a mixture of solid Bi and BiF_3 in a crucible to a temperature of ~ 700–1000 K.[69] We refer to this method as sublimation or vaporization. A microwave discharge in a favorable gas has been found useful in some cases, but the results are sometimes unpredictable. For example, a discharge in PF_3 was employed as a possible source of PF_2 and PF, but the only well-defined species that could be observed was P_2.[70] Laser-induced photolysis of a suitable precursor would appear to be an ideal source, but it has only recently begun to be exploited.[71] Most precursor molecules have large photoabsorption and photodissociation cross sections at wavelengths $\lambda < 2000$ Å. An intense laser source is required to obtain large fractional conversion. The ArF excimer laser ($\lambda \approx 1935$ Å) has a convenient wavelength and is intense, but it is pulsed. Its poor duty cycle is not well matched to a cw VUV source such as a laboratory discharge or a (quasi-cw) synchrotron light source. The chemical reaction method has been found to have broad application. Typical reactions are hydrogen abstractions, using atomic hydrogen or atomic fluorine as reagents. The atomic species are generated in a microwave discharge. An example of such an apparatus is shown in Figure 9. Often, several stages of reaction can be observed, enabling one to measure more than one transient species. For example, H atom reactions with PH_3 yield PH_2, PH, and even atomic phosphorus in sufficient abundance for photoionization experiments.[72,73] The reaction of F atoms with SiH_4 has been shown[74] to generate SiH_3, two electronic states of SiH_2 and SiH. Mass analysis, if not essential, is extremely helpful in such studies (compared to non-mass-selective PES). Needless to say, in all of these experiments with relatively low counting rates, it is very important to minimize backgrounds. Also, whenever transient species make collisions with surfaces, some coatings must be used to minimize loss by recombination.

TABLE VIII
Transient Species Studied by PIMS, and Their Methods of Preparation

A.	Sublimation, vaporization (I, Te, As_2, As_4, Sb_2, Sb_4, Bi, Bi_2, Bi_3, Bi_4, BiF, BiF_2, C_{60})
B.	Pyrolysis (NF_2, PF_2, BH_3, C_2H_3, CH_3, PH_2, CH_2S, N_2H_2)
C.	Microwave discharge (H, O, N, F, Cl, Br, P_2, SO, N_2H_2)
D.	Chemical abstraction reactions involving H and F atoms [S, Se, P, As, OH, SH, SeH, PH, PH_2, AsH, AsH_2, NH_2, SiH_3, SiH_2 , ($^3B_1$ and $^1A_1$), SiH, GeH_3, GeH_2, C_2H_3, C_2H_5, COOH, B_2H_5, B_2H_4, Si_2H_5, Si_2H_4, CH_2OH, CD_3O, CH_2SH, CH_3S, CH_2S, HCS]
E.	Laser photodissociation (CH_3S)

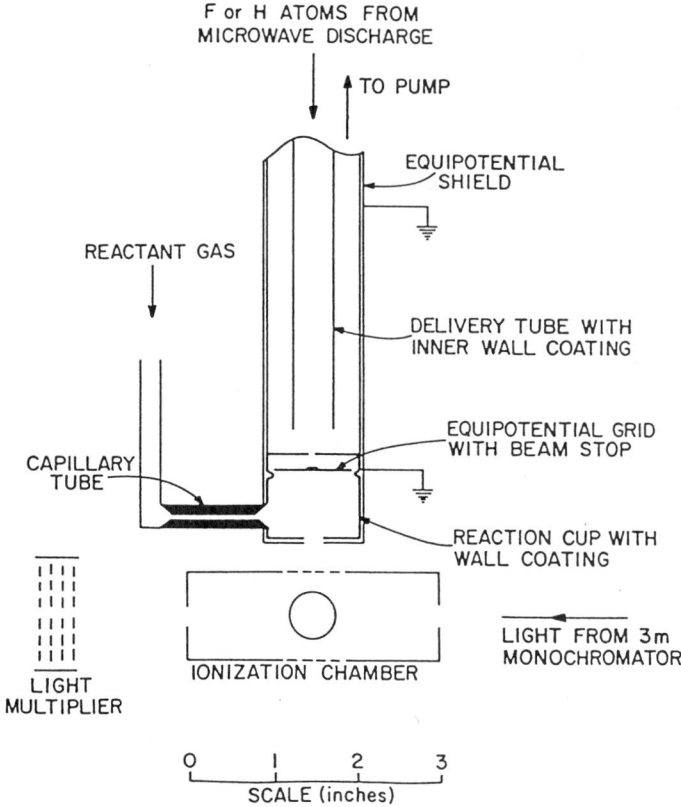

FIGURE 9. Schematic diagram of chemical reaction apparatus for generating transient species.

3.1. Open-Shell Atoms

The electron spectrometry of atoms is treated in detail by Krause and Caldwell in Chapter 6. In most of these studies to date, the atoms either are permanent gases (the noble gases) or are normally solids, and are generated as atomic vapors by sublimation or vaporization. Of the latter, some are closed shell systems (Cd, Ag, Pb, Yb) while others are open-shell systems (Mn, Cr). Recent work has involved the open-shell atoms O, Cl, and Br, prepared by microwave discharge through the corresponding diatomic molecules. We shall restrict our remarks here to valence shell photoionization studies of open-shell atoms prepared by the chemical reaction method.

Sublimation (vaporization) of elemental P, As, Sb, Se, Te produces complex molecules in the vapor phase—tetramers for P, As, and Sb, still more complex molecular species for S and Se, Te_2 and Te_5 for Te.[75] Either massive superheating with double oven techniques or reducing the activity of the element by suitable choices of precursors would be necessary to generate free atoms. The latter approach was successful for Te, using Cu_2Te.[75] A more general approach has been the use of the chemical reaction, hydrogen abstraction method. Thus, the reaction of atomic hydrogen with PH_3, AsH_3, H_2S, and H_2Se has produced a sufficient abundance of P, As, S, and Se for photoionization studies. The results of these studies, together with experiments on N, O, F, Cl, and Br produced by electric discharge, and atomic iodine by sublimation of AgI[75] provide an almost complete list of atomic species in the main groups V–VIII of the periodic chart. With this array of data, it is possible to elicit some generalizations.

For the interested reader, there exists a rather detailed analysis.[76] The atoms of Groups V–VIII can be described as having the generic valence shell configuration $...(ns)^2(np)^k$, where $n = 2$–4 and $k =$

3–6. Photoexcitation from the outer valence p orbitals is electric dipole allowed to ms and md Rydberg states. Some of these Rydberg levels lie above the lowest state of the ion. For example, in the noble gases (Ng) the lowest state of the ion Ng^+ is $^2P_{3/2}$, the next higher one $^2P_{1/2}$. Rydberg levels converging to $^2P_{1/2}$, and lying above $^2P_{3/2}$, are observed to autoionize to the ground state. The open-shell halogen cations with the configuration $...(ns)^2(np)^4$ can form $^3P_{2,1,0}$, $^1D_2$, and $^1S_0$ states. Rydberg levels lying higher than the ground state ($^3P_2$) are subject to autoionization. Similar considerations apply to the Group VI atoms.

It is observed that the ms and md quasidiscrete states (those lying above the lowest ionization limit) all give rise to sharp autoionization peaks in the first-row atoms (O, F, Ne). For the heavier members of Groups VI–VIII, the ms-like quasidiscrete states still manifest themselves as sharp autoionization peaks, but the md-like quasidiscrete states now display broad autoionization profiles. The interpretation that emerges from a detailed analysis[76] is as follows:

1. Autoionization of ms-like Rydberg states to available continua, when required to maintain parity, spin, and orbital angular momentum ℓ between quasidiscrete state and continuum, must undergo $\Delta\ell = +2$ transitions between Rydberg electron and continuum electron. The single electron operator governing these transitions $\langle ns | 1/r^3 | \varepsilon d\rangle$ is relatively weak. With weak configuration interaction between quasidiscrete state and continuum, the autoionization peak width will be sharp (narrow).

2. Autoionization of md-like Rydberg states to available continua, while maintaining the above selection rules, does not require a change in angular momentum between Rydberg electron and continuum electron, i.e., $\Delta\ell = 0$. The single electron operator is now of the form $\langle nd | 1/r | \varepsilon d\rangle$. This allows for a much stronger configuration interaction, and hence a broader resonance profile.

3. Among the first-row elements, even the md-like Rydberg states give rise to sharp autoionization peaks. This is not a consequence of selection rules, but is attributed to the inherently weak interaction between a "tight" ion core and the Rydberg electron. The tightness of a core can be described by its polarizability, which is significantly smaller for first-row ions than heavier ones. It has been shown[76] that the polarizability effect can reduce the peak widths of first-row atoms by a factor of ~20. The polarizability, in turn, can be related to a sum over transition probabilities. The most significant of these is the lowest p–d transition. In first-row elements, this transition must access the next higher principal quantum number, i.e., for Ne it is $...2p^6 \rightarrow ...2p^5\,3d$; for heavier members, this transition occurs within the same principal quantum number, i.e., for Ar it is $...3p^6 \rightarrow ...3p^5\,3d$. The intrashell transition typically has a higher oscillator strength, a lower excitation energy, and hence a larger contribution to the polarizability.

For the Group V elements, the above arguments are not applicable. The ground states of these atoms are $^4S_{3/2}$, and the configuration of the ion, $...ns^2np^2$, gives rise to the states ^3P, ^1D, and ^1S. Hence, as long as L–S coupling and spin conservation are maintained in the photoexcitation, ionization is permitted to $^3P + e^-$, but the combinations $(^1D)ms, md$ and $(^1S)ms, md$ do not lead to quartet states. Thus, the probability of forming doublet quasidiscrete states is very small, and hence they do not appear in photoabsorption or photoionization.

Photoexcitation of the inner valence shell ns electrons to mp Rydberg levels manifests itself in other ways. For the first-row elements N, O, Ne (F not yet studied), one observes a characteristic asymmetric peak shape in the autoionized Rydberg series. For the heavier elements in these groups, the corresponding autoionization resonances are more nearly windows, i.e., they manifest themselves as dips into the ionization continuum. In order to simplify the ensuing discussion, we refer to Fano's formula[77] for the peak shape (or profile index) resulting from the configuration interaction between a quasidiscrete state and a continuum

$$q = \frac{\langle \Phi | T | i \rangle}{\pi V_E^* \langle \psi_E | T | i \rangle}$$

Here, $\langle \Phi | T | i \rangle$ describes the transition moment from the initial state i to the quasidiscrete state Φ (T is a transition operator, which for the present purpose is the electric dipole operator), $\langle \psi_E | T | i \rangle$ describes the transition to a nearby ionization continuum, and V_E^* is the configuration interaction matrix element between quasidiscrete state and continuum referred to earlier, which governs the width of the resonance. A value of $q = 0$ corresponds to a window resonance, while $|q| > 2$ is the more typical situation of a peak surmounting the continuum. An intermediate value, $|q| \simeq 1-2$, describes an asymmetric resonance profile.

Low values of q can, in principle, be the result of a very small numerator or a very large denominator. The magnitude of the latter is limited to fully allowed transition probabilities to the continuum, and strong configuration interaction. Even under these conditions, q can have a value significantly larger than zero. Hence, more typically, a low q value can be related to a small numerator, i.e., a low transition probability to the quasidiscrete state.

This, in turn, implies a small oscillator strength (f) for that Rydberg transition. From the definition of q, all of the members of a Rydberg series will have about the same value of q. Consequently, the members of a Rydberg series characterized by a low q all have small oscillator strengths, and converge to a continuum with a small value of df/dE, proportional to the partial absorption cross section of that continuum. Detailed analysis[78] shows that the integrated oscillator strength for photoexcitation from the $(ns)^2$ subshell is significantly less than 2, the number of electrons in that shell, i.e., the partial sum rule is invalid. Some of the oscillator strength is being shifted into other channels.

Amusia and co-workers[79] have been able to reproduce the essential peak profiles of the asymmetric $2s \to 3p$, $4p$ resonances in Ne and the window resonances in Ar and Kr by the random phase approximation with exchange (RPAE) method. They show that the outer p shell has a strong shielding effect. By including the interaction of the $ns \to \varepsilon d$ transition, they show that the peak profile changes significantly and closely approaches the experimental shape. They also show that the intershell correlation can influence the partial oscillator strength, shifting some of the ns oscillator strength into other shells.

There are hints[78] that the shielding effect of the outer p electrons increases as the p shell fills (between Groups V and VIII). In addition, one may speculate that outer shielding may be less effective for first-row elements than heavier ones. The $2s$ and $2p$ wave functions are closer in energy than $3s$-$3p$, etc., and tend to occupy similar regions of configuration space. The lesser shielding might explain the somewhat higher values of q for first-row elements than for the heavier ones.

3.2. Group V and Group VI Hydrides

In 1978, Goddard and Harding[80] presented a semiempirical scheme for estimating the consecutive bond energies in the Pn–H$_n$ and Ch–H$_n$ systems (here, Pn = pnicogen = N, P, As, Sb; Ch = chalcogen = O, S, Se, Te). They assumed each bond pair to be covalent. The simple expectation might be that the three bond energies in the pnicogen hydrides (and the two in the chalcogen hydrides) would be comparable, and hence close to the average bond energy. If anything, one might anticipate a slight decrease in bond energies on successive hydrogenation, as a result of bond–bond repulsions. Goddard and Harding predicted that this was not the case, but rather that the dominant role was played by the stabilization related to p - p' exchange integrals. The stabilization related to these exchange interactions is greatest in Pn–H, least in PnH$_3$. This leads to D_0(Pn–H) < D_0(HPn–H) < D_0(H$_2$Pn–H), the difference in consecutive bond energies being equal to one-half the p - p' exchange integral. With two prescribed parameters (the experimental atomization energy, derived from calorimetry, and the magnitude of the p - p' exchange integral, derived from atomic spectra) they were able to predict bond energies "in most cases...probably more accurate than the current [1978] experimental data."

In recent years, it has become possible to test their predictions more stringently with the aid of photoionization studies on selected free radicals. These studies include NH$_2$, PH$_2$, PH, AsH$_2$, AsH, SH, and SeH by PIMS; NH$_2$, NH, PH$_2$, PH, SH, and OH by PES. In the PIMS experiments, the transient species were mostly generated in situ by reaction of hydrogen atoms with the various reagents. For example, AsH$_2$ and AsH were observed[81] in the H + AsH$_3$ reaction. The photoion yield curves of these species are shown in Figures 10 and 11. In most cases, there is good agreement between PIMS and PES in determining the adiabatic ionization potentials of these transient species. In one case (NH$_2$) there was significant disagreement, the PIMS value[82] for IP(NH$_2$) of 11.14 ± 0.01 eV being substantially lower than the PES value,[83] 11.46 eV. The discrepancy was attributed to an overlapping NH$_3$ band in the PES experiment, which necessitated a subtraction, and consequently an uncertain

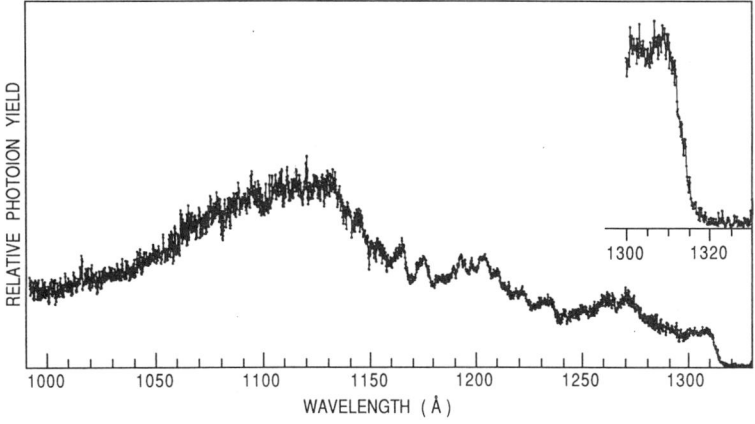

FIGURE 10. Photoion yield curve of AsH_2^+ (AsH_2). (From Ref. 14, with permission.)

residue near threshold. The adiabatic ionization potentials obtained for these transient species, together with appearance potentials of radical cations from the stable pnicogen trihydrides and chalcogen dihydrides, enable one to deduce consecutive bond energies, e.g.,

$$PnH_3 + h\nu \to PnH_2^+ + H + e \qquad AP$$

$$\underline{PnH_2 + h\nu \to PnH_2^+ + e \qquad IP}$$

$$PnH_3 \to PnH_2 + H \qquad AP - IP$$

Tables IX and X summarize these results for the pnicogen hydrides and chalcogen hydrides, respectively. Also included in these tables are the semiempirical predictions of Goddard and Harding, and the results of recent, high-quality *ab initio* calculations. It is clear from these tables that the qualitative conclusions of Goddard and Harding are correct. There is an *increase* in the bond energies of these hydrides on successive hydrogenation. There is even fairly good agreement quantitatively. The poorest

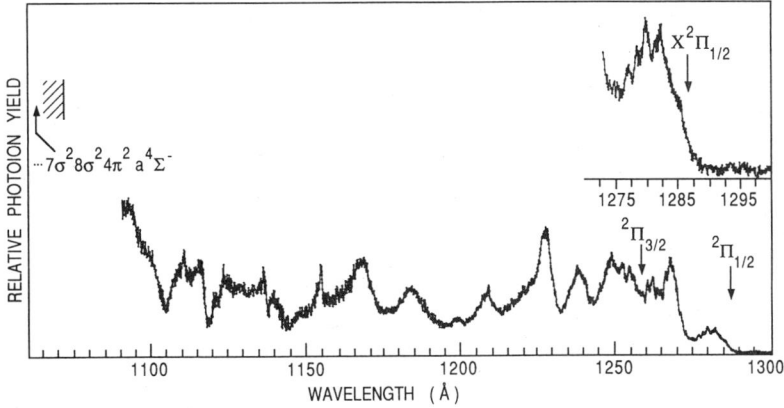

FIGURE 11. Photoion yield curve of AsH^+ (AsH). (From Ref. 14, with permission.)

TABLE IX
Comparison of Experimental Pn–H$_n$ Bond Energies (Pn = N, P, As) with Semiempirical Calculations and *Ab Initio* Calculations (in kcal/mol at 0 K)

	Ab initio[a]	Semiempirical[b]	PIMS[c]
N–H$_n$ compounds			
D_0(H$_2$N–H)	105.8	101.39	106.7 ± 0.3
D_0(HN–H)	91.2	92.23	91.0 ± 0.5
D_0(N–H)	77.2	83.07	79.0 ± 0.4
P–H$_n$ compounds			
D_0(H$_2$P–H)	82.4	81.11	82.5 ± 0.5
D_0(HP–H)	75.4	75.69	74.7 ± 0.5
D_0(P–H)	67.6	70.27	70.1 ± 0.5
As–H$_n$ compounds			
D_0(H$_2$As–H)	~73.0(74.6)[d]	74.89	74.9 ± 0.2
D_0(HAs–H)	~69.2(69.1)[d]	69.75	66.5 ± 0.2
D_0(As–H)	~62.3(62.4)[d]	64.60	64.6 ± 0.7

[a]N–H$_n$ from J. A. Pople, M. Head-Gordon, D. J. Fox, K. Raghavachari, and L. A. Curtiss, *J. Chem. Phys.* **90**, 5622 (1989); P–H$_n$ from J. A. Pople, B. T. Luke, M. J. Frisch, and J. S. Binkley, *J. Phys. Chem.* **89**, 2198 (1985); As–H$_n$ from R. C. Binning, Jr., and L. A. Curtiss, *J. Chem. Phys.* **92**, 1860 (1990).
[b]Ref. 80.
[c]N–H$_n$ from Ref. 15; P–H$_n$ from J. Berkowitz, L. A. Curtiss, S. T. Gibson, J. P. Greene, G. L. Hillhouse, and J. A. Pople, *J. Chem. Phys.* **84**, 375 (1986), slightly modified in J. Berkowitz and H. Cho, *J. Chem. Phys.* **90**, 1 (1989); As–H$_n$ from Ref. 81.
[d]Quantities in parentheses are described further by Binning and Curtiss (see footnote a).

agreement occurs for the first-row N–H$_n$ and O–H$_n$ bonds. One can show by electronegativity differences that these bonds have more ionic character than the heavier Pn–H$_n$ and Ch–H$_n$ bonds, and hence depart more significantly from the assumption of pure covalent character implicit in the Goddard–Harding treatment. The *ab initio* calculations agree with the experimental results to ± 2 kcal/mol in most cases.

TABLE X
Comparison of Experimental Ch–H$_n$ Bond Energies (Ch = O, S, Se) with Semiempirical Calculations and *Ab Initio* Calculations (in kcal/mol at 0 K)

	Ab initio[a]	Semiempirical[b]	Expt.[c]
O–H$_n$ compounds			
D_0(HO–H)	116.8	115.32	117.91 ± 0.29
D_0(O–H)	100.5	104.04	101.44 ± 0.29
S–H$_n$ compounds			
D_0(HS–H)	89.8	90.03	89.65 ± 0.69
D_0(S–H)	81.9	83.56	83.48 ± 0.69
Se–H$_n$ compounds			
D_0(HSe–H)	78.6	77.75	78.89 ± 0.18
D_0(Se–H)	72.7	71.58	74.27 ± 0.23

[a]O–H$_n$ from J. A. Pople, M. Head Gordon, D. J. Fox, K. Raghavachari, and L. A. Curtiss, *J. Chem. Phys.* **90**, 5622 (1989); S–H$_n$ from J. A. Pople, B. T. Luke, W. J. Frisch, and J. S. Binkley, *J. Phys. Chem.* **89**, 2198 (1985); Se–H$_n$ from R. C. Binning, Jr., and L. A. Curtiss, *J. Chem. Phys.* **92**, 1860 (1990).
[b]Ref. 80.
[c]O–H$_n$ is inferred from M. W. Chase, Jr., C. A. Davies, J. R. Downey, Jr., D. J. Frurip, R. A. McDonald, and A. N. Syverud, JANAF Thermochemical Tables, Third Edition, *J. Phys. Chem. Ref. Data* **14**, Suppl. 1 (1985); S–H$_n$ from R. E. Continetti, B. A. Balko, and Y. T. Lee, *Chem. Phys. Lett.* **182**, 400 (1991); Se–H$_n$ from S. T. Gibson, J. P. Greene, and J. Berkowitz, *J. Chem. Phys.* **85**, 4815 (1986).

The shapes of the photoion yield curves in PIMS and the breadth of the vibrational band envelopes in PES also provide some structural information. Thus, neutral NH_2 has a bond angle of 103.2°, while the calculated bond angles of NH_2^+ are 150.9° for $^3B_1$, 110.0° for $^1A_1$. Photoionization to $^1A_1$ can be expected to display a narrower Franck–Condon vibrational envelope than photoionization to $^3B_1$. It is observed that the excited state of NH_2^+ has the narrow Franck–Condon band, and hence the energy ordering in NH_2^+ is a $^3B_1$ ground state, and a $^1A_1$ excited state. This is the same ordering as in the isoelectronic CH_2 radical. By contrast, photoionization of PH_2 and AsH_2 displays a rather sharp onset (narrow Franck–Condon band), and hence the ordering here is $^1A_1$ ground state, $^3B_1$ excited state for PH_2^+ and AsH_2^+, as in the isoelectronic second- and third-row neutral species SiH_2 and GeH_2.

3.3. Group V Fluorides

The Group V fluorides present a contrast to the simple progression of bond energies encountered in the Group V and Group VI hydrides, where the influence of the exchange integrals appeared to dominate, and other factors (bond–bond repulsions and ionicity) were minor effects. For the nitrogen and phosphorus fluorides, the bond energies were evaluated from adiabatic ionization potentials of NF and PF measured by PES,[84,85] those of NF_2 and PF_2 determined by PIMS,[55] and appearance potentials of the cations also measured by PIMS.[55] The results for the nitrogen fluorides display a *diminution* of bond energy with increasing fluorination—$D_0(N–F) = 75.4 \pm 0.5$, $D_0(FN–F) = 65.7 \pm 0.5$, and $D_0(F_2N–F) = 57.0 \pm 0.2$, all in kcal/mol. This is the opposite direction from that described for the Group V hydrides, and presumably represents the increasing influence of bond–bond and F–F repulsions. However, this pattern does not persist for the phosphorus fluorides. Here, the strongest bond is the last one, $D_0(F_2P–F) = 131.7 \pm 0.5$ kcal/mol, while $D_0(FP–F) = 120 \pm 10$ kcal/mol and $D_0(P–F) = 106 \pm 10$ kcal/mol. The much larger bond energies in PF_n compared to NF_n, and the variation in the consecutive bond energies, have been attributed[55] to the relative importance of ionic bonding in the phosphorus–fluorine system. Information on two other members of this Group, AsF_n and SbF_n, is inconclusive and sparse at this time. Preliminary data on the BiF_n system[69] indicate that the last bond remains the strongest $[D_0(F_2Bi–F) \approx 104$ kcal/mol], with the middle bond $[D_0(FBi–F) \approx 81.5$ kcal/mol] weaker than the first bond $[D_0(BiF) \approx 87$ kcal/mol]. At the present time, no simple model has been proffered to rationalize these diverse observations.

3.4. Group IV Hydrides

3.4.1. SiH_n and GeH_n.
The heats of formation and the molecular structures of the $C–H_n$ species are rather well established.[70] The adiabatic ionization potential of CH_4 (12.615 eV)[86] corresponds to a transition from tetrahedral CH_4 to a Jahn–Teller distorted CH_4^+. The ionization potential of CH_3 (9.843 eV)[87] represents a transition from planar or near-planar CH_3 to planar CH_3^+. It is now generally accepted that the ionization potential of CH_2 (10.396 eV)[88] involves a $^3B_1$ ground state. The ionization potential of CH is 10.64 eV.[89]

For the corresponding $Si–H_n$ species, considerable controversy existed before a recent flurry of experimental and calculational investigations. Even less was known about the $Ge–H_n$ species.

The He I photoelectron spectra of SiH_4 and GeH_4 have been known for some time.[90] In both cases (as also in CH_4) the first band is broad, with some vibrational fine structure. It corresponds to electron emission from the uppermost occupied t_2 orbital, and leads to Jahn–Teller distortion in the ionic ground state. The "lowest detected I. P." reported by Potts and Price[90] was 11.60 eV for SiH_4, 11.34 eV for GeH_4. Although ionization of CH_4 creates a stable CH_4^+ readily seen in mass spectrometry, it was not clear from prior work whether SiH_4^+ and GeH_4^+ could be observed. At best they were weak ion signals, and the isotopic variants of Si and Ge confused the issue. Careful photoionization studies have now established that both SiH_4^+ and GeH_4^+ are observable, but their adiabatic ionization potentials are much lower than could be inferred from the photoelectron spectra—11.00 ± 0.02 eV for SiH_4,[74] ≤ 10.53 eV for GeH_4.[91] These low ionization potentials (relative to photoelectron spectra) and low abundances have implications for the geometric structures of SiH_4^+ and GeH_4^+, which have now been calculated.[92,93] Both are highly distorted from the tetrahedral structure of the neutral species, and are best described as $SiH_2^+ \cdot H_2$ and $GeH_2^+ \cdot H_2$. Thus, the Franck–Condon transitions are very unfavorable near threshold. When they do become favorable, the molecular ion dissociates.

Dyke et al.[94] prepared SiH_3 by the F + SiH_4 reaction, and obtained its He I photoelectron spectrum. A long vibrational progression was observed, and interpreted as excitation of the umbrella motion in the transition from pyramidal SiH_3 to planar or near-planar SiH_3^+, similar to the situation with NH_3. They extracted an adiabatic IP of 8.14 ± 0.01 eV after subtracting some contamination bands near threshold. Subsequently, Berkowitz et al.[74] examined SiH_3 (also prepared by the F + SiH_4 reaction) by PIMS. They observed a succession of round steps, whose derivative would look rather similar to the photoelectron spectrum of Dyke et al. This is the expected behavior if direct ionization (rather than autoionization) is the dominant process. They noted "a distinct rise above the background level at ~1527 Å ≡ 8.12 eV" but "an even smaller step (about 3–4 times smaller) which begins at ~1547 Å ≡ 8.01 eV." More recently, Johnson et al.[95] observed a Rydberg series in SiH_3 using multiphoton ionization, and obtained an extrapolated value of $8.135^{+0.005}_{-0.002}$ eV for the adiabatic IP of SiH_3. This is the preferred value. In hindsight, the "even smaller step" observed by Berkowitz et al. was very likely a hot band. Its magnitude is consistent with a Boltzmann vibrational population at a temperature of ~300 K.

A significant advantage of PIMS over PES is its ability to investigate species with inherently low abundance in the presence of large quantities of other species. Thus, in the F + SiH_4 experiments of Berkowitz et al.[74] they noted the presence of SiH_2^+, and even SiH^+, at photon energies much too low to be attributed to dissociative ionization. An analysis of the SiH_2^+ photoion yield curve revealed the presence of two components, attributed to photoionization of SiH_2 ($^3B_1$) and SiH_2 ($^1A_1$), the latter being the ground state. The difference in the adiabatic IP is the $^3B_1$–$^1A_1$ splitting in SiH_2, since the ionic ground state is the same for both. Berkowitz et al.[74] found IP(SiH_2, $^3B_1$) = $8.24_4 \pm 0.02_5$ eV, while IP(SiH_2, $^1A_1$) was ≤ 9.15 eV. This leads to a singlet–triplet splitting of 21.0 ± 0.7 kcal/mol. Several ab initio calculations[96–98] have now arrived at very nearly this value.

The photoion yield curve of SiH^+ (SiH) displays prominent autoionization structure.[74] The adiabatic IP, corresponding to formation of SiH^+ ($X^1\Sigma^+$), is found to be $7.91_2 \pm 0.01_0$ eV. The autoionization has been identified as Rydberg series whose convergence limit is SiH^+, $a^3\Pi$, with an ionization energy of 10.21 eV.

An analogous study of the F + GeH_4 system by PIMS was performed by Ruščić et al.[91] The GeH_3 species was observed to photoionize with steplike structure, similar to SiH_3, and indicative of a pyramidal to near-planar transition. The adiabatic IP inferred was ≤ $7.94_8 \pm 0.005$ eV. The GeH_2 species, produced in low abundance, was more difficult to study because of the multiplicity of germanium isotopes and some background related to atomic germanium. Only a crude value was obtained for this IP, ≤ 9.25 eV.

By combining the adiabatic ionization potentials of the species discussed above with appearance potentials of these ions from SiH_4 and GeH_4, it is possible to deduce values for the successive bond energies. For details, the reader is referred to the original papers.[74,91] It is instructive to compare these successive bond energies in the Si–H_n and Ge–H_n systems with the previously established C–H_n system. In Figure 12, the fractional bond energy is plotted versus the type of bond, i.e., M–H, HM–H, etc., where M = C, Si, Ge. A striking difference is seen between the bonding pattern in the C–H_n system, on the one hand, and the Si–H_n and Ge–H_n systems, on the other. The H_2M–H bond, which is the strongest bond in the C–H_n system, is the weakest one in both the Si–H_n and Ge–H_n systems. The latter two track one another closely. Thus, there is a characteristic difference between the bonding behavior of the first-row element (C) and the second- and third-row elements. In this instance, the difference in bonding pattern can be traced to the nature of the MH_2 ground state. In CH_2, the ground state is a triplet; an unpaired electron awaits bonding to the incoming H atom. With SiH_2 and GeH_2, the ground state is $^1A_1$; energy must be expended to uncouple the spins before bonding to an incoming H atom can occur.

3.4.2. C_2H_n and Si_2H_n. C_2H_6, C_2H_4, and C_2H_2 are kinetically stable gases, with well-known heats of formation. They represent the prototypical single, double, and triple C–C bonds. From well-established heats of formation of CH_3, CH_2, and CH (see above), we know that

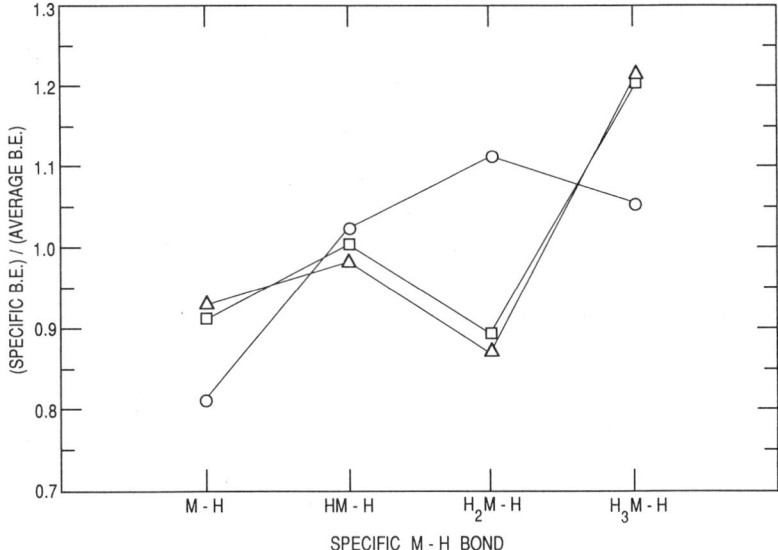

FIGURE 12. Fractional bond energy as a function of the specific M–H bond, where M = C (○), Si (□), and Ge (△). (From Ref. 91, with permission.)

$$C_2H_6 \rightarrow 2CH_3, \Delta H_0 = 87.6 \text{ kcal/mol}$$

$$C_2H_4 \rightarrow 2CH_2, \Delta H_0 = 171.9 \text{ kcal/mol}$$

$$C_2H_2 \rightarrow 2CH, \Delta H_0 = 229.5 \text{ kcal/mol}$$

These dissociation energies refer to ground-state products. For CH_3 ($^2A_2''$) and CH_2 ($^3B_1$), these products are also the "valence states." For CH, however, the ground state is $^2\Pi$, but the "valence state" is $^4\Sigma^-$, excited by 16.7 kcal/mol. With this correction, the dissociation energy of C_2H_2 into two $^4\Sigma^-$ states of CH is 262.9 kcal/mol. The ratio of single:double:triple bonds is 1:1.96:3.00. Hence, the π bonds in C are almost as strong as the σ bond. Rather similar results can be obtained by comparing the D_0 values of B_2, C_2, and N_2, where the corresponding ratios of bond energies are 1:2.057:3.23.

By contrast, in the disilicon hydride system, only Si_2H_6 is kinetically stable. Until 1991, all attempts to detect Si_2H_4 and Si_2H_2 spectroscopically had failed. There have been many *ab initio* calculations predicting structures and stabilities, but with no experimental verification. The transient Si_2H_n and $Si_2H_n^+$ species (n = 1–5) are believed to be important components of the plasma in chemical vapor deposition (CVD) methods for producing silicon-based devices.

Ruščić and Berkowitz[99] recently studied the F + Si_2H_6 system by PIMS. Not unexpectedly, the most abundant radicals observed were Si_2H_5 and Si_2H_4, for which photoion yield curves with good statistics were obtained. The adiabatic IP for Si_2H_5 was 7.60 ± 0.05 eV, and for Si_2H_4, 8.09 ± 0.03 eV. Both photoion yield curves increased gradually from threshold, suggesting a change in geometry between the neutral species and its cation. More surprisingly, some $Si_2H_2^+$ was observed at photon energies too low to be attributable to fragmentation. It is likely that the Si_2H_2 species generated was the result of a surface reaction, rather than successive hydrogen abstraction by F atoms. Unlike the photoion yield curves of Si_2H_5 and Si_2H_4, the $Si_2H_2^+$ (Si_2H_2) photoion yield curve displayed an abrupt onset (see Figure 13). This behavior, indicative of a favorable Franck–Condon factor near threshold, implies that Si_2H_2 and $Si_2H_2^+$ have very similar geometrical structures. *Ab initio* calculations[100,101] had shown that the most stable structures of Si_2H_2 and $Si_2H_2^+$ were the so-called "butterfly" structures (see Figure 14). Furthermore, the calculated ionization potential of Si_2H_2 (8.30 eV)[102] was quite close to

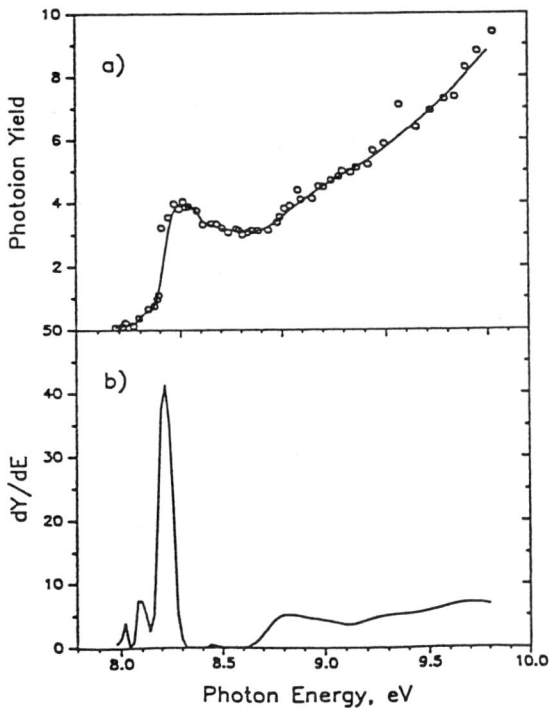

FIGURE 13. (a) Photoion yield curve of $Si_2H_2^+$ (Si_2H_2); (b) energy derivative of (a). (From Ref. 99, with permission.)

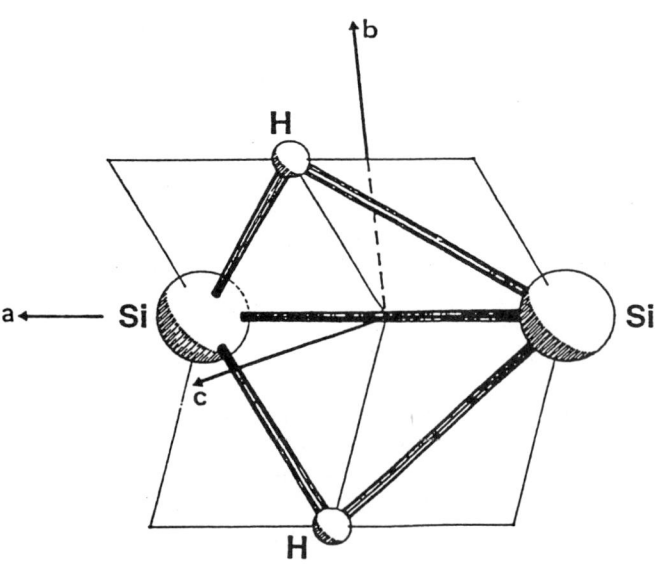

FIGURE 14. Geometrical structure of Si_2H_2. (From Ref. 103, with permission.)

the experimental value ($8.20^{+0.01}_{-0.02}$ eV). From this and other evidence, Ruščić and Berkowitz[99] concluded that they had generated and photoionized Si_2H_2 with this structure. At about this time, Bogey et al.[103] published their results on the submillimeter wave spectrum of Si_2H_2, from which they determined the structure shown in Figure 14.

The structure of Si_2H_4 has not yet been determined spectroscopically. However, the photoion yield curve and the adiabatic IP are consistent with *ab initio* calculations, i.e., C_{2h} symmetry, with the two SiH_2 units tilted from planarity, while $Si_2H_4^+$ has the planar ethylene structure.

The heats of formation of most of the Si_2H_n cations have been determined by appearance potential measurements.[104] These values, together with the measured adiabatic ionization potentials,[99] enable one to extract the heats of formation of the corresponding neutral Si_2H_n species. Various Si–H and Si–Si bond energies can be inferred from these quantities. We shall concentrate here on the Si–Si bond energies, and compare them with the aforementioned C–C bond energies.

A value for ΔH_f^{298} (Si_2H_6) = 17.1 ± 0.3 kcal/mol had previously been determined by a standard calorimetric technique.[105] At 0 K, ΔH_f (Si_2H_6) = 22.9 ± 0.3 kcal/mol. Recent values of ΔH_f (SiH_3) are ≤ 50.0 kcal/mol by PIMS[74] and 49.5 ± 1.2 by a chemical kinetics method.[106] Thus, the breaking of a Si–Si single bond, corresponding to the reaction

$$Si_2H_6 \rightarrow 2\ SiH_3$$

requires an expenditure of ≤ 77.1 (76.1±1.2) kcal/mol, somewhat smaller than the analogous C–C bond energy (87.6 kcal/mol). The value for ΔH_f° (Si_2H_4) obtained from PIMS is 67.9 ± 0.9 kcal/mol,[74] (while that for SiH_2 in its ground state ($^1A_1$) is 69.6 kcal/mol. Thus, for the reaction

$$Si_2H_4 \rightarrow 2\ SiH_2\ (^1A_1)$$

we obtain ΔH_0 = 63.3 ± 1.2 kcal/mol. Paradoxically, an apparent double bond energy appears to be weaker than the Si–Si single bond energy. This is related in part to the reversal in stability between the "valence" $^3B_1$ state and the $^1A_1$ state between methylene and silylene. By introducing the $^1A_1 \rightarrow\ ^3B_1$ excitation energy of SiH_2, we obtain for the dissociation reaction

$$Si_2H_4 \rightarrow 2\ SiH_2\ (^3B_1)$$

the value ΔH_0 = 105.3 kcal/mol. The nominal Si–Si double bond is now stronger than the single bond, and the reaction is analogous to the dissociation of ethylene into two CH_2 ($^3B_1$) moieties. However, in the latter case the double bond was indeed twice as strong as the simple C–C bond, whereas in the silicon case, the ratio of the nominal double bond to the single bond is ~1.37. This weakness of the double bond, attributed to the weakness of π-bonding, manifests itself in the nonplanar structure of Si_2H_4, and perhaps accounts for the prior difficulties in isolating this molecular species. It is exacerbated in the case of Si_2H_2. A structure analogous to acetylene involving two π-bonds has been calculated[107] to be much less stable than the "butterfly" structure, and may not even be a local minimum on the Si_2H_2 potential surface. Thus, the weakness of π-bonding in Si–Si compounds enables other structures to compete successfully. The calculated structure[102,108] of $Si_2H_3^+$, with a triple hydrogen bridge, provides another example for which some experimental support exists from PIMS.

Although experimental evidence does not exist for the corresponding Ge_2H_n, *ab initio* calculations[109] are beginning to show that these structures are analogous to the Si_2H_n structures. Thus, our familiar view of single, double, and triple bonds, based largely on extensive studies of carbon and other first-row compounds, may be the exception rather than the rule when due consideration is taken of the heavier congeners.

3.5. CH_2OH/CH_3O and CH_2SH/CH_3S

The two sets of isomeric transient species, CH_2OH/CH_3O and CH_2SH/CH_3S, provide interesting examples of

1. The relative merits of photoionization mass spectrometry and photoelectron spectroscopy, and
2. The advantages and disadvantages of several methods for generating certain transient species

Dyke et al.[110] initially obtained the He I photoelectron spectrum of CH_2OH. The CH_2OH species was generated by the reaction of F atoms with CH_3OH. In a later publication, Dyke[111] reported on the isotopic hydrogen variants. He obtained adiabatic ionization potentials of 7.56 ± 0.01 eV for CH_2OH and CD_2OD, and 7.55 ± 0.01 eV for CH_2OD and CD_2OH. It was known from kinetics experiments that both CH_2OH and CH_3O are formed in the $F + CH_3OH$ reaction, in roughly comparable amounts. If these species have similar ionization potentials, they would be difficult to distinguish in a PES experiment. Dyke therefore chose the pyrolysis of CH_3O-OCH_3 as a potential source of "pure" CH_3O. He obtained a spectrum, with a reported adiabatic IP of 7.37 ± 0.03 eV, attributed to CH_3O.

The PIMS approach of Ruščić and Berkowitz[112] utilized CD_3OH as the reagent to react with atomic fluorine. Hydrogen (deuterium) abstraction from the carbon end of the molecule would generate CD_2OH, and from the oxygen end, CD_3O. On ionization, these species are separable in a mass spectrometer. The resulting photoion yield curves are presented in Figures 15 and 16. The adiabatic IP of CD_2OH was found to be 7.540 ± 0.006 eV, in good agreement with Dyke's value (7.55 ± 0.01 eV). However, the adiabatic IP of CD_3O obtained by PIMS was very different—10.726 ± 0.008 eV. Surprisingly, CH_3O^+ could not be observed in the corresponding experiments with CH_3OH and CH_3OD, although CD_3O^+ was observed in the $F + CD_3OD$ reaction. It was surmised that CH_3O^+ is unstable with respect to $HCO^+ + H_2$ (and also with respect to CH_2OH^+), and decays in less than $\sim 10^{-5}$ s, the characteristic time between formation and detection, whereas CD_3O^+ survives. Although the decomposition process is exothermic, it involves a triplet–singlet surface crossing (CH_3O^+ has been calculated[113] to have a triplet bound ground state, whereas HCO^+ has a $^1\Sigma^+$ ground state). Manaa and Yarkony[114] have recently calculated the dynamics of this process, and conclude that there is a potential barrier to the decomposition. This allows for the possibility of a strong isotope effect.

Dyke[115] has noted that a sharp peak existed at 10.72 ± 0.01 eV in their original He I photoelectron spectrum from the $F + CH_3OH$ reaction. They thought at the time that it might be attributable to $CHOH^+$. In hindsight, it is very likely CH_3O. Its appearance in PES, and absence in PIMS, can be rationalized by the very different time scales characteristic of these events. The photoelectron can depart in $\sim 10^{-15}$ s. The photoelectron kinetic energy remains unaffected by the subsequent decay of CH_3O^+. The CH_2OH species, having a heat of formation $\Delta H_{f_0}^0 \leq -2.1 \pm 0.7$ kcal/mol, is more stable than CH_3O, with $\Delta H_{f_0}^0 = 5.9 \pm 1$ kcal/mol. On ionization CH_2OH^+ ($\Delta H_{f_0}^0 \leq 172.0 \pm 0.7$ kcal/mol) is much more stable than CH_3O^+ ($\Delta H_{f_0}^0 = 253.2 \pm 1.0$ kcal/mol). Earlier, Burgers and Holmes[116] had inferred $\Delta H_f^0(CH_3O^+) = 247 \pm 5$ kcal/mol, and Ferguson et al.[117] had deduced $\Delta H_f^0(CH_3O^+) = 245 \pm 6$

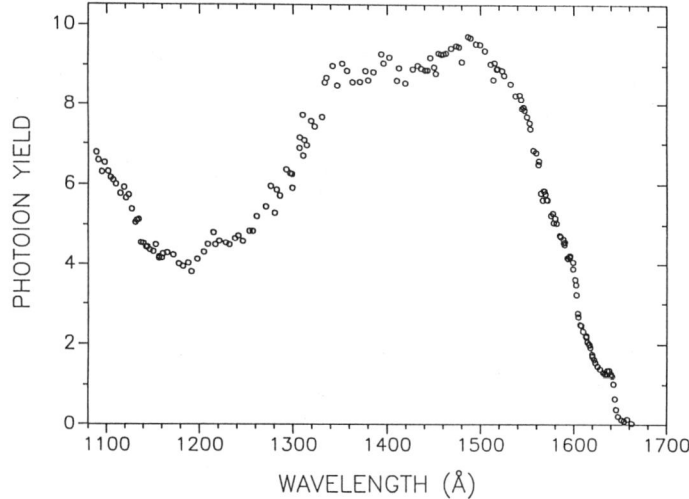

FIGURE 15. The photoion yield curve of CD_2OH^+ (CD_2OH). (From Ref. 112, with permission.)

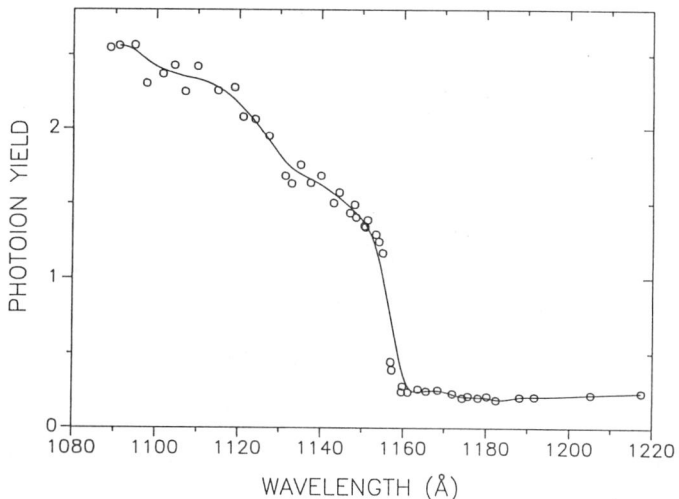

FIGURE 16. The photoion yield curve of CD_3O^+ (CD_3O). (From Ref. 112, with permission.)

kcal/mol, by indirect arguments. The adiabatic IP of CH_3O obtained by PIMS provides firm support for such a high value.

The relative stabilities of CH_2OH and CH_3O can be expressed in terms of the relative strengths of the C–H and O–H bonds in CH_3OH. Thus, from the values given above, $D_0(H–CH_2OH) \leq 95.1$ kcal/mol, $D_0(CH_3O–H) = 103.1 \pm 1$ kcal/mol. A major difference in the analogous CH_3SH system is that one expects the S–H bond to be considerably weaker than the O–H bond. Consequently, CH_3S may be expected to be more stable than CH_2SH.

Recently, three different experimental techniques have been brought to bear on this problem. Nicovich et al.[118] studied the kinetics of bromination of CH_3SH, CH_3SCH_3, and CH_3SSCH_3. One of their findings was that CH_3S, rather than CH_2SH, was formed on reacting bromine with CH_3SH, and hence (in agreement with expectations) CH_3S is more stable than CH_2SH. Nicovich et al.[118] also were able to establish that $\Delta H^0_{f_0}$ (CH_3S) = 31.44 ± 0.54 kcal/mol. Nourbakhsh et al.[71] prepared CH_3S by photolyzing CH_3SCH_3 with 193-nm laser light, and then obtained the photoion yield curve of CH_3S^+ (CH_3S) using PIMS. They obtained an adiabatic IP of 9.225 ± 0.014 eV. In addition, from time-of-flight measurements of the laser photofragments on photolysis of CH_3SH, CH_3SCH_3, and CH_3SSCH_3, they deduced $\Delta H^0_{f_0}$ (CH_3S) = 35.0 ± 1.0 kcal/mol. This latter heat of formation is equivalent to an S–H bond energy in CH_3SH of 89.6 ± 1 kcal/mol, whereas $\Delta H^0_{f_0}$ (CH_3S) from Nicovich et al. corresponds to an S–H bond energy of 86.1 ± 0.6 kcal/mol.

Ruščić and Berkowitz[119] succeeded in generating both CH_3S and CH_2SH by reacting atomic fluorine with CH_3SH. These species were clearly distinguishable by mass analysis when the reagent was CD_3SH (see Figures 17 and 18), although in this case the onset for CH_3S^+ (CH_3S) could be discerned superimposed on the photoion yield curve of CH_2SH^+(CH_2SH) at the same mass ($m/z = 47$). The adiabatic IP of CH_3S was found to be 9.262 ± 0.005 eV, somewhat higher than that of Nourbakhsh et al., while that of CH_2SH was 7.536 ± 0.003 eV. Thus, the IP of CH_2SH is very close to that of CH_2OH, while the IPs of both CH_3O and CH_3S are much higher.

The IP of CH_2SH could be used to infer the C–H bond energy in CH_3SH, if one had a reliable appearance potential for the photodissociative ionization process

$$CH_3SH + h\nu \to CH_2SH^+ + H + e$$

Kutina et al.[120] initially obtained $\leq 11.611 \pm 0.005$ eV at 0 K for this threshold. More recently, Nourbakhsh et al.[121] reported a value of 11.23 ± 0.05 eV. If we combine this latter quantity with

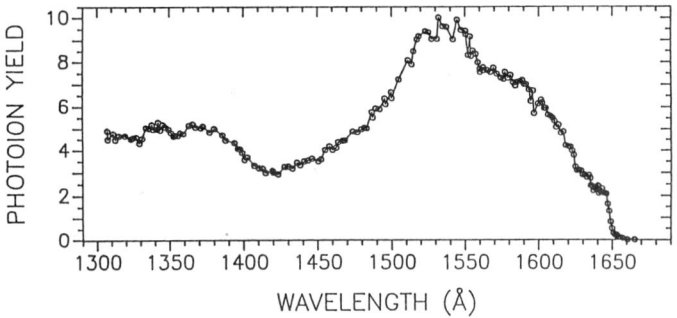

FIGURE 17. The photoion yield curve of CD_2SH^+ (CD_2SH). (From Ref. 119, with permission.)

IP(CH_2SH) = 7.536 eV, we obtain D_0(H–CH_2SH) = 3.69 ± 0.05 eV, or a C–H bond energy of 85.2 ± 1.2 kcal/mol. Thus, the C–H bond energy in CH_3SH would be less than the S–H bond energy, which seems unlikely. By using the appearance potential of Kutina et al. one arrives at D_0(H–CH_2SH) ≤ 4.075 eV, or a C–H bond energy of ≤ 93.97 kcal/mol. If we give weight to an alternative value for ΔH_f^0(CH_2SH^+) based on the proton affinity of CH_2SH, we arrive at a C–H bond energy of 92.4 ± 2.0 kcal/mol, 6–8 kcal/mol larger than the S–H bond energy.

There are other reasons for favoring the Kutina et al. value[120] for AP(CH_2SH^+)/CH_3SH to that of Nourbakhsh et al.[121] The most convincing one is based on quasiequilibrium theory. Both Kutina et al. and Nourbakhsh et al. observe a lower appearance potential for CH_2S^+(CH_3SH) than CH_2SH^+(CH_3SH). After an initial increase in intensity from threshold, the CH_2S^+ photoion yield abruptly diminishes at ~11.6 eV in the spectra of both groups of workers, leaving a cusp in the CH_2S^+ curve. The probable interpretation of this cusp is that the process yielding CH_2S^+ begins to suffer competition from a new channel—the one yielding CH_2SH^+. This is a plausible inference, since the

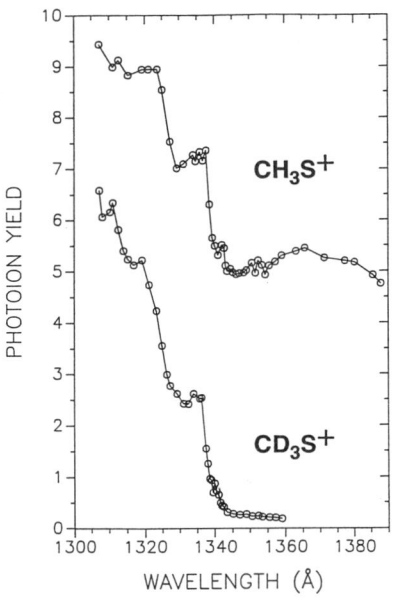

FIGURE 18. The photoion yield curve of CD_3S^+ (CD_3S). (From Ref. 119, with permission.)

latter is a simple bond rupture process, whereas the former (in which H_2 is a product) derives from a more constrained transition state.

In retrospect, we make two observations on this set of studies.

1. Although laser-induced photodissociation will undoubtedly prove to be a useful method for generating transient species, it has turned out to be less effective than F atom reaction in the CH_2SH/CH_3S study. Only CH_3S was produced by laser photolysis, whereas both transient species were formed by the F atom reaction.
2. The importance of mass analysis and preparation of isotopic variants was clearly demonstrated in the CH_2OH/CH_3O study.

4. PHOTOIONIZATION AND COINCIDENCE STUDIES OF SOME LARGE MOLECULES

4.1. Benzene Isomers

About 14 different $C_6H_6^+$ isomers are reported in the literature.[34,122] This includes the mono-, bi-, and tricyclic ring systems, and open-chain isomers. The heats of formation of these cations are found to be between 233.2 kcal/mol (benzene) and 327 kcal/mol (1,5 hexadiyne).

The low-energy fragmentations of benzene and its isomers, 1,5-hexadiyne and 2,4-hexadiyne, have been identified earlier by PEPICO measurements.[123] Photodissociation of benzene isomers above 25 eV excitation energy has been investigated by PEPICO and PEPIPICO spectroscopy.[124] In the low-energy regime,[123] the ratio of the fragment ion intensities of $C_3H_3^+$ and $C_4H_4^+$ remained the same, which led to the suggestion that a common, more stable isomer was formed prior to dissociation. Isotopic labeling in benzene has shown that significant scrambling of carbon atoms prior to ionic dissociation occurs.[125] The concept of noncommunicating isolated states[126] was therefore not verified by PEPICO measurements and rates derived from statistical theory.[123]

More recently, multiphoton ionization techniques have been applied together with reflectron time-of-flight mass spectrometry to measure the decay rate constants of two hydrogen-loss channels (H and H_2), and two C-loss channels (C_2H_2 and C_3H_3) in energy-selected benzene cation.[127] This technique overcomes the disadvantage of PEPICO where the limited mass resolution does not allow the observation of losses of light neutrals (e.g., H or H_2). The production rate was found to be identical for all fragmentations under investigation. This has been interpreted as evidence for competing fragmentations from the benzene cation ground state after fast nonradiative processes, since radiative deactivation has not been observed for benzene cations.[128] A crossover for the rate constants of the hydrogen and carbon losses is found experimentally and agrees with RRKM calculations,[127] based on different structures and the tightness of the transition states. This leads at low excitation energies to the energetically favored hydrogen losses, whereas at higher excess energy the entropically more favorable carbon losses dominate.

Fragmentation of both isomers 1,5-hexadiyne and 2,4-hexadiyne has been investigated recently by time resolved mass spectrometry.[129,130] The motivation for this work has been to find whether there is an isomerization barrier between the open-chain $C_6H_6^+$ isomers, 1,5- and 2,4-hexadiyne, and the benzene cation. The results indicate that direct dissociation of both open-chain isomers is not observed. This is in accordance with RRKM calculations, which have been used to fit the time-dependent photoion yield curves. The best agreement is obtained with an isomerization model rather than with direct dissociation mechanisms of the linear molecule.

Laser photodissociation experiments on $C_6H_6^+$ cation beams have been performed.[131] The results are contradictory to the above-mentioned isomerization model. Especially for 2,4-hexadiyne the large anisotropy of the photoproduct angular distribution indicates that the cation does not isomerize.

It has been concluded that more refined photodissociation experiments are needed.[129]

4.2. Naphthalene, Azulene

Polycyclic aromatic hydrocarbons (PAHs) have gained more attention recently, since it has been proposed that they may be carriers of infrared emission bands in interstellar media between 3 and 13 μm.[132–134] This so-called "unidentified infrared emission" has been specifically assigned to PAHs mainly in dehydrogenated and cationic form. Species with more than 20–25 atoms have been considered as resistant to interstellar radiation.[133] Naphthalene is proposed to be a major—if not exclusive—component of the PAHs responsible for the unidentified infrared emission besides other small PAHs such as anthracene ($C_{14}H_{10}$) and tetracene ($C_{18}H_{12}$).[135] This has motivated several studies on the photostability of singly and doubly charged PAHs.[136–138]

Naphthalene has been studied extensively since it is the simplest PAH. Its photostability is examined with the threshold PEPICO (T-PEPICO) technique,[136] where both charged particles (threshold photoelectrons and photoions) are detected in coincidence. The advantage of T-PEPICO with respect to simple PIMS is that exclusively threshold processes are selected. Monochromatized synchrotron radiation is well suited to scan the photon energy while detecting threshold photoelectrons. The bandwidth of these species is confined by using both angular discrimination with soft drawout fields (steradiancy effect) as well as the electron flight time. The electron flight time is easily accessible if the storage ring is operated in a single bunch mode. This enables suppression of fast electrons that hit the detector on the same trajectory as the true threshold photoelectrons. Electrons with kinetic energies between 0 and 20 meV are selected with this technique. Space focusing of the cations going in the opposite direction is achieved by a high-voltage pulse that is triggered by the arriving electrons. Further details concerning the experimental aspects are described in Ref. 139.

The branching ratios of fragmentation channels and the breakdown graph are obtained from T-PEPICO work.[136] A considerable blueshift of most fragmentation threshold energies is found for naphthalene-d_8 compared to PIMS[137] and older electron impact work[140] on naphthalene-h_8. A considerably lower detection efficiency in the coincidence experiment was discounted so that this observation was rationalized in terms of an isotope effect. Further support for this interpretation came from recent work on benzene where similar effects are observed.[141] The kinetic isotope shift ranges between 30 and 780 meV depending on the fragmentation reaction. The main fragmentations of both naphthalenes and their threshold energies are listed in Table XI.[137] These threshold values differ somewhat from the T-PEPICO values reported in Ref. 136. The reason for this may be the different sensitivities of both experimental setups, especially if the cation yield increases slowly above the threshold. With respect to isotope shifts it is important that an identical setup is used for a comparison of threshold energies. Another advantage comes from a comparison of the spectral shapes of the corresponding photoion yield curves above threshold. Higher precision in determination of isotope shifts is obtained if the displacement of the PIE curves is analyzed. This is shown in Figure 19 for the acetylene loss from both naphthalenes. The assumption that is made is that both cation yield curves are similar in growth in their threshold regions. The physicochemical factors for a kinetic isotope shift are discussed in Ref. 137. It is concluded that the existence of such an effect is the result of contributions from C–H (and C–D) vibrations to the density of the vibrational states of the reactants. This is consistent with the basic RRKM/QET assumptions of rapid internal conversion to the cation ground state and subsequent random flow of the molecular cation vibrational energy. Another important factor may be mass-dependent rates of isomerization. In experiments involving acetylene loss from naphthalene-d_8

TABLE XI
Appearance Potentials (AP) of Some Ionic Fragmentation Channels in $C_{10}D_8^+$ and $C_{10}H_8^+$

Cation	Neutral	AP($C_{10}H_8$) (eV)	AP($C_{10}D_8$) (eV)
$C_{10}D_7^+$	D	15.41	15.68
$C_8D_6^+$	C_2D_2	15.50	15.77
$C_7D_5^+$	C_3D_3	15.90	16.60
$C_6D_6^+$	C_4D_2	15.64	15.83
$C_6D_5^+$	C_4D_3	18.72	18.87
$C_5D_3^+$	C_5D_5	19.74	19.90
$C_3D_3^+$	C_7D_5	19.35	19.32

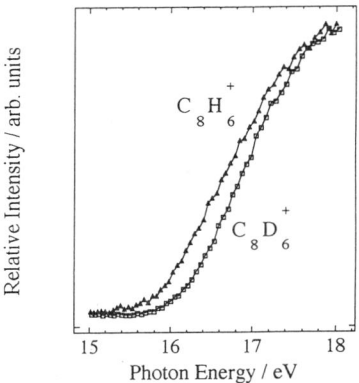

FIGURE 19. Photoion yield curves of $C_8H_6^+$ and $C_8D_6^+$ from naphthalene and naphthalene-d_8.

analyzed by RRKM calculations, it is found that a tight transition state with a negative activation entropy ($\Delta S^{\#} = -14.1$ J/mol K) can explain the hydrogen rearrangements apparently occurring.[136] The similarity in isotope shift has supported the conclusion that for the low-energy fragmentations (losses of hydrogen, acetylene, and diacetylene) a common intermediate may exist. The formation of this intermediate may be related to the rate-determining step. In contrast to this, "high-energy decomposition" mechanisms (formation of smaller fragment ions) involve the formation of open-chain intermediates and show consistently a different isotope shift.

No observations on deuterated PAHs are reported in connection with infrared emission in interstellar media. It may be possible that some of the emission bands interfere with features from atmospheric absorption. It is expected that isotopic fractionation of neutral and ionic PAHs will be influenced by the internal energy deposited in the molecules by photoexcitation and photoionization.

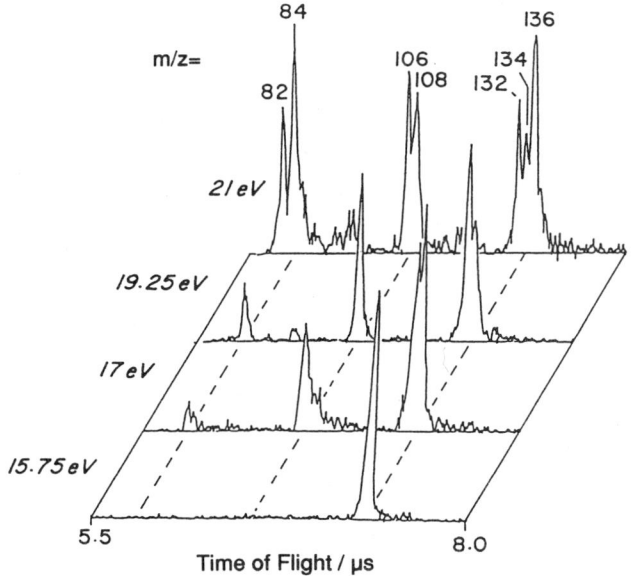

FIGURE 20. Threshold photoelectron–photoion coincidence (T-PEPICO) spectra of naphthalene-d_8 at various excitation energies.

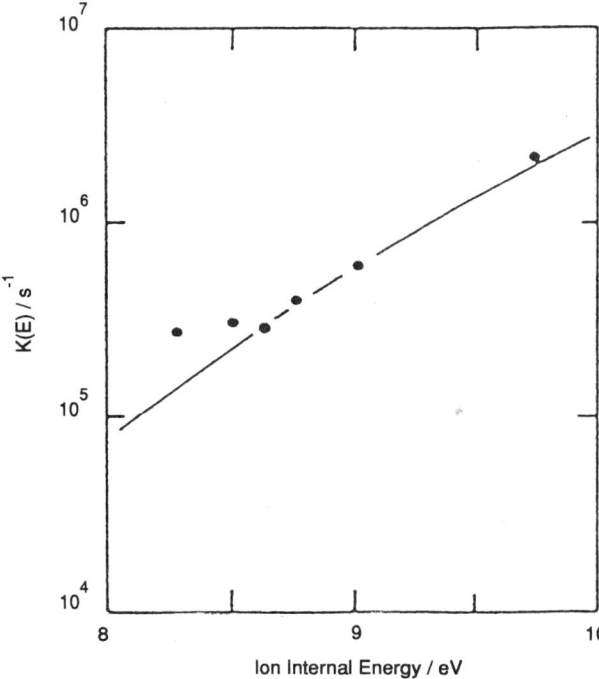

FIGURE 21. Rate of dissociation of the acetylene loss from ionic naphthalene-d_8. The dots represent the experimental data; the solid line is obtained from RRKM calculations.

The acetylene loss from naphthalene-d_8 shows in the T-PEPICO spectrum an asymmetric peak shape (m/z = 108) in the threshold regime. This is shown in Figure 20 for the T-PEPICO spectrum at 17 eV excitation energy. This behavior has been rationalized in terms of a metastable transition of $C_{10}D_8^+$. The dissociation rate of the reaction is determined from the peak shapes according to the method of Baer et al.[142] The results are shown in Figure 21 (dots). The experimental data are interpreted with RRKM calculations (solid line in Figure 21).

The extrapolated threshold value is 11.62 eV, which is considerably smaller than the experimental appearance potential ($C_{10}H_8$: 15.5 eV; $C_{10}D_8$: 15.77 eV). The thermodynamical threshold for acetylene loss is 11.34 eV, which is close to the activation energy obtained from the RRKM fit. This shows that low fragmentation rates can give rise to significant kinetic shifts.

From a mechanistic viewpoint the acetylene loss is interpreted via a primary ring opening followed by subsequent hydrogen and electron rearrangements. The most stable product that is expected is phenyl acetylene.[137]

Mechanisms have been considered for the cation decay of both isomers naphthalene and azulene. Already in early electron impact work[140] it was clear that the 70-eV electron impact mass spectra of both isomers are remarkably similar. The same is found for 20.65-eV photon impact mass spectra.[138] An almost constant difference in appearance potentials of the major fragments was observed by electron impact threshold spectroscopy.[140] This observation suggested that a common precursor cation is reached from both isomers. More recent work has verified this constant difference in appearance potentials by photoionization.[138] This difference is rationalized as the difference in heat of formation of both neutrals ($\Delta\Delta H_f$ = 1.43 eV).[34] The work of Jochims et al.[138] investigated the possibility of photoisomerization of both cations and their intermediates. Since the entire ionic hypersurface cannot be calculated, Jochims et al. have tried to adapt mechanisms which are known for the isomerization of the neutrals to the corresponding cations. These are the mechanisms proposed by Scott[143] and Dewar–Becker.[144,145] The Scott mechanism assumes that the transannular bond is opened and that the diradical that is formed undergoes a H 1,2-shift so that the other isomer can be formed. Since the calculated activation barrier (4.52 eV) for the formation of naphthalene is much higher than the experimental value (2.12 eV) an alternative mechanism is suggested by Dewar and Merz, which starts from neutral azulene, and involves a tricyclic intermediate that forms a carbene or a diradical. Both intermediates can collapse to form naphthalene. This mechanism has an activation barrier of only 3.41 eV which is still higher than the experimental value.

These reaction schemes have been exploited for the cation fragmentation of both isomers, where the common intermediate is analogous to those of the Dewar–Becker mechanism. All "low-energy fragmentations," i.e., the losses of atomic and molecular hydrogen, acetylene, and diacetylene, can be explained by the central role of the intermediate.[138]

The "high-energy fragmentations" are mostly those reactions with thresholds in the region 18.5–22.3 eV. These fragmentation paths are assumed to decay via open-chain cations into the various product channels.

4.3. Larger PAHs

Photoionization of larger PAHs has been investigated very recently.[146] This work is still in progress. The systems investigated are isomeric PAHs with up to seven condensed benzene rings: anthracene, phenanthrene, tetracene, triphenylene, tetraphene, perylene, and coronene. It is found in accordance with electron impact mass spectra[147] that the fragmentation pattern of these larger PAHs simplifies. Most of the cation intensity is found in the singly and doubly charged parent cation. It is interesting to note that the stability of dications increases with increasing number of benzene rings, which is equivalent to an increase in size of the molecules.

Threshold determination of singly charged parent cations by PIMS shows that the first ionization energies decrease slightly as the number of carbons increases. The hydrogen loss channel is found to increase in threshold energy in the same sense. This may be related to a decreasing rate of fragmentation as found for naphthalene cation fragmentation.[136] The threshold for the formation of dications decreases with increasing molecular size, so that in the case of coronene a lower threshold for dication formation is found compared to that of the hydrogen loss.

Based on earlier electron impact experiments, Tsai and Eland inferred a "rule of thumb" for the ratio between single and double ionization potentials of atoms and molecules $[I^{++}/I^+ = 2.8(1)]$.[148] The difference in ratio from twice the first ionization potential has been ascribed to electron correlation effects. It was reported by Tsai and Eland that for larger molecular systems this ratio may be smaller than the above-mentioned value. However, this finding was based on electron impact thresholds. Recent photoionization work[146] suggests that the ratio between double and single ionization potentials decreases from benzene to coronene continuously from 2.8 to 2.5. A ratio of 2.5 is also found for C_{60}.[149] A linear relationship between the size of the system and the double ionization potential is found in recent work on van der Waals clusters[150]: The double ionization thresholds decrease as a function of cluster size, so that the I^{++}/I^+ ratio approaches in the large cluster limit the value of 2, which is found for the solid. This difference can be rationalized in terms of two independent ionization steps at the same or at neighboring centers in the condensed phase,[151] such that no contribution of electron correlation is present.

The conditions in the H-I regions (regions with photon energies < 13.6 eV) in interstellar media can therefore contribute to double photoionization of PAHs via sequential ionization processes. Singly charged PAHs may absorb another photon which may lead either to fragmentation or to double ionization. From photoionization work on larger PAHs it seems to be more likely that double ionization is the more favorable process. However, in H-II regions ($E > 13.6$ eV) dissociative ionization can occur in an absorption process of a single more energetic photon.[152]

4.4. Buckminsterfullerene

Numerous studies on properties of fullerenes have been performed since the structure of the C_{60} fullerene was first proposed.[153] Single and double ionization of C_{60} as well as ionic fragmentation studies have been done in the VUV regime.

The first IP of C_{60} is found to be between 7.57 and 7.61 eV.[154–156] A difference of 40 meV is found between the values from PIMS and PES. Substantial heating of the sample may contribute to a considerable vibrational energy of 4.6 eV. A large energy shift caused by hot band excitation is not observed since most of the vibrational modes are inactive and even the symmetric normal modes (calculated to be at 610 and 1667 cm^{-1}, respectively[157]) will shift the first IP more than experimentally observed. This finding may also indicate that geometrical changes are quite small between the neutral and ionic ground states. The threshold behavior of the C_{60}^+ cation yield is nearly linear.[154] For a Franck–Condon transition without geometry changes one might expect a step-function onset of the cation yield. This difference is perhaps characteristic of the behavior of large systems.

The energy derivative of the C_{60}^+ photoion yield is fairly similar in shape to the photoelectron spectrum of C_{60}. This indicates that autoionization is not too important in the energy regime from 7 to 20 eV. However, differences have been rationalized in terms of ejection of low-energy electrons into high-angular-momentum waves, which follows from selection rule arguments.

The photoion yield of C_{60}^+ has been interpreted recently in terms of autoionization via a giant plasmon resonance,[158] because the energy position of this feature has been predicted from the RPA linear-response theory[159] to be at the same place as the maximum of the photoion yield of C_{60}^+.

The dication formation (C_{60}^{2+}) by one-photon ionization has been reported recently.[149,154] The threshold value is found at 19.00±0.03 eV. This agrees well with a value ≤ 19.5 eV that is deduced from discrete resonance light sources.[154] Since the adiabatic IP of C_{60} is 7.6 eV, the formation of C_{60}^{2+} from C_{60}^+ should require 11.8 eV.[160] Charge transfer bracketing has yielded 9.48 eV for this quantity,[161] whereas charge stripping experiments give 12.25 eV.[162] Possible reasons for these differences may be related to experimental problems in charge stripping experiments, which rely on the precision of reference data and calibration procedures, as well as in the threshold behavior of the direct double ionization potential, where a weak linear onset is found. On the other hand, charge transfer bracketing is sensitive to reaction barrier effects arising from the Coulomb repulsion between two singly charged cations that are formed as stable products.[160]

Fragmentation of C_{60}^+ is weak and occurs at higher photon energies than the threshold of double ionization, which is consistent with fragmentation of PAHs. Yoo et al., using line sources, have observed C_{58}^+ at 40.8 eV excitation energy, as well as marginal intensities of C_{56}^+, but not at 26.95 eV.[154] This implies the occurrence of a large kinetic shift which is on the order of 30 eV. RRKM calculations have been performed in order to model the C_2 loss reaction. The results are in agreement with the expected kinetic shift. Some disagreement exists regarding the threshold value for the C_2 loss. Studies of kinetic energy release measurements from laser ablation of C_{60}^+, together with classical phase space theory, have arrived at 4.6 eV.[163] The VUV PIMS measurements of Yoo et al.[154] have been interpreted by quasiequilibrium theory to imply a threshold energy ≥ 6 eV. More recently, C_{60}^+ was produced by electron impact (70 eV) on C_{60}, and the metastable ion attributed to the decomposition into $C_{58}^+ + C_2$ was studied.[164] From the kinetic energy release distribution of the metastable ion, the authors deduced 5.23 eV as the threshold for C_2 loss from C_{60}^+, using a form of unimolecular decay theory. Wurz and Lykke[165] observed both C_{58}^+ and delayed ionization of C_{60} from multiphoton excitation of C_{60}. They modeled both processes by forms of unimolecular decay theory and arrived at 5.6 eV for the activation energy to produce $C_{58}^+ + C_2$ from C_{60}^+. However, it should be mentioned that values as high as 11.1 eV have been reported.[166–169]

A more precise evaluation of this threshold requires better experiments such as PEPICO measurements. It would also help to know the fragmentation ratio at different times, to obtain an experimental measure of fragmentation rates. With such data, one can extrapolate to obtain the threshold value, and perhaps deduce the properties of the activated complex.

Interestingly, odd-carbon losses, such as C_3, are not observed. A plausible explanation involves the low stability of the remaining cation (C_{57}^+), which cannot form closed carbon cages. This is in contrast to loss of C_2, or at higher excitation energy multiples of C_2.

5. CONCLUSIONS

The topics covered in this chapter are somewhat arbitrary, and are not intended as a complete summary of valence ionization processes in the VUV. Rather, they are a kind of snapshot of current activities with certain species. In some cases, fairly complete information is available on total and partial cross sections, as well as the onset energies for various ionization processes. In most cases, e.g., transient species, basic information such as absolute photoabsorption and photoionization cross sections is lacking. With the advent of VUV lasers, higher-resolution studies will certainly be performed, yielding fine structure heretofore unobservable, as well as more precise threshold energies. PEPICO measurements, especially on large molecules such as C_{60}, should be quite informative about the applicability of QET and the magnitude of kinetic shifts. A few examples of delayed ionization in large clusters have been noted recently, but not yet adequately explained. Although much remains to be done, the present chapter has shown that considerable knowledge has been gained in the recent period. The radically different bonding between elements of the first complete row and their heavier congeners now has numerous experimental examples. Unexpected molecular structures are found, such as the nonplanar ring structure of Si_2H_2. Photoionization of isomers has been compared. In some cases, the evidence supports the view that isomerization to a common stable structure occurs on ionization, but in other cases it does not. In larger molecules, the proclivity toward double ionization exceeds that to fragmentation. Autoionization (which is particularly prominent in the VUV region) has been explored for several groups of atoms, and systematized. While not discussed in this chapter, much recent work has focused on rotational, as well as vibrational autoionization in molecules, using the ZEKE (zero electron kinetic energy) method. Future studies with VUV lasers, the new generation of synchrotron storage rings, and free electron lasers will no doubt provide a more detailed understanding of photoionization in the chemically important valence shells of atoms and molecules.

ACKNOWLEDGMENTS. This work was supported by the U.S. Department of Energy, Office of Basic Energy Sciences, under Contract No. W-31-109-ENG-38, the Bundesministerium für Forschung und Technologie, and the Freie Universität Berlin.

REFERENCES

1. J. BERKOWITZ, *Photoabsorption, Photoionization and Photoelectron Spectroscopy* (Academic Press, New York, 1979).
2. *Vacuum Ultraviolet Photoionization and Photodissociation of Molecules and Clusters*, edited by C. Y. Ng (World Scientific, Singapore, 1991).
3. *Synchrotron Radiation and Dynamic Phenomena*, edited by J. A. Beswick, *AIP Conference Proceedings* **258** (American Institute of Physics, New York, 1992).
4. I. NENNER AND J. A. BESWICK, in *Handbook on Synchrotron Radiation*, Vol. 2, edited by G. V. Marr (North-Holland, Amsterdam, 1987).
5. E. POLIAKOFF, in *Vacuum Ultraviolet Photoionization and Photodissociation of Molecules and Clusters*, edited by C. Y. Ng (World Scientific, Singapore, 1991), pp. 345–377.
6. J. H. D. ELAND, M. DEVORET, AND S. LEACH, *Chem. Phys. Lett.* **43**, 97 (1976); G. DUJARDIN, S. LEACH, AND G. TAIEB, *Chem. Phys.* **46**, 407 (1980); R. P. TUCKETT, E. CATELLUCCI, M. BONNEAU, G. DUJARDIN, AND S. LEACH, *Chem. Phys.* **92**, 43 **(1985)**.
7. H. ROTTKE AND K. H. WELGE, *Chem. Phys. Lett.* **99**, 456 (1983); D. J. HART AND J. W. HEPBURN, *Chem. Phys.* **129**, 51 (1989); H. H. FIELDING, T. P. SOFTLEY, AND F. MERKT, *Chem. Phys.* **155**, 257 (1991); F. MERKT AND T. P. SOFTLEY, *J. Chem. Phys.* **96**, 4149 (1992).
8. D. D. WAGMAN, W. H. EVANS, V. B. PARKER, R. H. SCHUMM, I. HALOW, S. M. BAILEY, K. L. CHURNEY, AND R. L. NUTTALL, *J. Phys. Chem. Ref. Data* **11**, Suppl. 2 (1982).
9. J. M. NICOVICH, K. D. KREUTTER, C. J. SHACKELFORD, AND P. H. WINE, *Chem. Phys. Lett.* **179**, 367 (1991).
10. *Gmelins Handbuch der Anorganischen Chemie, Chlor*, Ergänzungsband, Teil B, Lieferung 2 (Verlag Chemie, Weinheim, 1969).
11. A. ARKELL AND I. SCHWAGER, *J. Am. Chem. Soc.* **89**, 5999 (1967).
12. A. WAHNER, G. S. TYNDALL, AND A. R. RAVISHANKARA, *J. Phys. Chem.* **91**, 2734 (1987).
13. V. VAIDA, S. SOLOMON, E. C. RICHARD, E. RÜHL, AND A. JEFFERSON, *Nature* **342**, 405 (1989).
14. A. J. HILLS, R. J. CICERONE, J. C. CALVERT, AND J. W. BIRKS, *J. Phys. Chem.* **92**, 1853 (1988).
15. R. I. DERBY AND W. S. HUTCHINSON, *Inorg. Synth.* **4**, 152 (1953).
16. R. A. COX AND G. D. HAYMAN, *Nature* **332**, 796 (1988).
17. E. C. RICHARD AND V. VAIDA, *J. Chem. Phys.* **94**, 153 (1991); **94**, 163 (1991).

18. W. G. LAWRENCE, K. C. CLEMITSHAW, AND V. A. APKARIAN, *J. Geophys. Res.* **95**, 18591 (1990).
19. E. RÜHL, A. JEFFERSON, AND V. VAIDA, *J. Phys. Chem.* **94**, 2990 (1990).
20. E. BISHENDEN, J. HADDOCK, AND D. J. DONALDSON, *J. Phys. Chem.* **95**, 2113 (1991).
21. K. A. PETERSON AND H.-J. WERNER, *J. Chem. Phys.* **96**, 8948 (1992).
22. V. VAIDA, E. C. RICHARD, A. JEFFERSON, L. A. COOPER, R. FLESCH, AND E. RÜHL, *Ber. Bunsenges. Phys. Chem.* **96**, 391 (1992).
23. H. F. DAVIS AND Y. T. LEE, *J. Phys. Chem.* **96**, 5681 (1992).
24. C. M. HUMPHRIES, A. D. WALSH, AND P. A. WARSOP, *Discuss. Faraday. Soc.* **35**, 137 (1963).
25. N. BASCO AND R. D. MORSE, *Proc. R. Soc. London Ser. A* **336**, 495 (1974).
26. R. FLESCH, E. RÜHL, K. HOTTMANN, AND H. BAUMGÄRTEL, *J. Phys. Chem.* **97**, 837 (1993).
27. C. Y. R. WU AND D. L. JUDGE, *J. Chem. Phys.* **74**, 3804 (1981).
28. A. B. CORNFORD, D. C. FROST, F. C. HERRING, AND C. A. MCDOWELL, *Chem. Phys. Lett.* **10**, 345 (1971); *Discuss. Faraday Soc.* **54**, 56 (1972).
29. J. B. COON AND E. ORITZ, *J. Mol. Spectrosc.* **1**, 495 (1957).
30. K. KIMURA, S. KATSUMATA, Y. ACHIBA, T. YAMAZAKI, AND S. IWATA, *Handbook of Photoelectron Spectra of Fundamental Organic Molecules* (Japan Scientific Society, Tokyo, 1981).
31. L. S. CEDERBAUM, W. DOMCKE, W. VON NIESSEN, AND W. P. KRAEMER, *Mol. Phys.* **34**, 381 (1977).
32. K. P. HUBER AND G. HERZBERG, *Molecular Spectra and Molecular Structure. IV. Constants of Diatomic Molecules* (Van Nostrand–Reinhold, Princeton, NJ, 1979).
33. D. K. BULGIN, J. M. DYKE, N. JONATHAN, AND A. MORRIS, *Mol. Phys.* **32**, 1487 (1976); *J. Chem. Soc. Faraday Trans. 2* **75**, 456 (1978).
34. S. G. LIAS, J. E. BARTMESS, J. F. LIEBMAN, J. H. HOLMES, R. D. LEVIN, AND W. G. MALLARD, *J. Phys. Chem. Ref. Data* **17**, Suppl. 1 (1988).
35. U. ROCKLAND, H. BAUMGÄRTEL, E. RÜHL, O. LÖSKING, H. S. P. MÜLLER, AND H. WILLNER, *Ber. Bunsenges. Phys. Chem.*, (to be published).
36. J. L. GOLE, *J. Phys. Chem.* **84**, 1333 (1980).
37. K. A. PETERSON AND H. J. WERNER, (to be published).
38. V. H. DIBELER, J. A. WALKER, AND H. M. ROSENSTOCK, *J. Res. NBS* **70A**, 459 (1966).
39. M. J. WEISS AND G. M. LAWRENCE, *J. Chem. Phys.* **53**, 214 (1970).
40. P. J. BASSETT AND D. R. LLOYD, *Chem. Phys. Lett.* **6**, 166 (1970).
41. V. H. DIBELER AND J. A. WALKER, *Inorg. Chem.* **8**, 1728 (1969).
42. P. I. MANSELL, C. J. DANBY, AND I. POWIS, *J. Chem. Soc. Faraday Trans. 2* **77**, 1449 (1981).
43. E. A. LAWTON AND J. Q. WEBER, *J. Am. Chem. Soc.* **85**, 3595 (1963).
44. A. V. PANKRATOV, A. N. ZERCHENINOV, V. I. CHESNOKOV, AND N. N. ZHDANOVA, *Russ. J. Phys. Chem. (Engl. Transl.)* **43**, 212 (1969).
45. R. MINKWITZ, A. LIEDKE, AND R. NASS, *J. Fluorine Chem.* **35**, 307 (1987).
46. H. BAUMGÄRTEL, H.-W. JOCHIMS, E. RÜHL, H. BOCK, R. DAMMEL, R. MINKWITZ, AND R. NASS, *Inorg. Chem.* **28**, 943 (1989).
47. H. BOCK AND B. SOLOUKI, *Angew. Chem.* **93**, 425 (1981).
48. M. B. ROBIN, *The Higher Excited States of Polyatomic Molecules*, Vols. 1–3 (Academic Press, New York, 1974, 1975, 1985).
49. M. B. ROBIN, I. ISHII, R. MCLAREN, AND A. P. HITCHOCK, *J. Electron Spectrosc.* **47**, 53 (1988).
50. K. WITTEL AND H. BOCK, in *The Chemistry of Functional Groups*, edited by S. Patai and Z. Rappoport (Wiley, New York, 1983).
51. A. SCHMIEDEKAMP, S. SKAARUP, P. PULAY, AND J. E. BOGGS, *J. Chem. Phys.* **66**, 5769 (1977).
52. J. D. SWALEN AND J. A. IBERS, *J. Chem. Phys.* **36**, 1914 (1962).
53. D. COLBOURNE, D. C. FROST, C. A. MCDOWELL, AND N. P. C. WESTWOOD, *Chem. Phys. Lett.* **72**, 247 (1980).
54. H. HALIM, B. CIOMMER, AND H. SCHWARZ, *Angew. Chem.* **94**, 547 (1982); L. RADOM, W. J. BOUMA, R. H. NOBES, AND B. F. YATES, *Pure Appl. Chem.* **56**, 1831 (1984).
55. J. BERKOWITZ, J. P. GREENE, J. FOROPOULOS, JR., AND O. M. NESKOVIĆ, *J. Chem. Phys.* **81**, 6166 (1984).
56. K. WITTEL AND H. BOCK, *Chem. Ber.* **107**, 317 (1974).
57. D. REINKE, Ph.D. Thesis, DESY F41-73/6, Hamburg (1973).
58. R. KAUFEL, Ph.D. Thesis, Freie Universität Berlin, Berlin (1985).
59. D. P. STEVENSON, *Discuss. Faraday Soc.* **10**, 35 (1951).
60. E. RÜHL, H.-W. JOCHIMS, AND H. BAUMGÄRTEL, *Can. J. Chem.* **63**, 1949 (1985).
61. E. RÜHL, Diploma Thesis, Freie Universität Berlin, Berlin (1983).
62. G. FRENKING, W. KOCH, M. SCHAALE, AND H. BAUMGÄRTEL, *Int. J. Mass Spectrom. Ion Proc.* **61**, 305 (1984).
63. H.-W. JOCHIMS, W. LOHR, AND H. BAUMGÄRTEL, *Nouv. J. Chim* **3**, 109 (1979).

64. J. L. Franklin, J. G. Dillard, H. M. Rosenstock, J. T. Herron, K. Draxl, and F. H. Field, *Natl. Stand. Ref. Data Ser. Natl. Bur. Stand.* **26**, Washington, DC (1969).
65. G. Kauschka and L. Kodlitz, *Z. Chem.* **16**, 377 (1976).
66. R. M. Joshi, *J. Macromol. Sci. Chem.* **A4**, 1819 (1970); **A5**, 687 (1971); **A8**, 861 (1974).
67. H. M. Rosenstock, K. Draxl, B. W. Steiner, and J. T. Herron, *J. Phys. Chem. Ref. Data* **6**, Suppl. 1 (1977).
68. R. Kaufel and H. Baumgärtel, unpublished work.
69. R. K. Yoo, B. Ruščić, and J. Berkowitz, *Chem. Phys.* **166**, 215 (1992).
70. J. Berkowitz and B. Ruščić, in *Vacuum Ultraviolet Photoionization and Photodissociation of Molecules and Clusters* (World Scientific, Singapore, 1991), pp. 1–41.
71. S. Nourbakhsh, K. Norwood, G.-Z. He, and C. Y. Ng, *J. Am. Chem. Soc.* **113**, 6311 (1991).
72. J. Berkowitz and H. Cho, *J. Chem. Phys.* **90**, 1 (1989).
73. J. Berkowitz, J. P. Greene, H. Cho, and G. L. Goodman, *J. Phys. B* **20**, 2647 (1987).
74. J. Berkowitz, J. P. Greene, H. Cho, and B. Ruščić, *J. Chem. Phys.* **86**, 1235 (1987).
75. J. Berkowitz, C. H. Batson, and G. L. Goodman, *Phys. Rev. A* **24**, 149 (1981).
76. J. Berkowitz, *Adv. Chem. Phys.* **LXXII**, 1 (1988).
77. U. Fano, *Phys. Rev.* **124**, 1866 (1961).
78. J. Berkowitz, B. Ruščić, and R. K. Yoo, *Comments on Atomic and Molecular Physics*, **28**, 95 (1992).
79. M. Y. Amusia, Proc. of the 8th Intl. Conf. on the Physics of Electronic and Atomic Collisions, Belgrade, Yugoslavia (1973). Invited Lectures and Progress Reports, edited by B. C. Cobic and M. V. Kurepa (Institute of Physics, Belgrade, Yugoslavia, 1973), p. 172.
80. W. A. Goddard III and L. B. Harding, *Annu. Rev. Phys. Chem.* **29**, 363 (1978).
81. J. Berkowitz, *J. Chem. Phys.* **89**, 7065 (1988).
82. S. T. Gibson, J. P. Greene, and J. Berkowitz, *J. Chem. Phys.* **83**, 4319 (1985).
83. S. J. Dunlavey, J. M. Dyke, N. Jonathan, and A. Morris, *Mol. Phys.* **39**, 1121 (1980).
84. J. M. Dyke, N. Jonathan, A. E. Lewis, and A. Morris, *J. Chem. Soc. Faraday Trans. 2* **78**, 1445 (1982).
85. J. M. Dyke, N. Jonathan, and A. Morris, *Int. Rev. Phys. Chem.* **2**, 3 (1982).
86. See, for example, J. Berkowitz, *Photoabsorption, Photoionization and Photoelectron Spectroscopy* (Academic Press, New York, 1979), p. 275.
87. G. Herzberg and J. Shoosmith, *Can. J. Phys.* **34**, 523 (1956).
88. G. Herzberg, *Can. J. Phys.* **39**, 1511 (1961).
89. G. Herzberg and J. W. C. Johns, *Astrophys. J.* **158**, 399 (1969).
90. A. W. Potts and W. C. Price, *Proc. R. Soc. London Ser. A* **326**, 165 (1972).
91. B. Ruščić, M. Schwarz, and J. Berkowitz, *J. Chem. Phys.* **92**, 1865 (1990).
92. J. A. Pople and L. A. Curtiss, *J. Phys. Chem.* **91**, 155 (1987).
93. T. Kudo and S. Nagase, *Chem. Phys. Lett.* **148**, 73 (1988).
94. J. M. Dyke, N. Jonathan, A. Morris, A. Ridha, and M. J. Winter, *Chem. Phys.* **81**, 481 (1983).
95. R. D. Johnson III, B. P. Tsai, and J. W. Hudgens, *J. Chem. Phys.* **91**, 3340 (1989).
96. K. Balasubramanian and A. D. McLean, *J. Chem. Phys.* **85**, 5117 (1986).
97. C. W. Bauschlicher, Jr., S. R. Langhoff, and P. R. Taylor, *J. Chem. Phys.* **87**, 387 (1987).
98. L. A. Curtiss and J. A. Pople, *Chem. Phys. Lett.* **144**, 38 (1988).
99. B. Ruščić and J. Berkowitz, *J. Chem. Phys.* **95**, 2407 (1991).
100. J. S. Binkley, *J. Am. Chem. Soc.* **106**, 603 (1984).
101. B. T. Colegrove and H. F. Schaefer III, *J. Phys. Chem.* **94**, 5593 (1990).
102. L. A. Curtiss, K. Raghavachari, P. W. Deutsch, and J. A. Pople, *J. Chem. Phys.* **95**, 2433 (1991).
103. M. Bogey, H. Bolvin, C. Demuynck, and J. L. Destombes, *Phys. Rev. Lett.* **66**, 413 (1991).
104. B. Ruščić and J. Berkowitz, *J. Chem. Phys.* **95**, 2407 (1991).
105. S. R. Gunn and L. G. Green, *J. Phys. Chem.* **65**, 779 (1961).
106. J. A. Seetula, Y. Feng, D. Gutman, P. W. Seakins, and M. J. Pilling, *J. Phys. Chem.* **95**, 1658 (1991).
107. A. F. Sax and J. Kalcher, *J. Phys. Chem.* **95**, 1768 (1991).
108. B. T. Colegrove and H. F. Schaefer III, *J. Chem. Phys.* **93**, 7230 (1990).
109. R. S. Grev, B. J. DeLeeuw, and H. F. Schaefer III, *Chem. Phys. Lett.* **165**, 257 (1990).
110. J. M. Dyke, A. R. Ellis, N. Jonathan, N. Keddar, and A. Morris, *Chem. Phys. Lett.* **111**, 207 (1984).
111. J. M. Dyke, *J. Chem. Soc. Faraday Trans. 2* **83**, 69 (1987).
112. B. Ruščić and J. Berkowitz, *J. Chem. Phys.* **95**, 4033 (1991).
113. W. J. Bouma, R. H. Nobes, and L. Radom, *Org. Mass Spectrom.* **17**, 315 (1982).
114. M. R. Manaa and D. R. Yarkony, *J. Am. Chem. Soc.* **116**, 11444 (1994).
115. J. M. Dyke, private communication.
116. P. C. Burgers and J. L. Holmes, *Org. Mass Spectrom.* **19**, 452 (1984).

117. E. E. FERGUSON, J. RONCIN, AND L. BONAZZOLA, *Int. J. Mass Spectrom. Ion Proc.* **79**, 215 (1987).
118. J. M. NICOVICH, K. D. KREUTTER, C. A. VAN DIJK, AND P. H. WINE, *J. Phys. Chem.* **96**, 2518 (1992).
119. B. RUŠČIĆ AND J. BERKOWITZ, *J. Chem. Phys.* **97**, 1818 (1992).
120. R. E. KUTINA, A. K. EDWARDS, G. L. GOODMAN, AND J. BERKOWITZ, *J. Chem. Phys.* **77**, 5508 (1982). Although it was not clear at that time whether CH_2SH^+ or CH_3S^+ was more stable, the experimental threshold for mass 47 from CH_3SH is determined accurately, and is currently to be interpreted as CH_2SH^+.
121. S. NOURBAKHSH, K. NORWOOD, H.-M. YIN, C.-L. LIAO, AND C. Y. NG, *J. Chem. Phys.* **95**, 946 (1991).
122. H. M. ROSENSTOCK, J. DANNACHER, AND J. F. LIEBMAN, *Radiat. Phys. Chem.* **20**, 7 (1982).
123. T. BAER, G. D. WILLETT, D. SMITH, AND J. S. PHILLIPS, *J. Chem. Phys.* **70**, 4076 (1979).
124. O. BRAITBART, S. TOBITA, S. LEACH, P. ROY, AND I. NENNER, in *Synchrotron Radiation and Dynamic Phenomena*, edited by J. A. Beswick, *AIP Conference Proceedings* **258** (American Institute of Physics, New York, 1992), pp. 42–52.
125. J. H. BEYNON, R. M. CAPRIOLI, W. O. PERRY, AND W. E. BAITINGER, *J. Am. Chem. Soc.* **94**, 6828 (1972).
126. B. ANDLAUER AND C. OTTINGER, *J. Chem. Phys.* **55**, 1471 (1971); *Z. Naturforsch.* **27a**, 293 (1972).
127. H. KÜHLEWIND, A. KIERMEIER, AND H. J. NEUSSER, *J. Chem. Phys.* **85**, 4427 (1986).
128. O. BRAITBART, E. CASTELLUCCI, G. DUJARDIN, AND S. LEACH, *J. Phys. Chem.* **87**, 4799 (1983).
129. N. OHMICHI, Y. MALINOVICH, J. P. ZIESEL, AND C. LIFSHITZ, *J. Phys. Chem.* **93**, 2491 (1989).
130. C. LIFSHITZ AND N. OHMICHI, *J. Phys. Chem.* **93**, 6329 (1989).
131. R. E. KRAILLER, D. H. RUSSELL, M. F. JARROLD, AND M. T. BOWERS, *J. Am. Chem. Soc.* **107**, 2346 (1985).
132. A. LÉGER AND J. L. PUGET, *Astron. Astrophys.* **137**, L5 (1984); A. LÉGER AND L. D'HENDECOURT, *Astron. Astrophys.* **146**, 81 (1985).
133. A. LÉGER et al. (Eds.), *Polycyclic Aromatic Hydrocarbons and Astrophysics* (Reidel, Dordrecht, 1987).
134. L. J. ALLAMANDOLA, A. G. G. M. TIELENS, AND J. R. BARKER, *Astrophys. J.* **290**, L25 (1985).
135. W. W. DULEY AND A. P. JONES, *Astrophys. J.* **351**, L49 (1990).
136. E. RÜHL, S. D. PRICE, AND S. LEACH, *J. Phys. Chem.* **93**, 6312 (1989).
137. H. W. JOCHIMS, H. RASEKH, E. RÜHL, H. BAUMGÄRTEL, AND S. LEACH, *J. Phys. Chem.* **97**, 1312 (1993).
138. H. W. JOCHIMS, H. RASEKH, E. RÜHL, H. BAUMGÄRTEL, AND S. LEACH, *Chem. Phys.* **168**, 159 (1992).
139. M. RICHARD-VIARD, These d'Etat, Université de Paris-Sud, Paris (1988).
140. R. J. VAN BRUNT AND M. E. WACKS, *J. Chem. Phys.* **41**, 3195 (1964).
141. H. KÜHLEWIND, A. KIERMEIER, H. J. NEUSSER, AND E. W. SCHLAG, *J. Chem. Phys.* **87**, 6488 (1987).
142. T. BAER, O. DUTUIT, H. MESTDAGH, AND C. ROLANDO, *J. Phys. Chem.* **92**, 5674 (1988).
143. L. T. SCOTT AND M. A. KIRMS, *J. Am. Chem. Soc.* **103**, 5875 (1981).
144. J. BECKER, C. WENTRUP, E. KATZ, AND K.-P. ZELLER, *J. Am. Chem. Soc.* **102**, 5112 (1980).
145. M. J. S. DEWAR AND K. M. MERZ, JR., *J. Am. Chem. Soc.* **108**, 5142 (1986).
146. H. W. JOCHIMS, E. BILLER, H. BAUMGÄRTEL, AND S. LEACH, (to be published).
147. E. STENHAGEN, S. ABRAHAMSON, AND F. W. MCLAFFERTY, *Registry of Mass Spectral Data*, Vol. 1 (Wiley, New York, 1974).
148. B. P. TSAI AND J. H. D. ELAND, *Int. J. Mass Spectrom. Ion Phys.* **36**, 143 (1980).
149. H. STEGER, J. DE VRIES, B. KAMKE, W. KAMKE, AND T. DREWELLO, *Chem. Phys. Lett.* **194**, 452 (1992).
150. E. RÜHL, C. SCHMALE, H. C. SCHMELZ, AND H. BAUMGÄRTEL, *Chem. Phys. Lett.* **191**, 430 (1992).
151. H. W. BIESTER, M. J. BESNARD, G. DUJARDIN, L. HELLNER, AND E. E. KOCH, *Phys. Rev. Lett.* **59**, 1277 (1987).
152. S. LEACH, in *Interstellar Dust*, IAU/35, edited by L. Allamandola and A. G. G. M. Tielens (Kluwer, Amsterdam, 1989).
153. H. W. KROTO, J. R. HEATH, S. C. O'BRIEN, R. F. CURL, AND R. E. SMALLEY, *Nature* **318**, 162 (1985).
154. R. K. YOO, B. RUŠČIĆ, AND J. BERKOWITZ, *J. Chem. Phys.* **96**, 911 (1992).
155. D. L. LICHTENBERGER, K. W. NEBESNY, C. D. RAY, D. R. HUFFMAN, AND L. D. LAMB, *Chem. Phys. Lett.* **176**, 203 (1991).
156. J. A. ZIMMERMAN, J. R. EYLER, S. B. H. BACH, AND S. W. MCELVANY, *J. Chem. Phys.* **94**, 3556 (1991).
157. R. E. STANTON AND M. D. NEWTON, *J. Phys. Chem.* **92**, 2141 (1988).
158. I. V. HERTEL, H. STEGER, J. DE VRIES, B. WEISSER, C. MENZEL, B. KAMKE, AND W. KAMKE, *Phys. Rev. Lett.* **68**, 784 (1992).
159. G. F. BERTSCH, A. BULGAC, D. TOMÁNEK, AND Y. WANG, *Phys. Rev. Lett.* **67**, 2690 (1991).
160. S. PETRIE, G. JAVAHERY, J. WANG, AND D. K. BOHME, *J. Phys. Chem.* **96**, 6121 (1992).
161. S. W. MCELVANY AND S. B. H. BACH, *A.S.M.S. Mass Spectrom. Allied Top.* **39**, 422 (1991); S. W. MCELVANY, M. M. ROSS, AND J. H. CALLAHAN, *Proc. Matter Res. Soc. Symp.* **206**, 697 (1991).
162. C. LIFSHITZ, M. IRAQI, T. PERES, AND J. E. FISCHER, *Rapid Commun. Mass Spectrom.* **5**, 238 (1991).
163. P. P. RADI, M.-T. HSU, M. E. RINCON, P. R. KEMPER, AND M. T. BOWERS, *Chem. Phys. Lett.* **174**, 223 (1990).
164. P. SANDLER, C. LIFSHITZ, AND C. E. KLOTS, *Chem. Phys. Lett.* **200**, 445 (1992).

165. P. WURZ AND K. R. LYKKE, *J. Phys. Chem.* **96**, 10129 (1992).
166. R. C. MOWREY, D. W. BRENNER, B. I. DUNLAP, J. W. MINTMIRE, AND C. T. WHITE, *J. Phys. Chem.* **95**, 7138 (1991).
167. R. D. BECK, P. ST. JOHN, M. M. ALVAREZ, F. DIEDERICH, AND R. L. WHETTEN, *J. Phys. Chem.* **95**, 8402 (1991).
168. R. E. STANTON, *J. Phys. Chem.* **96**, 111 (1992).
169. J.-Y. YI AND J. BERNHOLC, *J. Chem. Phys.* **96**, 8634 (1992).

CHAPTER 8

PHOTOION-PAIR FORMATION

J. BERKOWITZ

1. BRIEF SUMMARY AND OVERVIEW OF ION-PAIR PROCESSES

The term *photoionization* usually implies the production of a positive ion–electron pair on interaction of sufficiently energetic photons with gaseous molecules. In the absence of collisional and field effects, and taking into account thermal Boltzmann effects, this process has a threshold which (if Franck–Condon effects are not too restrictive) is the adiabatic ionization potential. In some cases, ionization can arise from a quite different process—the formation of a positive ion–negative ion pair. Given favorable circumstances, this latter process may occur below the adiabatic ionization potential. Consider a diatomic metal halide, MX. Then, one can readily show that the ion-pair process may occur below the ionization potential if the electron affinity (EA) of X exceeds the dissociation energy (D_0) of MX^+. This condition is met for all of the thallium halides (see below). For TlI, TlBr, and (borderline) TlCl, the threshold for ion-pair formation occurs before the onset of the vacuum-ultraviolet region. This fortuitous circumstance enabled Terenin and Popov[1–2] to observe photoionization (into ion pairs) in the early 1930s well before the advent of vacuum-ultraviolet photoionization technology.

Vacuum-ultraviolet photoionization mass spectrometry commenced in earnest about three decades later, and with the passing of another three decades, approximately 50 examples of photoion-pair production in the valence region have been studied. Table I is a summary (almost certainly incomplete) of the molecules studied, the observed and calculated thresholds, the adiabatic ionization potential (IP), and (very approximately) the abundance of the ion-pair process relative to the major photoionization process. Several generalizations can be drawn from this list.

1. Only diatomic molecules display a cross section for ion-pair production which is comparable to that for photoionization.
2. Even among diatomic molecules, the relative cross section for ion pairs is large only when the corresponding threshold occurs below the adiabatic ionization potential.
3. For triatomic molecules, the relative cross section (compared to photoionization) is roughly 2 to 4 orders of magnitude lower. Among polyatomic molecules, CH_3F has $\sigma_{ionpair}/\sigma_{photoionization} \approx 1/20$; for all other species in Table I, this ratio is 10^{-2}–10^{-5}
4. Noteworthy by their absence in Table I are the most ionic molecules, the alkali halides. Noteworthy by their presence are the purely covalent homonuclear halogens, H_2 and O_2.
5. In many (but not all) cases, the observed threshold occurs at approximately the calculated thermochemical energy. Prominent departures occur when the polyatomic species has five or more atoms.

Let us first direct our attention to point 4. Parks *et al.*[3] have examined the relative probability of ion-pair formation in thallium halides and CsCl in single collisions with fast Xe or Kr atoms. Their

---

J. BERKOWITZ • Chemistry Division, Argonne National Laboratory, Argonne, Illinois 60439.

VUV and Soft X-Ray Photoionization. Edited by Uwe Becker and David A. Shirley. Plenum Press, New York, 1996.

TABLE I
Compilation of Observed Photoion-Pair Processes

Molecule	Anion observed	Threshold, eV Observed	Threshold, eV Calculated	Adiabatic IP, eV	Fractional abundance[a]	Reference
TlF	F^-	~7.5	7.28	10.52^b	~1:1	c
TlCl	Cl^-	~6.7	6.31	9.70	(~1/10)	c
TlBr	Br^-	6.17	6.17	9.14	~1:1	c
TlI	I^-	5.79	5.79	8.47	~1:1	d
HF	F^-	16.06	16.068	16.044^e	≲1:1	f
F_2	F^-	15.58–15.60	15.626	15.694^g	~1:1	f
Cl_2	Cl^-	~11.64	11.834	11.48	$~10^{-2}$	h
Br_2	Br^-	$10.44,^i$ 10.48 ± 0.02^j	10.4206	$10.52^{i,j}$	~1:1	i,j
I_2	I^-	8.79 ± 0.01^k 8.95 ± 0.02^j 8.946^l	8.9327	9.31	~1:1	j,k,l
ClF	F^-	12.04	12.18_5	12.65	0.12	m
H_2	H^-	17.322	17.322	15.4259	$~4 \times 10^{-3}$	n
O_2	O^-	17.272	17.272	12.071	9×10^{-3}	o
NO	O^-	19.56	19.54	9.26436	10^{-2}–10^{-3}	p
CO	O^-	20.91	20.898	14.0139	10^{-2}–10^{-3}	p
	C^-	23.45	23.442		10^{-3}–10^{-4}	p
H_2O	OH^-	~16.9	16.883	12.612	$~5 \times 10^{-4}$	q
HCN	CN^-	15.18	15.18	13.59^r	$~10^{-3}$	s
NO_2	O^-	10.92	10.930	$<9.62^t$	5×10^{-3}	q
N_2O	O^-	15.784	15.794	12.886	$~10^{-3}$	u
CO_2	O^-	17.990	18.006	13.773	1.3×10^{-4}	v
COS	S^-	15.10 ± 0.04^v $~15.09^q$	15.076	11.174	4.6×10^{-5v} $~10^{-3q}$	q,v
	O^-	17.08 ± 0.09^v $~16.8^q$	16.63		4.7×10^{-5v} $~10^{-4q}$	q,v
SO_2	O^-	14.5	14.52	12.32	$~10^{-3}$	q
	S^-	~16	15.89		$~10^{-4}$	q
KrF_2	$F^-(Kr^+ + F)$	11.52	11.57	13.34	10^{-1}–10^{-2}	w
XeF_2	$F^-(Xe^+ + F)$	11.48	11.43	12.35	10^{-1}–10^{-2}	x
C_2H_2	C_2H^-	$16.33_5 \pm 0.02$	16.325	11.398	$~2 \times 10^{-4}$	y
CH_4	H^-	13.37 ± 0.15	13.567	12.615	$~3 \times 10^{-4}$	z
CH_3F	F^-	12.56–12.59	11.26	12.50	$~5 \times 10^{-2}$	aa,bb
CH_3Cl	Cl^-	10.07	9.775	11.28	$~1.2 \times 10^{-2}$	aa,bb
CH_3Br	Br^-	9.48–9.55	9.485	10.53	10^{-2}–10^{-3}	cc
CF_4	F^-	13.25	≤11.186	~15	2–3×10^{-5}	dd
CF_3Cl	$F^-(CF_2Cl^+)$	15.90	~10.13	12.45	$~10^{-3}$	ee
	$Cl^-(CF_3^+)$	16.00	9.10		$~10^{-4}$	ee

TABLE I
(*Continued*)

Molecule	Anion observed	Threshold, eV Observed	Threshold, eV Calculated	Adiabatic IP, eV	Fractional abundance[a]	Reference
CF_2Cl_2	$Cl^-(CF_2Cl^+)$	10.60	~8.06	11.75	~1.8×10^{-3}	ee
	$F^-(CFCl_2^+)$	12.07	~9.5		~1×10^{-3}	ee
$CFCl_3$	$F^-(CCl_3^+)$	13.7 ± 0.5	8.76	11.57	~10^{-5}	ee
	$Cl^-(CFCl_2^+)$	15.6 ± 0.1	7.63 ± 0.03		~10^{-3}	ee
C_2H_6	H^-	12.00 ± 0.05	11.66	11.51	~3×10^{-3}	ff
C_3H_8	H^-	~11.0	10.84	11.0	~10^{-3}	ff
C_4H_{10}	H^-	10.9 ± 0.1	10.43	10.6 ± 0.1	~5×10^{-4}	ff
C_2H_5Cl	Cl^-	9.5 ± 0.1	8.10	10.97	10^{-3}–10^{-4}	gg
SF_6	$F^-(SF_5^+)$	12.98 ± 0.03	11.92	15.32	~2×10^{-4}	hh
cyclo-$C_6H_{11}Cl$	Cl^-	8.57 ± 0.05	6.90	10.10	$10^{-3} \times 10^{-4}$	gg
CH_3ONO_2	NO_2^-	8.2 ± 0.1	6.8	11.25	10^{-3}–10^{-4}	ii

[a] Estimated intensity in the ion-pair spectrum, relative to primary photoionization.
[b] J. L. Dehmer, J. Berkowitz, and L. C. Cusachs, *J. Chem. Phys.* **58**, 5681 (1973).
[c] J. Berkowitz and T. A. Walter, *J. Chem. Phys.* **49**, 1184 (1968); see also J. Berkowitz, *Adv. High Temp. Chem.* **3**, 147 (1971).
[d] J. Berkowitz and W. A. Chupka, *J. Chem. Phys.* **45**, 1287 (1966).
[e] T. E. H. Walker, P. M. Dehmer, and J. Berkowitz, *J. Chem. Phys.* **59**, 4292 (1973).
[f] Ref. 21.
[g] P. M. Guyon, R. Spohr, W. A. Chupka, and J. Berkowitz, *J. Chem. Phys.* **65**, 1650 (1976).
[h] Ref. 14.
[i] Ref. 17.
[j] J. D. Morrison, H. Hurzeler, M. G. Inghram, and H. E. Stanton, *J. Chem. Phys.* **33**, 821 (1960).
[k] Ref. 19.
[l] Ref. 20.
[m] V. H. Dibeler, J. A. Walker, and K. E. McCulloh, *J. Chem. Phys.* **53**, 4414 (1970).
[n] W. A. Chupka, P. M. Dehmer, and W. T. Jivery, *J. Chem. Phys.* **63**, 3929 (1975).
[o] P. M. Dehmer and W. A. Chupka, *J. Chem. Phys.* **62**, 4525 (1975).
[p] Ref. 25.
[q] Ref. 30.
[r] V. H. Dibeler and S. K. Liston, *J. Chem. Phys.* **48**, 4765 (1968).
[s] Ref. 64.
[t] Ref. 34.
[u] Ref. 31.
[v] Ref. 37.
[w] J. Berkowitz and J. H. Holloway, *J. Chem. Soc. Faraday Trans. II* **74**, 2077 (1978).
[x] J. Berkowitz, W. A. Chupka, P. M. Guyon, J. H. Holloway, and R. Spohr, *J. Phys. Chem.* **75**, 1461 (1973).
[y] B. Ruscic and J. Berkowitz, *J. Chem. Phys.* **93**, 5586 (1990).
[z] Ref. 54.
[aa] M. Kraus, J. A. Walker, and V. H. Dibeler, *J. Res. Natl. Bur. Std.* **72A**, 281 (1968).
[bb] Ref. 49.
[cc] Ref. 48.
[dd] Ref. 60.
[ee] Ref. 61.
[ff] W. A. Chupka and J. Berkowitz, *J. Chem. Phys.* **47**, 2921 (1967).
[gg] M. E. Akopyan, F. I. Vilesov, and Y. L. Sergeev, *Teor. Eksp. Khim.* **7**, 271 (1971).
[hh] Ref. 62.
[ii] M. M. Lipei and V. K. Potapov, *Khim. Vys. Energ.* **9**, 332 (1975).

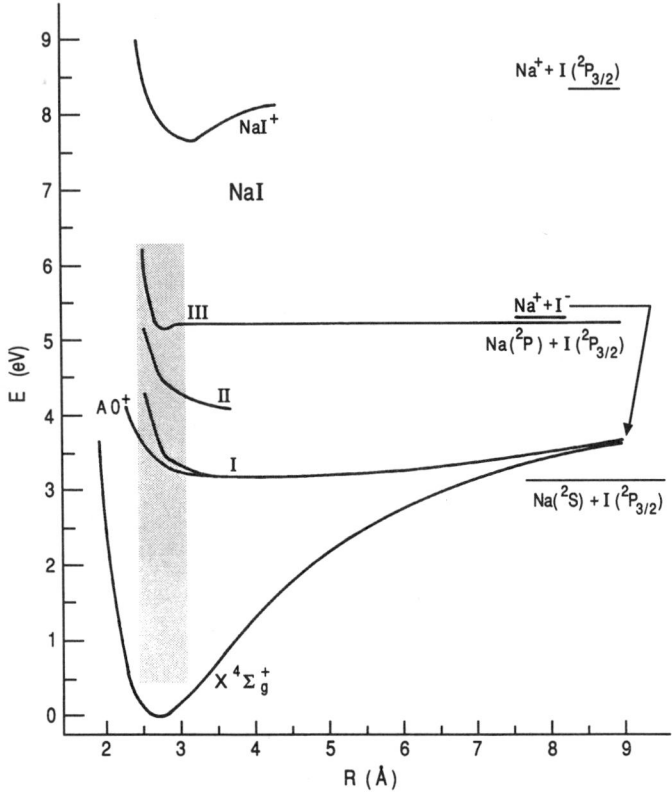

FIGURE 1. Relevant potential energy curves for NaI for examining ion-pair formation.

results for xenon-induced dissociation cross sections are: $\sigma(TlCl) \approx 1/3$ $\sigma(CsCl) \approx 10$ $\sigma(TlF) \approx 10$ $\sigma(TlBr) \approx 1000$ $\sigma(TlI)$. Tully et al.[4] have performed similar experiments with CsI, CsBr, RbI, and KI, and found the ion-pair cross sections of these molecules to be within a factor 2 of one another. We conclude that collisionally induced ion-pair formation is efficient for at least some alkali halides, but varies by three orders of magnitude among the thallium halides. By contrast, photoion-pair formation is efficient for the thallium halides, and has not been observed for the alkali halides. This disparity in behavior can be rationalized by reference to Figures 1 and 2, which describe relevant potential energy curves of NaI and TlCl.

Impulsive excitation with translational energies of ~ 6 eV will induce vibrational excitation of the electronic ground state. Near its minimum, this ground state has dominant ionic character, somewhat more so in NaI than in TlCl. With increasing vibrational energy, this ground-state potential curve approaches a more covalent curve, whose asymptote corresponds to ground-state atoms. These curves may cross (nonadiabatic) or undergo an avoided crossing (adiabatic). For alkali halides, these crossing points are typically at much larger internuclear distances than for thallium halides. For example, it is ~51.9 Å for CsCl and ~4.5 Å for TlCl. However, for NaI it is only ~6.9 Å. If curve crossing occurs, the nonadiabatic curve is followed, and ion pairs result. If the states interact, and there is an avoided crossing, there is still a possibility that the system will hop from the lower to the higher state, again resulting in ion pairs. Various mathematical formulations of this curve hopping probability can be qualitatively described by the Massey adiabatic criterion,[5] which relates the uncertainty principle time $\tau = \hbar/H_{ia}$ to the time of passage through this critical zone. Here, H_{ia} is the Hamiltonian connecting atomic and ionic states. The actual minimum energy gap between the two states turns out[6,7] to be $\Delta E° \cong 2 H_{ia}$. The time of passage through the critical zone may be roughly estimated from the

Coulomb energy at the crossing region, or from the vibrational energy spacing near this region. It does not vary as much, and hence is not as critical as $\Delta E°$. When $\Delta E°$ is relatively small, one can expect ion-pair formation, and conversely. Parks et al.[3a] have estimated $\Delta E° < 800$ cm^{-1} for TlCl, ~800 cm^{-1} for TlBr, and ~2800 cm^{-1} for TlI. Grice and Herschbach[8] have given a semiempirical formulation that enables one to estimate $\Delta E°$ from the crossing point. Using their expression, one obtains ~780 cm^{-1} for TlCl, ~1330 cm^{-1} for TlBr, and ~2200 cm^{-1} for TlI, in rough agreement with the estimates of Parks et al.[3] Among the alkali halides, one obtains ~0.07 cm^{-1} for CsI, $<<10^{-2}$ cm^{-1} for CsF, CsCl, and CsBr, ~30 cm^{-1} for NaCl, and ~400 cm^{-1} for NaI. These values of $\Delta E°$ rationalize the observation of relatively strong collisionally induced ion-pair formation for TlCl and the cesium halides, and successively weaker probability for TlBr and TlI.

By contrast, photoion-pair formation must involve photoexcitation to an electronically excited state, with subsequent evolution to ion-pair products. The excited state must be formed at an energy greater than or equal to the asymptotic energy for ion-pair formation. If the same states were involved as in the collisional problem, one might anticipate that a large value of $\Delta E°$ would enhance the prospect of observing ion pairs, since the photoexcited state would avoid the nonadiabatic behavior that would lead to atomic products, and might instead follow the ion-pair curve at large internuclear distances. The molecule NaI, which was observed to give neutral products in shock tube studies,[7] and has a relatively large value of $\Delta E°$, would, by this reasoning, appear to be a good candidate for photoion-pair formation. Figure 1, incorporating data from Davidovits and Brodhead,[9] and from Schaefer et al.,[10] provides a rationalization for the absence of photoion-pair formation. Davidovits and Brodhead observed three bands in absorption. The lowest of these probably involves two upper states, one of which (0^+) is very weakly bound,[10] and the other is probably purely repulsive. In any event, the Franck–Condon region does not access sufficiently high energy for ion-pair formation. The repulsive limb of the next higher state appears to barely reach the minimum energy necessary for the creation of ion pairs. However, using the Grice–Herschbach formula, $\Delta E°$ between this state and the ion-pair state is only ~3.9 cm^{-1}, and the kinetic energy at crossing is about 1.1 eV. Consequently, nonadiabatic

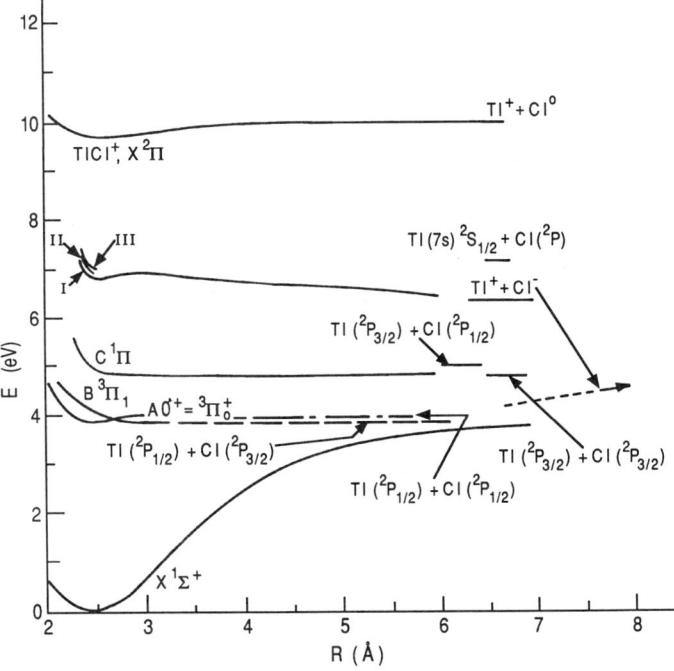

FIGURE 2. Relevant potential energy curves for TlCl for examining ion-pair formation.

behavior is expected here, and atomic products should result. For the uppermost curve, which has the requisite energy in the Franck–Condon region, there would be no crossing at all, since the products $Na(^2P) + I(^2P_{3/2})$ lie 0.02 eV above $Na^+ + I^-$. Hence, the absence of ion-pair formation can be understood, even in an alkali halide for which this process appeared favorable.

By contrast, TlCl has more known excited states, and more possible asymptotes near the ion-pair limit. Some of these are shown in Figure 2. In particular, Rao[11] has assigned two states observed in emission, one of which matches in energy a broad peak in the photoion-pair spectrum of TlCl (see Figure 3). The fact that peak structure appears in the photoion-pair spectrum implies the likelihood of curve crossing (or hopping) rather than direct photoabsorption into the ion-pair state. The close correspondence between the energy of the emitting state and the peak in the ion-pair spectrum is further evidence. However, the detailed evolution from initially excited molecular state to ion-pair products cannot be further elaborated with existing data. In fact (see below), even in cases where a great deal is known about many potential energy curves, from experiment and *ab initio* calculation, it is still difficult to establish unambiguously the path to ion-pair formation.

The other aspect of point 4 which warrants attention is the observation of ion pairs from homonuclear diatomic molecules. We are accustomed to assigning the states of such molecules as gerade or ungerade, implying that they have a center of symmetry. However, as they evolve along their ion-pair trajectory, they obviously must lose this character. Durup[6] has discussed this situation for the case of H_2. He shows that two appropriately chosen Born–Oppenheimer states having g and u character can be combined to give outgoing ion-pair states, i.e.,

$$|ij\rangle = \frac{\sqrt{2}}{2}(|\phi_g\rangle + |\phi_u\rangle)$$

$$|ji\rangle = \frac{\sqrt{2}}{2}(|\phi_g\rangle - |\phi_u\rangle)$$

where $|ij\rangle$ and $|ji\rangle$ are the channel, or ion-pair states.

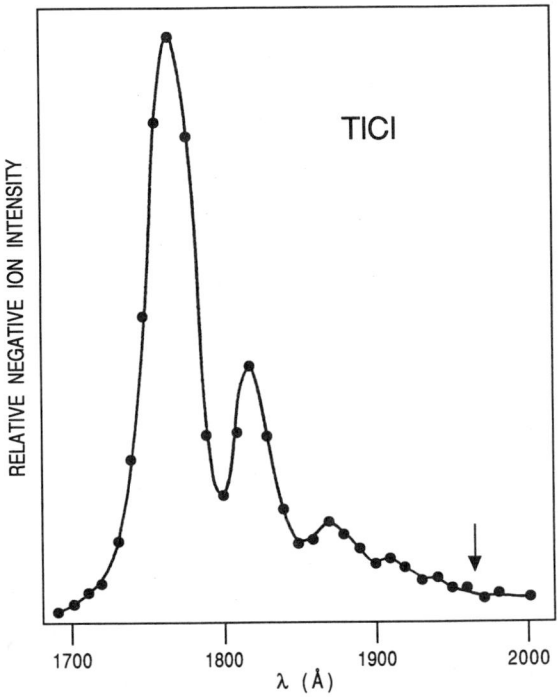

FIGURE 3. Ion yield curve of Cl^- from photoion-pair spectrum of TlCl. [From J. Berkowitz, *Adv. High Temp. Chem.* Vol. 3, p. 147 (1971), with permission from the publisher.]

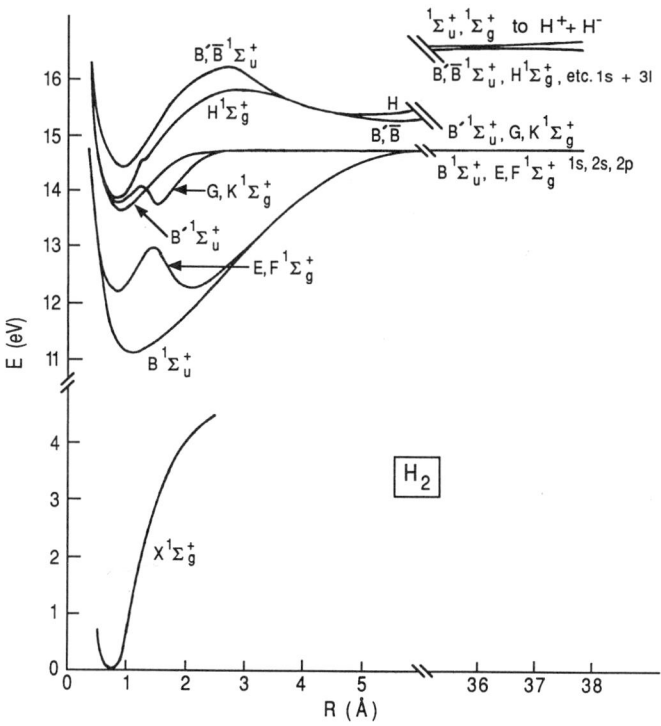

FIGURE 4. Relevant potential energy curves for photoion-pair formation in H_2.

The Born–Oppenheimer states are coupled by two operators, $\partial/\partial R$ and ∇_R^2. The first-order radial coupling $\langle \phi_g | \partial/\partial R | \phi_u \rangle$ vanishes when the center of mass coincides with the center of charge, and will not interest us here. Durup chose the $E,F\,^1\Sigma_g^+$ state and the $B\,^1\Sigma_u^+$ (see Figure 4) as the Born–Oppenheimer states in his detailed treatment, and was unable to account for a pronounced isotope effect in photoion-pair formation from HD (which we can take as a diagnostic for identifying the prominent ion-pair formation pathway). Durup briefly considers the $G,K\,^1\Sigma_g^+$ state and the combination of B'', $\bar{B}\,^1\Sigma_u^+$, and $H^1\,\Sigma_g^+$ states, discounting them in favor of Rydberg states which correlate with $H1s + H3p$ before proceeding to the coulombic path. To our knowledge, this issue is still not resolved.

Although the detailed evolution from initially excited state to ion pairs remains in doubt, the nature of the interaction seems to require that the coupling occur at an internuclear distance $R_x > R_e$, where R_e is the equilibrium internuclear distance. At R_e, the g and u character of individual states are well defined. At the united atom limit, they are also well defined. Therefore, it seems reasonable to expect that for $R < R_e$, the Born–Oppenheimer states g and u retain their character, and hence a transition to the ion-pair states is not expected on the repulsive limbs of the potential energy curves for homonuclear diatomic molecules.[6,12]

The interaction of the Coulomb curve with other states may give rise to the double-well structure seen in some of these H_2 curves. This is particularly noticeable also in the case of Cl_2, for which many potential curves have been calculated by Peyerimhoff and Buenker.[13] An eclectic selection of these curves appears in Figure 5. The experimental photoion-pair curve for Cl_2[14] (Figure 6) has a wealth of structure, which can be assigned to vibrational components of four Rydberg states. Three of these Rydberg states converge to Cl_2^+, $^2\Pi_u$, which is more weakly bound than the Cl_2 ground state. Hence, many vibrational levels in each of these Rydberg states are populated in the Franck–Condon region, accounting for the complex structure.

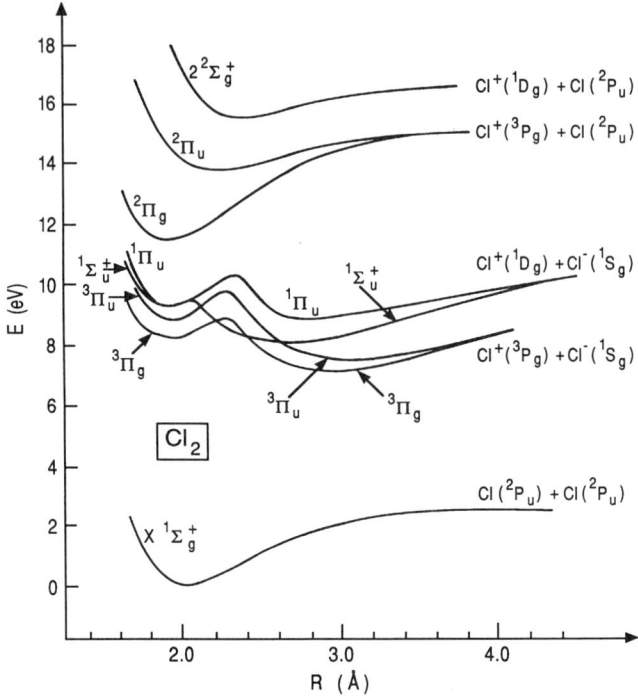

FIGURE 5. Selection of calculated potential energy curves for Cl_2. (From Ref. 13, with permission.)

Another clue to the preferred pathways to photo-ion pair formation is that at least some of the prominent processes tend to occur abruptly near onsets for ion-pair asymptotes. Thus, in Figure 6 there is an abrupt onset corresponding to Cl^+ ($^1D_2$) + Cl^-($^1S_0$), and another corresponding to Cl^+ ($^1S_0$) + Cl^-($^1S_0$). The relative probability of ion-pair formation (i.e., predissociation) compared to autoionization from these quasidiscrete Rydberg states varies somewhat from Rydberg to Rydberg, but not greatly,

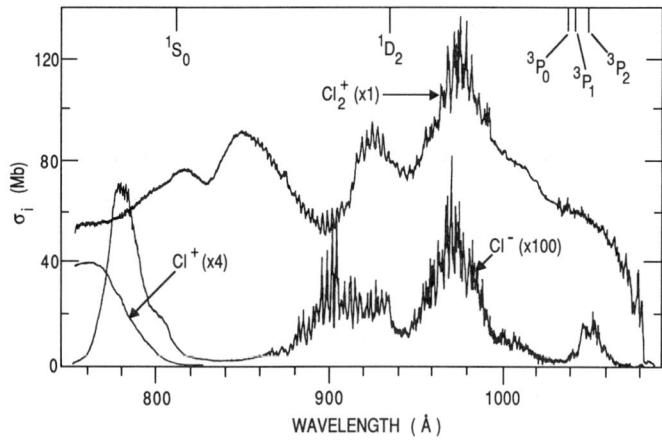

FIGURE 6. Photoion-pair spectrum and photoion yield curve of Cl_2 (overall). (From Ref. 14, with permission from the publisher.)

the ratio staying close to $\sim 10^{-2}$. Also, the vibrational intensity distribution within a Rydberg member does not depart substantially from the expected Franck–Condon distribution in absorption. This is surprising since predissociation is often associated with abrupt onsets at specific vibrational levels.

The adiabatic onset for ion-pair formation in Cl_2 [corresponding to asymptotic products Cl^+ ($^3P_2$) + Cl^- ($^1S_0$)] has been correlated with a Rydberg state converging to the ground state of Cl_2^+, $X\,^2\Pi_g$. This Rydberg state may be one of the double-well states observed and calculated for Cl_2^*. If so, the pathway here may be migration from the inner well to the outer well, and thence to ion pairs. One striking feature of this process is its sensitivity to temperature, which has been rationalized as an instance of heterogeneous coupling.[14] It may occur in other diatomic halogens (see below).

The molecules Br_2 and I_2 provide clear illustrations of point 2. In both cases, Rydberg series are implicated which converge to the ground state of the ion, $^2\Pi_g$. This ground state is substantially split by spin-orbit interaction into $^2\Pi_{3/2,g}$ and $^2\Pi_{1/2,g}$. In the He I photoelectron spectrum, these spin-orbit partners are formed in almost equal abundance. Hence, one would expect corresponding Rydberg members to be populated in approximately equal abundance by photoabsorption. However, the onset of photoionization (forming $X_2^+ + e$) straddles the region of corresponding Rydberg members converging to $^2\Pi_{3/2,g}$ and $^2\Pi_{1/2,g}$. As a consequence, the Rydberg(s) converging to the lower $^2\Pi_{3/2,g}$ state can predissociate into ion pairs without competition from autoionization, whereas those converging to the $^2\Pi_{1/2,g}$ state experience this competition. Although some ion-pair intensity is observed in the latter case, it is $\sim 1/10$ as strong. Thus, for Br_2 and I_2 near threshold, but above the ionization potential, predissociation is about $1/10$ as fast as autoionization. In Cl_2, it will be recalled, the corresponding ratio was $\sim 1/100$ over a substantial energy range. In that study,[14] the instrumental resolution was sufficient to establish a natural linewidth of ~ 0.5 Å for many of the peaks, corresponding to an autoionization rate of $\sim 10^{13}\,s^{-1}$, or a predissociation rate of $\sim 10^{11}\,s^{-1}$. Such a rate is considered a strongly allowed predissociation.[15]

The typical lifetime of states which do not autoionize or predissociate is $\gtrsim 10^{-8}$ s. Hence, predissociating states with the above rates would (in the absence of autoionization) be observed as a "breakoff" of absorption bands at particular values of v'. What seems surprising in the ion-pair spectrum of Cl_2 (and perhaps the other halogens) is that the rates of predissociation for many vibronic states appear to be closely comparable, since the envelope of autoionizing Rydberg states do not differ greatly from the corresponding envelope of predissociating vibronic states, which appear to be the same quasidiscrete states. It appears as if several Rydberg states, in several vibrational levels, are predissociating into ion pairs at comparable rates, which superficially seems contradictory to the expectation of specificity in the curve crossing process. Further theoretical work is required to elucidate the details of these processes.

Point 3 focuses on the relative decline of the ion-pair process as one proceeds to triatomic and polyatomic molecules. This observation can be rationalized within the assumptions of quasiequilibrium theory (QET). A necessary condition for the applicability of QET in its strong form is the prior occurrence of radiationless transitions from electronically excited states to the electronic ground state (internal conversion). The predissociative mechanism described above is an electronic transition in competition with internal conversion. Hence, if QET is perfectly obeyed, ion-pair formation would not be observed. In fact, QET appears to be more valid as the size of the molecule increases, while ion-pair formation exhibits the opposite trend. Ion-pair formation can thus be viewed as a diagnostic for the degree of applicability of QET.

The typical fragmentation product following formation of an excited photoion displays a gradual (usually linear) increase in intensity with excess energy above threshold. By contrast, photoion-pair formation often displays an abrupt onset. There are several cases (XeF_2, KrF_2, COS, CO_2, SO_2, CF_nCl_n) in which simultaneous production of three particles (e.g., $O^- + C^+ + O$, $Xe^+ + F^- + F$) occurs rather abruptly at threshold. In the conventional fragmentation, such processes usually appear sequentially, i.e., $ABC^+ \rightarrow AB^+ + C \rightarrow A^+ + B + C$, and then display very gradual onsets. These differences imply that ion-pair formation is favored by low excess kinetic energies, whereas normal fragmentation is favored by larger excess energies. The former implies curve crossing (or surface crossing).

It has been the conventional wisdom that photoion-pair production is confined to excitation processes involving the valence shells. Recently, Dujardin et al.[16] discovered that photoexcitation of

SO_2 in the vicinity of the $S2p$ edge (~ 175 eV) results *inter alia* in a weak channel ($\lesssim 10^{-5}$ of positive ions) forming $SO^{2+} + O^- + e^-$. A few examples of these higher-energy processes are discussed in the last section of this review.

2. RECENT STUDIES

2.1. Diatomic Molecules

2.1.1. The Diatomic Halogens. The experimental photoion-pair spectra of F_2, Cl_2, Br_2, and I_2 appear in Figures 7, 6, 8, and 9, respectively. In each instance, the photoion yield curve for parent ionization is displayed, for comparison. A more recent spectrum has appeared[17] for Br_2; we chose the earlier one[18] because it appears to have higher resolution, and it displays additional structure near the higher limits. For I_2, an earlier spectrum[19] appears superimposed on the more recent one,[20] in order to incorporate the higher-energy structure.

In each of the halogens, the lowest-energy structure corresponds to formation of X^+ (^3P) + X^- (^1S). In L-S coupling, the asymptotic molecular states would be $^3\Sigma^-$ and $^3\Pi$. Since the ground state of X_2 is $^1\Sigma_g^+$, conventional electric dipole selection rules would favor initial formation of $^1\Sigma_u^+$ and $^1\Pi_u$ Rydberg states, and curve crossing or mixing would involve triplet–singlet coupling. Yencha *et al.*[17] (Br_2) and Kvaran *et al.*[20] (I_2) overcame this difficulty by invoking Ω-ω coupling for these heavier halogens. In this coupling scheme, the lowest-energy atomic cation ($^3P_2$), together with X^- ($^1S_0$) can give rise to Ω = 2, 1, and 0^+. Yencha *et al.* favor photoabsorption from the $^1\Sigma_g^+$ ground state to a Rydberg having 0_u^+ character ($8p\pi$ for Br_2, $9p\pi$ for I_2), which can interact with the ion-pair state by homogeneous coupling, and lead to Br^+ (I^+) $^3P_2$ + Br^- (I^-) $^1S_0$. Furthermore, it is suggested by them that this crossing or near-crossing occurs "on the inner repulsive limbs of the two potentials."

While Ω-ω coupling may be appropriate for I_2 and Br_2, it is much less likely for Cl_2 and F_2. Yet these higher halogens apparently can overcome the obstacle represented by the triplet–singlet transition and form X^+ (^3P) + X^- (^1S). Also, we have previously discussed the g-u problem with regard to homonuclear diatomics and ion-pair formation, and shown that there is g-u mixing which is radially coupled at internuclear distances $> R_e$, and that it is unlikely on the repulsive limbs. Finally, we note that in all cases, the onset of ion-pair formation occurs at energies significantly *below* the ion-pair threshold. In the case of Cl_2,[14] it was shown that this below-threshold component was independent of pressure, but was strongly temperature dependent. In the case of F_2,[21] there is still a significant

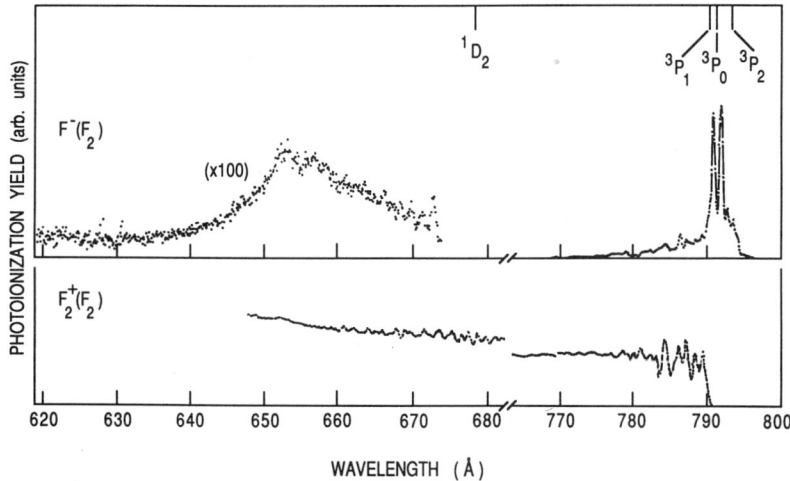

FIGURE 7. Photoion-pair spectrum and photoion yield curve of F_2. (From Ref. 21, with permission.)

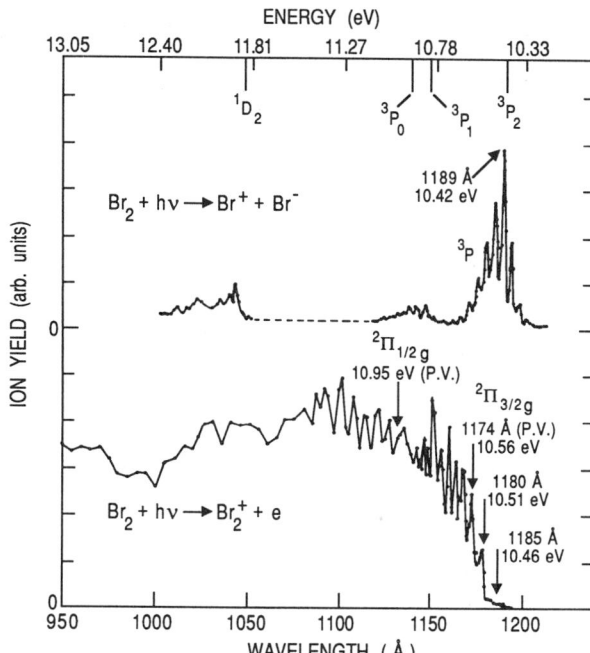

FIGURE 8. Photoion-pair spectrum and photoion yield curve of Br_2. (From Ref. 18, with permission.)

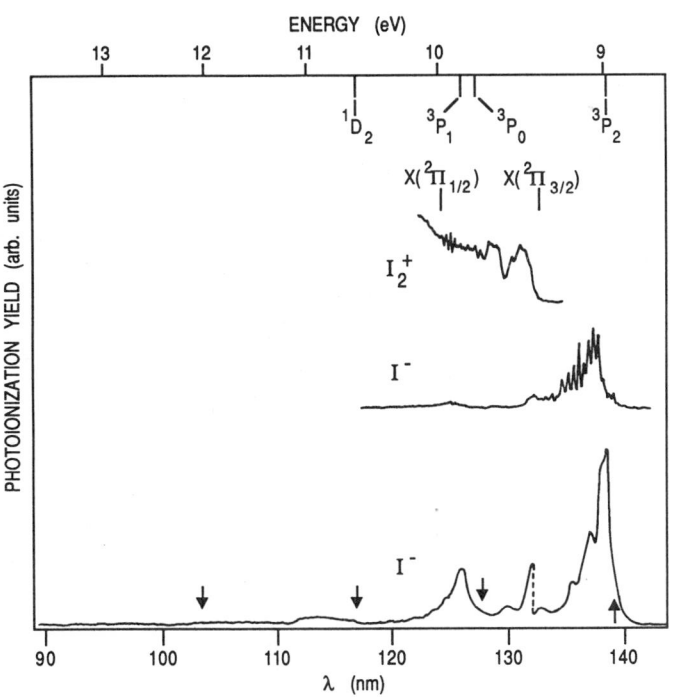

FIGURE 9. Photoion-pair spectrum and photoion yield curve of I_2. [From Refs. 20 and 19 (lowest I^-), with permission.]

below-threshold component, although the experiment was carried out at 80 °K. In the Cl_2 study,[14] it was suggested that heterogeneous predissociation, which makes efficient use of the rotational energy, might explain the below-threshold intensity. The rate of heterogeneous predissociation is proportional to $J(J + 1)$.[22] Thus, two different lines of reasoning—g-u mixing and below-threshold ionization—invoke the influence of rotation. Normally, the effective potential curve incorporating the centrifugal potential gives rise to a maximum, a consequence of the fact that the Born–Oppenheimer potential curve approaches the asymptote more quickly than $1/R^2$. However, when the asymptotic electrostatic potential is Coulomb-like, as in ion-pair formation, no potential maximum arises.[15]

In each of the halogens, the thermochemical onset for ion pairs occurs just below or not far above the first ionization potential. Hence, it is not surprising that high Rydberg states have been implicated in the initial absorption act. For these states, the Rydberg electron is far from the nuclei, and its angular momentum is more likely to couple to the rotational angular momentum vector, characteristic of Hund's case d. This is another indication that rotational angular momentum plays a role in photoion-pair formation in the halogens. Heterogeneous predissociation has already been shown to be the operative mechanism in the low-lying $B\,^3\Pi_0\,(0_u^+)$ states of Cl_2, Br_2, and I_2,[23] predissociated by a potential curve having 1_u character. In these low-lying states, it is not necessary to invoke case d coupling. It may be noteworthy that homogeneous coupling becomes permissible when the ion-pair asymptotes are $^1D + {}^1S$, or $^1S + {}^1S$. Yet, there is no indication in the ion-pair spectra above these asymptotes that the rate of predissociation is much more rapid than it is for the lowest $(^3P + {}^1S)$ asymptote.

2.1.2. O_2. Dadouch[24] has recently reexamined ion-pair formation in O_2 between 17 and 50 eV. There are two relatively intense regions, 17.2–18 eV and 20.8–24.5 eV. Both of these regions had previously been studied by Oertel et al.,[25] with somewhat lower resolution and signal strength. With $D_0(O_2) = 5.115_6$ eV,[26] IP(O) = 13.6181 eV,[27] and EA(O) = 1.4611103 eV,[28] the calculated threshold for

$$O_2 + h\nu \rightarrow O^+\,(^4S_u) + O^-\,(^2P_u)$$

is 17.272_6 eV. The threshold value of Dadouch is 17.26 eV, that of Oertel et al., 17.28 eV. The structure in the O^- yield curve in the region of 17.3–18.5 eV appears to be dominated by Rydberg series converging to the $b\,^4\Sigma_g^-$ state of O_2^+. The Rydberg series are assigned to

$$\ldots 3\sigma_g\,(1\pi_u)^4\,(1\pi_g)^2\,np\sigma_u,\,^3\Sigma_u^-$$

$$\ldots 3\sigma_g\,(1\pi_u)^4\,(1\pi_g)^2\,np\pi_u,\,^3\Pi_u$$

based on their quantum defects. The $np\sigma_u$ series is much the stronger one. Dadouch can assign Rydberg series converging to $v^+ = 0, 1, 2, 3$. Since the combination of $O^+(^4S_u)$ and $O^-(^2P_u)$ can give rise to $^3\Sigma_u^-$ and $^3\Pi_u$ states, there do not appear to be any restrictions to predissociation of the Rydberg states into ion pairs. A direct photoabsorption to the ion-pair states would have weak probability.

Between 18.5 and 20.5 eV, Dadouch observes structure previously undetected. It is weaker by about a factor of 30 from the adjoining features. It is attributed to series analogous to the above, but converging to $B^2\Sigma_g^-$.

Between 20.5 and 24.5 eV, Dadouch observes relatively intense, sharp structure in the O^- channel. Two apparent thresholds are seen, at 20.59 and 21.26 eV. The 20.59 eV onset is in good agreement with a calculated threshold for $O^+(^2D_u) + O^-(^2P_u)$. [The $^4S_u \rightarrow {}^2D_u$ excitation in O^+ is 3.3249 eV,[27] and hence the calculated threshold to $O^+(^2D_u) + O^-(^2P_u)$ is 20.597_5 eV.] Dadouch has no explanation for the increase at 21.26 eV, but sees a smaller, abrupt increase in O^- at 22.29 eV which matches the expected onset for $O^+(^2P_u) + O^-(^2P_u)$. The sharp structure is attributed to $ns\sigma_g$ and $nd\sigma_g$ Rydberg series emanating from the $2\sigma_u$ orbital, and converging to $c^4\Sigma_u^-$. Both $v^+ = 0$ and $v^+ = 1$ series are observed. Dadouch also sees a broad feature at ~21.5 eV, which is attributed to a shape resonance (i.e., excitation into an unoccupied σ_u orbital), with some fine structure superimposed, and attributed to two-electron excitation processes.

Some broad structure is observed between 26 and 30 eV, with fine structure superimposed. Dadouch tentatively assigns the lower-energy features to Rydberg series converging to $B^2\Sigma_g^-$ (IP = 27.25 eV), and some higher-energy features to early members of Rydberg series converging to a $^2\Pi_u$ state of O_2^+ at 33 eV.

There is a broad peak between 31 and 34 eV which Dadouch attributes to a direct transition to an ion-pair state with asymptotes $O^+(^4P) + O^-(^2P)$, where $O^+(^4P)$ has the configuration $2s2p^4$, and an excitation energy of 14.870 eV.[27] This assignment implies a calculated threshold of 32.14 eV for this process. This interpretation is possible, but not entirely convincing.

Between 34 and 40 eV, eight resonances are observed, four below the O_2^{2+} threshold (stated to be 36.2 eV), and four above. Dadouch does not assign these resonances, but presumes that the excited states predissociate to $O^+(^4P) + O^-(^2P)$ at 32.14 eV. In the region of 40–50 eV, Dadouch restricts himself to assigning a sharp onset at 40.31 eV, peaking at 40.62 eV. If this corresponds to a new O^+ channel, the likely candidate is the ^4P state of O^+ with the configuration $...2s^22p^2\,(^3P)3s$, which lies 22.987 eV[27] above the O^+ ground state. The calculated threshold would be 40.26 eV, slightly below the observed onset.

2.1.3. NO.

The calculated thermochemical thresholds for ion-pair formation in NO are: $N^+(^3P) + O^-(^2P)$, 19.585 eV; $N^+(^1D) + O^-(^2P)$, 21.484 eV; and $N^+(^1S) + O^-(^2P)$, 23.638 eV. All of these are much higher than the first ionization potential, 9.26436 eV. Not surprisingly, the cross section for ionization is described as "orders of magnitude larger than for the ion pair processes."[25] Oertel et al.[25] observe "steep onsets" at 19.56 and 21.46 eV, which they attribute to direct transitions to the ion-pair states. This does not seem likely, since the ion-pair curve, without configuration interaction, is likely to have a relatively large equilibrium internuclear distance. At the Franck–Condon absorption maximum, the curve is likely to be in the steeply rising repulsive limb, which would not lead to a steep onset. With their resolution (2 Å), the ion-pair spectrum has structure at higher energy. They comment that they can recognize members of a Rydberg series converging to $B'\,^1\Sigma^+$ of NO^+, which "should predissociate to ion-pair states leading to the dissociation limit at 19.56 eV or 21.46 eV," and further note that "a symmetry allowed transition is possible." Higher resolution studies may clarify the relative contributions of direct- and pre-dissociation.

2.1.4. CO.

The calculated thermochemical thresholds for ion-pair formation in CO are: $C^+(^2P) + O^-(^2P)$, 20.898 eV; $O^+(^3P) + C^-(^4S)$, 23.447 eV. Both of these thresholds are much higher than the adiabatic ionization potential of CO, 14.0139 eV. In fact, if we write the ground-state configuration of CO as

$$(1\sigma)^2\,(2\sigma)^2\,(3\sigma)^2\,(4\sigma)^2\,(1\pi)^4\,(5\sigma)^2,\; X^1\Sigma^+$$

then the adiabatic ion-pair thresholds are also well above $(1\pi)^{-1} = 16.54$ eV and $(4\sigma)^{-1} = 19.67$ eV, but well below $(3\sigma)^{-1} = 38.3$ eV. Hence, single electron excitations giving rise to Rydberg series in the vicinity of the ion-pair thresholds cannot involve $(5\sigma)^{-1}$, $(1\pi)^{-1}$, or $(4\sigma)^{-1}$ as limits, and are unlikely to involve $(3\sigma)^{-1}$. Nonetheless, experimental onsets very close to the predicted values are observed: 20.91[25] and 20.98[29] for $C^+ + O^-$, and 23.45[25,29] for $O^+ + C^-$. Dadouch et al.[29] divide their spectra into three energy regions:

1. 20.9–24 eV, where the O^- and C^- spectra display narrow resonances
2. 24–30 eV, characterized by very broad resonances
3. 30–40 eV, where narrow resonances appear once more

The narrow resonances appearing between 20.9 and 24 eV are assigned by Dadouch et al. to two-electron excitations (one electron excited to a nearby valence state, the other to a Rydberg state). The convergence limits of these states are believed to be the $D^2\Pi$ (VIP = 22.734 eV), $E^2\Sigma^+$ (VIP = 23.383 eV), and $E'\,^2\Sigma^+$ (VIP = 25.483 eV). The configurations of these ionic states are:

$D^2\Pi$: $\quad\ldots(3\sigma)^2\,(4\sigma)^2\,(1\pi)^2\,(5\sigma)^2\,2\pi$
$E^2\Sigma^+$: $\quad\ldots(3\sigma)^2\,(4\sigma)^2\,(1\pi)^3\,(5\sigma)\,2\pi$
$E'\,^2\Sigma^+$: $\quad\ldots(3\sigma)^2\,(4\sigma)^2\,(1\pi)^3\,(5\sigma)\,2\pi$

The broad resonances appearing in the region of 24–30 eV are believed[29] to be Rydberg series converging to $F^2\Sigma^+$ [VIP = 27.44 eV, configuration $...(5\sigma)^{-2}6\sigma$] and $G^2\Sigma^+$ [VIP = 28.086 eV, configuration ...$(4\sigma)^{-1}(1\pi)^{-1}2\pi$] but the large widths of the resonances preclude a definite assignment. In the region of 30–40 eV, data are lacking on CO^+ excited states. Hence, despite the observation of sharp resonances, no assignment is possible.

No evidence could be obtained by Dadouch et al.[29] for the known shape resonance in CO, although a coupling between shape resonances and ion-pair states may be expected to exist, and was observed for O_2. They also remark that the observed O^- resonances generally do not coincide with the C^- ones.

2.2. Triatomic Molecules

2.2.1. N_2O.
Photoion-pair formation in N_2O was examined by Kratzat[30] in 1984 at a resolution of ~2 Å, and more recently by Mitsuke et al.[31] at 0.8 Å. The results of the two studies are in good agreement. Both observe only the $O^- + N_2^+$ process, about a factor of 10^{-3} weaker than photoionization. Kratzat reports some NO^- about two orders of magnitude weaker. The thermochemical threshold for $O^- + N_2^+$ occurs at 15.794 eV, that for $NO^- + N^+$ at 19.444 eV. Mitsuke et al. observe a threshold at 15.784 ± 0.01 eV for O^- while Kratzat gives 15.8 eV. The adiabatic ionization potential of N_2O is much lower, 12.886 eV.[32] A wealth of structure is observed. In the near-threshold region it is mostly assigned to $nd\pi$ series converging on $\tilde{A}\,^2\Sigma^+$ (0,0,0), and (1,0,0) of N_2O^+. At higher energies, more intense peaks are observed, and assigned primarily to $nd\pi$ series converging to $\tilde{C}\,^2\Sigma^+$ (0,0,0), with less intense features attributed to $nd\sigma$, $np\sigma$, and $np\pi$. These series had been observed previously in autoionization, with varying peak profiles.[32] Between these regions, weaker predissociative structure is seen, with constant spacing, and assigned to a vibrational progression in the first Rydberg member converging on $\tilde{B}\,^2\Pi$.

2.2.2. NO_2.
Kratzat's[30] study of the negative ion spectrum on photoexcitation of NO_2 revealed primarily O^- (~5×10^{-3} of primary ionization), with NO^- about two orders of magnitude lower. The calculated thermochemical threshold for formation of $NO^+ + O^-$ is 10.930 eV, that for $O^+ + NO^-$, 16.72 eV. In Kratzat's spectrum (Figure 10) there is a weak onset at ~10.92 eV, and a stronger one at ~11.1 eV. Kratzat assigns the weak, sloping onset at 10.92 eV to a direct transition to the ion-pair state, which may be plausible in this case since it represents only a few percent of the ion-pair intensity, and it is not abrupt.

The ground-state configuration of NO_2 may be written as:

$$(1a_1^2)^2(1b_2)^2(2a_1)^2(3a_1)^2(2b_2)^2(4a_1)^2(3b_2)^2(5a_1)^2(1b_1)^2(1a_2)^2(4b_2)^2(6a_1),\,^2A_1$$

The photoion yield curve of $NO_2^+(NO_2)$ rises very gradually from threshold[33] presumably because of weak Franck–Condon factors. The adiabatic ionization potential, corresponding to $(6a_1)^{-1}$, is 9.586 ± 0.002 eV.[34] The most prominent features in Figure 10 appear to be Rydberg members emanating from $4b_2$. The doublet at ~12.45 eV can be assigned to $4b_2 \to 4d$, $v' = 5$ and 6 or $4b_2 \to 5d$, $v' = 1$ and 2. These energies are almost identical, according to Morioka et al.[35] What is striking about this doublet is that two vibrational members selectively predissociate, within a longer Franck–Condon absorption progression. The next most prominent grouping, between 11.500 and 11.834 eV, can be assigned to $4b_2 \to 4s\sigma$, $v' = 2$–6. Here again, the $v' = 0$ and 1 members predissociate much more weakly than the higher members of the Franck–Condon progression. The peaks near 12.9 eV, not prominent in absorption, have been assigned by Kratzat[30] to two $1a_2 \to np$ series. The broad band between ~13.5 and 15.3 eV with weak structure superimposed is most likely the result of unresolved Rydberg members emanating from $1b_1$ and $5a_1$ orbitals, since the corresponding states in the photoelectron spectrum[36] display a similar breadth. Finally, there is an isolated peak at 15.38 eV which is assigned to $3b_2 \to 3s\sigma$, the first member of this Rydberg series.

2.2.3. COS.
Photoion-pair formation in COS was studied by Kratzat[30] at a resolution of ~2 Å, and more recently by Mitsuke et al.[37] at 0.8 Å. Both groups found S^- and O^- to be the only significant negative ions. Kratzat estimated S^- to be about 10^{-3} of the primary ionization, with O^- about five times weaker, while Mitsuke et al. found both to be about 5×10^{-5} of primary ionization. Kratzat encountered background at the S^- position related to photoelectron dissociative attachment on COS, whereas

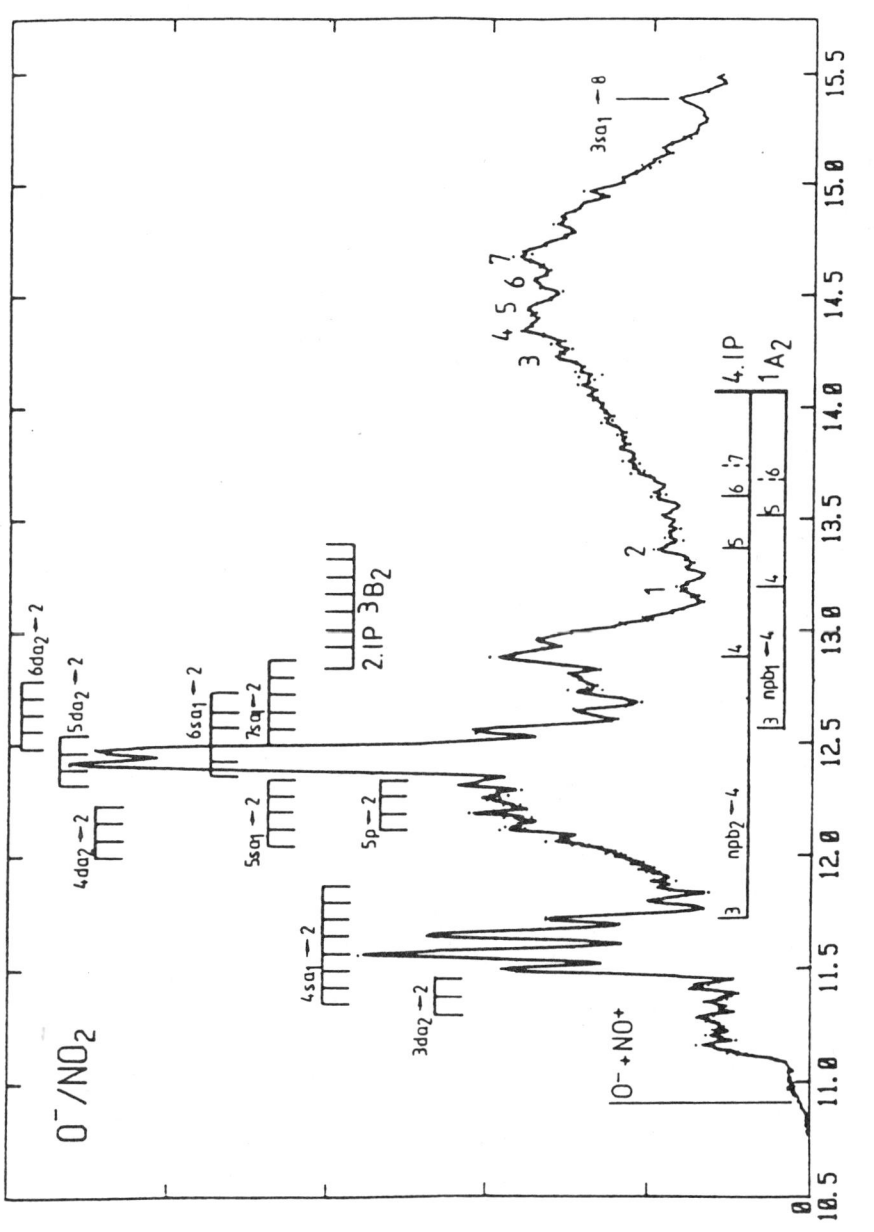

FIGURE 10. Photoion-pair spectrum of $NO_2 \rightarrow O^- + NO^+$. (From Ref. 30, with permission.)

Mitsuke et al., using a supersonic expansion molecular beam with a skimmer, had much less difficulty with this unwanted contribution. The calculated thermochemical thresholds are: $CO^+ + S^-$, 15.076 eV; $CS^+ + O^-$, 16.63 eV. The observed values are ~15.09 eV[30] and 15.10 ± 0.04 eV[37] for the first process, ~16.8 eV[30] and 17.08 ± 0.09 eV[37] for the second. Both ion-pair thresholds are much higher than the adiabatic ionization potential, 11.174 eV.[38] Asymptotically, both ion-pair products correlate with $^{1,3}\Sigma, ^{1,3}\Pi$. Since absorption from the $^1\Sigma$ ground state can yield upper states $^1\Sigma, ^1\Pi$, there are no symmetry restrictions to ion-pair formation. Neither group saw any evidence for abrupt onsets near excited cation states, but Mitsuke et al. (whose work extended to much higher energy) provide evidence for increases in ion yield at the onsets of three-body decomposition ($S^- + C^+ + O$; $O^- + C + S^+$).

In the S^- photoion yield curve, there are two broad regions, which Mitsuke et al. designate as F1 and F2. The F1 region, between threshold and ~16 eV, consists of peaks attributed to predissociation of Rydberg series converging to the $\tilde{B}\,^2\Sigma^+$ state of COS^+. Both Kratzat and Mitsuke et al. identify most of these features with $ns\sigma$ and $nd\sigma$ series previously assigned. Mitsuke et al. are able to find one feature attributable to an $np\sigma$ ($n = 5$) member. The interpretation of the F2 region (which is a broad peak between ~16 and 20 eV) is more controversial. Kratzat observes some weak structure on the rising (16–18 eV) region, which he interprets as members of two series ($ns\sigma$ and $np\pi$) converging to $\tilde{C}\,^2\Sigma^+$ of COS^+. Mitsuke et al. observe some weak structure, but cannot correlate it with previously observed Rydberg states converging to $\tilde{C}\,^2\Sigma^+$. In any event, the convergence limit of $\tilde{C}\,^2\Sigma^+$ (0,0,0) is 17.96 eV; the maximum of F2 occurs at ~18.4 eV, and it is quite broad, compared to the Franck–Condon breadth of the $\tilde{C}$ state. Instead, Mitsuke et al. attribute the F2 region to predissociation of an excited valence state. Taking the ground-state configuration of COS to be

$$(\text{core})^{14}(6\sigma)^2(7\sigma)^2(8\sigma)^2(9\sigma)^2(2\pi)^4(3\pi)^4$$

they identify the formation of the excited valence state with a $9\sigma \to 10\sigma$ transition. They cite an SCF calculation by Sheehy and Langhoff,[39] which finds a large oscillator strength for the $9\sigma \to 10\sigma$ transition, and an $(e,2e)$ experiment which finds a large partial cross section for formation of $\tilde{C}$, interpreted as autoionization from the $9\sigma \to 10\sigma$ resonance state.

The explanation offered by Mitsuke et al. has been utilized to rationalize a portion of the $O^- + CS^+$ spectrum. Here there is a first broad band, designated by them as F3, which has a maximum and breadth comparable to F2 above. They present some correlation diagrams to suggest that predissociation to both $O^- + CS^+$ and $S^- + CO^+$ should proceed readily from … 10σ. Kratzat finds some weak evidence for structure in this band, which he thinks might be attributable to $ns\sigma$ Rydbergs converging to $\tilde{C}$, but poor statistics prevent him from drawing definitive conclusions.

2.2.4 CO_2. Mitsuke et al.[37] have recently reported on the formation of $O^- + CO^+$ from CO_2, with a relative abundance of ~1 × 10^{-4} compared to primary photoionization. The thermochemical threshold for this process is 18.006 eV, while the observed onset is 17.990 ± 0.025 eV. The striking character of their photoion yield curve is a very weak, structured threshold region (where the structure can be assigned to $np\sigma_u$ and $nd\pi_g$ Rydberg series converging on $\tilde{C}\,^2\Sigma_g^+$ of CO_2^+ and a piling up of high Rydberg members near the $\tilde{B}\,^2\Sigma_u^+$ onset) and two broad, intense bands at higher energy which apparently cannot be explained by Rydberg predissociation. Mitsuke et al. try to explain the broad structure between ~20.6 and 23.4 eV in terms of an excited valence state. Taking the electronic configuration of the ground state of CO_2 as

$$(\text{core})^6(3\sigma_g^2)(2\sigma_u)^2(4\sigma_g)^2(3\sigma_u)^2(1\pi_u)^4(1\pi_g^4), \quad \tilde{X}^1\Sigma_g^+$$

they focus on the $3\sigma_u \to 5\sigma_g$ excitation. As support, they cite static-exchange calculations by McKoy and collaborators[40,41] which predict considerable oscillator strength for this transition, and some experimental observations.[42–44] There is a hint of structure in the partial cross section for formation of $(\tilde{A}^2\Pi_u + \tilde{B}^2\Sigma_u^+)$, unresolved, in the synchrotron-based[42] and $(e,2e)$ experiments[43] in the region of 20–21 eV. Perhaps the most convincing evidence is a peak at ~21.7 eV in the partial cross section of CO^+ from CO_2.[45] However, the other component at ~23 eV is not discernible. An alternative

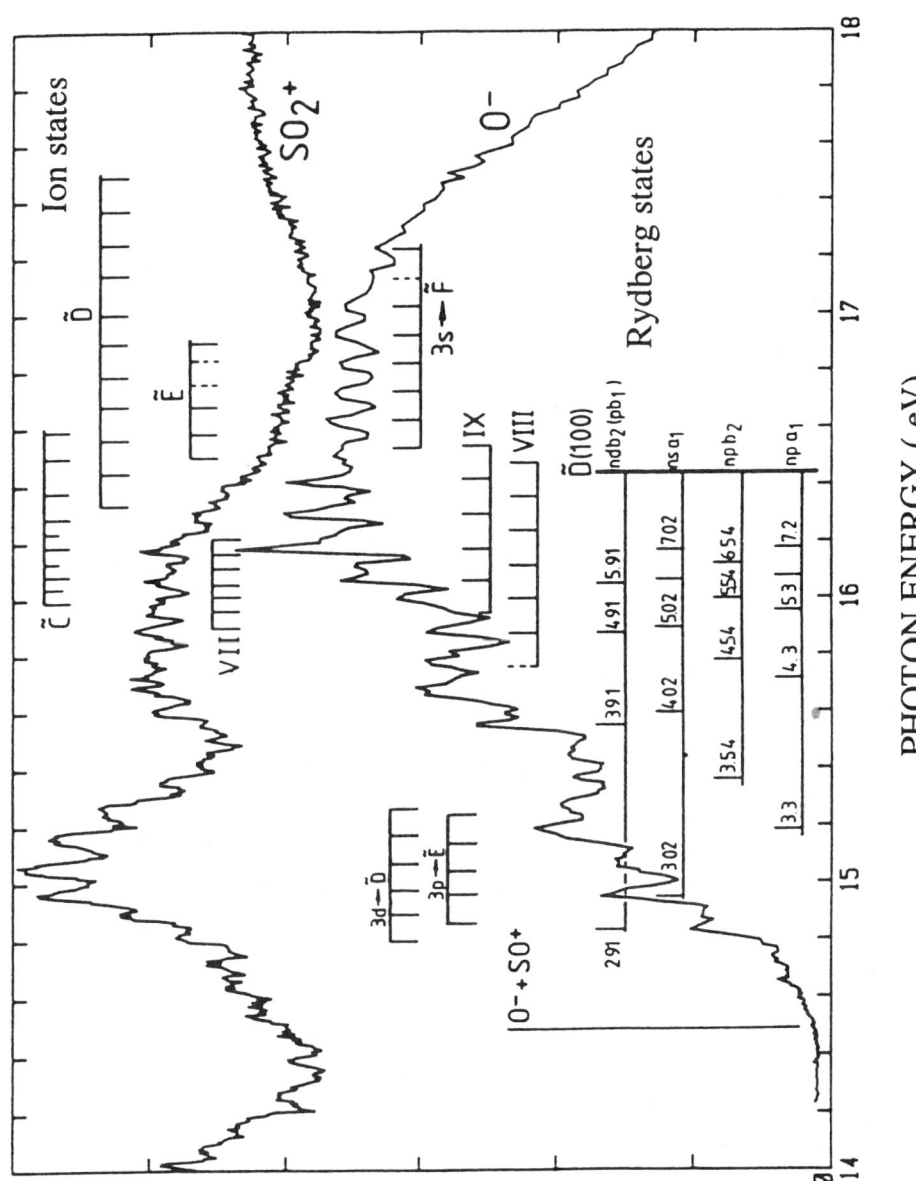

FIGURE 11. Photoion-pair spectrum of $SO_2 \rightarrow O^- + SO^+$. (From Ref. 30, with permission.)

explanation for the 21.7-eV peak could involve Rydbergs converging on the shake-up state of CO_2^+ at 22.8 eV.[42,46]

Mitsuke et al. observe another broad band at higher energy beginning abruptly at the threshold for fragmentation into $(O^- + C^+ + O)$ at ~26.34 eV ≡ 471 Å, and having a maximum at ~460 Å. There is a corresponding broad peak in CO^+ production from CO_2,[45] with a maximum ~450 Å. This feature may be related to another shake-up state at about 27.3 eV.[42] In any event, the major intensity in the O^- spectrum arises from minor excitations, not readily observed in other ionization and absorption channels.

2.2.5. SO_2. Kratzat[30] has examined the negative ion spectrum on photoexcitation of SO_2. The primary ion-pair channel is $O^- + SO^+$, with an abundance of ~10^{-3} of primary ionization. He also has observed $S^- + O_2^+$ (with some contribution from $O_2^- + S^+$), about an order of magnitude weaker. The calculated thermochemical threshold for $O^- + SO^+$ is 14.52 ± 0.02 eV. In Kratzat's ion-pair spectrum (see Figure 11) the onset is gradual, at about 14.5 eV. The calculated thresholds for $S^- + O_2^+$ and $O_2^- + S^+$ are 15.89 and 15.82 eV, respectively. Kratzat's spectrum for m/z = 64 near 16 eV is contaminated by photoelectron-induced dissociative attachment, making both the precise threshold and its assignment uncertain. Kratzat argues that $O_2^- + S^+$ must correlate with a triplet (or higher multiplicity), making it a less likely candidate for both direct and indirect (i.e., predissociative) ion-pair formation. Although SO^- is observed (the calculated threshold for $SO^- + O^+$ is higher, 18.19 eV), Kratzat also attributed this signal to dissociative attachment, since these products also correlate with a triplet or higher multiplicity, whereas the ground state is a singlet.

The electronic configuration of the ground state of SO_2 may be written

$$(1a_1)^2(1b_2)^2(2a_1)^2(3a_1)^2(2b_2)^2(4a_1)^2(1b_1)^2(5a_1)^2(3b_2)^2(6a_1)^2$$
$$(2b_1)^2(7a_1)^2(4b_2)^2(5b_2)^2(1a_2)^2(8a_1)^2, \quad {}^1A_1$$

The adiabatic ionization potential, corresponding to ejection of an $8a_1$ lone pair electron, is ~12.35 eV,[47] much lower than the ion-pair thresholds. The photoionization mass spectrum[47] displays weak fine structure as a result of autoionization, superimposed on a broad, undulating ion yield curve. The peak structure is somewhat more pronounced in the ion-pair spectrum (Figure 11). The spectrum is congested and difficult to assign unambiguously, but Kratzat finds at least some correlation with four Rydberg series (one ns, two np, and one nd) converging to v' = 1 of the $\tilde{D}$ ion state, corresponding to $(7a_1)^{-1}$. The intensities of the higher members of the series (n = 5–8) are anomalously high in the ion-pair spectrum, an observation attributed to stronger interaction between these members and the ion-pair state. Some distinct structure in the 17 eV energy range, which has no counterpart in the photoionization spectrum of SO_2^+, SO^+, or S^+, is assigned as the first Rydberg member of a series converging to the $\tilde{F}$ ion state, corresponding to $(6a_1)^{-1}$.

After a precipitous decline to ~18.5 eV, the O^- yield displays a rather abrupt increase at about 19.9 eV, which is attributed to the onset for three-particle formation, i.e., $O^- + S^+ + O$ (calculated threshold = 19.91 eV). In the S^- ion yield curve, there is also an abrupt onset at ~ 22.5 eV, identified with the appearance of $S^- + O^+ + O$ (calculated threshold = 22.5 eV).

2.3. Polyatomic Molecules

2.3.1. CH_4 and Methyl Halides. It is convenient to treat these molecules together, because there are many similarities. In Table II, we list the adiabatic ionization potentials, their spin-orbit partner potentials (where appropriate), and the calculated thresholds for ion-pair formation. Only in CH_4 is the ion-pair threshold at higher energy than the adiabatic ionization potential. Hence, there should be a region where ion-pair formation does not have competition from (auto)ionization. In fact, ion-pair formation has now been observed for all cases except CH_3I. For CH_3Br, previously unpublished, relatively crude data were obtained a number of years ago in this laboratory.[48] A quite recent study of CH_3Br[49] was unable to provide data.

The photoion-pair spectra of CH_3F, CH_3Cl, and CH_3Br near the respective thresholds for these processes are presented in Figures 12, 13, and 14, respectively. For CH_3F, there is no distinct peak

TABLE II

Observed and Calculated Ion-Pair Thresholds for CH_4 and CH_3X (X = F, Cl, Br, I), Adiabatic Ionization Potentials, and Spin-Orbit Excited State Energies

	Ionization potential, eV		Ion-pair threshold	
	Adiabatic	s/o excited state	Calculated	Observed
CH_3F	~12.50 (broad, structured band)[a]		11.36 ± 0.16 eV[b] (1092 ± 16 Å)	< 1020 Å[c]
CH_3Cl	11.28[a]	11.36[a]	9.788 ± 0.006 eV[b] (1266.7 ± 0.8 Å)	< 1250 Å[c]
CH_3Br	10.53[a]	10.85[a]	9.470 ± 0.005 eV[b] (1309.2 ± 0.7 Å)	~1310 Å[d]
CH_3I	9.94[a]	10.16[a]	9.198 ± 0.005 eV[b] (1348.0 ± 0.7 Å)	?
CH_4	12.615 (broad band)[e]		13.570 ± 0.002 eV[b]	13.37 ± 0.15 eV[f]

[a] Ref. 53.
[b] Heats of formation of CH_3, CH_4, and CH_3X (X = F, Cl, Br, I) taken from V. P. Glushko, L. V. Gurvich, G. A. Bergman, I. V. Veitz, V. A. Medvedev, G. A. Khachkuruzov, and V. C. Yungman, *Termodinamicheskie Svoistva Individual'nikh Veshchestv* (Nauka, Moscow, 1978 and 1979), Vols. 1 and 2; I. P. (CH_3) is from Ref. 44; E. A. (F, Br, I) from H. Hotop and W. C. Lineberger, *J. Phys. Chem. Ref. Data* **14**, 731 (1985); E. A. (Cl) from R. Trainham, G. D. Fletcher, and D. J. Larson, *J. Phys. B* **20**, L777 (1987).
[c] Ref. 49.
[d] Ref. 48.
[e] See, for example, J. Berkowitz, *Photoabsorption, Photoionization and Photoelectron Spectroscopy* (Academic Press, New York, 1979), p. 275.
[f] Ref. 54.

structure, and the onset of an observable signal is gradual, and occurs well above threshold. In the case of CH_3Cl, there is distinct peak structure, which becomes blurred or absent just above the ionization potential. The observed onset is weak, and occurs above the calculated ion-pair threshold. The situation is similar for CH_3Br, except that the observed onset is just about at the calculated ion-pair threshold. The structure in CH_3Cl and CH_3Br can be assigned to Rydberg series converging on the first ionization potential ($X^2E_{3/2}$) or its spin-orbit partner ($^2E_{1/2}$).

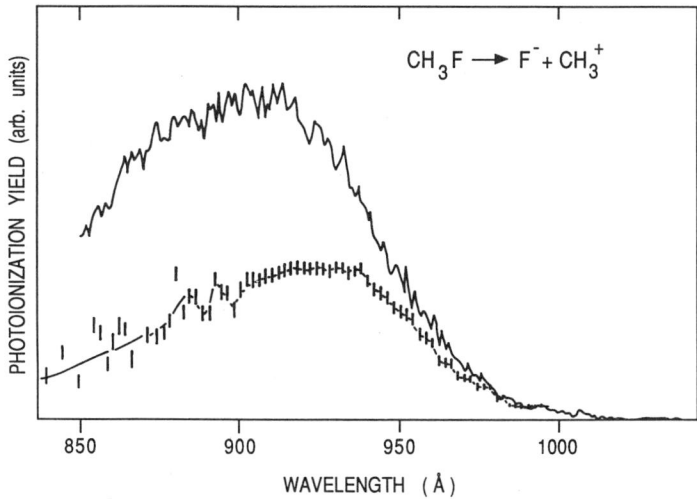

FIGURE 12. Photoion-pair spectrum of $CH_3F \rightarrow CH_3^+ + F^-$. [Smooth curve from Ref. 49, barred (lower) curve from M. Krauss, J. A. Walker, and V. H. Dibeler, *J. Res. Natl. Bur. Stand.* **72A**, 281 (1968), both with permission.]

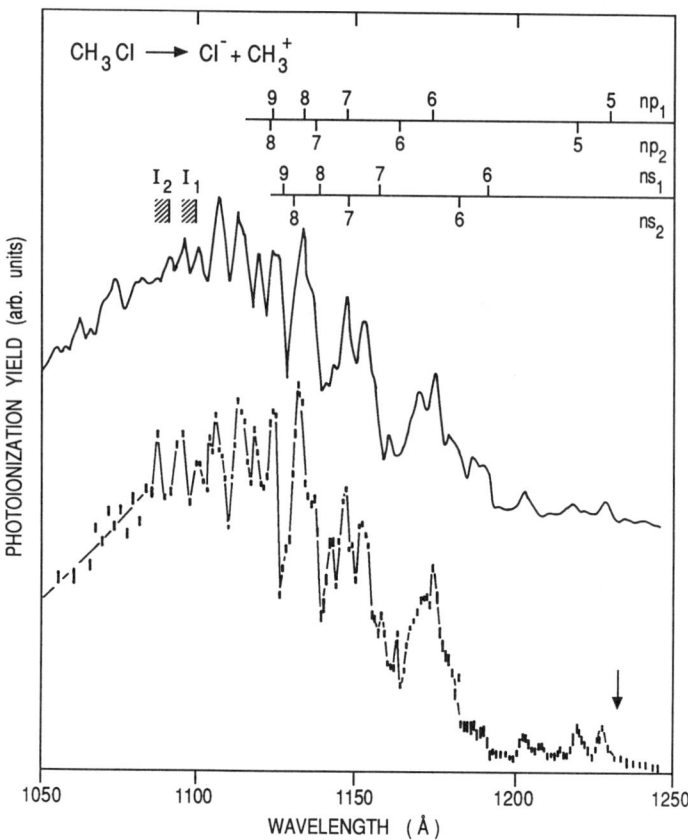

FIGURE 13. Photoion-pair spectrum of $CH_3Cl \rightarrow CH_3^+ + Cl^-$ (same references as for Figure 12).

A plausible explanation for this behavior can be inferred from the experiments of Munakata and Kasuya.[50] These authors used 1180-Å laser radiation impinging on CH_3Cl and CH_3Br, and examined ion-pair formation by TOF mass spectrometry. They found from the velocity distributions that only a fraction (17% for CH_3Br, 25% for CH_3Cl) of the excess energy appeared as fragment translational energy. They inferred that the missing energy resided in the out-of-plane bending vibration of CH_3^+. It is known[51,52] that the $\tilde{X}^1A_1'$ ground state of CH_3^+ is planar. The conformation of H atoms in methane and the methyl halides is near tetrahedral. To a significant extent, this is true of the corresponding ground-state cations. Hence, Rydberg series converging to these limits will be formed preferentially with a nonplanar CH_3 moiety, and will encounter Franck–Condon restrictions in order to predissociate into $CH_3(\tilde{X}^1A_1')$ in a planar, vibrational ground state. This argument rationalizes the declining ion-pair cross section as the yield curve approaches the thermochemical threshold, and the observation of Munakata and Kasuya.[50] The differing behavior of CH_3F, on the one hand, and both CH_3Cl and CH_3Br, on the other, can be rationalized from the corresponding photoelectron spectra.[53] The first ionization band in CH_3F is broad, with some vibrational fine structure, but no discernible spin-orbit splitting. Consequently, the Rydberg members converging to this ionic state should envelope a number of vibrations, which will result in a smeared ion-pair spectrum. By contrast, CH_3Cl and CH_3Br (and also CH_3I) display a sharp peak in PES, with very little vibration, and a distinct spin-orbit partner which is also sharp. Rydberg transitions approaching these limits will also be sharp, giving rise to the structured ion-pair spectra of CH_3Cl and CH_3Br.

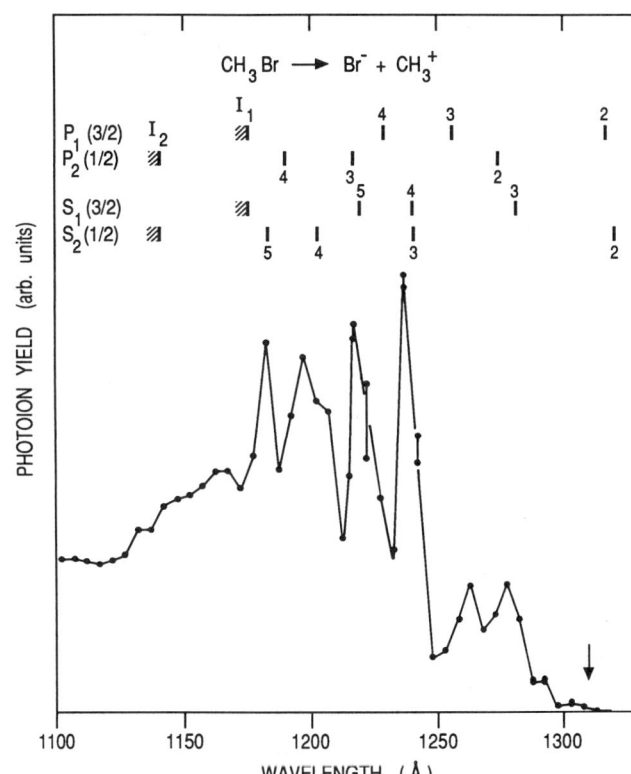

FIGURE 14. Photoion-pair spectrum of $CH_3Br \rightarrow CH_3^+ + Br^-$. [The Rydberg series assignments are from P. Hochmann, P. H. Templet, H.-t. Wang, and S. P. McGlynn, *J. Chem. Phys.* **62**, 2588 (1975). The pair formation plot is from early previously unpublished data of J. Berkowitz.] The resolution is about 1.7 Å (FWHM).

Recently, Koyano and co-workers[49,54] have recorded the photoion-pair spectrum of $CH_4 \rightarrow CH_3^+ + H^-$ over an extended energy region, and also examined ion-pair formation in CH_3X (X = F, Cl, Br) in a higher-energy region. The methane spectrum appears in Figure 15. One sees a lower-energy, unstructured band (designated by them as F1), and a structured higher-energy band with a relatively abrupt onset. The F1 band is reported by them to have an onset at 13.37 ± 0.15 eV, actually lower than

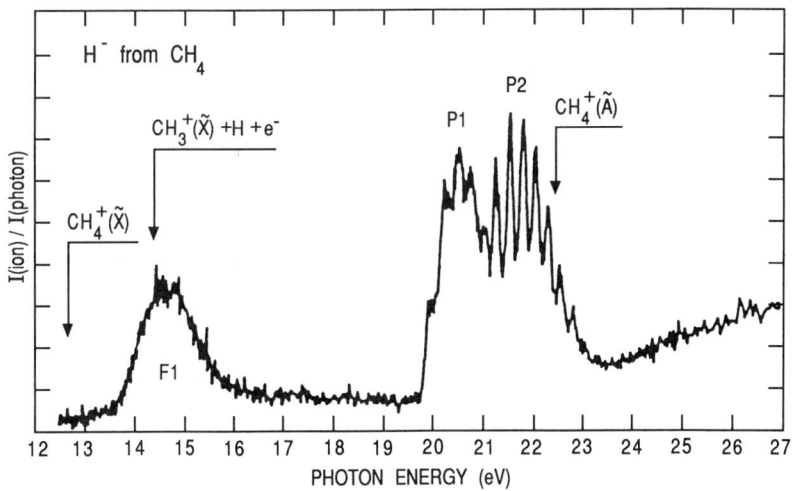

FIGURE 15. Photoion-pair spectrum of $CH_4 \rightarrow CH_3^+ + H^-$, in both threshold and higher-energy regions. (From Ref. 54, with permission.)

the thermochemical threshold, 13.570 ± 0.006 eV. The lack of structure can be understood in terms of the photoelectron spectrum of CH_4. The first band in PES, corresponding to electron ejection from the uppermost occupied $1t_2$ orbital, is substantially split and broadened by Jahn–Teller distortion in the cation. Rydberg series approaching the split cationic states will be similarly broadened.

The second band in the photoelectron spectra of CH_4, corresponding to electron ejection from the next deeper $2a_1$ orbital, gives rise to a structured spectrum, characteristic of a vibrational progression in the v_1 totally symmetric stretching mode.[55,56] The sharp structure in the ion-pair band in CH_4 between ~19.8 and 23 eV has been interpreted in terms of two members (P1 and P2) of a Rydberg series converging on $(2a_1)^{-1}$, $\tilde{A}^2A_1$, with their associated vibrational members. The abrupt onset of this ion-pair band at ~19.75–19.8 eV implies that a channel for new products has become available. Mitsuke et al.[54] assign this channel, quite plausibly, to CH_3^+ in its excited $\tilde{A}^1E'$ state, plus H^-. Dyke et al.[52] have observed broad features in the region of 13.4–16.5 eV in their photoelectron spectroscopic study of CH_3 generated by pyrolysis of azomethane. Although significant subtraction of impurity peaks was necessary to reveal this broad structure, they obtained good agreement between their observed peaks and vertical ionization potentials calculated by ΔSCF and INDO CI methods. In this way, they assigned a broad feature at 14.76 ± 0.03 eV to CH_3^+, $^3E'$, and the next broad feature, at 16.10 ± 0.03 eV, to $^1E'$. The adiabatic onset of the latter can be estimated from their spectrum to be about 15.8 eV, and hence it lies about 6 eV above the ground state (whose IP is 9.843 ± 0.001 eV).[51]

If the lower lying $^3E'$ state of CH_3^+ was to be formed, the products would correlate with a triplet ion-pair curve. Since excitation of CH_4 is far more likely to be to singlet Rydberg states, the interaction would require a singlet–triplet surface crossing (or avoided crossing), which may be retarded (although we have seen evidence for this type of interaction in the halogens, below the first ionization potential). If the new product were the $^1E'$ state of CH_3^+, this impediment would not arise. The abrupt onset may well be related to the geometrical structure of CH_3^+, $^1E'$. Although detailed calculations are not available, those reported by Blint et al.[57] imply that this state of CH_3^+ is nonplanar, which would facilitate its formation in its vibrational ground state from Rydberg states of CH_4. Consequently, this abrupt onset could be a good measure of the excitation energy of CH_3^+, $^1E'$, or equivalently, the ionization energy for the process $CH_3 \rightarrow CH_3^+$, $^1E'$. Taking the calculated threshold for the lower-energy ion-pair process (CH_3^+, $X^1A_1' + H^-$) to be 13.570 ± 0.006 eV, we obtain 6.18–6.23 eV for the excitation energy, or an ionization energy of 16.02–16.07 eV.

Suzuki et al.[49] (Figure 16) have observed broad-banded features resulting from higher-energy ion-pair formation in CH_3F, CH_3Cl, and CH_3Br. They attribute this higher-energy structure to formation of the aforementioned, excited CH_3^+, $^1E' + X^-$. They assign the broad features to Rydberg excitations emanating from inner ma_1 orbitals (m = 4, 6, and 8 for CH_3F, CH_3Cl, and CH_3Br). The photoelectron spectra[58,59] corresponding to electron ejection from these orbitals are indeed bereft of vibrational fine structure. If their attribution of these processes is correct, the thresholds they obtain (18.5 eV for CH_3^{+*} + F^-; 16.1 eV for CH_3^{+*} + Cl^-; 15.3 eV for CH_3^{+*} + Br^-) should provide a redundancy for establishing the excitation energy of CH_3^+, $^1E'$. From these reported thresholds, and the calculated onsets for the lower-energy processes given in Table II, we deduce 7.1_4, 6.3, and 5.8_3 eV for this excitation, from CH_3F, CH_3Cl, and CH_3Br, respectively. All of these reported thresholds are not abrupt, and the data have significant scatter. Nonetheless, we can conclude that in CH_3F the threshold is not reached, since the deduced excitation energy is higher than that deduced from the abrupt onset in CH_4. The CH_3Cl data are in fair agreement with those from CH_4. Only the CH_3Br data suggest a lower excitation energy, which may yet be within the scatter of the data. It must be kept in mind, however, that the adiabatic excitation energy extracted from the photoelectron spectrum is ~6.0 eV, which is lower than that from CH_4 ion-pair formation.

2.3.2. CF_4, CF_3Cl, CF_2Cl_2, $CFCl_3$. Mitsuke et al.[60] have recently reported on ion-pair formation from CF_4, and earlier, Schenk et al.[61] obtained data for CF_3Cl, CF_2Cl_2, and $CFCl_3$. The negative ion photoion yield curves of these halomethanes display similar patterns. The monatomic halogen anions (F^- or Cl^-) are the dominant ions. [Some CF_3^- (CF_3Cl) was observed, but was too weak for spectral dependence measurements and CF^-(CF_2Cl_2) had a very high appearance potential.] Structure is observed, but it is typically rather broad. The onset energies are all significantly higher than calculated

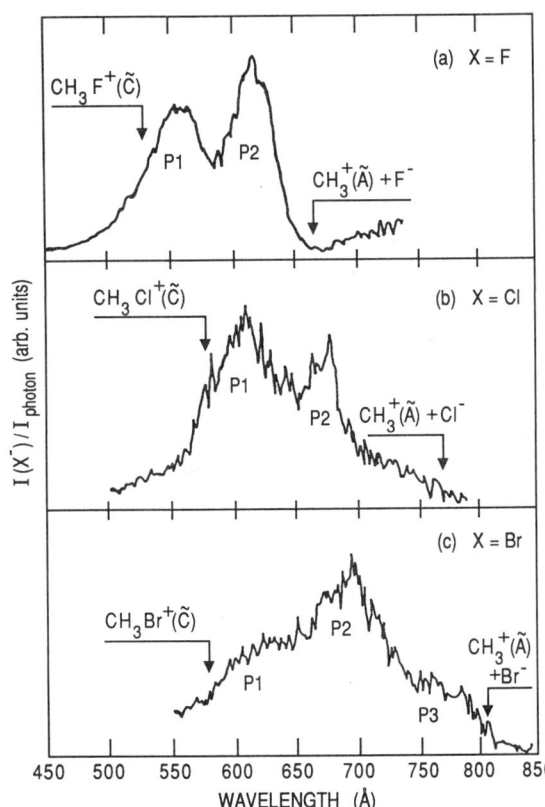

FIGURE 16. Photoion-pair spectrum of CH_3F, CH_3Cl, and CH_3Br in the higher-energy regions. (From Ref. 49, with permission.)

thermochemical thresholds. Table III summarizes the observed and calculated thresholds. For CF_4, CF_2Cl_2, and $CFCl_3$, where the lowest observed thresholds can be identified with formation of just two particles, the calculated thresholds are below the observed ones by at least 2 eV. For still higher thresholds, as in CF_3Cl, formation of three particles is assumed, but even here, the observed thresholds are higher than the calculated ones. Hence, the major interest in these studies is a dynamical one, i.e., assigning the initial absorptions which lead to the broad structures.

2.3.3. *SF_6 and C_2H_2.* These two polyatomic molecules, for which ion-pair processes have been examined recently, can be viewed in hindsight as representatives of two different classes. In both molecules, ion-pair formation is a small fraction (ca. 2×10^{-4}) of primary ionization. In SF_6, the only negative ion observed[62] that was uncontaminated by dissociative attachment processes was F^-. Its observed threshold was 12.98 ± 0.03 eV, more than 1 eV above the calculated threshold, ≤ 11.92 eV. About nine features were observed in the F^- yield curve, some of which could be identified with features in the absorption spectrum, but typically with drastically different relative intensities. Hence, photoion-pair formation in SF_6 is similar in behavior to that in CF_nCl_n.

The dominant negative ion on photoexcitation of C_2H_2 is C_2H^-.[63] It has been noted[63] that C_2H_2 is isoelectronic with HCN, where ion-pair formation had been observed to yield $H^+ + CN^-$. The electron affinity of CN, 3.82 eV,[64,65] is higher than any of the atomic halogens (a more recent value for EA(CN) = 3.862 ± 0.004 eV[69]), and in this sense CN can be regarded as a pseudohalogen. The electron affinity of C_2H, 2.969 ± 0.006 eV,[66] is also substantial. Thus, in a very simplified sense, both HCN and C_2H_2 can be viewed as pseudo hydrogen halides. One can also put the methyl halides in this category. However, as we have seen, excitation of the internal vibrations of the CH_3^+ moiety can make it difficult to observe a true thermochemical threshold. Thus, there is no guarantee that the observed thresholds for C_2H^- and CN^- are the true thermochemical thresholds. In both of these cases, the approach to

TABLE III
Observed and Calculated Ion-Pair Thresholds for Some Halomethanes

Species	Process	Observed threshold, eV	Calculated threshold, eV[c]
CF_4	$F^- + CF_3^+$	13.25 ± 0.03[a]	≤ 11.3
CF_3Cl	$Cl^- + CF_3^+$		≤ 9.1
	$Cl^- + CF_2^+ + F$	16.0 ± 0.1[b]	15.0
	$F^- + CF_2Cl^+$		10.1
	$F^- + CF_2^+ + Cl$	15.9 ± 0.3[b]	15.2
CF_2Cl_2	$Cl^- + CF_2Cl^+$	10.60 ± 0.02[b]	~8.0
	$F^- + CFCl_2^+$	12.07 ± 0.05[b]	~9.6
$CFCl_3$	$F^- + CCl_3^+$	13.7 ± 0.5[b]	8.76
	$F^- + Cl^+ + CCl_2$	17.5 ± 0.3[b]	16.03
	$Cl^- + Cl^+ + CFCl$	15.6 ± 0.1[b]	14.35

[a] Mitsuke et al.[(60)]
[b] Schenk et al.[(61)]
[c] Based on the compilation entitled "Gas Phase Ion and Neutral Thermochemistry," S. G. Lias, J. E. Bartmess, J. F. Liebman, J. L. Holmes, R. D. Levin, and W. G. Mallard, *J. Phys. Chem. Ref. Data* **17**, Suppl. 1 (1988).

threshold is quasilinear, with some diffuse fine structure, probably vibrational, superimposed. Other experiments[(65,67)] have since verified that the true thermochemical thresholds is indeed observed in $C_2H^-(C_2H_2)$ and $CN^-(HCN)$. Both studies yielded important information—the HCN experiment because it provided the first accurate value for EA(CN), the C_2H_2 investigation because it helped to resolve a dispute regarding the C–H bond energy in acetylene.

2.4. Ion-Pair Formation at Higher Energies

Until recently, it was thought that photoion-pair formation was confined to processes such as AB + $h\nu \rightarrow A^+ + B^-$, where typically A^+ could be in its ground state or in a few excited states. In 1989, Dujardin et al.[(16)] observed a new type of photoion-pair process in SO_2, i.e.,

$$SO_2 + h\nu \rightarrow SO^{2+} + O^- + \overline{e}$$

This process has an observed threshold (35.6 ± 0.3 eV) near the estimated thermochemical threshold (~ 33.0 eV). They believe that it can be explained as a two-step process, i.e.,

$$SO_2 + h\nu \rightarrow (SO_2^+)^* + \overline{e}$$

$$(SO_2^+)^* \rightarrow SO_2^{2+}\,(^1\Sigma^+) + O^-(^2P)$$

They have also observed O^- from SO_2 in the vicinity (above and below) of the S2p edge at ~ 175 eV, albeit with low intensity ($\leq 10^{-5}$ of the positive ion signal). Their explanation for this latter process, which displays structure in the O^- channel, is that the initial excitation is followed by an Auger process in which the excited electron (in a previously unoccupied valence or Rydberg state, or a shape resonance) acts as a spectator, thereby forming $(SO_2^+)^*$ which then autoionizes and decomposes to $SO_2^{2+} + O^-$. By comparing the structure in the O^- channel with the positive ion channels at $m/e = 16$ (O^+, S^{2+}, O_2^{2+}) and $m/e = 32$ (S^+, O_2^+) they show that the shape resonance is less important in the O^- channel, but that the valence or Rydberg excitations are more prominent. Whereas the excitation leading to the shape resonance tends to eschew the more electronegative O atom, and hence does not readily attach to it in the ensuing decomposition, excitation to the unoccupied valence $6b_2$ orbital enhances O^- production, perhaps because this $6b_2$ orbital has strong O3s character. Thus, attachment to the more electronegative O atom on dissociation is favored by valence excitation, but is disfavored in the shape resonance.

This type of study has been pursued[24] in other cases, including SiF_4 (near the Si2p edge at 111.6 eV) and SF_6 near the S2p edge. With SiF_4, they observe negative ions in the F^-, F_2^-, and Si^- channels, with similar structure in all three channels. These structures had previously been seen in photoabsorption[68] and assigned as excitations to valence or low Rydberg levels, as well as to two shape resonances. The presumed mechanism is again believed to proceed in two steps:

$$SiF_4 + h\nu \rightarrow [Si\,(2p^{-1})\,F_4]^* \rightarrow (SiF_4^+)^* + \bar{e}$$

where spectator Auger decay occurs, much more likely than x-ray emission, followed by

$$(SiF_4^+)^* \rightarrow SiF_3^{2+} + F^-$$

or

$$SiF^{2+} + F_2^-$$

The channel forming Si^- appears to be much weaker than that forming F^-, but shows relative enhancement in the shape resonance region. The explanation would follow that given for SO_2, i.e., the excited electron in the shape resonance remains closer to the central atom, and prefers to attach to it on dissociation.

In SF_6, where the observed ions are F^- and S^- in the vicinity of the S2p edge, similar behavior is seen in the corresponding structured spectra. Dadouch[24] reports the observation of both C^{2+} and O^- from CO, with an onset at 45.3 ± 0.2 eV, compared to a calculated threshold of 45.28 eV. He has examined N^{2+} and O^- from NO, and indeed observed these ions in the region calculated for the threshold (49.2–50 eV), but it is very difficult to see an experimental onset in his spectra.

3. CONCLUSIONS

If there were no interactions between the coulombic ion-pair curves (at large internuclear distances) with valence or Rydberg state potential energy curves at shorter distances, the minima of the ion-pair curves would occur at substantially larger internuclear distances than are characteristic of the neutral ground state. Hence, Franck–Condon transitions from the ground state would access the repulsive inner wall of the ion-pair curves, producing a continuum with a threshold significantly higher than the thermochemical threshold. By contrast, most ion yield curves associated with photoion-pair formation are structured, and often the observed onset occurs at the thermochemical threshold, within experimental uncertainty. The structure in the ion yield curve can frequently be assigned to excited valence and/or Rydberg states. Hence, the prevailing mechanism, in most cases, is predissociation.

Predissociation is a valuable spectroscopic method for establishing upper limits, and sometimes precise values, for dissociation energies. The same comment applies to observed onsets for photoion-pair formation. In the vast majority of cases, the heat of formation of the positive ion fragment is known. If the heat of formation of the negative ion is also known, the observed onset is an upper limit to the bond energy. Thus, the observed onset for the process

$$C_2H_2 + h\nu \rightarrow C_2H^- + H^+$$

has provided an upper limit (which turns out to be the true thermochemical threshold) for $D_0(HCC-H)$. If, alternatively, the dissociation energy is known but the heat of formation of the negative ion is not (i.e., the electron affinity is poorly known), then the observed onset provides a lower limit to the electron affinity. For example, the observed threshold for

$$HCN + h\nu \rightarrow CN^- + H^+$$

yielded an accurate value for EA(CN).

The excited states (valence and/or Rydberg) probed by the ion-pair process may be the same ones observed in autoionization, but they may also be states not readily apparent in other photoionization channels. The detailed mechanism of coupling between the initially excited states and the outgoing ion pairs is not well established in most cases. Especially in homonuclear diatomic molecules, which have g and u character at short internuclear distances but lose that character in the exiting channels, rotational coupling is implicated. The ion-pair process is most likely to be relatively intense when its threshold occurs below the adiabatic ionization potential, i.e., where competition from autoionization is absent. Photoion-pair formation is not observed in the ionic alkali halides, but is seen in covalently bound molecules. This is contrary to naive expectations, but is readily explainable.

Although the field of photoion-pair formation is about 60 years old, there has recently been an upsurge of interest, e.g., the work of Koyano and collaborators, and Dujardin and collaborators. This may be attributed, at least in part, to the more intense synchrotron light sources available. Weaker processes, previously unobservable or barely detected when at the 10^{-4}–10^{-5} level compared to total ionization, are now within range. This applies also to the higher-energy processes, previously unanticipated. With still brighter light sources on the horizon, the exploration of ion-pair formation processes is likely to experience a resurgence. Active involvement of theorists in helping to trace the ion-pair process from the initial photoabsorption act to the observed ion pairs is highly desirable.

ACKNOWLEDGMENT. This work was supported by the U.S. Department of Energy, Office of Basic Energy Sciences, under Contract No. W-31-109-ENG-38.

REFERENCES

1. A. TERENIN AND B. POPOV, *Phys. Z. Sowjetunion* **2**, 299 (1932).
2. R. L. KAGAN, *Zh. Eksp. Teor. Fiz.* **5**, 811 (1935).
3. (a) E. K. PARKS, N. J. HANSEN, AND S. WEXLER, *J. Chem. Phys.* **58**, 5489 (1973). (b) E. K. PARKS, J. G. KUHRY, AND S. WEXLER, *J. Chem. Phys.* **67**, 3014 (1977). (c) S. H. SHEEN, G. DIMOPLON, E. K. PARKS, AND S. WEXLER, *J. Chem. Phys.* **68**, 4950 (1978).
4. F. P. TULLY, N. H. CHEUNG, H. HABERLAND, AND Y. T. LEE, *J. Chem. Phys.* **73**, 4460 (1980).
5. N. F. MOTT AND H. S. W. MASSEY, *The Theory of Atomic Collisions* (Clarendon Press, Oxford, 1965), p. 665; H. S. W. MASSEY AND E. H. S. BURHOP, *Electronic and Ionic Impact Phenomena* (Clarendon Press, Oxford, 1952), pp. 441–442.
6. J. DURUP, *J. Phys. (Paris)* **39**, 941 (1978).
7. J. J. EWING, R. MILSTEIN, AND R. S. BERRY, *J. Chem. Phys.* **54**, 1752 (1971).
8. R. G. GRICE AND D. R. HERSCHBACH, *Mol. Phys.* **27**, 159 (1974).
9. P. DAVIDOVITS AND D. C. BRODHEAD, *J. Chem. Phys.* **46**, 2968 (1967).
10. S. H. SCHAEFER, D. BENDER, AND E. TIEMANN, *Chem. Phys.* **89**, 65 (1984).
11. P. T. RAO, *Indian J. Phys.* **23**, 393 (1949).
12. W. THORSON, *J. Mol. Spectrosc.* **37**, 199 (1971).
13. S. D. PEYERIMHOFF AND R. J. BUENKER, *Chem. Phys.* **57**, 279 (1981).
14. J. BERKOWITZ, C. A. MAYHEW, AND B. RUSCIC, *Chem. Phys.* **123**, 317 (1988).
15. G. HERZBERG, *Molecular Spectra and Molecular Structure I. Spectra of Diatomic Molecules* (Van Nostrand, Princeton, NJ, 1950), pp. 410, 419.
16. G. DUJARDIN, L. HELLNER, B. J. OLSSON, M. J. BESNARD-RAMAGE, AND A. DADOUCH, *Phys. Rev. Lett.* **62**, 745 (1989).
17. A. J. YENCHA, D. K. KELA, R. J. DONOVAN, A. HOPKIRK, AND A. KVARAN, *Chem. Phys. Lett.* **165**, 283 (1990).
18. V. H. DIBELER, J. A. WALKER, K. E. MCCULLOH, AND H. M. ROSENSTOCK, *Int. J. Mass Spectrom. Ion Phys.* **7**, 209 (1971).
19. M. E. AKOPYAN, F. I. VILESOV, AND Y. L. SERGEEV, *Opt. Spektrosk.* **35**, 812 (1973); Engl. transl. *Opt. Spectrosc.* **35**, 472 (1974).
20. A. KVARAN, A. J. YENCHA, D. K. KELA, R. J. DONOVAN, AND A. HOPKIRK, *Chem. Phys. Lett.* **179**, 263 (1991).
21. J. BERKOWITZ, W. A. CHUPKA, P. M. GUYON, J. H. HOLLOWAY, AND R. SPOHR, *J. Chem. Phys.* **54**, 5165 (1971).
22. R. DE L. KRONIG, *Z. Phys.* **50**, 347 (1928).
23. M. C. HEAVEN, *Chem. Soc. Rev.* **15**, 405 (1986).

24. A. DADOUCH, Thesis, Universite de Paris Sud, Centre d'Orsay, June, 1991, "Utilisation de la Detection d'Ions Fragments Negatifs pour v'Etude des Transferts de Charges Intramoleculaires a Haute Energie (25–200 eV)."
25. H. OERTEL, H. SCHENK, AND H. BAUMGÄRTEL, *Chem. Phys.* **46**, 251 (1980).
26. K. P. HUBER AND G. HERZBERG, *Molecular Spectra and Molecular Structure IV. Constants of Diatomic Molecules* (Van Nostrand Reinhold, New York, 1979).
27. C. E. MOORE, *Atomic Energy Levels*, Vol. I, NSRDS-NBS 35 (U.S. Government Printing Office, Washington, DC, 1971).
28. D. M. NEUMARK, K. R. LYKKE, T. ANDERSEN, AND W. C. LINEBERGER, *Phys. Rev. A* **32**, 1890 (1985).
29. A. DADOUCH, G. DUJARDIN, L. HELLNER, M. J. BESNARD-RAMAGE, AND B. J. OLSSON, *Phys. Rev. A* **43**, 6057 (1991).
30. M. KRATZAT, Ph.D. Thesis, Fachbereich Physik, Freie Universitat Berlin (1984), "Dissoziative Photoionisation und Ionenpaarbildung bei H_2O, SO_2, NO_2, N_2O and OCS."
31. K. MITSUKE, S. SUZUKI, T. IMAMURA, AND I. KOYANO, *J. Chem. Phys.* **92**, 6556 (1990).
32. J. BERKOWITZ AND J. H. D. ELAND, *J. Chem. Phys.* **67**, 2740 (1977).
33. P. C. KILLGOAR, JR., G. E. LEROI, W. A. CHUPKA, AND J. BERKOWITZ, *J. Chem. Phys.* **59**, 1370 (1973).
34. K. S. HABER, J. W. ZWANZIGER, F. X. CAMPOS, R. T. WIEDMANN, AND E. R. GRANT, *Chem. Phys. Lett.* **144**, 58 (1988).
35. Y. MORIOKA, H. MASUKO, M. NAKAMURA, M. SASANUMA, AND E. ISHIGURO, *Can. J. Phys.* **56**, 962 (1978).
36. O. EDQVIST, E. LINDHOLM, L. E. SELIN, L. ASBRINK, C. E. KUYATT, S. R. MIELCZAREK, J. A. SIMPSON, AND Y. FISCHER-HJALMARS, *Phys. Scr.* **1**, 172 (1970).
37. K. MITSUKE, S. SUZUKI, T. IMAMURA, AND I. KOYANO, *J. Chem. Phys.* **93**, 1710 (1990).
38. J. DELWICHE, M.-J. HUBIN-FRANSKIN, G. CAPRACE, P. NATALIS, AND D. ROY, *J. Electron Spectrosc. Relat. Phenom.* **21**, 205 (1980).
39. J. A. SHEEHY AND P. W. LANGHOFF, *Chem. Phys. Lett.* **135**, 109 (1987).
40. N. PADIAL, G. CSANAK, B. V. MCKOY, AND P. W. LANGHOFF, *Phys. Rev. A* **23**, 218 (1981).
41. R. R. LUCCHESE AND V. MCKOY, *Phys. Rev. A* **26**, 1406 (1982).
42. T. GUSTAFSSON, E. W. PLUMMER, D. E. EASTMAN, AND W. GUDAT, *Phys. Rev. A* **17**, 175 (1978).
43. C. E. BRION AND K. H. TAN, *Chem. Phys.* **34**, 141 (1978).
44. R. W. CARLSON, D. L. JUDGE, AND M. OGAWA, *J. Geophys. Res.* **78**, 3194 (1973).
45. T. MATSUOKA AND J. A. R. SAMSON, *J. Chim. Phys.* **77**, 623 (1980).
46. A. W. POTTS AND T. A. WILLIAMS, *J. Electron Spectrosc.* **3**, 3 (1974).
47. J. ERICKSON AND C. Y. NG, *J. Chem. Phys.* **75**, 1650 (1981).
48. J. BERKOWITZ, previously unpublished data.
49. S. SUZUKI, K. MITSUKE, T. IMAMURA, AND I. KOYANO, *J. Chem. Phys.* **96**, 7500 (1992).
50. T. MUNAKATA AND T. KASUYA, *Chem. Phys. Lett.* **154**, 604 (1989).
51. G. HERZBERG AND J. SHOOSMITH, *Can. J. Phys.* **34**, 523 (1956).
52. J. DYKE, N. JONATHAN, E. LEE, AND A. MORRIS, *J. Chem. Soc. Faraday Trans. 2* **72**, 1385 (1976).
53. K. KIMURA, S. KATSUMATA, Y. ACHIBA, T. YAMAZAKI, AND S. IWATA, *Handbook of He I Photoelectron Spectra of Fundamental Organic Molecules* (Halsted Press, New York, 1981), pp. 72, 75, 85; see also p. 86.
54. K. MITSUKE, S. SUZUKI, T. IMAMURA, AND I. KOYANO, *J. Chem. Phys.* **94**, 6003 (1991).
55. A. W. POTTS AND W. C. PRICE, *Proc. R. Soc. London Ser. A* **326**, 165 (1972).
56. G. BIERI AND L. ÅSBRINK, *J. Electron Spectrosc.* **20**, 149 (1980).
57. R. J. BLINT, R. F. MARSHALL, AND W. D. WATSON, *Astrophys. J.* **206**, 627 (1976).
58. G. BIERI, L. ÅSBRINK, AND W. VON NIESSEN, *J. Electron Spectrosc.* **23**, 281 (1981).
59. W. VON NIESSEN, L. ÅSBRINK, AND G. BIERI, *J. Electron Spectrosc.* **26**, 173 (1982).
60. K. MITSUKE, S. SUZUKI, T. IMAMURA, AND I. KOYANO, *J. Chem. Phys.* **95**, 2398 (1991).
61. H. SCHENK, H. OERTEL, AND H. BAUMGÄRTEL, *Ber. Bunsenges. Phys. Chem.* **83**, 683 (1979).
62. K. MITSUKE, S. SUZUKI, T. IMAMURA, AND I. KOYANO, *J. Chem. Phys.* **93**, 8717 (1990).
63. B. RUSCIC AND J. BERKOWITZ, *J. Chem. Phys.* **93**, 5586 (1990).
64. J. BERKOWITZ, W. A. CHUPKA, AND T. A. WALTER, *J. Chem. Phys.* **50**, 1497 (1967).
65. R. KLEIN, R. P. MCGINNIS, AND S. R. LEONE, *Chem. Phys. Lett.* **100**, 475 (1983).
66. K. M. ERVIN AND W. C. LINEBERGER, *J. Phys. Chem.* **95**, 1167 (1991).
67. D. P. BALDWIN, M. A. BUNTINE, AND D. W. CHANDLER, *J. Chem. Phys.* **93**, 6578 (1990).
68. H. FRIEDRICH, B. PITTEL, P. RABE, W. H. E. SCHWARZ, AND B. SONNTAG, *J. Phys. B* **13**, 25 (1980).
69. S. E. BRADFORD, E. H. KIM, D. W. ARNOLD, AND D. M. NEUMARK, *J. Chem. Phys.* **98**, 800 (1993).

CHAPTER 9

ELECTRONIC AND NUCLEAR RELAXATION OF CORE-EXCITED MOLECULES

IRÈNE NENNER AND PAUL MORIN

1. INTRODUCTION

Ionization and dissociation of core-excited molecules is a subject intimately related to the problem of the relaxation of the inner vacancy and the coupling to nuclear motion. For molecules built with light atoms, the Auger decay rate dominates largely the x-ray fluorescence rate and so we will concentrate on the effect of *ejection of several electrons* through Auger-like processes. In contrast to atoms, molecules which are stripped of several outer valence electrons through core excitation, lose their integrity and are efficiently destroyed because many electrons which form the cement of the chemical bonds are removed. In other words, the dissociation efficiency, defined as the number of broken bonds over the total number of bonds in the original molecule, is high and may reach unity (a full "atomization" of the system) for a number of molecules of small or medium size.

Fragmentation of a core-ionized molecule has been recognized[1–4] as one of the radiation damage phenomena besides the interaction of primary and secondary (Auger) electrons with the medium, usually called radiolysis. Notice that such studies appeared long before Siegbahn *et al.*[5] reported detailed information on Auger processes in molecules through high-resolution photoelectron spectroscopy. The interest in studying core-excited molecules appeared primarily because synchrotron radiation has developed widely throughout the world with more or less routine access to sophisticated beam lines, monochromators, and detection devices. Its large tunability in the far ultraviolet and soft x-ray radiation allows routinely to control the energy deposited in the molecule, in contrast to previous sources: electrons, laboratory x-ray tubes, discharge lamps. It is the purpose of this chapter to review the basic concepts and present understanding of the elementary processes describing the relaxation of core-excited molecules.

Such fundamental properties of isolated molecules have a more or less direct relevance to other fields. The first direction is very close to the subject of this review, that is, the study of core photoionization and photodissociation of free (semi)metallic (e.g., see Ref. 6) or van der Waals molecular clusters.[7] This could be considered as a natural extension of the present subject with another class of phenomena originating from environment or solvation effects and/or from the increase of the size of the system. A second close axis of research is the understanding of electronic relaxation processes of the core hole and associated photodesorption processes in molecules adsorbed on surfaces[8] or in the condensed phase (e.g., see Ref. 9 and references therein). In technological areas, the present subject is of direct relevance like in soft x-ray photoetching and photolithography processes[10–12] or the production of semiconductors through photoinduced chemical vapor deposition.[13] More recently, it has been recognized in radiobiology[14] that radiation damage is not always

---

IRÈNE NENNER AND PAUL MORIN • CEA/DSM/DRECAM, Service des Photons, Atomes et Molécules, Centre d'Études de Saclay, 91191 Gif sur Yvette Cedex, France, and Laboratoire pour l'Utilisation du Rayonnement Electromagnétique, Laboratoire mixte, CNRS, CEA et MESR, Centre Universitaire, 91405 Orsay Cedex, France.

VUV and Soft X-Ray Photoionization. Edited by Uwe Becker and David A. Shirley. Plenum Press, New York, 1996.

dominated by "indirect" effects (secondary reactions resulting from various radicals produced by water radiolysis, etc.) but by "direct" effects such as photodissociation of the core-excited molecule and dissociation induced by photoelectrons and Auger electrons. In other words, there is a direct correlation between the cell death and the photodissociation of the core-ionized system.[15] Furthermore, the hope to dissociate selectively a system such as the DNA backbone, by only tuning the energy of the incident photon near a selected atomic site (phosphorus), is not a vain idea.[16] In astrophysics, observations of soft x-ray emission in young stars[17] with a close spatial correlation with infrared radiation require laboratory experiments based on soft-ray light with large hydrocarbon molecules and solids (carbonaceous and silicate grains) to identify the nature of hot interstellar dust and various fluorescent small molecular species, which survive in this medium.[18]

Turning back to the main subject of this chapter, we show in Figure 1 a schematic and simplified outline of the various processes encountered after the creation of the core hole in a triatomic molecule in two situations, one after core ionization (Figure 1a) and the other after core excitation (Figure 1b). For the sake of simplicity, we have limited the decay up to triple ionization, but one should keep in mind that higher-order ionization is possible. The main idea of this scheme is to show that dissociation plays a major role even in the early stage of the electronic relaxation. It results in a complex manifold

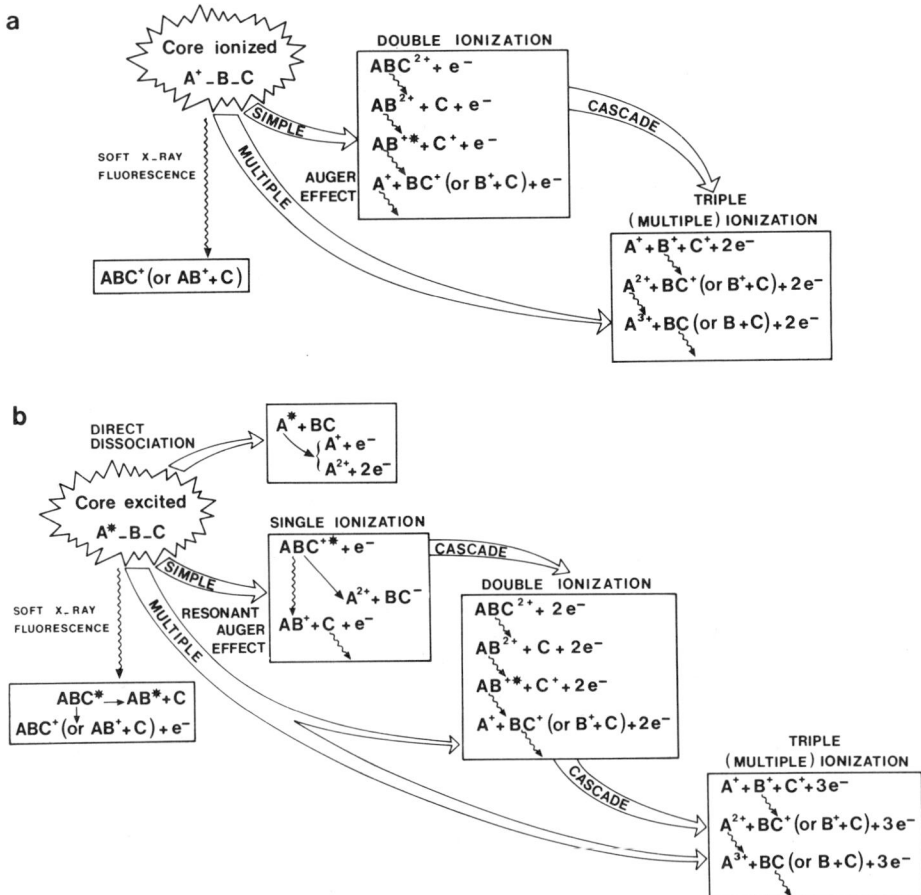

FIGURE 1. Basic photofragmentation processes following core ionization (a) and core excitation (b) of a triatomic molecule ABC built with chemically different atoms. The core vacancy is chosen on the A atomic site. The scheme is limited to double (b) and to triple (a) ionization for the sake of simplicity but higher-order ionization is possible (see text).

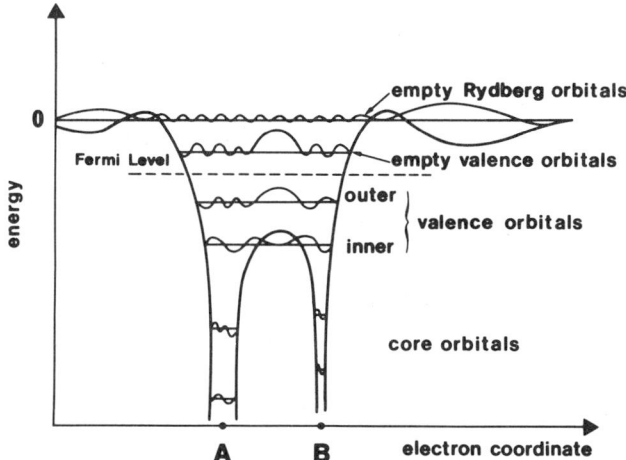

FIGURE 2. Schematic view of the electronic potential in a diatomic molecule.

of competitive or sequential events involving radiative decay, autoionization, Auger decay, as well as dissociation of the neutral, singly, doubly, and multiply charged ions.

The understanding of such processes is based on our knowledge of the properties of molecular orbitals. We show in Figure 2 a schematic representation of the potential of a diatomic molecule, A-B, with occupied and nonoccupied orbitals. Among outer valence orbitals, "bonding" orbitals are of special importance because they form the cement of the system. In contrast, core orbitals are essentially localized on each atomic site and core electrons do not participate in the bonding. For polyatomic molecules, the electronic structure (number of valence electrons, molecular orbitals, interaction between molecular orbitals) controls the geometry of the system and its reactivity. The reader should refer to the textbook by Jean and Volatron[19] for a comprehensive description of the electronic structure based on both simple and quantum-mechanical models and its connection to structure and correlation to fragments.

The (soft) x-ray photoabsorption spectrum of a molecule reveals resonances in the vicinity of a core ionization edge (several tenths of an electron volt) of each of its atomic components (for a review, see Ref. 20). The fact that there is a large energy gap between the ionization edge of an atom with a Z atomic number, with respect to a $Z + 1$ atom, allows one to choose the incident photon wavelength in order to *select an atomic site* for excitation, as long as there is no other chemically equivalent atoms in the molecule. In addition, the energy of such resonances depends on the nature of the chemical bonds between the initially excited atom and its neighbors, as well as on bond distances and angles. Again the choice of incident wavelength on a given resonance allows one to excite a specific electronic transition, e.g., a "core to antibonding valence orbital." One of the major interests in studying the decay of core-excited molecules is to learn how the initial energy and spatial localization is kept (or vanishes) throughout the many multiionization and multifragmentation processes.

We turn now to the Auger decay of a core vacancy (e.g., on the L shell of atom A) in a molecule AB. The formation (first step) and relaxation (second step) of the L vacancy is supposed to induce two holes in the valence shells V. The probability of the Auger decay, commonly called LVV, depends strongly on the A atomic components of the double-hole final states. This is because the Auger probability, as a particular case of electronic autoionization, is governed by a two-electron process and therefore depends on the product of two matrix elements $\langle c | i \rangle$ and $\langle f | c \rangle$ with i, c, and f representing the wave functions of the initial neutral ground state, the core-ionized state, and the final dicationic states. Intuitively, one can imagine that the nature of the bonding of valence orbitals is of importance in characterizing the intra- or interatomic character of the overall transition. In practice, numerous situations are found from a strong localization up to a full delocalization of the initial excitation.

The complete description of an Auger process in a molecule implies the knowledge of the spectroscopy of the final dicationic states. Unlike atoms, those states are poorly known except for some diatomics and triatomics, primarily because of the very high density of states and peculiarities of their potential energy curves. First, for stable cationic states, bond distances and angles and even atomic arrangement are quite different from those of neutral species in both small[21] and large organic[22] systems. As an example, the ground-state $C_2H_4^{2+}$ cation is nonplanar, with two CH_2 groups in perpendicular planes and a longer C–C bond distance than in the ground-state planar C_2H_4.[19] Second, cations as well as highly ionized systems are very often unstable because the number of bonding electrons has decreased. The potential surface of cationic states are often repulsive along specific coordinates and the systems dissociate (see Ref. 23 for a review). The simplest example is that of the hydrogen molecule. The ejection of its two electrons by one-photon excitation produces two H^+ atoms which are exposed to the pure Coulomb forces and depart from each other with a kinetic energy release defined by H–H distance in the neutral molecule. In H_2, Auger decay is not possible unlike for all other molecules for which the population of the dication proceeds from the ionization of a deeper shell thus inducing dissociation of the system by Coulomb forces (see Figure 3a for the simplest scheme of the concept of Coulomb explosion in a diatomic molecule other than H_2). The importance of Coulomb forces in a multiply ionized molecule, to produce ionized fragments with high kinetic energy, was shown in the 1960s.[24–31] The molecular effect of the "Coulomb explosion" appeared in 1951 in a study[32] of hot atom chemistry undergoing inner-shell ionization by means of internal conversion or electron capture. Later, several authors[1,3] investigated the Coulomb explosion in gaseous molecules containing a radioactive atom. Today, Coulomb explosion can be associated with the production of atomic ionized fragments after a molecule has been exposed to any ionizing agent such as x-ray or

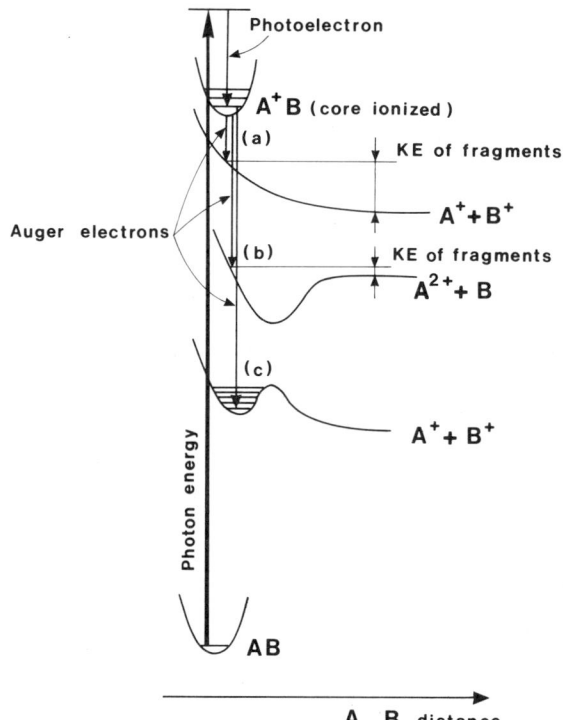

FIGURE 3. Schematic of dissociation of a dication after inner-shell ionization of a diatomic molecule other than hydrogen. The normal Auger process produces three types of double-ionization mechanisms: (a) a pure Coulomb explosion mechanism with charge separation; (b) dissociation with localization of the charges on one particle; (c) no dissociation. Cases (b) and (c) are governed by chemical forces related to the valence electrons.

γ-ray photons, fast electrons or ions (e.g., see Ref. 33) or even under high-power laser irradiation (e.g., see Ref. 34) after electron stripping through a beam foil.[35] In fact, Coulomb forces are not the unique factor controlling dissociation. In all molecules but hydrogen, dissociation processes are also governed by regular *chemical forces* originating from the residual valence electrons of the system, as we have mentioned above. The dication may be a bound state as shown in Figure 3b and c and the internal energy of the fragments, if produced through a direct dissociation mechanism, depends on the shape of the potential curve. In practice, the potential surface manifold of dicationic states is such that indirect predissociation mechanisms through electronic, vibrational, or rotational couplings are of importance up to the statistical limit for large systems, as one normally finds for neutral and singly charged positive ion species.[36] Furthermore, after electronic relaxation, the molecular cation is produced in a large manifold of states with very large internal energies as compared to the bonding energies and to most thermodynamical limits of two-, three-, and many-body dissociation processes. Consequently, the molecule dissociates efficiently in breaking several bonds.

The time scale of such processes brings another level of complexity to the decay of core-excited molecules. In a molecule, the nuclei can move as a whole in translational or rotational motion or with respect to each other in a vibrational motion. Because of the large mass of the nuclei relative to electrons, they move much more slowly than electrons and this justifies the Born–Oppenheimer adiabatic separation of electronic and nuclear motion. In other words, one should normally expect that ionization is much faster than dissociation. When a molecule is photoexcited into an electronic excited state (or relaxes from an excited state), the dipole electric transition probability is governed by a matrix element with an electronic and a nuclear term. Usually, one can separate the electronic from the nuclear terms, the latter being called the Franck–Condon factors and the transition is called "vertical." In the present situation of the core-excited decay processes, the short lifetime of the core vacancy, occurring on a femtosecond time scale, and comparable to vibrational motion and to direct dissociation along a repulsive surface, explains why departures from the classical vertical transition description are often found. Finally, the core hole localization in a molecule may induce in highly symmetric polyatomic molecules the excitation of non-totally symmetric vibrational modes because of coupling between electronic and nuclear motion, known as non-Born–Oppenheimer effects.

In the following, we analyze the spectroscopy of core-excited states in the photoabsorption (Section 2), the electronic relaxation (Section 3), then how the electronic and nuclear motion are coupled (Section 4) before examining the dynamics of the multifragmentation processes (Section 5) and future trends (Section 6).

2. PHOTOABSORPTION

The inner-shell photoabsorption spectrum of free molecules is a very powerful tool for investigating the internuclear potential and its electronic properties, through electronic transitions from the core to valence, Rydberg, and continuum orbitals schematically represented in Figure 2, for a diatomic molecule AB. The presence of a unique (for the $1s$ or $2s$ shells) or two (for the $2p$, $3d$, etc. shells) ionization limits within a large photon energy region explains the intrinsic simplicity of the spectroscopy compared to the valence region. Therefore, it is possible to identify rather easily the resonances originating from "one-electron" transitions which result from the excitation of a core electron into a valence unoccupied orbital, into a Rydberg orbital, or into the continuum, from those which are related to "two-electron" transitions, i.e., a core to bound plus a valence excitation or a core to bound associated with another core excitation. Such an analysis requires very high resolution and very good signal-to-noise ratio measurements. In other words, the spectral resolution should be better than the natural linewidth of the resonances. Besides photoabsorption spectra, one should mention the technique of resonant Raman scattering in which the spectral width of the exciting and secondary radiation is smaller than the natural width and only limited to the lifetime of the final state. Such a method, used first by Brown et al.[37] in atoms, has been applied only recently to molecules.[38] Turning back to photoabsorption methods and related techniques, the very first high-resolution results were obtained by electron energy loss spectroscopy (EELS), for a very large number of molecules from simple to

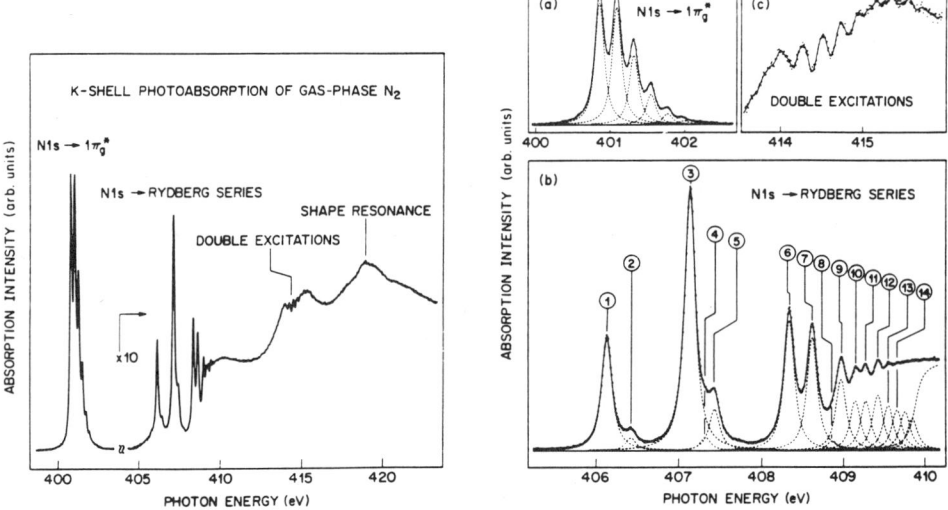

FIGURE 4. K-shell photoabsorption spectrum of N_2, with expanded view of fine structure. (From Chen et al.[44])

complex ones including organometallics. The reader may refer to the reviews of King and Read,[39] Hitchcock,[40] as well as the compilation of Hitchcock and Mancini[41] and the book by Stöhr,[20] which includes an extensive bibliography on the subject. More recently, photoabsorption spectra obtained using synchrotron radiation associated with advanced grating monochromators revealed unprecedented energy resolution in the soft x-ray range and statistical accuracy. Among the numerous works published since 1989, some are remarkable and are classified in the following according to the spectral ranges from the far ultraviolet up to the x-ray range.

From 150 to 250 eV, the S2p spectra of H_2S, CS_2, SO_2,[42,43] and SF_6[43] are now available. From 250 to 700 eV, the first high-resolution 1s spectrum of nitrogen has been produced by Chen et al.[44] Later on, Domke et al.[45,46] published very-high-resolution spectra of CO at the 1s edge of carbon and oxygen. Similar results were obtained on NO[47] at the N1s edge. In polyatomic systems, more results on formaldehyde H_2CO and D_2CO,[48] C_2H_4[49] at the C1s edge, SF_6[43] near the F1s edge, H_2O, NH_3,

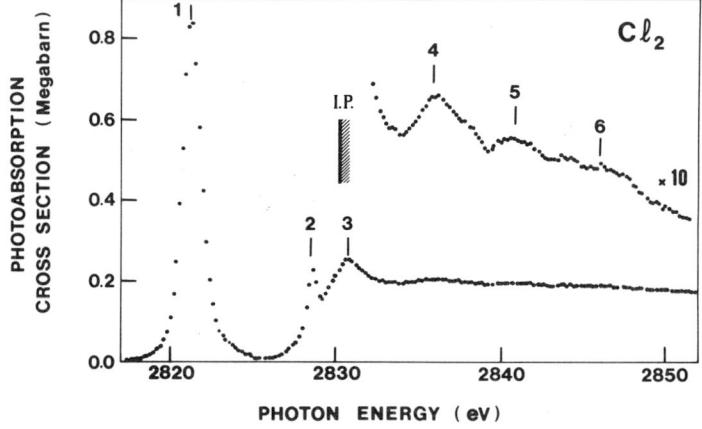

FIGURE 5. K-shell photoabsorption spectrum of Cl_2. (From Bodeur et al.[57])

CH_4[50] near the 1s edge, H_2S, D_2S near the sulfur 2p, 2s edges,[51] CH_4, C_2H_6, C_3H_8, and their deuterated analogues[52] provide better results than EELS.

At higher energies, i.e., from 1800 eV and up, EELS becomes very difficult and performances of photoabsorption with synchrotron radiation and crystal monochromators are excellent. Numerous results have been obtained in molecules built with third-row elements: SiX_4 (X = H, D, F, Cl, Br, CH_3, C_2H_5, OCH_3, OC_2H_5) near the Si1s edge,[53] phosphorus compounds such as PCl_3, PCl_5,[54] sulfur and chlorinated molecules, SF_5Cl,[55] H_2S,[56] HCl and Cl_2,[57] thiols RSH with R = CH_3, C_2H_5, C_6H_5 and thioethers RSHR', R = R' = CH_3 and R = R' = C_2H_5 or R = CH_3, R' = C_6H_5 or R' = CN, isocyanide CH_3NCS and SCl_2,[58,59] organometallics like $Mo(CO)_6$ near the molybdenum 2s, 2p edge,[60] or radicals like SiO near the silicon 1s edge.[61]

Taking the N_2[44] and Cl_2[57] 1s photoabsorption spectra (see Figures 4 and 5) as representative examples of high-resolution measurements on homonuclear molecules, we can describe the various resonances in terms of "one-electron" or single-excitation features and of "two-electron" or double-excitation features.

2.1. "One-Electron" Transitions

2.1.1. Discrete Spectrum.
In the discrete part of the spectrum, one observes strong and narrow resonances (typically a fraction of an electron volt) resulting from "one-electron" transitions, 1s to unoccupied valence orbital ($1\pi^*$ for N_2 and σ^* for Cl_2) and to Rydberg orbitals. Vibrational structure is observed for valence and Rydberg core-excited states of N_2, but not for Cl_2. This is related to the longer lifetime of the core hole vacancy of nitrogen (Γ = 132 meV) compared to chlorine (Γ = 400 meV) but also to the difference of the potential energy curve of the core-excited state. At this point, it is worth introducing the equivalent-core approximation, which allows one to account for the discrete resonances. This model was introduced by Hollander and Jolly[62] to describe core-excited ions in photoelectron spectroscopy, then used by Schwarz[63] for analyzing core-excited states in photoabsorption spectra. In this model, also called the "Z + 1" approximation, the valence electron cloud experiences the core as if it was replaced by the following atom in the periodic table. In other words, the picture of the core hole localization is valid. Then various properties such as term values, vibrational constants, equilibrium geometries, the bound or repulsive nature of the core-excited state, as measured in a core photoabsorption spectrum are directly comparable with those of the core equivalent species. The reader can refer for example to the theoretical work of Koch et al.[64] for appreciating the applicability of the model in tetrahedral compounds. We show in Figure 6 a schematic representation of the Z + 1 approximation model, applied for core to valence excited N_2 and Cl_2.

The N_2 core-excited states are bound since vibrational structures are observed (more details on vibrational effects are reported in Section 3) and resemble those of the NO molecule. The interpretation of Chen et al.[44] based on a comparison of vibrational constants, term values, and equilibrium distance of core-excited N_2 with those of the ground state of NO shows that the Z + 1 approximation holds very well for Rydberg states, but not totally for the N1s → π^*. The differences for the latter are larger than the experimental accuracy, showing that screening of the hole is not perfect because of the valence character of the excited electron. In contrast, there is no vibrational structure in the chlorine molecule spectrum; the Z + 1 approximation allows one to describe the Cl1s → σ^* resonance in terms of a repulsive state, because the σ^* unoccupied orbital is antibonding. Indeed the equivalent-core species is ClAr, which is a known excimer, unstable in its ground state. The very large Franck–Condon envelope of a core to repulsive state explains the larger width of the resonance compared to the atomic Γ value, as confirmed theoretically by Schimmelpfennig et al.[65]

Generally, there is no sharp distinction between valence and low-lying Rydberg states and one talks about valence–Rydberg mixing. Such effects are certainly very strong for hydrides compared to homonuclear molecules, because the overlap between the atomic orbitals which form the bonding valence orbital, is much stronger and the bond strength is larger (e.g., see Ref. 57). The Rydberg states manifold can be congested and series of different symmetry often overlap. The N_2 spectrum of Figure 4 shows a complex set of Rydberg states from 406 to 410 eV. Using the core-equivalent model, the peaks labeled 3, 7, 8, 10, 12, 14 are assigned[44] to $ns\sigma$ states, whereas the others are assigned to $np\pi$

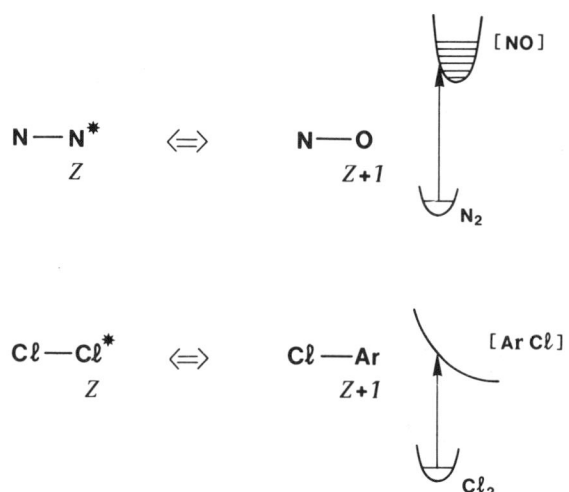

FIGURE 6. Schematic of the $Z + 1$ equivalent-core approximation for describing core-excited N_2 and Cl_2 molecules.

states, both sets being allowed by electric dipole selection rules. This assignment has been confirmed independently by angle-resolved photoion measurements.[66] Such experiments, introduced by Yagishita et al.,[67] allow one to measure "symmetry-resolved" photoabsorption spectra. The basic principle is the following: if one assumes that energetic fragments originate from fast dissociation processes compared to rotation, the fast fragment signal carries the initial orientation of the excited molecules and its angular distribution with respect to the electric vector of the incident light allows one to decompose degenerate ionization channels. Such a method has been used by several groups (Saito and Suzuki,[68–72] Lee et al.,[73] Shigemasa et al.,[66,74,75] Kosugi et al.,[76] Kim et al.,[77] Simon et al.,[78] Bozek et al.[79,80]). We present in Figure 7 the symmetry-resolved N_2 photoabsorption spectrum for two angles, as measured by Shigemasa et al.[66] The $1s \to \pi^*$ valence state at 401 eV and the $1s \to 3p\pi$ Rydberg state at 407 eV as well as the doubly excited states around 415 eV, dominate the spectrum for a 90° angle ($\sigma_\perp$). In contrast, for a 0° angle (σ_0), the peak at 406 eV as well as the shape resonance in the continuum show a σ symmetry, in excellent agreement with Chen et al.[44] Interestingly, the recent high-resolution symmetry-resolved measurements of Lee et al.[81] on N_2 revealed the presence of $s\sigma$, $p\sigma$, $p\pi$ as well as $d\sigma$, $d\pi$, and $d\delta$ Rydberg series in the spectrum. For the d series, they found that the $Z + 1$ approximation does not apply because of the d electron penetrating the valence cloud much more than the s and p ones.

Symmetry-resolved photoabsorption cross section can also be obtained by angular-resolved photoelectron spectroscopy (see Section 3). Several measurements were performed for N_2 and CO near the nitrogen and carbon $1s$ edges, respectively, by Truesdale et al.[82] and Becker et al.[83] Indeed, these authors found negative values of the β parameter for the π resonance and positive ones for the σ shape resonance. However, the information in the Rydberg state or doubly excited state region is still not available at least under the high-resolution conditions. Another interesting method for resolving the symmetry of core-excited states is to measure polarization of the direct x-ray fluorescence emission of the core hole. This has been successfully achieved by Lindle et al.[84] in CH_3Cl and other chlorinated compounds, photoexcited with linearly polarized synchrotron radiation near the chlorine $1s$ edge. The time scale of x-ray fluorescence lying in the picosecond time range, at a shorter rate than rotation, carries indeed the photoabsorption anisotropy at resonances.

In polyatomic molecules, the study of the discrete spectrum of core-level photoabsorption spectra is useful in studying the nature of the bonding. The examples of SF_6 and SF_5Cl following the measurements at the sulfur and chlorine $1s$ edges by Reynaud et al.[55] show that the lower symmetry of the substituted molecule induces the appearance of new peaks in the spectrum by breaking down

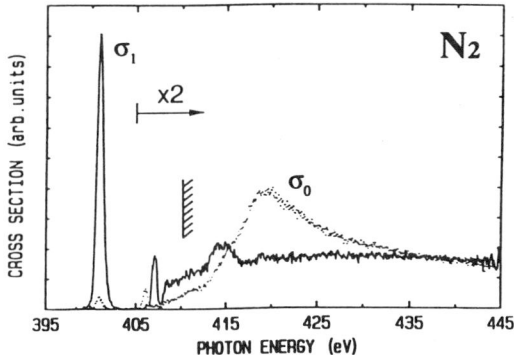

FIGURE 7. Symmetry-resolved photoabsorption spectra of N_2. (From Shigemasa et al.[66])

the degeneracy of some states. The discrete spectrum has been interpreted with *ab initio* multireference configuration calculations on the core equivalent molecules ClF_6 and ClF_5Cl. The very good agreement between theory and experiment[55] for the energy position of resonances in the discrete spectrum has shown conclusively that the old idea of the *spd* hybridization of the valence orbital is wrong, following the previous theoretical analysis of Magnusson.[85] Instead, the *d* orbitals have a true Rydberg character and do not participate to the σ valence orbitals. The CH_3SH molecule for which the $S1s$ core-to-bound excited states have been theoretically investigated[86] offers another example of the nature of the bonding and how the potential energy surface behaves along the normal coordinates. The two low-lying states are shown to have a strong valence and Rydberg mixed character. The first one carries a dominant $\sigma^*(S-C)$ character associated with a *p* (valence and Rydberg) contribution on sulfur. The second is described mostly as a $\sigma^*(S-H)$ with an *s* (valence and Rydberg) contribution on sulfur. Sevin et al.[86] also performed a detailed adiabatic and diabatic analysis of the potential surfaces of the first two core-excited states. Upon small geometrical changes simulating various vibrational modes, these states exhibit important energy variations along the S-C and S-H coordinates, which are associated with drastic changes in their relative *s* or *p* character.

2.1.2. Continuum Spectrum. In the ionization continuum, one generally observes both very wide and very narrow features. The first ones are intense near the ionization threshold in a 20-eV region and become much weaker at higher energy, i.e., several hundreds of electron volts above the edge. Such resonances are intrinsically due to one-electron effects. In other words, the ejected core electron scatters on the neighboring atoms and experiences the anisotropy of the molecular potential. The set of discrete resonances and multiple scattering resonances are called XANES (x-ray absorption near edge structure) or NEXAFS (near edge x-ray absorption fine structure) as reviewed in detail in Ref. 20. At higher excess energy (100 to 1000 eV typically) above the edge, one observes other undulations which are due to a single scattering regime of the core electron on its neighbors. Such oscillations are EXAFS (extended x-ray absorption fine structure) which provide by Fourier transform, local geometrical information such as atomic distances and coordination numbers. Second, there are narrow features in the continuum, which are due to multielectron excitation. They reveal that the molecule cannot be described with the independent electron model. The excitation of a single photon may induce the simultaneous excitation of a core and a valence electron, or even two core electrons. Those double excitations appear as resonances in the photoabsorption spectrum at quite different energies, as we will see below.

The first description of wide and intense resonances in the continuum of photoabsorption spectra was given by Dehmer.[87] They were called "shape" resonances because they depend on the shape of the molecular potential which is experienced by the outgoing electron. The reader may refer to the excellent review on this matter by Dehmer et al.[88] Such resonances have been described as transitions into quasi-discrete unoccupied orbitals.[89] A quasi-atomic approach based on the theory of localized

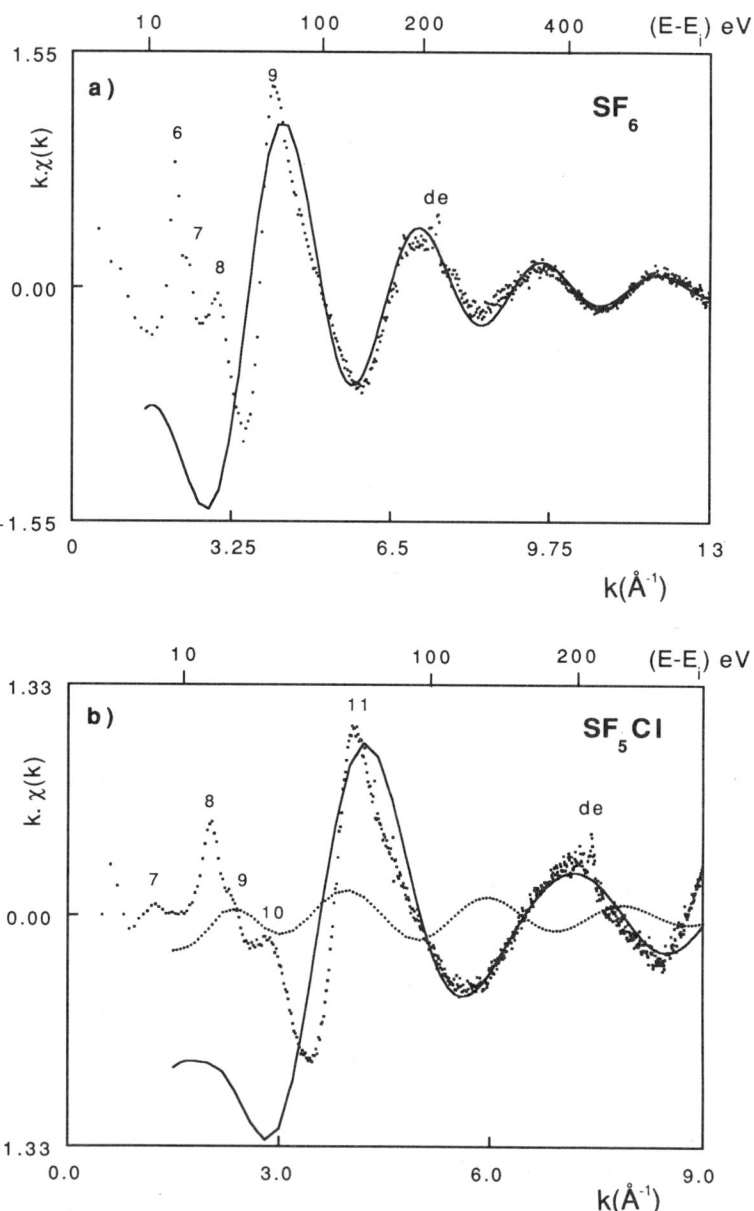

FIGURE 8. EXAFS oscillations in SF_6 (a) and SF_5Cl (b). Solid lines correspond to the calculated single scattering using tabulated phases and amplitudes. In the SF_5Cl case, the dotted line represents the contribution of the chlorine atom alone. "de" denotes the position of double vacancy excited states. (From Reynaud et al.[55])

orbitals in molecules was used successfully by Pavlychev et al.[90] In this theory, the x-ray excitation of a polyatomic system can be regarded as an atomic center modified by the surrounding potential (see also the review of Hallmeier et al.[91]). Recently, Stöhr and Bauchspiess[92] demonstrated a unified picture of near-edge and extended x-ray absorption fine structures in low-Z molecules. Such an approach allows considering shape resonances as multiple scattering, contributing to the strong and wide resonance and considered at the first enhanced EXAFS oscillation. The most recent EXAFS theories of Rehr et al.[93] as well as those of Tyson et al.[94] provide very encouraging results in a number of systems.

We illustrate the difficulty of analyzing XANES spectra in polyatomic systems using the SF_5Cl and SF_6 photoabsorption spectra, following the experimental and theoretical work of Reynaud et al.[55] We present in Figure 8 the EXAFS oscillations in SF_5Cl and SF_6 presented on the usual momentum scale, near the sulfur $1s$ edge. The calculated single scattering contribution is in good agreement with experiment. The small differences between theory and experiment, in the high-k region, are due to the significant contribution of triple scattering especially the diffusion of electrons in the forward direction like $S \to F_1 \to S \to F_2 \to S$ (or $S \to F_1 \to S \to Cl \to S$), in agreement with the General Multiple Scattering calculations of Tyson et al.[94] The similarity of single scattering spectra between both molecules shows that the contribution of the chlorine atom alone is small because the S–Cl distance is much longer (2.039 Å) than the S–F (1.57 Å) distance and more importantly the backscattering amplitude is much higher for fluorine than chlorine. In the low-k region, the spectrum is dominated by multiple scattering. This is the case of the structure labeled 8 in the SF_6 spectrum (or 10 for SF_5Cl), in agreement with the photoelectron spectroscopy measurements of Ferrett et al.[95] and the calculations of Tyson et al.[94] The other resonances, labeled 6 and 7 for SF_6 (7, 8, 9 for SF_5Cl) are fairly narrow compared to normal multiple scattering features, and originate from doubly excited states.[96,97] They interact with the nearby multiple scattering resonance (No. 8 for SF_6 and No. 10 for SF_5Cl) via continuum–continuum interaction providing an exceptional intensity to this feature.[55]

An interesting aspect of shape resonances (or the first multiple scattering structure) in molecules is that they are "fingerprints" of molecular structure, and particularly of bond length according to the original multiple scattering $X\alpha$ calculations of Dehmer et al.,[98] the Stieltjes–Tchebycheff MO calculations of Orel et al.[99] and Langhoff et al.,[89] or more recent multiple scattering calculations of Rehr et al.[93] and Tyson et al.[94] Various empirical correlations of resonance position and internuclear distance, R (in the form of R, $1/R^2$, or the sum of atomic numbers of the bonded atoms), were proposed by many authors (see Refs. 100, 101, 20 for a detailed bibliography and history of this subject) to measure bond length in molecules (free, physisorbed, or chemisorbed on a surface). After several years of controversy[102–105,20] it is clear that there is no unique answer because analysis of multiple scattering resonances is a difficult task, contrary to EXAFS. The spectra of diatomics and pseudo-diatomics (like R–X with R being a polyatomic radical and X an atom) are reasonably well understood and the correlation is reliable. For large molecules (conjugated, aromatic, cyclic nonaromatic molecules, etc.) there are severe difficulties. There is not always a one-to-one correspondence between resonances and a specific bond. Furthermore, strong continuum–continuum interaction due to multielectron effects (see below) contributes to the resonance pattern in the continuum. Further theoretical efforts are still needed to obtain quantitative understanding of shape resonances. A critical point of view on this matter can be found in the book by Stöhr.[20]

2.2. "Two-Electron" Transitions

Core photoabsorption spectra of molecules, like atoms, carry information on electron correlation, and therefore on the departure from the independent electron model. All of these effects are included in the inelastic part of the signal. Here, we deal with the specific aspect of correlation between one core electron and valence electrons, or between two core electrons. The results of Schaphorst et al.[106] on krypton $1s$ photoionization accompanied by $4p$, $3d$, $2s$, and $2s4p$ excitation illustrate the present limits of experiment and theory describing multielectron processes in atoms. The probability decreases in going from outer to inner shell-additional excitation. The calculated double-photoexcitation cross sections and energies are made in the framework of the multichannel multiconfiguration Dirac–Fock method and require taking into account relativistic, quantum electrodynamics and relaxation effects for a good accord with experiment. The theoretical situation in molecules built with the second- or third-row elements is not so advanced, because only energies have been calculated in a limited number of cases. On the experimental side, a significant number of results are now available and give the general spectroscopic and cross section trends and some specific molecular effects such as vibrational bands and chemical shifts. In the following, we will see that simple models can help to make a fairly good analysis of multielectron effects.

2.2.1. Double Core–Valence Excited States. The presence of double excitation resonances, originating from the simultaneous excitation of a core and a valence electron, in core photoionization spectra of atoms has been recognized for more than 30 years.[107–109] In molecules, such features are also seen but unlike the case for atoms, they are often found in the same region as multiple scattering resonances, raising difficulties in the spectroscopic assignments. The N_2 (Figure 4) spectrum of Chen et al.[44] and the Cl_2 (Figure 5) spectrum of Bodeur et al.[57] illustrate two situations. In the first case, double excitation features are overlapping the shape resonance and only high-resolution spectra or symmetry-resolved spectra (see Figure 7) can solve this problem. In the second, the situation is more simple, because the first shape resonance lies at higher energy. Nevertheless, a good resolution and good signal-to-noise ratio are necessary to reveal these weak features.

These sets of resonances should be seen as a manifold of doubly excited states converging to different ionization limits, for which one electron is excited, the other is ionized or two electrons (core and valence) are ionized. The excitation plus ionization limits are given experimentally by correlation satellites (or "shakeup" satellites) in core photoelectron spectra. The double ionization limit can only be given by theory. In analogy to photoelectron spectra, where satellites are described as "two hole–one electron" or "2h–1e," doubly excited states are often called "2h–2e." The first detailed attempt to analyze the fine structure of such states has been made by Domke et al.[45] in the high-resolution spectrum of CO near the carbon 1s edge. Considering that each of these electronic states has a vibrational structure, the analysis becomes more complicated, especially without theoretical calculations. Domke et al.[45] successfully obtained a reasonable assignment based on the knowledge of the spectroscopy of shakeup satellites in the C1s photoelectron spectrum of CO.[110–112] The authors[45] were able to compare the term values with those of the 2π excited core equivalent NO molecule. The good agreement shows that the $Z+1$ equivalent core approximation holds even for double core–valence excited states.

In polyatomic molecules, the first attempt to theoretically analyze these "2h–2e" states was made by Connerade and Hormes[113] on H_2S and SCl_2, near the sulfur 1s edge. In many others, the situation is much more complex[53] because the understanding of the one-electron features (multiple scattering resonances) is still incomplete. The SF_5Cl and SF_6 cases which have been analyzed in Section 2.1.2 (Figure 8) are representative of the difficulties of analyzing double core-valence excited states in polyatomics.

2.2.2. Double Core Excited States. Double core excited states refer to the simultaneous excitation of two core electrons into unoccupied orbitals. They are expected at very high energy excess above the core ionization edge, i.e., in the region of EXAFS oscillations. Up to now, there is scarce experimental evidence of such states in free atoms,[106,114] because of the extremely low cross sections (about 200 barns for argon $1s + 2p$ double excitation). In molecules, there are several examples available for systems built with second-row elements. The first experimental report of such states was made by Bodeur et al.[115] on silicon compound molecules, SiX_4 (X = H, F, Br, Cl, CH_3) in the Si 1s ionization continuum, in which two electrons ($1s + 2p$) were promoted into valence unoccupied orbitals. One distinguishes *single-center states* (the holes are created at a single atom) from *two-center states* (the holes are created at different atoms) following the theoretical self-consistent field (SCF) predictions of Ohrendorf et al.[116] The latter have never been evidenced experimentally. Similar single-center states were observed in other silicon compounds like $SiCl_x(CH_3)_{4-x}$,[117] phosphorus molecules[54] like PCl_3 and PCl_5, and sulfur molecules like SF_6, SF_5Cl, H_2S, SO_2 above the sulfur 1s edge[118,56] but not in the chlorine 1s continuum. More recently, D'Angelo et al.[119] showed, in HBr and Br_2, the existence of several double excitation channels of the bromine atom including $1s3p$. The excellent resolution and signal-to-noise ratio allowed the observation of resonances on the continuum, with an amplitude always inferior to 1% of the total photoabsorption cross section, as illustrated in the photoabsorption spectrum of Figure 8. There are no reports on double core–core vacancy states involving two 1s or 2p electrons excited from the same atom. The same comment is true for double core–core excited states in which the core electrons belong to two neighboring atoms.

We now analyze the H_2S case in more detail, because double $1s$–$2p$ core excited states have been studied both experimentally and theoretically. We show in Figure 9 the high-resolution sulfur 1s

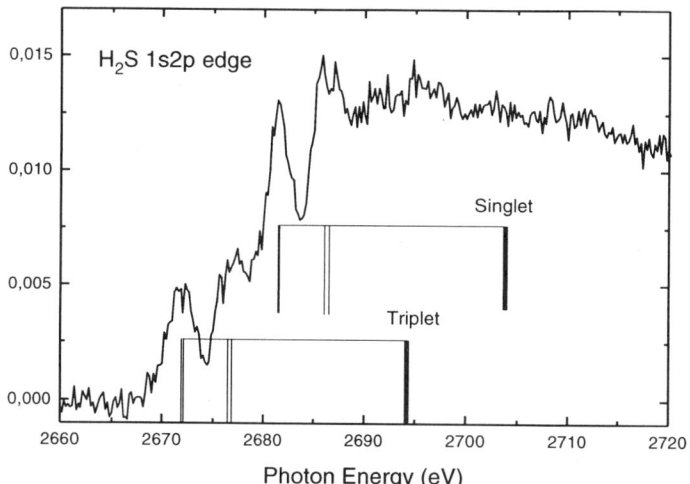

FIGURE 9. Sulfur K-shell photoabsorption spectrum of H_2S in the threshold and in the double core ($1s,2p$) excitation regions. Calculated energies of double core excited states and double ionized limits are indicated as vertical bars. The calculated energy of the first resonance has been adjusted to the experimental value. (From Reynaud et al.[56])

photoabsorption spectrum of H_2S[56] with an enlargement of the region of the double $1s$–$2p$ double core excitation. The simplest description of the energy E_{ij-kl} is given by a Hartree–Fock model,[115]

$$E_{ij-kl} = E_{i-k} + E_{j-l} + J + K$$

where i and j represent the core vacancies, k, l the unoccupied orbitals, E_{i-k} and E_{j-l} the single-core vacancy energies given by experiment, J and K the total Coulomb and total exchange terms, respectively. The J term is extremely large and dominates the exchange term. As an example, it amounts to 65.6 eV for a double hole $1s^{-1}2p^{-1}$ or 229.7 eV for a $1s^{-1}1s^{-1}$ in silicon. Therefore, one should search for these features at much higher energies than the sum of single-core excitation energies. An interesting characteristic of these states is that they appear as twin resonances, as the core holes may couple in triplet and singlet. The calculation of the energy of such states (including singlet–triplet splitting) with a high accuracy requires going beyond the simple Hartree–Fock formula. One difficulty is accounting for the relaxation energy of the electronic cloud in the presence of two core holes. The first successful SCF calculation,[116] in silane and tetrafluorosilane, showed the breakdown of the molecular-orbital picture of double ionization in core-vacancy situations, and found a good agreement with the observations of double core-vacancy states.[115] Very recently, Reynaud et al.[56] developed an original method to calculate directly this relaxation energy and the singlet–triplet splitting in excellent agreement with experiment, i.e., 9.6 eV for H_2S (the energy difference between the first two resonances) and the double core ionization energy. In addition, Reynaud et al.[56] introduced for the first time the $Z + 2$ core equivalent model to calculate term values, i.e., the energy difference between ArH_2 states and the ArH_2^{2+} limit. This molecule can be seen as a van der Waals system in which the argon atom is weakly bound to an elongated H_2 molecule, i.e., with the geometry of the original H_2S molecule. Ab initio multiconfiguration interaction calculations performed on this system[56] provide excellent agreement with experiment for the energies of the resonances (see Figure 9). Moreover, using the calculated ionization energy $1s^{-1}2p^{-1}$ including relativistic corrections, the absolute energy of the first resonance is predicted to equal 2674.4 eV and compared favorably with the 2672 eV experimental value.

Finally, double core-excited states, as observed with compounds built with third-row elements,[115,120] show a chemical shift 1.7 times larger than the corresponding $1s$ edge shift, a relative intensity which varies strongly with the electronegativity of the ligands, the most intense being found for fluorine and chlorine ligands.

3. ELECTRONIC RELAXATION

The electronic relaxation processes of a core hole in molecules are governed by soft x-ray fluorescence, electron emission as well as ionic and neutral fragmentation, and fluorescence of the fragments, as schematically represented in Figures 1a and 1b. Even if soft-ray spectrally resolved fluorescence spectroscopy has been done under high-resolution conditions, in some free molecules (see Ref. 121 and references therein), the fluorescence yield is very small (e.g., see Ref. 122) for species built with light elements and Auger-like processes dominate. Consequently, electron spectroscopy is the most common tool to investigate the relaxation of core holes. Molecular Auger spectra in diatomic and simple polyatomic molecules have been available with very high resolution for 25 years, using high-resolution electron analyzers associated with laboratory ionization (x-ray tubes or electron guns) sources.[5,123] The routine use of synchrotron radiation in the far ultraviolet and the soft-ray range with its polarization properties (linear or circular) as a photoionization source, associated with electron spectrometers, started much later, in the early 1980s, and triggered a wealth of new photoionization measurements using widely the tunability of the incident radiation. Indeed, it was possible to tune the photon energy on a given resonance, near a selected ionization edge, to prepare the molecule in a well-defined core-excited state and investigate the electronic relaxation channels. The ejection of two (or more) electrons being closely associated with fragmentation, the investigation of multiionization became possible starting in the mid-1980s, with time-of-flight mass spectrometry operated in the multicoincidence mode. Several aspects of this photoelectron and photoion spectroscopic work can be found in numerous reviews.[40,88,124–127]

3.1. The Core Level Photoelectron Spectrum: A General Description

The core level photoelectron spectrum of a molecule can be described at first sight, along the same lines as for atoms. It contains the basics on the electronic properties of core orbitals and provides an extremely rich variety of information on many-body effects (or electron correlations) necessary for a successful interplay between theory and experiment (the reader should refer to the recent reviews on photoionization of atoms[128–130] including the exhaustive bibliography of theoretical and experimental works on the subject). However, the status of understanding molecular spectra lags far behind that of atoms primarily because of the density of final states and their complex potential energy surfaces. Furthermore, as we have seen in the Introduction, there are specific questions relevant to molecules only, which broadens the interests of such studies. One is the variation of the binding energy of a core line as a function of the chemical environment of the excited atom. This is known as the chemical shift and widely used for analytical purposes.[5,131–134] A second is the variation of the core hole lifetime as a function of the chemical environment. This subject is more recent because it requires, in the absence of real-time experiments, the measurement of photoelectron, photoabsorption, or electron energy loss spectra under very high resolution (see Section 2). Several theoretical analyses have been made based on a one-center[135] or multicenter model[136] without converging on the same predictions. Coville and Thomas[135] found a large reduction of the 1s lifetime in carbonaceous molecules compared to the free carbon atom and in systems with ligands of increasing electronegativity, at variance with the results of Hartmann and Der[136] but in good agreement with experiment except for CO_2. Cutler et al.[137] established experimentally, in a series of iodine compounds, that the I4d linewidth decreases as the electronegativity of the ligand increases. This subject is still a matter of debate because vibrational and molecular field effects are not easily interpreted (see Section 4). A third question is the role of nuclear motion and dissociation in the photoelectron ejection. The extremely large number of vibrational, rotational, and dissociation continua populated in the decay of a core vacancy are bringing new fine structures or washing them out in photoelectron spectra thus complicating the interpretation of electronic relaxation processes. This important point is the main subject of this section.

The general features of a photoelectron spectrum of a molecule, from an experimental point of view, include the core lines and satellites, valence lines with satellites corresponding to the primary photoionization step and Auger-like lines (main lines and satellites) which correspond to secondary electron emission. Among the latter, one distinguishes the "first-order" decay lines which are the Auger (or resonant Auger and autoionization) and the "second-order" lines such as two-step autoionization, double resonant Auger, double Auger, and shakeoff. Notice that "shakeoff" or double Auger processes appear as a continuum in the spectrum, whereas the other processes lead in principle to well-defined lines in the spectrum. In practice, those features are always congested or even quite structureless because of the presence of satellite lines due to electron correlation, vibrational excitation, or the dissociative character of the final state. This complication explains why second-order processes are often investigated by studying the dissociation pathways of multiply charged ions (see Section 3.3).

The photoelectron spectrum varies considerably in line energy and intensity, with photon energy $E(h\nu)$, especially when one excites resonances near core levels. Despite the intrinsic interest of tuning the photon energy on resonance, the photon energy tunability allows one to easily identify the direct core and valence photolines from the others. The former have an energy ε directly related to the binding energy E_b by

$$E(h\nu) = E_b + \varepsilon$$

The latter have characteristic energies only related to the dynamics of the core hole relaxation and are, at first order, independent of the incident photon energy.

The dynamics of the photoemission process is very sensitive to the initial and final wave functions and to electron correlation and resonant processes. Each photoelectron line is characterized by the ionization probability or cross section (σ) and its angular distribution asymmetry parameter β. Other important properties of the dynamics of the photoionization process can be extracted from polarization of fluorescence of the residual ion and angular distribution of Auger electrons (see Section 3.2), and more importantly from all parameters defining *completely* the photoionization process, as one extracts from electron/Auger electron coincidence, Auger electron angular distribution, and spin polarization experiments (see Ref. 128 for a review). In molecules, such sophisticated methods (except angular distribution of Auger electrons) have not to our knowledge been applied, mainly because of the very high density of states and electron–nuclear motion interaction (see Section 4) but also because the first theoretical predictions regarding angular distribution and spin polarization of Auger electrons are very recent.[138] As mentioned above, angle-resolved photoelectron spectroscopy has been used on many occasions for approaching the photoionization dynamics of core electron in molecules.

Assuming a linearly polarized light and electric dipole transition, the differential cross section $d\sigma(h\nu,\theta)$ of each direct photoelectric photoline, and measured for a given photon energy $h\nu$, is

$$d\sigma(h\nu, \theta)/d\Omega = \sigma(h\nu)/4\pi[1 + \beta(h\nu) P_2 (\cos\theta)]$$

where θ is the angle between the photon polarization vector and the electron emission direction and $P_2 (\cos\theta)$ is the second Legendre polynomial. The β parameter contains important information on the transition dipole moments into the individual angular momentum channels, on the Coulomb phases shifts, and on the intra- and interchannel phase shifts. The β parameter associated with a specific photoline is expected to vary with electron energy, especially across a resonance.

High-resolution photoelectron spectroscopy with x-ray laboratory sources has provided high-quality spectra for various simple molecules.[5,123] In the 1980s, angle-resolved photoelectron spectroscopy associated with synchrotron radiation was used by many groups to measure partial photoionization cross sections and angular distributions. An extensive review can be found in Chapter 5 of this volume.[139] The ZEKE (zero-kinetic-energy-electron) photoelectron spectroscopy technique also appeared[140] offering a complementary tool to conventional PES methods for high-resolution spectroscopy. Meanwhile, theoretical efforts developed strongly.[88,89,99]

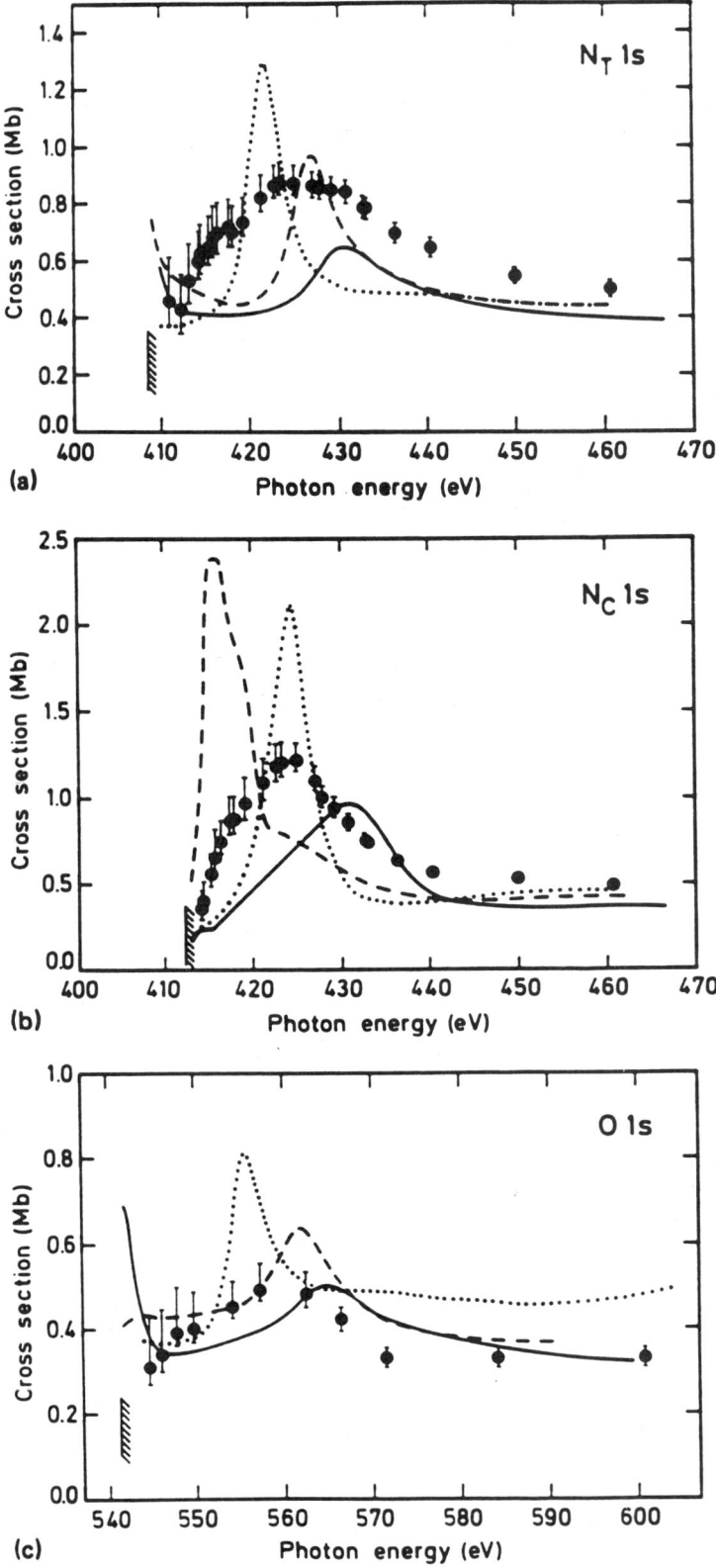

3.2. Shape Resonances

Shape resonances are typical of molecules because they are induced by the anisotropy of the molecular field. We use the term "shape resonance" rather than the more appropriate term "multiple scattering resonance" for historical reasons and also because in small molecules there is a coincidence between the two concepts. The main property of shape resonance is that it belongs to the dynamics of ejection of the core electron only. It is a "one-electron" effect in contrast to electronic autoionization which involves bielectronic matrix elements. This is a well-visited subject which has stimulated many theoretical studies. Experimentally, measurements of partial and differential photoionization cross section by angle-resolved photoelectron spectroscopy have been performed as soon as synchrotron radiation, soft-x-ray monochromators, and appropriate rotatable electrostatic analyzers were routinely accessible for gas-phase experiments: The first experimental efforts started in the early 1980s with Carlson et al.[141] on HBr, Br_2, CH_3Br molecules, and developed further with Keller et al.[142] on SiF_4, $Si(CH_3)_4$, Truesdale et al.[82,143] on CO, CO_2, CF_4, OCS, Lindle et al.[144] on N_2, CO, NO, Kanamori et al.[145] on BF_3, Becker et al.[83] on CO, de Souza et al.[146,147] on $Si(CH_3)_4$, SiF_4, Ferrett et al.[95,96,148] on SF_6, SiF_4, SO_2, H_2O, Piancastelli et al.[103,149] on C_6H_6, C_2H_4, Grimm et al.[150] on N_2O, Schmidbauer et al.[151,152] on CO, N_2O, CO_2, and Kilcoyne et al.[153] on H_2CO, C_2H_4.

Let us concentrate on the dynamics of the primary core photoionization process, on the basis of angle-resolved photoelectron spectroscopy performed in the triatomic N_2O molecule,[150,151] photoexcited near the nitrogen and oxygen 1s edges. This molecule is the prototypical example of polyatomic systems. It is a linear N–N–O triatomic species, in which the two nitrogen atoms have different binding energies because they have different chemical environments (the chemical shift between the central, N_c, and terminal, N_t, 1s core lines amounts to 4 eV). It is then possible to perform site-specific photoionization studies for each of the three atomic components. We show in Figure 10 partial 1s single-hole photoionization cross sections of the N_c, N_t, and O atoms of N_2O as reported from the work of Schmidbauer et al.[151] The figure includes the multiple scattering calculations (MSMXα) of Grimm et al.[150] and the Hartree–Fock calculations of Schmidbauer et al.[151] On a kinetic energy scale, the resonance position is found at 15.5, 11.5, and 17 eV for N_c, N_t, and O 1s ionization, respectively, showing a clear dependence on the specific ionization site. The Hartree–Fock calculations of Schmidbauer et al.,[151] which account for the relaxation of the core orbital, show a better agreement with experiment, compared to those in which the core orbital is kept frozen or to the earlier results of the multiple scattering model (MSMXα). The steep cross section theoretical curve (solid line in Figure 10) at the oxygen ionization edge suggests the existence of another shape resonance at very low kinetic energy. ZEKE photoelectron spectroscopy measurements[154] indicate this trend but lack absolute values for a definite conclusion. Asymmetry parameters, β, for the 1s single-hole photoionization cross sections of the N_c, N_t, and O atoms (experiment and theory) have been measured by the same authors.[151] Again, one observes quite different photon energy dependence with the atomic ionization site. Theory reproduces only qualitatively the existence of minima but the discrepancy in the energy position suggests that various effects which have not been accounted for by calculations, play a significant role. These are interference between the (degenerate) partial photoionization channels and discrete–continuum interaction with autoionizing doubly excited states lying in the same energy region. Therefore, the final state is not strictly speaking the core photoionization continuum but also excited configuration states of the core hole (e.g., see the SF_6 case in Ref. 95), or excited configuration valence ionic states (see the dispersed fluorescence experiments in Ref. 155 in the nitrogen molecule and Section 4.1 of this chapter).

FIGURE 10. Partial 1s single-hole photoionization cross sections of the N_t (a), N_c (b), and O (c) atoms of N_2O. The experimental points are scaled to the total photoionization cross section (Schmidbauer et al.[151]). Full lines represent relaxed-core Hartree–Fock calculations, dashed lines represent frozen-core Hartree–Fock calculations from Schmidbauer et al.[151] Dotted lines correspond to MSMXα calculations of Grimm et al.[150] The 1s ionization thresholds are indicated by hatched bars. (From Schmidbauer et al.[152])

More information on the alignment induced by the photoabsorption process at the energy of a shape resonance can be obtained by measured Auger angular distributions. The partial differential cross section is defined through the simple relation[156,157]

$$d\sigma(h\nu, \theta)/d\Omega = (\sigma(h\nu)/4\pi)[1 + \beta_m(h\nu) \, c_a P_2 (\cos\theta)]$$

The overall asymmetry parameter $\beta_A = \beta_m(h\nu)c_a$ depends on the molecular alignment $\beta_m(h\nu)$ and the intrinsic anisotropy of the Auger decay c_a and thus the final dicationic state. The alignment term includes only dipole transition matrix elements and not phase shift terms.[157] The CO molecule photoexcited at the carbon $1s \rightarrow \sigma^*$ shape resonance is the most illustrative example. The theoretical predictions of Dill et al.[157] of a strong alignment related to the symmetry of the excited electron were not reproduced by angle-resolved photoelectron spectroscopy[83,143] but were in good agreement with angle-resolved photoion spectroscopy.[67,68] Hemmers et al.[158] clarified this puzzling discrepancy by showing that the alignment parameter $\beta_m(h\nu)$ is fairly small at the shape resonance, at variance with the early predictions but in agreement with the most recent calculations of Schirmer et al.[159] and those of Lynch.[160] One important conclusion of this recent study is that the excited electron is indeed in the "continuum" and not in a bound state.

3.3. The Auger Process

The Auger process is usually described as a two-step process, the ejection of the core photoelectron followed by secondary Auger electrons. When the molecule relaxes through this second step, it "forgets" the way the core hole was formed. More than one electron can be ejected either simultaneously (shakeoff) or sequentially (Auger cascades). In the simplest picture, the Auger process is an electronic autoionization process, in which the core hole is filled by an outer electron and another external electron is ionized. The residual ion is primarily a doubly charged molecular ion (or a dication). The Auger electron is labeled as KLL, LVV, depending on the initial hole (K or L) and on the nature of the two final holes (LL, VV). The Auger electron energy spectrum is therefore independent of the incident photon energy and is often presented on a double ionization energy scale since the dication energy state $E(M^{2+}) = I_c - \varepsilon_A$ where I_c is the core ionization energy and ε_A the Auger electron energy. Contrary to atoms, the number of final dicationic states in molecules is enormous (10^2 to 10^3 for small polyatomics). Our knowledge as to the spectroscopy of dications is not very developed, except for the low-lying states of diatomics and triatomics for which a variety of experimental techniques[23,161] and accurate *ab initio* calculations are available (see Ref. 21 and references therein). Furthermore, it is difficult to calculate localized states, without breaking the symmetry of the wave function. All of these difficulties rule out accurate configuration interaction calculations of molecular Auger spectra, like for atomic spectra, and put in perspective the recent theoretical efforts on this matter (see Refs. 162–165 and references therein).

The hole localization of the initial ionized state *and* the final dicationic states is expected to strongly influence the Auger matrix element and has been invoked to explain qualitatively the features of molecular Auger spectra as well as fragmentation mechanisms (for a review see Ref. 128). The problem of localization in valence final states of Auger spectra has been discussed in homonuclear molecules[166,167] and more generally by several authors,[164,168–172] but although the *self-imaging* picture of Auger spectra of covalent molecules was accepted, the understanding of molecular Auger spectra was still limited to a qualitative or semiquantitative level. The recent spectacular theoretical progress (for a review, see Ref. 165) allows molecular spectra to be calculated quantitatively, as seen below.

The characteristics (intra- or interatomic) of the Auger decay and especially the spatial distribution of the two-hole final states are strongly dependent on the bonding (covalent or ionic) in the neutral molecule. First, we consider the CO molecule as a prototype of covalent (polar) species, because various measurements of the Auger spectrum are available.[123,162,163] We present in Figure 11 the carbon KVV Auger spectra of CO, as measured by Moddemann et al.[123] This spectrum is quite different from the oxygen KVV one (not shown) and one readily suspects that the holes are probably localized on the initially ionized site. They are also very complex and up to recently, they have been explained

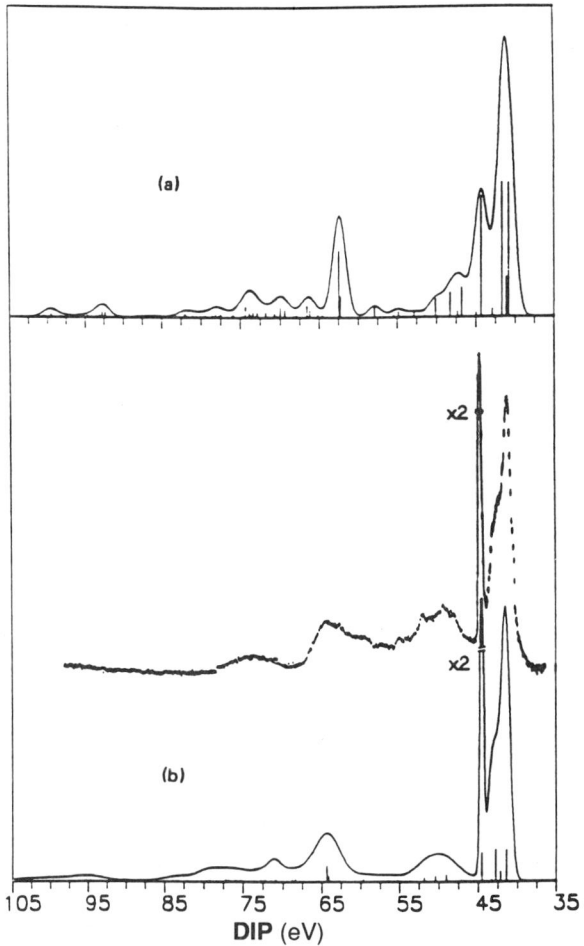

FIGURE 11. Carbon KVV Auger spectrum of CO. Ab initio, time-dependent calculations of Cederbaum et al.,[178] without (a) and with (b, lower spectrum) vibrational effects. The inherent broadening of the lines is included (Γ = 0.2 eV) to account for the finite lifetime of the core hole and experimental resolution. Experimental spectrum of Moddemann et al.[123] (b, upper spectrum). (From Cederbaum et al.[178])

only qualitatively.[162,165,173–177] Generally, the molecular Auger spectrum results in three distinct regions due to double-hole states involving ionization of two outer electrons, one outer and one inner electron, or two inner electrons, correlating strictly with atomic p^{-2}, $p^{-1}s^{-1}$, and s^{-2} valence holes, respectively. Cederbaum et al.[178] were the first to reproduce quantitatively (line position and widths) the carbon and oxygen KVV experimental spectra (see Figure 11 for the carbon KVV results) using an ab initio time-dependent theory based on the two-particle Green's-function algebraic diagrammatic construction (ADC) method. These authors demonstrated that the complexity of the spectra originated from the combined effect of the nuclear motion of both the short-lived intermediate (core ionized) state and the final state (dication), as well as electron correlation effects. They show that the correct ordering of the dicationic states with sufficiently accurate potential energy curves (especially the slope of the repulsive part of the potential curves, in the relevant internuclear distance region) is absolutely necessary to obtain vibrational broadening and energy shifts. In order to appreciate the importance of these effects, Cederbaum et al.[178] also calculated the Auger spectra without consideration of the vibrational effects (Figure 11). It is remarkable to observe that the resulting shifts of the lines compared

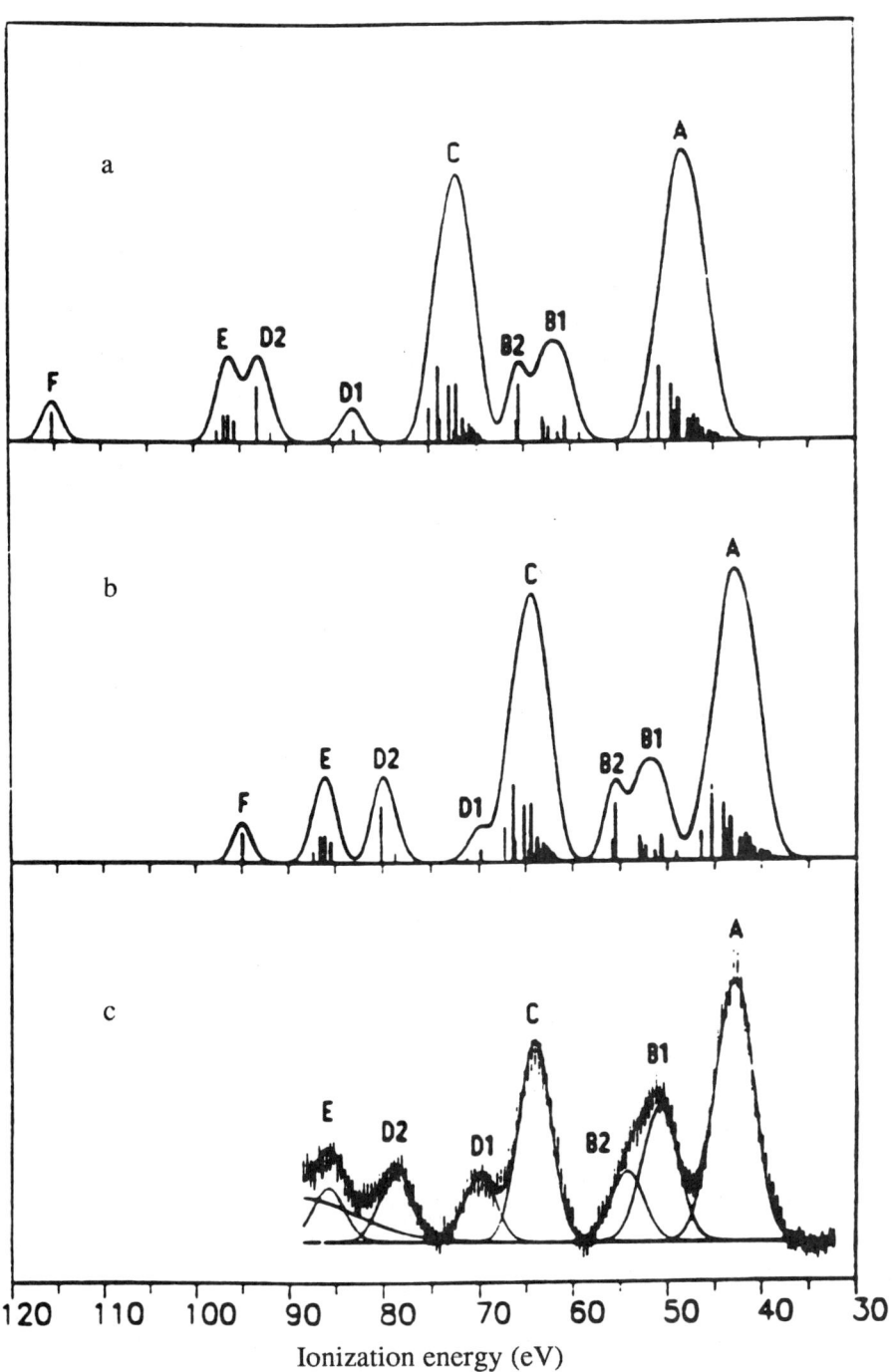

FIGURE 12. Silicon $2p$ Auger spectrum of SiF_4. Green's-function calculations of Tarantelli and Cederbaum[188] without (a) and with (b) relaxation energy shifts. Experimental spectrum (c) from Aksela et al.[190]

to those of the spectrum calculated with such effects, are very strong (1 to 3 eV for the carbon KVV spectrum) and this affects strongly the general shape of the spectrum. This shows that the description of the Auger process in terms of vertical transitions only is inadequate. In other words, one cannot apply the Franck–Condon principle and separate the electronic from the nuclear part in the matrix elements. The same authors also found that electron correlation induces the breakdown of the Auger intensity over many states above 60 eV, i.e., for double ionizations involving one inner valence and one outer valence electron or two inner valence electrons, thus increasing further the complexity of the spectrum. The authors also calculated the two-hole atomic population of each of the dicationic states and found that they are characterized by significant relative values of the one-site population on one atom, with a very small relative interatomic component on the other atom. In formaldehyde, which is also a covalent molecule, the carbon and oxygen Auger spectra of H_2CO of Correia et al.[179] have been quantitatively reproduced by Minelli et al.[180] Along the same lines, the experimental Auger spectra of HF[181] and LiF[182] have been interpreted by Zähringer et al.[183] Further theoretical developments based on time-dependent formulation[184,185] have treated more complex situations (see also Section 4).

The situation regarding Auger spectra of ionic compounds is drastically different, as demonstrated by recent *ab initio* theoretical calculations of Auger spectra, for the boron and fluorine KVV spectra of BF_3,[186] the potassium and fluorine KVV spectra of KF,[187] and the silicon LVV spectrum of SiF_4.[188] We take the SiF_4 case as typical of an ionic molecule which has been the subject of many experimental studies.[96,147,189,190] In contrast to the general description of the spectrum in three regions (see above), with the *self-imaging* picture of Auger spectroscopy, strongly ionic molecules display a *foreign-imaging* behavior.[188] This refers to the fact that the final two holes are found mostly on the ligands rather than on the initial atomic site. This description is opposite that of the classical picture of Larkins et al.,[191] who predicted, using a semiempirical independent particle model, only four bands out of the six bands observed experimentally. In Figure 12 the experimental Auger spectrum of Aksela et al.[190] is compared to the *ab initio* Green's-function calculations of Tarantelli and Cederbaum.[188] The six experimental bands are well reproduced in energy, width, and relative intensity and support the theoretical analysis. Indeed, taking the average two-hole populations of the group of states forming the contour of peak A in Figure 12, Tarantelli and Cederbaum[188] obtained relative populations distributed as follows: 0.0052 for Si^{-2}, 0.0028 for F^{-2}, 0.1330 for $Si^{-1}F^{-1}$, and 0.859 for F_1^{-1}, F_2^{-1}. This shows the importance of cationic states with two holes on different fluorine sites.

3.4. The Resonant Auger Process

This generic term refers to the Auger-like decay of core-excited molecules and in fact covers a manifold of processes in which the extra electron, which has been promoted from the core to the valence region, remains either as a spectator or participates in the relaxation process. The reader may refer to experimental results in atoms[192,193] and to theoretical aspects.[194] In the spectator resonant Auger decay, the core hole is filled, in the presence of the extra electron, with a valence electron and the energy released is enough to ionize another valence electron. The residual ion is a singly charged ion with an excited configuration (two hole–one particle). In the participator transition, the core hole is filled by the extra electron and the energy released is enough to ionize a valence electron. The residual ion is a singly charged ion with a single hole configuration. The main difference with the normal Auger process is found in the charge distribution of the final states which starts from single ionization rather than double ionization, due to the screening effect of this electron. Ungier and Thomas[195] with electron–electron coincidence spectroscopy in CO and Kanamori et al.[145] with synchrotron radiation in BF_3 reported the first experimental evidence for resonant Auger spectra in molecules. The main finding of these early reports is that in the photoelectron spectrum, the highest kinetic electrons are due to autoionization in which the extra electron participates in the decay, the low-energy part arising from deexcitation with the extra electron remaining as a spectator. The former lines correspond to a monocation with a single valence hole and cannot be distinguished from direct valence photolines (bottom spectrum of Figure 13). The latter form a set of lines similar to the Auger spectrum (top spectrum of Figure 13) but shifted by a fixed amount reflecting the Coulomb interaction with the spectator electron. The deexcitation spectrum of CO after the carbon $1s \rightarrow \pi^*$ core excitation[82,83,143,195–197] illustrates

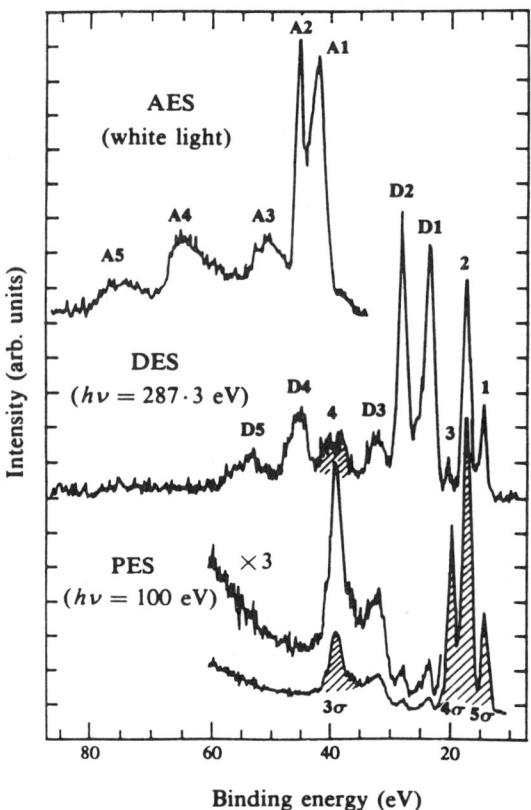

FIGURE 13. Experimental Auger (top), autoionization (or resonant Auger) of CO at 287.3 eV (middle), and photoelectron at 100 eV (bottom) from Eberhardt et al.[197]

quite well the main relaxation pathways. We show in Figure 13 the deexcitation spectrum of the $C1s \to \pi^*$ resonance compared to the normal Auger spectrum and to the photoelectron spectrum of CO obtained off resonance in the valence ionization continua, as measured by Eberhardt et al.[197] with synchrotron radiation. The deexcitation spectrum shows primarily D_n features which mimic the Auger spectrum with peaks labeled A_n, but shifted in energy. The D_n peaks represent the signature of spectator transitions and they dominate the autoionization spectrum. The deexcitation spectrum shows the enhancement of valence bands (one hole and two hole–one particle) labeled 1 to 4 in Figure 13. These are due to participator transitions in which the final states are identical (at least for the one-hole features) to those found for direct photoionization. The first theoretical analysis of the entire deexcitation spectrum of CO has been made by Freund and Liegener[198] within a Green's-function formalism. We show in Figure 14 the theoretical predictions for the carbon $1s$ KVV Auger spectrum of CO, and of Schirmer and Walter[199] for the direct photoelectron spectrum and of Freund and Liegener[198] for the deexcitation spectrum of the carbon $1s \to \pi^*$ excited molecule. Such early calculations do not include vibrational effects that are necessary for a quantitative fit with experiment. Nevertheless, several trends can be extracted. The intensity distribution of the deexcitation spectrum is quite different from the direct photolines because the intensity is governed by Coulomb matrix elements associated with few configurations. The 2π extra electron experiences the strongest Coulomb interaction, among other valence electrons, the 5σ electron dominates because it is located on the carbon atom. The calculations show the influence of screening the initial state of the deexcitation spectrum and validate the procedure of shifting the Auger spectrum by a fixed amount of energy. Above a binding energy of 30 eV, the one-to-one correspondence between the direct photolines and the autoionization spectrum is lost,

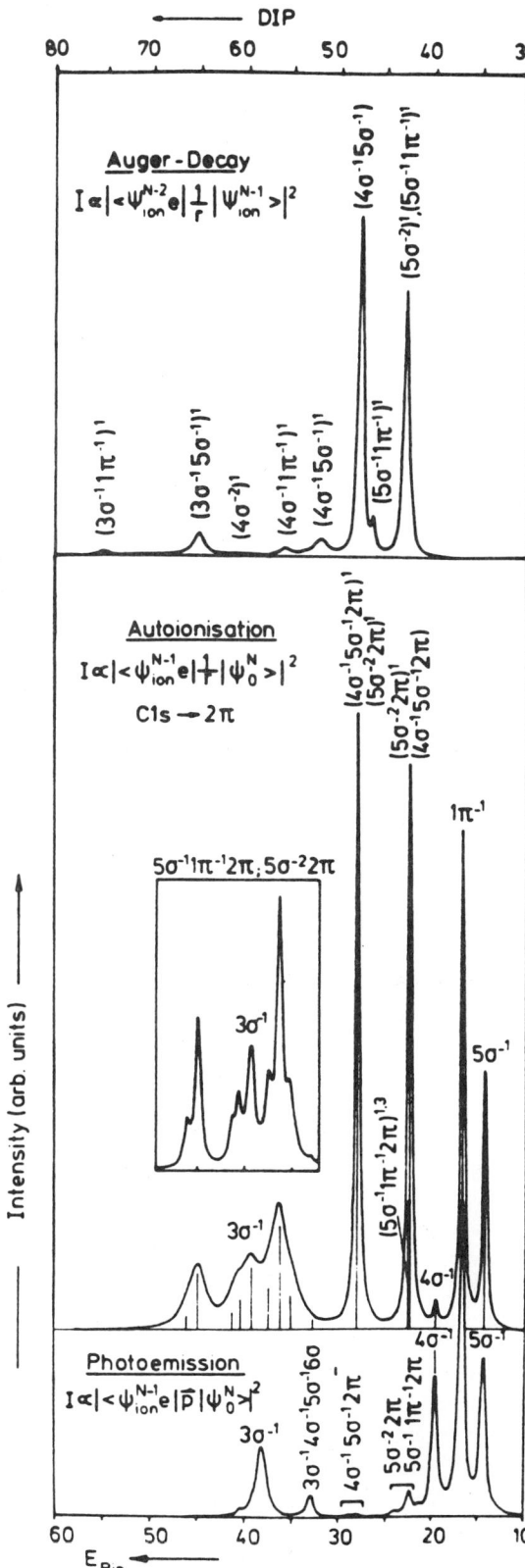

FIGURE 14. Theoretical Auger (top), autoionization (or resonant Auger) (middle), and photoelectron spectrum (bottom) of CO. The bottom binding energy scale refers to the photoelectron and autoionization spectrum; the top double-ionization potential energy scale refers to the Auger spectrum. (From Freund and Liegener.[198])

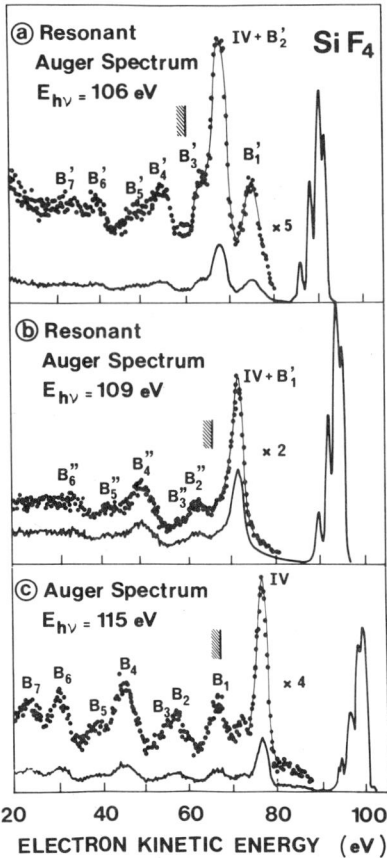

FIGURE 15. Experimental resonant Auger spectra (a,b) recorded at two photon energies (discrete resonances) and Auger spectrum of SiF_4 near the silicon $2p$ edge. (From de Souza et al.[147])

showing that the final states are totally different. However, more precise calculations are needed to confirm this last conclusion.

Similar behavior of the decay of core-excited molecules has been observed in a variety of other molecules: SiF_4,[96,190,200,147] SiH_4,[201] $SiCl_4$,[202] BF_3,[203] $Si(CH_3)_4$,[147,204] SF_6,[95,205] N_2O,[206] gas-phase N-heterocyclic molecules,[207] $SnCl_2$ and $SnCl_4$,[208,209] NO,[210] methyl formate.[211] Related results on adsorbates on surfaces are also reported for CO, CO_2[212] or azabenzenes[9] and benzene.[213]

Generally, the spectator Auger processes represent the preferred deexcitation pathways compared to participator transitions. The kinetic energy shift of the participator and spectator Auger lines relative to the normal Auger lines depends on the binding energy of the electron in the excited bound state orbital extracted from atomic nonrelativistic calculations.[205,164] We take the resonant Auger results in the SiF_4 molecule photoexcited near the $Si2p$ edge because this molecule has been the subject of many works[96,147,190,200] and the theoretical analysis of the Auger spectrum is now available (see Section 3.2). We show in Figure 15 the resonant Auger spectrum of SiF_4, measured on two discrete resonances below the silicon $2p$ edge, as obtained by de Souza et al.[147] In this energy range, the direct valence ionization continua are quite significant compared to the core hole photoionization cross section. The photoabsorption spectrum shows discrete resonances associated with the $Si2p \to \sigma^*(a_1)$, $\sigma^*(t_2)$, and Rydberg transitions. Spectator transitions (labeled B'_n and B''_n) are found shifted compared to the Auger lines (labeled B_n). They overlap the inner valence part (labeled IV in Figure 15) of the photoline lines. The assignment reported in Figure 15 allows one to show that the overall intensity distribution of the normal Auger spectrum is conserved and to measure a shift of 9.3 and 4.6 eV for the 106- and 109-eV resonant spectra, respectively. These values are decreasing as the binding energies of the σ^* electrons. Participator transitions appear with the enhancement of some outer valence bands in the high-electron-energy part of the spectrum. The enhancement of the inner valence part is much more difficult to evaluate because of the overlap with spectator transitions. Nevertheless, if one assumes that there is a significant contribution of participator transitions because of the Si3s contribution to the inner valence region, the spectator resonant Auger still dominates the deexcitation spectrum compared to participator transitions.

Up to now, we have implicitly considered spectator resonant Auger decay as a two-step process, i.e., the core excitation step and the relaxation of the core hole. There is another point of view which describes the resonant Auger as a one-step, two-electron excitation, very much like a shakeup process. This is known as the Auger analogue of resonant Raman scattering or the Auger resonance Raman effect following the pioneering experiments of Brown et al.[37] and Armen et al.[214] in atoms. The key point which differentiates the stepwise from the one-step picture is that in the former, the linewidth of the Auger lines is limited by the core hole lifetime, whereas in the latter, the linewidth is not and can be narrower. Liu et al.[215] using high-resolution photoelectron measurements in HBr, at the Br $3d \rightarrow 5p\pi$ Rydberg resonance, first reported the existence of Auger resonance Raman effect in a molecule. The resonant Auger lines are observed at energies normally shifted with respect to the normal Auger lines, but they are found to be narrower (0.15 eV compared to 0.5 eV). This Rydberg core-excited state is assumed not to be dissociative as the Br$3d \rightarrow \sigma^*$ resonance (see Section 4) and the resonant Auger decay is assumed not to be governed by the dynamical effects described above (Section 3.2). The Br $3d$ core hole is found to be broadened by the ligand field splitting on the basis of the $3d_{5/2}$ line shape measurements. The resonant Auger line shape limited only by the instrumental resolution, is found to be independent of this ligand field and even of the lifetime broadening (0.2 eV).

3.5. Double Auger Effects, Cascades, and Multiionization

The simultaneous emission of two Auger electrons, also called shakeoff, was recognized first by Carlson and Krause[216,217] and Krause et al.[218] in rare gases. The associated spectrum appears as a continuum in the low-energy side of the Auger spectrum and corresponds to the formation of a triply charged positive ion. There are other processes such as cascade Auger which should be seen as stepwise processes.[218–220] They are observed when an electron from a deep shell is ejected, leading to complex sequences of KLL, KLM, LMN, LNN, etc. pathways which produce multiply charged ions. The emission of two electrons has also been observed in resonant Auger decay of atoms (see Ref. 221 and references therein) which produces doubly charged ions. One distinguishes shakeoff processes which appear as a continuum in the low-energy part of the photoelectron spectrum from two-step autoionization, which is seen as two sets of well-defined peaks, superimposed on the shakeoff continuum.

In molecules, multiple ionization is just as efficient as in atoms.[222] As for atoms, one always observes low-kinetic-energy electrons in an Auger or resonant Auger spectrum. Using the threshold electron spectroscopy or the ZEKE method, one always observes a large signal at the energy of discrete resonances near a core ionization edge.[154,222] These reflect the existence of shakeoff processes. As for atoms, the double Auger ejection spectrum of a core-excited molecule extends in a larger range beyond zero. It is almost impossible to distinguish shakeoff from stepwise processes because of spectral congestion and more importantly because the final states are very often dissociative and Franck–Condon broadening smooths the lines into a broad lump. In contrast, cascade Auger from deep shell ionization can be observed[95] because the energies of electrons are in quite different ranges and do not overlap with direct photoionization processes. Nevertheless, fine details on the final states cannot be obtained because of the importance of dissociation. We illustrate this point using the photoelectron data of Ferrett et al.[95] on the SF_6 molecule, photoexcited in the region of the sulfur $1s$ edge. The deexcitation spectrum of the core-excited SF_6 molecule recorded at the $S1s \rightarrow 6t_{1u}$ discrete resonance is observed through the sulfur (LVV) and sulfur (LLV) bands, the former extending down to very low energy. The Auger intensity ratio LVV/LLV changes from 3 off resonance to 25 at 2486 eV on resonance showing that on resonance, the molecule is primarily formed via spectator transitions into highly excited $2s^{-2}v^*$, $2p^{-1}2s^{-1}v^*$, $2p^{-2}v^*$ states, where v^* represents the excited electron in the outer valence orbital. The dominant channel is found to be the $2p^{-2}v^*$ channel (75% of the resonant cross section) very much like the argon case. According to Ref. 95, the successive filling of the S$1s$, S$2p$, and S$2s$ holes leads to the ejection of two, three, four, and five electrons, i.e., the formation of SF_6^{2+}, SF_6^{3+}, SF_6^{4+}, and SF_6^{5+}, if one ignores the dissociation processes. The broadness of each Auger band prevents obtaining any further information on the dynamics of the ejection of these Auger electrons. Here the lifetime of the $2s$ and $2p$ holes are 1.4 and 5 fs, respectively. It is quite likely that the dissociation is direct and proceeds quite early after the core hole decay and explodes, but no experiment on the fragmentation of S$1s$ core-excited SF_6 has been reported.

The multiionization efficiency following core excitation or ionization can be studied by mass spectrometry operated in the coincidence mode (see Section 5 for more details). Such a method allows one to detect, for a single event, the intensity and kinetic energy release of two (or more) correlated ionic fragments. In the absence of information on the kinetic energy of the electron, this method does not provide spectroscopic information on the dissociative states. More sophisticated experiments offering the advantages of both photoelectron and photoion spectroscopy are needed for this purpose (see Section 5). Conventional TOF mass spectrometry and photoion–photoion coincidence measurements have been used for several molecules to investigate the dissociative ionization channels. Morin et al.[204] and Dujardin et al.[223] used this method for the first time to investigate dissociative double ionization in molecules, especially at energies of discrete resonances for which double ionization is already the signature of higher-order ionization processes. Subsequently, numerous studies on higher-order ionization processes leading to double, triple, quadruple (and higher order) ionization pathways were reported (see Section 5 for a detailed bibliography). We illustrate the interest of such a method using the CO results of Hitchcock et al.[224] for which the single bond breaking simplifies the description of the decay channels. We present in Figure 16 the partial ion yield oscillator strengths for C$1s$ photoionization of CO. Coincidence spectra show that the dominant channel is the symmetric dissociative double ionization producing $C^+ + O^+$, but other symmetric ($C^{2+} + O^{2+}$) or asymmetric ($C^{2+} + O^+$, $C^+ + O^{2+}$) channels. We have reported in Table I the relative importance of various dissociative

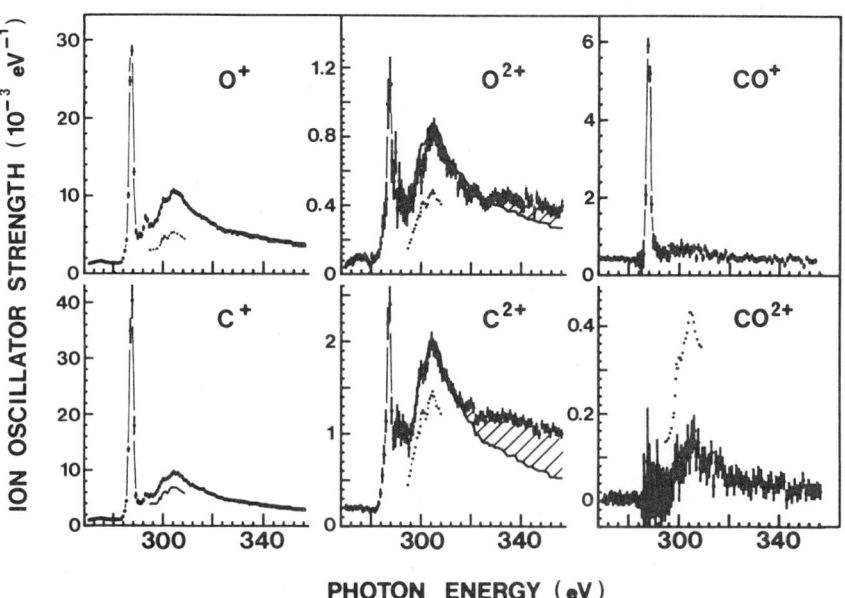

FIGURE 16. Partial ion yield oscillator strength for C$1s$ photoionization of CO. Dots are the (e,e + ion) results of Kay et al.[225] Solid lines are the total ion yield spectrum scaled to match each ion yield spectrum between 296 and 310 eV. The hatching highlights the difference between the shapes of the absorption and partial yield curves which are assigned to multielectron processes. (From Hitchcock et al.[224])

TABLE I
Relative Importance of CO Ionization Reactions following High-Energy Photoabsorption[a]

Reaction products	Ionization degree	v 280	C1sπ* 287	O1sπ* 534	C1s 305/516[b]	O1s 534
CO^+	one	0.18	0.09	0.02	0.00	0.01
$C^+ + O$	one	0.29	0.44	0.53	0.03	0.09
$C + O^+$	one	0.28	0.12	0.22	0.00	0.02
CO^{2+}	two	0.02	0.01	0.01	0.03	0.02
$C^{2+} + O$	two	0.06	0.04	0.02	0.11	0.07
$C + O^{2+}$	two	0.02	0.02	0.00	0.07	0.05
$C^+ + O^+$	two	0.13	0.25	0.19	0.69	0.66
$C^{2+} + O^+$	three	0.02	0.02	0.01	0.08	0.09

[a]From Ref. 224.
[b]The values are the average yields derived from the 305- and 516-eV data.

and nondissociative reactions following valence (v), $C1s \rightarrow \pi^*$, $O1s \rightarrow \pi^*$ photoexcitation, and $C1s$ and $O1s$ photoionization. Core excitation induces dissociative single-ionization channels in agreement with the spectator model. The corresponding branching ratios are quite different from those obtained for valence ionization. Dissociative double ionization is quite high at the resonances showing the efficiency of double resonant Auger. In the $C1s$ or $O1s$ continua, the Auger process is mainly found in the symmetric $C^+ + O^+$ channel. Higher-order processes reach some 8 to 9% of the total. Some remaining single-ionization pathways measured in the core level ionization continua are due to postcollision interaction.[225] Interestingly, dissociative double ionization leads to $C^{2+} + O$ and $C + O^{2+}$, showing that forces other than Coulomb repulsion initiate the dissociation processes. In fact, Hitchcock et al.[224] measured the kinetic energy release in the $C^+ + O^+$ and $C^{2+} + O^+$ processes. For the asymmetric dissociation process, the kinetic energy release amounts to 25 eV which is close to the Coulomb repulsion value. In contrast, for the symmetric dissociation process, there is a distribution of values ranging from 8 to 16 eV for resonances and from 4 to 28 eV for the continua. These values are consistent with the deexcitation spectrum of Eberhardt et al.[197] and the potential curves of CO^{2+} states which show a number of potential wells resulting from the competition between an ordinary chemical bond and electrostatic repulsion.[226–229]

Angle-resolved photoelectron spectroscopy reveals that core-excited relaxation can produce excited configurations, i.e., satellites lines in the Auger spectrum. Hemmers et al.[158] established in the CO molecule, on the basis of the photon energy dependence of the angular distribution parameter, that satellite Auger lines show unexpected large anisotropies because of the occurrence of a conjugate shakeup process (known in atoms but never observed in molecules) in the primary photoionization process leading to a strongly aligned intermediate state of the molecule. In other words, the core-excited $2\sigma^{-1}, 1\pi^{-1} 2\pi$ decays via a participator transition into $5\sigma^{-1} 1\pi^{-1}$ taking along the angular momentum deposited in the core-excited ion via the $2\sigma \rightarrow 2\pi$ core excitation. Along the same lines, measurement of the polarization of soft x-ray emission after photoexcitation of a molecule with linearly polarized synchrotron radiation is an attractive tool to investigate multivacancy effects during the radiative decay of core holes. For example, Mayer et al.[230] investigated the H_2S molecule near the sulfur $1s$ edge and established, above threshold, the existence of the simultaneous excitation of multiple vacancy final states, increasing in intensity with excess energy.

Finally, very high degrees of ionization can be found in molecules with a heavy element or in lighter systems photoexcited in a deep level, on the basis of mass spectrometry and/or photoion–photoion coincidence detection techniques. The early work of Carlson and White[31] on CH_3I excited at the iodine $2p$ edge with x-ray photons, shows that the molecule undergoes a complete atomization with the formation of ion fragments like H^+, C^{2+}, I^{n+} ($n = 1$ to 7) with a maximum loss of ten electrons. Similar results in other iodine and lead compounds were obtained by Carlson and White,[231] and on

1s core-excited N_2, O_2, NO, CO, CO_2, CF_4 molecules.[232] More recent works also illustrate the high efficiency of very high degrees of ionization in molecules especially when the vacancy is deeper inducing increasing cascade Auger events. The reader can refer to the articles of Lapiano-Smith et al.[233] on the carbon and silicon tetrafluoride molecules photoionized near the carbon 1s and fluorine 1s edges, of Nagaoka et al.[234] on $Pb(CH_3)_4$ near the Pb 5p, 4f, and C1s edges, of Ueda et al.[235] on $Ga(CH_3)_4$ near the Ga 3p, 3s edges, of Ueda et al.[236–238] on $Sn(CH_3)_4$ near the Sn 3d, 4p, and C1s edges, of Lindle et al.[239] for HCl and CF_3Cl near the chlorine 1s edge, H_2S near the sulfur 1s edge.

4. ELECTRON–NUCLEAR MOTION INTERACTION

It is well established that any photoionization process in a molecule always implies a consequence on nuclear motion, i.e., vibrational, rotational, and dissociation. In the first place, the relevant time scale (subpicosecond) of these phenomena is such that one intuitively understands that they can interfere with autoionization, Auger processes. This has some relevance only for specific core levels with a core lifetime comparable to a vibrational period, i.e., a few to several tenths of a femtosecond. The semiempirical calculated values of widths and core hole lifetimes[240] give, for example, 80 and 54 meV for the carbon 1s and sulfur 2p levels, respectively. The uncertainty principle gives ΔE (eV) Δt (fs) ≥ 0.659 and one obtains 8.2 and 12 fs for the C1s and S2p levels, respectively. These atomic values are just providing an order of magnitude compared to nuclear motion (see below) and do not represent the true core hole lifetime for a molecule since for a given atom, it is different when the atom is involved in a chemical bond and varies with its chemical environment (Ref. 136 and references therein). In any case, there is no pulsed subpicosecond soft or far-UV sources available as yet, and it is not possible to account for such phenomena on a real time scale but only on an energy scale. The tremendous progress brought by third-generation synchrotron radiation sources with ondulators and with monochromators with very high resolving power have brought a great deal of information regarding molecules. At this time, it is possible to resolve vibrational structure in photoabsorption and photoelectron spectra and search for interferences with stretching and some bending vibrational effects, i.e., on the subpicosecond time scale. In fact, this subject is a matter of controversy because in molecules, there are several reasons for the presence of fine structures and broadening of lines (vibrational state congestion, vibronic coupling, ligand field splitting) which complicate the interpretation and therefore the measurements of core hole widths and lifetimes. In order to illustrate this difficulty, it is worth mentioning the recent high resolution of the iodine 4d photoelectron spectrum of HI[241] and in other iodine molecules.[138] These authors showed the importance of ligand field splitting over vibrational effects and measured the inherent linewidth of the core level (100 meV) and reinterpreted the I4d NVV Auger spectrum of Karlsson et al.[242] More can be found in Ref. 214.

Apart from vibrational effects, rotational effects are out of reach in photoabsorption and photoelectron spectra since this is a slow motion (picosecond time scale) not only because experimentally they would require microelectronvolt resolution but more importantly because any autoionization process always occurs in less than 1 ps. Rotational effects, however, can be studied in fluorescence spectra of residual ions or fragments after core excitation because they emit light in the visible and adequate spectral resolution is easily obtained.[243,244] In contrast, the relevant time of dissociation scale extends from a very short time, i.e., a few femtoseconds for direct dissociation, up to the nanosecond scale. Such a short time scale for direct dissociation has been demonstrated[245,246] (see also Ref. 247 for a review), on the basis of real-time (femtosecond) pump-probe experiments performed on valence photoexcited molecules. Again, because autoionization processes rule out the longer time limit, it is clear that only direct dissociation is of relevance here. In addition, it is worthwhile considering other types of nuclear motion such as isomerization and atomic rearrangement. In other words, this refers to a concerted process in which at least one bond is broken and another is formed. Because we are restricted to very rapid phenomena, the transfer of H^+ from one atom to another, or the concerted dissociation of a polyhydride (e.g., SiH_4) with two Si–H breaks, the H–H bond formation giving H_2^+, has a chance to be of importance here. The other atomic rearrangement processes which are observed after core hole creation, are much slower, i.e., on a micro or submicrosecond time scale.

4.1. Vibrational Effects

4.1.1. Vibrational Effects in Photoabsorption Spectra.

Such effects arise because the potential surface of the core-excited molecule and of the final molecular ion is different from the ground-state surface. On the scheme of Figure 3, the difference in equilibrium bond distances between the ground and core-excited state leads to a vibrational structure as long as the core hole lifetime is long enough. Within the Born–Oppenheimer adiabatic separation of electronic and nuclear motion, one can separate the electronic matrix element from the nuclear terms, i.e., the Franck–Condon factors. In other words, the electronic motion is much faster than nuclear motion and when the electronic transition takes place, the molecule is essentially frozen in its initial geometry. At room temperature, a diatomic molecule is essentially in its ground vibrational state and one defines the amplitude of the internuclear distance as the Franck–Condon zone. The Franck–Condon factors provide information on the molecular geometry changes between the initial and final states. In photoabsorption spectra, vibrational structure has been revealed by electron energy loss spectroscopy (EELS) following the pioneering work of King and Read in the late 1970s.[248,249] (See Refs. 39, 250 for a review; a more exhaustive compilation has been made by Hitchcock and Mancini.[41]) Some years later, more synchrotron radiation facilities developed and soft x-ray monochromators with resolving powers $(E/\Delta E) \geq 5000$ allowed Chen et al.,[44] Heimann et al.,[140] and Domke et al.[46] to produce vibrationally resolved inner photoabsorption spectra in diatomic molecules N_2, NO near the nitrogen 1s edge, or in CO near the carbon edge. We show in Table II, the equilibrium distances R and vibrational energies $h\nu$ for N_2 in the ground and core-excited states, as measured by Chen et al.[44] and Lee et al.[251] These values are compared with those of the core-equivalent NO molecule.[252] The vibrational frequencies of the neutral molecule (ground and excited) have been corrected for the differences in reduced masses. This shows that the $Z + 1$ approximation holds very well, as recognized in the early EELS series of results.[253,254] Note that such verification is possible as long as the spectroscopy of the core-equivalent molecule is known. When the core-equivalent species is a radical and therefore unstable, the photoabsorption spectrum near the core edge may help to identify its vibrational frequencies of the radical and hopefully its geometry.

In polyatomic molecules, the vibrational structure in core-excitation spectra has been resolved for core-Rydberg bands in various systems such as SiH_4, near the Si2p edge,[255,256] H_2CO, D_2CO,[49] C_2H_4,[47] selected alkanes,[52] CH_4[257] near the carbon 1s edge, and SF_6,[43] H_2S, D_2S,[51] SO_2, CS_2[42] near the sulfur 2p edge. Similar high-resolution measurements revealed vibrational structures in the photoabsorption spectra of hydrocarbons adsorbed on surfaces (see Ref. 258 and references therein). Notice that for core-valence bands, the antibonding nature of the σ^* orbital makes the core excited repulsive and washes out any vibrational structure, as observed in many hydrides (HBr, HCl, H_2S, HCl, SiH_4) or other systems such as SF_6. This extreme situation is analyzed in detail in Section 4.2. Returning to the core-Rydberg bands, the situation is more complicated because of the vibrational progression including both stretching and bending vibrations and combination bands. Therefore, the vibrational spacing for bending (unlike stretching) may be inferior to the natural core hole lifetime and is likely not to be resolved. In H_2S and D_2S,[51] the molecular field splitting adds itself to vibrational

TABLE II
Equilibrium Distances, R, and Vibrational Energies, $h\nu$, for N_2 in the Ground and Core-Excited States (Ref. 44). Comparison with NO (Ground and Excited Neutral States)[a]

State	N_2		NO	
	R (Å)	$h\nu$ (meV)	R (Å)	$h\nu$ (meV)
Ground	1.09768[a]	292.42[a]		
$(N1s)^{-1}\pi^*$	1.164	235	1.1508	244
$(N1s)^{-1}3s\sigma$	1.077	293	1.0634	304
$(N1s)^{-1}3p\pi$	1.073	300	1.062	307

[a]From Ref. 44.

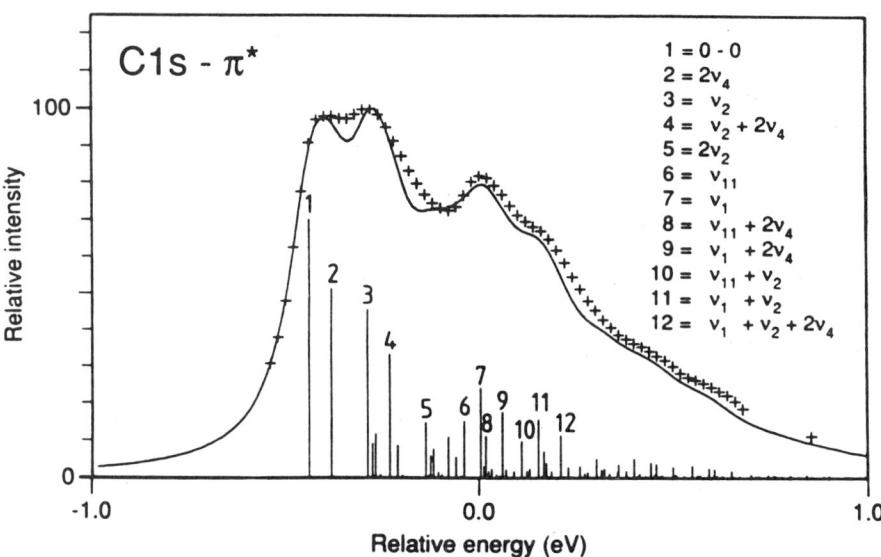

FIGURE 17. High resolution of the $1s \to \pi^*$ photoabsorption band in C_2H_4. Experiment (crosses) after Ma et al.,[47] theory (solid line) after Gadea et al.[259]

progressions of the S2p Rydberg states. Further complication arises from vibronic coupling due to the dynamic localization of the core hole. This is the case for the $C1s \to \sigma^*$ band in ethylene,[47] as reported in Figure 17. The *ab initio* calculation of Gadea et al.[259] performed with Green's-function formalism reported in Figure 17 shows that the fine structure cannot be interpreted as totally symmetric progressions. The localization of the core hole in a system with two equivalent core atoms, removes the degeneracy of the core-excited state and non-totally symmetric vibrational modes associated with each of the two electronic components via vibronic coupling are excited producing a congestion of closely spaced vibrational states which cannot be resolved. This is a typical case of nonadiabatic effects. Other *ab initio* but simpler calculations at the SCF level[260] were performed to reproduce the fine structures observed in the photoabsorption spectrum of more complicated conjugated hydrocarbons physisorbed on a surface. Using the $Z + 1$ approximation and vertical transitions, and the vibrational analysis in terms of normal modes, these authors were able to reproduce the positions of the maxima and established that only a few vibrations were active. They also obtained the first-order geometrical changes following the electronic excitation.

4.1.2. Vibrational Effects in Photoelectron Spectra. Vibrational structure can be observed only if the experimental resolution (photon + electron) is less than 0.1 eV as obtained from high-resolution monochromator and electron analyzers.[261–263] An exception is the high-resolution photoelectron spectrum of methane obtained with monochromatized AlK_α conventional laboratory source by Asplund et al.[264]; such a 1s spectrum exhibits a C–H stretching progression with a 390-meV spacing, in this favorable case. For other molecules, such as SiH_4, SiD_4, Si_2H_6, and various substituted silicon compounds like SiF_4, $Si(CH_3)_x(OCH_3)_{4-x}$, and numerous di- and trisilane molecules[261,262,265,266] near the Si2p edge, ICl, IBr, HI, I_2, CH_3I, CH_2I_2, CF_3I[241,267,137] near the I4d edge, or HBr near the Br3d edge,[215] the experimental energy resolution requirements are much severe. Taking the silane molecule as an example (see Figure 18), the photoelectron band shows a progression of three peaks for the two spin-orbit components $Si2p_{3/2}$ and $Si2p_{1/2}$ corresponding to the Si–H totally symmetric stretching mode v_1 of a_1 symmetry, with a 295-meV energy spacing. Such an interpretation is fully confirmed by the same measurements performed on deuterated silane. This value equals 2379 cm^{-1} and compares favorably with the stretching frequency mode (2327 cm^{-1}) of the core-equivalent species PH_4^+. There is also some evidence of another progression with a 110-meV energy spacing which has been attributed to the asymmetric v_2 bending frequency. This is unexpected in the Born–Oppenheimer approximation,

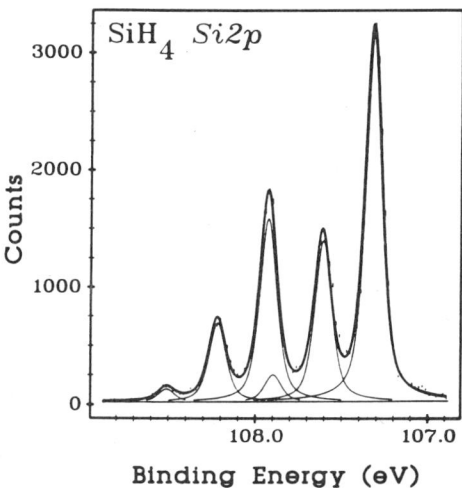

FIGURE 18. High-resolution Si2p photoelectron spectrum of SiH$_4$, after Bozek et al.[261]

but no theoretical calculations have as yet been performed to analyze this additional progression. Departure from this approximation has been suspected by Sutherland et al.[266] especially in disilane, where the vibrational structure cannot be analyzed with simple progressions of Si–Si or Si–H stretching progressions, rather with non-totally symmetric vibrational modes. This very first observation in photoelectron spectra of polyatomic molecules indicates that vibronic coupling[268] may exist here similarly to the ethylene photoabsorption spectrum (see Figure 17).

Other vibrational effects in photoelectron spectra have been investigated in the search for interference between the initial excitation and the decay process, due to the extremely short lifetime of the core hole excited state, following the prediction of Kaspar et al.[269] and Correia et al.[162] The first experimental attempt in the search for them was made by Correia et al.[162] and later by Carroll et al.[270] on the deexcitation of the core-excited N1s → $1\pi_g$ NO molecule. The peculiarity of the situation is that the core-excited molecule exists only 7 fs, which is roughly three vibrational periods. The photoabsorption and Auger decay should be considered coherently rather than as sequential events, and are expected to interfere. The amplitudes of the vibrational matrix elements for the excitation and decay have to be added rather than the amplitude squares. In this regard, recall the description of Cederbaum et al. (see Section 3) which showed the inadequacy of vertical transitions in normal Auger transitions. The short lifetime of the core-excited state is responsible for the coherent excitation of several vibrational levels. Such effects were evidenced first with high-resolution soft x-ray fluorescence[271] (see also Section 4.1.3) and later with high-resolution photoelectron spectroscopy allowing resolution of vibrational structure in deexcitation spectra.[272,273] We illustrate these interference effects using the recent results of Neeb et al.[274] on the 1s core excitation of the oxygen molecule after the predictions of Thomas and Carroll.[275] We show in Figure 19 (top) the shape of the first core to bound $(1s \rightarrow 1\pi_g^3)$ feature of the photoabsorption spectrum for which the core hole broadening amounts to 0.18 eV, i.e., slightly larger than the vibrational spacing (0.16 eV). The high resolution of the deexcitation spectrum corresponding to the formation of O_2^+ ($X^2\Pi_g$) presented in Figure 19 (bottom) has been recorded at four energies chosen along the core-excited resonance. Using the $Z + 1$ approximation for describing the core-excited state, and a two-step picture of the photoexcitation and resonant Auger decay, Neeb et al.[274] calculated the population probability of the final state using the formula

$$I_f(E_{exc}) \propto M(E_{exc}) \cdot \left| \sum_n \frac{\langle f|n\rangle\langle n|0\rangle}{E_{exc} - (E_n - E_0) + \frac{i\Gamma}{2}} \right|^2$$

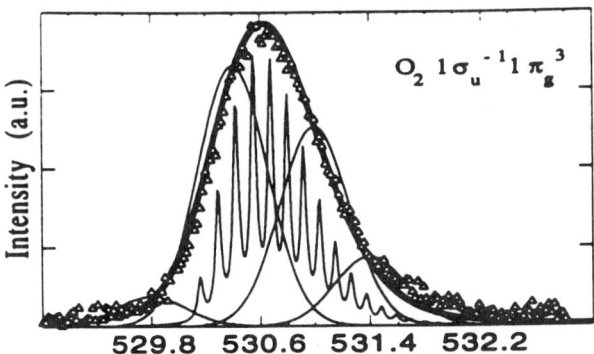

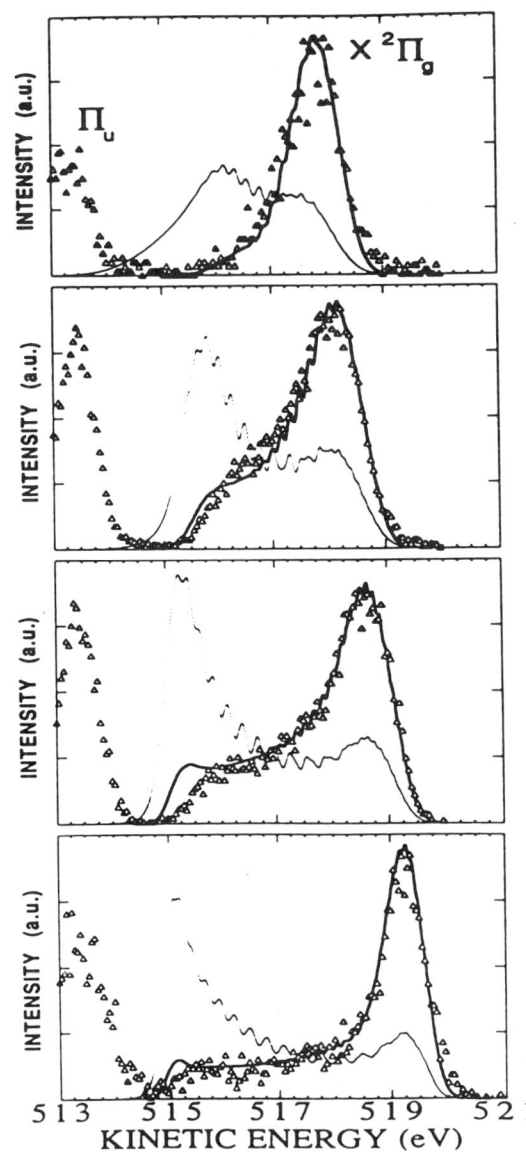

FIGURE 19. (Top) Experimental (triangles) and calculated (solid line) photoabsorption spectrum of the $(1s \rightarrow 1\pi_g^3)$ band in O_2. The calculations were performed using the $Z + 1$ approximation for the excited state, a 0.18-eV Lorentzian and a 0.5-eV Gaussian broadening: the width of the narrow peaks was arbitrarily set to 0.05 eV. The four Gaussian curves (FWHM0 = 0.6 eV, chosen to match the experimental conditions) indicate the monochromator function used to record the decay spectra (see below).
(Bottom) Experimental (triangles) and calculated (solid line) deexcitation spectrum in the region of the O_2^+ ($X^2\Pi_g$). The photon energies from top to bottom are 529.8, 530.4, 531.0, and 531.4 eV. The dotted lines show the results when lifetime vibrational interference is neglected. The curves were normalized arbitrarily. (From Neeb et al.[274])

where E_{exc} represents the excitation energy, $|0\rangle$, $|n\rangle$, and $|f\rangle$ refer to the vibrational eigenfunctions of the ground, intermediate, and final states, respectively. E_n and E_f are the vibrational energies of the intermediate and final states and Γ is the lifetime width of the intermediate state, the width of the final state being neglected. $M(E_{exc})$ in the monochromator function is assumed to be a Gaussian distribution. The results of the calculations are shown in Figure 19 in excellent agreement with experiment. The calculated curve obtained without interferences shows that this effect is especially strong in oxygen. Finally, Neeb et al.[274] performed the same study on N_2 and CO and found similar but smaller effects because of the longer lifetime of the intermediate core-excited state.

4.1.3. Vibrational and Rotational Effects in Dispersed Fluorescence Spectra. Dispersed fluorescence in the soft x-ray range[276,121] has been a tool with which to learn about several aspects of core vacancy relaxation phenomena, as a complement to the information provided by photoelectron spectra. One striking example has been given by Flores-Riveros et al.[271] who investigated in the CO molecule the carbon and oxygen radiative decay by dispersed soft x-ray fluorescence, leaving the molecule in monocationic states with one hole configuration in the outer valence shells. The high resolution of the spectra (0.15 eV for the carbon $1s$ emission and 0.21 eV for the oxygen $1s$ emission) allows resolution of the vibrational structures of the emission bands and established lifetime vibrational effects similar to effects observed in photoelectron spectra (see Section 4.1.2). Using large-scale multiconfiguration SCF calculations for core-excited and final ionic state potential curves, Flores-Riveros et al.[271] show that the fine and broad structures of the spectra result from constructive and destructive interference primarily because of the lifetime broadening of the core hole (0.1 eV for carbon and 0.15 eV for oxygen) but also because of the bonding, antibonding natures of the outer valence and core orbitals, respectively.

Dispersed fluorescence spectroscopy in the visible is a quite attractive method for understanding the core hole decay in a molecule because electronically excited molecular ions are sometimes produced during the course of the core hole decay and fluoresce and also because extremely high spectral resolution, compared to photoelectron spectroscopy, can be achieved. Poliakoff et al.[155,277] were able to measure the emission of N_2^+ ($B^2\Sigma_u^+ \to X^2\Sigma_g^+$), $v' = 0 \to v'' = 0,1$, in the region of the $1s$ ionization region of N_2. The excitation spectrum of the $v' = 0 \to v'' = 0$ transition ($\lambda_{fl} = 3915$ Å), recorded under medium spectral resolution, follows the general shape of the photoabsorption spectrum (see Figure 4). The vibrational branching ratios recorded at the $N1s \to \pi^*$ are quite different from those measured off resonance, in the valence ionization continuum, showing that the decay of the core-excited state couples strongly to molecular vibration. The experimental Franck–Condon factors are found to be different ($\sigma_1/\sigma_0 = 0.179$) from those calculated ($\sigma_1/\sigma_0 = 0.68$) from the shape of the three relevant potential curves, $N1s \to \pi^*$, and $N_2^+ B^2\Sigma_u^+, X^2\Sigma_g^+$ including interference terms due to the finite lifetime of the core-excited state. This discrepancy is not resolved at present and more experimental and theoretical studies are needed to investigate various hypotheses, such as cascade effects or the role of rotation.

4.2. Direct Dissociation and Competition with Electronic Relaxation

Photoionization of a molecule is intimately associated with its dissociation, because among the manifold of potential surfaces of excited or ionized molecules, there are many repulsive surfaces responsible for direct dissociation or predissociation of the system. This is especially true for core ionization. In the first place, the core-excited molecular state may be repulsive because the core electron is promoted into an antibonding orbital. Second, following the Auger process, the ionized molecule may also be found in a repulsive state, in the dissociation continuum of a bound state or in a predissociated state. This is true either for a molecular monocation in an excited state (satellite in a photoelectron spectrum) or for a doubly (or multiply) charged molecular ion.

4.2.1. Direct Dissociation of Core-Excited Molecules. Photoabsorption spectra near specific core edges of the elements bring important spectroscopic information regarding the occurrence of repulsive states for core-excited molecules. Indeed, when the discrete part of the spectrum shows a broad feature for the first core-to-bound transition and very narrow features in the Rydberg region, one should suspect that the broadening of this state originates from the wide Franck–Condon envelope as expected for a bound-to-repulsive transition. This situation is observed for hydrides like the three molecules HBr,[253] CH$_3$Br,[278] and SiH$_4$.[254] In these three molecules, the first feature is due to core to an antibonding σ^* orbital and carries a very large part of the oscillator strength. At higher energies, the core-to-Rydberg transitions are much narrower and their width is closer to the natural lifetime of the core vacancy. The equivalent core approximation may also be very helpful to predict a repulsive state. This was recognized very early by Shaw et al.[253,254] and Schwarz.[279] For example, the (chlorine 2p to σ^*) core-excited HCl is equivalent to HAr which is a well-known excimer compound with a very shallow potential well in its ground state. In other words, at room temperature, this system is fully unstable and dissociates through a direct process along a single potential surface.

We turn now to the competition of such direct and fast dissociation with electronic relaxation. Historically, the first experimental evidence for a dissociation process faster than autoionization was reported by Čermák[280] in the oxygen molecule photoexcited in the valence region, using Penning ionization. Thirteen years later, Morin and Nenner[281] showed in Br3d core-excited hydrogen bromide that direct dissociation could precede resonant Auger decay. In other words, the electronic decay of the core hole occurs after the atoms have separated. This demonstration has been made on the basis of photoelectron spectroscopy associated with tunable synchrotron radiation in the far ultraviolet. Since then, many other systems have shown similar behavior but with variable amplitudes using primarily photoelectron spectroscopy but in some occasions mass spectrometry: CH$_3$Br[282,283] near the Br3d edge, HI and CH$_3$I[283] near the I4d edge, SiH$_4$[201] near the Si2p edge, HCl[284,285] and CH$_3$Cl[286] near the Cl2p edge, HF[287] near the F1s edge, H$_2$S[288–290] near the S2p edge, Cl$_2$[291] near the Cl2p edge, O$_2$ near the O1s edge.[292] Apart from these gas-phase works, similar effects were reported on condensed layers, namely, H$_2$O,[293,294] NH$_3$,[8] and O$_2$.[295] It is worth noting that in the valence region, certain superexcited states dissociate faster than they autoionize (e.g., see Ref. 296). Examining the cases observed for core-excited states, it is possible to analyze the role of the core hole lifetime and the relative weight of the departing fragments on the competition between direct and electronic relaxation.

As the starting point we take the HBr case which shows the most dominant effect. Figure 20 shows the EELS spectrum of HBr in the region of the $3d_{5/2,3/2}$ core ionization thresholds. The wide and intense resonance around 70.6 eV has been interpreted as the $3d_{5/2} \rightarrow \sigma^*$ transitions. Photoelectron spectra recorded on and off resonance show striking differences which can be understood through a two-step process, as represented in Figure 20: (1) the core-excited HBr molecule dissociates into a ground-state atom and a core-excited bromine atom and (2) the latter decays through resonant Auger and autoionization into various Br$^+$ + e^- and Br^{2+} + 2e^- continua. The prominent features seen in Figure 20 are atomic bromine lines (participator resonant Auger decay) which are superimposed on the normal valence photoelectron spectrum of HBr. The fingerprint of the two-step process is obtained by scanning the photon energy along the resonance profile. The atomic lines are found to appear at a constant electron energy in the photoelectron spectrum (not shown here; see Ref. 281), very much like a normal Auger process. Using a simple kinematic model, one assumes that the kinetic energy release in the dissociation, which equals the resonance energy minus the energy of the dissociation limit, i.e., $E_R = 2.47$ eV, is entirely transferred to the H atom because of the high H/Br mass ratio of 80. After the 7 fs of the core hole lifetime τ, the interatomic distance is calculated to be about 3 Å, i.e., double the original distance, and the two-step process proposed is quite realistic. However, this picture with a pure atomic relaxation is indeed too simple. If we consider now the photon energy dependence of the molecular and atomic lines, the atomic lines resonate strongly at the position of the σ^* resonance.[281] But there is some residual signal of the molecular lines at the resonance showing clearly that molecular autoionization is also present. Coincidence mass spectrometry results obtained by LeBrun et al.[297] in HBr$^+$ confirmed that electronic relaxation occurs partly at short internuclear distances giving HBr$^+$, HBr^{2+}, and H$^+$ + Br$^+$ as residual ions.

We consider now the effect of changing the mass of the departing fragment. This is realized by studying the CH$_3$Br molecule near the 3d edge. The same photoelectron experiment has been performed.[282,283,298] When the molecule is photoexcited at the $3d_{5/2} \rightarrow \sigma^*$ resonance, one obtains atomic lines like for HBr but here, they are broader than in the previous case. The core-excited molecule dissociates into a ground electronic state CH$_3$ plus a core-excited bromine atom which relaxes at large interatomic distances. Working out the simple kinematic model mentioned above but also considering the reduced mass of the departing fragments, $m_1 m_2/(m_1 + m_2)$, one obtains that after 7 fs the C–Br

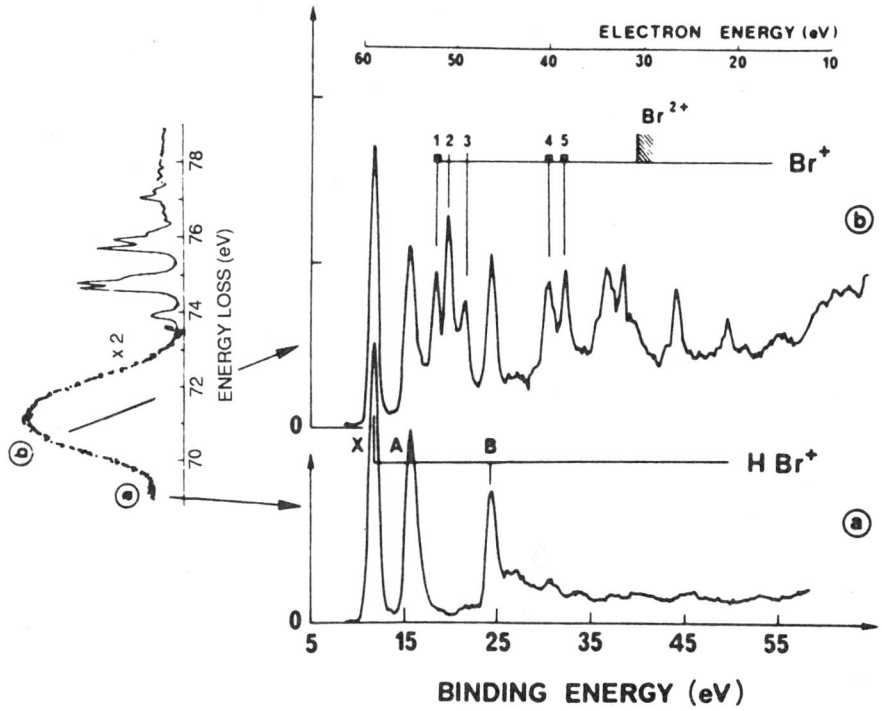
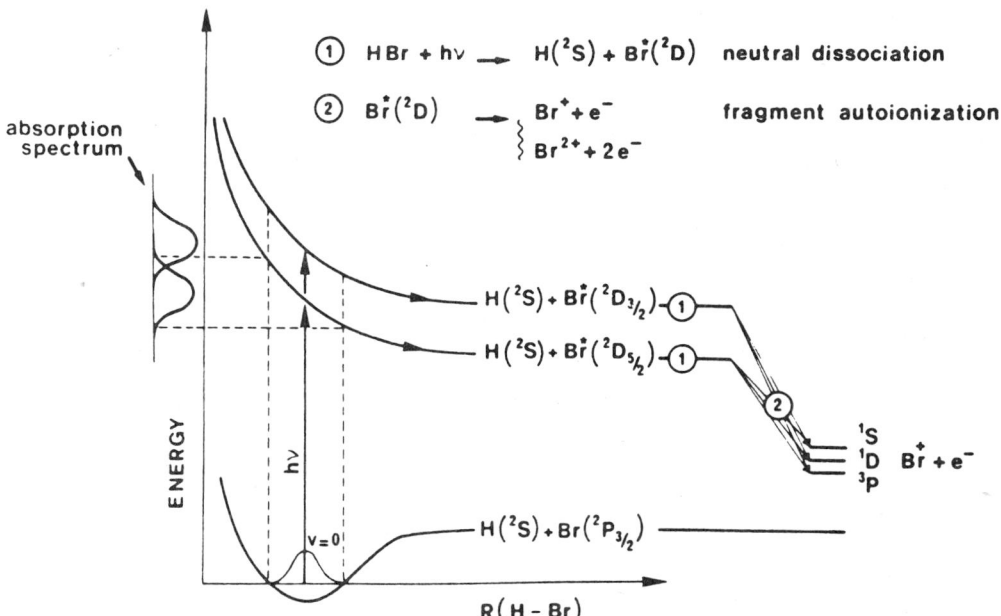

FIGURE 20. (A) Side: Electron energy loss spectrum of HBr near the 3d edge, after Shaw et al.[253]; top: photoelectron spectrum of HBr recorded off (a) and on (b) resonance at 68.2 and 70.6 eV photon energy, respectively. (B) Schematics of relevant potential curves of HBr describing the two-step relaxation process. (Both from Morin and Nenner.[281])

internuclear distance has increased only from 1.939 Å to 2.39 Å. In other words, the methyl group CH_3 departs more slowly than the H atom. Indeed, the reduced mass of the departing fragments amounts to 12.6 compared to 1 in HBr. Consequently, the resonant Auger decay occurs as soon as the C–Br stretches.

Let us consider the effect of the hole lifetime, by comparing the series of hydrides HCl,[284,287] HBr,[281] and HI[282] respectively excited from the Cl2p, Br3d, and I4d edge into the σ* antibonding orbital. The Cl2p, Br3d, and I4d lifetimes are 7.9, 7, and 3 fs, respectively. The core-excited HCl behavior decays through two competing pathways: a dominant direct dissociation and the normal molecular decay process. The competition seems to be more pronounced compared to HBr. The similarity of core hole lifetimes shows that other effects on the direct dissociation rate play a role here, as for instance the slope of the repulsive potential curve in the Franck–Condon zone. The decay of the core-excited molecule does not show a dominant stepwise process like HBr despite a favorable reduced mass of the departing fragments. In HI, the enhanced competition is clearly the effect of the core hole lifetime which amounts to a value about half those of the Br3d or Cl2p holes. The core-excited molecule undergoes direct dissociation but the core hole decay is so fast that the H–I distance stretches only a little. A quantitative modeling of the resonant Auger decay as a function of internuclear distance with time-dependent methods[184,185] would be of great interest, having in mind that accurate potential curves of HBr^{2+} and HCl^{2+} are available.[299,300]

4.2.2. Direct Dissociation of the Residual Molecular Ion and Competition with Further Autoionization.

Competition between dissociation and ionization is likely to persist after the first step of ionization, that is, for cascade events and only in cases where the overall core hole lifetime is long enough. Such cases are found for light elements or for some shallow core levels of heavier elements. Hitchcock et al.[224] suspected that in the relaxation of core-excited CO near the carbon 1s edge, double ionization was sequential rather than simultaneous, on the basis of the observation of two well-defined bands in the photoelectron spectrum of Eberhardt et al.[197] In other words, the spectator transition would give an excited CO^+ (first step) which then autoionizes into the $C^+ + O^+$ continuum (second step). In a different experiment, performed in the valence region, Lablanquie et al.[227] interpreted the low-lying threshold for the $C^+ + O^+$ pathway using the same two-step picture. Becker et al.[301] investigated in more detail by photoelectron spectroscopy at high resolution, the CO valence region and established that in addition to direct double ionization, there was evidence of atomic lines due to the same stepwise process but involving a fast dissociation in the first step and autoionization of a fragment in the second step:

$$CO + h\nu \rightarrow C^+ + O^* + e^-$$
$$\rightarrow C^+ + O^+ + 2e^-$$

Even if such a sophisticated experiment were performed at photon energies in the valence region, it brings the idea that double ionization and dissociation mechanisms should be considered on the same time scale.

Finally, it should be noted that dissociation and triple (or multiple) ionization represent limiting cases of the competition, that is, both are likely to be on a very short time scale. This is probably the case for systems built with heavy elements and for core levels with a short lifetime which undergo multiionization and multifragmentation on the femtosecond time scale (see Section 3.4). The early work of Carlson and White[31] on CH_3I excited at the iodine 2p edge with x-ray photons shows a complete atomization of the system with the formation of ion fragments like H^+, C^{2+}, I^{n+} ($n = 1$ to 7), and large recoil energies (40, 34, and 9 eV for C^{2+}, H^+, and I^{n+}, respectively). The recoil energies of the fragments were found to be significantly smaller than those expected from a pure Coulomb explosion model with a molecule frozen during the multiionization process. Carlson[302] recognized that in fact the molecule begins to dissociate during the vacancy cascade and the fragments do not receive the full recoil energy when the ionization is complete. Since then, several studies[231–233] have confirmed this point.

5. FRAGMENTATION OF CORE-EXCITED MOLECULES

Core excitation or ionization processes leave the molecule in highly excited states. The final result is the creation of very unstable ions (singly or multiply charged) which often undergo dissociation, as we have seen in Figure 1. The basic and simple question on the general rules describing the dissociation processes is difficult to answer because the data base is limited due to experimental difficulties and also because each system seems to behave specifically. Among the numerous dissociation pathways schematically depicted in Figure 1 for a triatomic molecule, it is of importance that the knowledge of the mass and charge of a fragment ion is not sufficient to identify one pathway from the other. The challenge is to select one dissociation channel by the simultaneous detection of *two or more* fragments that are produced from a single ionization event. This is the purpose of coincidence detection techniques that have been developed in the past 10 years for dissociative multiionization studies.

Meanwhile, the efficiency of fragmentation of core-excited molecules and the "localized" nature of the overall processes around an atomic site have triggered specific applications. The production of ionic fragments after soft-ray irradiation of a molecular gas appears as the very first step of a complex chain of reactions induced in a high-density medium (plasma, surface, or macromolecule). Soft x-ray photon interaction with a gas, like chemical vapor deposition, is a method to produce thin films or etching of specific semiconductors in microelectronics.[11,303–305] Cerrina et al.[304] used such techniques to deposit $Si_xN_yH_z$ from SiH_4 and N_2 mixture irradiation and Wen and Rosenberg[305] obtained nitration of silicon at room temperature. In radiobiology, the effect of soft x-ray radiation on biological compounds is also very attractive as an attempt to induce reversible or irreversible damage on cells for instance. Recently, Hieda et al.[306] established a direct correlation between the double-strand break of DNA and the photon energy chosen on and off the phosphorus P$1s \rightarrow \sigma^*$ strong resonance at 2153 eV. Kobayashi et al.[16] showed that on soft x-ray irradiation at the same phosphorus resonance, the survival of yeast cells was strongly affected, demonstrating the direct role of induced Auger cascade. More recently, Svensson et al.[307] measured the radiation damage (fragmentation) induced in free nucleotides and sulfur-containing amino acids after monochromatic x-ray absorption near the phosphorus and sulfur $1s$ edges. The molecular fragmentation pattern and the ionization degree are found to depend on the excitation site. More studies on these complex systems are under way but meanwhile there is a strong need to understand the rules of fragmentation in medium and large molecules to hope to bridge the gap between basic molecular physics and radiobiology.

Turning to the basic concepts of fragmentation processes, the relevant points of interest can be summarized as follows:

- What is the ionization degree of the molecule after Auger (resonant, normal) relaxation? The fragmentation pattern is highly dependent on the state of charge of the residual ion. The single, double, and multiple ionization branching ratios are expected to be very different (see Section 3) according to the nature of the photoexcitation (discrete or continuum). It is also very dependent on the nature of excited atomic site and on the depth of the core hole, responsible for weak or strong cascade effects.
- What is the dynamics of the dissociation? The multiply charged ions are usually unstable and undergo fragmentation where several chemical bonds are broken; are these fragmentation processes instantaneous or sequential, the time reference scale being rotation or vibration? In case of (quasi-) instantaneous dissociation, i.e., shorter than one vibrational period, can we describe the process as an explosion driven by Coulomb repulsion or do we have to invoke more sophisticated models? What is the role of the geometry of the molecule and the symmetry of intermediate (resonant) state?
- Can we take advantage of the initial hole localization near a specific atomic site in a complex molecule to induce a selective bond fission; in other words, are memory effects induced on a short time scale (femtoseconds) conserved at longer times (microseconds) and what is the spatial extension of the fragmentation?

So far, a rather limited number of reviews have been devoted to this subject. Fragmentation of doubly charged molecular ions produced near threshold has been extensively covered by Eland (see

Ref. 23 and references therein). Some aspects of dissociation of core-excited molecules have been reviewed by Nenner and Beswick,[124] Nenner et al.[222,308,309] Eberhard,[125] Hitchcock,[40] Hanson,[126] Crasemann,[127] and Morin et al.[290,310]

5.1. Experimental Techniques

The first attempt[311] to detect fragment ions in coincidence was done using the so-called photoion–photoion coincidence mass spectrometry (PIPICO) detection method, also known as charge separation mass spectrometry. In this technique, the ions are extracted from the ionization region toward the detector with an electric field, thus flying in a field-free region. The time of flight of the ion (i.e., the time it takes for the ion to reach the detector) is related both to its mass-to-charge (m/q) ratio and to the projection of its momentum along the detection axis, as given by, in the Wiley and McLaren[312] focusing conditions:

$$T = T^0 \left(\frac{m}{q}\right) + \frac{\alpha}{q} \cdot P \cdot \cos\theta \tag{1}$$

where θ is the ejection angle of the particle with respect to the detection axis and P its momentum. The advantages of the time-of-flight mass spectrometry are the very high detection efficiency (because of the extraction field), reasonably high mass resolution, and also some information on the momenta of the ejected particles. In the PIPICO technique, two fragment ions issued from the same ionization event are detected by measuring directly the difference $T_2 - T_1$ in their arrival times. This is obtained by keeping the ionization rate low enough to avoid false coincidences. Since the two ions do not have exactly the same time of flight, this can be done easily using just one detector. This technique is ideally suited for diatomics, for instance (see Ref. 224 and Section 3.4). The advantages are the simplicity and the high efficiency of the setup. Indeed, numerous studies on polyatomics have been reported using this method.[204,223,234–238,313–316] However, in some cases of very complex molecules, for which the number of dissociation channels are very high, the time difference spectrum is not enough to assign unambiguously the various ion pairs resulting from multiple bond breaking. This is the reason for using multicoincidence techniques.

Multicoincidence detection methods, known as the photoelectron–photoion–photoion coincidence method (PEPIPICO), were introduced by Frasinski et al.[317] and Eland et al.[318] in investigations of dissociative double ionization in the "threshold" region. Since then, Simon et al.[319] have used the same method to investigate these phenomena induced by core hole relaxation. In this operation mode, one measures separately the individual time of flight of each fragment. A schematic of the setup used by Simon et al.[319,320] is shown in Figure 21. The detection of the electron is used as a start, whereas the multihit TDC gives as many stops as fragment ions have been detected. Each ionization event is stored as a set of several times (i.e., $[T_1]$ or $[T_1,T_2]$ or $[T_1,T_2,T_3]$, etc.). This can be generalized to any higher coincidence level. Further analysis is done by examining the occurrence of the various events, as detailed later. This technique is very powerful because it allows determination of the branching ratio between all of the dissociation channels and kinetic energy release and also study of the dynamics, i.e., the sequentiality of bond breaking.

In order to obtain branching ratios between different ion pairs but also between different ionization channels (single, double, or triple ionization), it is required to use the PEPIPICO technique as soon as the molecule has more than two atoms. This is a real experimental difficulty of molecules as compared to atoms, for which similar information is readily obtained by conventional mass spectrometry since charge separation processes do not exist.

We show in Figure 22 examples of correlation diagrams measured in a triatomic molecule by the PEPIPICO method. The results are presented in a two-dimensional plot (the time of flight of the first ion T_1 versus the time of flight of the second ion T_2; the probability of the event is shown as various colors) in three different situations:

In the case of a two-body fragmentation the two particles are ejected with opposite momenta P_1 and P_2 ($P_1 + P_2 = 0$); one deduces easily from equation (1) that their times of flight also follow the relation $T_2 = -T_1$ neglecting the constant in equation (1). The case of a three-body fragmentation is

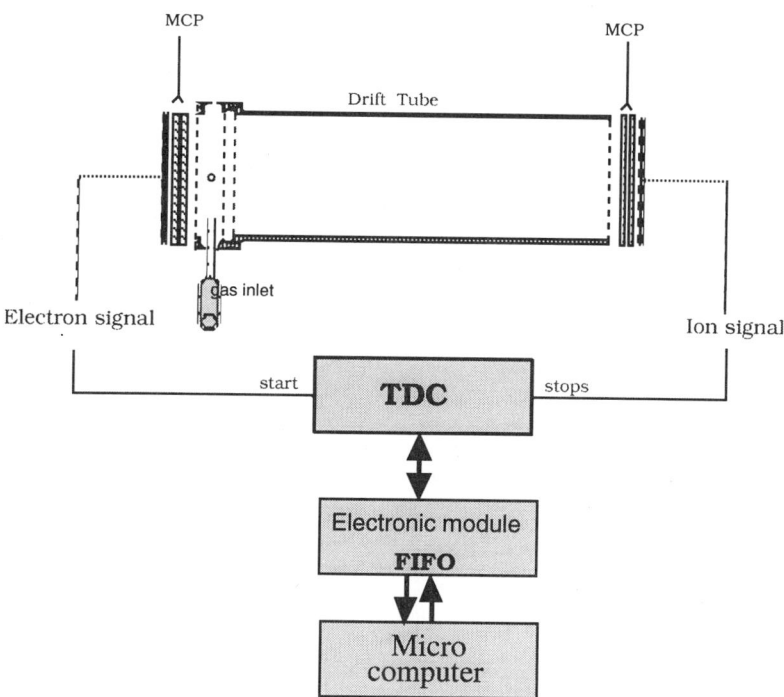

FIGURE 21. Schematic of the multicoincidence (electron–ion–ion) or PEPIPICO experimental setup.

more common and also more interesting. The momentum conservation implies now that $P_1 + P_2 + P_3 = 0$, or $P_1 + P_2 = -P_3$. This situation is depicted in Figure 22. The consequence is that the T_2 versus T_1 plot is no longer a "stick" with a -1 slope, but is significantly enlarged along the "$T_1 + T_2$" axis, due to the nonzero momentum of the third particle. This specific shape results from a superposition of ellipses.[321] Such measurements provide some information on the undetected particle, especially when it is a neutral. Generally speaking, a three-body fragmentation process is undetermined in the sense that the three departing fragments may share continuously the total available energy. Nevertheless, it is possible in particular situations, with some physical hypothesis, to evaluate the energy partitioning. For instance, when stepwise processes are invoked, one can apply the momentum conservation at each step of the dissociation and thus deduce the final momentum partitioning. Eland[23] showed first, for a stepwise process, that the final T_2 versus T_1 diagram shape should be a parallelogram of precise slope (see below). Later, Lavollée and Bergeron[322] gave an analytical demonstration of such shape and Simon et al.[323] proposed a generalization to the multifragmentation with more than three fragments.

In such PEPIPICO experiments, the internal energy of the dissociating ion is unknown. This problem can be overcome by analyzing the electron energy or by detecting zero-energy electron. In the latter technique[324] also called ZEKE (see Section 2.1), one selects the internal energy of the core-ionized molecule at the threshold. The power of this method is that in situations where there are two identical atoms in a molecule but with a different chemical environment, it is possible to select the two different sites because of the induced chemical shift. For instance, in the CH_3CF_3[325] molecule, significant changes in the fragmentation have been observed depending on the two carbon sites (C–F or C–H). The energy analysis of the Auger electron (of high kinetic energy) allows selection of the internal energy of the residual ion. The transmission of the electron analyzer is critical to obtain reasonable coincidence counting rates. In a pioneering experiment, Eberhardt et al.[197] succeeded in such an experiment on CO where coincidences between energy-analyzed Auger electrons and ions were performed. Similar measurements were reported recently by Ueda et al.[205] on BF_3 and Sato et

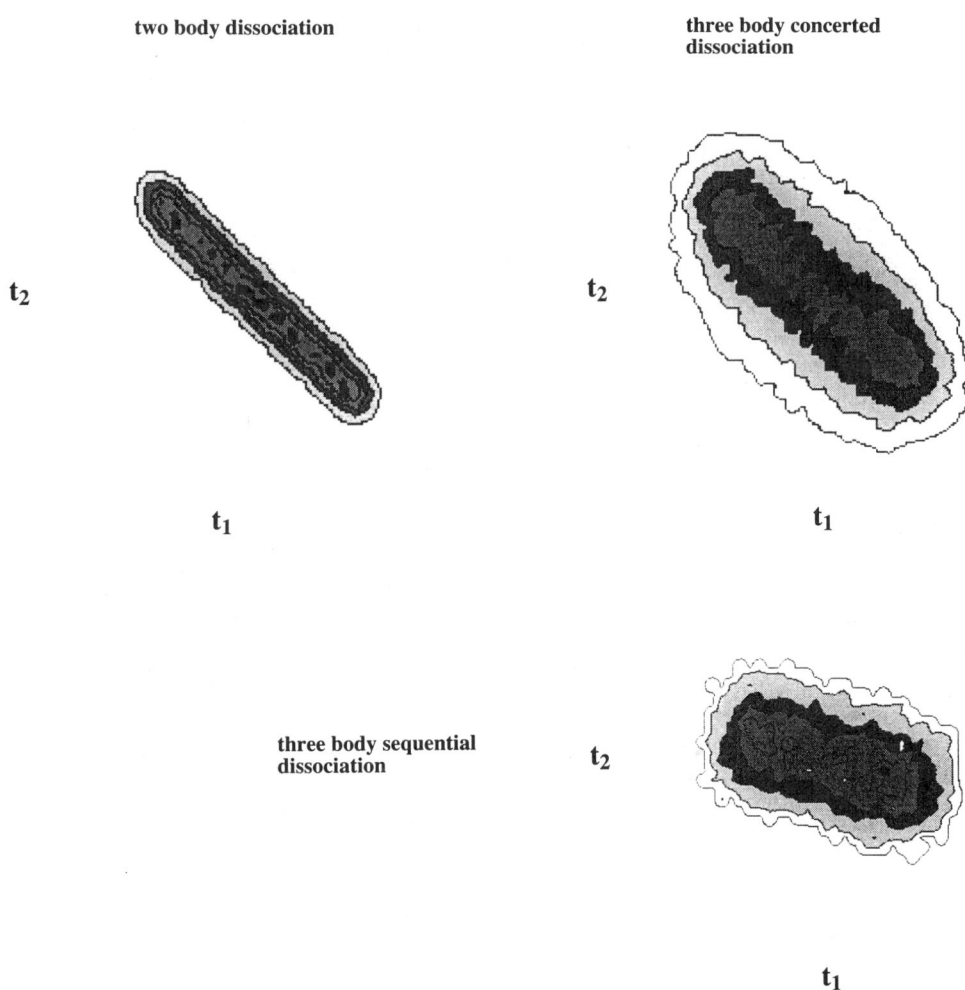

FIGURE 22. Typical coincidence PEPIPICO spectra represented as contour maps in a two-dimensional time-of-flight plot of a triatomic molecule dissociating in (left) two bodies and (right) three bodies with two extreme cases: concerted and sequential processes.

al.[326] on SF_6. Lindle et al.[239] were able to perform Auger electron–ion coincidence measurements on several atoms and molecules after deep core excitation (several keV), using a CMA facing a short time-of-flight spectrometer. Hanson et al.[327] Ma et al.[328] successfully performed energy-analyzed Auger electron ion–ion coincidences in N_2O. These very promising experiments nevertheless suffer from very low counting rates and the statistics achieved so far do not allow a precise contour determination of the coincidence peak shapes.

These energy-selected Auger electron experiments are all performed under constant extraction field to detect correlated electrons and ions, but the electron energy resolution is lowered and the ion mass resolution is not optimized because the electric field has to be kept low enough to minimize the previous effect. This problem can be overcome if one applies a pulsed extraction field, as done by Ferrand-Tanaka et al.[329] in a very recent experiment. Here, one has to face the problem of fortuitous coincidences which can be solved using a set of different pulsed electrodes. As a demonstrative example, we show in Figure 23 such mass spectra obtained by Ferrand-Tanaka et al.[329] on $^{14}N^{15}NO$ after $N_t 1s \to \pi^*$ excitation at 401 eV photon energy and for different Auger electron energies. The use

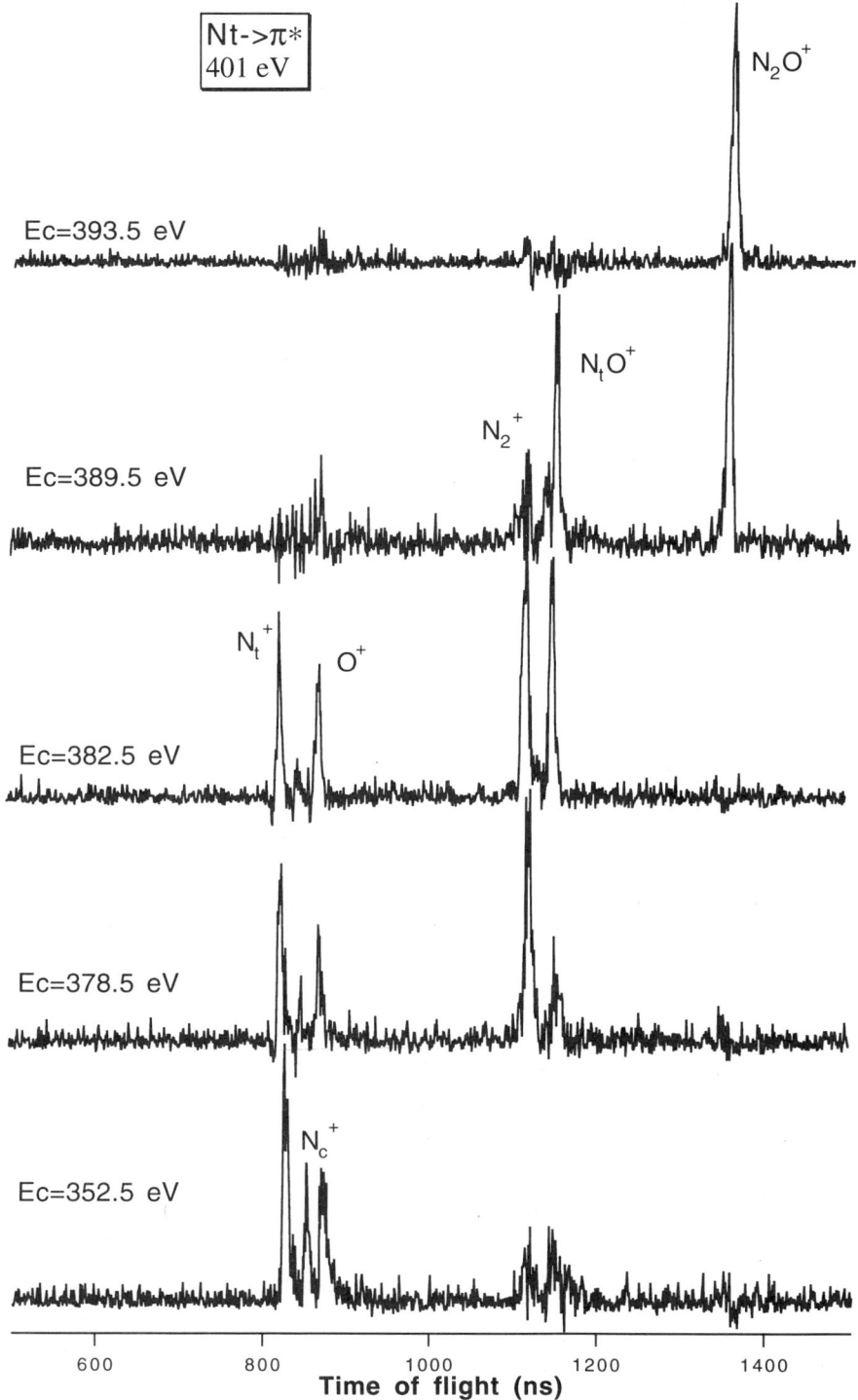

FIGURE 23. Coincidence spectra between energy-analyzed Auger electrons and fragment ions in ^{14}N^{15}NO, photoexcited at the N1s → π* resonance at 401 eV for different Auger electron energies. (From Ferrand-Tanaka et al.[329])

of the isotopically substituted molecule allows one to distinguish the terminal and central nitrogen atoms in the final products. The results show clearly the role of internal energy of the residual ion in the dissociation, produced after resonant excitation. The electrons are analyzed in a field-free regime through a commercial hemispherical analyzer operated with a broad band pass (~1.7 eV). The ions were extracted using a strong (2 kV/cm) pulsed field. The upper spectrum in Figure 23 was obtained by selecting electrons of 393.5 eV kinetic energy, which corresponds to the formation of an ion with small internal energy. Indeed, the observation of only the parent N_2O^+ ion shows that it does not have enough internal energy to dissociate. In contrast, when analyzing electrons of less than 382.5 eV kinetic energy, one observes a high ionic fragmentation. Here the residual ion carries a single and/or double positive charge and significant internal energy to produce the atomization of the molecule, like the dissociation channels:

$$N_2O^+ \rightarrow N_t + N_c^+ + O$$

$$N_2O^{2+} \rightarrow N_t^+ + N_c^+ + O$$

Further studies are under way to extend this experiment via detection of two correlated fragments and to separate out single from double dissociative ionization channels.

Other coincidence techniques such as photoion–photon of fluorescence coincidence detection methods (PIFCO) developed by Eland et al.[330] for the determination of lifetimes and quantum yields for the radiative decay of excited states of molecular ions are now used successfully[331] for investigating the internal energy and the relative abundance of excited photofragments.

5.2. Dissociation of Simple Polyatomic Molecules

5.2.1. Three-Body Dissociation: Concerted Processes. The N_2O molecule is of particular interest for a core excitation study, because of the presence of two chemically unequivalent nitrogen atoms: a terminal one (N_t) and a central one (N_c) bonded to different atoms (N or O). Due to the electronegativity of the oxygen atom, the N_c 1s energy level is shifted down about 4 eV in energy as compared to the N_t one. The photoabsorption spectrum near the N1s edge shows two distinct resonances $N_c \rightarrow \pi^*$ and $N_t \rightarrow \pi^*$ separated also by 4 eV, allowing a selective photon excitation of each nitrogen site. Numerous studies have been devoted to this prototype molecule: site-selective fragmentation has been studied by Murakami et al.[332] and the role of internal energy has been explored by Eberhardt et al.[197] Levasseur and Millié[333] calculated using *ab initio* quantum chemistry methods the low-lying potential surfaces of N_2O^{2+} and the unexpected barriers along N–N and N–O bond breaking pathways. Bozek et al.[79] investigated the anisotropy in the fragmentation due to the symmetry of the core-excited species. Using the PEPIPICO technique, LeBrun et al.[297,334] pointed out the role of the resonant state in the dissociation dynamics. The observation of angular distribution in the fragments showed clearly in that case that the dissociation takes place on a very short time scale (i.e., before significant rotation of the molecule) and rules out any interpretation invoking stepwise processes. Recently, Morin et al.[290,310] proposed an impulsive model to explain the observation that the central atom was always found with small kinetic energy, independently of its state of charge. In Figure 24 we show the PEPIPICO feature corresponding to the "atomization" of the molecule, i.e., when the two chemical bonds are broken. Using isotopically labeled molecules (i.e., $^{14}N^{15}NO$), LeBrun et al.[334] were able to distinguish in the final products the central and terminal nitrogen atoms. The upper, middle, and lower diagrams were obtained after $N_t 1s \rightarrow \pi^*$ (401 eV), $N_c 1s \rightarrow \pi^*$ (405 eV), and $N1s \rightarrow \sigma^*$ (430 eV) excitation, respectively. The right side contours correspond to N_t^+/N_c^+, N_t^+/O^+ and N_c^+, O^+ ion pairs, whereas on the left side six spectra are reported corresponding to triple ionization channels: N_t^{2+}/N_c^+, N_t^{2+}/O^+, N_c^{2+}/O^+, N_c^{2+}/N_t^+, O^{2+}/N_t^+, and O^{2+}/N_c^+. The most striking observations are the following:

- A significant change is observed in the shape of the diagrams depending on the excitation energy: this reveals the role of the initial resonant state in the final dissociation pattern.
- Above the core ionization threshold, the central nitrogen is ejected with very little kinetic energy (≤ 1 eV). This result is true for $N_c, N_c^+,$ and N_c^{2+}.

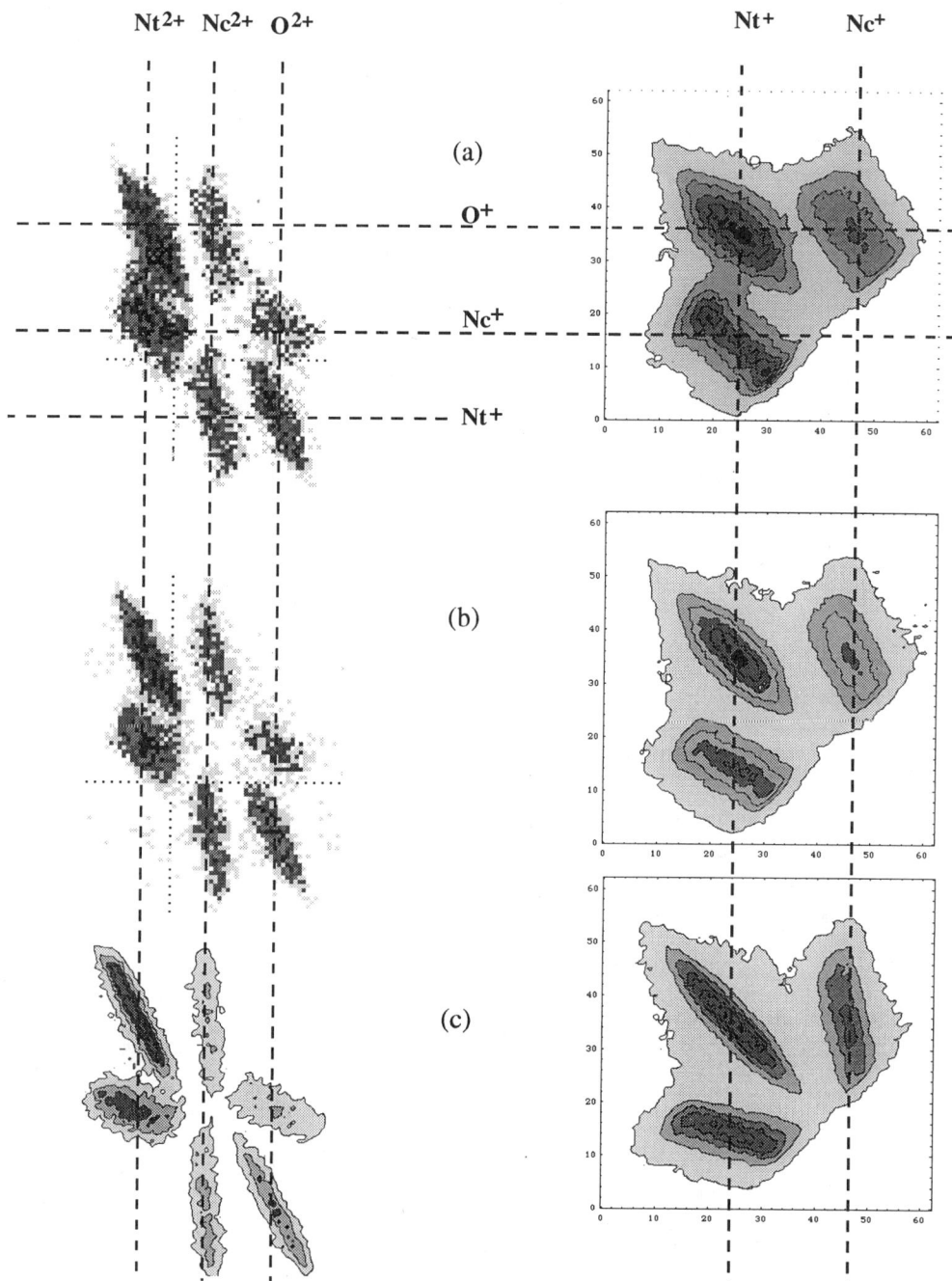

FIGURE 24. PEPIPICO spectra of $^{14}N^{15}NO$ for three-body dissociation obtained at the resonances energies (a) $N_t \to \pi^*$ at 401 eV, (b) $N_c \to \pi^*$ at 405 eV, and (c) in the continuum resonance σ^* at 430 eV. (From Morin et al.[290])

In the Coulomb explosion picture, charges are supposed to be localized around the atoms (just after the Auger effect) and the dissociation process is governed by the Coulomb repulsion forces between these charges. The final momentum partitioning is given by the choice of the initial conditions (distances and angles). Such a model was successfully applied in the pioneering work of Carlson and

White[31] on CH_3I after ionization of iodine in the L shell. In that case an average removal of 8 electrons is observed. This process occurs within the lifetime of the iodine $2p$ hole vacancy (subfemtosecond time scale) and this explains why the Coulomb explosion picture is valid. It should be noted, however, that in order to reproduce the (low) values of the kinetic energy of the fragments, Carlson and White[31,302] introduced a deformation of the molecule before applying a pure Coulomb model. In contrast, in a beam foil experiment,[335] in which all of the valence electrons of a molecule are rapidly removed by interaction of a fast ion beam (a few MeV/amu) through a thin film, the Coulomb explosion applies without restrictions and it is possible to obtain directly geometrical information regarding the projectile ion. Turning to the core photoionization but on "shallow" levels, far fewer electrons are removed and the Coulomb explosion model becomes questionable. In the N_2O example, the observation of low-kinetic-energy ions, and especially N_c^{2+} (from the $N_t^+ + N_c^{2+} + O$ and $N_t + N_c^{2+} + O^+$ channels), is a clear demonstration of the breakdown of the Coulomb explosion picture. An impulsive model was proposed[290,310] to explain the PEPIPICO results. The basic idea of this model is to emphasize the role of the direct breaking of the chemical bonds after removal of binding electrons, and to account for collisional effects or obstructed fragmentation due to the alignment of particles in case of quasilinear geometry of the molecule. This is the case of the central nitrogen (with or without a positive charge) in N_2O: it cannot escape freely without colliding with N_t or O atoms. The potential used to account for these points was the following:

$$V(r_1,r_2,\varphi) = D_1 \cdot e^{-\alpha_1(r_1-r_{e1})} + D_2 \cdot e^{-\alpha_2(r_2-r_{e2})} + \frac{q_1 q_2}{r} + D_3 \cdot e^{-(r_1-r_{e1}+r_2-r_{e2})} \cdot (\cos(\varphi) - \cos(\varphi_0))^2$$

where r_1, r_2 are the N–N and N–O distances, j is the N–N, N–O angle, and q_1 and q_2 are the state of charge of the atomic fragments. The other constants define the energetics of the dissociation. It can be divided into three terms:

- Two decreasing exponentials to account for the dissociative potential between the atoms (bond breaking)
- A Coulomb repulsion term which is dominant at long distances
- An angle-dependent term to account for eventual change in geometry (resonant state)

In the ionization continuum, this simple model was applied without any angular term (the linear geometry is conserved after core ionization) and a very good agreement was found with experimental results with the same set of parameters for all of the ion pairs. When discrete resonances are excited, the situation is different: the additional electron changes the electronic density around the atoms and induces a significant change in the geometry of the resonant state. A useful guide to appreciate this effect is the equivalent core model as described in Section 2. For instance, when the $N_t 1s \rightarrow \pi^*$ is excited, the system is equivalent to the ONO molecule which is known to be bent (134°). Of course, the molecule does not have time to totally bend within the 10^{-14} s of the Auger relaxation time, but strong repulsion forces between the electronic clouds located around each nucleus govern the dissociation pattern. This has been rationalized by de Miranda and Beswick[336] who used *ab initio* methods to calculate intramolecular forces after $N_t 1s$, $N_c 1s$, and $O 1s \rightarrow \pi^*$ or σ^* excitation. Including the angular term with $\varphi_0 = 134°$ in the impulsive model, Morin et al.[290,310] found an excellent agreement with experiment, namely, that the central nitrogen is ejected with a significant kinetic energy when discrete resonances are excited. The stronger effect is obtained when N_t is excited, as predicted by de Miranda and Beswick [336] calculations.

In order to confirm this description, a similar experiment was performed[310] on CO_2 around the $C1s$ edge. Indeed, this molecule can be core ionized in the $C1s$ shell and core excited into the first empty orbital in the 300 eV energy range. The interest of the comparison with N_2O is that in both cases the equivalent core molecule after π^* excitation is the same: $OC^*O \Leftrightarrow ONO \Leftrightarrow N^*NO$. The observations are very similar: the carbon is ejected with low kinetic energy in the continuum (σ^* resonance), producing "elongated" PEPIPICO peak shapes. On the contrary, along the π^* resonance, the carbon atom expends a significant amount of kinetic energy and the PEPIPICO peak shapes become

TABLE III

Measured and Corrected Slopes of the Coincidence Peak Shapes Associated with Various I^{p+}/C^{q+} and I^{p+}/N^{q+} Ion Pairs in ICN, at 700 eV Photon Energy[a]

Fragment ion pairs	Measured slope	Corrected slope	Fragment ion pairs	Measured slope	Corrected slope
I^+/C^+	−2.08	−2.1	I^+/N^+	−1.31	−1.3
I^{2+}/C^+	−1.15	−2.3	I^{2+}/N^+	−0.65	−1.3
I^{3+}/C^+	−0.71	−2.1	I^{3+}/N^+	−0.45	−1.3
I^{4+}/C^+	−0.50	−2.0	I^{4+}/N^+	−0.34	−1.4
I^{5+}/C^+	−0.44	−2.2	I^{5+}/N^+	−0.32	−1.6
I^+/C^{2+}	−6.67	−3.3	I^+/N^{2+}	−2.58	−1.3
I^{2+}/C^{2+}	−2.67	−2.7	I^{2+}/N^{2+}	−1.31	−1.3
I^{3+}/C^{2+}	−1.59	−2.4	I^{3+}/N^{2+}	−0.89	−1.3
I^{4+}/C^{2+}	−1.10	−2.2	I^{4+}/N_2^+	−0.66	−1.3
I^{5+}/C^{2+}	−0.85	−2.1	I^{5+}/N^{2+}	−0.52	−1.3

[a] From Ref. 319.

"oviform." Simulations using the impulsive model and relevant molecular constants gave an excellent agreement between calculated and measured shape and slope of the various PEPIPICO spectra.

Another interesting system is the linear ICN molecule because of the presence of a heavy atom and for which the chemical bonds I–C and C–N are quite different in nature. The dissociative ionization of this molecule has been studied both in the region of double ionization threshold[290,319,337] and in the region of I4d and I3d edges.[290,310] Of special interest for the present discussion is the three-body dissociation leading to the formation of $I^{p+} + C^{q+} + N^{r+}$. When the iodine 3d ionization is achieved (~700 eV), because of cascade processes (implying mostly 4d, 4s, and 4p iodine inner shells) one can reach a large set of p, q, r values: p = 0–5, q = 0–2, r = 0–2 with p + q + r = 2–7. The most striking observation is that all of the coincidence peaks appear as elongated cigars, with well-defined slopes. Interestingly, these data allow comparison of the situations in which the fragments are identical except the state of charge of one fragment. The state of charge q of a fragment influences the time of flight as $1/q$ [see equation (1)]. When comparing the slopes associated for instance with I^{p+}/N^{q+}, one has to introduce a multiplying factor of p/q. In doing so, most of the slopes, for any I^{p+}/C^{q+} or I^{p+}/N^{q+} pairs, are found as summarized in the Table III. Simon et al.[319] and Morin et al.[290] concluded that the dissociation dynamics are not governed by Coulomb repulsion after Auger decay but by the repulsion of electronic clouds around each atom within the linear core-excited molecule in the early stage of the photoexcitation. The same authors proposed an impulsive model to predict the momentum ratio: the C–I bond is first broken with an energy release E_1, the C atom then collides with the N atom (spectator in the first step). However, CN is not a rigid system and behaves like an oscillator and one ends up with an energy partitioning of E_1 into translational energy $k * E_1$ and internal energy $(1 − k) * E_1$ of CN. The CN fragment then dissociates rapidly, releasing its internal energy. The calculated slopes with $k = 0.85$ for I^{p+}/N^{q+} and $k = 1$ for I^{p+}/C^{q+} are in excellent agreement with the corrected slopes of Table III. This simple model, which weakens the role of Coulomb repulsion after the electronic relaxation in the final state, explains readily why the state of charge of the fragments does not influence the momenta of the fragments.

5.2.2. Comparison with Dissociative Ionization under Fast Ion Beam or Strong Laser Field. Other ionization sources like fast ion beam ionization and high-power laser multiphoton ionization are known to produce fragments issued from multiionization. However, the origin of multiionization is not obviously due to the creation of core vacancies but rather through valence charge stripping. Consequently, the final electronic states of the multiply charged ions are expected to be very different from the core excitation and subsequent Auger decay, for which the departing electrons are correlated.

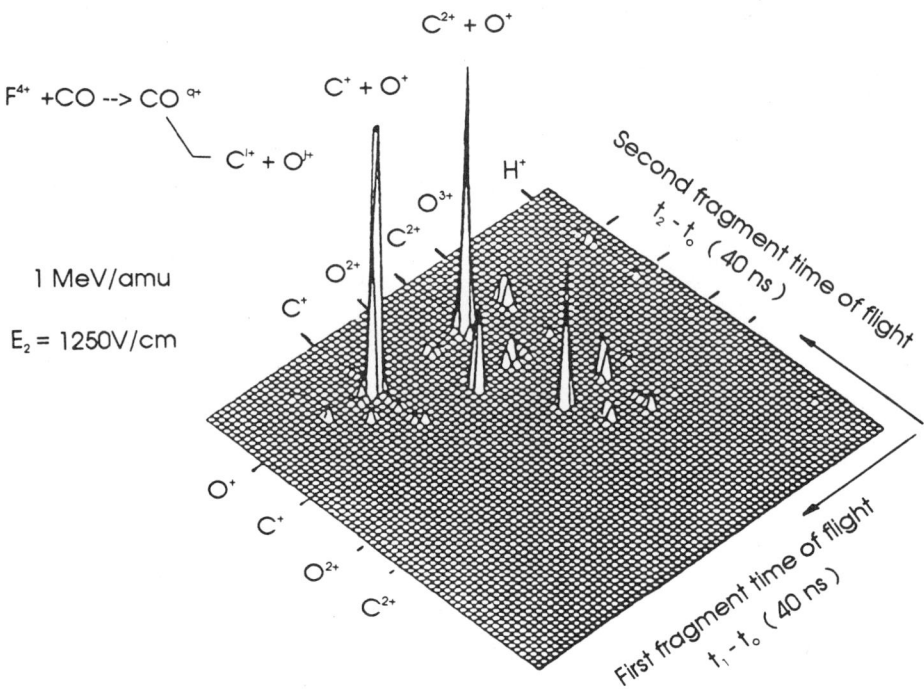

FIGURE 25. Coincidence spectrum of CO obtained by F^{4+} ion impact at 1 MeV/amu and strong extraction field conditions, represented in a three-dimensional plot. The $C^+ + O^+$ peak is truncated. (From Ben-Itzhak et al.[340])

Numerous studies using fast ion beam ionization have been devoted to the CO molecule.[338–340] They have shown, using the time-of-flight multicoincidence technique, that, similarly to photon impact, the process could be described as a two-step mechanism: fast removal of electrons followed by the electron cloud reorganization inducing rapid dissociation of the multiply charged ion. The resulting three-dimensional coincidence spectrum is shown in Figure 25. The symmetric breakup $CO^{q+} \rightarrow C^{i+} + O^{j+}$ with a distribution centered at $i = j = q/2$ is the dominant dissociation process, but nevertheless nonsymmetric dissociation, like breakup into a neutral and a charged fragment, is also observed with a decreasing probability on increasing the total number of removed electrons. Such a result can be compared, at least for $q = 2$ and 3 with the carbon and oxygen 1s photoionization data reported in Table I. An interesting point is the determination of kinetic energy released in the fragmentation, as it allows discussion of the dissociation dynamics. In a first approach, Sampoll et al.[338] used the time-of-flight technique with the following procedure: dissociation products were identified by gating the arrival time of flight of the first ion (C^+, O^+, C^{2+}, O^{2+}, etc.) and the time difference with the second ion was recorded event by event; an iterative matrix transformation was used to convert the time difference into total kinetic energy distributions. In this study, transient CO^{q+} ($q = 2–7$) were produced in collision with 97-MeV Ar^{14+} ions. Sampoll et al.[338] compared the average energy released with the results of Lablanquie et al.[227] for CO^{2+} and CO^{3+} dissociation, obtained by photon impact (35–150 eV) in the region of valence multiionization. Their higher average values were interpreted as the formation of ions in higher states of excitation than those produced by photoionization. This finding was corroborated by comparison with a point-charge Coulomb model which gives also systematically lower values than measured in the experiment. This surprising result is at variance with the results obtained by Mathur et al.[339] using 97-MeV F^{8+} collisions on CO. In this latter study, recoil ions were extracted into an electrostatic energy analyzer (0.3–0.5 eV energy resolution) followed by a quadruple mass spectrometer. The advantage of such an approach is the good energy determination but the drawback is that the ions are not detected in coincidence. Nevertheless, Mathur et al.[339]

compared the maximum kinetic energy released (KER) for a given ion with the one calculated (Coulomb model) for dissociation via the lowest energy coulombic potential energy curves. They found a reasonable agreement and concluded at variance with Sampoll et al.[338] that in general the observed KER was lower than predicted by pure Coulomb repulsion. This statement was corroborated by quantum chemical calculations over highly excited states. It appears that even for highly charged CO^{q+} ions ($q = 2–9$), most electronic states have noncoulombic potential curves and the kinetic energy release is systematically lower than the pure coulombic value. When all valence electrons are removed (i.e., $q = 10$), one expects the Coulomb picture to be correct. Despite the difficulty in understanding the apparent contradiction between the results of Sampoll et al.[338] and Mathur et al.,[339] it is clear that under ion impact the point-charge Coulomb model is not correct because many excited states are populated during the ionization process and branch into a complex set of dissociation channels.

Under strong field laser irradiation ($10^{13}–10^{16}$ W/cm^2) of molecules with subpicosecond pulses, it is possible to obtain multicharged fragments with relative probability depending on the laser parameters (wavelength, intensity). In a pioneering experiment, Frasinski et al.[341] showed that it was possible to investigate dissociation dynamics of multiply charged ions on a femtosecond time scale. Cornaggia et al.[34] have shown that irradiation of N_2 with a visible laser at 610 nm with a pulse duration of 2 ps and an intensity of 5×10^{15} W/cm^2 leads to fragments issued from the dissociation of N_2^{2+} up to N_2^{8+}. The kinetic energy analysis, as performed by conventional TOF technique, indicates that the fragments are produced by a sequential ionization occurring during the dissociation process. Later Cornaggia et al.[342] used a 100-fs pulse duration and concluded that the kinetic energy released is specific for a particular dissociation channel but does not depend on the laser intensity nor laser duration. A simple model involving Coulomb repulsion between the atomic ions could not explain the well-defined energies observed nor the independence with duration and intensity. In a very recent work on CO_2, Cornaggia et al.[342] found that generally, the measured kinetic energy released is 50% weaker than the Coulomb repulsion energy calculated at the equilibrium internuclear distances. Two possible explanations are proposed: (1) the ion–ion repulsion cannot be correctly represented by point charges—because of the electronic cloud distortion, the ions behave as they would repel each other with a reduced effective charge $Z_{eff} = 0.67\ Z$; (2) The internuclear distance does not remain constant during the laser interaction, resulting in an increased internuclear distance $R_{eff} = 2.2\ R_e$. The first explanation was also invoked[290,310] in the case of N_2O^{q+} (and $CO_2 q+$) dissociation induced by photon core excitation. The same conclusion was also made by Mathur et al.[339] in a multiple ionization induced by fast ion beam impact.

Cornaggia et al.[342] also investigated the three-body dissociation dynamics of CO_2^{q+} using the covariance mapping method introduced by Frasinski et al.[343] The covariance mapping technique is based on the correlated fluctuations of the ion signals coming from the same dissociation pathway. For instance, when two ions A^{p+} and B^{q+} arise from the same dissociation of $AB^{(p+q)+}$, there is an enhancement of the covariance coefficient $C(t_A, t_B) = \langle (t_A) \cdot s(t_B) \rangle - \langle s_A \rangle \cdot \langle s(t_B) \rangle$ where $s(t)$ is the ion signal intensity for time of flight t. The average value is obtained over a great number of laser shots (~ 20,000). Double ion/ion covariance maps can thus be obtained as shown in Figure 26 for the CO molecule. Both symmetric and asymmetric charge separation reactions are observed. It is interesting to compare at this point the photoion coincidence technique (PEPIPICO) as described in Section 5.1 and the covariance mapping method. Coincidence experiments are possible only when the time separating two successive ionization events is large enough to avoid false coincidences. Such a technique is suited for continuous or pseudocontinuous ionization sources like synchrotron radiation or laboratory discharge lamp which allow the counting rate to be kept low enough. When using ultrashort laser pulses, coincidence methods cannot be employed because a lot of ionization events are created simultaneously during one laser shot. Cornaggia et al.[342] were also able to extract third-order covariance coefficients which reveal triple dissociative events associated with three singly charged fragments. Similarly to synchrotron radiation results,[310] they show that the linear geometry is essentially preserved during the dissociation. The fragmentation process is symmetric with the central atom (C^{q+}) ejected with little kinetic energy. When the laser polarization is changed by 90° with the electric vector perpendicular to the spectrometer axis, the results show that there is a substantial bending around the linear geometry, as induced from the violent laser alignment of the molecule along the laser

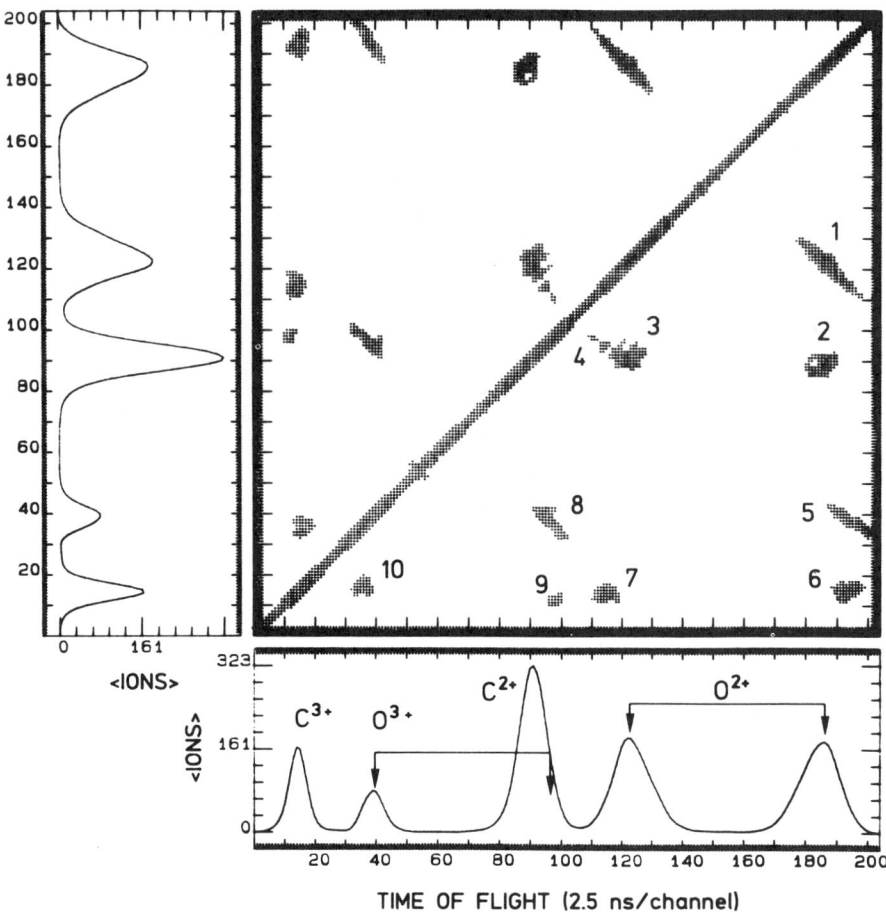

FIGURE 26. Double ion–ion correlation maps in CO_2 obtained under strong laser field conditions. (From Cornaggia et al.[342])

polarization direction before or in the course of the dissociation. It is interesting to note again that this strong torque applied to the molecule before atomization is similar to the one induced after core excitation when the resonant state is of different geometry (see above the N_2O and CO_2 cases).

5.2.3. Nature and Energy of the Fragments and Radiative Decay. The fragments produced after core excitation or ionization can be found mainly with one or several positive charges, because of the efficiency of multiionization. As discussed in Section 5.1, mass spectrometry operated in the conventional or in the multicoincidence modes is the ideal tool for investigating numerous decay channels leading to one or several correlated positively charged fragments (see Figure 1). A significant part of other fragments are neutral species and sometimes negatively charged particles. This is because the dissociation pattern is complex. After the electronic relaxation of the core hole, the molecule is produced with so much internal energy that several bonds are broken producing more fragments than the number of positive charges. One readily understands that some neutral particles are produced. This can be established indirectly by mass spectrometry in the multicoincidence mode (see Section 5.1) or by dispersed UV/visible fluorescence[344] at least for radiative species. The residual molecular ion can also be predissociated by electronic interaction and branches into more exotic channels such as the SO_2 sulfur $2p$ core excitation. Dujardin et al.[345] using negative ion mass spectrometry reported for

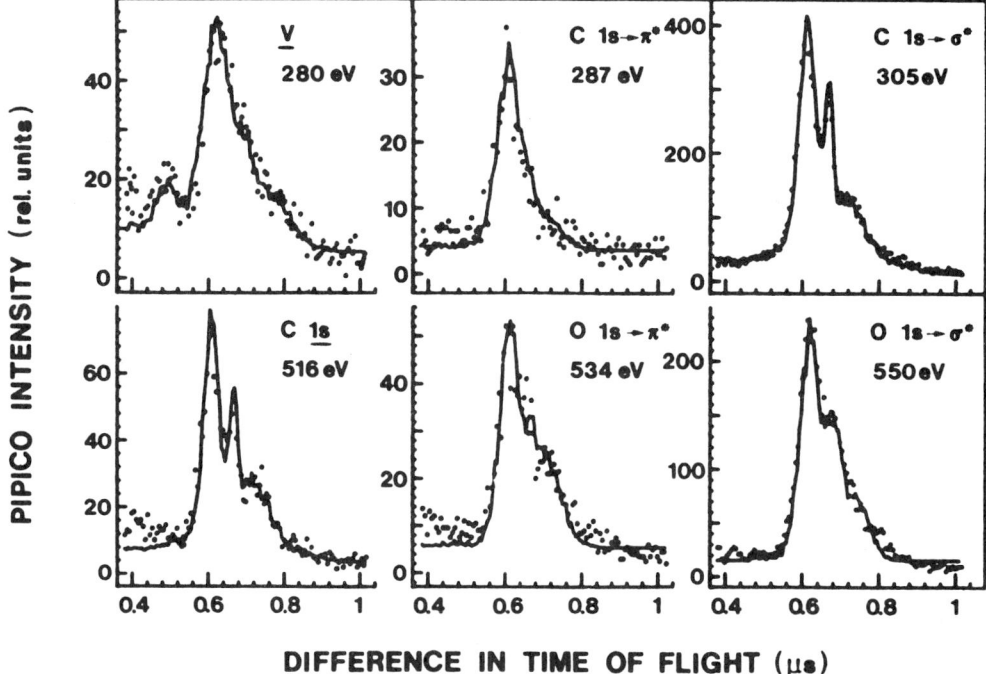

FIGURE 27. Photoion–photoion coincidence spectra of CO in the region of the $C^+ + O^+$ peak. The dots are the experimental data and the solid lines are the results of Monte Carlo simulations based on specific kinetic energy release distabutions. (From Hitchcock et al.[224])

the first time that this core-excited molecule relaxes by resonant Auger into an excited ion SO_2^{+*} which then decays into $SO^{2*} + O^-$. For ground-state neutrals, there is no report of any direct measurements although laser-induced fluorescence or double resonance mass spectrometry could be used in principle.[346]

The investigation of the internal energy of the fragments after dissociation of a core-excited or -ionized diatomic molecule is an important issue because it brings, along with the knowledge of the internal energy of the residual ion, a set of data necessary to approach a "state to state" description of the overall process. One way to obtain such information is the kinetic energy release in charge separation reactions after core excitation and ionization. This has been realized by Hitchcock et al.[224] in the CO molecule near the carbon and oxygen 1s edges, as seen in Figure 27 for the symmetric $C^+ + O^+$ pathway. According to Table I, the fragments are essentially ionized species. The combination of these results and the calculations of CO^{2+} potential energy curves[226] indicate that, after C1s or O1s ionization and in the absence of curve crossings, the C^+ and/or O^+ fragments are electronically excited in metastable states, i.e., with a lifetime long enough for an observable fluorescence. After core excitation in the C1s $\rightarrow \pi^*$ (or O1s $\rightarrow \pi^*$) states, the fragments are ions and/or neutrals (see Table I) with a great deal of electronic excitation.

Further insight on the fragmentation pathways of core-excited or -ionized molecules is given by a state selection of the residual ion. This is achieved by Auger electron–ion coincidence, as realized first in N_2 and CO by Eberhardt et al.[125,197] Other similar measurements were performed in N_2O,[327,347] CS_2 and C_4H_4S,[348] SF_6,[326] BF_3,[205] and HCl, CH_3Cl, CF_3Cl, and H_2S in Ref. 236. The information on the internal state energy is still poor because the energy resolution is not very good and generally, one learns the fragmentation pattern of all states contained within an Auger band rather than a single electronic state. Taking CO as an example following the work of Eberhardt et al.,[197] the relative yields of C^+, O^+, C^{2+}, O^{2+}, and CO^+ vary drastically with the Auger energy. So does the kinetic energy of the fragments. Interestingly, the C^{2+} ion only occurs for states with binding energies above

58 eV in the O1s hole decay. It is rather surprising that a fragment with two holes on the carbon atom is produced following the relaxation of a 1s hole in oxygen. Combining these results with those of Hitchcock et al.,[224] it is clear that the major route of these high-lying Auger bands is the dissociative triple ionization channel $C^{2+} + O^+$ rather than the $C^{2+} + O$ one.

In triatomic or polyatomic molecules, fragments can be molecular and may carry vibrational or rotational energy. Dispersed UV/visible fluorescence is a well-suited method to identify the internal energy of some fragments. This has been realized by Rosenberg et al.[349] in $SiCl_4$ and SiF_4 photoexcited in the vicinity of the silicon 2p edge. The fragments are found to be excited neutral and ionic such as Si*, F*, SiF*, Si^{+*} in the whole region of bound and continuum resonances. Interestingly, there is a large enhancement of the yield of excited-state fragments at the energy of the core-excited Rydberg state compared to the core-valence state. Rosenberg et al.[349] explain this difference by the diffuseness of the Rydberg electron (in contrast to a valence-excited electron) which favors, during the resonant Auger process (spectator transition), the formation of excited fragments. Very recently, Meyer et al.[331] using the photoion–photon of fluorescence coincidence method investigated the fluorescing fragments in core-excited N_2, N_2O, and ICN near the nitrogen 1s and iodine 4d edges. Generally they found that about 5% of the fragments are produced in an excited state. Taking the ICN molecule photoionized at 120 eV as an example, Meyer et al.[331] found that the I^{2+} and CN^+ fragments are associated with CN* and I* fluorescing species, respectively. In N_2, Meyer et al. clearly established that on N1s → σ* excitation, the fluorescing species in the 260–600 nm spectral region are atomic ion fragments rather than the ionized molecule in the $B\ ^2\Sigma_u^+$ state. This result is at variance with the results of Poliakoff et al.[155,277] and Mahalingham et al.[244] and further studies are needed to clarify this point.

5.3. Dissociation of Complex Molecules

5.3.1. Site-Selective Fragmentation. As mentioned earlier, one of the simple ideas that emerged in the last few years was to take advantage of the hole localization to induce a specific bond breaking in a molecule. The localized character of the hole appears clearly in the Auger spectrum (see Section 3.2) i.e., in the relative intensity of the lines and the overall shape of the spectrum: only orbitals with a significant overlap with the created hole participate in the Auger relaxation, at least for nonionic compounds. But this condition is not sufficient to ensure selective dissociation. Indeed there is a severe competition between the dissociation and a redistribution of the internal energy among the various vibration modes of the ion. We can imagine two extreme cases:

1. The dissociation is direct and very fast like in HBr after Br3d→σ* excitation (see Section 4.2.1) and the electronic relaxation (Auger effect) is achieved after fragmentation.
2. The dissociation is fairly slow and is ruled out by statistical laws. Generally, when the molecule is photoexcited in one electronic state, there is a fast internal conversion process from electronic energy into vibrational energy of the ground electronic state. This vibrational energy is redistributed statistically over the various degrees of freedom and the molecule dissociates. The fragmentation rate and branching ratio depend on the internal energy and the number of rovibrational states above the activation barrier along the reaction coordinate. Such branching ratios are always in favor of the rupture of the weakest bond and occur only after conversion of electronic internal energy into vibrational excitation of all the degrees of freedom of the system. These two limit cases are also featuring two extreme cases of site selectivity: The first case corresponds to a pure site-selective effect, because it is governed by the core-excited electronic state whereas the second corresponds to a total loss of memory of any initial localization of the initial excitation. The final result of fragmentation after core photoionization is thus a crucial balance between these two extreme behaviors.

Investigation of site-selective fragmentation has been reported for a large variety of molecules including halogeno-hydrocarbons,[282,324,350–353] organometallic compounds,[234–237,354–356] silicon compounds,[233,238,316,323,357] and other species.[358,359] We wish to focus here on two selected examples corresponding to the extreme cases discussed above: the Br(CH$_2$)$_n$Cl[352] and Fe(CO)$_2$(NO)$_2$[355] molecules.

Schmelz et al.[352] have reported a significant site-selective fragmentation of bromo-chloro-alkanes Br(CH$_2$)$_n$Cl ($n = 1$–3) using TOF mass spectrometry used in the multicoincidence mode. They observed differences between the fragmentation after bromine $3d$ (65–150 eV) or chlorine $2p$ (150–220 eV) excitation or ionization. In Figure 28 are reported mass spectra of the three molecules recorded at 70.67 eV (Br$3d \to \sigma^*$ resonance) and 212 eV (C1$2p \to$ continuum); the presented spectra have been corrected for the contribution of valence ionization at the bromine edge and for the Br$3d$ continuum at the chlorine edge) for a meaningful comparison. In spectrum a (upper), Br$^+$ clearly dominates the fragmentation pattern whereas the BrCH$_x^+$ fragment is almost negligible. On the contrary, as seen in spectrum a (lower), Cl$^+$ is now dominating whereas the BrCH$_x^+$ fragment is as intense as Br$^+$. Thus, after excitation of the Br$3d$ edge the bond with Br and the rest of the molecule is preferentially broken whereas at the Cl$2p$ edge, the bond with Cl is preferentially broken despite the fact that the C–Br bond strength is smaller than the C–Cl one. The effect is of course not total in the sense that some residual Cl bond breakage is observed at 70.67 eV and some remnant Br$^+$ and ClCH$_x^+$ fragments are also observed at 212 eV. Examination of ion pair formation showed that dissociative double ionization, especially at the Cl$2p$ edge, is faster than internal conversion, which may explain the observed selectivity. Schmelz et al.[352] have also investigated the size dependence of the selectivity by comparing molecules with increasing length of the carbon chain (C, C$_2$, C$_3$). Similar effects are observed as shown in Figure 28 (middle and right), but to a smaller extent. It is interesting to note that no general trends can be drawn because the intermediate molecule [Br(CH$_2$)$_2$Cl] is the one which shows the smallest site selectivity.

Another system of interest with respect to site selectivity is the Fe(CO)$_2$(NO)$_2$ molecule. Indeed in such a system one can selectively excite the carbon or the nitrogen 1s shells which are well separated

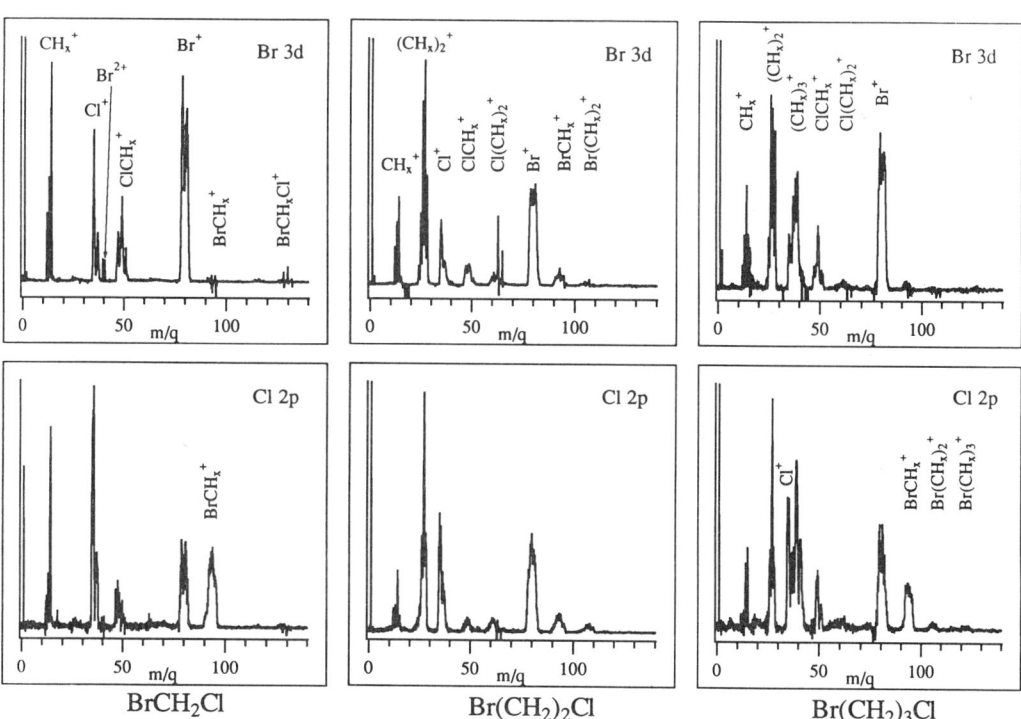

FIGURE 28. Mass spectra of Br(CH$_2$)$_n$Cl ($n = 1,2,3$) molecules near the bromine $3d$ and the chlorine $2p$ edges, after subtraction of the valence ionization continuum. (From Schmelz et al.[352])

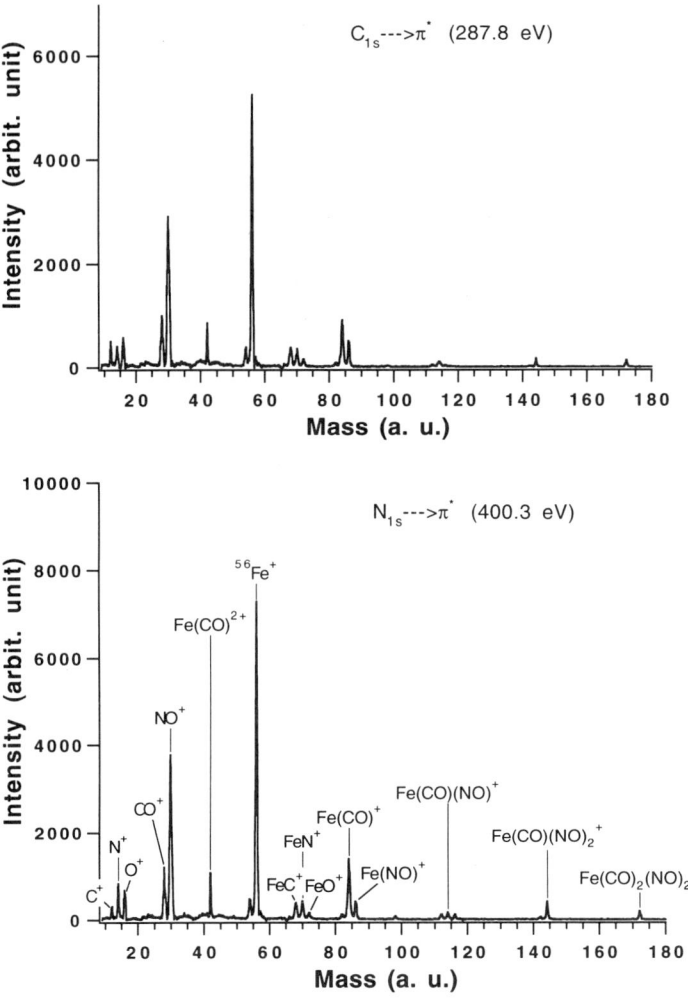

FIGURE 29. Mass spectra in Fe(CO)$_2$(NO)$_2$ recorded on the C1s → π* resonance (upper) and on the N1s → π* resonance (lower). These spectra correspond to the ejection of one ionic fragment only. (From Simon et al.[355])

in energy. Moreover, both CO and NO ligands exhibit strong and sharp π* resonances and broad σ* shape resonances that can be excited from the 1s shell. The fragmentation of this molecule has been investigated[319,355] using the TOF multicoincidence technique. Mass spectra recorded along the C1s →π* (287.8 eV) and N1s→π* (400.3 eV) resonances are reported in Figure 29. The most striking observation is that those spectra are very similar: at variance with the findings of Schmelz et al.[352] on bromo-chloro-alkanes, no memory effects due to the initial localization of the hole (selected site) and excitation (π* resonances) can be seen. Simon et al.[355] pointed out, however, some slight differences when comparing (at the same edge) π* and σ* excitation, which were interpreted as a change in the single/double ionization branching ratio along the two resonances: in the continuum, one expects only double ionization to be observed due to the dominant normal Auger effect, whereas on the π* resonance, resonant Auger relaxation favors single ionization channels (spectator model). The loss of memory effects was interpreted[319,354] as due to the statistical fragmentation of the residual ground-state multiply charged ion after the Auger decay. This is because the ground state of the ion is the same (localization of the charges, equilibrium geometry) regardless of the location of the initial vacancy. Furthermore, the fragmentation pattern which occurs after the redistribution of the electronic into vibration energy is governed by the internal energy of the ion and the electronic structure of the molecular ion populated by the core hole decay. Such a modeling has already been invoked for dissociation of $Fe(CO)_5$[360] and $Fe(CO)_5^+$[361] at low energy. Similar to Norwood et al.,[361] Simon et al.[355] concluded in favor of a sequential loss of CO (and NO) as deduced from ion pair analysis.

5.3.2. Sequential Processes. As stated previously, dissociation of core-excited complex molecules may occur on a (sub)picosecond time scale. This is enough for stepwise processes (rotation time scale) to take place. Eland and co-workers[23,362–364] have shown that the TOF multicoincidence technique was ideally suited for measuring dynamics on such a time scale, as far as two or more fragment ions are concerned. In the case of dissociation of a triatomic molecule, two extreme stepwise processes can be distinguished. The first one is the following:

$$ABC^{2+} \rightarrow A + BC^{2+} \quad \text{(kinetic energy released } U_1) \quad \text{(first step)}$$

$$BC^{2+} \rightarrow B^+ + C^+ \quad \text{(kinetic energy released } U_2) \quad \text{(second step)}$$

This mechanism is called "deferred charge separation" (DCS). The second one is the following:

$$ABC^{2+} \rightarrow AB^+ + C^+ \quad \text{(kinetic energy released } U_1) \quad \text{(first step)}$$

$$AB^+ \rightarrow B^+ + C \quad \text{(kinetic energy released } U_2) \quad \text{(second step)}$$

This mechanism is called "secondary decay" (SD). The SD mechanism is recognized if the intermediate survives even for a period much less than one rotation. When there is lack of both rotation and large-amplitude bending vibrations so as to retain fixed angular relations between the fragment trajectories, the mechanism is concerted like:

$$ABC^{2+} \rightarrow A^+ + B^+ + C$$

It is important to specify that the time scale involved in the various steps is large enough so that dissociation energies can be defined and that momentum conservation can be applied. Eland et al.[23,363] using Monte Carlo simulations and classical trajectory modeling and Lavollée and Bergeron[322] using analytical calculations found that in both cases the expected PEPIPICO contour should be a parallelogram of slope -1 (DCS) or $-m_B/m_{AB}$ (SD), as far as the molecular fragment has time to rotate. The width and length of the parallelogram are directly related to the kinetic energy released at each step (U_1 and U_2). Numerous molecules have been studied in the region of double ionization threshold, such as perfluoro compounds[364] for which SD processes are dominant. In the core ionization regime, Simon et al.[319,323] studied the fragmentation of hexamethyl disilane [$Si_2(CH_3)_6$] around the Si2p edge. Fragmentation mass spectra do not depend drastically on photon energy and are dominated by the $Si(CH_3)_3^+$ ion which corresponds to the Si–Si bond breaking. However, the presence of fragments such as $Si_2(CH_3)_x$ shows that the Si–C bond can be broken while keeping the Si–Si bond intact, despite the fact that the Si–C bond is stronger in the neutral molecule. Most of the ion pairs resulting from dissociative double ionization involve $Si(CH_3)_3^+$. Simon et al.[323] pointed out that the PEPIPICO contours for this ion in coincidence with SiH^+, $Si(CH_3)^+$, $SiH_2(CH_3)^+$, $Si(CH_3)_2^+$ were identified as parallelograms with specific slopes. The results of this measurement, as well as the calculated slope resulting from corresponding secondary decay are summarized in Table IV.

In the selected pairs, the same stepwise mechanism is observed, namely, the breaking of the Si–Si bond producing two $Si(CH_3)_3^+$ fragments, followed by the Si–C bond breaking. Simon et al.[323] pointed out that the dissociation energy released in the first step is found to be exactly the same in all of the mentioned channels which suggests that the first dissociation step is the same for all processes. It was finally concluded that the fragmentation of HMDS after Si2p excitation or ionization was dominated by a fast dissociation into $Si(CH_3)_3^+ + Si(CH_3)_3^+$, and each fragment undergoes a statistical decay as routinely observed for the singly charged ion,[365] in accordance with the photon-independent character

TABLE IV
Measured and Calculated Slopes of the Coincidence Peak Shapes in Core-Excited Hexamethyl Disilane with the Associated Fragmentation Mechanism[a]

Fragment ion pair	Slope	Mechanism	Calc. slope
$SiH^+ / Si(CH_3)_3^+$	−2.52	$Si(CH_3)_3^+ + Si(CH_3)_3^+$ $Si(CH_3)_3^+ \to SiH^+ + C_3H_8$	−2.5
$SiCH_3^+ / Si(CH_3)_3^+$	−1.65	$Si(CH_3)_3^+ + Si(CH_3)_3^+$ $Si(CH_3)_3^+ \to SiCH_3^+ + C_3H_8$	−1.7
$SiH_2CH_3^+ / Si(CH_3)_3^+$	−1.6	$Si(CH_3)_3^+ + Si(CH_3)_3^+$ $Si(CH_3)_3^+ \to$ $SiH_2CH_3^+ + C_3H_6$	−1.62
$Si(CH_3)_2^+ / Si(CH_3)_3^+$	−1.23	$Si(CH_3)_3^+ + Si(CH_3)_3^+$ $Si(CH_3)_3^+ \to Si(CH_3)_2^+ + CH_3$	−1.26

[a]From Ref. 323.

of the fragmentation. Further evidence of slow secondary decay is given by the observation of metastable ions like $SiCH_5^+$ and $SiC_3H_9^+$ with a lifetime of 49 and 45 ns, respectively. It should also be noted that several fragments like H_2^+, SiH^+, SiH_3^+ result from rearrangement reactions. The carbon substitution by hydrogen atoms around the silicon atom requires that several concerted fragmentation (two broken bonds and one formed) and isomerization reactions occur. This seems also the case to explain the formation of H_2^+ and H_3^+ during dissociative double ionization in core-excited silane near the silicon 2p edge.[313]

Finally, it should be noted that the recent attempt[366] to model the multifragmentation of a core-excited polyatomic system (CH_3CF_3) in terms of full statistical terms with energy, matter, and charge conservation has failed. The same authors introduced a site-specific constraint that distinguishes the two carbon atoms which provides a better agreement with experiment.[325] Surprisingly, the fragmentation of the singly charged ion is found more nonstatistical than that of the doubly charged ion.

6. FUTURE TRENDS

The most straightforward foreseeable progress in the near future for understanding the coupling of nuclear and electronic motion is based on the development of high-resolution spectroscopy for both photoexcitation and electron energy analysis associated with the use of third-generation synchrotron radiation sources. The challenge will be to understand the intricate interplay between the electronic and nuclear motion during the core hole decay. Experimentally, this will require one to select a vibrational level of a core-excited molecule (diatomics and triatomics), to analyze the photoelectron spectrum with a resolution sufficient to resolve the vibrational structure of electronic bands, and to obtain the corresponding angular distribution. The next step is to understand how the dynamics of the fragmentation can be described when the final state of the ion is selected. In other words, it should be particularly important to know whether the potential surface of the doubly charged ion also controls the dynamics, besides the impulse given in the early stage of the photoexcitation process. The ideal experiment will be to study state-selected ion fragmentation by Auger electron–ion–ion coincidence[126,327,328] with an investigation of the internal energy of the fragments and kinetic energy release with possibly the anisotropy of the process. This latter stage of experiments may require the use of two colors in a pump-probe arrangement in cases where the fragments do not fluoresce spontaneously.[346] Such ideal sequences of measurements, some of which have already been performed as seen in the preceding sections, are considered as a complete state-to-state study of the electronic and nuclear relaxation of a core-excited molecule and should be developed in order to offer the proper data bases to theoreticians.

So far, very few studies have been devoted to the investigation of fragmentation of state-selected ions produced after deep inner shell excitation. More should be learned through Auger electron–ion coincidence following the pioneering work of Lindle et al.[239] Other sophisticated experimental studies will be of great interest for studying rotational orientation and alignment of linear molecules thanks to angle- and/or spin-resolved Auger spectroscopy after the theoretical predictions of Chandra and Sen[138] and Zähringer et al.[183]

Further investigations of the relaxation of large molecules should be of great interest without, however, always seeking a very high resolution. Challenges are several: (1) understanding the size effects of the fragmentation efficiency (we do not have any idea of the size limit above which fragmentation is negligible); (2) understanding the role of the initial structure (aromatics, aliphatics, and substituted species) on the fragmentation; (3) investigating concerted multifragmentation with both bond formation and destruction; (4) clarifying the origin of site selectivity to make reliable predictions; and (5) investigating biological molecules.

Related studies on van der Waals, metallic, semimetallic, semiconductors or insulator clusters are discovering new classes of phenomena, i.e., interatomic processes (see Rühl et al.[7] for van der Waals clusters) by contrast to the intra-atomic ones. It would be especially interesting to investigate the effects of surrounding molecules (solvent effect) in heterogeneous clusters and the role of bonding on fragmentation as well as size effects (number of basic units) on the core hole electronic relaxation and fragmentation.

In conclusion, this review of the electronic and nuclear relaxation of core-excited molecules should be considered as a rich area of fundamental research which lies at the interface of many related fields such as surfaces and interfaces, technology, radiobiology, and astrophysics, showing that a great variety of applications are already foreseen in the near future.

ACKNOWLEDGMENTS. We are very grateful to C. Reynaud, A. Beswick, P. Lablanquie, M. Simon, and M. Meyer for a careful and critical reading of the manuscript and helpful discussions.

REFERENCES

1. S. WEXLER AND G. R. ANDERSON, *J. Chem. Phys.* **33**, 850 (1960).
2. S. WEXLER *J. Chem. Phys.* **36**, 1992 (1962).
3. T. A. CARLSON AND R. M. WHITE, *J. Chem. Phys.* **38**, 2930 (1963).
4. J. DURUP AND R. L. PLATZMAN, *Int. J. Radiat. Phys. Chem.* **7**, 121 (1975).
5. K. SIEGBAHN, C. NORDLING, G. JOHANSSON, J. HEDMAN, P. F. HEDÈN, K. HAMRIN, U. GELIUS, T. BERGMARK, L. O. WERME, R. MANNE, AND Y. BAER, *ESCA Applied to Free Molecules* (North-Holland, Amsterdam, 1969).
6. C. BRÉCHIGNAC, M. BROYER, P. CAHUZAC, M. DE FRUTOS, P. LABASTIE, AND J.-P. ROUX, *Phys. Rev. Lett.* **67**, 1222 (1991).
7. E. RÜHL, C. HEINZEL, A. P. HITCHCOCK, AND H. BAUMGÄRTEL, *Proceedings of the 10th International Conference on Vacuum Ultra-Violet and Radiation Physics*, Paris, July 27–31, 1992, edited by F. J. Wuilleumier, Y. Petroff, and I. Nenner (World Scientific, Singapore, 1993), pp. 226–235.
8. D. MENZEL, *Appl. Phys.* **A51**, 163 (1990); D. MENZEL, G. ROCKER, D. COULMAN, P. FEULNER AND W. WURTH. *Phys. Scr.* **41**, 588 (1990).
9. R. DUDDE, M. L. M. ROCCO, E. E. KOCH, S. BERNSTORFF AND W. EBERHARDT, *J. Chem. Phys.* **91**, 20 (1989).
10. H. KYURAGI AND T. URISU, *Appl. Phys. Lett.* **50**, 1254 (1987).
11. T. URISU AND H. KYURAGI, *J. Vac. Sci. Technol.* **B5**, 1436 (1987).
12. N. HAYASAKA, A. HIRAYA, AND K. SHOBATAKE, *J. Appl. Phys.* **26**, L1110 (1987).
13. H. OHASHI, K. INOUE, Y. SAITO, A. YOSHIDA, H. OGAWA, AND K. SHOBATAKE, *Appl. Phys. Lett.* **55**, 1644 (1989).
14. K. HIEDA AND T. ITO, *Handbook on Synchrotron Radiation*, Vol. 4, edited by S. Ebashi, M. Koch, and E. Rubenstein (Elsevier, Amsterdam, 1991), pp. 431–465.
15. Y. FURUSAWA, H. MAEZAWA, K. SUZUKI, K. KOBAYASHI, M. SUZUKI, AND K. HIEDA, *AAPM Symposium Series* **8**, p. 37, edited by R. W. Howell, V. R. Warra, K. S. R. Sastry, and D. V. Rao (1992).

16. K. KOBAYASHI, K. KIEDA, H. MAEZAWA, Y. FURUSAWA, M. SUSUKI, AND T. ITO, *Int. J. Radiat. Biol.* **59**, 643 (1991).
17. S. CASANOVA, T. MONTMERLE, E. D. FEIGELSON AND P. ANDRÉ, *Astrophys. J.* **439**, 752 (1994).
18. L. ALLAMENDOLA AND A. TIELENS (Eds.), *Interstellar Dust*, IAU Symp. 135 (Kluwer, 1989).
19. Y. JEAN AND F. VOLATRON, *An Introduction to Molecular Orbitals* (translated by J. K. Burdett) (Oxford University Press, London, 1993).
20. J. STÖHR, *NEXAFS Spectroscopy*, Springer Series in Surface Science No. 25 (Springer-Verlag, Berlin, 1992).
21. M. LARSSON, *Comments At. Mol. Phys.* **29**, 39 (1993).
22. K. LAMMERTSMA, P. VON RAGUÉ SCHLEYER, AND H. SCHWARZ, *Angew. Chem. Int. Ed. Engl.* **28**, 1321 (1989).
23. J. H. D. ELAND, *Vacuum Ultraviolet Photoionization and Photodissociation of Molecules and Clusters*, edited by C.Y. Ng (World Scientific, Singapore, 1991), pp 297–343.
24. T. TSUCHIYA, *J. Chem. Phys.* **36**, 568 (1962).
25. J. BRACHER, H. EHRHARDT, R. FUCHS, O. OSBERGHAUS, AND R. TAUBERT, *Adv. Mass Spectrom.* **2**, 285 (1963).
26. F. L. MOHLER, V. H. DIBELER, AND R. M. REESE, *J. Chem. Phys.* **22**, 394 (1954).
27. H. EHRHARDT AND T. TEKAAT, *Z. Naturforsch.* **19a**, 1382 (1964).
28. R. FUCHS AND R. TAUBERT, *Z. Naturforsch.* **19a**, 1181 (1964).
29. K. E. MCCULLOH, T. E. SHARP, AND H. M. ROSENSTOCK, *J. Chem. Phys.* **3**, 457 (1965).
30. J. APPELL, J. DURUP, AND K. HEITZ, *Adv. Mass Spectrom.* **3**, 457 (1965).
31. T. A. CARLSON AND R. M. WHITE, *J. Chem. Phys.* **44** 4510 (1966).
32. A. C. WAHL AND N. A. BONNER, *Radioactivity Applied to Chemistry* (Wiley, New York, 1951), p. 269.
33. C. DEUTSCH, *Laser Particle Beams* **10**, 217 (1992).
34. C. CORNAGGIA, J. LAVANCIER, D. NORMAND, J. MORELLEC, AND H. X. LIU, *Phys. Rev. A* **42**, 5464 (1990).
35. Z. VAGER, T. GRABER, E. P. KANTER, AND D. ZAJFMAN, *Phys. Rev. Lett.* **70**, 3549 (1993).
36. G. HERZBERG, *Molecular Spectra and Molecular Structure: III. Electronic Spectra and Electronic Structure of Polyatomic Molecules* (Van Nostrand Reinhold, Princeton, NJ, 1966).
37. G. S. BROWN, M. H. CHEN, B. CRASEMANN, AND G. E. ICE, *Phys. Rev. Lett.* **45**, 1937 (1980).
38. Z. F. LIU, G. M. BANCROFT, K. H. TAN, AND M. SCHACHTER, *Phys. Rev. Lett.* **72**, 621 (1994).
39. G. C. KING AND F. H. READ, in *Atomic Inner-Shell Physics*, edited by B. Crasemann (Plenum Press, New York, 1985), pp. 317–375.
40. A. P. HITCHCOCK *Phys. Scr.* **T31**, 159 (1990).
41. A. P. HITCHCOCK AND D. C. MANCINI, *J. Electron Spectrosc.* **67**, 1 (1994).
42. T. D. THOMAS, M. COVILLE, R. THISSEN, AND P. MORIN, *Synchrotron Radiat. News* **5**, 9 (1992).
43. E. HUDSON, D. A. SHIRLEY, M. DOMKE, G. REMMERS, A. PUSCHMANN, T. MANDEL, C. XUE, AND G. KAINDL, *Phys. Rev. A* **47**, 361 (1993).
44. C. T. CHEN, Y. MA, AND F. SETTE, *Phys. Rev. A* **40**, 6737 (1989).
45. M. DOMKE, C. XUE, A. PUSCHMANN, T. MANDEL, E. HUDSON, D. A. SHIRLEY, AND G. KAINDL, *Chem. Phys. Lett.* **173**, 122 (1990); **174**, 668 (1990).
46. M. DOMKE, T. MANDEL, A. PUSCHMANN, C. XUE, D. A. SHIRLEY, G. KAINDL, H. PETERSEN, AND P. KUSKE, *Rev. Sci. Instrum.* **63**, 80 (1992).
47. Y. MA, C. T. CHEN, G. MEIGS, K. RANDALL, AND F. SETTE, *Phys. Rev. A* **44**, 1848 (1991).
48. G. REMMERS, M. DOMKE, A. PUSCHMANN, T. MANDEL, C. XUE, G. KAINDL, E. HUDSON, AND D. A. SHIRLEY, *Phys. Rev. A* **46**, 3935 (1992).
49. Y. MA, F. SETTE, G. MEIGS, S. MODESTI, AND C. T. CHEN, *Phys. Rev. Lett.* **63**, 2044 (1989).
50. J. SCHIRMER, A. B. TROFIMOV, K. J. RANDALL, J. FELDHAUS, A. M. BRADSHAW, Y. MA, C. T. CHEN, AND F. SETTE, *Phys. Rev. A* **47**, 1136 (1993).
51. E. HUDSON, D. A. SHIRLEY, M. DOMKE, G. REMMERS, AND G. KAINDL, *Phys. Rev. A* **49**, 161 (1994).
52. G. REMMERS, M. DOMKE, AND G. KAINDL, *Phys. Rev. A* **47**, 3085 (1993).
53. S. BODEUR, P. MILLIÉ, AND I. NENNER, *Phys. Rev. A* **41**, 252 (1990).
54. S. BODEUR, D. BAZIN, C. REYNAUD, AND I. NENNER, Conference Proceedings Vol. 25, *2nd European Conference on Progress in X-ray Synchrotron Radiation Research*, edited by A. Balena, E. Bernieri and S. Mobilio (SIF, Bologna, 1990), p. 63.
55. C. REYNAUD, S. BODEUR, J. L. MARÉCHAL, D. BAZIN, P. MILLIÉ, I. NENNER, U. ROCKLAND, AND H. BAUMGÄRTEL, *Chem. Phys.* **166**, 411 (1992).
56. C. REYNAUD, M. A. GAVEAU, K. BISSON, P. MILLIÉ, I. NENNER, S. BODEUR, P. ARCHIREL, AND B. LEVY, *J. Electron Spectrosc.* (in press).
57. S. BODEUR, J. L. MARÉCHAL, C. REYNAUD, D. BAZIN, AND I. NENNER, *Z. Phys. D* **17**, 291 (1990).
58. C. DÉZARNAUD, M. TRONC, AND A. P. HITCHCOCK, *Chem. Phys.* **142**, 455 (1990).
59. C. DÉZARNAUD, M. TRONC AND A. MODELLI, *Chem. Phys.* **156**, 129 (1991).

60. M. Tronc and C. Dézarnaud-Dandine, *Chem. Phys. Lett.* **184**, 267 (1991).
61. E. Bouisset, J. M. Esteva, R. C. Karnatak, J. P. Connerade, A. M. Flank, and P. Lagarde, *J. Phys. B* **24**, 1609 (1991).
62. J. M. Hollander and W. L. Jolly, *Acc. Chem. Res.* **3**, 193 (1970).
63. W. H. E. Schwarz, *Angew. Chem. Int. Ed. Engl.* **13**, 454 (1974).
64. A. Koch, B. M. Nestmann, and S. D. Peyerimhoff, *Chem. Phys.* **161**, 169 (1992).
65. B. Schimmelpfennig, B. Nestmann, and S. D. Peyerimhoff, *J. Phys. B.* **25**, 1217 (1992).
66. E. Shigemasa, K. Ueda, Y. Sato, T. Sasaki, and A. Yagishita, *Phys. Rev. A* **45**, 2915 (1992).
67. A. Yagishita, H. Maezawa, M. Ukai, and E. Shigemasa, *Phys. Rev. Lett.* **62**, 36 (1989).
68. N. Saito and I. H. Suzuki, *Phys. Rev. Lett.* **61**, 2740 (1988).
69. N. Saito and I. H. Suzuki, *J. Phys. B* **22**, 3973 (1989).
70. N. Saito and I. H. Suzuki, *J. Chem. Phys.* **91**, 5329 (1989).
71. N. Saito and I. H. Suzuki, *J. Phys. B* **22**, L517 (1989).
72. N. Saito and I. H. Suzuki, *Phys. Rev. A* **43**, 3662 (1991).
73. K. Lee, D. Y. Kim, C. I. Ma, D. A. Lapiano-Smith, and D. M. Hanson, *J. Chem Phys.* **93**, 7936 (1990).
74. E. Shigemasa, K. Ueda, Y. Sato, T. Hayaishi, H. Maezawa, T. Sasaki, and A. Yagishita, *Phys. Scr.* **41**, 63 (1990).
75. E. Shigemasa, T. Hayaishi, T. Sasaki, and A. Yagishita, *Phys. Rev. A* **47**, 1824 (1993).
76. N. Kosugi, E. Shigemasa, and A. Yagishita, *Chem. Phys. Lett.* **190**, 481 (1992).
77. D. Kim, K. Lee, C. I. Ma, M. Mahalingam, D. M. Hanson, and S. H. Hulbert, *J. Phys. Chem.* **97**, 5915 (1992).
78. M. Simon, M. Lavollée, M. Meyer, and P. Morin, *J. Chim. Phys. (France)* **90**, 1325 (1993).
79. J. D. Bozek, N. Saito, and I. H. Suzuki, *J. Chem. Phys.* **98**, 4652 (1993).
80. J. D. Bozek, N. Saito, and I. H. Suzuki, *J. Chem. Phys.* **100**, 393 (1994).
81. K. Lee, S. H. Hulbert, P. Kuiper, D. Ji, and D. M. Hanson, *Nucl. Instrum. Methods* **347**, 446 (1994).
82. C. M. Truesdale, D. W. Lindle, P. H. Kobrin, U. Becker, H. G. Kerkhoff, P. A. Heimann, T. A. Ferrett, and D. A. Shirley, *J. Chem. Phys.* **80**, 2319 (1984).
83. U. Becker, R. Hölzel, H. G. Kerkhoff, B. Langer, D. Szostak, and R. Wehlitz, *Phys. Rev. Lett.* **56**, 1455 (1986).
84. D. W. Lindle, P. L. Cowan, R. E. La Villa, T. Jach, R. D. Deslattes, B. Karlin, J. A. Sheehy, T. J. Gil, and P. Langhoff, *Phys. Rev. Lett.* **72**, 621 (1988).
85. E. Magnusson, *J. Am. Chem. Soc.* **112**, 7940 (1990).
86. A. Sevin, C. Dézarnaud-Dandine, and M. Tronc, *Chem Phys.* **165**, 245 (1992).
87. J. L. Dehmer, *J. Chem. Phys.* **56**, 4496 (1972).
88. J. L. Dehmer, A. C. Parr, and S. H. Southworth, in *Handbook on Synchrotron Radiation*, Vol. II, edited by G. V. Marr (North-Holland, Amsterdam, 1987), pp. 241–353.
89. P. W. Langhoff, T. N. Rescigno, N. Padial, G. Csanak, and B. V. McKoy, *J. Chim. Phys.* **77**, 589 (1980).
90. A. A. Pavlychev, A. S. Vinogradov, A. P. Stepanov, and A. S. Schulakov, *Opt. Spectrosc. (USSR)* **75**, 554 (1993).
91. K.-H. Hallmeier, A. A. Pavlychev, R. Szargan, L. Beyer, C. Hennig, and F. Thiel, *Chem. Phys.* **178**, 349 (1993).
92. J. Stöhr and K. R. Bauchspiess, *Phys. Rev. Lett.* **67**, 3376 (1991).
93. J. J. Rehr, R. C. Albers, and S. I. Zabinsky, *Phys. Rev. Lett.* **69**, 3397 (1992).
94. T. A. Tyson, K. O. Hodgson, C. R. Natoli, and M. Benfatto, *Phys Rev. B* **46**, 5997 (1992).
95. T. A. Ferrett, D. W. Lindle, P. A. Heimann, M. G. Kerkhoff, U. E. Becker, and D. A. Shirley, *Phys. Rev. A* **34**, 1916 (1986).
96. T. A. Ferrett, D. W. Lindle, P. A. Heimann, M. N. Piancastelli, P. H. Kobrin, H. G. Kerkhoff, U. Becker, W. D. Brewer, and D. A. Shirley, *J. Chem. Phys.* **89**, 4726 (1988).
97. M. B. Robin, in *Higher Excited States of Polyatomic Molecules*, Vol. III (Academic Press, New York, 1985).
98. J. L. Dehmer, D. Dill, and S. Wallace, *Phys. Rev. Lett.* **43**, 1005 (1979).
99. A. E. Orel, T. N. Rescigno, B. V. McKoy, and P. W. Langhoff, *J. Chem. Phys.* **72**, 1265 (1980).
100. F. Sette, J. Stöhr, and A. P. Hitchcock, *J. Chem. Phys.* **81**, 4906 (1984).
101. J. Stöhr, F. Sette, and A. L. Johnson, *Phys. Rev. Lett.* **53**, 1684 (1984).
102. A. P. Hitchcock and C. E. Brion, *J. Phys. B* **14**, 4399 (1981).
103. M. N. Piancastelli, D. W. Lindle, T. A. Ferrett, and D. A. Shirley, *J. Chem. Phys.* **86** 2765 (1987); **87**, 3255 (1987).
104. A. P. Hitchcock and J. Stöhr, *J. Chem. Phys.* **87**, 3253 (1987).
105. J. A. Sheehy, T. J. Gil, C. L. Winstead, R. E. Farren, and P. W. Langhoff, *J. Chem. Phys.* **91**, 1796 (1989).

106. S. J. Schaphorst, A. F. Kodre, J. Ruscheinski, B. Crasemann, T. Åberg, J. Tulkki, M. H. Chen, Y. Azuma, and G. S. Brown, *Phys. Rev. A* **47**, 1953 (1993).
107. H. W. Schnopper, *Phys. Rev.* **131**, 2558 (1963).
108. F. J. Wuilleumier, *J. Phys.* **32**, Suppl. C4-88 (1971)
109. R. D. Deslattes, R. E. La Villa, P. L. Cowan, and A. Henins, *Phys. Rev. A* **27**, 923 (1983).
110. L. J. Medhurst, T. A. Ferrett, P. A. Heimann, D. W. Lindle, S. H. Liu, and D. A. Shirley, *J. Chem. Phys.* **89**, 6096 (1988).
111. L. J. Medhurst, P. A. Heimann, M. R. F. Siggel, D. A. Shirley, C. T. Chen, Y. Ma, S. Modesti, and F. Sette, *Chem. Phys. Lett.* **193**, 493 (1992).
112. G. Angonoa, O. Walter, and J. Schirmer, *J. Chem. Phys.* **87**, 6789 (1987).
113. J. P. Connerade and J. Hormes, *Z. Phys. D.* **4**, 3 (1986).
114. U. Kuetgens and J. Hormes, *Phys. Rev. A* **44**, 264 (1991).
115. S. Bodeur, P. Millié, E. Lizon à. Lugrin, I. Nenner, A. Filipponi, F. Boscherini, and S. Mobilio, *Phys. Rev. A* **39**, 5075 (1989).
116. E. M.-L. Ohrendorf, L. S. Cederbaum, and F. Tarantelli, *Phys. Rev. A* **44**, 205 (1991).
117. J. L. Ferrer, S. Bodeur, and I. Nenner, *J. Electron Spectrosc. Relat. Phenom.* **52**, 711 (1990).
118. S. Bodeur, C. Reynaud, K. Bisson, P. Millié and I. Nenner, in *Synchrotron Radiation and Dynamic Phenomena*, edited by A. J. Beswick (AIP, New York, 1992), pp. 300–308.
119. P. D'Angelo, A. Di Cicco, A. Filipponi, and N. V. Pavel, *Phys. Rev. A* **47**, 2055 (1993).
120. A. Filipponi, T. A. Tyson, K. O. Hodgson, and S. Mobilio, *Phys. Rev. A* **48**, 1328 (1993).
121. J. Nordgren and N. Wassdahl, *Phys. Scr.* **T31**, 103 (1989).
122. D. L. Walters and C. P. Bhalla, *Phys. Rev. A* **3**, 1919 (1971).
123. W. E. Moddemann, T. A. Carlson, M. O. Krause, and B. P. Pullen, *J. Chem. Phys.* **55**, 2317 (1971).
124. I. Nenner and J. A. Beswick in *Handbook on Synchrotron Radiation*, Vol. II, edited by G. V. Marr (North-Holland, Amsterdam, 1987), pp. 355–466.
125. W. Eberhardt, *Phys. Scr.* **T17**, 28 (1987).
126. D. M. Hanson, *Adv. Chem. Phys.* **77**, 1 (1990).
127. B. Crasemann in *The Physics of Electronic and Atomic Collisions*, edited by W. R. McGilliwray, and M. C. Strandage (Adam Hilger, Bristol, 1992), p. 69.
128. V. Schmidt, *Rep. Prog. Phys.* **55**, 1483–1659 (1992).
129. F. J. Wuilleumier, in *Fundamental Processes and Applications of Atoms and Ions*, edited by C. D. Lin (World Scientific, Singapore, 1993), pp. 283–356.
130. F. J. Wuilleumier, in *New Directions in Research with Third Generation Soft-X-ray Synchrotron Radiation Source* (Kluwer Academic, The Netherlands, 1994), pp. 47–102.
131. T. A. Carlson, *Photoelectron and Auger Spectroscopy* (Plenum Press, New York, 1975).
132. C. S. Fadley, in *Eletectron Spectroscopy: Theory, Techniques and Applications*, Vol. 2, edited by C. R. Brundle and A. D. Baker (Academic Press, New York, 1978), p. 1.
133. W. B. Perry and W. L. Jolly, *Inorg. Chem.* **13**, 1211 (1974).
134. T. D. Thomas, M. R. F. Siggel, and L. J. Saethre, *J. Electron Spectrosc. Relat. Phenom.* **51**, 417 (1990).
135. M. Coville and T. D. Thomas, *Phys. Rev. A* **43**, 6053 (1991).
136. E. Hartmann and R. Der, *J. Phys. B* **21**, 1751 (1988).
137. J. N. Cutler, and G. M. Bancroft, and K. H. Tan, *J. Chem. Phys.* **97**, 7932 (1992).
138. N. Chandra and S. Sen, *J. Chem. Phys.* **98**, 5242 (1993).
139. U. Becker and D. A. Shirley, Chapter 5 of this volume.
140. P. A. Heimann, U. Becker, H. G. Kerkhoff, B. Langer, D. Szostak, R. Wehlitz, D. W. Lindle, T. A. Ferrett, and D. A. Shirley, *Phys. Rev. A* **34**, 3782 (1986).
141. T. A. Carlson, M. O. Krause, F. A. Grimm, P. R. Keller, and J. W. Taylor, *Chem. Phys. Lett.* **87**, 552 (1982).
142. P. R. Keller, J. W. Taylor, F. A. Grimm, P. Senn, T. A. Carlson, and M. O. Krause, *Chem. Phys.* **74**, 247 (1983).
143. C. M. Truesdale, S. H. Southworth, P. H. Kobrin, U. Becker, D. W. Lindle, H. G. Kerkhoff, and D. A. Shirley, *Phys. Rev. Lett.* **50**, 1265 (1983).
144. D. W. Lindle, C. M. Truesdale, P. H. Kobrin, T. A. Ferrett, P. A. Heimann, U. Becker, H. G. Kerkhoff, and D. A. Shirley, *J. Chem. Phys.* **81**, 5375 (1984).
145. H. Kanamori, S. Iwata, A. Mikuni, and T. Sasaki, *J. Phys. B* **17**, 3887 (1984).
146. G. G. B. de Souza, P. Morin, and I. Nenner, *J. Chem. Phys.* **83**, 492 (1985).
147. G. G. B. de Souza, P. Morin, and I. Nenner, *J. Chem. Phys.* **90**, 7071 (1989).

148. T. A. Ferrett, P. A. Heimann, H. G. Kerkhoff, U. Becker, D. W. Lindle, and D. A. Shirley, *Chem. Phys. Lett.* **138**, 607 (1987).
149. M. N. Piancastelli, T. A. Ferrett, D. W. Lindle, L. J. Medhurst, P. A. Heimann, S. H. Liu, and D. A. Shirley, *J. Chem. Phys.* **90**, 3004 (1989).
150. F. A. Grimm, T. A. Carlson, J. Jiménez-Mier, B. Yates, J. W. Taylor, and B. P. Pullen, *J. Electron Spectrosc. Relat. Phenom.* **47**, 257 (1988).
151. M. Schmidbauer, A. L. D. Kilcoyne, K. J. Randall, J. Feldhaus, A. M. Bradshaw, M. Braunstein, and V. McKoy, *J. Chem. Phys.* **94**, 5299 (1991).
152. M. Schmidbauer, A. L. D. Kilcoyne, H.-M. Köppe, J. Feldhaus, and A. M. Bradshaw, *Chem. Phys. Lett.* **199**, 119 (1992).
153. A. L. D. Kilcoyne, M. Schmidbauer, A. Koch, K. J. Randall, and J. Feldhaus, *J. Chem. Phys.* **98**, 6735 (1993).
154. W. Habenicht, H. Baiter, K. Müller-Dethlefs, and E. W. Schlag, *Phys. Scr.* **41**, 814 (1990).
155. E. D. Poliakoff, L. A. Kelly, L. M. Duffy, B. Space, P. Roy, S. H. Southworth, and M. G. White, *Chem. Phys.* **129**, 65 (1989).
156. J. L. Dehmer and D. Dill, *Phys. Rev. Lett.* **35**, 213 (1975).
157. D. Dill, J. R. Swanson, S. Wallace, and J. L. Dehmer, *Phys. Rev. Lett.* **45**, 1393 (1980).
158. O. Hemmers, F. Heiser, J. Eiben, R. Wehlitz, and U. Becker, *Phys. Rev. Lett.* **71**, 987 (1993).
159. J. Schirmer, M. Braunstein, and B. V. McKoy, *Phys. Rev. A* **41**, 283 (1990).
160. D. L. Lynch, *J. Phys. Rev. A* **43**, 5176 (1991).
161. R. I. Hall, G. Dawber, A. McConkey, M. A. MacDonald, and G. C. King, *Phys. Rev. Lett.* **68**, 2751 (1992).
162. N. Correia, A. Flores-Riveros, H. Ågren, K. Helenlund, L. Asplund, and U. Gelius, *J. Chem. Phys.* **83**, 2035 (1985).
163. T. X. Carroll and T. D. Thomas, *J. Electron Spectrosc. Relat. Phenom.* **10**, 215 (1977).
164. F. P. Larkins, *J. Electron Spectrosc. Relat. Phenom.* **51**, 115 (1990).
165. F. Tarantelli, A. Sgamellotti, and L. S. Cederbaum, *Applied Many-Body Methods in Spectroscopy and Electronic Structure*, edited by D. Mukherjee (Plenum Press, New York, 1992), pp. 57–104.
166. L. S. Cederbaum, F. Tarantelli, A. Sgamellotti, and J. Schirmer, *J. Chem. Phys.* **85**, 6513 (1986).
167. L. S. Cederbaum, F. Tarantelli, A. Sgamellotti, and J. Schirmer, *J. Chem. Phys.* **86**, 2168 (1987).
168. J. A. D. Matthew and Y. Komninos, *Surf. Sci.* **53**, 716 (1975).
169. T. D. Thomas and P. Weightman, *Chem. Phys. Lett.* **81**, 325 (1981).
170. J. A. Kelber, D. R. Jennison, and R. R. Rye, *J. Chem. Phys.* **75**, 652 (1981).
171. M. Cini, F. Maracci, and R. Platania, *J. Electron. Spectrosc. Relat. Phenom.* **41**, 37 (1986).
172. F. P. Larkins, *J. Chem. Phys.* **86**, 3239 (1987).
173. I. H. Hillier and J. Kendrick, *Mol. Phys.* **31**, 849 (1976).
174. T. Ågren and H. Siegbahn, *Chem. Phys. Lett.* **72**, 498 (1980).
175. D. R. Jennison, *Phys. Rev. A* **23**, 1215 (1981).
176. G. E. Laramore, *J. Phys. Rev. A* **29**, 23 (1984).
177. C.-M. Liegener, *Chem. Phys. Lett.* **106**, 201 (1984).
178. L. S. Cederbaum, P. Campos, F. Tarantelli, and A. Sgamellotti, *J. Chem. Phys.* **95**, 6634 (1991).
179. N. Correia, A. Naves de Brito, M. P. Keane, L. Karlsson, S. Svensson, C.-M. Liegener, A. Cesar, and H. Ågren, *J. Chem. Phys.* **95**, 5187 (1991).
180. D. Minelli, F. Tarantelli, A. Sgamellotti, and L. S. Cederbaum, *J. Chem. Phys.* **99**, 6688 (1993).
181. R. W. Shaw and T. D. Thomas, *Phys. Rev. A* **11**, 1491 (1975).
182. M. Hotokka, H. Ågren, H. Aksela, and S. Aksela, *Phys. Rev. A* **30**, 1855 (1984).
183. K. Zähringer, H.-D. Meyer, L. S. Cederbaum, F. Tarantelli, and A. Sgamellotti, *Chem. Phys. Lett.* **206**, 247 (1993).
184. L. S. Cederbaum and F. Tarantelli, *J. Chem. Phys.* **98**, 9691 (1993).
185. L. S. Cederbaum and F. Tarantelli, *J. Chem. Phys.* **99**, 5871 (1993).
186. F. Tarantelli, A. Sgamellotti, and L. S. Cederbaum, *J. Chem. Phys.* **94**, 523 (1991).
187. F. Tarantelli, A. Sgamellotti, and L. S. Cederbaum, *Phys. Rev. Lett.* **72**, 428 (1994).
188. F. Tarantelli and L. S. Cederbaum, *Phys. Rev. Lett.* **71**, 649 (1993).
189. R. R. Rye and J. E. Houston, *Acc. Chem. Res.* **17**, 41 (1984).
190. S. Aksela, K. H. Tan, H. Aksela, and G. M. Bancroft, *Phys. Rev. A* **33**, 258 (1986).
191. F. P. Larkins, J. McColl, and E. Z. Chelkowska, *J. Electron Spectrosc. Relat. Phenom.* **67**, 275 (1994).
192. W. Eberhardt, G. Kalkoffen, and C. Kunz, *Phys. Rev. Lett.* **41**, 156 (1978).
193. D. W. Lindle, P. A. Heimann, T. A. Ferrett, M. N. Piancastelli, and D. A. Shirley, *Phys. Rev. A* **35**, 4605 (1987).

194. T. ÅBERG AND J. TULKKI, in *Atomic Inner-Shell Physics*, edited by B. Crasemann (Plenum Press, New York, 1985), Chapter 10.
195. L. UNGIER AND T. D. THOMAS, *Chem. Phys. Lett.* **96**, 247 (1983).
196. L. UNGIER AND T. D. THOMAS, *J. Chem. Phys.* **82**, 3146 (1985).
197. W. EBERHARDT, E. W. PLUMMER, C. T. CHEN, AND W. K. FORD, *Aust. J. Phys.* **39**, 853 (1986).
198. H.-J. FREUND, AND C.-M. LIEGENER, *Chem. Phys. Lett.* **134**, 70 (1987).
199. J. SCHIRMER AND O. WALTER, *Chem. Phys.* **78**, 201 (1983).
200. G. M. BANCROFT, S. AKSELA, H. AKSELA, K. H. TAN, B. W. YATES, L. L. COATSWORTH, AND J. S. TSE, *J. Chem. Phys.* **84**, 5 (1986).
201. G. G. B. DE SOUZA, P. MORIN, AND I. NENNER, *Phys. Rev. A.* **34**, 4770 (1986).
202. T. A. CARLSON, D. R. MULLINS, C. E. BEALL, B. W. YATES, J. W. TAYLOR, AND F. A. GRIMM, *J. Chem. Phys.* **89**, 4490 (1988).
203. G. M. BANCROFT, K. H. TAN, O.-P. SAIRANEN, S. AKSELA, AND H. AKSELA, *Phys. Rev. A* **41**, 3716 (1990).
204. P. MORIN, G. G. P. DE SOUZA, I. NENNER, AND P. LABLANQUIE, *Phys. Rev. Lett.* **56**, 131 (1986).
205. K. UEDA, H. CHIBA, Y. SATO, T. HAYAISHI, E. SHIGEMASA, AND A. YAGISHITA, *Phys. Rev. A* **46**, R5 (1992).
206. F. P. LARKINS, W. EBERHARDT, IN WHAN LYO, R. MURPHY, AND E. W. PLUMMER, *J. Chem. Phys.* **88**, 2948 (1988).
207. W. EBERHARDT, R. DUDDE, M. L. M. ROCCO, E. E. KOCH, AND S. BERNSTORFF, *J. Electron Spectrosc. Relat. Phenom.* **51**, 373 (1990).
208. S. STRANGES, M. Y. ADAM, C. CAULETTI, M. DE SIMONE, C. FURLANI, M. N. PIANCASTELLI, P. DECLEVA, AND A. LISINI, *J. Chem. Phys.* **97**, 4764 (1992).
209. S. STRANGES, M. Y. ADAM, M. DE SIMONE, P. DECLEVA, A. LISINI, C. CAULETTI, M. N. PIANCASTELLI, AND C. FURLANI, *J. Chem. Phys.* **102**, 3555 (1995).
210. T. X. CARROLL, M. COVILLE, P. MORIN, AND T. D. THOMAS, *J. Chem. Phys.* **101**, 998 (1994).
211. D. JI AND T. D. THOMAS, *J. Electron Spectrosc. Relat. Phenom.* **67**, 233 (1994).
212. G. ILLING, T. PORWOL, I. HEMMERICH, G. DÖMÖTÖR, H. KUHLENBECK, H.-J. FREUND, C.-M. LIEGENER, AND W. VON NIESSEN, *J. Electron Spectrosc. Relat. Phenom.* **51**, 149 (1990).
213. D. MENZEL, G. ROCKER, H.-P. STEINRUCK, D. COULMAN, P. A. HEIMANN, W. HUBER, P. ZEBISCH, AND D. R. LLOYD, *J. Chem. Phys.* **96**, 1724 (1992).
214. G. B. ARMEN, T. ÅBERG, J. C. LEVIN, B. CRASEMANN, M. H. CHEN, G. E. ICE, AND G. S. BROWN, *Phys. Rev. Lett.* **54**, 1142 (1985).
215. Z. F. LIU, G. M. BANCROFT, K. H. TAN, AND M. SCHACHTER, *J. Electron Spectrosc. Relat. Phenom.* **67**, 299 (1994).
216. T. A. CARLSON AND M. O. KRAUSE, *Phys. Rev. Lett.* **14**, 390 (1965).
217. T. A. CARLSON AND M. O. KRAUSE, *Phys. Rev. Lett.* **17**, 1079 (1966).
218. M. O. KRAUSE, T. A. CARLSON, AND R. D. DISMUKES, *Phys. Rev. A* **170**, 37 (1968).
219. M. O. KRAUSE AND T. A. CARLSON, *Phys. Rev.* **158**, 18 (1967).
220. T. A. CARLSON, in *Chemical Effects of Nuclear Transformation in Inorganic Systems*, edited by G. Harbottle, and A. G. Maddock (North-Holland, Amsterdam, 1979), p. 13.
221. U. BECKER AND R. WEHLITZ, *J. Electron Spectrosc. Relat. Phenom.* **67**, 341 (1994).
222. I. NENNER, P. MORIN, AND P. LABLANQUIE, *Comments At. Mol. Phys.* **22**, 51 (1988).
223. G. DUJARDIN, L. HELLNER, D. WINKOUN, AND M. J. BESNARD, *Chem. Phys.* **105**, 291 (1986).
224. A. P. HITCHCOCK, P. LABLANQUIE, P. MORIN, E. LIZON À. LUGRIN, M. SIMON, P. THIRY, AND I. NENNER, *Phys. Rev. A* **37**, 2448 (1988).
225. R. B. KAY, P. E. VAN DER LEEUW, AND M. J. VAN DER WIEL, *J. Phys. B* **10**, 2521 (1977).
226. R. W. WETMORE, R. J. LE ROY, AND R. K. BOYD, *J. Phys. Chem.* **88**, 6318 (1984).
227. P. LABLANQUIE, J. DELWICHE, M.-J. HUBIN-FRANSKIN, I. NENNER, P. MORIN, K. ITO., J. H. D. ELAND, J.-M. ROBBE, G. GANDARA, J. FOURNIER, AND P. G. FOURNIER, *Phys. Rev. A* **40**, 5673 (1989).
228. M. LARSSON, B. J. OLSSON, AND P. SIGRAY, *Chem. Phys.* **139**, 457 (1989).
229. N. LEVASSEUR, P. MILLIÉ, P. ARCHIREL, AND B. LEVY, *Chem. Phys.* **153**, 387 (1991).
230. R. MAYER, D. W. LINDLE, S. H. SOURTHWORTH, AND P. L. COWAN, *Phys. Rev. A* **43**, 235 (1991).
231. T. A. CARLSON AND R. M. WHITE, *J. Chem. Phys.* **48**, 5191 (1968).
232. T. A. CARLSON AND M. O. KRAUSE, *J. Chem. Phys.* **56**, 3206 (1972).
233. D. A. LAPIANO-SMITH, C. I. MA, K. T. WU, AND D. M. HANSON, *J. Chem. Phys.* **90**, 2162 (1989).
234. S. NAGAOKA, I. KOYANO, K. UEDA, E. SHIGEMASA, Y. SATO, A. YAGISHITA, T. NAGATA, AND T. HAYAISHI, *Chem. Phys. Lett.* **154**, 363 (1989).
235. K. UEDA, Y. SATO, S. NAGAOKA, I. KOYANO, A. YAGISHITA, AND T. HAYAISHI, *Chem. Phys Lett.* **170**, 389 (1990).

236. K. Ueda, E. Shigemasa, Y. Sato, S. Nagaoka, I. Koyano, A. Yaghishita, T. Nagata, and T. Hayaishi, *Chem. Phys. Lett.* **154**, 357 (1989).
237. K. Ueda, E. Shigemasa, Y. Sato, S. Nagaoka, I. Koyano, A. Yaghishita, and T. Hayaishi, *Chem. Phys. Lett.* **166**, 391 (1990).
238. K. Ueda, E. Shigemasa, Y. Sato, S. Nagaoka, I. Koyano, A. Yagishita, and T. Hayaishi, *Phys. Scr.* **41**, 78 (1990).
239. D. W. Lindle, W. Manner, L. Steinbeck, E. Villalobos, J. C. Levin, and I. A. Sellin, *J. Electron Spectrosc. Relat. Phenom.* **67**, 373 (1994).
240. M. O. Krause and J. H. Oliver, *J. Phys. Chem. Ref. Data* **8**, 329 (1979).
241. J. N. Cutler, G. M. Bancroft, D. G. Sutherland, and K. H. Tan, *Phys. Rev. Lett.* **67**, 1531 (1991).
242. L. Karlsson, S. Svensson, P. Baltzer, M. Carlsson-Göthe, M. P. Keane, A. Naves de Britos, N. Correia, and B. Wannberg, *J. Phys. B* **22**, 3001 (1989).
243. E. D. Poliakoff, *Vacuum Ultraviolet Photoionization and Photodissociation of Molecules and Clusters*, edited by C. Y. Ng (World Scientific, Singapore, 1991), pp. 345–377.
244. M. Mahalingham, K. Lee, and D. M. Hanson, *J. Chem. Phys.* **98**, 5239 (1993).
245. R. M. Bowman, M. Dantus, and A. H. Zewail, *Chem. Phys. Lett.* **161**, 297 (1989).
246. T. Baumert, M. Grosser, R. Thalweiser, and G. Gerber, *Phys. Rev. Lett.* **67**, 3753 (1991).
247. A. H. Zewail, in *Proceedings of the 10th Vacuum Ultraviolet Radiation Physics Conference*, Paris, July 27–31, 1992, edited by F. J. Wuilleumier, Y. Petroff, and I. Nenner (World Scientific, Singapore, 1993), pp. 20–28.
248. M. Tronc, G. C. King, and F. H. Read, *J. Phys. B* **12**, 137 (1979).
249. D. A. Shaw, G. C. King, F. H. Read, and D. Cvejanović, *J. Phys. B* **15**, 1785 (1982).
250. A. P. Hitchcock and C. E. Brion, *J. Electron Spectrosc. Relat. Phenom.* **10**, 317 (1977).
251. K. Lee, D. Y. Kim, C.-I. Ma, and D. M. Hanson, *J. Chem. Phys.* **100**, 8550 (1994).
252. K. P. Huber and G. Herzberg, *Molecular Spectra and Molecular Structure*, Vol. 4 (Van Nostrand Reinhold, New York, 1979).
253. D. A. Shaw, G. C. King, and F. H. Read, *J. Phys. B* **13**, L723 (1980).
254. D. A. Shaw, D. Cvejanović, G. C. King, and F. H. Read, *J. Phys. B* **17**, 1173 (1984).
255. W. Hayes and F. C. Brown, *Phys Rev. A* **6**, 21 (1972).
256. H. Friedrich, B. Sonntag, P. Rabe, W. Butscher, and W. H. E. Schwarz, *Chem. Phys. Lett.* **64** 360 (1979).
257. J. Shirmer, A. B. Trofimov, K. J. Randall, J. Feldhaus, A. M. Bradshaw, Y. Ma, C. T. Chen, and F. Sette, (unpublished) cited in A. M. Bradshaw, *Proceedings of the 10th International Conference on Vacuum Ultra-Violet and Radiation Physics*, Paris, July 27–31 1992, edited by F. J. Wuilleumier, Y. Petroff, and I. Nenner (World Scientific, Singapore, 1993), pp. 29–42.
258. H. Rabus, D. Arvanitis, M. Domke, and K. Baberschke, *J. Chem. Phys.* **96**, 1560 (1992).
259. F. X. Gadea, H. Köppel, J. Schirmer, L. S. Cederbaum, K. J. Randall, A. M. Bradshaw, Y. Ma, F. Sette, and C. T. Chen, *Phys. Rev. Lett.* **66**, 883 (1991).
260. M. P. de Miranda, J. A. Beswick, P. Parent, C. Laffon, G. Tourillon, A. Cassuto, G. Nicolas, and F. X. Gadea, *J. Chem. Phys.* **101**, 5500 (1994).
261. J. D. Bozek, G. M. Bancroft, J. N. Cutler, and K. H. Tan, *Phys. Rev. Lett.* **65**, 2757 (1990).
262. J. D. Bozek, G. M. Bancroft, and K. H. Tan, *Phys. Rev. A* **43**, 3597 (1991).
263. K. Randall, J. Feldhaus, W. Erlebach, A. M. Bradshaw, W. Eberhardt, Z. Xu, Y. Ma, and P. D. Johnson, *Synchrotron Radiat. News* **4**, 16 (1991).
264. L. Asplund, U. Gelius, S. Hedmann, K. Helenelund, K. Siegbahn, and P. E. M. Siegbahn, *J. Phys. B* **18**, 1569 (1985).
265. D. G. J. Sutherland, G. M. Bancroft, and K. H. Tan, *Surf. Sci. Lett.* **262**, L96 (1992).
266. D. G. J. Sutherland, G. M. Bancroft, and K. H. Tan, *J. Chem. Phys* **97**, 7918 (1992).
267. J. N. Cutler, G. M. Bancroft, and K. H. Tan, *J. Phys. B* **24**, 4897 (1991).
268. H. Köppel, L. S. Cederbaum, and W. Domcke, *J. Chem. Phys.* **89**, 2023 (1988).
269. F. Kaspar, W. Domcke, and L. S. Cederbaum, *Chem. Phys.* **44**, 33 (1979).
270. T. X. Carrol, S. E. Anderson, L. Ungier, and T. D. Thomas, *Phys. Rev. Lett.* **58**, 867 (1987).
271. A. Flores-Riveros, N. Correia, H. Ågren, L. Pettersson, M. Bäckström, and J. Nordgren, *J. Chem. Phys.* **83**, 2053 (1985).
272. R. Murphy, In-Whan Lyo, and W. Eberhardt, *J. Chem. Phys.* **88**, 6078 (1988).
273. W. Eberhardt, J.-E. Rubensson, K. J. Randall, J. Feldhaus, A. L. D. Kilcoyne, A. M. Bradshaw, Z. Xu, P. D. Johnson, and Y. Ma, *Phys. Scr.* **T41**, 143 (1992)
274. M. Neeb, J.-E. Rubensson, M. Biermann, and W. Eberhardt, *J. Electron Spectrosc. Relat. Phenom.* **67**, 261 (1994).
275. T. D. Thomas and T. X. Carroll, *Chem. Phys. Lett.* **185**, 31 (1991).

276. J. NORDGREN AND H. ÅGREN, *Comments At. Mol. Phys.* **14**, 203 (1984).
277. E. D. POLIAKOFF, L. A. KELLY, L. M. DUFFY, B. SPACE, P. ROY, S. H. SOUTHWORTH, AND M. G. WHITE, *J. Chem. Phys.* **89**, 4048 (1988).
278. A. P. HITCHCOCK AND C. E. BRION, *J. Electron. Spectrosc. Relat. Phenom.* **13**, 193 (1978).
279. W. H. E. SCHWARZ, *Chem. Phys.* **11**, 217 (1975).
280. V. ČERMÁK AND J. ŠRÁMEK, *J. Electron Spectrosc. Relat. Phenom.* **2**, 97 (1973).
281. P. MORIN AND I. NENNER, *Phys. Rev. Lett.* **56**, 1913 (1986).
282. I. NENNER, P. MORIN, M. SIMON, P. LABLANQUIE, AND G. G. G. DE SOUZA, in *Desorption Induced by Electronic Transitions*, DIET III, edited by R. H. Stulen and M. L. Knotek (Springer-Verlag, 1988), pp. 10–31.
283. P. MORIN AND I. NENNER, *Phys. Scr.* **T17**, 171 (1987).
284. H. AKSELA, S. AKSELA, M. ALA-KORPELA, O.-P. SAIRAINEN, M. HOTOKKA, G. M. BANCROFT, K. H. TAN AND J. TULKKI, *Phys. Rev. A* **41**, 6000 (1990); H. AKSELA, S. AKSELA, M. HOTOKKA, A. YAGISHITA, AND E. SHIGEMASA, *J. Phys. B* **25**, 3357 (1992).
285. A. MENZEL, O. HEMMERS, B. LANGER, R. WEHLITZ, AND U. BECKER, private communication (1994).
286. R. THISSEN, M. SIMON, AND M.-J. HUBIN-FRANSKIN, *J. Chem Phys.* **101**, 7548 (1994).
287. S. SVENSSON, L. KARLSSON, N. MÅRTENSSON, P. BALTZER, AND B. WANNBERG, *J. Electron Spectrosc. Relat. Phenom.* **50**, 1 (1990).
288. H. AKSELA, S. AKSELA, A. NAVES DE BRITO, G. M. BANCROFT, AND K. H. TAN, *Phys. Rev. A* **45**, 7948 (1992).
289. A. NAVES DE BRITO AND H. ÅGREN, *Phys. Rev. A* **45**, 7953 (1992).
290. P. MORIN, M. LAVOLLÉE, AND M. SIMON, in *Proceedings of the 10th International Conference on Vacuum Ultra-Violet and Radiation Physics*, Paris, July, 27–31, 1992, edited by F. J. Wuilleumier, Y. Petroff, and I. Nenner (World Scientific, Singapore, 1993), pp. 211–225
291. H. AKSELA, S. AKSELA, O.-P. SAIRAINEN, A. KIVIMÄKI, G. M. BANCROFT, AND K. H. TAN, *Phys. Scr.* **T41**, 122 (1992).
292. S. J. SCHAPHORST, C. D. CALDWELL, M. O. KRAUSE, AND J. JIMÉNEZ-MIER, *Chem. Phys. Lett.* **213**, 315 (1993).
293. D. COULMAN, A. PUSCHMANN, W. WURTH, H.-P. STEINRÜCK, AND D. MENZEL, *Chem. Phys. Lett.* **148**, 371 (1988).
294. D. COULMAN, A. PUSCHMANN, U. HÖFER, H.-P. STEINRÜCK, W. WURTH, P. FEULNER, AND D. MENZEL, *J. Chem. Phys.* **93**, 58 (1990).
295. P. KUIPER AND B. I. DUNLAP, *J. Chem. Phys.* **100**, 4087 (1994).
296. A. A. CAFOLLA, T. REDDISH, AND J. COMER, *J. Phys. B* **22**, L273 (1989).
297. T. LEBRUN, M. LAVOLLÉE, AND P. MORIN, in *X-ray and Inner-Shell Processes*, Knoxville, TN, edited by T. A. Carlson, M. O. Krause, and S. T. Manson, p. 846.
298. I. NENNER, P. MORIN, P. LABLANQUIE, M. SIMON, N. LEVASSEUR, AND P. MILLIÉ, *J. Electron. Spectrosc. Relat. Plenom.* **52**, 623 (1990).
299. A. BANICHEVICH, S. D. PEYERIMHOFF, M. C. VAN HEMERT, AND P. G. FOURNIER, *Chem. Phys.* **121**, 351 (1988).
300. A. BANICHEVICH, S. D. PEYERIMHOFF, B. A. HESS, AND M. C. VAN HEMERT, *Chem. Phys.* **154**, 199 (1991).
301. U. BECKER, O. HEMMERS, B. LANGER, A. MENZEL, R. WEHLITZ, AND W. B. PEATMAN, *Phys. Rev. A* **45**, R1295 (1992).
302. T. A. CARLSON, in *Desorption Induced Transitions*, DIET I, edited by N. H. Tolk, M. M. Traum, J. C. Tully, and T. E. Madey (Springer-Verlag, Berlin, 1983), p. 169.
303. T. URISU, H. KYURAGI, Y. UTSUMI, J. I. TAKAHASHI, AND M. KITAMURA, *Rev. Sci. Instrum.* **60**, 2157 (1989).
304. F. CERRINA, B. LAI, G. M. WELLS, J. R. WILEY, D. G. KILDAY, AND G. MARGARITONDO, *Appl. Phys. Lett.* **50**, 533 (1987).
305. C.-R. WEN AND R. A. ROSENBERG, *Surf. Sci.* **218**, L483 (1989).
306. K. HIEDA, A. AZAMI, M. SUZUKI, H. MAEZAWA, Y. FURUZAWA, AND K. KOBAYASHI, Photon Factory Activity Report, p. 286 (1987).
307. A. SVENSSON, J. BORDAS, E. A. HUGHES, AND G. MANT, Conference Proceedings "Biophysics and Synchrotron Radiation" (Tsukuba Oxford Press, 1992).
308. I. NENNER, in *Giant Resonances in Atoms, Molecules and Solids*, edited by J. P. Connerade, J. M. Esteva, and R. C. Karnatak (Plenum Press, New York, 1987), pp. 259–280.
309. I. NENNER, in *Electronic and Atomic Collisions*, edited by H. B. Gilbody, W. R. Newell, F. H. Read, and A. C. H. Smith (Elsevier, Amsterdam, 1988), pp. 517–532.
310. P. MORIN, M. LAVOLLÉE, M. MEYER, AND M. SIMON, in *Proceedings of the XVIII International Conference on the Physics of Electronic and Atomic Collisions*, ICPEAC, 1993, edited by T. Andersen, B. Fastrup, F. Folkmann, H. Knudsen and N. Andersen, pp. 139–151 (1994).
311. G. DUJARDIN, S. LEACH, O. DUTUIT, P.-M. GUYON, AND M. RICHARD-VIARD, *Chem. Phys.* **88**, 339 (1984).
312. W. C. WILEY AND I. H. MCLAREN, *Rev. Sci. Instrum.* **26**, 1150 (1955).

313. Y. Sato, K. Ueda, A. Yagishita, T. Sasaki, T. Nagata, T. Hayaishi, M. Yoshino, T. Koisumi, Y. Itoh, and A. A. MacDowell, *Phys. Scr.* **41**, 55 (1990).
314. S. Nagaoka, J. Ohshita, M. Ishikawa, T. Masuoka, and I. Koyano, *J. Phys. Chem.* **97**, 1488 (1993).
315. P. Lablanquie, A. C. A. Souza, G. G. B. de Souza, P. Morin, and I. Nenner, *J. Chem. Phys.* **90**, 7078 (1989).
316. R. Thissen, M.-J. Hubin-Franskin, M. Furlan, J.-L. Piette, P. Morin, and I. Nenner, *Chem. Phys. Lett.* **199**, 102 (1992).
317. L. J. Frasinski, M. Stankiewicz, K. J. Randall, P. A. Hatherley, and K. Codling, *J. Phys. B* **19**, L819 (1986).
318. J. H. D. Eland, F. S. Wort, and R. N. Royds, *J. Electron. Spectrosc. Relat. Phenom.* **41**, 297 (1986); J.H.D. Eland, *Mol. Phys.* **61**, 725 (1987).
319. M. Simon, T. LeBrun, P. Morin, M. Lavollée, and J. L. Maréchal, *Nucl. Instrum. Methods* **B62**, 167 (1991).
320. M. Simon, Thèse de Doctorat, Orsay University (unpublished, 1991).
321. T. LeBrun, Thèse de Doctorat, Université de Paris-sud, Orsay (1991).
322. M. Lavollée and H. Bergeron, *J. Phys. B* **25**, 3101 (1992).
323. M. Simon, T. Lebrun, R. Martins, G. G. B. de Souza, I. Nenner, M. Lavollée, and P. Morin, *J. Phys. Chem.* **97**, 5228 (1993).
324. K. Müller-Dethlefs, M. Sander, L. A. Chewter, and E. W. Schlag, *J. Chem. Phys.* **85**, 5755 (1986).
325. W. Habenicht, H. Baiter, K. Müller-Dethlefs, and E. W. Schlag, *J. Phys. Chem.* **95**, 6774 (1991).
326. Y. Sato, K. Ueda, H. Chiba, E. Shigemasa, and A. Yagishita, *Chem. Phys. Lett.* **196**, 475 (1991).
327. D. M. Hanson, C. I. Ma, K. Lee, D. Lapiano-Smith, and D. Y. Kim, *J. Chem. Phys.* **93**, 9200 (1990).
328. C. I. Ma, K. Lee, D. Ji, D. Y. Kim, and D. M. Hanson, *Nucl. Instrum. Methods A* **347**, 453 (1994).
329. L. Ferand-Tanaka, M. Simon, M. Lavollée, T. Thissen, and P. Morin, *Nucl. Instrum. Methods A* **66** 1587 (1995); *Rev. Sci. Instrum.* **66**, 1587 (1995).
330. J. H. D. Eland, M. Devoret, and S. Leach, *Chem Phys. Lett.* **43**, 97 (1976).
331. M. Meyer, J. Lacoursiére, M. Simon, P. Morin, and M. Larzillière, *Chem. Phys.* **187**, 143 (1994).
332. J. Murakami, M. C. Nelson, S. L. Anderson, and D. M. Hanson, *J. Chem. Phys.* **85**, 5755 (1986).
333. N. Levasseur and P. Millié, *J. Chem. Phys.* **92**, 2974 (1990).
334. T. LeBrun, M. Lavollée, M. Simon, and P. Morin, *J. Chem. Phys.* **98**, 2534 (1993).
335. Z. Vager, R. Naaman, and E. P. Kanter, *Science* **244**, 426 (1989).
336. M. P. de Miranda and J. A. Beswick, unpublished (1994).
337. J. H. D. Eland, *Chem. Phys. Lett.* **203**, 354 (1993).
338. G. Sampoll, R. L. Watson, O. Heber, V. Horvat, K. Wohrer, and M. Chabot, *Phys. Rev. A* **45**, 2903 (1992).
339. D. Mathur, E. Krishnakumar, K. Nagesha, V. R. Marathe, V. Krishnamurti, F. A. Rajgara, and U. T. Raheja, *J. Phys. B* **26**, L141 (1993).
340. I. Ben-Itzhak, S. G. Ginther, and K. D. Carnes, *Phys. Rev. A* **47**, 2827 (1993).
341. L. J. Frasinski, K. Codling, P. Hatherly, J. Barr, I. N. Ross, and W. T. Toner, *Phys. Rev. Lett.* **58**, 2424 (1987).
342. C. Cornaggia, M. Schmidt, and D. Normand, *J. Phys. B* **27**, L123 (1994).
343. L. J. Frasiński, K. Codling, and P. Hatherly, *Science* **246**, 1029 (1989).
344. R. A. Rosenberg, C.-R. Wen, J.-M. Chen, and K. Tan, *Phys. Scr.* **41**, 475 (1990).
345. G. Dujardin, L. Hellner, B. J. Olsson, M. J. Besnard-Ramage, and A. Dadouch, *Phys. Rev. Lett.* **62**, 745 (1989).
346. I. Nenner, P. Morin, M. Meyer, J. Lacoursière, and L. Nahon, in *New Directions in Research with Third Generation Soft-X-ray Synchrotron Radiation Sources*, edited by A. S. Schlachter and F. J. Wuilleumier, (Kluwer Academic, The Netherlands, 1994), pp. 129–160.
347. R. Murphy and W. Eberhardt, *J. Chem. Phys.* **89**, 4054 (1988).
348. R. G. Hayes and W. Eberhardt, *Chem. Phys.* **94**, 397 (1991).
349. R. A. Rosenberg, C.-R. Wen, K. Tan, and J.-M. Chen, *J. Chem. Phys.* **92**, 5196 (1990).
350. P. Morin, T. LeBrun, and P. Lablanquie, *J. Chim. Phys. (France)* **86**, 1833 (1989).
351. P. Morin, T. LeBrun, and P. Lablanquie, *Bull. Sté R. Sci. Liège* **58**, 135 (1989).
352. H. C. Schmelz, C. Reynaud, M. Simon and, I. Nenner, *J. Chem. Phys.* **101**, 3742 (1994).
353. I. Nenner, P. Morin, P. Lablanquie, M. Simon, N. Levasseur, and P. Millié, *J. Electron Spectrosc. Relat. Phenom.* **52**, 623 (1990).
354. E. Rühl, C. Heinzel, H. Bäumgartel, and A. P. Hitchcock, *Chem. Phys.* **169**, 243 (1993).
355. M. Simon, M. Lavollée, T. LeBrun, J. Delwiche, M. J. Hubin-Franskin, and P. Morin, in *Synchrotron Radiation and Dynamic Phenomena*, edited by J. A. Beswick, AIP, Conf. Proc. No. 258, pp. 323–331 (1992).
356. S. Nagaoka, S. Suzuki, and I. Koyano, *Phys. Rev. Lett.* **58**, 1524 (1987).
357. T. Imamura, C. E. Brion, I. Koyano, T. Ibuki, and T. Masuoka, *J. Chem. Phys.* **94**, 4936 (1991).

358. M. C. Nelson, J. Murakami, S. L. Anderson, and D. M. Hanson, *J. Chem. Phys.* **86**, 4442 (1987).
359. M. Simon, M. Lavollee, P. Morin, and I. Nenner, *J. Phys. Chem.* **99**, 1733 (1994).
360. I. Waller and J. Hepburn, *J. Chem. Phys.* **88**, 6658 (1988).
361. K. Norwood, A. Ali, G. D. Flesch, and C. Y. Ng, *J. Am. Chem. Soc.* **112**, 7502 (1990).
362. J. H. D. Eland and D. A. Hagan, *Int. J. Mass Spectrom. Ion Phys.* **100**, 489 (1990).
363. J. H. D. Eland, *Laser Chem.* **11**, 259 (1993).
364. J. H. D. Eland, L. A. Coles, and H. Bountra, *Int. J. Mass Spectrom.* **89**, 265 (1989).
365. L. Szepes and T. Baer, *J. Am. Chem. Soc.* **106**, 273 (1984).
366. I. Salman, J. Silberstein, and R. D. Levine, *J. Phys. Chem.* **95**, 6781 (1991).

CHAPTER 10

RESONANCES AND NEAR-THRESHOLD PROCESSES

G. C. KING AND K.-H. SCHARTNER

1. INTRODUCTION

Something interesting often happens when a reaction mechanism is excited close to its threshold. The process of photoionization is no exception. Here the incident photon excites the ionic state just above its threshold so that the outgoing electron has a very small, nominally zero energy, and this can have profound influence on the dynamics of the process. In particular, electron correlations can become dominant.

A good example of the effects of electron correlation is the formation of satellite states. These states occur in photoionization when one electron is ejected from the atom and another is raised to an unfilled orbital, e.g., Ar^+ $3s^2 3p^4 nl$. They only occur because of electron correlations and so have received much theoretical and experimental attention (e.g., Refs. 1–3 and references therein). It is of interest to understand, for example, the relative intensities of these satellite states compared to that of the main line (which corresponds to just the ejection of an electron) and how this ratio at threshold compares with its value for higher incident photon energies, allowing a discrimination between "intrinsic" and "dynamic" correlations.[4]

A further class of double excitation that is dominated by electron correlations occurs when two electrons are promoted to bound orbitals, e.g., Ar $3s^2 3p^4 nl n'l'$. These double excited neutral states are also called resonances. Each satellite state may be considered as the limit of a doubly excited Rydberg series, e.g., $3s^2 3p^4 3dnl$ in argon. As a consequence, photoionization near a threshold is closely related to these resonances. Numerous satellites occur in atoms and so there exists the possibility of rich resonance structure, where resonances of one Rydberg series may overlap with the threshold region of another ionization channel. While the decay of single excited neutral states like $3s 3p^6 nl$ in Ar produces the well-known interference effects in the direct ionization of the outer-shell p electrons, the double excited neutral states show a strong influence on the energy dependence of the outer-shell s-electron photoionization.

This resonance structure is one subject of the present chapter. Data obtained from recent applications of photoelectron spectroscopy (PES) in the constant ion state (CIS) mode and of photon-induced fluorescence spectroscopy (PIFS) are presented and discussed. One useful advantage of both techniques is that it will be possible to study selective decay of the resonances into different ionic states. Furthermore applying the PIFS technique, the thresholds of the rare-gas outer-shell s electrons could be scanned for the first time.

Single-photon double photoionization is a process which is extremely sensitive to the physics of electron correlations, especially at its threshold where the two electrons have to leave the remaining ion with very low energy. The physical concepts governing this process were first discussed in a pioneering paper by Wannier.[5] Since then, considerable effort has been put into, on the one hand,

---

G. C. KING • Department of Physics and Astronomy, Schuster Laboratory, The University of Manchester, Manchester M13 9PL, United Kingdom. K.-H. SCHARTNER • I. Physikalisches Institut der Justus-Liebig-Universität Giessen, 35392 Giessen, Germany,

VUV and Soft X-Ray Photoionization. Edited by Uwe Becker and David A. Shirley. Plenum Press, New York, 1996.

extending this theory or developing quantal analogues and on the other hand experimentally verifying the predictions and determining their range of validity (e.g., Refs. 6,7). Since the electron correlations are strongest near threshold, the study of near-zero-energy photoelectrons provides a most stringent test of the theory describing single-photon double photoionization. On the other hand, the cross section for this channel is zero at threshold and in consequence a technique with very high collection efficiency and sensitivity is needed. Both demands are fulfilled by threshold photoelectron spectroscopy (TPES) which also provides very high photoelectron resolution. The use of TPES to study both single-photon double photoionization and related coincidence techniques and also single photoionization is also discussed in this chapter.

2. THRESHOLD PHOTOELECTRON SPECTROSCOPY

2.1. Principles and Experimental Methods

2.1.1. Objectives of TPES. In TPES the outgoing electron of interest has a very low energy (<<1 eV), and it is required to collect these threshold photoelectrons and discriminate them from any higher-energy photoelectrons. It may be that angular discrimination of the ejected electrons is also required when, for example, the finer details of the process are being studied, e.g., in the measurement of the asymmetry parameter, β. Alternatively, it may be necessary to collect as many threshold electrons as possible. Clearly, if angular discrimination can be dispensed with, much higher collection yields are achievable and a collection angle of 4π sr is quite feasible.

The energy of the threshold photoelectron is never identically zero and it is more appropriate to talk in terms of a range of collection energy from zero to ΔE where ΔE is the threshold resolution of the analyzer. The value of ΔE can be as small as a few millielectron volts, even for fairly routine operation of the threshold analyzer. This situation may be contrasted with conventional PES where photoelectrons with energies between E and $E + \Delta E$ are detected, E being the collection energy and ΔE the energy resolution of the analyzer. Interestingly, so long as angular discrimination is not also required, it can be much more efficient to detect photoelectrons of energy between 0 and ΔE than for those with energies between E and $E + \Delta E$ where E is finite.

2.1.2. Threshold Photoelectron Spectrometers. A variety of ingenious experimental techniques have been used to collect and energy-select near-zero-energy photoelectrons. These include time-of-flight (TOF) techniques (e.g., Heimann et al.,[8] Heiser et al.[9]) penetrating field techniques (e.g., King et al.[10] Hall et al.[11]) techniques that depend on angular discrimination against energetic photoelectrons (e.g., Baer et al.,[12] Spohr et al.[13]) and ones that use electron attachment of the zero-energy photoelectrons to molecules, e.g., SF_6 (Ajello and Chutjian,[14] Chutjian and Ajello[15]). The two most commonly used methods are those based on the TOF and penetrating field techniques. TOF techniques exploit the relatively long transit times of the very-low-energy photoelectrons as they pass through the analyzer. Penetrating field techniques exploit the fact that very-low-energy photoelectrons can be readily trapped by applied electrostatic fields. In general, this means that penetrating field analyzers may have higher collection angles and therefore higher detection efficiencies than do TOF devices. On the other hand, TOF devices can more easily provide angular discrimination of the photoelectrons. The characteristics of a threshold electron analyzer will be exemplified by a penetrating field device.

Penetrating field threshold photoelectron spectrometer. In a conventional photoelectron spectrometer, the target/photon interaction region is placed in an electrostatically field-free region so that the emitted photoelectrons travel in straight trajectories to the energy analyzer. In sharp contrast to this, the penetrating field analyzer depends on the penetration of an electric field into the interaction region which draws out, preferentially, photoelectrons of near-zero energy. The penetrating field technique was first described by Cvejanović and Read[16] and an example of the action of the penetrating field is shown in Figure 1 (King et al.[10]). This corresponds to an interaction region surrounded by a cylindrical, high-transmission mesh of diameter 20 mm which contains a circular aperture of diameter 4 mm placed on one side. Outside of the mesh is placed the extracting electrode which also contains a 4-mm aperture placed at a distance of 3 mm and in line with the first. The mesh

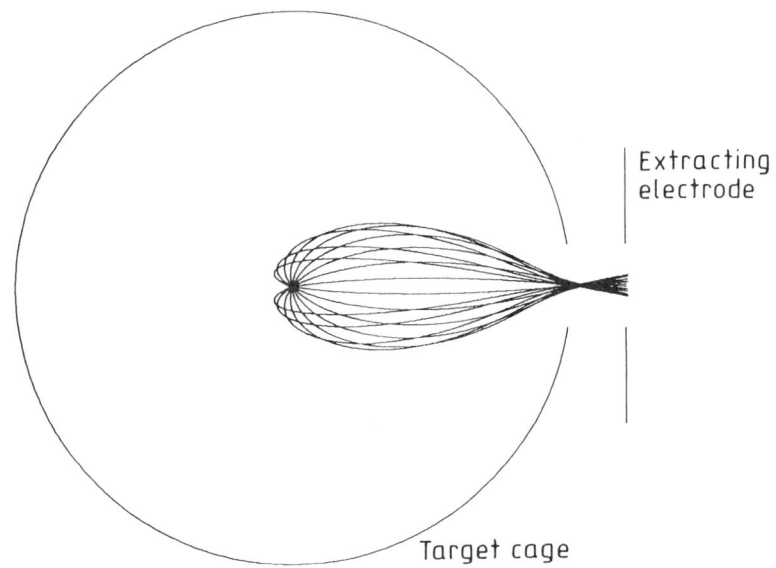

FIGURE 1. Computed trajectories of low-energy (2 meV) photoelectrons produced at the interaction region.

potential and the extractor at a potential of +10 V with respect to the mesh. Also shown in Figure 1 are the computed trajectories of 2-meV photoelectrons that emanate from the interaction region. The action of the penetrating field is to draw out these threshold photoelectrons over a solid angle of 4π sr. Similar computer simulations show that higher-energy photoelectrons, say, greater than 5 meV, are not drawn out in this way and leave the interaction region through the transparent mesh. Moreover, the extracting field produces a crossover in the threshold electron trajectories that is ideally suited for use as the object of an electrostatic lens system. The penetrating field extraction system thus provides very good threshold resolution (≤ 3 meV) and very high solid angle of collection ($\sim 4\pi$ sr).

A fraction of any energetic photoelectrons produced will be emitted into the solid angle subtended by the extracting electrode and this results in a high-energy tail in the transmission function of the extraction stage. This high-energy tail is removed by placing a conventional electrostatic analyzer after the penetrating field stage as illustrated in Figure 2 (Hall et al.[11]). Here a 127° cylindrical deflector analyzer (CDA) is used in conjunction with an input lens system. This lens system consists of two triple-aperture lenses with a real aperture placed between them and serves to image the crossover from the extraction stage onto the entrance slit of the CDA. The advantage of such a two-lens system is that spurious electrons are strongly suppressed which further enhances the sensitivity of the analyzer. The CDA discriminates against energetic electrons while the threshold electrons are transmitted and detected by a single-channel electron multiplier placed at the exit slit. A TPES spectrum is obtained by scanning the photon energy and measuring the yield of threshold electrons.

The performance of such a threshold analyzer is illustrated in Figure 3 which shows a threshold photoelectron spectrum obtained in the vicinity of the $^2P_{3/2,1/2}$ states of Ar$^+$. This spectrum was obtained using the 5-m normal-incidence monochromator at the Daresbury Laboratory Synchrotron Radiation Source (SRS) (Hall et al.[11]). There are several points of note about this spectrum. The widths of the threshold peaks are less than 3 meV indicating a threshold resolution of approximately 2 meV. Also the signal counting rate is high, e.g., 500 kHz in the $^2P_{3/2}$ threshold peak for $\sim 10^{10}$ photons/s and a target pressure of $\sim 10^{-3}$ mbarr, which demonstrates the high detection efficiency of the analyzer. Argon provides a stringent test of the threshold analyzer in that there exists a Rydberg series of autoionizing states converging to the Ar$^+$ $^2P_{1/2}$ state which gives rise to a series of photoelectrons of energy 3, 38, 62, 81, . . . meV (Peatman et al.[17]). The ion yield spectrum superimposed on the threshold photoelectron spectrum of Figure 3 indicates the positions and relative

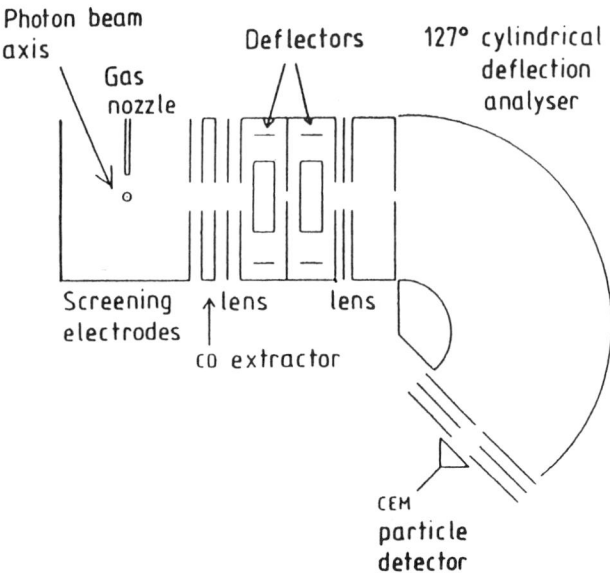

FIGURE 2. Schematic diagram of the coincidence spectrometer.

intensities of these autoionizing states. The absence of structure at these energy positions in the threshold spectrum demonstrates the successful suppression of these energetic photoelectrons.

Threshold coincidence techniques. In some photoionization studies, it can be very useful to detect the threshold photoelectron in coincidence with one or more of the other reaction products since this provides detailed information about decay pathways. This leads to, for example, threshold photoelectron–photoion coincidence (TPEPICO) spectroscopy for which several spectrometers have been

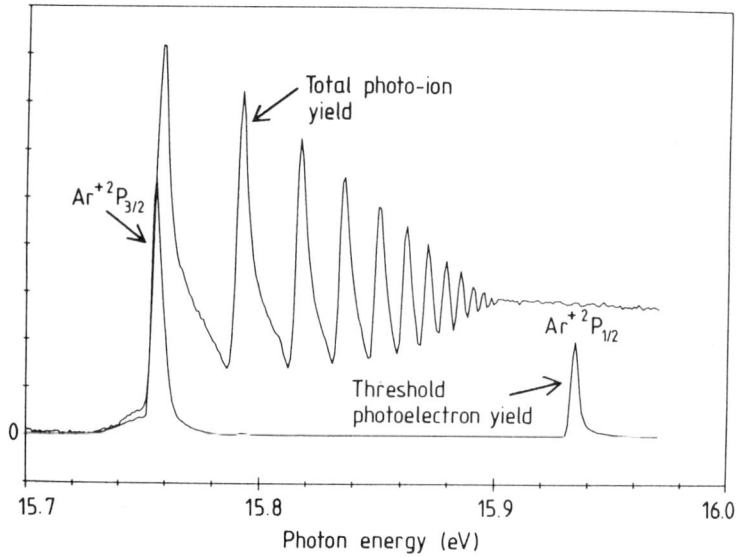

FIGURE 3. Threshold photoelectron spectrum superimposed on a photoion spectrum obtained in argon.

designed (e.g., see Nenner and Beswick[18] and Hall et al.[11] and references therein). Examples of the use of such coincidence techniques are given in the following sections.

2.2. Threshold Photoelectron Studies in Atoms

When an energetic (VUV) photon is incident on an atom, a valence electron may be emitted resulting in a peak in the photoelectron spectrum, which is called a main line. It is also possible that a second electron may be simultaneously excited to a bound orbital to produce an excited state of the ion which results in an additional peak in the photoelectron spectrum, which is called a satellite line. Such double electron excitations only occur because of electron correlations and because of this they have received a great deal of theoretical and experimental attention. Under some conditions the satellite structure can dominate the spectrum, and nowhere is this more true than in near-threshold photoionization where the relative intensities of the satellites can change dramatically. TPES provides a powerful technique for studying these satellites and identifying the dominant correlation effects that contribute to the intensity of a given satellite. The rare gases have received most of the theoretical and experimental attention and of these argon has been the best studied. The satellite states converge to double-ionization potentials and so single-photon double ionization may be seen as an extension of these states with electron correlation effects also playing a dominant role. Again, TPES provides a valuable experimental probe of the photo double-ionization process where it can provide valuable physical insights into the various kinds of electron correlations that may occur, e.g., angular and radial correlations.

2.2.1. Satellite Formation in Argon.
Photoelectron measurements of the argon satellite states have been performed from the x-ray domain (Svensson et al.[19]) down to threshold (Becker et al.,[1] Hall et al.,[20] Heiser et al.,[9] Lablanquie[21]). At high photon energies, final ion state correlation interaction (FISCI) feeds intensity from the p and s hole to satellite states with the same symmetry, P and S, respectively. The S symmetry dominates and the most intense line in the satellite spectrum is $(^1D) 3d\ ^2S$ reflecting the well-known strong coupling between the $3s$ and $3d$ orbitals. At intermediate photon energies, correlation in the initial state plays a role leading to the observation of states with other symmetries (Kossman et al.[22]). In the threshold region, Becker et al.[1] observed that many states, of all symmetries, are observed.

A threshold photoelectron spectrum obtained in argon, over the energy range from 32.0 to 40.0 eV, is shown in Figure 4 taken from Hall et al.[20] The widths of the observed peaks are typically 70 meV which is mainly due to the finite energy spread of the incident photon beam. The assignments of the peaks were deduced using the tables of Moore[23] or the spectroscopic studies of Minnhagen.[24] The numbering of certain peaks is that of Svensson et al.[19] corresponding to the energies of peaks observed at high photon energies (1.5 keV). The first thing to note about the spectrum is the large number of satellite states that occur in this energy region and the ability of the threshold photoelectron technique to see and resolve them. Excitation is observed of almost every known satellite state. This includes contributions from states with large angular momentum (i.e., F and G states) as well as quartet states which break the $\Delta S = \pm \frac{1}{2}$ rule indicating the presence of spin-orbit coupling. The spectrum also indicates that the most prominent states contain $3d$, $4s$, or $4p$ orbitals: peaks corresponding to states containing other orbitals such as $4d$, $5s$, or $5p$ are only visible because of the high instrumental sensitivity. The second point of note is the dramatic changes in relative peak intensities that have occurred for threshold excitation. The numbered satellites, which dominate the spectrum at high photon energies, make only a minor contribution to the total satellite intensity. The $(^1D)3d\ ^2S_{1/2}$ peak whose intensity is about 20% of the $3s$ main line at high energies is very weak at threshold with a relative intensity of about 3%. The most intense threshold peak is related to the $(^3P)3d(^2P)_{3/2}$ state with a relative intensity of over 20%. These dramatic changes in relative intensity indicate that there exist additional excitation mechanisms for the satellites in the near-threshold region.

These changes in relative intensity are highlighted in the near-threshold, constant photoelectron energy spectra of Figure 5 (Hall et al.[20]). For these spectra the collection energy of the spectrometer was held fixed at certain values (from 0.12 to 0.48 eV) as the photon energy was varied. This mode of operation is exactly analogous to that used for the TPES measurements except there the fixed value of collection energy was equal to zero. The explanation for the dramatic changes is that the satellite

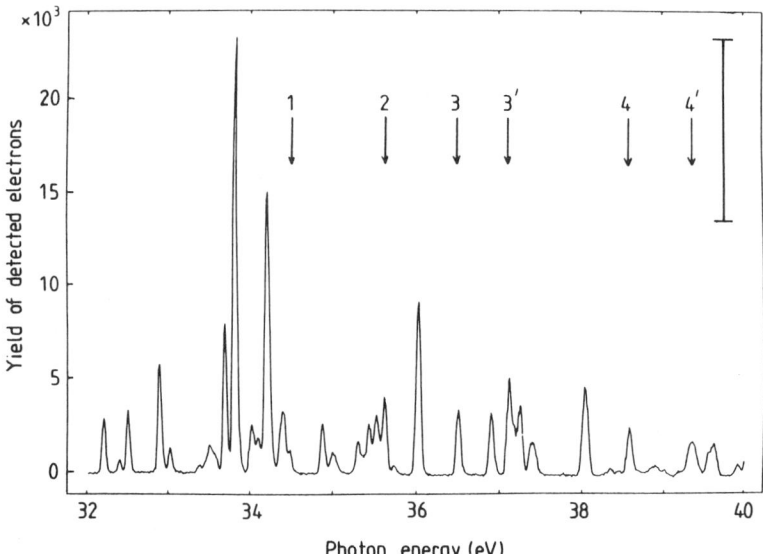

FIGURE 4. Threshold photoelectron spectrum in argon, obtained over the energy range 32.0–40.0 eV. The length of the vertical bar corresponds to 10% of the 3s threshold peak (not shown).

cross sections are dominated by resonance structure resulting from the decay of doubly excited neutral states into the satellites, i.e., the additional and dominant mechanism is the excitation of satellite states via the formation of doubly excited neutral states. When the photon energy is coincident with the energy of a doubly excited state, the satellite cross section may be considerably enhanced and this is reflected in the increased yield of photoelectrons of the appropriate energy. This resonance structure is also clearly evident when the partial cross sections of the satellites are measured by, for example, photoelectron or ion fluorescence spectroscopies which are discussed below in Sections 3 and 4, respectively.

Near-threshold excitation of satellite states in the other rare gases Ne, Kr, and Xe has also been studied (Hall et al.,[25] Wills et al.,[26,28] Hall et al.,[27] Heiser et al.[9]). The main conclusion of these studies was the same as for the argon case, namely, that the threshold region ionization proceeds entirely via the formation of doubly excited states. Moreover, essentially all of the possible satellite states are observed in the TPES spectra. On the instrumental side, the TPES technique provides both very high energy resolution (~a few millielectron volts) and excellent sensitivity. Indeed, the resolution of the TPES measurements in these energy regions is presently limited by the spread in the photon energy. The results also indicate the need for more theoretical work to understand the various configuration and interchannel interactions that are involved in satellite formation.

2.2.2. TPEPICO Studies of Satellites near Double IPs. Satellite states clearly play an important role in the photoionization of atoms. A further aspect of this role occurs in photodouble ionization when the satellite states lie above a double IP. There is then the possibility of decay to either singly or doubly charged ionic states. The decay paths to either species may be distinguished by measuring the threshold photoelectron (corresponding to excitation of the satellite state) in coincidence with the resultant product ions. Such a study has been performed by Hall et al.[29] who studied the decay of the satellite states that occur near the double IPs of the rare gases. These authors used a TOF analyzer in conjunction with their threshold photoelectron analyzer. The TOF analyzer consisted of an ion extractor stage and a triple-aperture electrostatic lens to collect and focus the product ions, a flight tube 60 mm long, and a channel electron multiplier to detect the ions. The ionic species, e.g., Ne^+ or Ne^{2+}, were discriminated by their flight times which differed by several microseconds. In the coincidence

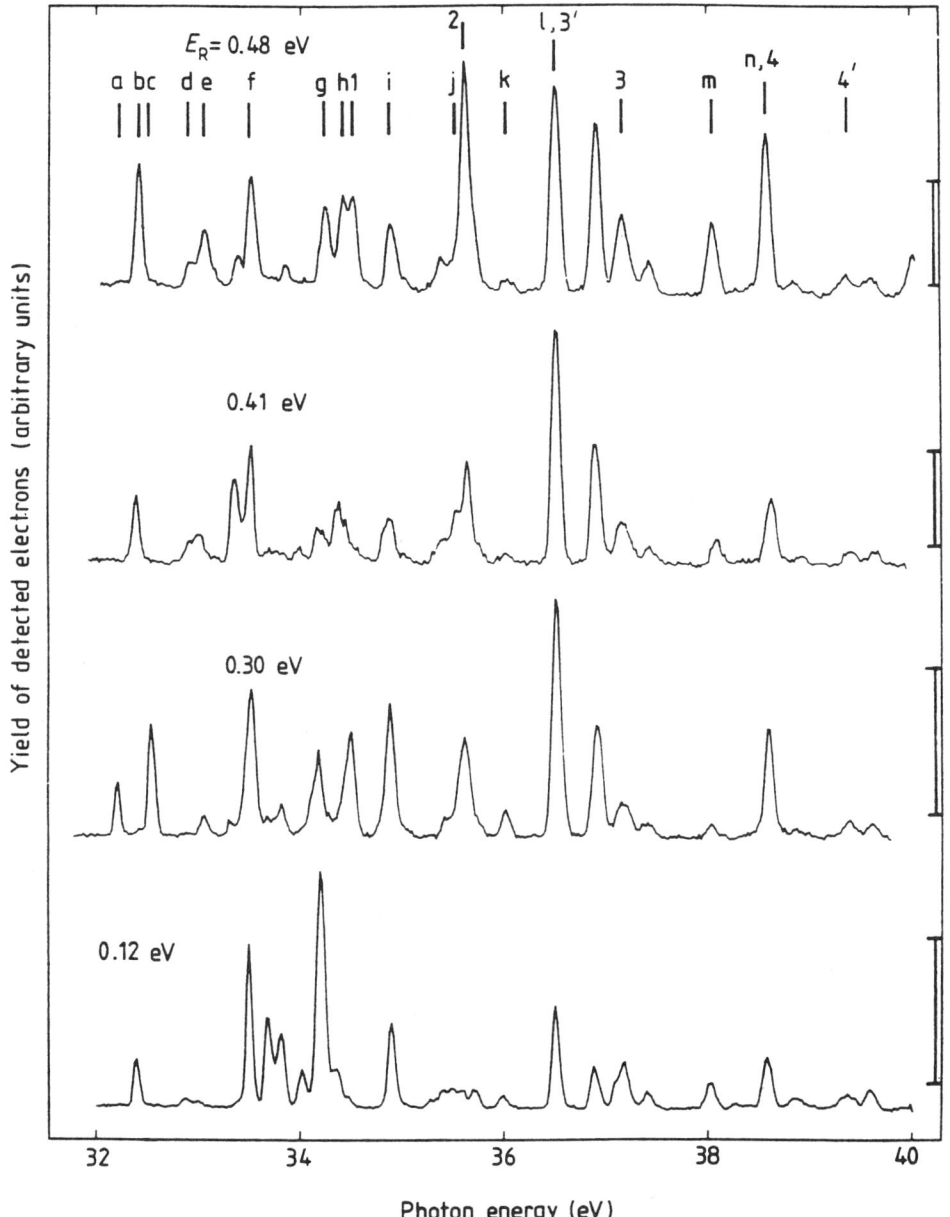

FIGURE 5. Constant photoelectron energy spectra obtained in argon for various values of photoelectron energy, E_R. The length of the vertical bars corresponds to 10% of the main line, $3s^{-1}(^2S_{1/2})$ photoelectron peak (not shown).

measurements, a time window was placed on either the Ne^+ or the Ne^{2+} signal. Since both the TOF and threshold analyzers used penetration field stages, detection efficiencies and true coincidence rates were high.

A threshold photoelectron spectrum of Ne is shown in Figure 6a and corresponds to the signal from the threshold photoelectron analyzer obtained during the coincidence measurements. The spectra of coincidences between threshold photoelectrons and Ne^{2+} and Ne^+ ions are shown in Figure 6b and 6c, respectively, on the same energy scale. The threshold spectrum reveals abundant structure converging to the ^3P, ^1D, and ^1S double-ionization thresholds, which are indicated in Figure 6b.

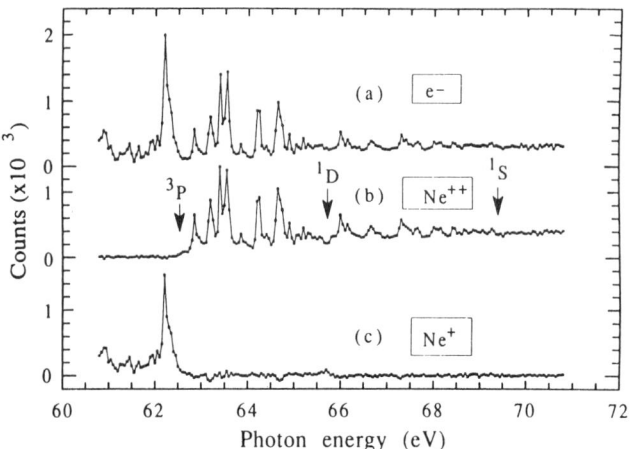

FIGURE 6. TPEPICO spectra recorded for doubly (b) and singly (c) charged ions of neon with a 50-meV channel width. The threshold electron spectrum is displayed in (a).

However, the most striking feature of the coincidence spectra is that once the Ne^{2+} channel is open, no further signal, to within experimental error is detected in the Ne^+ channel, i.e., all of the satellite states decay to Ne^{2+} in an autoionization process. Each of these satellite states, once formed, will emit an electron of fixed energy, irrespective of the photon energy at which it is excited. The energy of the emitted electrons is characteristic of the satellite state and equal to the difference between its excitation energy and the double IP. Furthermore, the electron intensity is directly indicative of the excitation probability. The photoelectron energy spectrum of the satellite states situated between the ^3P and ^1D double IPs has been obtained by Wills et al.[26] for photon energies corresponding to the production of these states within 1 eV of their respective thresholds. This spectrum revealed a strong similarity to the threshold spectrum in both the energy of the features and their intensity, thus providing further evidence for the strong coupling of these states to the double ion continuum. This two-step process leading to double ionization involves the sequential emission of two electrons which can interact and when the lifetimes are appropriate should produce the characteristic energy shifts and peak shapes associated with postcollision interaction (PCI) (e.g., see Avaldi et al.[30]). Wills et al.[26] observed no significant PCI effects and came to the conclusion that the satellite state lifetimes are sufficiently long to make these effects too small to be detected in their experiment. Another interesting observation is the low cross section at the double ion onset itself. This channel only becomes intense once satellite states are excited and in fact these states are the major contributors to the Ne^{2+} cross section. The direct double ionization process which would produce a continuous contribution to the spectrum is weak relative to the structure, at least up to the ^1D IP.

2.2.3. TPESCO Studies of Doubly Charged Ions. In a conventional TPES study, the yield of threshold electrons is measured as a function of photon energy and this maps out the various states of the singly charged ion. This principle has been extended to the study of doubly charged ions in experiments that use a combination of two threshold electron analyzers, and which measure coincidences between two essentially zero-energy photoelectrons (Hall et al.[31,32]). A true coincidence signal is then the signature of a double ionization event and will be recorded each time the scanned photon energy passes through the threshold of a double ion state. A spectrum obtained in this way maps out the various states of the doubly charged ion. This is proving to be a very valuable spectroscopic tool for double ion states since it measures their energies directly with a precision (~10 meV) that is limited only by the energy spread in the incident photon beam. It also measures partial double-ionization cross sections of the states at their thresholds.

RESONANCES AND NEAR-THRESHOLD PROCESSES 363

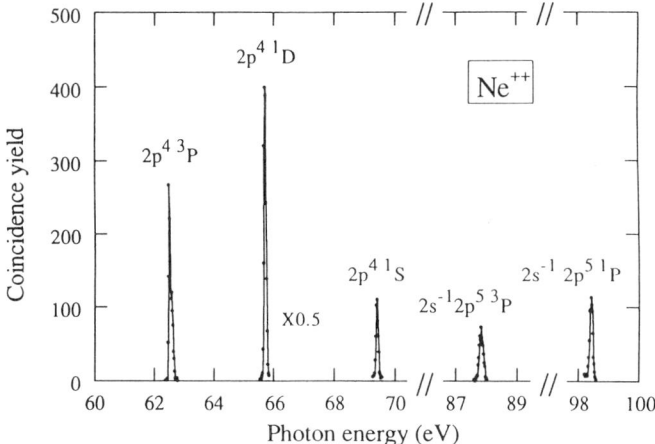

FIGURE 7. A threshold photoelectron coincidence spectrum in neon over the photon energy range 62 to 100 eV showing the $2p^4$ (^3P, ^1D, and ^1S) and $2s^2 2p^5$ (^3P and ^1P) states.

A TPESCO spectrum obtained in neon by Hall et al.[33] is shown in Figure 7. The energy range of this spectrum (62 to 100 eV) shows peaks corresponding to the $2p^4$ (^3P, ^1D, and ^1S) and $2s2p^5$ (^3P and ^1P) states of Ne^{2+}. The most important point to note is that all of the possible ionic states are observed and that they are seen with similar intensity. In near-threshold photodouble ionization the two emitted photoelectrons are strongly correlated and this electron pair has a wave function with associated values of orbital angular momentum, parity, and spin. Greene and Rau[34,35] have pointed out propensity rules, which indicate the favoredness of the various possibly double charged states which have subsequently been modified by Huetz et al.[7] The results of Hall et al.[33] in the rare gases neon, argon, krypton, and xenon were found to be in general agreement with the propensity rule of Huetz et al. except for the particular case of neon.

2.2.4. Threshold Photodouble Ionization in Helium. Near-threshold photoionization of helium is an archetypal system for studying electron–electron correlations. This is particularly the case for the region around the double IP. The processes in this region include satellite formation with high values of n and, in the limit of this, near-threshold double ionization which is one of the fundamental processes of atomic physics. TPES studies can provide valuable information about both of these reactions which can be written:

$$h\nu + \text{He} \rightarrow \text{He}^{+*} + e$$

$$\rightarrow \text{He}^{2+} + e + e$$

Since the original work of Wannier,[5] numerous theoretical studies of near-threshold photodouble ionization by electron or photon impact have been made (for reviews see Refs. 3, 6). The various theories yield predictions for (1) the energy dependence of the double-ionization cross section $\sigma^{2+}(E)$ where E is the excess energy above threshold, (2) the energy sharing of the two outgoing electrons, and (3) their angular correlations. For example, Wannier predicts that the cross section, $\sigma^{2+}(E)$, for single-photon double ionization of helium varies as $E^{1.056}$. Since the electron correlation effects are strongest close to threshold, TPES studies have become one of the main ways of studying them.

A threshold photoelectron spectrum obtained in helium is shown in Figure 8. It was obtained with a penetrating field analyzer used in conjunction with a toroidal grating monochromator on the Daresbury Laboratory SRS (Hall et al.[36]). The spectrum clearly reveals the helium satellite manifold up to $n = 11$ at 78.56 eV ($n = 2$ at 64.50 eV is shown as an insert). Above $n = 11$ the peaks merge to

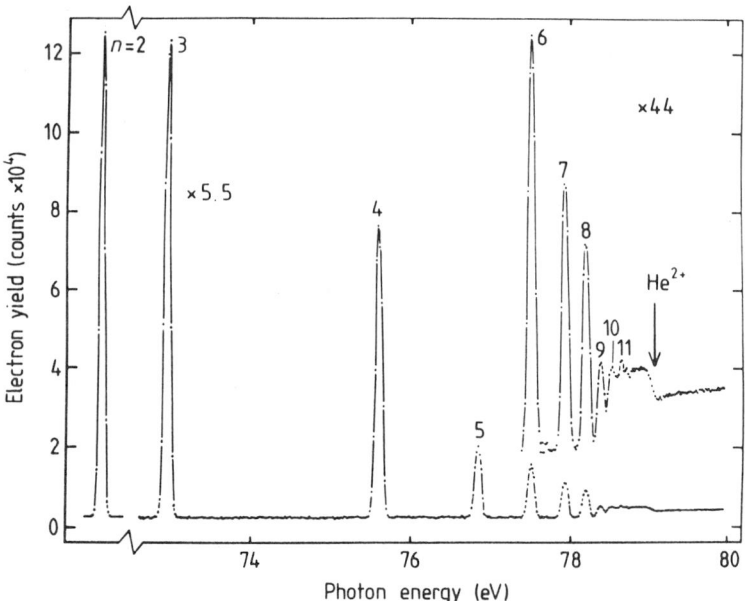

FIGURE 8. Threshold ($E_e = 0$ eV) photoelectron spectrum of helium obtained with a photon energy spread of 100 meV and photon energy increments of 20 meV. The peaks correspond to excitation of $He^{+*}(n)$ satellite states.

form a plateau which dips down to a minimum at the He^{2+} potential before rising slowly. The peak intensity in the threshold spectrum is expected to vary as n^{-3} (Fano and Cooper[37]), but when the electron correlation effects contained in the Wannier model are taken into account, the peak intensities would then vary as $n^{-3.11}$ (King et al.[38]). However as can be seen from Figure 8, the peak envelope is not smooth. This was previously pointed out by King et al.[38] who from a fit of the observed intensities determined a value of −3.9 for the exponent. These discrepancies can be explained by the presence of double excited neutral states. These contribute to the cross sections and correspond to processes of quantal origin not included in the Wannier mechanism. Recently, Zubek et al.[39] have reported large contributions from double excited states to the cross sections for satellite excitation up to $n = 4$. As n increases, the double excited states and satellite manifolds will overlap and threshold cross sections will consequently be modified.

The energy region around the double IP is shown in greater detail in Figure 9, obtained at an energy resolution of 60 meV (Hall et al.[36]). Satellite states up to $n = 12$ are clearly visible. Higher levels merge into an almost flat shoulder which falls off rapidly reaching a sharp minimum at the double IP. This singularity which has the form of a strongly asymmetric cusp can be seen more distinctly in the inset which is shown on an expanded scale. This singularity is a strong signature of electron–electron correlations which suppress the yield of low-energy electrons.

The energy differential form of the Wannier threshold law can be written

$$\sigma(E, E_e) = E^{\alpha-1} f(E_e/E)$$

where $f(E_e/E)$ is the energy partitioning function between the electrons and in the Wannier model is independent of E above the IP. This distribution is expected to be uniform and accurate calculations indeed show that this would be the case to within less than 10% (Read and Cvejanović[40]). The high degree of correlation between the electrons above the double IP which is responsible for the Wannier threshold phenomena is also present for near-threshold excitation of high-n satellite states. This has led to the threshold law being extended using classical reasoning to the high Rydberg excitation region

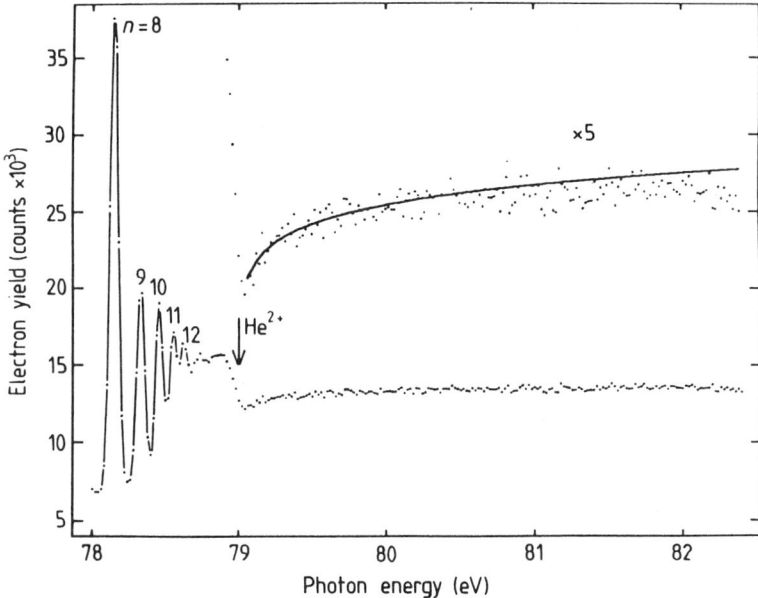

FIGURE 9. Threshold ($E_e = 0$ eV) photoelectron spectrum of helium over an energy range in the region of the double ionization potential.

below the ionization energy to accommodate negative values of E. The partial cross section is then written:

$$\sigma(E,E_e) = c|E|^{\alpha-1} \tag{1}$$

where c is a constant which would have a slightly smaller value (5%) for excitation than for ionization. It should be pointed out, however, that this relation would not account for the quantal behavior described above. Note that for measurement of the partial cross section, as obtained in threshold measurements where electrons of energy between 0 and ΔE are detected, the yield is proportional to $\alpha - 1$. Since α is predicted to be very close to unity, these threshold electron measurements provide a very sensitive test of theory.

The cusplike feature shown in Figure 9 was fitted with relation (1) using a least-squares fitting program after convolution with a Gaussian function 60 meV wide corresponding to the apparatus function. Clearly, the structure due to satellite excitation will prevent the Wannier law [equation (1)] from being fitted over an extended range below the double IP although reasonable fits were obtained over the energy range $-0.06 < E < +0.4$ eV and gave the value for the ratio c_2/c_1 of 1.43 where c_1 and c_2 are the coefficients in (1) above and below the double IP, respectively. This ratio reflects the strong asymmetry of the cusp and differs markedly from the value 0.95 determined using classical trajectories (Read and Cvejanović[40]). This discrepancy could be related to a particularly strong contribution from resonant processes to the satellite cross sections in the region of high n. When the fitting procedure was restricted to the energy region above the double IP $+0.08 < E < +0.48$ eV, the exponent was determined to be $\alpha = 1.060 \pm 0.007$ which is in excellent agreement with the Wannier prediction. The energy region over which good agreement was found indicated that the range of validity of the threshold law was about 2.0 eV.

In threshold photoelectron measurements, great care has to be taken to ensure that spurious backgrounds from other sources of low-energy electrons are eliminated. These would limit the accuracy of the above measurements especially since the shape of a very small cross section was being determined. In a subsequent study of threshold ionization of helium, Hall et al.[41] measured the threshold electrons in coincidence with singly or doubly charged ions. Since spurious electrons are not correlated with the ions, they give rise to a background of random coincidences that could be subtracted from the true coincidences with confidence. It was found that these subsequent measurements were in excellent agreement with the earlier ones of Hall et al.[36]

2.2.5. Angular Distribution Measurements in Near-Threshold Photodouble Ionization. The Wannier model makes predictions about the angular behavior of photoelectrons resulting from double ionization in helium. In this theory the distribution of the mutual angle between the two outgoing electrons, which is independent of the energy sharing between the two electrons, is represented by a Gaussian distribution peaked at 180° with a full width at half height of the distribution given by

$$\Theta^{1/2} = \Theta_o E^{1/4}$$

where Θ_o is the angular correlation coefficient, taking the value 91 or 103° (see Huetz et al.[7]). Huetz et al. showed that $\Theta_o = 91°$ leads to an asymmetry parameter, β, for the photoelectrons that rises rapidly from a value of –1 at threshold to zero at $E \approx 1.5$ eV. The extreme β value of –1 at threshold reflects the dominance of the electron–electron interaction compared to the electric field of the photon beam, which induces the electron to be emitted predominantly in a direction perpendicular to the polarization vector of the photon beam. The rise in β from threshold corresponds to the increasing influence of the photon electric field due to a weakening of the correlation effect as the electron energies increase.

Hall et al.[36] modified their threshold analyzer to make angular measurements of photoelectrons with energies as low as 0.14 eV. These authors measured β parameters for the region above the double IP where the detected photoelectrons correspond to those with a specific value of energy, equal to the collection energy of the spectrometer. The β parameters determined in this way, up to 2 eV above the double IP, are shown in Figure 10 for $E_e = 0.25$ eV.

Calculations of β have been performed for double ionization of helium by Huetz et al.[7] using the Wannier theory in the energy range $0 < E < 2$ eV. Their values of β increase from –1 to 0 approximately varying as $E^{1/2}$ as the photon energy goes from threshold to 1.4 eV above. The rate of increase of β depends on the angular correlation coefficient θ_0. Figure 10 reveals that β parameters

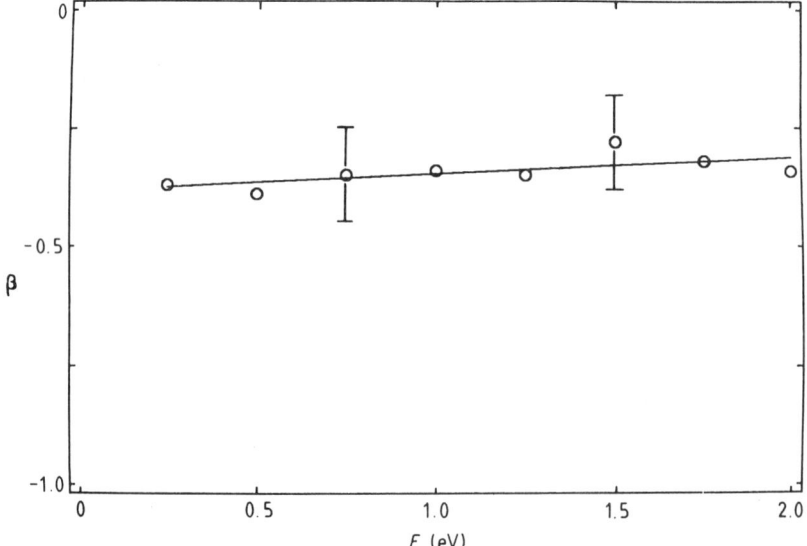

FIGURE 10. Asymmetry parameters β for the double ionization process in helium plotted against photon energy E above the He^{2+} potential. These parameters were obtained at a photoelectron energy $E_e = 0.25$ eV. The solid line is a quadratic fit through the data points to guide the eye.

measured above the double ionization threshold are lower and rise more slowly than predicted and appear to disagree with the detailed predictions of Wannier theory. Again because of the possible sources of systematic error arising from detection of spurious low-energy electrons, Hall et al. repeated their 1991 measurements but measured the photoelectrons in coincidence with the positive ions produced (Hall et al.[41]). These coincidence measurements were in agreement with the earlier ones and confirmed the conclusion that the behavior of the measured β parameter does not follow the trend predicted by Wannier theory and that the degree of angular correlation is underestimated in this model.

The first ab initio calculations of near-threshold double photoionization in helium were made recently by Maulbetsch and Briggs[42,43] and confirmed the dominant role of electron correlation in the final state. It was also shown that, contrary to the Wannier situation, in the first few electron volts above threshold the behavior of β is strongly dependent on the energy partitioning between the two electrons. In order to test these ideas, Hall et al.[44] measured β for the extreme case of unequal energy partitioning between the two electrons, where one electron has near-zero energy and the other carries off the entire excess energy. This is the situation where electron correlation is now expected to have the least effect and where this effect will diminish the most rapidly as the energy above threshold increases. These authors studied the $2p^{-2}$ ^1S state of neon rather than double ionization in helium. This state has the same attributes as those of helium, i.e., it has only one continuum state and it is also free of any indirect processes because of the absence of satellite states within several electron volts of threshold, but has the advantage of a larger cross section. In these observations coincidences were measured between near-zero-energy electrons detected by a threshold analyzer and finite-energy electrons detected by a second analyzer.

The measured asymmetry parameters for double photoionization of neon to the ^1S state in the energy range $0.2 < E < 2.0$ eV above threshold for the condition where one electron has near-zero energy and the other has energy E are shown in Figure 11. As can be seen, the predictions of Wannier theory (Huetz et al.[7]) which are independent of energy partitioning do not, as for helium, give a good account of the experimental observations. As noted above, the ab initio calculations of Maulbetsch and Briggs[42,43] reveal a strong dependence of the behavior of β on the energy partitioning between the electrons. These authors also find, as was predicted by Greene[45] from Wannier theory, that β should take the value -1 for $E = 0$. However, for unequal energy sharing β would be roughly flat over the present energy range and only turns down to -1 at lower energies. Only in the case of equal sharing is β expected to have considerable slope. These theoretical results are in accord with the results shown in Figure 11 although the values of β are generally somewhat lower than those predicted by theory

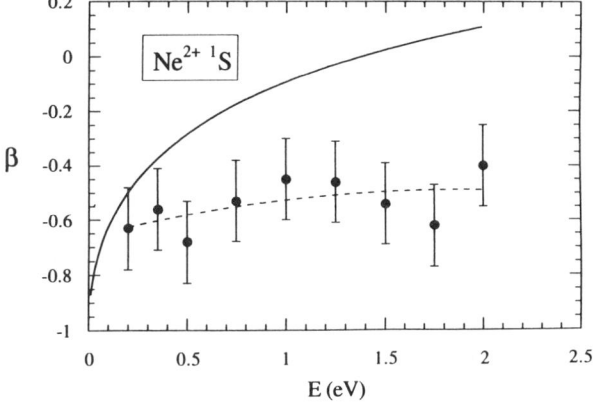

FIGURE 11. Asymmetry parameters β for double photoionization of neon to the ^1S state at energies E above threshold for the conditions where one electron has near-zero energy and the other has energy E. The dashed line is a quadratic fit to the data. The solid line represents a theoretical result (Huetz et al.[7]) obtained using Wannier theory and a value of the angular correlation coefficient $\Theta_e = 91°$.

assuming of course that the same calculations for neon would give the same results as for helium. This is a reasonable assumption as it was found that the final-state wave function was the determining factor in the calculations and the initial-state wave function had little influence on the angular behavior of the electrons. In any case, the negative values of β, which are still near –0.5 at 2 eV, indicate that even in this extreme situation of unequal energy sharing, the electron–electron interaction still plays a dominant role.

3. CONSTANT ION STATE SPECTROSCOPY

3.1. Principles and Experimental Methods

3.1.1. Principles. In conventional PES, the energy of the incident photons is kept fixed so that the photoelectron has a well-defined kinetic energy given by

$$\text{KE} = h\nu - E_{\text{spec}} \tag{2}$$

where E_{spec} is the spectroscopic energy of the ion state. The intensity of the photoelectron signal is a measure of the cross section for exciting the ion state at that particular value of photon energy. In the CIS technique the photoelectron yield is measured as the incident photon energy is varied continuously. Clearly, the kinetic energy of the photoelectron will also vary, according to equation (2), and so the collection energy of the photoelectron spectrometer ($\equiv$ kinetic energy of the photoelectrons) must be varied to account for this, i.e., the collection energy is maintained equal to the photon energy minus the spectroscopic energy of the ion state. The yield of the detected photoelectrons, measured as a function of photon energy, is the CIS spectrum, or equivalently, the partial ionization cross section of the ion state. The ion state may be excited by direct ionization or it may be populated by autoionization from doubly excited states. These two processes interfere and so the doubly excited states show up as sharp resonance structure in the measured CIS spectrum. The technique is a powerful way of studying these resonances giving, for example, their energies, widths, and shapes and their modes of decay into the various accessible ion states (e.g., Plummer et al.,[46] Morin et al.,[47] Zubek et al.[48]).

It is interesting to note that in the CIS mode, structure is observed with a resolution which is the better of the resolutions of the incident photon beam and the electron energy analyzer. Consider what happens, for example, when the photon energy spread is greater than the passband of the energy analyzer. The photon beam produces photoelectrons that have an energy spread equal to that of the photon beam. However, only those photoelectrons that have an energy within the bandpass of the analyzer are transmitted and detected. Thus, when the photon and electron collection energy are scanned synchronously, structure will be observed in the measured ionization cross section with a resolution that is equal to the bandpass of the analyzer, i.e., the better of the two resolutions. A similar result occurs when the photon energy spread is less than the analyzer bandpass. Of course, the resolution of the energy analyzer must be sufficient to resolve the particular ion state of interest if more than one exists.

3.1.2. Experimental Techniques. The CIS technique requires a photon source of tunable energy and a photoelectron spectrometer whose collection energy can be varied in synchronism with the photon energy according to equation (2). The photons are supplied by a synchrotron radiation source where, usually, a computer is used to step an optical monochromator to provide photons of variable energy. For CIS measurements it is convenient to step the photons linearly in energy rather than in wavelength and typical energy ranges would be 1 to 10 eV with step sizes of typically 5 meV, depending on the energy resolution of the measurements. The computer can also provide a linear voltage ramp that varies in synchronism with the changing photon energy and which can be used to control the photoelectron spectrometer. In the spectrometer an electrostatic lens system is placed between the interaction region and the energy dispersing element, e.g., a hemispherical deflection analyzer (HDA). The purpose of the lens system is to collect photoelectrons from the interaction region and focus them onto the entrance slit of the HDA, which is operated in the constant pass energy mode where the energy resolution remains constant across the spectrum. The collection energy of the photoelectron spectrometer is scanned by applying a ramp voltage between the interaction region and the spectrometer. This ramp voltage is the one provided by the computer so that the photon energy and collection energy are scanned synchronously. The CIS spectra obtained in this way will be affected by any nonuniformity in the transmission function of the spectrometer with respect to photoelectron kinetic energy. This transmission function may be determined and corrected for by measuring a well-known partial ionization cross section over the same range of kinetic energy, e.g., that of the He$^+$ (1s) state (Marr and West[49]). The spectrometer may be placed at the magic angle (about 54°, depending on the polarization of the photon beam) so that the measured cross sections are independent of any angular distribution effects. Alternatively, it can be used to obtain the variation of the asymmetry parameter, β, over the photon energy range by taking measurements at angles of 0 and 90° with respect to the plane of polarization of the photon beam (Zubek et al.[50]).

An example of a CIS spectrum, obtained by Zubek et al.[50] in helium, is presented in Figure 12. These authors used the spectrometer shown in Figure 12 but operated it in the CIS mode described above. The satellite states, He$^+$(N) form a Rydberg series converging to the double IP, i.e., to He^{2+} at 79 eV. In turn, the doubly excited neutral states form Rydberg series converging to the satellite states, e.g., He(2snl) converging to He$^+$(N = 2) (Madden and Codling,[51] Codling and Madden[52]). The CIS spectrum of Figure 12 corresponds to the partial ionization cross section of the He$^+$(N = 2) states over an energy range (68.8 to 76.9 eV), which includes the energies of the He$^+$(N = 3, 4, and 5) satellites. The CIS spectrum clearly shows strong structure related to the doubly excited, autoionizing states, which can be seen to converge to the higher-lying satellite states.

Comer and co-workers have developed a photoelectron spectrometer that employs a position-sensitive detector and have used it extensively for CIS measurements. The power of the apparatus is that a range of kinetic energies can be measured simultaneously. This shortens data accumulation times and also allows the unraveling of competing deexcitation and dissociation processes. This photoelectron spectrometer has been described in detail by Haworth et al.[53] The photoelectrons are energy analyzed using an HDA operated in constant pass energy mode with a resolution (typically 25–30 meV) that is sufficient to resolve the various final ion states. The energy-dispersed electron image at

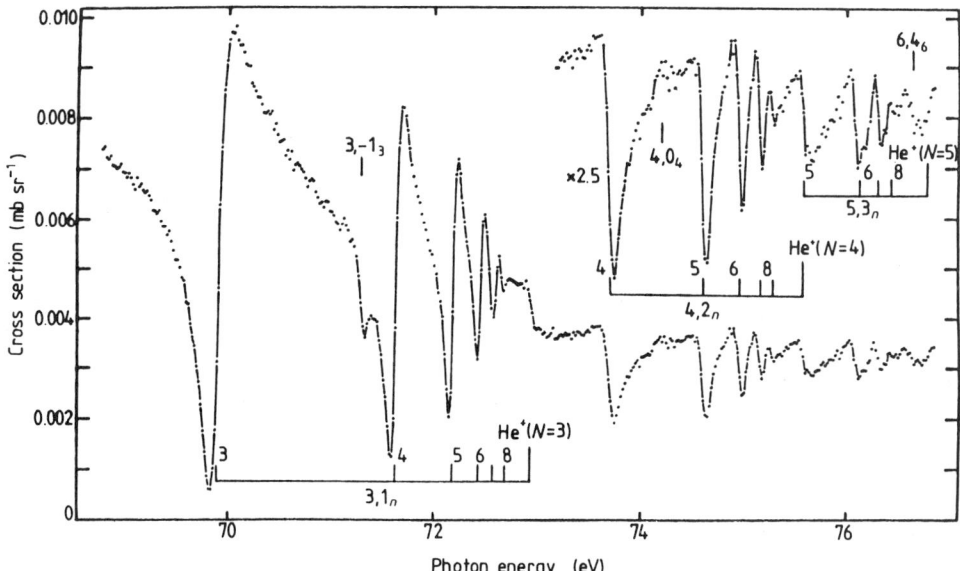

FIGURE 12. The differential cross section for photoionization into the He$^+$(N = 2) states obtained at an angle Θ = 90° over the photon energy range 68.8–76.9 eV. The energies of autoionizing states and potitions of photoionization thresholds of He$^+$ (N) are indicated by vertical bars.

the exit plane of the HDA is incident on the position-sensitive detector consisting of a pair of microchannel plates, a phosphor screen, and a charge-coupled device (Hicks et al.[54]). (More recently the phosphor screen/CCD combination has been replaced by a customized integrated circuit with an array of electrodes connected to individual amplifiers and counters on the integrated circuit.) The photon energy is scanned in small fixed energy steps (~5 meV) and at each photon energy a photoelectron spectrum is recorded by scanning the electron collection energy of the analyzer. In this way a two-dimensional surface of photoelectron yield as a function of photoelectron energy and electron collection energy is recorded. This two-dimensional spectrum can be used in various ways but for example one particular cut through the surface yields the CIS spectrum (see Section 2.2).

3.2. Constant Ion State Studies in Atoms

3.2.1. Helium. The doubly excited states of helium, where both electrons have a high value of principal quantum number, offer an ideal system in which to study electron–electron correlations. These states were first observed in a photoabsorption study by Madden and Codling[51] and since their discovery they have received a great deal of experimental and theoretical attention. The CIS technique has important advantages for the study of these doubly excited states. It allows resonance decay into different ion states to be distinguished and isolated, so that partial ionization cross sections can be determined and decay routes elucidated. This in turn means that a particular decay route may be chosen for study where most of the direct ionization can be avoided and the resonant structure becomes the dominant contribution. Zubek et al.[39,50] have made such CIS measurements and have studied photoionization into the $He^+(N)$ satellite states and demonstrated that in the satellite region photoionization is dominated by autoionization from the doubly excited states. (This has also been seen in the recent high-resolution measurements of Domke et al.[55])

The CIS spectrum of the $He^+(N = 2)$ states, measured by Zubek et al.[50] at an angle of 90°, is shown in Figure 12. It corresponds to the partial ionization cross section for that state. The vertical scale has been normalized to a value of 0.0071 Mb·sr^{-1} at 68.9 eV using the data of Lindle et al.[56] The resonance structure, which dominates the cross section, has been assigned using the simplified nomenclature N, K_n (Herick and Sinanoglu,[57]) where N and n are the principal quantum numbers of the inner and outer electrons, respectively, and K is the collective quantum number. Strong series corresponding to $3, 1_n$ for $n = 3$ to 8, $4, 2_n$ for $n = 4$ to 8, and $5, 3_n$ for $n = 5$ to 8 are clearly identified converging to the $He^+ (N = 3)$, $He^+ (N = 4)$, and $He^+ (N = 5)$ ionization thresholds, respectively. The energies of these states, recently determined by Zubek et al.,[50] are indicated in Figure 12. Also observed are members of weaker series, e.g., $3, -1_3$ and $4, 0_4$. Structures corresponding to the $3, -1_n$ series are clearly distinguishable up to $n = 8$ at which point the higher states merge to form a weak maximum in the measured yield, which then falls off rapidly, producing a step at the energy of the He^+ $(N = 3)$ threshold. This rapid change takes place within 60 meV, a value equal to the energy resolution of the spectrometer. It is due to the contribution of the doubly excited states, whose autoionization strength within the series is approximately constant, as predicted by recent calculations (Hayes and Scott,[58] Salomonson et al.[59]) and which disappears at their ionization limit. There is an indication of a similar steplike feature at the $He^+ (N = 4)$ threshold.

Zubek et al.[50] also obtained the β parameter as a continuous function of photon energy by measuring CIS spectra for the $He^+(N = 2)$ states at observation angles of 0 and 90°, and highlighted the dramatic influence of autoionization on the β parameter. The β parameter obtained as a function of photon energy in the 68.8–76.9 eV range is shown in Figure 13. As can be seen, β varies dramatically with energy, the oscillations being even more pronounced than the corresponding resonance features in the cross sections. This is particularly the case for the $3, -1_3$ structure which forms a strong peak. This fact allowed a weak structure at 72.06 eV to be tentatively assigned to the $3, -1_4$ state. Of note in the β variation with energy is the step increase at the $He^+ (N = 3)$ threshold, where it can be deduced that the total cross section decreases by approximately 19%. The rapid oscillations of β reflect the sharp variations of the relative contributions of the s- and d-wave channels in the ionization into the $He^+(2p)$ state. Ionization into the $He^+ (2p)$ state is a p-wave process for which the β parameter is constant and equal to 2. These oscillations are superimposed on a background that is increasing

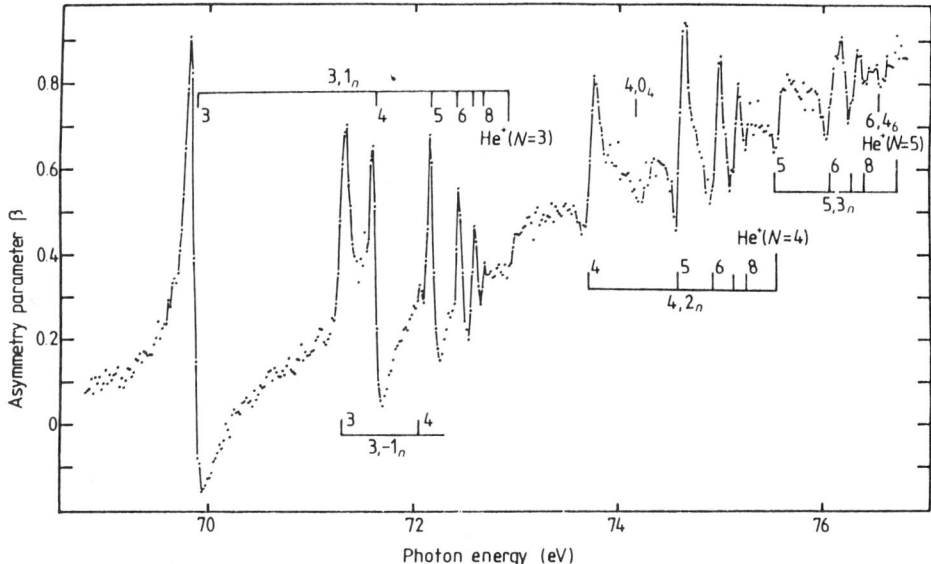

FIGURE 13. The measured asymmetry parameter β for photoionization into the $He^+(N=2)$ states obtained over the photon energy range 68.8–76.9 eV. The energies of autoionizing states and positions of photoionization thresholds of $He^+(N)$ are indicated by vertical bars.

monotonically with energy from 0.1 at 68.9 eV to 0.85 at 76.8 eV, which is in agreement with previous observations, e.g., Bizau et al.[60] and Lindle et al.[56] The recent calculations of β for energies up to 72.7 eV (Hayes and Scott,[58] Salomonson et al.[59]) are generally in good agreement with the present results.

3.2.2. Argon. Argon serves as a very useful case to illustrate the kinds of information that can be deduced from CIS measurements: it has many ion states and a wealth of resonance structure in its photoionization cross sections. Becker et al.[1] observed photoelectron spectra close to threshold and observed variations in the relative intensities of the satellite lines compared to the 3s main line. Hall et al.[20] and Wills et al.[61] made complementary measurements of CIS spectra for both main line and satellite states in argon. The measurements of Hall et al. extended over the range from about 0.1 to 1 eV above threshold, while Wills et al. covered the range from 0.3 to a maximum of 8 eV above threshold. The power of the CIS technique is that it provides detailed information about the ways in which the doubly excited states decay and complements photoabsorption measurements which relate to the excitation of the states. The measurements of Hall et al. and Wills et al. demonstrate the very selective decay of these doubly excited states.

3p and 3s main line CIS spectra. CIS spectra, obtained by Wills et al.,[61] for the 3p and 3s main lines are shown in Figure 14 and exhibit pronounced resonance structure. The 3p spectrum shows the well-characterized $3s3p^6np$ window resonances converging to the $3s3p^6(^2S)$ state threshold at 29.24 eV (Madden et al.[62]). Strong resonances are also observed in the 3s spectrum just above the ^2S ion threshold and correspond well to those seen by Madden et al. This dominance of the ^2S cross section by resonance structure near threshold has also been observed in the fluorescence measurements of Schartner et al.[63] As can be seen from Figure 14, these resonance structures are weak or absent in the ^2P channel. Main line CIS spectra have also been obtained by Hall et al.[20] for the region just above the ^2S threshold and these are presented in Figure 30 where detailed comparison is made with the available theoretical work. The main line CIS spectra also show structures due to Rydberg series of doubly excited states, labeled R2 and R4 in the 3s and 3p spectra, respectively, of Figure 14. These Rydberg series are also observed in the spectra of some satellites.

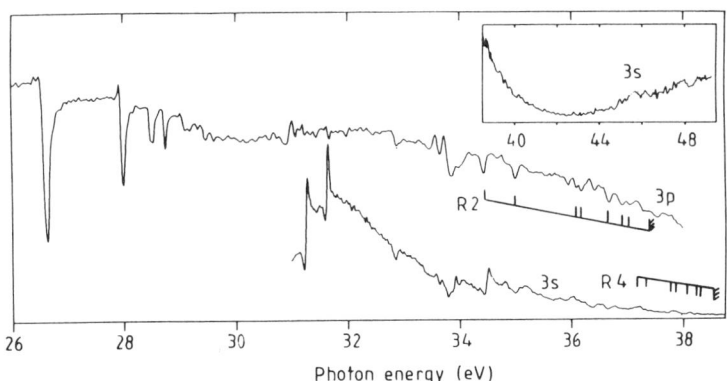

FIGURE 14. Constant ionic state spectra of the 3p and 3s main lines of argon. The inset shows the 3s main line in the region of its Cooper minimum.

Satellite state CIS spectra. Argon has a large number of satellite states as observed in the TPES spectra. Individual CIS spectra for each of these satellite states can be recorded separately (e.g., Hall et al.[20]). Alternatively, a number of satellite states may be observed simultaneously using a multidetector technique where the photoelectron yield is measured as a function of both photon and photoelectron energies (Wills et al.[61]). An overview of the data obtained by Wills et al. for the argon satellites between the ^2S threshold and the double IP is shown in the form of a contour plot in Figure 15. The data set was taken in about 60 hours of beam time and consists of about 60,000 data points. In Figure 15 each line at fixed photon energy corresponds to a photoelectron spectrum for that energy, with peaks due to satellite states. Following any of these peaks across the surface at a fixed value of binding energy yields a CIS spectrum. The variation in intensity of each of these as a function of photon energy shows clear resonance structure especially close to threshold. With such a surface it is possible to obtain a reliable estimate of the background signal by examining the data between any of the measured satellite lines. The data of Figure 15 can be displayed in a number of different ways. A composite photoelectron spectrum for the whole range of photon energies is obtained by summing all of the photoelectron spectra and the result is shown as the upper spectrum of Figure 16, which also shows a number of representative photoelectron spectra at selected photon energies. The composite photoelectron spectrum ($32.5 \leq h\nu \leq 42.3$ eV) displays all of the observed satellite peaks and is to some extent analogous to the spectrum obtained in the fluorescence experiment of Samson et al.[64] where a broad band of VUV radiation ($\sim 33 \leq h\nu \leq 125$ eV) was used to simultaneously excite all of the satellite states. The labels A–Z shown in Figure 16 refer to the assignments of Wills et al. which were taken from Moore[23] and Minnhagen.[24]

Examples of CIS spectra, corresponding to the satellite peaks indicated in Figure 16, are shown in Figure 17. Some of these peaks include a number of unresolved satellites. The spectra show that the satellite cross sections are dominated by resonance structure. In general, this is strongest just above the threshold of the satellite which suggests that the doubly excited states decay most strongly to satellites that lie energetically close.

Schartner et al.[63] have measured photon-induced fluorescence from the satellite $(^3P)4s(^4P)$, labeled B in Figure 16, at intervals in photon energy of between 0.15 and 0.5 eV (see also Figure 40). They observed features similar to those of the corresponding CIS spectrum B in Figure 17. Becker et al.[1] used PES to obtain branching ratios for selected satellites with respect to the 3s main line. For their satellite state 1, corresponding to A + B in Figure 16, they used photon energy steps of between 0.35 and 1 eV and observed a strong increase in signal around 1.5 eV above threshold which was identified as a resonance effect. This was seen in more detail in the work of Schartner et al. as well as in Figure 17. Becker et al. measured satellite 2, corresponding to I + J here, with energy steps of 1 eV

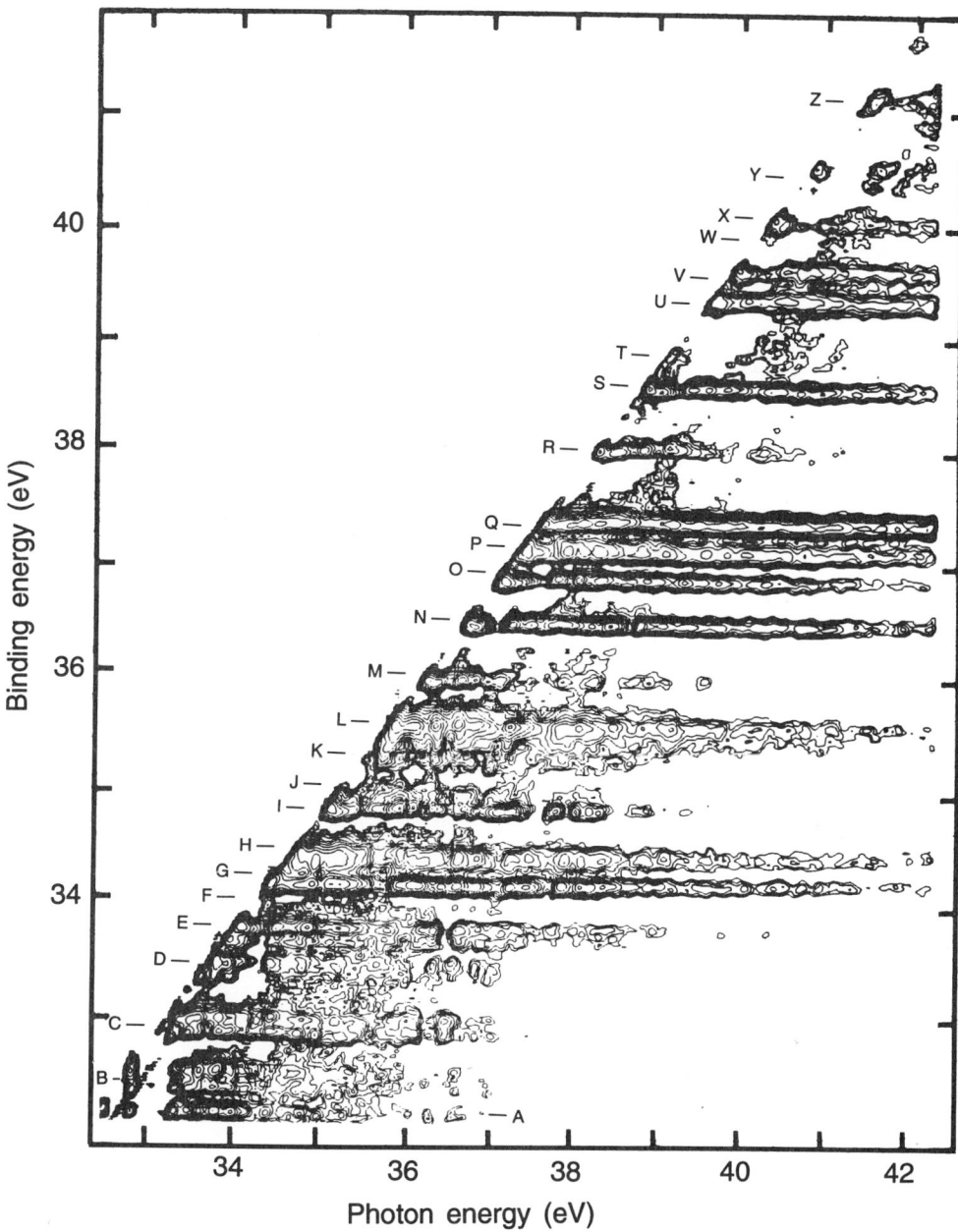

FIGURE 15. Contour map of the satellite data set in argon, showing the yield of photoelectrons as a functin of both photon energy and binding energy. For clarity the contours have been taken at equal intervals of (intensity)$^{1/3}$. The labels A–Z refer to the assignments of Wills et al.[61]

and this showed an increase in signal toward threshold, but this can also be explained by resonances observed in Figure 17. This demonstrates the importance of considering resonances when investigating the excitation mechanism in the threshold region.

One of the tasks in analyzing the CIS spectra is to assign the doubly excited states that give rise to the resonance structure. The first step is to compare the CIS spectra with the corresponding

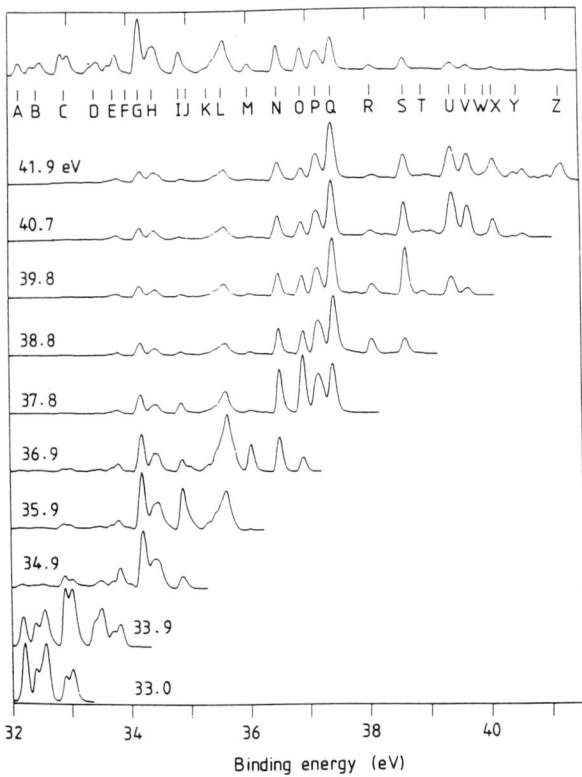

FIGURE 16. Argon photoelectron spectra obtained for the indicated values of photon energy. The top spectrum is a composite spectrum of the individual spectra shown below.

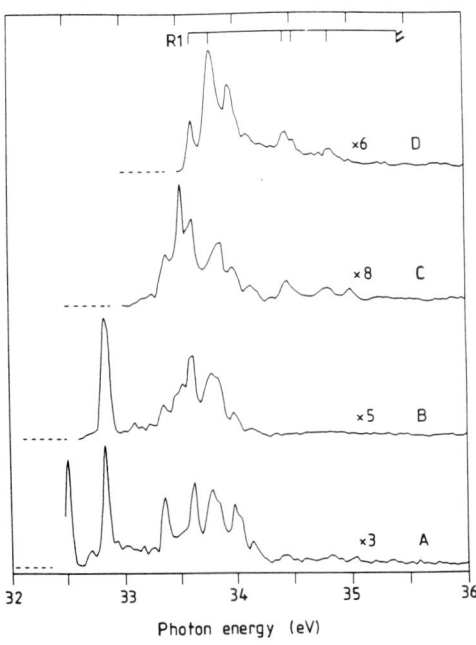

FIGURE 17. Constant ionic state spectra of the first four satellite states of argon. The letters above each spectrum refer to the assignments of Wills et al.[61]

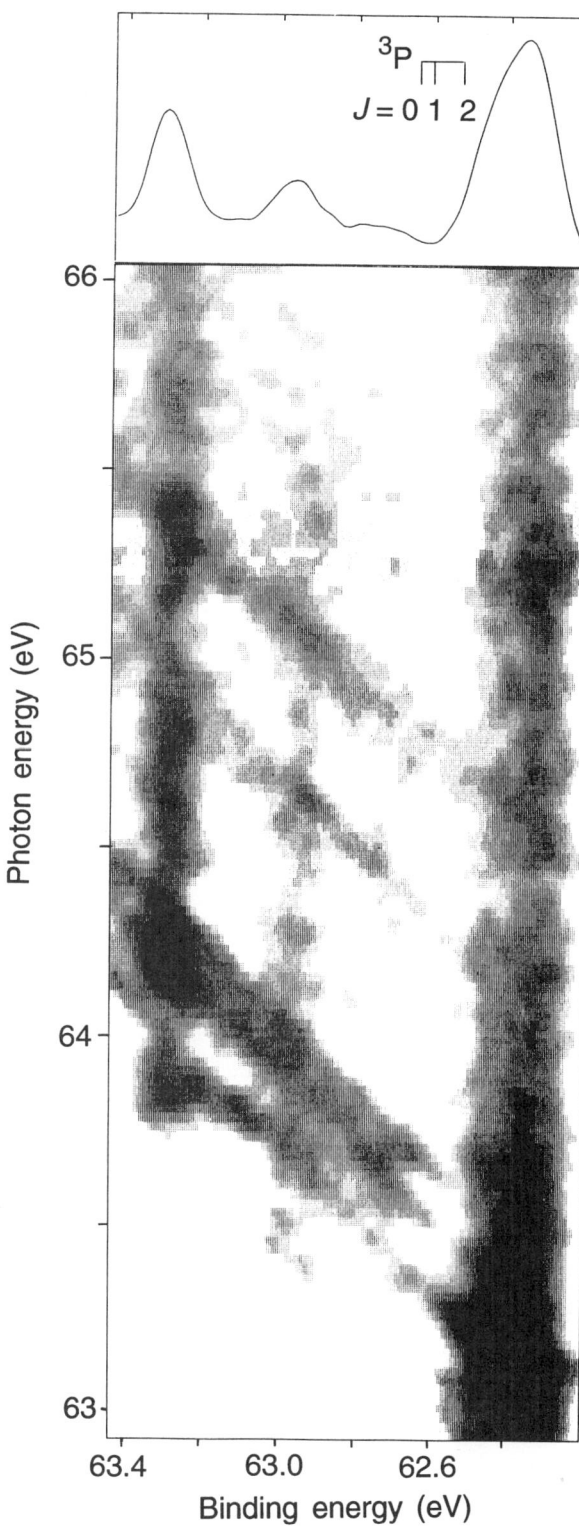

FIGURE 18. Gray scale plot of the electron yield as a function of binding energy and photon energy; the electron intensity is shown increasing from light to dark using a linear scale. The lines parallel to the photon energy axis correspond to states of Ne$^+$, which can be seen at the top of the figure in a composite photoelectron spectrum ($63 \leq h\nu \leq 66$ eV) which has been extracted from the data set shown below. The lines at 45° correspond to electrons with fixed kinetic energy as discussed in the text.

photoabsorption spectrum. Additional assignments can be made by fitting the structures to Rydberg series, taking into account parity considerations and by comparison with quantum defects known from Rydberg series in other atoms with similar configurations (Moore[23]). In these ways Wills et al.[61] were able to assign many of the resonances. There did, however, remain a number of resonances whose assignments were uncertain and this highlights the need for more theoretical work to understand the observed structure.

3.2.3. The Other Rare Gases. In analogy with helium and argon, the other rare gases show a wealth of resonance structure in their photoionization cross sections. They represent useful cases for study because, on the one hand, the heavier atoms, krypton and xenon, offer the opportunity to investigate the effects of the breakdown of L-S coupling on electron correlations. On the other hand, the lighter atom, neon, is expected to follow L-S coupling more closely and to produce much simpler spectra. The resulting, less complex CIS spectra offer the possibility of being able to assign all of the observed resonance structure and of being able to determine selection rules for the decay of these resonances to main line and satellite states.

Neon. For the above reasons Wills et al.[26] have investigated in some detail the excitation of the neon main line and satellite states. They observed a great deal of resonance structure and were able to assign much of this to doubly excited Rydberg series. This also enabled them to study changes in correlations and screening effects across a Rydberg series. They found a strong selectivity in the decay of the resonances and attempted to establish decay patterns for them. However, even for this relatively simple atom, it was not clear that general selection rules could be applied.

The study of Wills et al.[26] also demonstrated the ability of the multidetection technique to distinguish electrons from different emission processes. In addition to the photoelectron satellite peaks they observed several features related to autoionization of ionic states that lie energetically above the first double IP; a process that may be considered as an Auger decay. The multidetection technique can distinguish between the photoelectron and the Auger electron since the variation of their kinetic energies with photon energy is different. The photoelectron energy increases with photon energy, whereas the Auger electron has a fixed kinetic energy equal to the difference in energy between the energy of the excited ion and the doubly charged ion. The two types of behavior can be clearly seen from the spectrum of Figure 18 where a gray scale plot shows the electron yield as a function of binding energy and photon energy. In this spectrum photoelectrons produced by ionization into specific states of Ne^+ have a constant binding energy and so appear as lines parallel to the photon energy axis. The diagonal lines correspond to constant collected energy and are caused by autoionization of the excited ion. This ability to identify electron emission lines using two-dimensional PES has also been shown to be valuable in the photodissociation of molecules where autoionization fragments are produced

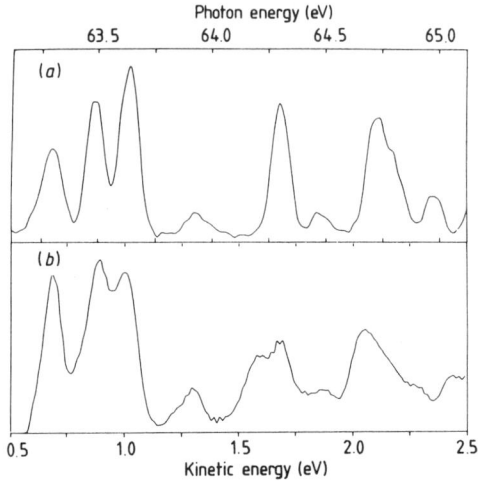

FIGURE 19. (a) Threshold spectrum (Hall et al.[27]) showing the Ne^{+*} states. (b) A spectrum showing the kinetic energy of electrons emitted from states of Ne^{+*}.

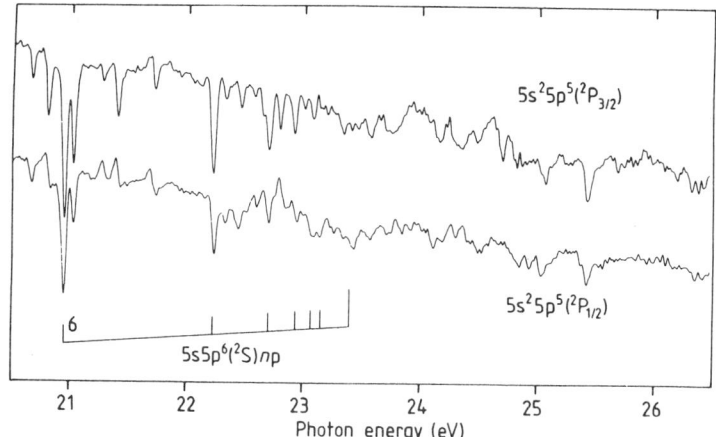

FIGURE 20. Constant ionic state spectra for the $5s^25p^5(^2P_{3/2})$ and $5s^25p^5(^2P_{1/2})$ states of xenon, taken at 0° relative to the direction of polarization of the radiation, displayed with the same zero. The singly excited series, $5s5p^6(^2S)np$, and many unassigned doubly excited resonances are observed.

(Cafolla et al.[65]). Wills et al. obtained spectra of the electrons emitted from the Ne^{+*} states by taking sections through the surface parallel to the photon energy axis. A spectrum obtained in this way is shown in Figure 19 where it is compared with a TPES spectrum of the Ne^{+*} states obtained by Hall et al.[25] In general, there is a good correspondence between these two spectra.

Xenon. In xenon, CIS spectra for main line and some satellite states have been obtained in complementary studies by Hall et al.[27] and Wills et al.,[28] while Schartner et al.[66] have made fluorescence measurements of the 5s main line and some satellite states. There has also been some recent theoretical work (Tulkki[67]) in which the photon energy dependence of ionization into the 5s and some satellite states was calculated but in which resonances were not included. CIS spectra for the $5s^25p^5$ ($^2P_{3/2}$) and ($^2P_{1/2}$) main lines, measured by Wills et al.[28] at 0° with respect to the direction of polarization of the radiation, are shown in Figure 20. These spectra exhibit complex and extensive structure resulting from one- and two-electron excitations which have been seen previously in photoabsorption measurements (Coaling and Madder[68]). These include the singly excited series $5s5p^6$ ($^2S)np$ and many unassigned doubly excited resonances. As may be seen, the resonance profiles displayed by the two spin-orbit components are significantly different. This is probably a reflection of the different angular dependencies exhibited by the two components caused by the strong spin-orbit interaction in xenon (Codling et al.[69]). Wills et al.[28] and Hall et al.[27] also obtained CIS spectra for a number of satellite states which again showed much resonance structure, and demonstrated that the decay of the resonant states to different satellites is very selective. However, because of the complex nature of the CIS spectra and the breakdown of L-S coupling, it was not possible to assign all of the structure that was observed.

4. PHOTON-INDUCED FLUORESCENCE SPECTROSCOPY

4.1. Principles

Fluorescence spectroscopy is based on a quantitative analysis of the fluorescence radiation emitted during the decay of the level of interest i, excited from a lower state 0, which is usually but not necessarily the ground state of an atom or molecule. Photon-induced fluorescence spectroscopy (PIFS) and PES can be regarded as alternative and complementary methods (Figure 21). When total cross sections are measured, they are equivalent. On the other hand, a number of basic differences exist for PIFS with respect to PES:

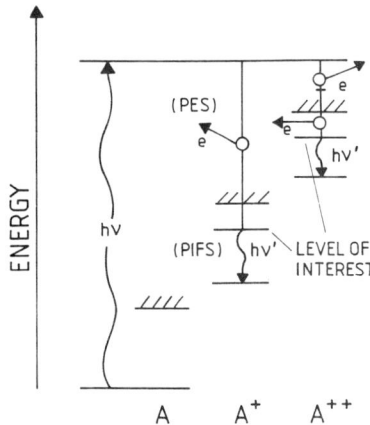

FIGURE 21. Schematic comparison of photoionization studied by photoelectron spectroscopy (PES) and by photon-induced fluorescence spectroscopy (PIFS).

1. The bandwidth of the exciting photons and the resolution of the fluorescence radiation are not coupled.
2. Thresholds are passed without transmission problems caused by the strongly varying photoelectron energies.
3. Double photoionization into excited states can be studied without coincidence techniques.

The photoionization cross sections, σ_i, follow from the fluorescence intensity corrected for any possible anisotropy of the fluorescence radiation. The latter is correlated with the alignment and orientation parameters of the excited ion.

Very often, the photoionization cross section σ_i has to be derived from an emission cross section σ_{ij}. The latter denotes the cross section for the photon-induced transition $i \to j$ with wavelength λ_{ij} (Figure 22).

The excitation cross section σ_i follows from σ_{ij} by subtraction of cascade processes, represented by σ_{ki} [equation (3)]. τ_i and A_{ij} are the lifetime of level i and the transition probability for the decay mode $i \to j$, respectively (Figure 22).

$$\sigma_i = \sigma_{ij} \cdot \frac{1}{A_{ij} \cdot \tau_i} - \sum_k \sigma_{ki} \qquad (3)$$

For a measurement of σ_{ij}, the signal rate S_{ij}, obtained from observing a length l of the photon beam within the solid angle Ω, has to be converted into an intensity by the quantum efficiency $k(\lambda_{ij})$ of the monochromator–detector system used for the analysis of the fluorescence radiation in the particular experiment [equation (4)]. The target density, n, and the rate of impinging photons, i_{ph}, also have to be measured.

$$\sigma_{ij} = \frac{S_{ij}}{k(\lambda_{ij})} \cdot \frac{1}{n \cdot i_{ph} \cdot l} \cdot \frac{\Omega}{4\pi} \cdot f(\theta,\Pi) \qquad (4)$$

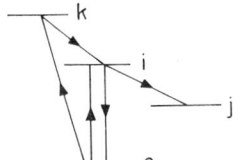

FIGURE 22. Principal level scheme and transition from the level of interest i to lower levels j or on the level i from higher levels k (cascading).

The function $f(\theta, \Pi)$ takes care of the anisotropic spatial intensity distribution of the fluorescence radiation:

$$f(\theta,\Pi) = \frac{3 - \Pi}{3(1 - \Pi \cos^2 \theta)} \tag{5}$$

For linear polarized synchrotron radiation, θ denotes the angle between the E vector of the electromagnetic field and the observation direction, the latter being oriented perpendicular to the photon beam. The linear polarization is defined by $\Pi = (I_\parallel - I_\perp)/(I_\parallel + I_\perp)$, with $I_\parallel$ and $I_\perp$ being the radiation components of the fluorescence radiation for $\theta = \pi/2$ with electric field vector parallel and perpendicular, respectively, to the E vector of the synchrotron radiation. Since $\Pi = 0$ for most of the following no further emphasis is given to alignment or even orientation features, which have been studied, e.g., by Jiménez-Mier et al.,[70] Greene and Zare,[71] Klar,[72] and Kronast et al.[73]

As an optical method, PIFS is in principle capable of a high resolution for all energies of the exciting photons, with values for $\lambda/d\lambda$ of a few thousand. This fundamental advantage is momentarily hampered by the comparatively small detection efficiency of optical spectrometers. For demonstration, Figure 23 shows a fluorescence spectrum obtained for photons with $h\nu = 100$ eV impinging on Ar[74] (see also Figure 27b). The fluorescence lines in Figure 23 are marked by the upper levels of the respective transitions. For the transitions within the ArIII $3s3p^5\,^3P_{2,1,0}$–$3s^23p^4\,^3P_{2,1,0}$ multiplet between 88.7 and 87.5 nm the fine-structure splitting is resolved. The FWHM of the lines in Figure 23 corresponds to $\lambda/d\lambda = 600$. A total instrumental bandwidth of 20 meV at $h\nu = 100$ eV would be necessary to resolve this fine-structure splitting by PES (see also Figure 1 of Chapter 6).

The sensitivity problem of PIFS mentioned above is the main reason for the comparatively small number of fluorescence studies of the photoionization of free atoms or molecules. It has partly been overcome by the introduction of multiplex detection techniques, although PIFS will profit from the increasing access to high-flux insertion devices. Without aiming at completeness, the x-ray fluores-

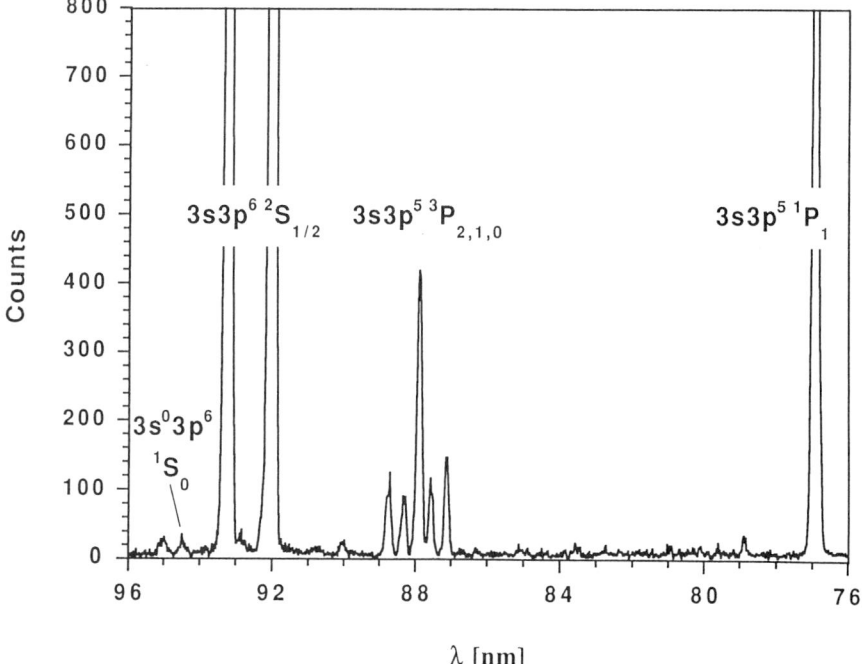

FIGURE 23. VUV-fluorescence spectrum for photons of 100 eV interacting with Ar (Möbus et al.[74]).

cence studies of the NIST group (Lindle et al.,[75] Deslattes[76]), the investigations of the satellite production in He by Woodruff and Samson[77] and of Jiménez-Mier et al.[70] studied by unresolved VUV fluorescence, the experiments with Cd and Zn using interference filters in the visible by Kronast et al.,[73] and the work of the Daresbury group studying resonance enhancement of satellite production in Ca$^+$ by registering the total fluorescence radiation in the spectral range of the visible (West[78]) are mentioned as previous studies.

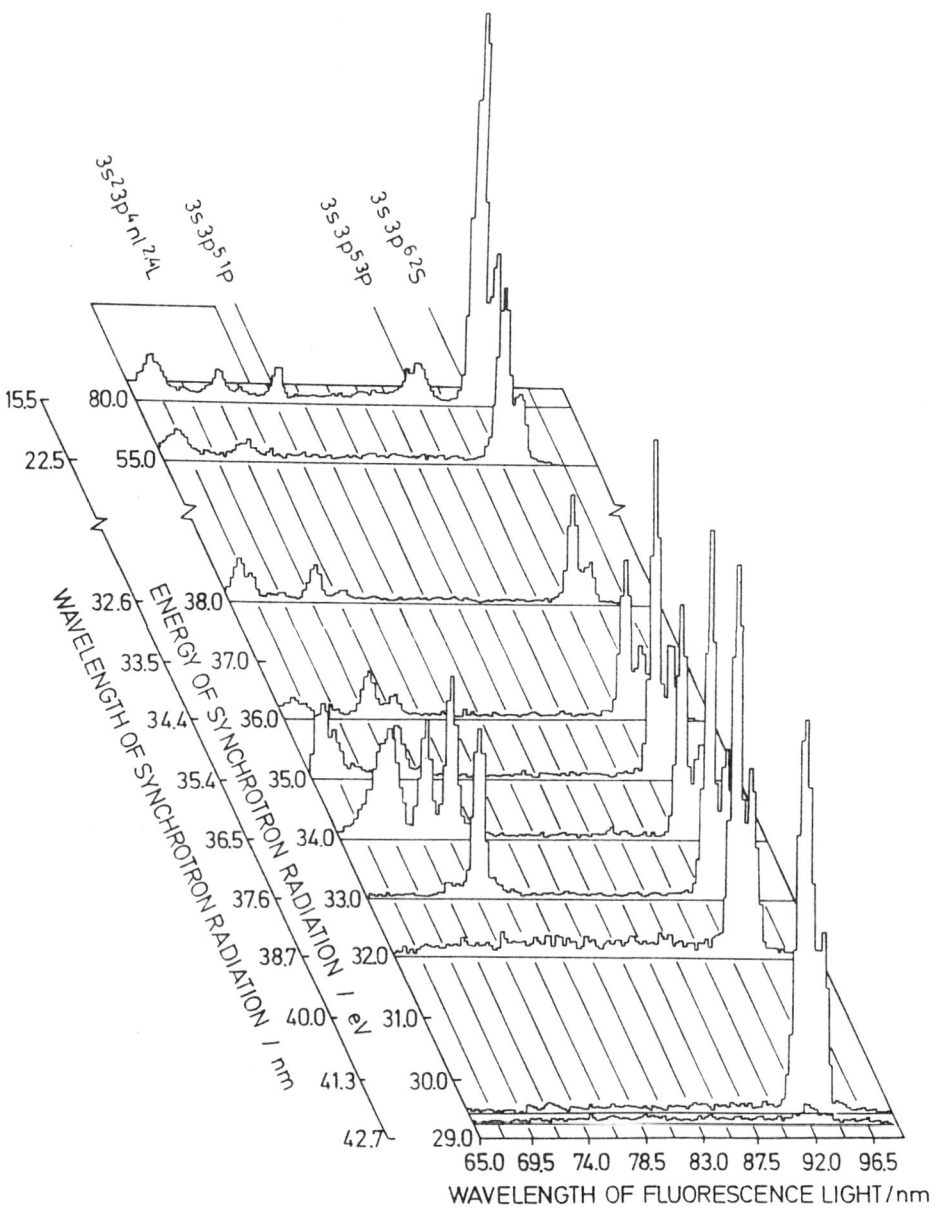

FIGURE 24. VUV-fluorescence spectra from Ar with the energy of the impinging photons as parameter (Schartner et al.[79]).

Spectroscopically resolved VUV fluorescence from rare gases was described by Samson et al.,[64] using "white" synchrotron radiation for the photoionization process, and simultaneously by Schartner et al.,[79] using monochromatized synchrotron radiation. Figure 24 shows a set of early VUV-fluorescence spectra from Ar, registered for photon energies increasing upwards from the Ar 3s-electron threshold. The collaboration of the group at the University of Giessen and the group at the University of Kaiserslautern has subsequently applied PIFS especially for investigations of photoionization processes near the thresholds of the rare-gas outer-shell s electrons. The results of these experiments as well as PES studies have initiated a number of theoretical investigations. No special references are given here since they will arise in the examples discussed below. PIFS has also been applied in studies of the single-photon double-ionization process (see Section 2.4). Figure 23 was reproduced from such a study.[74] These experiments are not included in the following sections on photon-induced fluorescence spectroscopy. Their results in the threshold range are closely related to Section 2.2.4.

4.2. Experimental Procedure

The experimental equipment necessary for PIFS is shown in schematic form in Figure 25. The ionizing photons are provided by a synchrotron radiation facility. They have to be monochromatized by an appropriate monochromator, e.g., by a toroidal grating monochromator. A mirror focuses the synchrotron radiation through a set of diaphragma into a target cell and a Faraday cup. The secondary electron currents from the diaphragma, especially from the one at the target-cell exit, have to be measured in order to ensure the complete detection of the photon beam. The experimental problems of a photon flux measurement are rather similar in PIFS and in PES. The photon flux can be monitored by the secondary electron emission from an Al cathode. Provided its electron yield as a function of the photon energy is known, a flux measurement is possible. Alternatively, a solid-state photodiode can be used. The target cell is differentially pumped. The target pressure is held constant by a regulated valve. For the pressure measurement a capacitance manometer is appropriate.

The optical treatment of the fluorescence radiation has to compromise between resolution and sensitivity. Interference filters were applied in the visible spectral range[73,80] with the intention of gaining sensitivity at the cost of a restriction in the number of fluorescence channels investigated. Schartner et al.[81] supplied a Pouey-type VUV monochromator of high luminosity[82] with a position-

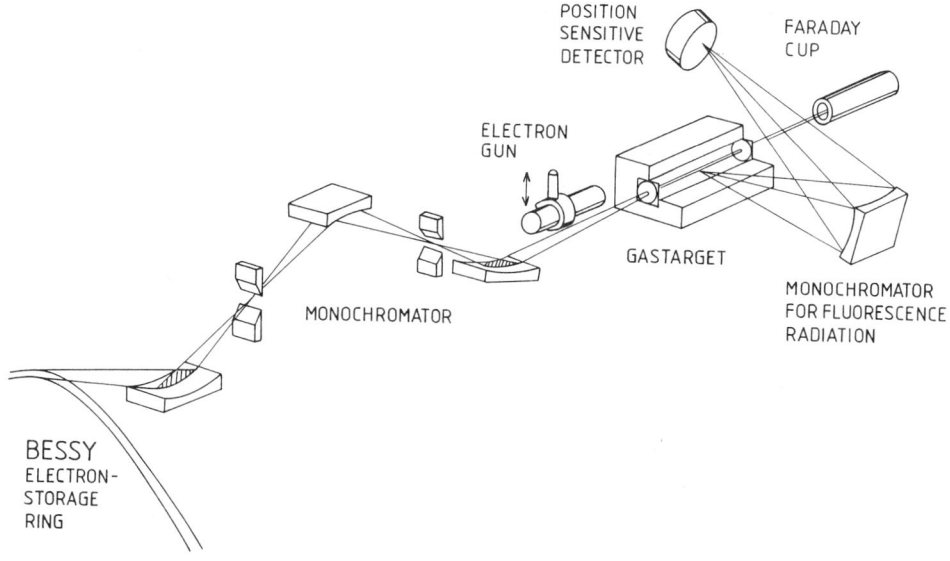

FIGURE 25. Scheme of an experimental setup used for PIFS.

sensitive channelplate detector. As indicated in Figure 25, the focus of the photon beam replaces the monochromator entrance slit, thereby optimizing the luminosity. In a later setup, a 1-m normal-incidence monochromator of the 225 McPherson type was used. Its entrance slit head was replaced by a special unit allowing two modes of operation. Either a variable entrance slit was used at a close distance of 15 mm to the photon beam or this slit was opened far enough to enable focusing of the photon beam onto the channelplate of the multiplex detector without limitation of the observed solid angle by the entrance slit. The detector position could be readjusted *in situ*. The spectrum of Figure 23 was registered with a 100 μm entrance slit.

A position-sensitive detector for the detection of the fluorescence radiation is essential for a broad application of the PIFS technique even with the presently increasing photon fluxes from insertion devices. Multiplex detectors are available for wavelengths from the visible to the x-ray range. They are based on photocathodes followed by channelplates and a resistive anode, by open channelplate chevrons with resistive or wedge and strip anode for readout,[82,83] and by position-sensitive proportional counters.[76] With the exception of detectors with photocathodes, noncommercial systems are mostly used. The spectra in Figures 23 and 24 were registered with a channelplate detector followed by a wedge and strip anode. The latter allows the determination, in two dimensions, of the position of the center of charge of the electrons leaving the final channelplate (Figure 26). The signals from the three parts of the anode are—after amplification and analog-to-digital conversion—coupled through an interface to a computer which calculates the position and is moreover used for the spectrum processing. Figure 27a displays a two-dimensional spectrum. The one-dimensional spectrum in Figure 27b follows from an integration within a rectangular part of the two-dimensional spectrum, variable in its dimensions. Two-dimensional spectra offer the advantage of a background reduction and of a control of the channelplate behavior, especially the manifestation of so-called hot spots having an increased background rate.

A second essential part of the experimental setup shown in Figure 25 is an electron gun. It is used to determine the relative detection efficiency of the monochromator/detector combination. The electron beam can be adjusted into the position of the photon beam and electron impact-induced line radiation from atoms or molecules can be registered under practically the same experimental and geometrical conditions as are realized for the photons from the storage ring. The method is based on

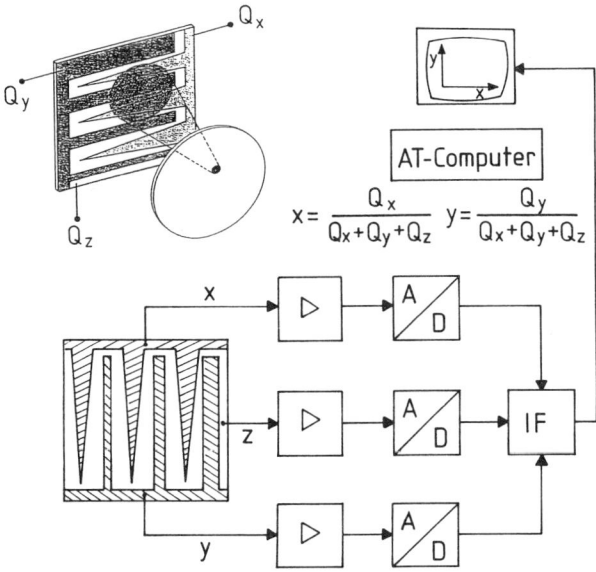

FIGURE 26. Principal scheme using a wedge and strip anode for a position-sensitive readout system.

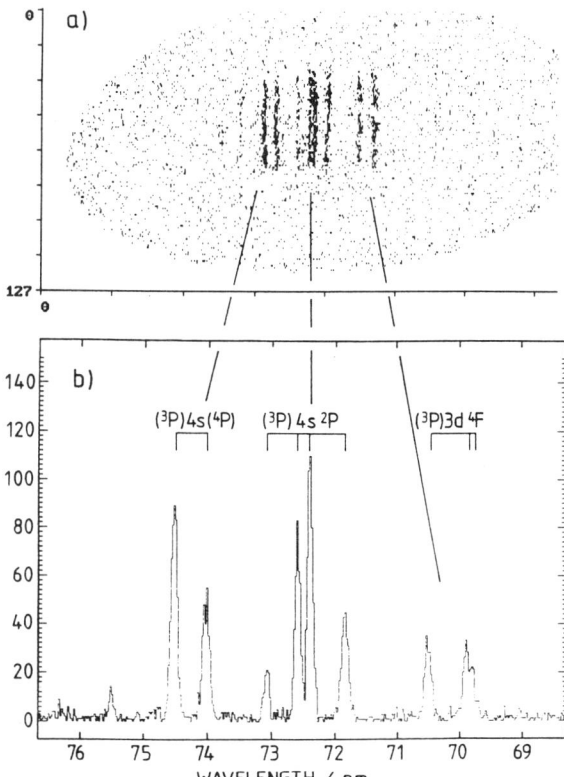

FIGURE 27. (a) Two-dimensional spectrum from a channelplate detector readout by a wedge and strip anode, registered on a resonance at 33.6 eV. (b) One-dimensional spectrum from integration of the two-dimensional spectrum within the rectangular area of panel (a).

a number of known line emission cross sections.[84–86] Electron currents of a few microamperes and target pressures of about 1 µbar yield signal rates comparable to the rates from photoionization of 10-µbar targets with flux values of about 10^{11} photons/s. Under these conditions a photoionization cross section of the order of 1 Mb corresponds to signal rates of the order of 10/s.

4.3. Photoionization of the Outer s Electrons of the Rare Gases near Threshold

The process of photoionization of the outer ns subshells ($n = 2,...,5$) of the rare gases is very suitable for study by PIFS. The s-electron vacancies decay exclusively by photon emission producing an $nsnp^6\ ^2S_{1/2}$–$ns^2np^5\ ^2P_{1/2,3/2}$ line doublet. The corresponding radiation is unpolarized, and the wavelengths belong to the spectral range of the VUV. They are listed in Table I. A principal term diagram showing schematically terms of the neutral, the singly and doubly ionized rare-gas valence shell is shown in Figure 28. It should be noted that the $nsnp^6\ ^2S_{1/2}$ level is the lowest excited level of the singly charged ion. The LS notation in Figure 28 is used for simplicity. With respect to the physics, the photoionization of the ns subshells of the rare gases is a so-called weak channel and has been found to be strongly determined by intershell correlations, described by a dipole polarization of the valence shell by the subvalence shell vacancy and by interchannel interaction (Amusia et al.,[87] Samson and Gardner[88]).

4.3.1. Ar 3s Electrons.
The Ar 3s photoionization cross section was favored in many investigations to demonstrate the breakdown of the independent particle model. During recent years it was shown, mainly by experiment, that the energy dependence of the 3s photoionization cross section is significantly structured near threshold. As observed by Schartner et al.[63] using PIFS and inde-

TABLE I
Wavelengths of the $nsnp^6\ ^2S_{1/2}-ns^2np^5\ ^2P_{1/2,3/2}$ ($n = 2,..., 5$) Transitions of the Rare Gases

	Transition		Wavelength (nm)
Ne II	$2s2p^6\ ^2S_{1/2} \longrightarrow$	$2s^22p^5\ ^2P_{1/2}$	46.2
		$2s^22p^5\ ^2P_{3/2}$	46.1
Ar II	$3s3p^6\ ^2S_{1/2} \longrightarrow$	$3s^23p^5\ ^2P_{1/2}$	93.2
		$3s^23p^5\ ^2P_{3/2}$	92.0
Kr II	$4s4p^6\ ^2S_{1/2} \longrightarrow$	$4s^24p^5\ ^2P_{1/2}$	96.5
		$4s^24p^5\ ^2P_{3/2}$	91.7
Xe II	$5s5p^6\ ^2S_{1/2} \longrightarrow$	$5s^25p^5\ ^2P_{1/2}$	124.4
		$5s^25p^5\ ^2P_{3/2}$	110.0

pendently by Adam et al.[89] using PES, later by Wills et al.[61] and by Hall et al.,[20] strong resonance structures appear at about 31 eV and higher photon energies. They result from the autoionization of doubly excited atomic states which were observed and identified previously in total absorption spectra by Madden et al.,[62] and reinvestigated recently by Baig et al.[90] A summary of the experimental and theoretical results for the Ar 3s photoionization cross section, combining data from the application of PES and of PIFS, was recently published by Möbus et al.[91]

Figure 29, which shows this summary, underlines one of the special features of PIFS, namely, that the threshold region can be scanned. The Ar 3s photoionization cross section is practically constant within 1 eV above threshold. This plateau is followed by strong resonance structure given in detail in

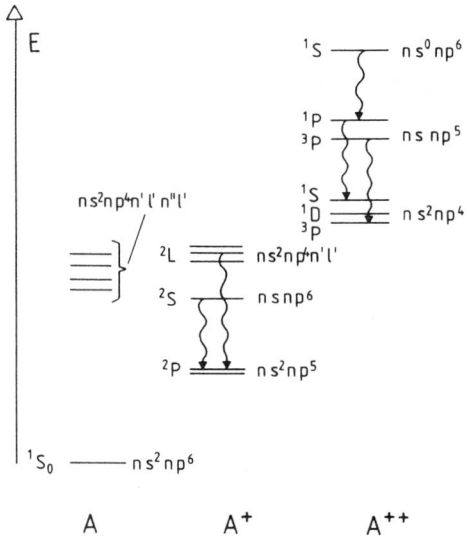

FIGURE 28. Schematic term diagram of rare-gas valence-shell levels ($n = 2, ..., 5$).

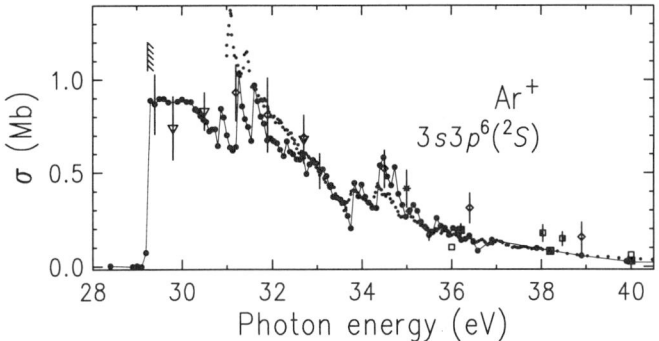

FIGURE 29. Partial cross sections of the Ar 3s → εp transition are shown as a function of the incident photon energy from threshold (29.24 eV) up to 40.5 eV. PIFS (only representative error bars are shown): ●●● (Möbus et al.[91]). PES: ■■■ (Möbus et al.[91]), ▽ (Samson and Gardner[88]), ◇ (Houlgate et al.[111]), ◨◼◻ (Adam et al.[112]), normalized to West and Marr[113] and Samson,[114] respectively; ○○○ (Adam et al.[89]), □ (Adam et al.[115]), ∗ (Kämmerling[116]). In order to guide the eye in the resonance region, the PIFS data points are connected by a solid line.

Figure 30 which contains PES and PIFS data and is reproduced from Ref. 20. Corresponding structures are also present in the energy dependence of the 3p photoionization cross section (Figure 30[20]). From absorption spectra, which were obtained with the very good resolution of 5 meV,[62] more details might be expected within these prominent resonances near threshold. They should be of special interest for a theoretical treatment, because they can only interfere with the direct photoionization of the 3s and the 3p electrons. No satellites can be influenced. Nevertheless a theoretical investigation of the interference has not been undertaken so far except for doubly excited states at higher photon energies. The result of this calculation by Wijesundera and Kelly,[92] based on the MBPT method, is compared in Figure 31 with experimental data from Figure 29 as well as with the earlier calculations (Kennedy

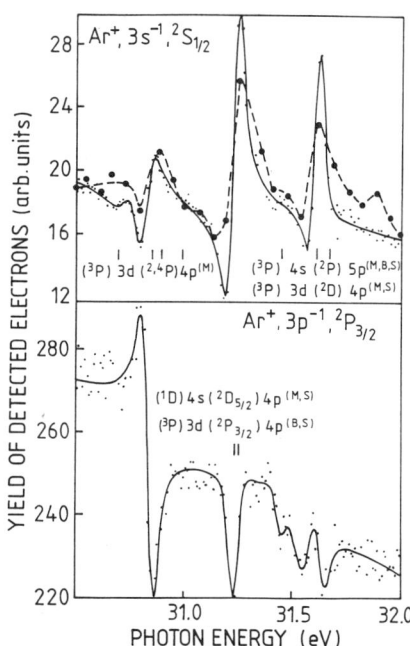

FIGURE 30. Ar 3s and 3p photoelectron spectra from Hall et al.[20] ●, PIFS data from Schartner et al.[63] Identification of doubly excited state resonances: M, Madden et al.[62]; B, Baig et al.[90]; S, Sukhorukov et al.[95]

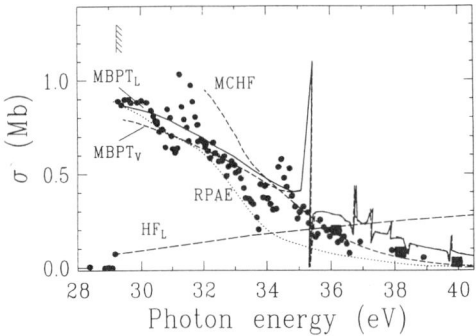

FIGURE 31. Data from Figure 29 compared with some selected theoretical calculations: Hartree–Fock (HF$_L$), Kennedy and Manson[93] (length form); random-phase approximation with exchange (RPAE), Amusia et al.[87]; many-body perturbation theory (MBPT$_{L,V}$), Wijesundera and Kelly[92] (length and velocity form, respectively); multiconfiguration Hartree–Fock (MCHF), Saha.[94]

and Manson,[93] Amusia et al.,[87] Saha[94]). The MBPT data agree very well with experiment as long as the resonances are excluded in the comparison. There is also a difference directly above threshold where the flat part of the cross section is not reproduced by the calculations. In a recent theoretical investigation, Sukhorukov et al.[95] calculated energy positions and cross sections for a considerable number of resonances of the type $3s^2 3p^4$ $nsnp$ $^1P_1$ and $3s^2 3p^4$ $ndnp$ $^1P_1$. For the calculation of the energies of these doubly excited states, the configuration interaction technique was applied. Transition amplitudes were determined in the frame of the second-order perturbation theory. Interference between the direct 3s ionization and autoionization was not taken into account. Figure 32a shows the calculated excitation cross sections for the doubly excited states represented by Lorentzian curves with 0.2 eV FWHM. In Figure 32b the sum of the cross sections is compared with the experimental Ar 3s photoionization cross section. The different vertical scales should be noted. Since the interference of the autoionization process with the direct ionization was not considered—and also no branching ratios for the decay of the doubly excited states were given—only the identification and energy positions from Ref. 95 can be compared with identifications based on the quantum defect approach from Madden et al.[62] and from Baig et al.[90]

This comparison is presented for the limited energy range between 29 and 32 eV in Table II. As mentioned before, satellites are produced only above 32.18 eV. All three studies attribute the structures around 31.6 eV to interference of the $3p^4(^3P)$ $4s(^2P)$ $5p$ and $3p^4(^3P)$ $3d(^2D)$ $4p$ resonances with the direct ionization of a 3s electron and/or a 3p electron. From the energy position at 31.6 eV and the comparison with the shape of absorption spectra,[62] the $3p^4(^3P)$ $4s(^2P_{1/2})$ $5p$ level should predominantly influence the 3s channel, while both resonances produce a more complicated structure in the 3p channel (Figure 30). There is also agreement between Madden et al.[62] and Sukhorukov et al.[95] for the $3p^4(^1D)$ $4s(^2D)$ $4p$ resonance causing interference around 31.25 eV. This resonance presents itself in the absorption spectra as a clear doublet with a window profile. The main difference in the interpretation of the absorption spectra concerns the position of the $3p^4(^3P)$ $3d(^2P)$ $4p$ doubly excited configuration. While Madden et al.[62] attribute the structures around 30.8 eV to this configuration, Baig et al.[90] quote higher energies. The calculated value from Ref. 95 does not help in this situation. The energy differences are within 0.2 eV which is the order of accuracy of the calculation and moreover the fine-structure splitting was not included. Probably also the $3p^4(^3P)$ $3d(^4P)$ $4p$ configuration has to be considered for a description of the processes around 30.8 eV. The largest difference between values from Sukhorukov et al.[95] and Madden et al.[62] is observed for the $3p^4(^3P)$ $4s(^2P)$ $4p$ configuration. At the calculated position of 30.15 eV no clear feature is seen in the absorption spectra or in the energy dependence of the 3s photoionization cross section. In conclusion, the existing experimental data certainly present prominent structures close to the threshold of the 3s photoionization cross sections which need more theoretical efforts for their final interpretation. In these efforts the autoionization process of the doubly excited states $3s^2 3p^4$ $nln'l'$ is of special interest. Their interference with Ar 3s photoionization is an indirect observation of a new three-electron Auger transition first observed in Ref. 96 for the decay of double L vacancies in Ar. Here two outer-shell vacancies decay jointly ejecting a third electron from the outer subvalence shell.[97]

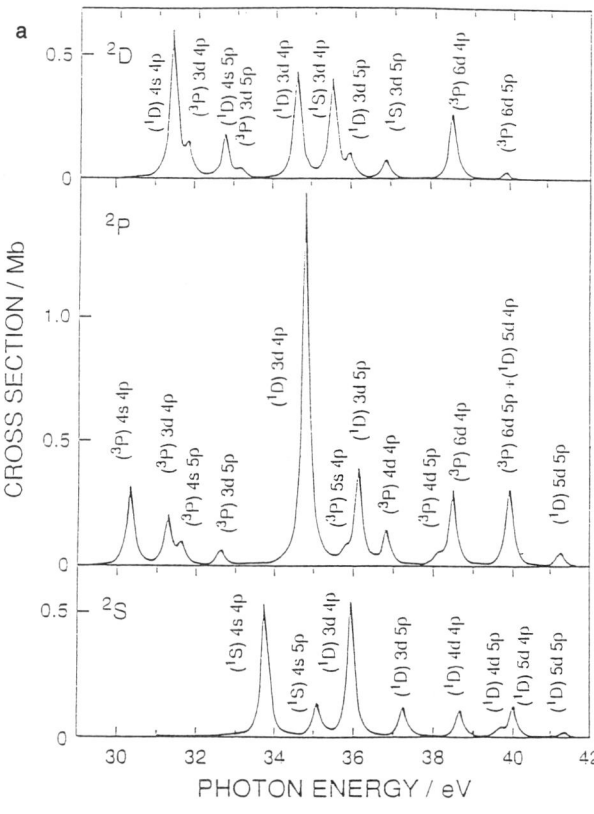

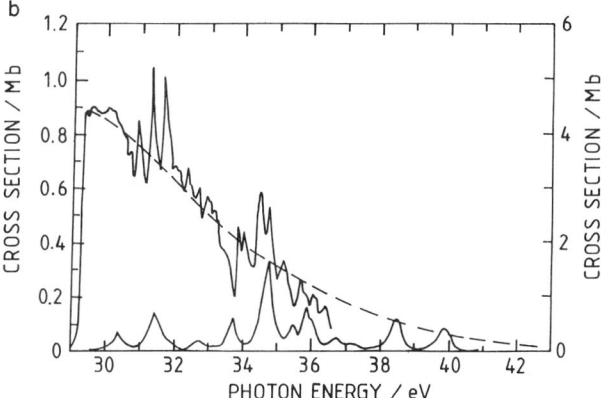

FIGURE 32. (a) Calculated excitation cross sections for doubly excited states of Ar, represented by Lorentzian curves of 0.2 eV FWHM (Sukhorukov et al.[95]). (b) Cross sections for 3s photoionization (left ordinate) and double excitation (right ordinate). Experiment: ———, Möbus et al.[91] Theory: ———, Sukhorukov et al.,[95] solid structured curve: cross sections of doubly excited states, Lorentzian curves with 0.2 eV FWHM superimposed (Sukhorukov et al.[95]), ---, Wijesundera and Kelly.[92]

4.3.2. Ne 2s Electrons. As for Ar, the near-threshold range of the Ne 2s photoionization cross section has been investigated experimentally only by PIFS. Applying a multiscaling scan technique, Schartner et al.[98] observed structures within 0.8 eV above threshold (Figure 33). In this energy range absorption spectra also show a series of narrow resonances (10 meV FWHM[99,100]). They were identified as belonging to the $2p^4(^3P)\,3s\,(^2P_{3/2,1/2})\,np$ series. The strongest absorption peak was attributed to the $2p^4(^1D)\,3s(^2D)\,3p$ resonance. The positions of these doubly excited states are marked in Figure 33 by the vertical lines. Their lengths represent the relative heights of the absorption resonances. A clear correlation is evident although the bandwidth of 40 meV of the ionizing photons for the fluorescence measurements prohibits a detailed interpretation. At present it is not known how

TABLE II
Comparison of the Identification of Doubly Excited $3p^4nln'l'$ ^1P States in Ar

E (eV)	Madden et al.[62]	Baig et al.[90]	Sukhorukov et al.[95]
29.03	$(^3P)4s(^2P_{3/2})4p$		
29.22	$(^3P)4s(^2P_{1/2})4p$		
30.15			$(^3P)4s(^2P)4p$
30.70	$(^3P)3d(^2P_{1/2})4p$		
30.84	$(^3P)3d(^2P_{3/2})4p$		
30.88	$(^3P)3d(^4P_{3/2,1/2})4p$		
30.99	$(^3P)3d(^2P_{3/2})4p$		
31.12			$(^3P)3d(^2P)4p$
31.23	$(^1D)4s(^2D_{3/2})4p$		
31.24	$(^1D)4s(^2D_{5/2})4p$		
31.25		$(^3P)3d(^2P_{3/2})4p$	
31.29			$(^1D)4s(^2D)4p$
31.45	$(^3P)4s(^2P_{3/2})5p$		
31.54		$(^3P)3d(^2P_{1/2})4p$	
31.57		$(^3P)4s(^2P_{3/2})5p$	$(^3P)4s(^2P)5p$
31.62	$(^3P)4s(^2P_{1/2})5p$		
31.68	$(^3P)3d(^2D_{3/2,5/2})4p$		$(^3P)3d(^2D)4p$

the above doubly excited states manifest themselves in the energy dependence of the Ne $2p$ photoionization cross section. A wider energy range with respect to Figure 33 was studied by PIFS (Figure 34)[81] and by PES.[25,26] The PES studies investigated mainly the satellite production which is also strongly influenced by doubly excited atomic states as demonstrated previously by Becker et al.[101] and by Heimann et al.[8] Figure 34 shows, for example, that the Ne$^+$ $2p^4(^3P)$ $3s$ ^2P satellite is predominantly populated by autoionization. The influence of the higher $2p^4(^1D)$ $3s(^2D)$ np series on the Ne $2s$-electron

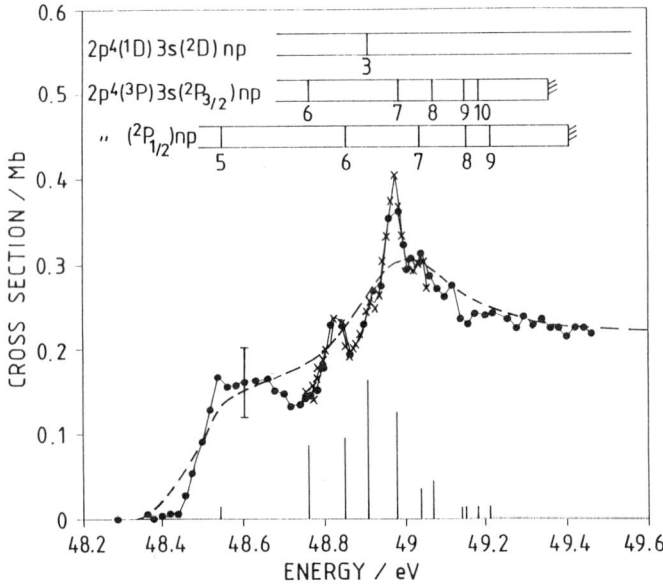

FIGURE 33. Photoionization cross section for the Ne $2s$ electron as a function of the exciting photon energy. Statistical uncertainty in the order of the dot size. ●, Schartner et al.[98]; ---, Schartner et al.[81]; vertical solid lines indicate positions and relative heights of resonances in arbitrary units from Codling et al.[99]

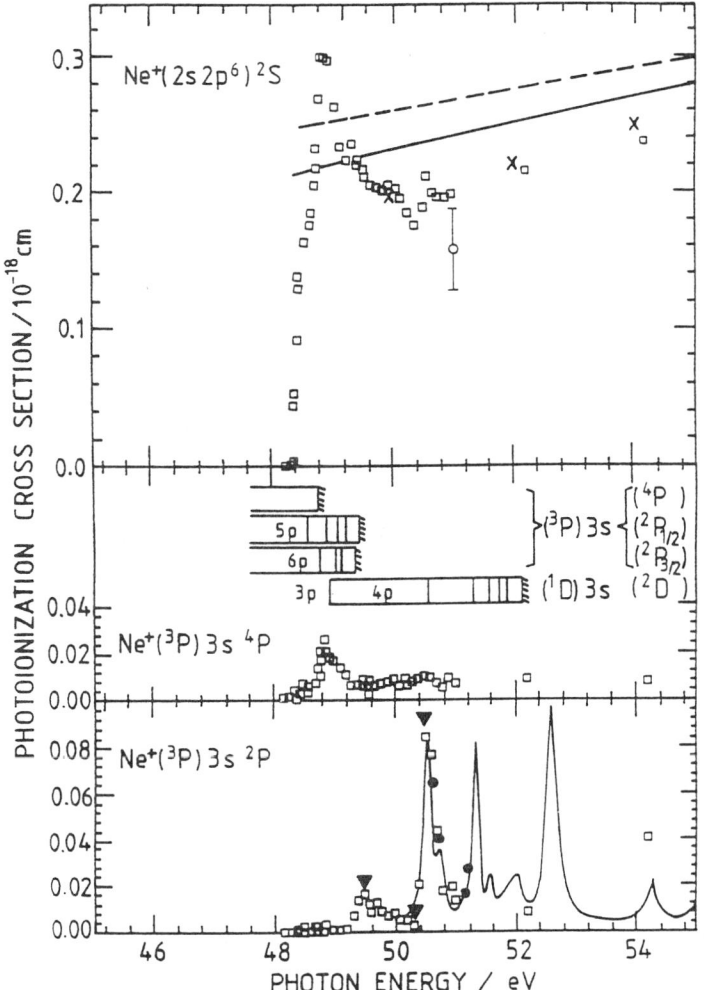

FIGURE 34. Photoionization cross section as a function of photon energy for: (Top) The Ne 2s electron: data □, Schartner et al.[81]; ○, Samson and Gardner[88]; ×, Wuilleumier and Krause[117]; calculations ---, Amusia et al.[87]; ——, Burke and Taylor.[118] (Middle) Production of the $2p^4(^3P)3s\ ^4P$ satellite: □, Schartner et al.[81] (Bottom) Production of the $2p^4(^3P)3s\ ^2P$ satellite: □, Schartner et al.[81]; ▼, Heimann et al.[8]; ●, ——, Becker et al.[101]

photoionization process was investigated in detail by Hall et al.[25] using PES. These authors verified (Figure 35) that the resonance features disappear in the energy dependence of the β parameter, the latter having a value of 2 as expected. The dip in β at the $2p^4(^3P)\ 3d(^2P)\ 3p$ resonance was attributed to a component of the Ne$^+$ $2p^4(^3P)\ 3s(^4P)$ state which is close in energy to the 2s vacancy. In contrast to Ar, no theoretical investigations of the resonance structures near the 2s-electron threshold have been published. From Figure 33 one can derive a threshold cross section of 0.16 Mb which deviates from the calculated values of 0.2 and 0.24 Mb (see Figure 34). No experimental evidence of $2p^4(^3P)\ 3s(^4P)$ np resonances which would converge toward 48.78 eV was given, but further measurements with reduced bandwidth should be carried out.

As for the $3p^4(^1D)\ 4s(^2D)\ 4p$ doubly excited state of Ar, the $2p^4(^1D)\ 3s(^2D)\ 3p$ state of Ne also seems to be a good candidate for a theoretical investigation of branching and interference. It is well

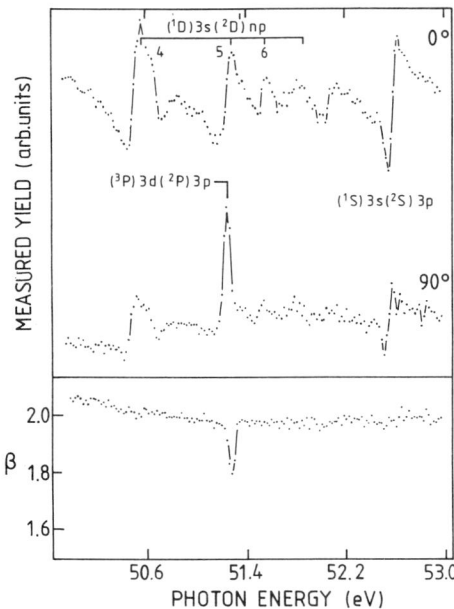

FIGURE 35. Measured differential ionization cross sections of the Ne$^+$ $2s^{-1}(^2S)$ state at 0 and 90°, and the deduced β parameter variation for the energies between 50 and 53 eV. The energy interval is 20 meV per channel (Hall et al.[25]).

separated and interferes only with the direct 2s-electron ionization, with the $2p^4(^3P)\ 3s(^2P)$ satellite production, and probably with the direct 2p-electron ionization. For Ar, the population of a satellite via the decay of the $3p^4(^1D)\ 4s(^2D)\ 4p$ resonance is energetically not possible.

4.3.3. Kr 4s Electrons and Xe 5s Electrons. Three data points from Samson and Gardner[88] were for many years the only experimental information about the Kr 4s photoionization cross section close to threshold. Schmoranzer et al.[102] applied PIFS to a study of the Kr 4s photoionization cross section across the threshold region. Not unexpectedly, in this measurement prominent structures were observed at energies which coincide with the energies of structures in the absorption spectra investigated by Codling and Madden.[68] Because of the complexity of the spectra, most of the doubly excited state resonances could not be identified. In a continuation of their measurements, Schmoranzer and co-workers increased the resolution of the fluorescence analysis, reduced the bandwidth of the exciting photons, and carried out absolute cross section measurements.[103] Figure 36a reproduces their fluorescence spectrum while panel (b) shows the term diagram of the lowest excited states of Kr$^+$ and the wavelengths of the observed transitions. The measured 4s photoionization cross section is displayed in Figure 37 together with the energetic positions of absorption resonances from Ref. 68. It is compared with the data from Samson and Gardner[88] and with calculations by Amusia et al.,[87] by Schmoranzer et al.,[104] and by Sukhorukov et al.[105] All of these calculations include electron correlations.

From Figure 37 the correspondence with the absorption features is evident, and the agreement with the earlier data[88] is very good. The comparison with the calculated results is hampered by the unknown influence of doubly excited states. The best overall agreement seems to be presented by the most recent calculations of Sukhorokov et al.[105] At energies where no absorption resonances were observed, a very good agreement with the experiment is obtained. The calculations by Schmoranzer et al.,[104] and by Sukhorukov et al.[105] differ by the inclusion of spin-orbit interaction and higher orders of perturbation theory in the latter work. Calculations by Tulkki et al.[106] give cross sections that are too low at 30 eV but describe the cross section minimum around 42 eV rather well as can be seen from a comparison carried out by Ehresmann et al.[103] for energies up to 90 eV. In contrast to the Ne 2s and

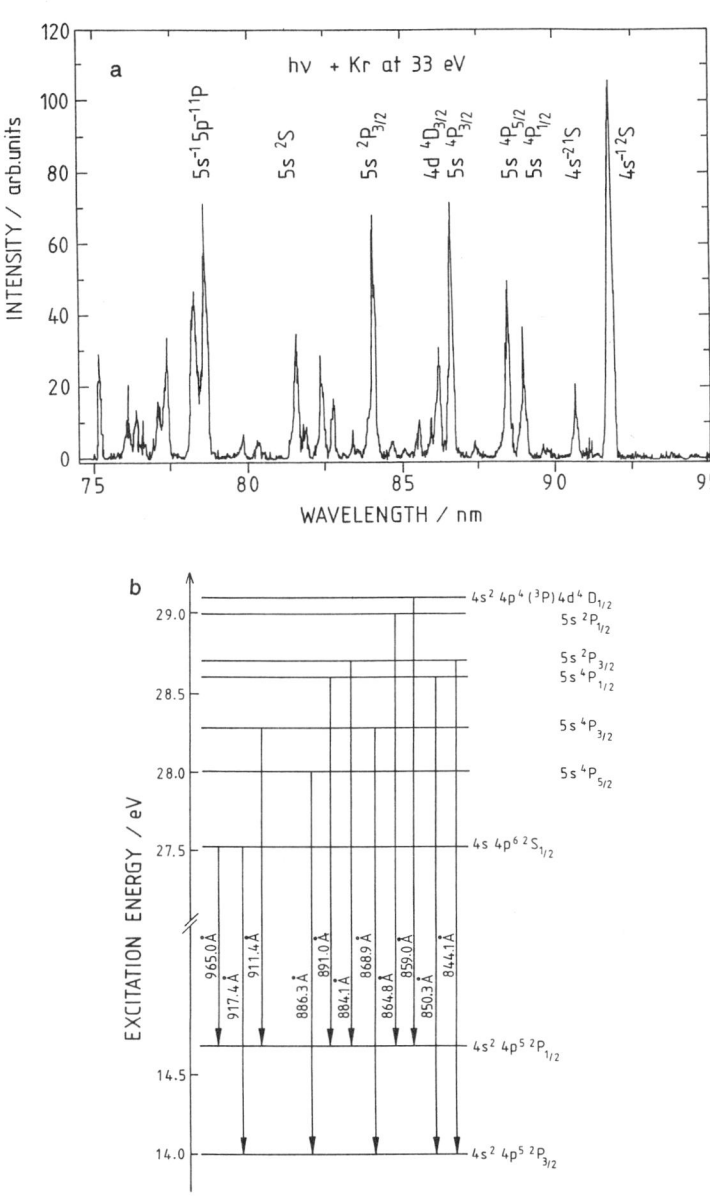

FIGURE 36. (a) Fluorescence spectrum from Kr at 33 eV exciting photon energy. The upper level of the most intense lines is specified. (b) Term diagram of the lowest states of Kr$^+$ with wavelengths of transitions between these states.

the Ar 3s photoionization, no "resonance free" cross section value very close to threshold could be derived from the experiments so far. The threshold peak in Figure 37 is still structured within 100 meV above threshold as a result of resonances at 27.55, 27.6, and 27.66 eV.[103]

The most detailed experimental knowledge about the Xe 5s-photoionization cross section close to threshold results from a threshold scan of the undispersed VUV-fluorescence radiation (Figure 38).[66] The bandwidth of the exciting photons was 15 meV and a range from 300 meV above threshold

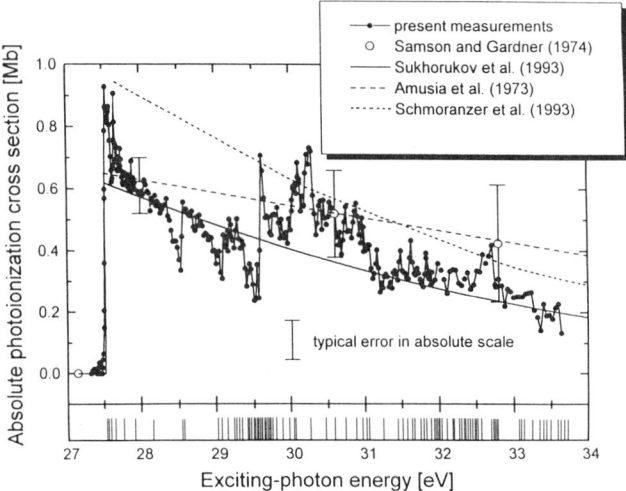

FIGURE 37. Kr 4s photoionization cross section as a function of the photon energy. Experiment: ●, Ehresmann et al.[103]; ○, Samson and Gardner.[88] Positions of photoabsorption resonances according to Codling and Madden.[68] Calculations (without inclusion of resonance excitation): ·····, Schmoranzer et al.[104]; ———, Sukhorukov et al.[105]; – – –, Amusia et al.[87]

(at 23.39 eV) upwards was covered (the first satellite has its energetic threshold at 23.67 eV). The bandwidth had to be increased to 70 meV for the applications of PIFS at 110.0 nm (see Table I). The main two peaks in Figure 38 were confirmed in the PIFS study but the double peak just above threshold was not resolved. The energetic positions of doubly excited states observed as absorption resonances were taken from Ref. 68. The authors attributed the resonances at 23.42, 23.46, 23.48, and 23.58 eV to the $5p^4$ (^3P) 6s ($^4P_{5/2}$) 11p, - ($^4P_{3/2}$) 9p, - ($^4P_{5/2}$) 12p, and - ($^4P_{3/2}$) 10p states, respectively. The remaining resonances in Figure 38 were not classified. With regard to the results for the Kr 4s photoionization described above and the even stronger breakdown of the LS-coupling scheme, a rich resonance structure would be expected from further studies of the Xe 5s photoionization process.

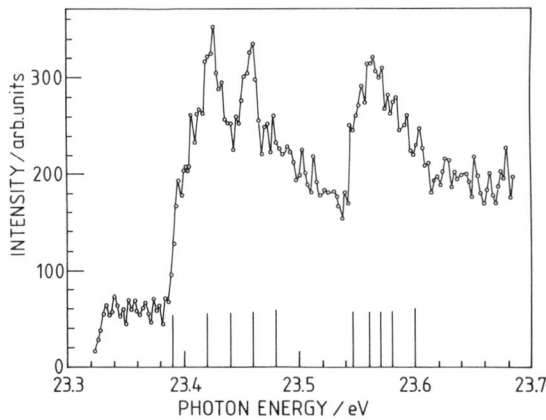

FIGURE 38. Threshold scan of the Xe 5s-electron photoionization cross section (Schartner et al.[66]).

4.4. Examples of Autoionization Resonances in Satellite Production, Comparison with Calculations

Photoionization satellites are a direct manifestation of electron correlations. The energy dependence of cross sections for their production is particularly interesting in the threshold range, revealing characteristics of the correlation dynamics (Becker and Shirley[4]). In general, the direct satellite production is a weak channel. However, resonance enhancement from the autoionization of doubly excited states can produce intense satellite lines for the heavier rare gases.[1,101] This happens in the threshold range since series of doubly excited states converge to the respective satellite states. The large number of doubly excited states dominates over the direct satellite production, complicating the comparison between calculated results and data, especially since practically no calculations have been carried out for the heavier rare gases which include the interference between the direct satellite production and the resonance effect. The exception is an MBPT calculation by Wijesundera and Kelly[92] for Ar. Unfortunately, the included doubly excited states have energies above 34 eV. Thus, they can influence a large number of satellites and not only the lowest, the ones which show the clearest features.

Extensive experimental studies of the resonance enhancement applying PES were undertaken by the Manchester groups (Hall et al.,[20,25] Wills et al.[26,61]) as described in Section 3.2. The application of PIFS is advisable only in a cascade-free energy range, otherwise detailed cascade corrections are necessary. As a consequence, the threshold range of only the lowest-energy satellites for Ne, Ar, and Kr were studied (Schartner et al.,[63,81] Schmoranzer et al.,[102] Sukhorukov et al.[105]).

4.4.1. Ne Satellites.

The $2p^4$ (^3P) $3s\,^4P$ and ^2P states produce the lowest-energy Ne$^+$ satellites, and have cross section energy dependences as shown in Figure 34 in connection with the Ne 2s cross section. No resonance enhancement is observed by PIFS for the ^4P state, which needs a spin flip for its population. The PES results of Wills et al.[26] show a weak effect at the $2p^4$ (^3P) $3s$ (^2P) $4p$ resonance. Since these data are on a relative cross section scale—and were not taken from threshold on—they cannot be compared with the PIFS data. Wills et al.[26] note that the ^4P state, with respect to the ^2P state, is enhanced with comparable intensity only by members of the $2p^4$ (^3P) $3p$ (^2D) nd series at 51.3 and 52.11 eV. On the other hand, the ^2P state—a prominent example for resonance enhancement—is populated by nearly all of the resonances observed in absorption spectroscopy.[99]

Figure 39 contains most of our present knowledge regarding the threshold range of the ^2P satellite.[107] The absolute values from the different measurements agree very well with each other. When normalized, the relative data from Ref. 26 could also be added showing more details especially around 52 eV. The calculation by Luke et al.[108] based on perturbation theory in lowest order seems to give the correct order of magnitude for the nonresonant mechanism. Luke et al. do not reproduce the cross section increase within 0.5 eV near threshold. Their cross section at threshold amounts to 0.005 Mb. Langer[107] determined the β parameter on the strong resonances, finding values close to −1, as expected for a parity-unfavored satellite. Hall et al.[25] found deviating β values between 1.9 and −0.4, varying dramatically at the 50.54- and 51.26-eV resonances.

4.4.2. Ar Satellites.

The lowest-binding-energy Ar$^+$ satellites are denoted in LS-coupling as $3p^4$ (^3P) $3d\,^4D$, and $3p^4$ (^3P) $4s\,^4P$ and $3p^4$ (^3P) $4s\,^2P$. Following Wills et al.[61] and Hall et al.,[20] the ^4D state shows a strong enhancement. In the VUV fluorescence spectra, the transitions from the ^4D state are of very weak intensity (Figure 27). The reason for this might be a strong population of the $J = \frac{7}{2}$ component which cannot decay to the Ar$^+$ ground state. The $3p^4$ (^3P) $4s\,^4P$ and ^2P satellites display complicated resonance structure. In contrast to the Ne $2p^4$ (^3P) $3s\,^4P$ and ^2P satellites, in Ar both states are populated with comparable magnitude (Figure 40). This is certainly related to the breakdown of LS-coupling for Ar. Madden et al.[62] classified many absorption resonances which are included in Figure 40 together with the identification given by Baig et al.[90] Most published cross sections for the production of the ^4P and ^2P satellites are included in Figure 40: PES data are from Becker et al.[1] and from Wills et al.[61] and Hall et al.,[20] normalized at an intense resonance on the PIFS data from Schartner et al.[63] Experimental cross sections at threshold are from Heiser et al.,[9] Heimann,[109] and Medhurst et al.[110] The reason for the disagreement of the PES data and the PIFS data for the ^2P satellite

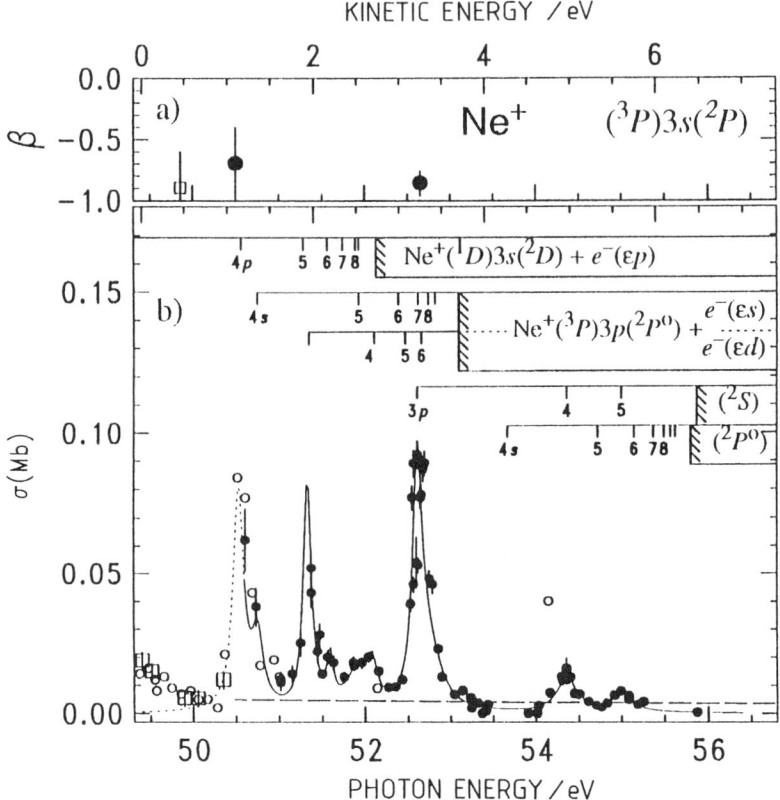

FIGURE 39. β parameter and cross section for the Ne$^+$ $2p^4$ (^3P) $3s$ (^2P) satellite as a function of photon energy. PES: —●—, Becker et al.[101]; □, Heimann et al.[119] PIFS: ○, Schartner et al.[81] Theory: ———, Luke et al.[108]

at threshold is unclear. The differences between the results obtained by PES and PIFS for the ^4P and the ^2P states at higher energies might well result from cascades from higher levels.

No theoretical investigations of the resonance structure of the lowest-binding-energy satellites near threshold have been performed to date. The direct ^2P satellite production was studied by Sukhorukov et al.[95] using a combination of the CI technique and second-order perturbation theory. Comparison to the data is not possible because of the rich resonance structure. Nevertheless, the calculation probably gives cross sections that are too large, as can be concluded from Figure 40.

A similarity between the Ne$^+$ ^2P and Ar$^+$ ^4P satellites (assuming that only the ^2P component of the so-called ^4P state is resonance enhanced) are the corresponding, strong $2p^4$ (^1D) $3s$ (^2D) $4p$ and $3p^4$ (^1D) $4s$ (^2D) $5p$ resonances. For both rare gases they are the lowest-energy resonances. The second strong resonance in Ne is attributed to the $2p^4$ (^3P) $3p$ (^2P) $3d$ resonance. The corresponding resonance for Ar was not identified by Madden et al.[62] From the identification of the $3p^4$ (^3P) $4p$ (^2D) $3d$ resonance by Baig et al.[90] and the ^2D–^2P shift, the resonance enhancement of the Ar$^+$ ^4P and ^2P population at 33.5 eV might result from the $3p^4$ (^3P) $4p$ (^2P) $3d$ resonance (Figure 40). The third strong Ne$^+$ ^2P resonance belongs to the $2p^4$ (^1S) $3s$ (^2S) $3p$ state. According to Madden et al.[62] the corresponding $3p^4$ (^1S) $4s$ (^2S) $4p$ resonance has an energy of 33.5 eV while Baig et al.[90] give a value of 33.85 eV. The assumption of a similar behavior for Ne and Ar would support the identification by Baig et al.[90] Two of the three dominating resonances discussed have a $3s$ and a $4s$ electron, respectively, in common. Becker et al.[101] suggested that resonance enhancement is strong for satellites and doubly excited states having an excited electron in common, or having at least the same symmetry of the ion core, e.g., a ^3P core for the ^2P and ^4P states discussed here.

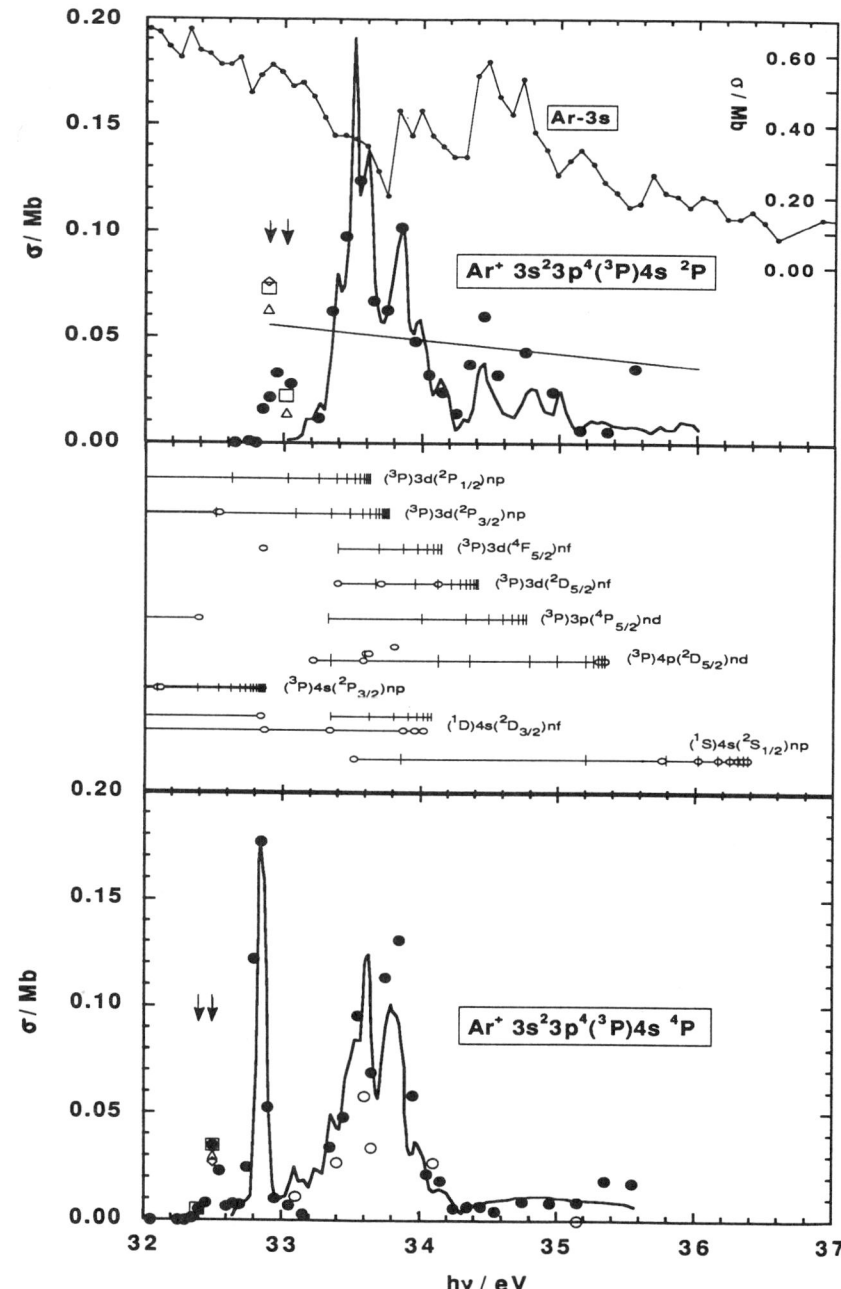

FIGURE 40. Cross sections for production of the Ar$^+$ $3p^4$ (^3P) $4s$ ^4P and ^2P satellites as a function of photon energy. PES: ○, Langer[107]; ◇, Heimann[8,109]; △, Medhurst et al.[110]; □, Heiser et al.[9]; ——, Wills et al.[61] normalized. PIFS: ●, Schartner et al.[63,79] Positions of absorption resonances: ○, Madden et al.[62]; +, Baig et al.[90]; ——, Sukhorukov et al.[95]

Many more experimental investigations with reduced bandwidth of the ionizing photons are needed to disentangle the complicated enhancement pattern, and to eliminate the discrepancies mentioned. On the other hand, the existing data for the lowest-energy satellites for Ne and Ar, in connection with the data for the main lines presented in the previous section, offer sufficient interesting material for increased theoretical effort. A step in the direction of including resonance effects in the calculation of the satellite production is the application of the MBPT by Wijesundera and Kelly,[92] e.g., for the $3p^4\,(^3P)\,4p\,^2P$ satellite. Samson et al.,[80] using PIFS in the visible, succeeded in measuring the resonance enhancement for this satellite with a 19 meV bandwidth but could not confirm an enhancement by certain doubly excited states which were quoted by Wijesundera and Kelly as dominating.

4.4.3. Kr Satellites. The lowest-binding-energy satellites in Kr$^+$ belong to the so-called (in *LS* notation) $4p^4\,(^3P)\,5s\,^4P_{5/2,3/2,1/2}$ and $^2P_{3/2,1/2}$ states. Their production cross sections as a function of photon energy were studied using PES by Hall et al.[27] and Wills et al.,[28] and using PIFS by Schmoranzer et al.[102] and by Sukhorukov et al.[105] Codling and Madden[68] observed a strongly structured absorption spectrum but gave no identification of resonances. Data from Ref. 105 are reproduced in Figure 41 for the $J = \tfrac{3}{2}$ component of the ^4P and the ^2P states. Below the abscissa the energy positions of resonances from Ref. 68 are marked. The calculated cross sections in Figure 41

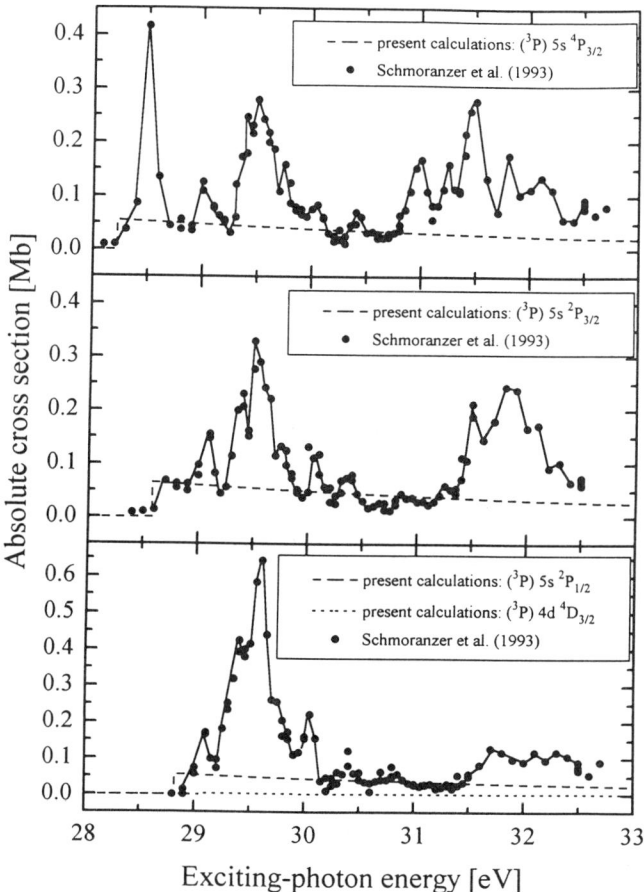

FIGURE 41. Cross sections for production of the Kr$^+$ $4p^4\,(^3P)\,5s\,^4P_{3/2}$, $^2P_{3/2}$, and $^2P_{1/2}$ satellites as a function of photon energy.[105]

TABLE III
Calculated Energies of Doubly Excited States in Kr

E (eV)	Sukhorukov et al.[105]
28.5	$4p^4\,(^3P)\,5s\,(^2D)\,5p$
	$4p^4\,(^1S)\,5s\,(^2S)\,5p$
	$4p^4\,(^1D)\,5s\,(^2D)\,6p$
29.1	$4p^4\,(^1D)\,5s\,(^2D)\,7p$
	$4p^4\,(^3P)\,4d\,(^2P,\,^2D)\,6p$

were obtained by Sukhorukov et al.[105] who refined their calculations in Ref. 103 by the introduction of spin-orbit coupling. A comparison at energies free of resonances, i.e., between 28.5 and 29 eV, yields a very good agreement between data and calculation. Sukhorukov et al.[105] additionally calculated energies of a number of doubly excited states, quoting an accuracy of 0.3 eV. In connection with the similarities discussed above for the Ne and Ar satellites, this might be sufficiently accurate to suggest a tentative classification for the isolated resonances between 28 and 29 eV (Table III).

5. SUMMARY AND FUTURE PROSPECTS

Selected examples have been presented of recent studies of threshold phenomena and resonance formation in the valence shell photoionization of the rare gases. The results illustrate that new experimental developments such as the combination of threshold photoelectron spectroscopy and coincidence techniques and the introduction of the multiplex detection mode in CIS and fluorescence spectroscopy have succeeded in revealing many interesting details of the photoionization process. This holds especially for the influence of electron correlations on the so-called weak processes like ns-subshell photoionization, satellite production, or single-photon double photoionization which is completely due to the electron–electron interaction.

Though the theoretical principles dealing with the many-electron nature of the photoionization process have been developed for many years, theoreticians are beginning just now, motivated by the rich and detailed experimental data, to increase their efforts at reproducing quantitatively the data for these weak processes, e.g., the angular correlation of the electrons emitted in single-photon double photoionization. The resonances of doubly excited atomic states offer many possibilities for testing structure calculations with respect to not only energy positions but also decay modes and branching ratios.

Further experiments will have to concentrate on specific details of the photoionization process trying to carry out experiments as completely as possible. That is, the activities in multiparameter experiments will increase. Many of them will become feasible only with the availability of the new generation of powerful synchrotron radiation facilities. These high-brightness sources will provide much higher photon fluxes than are available at present which will be particularly important for coincidence measurements. They will also provide the energy resolution in the photon beam that is required to fully resolve and characterize the wealth of structure that occurs in the valence shell photoionization of the rare gases.

REFERENCES

1. U. BECKER, B. LANGER, H. G. KERKHOFF, M. KUPSCH, D. SZOSTAK, R. WEHLITZ, P. A. HEIMANN, S. H. LIU, D. W. LINDLE, T. A. FERRETT, AND D. A. SHIRLEY, Phys. Rev. Lett. **60**, 1490 (1988).
2. H. SMID AND J. E. HANSEN, J. Phys. B **16**, 3339 (1983).
3. V. SCHMIDT, Rep. Prog. Phys. **55**, 1483 (1992).
4. U. BECKER AND D. A. SHIRLEY, Phys. Scr. **T31**, 56 (1990).

5. G. H. WANNIER, *Phys. Rev.* **90**, 817 (1953).
6. F. H. READ, in *Electron Impact Ionization*, edited by T. D. Mark and G. H. Dunn (Springer-Verlag, Berlin, 1985), p. 42.
7. A. HUETZ, P. SELLES, D. WAYMEL, AND J. MAZEAU, *J. Phys. B* **24**, 1917 (1991).
8. P. A. HEIMANN, U. BECKER, H. G. KERKHOFF, B. LANGER, D. SZOSTAK, R. WEHLITZ, D. W. LINDLE, T. A. FERRETT, AND D. A. SHIRLEY, *Phys. Rev. A* **34**, 3782 (1986).
9. F. HEISER, U. HERGENHAHN, J. VIEFHAUS, K. WIELICZEK, AND U. BECKER, *J. Electron Spectrosc. Relat. Phenom.* **60**, 337 (1992).
10. G. C. KING, M. ZUBEK, P. M. RUTTER, AND F. H. READ, *J. Phys. E* **20**, 440 (1987).
11. R. I. HALL, A. G. MCCONKEY, K. ELLIS, G. DAWBER, L. AVALDI, M. A. MACDONALD, AND G. C. KING, *Meas. Sci. Tech.* **3**, 316 (1992).
12. T. BAER, W. PEATMAN, AND E. W. SCHLAG, *Chem. Phys. Lett.* **4**, 243 (1969).
13. R. SPOHR, P. M. GUYON, W. A. CHUPKA, AND J. BERKOWITZ, *Rev. Sci. Instrum.* **42**, 1872 (1971).
14. J. M. AJELLO AND A. CHUTJIAN, *J. Chem. Phys.* **65**, 5524 (1976).
15. A. CHUTJIAN AND J. M. AJELLO, *J. Chem. Phys.* **66**, 4544 (1977).
16. S. CVEJANOVIĆ AND F. H. READ, *J. Phys. B* **7**, 1180 (1974).
17. W. B. PEATMAN, G. B. KASTING, AND D. J. WILSON, *J. Electron Spectrosc. Relat. Phenom.* **7**, 233 (1975).
18. I. NENNER AND J. A. BESWICK, in *Handbook on Synchrotron Radiation*, Vol. 2, edited by G. V. Marr (Elsevier, Amsterdam, 1987).
19. S. SVENSSON, K. HELENELUND, AND U. GELIUS, *Phys. Rev. Lett.* **58**, 1624 (1987).
20. R. I. HALL, L. AVALDI, G. DAWBER, P. M. RUTTER, M. A. MACDONALD, AND G. C. KING, *J. Phys. B* **22**, 3205 (1989).
21. P. LABLANQUIE, Doctoral Thesis, Université Paris Sud, Orsay (1989).
22. H. KOSSMAN, B. KRÄSSIG, H. SCHMIDT, AND J. E. HANSEN, *Phys. Rev. Lett.* **58**, 1620 (1987).
23. C. E. MOORE, *Atomic Energy Level*, Vol. 1 (NBS Ref. Data Series 35). (U.S. Government Printing Office, Washington, DC, 1971).
24. L. MINNHAGEN, *J. Opt. Soc. Am.* **61**, 1257 (1971).
25. R. I. HALL, G. DAWBER, K. ELLIS, M. ZUBEK, L. AVALDI, AND G. C. KING, *J. Phys. B* **24**, 4133 (1991).
26. A. WILLS, A. A. CAFOLLA, A. SVENSSON, AND J. COMER, *J. Phys. B* **23**, 2013 (1990).
27. R. I. HALL, L. AVALDI, G. DAWBER, M. ZUBEK, AND G. C. KING, *J. Phys. B* **23**, 4469 (1990).
28. A. WILLS, A. A. CAFOLLA, AND J. COMER, *J. Phys. B* **23**, 2029 (1990).
29. R. I. HALL, K. ELLIS, A. G. MCCONKEY, G. DAWBER, L. AVALDI, M. A. MACDONALD, AND G. C. KING, *J. Phys. B* **25**, 377 (1992).
30. L. AVALDI, R. I. HALL, G. DAWBER, P. M. RUTTER, AND G. C. KING, *J. Phys. B* **24**, 427 (1991).
31. R. I. HALL, A. MCCONKEY, L. AVALDI, M. A. MACDONALD, AND G. C. KING, *J. Phys. B* **25**, 441 (1992).
32. R. I. HALL, G. DAWBER, A. G. MCCONKEY, M. A. MACDONALD, AND G. C. KING, *Phys. Rev. Lett.* **68**, 275 (1992).
33. R. I. HALL, G. DAWBER, A. MCCONKEY, M. A. MACDONALD, AND G. C. KING, *Z. Phys. D* **23**, 377 (1992).
34. C. H. GREENE AND A. R. P. RAU, *Phys. Rev. Lett.* **48**, 533 (1982).
35. C. H. GREENE AND A. R. P. RAU, *J. Phys B* **16**, 99 (1983).
36. R. I. HALL, L. AVALDI, G. DAWBER, M. ZUBEK, K. ELLIS, AND G. C. KING, *J. Phys. B* **24**, 115 (1991).
37. U. FANO AND J. W. COOPER, *Rev. Mod. Phys.* **40**, 441 (1968).
38. G. C. KING, M. ZUBEK, P. M. RUTTER, F. H. READ, A. A. MCDOWELL, D. M. P. HOLLAND, AND J. B. WEST, *J. Phys. B* **21**, L403 (1988).
39. M. ZUBEK, G. C. KING, P. M. RUTTER, AND F. H. READ, *J. Phys. B* **22**, 3411 (1989).
40. F. H. READ AND S. CVEJANOVIĆ, *J. Phys. B* **21**, L371 (1988).
41. R. I. HALL, A. MCCONKEY, L. AVALDI, K. ELLIS, M. A. MACDONALD, G. DAWBER, AND G. C. KING, *J. Phys. B* **25**, 1195 (1992).
42. F. MAULBETSCH AND J. S. BRIGGS, *Phys. Rev. Lett.* **68**, 2004 (1992).
43. F. MAULBETSCH AND J. S. BRIGGS, *J. Phys. B* **26**, 1679 (1993).
44. R. I. HALL, A. MCCONKEY, L. AVALDI, M. A. MACDONALD, AND G. C. KING, *J. Phys. B* **26**, L653 (1993).
45. C. H. GREENE, *J. Phys B* **20**, L357 (1987).
46. E. W. PLUMMER, T. GUSTAFSSON, W. SUDAT, AND D. E. EASTMAN, *Phys. Rev. A* **15**, 2339 (1977).
47. P. MORIN, I. NENNER, M. Y. ADAM, J. DELWICHE, M. J. HUBIN-FRANSKIN, H. LEFEBVRE-BRION, AND A. GIUSTI, *Chem. Phys. Lett.* **92**, 609 (1982).
48. M. ZUBEK, G. C. KING, AND P. M. RUTTER, *J. Phys. B* **21**, 3585 (1988).
49. G. V. MARR AND J. B. WEST, *At. Data Nucl. Data Tables* **18**, 497 (1976).
50. M. ZUBEK, G. DAWBER, R. I. HALL, L. AVALDI, K. ELLIS, AND G. C. KING, *J. Phys. B* **24**, L337 (1991).

51. R. P. Madden and K. Codling, *Phys. Rev. Lett.* **10**, 516 (1963).
52. K. Codling and R. P. Madden, *Appl. Opt.* **4**, 1431 (1965).
53. A. Haworth, D. G. Wilden, and J. Comer, *J. Electron Spectrosc. Relat. Phenom.* **37**, 291 (1985).
54. P. J. Hicks, S. Daniel, B. Wallbank, and J. Comer, *J. Phys. E* **13**, 713 (1980).
55. M. Domke, C. Xue, A. Puschmann, T. Mandel, E. Hudson, D. A. Shirley, G. Kaindl, C. H. Greene, H. R. Sadeghpour, and H. Petersen, *Phys. Rev. Lett.* **66**, 1306 (1991).
56. D. W. Lindle, T. A. Ferrett, U. Becker, P. H. Kobrin, C. M. Truesdale, H. G. Kerkhoff, and D. A. Shirley, *Phys. Rev. A* **31**, 714 (1985).
57. D. R. Herrick and O. Sinanoglu, *Phys. Rev. A* **11**, 97 (1975).
58. M. A. Hayes and M. P. Scott, *J. Phys. B* **21**, 1499 (1988).
59. S. Salomonson, S. L. Carter, and H. P. Kelly, *Phys. Rev. A* **39**, 5111 (1989).
60. J. M. Bizau, F. Wuilleumier, P. Dhez, D. L. Ederer, T. N. Chang, S. Krummacher, and V. Schmidt, *Phys. Rev. Lett.* **48**, 588 (1982).
61. A. Wills, A. A. Cafolla, F. J. Currell, J. Comer, A. Svensson, and M. A. MacDonald, *J. Phys. B* **22**, 3217 (1989).
62. R. P. Madden, D. L. Ederer, and K. Codling, *Phys. Rev.* **177**, 136 (1969).
63. K. H. Schartner, B. Möbus, P. Lenz, H. Schmoranzer, and M. Wildberger, *Phys. Rev. Lett.* **61**, 2744 (1988).
64. J. A. R. Samson, Y. Chung, and E. Lee, *Phys. Lett.* **127A**, 171 (1988).
65. A. A. Cafolla, T. Redish, and J. Comer, *J. Phys. B* **22**, L273 (1989).
66. K. H. Schartner, P. Lenz, B. Möbus, H. Schmoranzer, and M. Wildberger, *J. Phys. B* **22**, 1573 (1989).
67. J. Tulkki, *Phys. Rev. Lett.* **62**, 2817 (1989).
68. K. Codling and R. P. Madden, *J. Res. NBS A* 761 (1972).
69. K. Codling, J. B. West, A. C. Parr, J. L. Dehmer, and R. L. Stockbauer, *J. Phys. B* **13**, L693 (1980).
70. J. Jiménez-Mier, C. D. Caldwell, and D. L. Ederer, *Phys. Rev. Lett.* **57**, 2260 (1986).
71. C. H. Greene and R. N. Zare, *Phys. Rev. A* **25**, 2031 (1982) [with corrected equation (13) in L. Kim and C. H. Greene, *Phys. Rev. A* **36**, 4272 (1987)].
72. H. Klar, *J. Phys. B* **15**, 4535 (1982).
73. W. Kronast, R. Huster, and W. Mehlhorn, *Z. Phys. D* **2**, 285 (1986).
74. B. Möbus, G. Mentzel, K.-H. Schartner, A. Ehresmann, and H. Schmoranzer, *Z. Phys. D* **30**, 285 (1994).
75. D. W. Lindle, P. L. Cowan, R. E. LaVilla, T. Jach, R. D. Deslattes, R. C. C. Perera, and B. Karlin, *J. Phys.* **48** (C9), 761 (1987).
76. R. D. Deslattes, *J. Phys.* **48** (C9), 579 (1987).
77. P. R. Woodruff and J. A. R. Samson, *Phys. Rev. A* **25**, 848 (1982).
78. J. B. West, *Z. Phys. D* **5**, 265 (1987).
79. K.-H. Schartner, P. Lenz, B. Möbus, H. Schmoranzer, and M. Wildberger, *Phys. Lett. A* **128**, 374 (1988).
80. J. A. R. Samson, E.-M. Lee, and Y. Chung, *Phys. Scr.* **41**, 850 (1990).
81. K.-H. Schartner, B. Magel, B. Möbus, H. Schmoranzer, and M. Wildberger, *J. Phys. B* **23**, L527 (1990).
82. H. Schmoranzer, K. Molter, T. Noll, and J. Imschweiler, *Nucl. Instrum. Methods A* **246**, 485 (1986).
83. B. Kraus, K.-H. Schartner, F. Folkmann, A. E. Livingston, and P. H. Mokler, *Proc. SPIE* **1159**, 217 (1989).
84. K.-H. Schartner, B. Kraus, W. Pöffel, and K. Reymann, *Nucl. Instrum. Methods B* **27**, 519 (1987).
85. P. J. M. van der Burgt, W. B. Westerveld, and J. S. Risley, *J. Phys. Chem. Ref. Data* **18**, 1757 (1989).
86. W. Jans, B. Möbus, M. Kühne, G. Ulm, M. Anton, and K.-H. Schartner, *Appl. Opt.* **34**, 3671 (1995).
87. M. Y. Amusia, V. K. Ivanov, N. A. Cherepkov, and L. V. Chernysheva, *Phys. Lett. A* **40**, 361 (1972).
88. J. A. R. Samson and J. L. Gardner, *Phys. Rev. Lett.* **33**, 671 (1974).
89. M. Y. Adam, P. Morin, and G. Wendin, in LURE Annual Activity Report No. 37 (1985).
90. M. A. Baig, S. Ahmad, J. P. Connerade, W. Dussa, and J. Hormes, *Phys. Rev. A* **45**, 7963 (1992).
91. B. Möbus, B. Magel, K.-H. Schartner, B. Langer, U. Becker, M. Wildberger, and H. Schmoranzer, *Phys. Rev. A* **47**, 3888 (1993).
92. W. Wijesundera and H. P. Kelly, *Phys. Rev. A* **39**, 634 (1989).
93. D. J. Kennedy and S. T. Manson, *Phys. Rev. A* **5**, 227 (1972).
94. H. P. Saha, *Phys. Rev. A* **39**, 2456 (1989).
95. V. L. Sukhorukov, B. M. Lagutin, H. Schmoranzer, I. D. Petrov, and K.-H. Schartner, *Phys. Lett. A* **169**, 445 (1992).
96. V. V. Afrosimov, D. K. Rasulov, A. P. Shergin, Y. S. Gordeev, and A. N. Zinovév, *JETP Lett.* **21**, 249 (1975).

97. V. Kilin, A. Ehresmann, H. Schmoranzer, and K.-H. Schartner, in *Europhysics Conference Abstracts, Fourth European Conference on Atomic and Molecular Physics, Riga*, Part I, p. 167 (1992).
98. K.-H. Schartner, B. Möbus, G. Mentzel, A. Ehresmann, F. Vollweiler, and H. Schmoranzer, *Phys. Lett. A* **169**, 393 (1992).
99. K. Codling, R. P. Madden, and D. L. Ederer, *Phys. Rev.* **155**, 26 (1967).
100. M. Domke, G. Remmers, and G. Kaindl, *Verh. Dtsch. Phys. Ges.* 1211 (1992).
101. U. Becker, R. Hölzel, H. G. Kerkhoff, B. Langer, D. Szostak, and R. Wehlitz, *Phys. Rev. Lett.* **56**, 1120 (1986).
102. H. Schmoranzer, M. Wildberger, K.-H. Schartner, B. Möbus, and B. Magel, *Phys. Lett. A* **150**, 281 (1990).
103. A. Ehresmann, F. Vollweiler, H. Schmoranzer, V. L. Sukhorukov, B. M. Lagutin, I. D. Petrov, G. Mentzel, and K.-H. Schartner, *J. Phys. B* **27**, 1489 (1994).
104. H. Schmoranzer, A. Ehresmann, F. Vollweiler, V. L. Sukhorukov, B. M. Lagutin, I. D. Petrov, K.-H. Schartner, and B. Möbus, *J. Phys. B* **26**, 2795 (1993).
105. V. L. Sukhorukov, B. M. Lagutin, I. D. Petrov, H. Schmoranzer, A. Ehresmann, and K.-H. Schartner, *J. Phys. B* **27**, 241 (1994).
106. J. Tulkki, S. Aksela, H. Aksela, E. Shigemasa, A. Yagishita, and Y. Furusawa, *Phys. Rev. A* **45**, 4640 (1992).
107. B. Langer, Thesis, FU Berlin (1993).
108. T. M. Luke, J. D. Talman, H. Aksela, and M. Levasalmi, *Phys. Rev. A* **41**, 1350 (1990).
109. P. A. Heimann, Thesis, University of California, Berkeley (1986).
110. L. J. Medhurst, A. Schach von Wittenau, R. P. van Zee, S. Zhang, S. H. Liu, D. A. Shirley, and D. W. Lindle, *J. Electron Spectrosc. Relat. Phenom.* **52**, 671 (1990).
111. R. G. Houlgate, J. B. West, K. Codling, and G. V. Marr, *J. Electron Spectrosc. Relat. Phenom.* **9**, 205 (1976).
112. M. Y. Adam, F. Wuilleumier, S. Krummacher, N. Sandner, V. Schmidt, and W. Mehlhorn, *J. Electron Spectrosc. Relat. Phenom.* **15**, 211 (1979).
113. J. B. West and G. V. Marr, *Proc. R. Soc. London Ser. A* **349**, 397 (1976); G. V. Marr and J. B. West, *At. Data Nucl. Data Tables* **18**, 497 (1976).
114. J. A. R. Samson, private communication.
115. M. Y. Adam, P. Morin, and G. Wendin, *Phys. Rev. A* **31**, 1426 (1985).
116. B. Kämmerling, Diplomarbeit, Universität Freiburg, unpublished (1986).
117. F. Wuilleumier and M. O. Krause, *J. Electron Spectrosc.* **15**, 15 (1979).
118. P. G. Burke and K. T. Taylor, *J. Phys. B* **8**, 2620 (1975).
119. P. A. Heimann, C. M. Truesdole, H. G. Kerkhoff, D. W. Lindle, T. A. Ferrett, V. C. Bahr, W. D. Bremer, U. Becker, and D. A. Shirley, *Phys. Rev. A* **31**, 2260 (1985).

CHAPTER 11

RESONANT AND NONRESONANT AUGER RECOMBINATION

H. AKSELA, S. AKSELA, AND N. KABACHNIK

1. INTRODUCTION

In the simplest picture the Auger process, named after its discoverer the French physicist Pierre Auger, can be considered as a two-step process. In the initial stage of the normal Auger process the atom is singly ionized. In the high-energy limit of the photoionization process the atom first relaxes in its core hole state. In a practically distinct second step, the core hole is then filled by an outer electron and the excess energy is released by means of photon emission or by emission of a second electron which is called the Auger electron (Figure 1a). When the binding energies of the initial core hole states are in the soft x-ray region (≤ 1 keV), the Auger process is a strongly dominating decay channel. There are commonly used symbols for the emitted Auger electrons which are based on x-ray energy notations and indicate the initial- and the two final-state core holes, e.g., $K - L_1L_{2,3}$. Alternatively, the orbital notations for the vacancies are used, e.g., $1s^{-1} \to 2s^{-1}2p^{-1}$. Auger transitions, where the initial-state hole is in the same main shell as one or both of the final-state holes, are called Coster–Kronig (CK) or super Coster-Kronig (SCK) transitions, respectively. The energy of the outgoing Auger electron is given by the difference of the total energy of the initial state of the system (atom, molecule, solid state) with a hole in an inner subshell and the final state with two holes in the outer subshells, e.g., $E(K - L_1L_{2,3}) = E[(K)^+] - E[(L_1L_{2,3})^{2+}]$. The Auger emission is very sensitive in testing the quantum theory of the structure of the emitting system. Theoretical background and calculation methods will be discussed in more detail below. For recent review papers the reader is referred to Refs. 1–6.

Because of its double hole final-state property, Auger spectra are always characterized, even in the case of closed-shell atoms, by several line components in comparison with the usually single photoelectron line of a core level. The splitting between numerous multiplet components is often so small that a very high electron energy resolution is needed to obtain reliable positions and intensities for more or less overlapping lines. Since molecular and solid-state effects always cause considerable broadening to the Auger lines, free atoms are most suitable for detailed studies of Auger spectra. Solid-state effects also cause a very high background to the spectra which is difficult to estimate accurately, especially below the extended Auger line groups. Therefore, we will concentrate in this chapter on the gas-phase spectra. On the other hand, atomic spectra serve as useful references when the influences of the molecular or solid-state surroundings as line shapes, chemical shifts, and extra-atomic relaxation effects are studied by means of Auger spectroscopy.

The simplest atomic samples and the only ones which appear as free atoms under normal conditions are the rare gases which have been studied first and most carefully. The number of atomic species can be essentially increased using the vaporization of solid samples. Most solid elements vaporize mainly as free atoms. This provides a very important method to also obtain open-shell atomic

---

H. AKSELA AND S. AKSELA • Department of Physics, University of Oulu, 90570 Oulu 10, Finland. N. KABACHNIK • Institute of Nuclear Physics, Moscow State University, 119899 Moscow, Russia.

VUV and Soft X-Ray Photoionization. Edited by Uwe Becker and David A. Shirley. Plenum Press, New York, 1996.

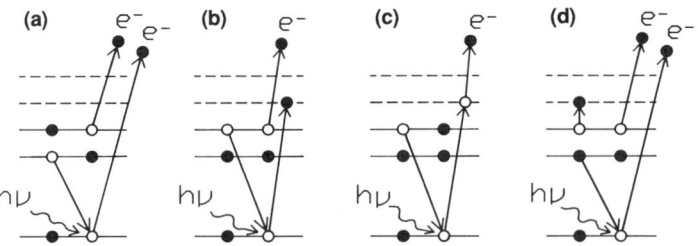

FIGURE 1. Schematic representation of (a) normal, (b) spectator, (c) participator, and (d) satellite Auger transitions.

species. Auger studies of metal vapors have been done for several atoms mainly by the research groups in Oulu and Freiburg. The other methods to generate atomic species as chemical reactions and discharge processes have not yet been applied significantly in the studies of Auger spectra, but they are very interesting and will certainly get more attention in the future.

The experimentally observed spectrum is always a convolution of the inherent spectrum emitted by the sample and the instrument function representing the broadening caused by the used spectrometer. For a reliable extraction of line energies, intensities, and widths, the resolution of energy analyzers should be such that the line broadening caused by the analyzer is smaller than the inherent lifetime width of the Auger lines. In practice this usually means that the energy resolution should be better than 0.05% of the initial kinetic energy of Auger electrons. The inherent linewidths can also be determined reliably if the instrument function of the spectrometer is accurately known or can be measured simultaneously.

Within the simplified stepwise model of the Auger process, the Auger transitions are independent of the primary ionization process and thus they can equally be created by photons or electrons. Photon excitation produces considerably less background from inelastically scattered electrons but the intensity of the usual laboratory soft x-ray sources is much lower than what can be easily obtained by simple electron guns. Especially in the case of gaseous targets, the atomic density in the interaction volume is so low that a high-intensity electron beam excitation is usually a very practical solution. Therefore, most of the published gas-phase Auger spectra have been excited by electron beams. However, recent progress[7,8] in the development of high-intensity soft x-ray sources equipped with rotating anodes, effective differential pumping systems, and position-sensitive detectors has also made the x-ray excited Auger spectroscopy a very powerful method.

Synchrotron radiation also provides a very elegant and unique excitation method for normal Auger spectroscopy. The photon flux from modern storage rings (especially those with insertion devices) is, even after high-resolution monochromators (e.g., toroidal or spherical grating monochromators), so high that Auger spectra from gaseous samples can be successfully studied. The unique advantage of synchrotron radiation is that the energy of the exciting radiation can be selected to fulfill optimal conditions. Thus, excitation of deeper core level than that considered can be avoided. High-energy (1–5 keV) electrons often cause ionization with rather high probability also in deeper levels. This causes Auger cascades and, as a result, a more complicated Auger spectrum. As an example of this effect the $L_{2,3}$–MM Auger spectra of Ar excited by synchrotron radiation at two photon energies are displayed in Figure 2. The 350-eV photons can also ionize $2s$ electrons causing additional structure in the spectrum and a higher continuous background.

Additional structure in the Auger spectrum exemplified in Figure 2 is the so-called satellite lines or simply satellites. Usually all lines, which appear in Auger spectra besides the normal or diagram lines and which arise from multiple ionizations or excitations or other many-electron effects, are called satellites (Figure 1d). In Auger spectra produced by photoionization or by electron impact the satellites are considerably weaker than the diagram lines. However if protons or heavier ions are used for initial ionization, then multiple ionizations or/and excitations are produced very effectively in addition to the considered single core hole ionization. Therefore, the typical spectra excited by ions are very rich in

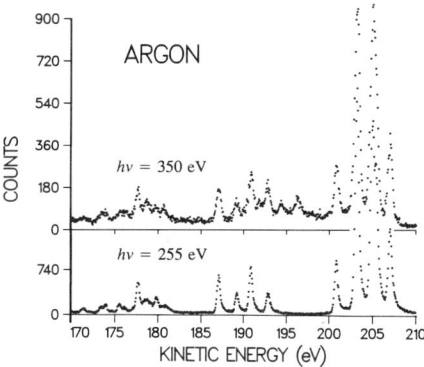

FIGURE 2. L–MM Auger spectra of Ar excited by 350- and 255-eV photons.

satellite structure and as a rule the satellites are more intensive than the diagram lines. This is, of course, a big advantage if one is interested in the satellite Auger processes but complicates the studies of normal Auger spectra.

The most powerful use of synchrotron radiation is when it is used not to ionize a given core level but to excite its electrons selectively to different unfilled orbitals (Figure 1b). These resonantly excited states, with one core hole and an extra electron in some outer level, decay again most probably via Auger recombination which can be classified into two major types. First, the excited electron may remain as a passive spectator when the core hole is filled by another outer-shell electron and a second electron is emitted. This is called the spectator resonant Auger transition (Figure 1b). The second possibility is that the excited core electron participates directly in the emission process which is then called the participator resonant Auger or the autoionization process (Figure 1c), because it is identical to the autoionization phenomenon appearing commonly in outer-shell excitations. As has been shown in many recent studies[9–14] the spectator electron can also shake up or down during the decay process, complicating this simple classification. The resonant Auger recombination can provide very useful complementary information to the normal Auger spectra, and their studies have increased considerably during the last few years. This is very much a consequence of the recent developments of high-brilliance synchrotron radiation sources, high-resolution monochromators and electron spectrometers. It is not difficult to predict that in the future, more and more experimental activities will be devoted to synchrotron radiation excited Auger and especially resonant Auger recombination studies.

Besides the conventional experiments with a fixed emission angle, the measurements of Auger electron angular distributions using the methods of angular resolved electron spectrometry are becoming more and more common, especially in the case of resonant Auger spectra where very strong asymmetry of the angular distribution of different lines has been observed recently.[15,16] Measurements of the angular distribution anisotropy provide complementary and more detailed information about the Auger recombination as well as the initial photoionization processes as compared to the conventional experiments.

Even more refined information can be obtained from coincidence measurements of the two outgoing electrons (photo and Auger electrons or two Auger electrons). Recently, this powerful method has been applied in studying Auger and resonant Auger transitions.[17,18] First measurements demonstrate advantages of the method in the efforts to gain better understanding of these phenomena.

2. INSTRUMENTATION AND METHODS

2.1. Conventional Auger Spectroscopy

In the gas phase the atom density in the source volume is always very low, several orders of magnitudes lower than in the solid targets. Therefore, rather intensive exciting beams are usually needed. Electron beam excitation is one of the most useful methods. Typical electron beam currents are from 100 µA to 1 mA compared to nanoamperes used in solid-state Auger spectroscopy. The electron beam accelerated usually by a potential difference of 1–3 kV crosses the atom beam or hits the effusive gas target inside of the gas cell or the oven. The Auger electrons created are then analyzed in a given direction (e.g., at 90 or 54.7°) relative to the direction of the primary beam. The electrons can be analyzed either directly with their full kinetic energy or after a given retardation caused by some electron lens or grid system.

The electron spectrometers which are used at present in Auger and resonant Auger spectroscopy are in most cases electrostatic analyzers, although a lot of important work has been done by Shirley and Becker and their co-workers[19–21] using time-of-flight spectrometers. The resolution of time-of-flight electron spectrometers deteriorates fast with increasing kinetic energy. Therefore, they are usually limited to studies of low kinetic energies. The most common basic types of electrostatic analyzers are cylindrical mirror analyzers (CMA) and hemispherical analyzers (HSA). A CMA has the advantage that its transmission is principally higher because of the higher usable solid angle related to cylindrical geometry. Electrons can also be detected conveniently at the magic angle of 54.7°, which means then that the measured intensity is independent of the possible angular anisotropy of the emitted electrons. This kind of setup was used e.g. by Mehlhorn and co-workers in their pioneering gas-phase Auger studies.[22,23] In order to measure the angular independent spectra of emitted electrons the HSA should be mounted at the magic angle of 54.7° with respect to the primary electron beam instead of the very commonly used angle of 90°. In the case of synchrotron radiation excitation, the HSA should be mounted at the double magic angle fulfilling the magic angle requirement for both the horizontal and vertical components of the electric vector of the radiation.[24,25]

2.2. Angular Resolving Spectrometers

The straightforward method for measuring the angular distributions of the emitted electrons is to mount the analyzer on a turntable and measure the intensity at different angles with respect to the reference direction. More advantageous is to use two or even three analyzers[26,27] looking at exactly the same source point and either keeping them at fixed angles (e.g., at 0 and 90°) or keeping one analyzer at a fixed position to measure the reference signal and moving the other ones. HSAs have been used most commonly as the energy analyzers[26,27] but also sector CMAs have been used very successfully, e.g., by V. Schmidt and his group.[28] A very interesting and promising method is to use the CMA equipped with several detectors or also with position-sensitive detectors as the angular resolving instrument by detecting the azimuthal angles of the electrons. This method has been used by the Debrecen group[29] and very recently by Feldhaus et al.[30] at BESSY. The essential advantage of this method is that a large range of emission angles are detected simultaneously and the angular independent total intensity is also obtained at the same time.

Experimental results obtained so far by different instruments show very large scatter. This clearly indicates the need for very careful calibration of the systems with well-known reference spectra. There also clearly exists a need for angular resolved measurements of normal Auger lines, because almost all available results are only for rare gases. Very interesting will be accurate measurements of angular distributions of open-shell metal atoms.

2.3. Synchrotron Radiation Excited Auger and Resonant Auger Spectroscopies

The recent progress in the development of intensive synchrotron radiation sources has also created completely new possibilities for gas-phase Auger and resonant Auger spectroscopy. The fundamental problem in the synchrotron radiation excited gas-phase photoelectron and Auger electron spectroscopy is the low intensity. With a given photon flux from the storage ring the monochromator resolution (slit widths) and the electron spectrometer resolution (slit widths or pass energy) must be adjusted so that a reasonable count rate of electrons can be obtained. Therefore, most of the experimental gas-phase studies have been done until now only with moderate total resolution especially for higher kinetic energies. Only very recently have the first really high-resolution photoelectron or resonant Auger studies been achieved, mainly by Bancroft's group[31–35] at the Canadian Synchrotron Radiation Facility and by Krause et al.[36] at the Aladdin storage ring in Madison, Wisconsin.

As an example of the needed photon flux to the sample, let us consider the case that we have a pressure of 1×10^{-4} mbar in the target volume. Then the number of atoms or molecules in 1 mm^3 is about 2.7×10^9. The photon spot size is typically 1–5 mm. Therefore, a cubic millimeter is a proper unit also for the effective source volume of the electron spectrometer. If the cross section for the primary photoionization is 1 Mb = 10^{-18} cm^2, the total cross section for the atoms in 1 mm^3 is 2.7×10^{-9} cm^2. With the flux of monochromatized photons of 10^{10} photons/mm^2·s to the source, then 2.7×10^3 photoionizations/mm^3 take place. The transmission of the spectrometer can be at best of the order of 0.1%, which corresponds with the spherical analyzer to an effective angular opening of $2° \times 20°$ without retardation. Thus, under these conditions we should count less than 3 electrons/s from 1 mm^3 effective source volume.

This estimation can be compared with some experimentally observed intensities. Bozek et al.[31] estimated that the intensity at the Canadian beamline with the Grosshopper IV monochromator to the source point is 10^{10} photons/s with a good resolution (0.08 eV) at the photon energy of 90 eV. Richter et al.[37] estimated their photon flux to be ~10^{11} photons/s at the TGM beamline at BESSY with a moderate resolution $E/\Delta E = 330$. In another study Bozek et al.[31] observed from the Xe $4d_{5/2}$ photoline a peak intensity of about 160 counts/s. The total cross section for Xe $4d$ absorption at 94 eV is ~ 25 Mb giving ~ 15 Mb for the $4d_{5/2}$ subshell. They applied a very high pressure of 0.1 mbar in the gas cell. From these values we get for the effective transmission of their spectrometer for 1 mm^3 effective source 4×10^{-6}. In the same study Bozek et al. obtained for Kr MNN Auger lines with experimental linewidths ≈ 0.17 eV the peak intensities of about 10^3 counts/s. For the Auger measurements the monochromator slits can be made much wider and thus the photon flux in the source volume can be of the order of 10^{12}–10^{13} photons/s. Also in this case the effective transmission seems to be of the order of $1-10 \times 10^{-5}$. These numbers are just rough estimates. It is difficult to know the actual pressure in the source region and the synchrotron radiation beam is probably not well matched to the slit when the analyzer is at the magic angle, causing too low transmission. In order to partially overcome these intensity problems, one very useful method is to use position-sensitive detectors, and then intensity gains of 10–100 can be easily obtained.

At present, the best numbers of monochromatized photons/s to the source volume at photon energies above 50 eV are of the order of 10^{10}–10^{11}. These fluxes under optimal target pressure and spectrometer transmission conditions with position-sensitive detectors make possible total experimental linewidths which are equal or better than 1.5 × the inherent lifetime width of lines. This goal was very recently reached[35] in the resonant Auger study of Kr M–NN spectra using the radiation from the undulator source at Aladdin.

Gases are the easiest samples because relatively high target pressures can be conveniently created. In the case of metal vapor targets, not so high target densities can be obtained. This in turn means that even higher photon fluxes are needed for very high-resolution measurements. Therefore, the coming so-called third-generation synchrotron radiation sources based on the use of insertion devices are very interesting. Such sources will be MAX II (Lund), ALS (Berkeley), APS (Argonne), Elettra (Trieste), and BESSY II (Berlin).

Besides the electron spectrometers, very high resolution is also needed for the monochromators in order to separate the adjacent resonances in the resonant Auger measurements. Typically a photon bandwidth less than 100 meV at 100 eV is necessary for high-resolution Kr M–NN resonant excitations. Such resolutions can be achieved by modern spherical[38–40] or toroidal grating monochromators with movable slits or with high-resolution plane grating monochromators.[41]

2.4. Metal Vapor Ovens

The most common method for producing free atoms other than rare gases is the vaporization of solid samples. Most solid elements vaporize mainly as free atoms and only some as molecular clusters (e.g., Sb_4, Te_2). The typical vapor pressure used in the source volume is about 10^{-4} mbar. The temperatures needed to generate these vapor pressures vary from some hundred degrees up to 3000 °C. Resistive heating applying bifilar heating wires can be done conveniently until ~ 1000 °C. For higher

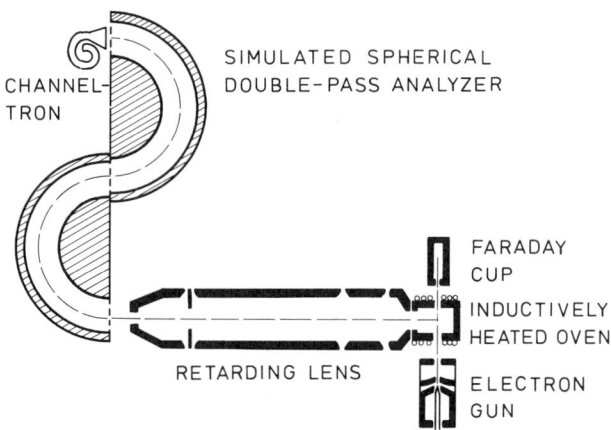

FIGURE 3. The experimental setup used to measure Auger electron spectra at the University of Oulu.

temperatures mainly two types of heating methods are used, namely, inductive heating[42–46] and electron bombardment heating.[47] Inductive heating is technically a rather simple method to realize. Its main problem is the radio frequency noise which is often induced in the electronic circuits, especially in data collection systems. Therefore, gating must usually be applied so that during the short heating periods (10–100 ms) no data are collected and during the data collection periods (10–100 ms) heating is off. This of course causes reduction in the data collection efficiency, which increases with needed temperatures because the heating to data collection time ratio must be increased. Electron bombardment heating has been applied[47] successfully by Sonntag's group at DESY. It is much more noise-free and allows continuous data collection which is important especially in measurements with synchrotron radiation because the beam time is very valuable.

The main problem in very high-temperature evaporation is extremely high reactivity of many metals as liquids and vapors. For several elements graphite or ceramic oven materials are best suited but for some Ta, Mo, or W crucibles have to be used.

As an example of high-temperature Auger electron spectrometers, the measuring system[43,44] used by our research group at the University of Oulu is shown in Figure 3. It comprises an inductively heated sample oven, a four-element retarding lens between the vapor sample and a double-pass simulated spherical-field electron energy analyzer. The system has been used to measure electron excited Auger spectra, e.g., from vapors of Au,[43] Pd,[44] and 3d transition metals.[45,46]

3. BASIC IDEAS OF THEORETICAL TREATMENT

3.1. Auger Transition Energies

Normal Auger electron lines result from transitions in atoms (molecules, solids) having a single initial vacancy and two final vacancies in a ground-state configuration. The energies of Auger transitions can be obtained by the energy difference between separate self-consistent-field (SCF) calculations for the total energies of the singly ionized initial- and doubly ionized final-state levels of the emitting system (the ΔE_{SCF} approach). For free atoms the total energies of electronic states can be calculated, e.g., with the multiconfiguration Dirac–Fock (MCDF) approach[48] which will be briefly discussed here. For the calculation of Auger transition energies using nonrelativistic or semiempirical methods, the reader is referred to the review articles by Åberg and Howat[3] and by Larkin.[49]

Atomic-state wave function (ASF) for the MCDF method is presented[48] as a linear combination of configuration-state functions (CSF). The ASF of the νth electronic state of an atom with total angular momentum JM is thus given by

$$\psi_v(JM) = \sum_{\beta=1}^{n} c_{v\beta}^J \phi_\beta(JM) \tag{1}$$

where n is the number of CSF included in the expansion and $c_{v\beta}^J$ are the mixing coefficients for state v. The CSF are constructed from antisymmetric products of Dirac central-field four-spinors.

For the MCDF method the eigenvectors (mixing coefficients) and the spinors are determined by applying the variational method to the expectation value of the Hamiltonian with respect to ASF (1) subject to orbital orthogonality constraints. The Hamiltonian for an N-electron atom or ion used in the variational principle is taken to consist of a sum of single-particle Dirac Hamiltonians plus the purely coulombic interelectronic repulsion

$$H = \sum_{i=1}^{N} H_i + \sum_{\substack{i,j \\ i<j}}^{N} \frac{1}{r_{ij}} \tag{2}$$

The Breit interaction, along with other quantum-electrodynamic (QED) contributions, are added to the Hamiltonian matrix once the orbitals have been determined. The complete matrix of the Hamiltonian is diagonalized to determine the corrected energy levels and mixing coefficients.

3.2. Transition Probabilities

The Auger transition probabilities are usually calculated from perturbation theory. This implies a two-step model in which the primary ionization is assumed to take place independently of the Auger decay. This excludes resonant Raman and postcollision interaction (PCI) effects, which are manifestations of interactions between the primary and Auger electrons close to ionization thresholds.[50,51]

In the case that the Auger process can be treated as a two-step process, the decay rate for the Auger part is given by

$$T = \frac{2\pi}{\hbar} |\langle \psi_i | H - E | \psi_f \rangle|^2 \tag{3}$$

where ψ_i and ψ_f are the many-electron wave functions of the initial and final states of the Auger decay, respectively. The final state also contains the continuum electron wave function which is assumed to be normalized per unit energy range. In equation (3) H is the full electronic Hamiltonian and E is the total energy of the final state.

In the frozen orbital approximation the E term can be dropped from equation (3) since $<\psi_i|\psi_f> = 0$. The transition probability thus reduces to a two-electron matrix element

$$T = \frac{2\pi}{\hbar} |\langle \psi_i | \sum_{\mu,\sigma} V_{\mu\delta} | \psi_f \rangle|^2 \tag{4}$$

Equation (4), which is equivalent to Wentzel's Ansatz,[52] is often used as the starting point for the theory of the Auger effect. In the MCDF method $V_{\mu\delta}$ in equation (4) is taken to be the sum of the Coulomb and transverse Breit operators

$$V_{\mu\delta} = \frac{1}{r_{\mu\delta}} - (\vec{\alpha}_\mu \cdot \vec{\alpha}_\delta) \frac{\exp(i\omega r_{\mu\delta})}{r_{\mu\delta}} + (\vec{\alpha}_\mu \cdot \nabla_\mu)(\vec{\alpha}_\delta \cdot \nabla_\delta) \frac{\exp(i\omega r_{\mu\delta}) - 1}{\omega^2 r_{\mu\delta}} \tag{5}$$

where $\vec{\alpha}_\mu$ are Dirac matrices and ω is the frequency of the virtual photon.[53]

The wave function of the vth singly ionized initial state is given by equation (1) and a similar expression can be written for the wave function of the ηth doubly ionized final state $\psi_\eta(J'M')$. By coupling the final ionic state wave function to the continuum orbital $|\varepsilon l j m \rangle$ of the emitted Auger electron one can obtain the Auger transition probability from an initial state (1) to any of the doubly ionized final states $\psi_\eta(J'M')$

$$T_{\nu\eta} = \frac{2\pi}{\hbar} \sum_{lj} |\sum_\alpha \sum_\beta c^{J'}_{\eta\alpha} c^{J}_{\nu\beta} \langle \phi_\beta(JM) | \sum_{\gamma,\delta} V_{\gamma\delta} | \phi_\alpha(J'M')\varepsilon lj; JM \rangle|^2 \qquad (6)$$

The matrix elements of the two-electron operator between two CSFs reduce to a sum over two-electron Slater integrals multiplied by angular momentum coefficients. The angular factors for any closed- or open-shell case can be evaluated by using available computer codes (e.g., the MCP code of Grant[48]).

The radial integrals can be calculated using different sets of bound and continuum orbitals. Bound orbitals have been optimized with respect to the initial state (I) or to the final state (F) of the ion. The continuum orbitals have usually been generated in the average potential of the final double-hole ion constructed from either set (I) or (F). Recently the continuum orbitals have been obtained by including fully the exchange interaction[54,55] (DF continuum wave function) whereas in earlier works[56] the exchange contribution was often approximated by a local model potential [Dirac–Slater (DS) approximation]. The Lagrangian multiplier technique ensures the orthogonality between the bound and DF continuum orbitals.

The effect of relativity on Auger transition rates consists of shifts in transition energies, relativistic corrections in one-electron orbitals, and generalizations of the electron–electron interaction to include retardation and magnetic interactions. The effect of nonlocality on the exchange interaction can be estimated from the difference between the rates calculated with DF and DS continuum wave functions.

3.3. Configuration Interaction, Channel Interaction, and Relaxation

The MCDF model given by equation (6) regards only the configuration interaction (CI) of bound states in atomic systems. This CI is usually divided into initial-ionic-state configuration interaction (IISCI) and final-ionic-state configuration interaction (FISCI).

Pure *LS* coupling or intermediate coupling (IC) states are in the MCDF model given as superpositions of *jj*-coupled states. In such cases the mixing coefficients can be obtained using the CI technique. The generalization to the cases of IISCI or FISCI is straightforward: summation over the *jj*-coupled states resulting from one nonrelativistic configuration is replaced by a more general summation over the states resulting from several nonrelativistic configurations in accordance with a parity rule.

The main extensions beyond the MCDF approximation given by equation (6) are to allow for nonorthogonal basis sets (relaxation) and to include the interactions between the final continuum states [final-continuum-state configuration interaction (FCSCI)]. The derivation of nonrelativistic multichannel Auger rates from the first principles of scattering theory has been presented by Åberg and Howat[3] on the basis of the CI approach of Fano.[57] Its relativistic extension, the multichannel multiconfiguration Dirac–Fock (MMCDF) method used by Tulkki et al.,[55] can account for IISCI, FISCI, FCSCI, and relaxation. Relaxation is taken into account by using separately optimized initial- and final-state orbitals and by including all of the overlap integrals resulting from the nonorthogonality between the initial and final orbital sets in the transition amplitude.

The multichannel Auger transition amplitude is given by[55]

$$\langle \Phi_0 | H - E | \Phi^-_{\Gamma E} \rangle = \sum_{\alpha=1}^{N_c} [\langle \Phi_0 | H - E | \Phi_{\alpha E} \rangle$$

$$+ \sum_{\beta=1}^{N_c} \mathcal{P} \int dE' \frac{\langle \Phi_0 | H - E | \Phi_{\beta E'} \rangle \langle \Phi_{\beta E'} | K | \Phi_{\alpha E} \rangle}{E - E'}] Z^-_{\alpha \Gamma} \qquad (7)$$

where Φ_0 denotes the initial-state wave function and $\Phi_{\Gamma E}^-$ the multichannel multiconfiguration final-state wave function. The function $\Phi_{\alpha E}$ stands for the multiconfiguration single channel state which is a properly coupled antisymmetrized product of ASF (1) and the continuum orbital. The K matrix describes the coupling between the N_c ionization channels. The incoming-wave boundary condition is taken into account by the phase shifts $Z_{\alpha \Gamma}^-$. For further details of the MMCDF approach the reader is referred to Ref. 55.

3.4. Molecular Auger Transitions

Theoretical analysis of Auger spectra of molecules requires computation of both the Auger transition energies and intensities. For energy analysis the potential energy curves/surfaces can be predicted for the initial and final states of the decay by employing various methods of quantum chemistry (e.g., see Refs. 58–61 and references therein). Transitions are assumed to take place within the Frank–Condon regime. Computations are commonly performed using the nonrelativistic approach within the $\Lambda\Sigma$ coupling scheme which is unable to predict level splittings and crossings due to spin-orbit interaction.

The calculation of Auger decay rates requires knowledge of the molecular continuum wave function. Fairly good agreement with experiment has been achieved by approximating the continuum function by an atomic continuum wave function in the case of atomic-like molecules, e.g., HF, H_2O, CH_4.[62–64] Several attempts to generate a molecular continuum wave function have been reported recently. Larkins and Richards[65] computed the continuum wave of linear molecules by a numerical integration method. Colle and Simonucci[66] determined the continuum wave by solving the Lippman–Schwinger equation within a Gaussian basis set. Carravetta and Ågren[67] adopted the Stieltjes imaging method. Zähringer et al.[68] used the numerical close-coupling method. For further details of the methods the reader is referred to Refs. 65–68.

4. NORMAL AUGER ELECTRON SPECTRA: EXPERIMENT VERSUS THEORY

4.1. Auger Transition Energies of Free Atoms

Experimental Auger transition energies from some recent free-atom measurements in the kinetic energy range 20–1500 eV, as well as binding energies of initial and final states, are tabulated together with calculated values in Table I.[69] The calculations were carried out within the ΔE_{SCF} approach using the MCDF code of Grant.[48] The discrepancies between the experimental and calculated energies arise from inadequate treatment of electron correlation. The common discrepancies (0–2 eV) can generally be removed by including CI with the near-lying bound states. If the hole states lie close to the continuum, the interaction with continuum may also affect the energy both in the initial and in the final state of the decay.[70]

Multiplet splitting of initial and final states manifests itself in the fine structure of Auger spectra. IC with CI is commonly needed to predict the energy splittings satisfactorily. In some cases the experimental spacings are available from optical data[71] and they can be used as a separate test. Note that attention should be paid to the accuracy of the energy calculations, which also provide the eigenvectors $c_{\eta\alpha}^{J'}$ and $c_{\nu\beta}^{J}$ of equation (6). Even small changes in the eigenvectors result in noticeable redistribution of the intensity between the multiplets.

4.2. Spectra of Rare Gases and Other Closed-Shell Atoms

Most extensively studied Auger spectra are the spectra where the outer-shell electrons take part in the decay process in rare gases Ne, Ar, Kr, and Xe. The spectra have been recorded at the magic angle using both the high-energy electron and photon impact.[54,72–79] The results agree well with each other thus supporting the validity of the two-step model. We shall compare the overall features of the rare gas spectra with each other since the transitions are frequently used to test various theoretical approaches. In the final state of the transitions, two electrons are missing from s and/or p outer orbitals ($2s$, $2p$ in Ne, $3s$, $3p$ in Ar, $4s$, $4p$ in Kr, and $5s$, $5p$ in Xe). FISCI between the ns^{-2}, $J = 0$ and np^{-2}, $J =$

TABLE I
Experimental and Calculated Auger and Binding Energies in eV[a]

		Auger energy		Binding energy			
				Initial state		Final state	
Element	Transition	Experiment	Theory	Experiment	Theory	Experiment	Theory
^{10}Ne	$K–L_{2,3}L_{2,3}(^1D_2)$	804.46	806.78	870.21	869.33	65.73	62.55
^{11}Na	$K–L_{2,3}L_{2,3}(^2D_{5/2})$	977.2	979.12	1078.9	1078.29	101.92	99.17
^{12}Mg	$K–L_{2,3}L_{2,3}(^1D_2)$	1167.0	1168.92	1311.4	1310.85	143.9	141.93
^{18}Ar	$K–L_{2,3}L_{2,3}(^1D_2)$	2660.5	2660.78				
	$L_1–L_{2,3}M_1(^3P_1)$	30.5	28.04	326.0	327.02	295.5	298.98
	$L_3–M_{2,3}M_{2,3}(^1D_2)$	203.50	205.01	248.63	248.18	45.13	43.15
^{19}K	$L_3–M_{2,3}M_{2,3}(^2D_{5/2})$	236.6	238.16	300.5	300.38	63.87	62.22
^{29}Cu	$L_3–M_{4,5}M_{4,5}(^2G_{7/2})$	900.7	905.33	939.7	938.88	39.05	33.55
^{30}Zn	$L_3–M_{2,3}M_{4,5}(^1F_3)$	886.7	886.31	1028.9	1028.56	142.2	142.24
	$L_3–M_{4,5}M_{4,5}(^1G_4)$	974.4	978.60	1028.9	1028.56	54.5	49.96
^{36}Kr	$L_3–M_{4,5}M_{4,5}(^1G_4)$	1460.4	1463.72				
	$M_5–N_{2,3}N_{2,3}(^1D_2)$	53.7	53.85	93.8	92.18	40.1	38.33
^{46}Pd	$M_5–N_{4,5}N_{4,5}(^1G_4)$	311.2	313.59	341.1	339.93	29.97	26.34
^{47}Ag	$M_4–N_{4,5}N_{4,5}(^2G_{7/2})$	335.8	338.06	375.5	374.64	39.69	36.58
^{48}Cd	$M_5–N_{4,5}N_{4,5}(^1G_4)$	361.2	363.16	412.1	411.39	50.82	48.22
^{49}In	$M_5–N_{4,5}N_{4,5}(^2H_{9/2})$	386.3	388.24				
^{54}Xe	$M_5–N_{4,5}N_{4,5}(^1G_4)$	520.0	521.83	676.7	676.49	156.7	154.66
	$N_5–O_{2,3}O_{2,3}(^1D_2)$	32.32	33.01	67.55	66.52	35.23	33.51
^{55}Cs	$M_5–N_{4,5}N_{4,5}(^2G_{9/2})$	543.9	545.56	731.6	732.02	187.7	186.46
	$N_5–O_{2,3}O_{2,3}(^2D_{5/2})$	36.3	36.88				
^{70}Yb	$M_5–N_{6,7}N_{6,7}(^1I_6)$	1494.1	1495.91				
	$N_5–N_{6,7}N_{6,7}(^1I_6)$	152.1	158.89				
^{79}Au	$N_7–O_{4,5}O_{4,5}(^2G_{9/2})$	54.9	55.25	91.6	88.76	36.89	33.51
^{80}Hg	$N_5–N_{6,7}N_{6,7}(^1I_6)$	119.3	124.47	366.0	368.98	246.7	244.51
	$N_7–O_{4,5}O_{4,5}(^1G_4)$	62.4	62.80	107.06	104.84	44.85	42.04

[a]Ref. 69 and references therein.

0 states plays an important role in all of the spectra as first pointed out by Asaad.[80] Furthermore, FISCI shifts the energies and redistributes the intensities in the groups with ns^{-2} and $ns^{-1}np^{-1}$ final states in Ar, Kr, and Xe.[76–79,81] As demonstrated for $M_{4,5}–N_1N_1$ and $M_{4,5}–N_1N_{2,3}$ spectra of Kr in Figure 4, the energies in a multiconfiguration calculation shift from the single-configuration positions and extra structure appears at lower energy by ~ 10 eV in fairly good agreement with experiment.[77]

The kinetic energies differ and the intensity is distributed between the Auger groups in a very different way in various rare gas spectra. Note that initial vacancy has different l quantum number in addition to the principal quantum number. Ne $K–LL$ transition energies lie around 800 eV and the $K–L_{2,3}L_{2,3}$ group is clearly more intense than the other groups. The $K–LL$ spectrum of Ne is well understood both from theory and from experiment.[55,72,73,81] Ar $L_{2,3}–MM$ transition energies are around 200 eV and the $L_{2,3}–M_{2,3}M_{2,3}$ group gains most of the intensity of the $L_{2,3}–MM$ transitions (see Figure 2) and it is also most carefully studied.[55,74,82,83] Because of the low intensity of the Ar $L_{2,3}–M_1M_1$ and

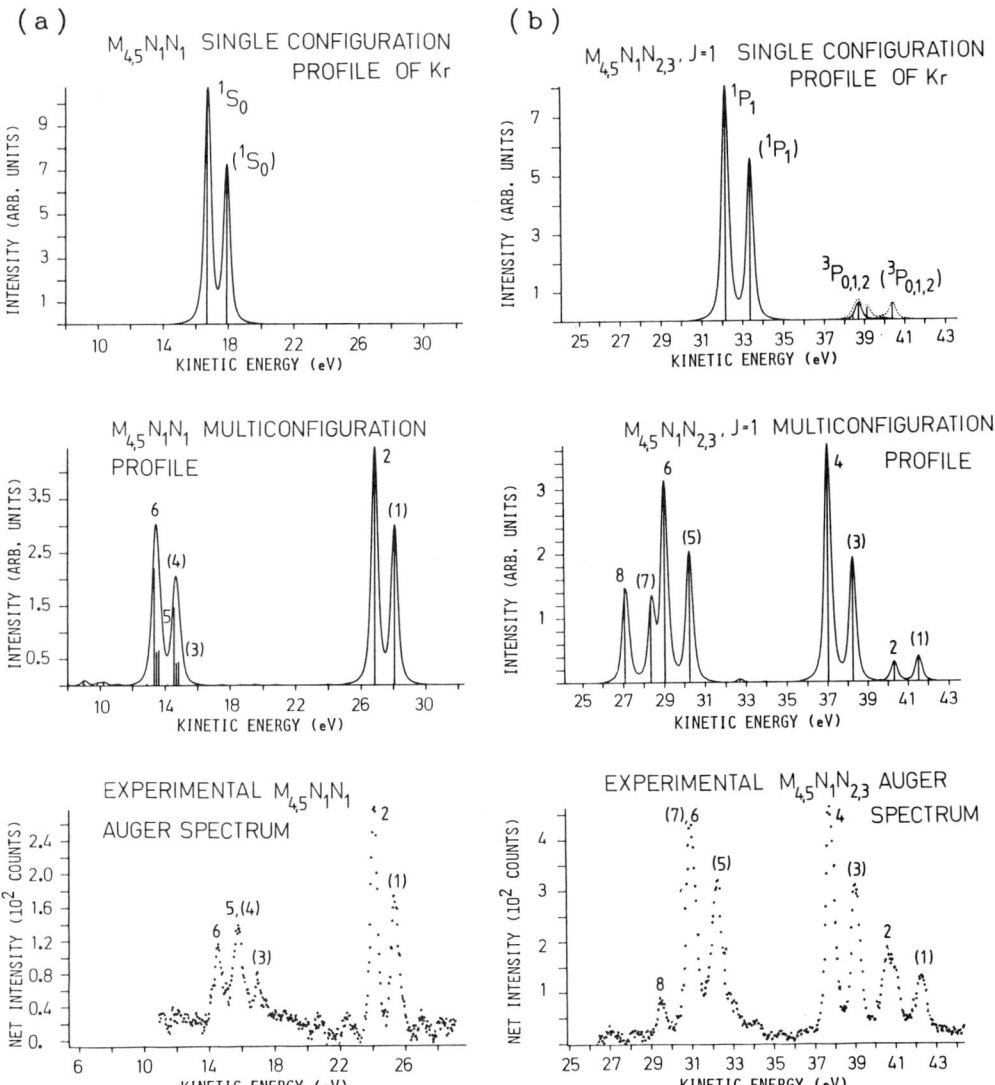

FIGURE 4. Comparison between calculated single-configuration (upper panels) and multiconfiguration (middle panels) spectra with experimental (lower panels) $M_{4,5}$–N_1N_1 (a) and $M_{4,5}$–$N_1N_{2,3}$ (b) Auger spectra of Kr.[77] The multiconfiguration calculations take into account FISCI between the $4s^04p^6$, $4s^14p^44d^1$, and $4s^24p^24d^2$ configurations in (a) and between the $4s^14p^5$ and $4s^24p^34d^1$ configurations in (b).

$L_{2,3}$–$M_1M_{2,3}$ groups and their overlap with an enormous number of satellite lines, their full understanding is still lacking. Kr $M_{4,5}$–NN and Xe $N_{4,5}$–OO spectra behave in a similar way. Their energies are well below 100 eV and intensity is distributed between all of the groups.[76–79] Because of the very low kinetic energy of the $M_{4,5}$–N_1N_1 transitions of Kr and the $N_{4,5}$–O_1O_1 transitions of Xe, experimental results are still rare and more detailed theoretical work is also needed to include both CI with higher $4s^14p^4(n+1)s, nd$ and $4s^24p^2(n+1)s, nd^2$ terms and FCSCI.

Inner-shell Auger spectra of rare gases, like the K–$L_{2,3}L_{2,3}$ spectrum of Ar, the $L_{2,3}$–$M_{4,5}M_{4,5}$ spectrum of Kr, and the $M_{4,5}$–$N_{4,5}N_{4,5}$ spectrum of Xe, have also been studied extensively, but the measurements are done using electron beam excitation only.[84,85] IC was found to be well suited to predict the spectral structures. The lines in the $M_{4,5}$–$N_{4,5}X, X = N_1, N_{2,3}$ spectra of Xe were found to be

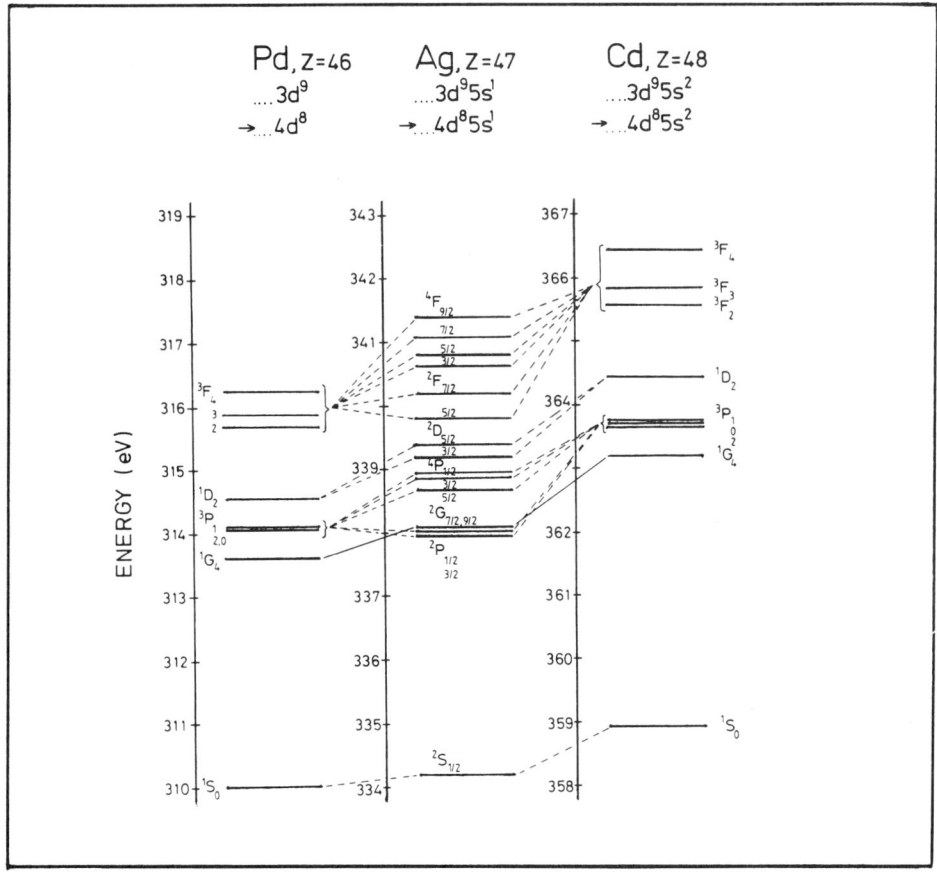

FIGURE 5. (a) Energy levels of Pd, Ag, and Cd with two $4d$ vacancies predicted by the MCDF calculations.

broadened and shifted because of the interaction with SCK continua.[85] The $L_{2,3}$–$M_{4,5}M_{4,5}$ transitions in Zn and the $M_{4,5}$–$N_{4,5}N_{4,5}$ transitions in Pd and Cd display Auger spectra which are very similar to those of rare gases, only the Z-dependence shows up.[44,86,87] The spectra of closed-shell metal atoms[44,86–89] give valuable additional information to the studies of Auger processes, since the rare gases cover only a very small part of periodic table.

4.3. Spectra of Open-Shell Atoms

In a closed-shell atom the fine structure of Auger spectra arises from the interaction of vacancies created by the Auger decay itself. The spectrum of an open-shell atom is more complicated and reveals more fine structure as a result of coupling of the participating hole states to the outermost open shell where the electrons stay passive during the decay.[90] This causes the splitting of the parent levels to daughter levels, as can be seen from Figure 5, where the final-state energy level structures in IC are presented for ^{46}Pd, ^{47}Ag, and ^{48}Cd. The energy splitting of the $4d^{-2}$ hole state increases slightly in going from Pd to Cd which has an effect that peaks are better resolved in the $M_{4,5}$–$N_{4,5}N_{4,5}$ spectrum of Cd[44,87,91] (see Figure 5). The spectrum of Ag differs clearly from those of Pd and Cd, not only because of the extra splitting but also because the IC conditions are dramatically changed when the number of levels with the same J values increases. In passing to the Auger spectra of complicated open-shell atoms with several outer electrons, the fine structure of the spectra smears out because of the overlap of transitions between numerous initial- and final-state levels.[92]

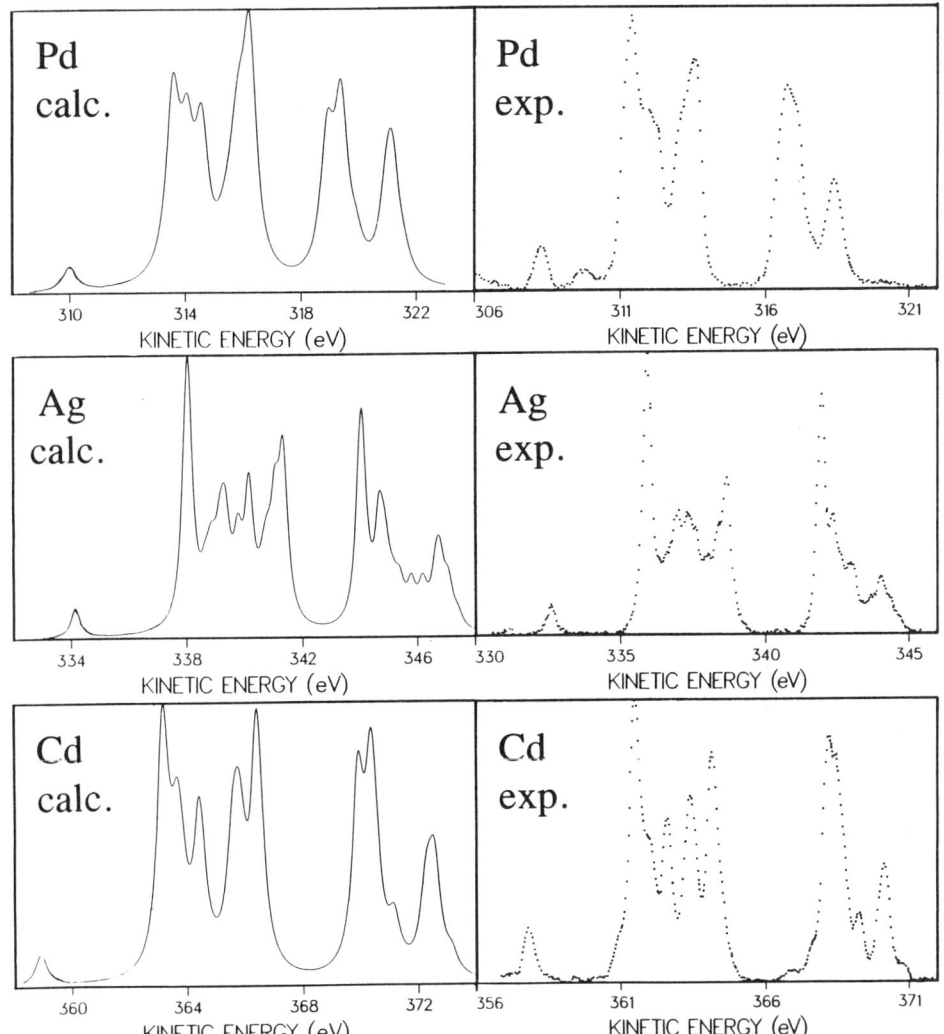

FIGURE 5. (*Continued*) (b) Intensity distribution of the $M_{4,5}$–$N_{4,5}N_{4,5}$ Auger transitions in Pd, Ag, and Cd from experiment[44,87,91] (dotted curve) and theory (solid curve). The calculations were done in IC with DF bound and continuum wave functions taking the coupling of the 5s electron fully into account for Ag.

Since the open-shell elements make up the larger part of the periodic table, better understanding of their spectra is important for fruitful progress in Auger electron spectroscopy. With the use of the tunability of the synchrotron radiation, it becomes possible to experimentally resolve the spectral contributions related to different initial-state multiplets in an open-shell atom.[93]

The same Auger transitions take place in an ionized open-shell atom and in a core to open-shell excited atom, provided that core holes and spectator electrons have the same quantum numbers. Comparisons between open-shell Auger spectra and resonance Auger spectra are thus useful in testing the dependence of Auger decay on Z, ionization degree, and population of initial-state levels.

4.4. Total Auger Decay Width, Group Rate, and Multiplet Structure

Most experiments are devoted to study of the fine structures of Auger groups. Both the energy splitting and relative intensities of the multiplets are then the quantities which can be compared with

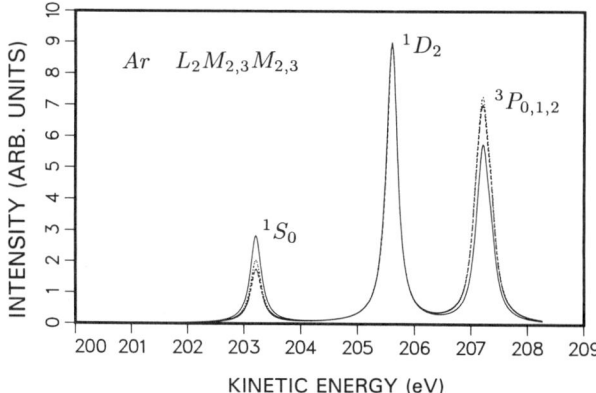

FIGURE 6. Experimental (solid line) and theoretical $L_2-M_{2,3}M_{2,3}$ Auger spectra of Ar.[55] Experimental linewidths and positions were also used for the calculated profiles. The calculations were done in the DF approximation with the final ionic state basis set as follows: Dotted line, calculation with FISCI; dashed line, calculation with FCSCI (21 channels included).

calculated ones. Calculations including IC and FISCI often reproduce the relative intensities of high-kinetic-energy Auger lines surprisingly well. This is demonstrated in Figure 5 for the $M_{4,5}-N_{4,5}N_{4,5}$ spectra of Pd, Ag, and Cd. Figure 6 compares the Ar $L_2-M_{2,3}M_{2,3}$ spectrum by plotting the experimental spectrum with theoretical profiles obtained using DF bound and continuum orbitals with FISCI and FCSCI.[55] FCSCI plays only a minor role in the Auger spectra when the kinetic energy of the Auger electrons exceeds about 200 eV.[55] In the low-kinetic-energy spectra, FCSCI is of major importance as recently demonstrated theoretically for the $N_{4,5}-OO$ spectrum of Xe.[94] Xe $N_{4,5}-O_{2,3}O_{2,3}$ transitions were found to be well reproduced by MMCDF calculations but for the low-kinetic-energy groups the comparison between experiment and theory is still lacking since FISCI was not included properly.

The influence of FCSCI has been studied most extensively for the $K-LL$ spectrum of Ne.[55,81,95,96] Since the Ne spectrum is one of the few spectra which also have been studied accurately in experiment, it is well suited to test different theoretical approaches. FCSCI seems to affect the distribution of the intensity between the different Auger groups. Some intensity from the $K-L_1L_1$ and $K-L_1L_{2,3}$ groups is transferred to the $K-L_{2,3}L_{2,3}$ group, resulting in a better agreement with experiment. This is demonstrated in Figure 7 where we show theoretical $K-LL$ spectra of Ne computed with DF wave functions by taking FISCI and FCSCI into account.[55] Energies are also obtained by including FISCI. The lower panel displays the experimental spectrum. As will be discussed in Section 5, satellite transitions make the interpretation of the experiment very complicated. Such satellite structures appear in the spectra in general, and they are the main reason why comparison between experiment and theory is difficult, especially if ratios of group intensities are considered. Another source of errors in the experimental data is the correction of the experiment for the variation of the transmission of the spectrometer as a function of kinetic energy. For example, the $M_{4,5}-NN$ spectra of Kr recorded by different experimental setups[75-77,97] differ considerably from each other, demonstrating the difficulty in transmission correction especially at low kinetic energies.

The natural width of Auger lines is determined by the total decay widths of the initial and final states of the Auger process. If the radiative decay is insignificant and the final state is not allowed to decay further, the width of the Auger lines is determined by the total Auger decay rate of the photoexcited state. The absolute rates show a clear dependence on the choice of one-electron orbitals. The total rates increase by 20–30% for Ne $K-LL$ and Ar $L_{2,3}-MM$ transitions if predicted by final- or initial-state orbitals, respectively.[55] The difference between similar predictions is only 7% for the $N_{4,5}-OO$ transitions of Xe.[94] FCSCI does not affect the total rates of Ne and Ar transitions considerably[55] whereas the total rates of Xe are more sensitive[94] to FCSCI. The exchange interaction may

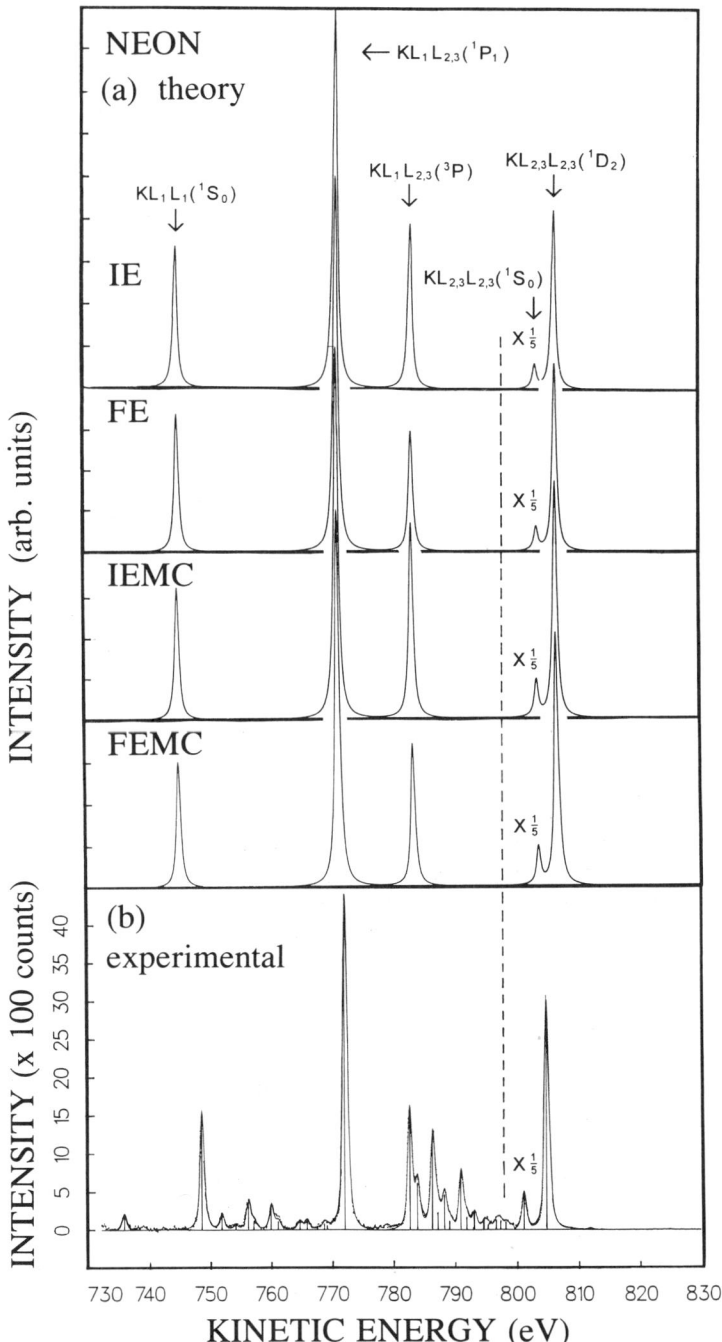

FIGURE 7. (a) Theoretical K–LL Auger spectra of Ne.[55] The energies are obtained with the ΔE_{SCF} approach taking FISCI into account in the final state. IE indicates single-channel DF results obtained using initial-state bound orbitals. FE is the same as IE but based on the use of final-state bound orbitals. IEMC and FEMC are the same as IE and FE but include FCSCI. (b) Experimental K–LL Auger spectrum of Ne.[73]

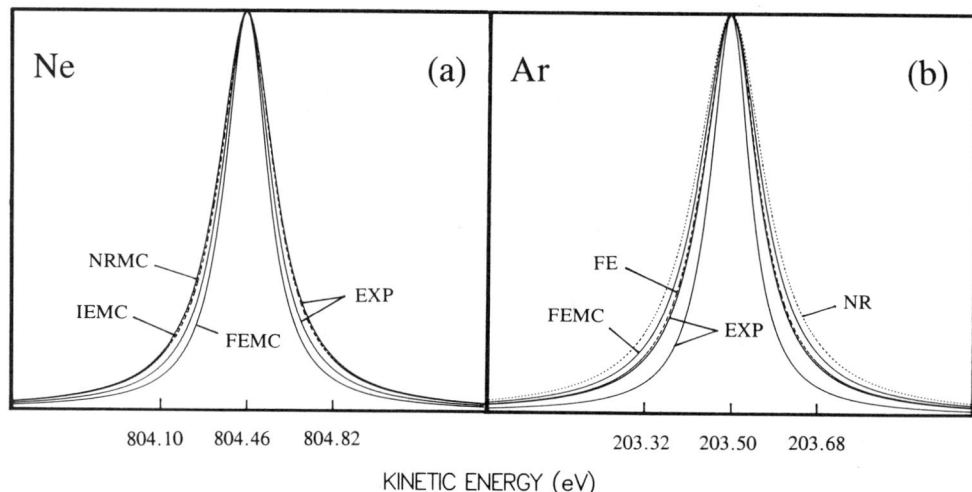

FIGURE 8. Calculated linewidths of (a) Ne $KL_{2,3}L_{2,3}, {}^1D_2$ and (b) Ar $L_2M_{2,3}M_{2,3}, {}^1D_2$ transitions in comparison with experiment. Upper and lower limits for the experiment are shown by solid lines (EXP). FE, IEMC, and FEMC are the same as in the caption of Figure 7. NRMC, nonrelativistic multichannel calculation (see Ref. 3 for details); NR, nonrelativistic single-channel calculation.[83]

also be of importance for low-energy transitions. Comparison of calculated decay widths with experiment is by no means straightforward. Experimental peaks always contain the spectrometer broadening contribution. This is often of about the magnitude of the natural width of the lines. It should, however, be pointed out that Auger electron spectra provide more reliable experimental values for the level widths than photoelectron spectra since the bandwidth of the exciting photon beam is the third contribution in the measured photoelectron line whereas the Auger line is a convolution of spectrometer and natural widths only. Figure 8 compares the calculated and experimental Auger linewidths of the most intense Ne K–LL and Ar $L_{2,3}$–MM transitions which are best known by experiment. Inspection of Figure 8 shows that various theoretical approaches differ from each other less than the experimental error limits but for Ar they all tend to overestimate the experiment.

4.5. Coster–Kronig Transitions

CK transitions involve small transition energies and their rates are exceedingly sensitive to the calculation model. When energetically possible, the CK transitions are the principal channels by which the core hole states relax. Because of low energies, the detection of CK spectra often becomes difficult. The determination of multiplet or group intensities is hampered because of the heavy overlap of lines characterized by large linewidths. For these reasons only few comparisons with experiment are available.[98]

An important effect that is associated with CK transitions is the interaction with continuum causing shifts in energy levels.[70] In a case where atomic CK decay is forbidden, extra-atomic relaxation in the solid state can shift the near-threshold CK energies just above the threshold.[99,100]

4.6. Molecular Auger Spectra

Because of the dependence of the electronic energy on the bond distance and also the appearance of additional energy modes (vibrational, rotational), the Auger spectra of molecules have extra features relative to the spectra of free atoms. High-resolution Auger spectra of molecules are very useful in predicting potential energy curves and spacings of vibrational levels. As an example the high-resolution $L_{2,3}$–MM spectrum of HCl studied by Svensson and co-workers[101,102] will be discussed shortly.

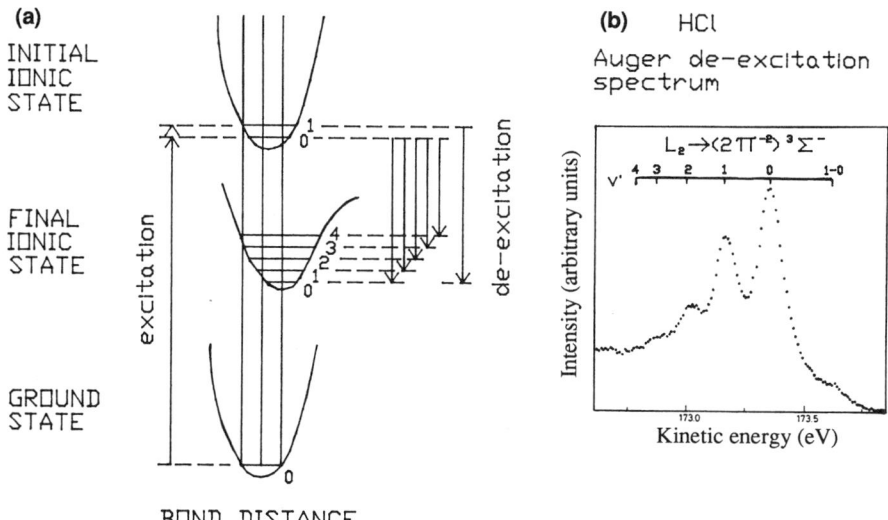

FIGURE 9. (a) Schematic representation of electronic transitions in HCl. (b) Experimental high-resolution spectrum of HCl[101,102] resulting from transitions shown in (a).

Several high-resolution studies for other small molecules were also reported recently by the Uppsala group.[103–107] HCl is isoelectronic with Ar which implies that their Auger spectra are to some extent similar to each other, e.g., the width of the lines in both spectra is observed to be 0.12 eV. Since the equilibrium distances in the ground state and singly ionized initial state of HCl are about the same, electronic excitation takes place between the $v = 0$ vibrational states only. The potential energy minima of final-state configurations appear at slightly larger bond lengths thus resulting in vibrational fine structures of the peaks when in the final state of the Auger decay the vibrational levels with $v = 0,1,2,3,\ldots$ become populated. The transitions are illustrated in Figure 9. Final states may also be repulsive resulting in the dissociation of the molecule.

4.7. Near-Threshold Phenomena

Near threshold the photoionization process followed by deexcitation has different characteristics from that in the high-energy limit, where the two-step model is valid (e.g., see Refs. 108–119 and references therein). In the vicinity of threshold, photoexcitation and Auger recombination can occur as a single-step process, the resonant Raman effect.[50,111] When the photon energy is tuned through a broad core level resonance, the width of emitted Auger electron line reflects the bandwidth of the photon beam. The energy of the line exhibits linear Raman dispersion.

The bridge which links the two extremes, the threshold Raman and the high-energy two-step regimes, is the postcollision interaction (PCI).[51,111–113] When near-threshold photoionization is followed by Auger recombination, the Auger electron screens the ionic field felt by the photoelectron. The photoelectron is decelerated, even recaptured[114,115] by the atom, and the energy it loses is transferred to the Auger electron. This results in energy shift and anomalous line shape of Auger electron lines. When the photoelectron becomes faster than the Auger electron, the PCI vanishes.[116–118]

In the theoretical resonant-scattering formulation the PCI and resonant Raman effect is described as a one-step process which appears as a property of a resonance in the double-photoionization cross section.[115,119,120] The unified theory of Auger electron emission describing the evolution of shake-modified resonant Auger decay into PCI as a function of photon energy, was discussed recently by Åberg.[120]

5. SATELLITE TRANSITIONS

5.1. Satellites Produced during Auger Decay

Satellites in the Auger spectra which result from the FISCI were discussed in the context of rare gas spectra. Those satellites which gain intensity through the mixing of Auger final ionic state configuration with other hole-filling configurations are commonly called CI satellites. The same nomenclature exists in photoelectron spectroscopy, e.g., for the satellites accompanying the outer s-shell photoelectron lines of the rare gases.

Besides the CI satellites Auger spectra may also contain shakeup satellites. The outer electron may shake up during the Auger decay. Shake probabilities, obtained from the square of the overlap integral $\langle nlj | n'lj \rangle$, where $|nlj\rangle$ stands for the orbital of the shaking electron before and $|n'lj\rangle$ after the decay, are usually considerably lower for Auger decay than for photoionization. For example, for the Ar $3p \rightarrow 4p$ shakeup during $2p$ ionization and Kr $4p \rightarrow 5p$ shakeup during $3d$ ionization the shake probabilities are a few percent but decrease distinctly if predicted for Auger decay. Relaxation of the orbitals is thus more pronounced when passing from neutral atom to singly ionized than from singly to doubly ionized.

5.2. Satellites Caused by the Initial-State Effects

Auger spectra may contain satellites where the initial state of the Auger decay is the satellite state in photoionization. An extra outer electron can shake up or off during photoionization. Shakeoff in photoionization gives rise to a continuous energy distribution whose contribution is hard to distinguish in experimental spectra. It can be obtained indirectly from Auger spectra where the satellites appear as discrete lines.[72,73] The satellite contribution in Auger spectra, which is the result of shake transitions in the course of photoionization, also varies as a function of exciting energy as the photoelectron satellite contribution. The shake probability may be as high as 40% in photoionization of rare gases.[72,73,121,122] This means that Auger spectra contain a rich satellite structure which often overlaps with the main Auger lines. The shaken electron may also further shake down or up during the Auger decay. Fine structure of satellites is complicated because of open-shell electronic structure. Multiply excited/ionized initial states produce their own satellite contribution. Satellites in the spectra created by particle impact are of this origin.

CI satellites of photoelectron spectra give rise to their own contribution in the Auger spectra as well. CI satellites may appear in addition to the satellites as a result of shake transitions and they may be remarkably populated in photoionization, e.g., up to 35% in the case of the Ba $4d$ ionized state.[123] The CI satellite contribution in photoelectron spectra is related to the eigenvectors of the ionized states that have the same passive electron configuration as the ground state. The Auger decay part, when IISCI is included, probes the eigenvectors of the mixed states in the ion as well. This may cause dramatic effects if admixed states have large Auger amplitudes. Branching ratios of the $4d_{3/2}$ and $4d_{5/2}$ transitions in Ba were therefore completely changed in Auger decay compared to photoionization[124] as shown in Figure 10. Auger electron spectra of transition metals and K, Ca, Rb, Sr, Cs, and Ba atoms are strongly affected by IISCI and FISCI.[2,125–130]

In addition to CI and shake processes in the initial state, the thermal population of the energy levels lying in close proximity to each other in the ground state of the atom may also result in a complicated structure of Auger spectra. The free iron group atoms at the end of the $3d$ transition elements are good examples.[45,46] Ground-state configuration interaction (GSCI) may also be a source for anomalous population of initial states which, furthermore, reflects the complexity of the Auger spectra.

5.3. Cascades

When deeper core levels are ionized, Auger cascades are possible and corresponding satellite contributions appear in the spectra.[5] The satellites are similar to those caused by shakeoff in photoionization in the sense that the initial state consists of two holes. The state is, however, populated via former Auger process. Auger spectra of open-shell atoms are often accompanied by cascade

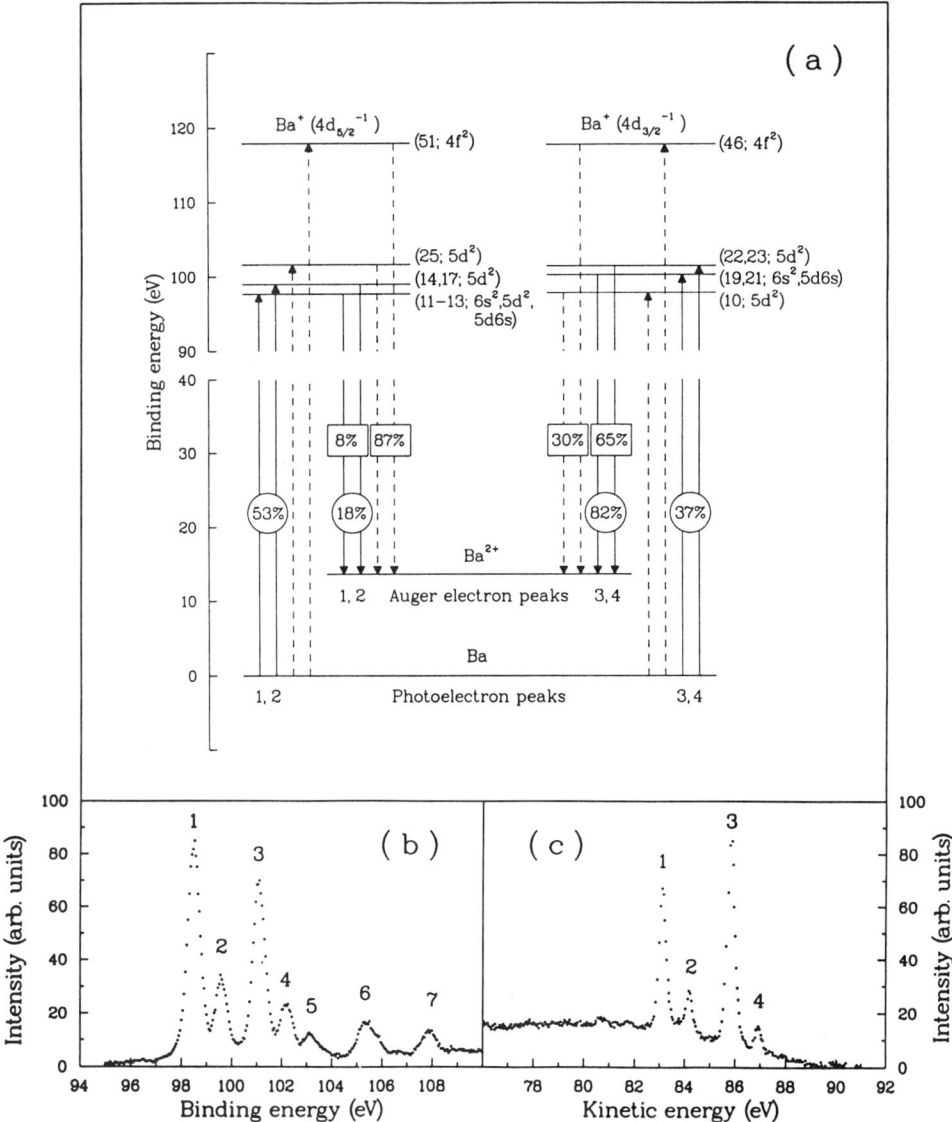

FIGURE 10. (a) Schematic diagram for the creation of Ba 4d-hole states and their subsequent Auger decay into the Ba^{2+} ground state.[124] Solid arrows represent the pathways which give rise to the peaks 1–4 in the experimental photoelectron (b) and Auger electron (c) spectra. Dashed arrows indicate pathways which do not populate 4d-hole states but have high Auger rates. Calculated relative numbers of emitted electrons in both processes and relative Auger rates are indicated by circles and squares, respectively. Numbers in parentheses refer to ASFs (see Ref. 124) followed by the leading nonrelativistic configuration.

satellites because the existing extra outer electron makes the second-step Auger decay energetically possible. The unique advantage of synchrotron radiation is that the energy of the ionizing radiation can be selected to ionize the given level. Satellites caused by cascades can thus be studied in more detail (see also Figure 2).

A wide variety of existing satellites, caused by shake transition (before and during the Auger decay), CI (ground, initial, and final state), and cascades, make the experimental Auger spectra often extremely complicated with many overlapping lines. When extracting experimental values for the energies and intensities of the main lines to be compared with some specific calculations, the satellite contribution should also be kept in mind. On the other hand, satellite production is a very useful probe of many-electron processes in free atoms and molecules which is why its detailed studies are important.

6. RESONANT AUGER RECOMBINATION

6.1. Decay Spectra of Core-Excited Atoms

A core electron can be excited into an unoccupied orbital at a certain photon energy below its ionization threshold. In their pioneering work Eberhardt et al.[131] for the first time presented the decay spectra of Kr at the $3d \to 5p$ resonance and of Xe at the $4d \to 6p$ resonance. Since then, decay spectra of core-excited rare gases have been studied extensively (e.g., Refs. 9, 10, 12, 13, 15–17, 35, 97, 132–134 and references therein). For other atoms, in reference to their number in the periodic table, only a few studies have been reported (see Refs. 11, 12, 14, 37, 135–148 and references therein).

The core hole is filled via Auger recombination in which the electron initially excited to the unoccupied orbital either remains as a spectator electron during the Auger decay or participates in the process. A remarkable probability for a process where the spectator electron shakes up or down during the resonant Auger decay has also been observed. In the decay of core-excited states the spectator Auger transitions gain most of the intensity. Participator Auger decay (autoionization) is the prominent decay channel only if the spectator decay is forbidden which is the case when outer electrons are excited.[149,150] Participator Auger spectra are characterized by interference effects between the participator Auger and the direct photoionization channels. Outer-shell autoionization resonances thus give special features in the photoelectron spectra. In this chapter we, however, restrict ourselves to the discussion about resonant Auger recombination associated with core electron excitations. Participator Auger decay becomes important also in a few cases when a core electron is excited. The participator Auger decay channel competes with the spectator decay channel in the deexcitation of Ba $4d^{-1}4f$ excited states manifesting itself as a reversed branching ratio of the spin-orbit components of the valence photoelectron lines.[37]

The simple classification into spectator and participator processes becomes meaningless when the electron is excited to an open shell. This is demonstrated in the resonant spectra of bromine and iodine which were obtained after producing atomic samples by laser-induced photodissociation of molecules.[141–143] The spectra show great similarities to the normal Auger spectra of krypton[76] and xenon[78] since electron configurations of initial and final states of the transitions are the same. Z, charge (transitions take place in neutral excited atoms), and kinetic energies, however, differ from those of noble gases. This offers an interesting case to test the ability of the theoretical calculations to reproduce experiment. Because the direct photoionization process also populates the same final states as the resonant Auger decay, the interference between them needs to be taken into account.[141–143]

In closed-shell atoms, the number of energy levels in the final state of the spectator Auger decay increases from that of the normal Auger decay because of the coupling between the core and the spectator electron. The lines corresponding to the transitions to the daughter levels of a given parent multiplet are, however, usually so close to each other that they cannot be resolved with the normal resolution of ≥ 0.2 eV. In addition, the photon resolution of generally ≥ 0.3 eV often makes it impossible to avoid exciting more than one resonance leading to even more complicated overlapping resonance Auger spectra. Because of these problems, simplifying assumptions, such as the intensity distributions in normal, spectator, and shakeup spectra are the same and that the spectator spectrum only shifts from the normal Auger spectrum (strict spectator model), have been made. However, with the use of high-resolution synchrotron radiation from an undulator beamline and high-resolution electron spectrometers, high-quality experimental results can be obtained. The strict spectator model breaks down in reproducing such data. This is demonstrated by the high-resolution results for Kr which indicate a considerable redistribution in the intensity of parent multiplets in passing from the normal Auger spectrum to the resonance spectrum.[35] The eigenvectors obtained from IC calculations indicate strong mixing of some parents, the strength of the mixing depending strongly on whether the spectator electron stays on the first or second Rydberg orbital. FCSCI may also affect the low-kinetic-energy spectra considerably but no detailed studies are yet available.

The spectator Auger spectra of Ar, K, and Ca show a pronounced difference at the $2p^{-1}3d$ resonance as compared with the normal Auger spectra.[11–13,74,125,130] This is related to the strength of the $3p$–$3d$ interaction in the presence of the collapsed $3d$ orbital which has a strong influence on the IC conditions. Spectator Auger spectrum of Ar also serves as an excellent test for CI predictions. Tuning the photon energy through the $2p^{-1}(n+1)s, nd, n = 3, 4, 5 \ldots$ resonances, the $3p^{-2}(n+1)s, nd$ states become populated via Auger recombination. These states with $J = \frac{1}{2}$ are strongly mixed with the $3s^{-1}, J = \frac{1}{2}$ ionic state, which manifests itself also as a strong satellite structure in the $3s$ photoelectron spectrum. The resonant Auger spectrum thus gives valuable additional information from the energy splitting and eigenvector composition of the $3p^{-2}(n+1)s, nd$ states.

6.2. Shake Modified Resonant Auger Decay

Shake transitions of the spectator electron accompanying the resonance Auger recombination have been observed to be very important in recent studies.[9,10,13,14,35,97,133] The collapse of the Rydberg orbitals on going from the initial state to the final state of the resonant Auger decay makes it possible that the electrons may shake to neighboring orbitals. Observed shake contributions are found to be reproduced fairly well with simple shake calculations. Shake contributions at first resonances of Ne, Ar, Kr, and Xe are found to be 20–30% from the spectator decay probability.[10,97] In the decay of the $2p^{-1}3d$ resonance of Ar, a very strong $3d \rightarrow 4d$ shakeup channel was reported and explained by the collapse of the $3d$ orbital.[9,13] Shake calculations predict an increasing shake probability with increasing n. Shakeoff is not, however, of major importance even for high n. Instead the shakedown starts to play an important role. Whitfield et al.[14] recently studied shake transitions in the decay of resonantly excited Mg $2p^5 ns, d$ states. Their study covered a wide range of principal quantum numbers of the initial and final states of the shaking electron. They found that the total shake probability oscillates as a function of n and that for some n values the total shake probability was very close to unity. The shakeoff contribution in the total shake probability was less than 2%.

Multiplet dependence of shake structures was recently observed for Kr and explained with the use of IC calculations.[35] Mixing of wave functions in IC shows a strong dependence on n. This gives rise to large differences in shake to spectator probability ratios if obtained for each parent multiplet separately. The effect is most pronounced in passing from the $3p^{-2}3d$ state of Ar to the $3p^{-2}4d$ state, but it is not negligible in more general cases either.

6.3. Auger Cascades and Their Role in Producing Multiply Charged Ions

Final states of the resonant Auger transitions which are above the second ionization threshold may further decay via second-step Auger process. This is demonstrated for Kr with the energy level diagram in Figure 11. The lines related to the second-step Auger decay fall in the low-kinetic-energy part of the spectrum where they strongly overlap with the low-kinetic-energy groups of the first-step Auger transitions and their satellites. Using electron–electron coincidence technique, Sonntag and co-workers were able to identify the lines related to the most intense first-step transitions.[17]

The cascade Auger decay is a prominent decay channel in producing multiply charged ions at the first resonances of noble gases. Ion yield measurements have shown a high yield of doubly charged Kr ions at the first resonance, whereas the Ar ions have been observed to remain mostly as singly charged. This is related to the intensity distribution between different groups in Ar and Kr resonance spectra.[97] According to the recent calculations, the low-kinetic-energy groups in Kr gain a remarkable part of the intensity. In Ar, however, the high-energy group whose final levels lie below the double-ionization threshold is the most intense one (see Table II for details). Calculated Auger rates were used to investigate the roles of the cascade Auger decay in producing multiply charged ions.[97] Comparison with ion yield measurements indicates that most of the production of multiply charged ions is the result of Auger cascades. The role of CI with continuum related to the final state of the first-step Auger transition with one or two s holes has not yet been fully studied.

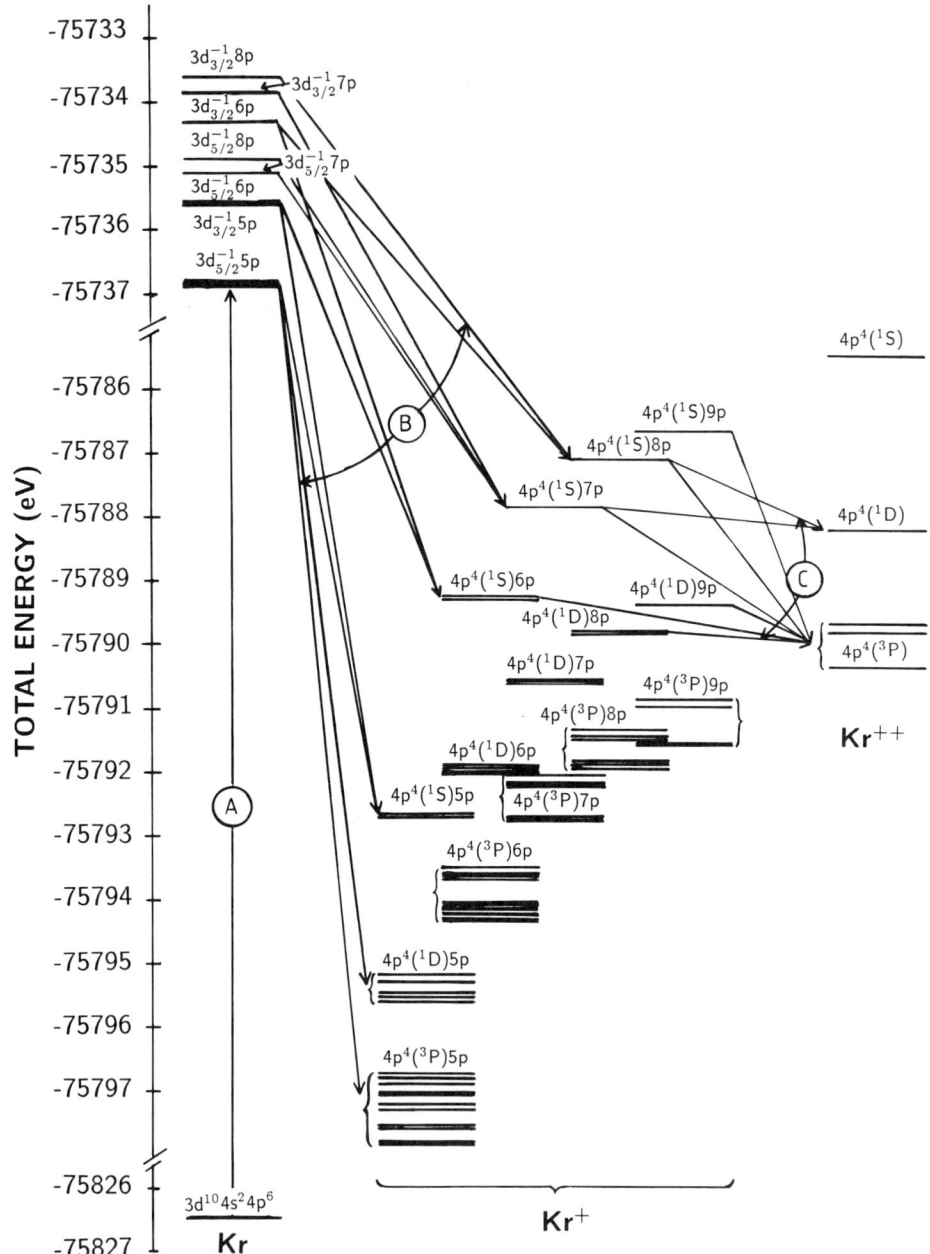

FIGURE 11. Energy level diagram of Kr showing the excitation and deexcitation processes.[133] First-step Auger transitions are indicated by B. Second-step Auger transitions C are possible from all of the final states of the $4s^{-2}$ and $4s^{-1}4p^{-1}$ groups in the first step. The second step becomes energetically possible also from the $4p^{-2}(^1S)np(n = 6,7,8,9)$ and the $4p^{-2}(^1D)np(n = 8,9)$ states.

TABLE II
Contribution (in %) of Each Auger Transition to the Total Transition Rate in the Decay of the $3d^9 5p$ and $3d^9 6p$ States of Kr and the $2p^5 4s$ and $2p^5 3d$ States of Ar[97]

	Kr			Ar	
	Contribution			Contribution	
Transition	$n = 5$	$n = 6$	Transition	$n = 4$	$m = 3$
$3d^9 np \to 4s^2 4p^5$	0.3	0.2	$2p^5 ns,md \to 3s^2 3p^5$	0.5	0.2
$3d^9 np \to 4s^1 4p^6$	0.0	0.0	$2p^5 ns,md \to 3s^1 3p^6$	1.7	
				75^a	
$3d^9 np \to 4s^2 4p^4 np$	24.6	24.7	$2p^5 ns,md \to 3s^2 3p^4 ns,md$	73.1	
$3d^9 np \to 4s^1 4p^5 np$	52.3	52.3	$2p^5 ns,md \to 3s^1 3p^5 ns,md$	22.7	22.7
$3d^9 np \to 4s^{0.} 4p^6 np$	22.7	22.7	$2p^5 ns,md \to 3p^0 3p^6 ns,md$	2.0	2.1

^{a}Because of the CI between final ionic states the transitions have not been separated from each other.

6.4. Decay of Two-Electron Excitations

Photoabsorption and total ion yield spectra of atoms are affected by discrete two-electron excitations. Such doubly excited states may decay via Auger-like transitions leaving the ion in an excited state. Decay spectra of two-electron excitations of outer shells of noble gas atoms have been studied recently.[149,150] Similar transitions may also take place at resonance energies where one core electron and one valence electron are excited. The peak structure appears in threshold electron spectra at higher photon energies than in total ion yield spectra of noble gases[151–153] indicating that doubly excited states decay producing Auger electrons with kinetic energies different from zero.

6.5. Decay Channels of Core-Excited Molecules

Decay processes of core-excited resonances in molecules have been studied actively in recent years (see Refs. 154–162 and references therein). The core-excited state may decay through direct dissociation followed by Auger decay in the dissociation products or alternatively through electronic recombination with molecular Auger electron emission. A competition between the decay channels may arise if the electronic decay and dissociation occur on overlapping time scales.

Neutral dissociation followed by Auger decay of atomic fragments was found to be the dominating decay channel in hydrogen halide molecules at the core to valence resonance.[156,157,160] In the H_2S molecule, the presence of a channel of fast dissociation into excited HS^* followed by Auger decay to HS^+ was recently reported.[158–160] Further information was obtained by comparing the synchrotron radiation-induced resonant Auger spectra with those obtained using more standard excitation sources such as monochromatized Al K_α x radiation and electron beam.[159] The decay was suggested to occur in a dissociating excited molecular fragment. Vibrational spacing in the fragment was found to be smaller than calculated for dissociated HS. Thus, the fragment may emit the electron before attaining complete dissociation.

7. ANGULAR DISTRIBUTION OF AUGER ELECTRONS

When a highly excited state of an ion or atom with a vacancy in an inner shell decays via an Auger transition, the angular distribution of the emitted Auger electrons may be nonisotropic.[163] The anisotropy is the result of the nonstatistical population of the magnetic substates in the process of photoionization or photoexcitation[163,164] which reflects asymmetry of the photon–atom interaction. Study of the Auger electron angular distribution provides additional and more refined information as compared to the conventional measurements of intensities of the Auger electron spectra. This information concerns both the Auger decay process and the primary photoexcitation or photoionization processes which created the vacancy. Until recently, experimental investigations of the anisotropy of Auger emission produced by photoionization were rather rare.[20] However, the progress in synchrotron radiation sources, monochromators, and energy analyzers has led to a rapid increase in the number of such experiments.[15,16,110,165–171] The results of these measurements and the problems connected with their interpretations as well as theoretical approaches to the Auger emission anisotropy have been discussed in several recent reviews.[1–3,169,172]

Within the framework of a two-step model the angular distribution of Auger electrons produced in photoabsorption of unpolarized or linearly polarized light by an unpolarized target atom may be written in the following form[173,174]:

$$W(\Theta) = \frac{W_0}{4\pi}(1 + \alpha_2 \mathcal{A}_{20} P_2(\cos\Theta)) \tag{8}$$

Here Θ is the angle between the z-axis and the direction of Auger electron emission, $P_2(\cos\Theta)$ is the second Legendre polynomial, and W_0 is the total probability of Auger emission per unit time. The z-axis is chosen along the direction of the photon beam (for unpolarized radiation) or along the pogarization vector of photons (for linearly polarized radiation). The angular distribution (8) is similar, naturally, to the angular distribution of photoelectrons in dipole photoionization

$$\frac{d\sigma}{d\Omega} = \frac{\sigma}{4\pi}(1 + \beta_2 P_2(\cos\Theta)) \tag{9}$$

However, for Auger electrons the anisotropy of the angular distribution is determined by the product of two factors $\beta_2 = \alpha_2 \mathcal{A}_{20}$. One of them, $\mathcal{A}_{20}$, is an alignment parameter which describes the anisotropy of the decaying state. It is determined by the photoionization amplitudes describing the population of the magnetic sublevels of the state. The coefficient α_2 is the Auger decay anisotropy parameter, which depends on the contributing Auger amplitudes. The factorization of the anisotropy coefficient reflects the two-step character of the Auger emission.

7.1. Alignment of Atoms in Photoionization and Photoexcitation

The alignment parameter $\mathcal{A}_{20}$ is a zero-projection component of the second rank normalized statistical tensor. A set of statistical tensors $\mathcal{A}_{kq} = \rho_{kq}/\rho_{00}$, which is fully equivalent to the density matrix, completely determines an atomic state produced in some process.[175,176] As follows from the general theory,[177] if the atom undergoes a decay the angular distribution of decay products is described only by the even rank statistical tensors because of parity conservation in the decay. Axial symmetry of the excitation process (only the direction of the beam is singled out, the photoelectron is not detected) leads to the conclusion that only zero-projection components of statistical tensors survive. Finally, in the photoexcitation and photoionization processes due to the dipole character of the photon–atom interaction only the component $\mathcal{A}_{20}$ (except trivial $\mathcal{A}_{00} = 1$) can be nonzero. The physical cause of the alignment of the atomic state produced by photoabsorption is the difference in the excitation cross sections $\sigma(J,M)$ for different magnetic substates M, while $\sigma(J,M) = \sigma(J,-M)$. The alignment parameter $\mathcal{A}_{20}$ can be expressed in terms of the substate cross sections. For example, for $J = \frac{3}{2}$ ionization one has

$$\mathcal{A}_{20}(J = 3/2) = \frac{\sigma(3/2,3/2) - \sigma(3/2,1/2)}{\sigma(3/2,3/2) + \sigma(3/2,1/2)} \tag{10}$$

The alignment in photoionization was theoretically considered in an early paper by Berezhko et al.[173] It was shown that for dipole photons the alignment depends on the relative strength of the excitation from a given orbital into two partial waves allowed by parity and momentum conservation. Usually one partial wave corresponding to $l \rightarrow l + 1$ transition dominates. This leads to a rather small alignment of the value 0.2, 0.3 which is typical for photoionization of any atomic subshell. However, the calculations[174] pointed out that, for ionization near threshold, it is to be expected that only the

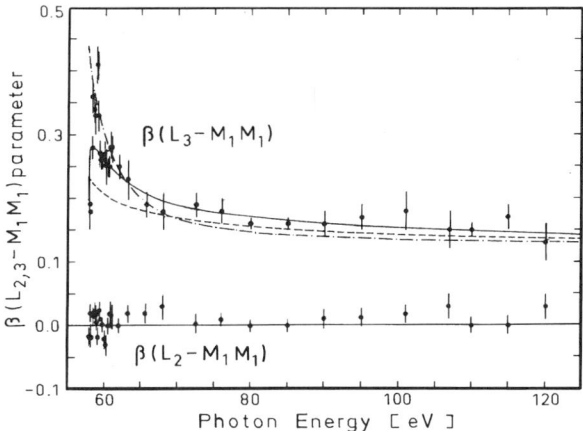

FIGURE 12. Anisotropy coefficient for the MgL_3–$M_1M_1(^1S_0)$ Auger transition (upper points) $\beta_{Auger} = -\mathcal{A}_{20}$. Experimental data are from Ref. 179 and curves represent the theoretical results obtained in HS approximation (dashed curve) and RRPA (solid curve).

wave with the lowest angular momentum ($l \to l - 1$ transition) will participate because of potential barrier effects. The result of this will be a large value of the alignment. The increase of the alignment near threshold was confirmed by experiments.[20,165,169,178,179] Similarly, in the region of the Cooper minimum of photoabsorption the $l \to l - 1$ transition dominates and the alignment is expected to be large. More refined calculations allowing for many-electron correlations and relativistic corrections confirm the general features of the alignment in photoionization discussed above.[172,180]

An experimental study of the alignment is possible by measuring either the angular distribution of Auger electrons or the polarization or angular distribution of fluorescence radiation.[2] If Auger electrons are detected, the transition is chosen for which the anisotropy coefficient α_2 may be calculated in a model-independent way[181,182] (see discussion below). Then the anisotropy for this line provides directly the alignment. For fluorescence emission of dipole photons the coefficient relating the alignment to the observed polarization (or anisotropy) is always a geometrical quantity only.[173] Therefore, both methods have been used for experimental determination of the alignment. Figure 12 shows a typical result of measurements[179] and a comparison with calculations. A more detailed discussion of the problem may be found in recent reviews.[169,172]

Measurements of the photoion alignment provide information about the photoionization amplitudes which is complementary to that obtained in the conventional measurements of the ionization cross sections and the photoelectron angular distributions. In some cases these measurements are sufficient for realizing the so-called "complete" experiment in photoionization[165,170] (see also discussion in Ref. 172 and in Chapter 15 of this volume).

In photoexcitation the calculation of the excited-state alignment is an easy problem because of the angular momentum conservation law. In the practically important particular case of the atom excited by linearly polarized light from the $J_0 = 0^+$ ground state to the $J = 1^-$ excited state, the substates with only one projection $M = 0$ may be populated (the quantization axis is chosen along the polarization vector). In this case according to the definition of the statistical tensors (see Ref. 173 for example) the alignment parameter $\mathcal{A}_{20}$ has an energy-independent value $\mathcal{A}_{20}(J = 1) = -\sqrt{2}$. Large alignment causes a strong anisotropy of the resonant Auger lines[15,16] found experimentally.

7.2. Anisotropy of Normal Auger Transitions

If one knows the alignment of an excited state from experiments, as described above, or from theoretical considerations, then measurements of the angular distribution of Auger electrons for different transitions from this state give the values of the α_2 parameters which provide information on the Auger decay process.

In general the anisotropy parameter α_2 can be expressed in terms of Auger amplitudes as follows[183]:

$$\alpha_2 = \sum_{lj,l'j'} \alpha_{lj,l'j'} \text{Re}(M_{lj}M^*_{l'j'}) / \sum_{lj} |M_{lj}|^2 \qquad (11)$$

Here $M_{lj} = \langle \Psi_f(J')\varepsilon lj; J | V | \Psi_i(J) \rangle$ is an Auger decay amplitude corresponding to a definite channel characterized by the final ionic state and the total and orbital angular momenta of the emitted electron. In this connection three different categories of Auger transitions can be distinguished[166]:

1. Transitions with only one partial wave (one channel) allowed by the selection rules independent of the model. For example, if the final ionic state has total angular momentum $J' = 0$, then the emitted Auger electron should have the same momenta as the primary vacancy. In this case α_2 is only a geometrical quantity independent of the Auger amplitudes.[181,182] These transitions have been used in earlier experiments to determine the alignment $\mathcal{A}_{20}$ of the ionic state.[2]
2. Transitions which are single channel in the LS-coupling approximation. Here, if the ionic wave functions are close to the LS-coupling ones, the anisotropy should be practically independent of the model.
3. Transitions with several partial waves in any coupling. These transitions are especially interesting for gaining additional information on the Auger decay.

Until now the angular distributions of Auger electrons have been studied for noble gas atoms only. There are several theoretical calculations in various approximations[183–187,189–193] and a number of measurements[16,110,165,166,168,169,171] for normal Auger transitions. A comparison of theoretical and experimental anisotropy coefficients shows a high sensitivity of these values to the models and approximations, especially for transitions of the third category. As an example in Table III the experimental and some theoretical data for several transitions in Kr and Xe are compared.* Transitions to the $^1S_0$ and $^3P_0$ final states are typical examples of the transitions where only one partial wave is possible in any coupling scheme. These lines are usually used for extraction of the alignment parameter $\mathcal{A}_{20}$ from the experimental β_2 values with the help of the known α_2 parameter. The transitions to the $^3P_1$ and $^3P_2$ states belong to the second category with only one contributing Auger amplitude in LS-coupling. Especially interesting is the transition M_4–$N_{2,3}N_{2,3}(^3P_2)$ in Kr which is predicted to be isotropic within LS-coupling approximation. Experiment[16,166] shows definitely positive anisotropy. It was shown in Ref. 193 that IC scheme can change considerably the anisotropy coefficient. A recent calculation[191] within the MCDF model with IC gives small positive anisotropy for this transition. However, the discrepancy with the experiment is still large. The effect of IC is even larger in the corresponding N_4–$O_{2,3}O_{2,3}(^3P_2)$ transition in Xe. Here also the experimental anisotropy is larger than predicted. The most striking example of the sensitivity of the anisotropy to the models is represented by the transitions $M_{4,5}$–$N_{2,3}N_{2,3}(^1D_2)$ in Kr which belong to the third category. While the LS-coupling calculations with Hartree–Slater wave functions[186] give a small negative anisotropy, the experiment[16,166] and recently more refined calculations[191] give considerable positive anisotropy. A similar difference of the results of two calculations can be seen for the $N_{4,5}$–$O_{2,3}O_{2,3}(^1D_2)$ transitions in Xe. Here also recent experiments confirm the results of MCDF calculations with IC. The main cause of the difference seems to be the inadequacy of the HS description of the continuum electron wave function.

---

*Results obtained by M. H. Chen[192] within the MCDF approach are very close to those of Kabachnik et al.[191] shown in Table III. Recently, more refined calculations including relaxation, channel coupling, and exchange effects were done.[193] See this reference for discussion.

TABLE III

Theoretical and Experimental Data for the Anisotropy Coefficients α_2 of Different Auger Lines in Kr and Xe

	Theory		Experiment	
Final state	(a)	(b)	(c)	(d)
Kr M_4–$N_{2,3}N_{2,3}$				
$^1S_0$	−1	−1		
$^1D_2$	−0.093	0.240		
$^3P_0$	−1	−1	−0.77(10)	
$^3P_1$	−0.8	−0.818		0
$^3P_2$	0	0.017	0.21(9)	1.45
Kr M_5–$N_{2,3}N_{2,3}$				
$^1S_0$	−1.07	−1.07		
$^1D_2$	−0.099	0.330	0.18(4)	0.50
$^3P_0$	−1.07	−1.07		
$^3P_1$	−0.748	−0.739		
$^3P_2$	−0.382	−0.303	−0.31(6)	0.14

	Theory				Experiment	
Final state	(a)	(e)	(f)	(g)	(h)	(i)
Kr M_4–$N_1N_{2,3}$						
$^1P_1$	−0.81					−0.77(4)
$^3P_0$	−1.00					
$^3P_1$	−0.82	−0.82	−0.92			−1.08(7)
$^3P_2$	−0.81	−0.82	−0.89	−0.90	−0.83(44)	−1.02(7)
Kr M_5–$N_1N_{2,3}$						
$^1P_1$	−0.86					−0.72(4)
$^3P_0$	−1.07					−1.04(13)
$^3P_1$	−0.99	−1.00	−1.03	−0.69	−0.77(25)	−1.20(5)
$^3P_2$	−0.62	−0.63	−0.85			−0.96(7)

	Theory			Experiment		
Final state	(a)	(b)	(d)	(j)	(k)	(l)
Xe N_4–$O_{2,3}O_{2,3}$						
$^1S_0$	−1	−1				
$^1D_2$	−0.554	0.055		0.05(6)	−0.7(1)	
$^3P_0$	−1	−1				
$^3P_1$	−0.8	−0.835	−0.43	−0.73(11)		
$^3P_2$	0	0.156	0.27	0.72(11)	1.2(2)	
Xe N_5–$O_{2,3}O_{2,3}$						
$^1S_0$	−1.07	−1.07				
$^1D_2$	−0.592	0.139		0.23(4)	0.3	0.30(12)
$^3P_0$	−1.07	−1.07	−0.79	−1.07(10)	−1.2(2)	−0.69(10)
$^3P_1$	−0.748	−0.734		−0.77(17)	−1.0(2)	
$^3P_2$	−0.382	−0.227		−0.47(3)		−0.09(10)

Theoretical calculations:
(a) HS approximation with Herman–Skillman potential and LS-coupling[186]
(b) Relativistic DF approximation with IC[191]
(e) HF approximation with LS-coupling[187]
(f) Many-body perturbation theory[187]
(g) Relativistic DF approximation[189]

Experimental data:
(c) Ref. 166
(d) Ref. 16
(h) Ref. 209
(i) Ref. 171
(j) Ref. 168
(k) Becker et al., preliminary results cited in Ref. 191
(l) Ref. 110

In the transitions Kr $M_{4,5}-N_1N_{2,3}(^3P_1, {}^3P_2)$ more than one partial wave is also involved. This case was treated theoretically in detail for different kinds of coupling schemes as well as for different approximations concerning the influence of many-electron correlations.[184,187,189] Some of the results are presented in Table III. Recently, accurate experimental data were obtained for these transitions excited by synchrotron radiation.[171] Experimental results favor the calculations which take into account many-electron effects. On the other hand, the coupling scheme in this case has negligible influence on the α_2 parameters.

7.3. Anisotropy of Resonant Auger Decay

The Auger decay of resonantly excited atomic states shows a remarkable angular distribution of electrons as compared with normal Auger lines. An unusually large degree of angular anisotropy has been discovered by Carlson et al.[15] and confirmed in later experiment (see Refs. 16, 167, 168, and preliminary results by Becker et al., cited in Ref. 195). A theoretical explanation of this phenomenon applying the angular momentum transfer theory was made by Cooper.[188] Another approach based on the two-step model has been developed in several papers.[168,194,195] Partly, the large anisotropy of resonant Auger spectra is explained by a large value of the alignment produced in photoexcitation as compared with photoionization. However, for a detailed explanation of the experimental data including the strong variation of the asymmetry parameter β_2 from line to line, one needs a systematic calculation of the anisotropy coefficients α_2. In this connection a spectator model of the resonant Auger decay was exploited. Within a spectator model the excited electron is treated as a spectator which does not take part in the decay. Then, it is possible to calculate the corresponding Auger amplitudes and the asymmetry parameter using the MCDF approach as it was discussed above for open-shell atoms. Such work is in progress. However, to clarify the physical picture of the process it is instructive to consider a simpler model, which has been suggested recently, the so-called strict spectator model, which relates the anisotropy parameters of the resonant and normal Auger transitions. Within the strict spectator model the interaction of the spectator electron with core electrons is supposed to be so weak that it can be ignored. In this approximation one can easily relate the amplitude of a resonant Auger transition with that of a corresponding normal transition[168,194,195]:

$$\langle \Psi_f((J'_c,j_s)J')\varepsilon l j;J | V | \Psi_i((J_c,j_s)J)\rangle = C(j) \langle \Psi_f(J'_c)\varepsilon l j;J_c | V | \Psi_i(J_c)\rangle \quad (12)$$

where j_s is the angular momentum of the spectator electron. The coefficient $C(j)$ is an angular momentum recoupling coefficient. In some particular cases, e.g., when the total angular momentum of the core electrons $J'_c = 0$, it is possible to relate not only the amplitudes but even the angular anisotropy coefficients for the normal α_2(norm) and the resonant α_2(res) Auger transitions. Remember that for normal Auger transition α_2(norm) is model independent in the case $J'_c = 0$, so one can predict the anisotropy of the corresponding resonant transition. For example, the anisotropy coefficient of the normal Auger transition $L_3-M_{2,3}M_{2,3}({}^1S_0)$ in Ar is α_2(norm) $= -1$.[173] The corresponding resonant transition $2p_{3/2}^{-1}4s \rightarrow 3p^{-2}({}^1S_0)4s$ should have anisotropy α_2(res) $= -1/\sqrt{2}$ and the asymmetry parameter $\beta = \alpha_2 A_{20} = +1$. This result was obtained from general considerations by Becker[167] using the spectator model and was also verified experimentally by Becker et al. (preliminary results cited in Ref. 195). Note the important difference between the predictions for normal and resonant Auger transitions at $J'_c = 0$. In the former case the predicted anisotropy is a completely model-independent result. On the contrary, the prediction for resonant transitions based on the relation (12) is valid only within the framework of the strict spectator model, i.e., model dependent.

The interaction of the spectator electron with core electrons splits the energy levels of the final ionic state in resonant Auger transitions. However, one can average the angular anisotropy over the final energy levels belonging to the same parent double-hole ionic core state (so-called gross spectator model[168]):

$$\alpha_2^{av} = \sum_{J'} \alpha_2 W_0 / \sum_{J'} W_0 \quad (13)$$

TABLE IV
Angular Anisotropy of the Resonant Auger Process for Ar $2p_{3/2} \to 4s$ Excitation[a]

Peak	Final State	β_{theor}	$\beta_{exp}(I)$	$\beta_{exp}(II)$	$\beta_{exp}(III)$
1	$3p^4(^3P)4s$	−0.496	−0.53		−0.46(10)
2	$3p^4(^1D_2)4s$	0.417	0.36, 0.48	0.6	0.55(10)
3	$3p^4(^1S_0)4s$	1.0	0.39, 0.66	1.0[b]	1.09(10)
1a	$3p^4(^3P)4s^4P$	0.525	0.23	0.3	
1b	$3p^4(^3P)4s^2P$	−0.941	−0.69	−0.9	

[a]Experimental data: I, Carlson et al.[16] (two sets of data); II, Becker[167]; III, Becker et al. (preliminary results cited in Ref. 195).
[b]By assumption.

Here J' is the total angular momentum of the final state of the core + spectator electron system. The averaged anisotropy coefficient for resonant transitions is simply related to the corresponding coefficient for the normal transition.[168,195] For example, for closed-shell atoms in jj coupling one has:

$$\alpha_2^{av}(\text{res}) = \alpha_2(\text{norm}) * \begin{cases} 1/\sqrt{2} & p_{3/2} \to s_{1/2} \\ -2\sqrt{2}/5 & d_{3/2} \to p_{3/2} \\ \sqrt{7}/5 & d_{5/2} \to p_{3/2} \end{cases} \quad (14)$$

This relation is based on the strict spectator model. However, it connects two measurable quantities and its experimental verification could serve as a rigit test of the model.

A detailed calculation of the anisotropy coefficients for resonant Auger transitions in noble gases was performed by Hergenhahn et al.[195,196] Table IV shows some of the results for Ar $2p_{3/2} \to 4s$ excitation. In general the agreement with experiments is satisfactory. For heavier atoms the agreement is worse. This may be caused by the roughness of the strict spectator model. However, because the existing experimental data are not sufficiently accurate, more experimental and theoretical efforts are necessary to clarify the situation.

8. PHOTOELECTRON–AUGER ELECTRON ANGULAR CORRELATION

In the previous section we discussed inner-shell photoionization experiments with only one electron (namely, Auger electron) detected. No information about the knocked-out photoelectron has been obtained in these measurements. In earlier papers[197–199] coincidence measurements were suggested for inner-shell study with two electrons in the final state (i.e., photoelectron and Auger electron) detected in coincidence. It was shown that experimental investigation of the angular correlation between two escaping electrons can provide more detailed information on both stages of the process, ionization and decay, as compared with separate noncoincidence measurements of angular distributions of photoelectron and Auger electron. Difficulties and advantages of the coincidence experiments exploring the photon-induced double-ionization processes were discussed recently in a review paper by V. Schmidt.[200]

8.1. Theoretical Background of Coincidence Measurements

A general formalism describing coincidence measurements in application to the two-step inner-shell ionization was considered by Berezhko et al.[198] Here we apply this formalism for discussing the two-step double photoionization. At the first step, photoionization of an inner atomic shell occurs and a photoelectron is detected in a direction $\mathbf{n}_1 = (\Theta_1, \Phi_1)$. We consider two cases of photoionization by a linearly polarized and unpolarized photon beam. In the former case we choose the coordinate frame as shown in Figure 13a (z-axis along the photon polarization vector, y-axis along the photon beam). In the latter case the coordinate frame is shown in Figure 13b (z-axis along the

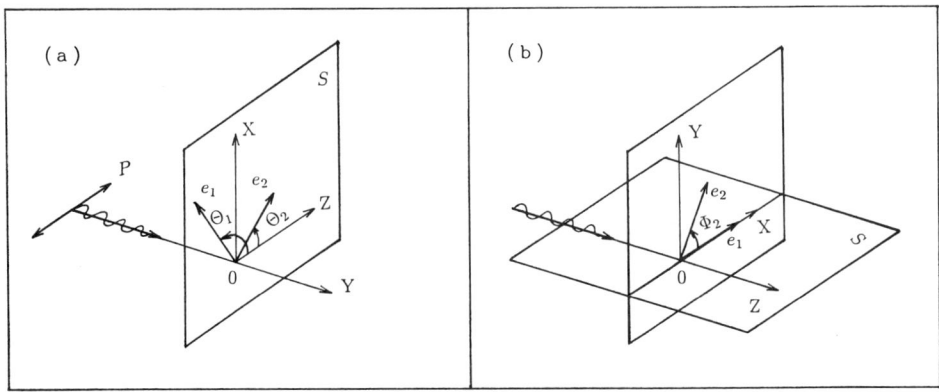

FIGURE 13. The frames chosen for the description of the photoelectron–Auger electron coincidence measurements: (a) linearly polarized photon, (b) unpolarized photon.

photon beam, x-axis along the photoelectron emission direction). As a result of photoionization, a highly excited state of the ion with a vacancy in its inner shell is formed. As we have already discussed in previous sections the ionic state produced is nonisotropic. In general, the polarization can be described by a set of statistical tensors $\mathcal{A}_{kq}(J,\mathbf{n}_1)$ which are functions of the photoelectron emission angles. The values of the tensors are determined by the photoionization amplitudes and their relative phases. The maximal rank of the tensors is limited by the total angular momentum of the ion $k \leq 2J$ and $-k \leq q \leq k$. If Auger decay of the excited state occurs, the angular distribution of emitted electrons is completely determined by the tensors $\mathcal{A}_{kq}(J,\mathbf{n})$[198]:

$$W(\Theta_2, \Phi_2) = \frac{W_0}{4\pi}\left[1 + \sum_{k=2,4,\ldots} \alpha_k \sum_q \mathcal{A}_{kq}(J,\mathbf{n}_1)\left(\frac{4\pi}{2k+1}\right)^{1/2} Y_{kq}(\Theta_2, \Phi_2)\right] \quad (15)$$

where α_k are the same anisotropy coefficients, determined by expression (11). The symmetry conditions of a particular experiment impose some restrictions on the number of independent nonzero tensor components. In the noncoincidence measurements discussed above the experimental conditions are axially symmetric. Therefore, only the component $\mathcal{A}_{20}$ turns out to be nonzero, i.e., the ionic state is aligned and the angular distribution is axially symmetric.

In coincidence measurements the experimental conditions are symmetric about reflection in the plane marked by an "S" in Figure 13. In this case the even rank statistical tensors are real and satisfy the equation $\mathcal{A}_{kq} = (-1)^q \mathcal{A}_{k-q}$.[198] Thus, for example, there are three independent real components of the second-rank tensor $\mathcal{A}_{20}$, $\mathcal{A}_{21}$, and $\mathcal{A}_{22}$ which can be obtained from the angular correlation coincidence measurements. For subshells with $J = \frac{3}{2}$ these components are the only quantities which give additional information on photoionization as compared to the measurements of cross section and photoelectron angular distribution. For $J = \frac{5}{2}$ subshells the components of a fourth-rank tensor contribute. The explicit expression for the $\mathcal{A}_{kq}(J,\mathbf{n}_1)$ tensor in terms of dipole photoionization amplitudes was obtained in an earlier paper[199] in the nonrelativistic case. More general expressions have been presented recently.[201]

8.2. Experimental Investigation

Recently the first measurement of the angular correlation between two escaping electrons in two-step double photoionization of a Xe atom was reported.[170] The yield of the $4d_{5/2}$ photoelectrons produced by a linearly polarized photon beam was measured in coincidence with the N_5-$O_{2,3}O_{2,3}(^1S_0)$ Auger electrons. The analyzer for the $4d_{5/2}$ photoelectron was placed in a fixed position (azimuthal angle $\Phi_1 = 0°$, polar angle $\Theta_1 = 150°$ see Figure 13a). The position of the second

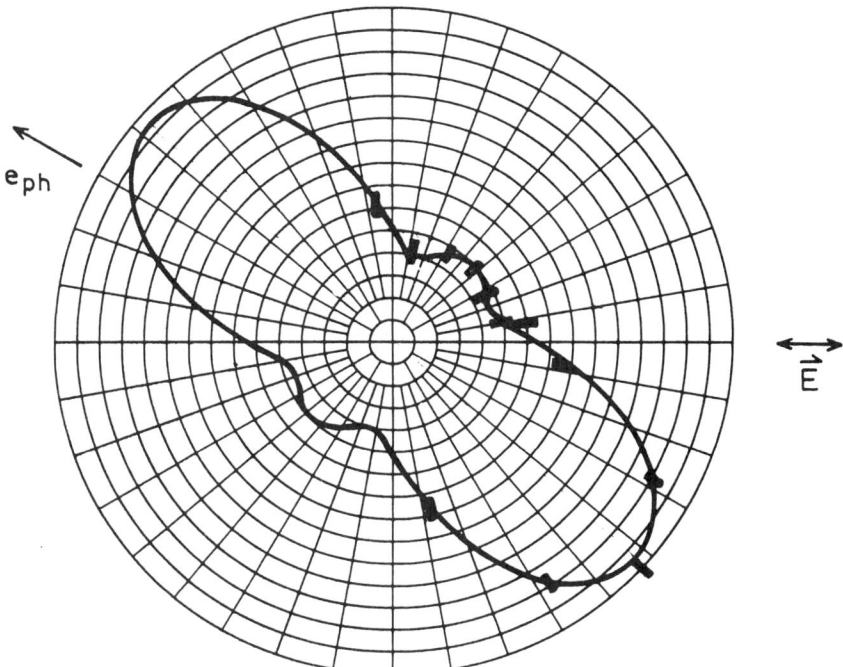

FIGURE 14. Polar plot of coincidence intensities (bars) between $4d_{5/2}$ photo- and $N_5-O_{2,3}O_{2,3}(^1S_0)$ Auger electrons in xenon. The dark curve represents the result of a least-squares fit according to theoretical expression (see text).

analyzer which scans the Auger lines was varied in the plane perpendicular to the photon beam ($\Phi_2 = 0°$, Θ_2 variable). For this particular geometry the theoretical expression (15) can be reduced to the following form[170]:

$$W(\Theta_2, \Phi_2) = A_0 + A_2\cos2\Theta_2 + B_2\sin2\Theta_2 + A_4\cos4\Theta_2 + B_4\sin4\Theta_2 \tag{16}$$

The selected Auger line has the advantage that only one partial wave is possible for the emitted Auger electron. Hence, the anisotropy coefficients α_k are known and the ratios of the coefficients A_i/A_0 and B_i/A_0 are known combinations of the tensors $\mathcal{A}_{2q}$ and $\mathcal{A}_{4q}$. The results of the experiment are shown in Figure 14 as points with error bars, and the solid line represents a least-squared fit according to expression (16). As could be expected from (16), the experimental angular distribution is not symmetric and exhibits the presence of at least two loops which confirm the existence of higher terms in the angular distribution expansion. Combining the experimental values of the ratios A_i/A_0 and B_i/A_0 with the experimental data for the photoionization cross section, the angular asymmetry parameter of the photoelectron and the alignment parameter in noncoincident measurements, one can derive the values of the dipole matrix elements and their relative phases for all ionization channels.[170] The experimental values of the ionization amplitude represent the most complete information about the process and can serve as a stringent test for theories.

It is interesting that for a $J = \frac{3}{2}$ vacancy the angular correlation function (16) contains only two coefficients A_2/A_0 and B_2/A_0. Therefore, the angular distribution is symmetric with respect to the direction $\widetilde{\Theta}$ which is determined by the relation $\tan\widetilde{\Theta} = A_2/B_2$. This case was considered in detail in Refs. 198 and 199 where the parameters of angular distribution were expressed in terms of statistical tensors.

Note that for the case of ionization by nonpolarized light for the same geometry of the experiment (Figure 13b) the angular correlation function is always symmetric[201]:

$$W(\Theta_2,\Phi_2) = \sum_{k-\text{even}} C_k \cos k\Phi_2$$

The ratios of coefficients C_k/C_0 give again new pieces of information on the photoionization amplitudes, which can increase the accuracy and check self-consistency of the derived amplitudes.

9. SPIN POLARIZATION OF AUGER ELECTRONS

It is well known that spin-polarized electrons can be produced in photoionization of atoms and molecules (e.g., see Chapter 15 of this volume). Much less is known about the spin polarization of Auger electrons. It was only recently realized that as a result of the spin-orbit interaction the Auger electrons arising from ionization of a free atom by a beam of photons or particles can be polarized even in the case when the target atom is unpolarized.[202,203]* One can distinguish two different causes of the spin polarization of Auger electrons. (1) In the photoionization of unpolarized atoms by unpolarized or linearly polarized photons a vacancy in the inner atomic shell turns out to be aligned (see Section 7). Owing to the spin-orbit interaction, the Auger decay of an aligned vacancy gives rise to spin-polarized electrons. This mechanism is called dynamic polarization.[202] (2) If atoms are ionized by circularly polarized photons, the photons are aligned and oriented,[202,206,207] and orientation is partially transferred to the spin polarization of Auger electrons. This mechanism is referred to as polarization transfer.[202] Below we discuss these two cases in more detail.

9.1. Dynamic Polarization in Photoionization

Consider the production of Auger electrons in photoionization of an unpolarized atom by unpolarized or linearly polarized light. If the xz-plane is determined by the quantization axis of the incoming photon beam (beam direction for unpolarized photons, polarization vector for linearly polarized photons) and the direction of emission of the Auger electron, then it follows from simple symmetry considerations that the spin polarization can have only one component perpendicular to the xz-plane, so that in this case the Auger electrons are transversely polarized. The value of the polarization vector may be presented in the form[203]:

$$P_y = \frac{\xi_2 \mathcal{A}_{20} \sin 2\Theta}{1 + \alpha_2 \mathcal{A}_{20} P_2(\cos\Theta)} \tag{17}$$

where Θ is the angle between the direction of emission of an Auger electron and the quantization axis. The polarization coefficient ξ_2 is determined by Auger amplitudes

$$\xi_2 = \sum_{lj,l'j'} b_{lj,l'j'} \text{Im}(M_{lj} M^*_{l'j'}) / \sum_{lj} |M_{lj}|^2 \tag{18}$$

The denominator of equation (17) is the angular distribution of Auger electrons. Now we point out the main properties of Auger electron polarization, which follow from the general formulas (17) and (18):

1. The dynamic polarization arises only when the alignment is nonzero. It thus appears that the Auger electrons of the K spectrum are unpolarized because the $1s_{1/2}$ shell vacancy cannot be aligned. Of all lines of the L spectrum only the lines corresponding to the L_3 vacancy decay can be polarized. The degree of spin polarization is proportional to the degree of alignment.

---

*The spin polarization of Auger electrons from oriented atoms in ferromagnetic solids was observed and used to advantage for studying intricate details of the magnetic solid band structure and local surface magnetization (e.g., see Refs. 204, 205 and references therein). Here we discuss only processes in free atoms and molecules.

2. At least two decay channels should contribute for arising polarization. In the single-channel case (i.e., for $J' = 0$) the Auger electrons are unpolarized.
3. Polarization of Auger electrons vanishes if they are ejected at an angle of $\Theta = 90°$ to the quantization axis. The maximum value of the degree of polarization is reached near $\Theta \simeq 45°$. (This angle depends on the values of α_k.) The net dynamic polarization of Auger electrons (integrated over all emission angles) is zero.
4. The polarization value depends on the ratio of the Auger amplitudes for different channels and on their relative phases. Hence, the polarization measurements can provide, in principle, nontrivial information on Auger amplitudes complementary to that obtained from the study of the line intensities and angular distributions.

There are several calculations of the polarization parameters ξ_2 for various Auger transitions in noble gas atoms.[183–187,189,191] However, practically for all transitions studied to date the values of ξ_2 are very small ($\xi_2 \leq 0.1$). Recalling that the alignment for photoionization is also not large one can conclude that the dynamic polarization for normal Auger lines excited in photoionization is so small that it is at the limit of sensitivity of modern experiments.

Until presently only one group attempted to measure the dynamic polarization of the Auger electrons.[208,209] Merz and co-workers have measured the angular distribution and spin polarization of the krypton $M_{4,5}-N_1N_{2,3}$ ($^3P_2$) Auger transitions excited by electron[208] and proton[209] impact. The experimental value of the ξ_2 parameter for $M_4-N_1N_{2,3}$ ($^3P_2$) line 0.06 (5) agrees with theoretical predictions of very small polarization.

Quite recently the first measurement of spin polarization of Auger electrons produced by photoionization has been done (see below). This measurement demonstrates the feasibility of such experiments and the advantages of the photon beam as an ionizing agent. One of the interesting possibilities for the study of the Auger electron polarization would be the resonant Auger transitions excited by linearly polarized light.[1] Big alignment of the excited state ($\mathcal{A}_{20} = -\sqrt{2}$) is an advantage of this case as compared with photoionization. So far no ξ_2 values have been calculated for resonant transitions although theoretical expressions for ξ_2 in the spectator model have been discussed in Ref. 194.

9.2. Polarization Transfer in Photoionization by Circularly Polarized Light

The second mechanism which can yield spin-polarized Auger electrons, polarization transfer, in particular, in the process of photoionization by circularly polarized photons, was considered in theoretical papers by Klar[202] and Huang.[210] Detailed calculations for some noble gas Auger transitions and analysis of possible experiments were performed in Ref. 211.

Within the framework of the standard two-step model the components of the Auger-electron polarization vector can be written in the following form[211] (we use the same reference frame as described in Section 9.1, Θ is the angle between the direction of emission of an Auger electron and the quantization axis):

$$P_x = \frac{\frac{3}{4}\gamma_1 \mathcal{A}_{10}\sin 2\Theta}{1 + \alpha_2 \mathcal{A}_{20} P_2(\cos\Theta)} \tag{19}$$

$$P_y = \frac{\xi_2 \mathcal{A}_{20}\sin 2\Theta}{1 + \alpha_2 \mathcal{A}_{20} P_2(\cos\Theta)} \tag{20}$$

$$P_z = \frac{\beta_1 \mathcal{A}_{10} + \gamma_1 \mathcal{A}_{10} P_2(\cos\Theta)}{1 + \alpha_2 \mathcal{A}_{20} P_2(\cos\Theta)} \tag{21}$$

The denominator in expressions (19)–(21) determines, as earlier, the angular distribution of the Auger electron. The tensor $\mathcal{A}_{10}$ describes the orientation of the photoion, and the tensor $\mathcal{A}_{20}$ describes its alignment. The presence of only zero tensor projections is a consequence of the axial symmetry of the process of vacancy production by a circularly polarized photon. Both tensors depend on photoionization amplitudes. The expressions for $\mathcal{A}_{10}$ and $\mathcal{A}_{20}$ obtained in Refs. 174 and 211 show that the alignment parameter $\mathcal{A}_{20}$ is independent of the degree of circular polarization and is identical with the alignment induced by an unpolarized photon. Therefore, the component P_y of the polarization vector is equal to the dynamic polarization arising as a result of photoionization by an unpolarized photon. On the contrary, the orientation parameter $\mathcal{A}_{10}$ is proportional to the degree of circular polarization. The value of $\mathcal{A}_{10}$ depends on the ratio of the absolute values of the photoionization amplitudes for transitions $l \rightarrow l \pm 1$. The calculations[211] show that, in contrast to the alignment, $\mathcal{A}_{20}$, which is large only in the near-threshold region and in the vicinity of the Cooper minimum, the orientation parameter $\mathcal{A}_{10}$ is large ($\sim 0.6, 0.7$) in a wide range of photon energies.

The parameters β_1, γ_1, and ξ_2 in (19)–(21) depend only on the amplitudes of Auger decay. We have already mentioned that ξ_2 determines the dynamic polarization, connected with the alignment of vacancies. The parameter β_1 determines the net polarization (integral with respect to the angle Θ) arising exclusively from the polarization transfer. The parameter γ_1 determines the differential polarization arising from orientation of a vacancy by a circularly polarized photon. It is interesting that β_1 and γ_1 are nonzero even for a one-channel Auger transition. Measurements of the spin polarization components of an Auger electron together with the angular anisotropy thus yield four parameters, four independent relations connecting the Auger amplitudes.

Recently, the first experimental study of the polarization transfer was reported. The spin polarization of Auger electrons from a Ba $5p$ hole state oriented by circularly polarized synchrotron radiation has been measured.[215] The experiment provides all four parameters characterizing the spin polarization components. Comparing the two mechanisms yielding polarized Auger electrons in photoionization, one can note that in both cases the spin polarization contains valuable information about both the Auger decay and photoionization. However, the polarization transfer has the advantage that in a wide region of photon energies the polarization of Auger electrons has a fairly large value and can be easily measured in experiment. Another advantage is that the polarization transfer gives nonzero polarization even for those lines which cannot be dynamically polarized (single-channel transitions and transitions from $J = \frac{1}{2}$ initial state, e.g., a $p_{1/2}$-vacancy, which cannot be aligned but can be oriented).

9.3. On the Possibility of a "Complete" Experiment in Auger Recombination

From the above discussion it is clear that vacancy production by photoionization and photoexcitation is a unique tool for studying the process of Auger recombination. Combining the measurements of line intensities, angular distribution, and spin polarization of Auger electrons one receives a wealth of information on Auger process dynamics. This raises the question whether these measurements provide a complete description of the Auger process, or in other words what types of measurements are necessary in order to realize the so-called "complete" or "perfect" experiment.[214] A set of measurements is referred to as realizing a "complete" experiment if its analysis provides the theoretical parameters (usually transition matrix elements), which give a complete quantum mechanical description of the process. In atomic photoionization this problem was discussed for a rather long time[212,213] and several types of the complete experiment were realized.[165,170,212] Here we discuss this problem in the context of photoionization or photoexcitation experiments.

The complete experiment should provide the total set of the complex Auger transition amplitudes $M_{lj} \equiv \langle \Psi_f(J') \varepsilon lj; J | V | \Psi_i(J) \rangle$ for all possible values of l and j consistent with the conservation laws. Using the angular momentum and parity selection rules one can easily calculate that in the general case the total number of the amplitudes describing the transition to the definite final ionic state is $2J + 1$. The moduli of the amplitudes and relative phase shifts form the set of the $4J + 1$ parameters to be determined experimentally for a complete characterization of the Auger process. It is worth noting that there are many transitions for which the number of possible decay channels is less than the maximum number.

Now we discuss the possible experiments and the number of nonredundant measurable quantities. Although, in principle, it is possible to detect the ion in the final state, practically in all experimental studies of the Auger decay, only electrons have been detected and their characteristics have been measured. If the electron detector is not sensitive to the spin polarization, then a measurement of the photoinduced Auger electron angular distribution gives only two values: W_0, the total intensity of the line, and α_2, the anisotropy coefficient [see expression (8), we assume that the alignment parameter is known either from theoretical considerations or from experiment]. For obtaining higher-order anisotropy coefficients (α_4, etc.) one has to perform photoelectron–Auger electron coincident measurements.

Now, let us consider the case where a detector of electrons is sensitive to their spin polarization. If the photon beam is unpolarized or linearly polarized, then a measurement of the angular distribution of spin polarization yields the coefficient ξ_2 which determines the dynamic polarization [see expression (17)]. Finally, if the photon beam is circularly polarized, then a measurement of the net spin polarization, integrated over all directions of electron emission, yields the coefficient β_1 and a measurement of the spin polarization angular dependence gives additionally the coefficient γ_1 [see expressions (19) and (21)]. In general, this information is still insufficient for a complete experiment.

Missing information might be obtained from more elaborate coincidence experiments with spin analysis of the fragments. As was shown recently[214] the total number of measurable parameters determining the angular correlation of the two spin-analyzed electrons in the final state is just $4J + 1$, i.e., it is equal to the number of independent variables (amplitudes and phase shifts) describing the Auger decay. Therefore, the complete experiment is possible.

ACKNOWLEDGMENTS. The authors are grateful to S. Whitfield for careful reading of the manuscript and useful suggestions. The financial support of the Research Council for the Natural Sciences of the Academy of Finland is acknowledged. One of the authors (N.M.K.) acknowledges the hospitality of the University of Oulu extended to him during his visit.

REFERENCES

1. W. MEHLHORN, in *X-ray and Inner-shell Processes*, edited by T. A. Carlson, M. O. Krause, and S. T. Manson (American Institute of Physics, New York, 1990), p. 465.
2. W. MEHLHORN, in *Atomic Inner-shell Physics*, edited by B. Crasemann (Plenum Press, New York, 1985), p. 119.
3. T. ÅBERG AND G. HOWAT, in *Encyclopedia of Physics*, Vol. 31, edited by S. Flngge and W. Mehlhorn (Springer-Verlag, Berlin, 1982), p. 469.
4. H. AKSELA AND S. AKSELA, *J. Phys.* **48**, C9-565 (1987).
5. S. AKSELA, in *X-ray Spectroscopy in Atomic and Solid State Physics*, edited by J. C. Ferreira and M. T. Ramos (Plenum Press, New York, 1988), p. 1.
6. H. AKSELA, in *X-ray Spectroscopy in Atomic and Solid State Physics*, edited by J. C. Ferreira and M. T. Ramos (Plenum Press, New York, 1988), p. 15.
7. U. GELIUS, L. ASPLUND, E. BASILIER, S. HEDMAN, K. HELENELUND, AND K. SIEGBAHN, *Nucl. Instrum. Methods B* **1**, 85 (1984).
8. P. BALTZER, B. WANNBERG, AND M. CARLSSON GÖTHE, *Rev. Sci. Instrum.* **62**, 643 (1991).
9. H. AKSELA, S. AKSELA, H. PULKKINEN, G. M. BANCROFT, AND K. H. TAN, *Phys. Rev. A* **37**, 1798 (1988).
10. H. AKSELA, S. AKSELA, H. PULKKINEN, A. KIVIMÄKI, AND O.-P. SAIRANEN, *Phys. Scr.* **41**, 425 (1990).
11. M. MEYER, E. V. RAVEN, M. RICHTER, B. SONNTAG, R. D. COWAN, AND J. E. HANSEN, *Phys. Rev. A* **39**, 4319 (1989).
12. M. MEYER, E. V. RAVEN, M. RICHTER, B. SONNTAG, AND J. E. HANSEN, *J. Electron Spectrosc. Relat. Phenom.* **51**, 407 (1990).
13. M. MEYER, E. V. RAVEN, B. SONNTAG, AND J. E. HANSEN, *Phys. Rev. A* **43**, 177 (1991).
14. S. B. WHITFIELD, J. TULKKI, AND T. ÅBERG, *Phys. Rev. A* **44**, R6983 (1991).
15. T. A. CARLSON, D. R. MULLINS, C. E. BEALL, B. W. YATES, J. W. TAYLOR, D. W. LINDLE, B. P. PULLEN, AND F. A. GRIMM, *Phys. Rev. Lett.* **60**, 1382 (1988).
16. T. A. CARLSON, D. R. MULLINS, C. E. BEALL, B. W. YATES, J. W. TAYLOR, D. W. LINDLE, AND F. A. GRIMM, *Phys. Rev. A* **39**, 1170 (1989).
17. E. V. RAVEN, M. MEYER, M. PAHLER, AND B. SONNTAG, *J. Electron Spectrosc. Relat. Phenom.* **52**, 677 (1990).

18. J. H. D. ELAND, in *X-ray and Inner-Shell Processes*, edited by T. A. Carlson, M. O. Krause, and S. T. Manson (American Institute of Physics, New York, 1990), p. 549.
19. U. BECKER, D. SZOSTAK, M. KUPSCH, H. G. KERKHOFF, B. LANGER, AND R. WEHLITZ, *J. Phys. B* **22**, 749 (1989).
20. S. H. SOUTHWORTH, U. BECKER, C. M. TRUESDALE, P. H. KOBRIN, D. W. LINDLE, S. OWAKI, AND D. A. SHIRLEY, *Phys. Rev. A* **28**, 261 (1983).
21. U. BECKER AND D. A. SHIRLEY, *Phys. Scr.* **T31**, 56 (1990).
22. W. MEHLHORN, *Z. Phys.* **160**, 247 (1960).
23. H. KÖRBER AND W. MEHLHORN, *Z. Phys.* **191**, 217 (1966).
24. B. W. YATES, K. H. TAN, G. M. BANCROFT, L. L. COATSWORTH, AND J. S. TSE, *J. Chem. Phys.* **83**, 4906 (1985).
25. B. W. YATES, K. H. TAN, L. L. COATSWORTH, AND G. M. BANCROFT, *Phys. Rev. A* **31**, 1529 (1985).
26. M. O. KRAUSE, T. A. CARLSON, AND A. FAHLMAN, *Phys. Rev. A* **30**, 1316 (1984); M. O. KRAUSE, F. CERRINA, AND A. FAHLMAN, *Phys. Rev. Lett.* **50**, 1118 (1983).
27. C. D. CALDWELL AND M. O. KRAUSE, *J. Phys. B* **23**, 2233 (1990).
28. H. DERENBACH, C. FRANKE, R. MALUTZKI, A. WACHTER, AND V. SCHMIDT, *Nucl. Instrum. Methods A* **260**, 258 (1987).
29. D. VARGA, I. KÁDÁR, S. RICZ, J. VÉGH, Á. KÖVÉR, B. SULIK, AND D. BERÉNYI, *Nucl. Instrum. Methods A* **313**, 162 (1992).
30. J. FELDHAUS, W. ERLEBACH, A. L. D. KILCOYNE, K. J. RANDALL, AND M. SCHMIDBAUER, *Rev. Sci. Instrum.* **63**, 1454 (1992).
31. J. D. BOZEK, J. N. CUTLER, G. M. BANCROFT, L. L. COATSWORTH, K. H. TAN, D. S. YANG, AND R. G. CAVELL, *Chem. Phys. Lett.* **165**, 1 (1990).
32. J. D. BOZEK, G. M. BANCROFT, J. N. CUTLER, AND K. H. TAN, *Phys. Rev. Lett.* **65**, 2757 (1990).
33. J. N. CUTLER, G. M. BANCROFT, AND K. H. TAN, *J. Phys. B* **24**, 4897 (1991).
34. J. N. CUTLER, G. M. BANCROFT, D. G. SUTHERLAND, AND K. H. TAN, *Phys. Rev. Lett.* **67**, 1531 (1991).
35. H. AKSELA, G. M. BANCROFT, AND B. OLSSON, *Phys. Rev. A* **46**, 1345 (1992).
36. M. O. KRAUSE, C. D. CALDWELL, S. B. WHITFIELD, S. J. SHAPHORST, AND Y. AZUMA, *Synchrotron Radiation News* **4**, 25 (1992).
37. M. RICHTER, M. MEYER, M. PAHLER, T. PRESCHER, E. V. RAVEN, B. SONNTAG, AND H. E. WETZEL, *Phys. Rev. A* **39**, 5666 (1989).
38. C. T. CHEN, *Nucl. Instrum. Methods A* **256**, 595 (1987).
39. C. T. CHEN AND F. SETTE, *Rev. Sci. Instrum.* **60**, 1616 (1989).
40. P. A. HEIMANN, F. SENF, W. MCKINNEY, M. HOWELLS. R. D. VAN ZEE, L. J. MEDHURST, T. LAURITZEN, J. CHIN, J. MENEGHETTI, W. GATH, H. HOGREFE, AND D. A. SHIRLEY, *Phys. Scr.* **T31**, 127 (1990).
41. M. DOMKE, T. MANDEL, A. PUSCHMANN, C. XUE, D. A. SHIRLEY, AND G. KAINDL, *Rev. Sci. Instrum.* **63**, 80 (1992).
42. D. BULGIN, J. DYKE, F. GOODFELLOW, N. JONATHAN, E. LEE, AND A. MORRIS, *J. Electron Spectrosc. Relat. Phenom.* **12**, 67 (1977).
43. S. AKSELA, M. HARKOMA, M. POHJOLA, AND H. AKSELA, *J. Phys. B* **17**, 2227 (1984).
44. S. AKSELA, M. HARKOMA, AND H. AKSELA, *Phys. Rev. A* **29**, 2915 (1984).
45. H. AKSELA, S. AKSELA, T. PEKKALA, AND M. WALLENIUS, *Phys. Rev. A* **35**, 1522 (1987).
46. S. AKSELA, T. PEKKALA, H. AKSELA, M. WALLENIUS, AND M. HARKOMA, *Phys. Rev. A* **35**, 1426 (1987).
47. T. PRESCHER, M. RICHTER, B. SONNTAG, AND H. E. WETZEL, *Nucl. Instrum. Methods A* **254**, 627 (1987).
48. I. P. GRANT, B. J. MCKENZIE, AND P. H. NORRINGTON, *Comput. Phys. Commun.* **21**, 207 (1980); **21**, 233 (1980).
49. F. P. LARKINS, in *Atomic Inner-Shell Processes I*, edited by B. Crasemann (Academic Press, New York, 1975), p. 377.
50. G. B. ARMEN, T. ÅBERG, J. C. LEVIN, B. CRASEMANN, M. H. CHEN, G. E. ICE, AND G. S. BROWN, *Phys. Rev. Lett.* **54**, 1142 (1985).
51. G. B. ARMEN, J. TULKKI, T. ÅBERG, AND B. CRASEMANN, *Phys. Rev. A* **36**, 5606 (1987).
52. G. WENTZEL, *Z. Phys.* **43**, 524 (1927).
53. K.-H. HUANG, *J. Phys. B* **11**, 787 (1978).
54. H. AKSELA, S. AKSELA, J. TULKKI, T. ÅBERG, G. M. BANCROFT, AND K. H. TAN, *Phys. Rev. A* **39**, 3401 (1989).
55. J. TULKKI, T. ÅBERG, A. MÄNTYKENTTÄ, AND H. AKSELA, *Phys. Rev. A* **46**, 1357 (1992).
56. M. H. CHEN AND B. CRASEMANN, *Phys. Rev. A* **28**, 2829 (1983).
57. U. FANO, *Phys. Rev.* **124**, 1866 (1961).
58. H. ÅGREN, *J. Chem. Phys.* **75**, 1267 (1981).
59. P. E. M. SIEGBAHN, *Chem Phys.* **66**, 443 (1982).
60. C.-M. LIEGENER, *J. Chem Phys.* **79**, 2924 (1983); **106**, 201 (1984).
61. F. TARANTELLI, A. SGAMELLOTTI, AND L. S. CEDERBAUM, *J. Chem. Phys.* **94**, 523 (1991).

62. H. Siegbahn, L. Asplund, and P. Kelfve, *Chem. Phys. Lett.* **35**, 330 (1975).
63. D. R. Jennison, *Chem. Phys. Lett.* **69**, 435 (1980).
64. K. Faegri, Jr., and H. P. Kelly, *Phys. Rev. A* **19**, 1649 (1979).
65. F. P. Larkins and J. A. Richards, *Aust. J. Phys.* **39**, 809 (1986).
66. R. Colle and S. Simonucci, *Phys. Rev. A* **39**, 6247 (1989).
67. V. Carravetta and H. Ågren, *Phys. Rev. A* **35**, 1022 (1987).
68. K. Zähringer, H.-D. Meyer, and L. S. Cederbaum, *Phys. Rev. A* **45**, 318 (1992).
69. H. Aksela, S. Aksela, and H. Patana, *Phys. Rev. A* **30**, 858 (1984).
70. M. H. Chen, B. Crasemann, and H. Mark, *Phys. Rev. A* **24**, 1158 (1981).
71. C. E. Moore, Atomic Energy Levels, *Nat. Stand. Ref. Data Ser.* 467 (National Bureau of Standards, Washington, DC, 1971), Vols. 1–3.
72. A. Albiez, M. Thoma, W. Weber, and W. Mehlhorn, *Z. Phys. D* **16**, 97 (1990).
73. M. Leväsalmi, H. Aksela, and S. Aksela, *Phys. Scr.* **T41**, 119 (1992).
74. J. Nordgren, H. Ågren, C. Nordling, and K. Siegbahn, *Phys. Scr.* **19**, 5 (1979).
75. L. O. Werme, T. Bergmark, and K. Siegbahn, *Phys. Scr.* **6**, 141 (1972).
76. H. Aksela, S. Aksela, and H. Pulkkinen, *Phys. Rev. A* **30**, 2456 (1984).
77. H. Aksela, S. Aksela, H. Pulkkinen, G. M. Bancroft, and K. H. Tan, *Phys. Rev. A* **33**, 3876 (1986).
78. H. Aksela, S. Aksela, and H. Pulkkinen, *Phys. Rev. A* **30**, 865 (1984).
79. H. Aksela, S. Aksela, G. M. Bancroft, K. H. Tan, and H. Pulkkinen, *Phys. Rev. A* **33**, 3867 (1986).
80. W. N. Asaad, *Nucl. Phys.* **66**, 494 (1965).
81. J. Bruneau, *J. Phys. B* **20**, 713 (1987).
82. K. G. Dyall and F. P. Larkins, *J. Phys. B* **15**, 2793 (1982).
83. J. Bruneau, *J. Phys. B* **16**, 4135 (1983).
84. H. Aksela, S. Aksela, J. Väyrynen, and T. D. Thomas, *Phys. Rev. A* **22**, 1116 (1980).
85. S. Aksela, H. Aksela, and T. D. Thomas, *Phys. Rev. A* **19**, 721 (1979).
86. S. Aksela, J. Väyrynen, and H. Aksela, *Phys. Rev. Lett.* **33**, 999 (1974).
87. H. Aksela and S. Aksela, *J. Phys. B* **7**, 1262 (1974).
88. S. Aksela and H. Aksela, *Phys. Rev. A* **27**, 3129 (1983).
89. H. Aksela and S. Aksela, *J. Phys. B* **16**, 1531 (1983).
90. H. Aksela, T. Mäkipaaso, V. Halonen, M. Pohjola, and S. Aksela, *Phys. Rev. A* **30**, 1339 (1984).
91. J. Väyrynen, S. Aksela, M. Kellokumpu, and H. Aksela, *Phys. Rev. A* **22**, 1610 (1980).
92. H. Aksela, T. Mäkipaaso, and S. Aksela, *Phys. Rev. A* **31** 499 (1985).
93. R. Malutzki, A. Wachter, V. Schmidt, and J. E. Hansen, *J. Phys. B* **20**, 5411 (1987).
94. A. Mäntykenttä, *Phys. Rev. A* **47**, 3961 (1993).
95. G. Howat, T. Åberg, and O. Goscinski, *J. Phys. B* **11**, 1575 (1978).
96. H. P. Kelly, *Phys. Rev. A* **11**, 556 (1975).
97. H. Aksela, S. Aksela, A. Mäntykenttä, J. Tulkki, E. Shigemasa, A. Yagishita, and Y. Furusawa, *Phys. Scr.* **T41**, 113 (1992).
98. H. Aksela, S. Aksela, and R. Lakanen, *Phys. Rev. A* **42**, 1791 (1990).
99. N. Mårtensson and B. Johansson, *Phys Rev. B* **28**, 3733 (1983).
100. S. L. Sorensen, S. J. Schaphorst, S. B. Whitfield, B. Crasemann, and R. Carr, *Phys. Rev. A* **44**, 350 (1991).
101. L. Karlsson, P. Baltzer, S. Svensson, and B. Wannberg, *Phys. Rev. Lett.* **60**, 2473 (1988).
102. S. Svensson, L. Karlsson, P. Baltzer, M. P. Keane, and B. Wannberg, *Phys. Rev. A* **40**, 4369 (1989).
103. B. Wannberg, S. Svensson, M. P. Keane, L. Karlsson, and P. Baltzer, *Chem. Phys.* **133**, 281 (1989).
104. L. Karlsson, S. Svensson, P. Baltzer, M. Carlsson-Göthe, M. P. Keane, A. N. de Brito, N. Correia, and B. Wannberg, *J. Phys. B* **22**, 3001 (1989).
105. M. Larsson, P. Baltzar, S. Svensson, B. Wannberg, N. Mårtensson, A. N. de Brito, N. Correia, M. P. Keane, M. Carlsson-Göthe, and L. Karlsson, *J. Phys. B* **23**, 1175 (1990).
106. A. Cesar, H. Ågren, A. N. de Brito, S. Svensson, L. Karlsson, M. P. Keane, B. Wannberg, P. Baltzer, P. G. Fournier, and J. Fournier, *J. Chem. Phys.* **93**, 918 (1990).
107. L. G. M. Pettersson, L. Karlsson, M. P. Keane, A. N. de Brito, N. Correia, M. Larsson, L. Broström, S. Mannervik, and S. Svensson, *J. Chem. Phys.* **96**, 4884 (1992).
108. B. Crasemann, *J. Phys.* **48**, C9-389 (1987).
109. M. O. Krause, *Phys. Scr.* **T17**, 146 (1987).
110. S. B. Whitfield, C. D. Caldwell, D. X. Huang, and M. O. Krause, *J. Phys. B* **25**, 4755 (1992).
111. B. Crasemann, *Proc. ICPEAC* **XVII**, 69 (1991).
112. V. Schmidt, *J. Phys.* **48**, C9-401 (1987).

113. A. Russek and W. Mehlhorn, *J. Phys. B* **19**, 911 (1986).
114. W. Eberhardt, S. Bernstorff, H. W. Jochims, S. B. Whitfield, and B. Crasemann, *Phys. Rev. A* **38**, 3808 (1988).
115. J. Tulkki, T. Åberg, S. B. Whitfield, and B. Crasemann, *Phys. Rev. A* **41**, 181 (1990).
116. M. Borst and V. Schmidt, *Phys. Rev. A* **33**, 4456 (1986).
117. V. Schmidt, *Z. Phys. D* **2**, 275 (1986).
118. G. B. Armen, S. L. Sorensen, S. B. Whitfield, G. E. Ice, J. C. Levin, G. S. Brown, and B. Crasemann, *Phys. Rev. A* **35**, 3966 (1987).
119. J. Tulkki, G. B. Armen, T. Åberg, B. Crasemann, and M. H. Chen, *Z. Phys. D* **5**, 241 (1987).
120. T. Åberg, *Phys. Scr.* **T41**, 71 (1992).
121. S. Svensson, B. Eriksson, N. Mårtensson, G. Wendin, and U. Gelius, *J. Electron Spectrosc. Relat. Phenom.* **47**, 327 (1988).
122. J. Tulkki, T. Åberg, S. Schaphorst, and B. Crasemann, Program book of abstracts of X-10, E03 (1990).
123. A. Mäntykenttä, H. Aksela, S. Aksela, A. Yagishita, and E. Shigemasa, *J. Phys. B* **25**, 5315 (1992).
124. A. Mäntykenttä, H. Aksela, S. Aksela, J. Tulkki, and T. Åberg, *Phys. Rev. A* **47**, 4865 (1993).
125. S. Aksela, M. Kellokumpu, H. Aksela, and J. Väyrynen, *Phys. Rev. A* **23**, 2374 (1981).
126. H. Aksela, S. Aksela, R. Lakanen, J. Tulkki, and T. Åberg, *Phys. Rev. A* **42**, 5193 (1990).
127. H. Aksela and S. Aksela, *Phys. Rev. A* **28**, 2851 (1983).
128. H. Aksela, S. Aksela, A. Mäntykenttä, A. Kivimäki, A. Yagishita, and E. Shigemasa, *J. Phys. B* **26**, 1435 (1993).
129. M. Kellokumpu and H. Aksela, *Phys. Rev. A* **31**, 777 (1985).
130. W. Weber, B. Breuckmann, R. Huster, W. Menzel, W. Mehlhorn, M. H. Chen, and K. G. Dyall, *J. Electron Spectrosc. Relat. Phenom.* **47**, 105 (1988).
131. W. Eberhardt, G. Kalkoffen, and C. Kunz, *Phys. Rev. Lett.* **41**, 156 (1978).
132. T. A. Carlsson, D. R. Mullins, C. E. Beall, B. W. Yates, J. W. Taylor, D. W. Lindle, and F. A. Grimm, *Phys. Rev. A* **39**, 1170 (1989).
133. H. Aksela, S. Aksela, H. Pulkkinen, and A. Yagishita, *Phys. Rev. A* **40**, 6275 (1989).
134. P. Lablanquie and P. Morin, *J. Phys. B* **24**, 4349 (1991).
135. M. Meyer, E. v. Raven, M. Richter, B. Sonntag, R. D. Cowan, and J. E. Hansen, *Phys. Rev. A* **39**, 4319 (1989).
136. B. Sonntag, *Phys. Scr.* **T34**, 93 (1991).
137. B. Sonntag and P. Zimmermann, *Rep. Prog. Phys.* **55**, 911 (1992).
138. A. Yagishita, S. Aksela, T. Prescher, M. Meyer, M. Richter, E. v. Raven, and B. Sonntag, *J. Phys. B* **21**, 945 (1988).
139. L. Nahon, L. Duffy, P. Morin, F. Combet-Farnoux, J. Tremblay, and M. Larzilliere, *Phys. Rev. A* **41**, 4879 (1990).
140. L. Nahon, P. Morin, and F. Combet-Farnoux, *Phys. Scr.* **T41**, 104 (1992).
141. L. Nahon, L. Duffy, P. Morin, F. Combet-Farnoux, J. Tremblay, and M. Larzilliere, *Phys. Rev. A* **41**, 4879 (1990).
142. L. Nahon, P. Morin, and F. Combet-Farnoux, *Phys. Scr.* **T41**, 104 (1992).
143. L. Nahon and P. Morin, *Phys. Rev. A* **45**, 2887 (1992).
144. M. Meyer, T. Prescher, E. v. Raven, M. Richter, E. Schmidt, B. Sonntag, and H.-E. Wetzel, *Z. Phys. D* **2**, 347 (1986).
145. C. D. Caldwell, M. G. Flemming, M. O. Krause, P. van der Meulen, C. Pan, and A. F. Starace, *Phys. Rev. A* **41**, 542 (1990).
146. S. B. Whitfield, C. D. Caldwell, and M. O. Krause, *Phys. Rev. A* **43**, 2338 (1991).
147. J. M. Bizau, P. Gérard, F. J. Wuilleumier, and G. Wendin, *Phys. Rev. A* **36**, 1220 (1987).
148. J.M. Bizau, P. Gérard, F. J. Wuilleumier, and G. Wendin, *Phys. Rev. Lett.* **53**, 2083 (1984).
149. U. Becker and R. Wehlitz, *Phys. Scr.* **T41**, 127 (1992).
150. G. B. Armen and F. P. Larkins, *J. Phys. B* **25**, 931 (1992).
151. T. Hayaishi, Y. Morioka, Y. Kageyama, M. Watanabe, I. H. Suzuki, A. Mikuni, G. Isoyama, S. Asaoka, and M. Nakamura, *J. Phys. B* **17**, 3511 (1984).
152. T. Hayaishi, E. Murakami, Y. Morioka, H. Aksela, S. Aksela, E. Shigemasa, and A. Yagishita, *Phys. Rev. A* **44**, R2771 (1991).
153. T. Hayaishi, E. Murakami, Y. Morioka, H. Aksela, S. Aksela, E. Shigemasa, and A. Yagishita, *J. Phys. B* **25**, 4119 (1992).
154. I. Nenner, P. Morin, P. Lablanquie, M. Simon, N. Levasseur, and P. Millie, *J. Electron Spectrosc. Relat. Phenom.* **52**, 623 (1990).

155. W. Eberhardt, E. W. Plummer, C. T. Chen, and W. K. Ford, *Aust. J. Phys.* **39**, 853 (1986).
156. P. Morin and I. Nenner, *Phys. Rev. Lett.* **56**, 1913 (1986).
157. H. Aksela, S. Aksela, M. Ala-Korpela, O.-P. Sairanen, M. Hotokka, G. M. Bancroft, K. H. Tan, and J. Tulkki, *Phys. Rev. A* **41**, 6000 (1990).
158. H. Aksela, S. Aksela, A. N. de Brito, G. M. Bancroft, and K. H. Tan, *Phys. Rev. A* **45**, 7948 (1992).
159. A. N. de Brito, N. Correia, B. Wannberg, P. Baltzer, L. Karlsson, S. Svensson, M. Y. Adam, H. Aksela, and S. Aksela, *Phys. Rev. A* **46**, 6067 (1992).
160. H. Aksela, S. Aksela, O.-P. Sairanen, A. Kivimäki, G. M. Bancroft, and K. H. Tan, *Phys. Scr.* **T41**, 122 (1992).
161. U. Becker, R. Hölzel, H.G. Kerkhoff, B. Langer, D. Szostak, and R. Wehlitz, *Phys. Rev. Lett.* **56**, 1455 (1986).
162. T. X. Carroll and T. D. Thomas, *J. Chem. Phys.* **90**, 3479 (1989).
163. W. Mehlhorn, *Phys. Lett. A* **26**, 166 (1968).
164. S. Flügge, W. Mehlhorn, and V. Schmidt, *Phys. Rev. Lett.* **29**, 7 (1972).
165. A. Hausmann, B. Kämmerling, H. Kossmann, and V. Schmidt, *Phys. Rev. Lett.* **61**, 2669 (1988).
166. B. Kämmerling, V. Schmidt, W. Mehlhorn, W. B. Peatman, F. Schaefers, and T. Schroeter, *J. Phys. B* **22**, L597 (1989).
167. U. Becker, in *The Physics of Electronic and Atomic Collisions* (AIP Conf. Proc. 205), edited by A. Dalgarno, R. S. Freund, P. M. Koch, M. S. Lubell, and T. B. Lucatorto (AIP, New York, 1989), p. 162.
168. B. Kämmerling, B. Krässig, and V. Schmidt, *J. Phys. B* **23**, 4487 (1990).
169. C. D. Caldwell, in *X-ray and Inner-Shell Processes* (AIP Conf. Proc. 215), edited by T. A. Carlson, M. O. Krause, and S. T. Manson (AIP, New York, 1990), p. 685.
170. B. Kämmerling and V. Schmidt, *Phys. Rev. Lett.* **67**, 1848 (1991).
171. B. Kämmerling, B. Krässig, O. Schwarzkopf, J. P. Ribeiro, and V. Schmidt, *J. Phys. B* **25**, L5 (1992).
172. W. Mehlhorn, in *Proceedings of the UK/USSR seminar on Today and tomorrow in photoionization*, edited by M. Y. Amusia and J. B. West, Science and Engineering Council, Daresbury Lab., Daresbury, Warrington WA4 4AD (1991).
173. E. G. Berezhko and N. M. Kabachnik, *J. Phys. B* **10**, 2467 (1977).
174. E. G. Berezhko, N. M. Kabachnik, and V. Rostovsky, *J. Phys. B* **11**, 1749 (1978).
175. U. Fano and G. Racah, *Irreducible Tensorial Sets* (Academic Press, New York, 1959).
176. K. Blum, *Density Matrix Theory and Applications* (Plenum Press, New York, 1981).
177. S. Devons and L. J. B. Goldfarb, *Handbuch der Physik* **42** (Springer, Berlin, 1957), p. 362.
178. W. Kronast, R. Huster, and W. Mehlhorn, *Z. Phys. D* **2**, 285 (1986).
179. B. Kämmerling, A. Hausmann, J. Läuger, and V. Schmidt, *J. Phys. B* **25**, 4773 (1992).
180. E. G. Berezhko, V. K. Ivanov, and N. M. Kabachnik, *Phys. Lett. A* **66**, 474 (1978).
181. B. Cleff and W. Mehlhorn, *J. Phys. B* **7**, 605 (1974).
182. J. Eichler and W. Fritsch, *J. Phys. B* **9**, 1477 (1976).
183. N. M. Kabachnik and I. P. Sazhina, *J. Phys B* **17**, 1335 (1984).
184. K. Blum, B. Lohmann, and E. Taute, *J. Phys. B* **19**, 3815 (1986).
185. N. M. Kabachnik and I. P. Sazhina, *Opt. Spectros. (USSR)* **60**, 683 (1986).
186. N. M. Kabachnik and I. P. Sazhina, *J. Phys. B* **21**, 267 (1988).
187. N. M. Kabachnik, I. P. Sazhina, I. S. Lee, and O. V. Lee, *J. Phys. B* **21**, 3695 (1988).
188. J. W. Cooper, *Phys. Rev. A* **39**, 3714 (1989).
189. B. Lohmann, *J. Phys. B* **23**, 3147 (1990).
190. N. M. Kabachnik and O. V. Lee, *Z. Phys. D* **17**, 169 (1990).
191. N. M. Kabachnik, B. Lohmann, and W. Mehlhorn, *J. Phys. B* **24**, 2249 (1991).
192. M. H. Chen, *Phys. Rev. A* **45**, 1684 (1992).
193. J. Tulkki, N. M. Kabachnik, and H. Aksela, *Phys. Rev. A* **48**, 1277 (1993).
194. B. Lohmann, *J. Phys. B* **24**, 861 (1991).
195. U. Hergenhahn, N. M. Kabachnik, and B. Lohmann, *J. Phys. B* **24**, 4759 (1991).
196. U. Hergenhahn, B. Lohmann, N. M. Kabachnik, and U. Becker, *J. Phys. B* **26**, L117 (1993).
197. S. C. McFarlane, *J. Phys. B* **8**, 895 (1975).
198. E. G. Berezhko, N. M. Kabachnik, and V. V. Sizov, *J. Phys. B* **11**, 1819 (1978).
199. E. G. Berezhko and N. M. Kabachnik, *J. Phys. B* **12**, 2993 (1979).
200. V. Schmidt, in *X-ray and Inner-Shell Processes* (AIP Conf. Proc. 215), edited by T. A. Carlson, M. O. Krause, and S. T. Manson (AIP, New York, 1990), p. 559.
201. N. M. Kabachnik, *J. Phys. B* **25**, L389, (1992).
202. H. Klar, *J. Phys. B* **13**, 4741 (1980).

203. N. M. KABACHNIK, *J. Phys. B* **14**, L337 (1981).
204. M. LANDOLT, in *Polarized Electrons in Surface Physics*, edited by R. Feder (World Scientific, Singapore, 1985).
205. M. LANDOLT, *Appl. Phys. A* **41**, 83 (1986).
206. H. KLAR, *J. Phys. B* **13**, 2037 (1980).
207. W. BUßERT AND H. KLAR, *Z. Phys. A* **312**, 315 (1983).
208. U. HAHN, J. SEMKE, H. MERZ, AND J. KESSLER, *J. Phys.* **18**, L417 (1985).
209. H. MERZ AND J. SEMKE, in *X-ray and Inner-Shell Processes* (AIP Conf. Proc. 215), edited by T. A. Carlson, M. O. Krause, and S. T. Manson (AIP, New York, 1990), p. 719.
210. K.-N. HUANG, *Phys. Rev. A* **26**, 2274 (1982).
211. N. M. KABACHNIK AND O. V. LEE, *J. Phys. B* **22**, 2705 (1989).
212. U. HEINZMANN, *J. Phys. B* **13**, 4353, 4367 (1980).
213. J. KESSLER, *Comments At. Mol. Phys.* **10**, 47 (1981).
214. N. M. KABACHNIK AND I. P. SAZHINA, *J. Phys. B* **23**, L353 (1990).
215. R. KUNTZE, M. SALZMANN, N. BÖWERING, AND U. HEINZMANN, *Phys. Rev. Lett.* **70**, 3716 (1993).

CHAPTER 12

LASER-BASED UV AND VUV SPECTROSCOPY OF DOUBLY EXCITED ATOMS

J.-P. CONNERADE

1. INTRODUCTION

This chapter is intended as a brief introduction to recent experiments in which tunable lasers have been used instead of synchrotron radiation to probe highly excited states in the photoionization continuum of free atoms. The methods involved, which are comparatively new, are described, and the relative advantages and disadvantages of tunable laser versus synchrotron radiation sources are explained. Examples of recent spectra are then given, drawn from our program at Imperial College, and this is followed by some consideration of specific problems for which laser-based experiments are most fruitful.

The structure which appears in atomic spectra above the first ionization continuum is related to either the simultaneous excitation of two electrons or to the excitation of inner shells. Tunable lasers are not yet capable of reaching very high energies, and it is therefore still necessary to choose the spectra for study according to the range available: at the time of writing, it remains difficult to excite inner shells by using tunable laser radiation, and it is still necessary to select atoms with very low double ionization potentials if one wishes to scan through doubly excited resonances. Currently, this represents the main limitation of laser-based VUV spectroscopy.

In order to overcome the limitations of wavelength range and extend the method to more interesting cases, which tend to lie toward higher energies, one often makes use of more than one photon in the excitation scheme, by exploiting either stepwise or multiphoton excitation, or indeed by a combination of both. Thus, the laser-based methods divide naturally into two classes, according to whether a tunable photon of adequately high energy is available for single-photon excitation, or whether it is necessary to "bounce" off a suitable intermediate state of the sample, in which case more than one laser may prove necessary. Another option which sometimes exists is that of populating an excited metastable state by using collisions, for example by running a discharge in a gaseous sample, and then achieving a further excitation by using a tunable laser.

Despite considerable progress with all of these techniques, much remains to be done in the development of suitable sources, and there are fundamental limitations which are not easy to overcome: indeed, it is probable that the need for very-high-resolution spectroscopy will remain confined by the nature of atomic physics to a certain range of problems. Even though autoionization broadening occurs above the first ionization thresholds of atoms, very sharp lines are still encountered in the autoionization range. Indeed, provided the resonances belong to series converging on a metastable state of the ion, they will become very sharp as the series limit is approached, and high-resolution spectroscopy will be necessary to probe the details of their structure.

J.-P. CONNERADE • The Blackett Laboratory, Imperial College, London SW7 2BZ, United Kingdom.

VUV and Soft X-Ray Photoionization. Edited by Uwe Becker and David A. Shirley. Plenum Press, New York, 1996.

As one crosses above the double ionization potential of the atom, the parent ion state in turn may autoionize. In the soft x-ray range, this is referred to as Auger broadening, but when the process first becomes possible, i.e., near the double ionization threshold, one tends to refer to it as two-step autoionization. This source of broadening results in lines of a finite width which does not reduce as resonances approach their series limit. Consequently, autoionizing series are terminated before the limit is reached, and very high resolution ceases to be necessary, since very high Rydberg states cannot be observed.

As a rough guide, we may say that the double ionization limit of helium, which occurs at about 54 eV, is probably a practical limit above which high-resolution laser spectroscopy ceases to be useful for neutral species. Above this energy, there is still a wide open field in the study of ions, which is beyond the scope of the present chapter.

2. LASER SOURCES FOR VUV SPECTROSCOPY

We embrace in VUV spectroscopy both single-photon techniques which require a VUV photon and excitation techniques involving more than a single photon which achieve equivalent transition energies.

The basic problem of making a VUV laser is readily understood if one examines the Einstein–Milne relations:

$$g_i B_{ij} = g_j, \quad B_{ji} A_{ji} = \frac{2h\nu^3}{c^2} B_{ji} \tag{1}$$

where the g_i, g_j are statistical weights, h is Planck's constant, ν is the frequency, and A, B are the Einstein coefficients.

While the B_{ji} coefficient is essentially determined by atomic size, and therefore decreases reasonably slowly as the transition energy increases, the A_{ij} spontaneous emission coefficient increases much faster, because of the ν^3 dependence. The consequence is that laser action, which depends on B_{ji} and is spoiled by spontaneous emission, becomes increasingly difficult to sustain as the energy is increased. For this reason, much effort has been devoted to nonlinear mixing schemes, in which the energies of several photons are added together, by combining them in such a way that a single photon of higher energy is produced. There is the further advantage that tunable sources may be realized by this method, if one combines radiation from tunable lasers which already exist at longer wavelengths.

The techniques of frequency upconversion and nonlinear mixing are a subject in themselves which will not be reviewed in detail here. The interested reader is referred to earlier reviews (e.g., Bjorklund[1]). We content ourselves with a brief outline of the basic principles involved, and give an example of the current practice.

A simple scheme for frequency upconversion by nonlinear four-wave mixing is illustrated in Figure 1. A tunable laser held at a constant frequency is used to excite a two-photon transition of rather high energy in a rare gas. The choice of a rare gas is important, because the ionization potential is high, and short wavelengths thus become accessible. A second laser, which is tuned in order to scan the VUV source, then further excites the atoms by a third participating photon, and the nonlinear process produces the fourth, tunable VUV photon, which may be produced either by sum–sum or by sum–difference upconversion, depending on whether the fourth photon lies above or below the two-photon energy difference.

There exist many schemes of this type, whose use was advocated by a number of authors, including in particular Hilbig and Wallenstein,[2] who demonstrated continuous coverage of the entire range 1100 to 2100 Å.

Of course, continuous coverage must be understood as an overlap of a number of tuning ranges in which the same apparatus is used, but different dyes are employed. Changing the dye in a laser is not a trivial operation, and requires the system to be carefully flushed out. Indeed, it is usually preferable to move in a given direction (toward longer wavelengths) to minimize poisoning of the dyes. Thus, lasers are more complicated to tune than a synchrotron radiation source when the wavelengths cover

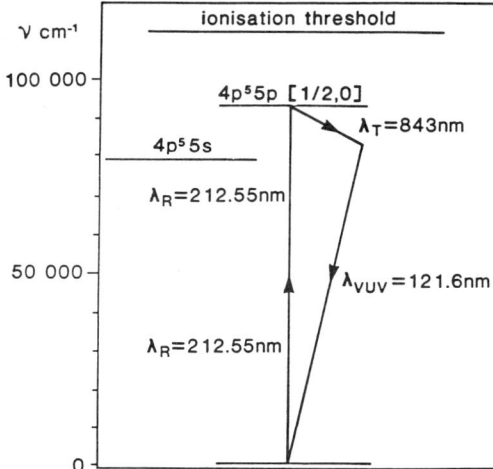

FIGURE 1. Schematic term diagram showing the sum–difference four-wave mixing scheme in krypton used to cover the range down to 120 nm.

a large range. On the other hand, accelerators are not always available either, and are not usually under the direct control of the experimenter, whereas a laser system can be operated in-house.

There are a number of other experimental advantages of laser sources which will be further discussed below. In general, a laser-based source is superior to synchrotron radiation for high-resolution spectroscopy over the limited wavelength range within which it can be conveniently tuned.

Consequently, frequency upconversion of laser radiation by nonlinear mixing in gaseous media is fast becoming a standard technique for generating bright, pulsed, tunable radiation in the VUV. Sources of coherent, narrow-bandwidth VUV light find wide application in atomic and molecular spectroscopy. Advances continue to be made in increasing the brightness and tunability of such sources (Vidal,[3] Reintjer[4]).

As an example, we will describe a system we have set up and operated at Imperial College in which VUV light is generated in Kr by using a sum–difference four-wave mixing scheme two-photon resonantly enhanced by the $5p$ $[0, 1/2]_0$ state (Marangos et al.[5]). Such techniques have previously been exploited by Hilbig and Wallenstein,[2] Hutchinson and Thomas,[6] and Hilber et al.[7]

It is an intrinsic feature of the sum scheme that the polarization properties of the generated field are determined solely by the polarization of the input fields. For instance, if both input lasers are linearly polarized in the same direction (as was the case in our experiment), then the generated beam will be polarized likewise. This property can be important for many experiments. For example, we have successfully combined tunable laser sources and polarization spectroscopy with pulsed magnetic fields, so as to extend the measurement of f values to high Rydberg members lying in the UV and VUV. Some examples will be given later in the chapter.

Our own experimental setup is shown in Figure 2. The system employs two identical dye lasers (Lambdaphysik 3002) synchronously pumped by a single XeCl excimer (Lambdaphysik EMG 201 MSC). Typically, a dye solution of stilbene 3 in methanol is used in dye laser A, giving output energies of ~10 mJ per pulse at 4250 Å and as much as 1 mJ after frequency doubling to 2125 Å in a β-barium borate crystal. A quartz prism is used to separate the second harmonic, at 2125 Å, from the fundamental. The wavelength of the first laser A is tuned exactly to the two-photon resonance, $5p[0,1/2]_0$, by optimization of the two-photon resonantly enhanced three-photon ionization signal from Kr in an ionization chamber. The second laser B can then be tuned with continuous coverage from 5880 to 6440 Å (using RhB in methanol as the dye), giving a UV wavelength coverage of 2940 to 3220 Å after frequency doubling.

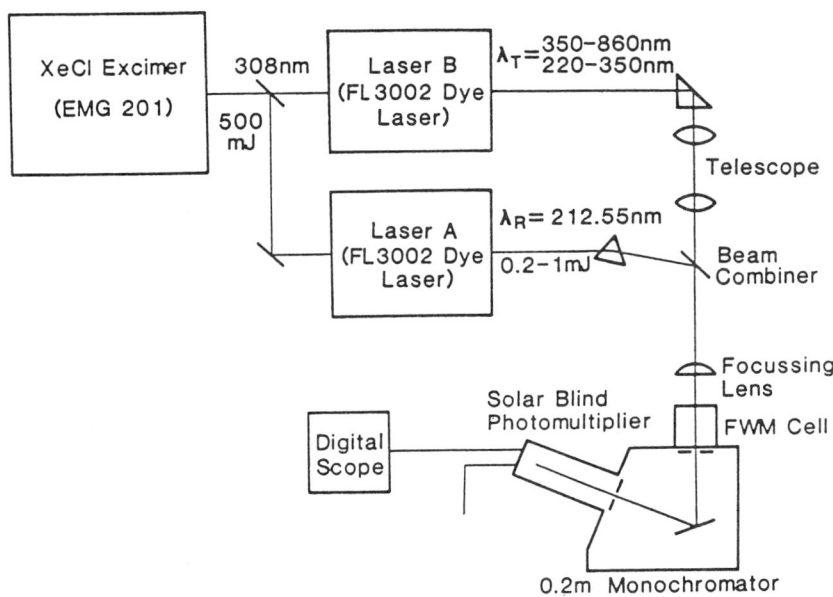

FIGURE 2. Block diagram of the experimental arrangement for a tunable four-wave mixing coherent VUV source.

The two laser beams from laser A and B are then merged using a dichroic mirror. A quartz lens of 225 mm focal length brings the colinear beams to a common focus in the mixing cell, as shown in Figure 2. The common focus is achieved by inserting a telescope to compensate for the effects of chromatic aberration in the quartz focusing lens.

The four-wave mixing cell is a 100-mm-long steel cylinder with a quartz window on the input side and an LiF exit window. It was filled with Kr at 20–50 mbar. From the standard treatment of four-wave mixing in focused Gaussian beams (Bjorklund[1]), the output power of VUV light generated by sum–difference frequency mixing is given by the relation

$$P_{VUV} \propto P_R^2 P_T N^2 \chi_{(3)}^2 F^2(b\Delta k, b/L) \qquad (2)$$

where P_R, P_T are the powers of laser A and tunable laser B, N is the number density of atoms in the nonlinear medium, and $\chi_{(3)}$ is the third-order susceptibility, while F is a dimensionless phase-matching function which depends on the wave-vector mismatch $\Delta k = k_{VUV} - (2k_R - k_T)$, on the confocal parameters of the incident beams, and on the focusing conditions (principally, on b/L, the ratio of the confocal parameter to the cell length).

The resonant enhancement of the susceptibility $\chi_{(3)}$ and the possibility of nonvanishing values of $F(b\Delta k, b/L)$ for a sum–difference scheme, regardless of the sign of dispersion in the medium, lead to our source having both a wide tunability and a high conversion efficiency ($\sim 10^{-5}$–10^{-4}). The input wavelengths in this experiment generated VUV radiation in the range 1664 to 1586 Å.

The second-harmonic generation in lasers A and B leads to, effectively, a 100% linear polarization of both beams in a mutually parallel direction. As a consequence, the generated VUV will also be effectively 100% linearly polarized. Any perpendicularly polarized components in the fundamental laser wavelengths will be strongly discriminated against by the nonlinearity of the frequency doubling and four-wave mixing processes. Any residual ellipticity that might remain is also accounted for in the fitting routines used for the analysis of the magneto-optical spectra.

A collimating MgF_2 lens is used after the mixing cell.

This kind of system and its use for high-resolution spectroscopy have been described in detail by Marangos et al.,[5] who used a grating spectrometer to select the VUV beam and separate it from the fundamental.

In the present arrangement, VUV radiation is separated from the laser beams by a prism monochromator and led into the interaction region. The prism separator is a useful improvement: it consists of two LiF prisms, a high-reflectivity VUV mirror (reflectivity ~80% for $\lambda > 1500$ Å), a slit edge and a beam block to stop the unwanted dispersed wavelengths as well as any light which might leak through in the zero order. The two 60° LiF prisms have a height of 10 mm. To minimize reflection losses related to the high refractive index at short wavelengths they are mounted with their axes of symmetry inclined at 10° to the incident beam. Without this precaution, incoming radiation at wavelengths below 1400 Å would be totally internally reflected in the system.

The entire assembly is contained inside a cylindrical vacuum vessel of 280-mm outer diameter, making the whole instrument compact and easily supported. Tuning of the transmitted wavelength is achieved by a simple translation of the mirror.

A prism instrument transmits VUV radiation much more efficiently than a grating spectrometer, and the prisms are less susceptible to radiation damage than the coating and surface of a reflection grating. The improved throughput turns out to be important in any experiments in which high intensity is desirable, for example in compensating for the losses caused by insertion of a VUV polarizer.

Under optimum conditions, a source of the kind described can deliver 1 kW of coherent VUV light into a linewidth of about 5 GHz.

Because the source is pulsed, there are problems of shot-to-shot reproducibility, and it is desirable to monitor the intensity of the beam and perform on-line division of the data. When tuning the dye laser cavities mechanically, one should bear in mind that a 5-GHz bandwidth contains many modes. This means that, at a 5-GHz bandwidth, variations of intensity related to changes in the mode structure are not too obtrusive. If, however, one attempts to improve the linewidth by the insertion of an intracavity Fabry–Pérot etalon, then mode-hopping due to mechanical tuning becomes much more of a problem. The most convenient solution to this problem is to build the cavity into a pressurized vessel, so that the dye laser may be optimized and then tuned without moving any parts. The linewidth in this case can be reduced to 1.5–2 GHz.

3. DETECTION

Experiments on the total cross section of atoms are usually performed in photoabsorption when using a synchrotron radiation source. In fact, this is not the only option; while the experimental arrangement is a simple one, the method possesses a number of disadvantages, which can be briefly stated as follows. First, absorption is very insensitive: at least a 2% dip in intensity is required for detection, and usually one needs more. Second, one cannot tolerate a high level of absorption, because opacity effects intervene which do not act linearly. Third, the dynamic range is therefore very poor. Fourth, detection must be performed on a very bright background, so that the signal-to-noise conditions are very unfavorable.

Even with synchrotron radiation, better experimental conditions can be obtained by using a thermionic diode detector, in which a column of sample is produced in a tube with a hot wire cathode and a collecting anode, biased in such a manner that the production of an ion neutralizes the space charge and produces a large gain in the anode current. The advantage of such a device is that high sensitivity is achieved, so that it can be operated at low vapor pressure, i.e., for low absorption levels at which opacity is not experienced. The dynamic range is therefore many orders of magnitude greater than in photoabsorption. Finally, the ionization signal is proportional to the measured cross section, so that detection occurs on a dark background.

An example of a spectrum obtained by this method using a synchrotron radiation source is shown in Figure 3. Thermionic detection is, however, not generally very convenient for synchrotron radiation spectroscopy. The ionization signal is only proportional to the total cross section below the double ionization potential. Usually also, such cells are closed and operated with a buffer gas, so that

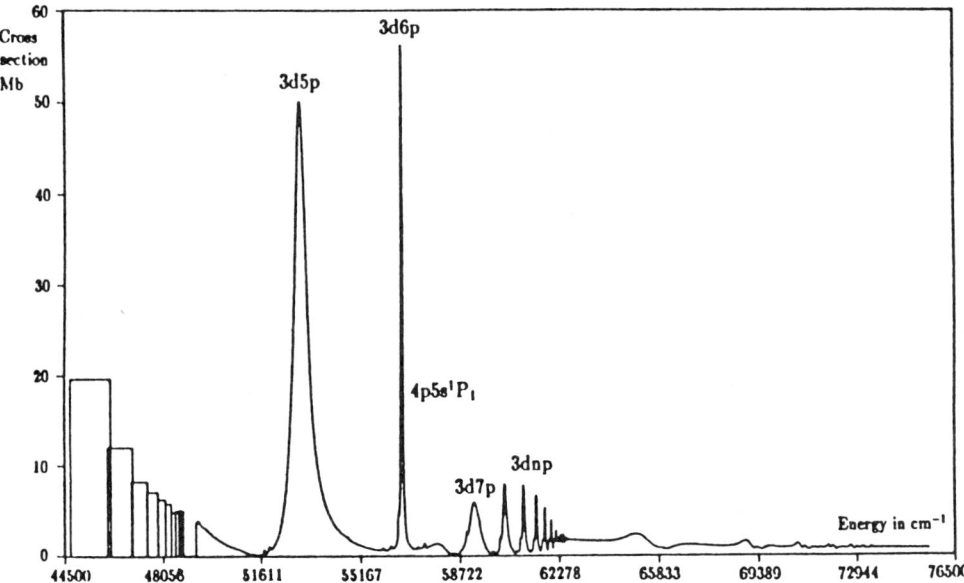

FIGURE 3. The doubly excited spectrum of Ca I just above the first ionization threshold, as obtained by Griesmann et al.,[8] using thermionic diode spectroscopy combined with synchrotron radiation from the undulator beamline at Hasylab.

differential pumping is needed to extend observations below the LiF cutoff. Both reasons limit the accessible wavelength range. For sensitive detection, a lock-in method is also preferred, and this cannot be done at the characteristic frequencies of most synchrotron radiation sources: a low-frequency chopper is therefore necessary.

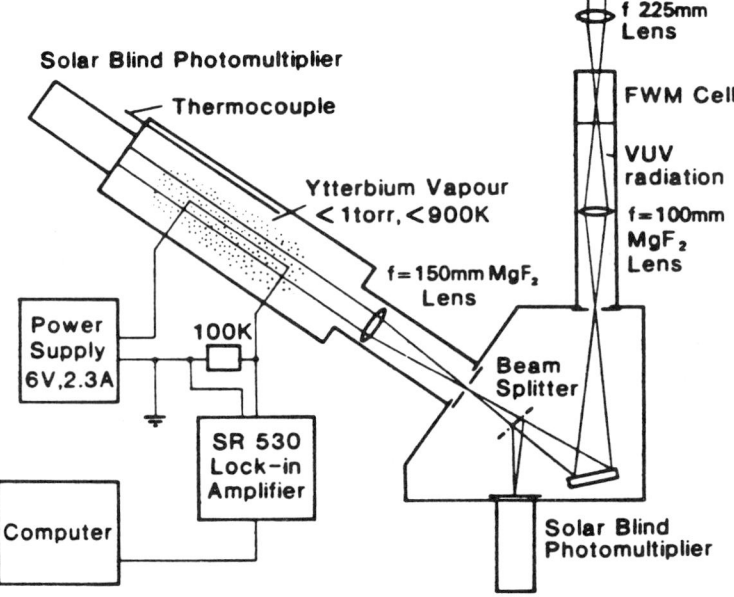

FIGURE 4. Diagram of the experimental arrangement for VUV spectroscopy using tunable lasers and a thermionic diode detector.

With laser-based sources, on the other hand, the thermionic diode is a natural detector to use. The lock-in detector can be operated at the repetition frequency of the laser source, and ions are produced in large numbers by the laser light, so that one tends anyway to turn down the pressure of the sample. Indeed, one can monitor the laser beam at the exit of the diode, and also check directly that no absorption has occurred in the experiment.

A block diagram of a typical experimental arrangement for thermionic diode spectroscopy using a coherent four-wave mixing VUV source is shown in Figure 4. The limitations of wavelength range which are intrinsic to thermionic diode detection are not too much of a problem at least down to 1200 Å, and the gain available using such detectors is a considerable advantage. It is also possible to study the influence of collisions with various noble gases using thermionic diodes by varying the buffer gas pressure.

When the laser range is extended to shorter wavelengths, it is possible to exploit atomic beams and detect photoionization directly so as to preserve the advantage of low sample pressure, as long as one remains below the double ionization limit.

4. WAVELENGTH CALIBRATION

Whenever one describes high-resolution spectroscopy experiments, one should not forget the problem of wavelength calibration, which is especially acute in the VUV. Of course, many atomic spectra have been studied by conventional spectroscopy of very high dispersion (using, for example, spectrographs of 6.6-m focal length), and reliable superpositions of standards have been accomplished up to the limit of resolution thus available, using well-established methods of term differences to provide good references (Edlén[9]). However, such painstaking work must be repeated afresh whenever any improvement in resolution is claimed, and in this respect much remains to be done by laser-based VUV methods.

The superposition of spectra cannot be performed in the same way as in conventional spectroscopy, but there are some attractive new features of nonlinear mixing methods which are worth mentioning. Over some wavelength ranges, one can, for example, keep track of the tuned laser in a four-wave mixing source at the fundamental frequency by passing a portion of the beam through an iodine cell (the I_2 spectrum contains many closely spaced and very well measured absorption lines) and measuring the attenuation with a photodiode. However, the wavelength range thus accessible is restricted, and one sometimes wishes to measure more than just one contributing laser frequency. Another possibility is to peel off part of the mixed beam and pass it through a hollow cathode source. The optogalvanic effect then provides a useful signal for wavelength calibration. The latter is a useful technique because it considerably broadens the range of available lines to use as standards, and because ions provide sharp lines in the UV and VUV. One may use either sealed lamps (which are commercially available with quartz windows) or open, differentially pumped sources, depending on the wavelength range. Hybrid fillings (rare gas–metal combinations) to suit the user's requirement are available with the sealed units.

Over some ranges where there is a paucity of known lines, a pulsed wavemeter of high resolution is attractive. However, any self-respecting spectroscopist will not believe purely instrumental readings unless they are referred to atomic standards at a sufficient number of points over the wavelength range. This imposes the need to determine standards at a precision comparable with laser resolution in the UV and VUV. The problem is incompletely resolved at the time of writing, but important steps have been made toward this goal. In particular, Learner and Thorne[10] have developed a VUV Fourier transform spectrometer which is yielding standards of interferometric quality through the UV and into the VUV, down to a present limit of 1750 Å. More work is needed to extend the range even further.

5. FARADAY ROTATION

Another method of detection which often proves useful to study perturbations related to doubly excited states occurring just below the first ionization limit is Faraday rotation spectroscopy. Again,

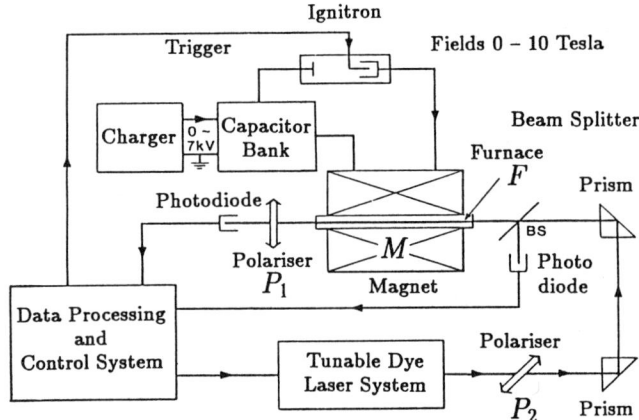

FIGURE 5. Diagram of the experimental arrangement for Faraday rotation spectroscopy using pulsed Bitter coils synchronized with excimer-pumped pulsed dye lasers.

this technique has been used with synchrotron radiation, but is best exploited using laser-based sources because it is particularly suited to very-high-resolution spectroscopy.

In the thermionic diode, the pressure of the sample is kept low so as not to disturb the measurement by opacity effects. This consideration is far less important when dealing with Faraday rotation, because the rotation angle is cumulative down a vapor column, and essentially unaffected by opacity. One can therefore detect very weak and very strong transitions without losing accuracy by increasing the vapor pressure, i.e., the method possesses a good dynamic range. The experiment is also performed using crossed polarizers, so that detection can take place on a dark background.

Faraday rotation experiments are best performed with fairly high magnetic fields, so that substantial rotation angles occur (several π), which increases the accuracy of the measurement and also provides intrinsic calibration through the intensity oscillations as a function of wavelength.

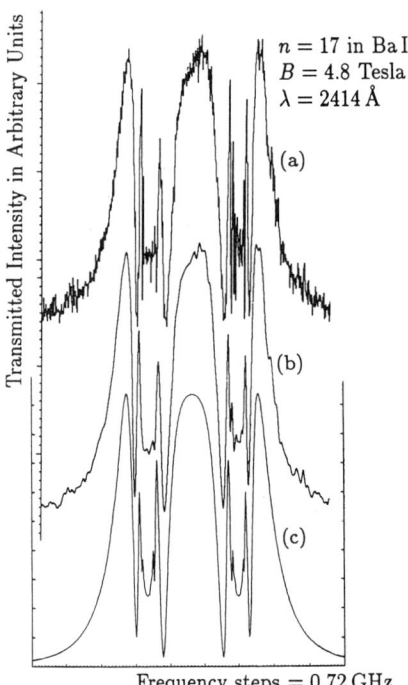

FIGURE 6. Illustrating how experimental and synthesized magneto-optical patterns are matched: (a) experimental raw data; (b) 3-5-3 smoothed data; (c) computed pattern for the $n = 17$ line in Ba I.

In such experiments, the pulsed nature of the laser-based sources actually turns out to be an advantage: with synchrotron radiation, large superconducting magnets are required, and consume much liquid helium. With pulsed magnets, on the other hand, no liquid helium is needed, and the typical time characteristic of a pulsed magnet is in the millisecond range, which can be regarded as a constant field for the duration of a laser pulse (in the 20 ns range). With appropriate synchronization, a fairly simple experiment can be devised; a block diagram of this arrangement is shown in Figure 5.

The analysis of magneto-optical rotation patterns is quite an intricate affair if one wishes to extract the maximum information from the data. The best approach is the magneto-optical vernier (MOV) method, in which the effects of both magnetic circular birefringence, responsible for magneto-optical rotation, and magnetic circular dichroism, are computed point by point across the full profile of the transition under study. This procedure has been described in some detail by Connerade[11] and has now been further refined and applied to a wide range of situations (e.g., Connerade,[12] Connerade et al.[13,14]). An example of the fitting of an experimental pattern is shown in Figure 6. Accuracies at the percent level or better in the determination of relative oscillator strengths can be achieved in favorable cases, provided the shot-to-shot variations in magnetic field strength and synchronization are not too large.

Recently, the MOV technique has been extended *above* the ionization threshold and into the continuum, where Faraday rotation is usually very small—which is perhaps why this possibility had been neglected in the past. It was shown by Connerade[12] that Faraday rotation does occur in sufficiently narrow autoionizing resonances, and that the analytic Fano formula[15] has an analytic counterpart to describe the rotation angle, through a Kramers–Kronig transformation to recover the corresponding refractive index variation. This theoretical development opened the way for experiments[13,16] and the method appears suitable for the study of very narrow autoionizing transitions. An example of an asymmetric pattern for an autoionizing line is shown in Figure 7.

Note the pronounced asymmetry in the observed rotation pattern.

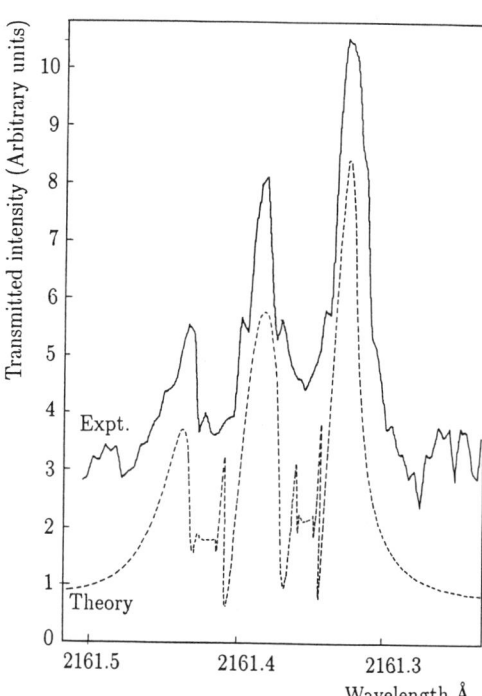

FIGURE 7. Magneto-optical pattern for an autoionizing line. Solid curve: experimental data; dashed curve: calculated.

6. STUDIES OF DOUBLY EXCITED SERIES

Laser-based experiments are very suitable for the study of doubly excited series in a number of heavy elements, prime candidates being the alkaline earths, whose doubly excited spectra lie quite low in energy, but nevertheless contain just two outer electrons, so that the doubly excited configurations remain simple. Consequently, they come very close in energy to the first Rydberg spectrum, and a number of interchannel perturbations occur which can be opened up for examination at high resolution.

Although the situation is more complex in such elements than in helium, where the doubly excited spectrum is far removed from single excitations and can therefore be considered as a relatively pure example, it is perhaps worth remembering that the overwhelming majority of doubly excited spectra are of the kind found in the heavy elements. Indeed, the very first doubly excited transitions were originally recognized in the spectra of the alkaline earths, in the early days of quantum mechanics, when the properties and origin of such excitations were rather mysterious. Recently, these spectra have been reexamined at very high resolution, and the new investigations have revealed a number of unexpected features or novel effects which do not seem to occur in the spectrum of helium.

6.1. Perturbations in the Principal Series of Barium

An excellent example of the energy overlap just described occurs in the photoabsorption spectrum of barium in the vicinity of the first series limit. This spectrum was investigated by classical high resolution spectroscopic techniques using conventional laboratory sources as early as 1960 by Garton and Codling.[17] Figure 8 shows early data in which a number of interesting features are apparent. First, one notes the presence of many asymmetric, autoionizing resonances in the continuum which are related to double excitations. The most dramatic one of these is the $5d8p\,^1P_1$ resonance which sits almost exactly on the series limit and actually straddles the threshold: it is responsible for an intensity anomaly among the high series members of the principal series. Around $n \sim 25$ a pronounced intensity minimum can be seen, although the data in Figure 8 are actually photographic, so that quantitative statements cannot be made. It was inferred by Garton[18] that this minimum corresponds to the window on the low-energy side of the Beutler–Fano $5d8p\,^1P_1$ resonance.

This is at first sight a rather surprising effect: a local reduction in the f values of the transitions implies that, around $n \sim 25$, the Rydberg state is stabilized against radiative decay, i.e., that its lifetime

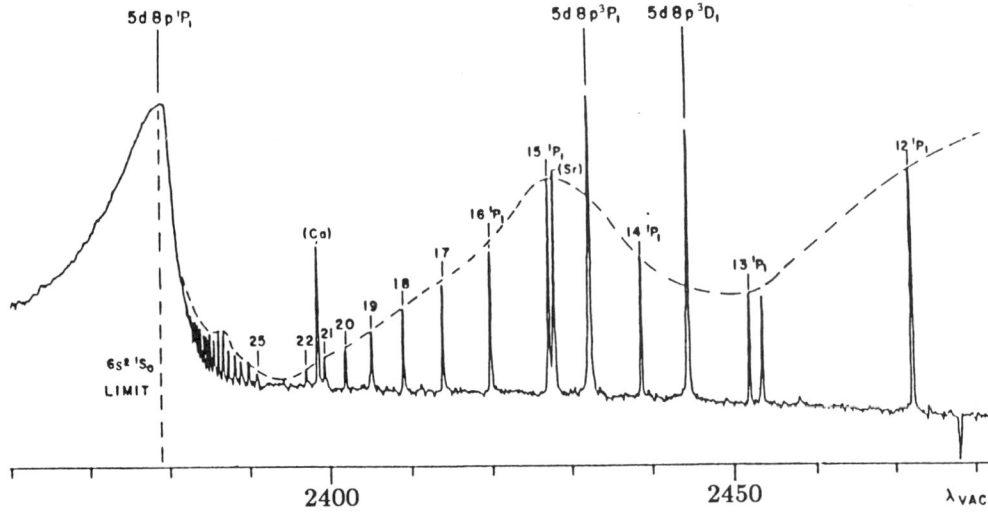

FIGURE 8. The photoabsorption spectrum of barium near the first ionization limit as obtained by classical dispersive grating spectroscopy using conventional sources. (After Ref. 18.)

is increased by the perturbation. This phenomenon was discussed theoretically by Connerade and Lane[19] in the context of Wigner's scattering theory.[20] In scattering theory, radiative and particle decay channels are treated on an equal footing in the scattering matrix, and it is therefore easy to understand how they can interfere. If a Rydberg state occurs at a specific energy where there is cancellation, then the interference can be completely destructive, i.e, stabilization against radiation occurs as a result of the perturbation.

This phenomenon is intrinsically so interesting that it was desired to acquire quantitative, high-resolution data to confirm the effect. Classical high-resolution spectroscopy turns out not to be very appropriate for this purpose, because the highest resolution is achieved by using nonlinear (photographic) detection, and because of the problems associated with opacity in photoabsorption experiments.

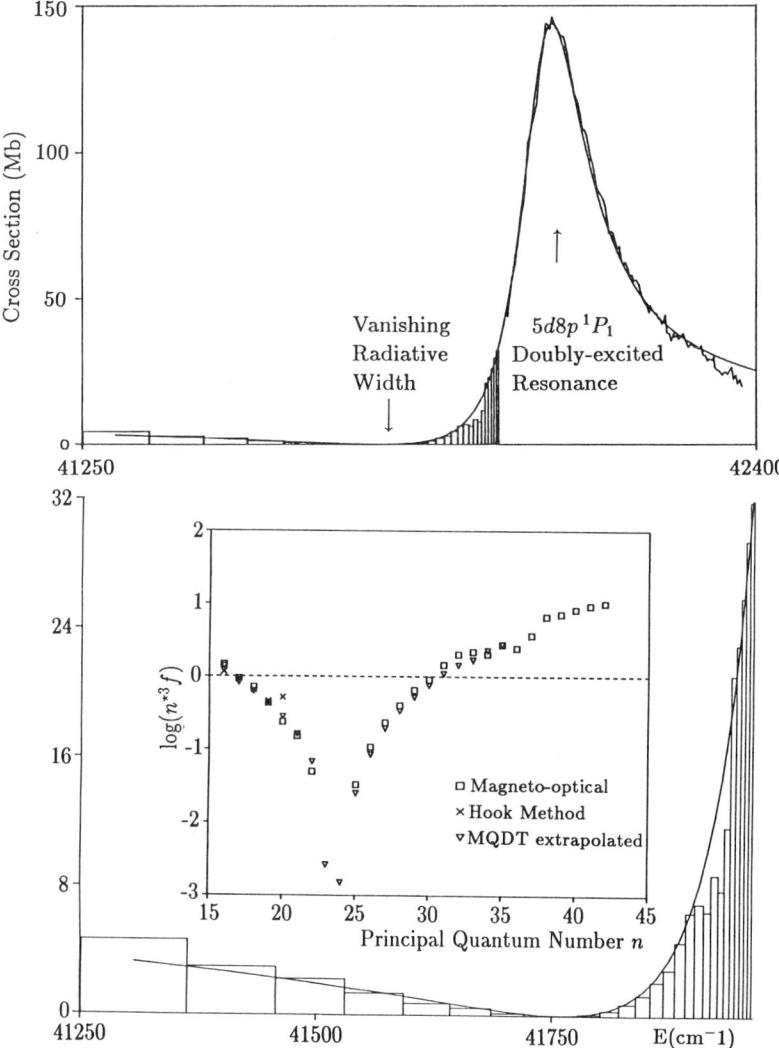

FIGURE 9. The vanishing radiative width in Ba as determined by high-resolution, laser-based Faraday rotation spectroscopy near the first ionization threshold, and by direct photoionization measurements above the threshold. Curves: ———, many body calculations; ---, experiment by Newsom (1966) from 6.11 to 7.45 eV and by Connerade et al. (1980) from 7.45 to 9.2 eV (normalized to the Newsom (1966) result at 7.45 eV).

It turns out, on the other hand, that Faraday rotation is an extremely sensitive technique, which combines independence from opacity with excellent spectral resolution when used in combination with laser-based sources. Thus, Connerade et al.[14,21] were able to perform f-value determinations by magneto-optics up to $n = 45$ in barium (see Figure 9), and to provide the first quantitative determination of what has been called the *vanishing radiative width* effect in barium.

An additional, interesting feature of the spectrum in Figure 9 is the continuity between the df/dE plot below threshold and the photoionization current above threshold, measured using an ionization detector. Continuity theorems are of course a well-established feature of atomic spectra, this particular kind being known since the work of Sugiura,[22] but it is interesting to consider an example in which a resonance occurs at the threshold: one has an intuitive feeling that if any breakdown were possible, it would be more likely in such a case.

We return in a later section to some further aspects of the doubly excited spectrum of barium.

6.2. Double Excitations in Calcium

An interesting counterpart of the vanishing radiative width effect in barium is the vanishing particle width which occurs in the double excited spectrum of calcium. Since, as previously remarked, photon and particle decay channels are treated on an equal footing in Wigner's theory, there is of course the more obvious interference effect in which two or more autoionizing channels interfere to produce stabilization.

A beautiful example of this effect occurs in the $3dnp$ doubly excited spectrum of calcium: the series involved is interesting in that it provided among the first recorded examples of doubly excited resonances. Remarkably, as the experimental resolution has increased, this spectrum and others directly related to it have provided many new insights into interchannel interactions involving double excitations.

We begin with an example obtained using a linear detector (thermionic diode) and together with synchrotron radiation. There are a number of practical problems in harnessing thermionic diode detection to synchrotron radiation sources, mainly because of the different characteristic time structures. To achieve a good signal-to-noise ratio, it is necessary to use either a box-car or lock-in detection. The latter is feasible with synchrotron radiation, but only if it is chopped at low frequency, as was first done by Griesmann et al.[8] who obtained the spectrum shown in Figure 3.

This spectrum contains a prominent doubly excited series, labeled $3dnp\,^1P_1$, which exhibits a remarkable series perturbation. Indeed, if one examines the spectrum of Figure 3, it is not at once obvious that all of the strong lines apparent at the resolution of this experiment belong to the same Rydberg system. One normally thinks of Rydberg series possessing smoothly decreasing intensities and widths, but this is not true in the present instance. In particular, one notes the extremely narrow autoionization width of the $6p$ member, as compared not only to the very broad $5p$ resonance but even to the $7p$ line, which is quite a remarkable departure from the normal course of linewidths. A closer scrutiny reveals that the anomalously narrow $6p$ line is close to the $4p5s$ perturber, apparently a very weak feature, which is nevertheless responsible for a profound transformation of the $3dnp\,^1P_1$ excitation channel.

This stabilization of one line against autoionization can be traced to an interference between two particle decay channels. It has also been computed theoretically. The first calculation to show it clearly seems to have been that of Altun et al.,[23] shown in Figure 10, which may not have represented the profiles of all of the surrounding resonances correctly, but did show the stabilization clearly. A more recent calculation (Greene and Kim[24]), shown in Figure 11, is the one which agrees best with the data of Figure 3.

Another, alternative explanation for the narrowing of the $6p$ resonance has also been advanced: it was suggested by Scott et al.[25] that a Cooper minimum in the underlying continuum, by forcing a sign change in the integrand of the matrix element which determines the autoionization rate, would lead to cancellation and thereby reduce the rate dramatically. This is a quite different mechanism, which was apparent in *ab initio* calculations for calcium performed by the Belfast group. The question therefore arises: which of the two mechanisms is correct? Furthermore, the calculations of Greene and Kim[24] suggested that much sharper structure (hardly visible except as tiny "blips" in the data of Figure

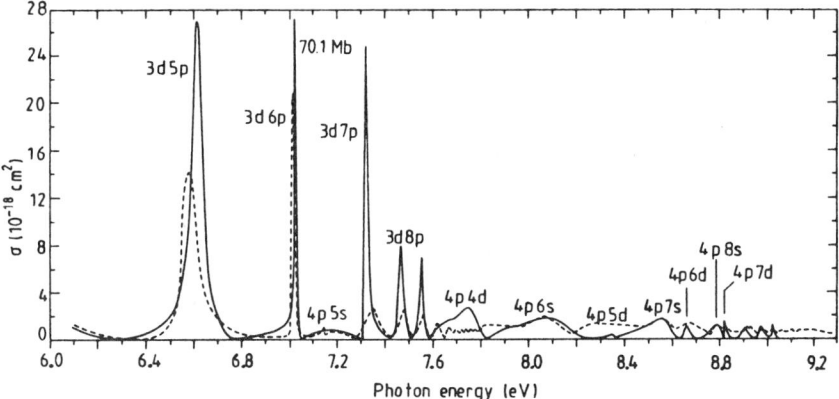

FIGURE 10. The doubly excited spectrum of Ca I just above the first ionization threshold as calculated by Altun et al.[23] using many-body perturbation theory.

3) played an important role in the latter calculation, so that a two-channel description of the influence of the 4p5s perturber would be insufficient, i.e., the interference mechanism involving this perturber is more complex and involves other configurations.

In order to investigate the matter further, it was therefore decided to embark on experimental studies at much higher resolving power, using laser-based methods, and in particular the four-wave mixing source described above, so as to reveal the finer lines and study their intensities and widths as well.

This new investigation reveals a profusion of structure, of which some idea can be gained from Figures 12 and 13. It will now be clear that the spectrum of Figure 3 was far from giving a complete

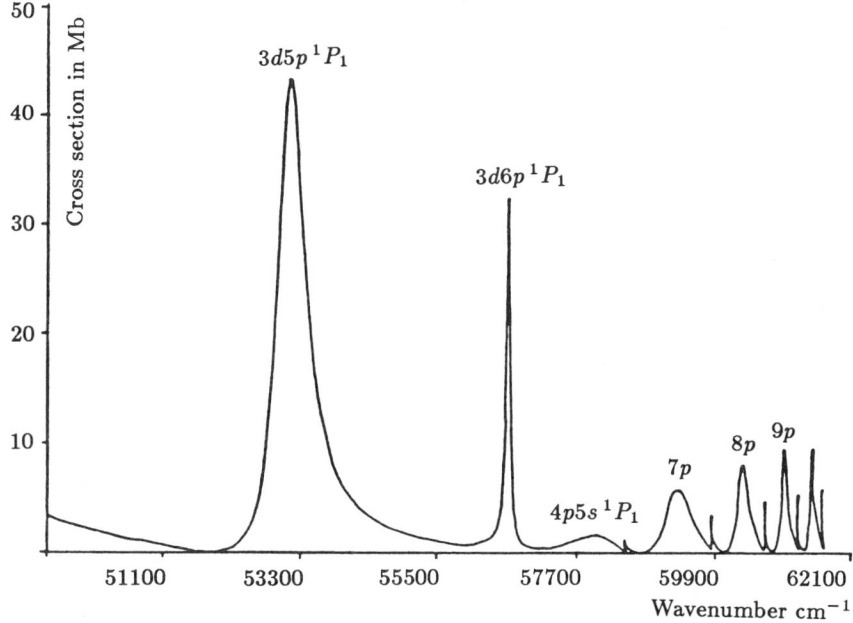

FIGURE 11. The doubly excited spectrum of Ca I just above the first ionization threshold as calculated by Greene and Kim[24] using the eigenchannel R-matrix approach. This figure should be compared with Figure 3.

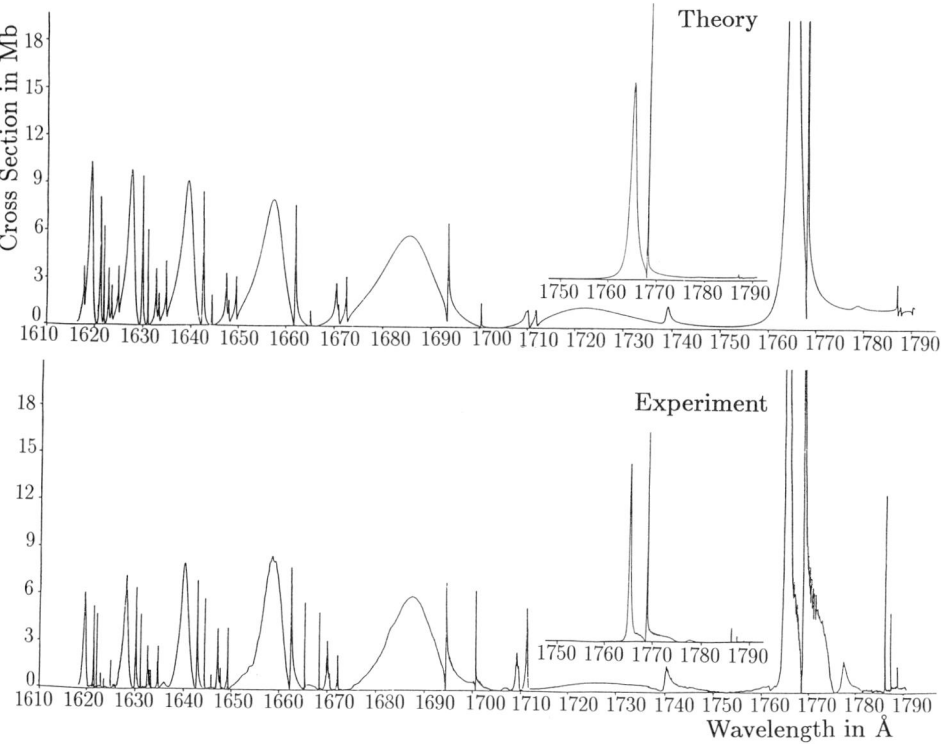

FIGURE 12. The doubly excited spectrum of Ca I at high resolution. Top spectrum calculated by Greene using the eigenchannel R-matrix approach and bottom spectrum as measured by thermionic diode spectroscopy using a VUV coherent four-wave mixing source.

picture: all of the very sharp features were lost, owing to the low resolution, and the real complexity of the spectrum emerges as the resolution is increased. In particular, many sharp lines are found some of which exhibit reversals of symmetry reminiscent of those found by Altun et al.,[23] but for different transitions.

These reversals of symmetry or *q-reversals*[26] are a well-known feature of interchannel interactions involving relatively sharp Rydberg series which are weakly coupled to broader interlopers (Connerade and Lane[19]). Further experimental examples will be given below. The presence of these finer lines is important as they perturb and shift the broad features, which has a profound influence on the phase cancellation in autoionization matrix elements.

In the light of these new experiments, the calculations of Greene and Kim[24] were updated, and performed on a finer energy scale, so as to be more directly comparable with the data (Farooqi et al.[27]). These are the theoretical curves in Figures 12 and 13.

Overall, the agreement is now very good, although the calculations begin to break down close to the series limit, where the density of excited states becomes large. This part of the experimental data is shown at high magnification in Figure 14.

The theory is an implementation of the scattering or R-matrix approach, adapted to an asymptotic Coulomb field. However, the states involved are so heavily mixed that, after the calculation has been performed, it is no longer possible to attach meaningful configuration labels to each individual state. Thus, the theory provides a global interpretation of the spectrum rather than individual assignments, a point to which we return later.

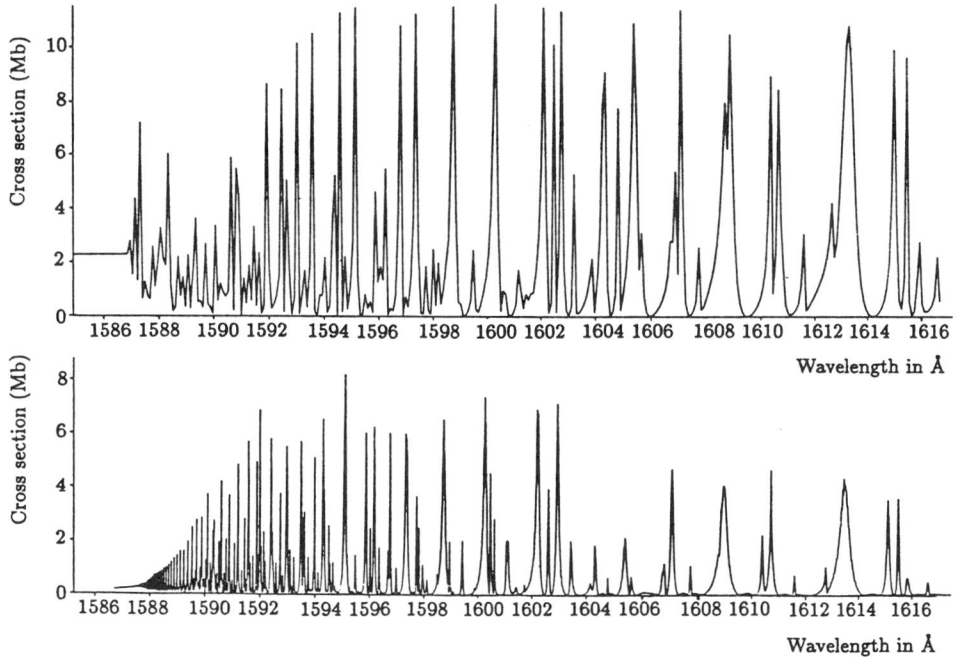

FIGURE 13. The doubly excited spectrum of Ca I at high resolution. Top spectrum calculated by Greene using the eigenchannel *R*-matrix approach and bottom spectrum as measured by thermionic diode spectroscopy using a VUV coherent four-wave mixing source.

6.3. The Doubly Excited Spectrum of Strontium

Similar calculations have been performed for the strontium spectrum by Aymar[28] in order to analyze conventional high-resolution classical photoabsorption data obtained by Ginter et al.[29] One region was found to be somewhat ambiguous in its interpretation (see Figure 15), and it was therefore decided to reinvestigate it at high resolution by laser-based spectroscopy, using a tunable, coherent VUV source.

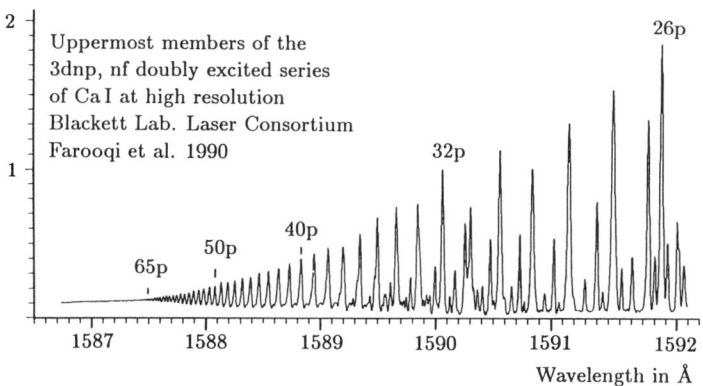

FIGURE 14. The doubly excited spectrum of Ca I at high resolution (experimental) close to the 3*p* ionization limit at high magnification, as measured by thermionic diode spectroscopy using a VUV coherent four-wave mixing source.

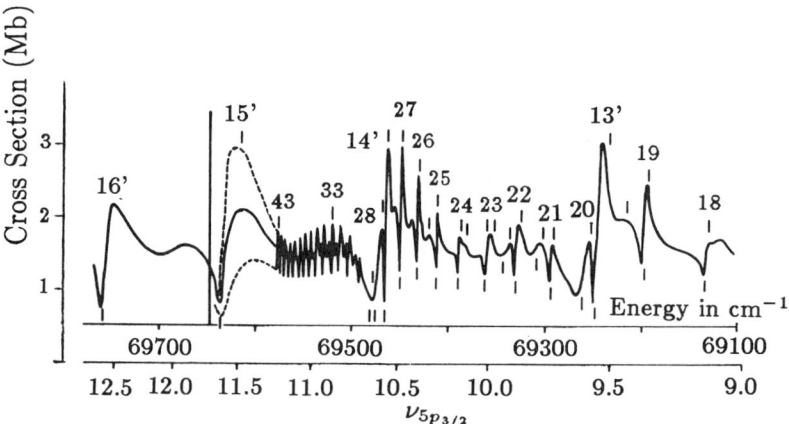

FIGURE 15. A portion of the doubly excited spectrum of Sr I in a region where early calculations presented some ambiguities.

The new data obtained are illustrated in Figure 16: Aymar repeated the calculation with some improvements, and the results of Figure 17 were obtained (both Figures 16 and 17 are from Farooqi et al.[30]).

A very interesting feature of the experimental plot is the disappearance of spectral fluctuations at the point marked "X" in the figure: a quantitative measure of spectral fluctuations is provided by the variance, which is itself an observable, and can be shown to possess vanishing points, which do not necessarily coincide with Rydberg series limits.

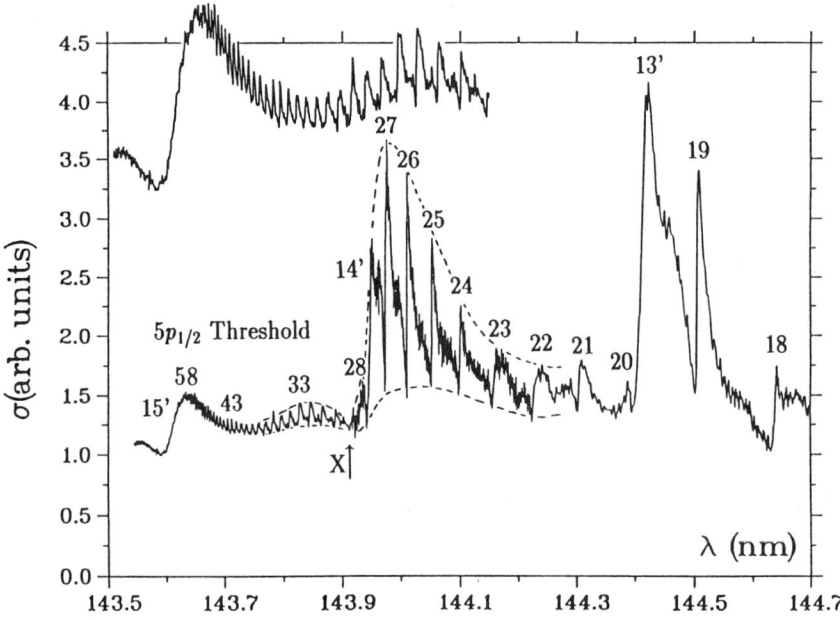

FIGURE 16. The same region of the doubly excited spectrum of Sr I as in the previous figure, measured using a tunable laser-based VUV coherent source. The inset shows the region near the series limit in greater detail.

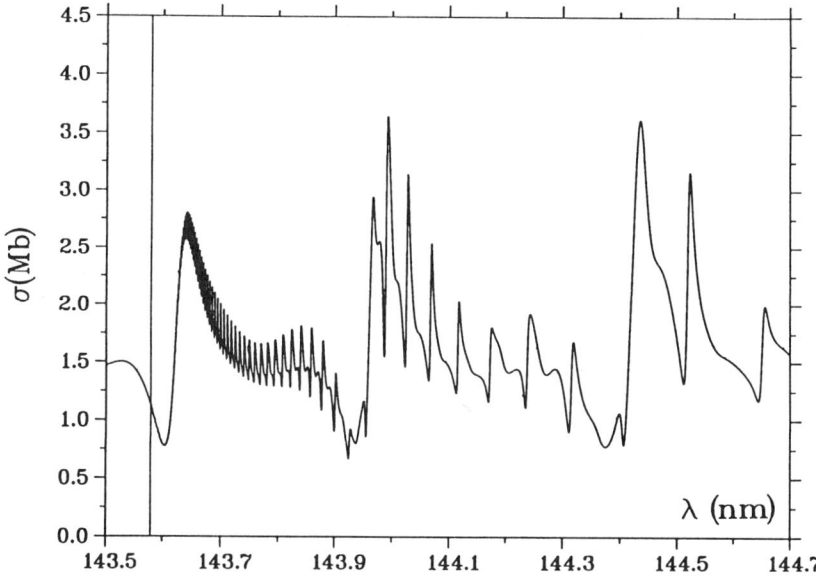

FIGURE 17. The same region of the doubly excited spectrum of Sr I as in the previous two figures, as recalculated by Aymar.[30]

6.4. The Doubly Excited Spectrum of Barium

An example of vanishing fluctuations which is perhaps more conspicuous than the one in the previous spectrum occurs in the case of Ba I. It is found that *for the corresponding series* of double excitations, a perturbation occurs in which there is a pronounced local disappearance of the spectral structure well before the series limit is reached (see Figure 18).

The origin of vanishing fluctuations can readily be understood once it is accepted that a vanishing radiative width can occur in one channel: if several channels are open, but responsible only for fairly constant or monotonically changing particle widths as a function of excitation energy, while the radiative width of the unperturbed Rydberg channel goes through a zero at a specific energy, then after coupling to the open channels, instead of a zero in the cross section, one expects a zero in the spectral fluctuations.

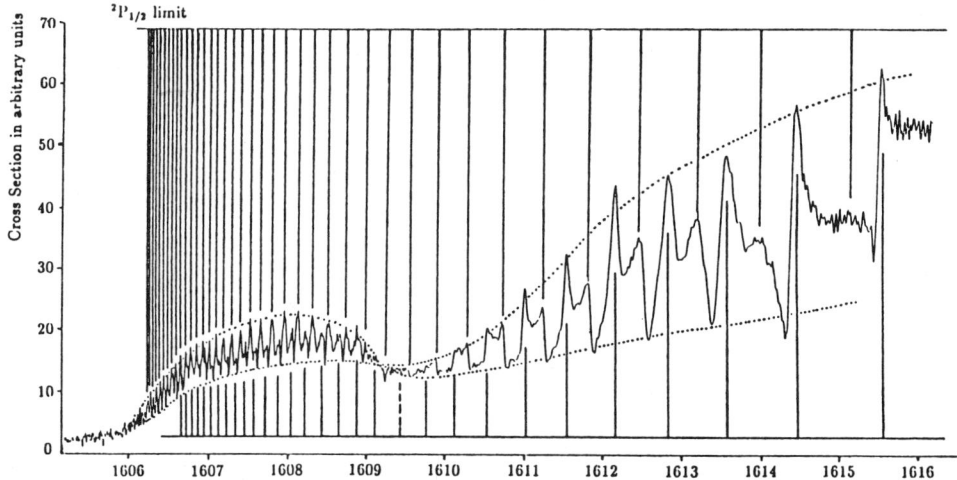

FIGURE 18. Vanishing fluctuations in the doubly excited spectrum of Ba I close to the 5p thresholds.

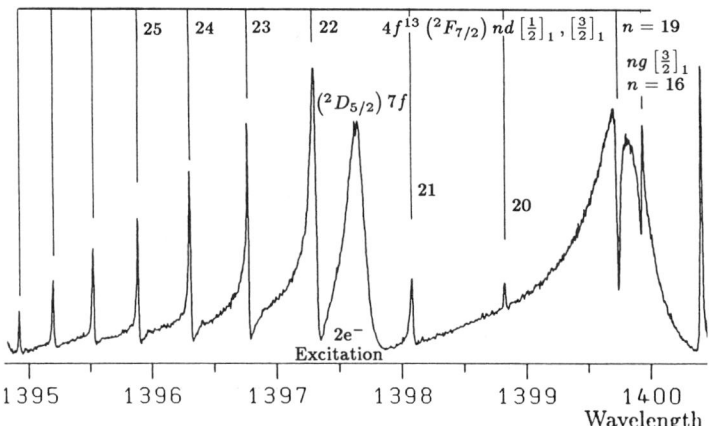

FIGURE 19. Part of the double-excitation spectrum of Yb, showing overlapping resonances caused by single excitation from the 4f subshell and double excitation of 6s electrons.

Experimentally, it is important that the instrumental resolution be high enough that one can exclude optical interference phenomena which sometimes occur[31] when the instrumental resolution becomes comparable to the spacing between spectral lines.

6.5. Energy Degeneracies between Single and Double Excitation Spectra

As was mentioned in the introduction, an energy degeneracy between single and double excitation spectra can significantly enhance double excitation spectra. This does not necessarily occur simply because the doubly excited spectrum comes close to the first Rydberg spectrum as in the examples just given. Another possibility is that inner-shell excitation spectra, which carry significant oscillator strength since they are allowed by one-electron dipole selection rules, can be degenerate in energy with double excitations.

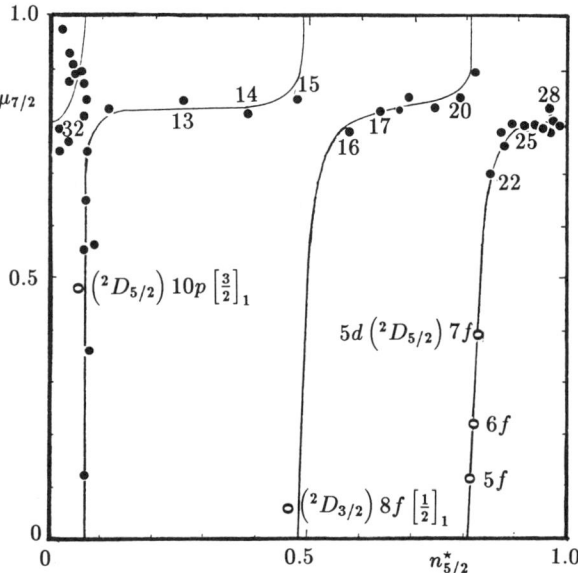

FIGURE 20. Lu–Fano graph, showing the rather narrow avoided crossing associated with interchannel interactions between singly and doubly excited channels of Yb I.

A general discussion of the conditions affecting such degeneracies (see Connerade[32]) would be beyond the scope of the present chapter. However, it is useful to illustrate by an example the fact that energy degeneracy alone is not sufficient to guarantee a strong interaction between single and double excitation spectra.

The spectrum of Yb I presents the interesting situation that because Yb I is the lightest element with a closed $4f$ subshell, it is easily broken open, i.e., the spectrum related to the excitation of a single $4f$ electron lies not far above the first ionization potential, and is embedded among doubly excited resonances.

Despite this energy degeneracy, the interaction between these different types of channel is not as strong as one might expect, as is revealed by the fact that the lines of the inner-shell spectrum remain sharp even when they are close to strong doubly excited resonances (see Figure 19), or alternatively by the narrow avoided crossings of the Lu–Fano graphs (Figure 20) obtained from high-resolution experimental data.

The reason for this behavior, which at first sight seems surprising, is that angular momentum barriers related to the centrifugal term in the Schrödinger equation reduce the spatial overlap between wave functions, and hence the interchannel matrix elements. Thus, it is not sufficient that energy degeneracies should occur. It is also necessary that the spatial overlap between the states should be large for strong interchannel coupling to occur.

7. INNER-SHELL EXCITATION

Most inner-shell excitation spectra are beyond the reach of laser spectroscopy using the four-wave mixing source described above, which depends on a sum–difference scheme. Working with a sum–sum frequency mixing scheme allows the range of the source to be extended toward higher energies. Another option is to perform the photon addition within the sample. For example, if the atom has an $ms^2\ {}^1S_0$ ground state, then the first stage can be a two-photon excitation to some $msns\ {}^1S_0$ state below the first ionization threshold, followed by a further photoexcitation using another synchronously pumped tunable laser. Since the terms are the same, the final states accessed in this way are also the same as one excited by using a single rather-high-energy photon. Such experiments can bring the lowest of the subvalence shells within the range of laser spectroscopy. An example is shown in Figure 21.

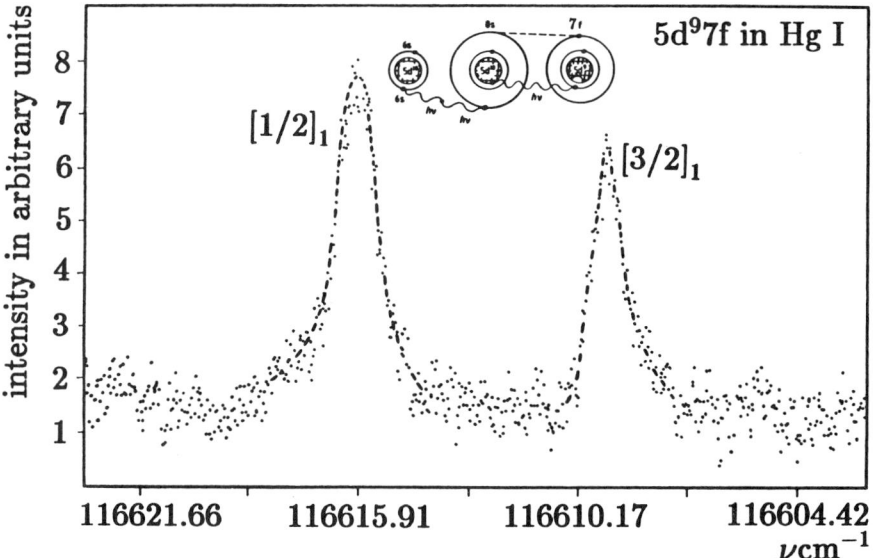

FIGURE 21. Excitation of the outermost inner-shell spectrum of Hg I by multiphoton excitation. The scheme is shown in the inset.

This approach does not actually yield the same spectrum as direct excitation from the ground state, because of the influence of the intermediate *msns* state, which can have a large radius if *n* is large, and therefore samples the outer reaches of the atom.

8. COLLISIONS AND EXTERNAL FIELDS

One should not forget that high resolution allows one to study the influence of collisions on spectra, especially when high Rydberg states are involved, because their large volume makes them sensitive to external fields. This is a very large subject, which cannot be covered in such a short chapter. Staying within the general subject area of doubly excited states which is the main thread of our argument, we show, in Figure 22, the manner in which a doubly excited level $3d^2\ ^1G_4$, which is not

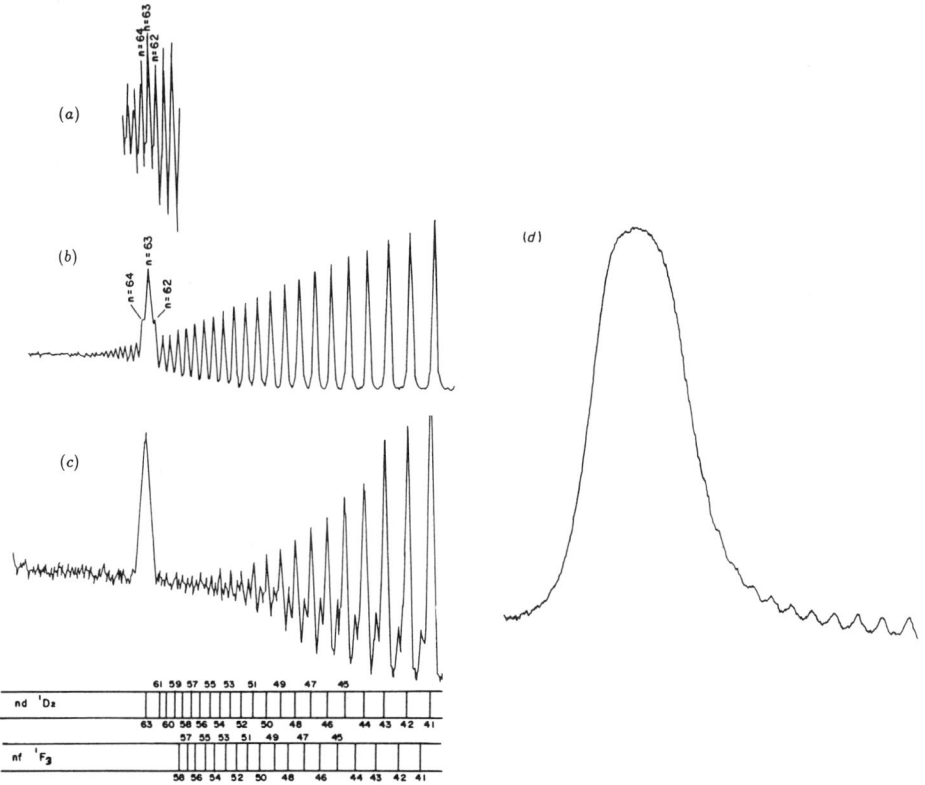

FIGURE 22. Influence of collisions with a rare gas on the two-photon spectrum of Ca I near the series limit. Note the strong perturbation around $n = 62$ which varies as a function of pressure and of the gas. (a) A portion of the two-photon spectrum of Ca I at zero pressure of foreign gas in the region of the $3d^2\ ^1G_4$ energy, showing that there is no intruding feature in this portion of the $4snd\ ^1D_2$ series. (b) The two photon spectrum of Ca I with a 10 ± 1 mbar pressure of argon, showing the depression of high series members and emergence of a strong feature described as $3d^2\ ^1G_4$. Note that, in this case, three members of the two-photon Rydberg series are enhanced. (c) The two-photon spectrum at a 100 ± 1 mbar pressure of argon, showing a yet faster fall off in the intensity of the Rydberg members with increasing n than in (b), despite which the $^1B_4$ becomes even more prominent. A new series, attributed in the present paper to collisional mixing between the $4snd\ ^1D_2$ series and both $4snp\ ^1P_1$ and $4snp\ ^1F_3$ with possible admixtures of higher angular momenta, emerges as weak satellite lines in the higher energy wings of the main two photon series. (d) The two-photon spectrum of Ca I at 400 ± 10 mbar of xenon, showing how the Rydberg series is almost completely quenched, while the $^1G_4$, feature dominates the spectrum, as a very broad, but fairly symmetric line. The survival of this feature in spite of the quenching of the Rydberg structure is a clear indication that the wavefunction of the excited state is much more compact than a Rydberg orbital. (After Bhatia et al.[33])

accessible from the ground state other than through collisions, can act as a strong perturber of a two-photon excited Rydberg series in the presence of a foreign gas which allows nearby states to mix by breaking the ΔJ rule.

9. THE USE OF COLLISIONS TO EXTEND THE WAVELENGTH RANGE

Since tunable lasers are still limited in their wavelength coverage, there exists yet another approach, which is to run a discharge in a gaseous sample when the atom of interest possesses a well-placed metastable state which can readily be populated. This provides a useful energy step, although the resolution available by this technique is of necessity limited, since all of the levels suffer Stark broadening by the microfields. One might also suspect that such a method would not be suitable for the observation of high Rydberg states, since these are particularly sensitive to collisions.

However, it turns out that this is not the case. Experiments have been done using both molecular dissociation (Gay et al.[34]) to populate metastable states of alkalis with low ionization potentials and energetic discharges to populate metastable states in helium[35] the highest ionization potential for a neutral atom, and in both cases high Rydberg states were reached.

In the latter instance, optogalvanic methods are the most appropriate means of detection, since a gaseous discharge is already running, and this is easily combined with laser irradiation of the cell. Such experiments have also been performed by Baig and co-workers in Islamabad: a typical example is shown in Figure 23.

10. FROM ORDER TO CHAOS IN MANY-ELECTRON SPECTRA

With high resolution, it also becomes possible to tackle another interesting problem. Any many-electron atom, beginning with helium, does not necessarily exhibit regularity in its spectrum. Indeed, the classical three-body problem contains chaotic trajectories. As the complexity increases, the situation worsens, and one may well ask whether and when correlations are sufficient to trigger an order-to-chaos transition.

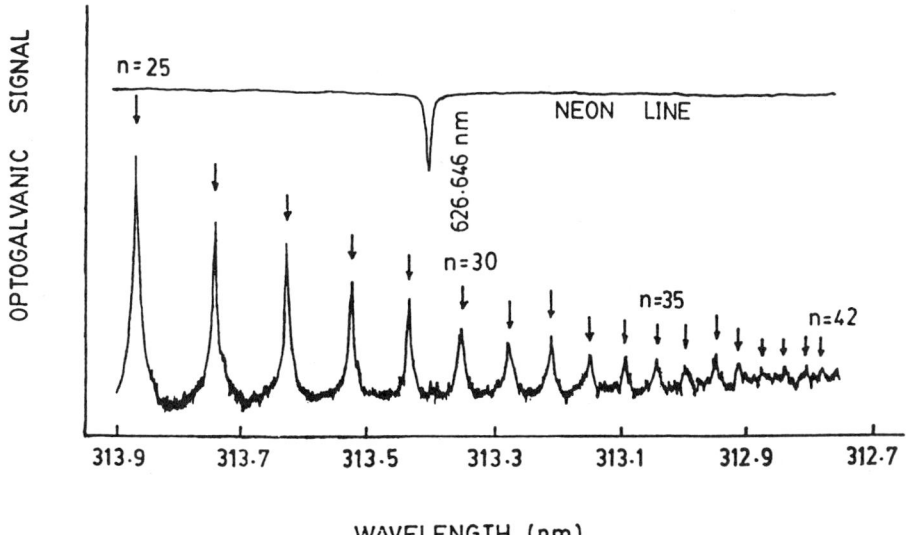

FIGURE 23. High Rydberg members of the helium spectrum observed by optogalvanic spectroscopy using a gaseous discharge to populate a metastable level. [After Baig et al. (unpublished).]

A more limited, but related, question is whether Wigner statistics appear in the nearest-neighbor spacings as well as a Thomas–Porter distribution in autoionization widths. The former question was considered by Rosenzweig and Porter,[36] but the latter seems never to have been addressed. Unfortunately also, the experimental data used in Ref. 36 seem to be unreliable for this purpose. There are several reasons for this. First, the experimental data originate from different sources for which instrumental resolution may differ and observations are probably incomplete. Second, the levels were binned by J value, and some of the J assignments may well be unreliable. Third, although the configurations are very complex, they are so for reasons connected with angular momentum symmetry groups and not necessarily because of electron correlations.

A much more satisfactory situation arises when dipole transitions from a $^1S_0$ ground state to a dense set of doubly excited levels are considered. Because of selection rules, the excited levels all have $J = 1$ so that binning is not required. Second, double excitations are necessarily dominated by correlations. Third, we may verify that the complexity of the spectrum is not merely related to angular momentum.

An important criterion is that one-electron labels for the excited states should not exist, i.e., that an independent electron assignment of the states is unsuitable. This is indeed the case discussed above for the spectra of Figures 12 and 13, and therefore it is interesting that the accumulated density of states in wide sections of these spectra rise nearly linearly with energy, while the level spacings do exhibit a Wigner-type behavior. Work is still in hand on the statistics of the level widths. Preliminary indications are that the distribution is double-peaked, perhaps because of a survival of singlet and triplet character in the spectrum.

11. CONCLUSION

The spectroscopy of atoms which have been raised into excited states, and which therefore present modified cross sections, is a rapidly expanding field. There are also many possibilities of combining lasers with other sources, such as synchrotrons. Multiphoton excitation using high-power pulses also offers new opportunities. In the interest of brevity, I have not attempted to cover these areas, although they are all in some way related to my theme.

In conclusion, it is hoped that the present chapter gives some convincing examples to show that laser-based UV and VUV spectroscopy of atoms, by providing higher spectral resolution and also more photons within a narrow bandwidth than any other known light source, can be applied effectively to yield new insights into the influence which many-electron correlations have on the atomic response.

ACKNOWLEDGMENTS. I wish to thank my numerous colleagues, co-workers, and former students from both the Physics Department at Imperial College and the Physikalisches Institut der Universität Bonn who have done most of the work described in this chapter.

REFERENCES

1. G. C. BJORKLUND, *IEEE J. Quantum Electron.* **QE11**, 287 (1975).
2. R. HILBIG AND R. WALLENSTEIN, *IEEE J. Quantum Electron.* **QE19**, No. 2 (1983); see also R. Hilbig, G. Hilber, A. Timmermann, and R. Wallenstein, *Ann. Isr. Phys. Soc.* **6**, 457 (1983).
3. C. R. VIDAL, in *Topics in Applied Physics*, Vol. 59, edited by Mollenauer and White (Springer-Verlag, Berlin, 1987).
4. J. F. REINTJER, *Nonlinear Parametric Processes in Liquids and Gases* (Academic Press, New York, 1984).
5. J. P. MARANGOS, N. SHEN, H. MA, M. H. R. HUTCHINSON, AND J. P. CONNERADE, *J. Opt. Soc. Am. B* **7**, 1254 (1990).
6. M. H. R. HUTCHINSON AND C. THOMAS, *J. Quantum Electron*, **QE19**, 1823 (1983).
7. G. HILBER, A. LAGO, AND R. WALLENSTEIN, *J. Opt. Soc. Am B* **4**, 1753 (1987).
8. U. GRIESMANN, N. SHEN, J. P. CONNERADE, K. SOMMER, AND J. HORMES, *J. Phys. B* **21**, L83 (1988).
9. B. EDLÉN, *Rep. Prog. Phys.* **26**, 181 (1963).
10. R. C. M. LEARNER AND A. P. THORNE, *J. Opt. Soc. Am.* **5**, 2045 (1988).

11. J. P. CONNERADE, *J. Phys. B* **16**, 399 (1983).
12. J. P. CONNERADE, *J. Phys. B* **21**, L551 (1988).
13. J. P. CONNERADE, W. A. FAROOQ, AND M. NAWAZ, *J. Phys. B* **25**, L175 (1992).
14. J. P. CONNERADE, W. A. FAROOQ, H. MA, M. NAWAZ, AND N. SHEN, *J. Phys. B* **25**, 1405 (1992).
15. U. FANO, *Phys. Rev.* **124**, 1866 (1961).
16. J. P. CONNERADE, *J. Phys. B* **24**, L51 (1991).
17. W. R. S. GARTON AND K. CODLING, *Proc. Phys. Soc.* **75**, 87 (1960).
18. W. R. S. GARTON, *Autoionisation Effects in Ultraviolet Absorption of Hot Gases*, Harvard College Observatory Scientific Report (November 1965).
19. J. P. CONNERADE AND A. M. LANE, *Rep. Prog. Phys.* **51**, 1439 (1988).
20. A. M. LANE AND R. G. THOMAS, *Rev. Mod. Phys.* **30**, 257 (1958).
21. J. P. CONNERADE, H. MA, N. SHEN, AND T. A. STAVRAKAS, *J. Phys. B* **21**, L241 (1988).
22. M. Y. SUGIURA, *J. Phys. Radium* Ser. VI **VIII**, No. 3, 8 (1927).
23. Z. ALTUN, S. L. CARTER, AND H. P. KELLY, *J. Phys. B* **15**, L709 (1982).
24. C. H. GREENE AND L. KIM, *Phys. Rev. A* **36**, 2706 (1987).
25. P. SCOTT, A. E. KINGSTON, AND A. HIBBERT, *J. Phys. B* **16**, 3945 (1983).
26. J. P. CONNERADE, *Proc. R. Soc. London Ser. A* **362**, 361 (1978).
27. S. M. FAROOQI, J. P. CONNERADE, C. H. GREENE, J. P. MARANGOS, M. H. R. HUTCHINSON, AND N. SHEN, *J. Phys. B* **24**, L179 (1991).
28. M. AYMAR, *J. Phys. B* **20**, 6507 (1987).
29. M. L. GINTER, D. S. GINTER, AND C. M. BROWN, *Appl. Opt.* **19** 4015 (1980).
30. S. M. FAROOQI, J. P. CONNERADE, AND M. AYMAR, *J. Phys. B* **25**, L219 (1992).
31. M. L. GINTER, D. S. GINTER, AND C. M. BROWN, *Appl. Opt.* **19**, 4015 (1980).
32. J. P. CONNERADE, *J. Phys. B* **10**, L239 (1977).
33. K. S. BHATIA, J. P. CONNERADE, AND Y. Y. MAKDISI, *J. Phys. B* **23**, 3475 (1990).
34. J. C. GAY, D. DELANDE, AND F. BIRABEN, *J. Phys. B* **13**, L729 (1980).
35. D. H. KATAYAMA, J. M. COOK, V. E. BONDYBEY, AND T. A. MILLER, *Chem. Phys. Lett.* **62**, 542 (1979).
36. N. ROSENZWEIG AND C. E. PORTER, *Phys. Rev.* **120**, 1698 (1960).

CHAPTER 13

ION YIELD SPECTROSCOPY WITH SOFT X RAYS

T. HAYAISHI AND P. ZIMMERMANN

1. INTRODUCTION

The soft x-ray region covers the photon energy range of approximately 40 eV to 6 keV.[1] This range corresponds to the region of atomic inner-shell excitation or ionization. Once the photon energy of soft x rays is absorbed by an atom, an inner-shell electron of the atom is excited or ejected. The corresponding inner-shell hole originated in this photoabsorption process is refilled through the rearrangement of the residual electrons in the atom such as Auger decays. As the final result of the photoabsorption, usually several electrons are ejected from the atom and multiply charged ions are produced. These decays following the production of inner-shell holes are studied by measurements of the electrons and ions. The measurement of the electrons, photoelectron spectroscopy, is a powerful technique in the study of the different decay channels when they appear as discrete lines in the photoelectron spectrum. If caused by multielectron ejection, however, the photoelectron spectrum shows continuous structures; it is hard to unravel the various routes of the decays by means of photoelectron spectroscopy alone. Here, the measurement of the ions, ion yield spectroscopy, makes up the deficit of photoelectron spectroscopy. Species and yields of the ions measured using mass spectrometers reflect the different routes of the decays. Hence, ion yield spectroscopy allows us to discuss the mechanism of the decays. Thus, the main aim of ion yield spectroscopy with soft x rays is to study what happens to the atom after the creation of an inner-shell hole.

After a short description of the time-of-flight and the atomic beam technique, this chapter will present results for the rare gases and metal vapors obtained recently by ion yield studies.

2. EXPERIMENTAL TECHNIQUES

2.1. Time-of-Flight Techniques

2.1.1. Characteristics of the Time-of-Flight Mass Spectrometer. Mass spectrometers which are designed to distinguish gaseous ions according to their mass–charge ratio are classified into conventional magnetic field types, quadrupole mass filter types, time-of-flight (TOF) types, and others.[2] The TOF mass spectrometer is widely used in atomic and molecular physics at present. The principle of the TOF is based on the idea that the velocity of ions depends on the mass–charge ratio under the condition of a constant electric field.[3] The TOF mass spectrometer can be characterized as follows: the construction of the apparatus is very simple, ions of all masses can be simultaneously observed, and the collection efficiency of ions is very high. Other types except for the TOF have slits to lead ions into the analyzers. Those types with slits only take a part of the ions produced by photoionization while the TOF mass spectrometer without any slit enables basically to take all of the ions produced by photoionization.

T. HAYAISHI • Institute of Applied Physics, University of Tsukuba, Tsukuba, Ibaraki 305, Japan.   P. ZIMMERMANN • Institut für Strahlungs- und Kernphysik, Technische Universität Berlin, 10623 Berlin, Germany.

VUV and Soft X-Ray Photoionization. Edited by Uwe Becker and David A. Shirley. Plenum Press, New York, 1996.

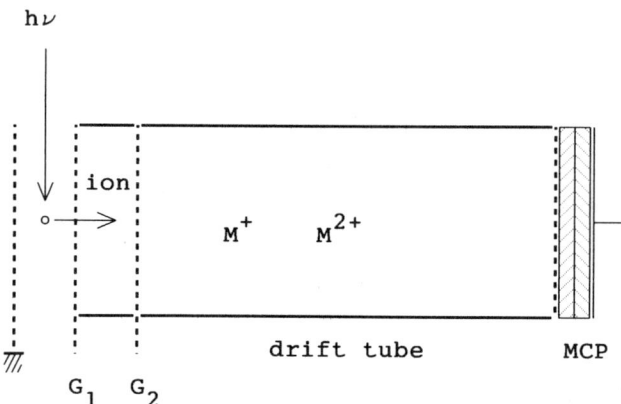

FIGURE 1. Schematic diagram of the TOF mass spectrometer; G_1 and G_2, grids; MCP, multichannel plates.

A schematic diagram of a typical TOF mass spectrometer is shown in Figure 1. The construction is based on the original feature of Wiley and McLaren.[4] The TOF mass spectrometer consists of three sections, namely, an ionization section, an acceleration section, and a field-free drift section. In the ionization section, sample gases are ionized by photon beams of monochromatized synchrotron radiation. Ions produced by the photoionization event are accelerated through grid G_1 and further accelerated through G_2. G_1 and G_2 are operated to increase the mass resolution under the double-field space focusing conditions described by Wiley and McLaren.[4] After acceleration the ions enter into the drift section and then are detected by a multichannel plate (MCP) detector following flights of the section. Mass analyses of the ions are performed by the flight times of the ions from the ionization region to the MCP detector, i.e., in coincidence with the initial photoionization and final detection events. In the coincidence mode it is necessary to settle the initial time of the photoionization event. The settlement of the initial event is attained by various techniques: detection of electrons ejected by photoionization,[5,6] the pulsed acceleration field applied through grid G_1,[7,8] or pulsed synchrotron radiation in the single bunch operation.[5]

Ion yields are accumulated as a function of the flight time in the TOF mass spectrometer. The ion yields give the so-called mass spectrum distributed into charge states. A mass spectrum can be obtained by an accumulation in this TOF technique, i.e., ions of all species caused by photoionization can be simultaneously detected. Thus, the TOF technique is more effective for simultaneous detection for ions of many species.

2.1.2. Precautions for the Collection Efficiency of Ions. The TOF mass spectrometer is one of the powerful tools in atomic and molecular physics. However, we should take careful precautions with respect to the collection efficiency of ions, because the efficiency is inconstant for different charge states in certain experimental conditions.

The monochromatized synchrotron radiation must be free from stray and second-order light. In the stray light including long wavelengths, detection of singly charged ions increases unexpectedly. In the second-order light, detection of multiply charged ions increases. To eliminate these drawbacks, spectroscopic filters are effective, such as polypropylene films[5] and beryllium windows.[7]

The gas pressure must be kept sufficiently low (e.g., 10^{-3}–10^{-5}Pa), because other processes such as recombination and electron impact ionization are pressure dependent. Therefore, it is necessary to check the pressure dependence of the M^{2+}/M^+ ratio.[5,9]

The kinetic energies of the ions incident on the MCP must be high enough to avoid the different collection efficiencies for different charge states. To determine the optimal voltage applied to the front of the MCP, the M^{2+}/M^+ ratio has been measured in previous research.[5,9,10] Voltages over -4 kV are recommended.

2.2. Atomic Beam Technique

There is a large number of methods for the production of atomic or molecular beams. Therefore, only some general principles and a few sources will be described. For a more detailed description the reader is referred to the literature (e.g., *Atomic and Molecular Beam Methods*, Vol. I, edited by G. Scoles, Oxford University Press, 1988).

2.2.1. The Effusive Source.

The simplest beam source still very often used is the effusive source. The principle of this source is that the particles effuse from a source chamber through a small hole into the vacuum chamber without undergoing collisions (molecular flow conditions). These conditions impose certain restrictions on the vapor pressure and the dimensions of both source and vacuum chamber which may be estimated considering the mean free path of the particles. From gas kinetics the product of the mean free path l and the pressure p is given by

$$l \cdot p = \frac{kT}{\sqrt{2\pi d^2}}$$

where d is the particle diameter. For thermal atomic beams typical values are $lp \approx 10^{-2}$ m · Pa (1 Pa = 10^{-5} bar) and the dimensions of the vacuum chamber and the orifice of the source chamber usually are on the order of 1 m and 1 mm, respectively. Therefore, the upper limits of the pressure in the vacuum chamber should be $p < 10^{-2}$ Pa (typical values: 10^{-3}–10^{-5} Pa) and in the source chamber $p < 10$ Pa (typical values: 10^{-1}–1 Pa).

2.2.2. Gases.

For experiments with gases the source chamber may be just a tube with an appropriate orifice and a connection to a gas reservoir with a dosaging valve for the control of the gas flow. For effusive beams the flow rate is very low and, therefore, the drop in the pressure of the reservoir during a measurement usually can be neglected. Otherwise, flow control valves or pressure-controlled reduction valves may be used.

Often one is interested in increasing the intensity of the beam in the interaction region of the vacuum chamber. For a given source pressure this can be achieved by using a channel or a multichannel array to collimate the beam into the forward direction instead of using a thin-walled orifice with its cosine-dependence of the beam intensity. Another possibility for a gain in the beam intensity is the increase of the source pressure but normally there are certain restrictions related to the pumping speed in the vacuum chamber. If the pressure in the vacuum chamber increases, the smaller mean free path can destroy the conditions of a collision-free beam. This difficulty can be overcome by increasing the pumping speed and by decreasing the orifice. Thus, one gradually leaves the conditions of an effusive source into the direction of a jet beam where the gas dynamic of the supersonic expansion from a high-pressure source ($p > 1$ bar) through a small nozzle ($d < 100$ μm) into the vacuum chamber is used.

2.2.3. Metal Vapors.

For experiments with metal vapors the source chamber is an oven in which the vapor pressure is produced by thermal evaporation of the element. The first consideration for the construction of such a source is the temperature that is necessary for the vapor pressure inside the source chamber. If one considers $p = 10^{-2}$ torr (1.3 Pa) as a typical value for the vapor pressure inside the source chamber, then the following list gives the corresponding temperatures (in °C):

- Alkali metals: Li (535), Na (290), K (210), Rb (170), Cs (150)
- Alkali-earth metals: Be (1205), Mg (430), Ca (605), Sr (540), Ba (680)

- 3d transition metals (iron group): Sc (1390), Ti (1730), V (1850), Cr (1390), Mn (970), Fe (1470), Co (1540), Ni (1530)
- 4d transition metals (palladium group): Y (1630), Zr (2440), Nb (2700), Mo (2510), Ru (2380), Rh (2040), Pd (1480)
- 5d transition metals (platinum group): Lu (1660), Hf (2440), Ta (3060), W (3230), Re (3060), Os (2960), Ir (2480), Pt (2100)
- 4f elements (rare earths): La (1750), Ce (1730), Pr (1520), Nd (1340), Sm (740), Eu (615), Gd (1580), Tb (1530), Dy (1120), Ho (1170), Er (1250), Tm (850), Yb (470)
- $d^{10}s^n$ elements (atomic noble metals): Cu (1260), Zn (345), Ag (1030), Cd (265), Au (1410)
- p^n elements (third and fourth group of the periodic table)
  p: B (2100), Al (1220), Ga (1045), In (940), Tl (630)
  p^2: C (2470), Si (1630), Ge (1390), Sn (1240), Pb (715)

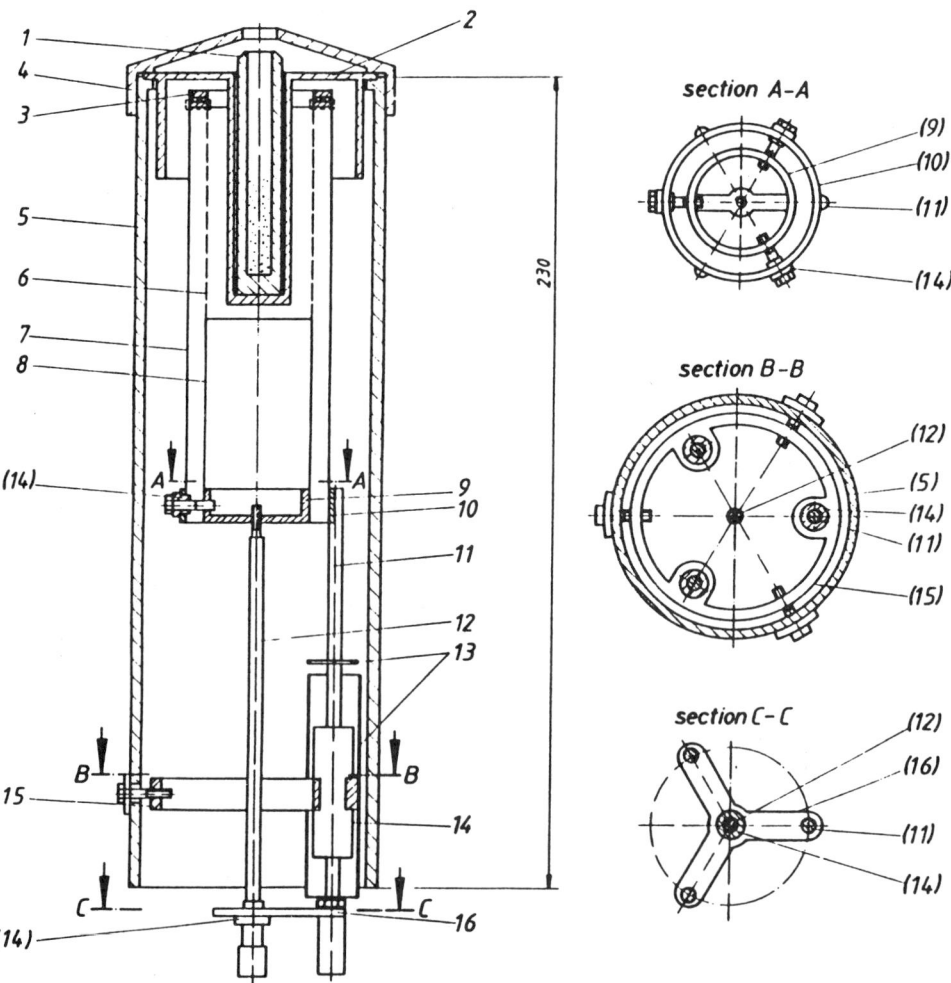

FIGURE 2. Sectional views of the electron-beam-heated atomic source. 1, crucible; 2, lid; 3, Nb ring; 4, Cu cap; 5, Cu cylinder; 6, W filaments; 7, outer Ta sheet cylinder; 8, inner Ta sheet cylinder; 9, 10, stainless steel rings; 11, 12, stainless steel rods; 13, metal shields; 14, ceramic insulators; 15, stainless steel ring; 16, stainless steel support.[41]

From a practical view one may divide the range of temperature into three regions: I ($\to$ 1000 K), II ($\to$ 2000 K), III ($\to$ 3500 K).

I. *Operation at medium temperatures to about 1000 K.* As direct heating in this region is possible, the construction of the source may be quite simple. The oven, for example, can consist of a metal tube with one closed end which is surrounded by heating wires. The orifice should be (separately) heated to a higher temperature as the rest of the oven to prevent clogging. Depending on the chemical aggressiveness of the beam material, different oven materials such as stainless steel, copper, or nickel may be used. Typical elements for this type of oven are the alkali and alkali-earth metals (except Be) or special elements from other groups like Zn, Cd, Sm, Eu, Yb, Tl, Pb.

II. *Operation at high temperatures between 1000 and 2000 K.* Although heating wires or direct heating of thin-walled tubes by high currents can be applied, this region of operation usually uses heating by electron bombardment. An example of such a source is given in Figure 2[41] where the oven is surrounded by a tungsten filament (cathode) and a dc voltage ($\approx$1 kV) is applied between the oven and the cathode; typical electron currents are 100 mA–1 A. The increasing heat loss by radiation ($\sim T^4$) has to be taken into account by a careful shielding of the source with molybdenum or tantalum metal sheets. Other severe problems may arise by the aggressiveness of the melting charge which can react or form an alloy with the oven material (usually molybdenum, tantalum, tungsten, or graphite). In these cases the inner part of the oven should be fabricated using materials such as tantalum carbide. Typical elements in this region up to 2000 K are the $3d$ transition metals (iron group) or the $4f$ elements (rare earths) except Sm, Eu, Yb.

III. *Operation at very high temperatures between 2000 and 3500 K.* This region of operation is characterized by increasing problems already described in region II which are inherently connected with the radiation losses and the choice of the oven material. For photoion spectroscopy there is also the special problem that at these temperatures the oven produces a large number of ions which can severely disturb the photoion detection. To obtain these high temperatures usually the dimensions of the oven heated by electron bombardment are reduced in order to minimize the radiation losses. In extreme cases the electron beam may be focused on the beam material itself.[42] Another possibility for producing high temperatures in small areas is the focusing of high-power laser beams on the beam material or the direct heating of wires by metals such as tungsten which have an efficient sublimation rate below their melting point. Typical elements in this region are the elements of the $4d$ and $5d$ transition metals (except Y, Pd, Lu).

3. RESULTS

3.1. Rare Gases

3.1.1. Partial Ion Yields. Early on, the measurements of ion yield were performed by total yields of ions acquired without any mass analysis, which are often called total ion yields. The yields measured as a function of photon energy correspond to the photoabsorption spectra. These spectra have provided us with certain phenomena in the soft x-ray region, such as autoionization[11] or shape resonances.[12] In inner-shell ionization regions such a phenomenon is entangled by various processes such as Auger processes following inner-shell ionization and multielectron ejection of outer-shell electrons. It is difficult to find distinctly the individually unraveled processes from the measurements of total ion yields alone. The finding of the individual processes is feasible by means of measurements of the partial ion yields discriminating the charge states of the ions, because a decay channel from a distinct phenomenon ends in a certain charge state.

The partial ion yields of rare gases have been measured by van der Wiel and Wiebes[13] El-Sherbini and van der Wiel,[14] and Holland *et al.*[9] by means of the TOF techniques. van der Wiel and co-workers measured partial ion yields for Ar, Kr, and Xe with electron impact (pseudo photon impact). Holland *et al.* also measured ions yields covering the wide range of photon energies for Ne, Ar, Kr, and Xe by the use of synchrotron radiation. These measurements revealed that Auger processes play an important role in the formation of charge states from inner-shell ionization; for instance, doubly and triply charged ions are produced from the $4d$ ionization of Xe. The formation of doubly charged

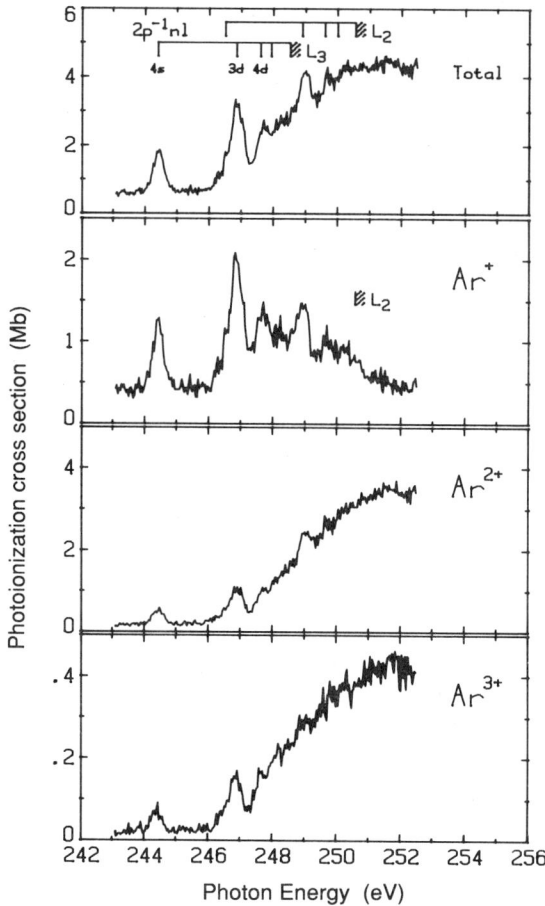

FIGURE 3. Photoionization cross sections of multiply charged ions taken by a TOF mass spectrometer in the vicinity of the Ar $2p$ ionization limits.[5]

ions is caused by Auger processes replacing the $4d$ inner-shell hole by two outer-shell holes. The formation of triply charged ions is caused by double Auger (Auger shakeoff) processes replacing the inner-shell hole with three outer-shell holes.

The measurements of van der Wiel and co-workers[13,14] and Holland et al.[9] were performed with low spectral resolution. For the last decade, measurements with higher spectral resolution have been permitted by the use of intense and stable synchrotron radiation. Much valuable information has been obtained from the measurements of partial ion yields: multiple ionization following inner-shell photoexcitation,[5,8,15] postcollision interaction closely above inner-shell ionization thresholds (the enlargement of the Ar^+ yield as a result of recapture of the receding photoelectron),[5,16–18] and partial cross sections (for subshells participating in ionization) derived from partial ion yields in inner-shell ionization regions.[6,7,14] Here, we will concentrate on the partial ion yields in the inner-shell resonance regions.

Hayaishi et al.[5] have measured partial ion yields in the Ar $2p$, Kr $3d$, and Xe $4d$ inner-shell resonance regions with high resolution of 0.008 nm using the TOF mass spectrometer in coincidence with electrons. Figures 3, 4, and 5 show the partial ion yields for singly to triply charged ions of Ar, Kr, and Xe, respectively. The ion yields have been verified by means of the pulsed synchrotron radiation technique to avoid counting errors in coincidence with electrons. The sum yield of singly to triply charged ions, which is shown in the upper part of each figure, corresponds to the total ion yield. Those

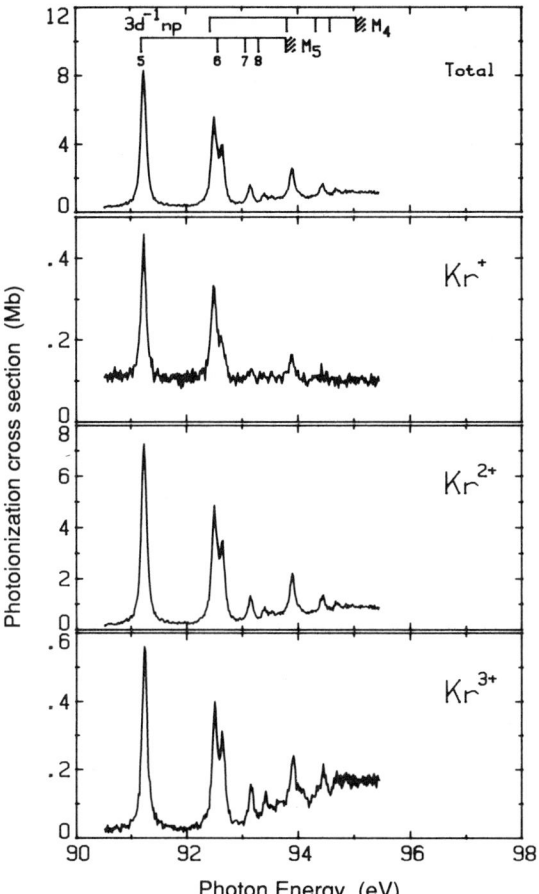

FIGURE 4. Photoionization cross sections of multiply charged ions taken by a TOF mass spectrometer in the vicinity of the Kr $3d$ ionization limits.[5]

ion yields were converted to absolute photoionization data of Marr and West[19] and West and Morton.[20]

The sum yields exhibit discrete structures related to the Rydberg series of Ar $2p^{-1}nl$, Kr $3d^{-1}nl$, and Xe $4d^{-1}nl$. Those structures can also be seen in the respective partial ion yields, in particular in doubly charged ions of Kr and Xe. Notice that comparing the ion yields of Ar, Kr, and Xe, the signals show different ratios and profiles for the different charge states. This fact suggests that they reflect different decay channels to these charge states.

3.1.2. Inner-Shell Resonance Decays. The findings obtained by ion yield measurements are the final information on the charge states from the resonance decays. In order to discuss the mechanisms of the decays in detail, however, more information on the decay routes is required. Such information can be supplied by photoelectron spectroscopy which analyzes the kinetic energies of the electrons ejected after photoexcitation. The first photoelectron measurements of the Kr $3d$ and Xe $4d$ resonances were performed by Eberhardt et al.[21] Next, Southworth et al.[22] measured minutely the Xe $4d$ resonant photoelectron spectrum. Those measurements revealed the following facts: Two major decay processes are possible in inner-shell resonance states. One is the case where the excited electron participates in the decay. The inner-shell hole is replaced by one outer-shell hole and the excited electron is ejected. This process is called the participant Auger process (autoionization). The final state of this process is

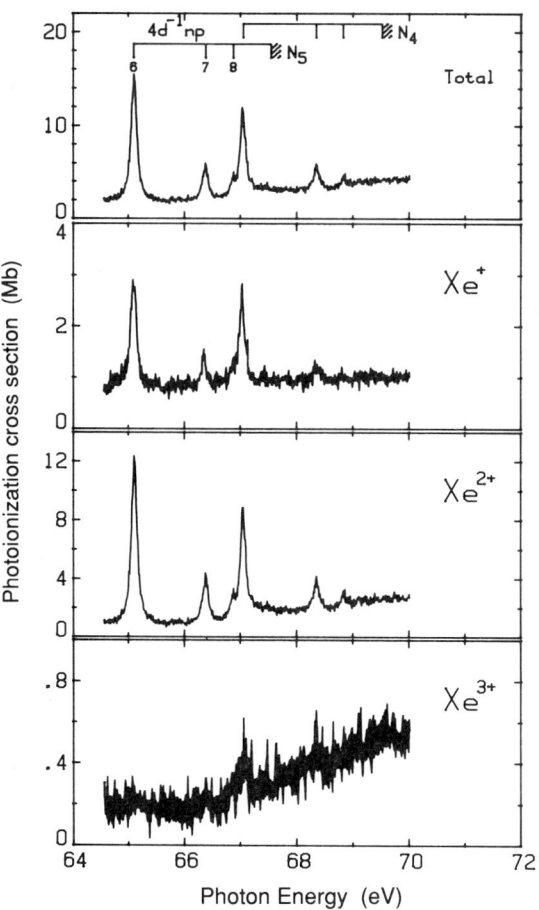

FIGURE 5. Photoionization cross sections of multiply charged ions taken by a TOF mass spectrometer in the vicinity of the Xe 4d ionization limits.[5]

the same as the direct ionization of the outer-shell electron. The other is the case where the excited electron acts as a spectator in the resonance decay. The inner-shell hole is replaced by two outer-shell holes and the excited electron remains in its orbital. This process is called the spectator Auger process. If the overlap of the core electrons with the excited electron is small, the spectator Auger process is more dominant than the participant Auger process. In some occasions, shake-up, -down, or -off transitions of the spectator electron can be important in which the spectator electron is promoted to a higher orbital, declined to a lower orbital, or ejected to an unbound state, respectively.

On the basis of these findings in resonant photoelectron spectroscopy, Hayaishi et al.[5] proposed that the formation of charge states from resonance states is attributed to two-step decays via spectator Auger processes including shakeup of the spectator electron. In the first step, singly charged states are caused by the spectator Auger processes. In the second step the singly charged states are relaxed by emission of photons and/or electrons. If the singly charged states are located between the single and double ionization thresholds, the decay ends in the formation of singly charged ions. This case can be seen in the Ar $2p^{-1}nl$ states. If the singly charged states are located between the double and triple ionization threshold, the decay predominantly ends in the formation of doubly charged ions. This case can be seen in the Kr $3d^{-1}nl$ and Xe $4d^{-1}nl$ states. Thus, this model of the two-step decay enables one to explain the dominant formation of charge states from the resonant states.

Ion Yield Spectroscopy with Soft X Rays

After the presentation of the formation of charge states from resonance states, resonant photoelectron studies with respect to the formation of doubly charged ions have been performed by many workers.[23–32] For the most part the two-step decay model was supported. However, Becker et al.[30,31] and Heimann et al.[32] proposed the resonant shakeoff (Auger shakeoff) model as the main reason for the formation of doubly charged ions from resonance states where the excited electron is shaken off during the resonant Auger processes. Becker et al.[31] estimated the shakeoff distributions from Xe $4d^{-1}np$ according to the partitions of continuous and discrete structures in their resonant photoelectron spectrum: resonant shakeoff leads to continuous structures in the spectrum because two electrons ejected simultaneously from the decay give a continuous distribution of kinetic energies in the spectrum.[33] They reported that the resonant shakeoff decay is important in the Xe^{2+} formation. In succession, Heimann et al.[32] measured threshold electrons having near zero kinetic energies and resonant photoelectrons from the Ar $2p^{-1}nl$, Kr $3d^{-1}nl$, and Xe $4d^{-1}nl$ resonances. From the partitions of the spectral structures in their resonant photoelectron spectra in the same manner as Becker et al.,[31] they concluded that the resonant shakeoff intensity is of comparable or greater strength than the two-step one in those first resonance states.

It is therefore of great interest to determine which decay (two-step or resonant shakeoff) plays a major role in the formation of charge states from resonance states. In the following sections the decays to the formation of multiply charged ions will be discussed in more detail.

Decay channels from resonance states. Before further discussion on resonance decays, it is more effective to classify the different decay channels from the resonance states. In all cases decays following inner-shell photoexcitation are directed by Auger processes filling the inner-shell hole. Figure 6 shows the typical decays from Xe $4d^{-1}np$. The first path from the 4d excited state is divided into the Auger, double Auger, and triple Auger processes where one, two, and three electrons are ejected. Needless to say, the Auger process is more dominant than the others. In the second path, each process is partitioned into participant and (dominant) spectator transitions. Here, for example, the notation of $(5s + 5p)^{-2}$ signifies $5s^{-2}$, $5s^{-1}5p^{-1}$, or $5p^{-2}$ holes. In the third path, the participant and spectator Auger processes end in the formation of the charge states. Numbers in parentheses in the path indicate the different decay channels to the formation of the corresponding charge states. Information on the decays in the first and second paths can be obtained by resonant photoelectron spectroscopy and in the third path by ion yield spectroscopy.

Formation of singly charged ions. As shown in Figure 6, singly charged ions are produced through the (1) participant Auger process and the (2) spectator Auger process. The (1) channel corresponds to the so-called autoionization which ends in one 5p or 5s outer-shell hole. If autoionization plays an important role in the formation of singly charged ions, the spectral widths of autoionizing

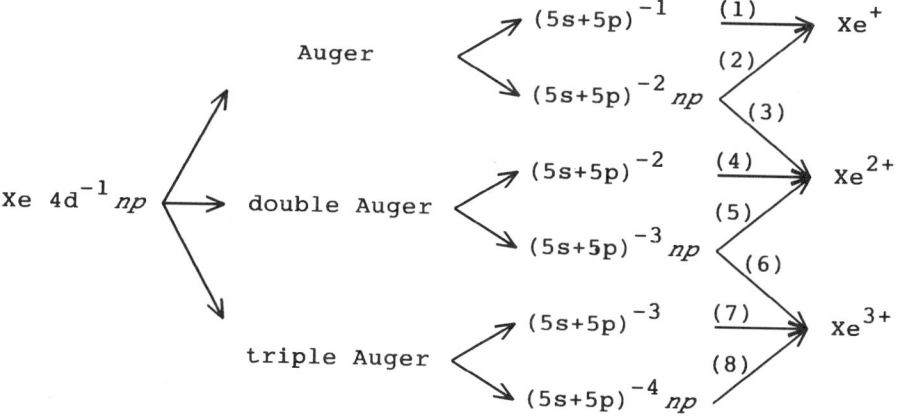

FIGURE 6. Classification of decay channels from Xe $4d^{-1}np$. Numbers in parentheses represent decay channels to the different charge states.

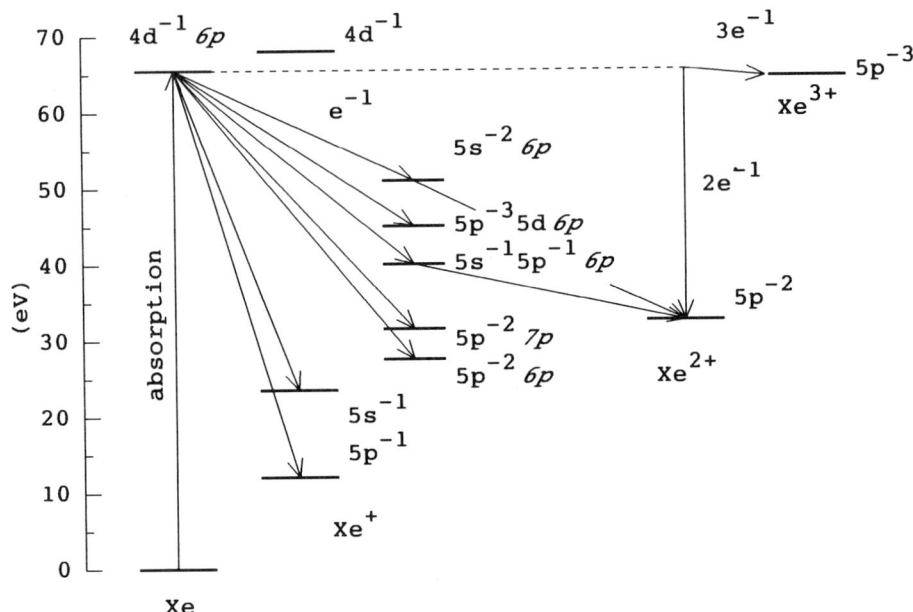

FIGURE 7. Decay channels from Xe $4d^{-1}6p$.[5]

resonance lines decrease with quantum number n^3 in the Rydberg series.[34] However, the spectral widths of the resonance lines observed as seen in Figures 3–5 are essentially constant irrespective of n. From this fact it is recognized that autoionization should be a comparatively weak process in these inner-shell resonance decays. This is consistent with the findings of Southworth et al.[22] for Xe and Lindle et al.[27] for Kr by resonant photoelectron spectroscopy. In the later discussion on the decay channels the contribution of (1) is therefore neglected.

The (2) channel is the spectator Auger process under the condition that the intermediate states caused by the first-step decay are located below the double ionization thresholds. Since the spectator Auger processes are the dominant decays from the resonance states, the formation of singly charged ions is mainly related to the (2) channel.

Formation of doubly charged ions. The decay channels to the formation of doubly charged ions from the resonance states are the (3) spectator Auger process including shakeup and -down, the (4) participant double Auger process, and the (5) spectator double Auger process. The (3) channel is expected to be the dominant channel. The (4) channel is the one-step decay which corresponds to the resonant shakeoff decay. The (3) and (5) channels are two-step decays. The intermediate excited states caused by the first step of the (3) channel are represented by $(5s + 5p)^{-2}np$. According to the resonant photoelectron spectrum for Xe $4d^{-1}6p$ of Becker et al.,[30] these excited states, e.g., $5s^{-1}5p^{-1}6p$, $5p^{-3}5d6p$, and $5s^{-2}6p$, are produced prominently from the resonance state. Those intermediate states are located above the double ionization thresholds as shown in Figure 7. Hence, these intermediate states finally can decay to doubly charged ions. On the other hand, it is hard to acquire information on the (4) and (5) channels by means of resonant photoelectron spectroscopy because of the continuous structures resulting from simultaneous two-electron ejection. However, double Auger decays are able to take place in the decays from the resonance states. The double Auger processes have probabilities of 20–30% in the Xe $4d$ and Kr $3d$ ionization.[6,7,14] If the intermediate states caused by the (5) channel are located between the double and triple ionization thresholds, the decays lead to the formation of doubly charged ions.

Formation of triply charged states. Since three electrons are ejected in the formation of the triply charged ions, the formation has a lower probability than the formation of singly and doubly charged

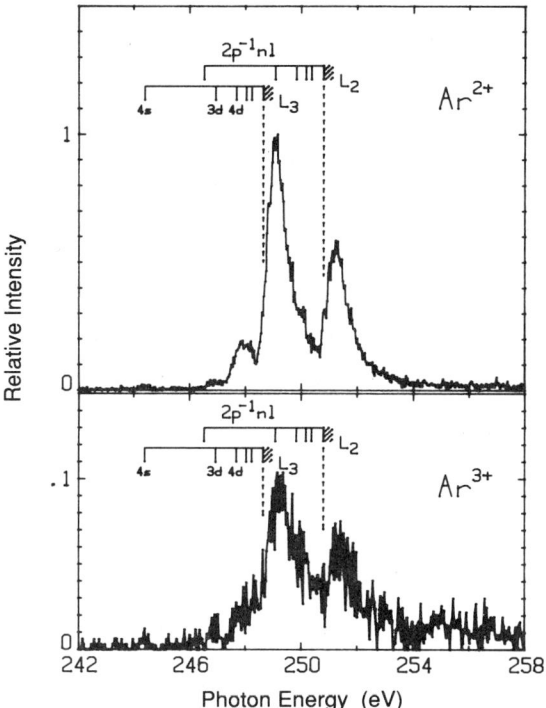

FIGURE 8. Relative yield spectra of Ar^{2+} and Ar^{3+} ions near the Ar $2p$ ionization limits taken with the threshold electron–ion coincidence method.[35]

ions. This fact can be seen in Figures 3–5. The decay channels related to the formation of triply charged ions are the spectator double Auger process of (6), the participant triple Auger process of (7), and the spectator triple Auger process of (8), as seen in Figure 6. The (7) channel is a one-step decay, the (6) and (8) channels are two-step decays. It is difficult to estimate the contributions of these decay channels, because the contributions give continua in the resonant photoelectron spectra. Therefore, the study of the formation of the triply charged ions must be combined with other experimental techniques, e.g., coincidence between threshold electrons and ions.[8,35–37] The coincidence technique enables one to find the decay channel accompanied by the threshold electron ejection. We will only give the coincidence spectra of Hayaishi et al.[35–37] here. Figures 8–10 show the coincidence spectra in the Ar $2p^{-1}nl$, Kr $3d^{-1}nl$, and Xe $4d^{-1}nl$ regions.

Decay channels following photoexcitation were distinguished as shown in Figure 6. In the next section, we report the quantitative analysis of the decay channels from the Ar, Kr, and Xe resonance states.

Multiple ionization from Ar $2p^{-1}4s$. Table I shows the ion yield ratios (%) from the first resonance state $2p^{-1}4s$ of Ar. The ratios were obtained by two coincidence modes with electrons[5] and with threshold electrons.[35] Numbers in parentheses indicate the decay channels classified in Figure 6. Since the ejection of threshold electrons from this state is mainly due to multielectron ejection,[35] the Ar^{2+} formation is due to (4), and the Ar^{3+} formation is due to (7) and (6) + (8).

From Table I we can estimate the branching ratios of the different decay channels for the formation of Ar^{2+} and Ar^{3+} ions. Since the ratio of (4) to (7) + (6) + (8) in coincidence with electrons is 67/33, the branching ratio of (4) is calculated to be 10% (= 5% × 67/33) for 5% of (7) + (6) + (8) in coincidence with electrons (total decays). Therefore, the branching ratio of (3) + (5) is estimated to be 20% by the subtraction of the (4) fraction (10%) from the (4) + (3) + (5) fractions (30%) in the total decays. Table II shows the branching ratios obtained from this estimation. The branching ratios for the

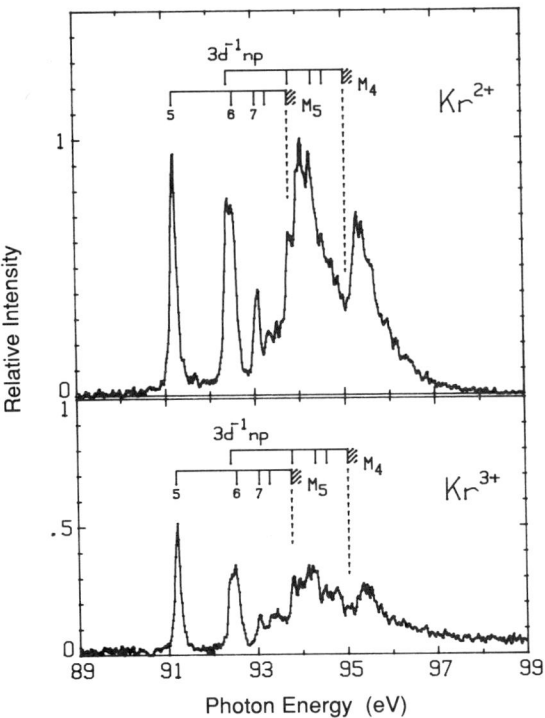

FIGURE 9. Relative yield spectra of Kr^{2+} and Kr^{3+} ions near the Kr $3d$ ionization limits taken with the threshold electron–ion coincidence method.[36]

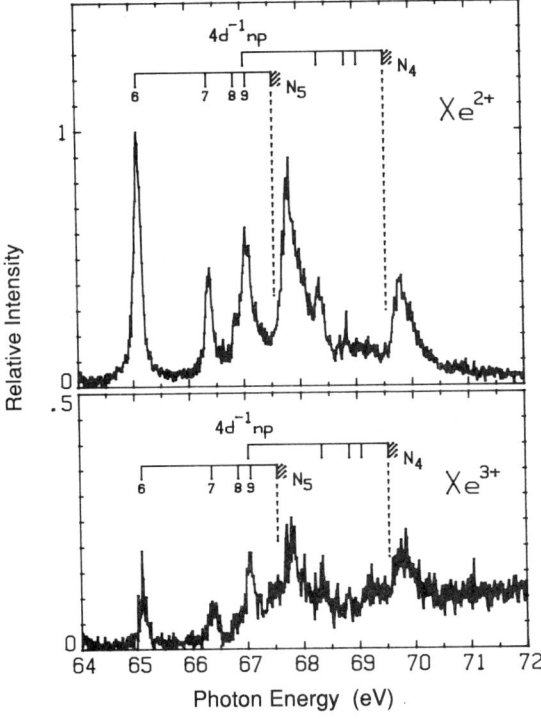

FIGURE 10. Relative yield spectra of Xe^{2+} and Xe^{3+} ions near the Xe $4d$ ionization limits taken with the threshold electron–ion coincidence method.[37]

TABLE I
Ion Yield Ratios (%) from Ar $2p^{-1}4s$

Coincidence Mode	Ar^+	Ar^{2+}	Ar^{3+}	Decay
With electrons	65	30	5	
		$(4)^a$	(7)	One-step
	(2)	(3) + (5)	(6) + (8)	Two-step
With threshold electrons		67	33	
		(4)	(7)	One-step
			(6) + (8)	Two-step

^{a}Numbers in parenthesis indicate the decay channels in Figure 6.

one-step decay of (4) + (7) and the two-step decay of (6) + (8) + (3) + (5) cannot be estimated fully, because (7) and (6) + (8) are further inseparable from the Ar^{3+} formation. From the photoelectron spectroscopic point of view it is possible to distribute to decay channels causing continuous and discrete structures in the spectrum. The continuous structures are given by (4), (7), (6), and (8), whereas the discrete structures are given by (3) and (5). Hence, branching ratios for decays to continuous and discrete structures are estimated to be 15% : 20%. For comparison the data of Heimann et al.[32] are included in Table II. They assigned continuous structures to the resonant shakeoff decays and discrete structures to the two-step decays. The branching ratios were 16% : 18%. Their result is in good agreement with the finding from ion yield spectroscopy, although the (6) + (8) interpretation is in disagreement. Thus, the two-step contribution (20%) is more dominant than the resonant shakeoff contribution (10%), at least for the Ar^{2+} formation.

Multiple ionization from Kr $3d^{-1}5p$. Table III shows partial ion yields (%) from the $3d^{-1}5p$ resonance state of Kr. Column (a) presents data of the ion yields measured by Hayaishi et al.[5] The formation of Kr^{2+} ions is surprisingly prominent even below the ionization limits. Hayaishi et al. proposed that the origin of the formation is the two-step decay via the spectator Auger process including shakeup transitions of the spectator electron; $3d^{-1}5p \to 4s^{-1}4p^{-1}5p$, $4p^{-3}4d5p$, and $4p^{-2}6p$. These transitions give intense lines in the resonance photoelectron spectra.[26,27] Recently, Lablanquie and Morin[8] have studied minutely the decay processes from the Kr $3d^{-1}5p$ resonance state by combining ion yield, photoelectron spectroscopy, and threshold electron–ion coincidence experiments. Column (b) shows the data obtained by their ion yield experiment. The formation of Kr^{2+} ions is likewise prominent, but the fractional ratios for the charge states are different from those in column (a). The partial ion yields of Hayaishi et al. were measured under the conditions of the coincidence mode with electrons and high spectral resolution of 0.008 nm (0.05 eV at 90 eV), whereas Lablanquie and Morin obtained their results under the conditions of the coincidence mode with the pulsed acceleration field and a photon bandpass of 0.03 nm (0.2 eV at 90 eV). It seems that the difference of columns (a) and

TABLE II
Branching Ratios (%) of One- and Two-Step Decays for the Formation of Ar^{2+} and Ar^{3+} Ions from Ar $2p^{-1}4s$

		This Study		Heimann et al.[32]	
One step	$(4)^a$	10			
	(7)		15	16	Shakeoff
		5			
Two-step	(6) + (8)				
	(3) + (5)	20	20	18	Two-step

^{a}Numbers in parenthesis indicate the decay channels in Figure 6.

TABLE III
Partial Ion Yields (%) from Kr $3d^{-1}5p$

	Hayaishi et al.[5]	Lablanquie and Morin[8]		
	(a)	(b)	(c)	
Kr^+	5	17	11	Spectator
			6	Shakeup
			33	Spectator
Kr^{2+}	88	73	18	Shakeup
			22	One-step
Kr^{3+}	7	10	10	One-step

(b) for the partial ion yields is related to different ion-detection efficiencies caused by different experimental conditions.

Column (c) of Lablanquie and Morin[8] from their resonant photoelectron spectrum indicates branching ratios for the different decay channels from the resonance state. On the basis of partitions of discrete structures in the spectrum, the branching ratios for the Kr^+ and Kr^{2+} formation were estimated to be 17% : 51% which were divided further into spectator and shakeup contributions. The remaining Kr^{2+} formation is 22% (= 73%–51%) which corresponds to the branching ratio of the Kr^{2+} formation due to one-step decays. On their assumption that the Kr^{3+} formation is mainly due to one-step decays, the branching ratio of the Kr^{3+} formation due to one-step decays was estimated to be 10%.

Hence, branching ratios of one- and two-step decay channels for the formation of Kr^{2+} and Kr^{3+} ions are estimated to be 51% : 32%. These facts mean that spectator Auger processes are dominant in the Kr $3d^{-1}5p$ resonance state. This finding is in disagreement with the resonant shakeoff model of Heimann et al.[32] They estimated the branching ratios of the resonant shakeoff (one-step) and two-step decays to be 37% : 16% from partitions of continuous and discrete structures in their resonant photoelectron spectrum. One reason for the disagreement is considered to arise from lower energy resolution in the resonant photoelectron spectrum of Heimann et al.[32] In their spectrum, many discrete lines due to the second step of the two-step decays are not resolved. Therefore, the two-step contributions are regarded as the continuous part of the spectrum. Consequently, the two-step contributions may be underestimated.

Multiple ionization from Xe $4d^{-1}6p$. Fractional ratios (%) for the ion yields and electron intensities from the $4d^{-1}6p$ resonance state of Xe are summarized in Table IV. Partial ion yields in columns (a) and (b) were measured by Hayaishi et al.[5] and Eland et al.,[15] respectively. Notice that the Xe^{2+} formation is prominent. Hayaishi et al.[5] proposed that the prominent formation is mainly due to two-step decays such as spectator Auger processes. On the contrary, Becker et al.[30] and Heimann et al.[32] reported that the formation is due to one-step decays such as resonant shakeoff.

TABLE IV
Fractional Ratios (%) for Ion Yield and Electron Intensities from Xe $4d^{-1}6p^a$

	(a)	(b)		(c)	
Xe^+	16	12	23	23	Two-step
				34	Two-step
Xe^{2+}	83	87	77		
				43	One-step
Xe^{3+}	< 1	< 1			One-step

^{a}Columns (a), (b), and (c) are data of Refs. 5, 15, and 31, respectively

Becker et al.[31] further studied the decays from the $4d^{-1}np$ states using resonant photoelectron spectroscopy. The fractional electron intensities from the $4d^{-1}6p$ state are indicated in column (c). On the basis of the structures in their resonant photoelectron spectrum, the contributions to the Xe^+ and Xe^{2+} formation were estimated to be 23 and 77%, respectively. Furthermore, the Xe^{2+} formation was separated into 34% for two-step contributions and 43% for resonant shakeoff contributions. Thus, the shakeoff decay is somewhat dominant in the Xe^{2+} formation from the resonance state $4d^{-1}6p$.

3.1.3. Summary and New Trends. In the first resonance states of Ar $2p^{-1}4s$, Kr $3d^{-1}5p$, and Xe $4d^{-1}6p$, two-step and one-step decays compete with each other in the formation of doubly charged ions. To be more exact, two-step decays are important in Ar and Kr, and one-step decays predominate somewhat in Xe.

In addition to these experimental studies, theoretical approaches for resonance decays also have been performed by some workers. Heimann et al.[32] calculated the shake probabilities for ejection (or excitation) of the Rydberg electron from the Ar $2p^{-1}ns$, nd, Kr $3d^{-1}np$, and Xe $4d^{-1}np$ resonance states. They reported that shakeoff is the main component in the shake probabilities and the shakeoff contribution increases with increasing n in the Rydberg series. On the contrary, Aksela et al.[38] found that shakeup and -down contributions increase in going to higher Rydberg orbitals in the Kr $3d^{-1}np$ decays. Furthermore, Aksela et al. showed the same n dependence in the decays of Ne $1s^{-1}np$, Ar $2p^{-1}ns$, Kr $3d^{-1}np$, and Xe $4d^{-1}np$. Thus, the n dependence for contributions of two- and one-step decays is not in agreement at the present time.

It is desirable to study the problems concerning the branching ratios of the decay channels to the charge states and the n dependence of the shakeup, -down, and -off probabilities in the Rydberg states in more detail. Recently, new experimental techniques have been applied to study the resonance decay. Electron–electron coincidence studies were performed by Okuyama et al.[39] and Raven et al.[40] Okuyama et al. estimated the branching ratios for the different Xe^{2+} states caused finally by the decays from the Xe $4d^{-1}nl$ resonance states. Raven et al. also identified Auger lines due to the first and second steps in two-step decays from the Ar $2p^{-1}4s$, Kr $3d^{-1}5p$, and Xe $4d^{-1}6p$ resonance states. However, the decay processes have not been defined fully because of low resolution and low detection efficiency. Nevertheless, there are high potentialities in the new techniques and it is necessary to extend these studies using the coincidence techniques.

Besides, in the extension of the present techniques, more accurate information on the resonance decays can be obtained by the following experimental conditions. The measurements of resonant photoelectrons must be performed with higher energy resolution for sufficient partitions of the spectral structures in the photoelectron spectra. The measurements of ion yields must pay full attention to the collection efficiencies for the different charge states. Both should be measured with higher spectral resolution for the explanation of the n dependence of shake probabilities in the Rydberg states. In the near future, ion–electron coincidence methods can be considered as new techniques: measurements of ions in coincidence with electrons having a specific kinetic energy, and measurements of photoelectrons in coincidence with ions of a specific charge state. The former gives information on the charge states produced from a specific decay channel, and the latter gives information on the photoelectron spectrum participating in the decays to a specific charge state. These new techniques will provide useful information on decay channels from resonance states.

3.2. Alkali and Alkali-Earth Metals

The production of atomic beams from alkali or alkali-earth metals (except Be) requires only moderate temperatures in the range of 400–1000 K (see Section 2.2). Therefore, these elements are favorite candidates for photoionization studies. For the alkali-earth metals there is the additional advantage of the simple atomic structure of closed subshells in the ground state.

3.2.1. The Region of the np Excitation/Ionization. The first systematic measurements for the production of singly and doubly charged photoions of Ca, Sr, and Ba around 20–35 eV were performed by Holland et al.[43] and Holland and Codling[44] using the synchrotron radiation of the Bonn 2.3-GeV synchrotron and a 1-m Seya monochromator with a bandwidth of 0.2 nm. Later this region was also studied with higher resolution via measurements of the X^+ and X^{2+} ion yield for Ca by Sato et al.[45]

and for Sr by Nagata et al.[46] using the synchrotron radiation of the 2.5-GeV storage ring of the Photon Factory in Tsukuba and a 1-m Seya monochromator with a bandwidth of 0.15 nm (Ca) and 0.08 nm (Sr), respectively. The region of 20–23 eV for the measurements of the Ba^+ and Ba^{2+} ion yield were performed by Lewandowski et al.[47] with the synchrotron radiation of the DORIS storage ring in Hamburg and a 3-m normal-incidence monochromator with a bandwidth of 0.09 nm and in a limited range around 21.4 eV with a bandwidth of 0.015 nm.

The range of 20–35 eV photon energy is above the limit of double ionization, i.e., the energy required to remove both valence electrons (Ca: 18.0 eV, Sr: 16.7 eV, Ba: 15.2 eV) and in the region of the excitation/ionization of the np electrons. The corresponding ionization limits are Ca $3p^54s^2\ {}^2P_{3/2,1/2}$: 34.3 and 34.7 eV, respectively; Sr $4p^54s^2\ {}^2P_{3/2,1/2}$: 28.2 and 29.2 eV; Ba $5p^56s^2\ {}^2P_{3/2,1/2}$: 22.7 and 24.7 eV.

There are essentially three regions of interest:

1. The region above the double ionization limit, but below the excitation of the np electrons. In this region only the *direct* double ionization process is energetically possible:

$$X\ np^6(n+1)s^2 \to X^{2+}\ np^6 + e_1 + e_2$$

By the detection of X^{2+} one can study this interesting process in which both electrons e_1 and e_2 share the excess energy forming a continuum in the photoelectron spectrum. Therefore, it is difficult to deduce the cross section of direct double photoionization from photoelectron spectra because one has to distinguish these electrons from the background.

2. The region above the ionization limit of the np electrons. Here a large increase in the production of doubly charged photoions is expected because now the X^{2+} ions can be produced in a two-step process in which the ejection of the np photoelectron is followed by the Auger decay

$$X\ np^6(n+1)s^2 \to X^+\ np^5(n+1)s^2 + e_1 \to X^{2+}\ np^6 + e_1 + e_2$$

3. The region of the excitation of the np electrons. This region is characterized by strong *resonances* in the photoabsorption spectra. In the case of Ca for example, the absorption spectrum[48] is dominated by the peak of the transition Ca $3p^64s^2\ {}^1S_0 \to$ Ca $3p^54s^23d\ {}^1P_1$ at 31.4 eV. The reason for such a strong transition is the overlap of the $3p$ with the $3d$ orbital. This $3p$–$3d$ overlap also controls the autoionization decay of this resonance by the $3d \to 3p$ recombination with the ejection of a valence electron like

$$\text{Ca } 3p^54s^23d \to \text{Ca}^+\ 3p^64s + e_1$$

giving rise to Ca^+ photoions. But the large perturbation of the $3p$ core-hole on the $3d$ orbital which lowers its energy relative to the $4s$ orbital is responsible for the possibility of a decay by a two-step process[49,50] like

$$\text{Ca } 3p^54s^23d \to \text{Ca}^+\ 3p^53d4s + e_1 \to \text{Ca}^{2+}\ 3p^6 + e_1 + e_2$$

into Ca^{2+} photoions.

Therefore, it is quite interesting to investigate these resonances by photoion spectroscopy especially in combination with the results of the corresponding photoelectron spectra. See for example the study of the decay channels of the $4p$ resonances in Sr by Yagishita et al.[51]

Photoion yield measurements in this energy region for the alkali metals were performed for Rb between 18 and 30 eV by Itoh et al.[52] and for Cs between 16 and 19 eV by Yoshino et al.[53] The energy range in both experiments were lower than the corresponding limits of double ionization (Rb: 31.5 eV, Cs: 29.1 eV). Therefore, only the photoion yield of Rb^+ and Cs^+, respectively, were measured.

In both papers a rich structure of resonances, due to Rydberg series converging to the limits Rb $4p^55s$ and $4p^54d$ and Cs $5p^56s$, respectively, was reported.

3.2.2. The Region of the nd Excitation/Ionization. The extension of the photon energy into the soft x-ray region makes it possible that electrons of deeper inner-shells can be excited or ejected. Especially the inner d shells have been the focus of photoionization studies. The corresponding ionization limits are:

$$3d^9\,{}^2D_{5/2,3/2}: 117.3 \text{ eV resp. } 118.9 \text{ eV (Rb)}$$

$$142.2 \text{ eV resp. } 144.1 \text{ eV (Sr)}$$

$$4d^9\,{}^2D_{5/2,3/2}: 82.9 \text{ eV resp. } 85.1 \text{ eV (Cs)}$$

$$98.3 \text{ eV resp. } 101.0 \text{ eV (Ba)}$$

Therefore, the photoion yield measurements of Koizumi et al.[54,55] for Rb and Sr and of Nagata et al.[56] for Cs and Ba using the 2.5-GeV storage ring in Tsukuba and a 2-m Grasshopper monochromator with a bandwidth of 0.04–0.02 nm are located in this region.

The creation of a d hole by the excitation or ionization of such an inner d-shell electron forces the atom to rearrange its residual electrons usually by the emission of one or several electrons. Therefore, in most cases a multiply charged photoion is produced. If for example in Sr a $3d$ electron is removed

$$\text{Sr } 3d^{10}4s^24p^65s^2 \rightarrow \text{Sr}^+ \, 3d^94s^24p^65s^2 + e_1$$

the filling of the $3d$ hole can take place under the participation of the $4s$, $4p$, and $5s$ electrons in many different ways like

$$\text{Sr}^+ \, 3d^94s^24p^65s^2 \rightarrow \text{Sr}^{2+} \, 3d^{10}4s^24p^55s + e_2 \qquad (a)$$

$$4s^24p^45s^2 + e_2 \qquad (b)$$

$$4s4p^55s^2 + e_2 \qquad (c)$$

Energetically the different routes can lead to differently charged photoions; for example, route (a) leads to Sr^{2+} as the final product of the photoionization process, whereas routes (b) and (c) open the way for the production of Sr^{3+} by the following Auger decay:

$$\text{Sr}^{2+} \, 4s^24p^45s^2 \rightarrow \text{Sr}^{3+} \, 4s^24p^5 + e_3 \qquad (d)$$

The different decay routes to Sr^{2+} and Sr^{3+} are schematically depicted in Figure 11.

The ratio of $\text{Sr}^{2+}/\text{Sr}^{3+}$ in these multistep processes depends on the branching ratios of the corresponding Auger decays which may be obtained from the Auger electron spectra. The analysis of the ion yield spectra, however, reveals that the ionization and decay process due to strong electron correlations can also take place in a more complicated way. The ionization of the nd electron, for example, can be accompanied by a simultaneous ionization of another electron (shakeoff process) such as

$$\text{Sr } 3d^{10}4s^24p^65s^2 \rightarrow \text{Sr}^{2+} \, 3d^94s^24p^65s + e_1 + e_2$$

with the subsequent Auger decay of the $3d$ hole state of Sr^{2+} into Sr^{3+} or Sr^{4+}.

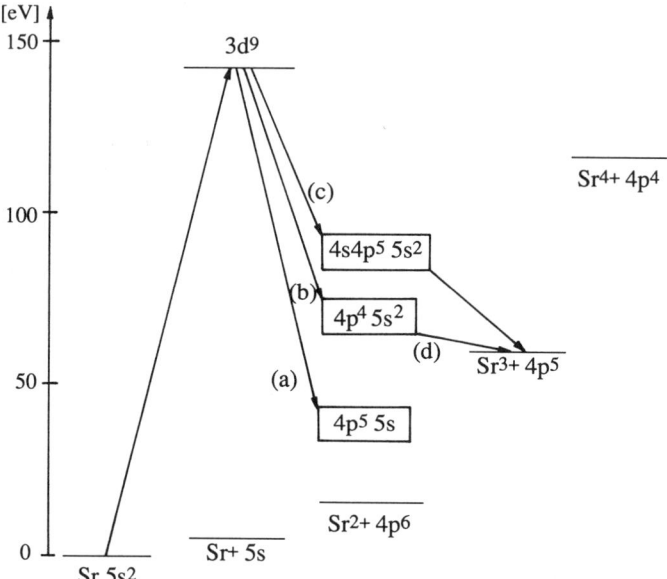

FIGURE 11. Simplified energy-level diagram for Sr showing some decay routes to Sr^{2+} or Sr^{3+} after ejection of a 3d electron.

Similar to the double ionization of the valence electrons the excess energy is shared by the two electrons e_1 and e_2 forming a continuum in the electron spectrum which is difficult to distinguish from the background. Another "one-step double ionization" process is the Auger shakeoff decay when simultaneously two electrons are emitted.

Below the nd ionization limits the excitation leads to resonance states X $nd^9n'l'$. Similar to the analysis of the decay of the resonances for the noble gases (see section 3.1.2), generally one may distinguish two cases:

1. The interaction of the excited electron with the residual electrons is large. Then the excited electron participates in the decay process. A typical decay may be the filling of the nd hole by the participation of the excited electron and a valence electron such as

$$Sr\ 3d^9 4s^2 4p^6 5s^2 nl \rightarrow Sr^+\ 3d^{10} 4s^2 4p^6 5s + e_1$$

In this case the final product of the photoionization process is a singly charged photoion Sr^+. The investigations of Rb, Sr, Cs, and Ba, however, show that this process for the nd resonances is very weak.

2. The interaction of the excited electron with the residual electrons is small. In this case the excited electron acts as a "spectator" in the subsequent decay process of the X $ns^9n'l'$ resonances and the similarity to the "normal" Auger decay of the $X^+\ nd^9$ states should lead to multiply charged photoions.

3.3. 3d Transition Metals (Iron Group)

3.3.1. The Region of the 3p Excitation/Ionization. The elements of the iron group ($Z = 21$–28) are characterized by the successive filling of the $3d$ subshell. These $3d$ electrons have a strong influence on the VUV absorption spectra in the region of the $3p$ excitation/ionization around 30–80 eV. The reason is the large overlap of the $3p$ and $3d$ orbitals resulting in strong transitions to resonance states

$$X\,3p^63d^n \to X\,3p^53d^{n+1}$$

The localization of the 3d electrons in the vicinity of the 3p electrons also produces a strong coupling in the decay processes of these resonances which, therefore, preferably decay via autoionization

$$X\,3p^53d^{n+1} \to X^+\,3p^63d^{n-1} + e_1$$

The interference of the discrete transitions with the direct 3d ionization

$$X\,3p^63d^n \to X^+\,3p^63d^{n-1} + e_1$$

gives rise to asymmetric Beutler–Fano profiles of the resonances.

The 3p resonances of the iron group elements were very intensively investigated in photoabsorption and photoelectron spectroscopy by Sonntag and co-workers (e.g., see Meyer et al.[57] and references therein) using the synchrotron radiation of the storage ring DORIS in Hamburg and BESSY in Berlin and toroidal grating monochromators. Figure 12[57] shows the absorption spectra of Sc, Ti, Cr, Mn, Fe, Co, and Ni in the region of the 3p excitation/ionization. The vertical bars mark the position of the 3p ionization limits split by the multiplet structure. One realizes that the resonances which, due to the multiplet splitting, may also have several absorption maxima gradually change from Sc to Ni. For the lighter elements the resonances are in the middle of the 3p ionization limits and exhibit rather

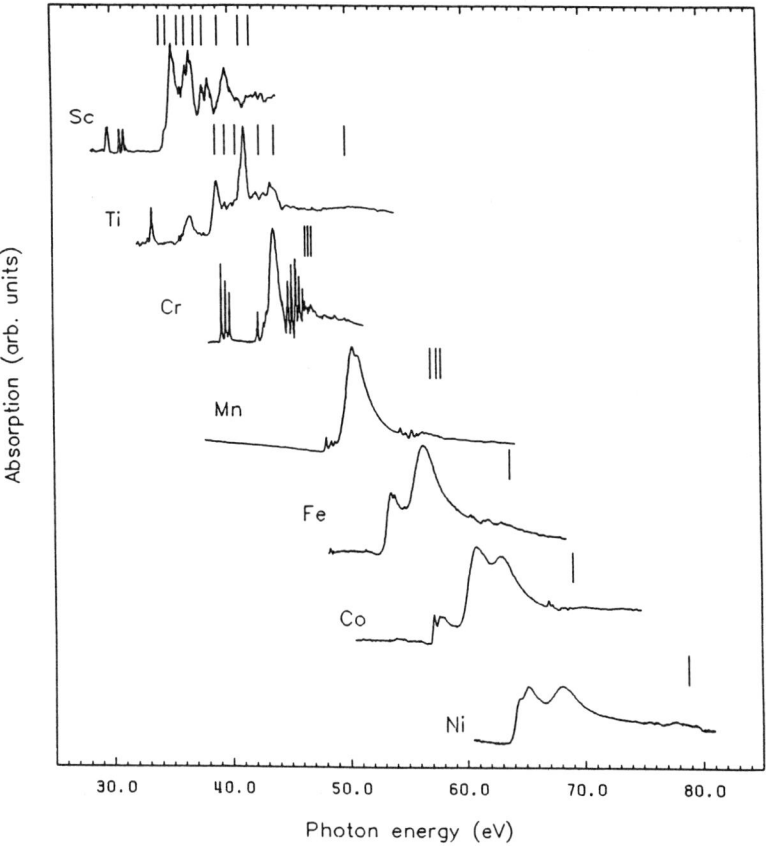

FIGURE 12. Photoabsorption spectra of atomic 3d transition metals. The spectra have not been normalized to one another. The vertical bars mark the energy positions of the 3p ionization thresholds.[57]

symmetric and sharp lines, but with increasing nuclear charge the resonances are shifted well below the 3p ionization limits and their profiles get broader and more asymmetric. These trends may be explained by the successive filling of the 3d subshell with an increasing amplitude of the direct 3d ionization which results in an increasing asymmetry of the Beutler–Fano profiles. The same arguments hold for the coupling between the 3d and 3p electrons which results in increasing linewidths of the resonances.

A more careful analysis of the spectra, however, reveals that there are also other processes which have a strong influence on the photoionization process. In the case of Sc and Ti, for example, one observes that the resonances are on top of a rather broad and large background. As the resonances here are mainly situated above the lowest 3p ionization limit, this background can be attributed to the 3p ionization

$$X\ 3p^6 3d^n \rightarrow X^+\ 3p^5 3d^n + e_1$$

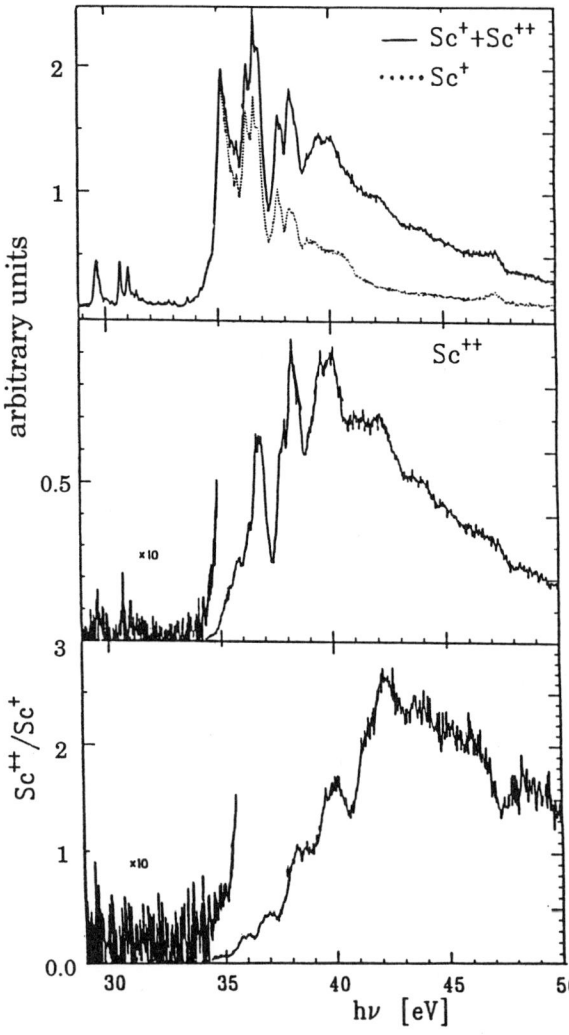

FIGURE 13. Photoion yield spectra of Sc^+ and Sc^{2+} in the region of the 3p excitation/ionization.

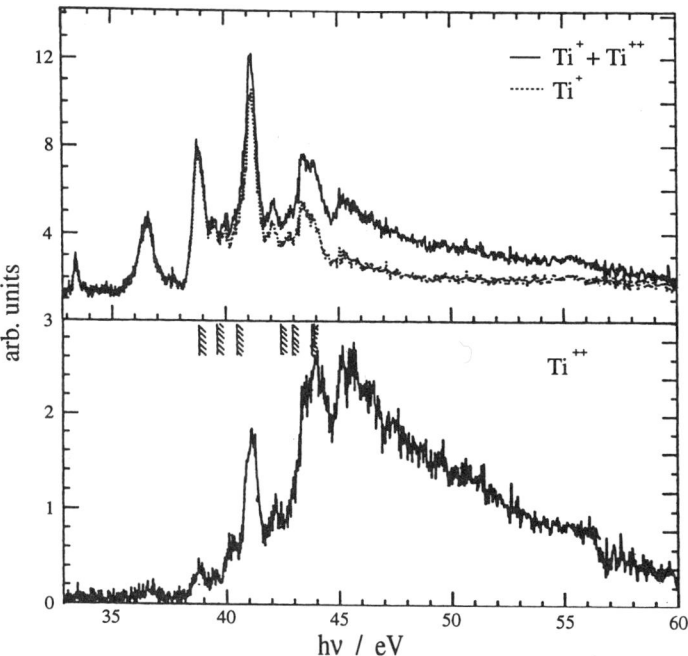

FIGURE 14. Photoion yield spectra of Ti^+ and Ti^{2+} in the region of the 3p excitation/ionization.

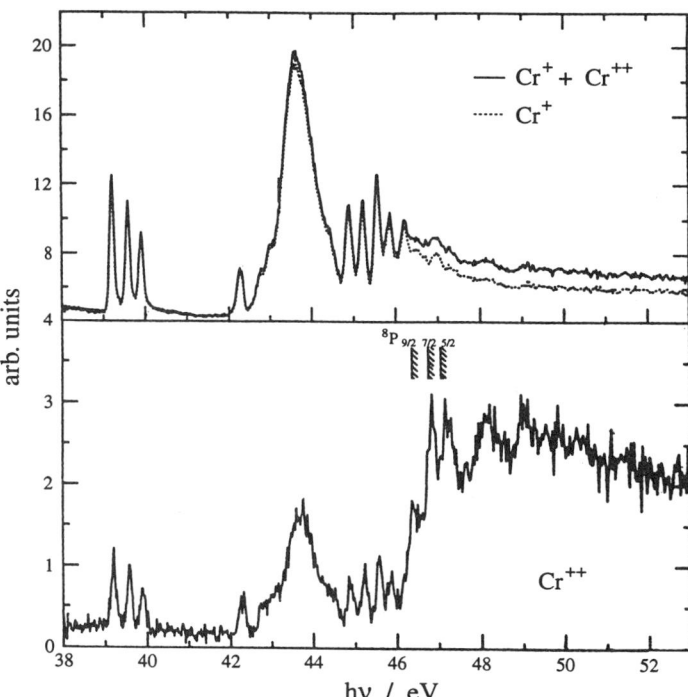

FIGURE 15. Photoion yield spectra of Cr^+ and Cr^{2+} in the region of the 3p excitation/ionization.

This interpretation is corroborated by the results of the photoion yield spectra. The photoion spectra were obtained by the group of one of the authors (P.Z.) using the synchrotron radiation of the storage ring BESSY in Berlin and TOF technique for the detection of the singly and doubly charged photoions. Figures 13 and 14 show the yield spectra of Sc^+ and Sc^{2+} and Ti^+ and Ti^{2+} in the region of the $3p$ excitation/ionization. Comparing the signals with the photoabsorption curves in Figure 12 one can see that the singly charged photoions essentially reproduce the $3p$ resonances which decay by autoionization to X^+, whereas the yield of the doubly charged photoions reflects the broad background of the $3p$ ionization because this process together with the subsequent Auger decay of the $3p$ hole produces X^{2+} ions such as

$$X\, 3p^6 3d^n \to X^+\, 3p^5 3d^n + e_1 \to X^{2+}\, 3p^6 3d^{n-2} + e_1 + e_2$$

But obviously also the resonances can decay to doubly charged photoions. Although this is already noticeable in Figures 13 and 14 for certain resonances, it is more convincingly demonstrated by the

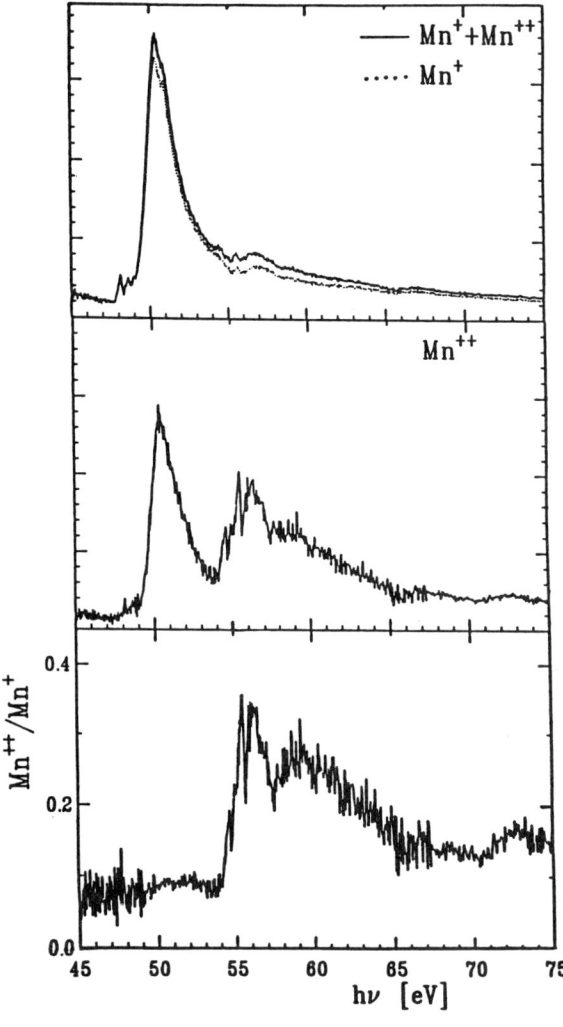

FIGURE 16. Photoion yield spectra of Mn^+ and Mn^{2+} in the region of the $3p$ excitation/ionization.

photoion yields for Cr^+, Cr^{2+} and Mn^+, Mn^{2+} in Figures 15 and 16. Clearly the resonances in the X^+ signals are reproduced by the X^{2+} signals. This decay mode of the $3p$ resonances into X^{2+} may be explained by the increasing binding energies of the $3d$ orbitals relative to the $4s$ orbitals. In Sc (like Ca) the $3d$ and $4s$ orbitals are nearly degenerate; the perturbation by the $3p$ hole with the contraction of the $3d$ orbital can lower $3d^24s$ relative to $3d4s^2$ and enables the autoionization of Sc $3p^53d^24s^2 \rightarrow Sc^+ 3p^53d^24s + e_1$ with the subsequent Auger decay $Sc^+ 3p^53d4s \rightarrow Sc^{2+} 3p^6(3d, 4s)^1 + e_2$. For increasing nuclear charge the binding energy of the $3d$ orbitals increases relative to the $4s$ orbitals and enables decays such as $Mn^+ 3p^63d^44snl \rightarrow Mn^{2+} 3p^63d^5 + e_2$. For a quantitative analysis of the X^{2+} photoions, shakeoff processes with the simultaneous emission of two electrons must also be taken into account.

Cr plays a unique role in the iron group because of its unfilled $4s$ subshell in the ground state Cr $3d^54s$ ^7S. Here, the $3p$ excitation can also be realized by strong $3p \rightarrow 4s$ transitions. Those transitions like Cr $3p^63d^54s$ $^7S \rightarrow$ Cr $3p^53d^54s^2$ ^7P are responsible for the three sharp lines at 39–40 eV. For Cr there is also a well-developed set of Rydberg series Cr $3p^63d^54s \rightarrow$ Cr $3p^53d^54snd$ (see Figure 12 or 15 at 45–47 eV) which may be explained by strong mixing with a nearby Cr $3p^53d^64s$ ^7P resonance.[58]

3.4. 4f Elements (Rare Earths)

3.4.1. The Region of the 5p Excitation/Ionization.

The first absorption spectra of the rare earths in the region of the $5p$ excitation/ionization between 20 and 40 eV were measured by Tracy[59] who investigated the heavier elements Sm, Eu, Dy, Ho, Er, Tm, and Yb with the synchrotron radiation of the Bonn 0.5-GeV synchrotron. He found strong resonances which for the special case of Yb $5p^64f^{14}6s^2$ with its closed $4f$ subshell could be attributed to discrete transitions of four series $5p \rightarrow nd$ converging to the $5p$ ionization limits $5p^5$ $^2P_{3/2}$ at 31.35 eV and $^2P_{1/2}$ at 37.52 eV. Photoion yield measurements in the same energy range were carried out at the Bonn electron synchrotron by Holland et al.[43] for Yb and by Holland and Codling[60] for Sm, Eu, and Tm. A systematic study of all elements of the rare earths (except Pm) was performed by Dzionk et al.[61,62] using the synchrotron radiation facility BESSY in Berlin.

Figure 17[61,62] shows the photoion yield of singly and doubly charged photoions for every second member of the rare earths group: Ce, Nd, Eu, Tb, Ho, and Tm. Comparing the spectra of the different elements, one sees distinct differences in the relation of the X^+ and X^{2+} signals which change gradually between the lighter and heavier elements: Whereas for the lighter elements the X^+ signals show rather sharp resonances with practically no correlations to the corresponding X^{2+} signals, for growing nuclear charge there is the increasing tendency that the resonances in the X^+ signals are reproduced by the X^{2+} signals.

Although the term-splitting of the unfilled $4f$ subshell and the broadening of the lines by autoionization prevent a detailed analysis of the spectra at the present time, the gross features can be explained by concentrating on the $5p$ electrons. For these electrons one can distinguish between discrete and continuum transitions like $5p \rightarrow nd$ and $5p \rightarrow \varepsilon d$, respectively. The continuum transitions

$$X\ 5p^6 \rightarrow X^+\ 5p^5 + e_1$$

with the subsequent Auger decay of the $5p$ hole produce doubly charged photoions and can therefore be studied by the X^{2+} signals. The X^{2+} signals of the lighter elements like Ce show the typical features of a continuum transition with a delayed onset[63] due to the centrifugal effects of the εd wave function. The rather sharp resonances, however, which are observed in the X^+ signals and also in the corresponding X^{2+} signals of the heavier elements, consequently must be attributed to discrete transitions such as

$$X\ 5p^64f^n6s^2 \rightarrow X\ 5p^54f^n6s^25d$$

The autoionization decay of these resonances essentially may proceed in two different ways where in the next step the $5p$ hole is either filled or remains unfilled. If we denote all electrons of the subshells $4f$, $6s$, nd ($n \geq 5$) by V^x (also to cover strong mixing between these subshells caused by the perturbation

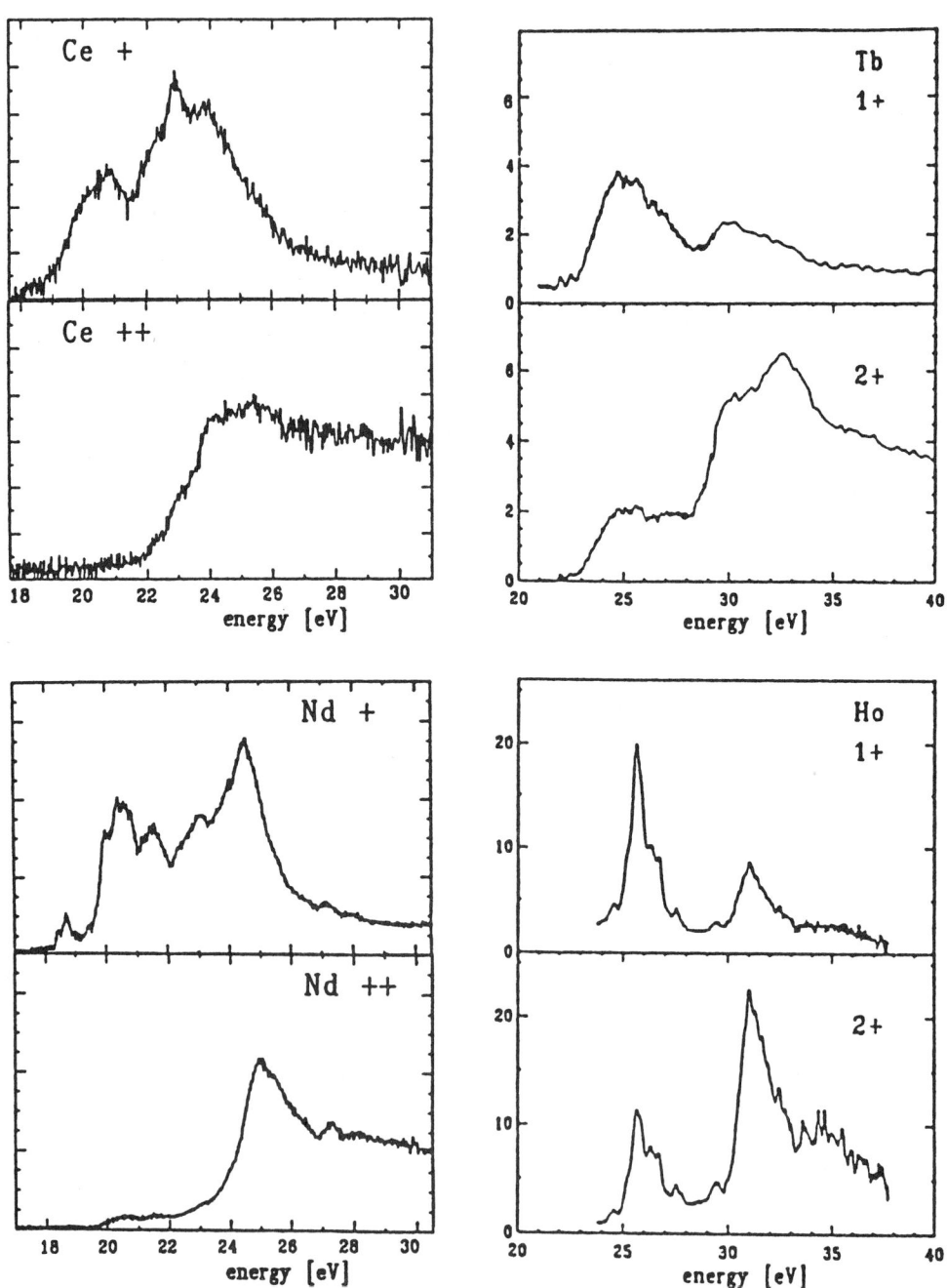

FIGURE 17. Photoion yield spectra of singly and doubly charged ions for Ce, Tb, Nd, and Ho (above) and Ev and Tm (facing page) in the region of the 5p excitation/ionization.[61,62]

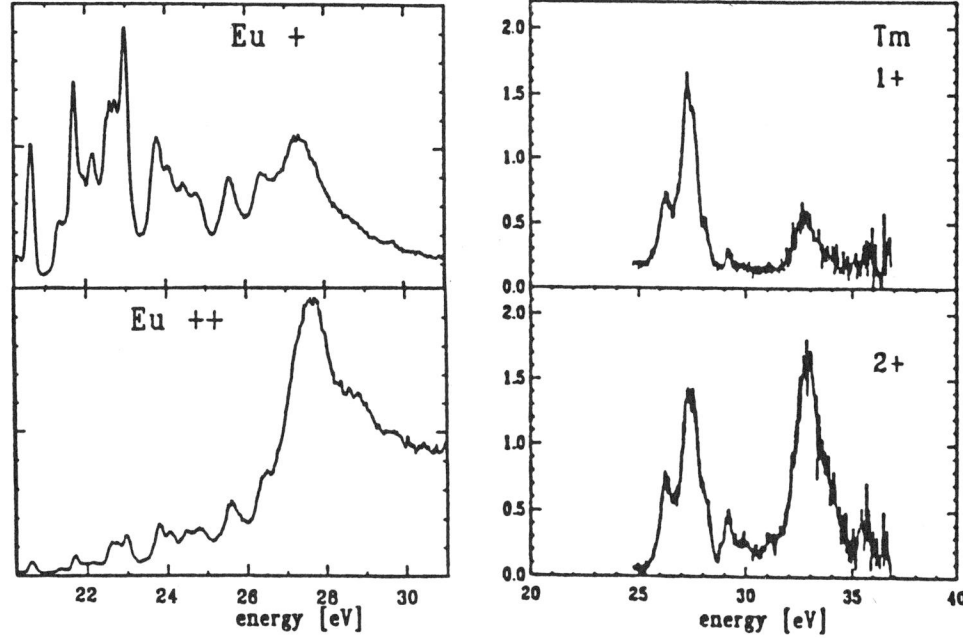

FIGURE 17. (*Continued*)

of the 5p hole and experimentally by the occupation of different configurations due to the high temperature for the production of the atomic beam), the excitation can be written as

$$X\ 5p^6 V^x \to X\ 5p^5 V^{x+1}$$

and the different ways of the decay as

$$X\ 5p^5 V^{x+1} \to X^+\ 5p^6 V^{x+1} + e_1 \tag{1}$$

or

$$X\ 5p^5 V^{x+1} \to X^+\ 5p^5 V^x + e_1 \to X^{2+}\ 5p^6 V^{x-2} + e_1 + e_2 \tag{2}$$

As the process (1) preferentially leads to singly charged photoions and the process (2) with the subsequent filling of the 5p hole to doubly charged photoions, one may discuss (1) with respect to the X^+ signals and (2) with respect to the X^{2+} signals. The increasing tendency of the heavier elements to reproduce the X^+ resonances by X^{2+} then can be attributed to the growing influence of processes of type (2) which may be seen as the "resonant" 5p ionization. For a quantitative analysis, other processes like the direct ionization of the valence electrons in connection with the X^+ signals or shakeoff processes in connection with the X^{2+} signals must also be considered.

3.4.2. The Region of the 4d Excitation/Ionization. Since the first solid-state measurements of Zimkina *et al.*[64] the region of the 4d excitation/ionization between 100 and 200 eV has attracted the interest of many experimental and theoretical studies.[65] The reason is the existence of strong resonances which cannot be explained by a simple one-electron picture. Figure 18[66] shows the absorption spectra of Ba and eight elements of the rare earths in this region. One can see that for every element the region is dominated by a broad asymmetric resonance (the "giant resonance") with a striking similarity between the different elements. The term "giant resonance" was adopted from

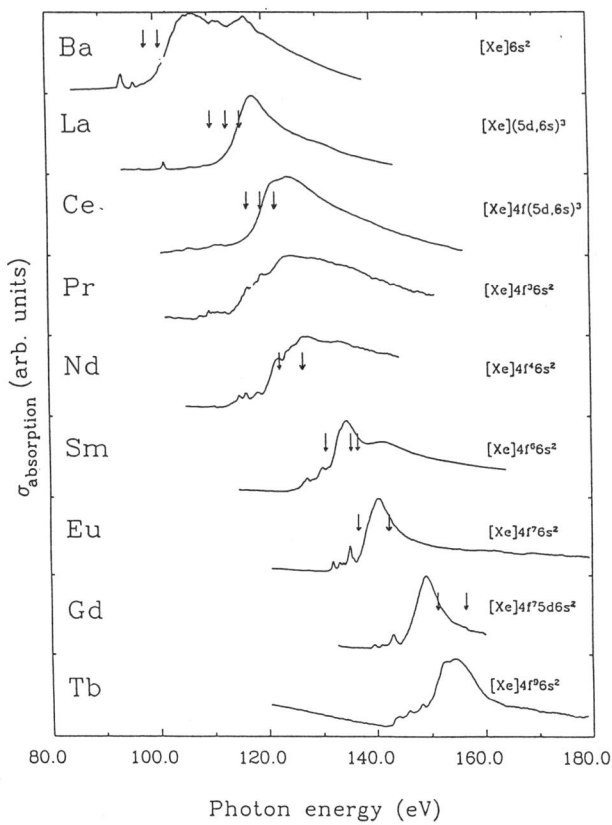

FIGURE 18. Photoabsorption spectra of Ba and eight elements of the rare earths in the region of the 4d excitation/ionization.

nuclear physics where in the photoneutron cross section of the heavier elements the giant electric dipole resonances at about 14–18 MeV with widths of some megaelectron volts are observed. These giant E1 resonances in the nucleus can be explained by a model in which the protons move collectively with a damped harmonic motion relative to the neutrons like two interpenetrating, incompressible fluids. Transferring the main ideas of a many-body motion to the atom, one may think of a full electronic subshell moving collectively under the influence of the photon field.

For a more detailed discussion of these resonances, however, it is necessary to realize that apart from the similarities there are also distinct differences between the resonances of these elements. A closer look at Figure 18 reveals that for light elements the resonance is above the 4d ionization limits (indicated by the vertical arrows), whereas for increasing nuclear charge the position of the resonance is shifted in the direction of the ionization limits, passes them, and for heavier elements lies below these limits. Therefore, the analysis of the resonances must also concentrate on this competition between continuum transitions $4d \rightarrow \varepsilon f$ and discrete transitions $4d \rightarrow 4f$.

A decisive part in this competition is played by the 4f electrons which move in an effective double-well potential[67] caused by a delicate balance of attractive coulombic forces and repulsive centrifugal forces. Changes in the potential by the increase of nuclear charge, by ionization, inner-shell excitation, or term-dependent effects can force the 4f wave function to collapse from its Rydberg-like orbital ($r > 10$ a.u.) in the outer well into a corelike orbital ($r > 1$ a.u.) in the inner well. If the 4f electron is localized in the outer well, the overlap between 4d and 4f is poor and only very weak discrete $4d \rightarrow 4f$ transitions are expected. For an increasing number of 4f electrons with corelike orbitals, however, strong discrete transitions may occur.

In the absorption spectrum the common distinction between broad resonances of continuum transitions and sharp resonances of discrete transitions can be obscured by the numerous term-splitting of an open 4f-subshell configuration. But these transitions show distinct differences in the following decay processes. In the case of the continuum transitions $4d \to \varepsilon f$

$$X\, 4d^{10} 4f^n 5s^2 5p^6 6s^2 \to X^+ \, 4d^9 4f^n 5s^2 5p^6 6s^2 + e_1$$

the subsequent filling of the 4d hole with electrons of the 5s and 5p subshells and the following filling of the corresponding 5s and 5p holes create additional Auger electrons so that multiply charged photoions (predominantly X^{3+} and X^{4+}) are produced. In the case of discrete transitions $4d \to 4f$

$$X\, 4d^{10} 4f^n 5s^2 5p^6 6s^2 \to X\, 4d^9 4f^{n+1} 5s^2 5p^6 6s^2$$

the autoionization decay is dominated by the $4f \to 4d$ recombination with the emission of a 4f electron

$$X\, 4d^9 4f^{n+1} 5s^2 5p^6 6s^2 \to X^+ \, 4d^{10} 4f^{n-1} 5s^2 5p^6 6s^2 + e_1$$

Therefore, the final products of discrete transitions will predominantly be singly charged photoions.

Photoion spectroscopy obviously is a very appropriate method to study the different decay processes. Photoion yield measurements of the rare earths in the region of the 4d excitation/ionization were reported by Dzionk et al.[68] Zimmermann,[69,70] and Nagata et al.[71] Figures 19 and 20[69] show as representative examples of light and heavier elements the photoion yield spectra of Ce and Dy. Comparing both spectra one can see that for Ce the dominant signal is the contribution of Ce^{3+}, whereas for Dy the yield spectra are dominated by Dy^+. Also the profiles of both signals differ widely. Whereas the Ce^{3+} signal shows the typical profile of a continuum resonance with the characteristic feature of a delayed onset due to centrifugal effects of the εf wave function, the Dy^+ signal has the typical asymmetric form of a Beutler–Fano profile, which can be explained by the interference of the amplitude

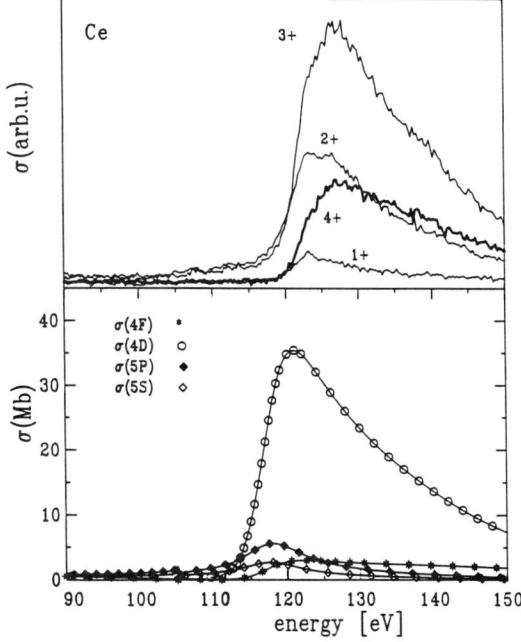

FIGURE 19. (Upper) Photoion yield spectra of Ce^{n+} ($n = 1$–4) in the region of the 4d excitation/ionization. (Lower) Calculated atomic partial photoionization cross sections $\sigma(nl)$ for the subshells 4d, 5p, 5s, and 4f using TDLDA.[72]

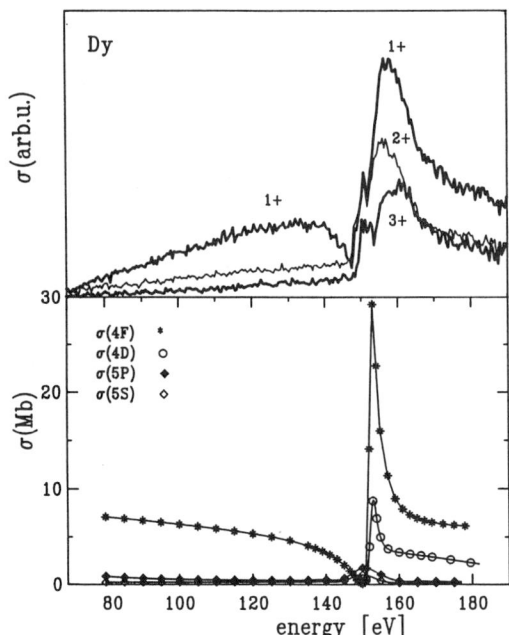

FIGURE 20. (Upper) Photoion yield spectra of Dy^{n+} ($n = 1-3$) in the region of the 4d excitation/ionization. (Lower) Calculated atomic partial photoionization cross sections $\sigma(nl)$ for the subshells 4d, 5p, 5s, and 4f using TDLDA.[72]

of the discrete transition X $4d^{10}4f^n \to X\ 4d^9 4f^{n+1} \to X^+\ 4d^{10}4f^{n-1} + e_1$ with the amplitude of the direct 4f ionization X $4d^{10}4f^n \to X\ 4d^{10}4f^{n-1} + e_1$.

For a qualitative interpretation of the spectra the atomic partial photoionization cross section $\sigma(nl)$ for the different subshells 4d, 5s, 5p, and 4f may be used. Figures 19 and 20 give in the lower panel these cross sections which were obtained with a relativistic version[72] of the time-dependent local density approximation. The cross sections reproduce the main trends of the dominance of the 4d ionization in Ce and of the 4f ionization with a Beutler–Fano profile in Dy. These theoretical results for the cross sections $\sigma(nl)$ are corroborated by the intensities and profiles of the corresponding photoelectron spectra with the dominance of the 4d line for the light elements and of the 4f line in the heavier elements.[73]

REFERENCES

1. J. A. R. SAMSON, *Techniques of Vacuum Ultraviolet Spectroscopy* (Wiley, New York, 1967), p. 2.
2. J. B. FARMER, in *Mass Spectrometry*, edited by C. A. MCDOWELL (McGraw–Hill, New York, 1963), pp. 7–44.
3. A. E. CAMERON AND D. F. EGGERS, JR., *Rev. Sci. Instrum.* **19**, 605 (1948).
4. W. C. WILEY AND I. H. MCLAREN, *Rev. Sci. Instrum.* **26**, 1150 (1955).
5. T. HAYAISHI, Y. MORIOKA, Y. KAGEYAMA, M. WATANABE, I. H. SUZUKI, A. MIKUNI, G. ISOYAMA, S. ASAOKA, AND M. NAKAMURA, *J. Phys. B* **17**, 3511 (1984).
6. E. MURAKAMI, T. HAYAISHI, A. YAGISHITA, AND Y. MORIOKA, *Phys. Scr.* **41**, 468 (1990).
7. U. BECKER, D. SZOSTAK, H. G. KERKHOFF, M. KUPSCH, B. LANGER, R. WEHLITZ, A. YAGISHITA, AND T. HAYAISHI, *Phys. Rev. A* **39**, 3902 (1989).
8. P. LABLANQUIE AND P. MORIN, *J. Phys. B* **24**, 4349 (1991).
9. D. M. P. HOLLAND, K. CODLING, J. B. WEST, AND G. V. MARR, *J. Phys. B* **12**, 2465 (1979).
10. M. J. VAN DER WIEL, T. M. EL-SHERBINI, AND L. VRIENS, *Physica* **42**, 411 (1969).
11. R. P. MADDEN AND K. CODLING, *Phys. Rev. Lett.* **10**, 516 (1963).
12. U. FANO AND J. W. COOPER, *Rev. Mod. Phys.* **40**, 441 (1968).

13. M. J. VAN DER WIEL AND G. WIEBES, *Physica* **53**, 225 (1971).
14. T. M. EL-SHERBINI AND M. J. VAN DER WIEL, *Physica* **62**, 119 (1972).
15. J. H. D. ELAND, F. S. WORT, P. LABLANQUIE, AND I. NENNER, *Z. Phys. D* **4**, 31 (1986).
16. M. J. VAN DER WIEL, G. R. WIGHT, AND R. R. TOL, *J. Phys. B* **9**, L5 (1976).
17. W. EBERHARDT, S. BERNSTORFF, H. W. JOCHIMS, S. B. WHITFIELD, AND B. CRASEMANN, *Phys. Rev. A* **38**, 3808 (1988).
18. J. TULKKI, T. ÅBERG, S. B. WHITFIELD, AND B. CRASEMANN, *Phys. Rev. A* **41**, 181 (1990).
19. G. V. MARR AND J. B. WEST, *At. Data Nucl. Data Tables* **18**, 497 (1976).
20. J. B. WEST AND J. MORTON, *At. Data Nucl. Data Tables* **22**, 103 (1978).
21. W. EBERHARDT, G. KAIKOFFEN, AND C. KUNZ, *Phys. Rev. Lett.* **41**, 156 (1978).
22. S. SOUTHWORTH, U. BECKER, C. M. TRUESDALE, P. H. KOBRIN, D. W. LINDLE, S. OWAKI, AND D. A. SHIRLEY, *Phys. Rev. A* **28**, 261 (1983).
23. H. AKSELA, S. AKSELA, H. PULKKINEN, G. M. BANCROFT, AND K. H. TAN, *Phys. Rev. A* **37**, 1798 (1988).
24. M. MEYER, E. V. RAVEN, M. RICHTER, B. SONNTAG, AND J. E. HANSEN, *J. Electron Spectrosc. Relat. Phenom.* **51**, 407 (1990).
25. M. MEYER, E. V. RAVEN, B. SONNTAG, AND J. E. HANSEN, *Phys. Rev. A* **43**, 177 (1991).
26. H. AKSELA, S. AKSELA, H. PULKKINEN, G. M. BANCROFT, AND K. H. TAN, *Phys. Rev. A* **33**, 3876 (1986).
27. D. W. LINDLE, P. A. HEIMANN, T. A. FERRETT, M. N. PIANCASTELLI, AND D. A. SHIRLEY, *Phys. Rev. A* **35**, 4605 (1987).
28. H. AKSELA, S. AKSELA, H. PULKKINEN, AND A. YAGISHITA, *Phys. Rev. A* **40**, 6275 (1989).
29. H. AKSELA, S. AKSELA, G. M. BANCROFT, K. H. TAN, AND H. PULKKINEN, *Phys. Rev. A* **33**, 3867 (1986).
30. U. BECKER, T. PRESCHER, E. SCHMIDT, B. SONNTAG, AND H.-E. WETZEL, *Phys. Rev. A* **33**, 3891 (1986).
31. U. BECKER, D. SZOSTAK, M. KUPSCH, H. G. KERKHOFF, B. LANGER, AND R. WEHLITZ, *J. Phys. B* **22**, 749 (1989).
32. P. A. HEIMANN, D. W. LINDLE, T. A. FERRETT, S. H. LIU, L. J. MEDHURST, M. N. PIANCASTELLI. D. A. SHIRLEY, U. BECKER, H. G. KERKHOFF, B. LANGER, D. SZOSTAK, AND R. WEHLITZ, *J. Phys. B* **20**, 5005 (1987).
33. G. H. WANNIER, *Phys. Rev.* **90**, 817 (1953).
34. U. FANO AND J. W. COOPER, *Phys. Rev.* **137**, A1364 (1965).
35. T. HAYAISHI, E. MURAKAMI, A. YAGISHITA, F. KOIKE, Y. MORIOKA, AND J. E. HANSEN, *J. Phys. B* **21**, 3203 (1988).
36. T. HAYAISHI, A. YAGISHITA, E. MURAKAMI, E. SHIGEMASA, Y. MORIOKA, AND T. SASAKI, *J. Phys. B* **23**, 1633 (1990).
37. T. HAYAISHI, A. YAGISHITA, E. SHIGEMASA, E. MURAKAMI, AND Y. MORIOKA, *J. Phys. B* **23**, 4431 (1990).
38. H. AKSELA, S. AKSELA, H. PULKKINEN, A. KIVIMAKI, AND O. P. SAIRANEN, *Phys. Scr.* **41**, 425 (1990).
39. K. OKUYAMA, J. H. D. ELAND, AND K. KIMURA, *Phys. Rev. A* **41**, 4930 (1990).
40. E. V. RAVEN, M. MEYER, M. PAHLER, AND B. SONNTAG, *J. Electron Spectrosc. Relat. Phenom.* **52**, 677 (1990).
41. T. PRESCHER, M. RICHTER, B. SONNTAG, AND H. E. WETZEL, *Nucl. Instrum. Methods A* **254**, 627 (1987).
42. S. BÜTTGENBACH, G. MEISEL, S. PENSELIN, AND K. H. SCHNEIDER, *Z. Physik* **230**, 329 (1970).
43. D. M. P. HOLLAND, K. CODLING, AND R. N. CHAMBERLAIN, *J. Phys. B* **14**, 839 (1981).
44. D. M. P. HOLLAND AND K. CODLING, *J. Phys. B* **14**, 2345 (1981).
45. Y. SATO, T. HAYAISHI, Y. ITIKAWA, Y. ITOH, J. MURAKAMI, T. NAGATA, T. SASAKI, B. SONNTAG, A. YAGISHITA, AND M. YOSHINO, *J. Phys. B* **18**, 225 (1985).
46. T. NAGATA, J. B. WEST, T. HAYAISHI, Y. ITIKAWA, Y. ITOH, T. KOIZUMI, J. MURAKAMI, Y. SATO, H. SHIBATA, A. YAGISHITA, AND M. YOSHINO, *J. Phys. B* **19**, 1281 (1986).
47. B. LEWANDOWSKI, J. GANZ, H. HOTOP, AND M.-W. RUF, *J. Phys. B* **14**, L803 (1981).
48. M. W. D. MANSFIELD AND G. H. NEWSOM, *Proc. R. Soc. London Ser. A* **357**, 77 (1977).
49. J. E. HANSEN, *J. Phys. B* **8**, 6403 (1975).
50. J. P. CONNERADE, S. J. ROSE, AND I. P. GRANT, *J. Phys. B* **12**, 653 (1979).
51. A. YAGISHITA, S. AKSELA, T. PRESCHER, M. MEYER, M. RICHTER, E. VON RAVEN, AND B. SONNTAG, *J. Phys. B* **21**, 945 (1988).
52. Y. ITOH, T. HAYAISHI, Y. ITIKAWA, T. KOIZUMI, T. NAGATA, Y. SATO, H. SHIBATA, A. YAGISHITA, and M. YOSHINO, *J. Phys. B* **21**, L727 (1988).
53. M. YOSHINO, T. HAYAISHI, Y. ITIKAWA, T. KOIZUMI, T. NAGATA, Y. SATO, H. SHIBATA, AND A. YAGISHITA, *J. Phys. B* **19**, L849 (1986).
54. T. KOIZUMI, T. HAYAISHI, Y. ITIKAWA, T. NAGATA, Y. SATO, AND A. YAGISHITA, *J. Phys. B* **20**, 5393 (1987).
55. T. KOIZUMI, T. HAYAISHI, Y. ITIKAWA, Y. ITOH, T. MATSUO, T. NAGATA, Y. SATO, E. SHIGEMASA, A. YAGISHITA, AND M. YOSHINO, *J. Phys. B* **23**, 403 (1990).
56. T. NAGATA, Y. ITOH, T. HAYAISHI, Y. ITIKAWA, T. KOIZUMI, T. MATSUO, Y. SATO, E. SHIGEMASA, A. YAGISHITA, AND M. YOSHINO, *J. Phys. B* **22**, 3865 (1989).

57. M. Meyer, T. Prescher, E. v. Raven, M. Richter, E. Schmidt, B. Sonntag, and H.-E. Wetzel, *Z. Phys. D* **2**, 347 (1986).
58. J. W. Cooper, C. W. Clark, C. L. Cromer, T. B. Lucatorto, B. F. Sonntag, E. T. Kennedy, and J. T. Costello, *Phys. Rev. A* **39**, 6074 (1989).
59. D. H. Tracy, *Proc. R. Soc. London Ser. A* **357**, 485 (1977).
60. D. M. P. Holland and K. Codling, *J. Phys. B* **14**, L359 (1981).
61. C. Dzionk, W. Fiedler, M. v. Lucke, and P. Zimmermann, *Phys. Rev. A* **39**, 1780 (1989).
62. C. Dzionk, W. Fiedler, M. v. Lucke, and P. Zimmermann, *Phys. Rev. A* **41**, 3572 (1990).
63. J. W. Cooper, *Phys. Rev. Lett.* **13**, 762 (1964).
64. T. M. Zimkina, V. A. Fomichev, S. A. Gribovskii, and I. I. Zhukova, *Sov. Phys. Solid State* **9**, 1128, 1163 (1967).
65. J. P. Connerade, J. M. Esteva, and R. C. Karnatak (Eds.), *Giant Resonances in Atoms, Molecules and Solids* (Plenum Press, New York, 1987).
66. M. Richter, M. Meyer, M. Pahler, T. Prescher, E. v. Raven, B. Sonntag, and H.-E. Wetzel, *Phys. Rev. A* **40**, 7007 (1989).
67. M. Goeppert Mayer, *Phys. Rev.* **60**, 184 (1941).
68. C. Dzionk, W. Fiedler, M. v. Lucke, and P. Zimmermann, *Phys. Rev. Lett.* **62**, 878 (1989).
69. P. Zimmermann, *Comments At. Mol. Phys.* **23**, 45 (1989).
70. P. Zimmermann, in *Today and Tomorrow in Photoionization*, edited by M. Y. Amusia and J. B. West (Science and Engineering Research Council, Daresbury Laboratory, Daresburg, 1991), p. 101.
71. T. Nagata, M. Yoshino, T. Hayaishi, Y. Itikawa, Y. Itoh, T. Koizumi, T. Matsuo, Y. Sato, E. Shigemasa, T. Takizawa, and A. Yagishita, *Phys. Scr.* **41**, 47 (1990).
72. D. A. Liberman and A. Zangwill, *Comp. Phys. Commun.* **32**, 75 (1984).
73. M. Richter, Thesis Hamburg University (1988).

CHAPTER 14

COINCIDENCE MEASUREMENTS ON IONS AND ELECTRONS

J. H. D. ELAND AND V. SCHMIDT

1. INTRODUCTION

Coincidence methods are among the most powerful tools in physics for experiments at the microscopic level, because they select for study single *events* of reaction by individual atoms or molecules. Coincidence detection of at least two products is essential for complete characterization of any process that produces three or more particles, such as double ionization; it also allows particular processes to be studied selectively in experimental situations where events of many different types happen concurrently. Ionization processes, for example, often produce both electrons and ions of many different energies; a coincidence experiment makes it possible to examine the ions that are formed together with electrons of a single chosen energy. The electron and the ion from one event arrive at their respective detectors at different times after their formation, and although the signals at the detectors are not simultaneous (coincident in the normal dictionary sense) they do have a definite temporal relationship. All coincidence experiments involve the search for time-correlated signals such as these from the arrivals of single particles at suitable detectors. Electrons and ions are particularly easy to detect as single particles, but photons and high-energy neutral atoms or molecules can be detected in coincidence too. The majority of experiments in atomic and molecular physics involve coincidences between just two particles, and are called twofold coincidence techniques; triple (threefold) and higher-order coincidence techniques are in use, but are considerably more difficult. As primary exciting particles, photons rather than electrons have a distinct advantage; because they are absorbed, not scattered, the total count of final particles is reduced by one for the same final state of the target. They also have precisely controllable energy and polarization, which are highly advantageous in many experiments.

By analogy with the popular *pump-probe* laser technique, selective coincidence experiments could be called *select-detect* experiments. In ionization by a photon of energy $h\nu$, below the threshold for double ionization, detection of an electron of energy E shows that an energy $h\nu - E$ has been transferred to the target. If the lowest ionization energy is I, an ion of internal energy $h\nu - I - E$ has been formed. Its internal quantum state will then be known from the energy, and the second particle in coincidence will probe the evolution of that selected ionic state. The wavelength of coincident photons, for instance, would give the identity of the final state of the decay. Above the double ionization threshold two energy-analyzed photoelectrons must be detected in coincidence to fix the energy transfer in double photoionization as $h\nu - (E_1 + E_2)$; measurement of a third particle in coincidence then allows the evolution of selected doubly charged ions to be investigated.

J. H. D. ELAND • Physical and Theoretical Chemistry Laboratory, Oxford University, Oxford OX1 3QZ, United Kingdom.   V. SCHMIDT • Fakultät für Physik, Albert-Ludwigs-Universität Freiburg, 79104 Freiburg, Germany.

VUV and Soft X-Ray Photoionization. Edited by Uwe Becker and David A. Shirley. Plenum Press, New York, 1996.

In the rest of this Introduction the objectives of coincidence work with ions and electrons are reviewed in a survey of the most important types of coincidence experiment. In accordance with the overall subject of this book, the emphasis both here and throughout the chapter is on photoionization using vacuum-ultraviolet radiation, though the principles of the coincidence technique, which are the next topic, apply equally to experiments using other forms of excitation. Both the techniques and the objectives of coincidence work are illustrated in the last two main sections by examples drawn from work on photoionization of atoms and of molecules, respectively. The choice of particular examples has inevitably led to the exclusion of some other topics; the authors have tried to include what they regard as the most active and exciting areas of current work and also to cover a representative range.

Survey of Coincidence Experiments on Photoionization

Since the particles detected may be electrons, ions (positive or negative), photons, or high-energy neutrals and because up to three of the same or of different sorts may be detected in coincidence, many distinct coincidence experiments are theoretically possible. In practice, not all imaginable experiments are useful; the most valuable ones and their objectives are described briefly in the following paragraphs. It is convenient to introduce a shorthand notation of the form $C(p_1, p_2 \ldots)$ to indicate that product particles p_1, p_2 (or more) are detected in coincidence. The symbols for different particles—e for an electron, $h\nu$ for a photon, m for a molecule or atom, and so on—should be self-explanatory.

Electron–Electron Coincidences, C(e,e). By measuring the energies of two photoelectrons in coincidence, the experimenter determines the state of the doubly charged ion at the instant of ionization and also learns about the dynamics of two-electron ejection. As the energy transfer on ionization is unambiguously determined, the intensity of the coincidence signal as a function of energy transfer is equivalent to a photoelectron spectrum of the doubly charged ion produced. The first technique implemented on these lines was to detect one electron of near-zero energy in coincidence with an energy-analyzed electron $[C(e_0,e)]$.[59] The most promising recent method for such spectroscopy is a $C(e_0,e_0)$ technique in which two electrons of near-zero energy are detected in coincidence as the photon energy is varied;[39,40,57] this has the advantage that electrons of near-zero energy can be detected with very good efficiency and with relatively good resolution. Even higher resolution, which may be applicable to such coincidence experiments, has been demonstrated in the closely related ZEKE (zero kinetic energy) technique.[73]

When two electrons are energy analyzed at energies away from zero [$C(e,e)$, PEPECO (photo-electron–photoelectron coincidence)[80]], the experiment also shows how the excess energy of the photon above the double-ionization energy is distributed between the two electrons; the distribution is characteristic of the mechanism of the reaction, particularly of whether it is single- or multi-step. The angular distributions of the two electron directions also provide detailed diagnostics of the ionization mechanism and tests of theoretical models.[50,68]

Electron–Ion Coincidences, $C(e,m^+)$. If photoelectron energies are selected by an energy analyzer, the energy transfer on single photoionization is fixed and ions in a defined initial state are selected. Either the electron energy is chosen to be near zero and the photon energy is varied [$C(e_0,m^+)$, TPEPICO (threshold photoelectron–photoion coincidence)[3,4]] or the photon energy is fixed and different electron energies are selected (PEPICO)[11,20] to control the choice of initial internal energy in the ion. The ions may be analyzed by time-of-flight (TOF) or by other types of mass spectrometer; branching ratios, kinetic energy releases, and reaction rate constants can all be measured as a function of the initial internal energy. These experiments offer a powerful tool for state-selective investigation of singly charged cations. In a variant of this technique the photon energy is set above an inner-shell threshold and the electrons selected are Auger electrons. This provides state-selected doubly charged ions whose chemistry can be probed, even though only one electron is energy analyzed,[18,72] as the other electron is of constant and known energy. In fact, the energy of an Auger electron from a known hole state identifies the energy of the doubly charged ion irrespective of the photon energy or the photoelectron energy.

In a simpler $C(e, m^+)$ technique[58] the electrons are not intentionally energy selected, and efforts may be made to ensure that all electrons are equally likely to be detected whatever their energy or initial direction.[67] If this is achieved, a true mass spectrum for single ionization below the double-ionization threshold can be measured by counting ions in coincidence with the electrons using a TOF mass spectrometer where electron detections serve as a time zero from which the ions' flight times are measured. The initial kinetic energies of the ions are also very conveniently measured by this technique at each photon wavelength, as explained later. At wavelengths where double or higher photoionization is possible, care is needed, as the contributions of processes giving two or more electrons, for which the sensitivity is higher, are difficult to calculate.

Coincidences of Electrons or Ions with Photons, $C(e,h\nu)$ and $C(m^+,h\nu)$. Photons and ions can be generated in the same events in several ways. Superexcited neutral states may emit light before autoionization, or ions produced in excited states may decay radiatively afterwards. These processes are separated from nonionizing emissions and are characterized by detecting photons in coincidence with the energy-analyzed electrons. This $C(e,h\nu)$ experiment [PEFCO (photoelectron–fluorescence coincidence)[37,66]] identifies the emitting state from the energies of the incident photon and outgoing electron (whether of near-zero energy or not) and allows determination of the lifetime, wavelength, and quantum yield of the emission. Quantum yield determination is a special strength of coincidence experiments in general, as we shall see later.

Photons emitted in the course of photoionization can also be distinguished from photons produced in other processes by detection in coincidence with the photoions. The PIFCO technique [photoion–fluorescence coincidence, $C(m^+,h\nu)$][27] is very sensitive because ions can be detected with high efficiency; it has recently been used to discover emission from a doubly charged molecular ion[10] and from many monocation fragments at energies just below double-ionization thresholds.[33] Quantum yields of monocation emission can be determined if the photoelectron spectrum is known at the photoionization wavelength.

Ion–Ion Coincidences, $C(m^+,m^+)$. At photon energies below the triple ionization energy, the detection of two cations in coincidence with each other, PIPICO (photoion–photoion coincidence),[14,16] singles out dissociative double ionization of molecules. It identifies ion pairs produced by charge separations such as

$$m + h\nu \rightarrow m^{2+}: \quad m^{2+} \rightarrow m_1^+ + m_2^+ + \text{other products}$$

In modern ion–ion coincidence experiments, both ions travel along a flight path to a single detector, where the overall difference in arrival times encodes the difference in the two ions' masses: fine details of the arrival time difference distribution reflect the initial velocities (momenta) of the ions and their angular distributions. For heteronuclear diatomic molecules the method is very effective for determination of energy releases at different wavelengths in dication decays. In general, however, PIPICO peak shapes in spectra of polyatomic molecules represent an inextricable convolution of the effects of several factors and in addition, all ion pairs where the two masses are the same are centered at zero time difference, and cannot be distinguished.

Threefold and Higher-Order Coincidences, $C(e,m^+,m^+)$ and $C(e,e,m^+)$. The threefold coincidence experiment most widely used hitherto in molecular photoionization is $C(e,m^+,m^+)$ (PEPIPICO),[30,35] which was developed to remove some of the ambiguity of PIPICO. The electrons are not usually energy analyzed on purpose, though some electron energy selectivity is accidentally present in most existing instruments.[29] As in the simplest $C(e,m^+)$ experiment the electron serves to define a time zero for the TOF spectra of the two cations, which can be detected with high efficiency. The individual ion masses are measured directly, even in equal mass pairs, and the peak shapes are two-dimensional distributions containing information on the dynamics of charge separation reactions (Section 3.2.5). The spectra also show distinct features for nanosecond to microsecond decay times which allow doubly charged ion lifetimes toward charge separation to be determined.[38]

In another form of this experiment the electrons are energy analyzed, and Auger electrons are chosen. This allows the initial energy (state) of the doubly charged ions to be selected,[43] just as in the related experiments of Eberhardt[18] and Murphy and Eberhardt.[72] It should be well able to address the interesting question of whether the choice of decay pathway is controlled by the initial localization of charge[74] in dications.

Another way to achieve dication state selectivity is to detect two energy-analyzed photoelectrons in coincidence with the ion or ions, but the low collection efficiency of energy analyzers makes this very difficult if the electrons are of high energy. Electrons of near-zero energy can be detected with very good efficiency, however, and with narrow bandwidth. By taking threshold electron pairs with ions in coincidence, we get the $C(e_0,e_0,m^+)$ and $C(e_0,e_0,m^+,m^+)$ techniques, for which the general acronym EPIC (electron pair ion coincidence) has been coined. These techniques have been tried out,[29] but have not yet been fully developed; they are extremely promising for the future.

The study of higher multiple ionization calls for other multiple coincidence techniques; dissociative triple ionization is being investigated in detail by $C(e,m^+,m^+,m^+)$ experiments using photons[70] or electrons[28] as the ionizing particles.

2. EXPERIMENTAL METHODS

Because the essence of coincidence experiments is the detection of temporal correlations, detectors with good time resolution for single particles are always required. For electrons and ions of energies below about 100 keV, this makes the use of fast electron multiplier detectors such as microchannel plates mandatory, and photomultipliers must be used for photon detection. Solid-state detectors and Geiger–Müller tubes, which can be used for high-energy particles, have some advantages, but are not widely used in atomic and molecular physics. The rest of the physical apparatus is specific to each experiment, but there are several general characteristics of coincidence techniques, a knowledge of which is vital for the design of experiments and for the evaluation of the results. These general characteristics are the main topic of this section.

General Aspects of Coincidence Techniques

The measurement of at least two fragmentation partners "in coincidence" requires that the fragmentation process itself occurs on a time scale which is short compared with the resolving time δt of the coincidence counting device. The finite resolving time, however, gives rise not only to the desired coincidences, usually named true coincidences, but also to accidental coincidences (also named random or false coincidences). True and accidental coincidence counting rates will be discussed first in the lowest approximation and for the case of just two coincident partners. This elementary discussion points the way to the optimization of apparatus for coincidence measurements. The extraction of the true coincidence signal from the observed coincidence counting rate in the more general case of nonnegligible higher-order contributions to the counting statistics is treated next, and finally we discuss data reduction in higher multiple coincidence experiments.

True and Accidental Coincidences. In an experimental setup for coincidence measurements it is always possible to retard the arrival time of one signal with respect to the other one by means of an electronic or cable delay. The first signal can be designated as the START and the second as a STOP. These signals are fed into the corresponding inputs of a time-measuring device, usually a time-to-amplitude converter (TAC) or a time-to-digital converter (TDC), where the time difference between the STOP and START signal is measured. After a certain dead time the time-measuring device is reset and waits for the next signal. Modern TDCs, unlike TACs, can respond to many STOP signals after a single START; this improves the counting rate, simplifies the data reduction, and makes multiple coincidence experiments possible using a single detector.

Even if the particles of a pair are emitted at exactly the same moment and with the same energies, the resulting START–STOP time differences vary from pair to pair. Two reasons for this are different traveling times of the coincident particles from their place of birth toward the detectors, and different electronic processing times of the signals depending on the exact forms of the electronic pulses. Hence, the coincidence signal ("true coincidence peak") always has a certain time width as illustrated by the tall peak in Figure 1. The coincidence resolving time δt has to be long enough to accept the true

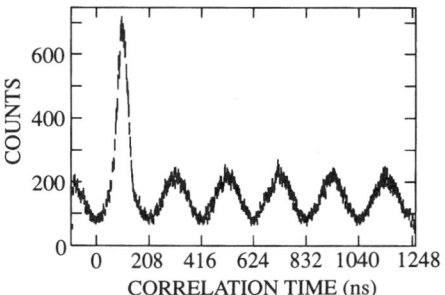

FIGURE 1. Example of a C(e,e) spectrum registered using a time-to-digital converter. The tall peak contains both true and accidental coincidences, while the wavy background is due to accidental coincidences only and reflects the time structure of radiation from the electron storage ring BESSY (Berlin), with a repetition time of 208 ns under the conditions used.

coincidences, but any finite resolving time also allows the registration of accidental coincidences. The rate of accidental coincidences can be calculated from the standard formula[81]

$$C_{acc} = I_1 I_2 \, \delta t \qquad (1)$$

where I_i are the single counting rates in each of the two detectors and δt is the coincidence resolving time, provided the rate of true coincidences, C_{true}, is much less than either I_1 or I_2. In the lowest approximation, true and accidental coincidences can be considered to be independent of each other, within the time range where both are present, and are additive:

$$C_{obs} = C_{true} + C_{acc} \qquad (2)$$

If the ionization events are randomly distributed in time (continuous photon beam) the accidental coincidences just form a constant "background" in the time spectrum, and their intensity can be obtained easily from a suitably selected range of the measured time spectrum by delaying one of the two detector signals. A modification occurs if ionization is possible only during finite production intervals of duration T_d; the time structure of an electron storage ring as radiation source is an important example. Depending on the operation mode, light flashes of duration T_d are delivered with repetition times t_{rep}. If T_d together with any instrumental broadening is smaller than t_{rep}, the observed time distribution will reflect the time structure of the storage ring in the accidental coincidences. Figure 1 shows an example of this; the large peak shows the sum of true and accidental coincidences, while the wavy structure comes from false coincidences only. To obtain C_{true} one coincidence resolving time δt is defined at the position of the main peak, at 100 ns in the figure, and several separate coincidence times displaced from it by multiples of the repetition time t_{rep} are used to determine C_{acc}.

One objective of determining C_{true} in experiments of the *select-detect* type is to measure quantum yields. Let the rate of detection of selected events be designated S counts per second; in single ionization this is the rate of detection of electrons of a chosen energy, in double ionization it could be the rate of detection of electron pairs with a chosen energy sum. Then the rate of true coincidences with another particle produced in the final state, C_{true}, is

$$C_{true} = S \phi f_2 \qquad (3)$$

where ϕ is the quantum yield for formation of particle 2 and f_2 is the detection efficiency for that particle. This expression shows the advantage of the coincidence method for measuring quantum yields, since the only apparatus parameter to be determined is the detection efficiency; there is no need to know the target gas density or the light flux.

In order to evaluate the importance of true and accidental coincidences for a given experiment, relations have to be derived which connect the counting rates with individual experimental parameters. The exact relations are specific to each experiment, but the principles behind their calculation are quite simple. To explain them, we first introduce the concept of the *source strength* $\mathbb{N}$, which is the rate of events of the specific chosen type occurring in the source volume from which particles are accepted. Formally the source strength for detector j is

$$\mathbb{N}_j = N_{hv} n_v \Delta x_j \sigma_j \tag{4}$$

Here N_{hv} is the number of incident photons per second per unit area, n_v is the target density, σ_j is the cross section for the process selected by detector j, and Δx_j is the source volume to which that detector is sensitive.

The source strength is used below to derive detailed rate expressions for the particular case of two-electron emission following photon-induced double ionization. The two electron branches, selected by two analyzers with spectrometer voltage settings U_1 and U_2, respectively, lead to single counting rates I_1 and I_2 and to the coincidence rates C_{true} and C_{acc}:

$$I_1 = \mathbb{N}_1 f_1 F(U_1, E_1) \tag{5a}$$

$$I_2 = \mathbb{N}_2 f_2 F(U_2, E_2) \tag{5b}$$

$$C_{true}(U_1, U_2) = \mathbb{N}_{12} f_1 f_2 F(U_1, E_1, U_2, E_2) \tag{6}$$

$$C_{acc}(U_1, U_2) = I_1 I_2 \delta t \tag{7}$$

In these expressions the source strengths $\mathbb{N}_1$ and $\mathbb{N}_2$ contain all processes at the chosen wavelength producing electrons of energy E_1 and E_2, respectively. The parameters f_i are overall detection efficiencies of the two spectrometers including transmission and detector effects, while the functions $F(U_i, E_i \ldots)$ take into account the influence of the different energy-dependent distributions (photon bandwidth, level width of any intermediate state and spectrometer function). In particular, $F(U_1, E_1, U_2, E_2)$ must also include the interdependence of the kinetic energies E_1 and E_2, which is required by energy conservation. The source strength $\mathbb{N}_{12}$ for the true coincidence signal contains the partial cross section only for double ionization giving a pair of electrons of energies E_1 and E_2, and contains the source volume Δx_{12} seen by both spectrometers in common. The most favorable situation is evidently for both analyzers to accept particles from this common overlap volume only.

From these relations, and assuming that all of the volume elements are the same and the $F()$ factors are unity in a first approximation, the ratio r of true to accidental coincidences is

$$r = \mathbb{N}_{12}/(\mathbb{N}_1 \mathbb{N}_2 \delta t) \tag{8a}$$

If there is only a single doubly charged ion state accessible, and there are no other processes producing single electrons, the source strengths are all identical and this simplifies to

$$r' = 1/(\mathbb{N}^+ \delta t) \tag{8b}$$

Quite often the ratio r' is used as a criterion for the quality or feasibility of coincidence experiments. From the above relations it follows that a large ratio of true to accidental coincidences requires a small coincidence resolving time δt and a small source strength $\mathbb{N}$. However, in most cases such a favorable situation can be realized only at the expense of an extremely low source strength which then leads to an unacceptably low coincidence rate C_{obs}. Therefore, the ratio r or r' is not a good criterion. A more appropriate figure of merit is the total data collection time T_{coll} necessary for obtaining a preselected relative error α of true coincidences (I_1 and I_2 are large; see Wapstra[94] and Völkel and Sandner[91]):

$$T_{coll} = \frac{1}{\alpha^2} \frac{1}{f_1 f_2} \frac{1}{\sigma_{12} N_{hv} n_v \Delta x} \left(1 + \frac{\sigma_1 \sigma_2}{\sigma_{12}} N_{hv} n_v \Delta x \delta t \right) \tag{9}$$

Optimization of Coincidence Apparatus. From the expressions given in the last section, general criteria for coincidence experiments can be stated.

1. The source volumes from which particles are detected by the two or more analyzing systems should coincide.
2. The coincidence resolving time δt should be as small as possible, but large enough to collect coincident events appropriately.
3. The product of transmission and detection efficiency, in particular the solid angle accepted by each analyzer, should be as large as possible.
4. The source strength should be adjusted to obtain rates of true and accidental coincidences of the same order of magnitude.
5. Because true and accidental coincidence rates depend differently on the source strength, for changing strength they should be measured simultaneously.

These criteria and their application in the design of coincidence experiments are discussed in this section. Often the counting rate in electron–electron coincidence experiments is extremely low, and rather long times are needed to accumulate data with sufficient statistics. This point is particularly important for experiments using synchrotron radiation where only limited beam time is available and it may be difficult to ensure stability of the electron storage ring over a long period. Hence, a careful optimization of the experimental setup is required.

It is a difficult but most important task to ensure that the source volumes seen by the different analyzers in a coincidence experiment are coterminous. Where the source is field free, it is sometimes possible to use aperture stops to define a single volume precisely, but where a uniform field is required the source must be defined by the intersection of the light beam and a jet of target gas. In either case the analyzers should be designed to accept particles from a larger volume than the interaction volume, especially if it is intended to determine absolute cross sections. Both the active volume and the analyzer functions may depend in real experiments on such variables as analysis energy, lens potentials, steering potentials, and angle settings. The manual optimization of all possible variables at every setting is difficult and tedious; a solution for experiments under full computer control is real-time optimization using a simplex algorithm.[71]

The coincidence resolving time δt should be as small as possible. However, the lower limit for δt follows from the inherent spread of arrival times t of the coincident partners in the time-measuring device. Differences in these times are mainly due to the time spread from different geometrical paths of, for instance, electrons in electron spectrometers.[46,91] In order to reduce the geometrical time spread, several devices have been employed: one is a geometrical compensation where a large-size detector (microchannel plate) is aligned parallel to the surface of equal flight time.[91] In another geometrical compensation method, two analyzers are combined such that longer/shorter electron paths in the first analyzer become the shorter/longer paths in the second analyzer.[86] In an electronic method the position sensitivity of a microchannel plate detector with discrete multianodes is used to compensate for different position-dependent flight times by properly adjusted time delays.[91] These devices can be used for ions as well as for electrons; it is vital in the case of ions to ensure that the time–focus conditions[95] are met to eliminate time differences due to different positions of ionization in the source. The major cause of ion time spread, however, is the initial velocity of the ions, either from random thermal motion or from energy release in fragmentation. The resulting time spread can be reduced only by using a strong drawout field in the source region.

A prerequisite for successful coincidence experiments is high collection efficiency, which demands a large solid angle Ω accepted by each detector since the maximum observable fraction from isotropic emission is $\Omega/4\pi$. However, large acceptance angles allow the registration of particles at angles which differ widely from those of the principal rays, obviously restricting the angular resolution. Where differential cross sections are to be measured, the influence of finite acceptance angles of all analyzers on the coincident angular distribution has to be worked out quantitatively and incorporated in the correct interpretation of the experimental data. The solid-angle effect can be treated easily for a pointlike source, since general transformation properties of the spherical harmonics can be used to derive a closed form for the angular distribution pattern observed by detectors with finite acceptance angles.[34,62] It should be mentioned that the essential requirement is not so much that the source should be small compared with a characteristic length of the detection device, but that the acceptance should be independent of the actual starting point of particles within the finite source volume.

Finally, the correct tuning of analyzers for the different particles has to be considered, particularly where the optimum setting of one analyzer may depend on a property of the particle detected in the other, even within the instrumental resolution. For instance, in double ionization, energy balance between the two electrons is ruled strictly by energy conservation. For proper registration of the coincidence intensity, both electron spectrometer voltages must be set accurately to the correct nominal energies, at which the transmission functions are known. However, from noncoincident electron spectrometry it is known that there may be shifts of electron line positions depending on the angular setting, or even depending (via penetrating fields) on other unrelated potentials or settings in the apparatus; any such shifts will, of course, affect the transmission function. For direct double photoionization with its continuous energy distribution, such "line" shifts may have less effect on measured coincident intensities than in the case of two-step double ionization where the electron energies are discrete; nevertheless, it is always better, and essential where discrete lines are involved, to scan the spectrometer voltages over both nominal electron energies. Such time-consuming procedures can be avoided by detection of all electrons within a suitable range of energies simultaneously using large aperture detectors. Similar precautions are seldom needed for ions, since the commonly used TOF mass spectrometry technique is fully multiplex, being always sensitive to ions of all masses. TOF spectrometry of electrons also obviates the problem, but can be applied only where the ionizing radiation has an appropriate time structure and the energy resolution of electron TOF is sufficient.[23]

Extraction of the True Coincidence Signal. The standard procedure to obtain true coincidences, C_{true}, from observed ones, C_{obs}, is subtraction of accidental coincidences, C_{acc} [equation (2)] which have ideally been measured simultaneously, but which are sometimes measured separately or calculated [equation (1)]. However, simple subtraction is valid only as a first approximation. When the rate of particle detections gets large, either because of increased source strength or because of increased transmission and detection efficiency, higher-order contributions must be allowed for. Complications arise frequently where coincidences with thermal ions are involved, because they can be collected, analyzed, and detected with extremely high efficiency. The data reduction can become complicated, and detailed prescriptions have been given for the usual case of a continuous source.[89] With multihit TDCs the problem of paralysis is minimized, but even then great care must be taken if C_{acc} is calculated rather than measured. The principle of the calculation is to consider each time channel as a coincidence gate of width δt; each START opens the gate, through which noncorrelated STOP signals can pass to contribute to C_{acc}. Thus, in most cases equation (1) must be modified to

$$C_{acc} = I_1(I_2 - C_{true})\delta t \tag{10}$$

which can always be solved recursively.

In some cases coincidence measurements must be made in a pulsed regime, and this exacerbates the problem of data reduction. An approach is given here for an example of electron–ion coincidences with electron energy analysis, but is quite general and applies to all situations where coincident and noncoincident processes are produced with high efficiency within a finite ionization time interval ΔT_i and at least one fragmentation product has a high detection probability.

For the optimum registration of coincidences between energy-analyzed electrons and thermal ions, two conditions have to be met. First, ions must be extracted efficiently from the source volume into the direction of a suitable ion analyzer, and second, the electrons have to reach their electron spectrometer without being disturbed by electric or magnetic fields. Since a field is needed to extract ions, these requirements are mutually incompatible under continuous field conditions so a pulsed ion extraction scheme is used in which electrons are observed during a field-free period, and an ion extraction field is applied once an electron is detected. A third requirement is then that uncorrelated ions should not be allowed to accumulate in the source before the extraction pulse. The solution is a pulse scheme with two regular parts, an ionization time interval ΔT_i where photoelectrons and ions are produced in a field-free region, and a clearance time interval ΔT_c during which unwanted photoelectrons and ions are pushed out of the source volume and away from their respective detectors. If an electron is detected, this sequence is interrupted for the analysis of the ions, and for a time interval ΔT_e an electrical field is put across the source volume to draw the ions out and accelerate them toward the ion analyzer. Depending on their masses (strictly mass-to-charge ratios) the accelerated ions have different flight times and can be separated by TOF analysis. When the analysis period is over, the external periodic time structure starts again with the electrical field pushing the ions out of the source region.

The key point for interpretation of the observed electron–ion coincidences is the following. After electron detection the coincident partner-ion and any other ions produced during ΔT_i are pushed with high efficiency into the TOF analyzer, and all ions with the same mass arrive at the ion detector at practically the same time. Hence, depending on how the detection system responds to such a bunch of events produced during ΔT_i the observed coincidence signal may be nonlinearly related to the source strength. Two special solutions exist. One is to use such a low source strength that within the time interval ΔT_i only one event can be produced at most, allowing for the Poisson counting statistics. This requires an average event rate at least ten times less than the repetition rate of the pulse scheme; as a consequence, low counting rates are observed. The other solution is based on the use of "zero-one" detectors which increment their attached counter only by one unit for any nonzero event produced during the measurement interval. In this case a general relation between the registered rate of coincidences, C_{obs}, and the average rate of initially produced correlated events, i.e., C_{true}, can be worked out which is valid for any source strength and/or particle transmission and detection efficiency (Kossmann[55]; compare also Kämmerling et al.[48]).

Multiple Coincidence Experiments. The data from an n-fold coincidence experiment form an intensity distribution in an $(n-1)$-parameter space, and the accidental coincidences are obviously of the same dimensionality. Unlike twofold coincidence experiments, where the background is a flat distribution in the simplest case, accidental coincidences always form a structured background in threefold and higher coincidence work. Either the source strength must be kept so low that the background is negligible, or great care must be taken in calculation of the accidental coincidence spectrum and its subtraction; simultaneous measurement of the accidental coincidences is certainly very difficult, and has not yet been attempted. Furthermore, the high-order events that give rise to true n-fold multiple coincidences also make disproportionately large contributions to measured $(n-1)$-fold and lower coincidence measurements on the same system.

We shall take a $C(e,m_1^+,m_2^+)$ experiment as an example, where natural choices for the two parameters are the arrival time differences $t_1 = t_{m1} - t_e$ and $t_2 = t_{m2} - t_e$. In this parameter space the true coincidences produce localized peaks, while the accidental coincidences form ridges; if both ions are accepted by a single detector, all coincidences are restricted to the segment of a rectangle above its leading diagonal, as illustrated in Figure 2, because m_2^+ is defined as the ion arriving after m_1^+. There are three types of accidental coincidence in the spectrum.[17,36] First, three entirely uncorrelated particles may be registered, producing a uniform flat distribution; this part is usually extremely weak. Second, the electron and one ion may be correlated, while the other ion, arriving either earlier or later, is uncorrelated. These signals produce ridges which form a rectangular grid pattern parallel to the axes at times which match ion peaks in the mass spectrum. Third, the electron may be uncorrelated while the ions are the members of a correlated pair; this produces a diagonal ridge in the spectrum joining

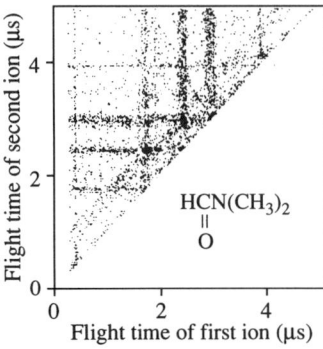

FIGURE 2. $C(e,m_1^+,m_2^+)$ or PEPIPICO spectrum of dimethylformamide at 30.4 nm, showing the accidental coincidence background. The major true coincidence peaks (solid black blobs) are more than 100 times stronger than the background and have been suppressed; the display scale is so coarse that the resolution into individual mass peaks is lost; the broad horizontal and vertical bands correspond to groups of close masses. All three types of accidental coincidence discussed in the text can be seen.

on to each true coincidence peak, parallel to the principal diagonal. Where the various ridges cross, their intensities are additive, provided the count rates are sufficiently low, as in Figure 2; of course the appearance would be entirely different if another parameterization was chosen, such as $t_{m1} - t_e$ and $t_{m2} - t_{m1}$.

The accidental coincidence count rate within an area δt^2 of the data space at (t_1, t_2) can be expressed approximately as a sum of the three contributions:

$$C_{acc} = E\,I^2\,\delta t^2 + \{C_{em}(t_1) + C_{em}(t_2)\}\,I\,\delta t + C_{mm}(\Delta t)\,E\,\delta t \qquad (11)$$

Here $C_{em}(t)$ is the rate of electron–ion coincidences within an interval δt at time difference t, $C_{mm}(\Delta t)$ is the rate of ion–ion coincidences for the given time difference, while E and I represent the counting rates for single electrons and single ions. $C_{em}(t)$ and $C_{mm}(\Delta t)$ are single points from the twofold electron–ion and ion–ion coincidence spectra, which should be measured concurrently with the threefold one and should have had their accidental coincidence backgrounds already subtracted. If the count rates or collection efficiencies are high, a more complex calculation is needed as in the preceding section. The equation given above can be modified by subtracting from I and E the rates of measurement of correlated signals where these are a significant fraction of the total; this correction is sufficiently accurate for practical needs. A serious difficulty is that the count in any individual element of the data array is either zero (the majority) or an integer, while the calculated accidental coincidence count is fractional and less than unity for many elements of the array. A numerical strategy is needed to clump together areas of low and uniform calculated C_{acc} so that statistically significant numbers of counts can be subtracted in an unbiased way. Even within the true coincidence peaks some clumping may be needed, and to avoid distortion of the peak shapes a random (Monte Carlo) method for selecting the individual events to be subtracted can be implemented.

In experimental situations where multiple coincidences are measured, we often want to determine the relative abundances of different types of event. Inner-shell photoionization and excitation of light molecules, for instance, produces single, double, and triple ionization; double and triple ionization are mostly followed by dissociation into singly charged fragment ions. To determine the relative cross sections for the different primary processes we must remove not only accidental coincidences but also the triple ionization contribution to $C(e,m_1^+,m_2^+)$ or $C(m_1^+,m_2^+)$ spectra, and both the triple and double ionization contributions to $C(e,m^+)$ spectra. These contributions can be very significant.

To illustrate the principle of the calculation, let us consider an experiment on triple ionization of a molecule where the highest-order coincidences to be measured are $C(e,m_1^+,m_2^+,m_3^+)$, and there is no electron energy analysis. We assume that the effective collection and detection efficiency for all ions is the same, f_i, and for all electrons it is f_e. This is unrealistic as the actual efficiencies will be at least energy and mass dependent, and would need to be measured. We further assume that all triple and double ionizations are purely dissociative to three and two singly charged ions, respectively, though in reality charge-retaining fragmentation would have to be considered. Let the true rates of triple, double, and single ionization be N_3, N_2, and N_1, respectively, then for the true coincidence rates we have

$$C_{\text{true}}(e,m_1^+,m_2^+,m_3^+) = N_3 \, 3 f_e f_i^3$$

$$C_{\text{true}}(e,m_1^+,m_2^+) = \{N_2 \, 2f_e + 2\, N_3 \, 3f_e\} f_i^2 \qquad (12)$$

$$C_{\text{true}}(e,m_1^+) = \{N_1 f_e + 2\, N_2 \, 2f_e + 3\, N_3 \, 3f_e\} f_i$$

The factors nf_e are the probabilities of a single START being produced by n-fold ionization; if the f_e are not much smaller than unity, nf_e must be replaced by the exact expressions $(1 - (1 - f_e)^n)$. Because f_e and f_i normally are small, the great majority of triple ionization events are detected as single $C(e,m^+)$ events or as $C(e,m_1^+,m_2^+)$ ion pairs rather than as triples. If f_e is 25%, for instance, a triple ionization event is 48 times more likely to be detected as a single ion than as a triple. The collection efficiencies must be known with good precision if the necessary subtractions for determination of N_3, N_2, and N_1 are to be accomplished accurately.[63]

3. ILLUSTRATIVE EXAMPLES

There have been so many applications of the coincidence technique, even in molecular and atomic photoionization, that only a few can be mentioned here. We have attempted to gather significant results that illustrate particular aspects of the technique as well as recent examples from current work. Much of the latter involve the use of synchrotron radiation, which provides an unrivaled source of photons in the extreme ultraviolet with its tunability and advantageous polarization characteristics. Several examples are concerned with double or multiple photoionization, which can be studied only with great difficulty without the aid of such a source.

3.1. Photoionization of Atoms

Photoionization in atoms will yield in general not just one electron but, according to the specific ionization process, several electrons, photons, and an ion in the complementary charge state. Hence, when considering twofold coincidences, the following possibilities exist: $C(e,e)$, $C(e,m^{n+})$, $C(h\nu,e)$, $C(h\nu,h\nu)$, and $C(h\nu,m^{n+})$. Among these only $C(e,e)$ and $C(e,m^{n+})$ experiments have been carried out so far, and examples are given below. The interest in these experiments attaches mainly to multiple ionization, of which double ionization is the most important representative; two clearly defined limiting forms of double-ionization mechanism can be distinguished. First, in direct double photoionization the primary photon interaction leads immediately to the emission of two electrons coming from the same or different orbitals. Second, in the best understood form of two-step double ionization, photoionization of an inner-shell electron takes place in a first step (photoelectron emission), and after a certain lifetime the inner-shell hole is filled while a second electron is emitted (Auger electron emission). By comparison with direct double photoionization this two-step process has several simplifying features. Instead of the available excess energy (photon energy minus the double-ionization energy) being shared between the two electrons in a continuous distribution, each electron has a definite energy; furthermore, instead of many orbital angular momenta of the two electrons being combined to form the orbital angular momentum of the two-electron pair function, the orbital angular momenta are restricted to fewer values.

Double ionization also occurs in other ways. At any energy above the first double-ionization threshold, one electron can be emitted while another is excited to a previously unoccupied orbital, after which a second electron can be ejected in an autoionization step. Both Auger electron spectra[7] and $C(e,e)$ spectra[77,78] have shown that this mechanism is important in the valence double photoionization of atoms. Three-step mechanisms may be involved when the incident photon excites a high-energy resonance. Coincidence experiments are essential for the elucidation of these more complex mechanisms; nevertheless, we concentrate here on the simple prototypes, direct double ionization and the two-step Auger process.

In C(e,e) coincidence experiments with atoms, the ultimate objective is to determine the energy-, angle-, and spin-resolved differential cross section for two-electron emission, represented by the sixfold differential cross section. This quantity shows separately the dependences on different parameters: on the solid angles $d\Omega_1$ and $d\Omega_2$ about the emission directions κ_1 and κ_2, respectively, on the energies E_1 and E_2 for each of the two electrons (with account taken of energy conservation), on the spin projections m_{s1} and m_{s2} of both electrons as measured in the individual detector frames, and on the polarization of the incident light which is described by the Stokes parameters $S = (S_1, S_2, S_3)$. Of course, the origin of the observable dependence of the differential cross section on these parameters lies in the inherent properties of the photoprocess, described by the matrix elements responsible for photon-induced double ionization. In order to simplify the discussion, the spin dependence can be omitted (for an example with spin analysis see Kämmerling and Schmidt.[51] In this case, a general parameterization based on invariance properties can be given where the "geometry" is separated from the "dynamics."[53] There are also alternative parameterizations where the dynamical parameters appear explicitly; these allow quantitative interpretation of two-electron emission (for direct double photoionization see Yudin et al.,[96] Selles et al.,[84] Huetz et al.,[45] and the treatment of continuous double Auger decay by Amusia et al.[2] which can be transferred to this case; for photon-induced two-step double ionization see Berezhko and Kabachnik,[9] Kabachnik,[47] Kämmerling and Schmidt,[51] Vegh and Becker[92]).

$C(e,m^{n+})$ coincidence experiments in atoms are an important tool for disentangling the different decay pathways that follow core excitation or ionization because the final ionic charge is determined directly. In particular, coincidences between threshold electrons and thermal ions have found wide application, because high efficiency for detection of both particles can be achieved by placing a small electric field across the source volume.

3.1.1. Near-Threshold Double Photoionization of Krypton.

Double photoionization in the outer shell of rare gases has long been and still is an extremely interesting subject, in particular with regard to the region close to threshold (Wannier region).[93] Its attractiveness is due to the possibility of deriving certain threshold laws without a full solution of the complicated three-body problem. Naturally, helium, as the simplest correlated system where direct double photoionization is not disturbed by other competing processes, is the ideal case for such studies; however, the cross section approaches its zero value at threshold with an extremely low constant of proportionality (unfavored structure of the two-electron wave function in the continuum) and the two electrons share continuously the available excess energy, and hence $C(e,e)$ coincidence measurements have not been feasible until recently.[87] Instead, the first experimental study of the correlation pattern in two-electron emission in near-threshold double photoionization was carried out on outer-shell ionization of krypton by HeII light, leading to the $^3P^e$ ionic state with an excess energy of 2.26 eV.[68] Because the symmetry of the full final state must be $^1P^o$, the two-electron wave functions $^3P^o$ and $^3D^o$ are both possible and will interfere. However, in the threshold region the theory of direct double ionization indicates that $^3P^o$ is favored (antinode in the two-electron wave function at $\theta_{12} = 180°$, θ_{12} being the angle between the two emitted electrons) while $^3D^o$ is unfavored (node at $\theta_{12} = 180°$). These properties together with experimentally selected energies $E_1 = E_2 = 1.13$ eV and Wannier geometry, $\theta_{12} = 180°$, lead to an important simplification that only the $^3P^o$ wave function contributes. This function has a component which is symmetric with respect to radial interchange, and for θ_{12} close to 180° it will still dominate. Hence, the angular correlation pattern between the two electrons shown in Figure 3 can be analyzed with respect to the angular correlation function $G(180 - \theta_{12})$ and the interference between the $^3P^o$ and $^3D^o$ contributions. The results are compatible with the angular dependence of the favored $^3P^o$ electron-pair wave function, and the angular correlation appears to be in agreement with the predictions of Wannier theory.[68]

3.1.2. State Selectivity in Double Photoionization of Argon.

For rare gases other than helium, double photoionization of p electrons in the outer shell leads to three final ionic states, $^3P^e$, $^1D^e$, and $^1S^e$. Because the total symmetry of the final state (from the dipole selection rule) is $^1P^o$, the associated electron-pair wave functions are $(^3P^o, ^3D^o)$, $(^1P^o, ^1D^o, ^1F^o)$ and $(^1P^o)$, respectively, and they are subject to certain symmetry requirements which can be formulated in the threshold region.

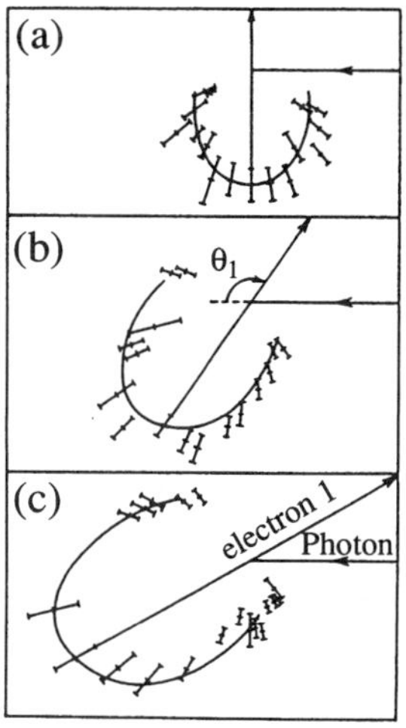

FIGURE 3. Energy- and angle-resolved $C(e,e)$ coincidences for near-threshold direct double photoionization of krypton to $Kr^{2+}(^3P^e)$. The unpolarized incoming photon beam direction and the first electron direction (polar angle θ_1) are indicated by arrows: (a) $\theta_1 = 90°$; (b) $\theta_1 = 125°$; (c) $\theta_1 = 150°$. Error bars give the intensities of the coincident second electron signal at different angles. The solid curve represents a fit of the relative differential cross section based on dominance of the $^3P^o$ electron-pair function, using the extended Wannier theory.[68]

Because of the above-mentioned favored character of the $^3P^o$ pair function, the processes leading to the $^3P^e$ ionic states are expected to be the strongest, and according to some propensity rules, $^1D^e$ should be more probable than $^1S^e$ (Huetz et al.[45] and references therein). In order to test this prediction, $C(e,e)$ coincidences are needed which allow the distribution among final ionic states to be determined from the kinetic energies of the two electrons. While the first $C(e_0,e)$ study on argon showed population of the argon dication $^3P^e$ ionic state only,[59] a strong population of the $^1D^e$ state was also observed in a later study.[77] Both experiments were performed at rather high excess energies, probably too high for the application of threshold predictions since indirect ionization may play a part as other experiments suggest[78] and the measured part of the triply differential cross section may not be related simply to the overall strength of the process as described by the total cross section. Hence, it was highly desirable to study the state selectivity in the vicinity of the ionization threshold and to collect all emitted electron pairs which belong to the selected threshold. In spite of the experimental difficulties, such $C(e_0,e_0)$ coincidence experiments have become possible now.[39,40,57] The spectrum of argon is shown in Figure 4; analysis of the data yields the constants of proportionality σ_0 of the Wannier cross section from which the validity of the expected propensity rule, $\sigma_0(^3P^e) > \sigma_0(^1D^e) > \sigma_0(^1S^e)$, can be inferred.

For the investigation of phenomena at double-ionization thresholds $C(e_0,m^{n+})$ experiments have also proved to be useful. This is demonstrated in Figure 5[41] where the intensities of $N(e_0)$, $C(e_0,Ar^+)$, and $C(e_0,Ar^{2+})$ are shown as functions of photon energy. At the openings of the respective double-ionization thresholds (arrows in Figure 5) a cusplike feature, expected from Wannier theory, can be seen clearly only at the $^1D^e$ threshold in the $N(e_0)$ threshold electron spectrum. In the $C(e_0,Ar^{2+})$ curve it is obscured, probably because of fluorescence decay of high-lying Rydberg states. It is not visible for the $^1S^e$ state, and at the $^3P^e$ channel threshold it is masked by the presence of strong satellite states. These satellite Ar^{+*} states are produced by photoionization with simultaneous excitation of another electron or possibly via doubly excited states of the neutral, and will decay afterwards. Because the $C(e_0,Ar^+)$ signal is practically zero when the photon energy is larger than the lowest Ar^{2+} ionization potential at 43.36 eV, one can conclude that radiative decay of these satellites is a rather weak process compared with autoionization, which will produce $C(e_0,Ar^{2+})$ coincidences when the photon energy matches the satellite excitation energy [compare the $N(e_0)$ curve too]. The kinetic energy of the electron emitted in the autoionization decay of a satellite is equal to the difference between the energy of the satellite state and the double-ionization potential and will be larger than zero in most cases. However, a satellite just at the onset of the $^3P^e$ ionization potential gives practically zero-kinetic-energy electrons if it decays to $Ar^{2+}(^3P^e)$, and in this special case the production of the satellite

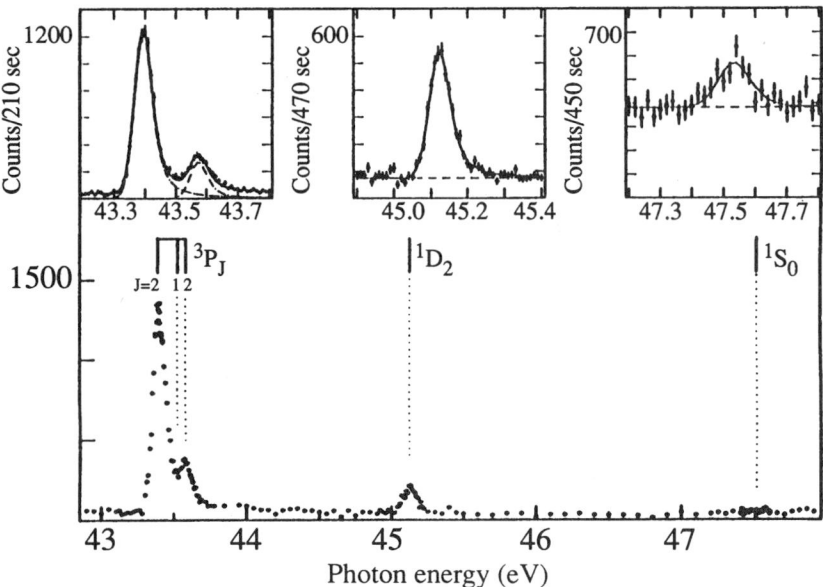

FIGURE 4. $C(e_0,e_0)$ coincidence spectrum of argon with enlarged insets covering the three double-ionization thresholds. The peak at the $^3P^e$ threshold also contains a disturbance from a two-step satellite decay. From Krässig and Schmidt,[57] to be compared with the similar spectrum.[40]

and its decay both yield electrons of near-zero energy which disturb the direct double photoionization at the $^3P^e$ threshold (for details see Hall et al.,[39,40] Krässig and Schmidt[57]).

3.1.3. Decay Paths in Core-Excited or Core-Ionized Xenon. Many processes of photon-induced double ionization involve inner-shell ionization or excitation followed by a subsequent Auger decay or autoionization. As an example, the electron spectrum of xenon following $4d_{5/2} \rightarrow 6p$ resonance excitation and autoionizing decay is shown in Figure 6 in the range of low kinetic energies. Here the autoionizing decay can lead to final ionic states which are subject to a second-step Auger decay. The complex overlapping structure of such first- and second-step electron lines could be identified unambiguously by a $C(e,e)$ coincidence experiment, and relative intensities for the second-step decay branches could be determined. The matching decays are collected in the upper part of the figure where

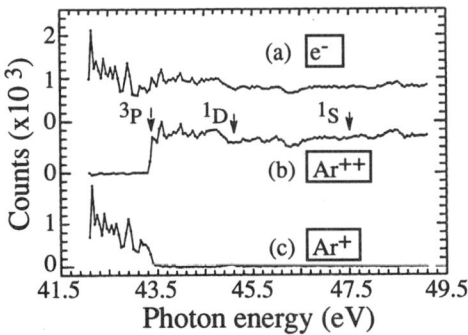

FIGURE 5. Threshold photoelectron spectrum, $N(e_0)$, and the coincidence spectra $C(e_0,Ar^{2+})$ and $C(e_0,Ar^+)$ of argon as functions of photon energy. The thresholds for direct double photoionization in the $3p$ shell are indicated by arrows.[41]

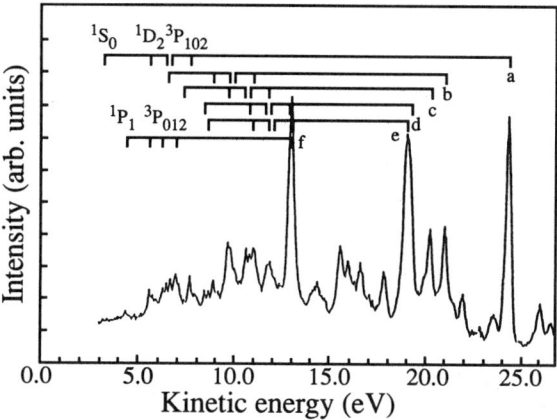

FIGURE 6. Low-energy part of the resonant Auger spectrum of xenon taken at the $4d_{5/2} \to 6p$ resonance. The letters (a) to (f) identify different states of Xe$^+$ produced in the first-step resonant Auger decay: (a) $5s5p^5(^1P)6p$; (b) $5s5p^5(^1P)7p$; (c)–(e) $5s^25p^35d6p$; (f) $5s^{-2}5p^6(^1S)6p$. The marks connect these transitions to the corresponding second-step Auger decay final states [electron configurations $5s^25p^4$ for (a) to (e); $5s5p^5$ for (f)].[82]

the peaks (a) to (f) from first-step transitions are connected with peaks produced in the second-step decay.[82]

In addition to the sequential decay of the $4d_{5/2} \to 6p$ resonance in xenon which leads to two-electron emission, there exists also the possibility of two-electron emission as part of the resonance decay itself (often called resonance shakeoff). The two processes can be distinguished by the energy distribution of the electrons, since sequential decay will produce discrete lines, while resonance shakeoff will give a continuum. In order to reveal both contributions, $C(e,e,)$ coincidences were studied, making no distinction between electrons from the first or second step, but collecting all of their energies at once and interpreting the coincidences in terms of selected total energies E_{tot} of the electron pair. Results for $E_{tot} = 30 \pm 5$ eV are shown in Figure 7. At this energy, resonance shakeoff decay to Xe$^{2+}(4d^{10}5s^25p^4)$ is possible, but combinations of sequential decays are possible also. Both features can be seen clearly in Figure 7: the solid line represents (on the linear time scale which is appropriate to the TOF analysis used) events that originate on the energy scale from a continuous and flat energy

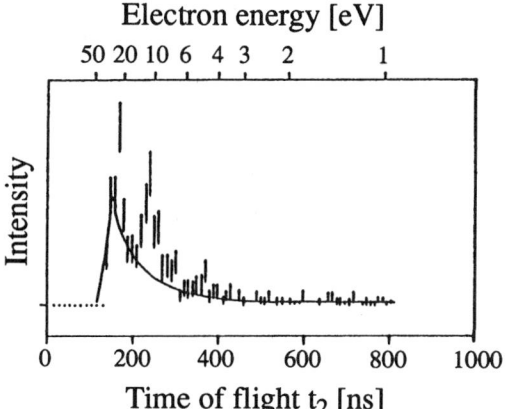

FIGURE 7. Energy distribution of a coincident electron pair $C(e,e)$ with 30 ± 5 eV total energy, measured for the $4d_{5/2} \to 6p$ resonance decay in xenon. The solid line describes resonance shakeoff decay; the remaining peaks represent sequential resonance decay. (From Okuyama et al.[76])

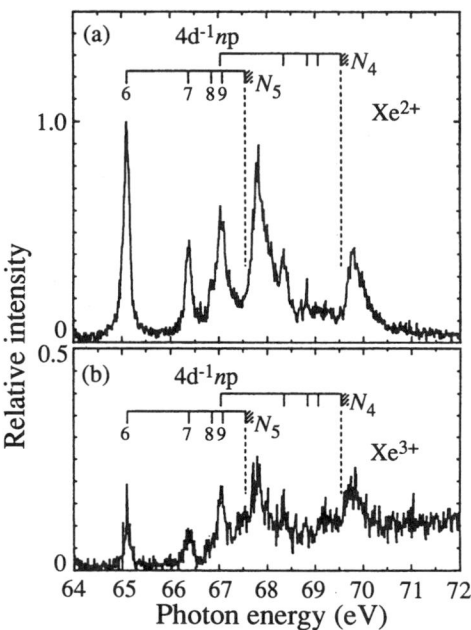

FIGURE 8. Relative yields of (a) $C(e_0,\mathrm{Ar}^{2+})$ and (b) $C(e_0,\mathrm{Ar}^{3+})$ coincidences in the range of the resonance excitations $4d^{-1} \to np$ and above the ionization thresholds N_5 and N_4 for $4d$ ionization in xenon.[44]

distribution, and the two peaks at 21 ± 3 and 9.5 ± 1.5 eV can be explained by combinations of the two-step decays seen in Figure 6. The analysis of several such spectra leads to the result that resonance decay to $\mathrm{Xe}^{2+}(5p^{-2})$ is predominantly due to a two-step resonant Auger process while one-step shakeoff is the dominant process leading to $\mathrm{Xe}^{2+}(5s^{-1}5p^{-1})$ and $\mathrm{Xe}^{2+}(5s^{-2})$.

$C(e_0,m^{n+})$ coincidences can also provide information on the decay branches, as shown in Figure 8. It can be seen that the $4d_{5/2} \to 6p$ resonance at 65.11 eV photon energy decays to doubly (and even triply) charged ions and, as selected by the coincidence technique, a threshold electron. A detailed analysis[44] yields the result that threshold-electron emission is enhanced in the $4d \to 6p, 7p, 8p$ resonance excitations through two-step decay. Xe^{3+} formation is presumably due to continuous double Auger decay (two-electron emission during the Auger transition), plus shakeoff. Above the two ionization thresholds indicated as N_5 and N_4, respectively, normal Auger decay (two-step process modified by postcollision interaction) and discrete double Auger decay (electron emission accompanied by excitation of another electron during the Auger transition) is responsible for the production of Xe^{2+}, while continuous double Auger decay is responsible for the formation of Xe^{3+}. From the $C(e_0,m)$ spectra it can be seen that threshold electrons are produced not only at these thresholds but also for appreciably higher energies, and moreover with higher intensity. This behavior is due to the combined action of an increasing cross section for $4d$ photoionization (shape resonance) and the effect of postcollision interaction between the slow photoelectron and the faster Auger electron; there is an energy exchange between the two electrons making the slower photoelectron even slower, leading to the production of zero-kinetic-energy photoelectrons with some probability. In addition, the two electrons emitted in the continuous double Auger process have an unknown but continuous energy distribution which will also contribute zero-energy electrons.

For rare gas atoms, the existence and relative strength of continuous double Auger transitions has often been inferred experimentally by interpreting the yields of differently charged ions. However, the natural way to study continuous double Auger decay and competing processes would be $C(e_0,m^{n+})$ and related coincidence experiments which directly yield the strengths of pathways to differently charged ions. Such experiments have become feasible recently (Kämmerling et al.[48]; see

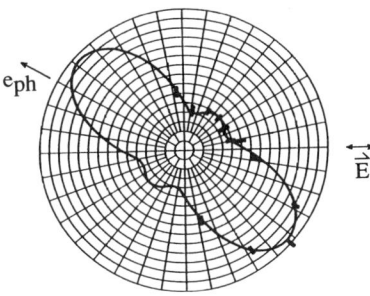

FIGURE 9. Energy- and angle-resolved C(e,e) coincidences for two-step double ionization induced by 94.5 eV photons in xenon. The photon beam goes through the origin of the figure and is perpendicular to the plane of the drawing. Its main component of linear polarization (Stokes parameter $S_1 = 0.957$) is indicated by E. The $4d_{5/2}$ photoelectrons are observed in the direction e_{ph}, the intensities of true coincident N_5-$O_{23}O_{23}$ $^1S_0$ Auger transitions are given as points with error bars. The solid line represents the result of a least-squares fit to the experimental data that follows from the theoretical treatment of this two-step process.[50]

also related coincidence studies by Chaudry et al.[13] and Levin et al.[65]) and offer exciting prospects for future work.

3.1.4. Complete Characterization of a Two-Electron Emission in Xenon. The $4d_{5/2}$ photoionization in xenon with subsequent N_5-$O_{23}O_{23}$ $^1S_0$ Auger decay, where the $4d_{5/2}$ hole is filled by one $5p$ electron while another $5p$ electron is emitted leaving a $^1S_0$ ion, represents a special case of photon-induced two-electron emission for two reasons. First, the condition for a two-step formulation can be shown to be fulfilled and a small perturbation by postcollision interaction can be taken into account.[49] Second, only one partial wave, $\varepsilon d_{5/2}$, contributes for the selected Auger transition and as a result the dynamics of this process depend on the photoionization matrix elements only. Therefore, a measurement of the angular correlation between the electrons gives additional information on the photoionization process (within the dipole approximation which is well fulfilled here). Experimental results for such a pattern of two-electron emission, measured at 94.5 eV photon energy and in a plane perpendicular to the photon beam direction, for a fixed position of the emitted photoelectron and for several angles of the coincident Auger electron are shown in Figure 9. The analysis of this pattern together with data on the partial cross section $\sigma(4d_{5/2})$, on the angular distribution parameter $\beta(4d_{5/2})$, and on the alignment parameter $A_{20}(4d_{5/2})$ provides a sufficient number of observables for the evaluation of the dipole matrix elements and their relative phases. In other words, a complete experiment can be performed which completely determines the energy-, angle, and spin-resolved two-electron emission patterns for any selected experimental setup.[50,51,85]

3.2. Photoionization of Molecules

Because of the possibilities of dissociation into atomic or molecular fragments the choice of decay pathways for an excited molecule is much richer than for an atom. For doubly charged and multiply charged ions the dissociation pathways are even more numerous, with many permutations of charges among the fragments. Coincidence experiments are essential to sort out this complexity, and some must inevitably be multiple rather than twofold coincidence experiments. There is no reason to think that the electronic process of double ionization by single photons is different in molecules as compared with atoms. There are both direct and indirect (multistep) routes to double ionization in both cases, but the consequences of indirect double ionization in molecules are more noticeable because of the extra richness of molecular energy levels among which intensity can be shared at each step. A molecular property absent in the decay of atoms is kinetic energy release. Happily the TOF method allows the initial kinetic energies of ionic species from dissociations of molecular ions to be determined easily. Under the proper time–focus conditions the time of flight t for any ion is

$$t = t^0 - p\cos\theta/eE \qquad (13)$$

where t^0 is the flight time for an initially stationary ion, p is the initial momentum, θ is its inclination to the spectrometer axis, E is the electric field strength in the source, and e is the elementary charge. Thus, for a single-valued initial kinetic energy the maximum excursion of t from its central value is p/eE, and the width of the TOF peak is just $2p/eE$, corresponding to forward and backward directed

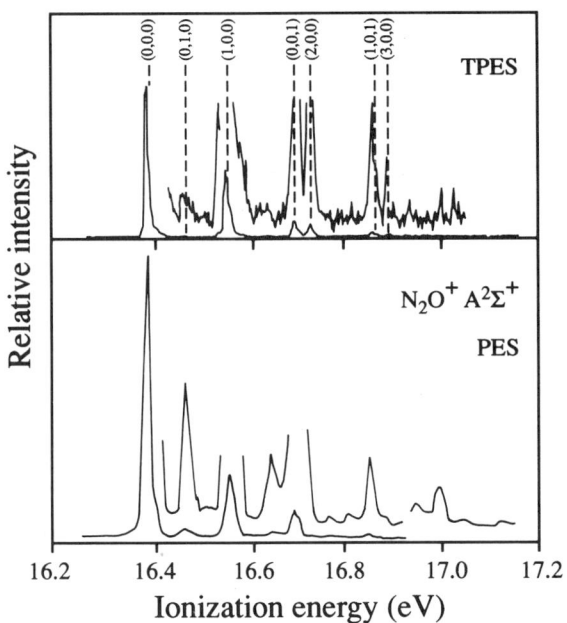

FIGURE 10. Threshold photoelectron spectrum (above) and HeI (58.4 nm) photoelectron spectrum (below) of N_2O showing the $A^2\Sigma^+$ state of N_2O^+. The differences between the two spectra are probably due to autoionization. (After Refs. 97 and 83.)

momenta. If the angular distribution of the initial momentum is isotropic, the TOF peak shape is flat-topped, and it is hollow or peaked, and easily calculable, for different angular distributions.

3.2.1. Predissociation from Vibrational Levels of N_2O^+ $A^2\Sigma^+$. The nitrous oxide molecule has the linear form NNO, and this linearity is preserved in all of the ionic states populated readily by photoionization. The first excited state of the ion, N_2O^+ $A^2\Sigma^+$, lies above the dissociation limits for formation of the products $NO^+ + N$ and $O^+ + N_2$. For every level within the A electronic state of N_2O^+ there are three possible fates: decay to the ground state of N_2O^+, dissociation to $NO^+ + N$, or dissociation to $O^+ + N_2$. Decay to the ground state might be either by emission of a photon or by nonradiative conversion from A to a high vibrational level of the ground state (internal conversion) followed by infrared emission terminating in a stable level. The actual decay pathways of N_2O^+ from the A state have been investigated by both coincidence techniques and laser techniques, and illustrate the relationship between the two groups of methods.

The accessible energy levels in N_2O^+ $A^2\Sigma^+$ are shown in Figure 10 in the form of a photoelectron spectrum taken at an excitation energy of 21.22 eV, and a threshold photoelectron spectrum. The strongest peak in both spectra represents formation of the vibrationless (0,0,0) level; the first coincidence experiments on N_2O^+ were measurements of product ions in coincidence with energy-analyzed photoelectrons, $C(e,m^+)$ and showed[12,21] that all the ions formed initially in $A(0,0,0)$ remain as N_2O^+ molecules, and do not dissociate. That their fate is actually to emit light in falling down to the ground state of N_2O^+, rather than to undergo internal conversion, is proved by two later experiments in which the emitted photons were detected in coincidence with ions or with electrons. In the $C(h\nu,m^+)$ or PIFCO experiment,[27] the total yield of photons per ion formed was measured. By comparison with the photoelectron spectrum, which gives the fraction of the ions formed in each initial state and level, the quantum yield of emission was found to be 1.0 ± 0.03. The same quantum yield was determined more directly later by a $C(h\nu,e)$ or PEFCO measurement in which the emission of each initial vibrational level was studied.[66]

A unit quantum yield of fluorescence from the (0,0,0) level of the A state of N_2O^+ is consistent with the well-studied emission spectrum of N_2O^+, which also contains emission from several other vibrational levels for which the coincidence measurements show that there is competition between radiative and dissociative decay pathways. The original $C(e,m^+)$ experiment[21] indicated an approximately 60:40 split between emission and dissociation to $NO^+ + N$ for all of the vibrationally excited levels, which were only partially resolved. In the first $C(e_0,h\nu)$ experiment on this system,[37] the emission quantum yield of (1,0,0) was determined as 66% and the emission lifetimes were also measured. The quantum yields and lifetimes for several more vibrational levels were determined in the $C(e,h\nu)$ experiment of Maier and Thommen,[66] with similar results. Knowledge of the lifetimes and quantum yields allows the rate constants for predissociation and radiative decay to be determined from each initial quantum state; in practice the radiative decay rate is found to be essentially independent of vibrational level. The predissociation rates from individual levels have been augmented by more precise lifetime determinations by a noncoincidence method[52] and by high-resolution $C(e_0,m^+)$ experiments giving dissociation quantum yields for all of the levels shown in Figure 10.[83]

In addition to the rates of radiative and dissociative decay from each vibrational level, the early $C(e,m^+)$ experiments[12] and the latest $C(e_0,m^+)$ experiments[83] also give the branching ratios to the different possible products, NO^+ or O^+, and the initial kinetic energies of the ions formed. Because the N_2O ion is linear, strong rotational excitation of the diatomic products is not expected and the kinetic energy distributions can be translated into vibrational state distributions for the NO^+ ion. These show a sharply peaked population distribution centered at NO^+ $v = 4$, with marked population inversion, when decay is from N_2O^+ (1,0,0). Higher initial vibration levels also give sharply peaked distributions, with intensity moving to even higher levels in NO^+.

These coincidence results provide the richest set of state-to-state reaction dynamics information in existence on any polyatomic molecular ion. They have recently been confirmed and amplified by laser techniques which have provided much higher resolution. The latest work, where references to other laser experiments will be found, involved excitation of N_2O^+ in a fast-ion-beam spectrometer and detection of the products with an electrostatic analyzer.[64] The remarkable resolution of this technique allowed selection not only of individual rotational levels but even of hyperfine sublevels.[1] There has also been theoretical activity; according to the currently accepted model,[54,83] the $A^2\Sigma^+$ state of N_2O^+ is predissociated weakly by a $^4\Pi$ state which crosses it above (0,0,0). The $^4\Pi$ state is strongly predissociated in its turn by a $^4\Sigma^-$ state that leads to the observed products. The position at present is that the experimental results from coincidence and other experiments contain much detail still unmatched by theoretical explanation.

3.2.2. Tests of Statistical Theory of Ion Dissociation. As the number of atoms in a molecule increases, so does the number of vibrational modes; at a given excitation energy the density of states increases roughly exponentially with the number of atoms. If the density of states is high enough, any individual vibrational level can be in communication with many other levels of the same symmetry and at the same energy. Free flow of energy between all of the accessible states [internal vibrational redistribution (IVR)] defines the "statistical limit" of nonradiative transition theory, and is assumed in the standard forms of reaction rate theory of unimolecular reactions. Coincidence methods have played an important part in testing this theory, particularly as it is applied in mass spectrometry where the name QET (quasi-equilibrium theory) is used.

Coincidence tests of the QET theory are of two types, both based on $C(e,m^+)$ or $C(e_0,m^+)$ experiments in which reactant ions of defined energy content are selected. The mass spectrum, measured in coincidence, represents the decay of the state-selected ions; details of the peak shapes give the kinetic energy releases and rates of reaction in favorable cases. First, the fundamental QET assumption of free flow of energy can be tested directly, because it leads to the predictions that rates, branching ratios, and energy releases should vary smoothly as functions of energy only, and that the lifetime and kinetic energy release distributions should be monotonic functions of their parameters. Furthermore, all decays of the same state must be in competition, so the same lifetime distribution of parent molecules must be measured in all decay channels at the same energy. Second, measured lifetime distributions, kinetic energy release distributions, and branching ratios can be compared with numerical predictions of the detailed theory, assuming free internal energy flow. The rate constant for a single internal energy, for instance, is calculated as

$$k(E) = \frac{\int_0^{E-E_0} \rho(E^+) dE^+}{h \rho(E)} \quad (14)$$

where E_0 is the reaction threshold, a superscript "+" denotes the transition state, h is Planck's constant, and $\rho(E)$ is the density of states at energy E. The difficulty with this approach is that the properties of the transition state, which are needed for the calculation and may be different in different decay channels, are not open to experimental measurement and so offer adjustable fitting parameters. Coincidence studies do show, however, that the QET theory accounts for the overall form of the decay curves in large molecular ions, and can fit many results in some detail with a reasonable choice of parameters.[5]

As an example of a case where the QET model is not applicable, Figure 11 shows part of the breakdown diagram of formic acid, HCOOH, determined by Nishimura et al.[75] The intensities of the different ions in the mass spectrum are shown as a function of the initial internal energy, with chemical identities of the products as determined by isotopic labeling. The strange oscillations of the CHO^+ and $COOH^+$ intensities are completely inexplicable on any statistical basis; instead they correlate strongly with different electronic states of formic acid seen as bands in the photoelectron spectrum. In the same experiments the kinetic energy release distributions in CHO^+ formation were shown to be bimodal, which also denotes a nonstatistical origin, probably from noninterconverting electronic states.

3.2.3. Double Photoionization of O_2 and CO. Double ionization of small molecules significantly affects the bonding, so according to the Franck–Condon principle we expect molecular double-ionization spectra to contain broad bands or long vibrational progressions. Two-step double ionization is even more complicated because the first step can populate a range of vibrational levels of an intermediate singly charged ion whose equilibrium geometry may be different; in the second step each of these levels has its own set of Franck–Condon factors to levels of the final doubly charged ion. The resulting distribution over final vibrational levels is likely to be very broad and possibly bi- or multimodal, like the vibrational distributions seen in photoelectron spectra at wavelengths where there is strong autoionization.[88] The effect of both direct and indirect ionization pathways is evident in the case of double photoionization of oxygen, for which examples of cuts from the $C(e,e)$ spectrum at a wavelength of 30.4 nm[79] are given in Figure 12. They are contrasted with a $C(e_0,e_0)$ spectrum of the same molecule using variable-wavelength light.[42] The most obvious contrast is the much better

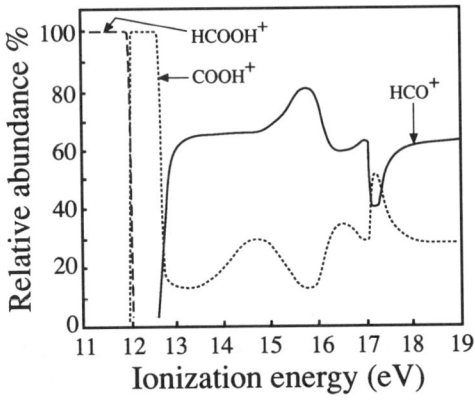

FIGURE 11. Part of the corrected breakdown diagram of formic acid ions from the $C(e_0,m^+)$ results of Ref. 75 showing branching ratios to the principal product ions as a function of the initial energy.

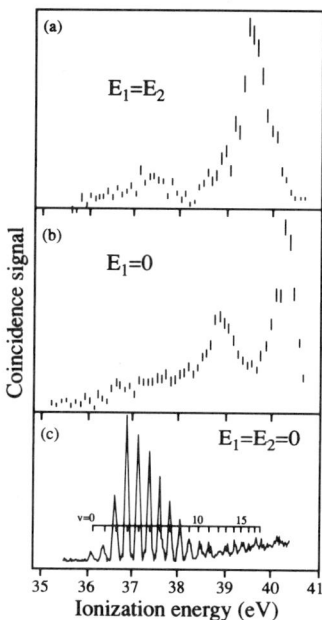

FIGURE 12. $C(e,e)$ spectra of oxygen in the region of the lowest state of O_2^{2+}. Spectra (a) and (b) were taken with fixed-wavelength light at 30.4 nm (40.8 eV) at a resolution of 0.4 eV, while (c) was measured with variable-wavelength light at a resolution of 0.08 eV. (After Refs. 42 and 79.)

resolution of the $C(e_0,e_0)$ spectrum, but the lower resolution cuts also exhibit structure. The difference between them in intensity distribution and their multihumped nature both stem from indirect double-ionization processes, because in the available energy range, only one electronic state of O_2^{2+}, the ground state, $^1\Sigma_g^+$, exists. In this case an intermediate state of O_2^* is formed and can partially dissociate before emitting a second electron in an autoionization to produce either the molecular dication O_2^{2+} by transition to the ground state in a bound region, or a dissociating pair $O^+ + O^+$ by transition to the same ground state in an unbound region. The photoelectron spectrum at this wavelength confirms the mechanism by showing line structure characteristic of autoionization of atomic oxygen atoms. New $C(h\nu,m^+)$ experiments indicate that the intermediate singly charged oxygen molecule dissociates partly to an atomic ion plus an excited atom, which can emit light in the process of decay.[33]

The formation of cation pairs from dication dissociation is observed directly in the $C(m^+,m^+)$ experiment (PIPICO), where two ions are detected in coincidence at a single detector. The PIPICO spectra of O_2 at different wavelengths[31] confirm the above model of indirect dissociative double ionization, by showing formation of $O^+ + O^+$ pairs of low kinetic energy at photon energies below the dissociation barrier in the O_2^{2+} ground state. A similar process of indirect dissociative double ionization below the direct double-ionization threshold has been found for CO in a particularly detailed PIPICO study (Lablanquie et al.[60]), where $C^+ + O^+$ appears at longer wavelengths than CO^{2+}. Auger electron–ion coincidence experiments of Eberhardt et al.[19] had already shown directly that the singlet states of CO^{2+} at energies near threshold do not dissociate at all, while higher states form cation pairs. Lablanquie et al. attributed the formation of the $C^+ + O^+$ ion pair at low energy to a two-step process where excited CO^+ dissociates before one of the atomic products autoionizes, releasing a second electron. This suggestion has been confirmed by the observation of discrete atomic autoionization lines at low kinetic energies in the photoelectron spectrum of CO at several wavelengths.[6] When the kinetic energies of C^+ and O^+ cation pairs from the higher two-hole valence states of CO^{2+} formed by Auger decay are compared in detail with energies of the same ion pairs formed by pure valence-shell double photoionization, they are found to be distinctly different.[60] This important result demonstrates that the strong electronic polarization and geometry changes induced in molecules by localized inner-shell holes play a major role in determining the final states of molecular Auger decay. In molecules, valence and inner-shell excitation do not, in general, populate the same levels of the doubly charged ions.

3.2.4. Angular Distribution in Double Photoionization of H_2. Hydrogen, being the simplest molecular two-electron system, frequently serves as a model for more complicated species and it allows comparison with its atomic counterpart, helium. In the present context the complete fragmentation pattern of H_2 by double photoionization to yield four charged particles in the continuum, subject to their mutual Coulomb interactions, is of special interest. For energies not too far from threshold, the photon operator is well described by the dipole approximation which allows only two transition amplitudes, D_Σ and D_Π, from the molecular ground state to the double-ionization continuum. These amplitudes belong to photoionization with molecular orientation along (Σ) or perpendicular (Π) to the electric vector of the incident light. Within the axial-recoil model which applies well in the present case, the angular distribution parameter β_m of the fragment ions is given by

$$\beta_m = 2\frac{D_\Sigma^2 - D_\Pi^2}{D_\Sigma^2 + 2D_\Pi^2} \tag{15}$$

and will provide information on the spatial orientation of the fragmentation process.

In measurements of $C(H^+,H^+)$ coincidences a pronounced angular distribution for the recoiling proton-pair has been found where the β_m values decrease toward $\beta_m = -0.75$ in the threshold region (see Figure 13). This value implies a predominance of the D_Π amplitude in the threshold region, so that the fragment ions show a preference for sideways emission with respect to the electric vector of the light. A similar observation slightly below threshold was made by Dehmer and Dill.[15] This emission pattern may be a signal that the electron pair is also emitted preferentially sideways and both pairs are emitted at right angles to each other as predicted theoretically.[32]

3.2.5. Mechanisms of Decay of Molecular Dications; NF_3^{2+}. The decay of a molecule into three fragments is analogous to the double ionization of an atom; the way the products share any excess energy is not determined by kinematics, but depends on the intrinsic dynamics of the dissociation reaction. Conventional experiments which examine reaction cross sections or angular distributions for one fragment at a time cannot easily determine even whether the overall mechanism is stepwise or simultaneous.[90] This problem is solved for dication decays by the threefold coincidence technique $C(e,m^+,m^+)$, PEPIPICO, where the detection of a photoelectron defines a START time from which the

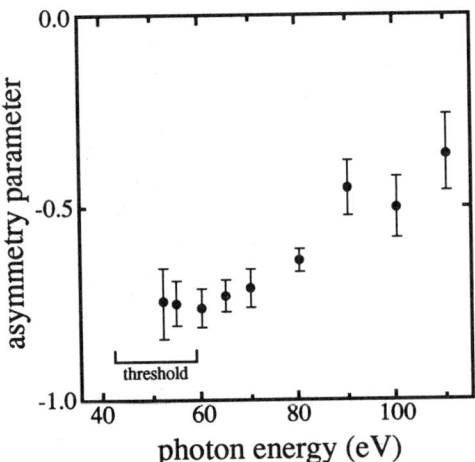

FIGURE 13. Angular distribution parameter β_m for $C(H^+,H^+)$ coincidences between the fragment ions from double photoionization of H_2. Nuclear motion in the molecular ground state produces an extended threshold range, which is also indicated.[56]

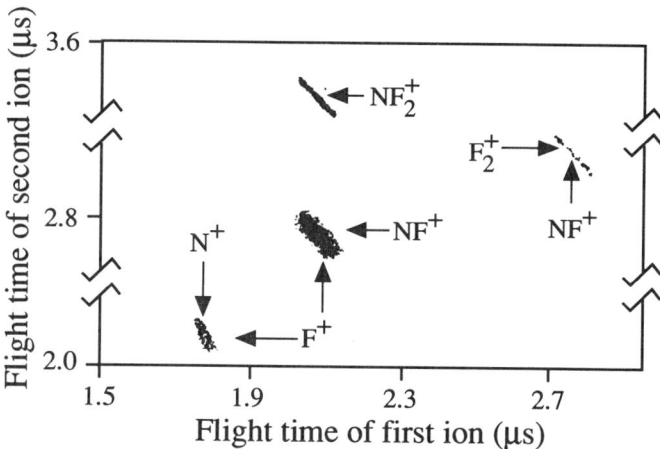

FIGURE 14. PEPIPICO or $C(e, m_1^+, m_2^+)$ spectrum of NF_3 at 25.6 nm exciting wavelength and with an extraction field of 300 V cm^{-1}. The accidental coincidence background and some interfering ion–neutral coincidences have been removed for clarity.

flight times of two ions are measured. The PEPIPICO experiment gives a two-parameter spectrum with intensity as a function of two ion flight times. Each distinct reaction produces a coincidence peak, as illustrated in Figure 14, and each peak has a distinctive shape.

For each ion of a pair the measured time of flight (under the correct focus conditions) depends linearly on its initial momentum component along the spectrometer axis, [equation (13)], which can be deduced directly. If the ions are formed in a two-body reaction, they must start with equal but opposed momenta, to obey the conservation law. If one ion arrives early, the other must arrive equally late; thus, the shape of the PEPIPICO peak should be a straight line of unit negative slope; real two-body peaks are indeed like this. For three-body reactions, by contrast, many different peak shapes are found, because the initial momentum distributions are determined by the reaction dynamics. In most cases the peak shapes show negative nonunit slopes (partial anticorrelation of momenta) and more or less enlarged perpendicular widths. The detailed mechanism of the decay reactions can be deduced from these shapes.[24,25]

Because the relation between peak shape and the initial vectors is complex, the analysis is based on model reaction mechanisms. Three models give characteristic and recognizable types of shape,[25] which have been well observed experimentally. The first is *instantaneous ("coulomb") explosion*, in which three or more particles fly apart with significant momenta at fixed relative angles. This mechanism produces peaks with curved outlines; a few examples have been found among doubly charged ions, and many more where triply charged ions dissociate. A second model mechanism is a two-step reaction in which a neutral fragment is emitted from a doubly charged parent well before the latter dissociates into two ions. This mechanism, called *deferred charge separation*, is very well attested both by the accurate match between simulated and observed peak shapes and by the observation of characteristic "metastable" phenomena in PEPIPICO spectra. The third model mechanism also has two steps. First the parent dication breaks into two ions, then one of the ions breaks again into an ion and a neutral. This *secondary dissociation* mechanism gives as its peak shape a broadened line with slope equal to the ratio of two masses, if the intermediate has time to rotate freely to a random angle and to escape from the other ion before its own decay. The mathematical basis of a method to identify peaks from these two-step mechanisms and to determine the kinematic parameters has been given recently.[61]

The application of these ideas can be illustrated by the PEPIPICO spectrum of NF_3 at 25.6 nm in Figure 14. There are two two-body peaks representing $NF_2^+ + F^+$ and $F_2^+ + NF^+$ pairs, with relative intensities of 39 and 1%, respectively. Both are narrow lines of slope -1, as required; from the length of the stronger peak, the initial kinetic energy of the ions is found to be 5.5 ± 0.5 eV. This corresponds to an intercharge distance of 2.6 Å in the parent dication, compatible with the structure of the neutral molecule. The third peak is the strongest in the spectrum (100%), and shows three-body formation of $NF^+ + F^+ + F$. It is broad and has a slope near -1; this suggests that the basic mechanism is deferred charge separation in which a neutral F is emitted from the parent NF_3^{2+} leaving NF_2^{2+} which later undergoes charge separation to the products. The ion NF_2^{2+} is observed in the mass spectrum at this wavelength whereas NF_3^{2+} is not observed, confirming this idea. The final peak, for formation of $N^+ + F^+$, is weak (3.7%), broad, and of slope -1.5 ± 0.1; this suggests that the N^+ ion is of secondary origin, since if it were formed from NF^+ the peak slope should be -2.36 (i.e, m_{NF}/m_N). The value of the slope evidently does not match the prediction for the pure secondary decay model; the discrepancy can be explained within the same basic mechanism, if the reaction is so fast that secondary decay of the NF^+ occurs within the Coulomb field of the other F^+ ion.[22,26] On closer examination it is found that the detailed shape of the $NF^+ + F^+$ peak does not exactly fit the deferred charge separation model either, but requires F ejection partly before but partly during charge separation. This and many other cases show that the model mechanisms must be considered as limiting cases, which are only approached in real dication decays. Nevertheless it is clear that the decay pathways in NF_3^{2+} to three or four fragments are mainly stepwise, though fast enough to compete with charge separation.

4. CONCLUSIONS AND OUTLOOK

Coincidence methods in atomic and molecular physics began 30 years ago[69]; what we have chronicled here has been the result of a great expansion in the use of such methods in photoionization work over the last decade. The coincidence techniques applied to atomic and molecular physics stand at an interesting threshold, mainly as a result of technical advances. Commercial equipment of great sophistication is now available for multiparameter coincidence spectroscopy "off the shelf." At the same time a detailed understanding of coincidence capabilities, statistics, and data reduction, already long available among the nuclear and particle physics communities, is becoming more and more widespread in small-scale physics and chemistry groups. Some of the early twofold coincidence techniques have perhaps passed their heydays and been overtaken by laser pump-probe methods, but others, particularly based on synchrotron sources, are even now being pioneered. Many flagship experiments of today and tomorrow seem likely to involve threefold and higher coincidences, as questions beyond the range of the lower-order coincidence techniques are addressed. The new generation of much brighter synchrotron radiation sources now coming on-line will allow experiments to be done on inherently weak processes, such as the direct double ionization of helium[87], and will also allow current experiments to be done with higher energy resolution. More demanding and sophisticated experiments such as determination of the recently predicted circular dichroism in double photoionization (Berakdar and Klar[8]) will become possible. For very-high-resolution spectroscopy on ions and ionization, however, even the new generation of synchrotrons seem unlikely to be able to compete with frequency-shifted lasers. Lasers and the ZEKE technique, which allows complementary high resolution, are sure to be combined with coincidence methods to engender a group of new high-resolution experiments. It must remain an open question whether fundamental physical insights into high-energy phenomena of ionization and dissociation will in the future come more from synchrotron sources, from laser experiments, or from particle impact. Whichever source of input energy is chosen, all will unquestionably make more and more use of coincidence methods.

Note added in proof: This manuscript was originally completed in November 1992.

REFERENCES

1. S. ABED, M. BROYER, M. CARRÉ, M. L. GAILLARD, AND M. LARZILLIÉRE, *Phys. Rev. Lett.* **49**, 120 (1982).
2. M. Y. AMUSIA, I. S. LEE, AND V. A. KILIN, *Phys. Rev. A* **45**, 4576 (1992).
3. T. BAER, W. B. PEATMAN, AND E. W. SCHLAG, *Chem. Phys. Lett.* **4**, 243 (1969).

4. T. Baer, P. M. Guyon, I. Nenner, A. Tabche-Fouhaille, R. Botter, L. F. A. Ferreira, and T. R. Govers, *J. Chem. Phys.* **70**, 1585 (1979).
5. T. Baer, J. Booze, and K.-M. Weitzel, in *Vacuum Ultraviolet Photoionization and Photodissociation of Molecules and Clusters*, edited by C. Y. Ng (World Scientific, Singapore, 1991), p. 259.
6. U. Becker, O. Hemmers, B. Langer, A. Menzel, and R. Wehlitz, *Phys. Rev. A* **45**, R1295 (1992).
7. U. Becker, T. Prescher, E. Schmidt, B. Sonnatag, and H.-E. Wetzel, *Phys. Rev. A* **33**, 3867 (1986).
8. J. Berakdar and H. Klar, *Phys. Rev. Lett.* **69**, 1175 (1992).
9. E. G. Berezhko and N. M. Kabachnik, *J. Phys. B* **12**, 2993 (1979).
10. M. J. Besnard, L. Hellner, Y. Malinovich, and G. Dujardin, *J. Chem. Phys.* **85**, 1316 (1986).
11. B. Brehm and E. von Puttkamer, *Adv. Mass Spectrom.* **4**, 591 (1967).
12. B. Brehm, R. Frey, A. Küstler, and J. H. D. Eland, *Int. J. Mass Spectrom. Ion Phys.* **13**, 251 (1974).
13. M. A. Chaudry, A. J. Duncan, R. Hippler, and H. Kleinpoppen, *Phys. Rev. Lett.* **59**, 2036 (1987).
14. D. M. Curtis and J. H. D. Eland, *Int. J. Mass Spectrom. Ion Proc.* **63**, 241 (1985).
15. J. L. Dehmer and D. Dill, *Phys. Rev. A* **18**, 164 (1978).
16. G. Dujardin, S. Leach, O. Dutuit, P. M. Guyon, and M. Richard-Viard, *Chem. Phys.* **88**, 339 (1984).
17. C. Dupré, A. Lahmann-Bennani, and A. Duguet, *Meas. Sci. Technol.* **2**, 327 (1991).
18. W. Eberhardt, *Phys. Scr.* **T17**, 28 (1987).
19. W. Eberhardt, E. W. Plummer, I. W. Lyo, R. Reininger, R. Carr, W. K. Ford, and D. Sondericker, *Aust. J. Phys.* **39**, 633 (1986).
20. J. H. D. Eland, *Int. J. Mass Spectrom. Ion Phys.* **8**, 143 (1972).
21. J. H. D. Eland, *Int. J. Mass Spectrom. Ion Phys.* **12**, 389 (1973).
22. J. H. D. Eland, *Mol. Phys.* **61**, 725 (1987).
23. J. H. D. Eland, *AIP Conf. Proc.* **215**, 549 (1990).
24. J. H. D. Eland, in *Vacuum Ultraviolet Photoionization and Photodissociation of Molecules and Clusters*, edited by C. Y. Ng (World Scientific, Singapore, 1991), p. 297.
25. J. H. D. Eland, *Laser Chem.* **11**, 259 (1991).
26. J. H. D. Eland and B. J. Treves-Brown, *AIP Conf. Proc.* **258**, 100 (1992).
27. J. H. D. Eland, M. Devoret, and S. Leach, *Chem. Phys. Lett.* **43**, 97 (1976).
28. J. H. D. Eland and D. A. Hagan, *Int. J. Mass Spectrom. Ion Proc.* **100**, 489 (1990).
29. J. H. D. Eland and D. Mathur, *Rapid Commun. Mass Spectrom.* **5**, 475 (1991).
30. J. H. D. Eland, F. S. Wort, and R. N. Royds, *J. Electron Spectrosc. Relat. Phenom.* **41**, 297 (1986).
31. J. H. D. Eland, S. D. Price, J. C. Cheney, P. Lablanquie, I. Nenner, and P. G. Fournier, *Philos. Trans. R. Soc. London Ser. A* **324**, 247 (1988).
32. J. M. Feagin and R. D. Filipczyk, *Phys. Rev. Lett.* **64**, 384 (1990).
33. T. Field and J. H. D. Eland, *Chem. Phys. Lett.* **211**, 436 (1993).
34. S. Frankel, *Phys. Rev.* **83**, 673 (1951).
35. L. J. Frasinski, M. Stankiewicz, K. J. Randall, P. A. Hatherley, and K. Codling, *J. Phys. B* **19**, L819 (1986).
36. L. J. Frasinski, M. Stankiewicz, P. A. Hatherley, and K. Codling, *Int. J. Mass Spectrom Ion Proc.* **116**, 37 (1992).
37. R. Frey, B. Gotchev, W. B. Peatman, H. Pollak, and E. W. Schag, *Chem. Phys. Lett.* **54**, 411 (1978).
38. D. A. Hagan and J. H. D. Eland, *AIP Conf. Proc.* **225**, 163 (1990).
39. R. I. Hall, A. McConkey, K. Ellis, G. Dawber, M. A. MacDonald, and G. C. King, *J. Phys. B* **25**, 799 (1992).
40. R. I. Hall, G. Dawber, A. G. McConkey, M. A. MacDonald, and G. C. King, *Z. Phys. D* **23**, 377 (1992).
41. R. I. Hall, K. Ellis, A. McConkey, G. Dawber, L. Avaldi, M. A. MacDonald, and G. C. King, *J. Phys. B* **25**, 377 (1992).
42. R. I. Hall, G. Dawber, A. McConkey, M. A. MacDonald, and G. C. King, *Phys. Rev. Lett.* **68**, 2751 (1992).
43. D. M. Hanson, C. I. Ma, K. Lee, D. Lapiano-Smith, and D. Y. Kim, *J. Chem. Phys.* **93**, 9200 (1990).
44. T. Hayaishi, A. Yagishita, E. Shigemasa, E. Murakami, and Y. Morioka, *J. Phys. B* **23**, 4431 (1990).
45. A. Huetz, P. Selles, D. Waymel, and J. Mazeau, *J. Phys.* **24**, 1917 (1991).
46. R. E. Imhof, A. Adams, and F. H. Read, *J. Phys. E* **9**, 138 (1976).
47. N. M. Kabachnik, *J. Phys. B* **25**, L389 (1992).
48. B. Kämmerling, B. Krässig, and V. Schmidt, *J. Phys. B* **25**, 3621 (1992).
49. B. Kämmerling, B. Krässig, and V. Schmidt, *J. Phys. B* **26**, 261 (1993).
50. B. Kämmerling, B. Krässig, and V. Schmidt, *Phys. Rev. Lett.* **67**, 1848 (1991).
51. B. Kämmerling and V. Schmidt, *J. Phys. B* **26**, 1141 (1993).
52. D. Klapstein and J. P. Maier, *Chem. Phys. Lett.* **83**, 590 (1981).

53. H. KLAR AND M. FEHR, *Z. Phys. D* **23**, 295 (1992).
54. N. KOMIHA, Thèse de troisième cycle, University of Paris Sud (1981).
55. H. KOSSMANN, *Meas. Sci. Technol.* **4**, 16 (1993).
56. H. KOSSMANN, O. SCHWARZKOPF, B. KÄMMERLING, AND V. SCHMIDT, *Phys. Rev. Lett.* **63**, 2040 (1989).
57. B. KRÄSSIG AND V. SCHMIDT, *J. Phys. B* **25**, L327 (1992).
58. M. O. KRAUSE, M. L. VESTAL, W. H. JOHNSTON, AND T. A. CARLSON, *Phys. Rev.* **133**, A385 (1964).
59. P. LABLANQUIE, J. H. D. ELAND, I. NENNER, P. MORIN, J. DELWICHE, AND M. J. HUBIN-FRANSKIN, *Phys. Rev. Lett.* **58**, 992 (1987).
60. P. LABLANQUIE, J. DELWICHE, M.-J. HUBIN-FRANSKIN, I. NENNER, P. MORIN, K. ITO, J. H. D. ELAND, J. M. ROBBE, G. GANDARA, J. FOURNIER, AND P. G. FOURNIER, *Phys. Rev. A* **40**, 5673 (1989).
61. M. LAVOLLÉE AND H. BERGERON, *J. Phys. B* **25**, 3101 (1992).
62. J. S. LAWSON AND H. FRAUENFELDER, *Phys. Rev.* **91**, 649 (1953).
63. T. LEBRUN, Doctoral thesis, Université Paris Sud (1991).
64. J. LERMÉ, S. ABED, R. A. HOLT, M. LARZILLIÈRE, AND M. CARRÉ, *J. Chem. Phys.* **84**, 2167 (1986).
65. J. C. LEVIN, C. BIEDERMANN, N. KELLER, L. LILJEBY, C.-S. O, R. T. SHORT, I. A. SELLIN, AND D. W. LINDLE, *Phys. Rev. Lett.* **65**, 988 (1990).
66. J. P. MAIER AND F. THOMMEN, *Chem. Phys.* **51**, 319 (1980).
67. T. MASUOKA AND J. A. R. SAMSON, *J. Chim. Phys.* **77**, 623 (1980).
68. J. MAZEAU, P. SELLES, D. WAYMEL, AND A. HUETZ, *Phys. Rev. Lett.* **67**, 820 (1991).
69. K. E. MCCULLOH, T. E. SHARP, AND H. M. ROSENSTOCK, *J. Chem. Phys.* **42**, 3501 (1965).
70. P. MORIN, personal communication (1992).
71. A. J. MURRAY, B. C. H. TURTON, AND F. H. READ, *Rev. Sci. Instrum.* **63**, 3346 (1992).
72. R. MURPHY AND W. EBERHARDT, *J. Chem. Phys.* **89**, 4054 (1988).
73. K. MÜLLER-DETHLEFS AND E. W. SCHLAG, *Annu. Rev. Phys. Chem.* **42**, 109 (1991).
74. K. MÜLLER-DETHLEFS, M. SANDER, L. A. CHEWTER, AND E. W. SCHLAG, *J. Phys. Chem.* **88**, 6098 (1984).
75. T. NISHIMURA, G. G. MEISELS, AND Y. NIWA, *J. Chem. Phys.* **91**, 4009 (1989).
76. K. OKUYAMA, J. H. D. ELAND, AND K. KIMURA, *Phys. Rev. A* **41**, 4930 (1990).
77. S. D. PRICE AND J. H. D. ELAND, *J. Electron. Spectrom.* **52**, 649 (1990).
78. S. D. PRICE AND J. H. D. ELAND, *J. Phys. B* **23**, 2269 (1990).
79. S. D. PRICE AND J. H. D. ELAND, *J. Phys. B* **24**, 4379 (1991).
80. S. D. PRICE AND J. H. D. ELAND, *Meas. Sci. Technol.* **3**, 306 (1992).
81. J. RADELOFF, N. BUTTLER, W. KESTERNICH, AND E. BODENSTEDT, *Nucl. Instrum. Methods* **47**, 109 (1967).
82. E. VON RAVEN, M. MEYER, M. PAHLER, AND B. SONNTAG, *J. Electron Spectrosc. Relat. Phenom.* **52**, 677 (1990).
83. M. RICHARD-VIARD, O. ATABEK, O. DUTUIT, AND P. M. GUYON, *J. Chem. Phys.* **93**, 8881 (1990).
84. P. SELLES, J. MAZEAU, AND A. HUETZ, *J. Phys. B* **20**, 5183 (1987).
85. V. SCHMIDT, 1992 X. Int. Conf. on VUV Radiat. Phys., Paris, Conference Proceedings, edited by F. J. Wuilleumier, Y. Petroff, and I. Nenner (World Scientific, Singapore, 1993), p. 154.
86. M. SCHNETZ, Ph.D. thesis, Universität Freiburg (1992).
87. O. SCHWARZKOPF, B. KRÄSSIG, J. ELMIGER AND V. SCHMIDT, *Phys. Rev. Lett.* **70**, 3008 (1993).
88. A. L. SMITH, *Philos. Trans. R. Soc. London* **268**, 169 (1970).
89. D. SMITH AND J. W. MÜLLER, *Rev. Sci. Instrum.* **60**, 143 (1989).
90. C. E. M. STRAUSS AND P. L. HOUSTON, *J. Chem. Phys.* **94**, 8751 (1990).
91. M. VÖLKEL AND N. SANDNER, *J. Phys. E* **16**, 456 (1983).
92. L. VEGH AND R. L. BECKER, *Phys. Rev. A* **46**, 2445 (1992).
93. G. H. WANNIER, *Phys. Rev.* **90**, 817 (1953).
94. A. H. WAPSTRA, in *Alpha-, Beta- and Gamma-Ray Spectroscopy*, Vol. I, edited by K. Siegbahn (North-Holland, Amsterdam, 1966), p. 539.
95. W. C. WILEY AND H. MACLAREN, *Rev. Sci. Instrum.* **26**, 1150 (1955).
96. N. P. YUDIN, A. V. PAVLICHENKOV, AND V. G. NEUDATCHIN, *Z. Phys. A* **320**, 565 (1985).
97. P. M. DEHMES, J. L. DEHMES, AND W. A. CHÚPKO, *J. Chem. Phys.* **73**, 126 (1980).

CHAPTER 15

SPIN POLARIZATION IN PHOTOIONIZATION

U. HEINZMANN AND N. A. CHEREPKOV

1. INTRODUCTION

Photoionization has been commonly used for a long time for testing atomic and molecular structure. Since electrons possess a spin, the complete characterization of the photoionization process should also include the description of the photoelectron spin orientation. Sauter already pointed out that atomic photoelectrons can be spin-polarized,[1] but the first systematic investigation was not carried out until 30 years later by Nagel[2] and Nagel and Olsson.[3] They considered the K-shell photoionization in the Coulomb approximation with relativistic corrections taken into account up to the second-order terms in αZ, where $\alpha = e^2/\hbar c$ is the fine-structure constant and Z is the nuclear charge. It was shown that at the low-energy limit $v/c \ll 1$, v being the photoelectron velocity, the spin polarization is proportional to $(\alpha Z)^2$.

So, up to the famous paper of Fano,[4] it was believed that the spin polarization of photoelectrons is a relativistic effect, which is important at high photon energies when either photoelectron velocities are high ($v/c \sim 1$), or the charge of the nucleus is large ($\alpha Z \sim 1$).

A review of relativistic theory is given in Ref. 5, while in Ref. 6 the results of extensive calculations of photoelectron polarization parameters for all ns subshells of uranium are presented. The only experiment at the high-energy limit performed to date for the K shell of gold with 662-keV photons[7] shows that the theory accurately describes the experimental results.

Fano for the first time mentioned[4] that in the nonrelativistic energy region photoelectrons can be highly polarized too. Previously it was observed that in low-energy elastic electron scattering on atoms, the maxima for spin polarization of scattered electrons occur at the cross section minima.[8] In the photoionization cross section of alkali atoms near thresholds there are also minima[9] caused by the passing through zero of dipole matrix elements, which are called the Cooper minima.[10] Because of the spin-orbit interaction in the continuous spectrum, the dipole matrix elements corresponding to the $ns \to \varepsilon p_{1/2}$ and $ns \to \varepsilon p_{3/2}$ transitions change signs at slightly different points in energy. In the vicinity of these points the difference between matrix elements is not small, so that at some point the degree of spin polarization of the total photoelectron flux integrated over electron ejection angles reaches 100% when spin-polarized, i.e., circularly polarized, radiation is used. This phenomenon was called the Fano effect. Smallness of the order of $(\alpha Z)^2$ disappears from the expression for the degree of polarization, but defines the cross section magnitude. Experiments[11,12] fully supported the prediction of Fano.[4]

The most striking results have been obtained for subshells with $l \neq 0$.[13] Here, because of the spin-orbit splitting of an initial atomic and/or a final ionic state, photoelectrons corresponding to a definite transition between these states always have a high degree of spin polarization which does not contain any small parameter and is of the order of unity at all photon energies. In some cases it may reach 100% in the cross section maxima as in autoionization resonances.[14] Smallness of the order of

---

U. HEINZMANN • Fakultät für Physik, Universität Bielefeld, 33615 Bielefeld, Germany.   N. A. CHEREPKOV • State Academy of Aerospace Instrumentation, 190000 St. Petersburg, Russia.

VUV and Soft X-Ray Photoionization. Edited by Uwe Becker and David A. Shirley. Plenum Press, New York, 1996.

$(\alpha Z)^2$ defines in this case a magnitude of the fine-structure splitting, but not the degree of polarization. Photoelectrons ejected at a definite angle are spin polarized even if they have been ejected by unpolarized light.[15] Therefore, the spin polarization of photoelectrons ejected at a definite angle is not an exception but rather a rule. Experimental investigations of subshells with $l \neq 0$[16,17] supported the theoretical prediction and revealed new interesting features.[18]

Analogous effects appear also in molecular photoionization.[19-25] Here the degree of spin polarization of photoelectrons is usually lower than in atomic cases because of a larger number of ionization channels contributing to the process.

Spin polarization measurements give new information on atomic and molecular structure and, being combined with the partial photoionization cross section and the angular asymmetry parameter usually measured, allow performance of the complete quantum-mechanical experiment for atoms[15,26,27] and making an essential step toward it for molecules. This fact shows the fundamental importance of spin polarization measurements and justifies the construction of quite sophisticated experimental equipment required for this kind of investigation. An alternative way to gain additional information from investigations of polarized atoms or oriented molecules without analysis of spin polarization of photoelectrons will also be discussed here.

A high degree of spin polarization of atomic photoelectrons can be used in constructing the sources of polarized electrons required for a vast variety of experiments in different branches of physics. The first idea of producing a polarized electron beam by photoionization of polarized alkali atoms proposed in Ref. 28 was not realized until 40 years later.[29,30] The Fano effect in alkali atoms has also been used to construct the sources of polarized electrons[31-34] which proved to be quite efficient. At last there is the possibility of obtaining an electron beam of nearly 100% polarization by irradiating atoms of the Al group by circularly polarized light of the wavelength corresponding to one of the autoionization resonances $ns^2 np \to nsnp^2$,[14,35] where the cross section can reach 10^{-15} cm^2.[36] It leads to a high quantum yield, i.e., the yield of electrons per incident photon. Thus, all sources of polarized electrons based on the photoionization of atoms have the advantage of giving almost complete polarization. Unfortunately, these methods could not give high electron flux. Therefore, at the moment practically all sources of polarized electrons are based on photoemission from GaAs crystals[37] which give very high intensity of the electron beam although a relatively low degree of polarization around 40% (the theoretical upper limit is 50%). But recent investigations[38] have shown the way to overcome this problem and to obtain a degree of polarization quite close to 100%.

In this review we will restrict our consideration to nonrelativistic photon energies where the electric dipole approximation is valid. Multiphoton processes are also excluded (see Ref. 39 and references therein).

2. GENERAL DESCRIPTION OF PHOTOELECTRON SPIN POLARIZATION

2.1. Theoretical Description

Let us consider absorption of light of a given polarization by unpolarized atoms. For the complete quantum-mechanical description of the photoionization process it is necessary to find the probability of ejection of electrons in a given direction $\vec{q}$ with the spin orientation along some other direction $\vec{s}$ ($\vec{q}$ and $\vec{s}$ are the unit vectors) by light of a given polarization. Linearly polarized light will be characterized by the polarization vector $\vec{e}$, while circularly polarized and unpolarized light will be characterized by the unit vector $\vec{k}$ in the direction of the photon beam (see Figure 1). The operator of interaction between an atomic electron and electromagnetic field in the electric dipole approximation is $(\vec{e}_\lambda \cdot \vec{r})$, where $\lambda = 0, \pm 1$ is the helicity, $\vec{e}_0 = \vec{e}$ for linearly polarized light, and $\vec{e}_{\pm 1} = \mp(\vec{e}_x \pm i\vec{e}_y)/\sqrt{2}$ for circularly polarized light (atomic units $\hbar = m = e = 1$ are used in this chapter). Since this operator does not act on the electron spin variables, the spin polarization of photoelectrons can appear due only to the spin-orbit interaction.

The spin-orbit interaction manifests itself in the fine-structure splitting of atomic levels with $l \neq 0$ and in the fact that each partial wave of continuous spectrum wave function with a given orbital angular momentum l_1 is split into two waves with the total angular momenta j_1 equal to $l_1 + 1/2$ and

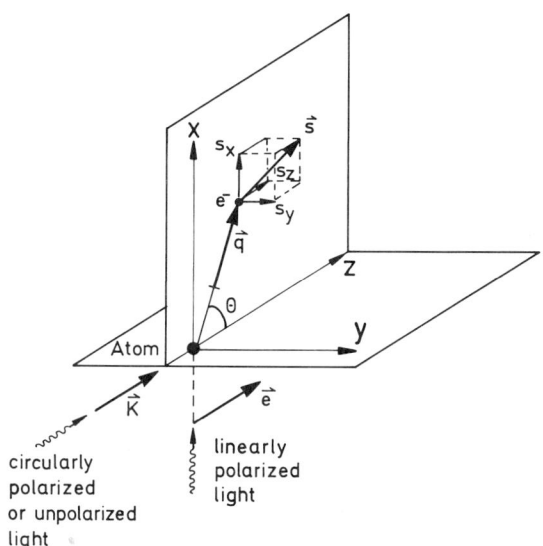

FIGURE 1. Definition of the unit vectors $\vec{k}, \vec{e}, \vec{q}, \vec{s}$ in the Cartesian frame to describe the angle- and spin-resolved photoelectron emission.

$l_1 - 1/2$. In the case of a closed-shell atom, the ground state is not split, but there is fine-structure splitting of the final ionic state, which can be resolved by photoelectron spectroscopy.

The spin-orbit splitting of atomic levels with $l > 0$ leads to the spin polarization of photoelectrons of the order of unity. To observe it, one has either to prepare the target atoms in the initial state with only one fine-structure component populated, or to select the photoelectrons corresponding to a definite fine-structure component of the final ionic state (or to do both for the general case of an open-shell atom). Strictly speaking, there is also the third possibility in the narrow energy regions between two thresholds, where only one fine-structure component can be ionized and where one need neither prepare the target nor analyze the photoelectron in energy. In practice it means that the spin polarization of photoelectrons can be readily observed in atoms with $Z \geq 10$ where the spin-orbit splitting is large enough to be resolved in experiment.

The spin-orbit interaction in the continuous spectrum normally leads to the effect of the order of $(\alpha Z)^2$ which is the order of magnitude of the spin-orbit interaction. But near the cross section minima, as was pointed out by Fano,[4] the effect is greatly enhanced since the dipole matrix element can become smaller than the difference between matrix elements corresponding to the $ns \to \varepsilon p_{1/2}$ and $ns \to \varepsilon p_{3/2}$ transitions, caused by the spin-orbit interaction in the continuous spectrum.

The derivation of the expression for the angular distribution of photoelectrons with defined spin polarization ejected from unpolarized atoms by light of a given polarization was outlined in detail in Ref. 35 without use of the density matrix formalism, and in Refs. 15, 40, and 41 using the density matrix formalism. It is convenient to present the result in the vector form independent of the particular choice of the coordinate frame as follows[42]:

$$I_j^\lambda(\vec{q}, \vec{s}) = \frac{\sigma_{nlj}(\omega)}{8\pi} \{1 + \frac{(2-3\lambda^2)}{2} \beta^j P_2(\vec{k}\,\vec{q}) + \lambda A^j \cdot (\vec{s}\,\vec{k})$$

$$- \lambda \alpha^j \left[\frac{3}{2}(\vec{q}\,\vec{k})(\vec{q}\,\vec{s}) - \frac{1}{2}(\vec{s}\,\vec{k})\right] + 2(2-3\lambda^2)\xi^j(\vec{s}\,[\vec{q}\,\vec{k}])(\vec{q}\,\vec{k})\} \quad (1)$$

where $\lambda = \pm 1$ for circularly polarized light and $\lambda = 0$ for unpolarized light, with $\vec{k}$ being the unit vector in the direction of the photon beam, $\lambda = 0$ for linearly polarized light with the replacement of $\vec{k}$ by the photon polarization vector $\vec{e}$ (see Figure 1), and $\sigma_{nlj}(\omega)$ is the partial photoionization cross section of the subshell with the quantum numbers nlj. The angular asymmetry parameter β^j and the polarization parameters A^j, α^j, ξ^j have been derived for different couplings in a number of papers.[42–45,15,35] Like β^j, the polarization parameters are dimensionless values of the order of 1 expressed through the dipole matrix elements and phase shift differences.

The degree of spin polarization of photoelectrons in the direction $\vec{s}$ is usually defined as

$$P_j^\lambda(\vec{s}) = \frac{I_j^\lambda(\vec{s}) - I_j^\lambda(-\vec{s})}{I_j^\lambda(\vec{s}) + I_j^\lambda(-\vec{s})} \tag{2}$$

From (1) and (2) we can easily find the spin polarization components along the axes of the laboratory frame defined in Figure 1. The XOZ plane containing the vectors $\vec{q}$ and $\vec{k}$ (or $\vec{e}$ for linearly polarized light) is called the reaction plane. For linearly polarized and unpolarized light there is only one nonzero component of spin polarization perpendicular to the reaction plane; this means that photoelectrons are transversely polarized. For example, for linearly polarized light we find

$$P_{jy}^0 = \frac{-4\xi^j \sin\Theta \cos\Theta}{1 + \beta^j P_2(\cos\Theta)} \tag{3}$$

where Θ is the angle between $\vec{e}$ and $\vec{q}$. Two components in the reaction plane for right circularly polarized light ($\lambda = +1$) are

$$P_{jx}^{+1} = -\frac{\frac{3}{2}\alpha^j \sin\Theta \cos\Theta}{1 - \frac{\beta^j}{2} P_2(\cos\Theta)} \tag{4}$$

$$P_{jz}^{+1} = \frac{A^j - \alpha^j P_2(\cos\Theta)}{1 - \frac{\beta^j}{2} P_2(\cos\Theta)} \tag{5}$$

So we can say that the parameters ξ^j and α^j define the degree of polarization along the y and x axes, respectively. If we integrate the photoelectron intensity (1) over electron ejection angles, then the degree of spin polarization of the total photoelectron flux turns out to be equal to zero for linearly polarized and unpolarized light, while for circularly polarized light it is different from zero and is defined by the parameter A^j:

$$P_j^\lambda(\vec{s}) = \lambda A^j \cdot (\vec{s} \cdot \vec{k}) \tag{6}$$

In the nonrelativistic approximation, that is, neglecting the spin-orbit interaction in both discrete and continuous spectrum wave functions, the polarization parameters A^j, α^j, ξ^j appear to be inversely proportional to the statistical weights of the states with a given l and different $j = l \pm \frac{1}{2}$, and have the opposite signs[42]:

$$\frac{A^{l+1/2}}{A^{l-1/2}} = \frac{\alpha^{l+1/2}}{\alpha^{l-1/2}} = \frac{\xi^{l+1/2}}{\xi^{l-1/2}} = -\frac{l}{l+1} \tag{7}$$

The same condition is valid for the degree of spin polarization. Corresponding photoionization cross sections in the same approximation are proportional to the statistical weights:

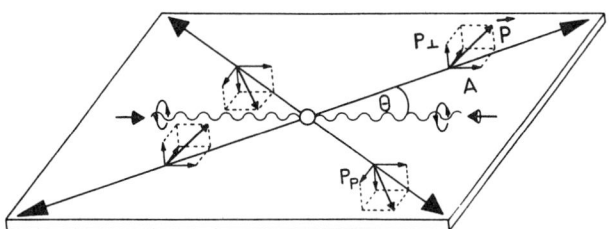

FIGURE 2. Photoionization reaction plane using circularly polarized radiation.

$$\frac{\sigma_{nl,l+1/2}(\omega)}{\sigma_{nl,l-1/2}(\omega)} = \frac{l+1}{l} \qquad (8)$$

Therefore, if the fine-structure levels are populated statistically and the photoelectrons corresponding to different j's are not separated, then the spin polarization of photoelectrons in this approximation will be equal to zero. Actually due to the j-dependence of the dipole matrix elements, the spin polarization will be different from zero but small, of the order of $(\alpha Z)^2$.

The spin polarization of photoelectrons ejected from subshells with $l \neq 0$ is a consequence of the dipole selection rules, which explains the absence of any small parameter. For example, if the $p_{1/2}$ subshell is ionized by right circularly polarized light, s-photoelectrons will be totally polarized. Indeed, in this case there is only one transition allowed: $np_{1/2}(m = -1/2) \rightarrow ns_{1/2}(m = +1/2)$, and the projection of the total angular momentum m of the s-state is the projection of spin. For other transitions the consideration is more complicated,[46] but the qualitative answer remains the same.

2.2. Experimental Examples of the Angular Dependence of Spin Polarization

The reaction plane for an angle- and spin-resolved photoionization process of an unpolarized atom or an unoriented molecule using circularly polarized radiation defined by the momenta of incoming photon and photoelectron ejected is shown in Figure 2. Since the momentum of the photon

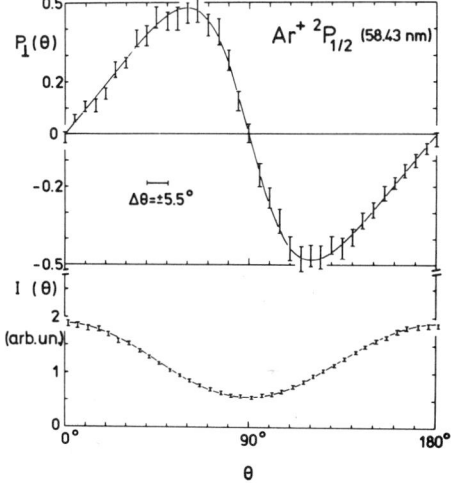

FIGURE 3. Angular distribution of photoelectron polarization $P_\perp(\Theta)$ (upper curve) and intensity $I(\Theta)$ (lower curve)[47] for the photoionization of argon atoms with linearly polarized radiation of 21.22 eV. The curves are least-squares fits according to equations (3) and (1).

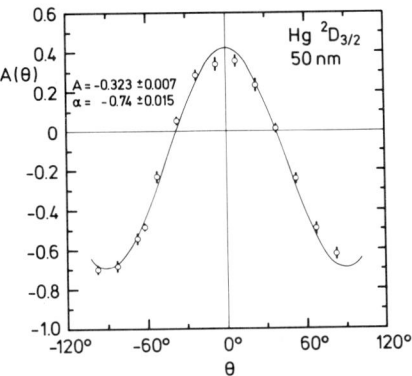

FIGURE 4. Angular dependence of the spin polarization component $A(\Theta)$ at 50 nm for the Hg^+ $^2D_{3/2}$ final ionic state. A least-squares fit yields the parameters A and α as indicated.[48]

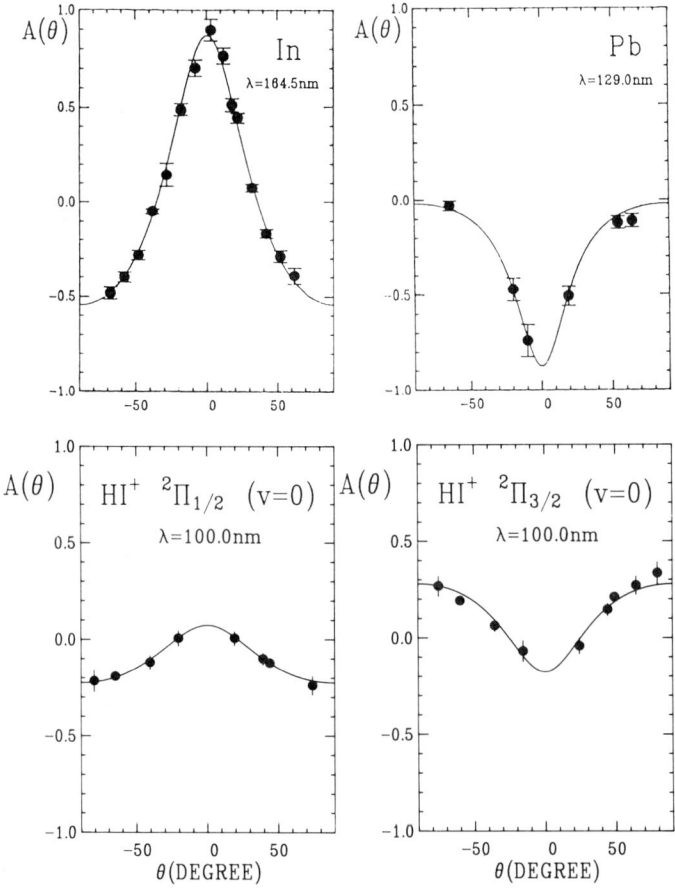

FIGURE 5. Angular dependences of the spin polarization component $A(\Theta)$ for the In^+ $^1S_0$,[49,50] Pb^+ $^2P_{3/2}$,[50,51] HI^+ $^2\Pi_{1/2}$ $v = 0$ and HI^+ $^2\Pi_{3/2}$ $v = 0$[25] ionic states. The curves are the least-squares fit according to equation (5).

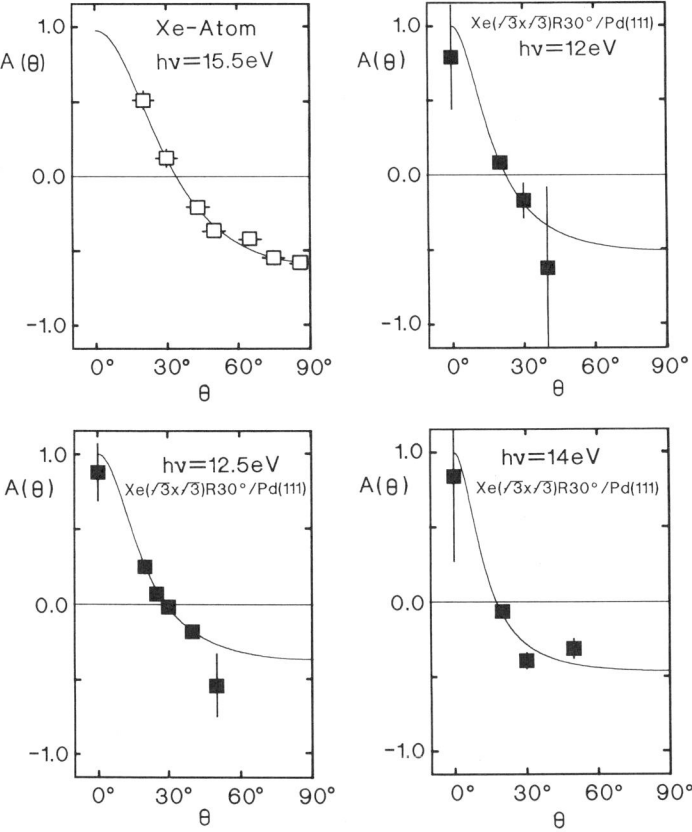

FIGURE 6. Spin polarization component $A(\Theta)$ (parallel to the light helicity) as a function of the polar angle Θ. Uppermost left: photoionization of Xe atoms, final ionic state Xe^+ $^2P_{1/2}$, photon energy 15.5 eV[53,54]; others: photoemission from adsorbed Xe $(p_{1/2})[\sqrt{3} \times \sqrt{3} R30° Pd(111)]$ at different photon energies.[55]

is negligibly small compared with the momentum of the photoelectron (valid in nonrelativistic approximation if the photon energy is smaller than 1000 eV) there is a forward–backward symmetry in the reaction plane (Figure 2). It makes no difference whether right-handed circularly polarized radiation comes from the left or left-handed comes from the right. This symmetry behavior and the rotational symmetry around the direction of the photon momentum causes both electron spin polarization components P_p and $P_\perp$ perpendicular to the photon spin to vanish for photoelectron emission angles $\Theta = 0, \pi/2, \pi$. This behavior is indeed in accordance with the proportionality of $P_\perp(\Theta) = P_{jy}$ [equation (3)] and $P_p(\Theta) = P_{jx}$ [equation (4)] to $\sin\Theta \cdot \cos\Theta$. The experimental verification[47] of this angular dependence of the spin polarization according to (3) is given in Figure 3, showing excellent agreement between the experimental data (error bars) and a fit following (3). It is worth noting that the component $P_\perp(\Theta)$ gained, regardless of the use of right or left circularly polarized light or even unpolarized or linearly polarized light, is slightly different for the experiment with circularly or unpolarized light compared with the case of linearly polarized light. In (3) the parameter -4ξ in the numerator and β in the denominator have to be replaced by 2ξ and $-\beta/2$, respectively, in the case of circularly and unpolarized radiation.

Unlike the spin polarization components perpendicular to the photon spin, the component $A(\Theta)$ parallel to the photon spin always has its positive and negative maxima and minima at the emission angles $\Theta = 0$ and $\Theta = 90°$: the experimental data (error bars) fit in very well with the theoretical angular dependence as given in equation (5) $A(\Theta) = P_{jz}^{+1}$ by means of the second Legendre polynomial

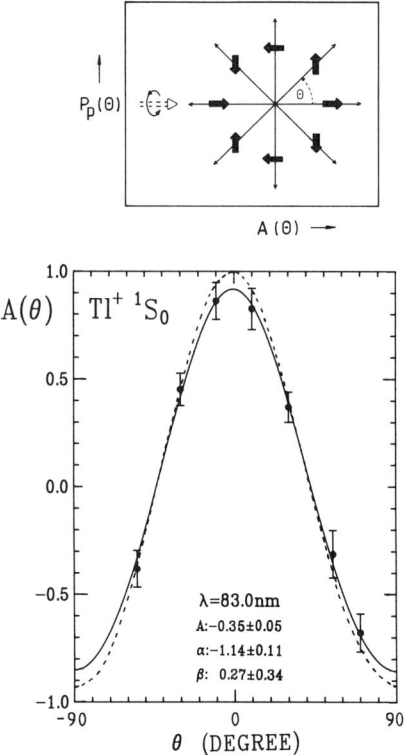

FIGURE 7. Spin polarization vector for complete polarization at all angles of emission, lying in the reaction plane for photoionization of thallium atoms using circularly polarized radiation of $\lambda = 83$ nm. The angular dependence for the spin polarization component $A(\Theta)$ and its least-squares fit according to equation (5) yielding the dynamical parameters given are shown in the lower panel.[18]

$P_2(\cos\Theta)$: this is shown for photoionization of a d-subshell in Figure 4,[48] leaving behind the mercury ion in the ionic state $^2D_{3/2}$ and for atomic p or molecular π orbitals in Figure 5 for indium[49,50] and lead[50,51] atoms as well as for HI molecules (vibrationally resolved).[25] Whether $A(\Theta=0)$ is positive or negative depends respectively on whether a $p_{1/2}$ or $p_{3/2}$ electron is removed, which clearly indicates that photoelectron spin polarization exhibits the quantum numbers of the bound states from which the photoelectron originates (or of the corresponding ionic states). As predicted by theory (see Section 2.1) and shown in Figure 5 for In within the experimental uncertainty, the ionization of an atomic $p_{1/2}$ subshell yields a complete spin polarization in forward (and backward) direction $\Theta = 0$ since there are only $s_{1/2}$ continuum states occupied due to the dipole selection rules $\Delta l = \pm 1$ and the fact that the d partial waves vanish for $\Theta = 0$.

This complete spin polarization for $\Theta = 0$ in creating a $p_{1/2}$ hole in photoionization has not only been shown for atomic photoionization but also for photoemission of a xenon monolayer adsorbate on different substrates (for a review see Ref. 52). Figure 6 shows the corresponding cross comparison in the angular dependence of the spin polarization component $A(\Theta)$ between atomic photoionization of xenon[53,54] and adlayer photoemission Xe/Pd(111).[55] Although the atomic and the monolayer target have a completely different geometrical symmetry, the spin polarization angular distribution shows, in contrast to the corresponding differential cross section, an atomic-like behavior in the condensed matter, since the spin polarization is primarily based on the local spin-orbit interaction rather than on the far geometrical order. It demonstrates that spin polarization spectroscopy may serve as a tool to study "atomic influences" in condensed matter photoemission.

The complete spin polarization at $\Theta = 0$ observed gave rise to the question of whether there are cases of atomic photoionization where the limit of complete polarization of the photoelectrons at all emission angles is approached.[18] From equations (3)–(5) there are two cases where the polarization vector can have a length of unity at all emission angles: case one (the trivial one), where one component $A(\Theta)$ does not vary with Θ, i.e., $\alpha = \beta = 0$ with $A(54°) = A = 1$, in other words where the spin polarization of the angle integrated flux is one; this has indeed been predicted and observed at certain wavelengths in the Fano effect of cesium[12] and for resonances in the photoionization of thallium.[14,56] Case two provides a polarization vector length of one where the components are strongly dependent on the emission angle Θ. Figure 7 demonstrates how the spin polarization vector with the components $A(\Theta)$ and $P_p(\Theta)$ rotates in the reaction plane as a function of Θ[18] and how the spin polarization component $A(\Theta)$ parallel to the photon momentum varies between +1 and −1.

In all cases shown in this section the fit of the experimental data points according to the predicted angular dependences [equations (3)–(5)] yields the values of the dynamical parameters A, α, β, ξ and their uncertainties.

2.3. The Possibility of Complete Experiments

The atomic photoionization process is determined by three complex dipole matrix elements according to the selection rules $\Delta J = \pm 1, 0$ for dipole transitions. Thus, it is characterized by not more than three real matrix elements and two phase shift differences, since the absolute value of the phase is not accessible. The five dynamical parameters parameterizing the spin- and angle-resolved photoelectron spectroscopy experiment, cross section σ, asymmetry parameter β, and the three dynamical spin polarization parameters A, α, ξ are analytically well-known functions (see Section 3.2) of the three real matrix elements and the two phase shift differences and thus form a complete set of experimental data to characterize the photoionization process completely according to quantum mechanics.[57] This means that the combination of the experimental data mentioned (five independent quantities) allows separate determination of all transition matrix elements and phase shift differences (including error bars) which appear in the theoretical description, for example of the $5p$ photoionization of xenon. This has indeed been done for atomic xenon in the autoionization range as well as in the continuous photoionization range.[54] Figure 8 shows a combined data set of experiments of different authors for the $5p$ autoionization of xenon between the two first ionization thresholds at 102.2 and 92.2 nm where only the ionization channels belonging to the Xe^+ $^2P_{3/2}$ final state are open. Not only the total cross section shows a pronounced resonance structure[58] in the well-known Beutler–Fano shape, but also the differential cross section characterized by the asymmetry parameter β[59] and the spin polarization vector characterized by the three spin parameters α, A, ξ.[60,27] The detailed analysis of the thus obtained "experimental" matrix elements and phase shift differences[54] in the autoionization range (not reviewed in this chapter in detail) clearly demonstrates that the broad resonance between 99 and 100 nm arises from a $5p_{3/2} \to \varepsilon d_{5/2}$ contribution only, whereas the sharp resonance at around 98.5 nm clearly shows up an amplitude for the $5p_{3/2} \to \varepsilon s_{1/2}$ transition with some weak $5p_{3/2} \to \varepsilon d_{3/2}$ contributions. Pronounced phase shift contributions between $\varepsilon d_{5/2}$ and $\varepsilon d_{3/2}$ are observed especially at the deep minimum between the resonances at about 97 nm.

Figure 8 also shows the excellent agreement between the experimental results and corresponding RRPA[61] and MQDT[15] calculations convoluted with the experimental resolutions.

Figures 9 and 10 show the corresponding experimentally obtained matrix elements and phase shift differences for the photoionization into the open continuum beyond the autoionization range (shown as vertical dashed lines) for both ionic states Xe^+ $^2P_{1/2}$ and $^2P_{3/2}$,[54] respectively.

The experimental data shown can be compared not only with theoretical calculations in RPAE[62,35] and RRPA–MQDT[61] but also with experimentally and theoretically obtained MQDT parameters in the discrete spectral range[27,63–65] given by oscillator strengths and quantum defects. Figures 9 and 10 clearly demonstrate the excellent quantitative agreement in this cross comparison. Figure 10 also exhibits a sign change of the $5p_{3/2} \to \varepsilon d_{3/2}$ matrix element at 20 eV (Cooper minimum) and a pronounced phase shift difference between the $\varepsilon d_{5/2}$ and $\varepsilon d_{3/2}$ continuum waves, increasing from 0 at threshold to $\pi/2$ at 10 eV higher energies. Both results demonstrate an important and nonnegligible influence of the spin-orbit interaction to the continuum states in xenon. Ref. 54 discusses and interprets

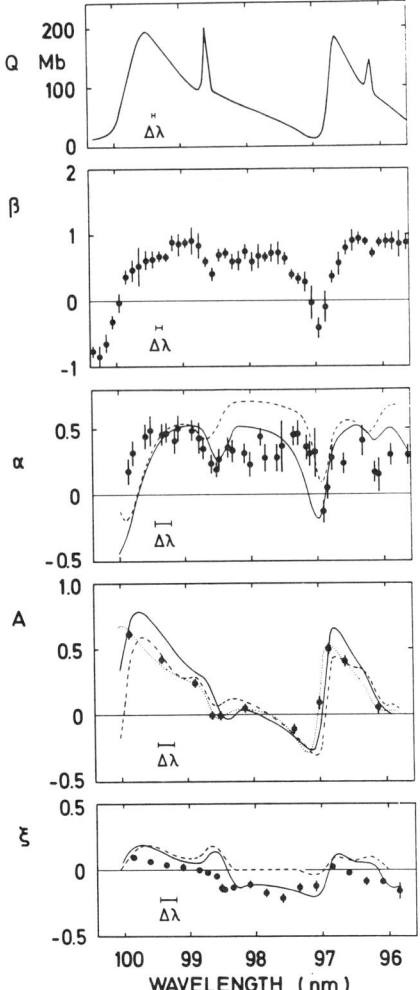

FIGURE 8. Cross section Q,[58] asymmetry parameter β,[59] and spin parameters α, A, and ξ for the $5p$ autoionization of xenon (from Ref. 60). For the spin parameters: circles, experimental results; full curves, RRPA calculation[61] (convoluted with $\Delta\lambda = 0.25$ nm); dashed curves, MQDT calculation[15] (convoluted with $\Delta\lambda = 0.25$ nm); dotted curve for A, based on experimental spin-polarization data of the angle-integrated photoelectron flux,[27] convoluted to correspond to the resolution of $\Delta\lambda = 0.25$ nm.[54]

this behavior in another angular momentum transfer classification scheme: the parity-favored and -unfavored transitions and the phase shifts in between.[63,66,67]

2.4. Other Types of Complete Experiments

Measuring the spin polarization of photoelectrons ejected from unpolarized atoms is far from being the only possibility of performing the complete experiment. The other evident way is to perform measurements of the angular distribution of photoelectrons ejected from polarized atoms, if, of course, atoms are polarizable. In this case one can extract as many as $(9J + 1)$ or $(9J + 1/2)$ parameters for integer or half-integer values of the total angular momentum J of the initial state, respectively, if circularly polarized light is absorbed[68] and $(6J + 1)$ or $6J$ parameters, if linearly polarized light is absorbed. The number of parameters will be further increased if the spin polarization of photoelectrons is also measured.[69] Since not so many parameters are usually necessary, the spin polarization of photoelectrons does not have to be measured if target atoms are polarized. Moreover, even the angular distribution of photoelectrons can be measured for some particular geometry, when the contributions of some terms disappear.

There is a more radical way to simplify the geometrical structure of the angular distribution, namely, by subtracting from the electron current ejected at a definite angle by right circularly polarized

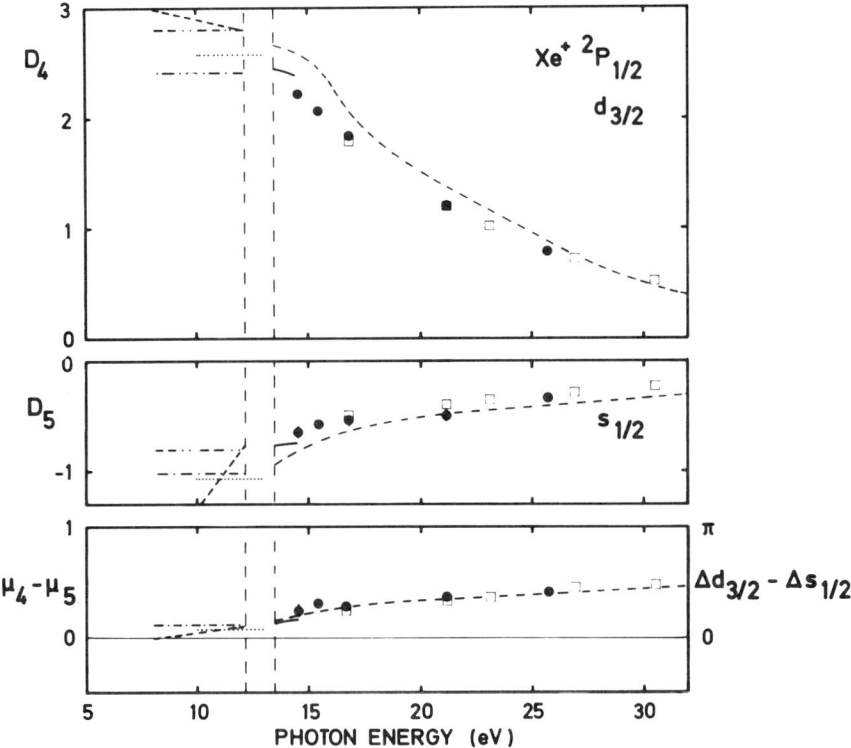

FIGURE 9. Matrix elements D_i and quantum-defect differences $\mu_i - \mu_j$ for photoionization with the final ionic state $Xe^+ \, ^2P_{1/2}$ as a function of the photon energy in the continuum region. The vertical dashed lines indicate the ionization thresholds. In the continuum region are shown: circles[54]; squares[27]; dashed curve, RPAE calculation[62,35]; full curve, calculated on the basis of RRPA–MQDT parameters from Ref. 61. The curves in the discrete region are calculated from MQDT parameters: dashed,[63] dotted,[64] double-dashed-dotted,[54] dashed-dotted.[65]

light the analogous current ejected by left circularly polarized light. This difference is called circular dichroism in the angular distribution of photoelectrons (CDAD), and is defined by three parameters[70] (if the second rank alignment tensor is the highest rank tensor characterizing the atomic polarization), which in the case of a one-electron subshell are proportional to the spin polarization parameters introduced above. In the same way one can define the linear dichroism in the angular distribution of photoelectrons (LDAD), which is a difference between photoelectron currents, ejected at a definite angle by linearly polarized light with two mutually perpendicular polarizations.[71] LDAD is characterized by more than three parameters. Both CDAD and LDAD will be discussed later. Here it is important to mention that the measurements of CDAD or LDAD, or the angular distribution of photoelectrons ejected from polarized atoms, can be used to perform the complete quantum-mechanical experiment.

There is an alternative possibility of investigating the photoionization process by small-angle inelastic scattering of fast electrons on atoms and molecules.[72,73] This $(e, 2e)$ simulation of photoionization processes was used for measurements of the absolute photoionization cross sections with high precision and for the angular distribution measurements.[74] The next evident step is to measure the spin polarization of slow electrons, which enables performance of the complete experiment. But there is another way to reach the same goal by measuring the angular distribution of slow electrons ejected from polarized atoms.

Usually in the $(e, 2e)$ simulation of photoionization processes the polar scattering angle θ of fast electrons is restricted by small values $\theta \ll 1$, while there is integration over all azimuthal angles φ.

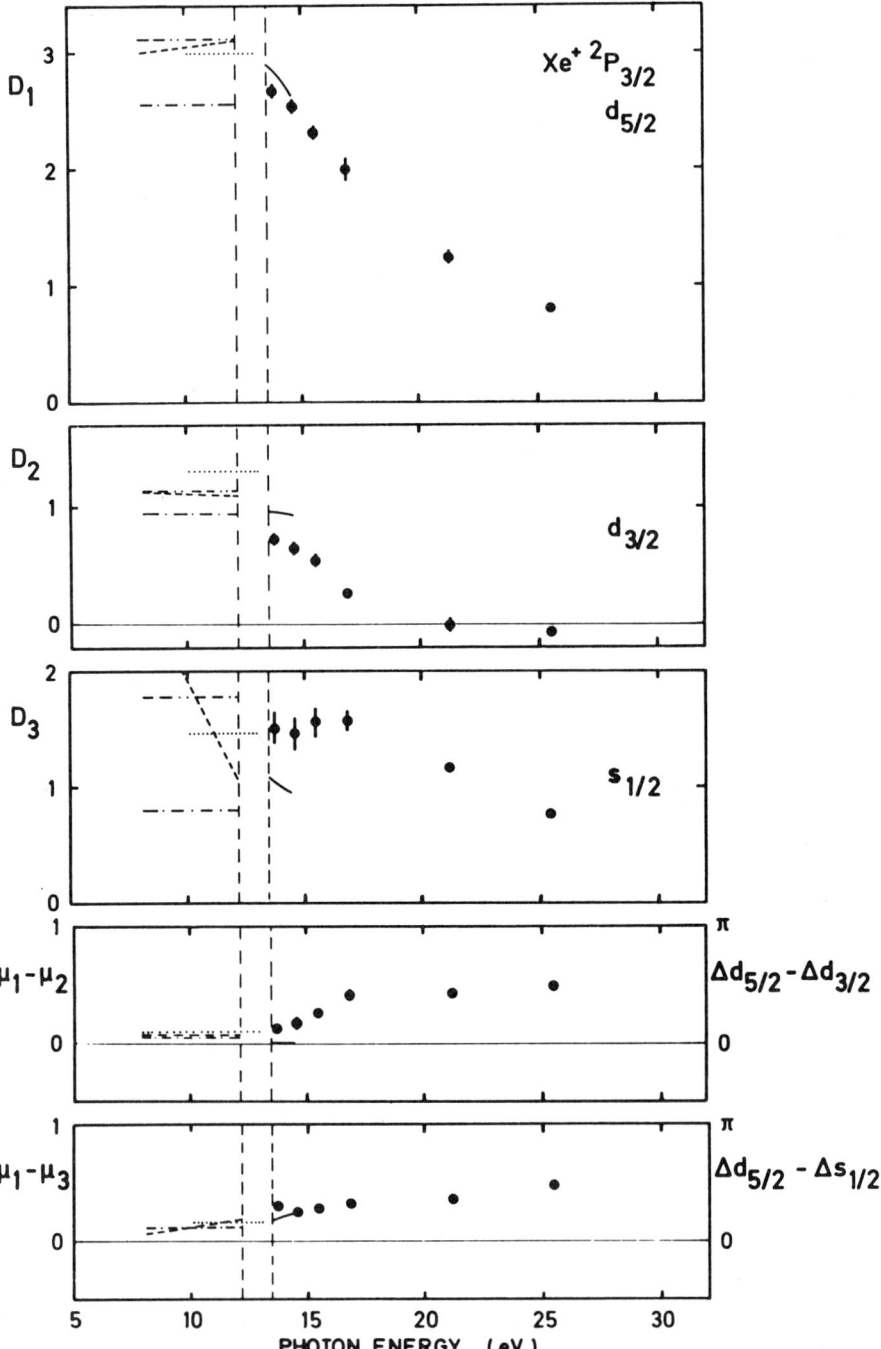

FIGURE 10. Matrix elements D_i and quantum-defect differences $\mu_i - \mu_j$ for photoionization with the final ionic state $Xe^+\ ^2P_{3/2}$ as a function of the photon energy in the continuum region.[54] For details see text and caption of Figure 9.

Here, the $(e, 2e)$ scattering is equivalent to the photoionization of the target by unpolarized light propagating in the direction of the fast electron beam. Measurement of the angular distribution of slow electrons in this case enables one to perform the complete experiment. However, since it again has a rather complicated geometrical structure, it is worthwhile finding some difference of two angular distributions. The only direction which can be changed now is the direction of atomic polarization. Therefore, one should find the difference between two intensities corresponding to two mutually perpendicular atomic polarizations, or, equivalently, the difference between the angular distribution measured for some fixed atomic polarization and the same distribution rotated by the angle $\pi/2$ around the proper direction. The result will be equivalent to the CDAD or LDAD measurements.[75]

So far we have discussed possible complete experiments based on the observation of photoelectrons. But the information on the photoionization process can also be extracted from the subsequent decay of the residual ion if it is left in an excited state. This is based on the fact that in the photoionization process not only photoelectrons but also residual ions are polarized (aligned or oriented depending on the light polarization).[76,77] The subsequent radiative or Auger decay process can be used for testing the polarization state of the ion. If the atoms in the initial state are unpolarized, the final ionic state can be described by the state multipoles of the rank not higher than two,[76] so that not more than two parameters could be extracted from investigations of the decay process. It is generally not sufficient for the complete experiment, but in combination with other measurements can be used successfully. It is also possible to use this kind of complete experiment if one can restrict the consideration to the nonrelativistic approximation where only three parameters are required. Here detailed discussion about this question is given in Ref. 78 (see also Chapter 11 of this volume).

Hence, there are many ways to perform the complete quantum-mechanical experiment in photoionization, and they are not necessarily connected with the photoelectron spin polarization measurement.

3. TECHNIQUES AND PARTICULAR RESULTS

3.1. Experimental Techniques

Most studies of photoionization with circularly polarized radiation are hampered by the fact that usually atoms and molecules have their ionization threshold in the VUV, where conventional methods for producing circularly polarized radiation by use of a quarter-wave plate and a prism break down (photon energy > 10 eV). They can be performed, however, with synchrotron radiation, which is linearly polarized when emitted in the plane of the storage ring but is elliptically polarized with a high degree of right (left)-handed circular polarization when emitted above (below) the plane. This "source" of circularly polarized VUV radiation has been used in the energy range up to 35 eV since 1978 in Bonn[79,27] and since 1982 at BESSY in Berlin.

The synchrotron radiation of BESSY is monochromatized by a 6.5-m normal-incidence UHV monochromator (Figure 11) of the Gillieson type[80] with the electron beam in the storage ring being the virtual entrance slit. A spherical mirror and a plane holographic grating (1200 or 3600 lines/mm) form a 1:1 image of the tangential point in one of the two exit slits. With a slit width of 2 mm, a bandwidth of 0.5 nm and 0.1 nm has been achieved for the 1200 grating (first diffraction order) and for the 3600 grating (second order), respectively. The optical degrees of polarization of the synchrotron radiation have been measured[53] by means of a rotatable four-mirror analyzer.[79,27] Figure 12 shows the results of the circular polarization P_{circ} and the linear polarization P_{lin} as functions of the vertical angle ψ (± 0.1 mrad) in comparison with theoretical predictions. Note that all radiation emitted above the storage ring plane (half-cone) has an average degree of circular polarization of 85% at 100 nm. The spin-resolved photoemission experiments are usually performed with the apertures set to accept radiation above and below ± 1 mrad. Thus, a photon flux of up to 5×10^{11} photons s^{-1} with a degree of circular polarization of $\pm 93\%$ passes the monochromator, which is about 30% of the intensity emitted in the full vertical cone.

Behind the monochromator exit slits, two sets of apparatus are in operation, which are very similar in their setup: one for photoionization of free atoms and molecules (high-vacuum system), the

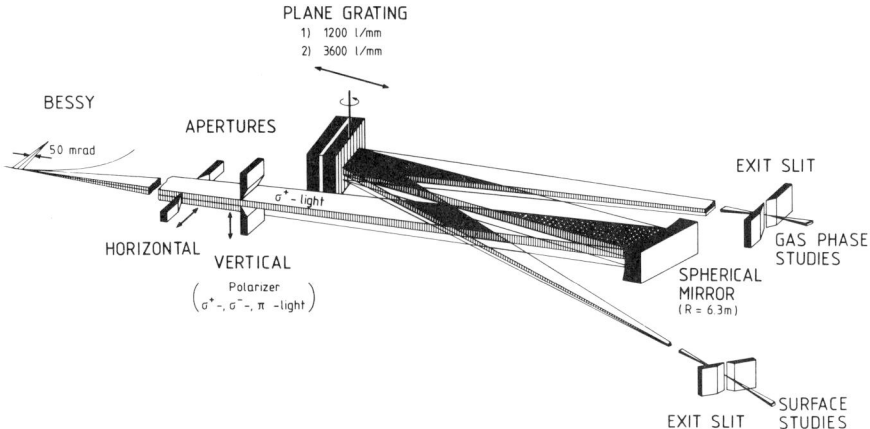

FIGURE 11. Schematic diagram of the 6.5-m normal-incidence monochromator for circularly polarized synchrotron radiation at BESSY.[80]

other for photoemission of solids and adsorbates (UHV). As shown in Figure 13 (schematic diagram of the HV system), the elliptically polarized VUV radiation hits the atomic or molecular beam under 90° producing photoelectrons in a region free of electric or magnetic fields.

The photoelectrons emitted into a cone of ±3° are analyzed with respect to their kinetic energy by a rotatable simulated hemispherical electron spectrometer[81] and are directed by a 90° electrostatic deflection along the axis of rotation of the spectrometer. This direction, which is the normal of the reaction plane spanned by the momenta of the incoming radiation and the outgoing photoelectrons, is rotated by 45° with respect to the major (and the minor) axis of the light polarization ellipse. Thus, the formulas describing the angular dependences of the spin polarization components of atomic photoelectrons become simple trigonometric relations.[54,44] After a second deflection the electrons are accelerated to 120 keV and scattered at the gold foil of the Mott detector for the spin polarization analysis.[46]

It is worth noting that in the near future a new type of circularly polarized synchrotron radiation source will be available at BESSY[82] to overcome the energetic restriction of 35 eV. A crossed field double undulator[83,84] generates radiation up to 200 eV of arbitrary polarization, depending on the selectable coherent superposition of the radiation of the two undulators, with a degree of circular polarization up to 50%.[82]

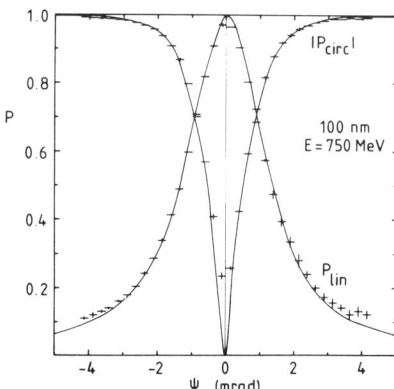

FIGURE 12. Degree of circular and linear polarization measured behind the monochromator as a function of the vertical angle with respect to the storage ring plane.[53]

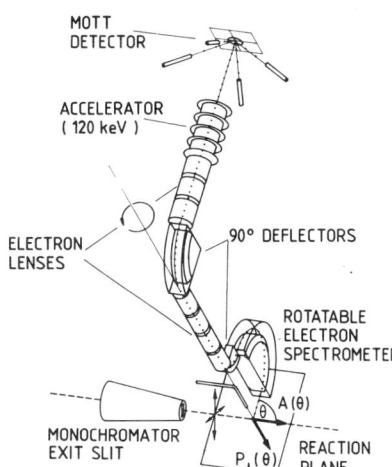

FIGURE 13. Schematic diagram of the apparatus for angle- and spin-resolved photoelectron spectroscopy built up at BESSY.[81]

Another light source in the VUV has become increasingly more important over the last few years, especially for molecular photoionization: the laser-based four-wave mixing in a mercury gas cell or a rare gas beam giving pulsed coherent VUV radiation in the energy range between 8 and 20 eV[85] of about 10^{11} photons/s at an energetic resolution of 250,000 (bandwidth 10^{-3}–10^{-4} nm) and a degree of circular polarization of 96%.[86]

This light source has been used for rotationally resolved measurements of the Fano effect in HI molecules[86] as well as of angular distribution measurements in molecular photoelectron spectroscopy.[87]

Figure 14 shows the experimental setup for angle-integrated spin polarization analysis using coherent VUV radiation.[86] It uses two Nd-doped yttrium-aluminum-garnet-pumped dye lasers (11-Hz repetition rate). The frequency of one laser is doubled by use of potassium dihydrogen phosphate crystal and tuned to match the Hg6s $^1D_2$ resonance. The other laser can be tuned throughout the visible part of the spectrum. Both beams are spatially overlapped by a dichroic mirror and focused ($f = 100$ cm) into the mercury cell. The homogeneous zone of mercury vapor ($p = 10$ mbar), used as conversion medium, is prepared in a heat-pipe construction[85] buffered with 13 mbar of neon. A LiF prism separates the sum-frequency from the intense input beams and unwanted by-products of the conversion.[85] After passing the target region, the intensity of the beam is monitored with an open photodiode; alternatively, the polarization can be checked by an analyzer (reflection type, using four gold mirrors). This radiation is used to ionize a supersonic beam of HI molecules that is rotationally cool, with only the ground-state rotational level populated. Regardless of their direction of emission, all photoelectrons produced are extracted by an electrostatic quadrupole field[88] and accelerated to 100 keV for an analysis of their intensity and their spin polarization in a Mott detector.[88]

3.2. Theoretical Calculations

In contrast to the experimental situation, practically every theoretical calculation of the photoionization cross section and the angular asymmetry parameter can be easily extended to include also the spin polarization parameters, since all that is needed in all cases are the dipole matrix elements and phase shift differences.

It is well known now that the photoionization cross section calculations in the Hartree–Fock approximation may differ substantially from the experiment[89] (see also Chapters 1 and 2 of this volume). Therefore, one needs to take into account the multielectron correlations. The first systematic calculations of polarization parameters for the rare gas atoms have been performed using both nonrelativistic[26] and relativistic[90] versions of the random phase approximation with exchange (RPAE).

In the nonrelativistic approximation when the spin-orbit interaction is neglected, the equations for the polarization parameters are especially simple[35]

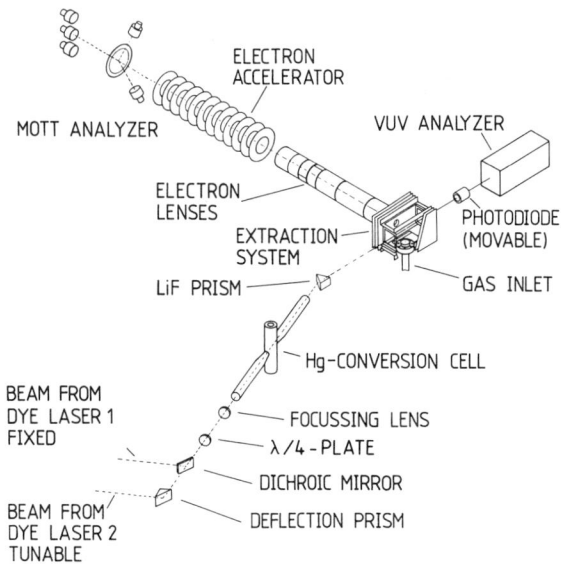

FIGURE 14. Schematic diagram of the experimental setup to measure the A parameter in rotationally resolved molecular photoionization by use of circularly polarized coherent VUV radiation.[86]

$$A^j(\omega) = \frac{(-1)^{j-l-1/2}}{(2j+1)} \cdot \frac{l|d^j_{l+1}|^2 - (l+1)|d^j_{l-1}|^2}{|d^j_{l+1}|^2 + |d^j_{l-1}|^2} \tag{9}$$

$$\xi^j(\omega) = \frac{(-1)^{j-l-1/2}}{2(2j+1)} \cdot \frac{3\sqrt{l(l+1)}\,\mathrm{Im}[d^j_{l+1} \cdot d^{*j}_{l-1} e^{i(\delta_{l+1}-\delta_{l-1})}]}{|d^j_{l+1}|^2 + |d^j_{l-1}|^2} \tag{10}$$

$$\alpha^j(\omega) = \frac{2(-1)^{j-l-1/2}}{(2j+1)(2l+1)}\{l(l+2)|d^j_{l+1}|^2 - (l^2-1)|d^j_{l-1}|^2$$
$$- 3\sqrt{l(l+1)}\mathrm{Re}(d^j_{l+1} \cdot d^{j*}_{l-1} e^{i(\delta_{l+1}-\delta_{l-1})})\}[|d^j_{l+1}|^2 + |d^j_{l-1}|^2]^{-1} \tag{11}$$

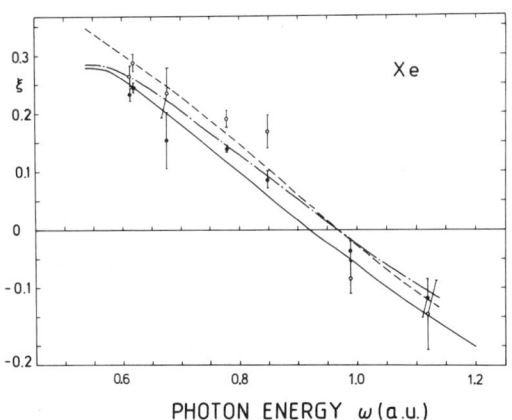

FIGURE 15. Spin polarization parameter ξ^j for the $5p^6$ subshell of Xe: RPAE[26] (full), RRPA[90] for $\xi^{1/2}$ (dash-dotted) and $-2\xi^{3/2}$ (dashed) in comparison with experimental results[27,91] for $^2P_{1/2}$ (full circles) and $^2P_{3/2}$ (multiplied by -2; open circles) ionic states.

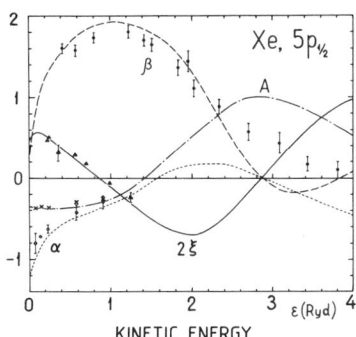

FIGURE 16. The parameters β, A, α, and 2ξ versus photoelectron energy for the $5p^6(p_{1/2})$ subshell of Xe, RPAE curves[35] in comparison with experimental results for A, α, ξ,[54] and β.[92,93]

where $d^j_{l\pm1}$ is the reduced dipole matrix element and $\delta_{l\pm1}$ is the phase shift. The spin-orbit interaction in this approximation manifests itself indirectly through the use of the linear combination of wave functions corresponding to a definite value of the total angular momentum j of the initial atomic and/or final ionic states, while the radial parts of the wave functions are calculated in the LS-coupling approximation. The parameters (9)–(11) satisfy the equation (7) exactly.

In Figure 15 we compare the results of the nonrelativistic calculation of the ξ^j parameter for the $5p^6$ subshell of Xe in RPAE[26] (full curve) with the analogous calculation in the relativistic random phase approximation (RRPA)[90] based on the Dirac–Fock wave functions. Results for the $\xi^{3/2}$ parameter are multiplied by -2 to visualize the deviation from equation (7) which in this case gives $\xi^{1/2} = -2\xi^{3/2}$. It is seen from Figure 15 that the difference between $\xi^{1/2}$ (chain) and $-2\xi^{3/2}$ (dashed) in the RRPA,[90] which appears due solely to the spin-orbit interaction, is quite small and comparable with experimental uncertainties.[27,91] Some systematic discrepancy between the RRPA and RPAE curves is connected with other relativistic effects. In Ar the relations (7) are practically fulfilled even in RRPA.[35,90] Therefore, we can conclude that the nonrelativistic theories like RPAE can be used for both prediction and quantitative description of polarization phenomena.

Figure 16 shows the parameters $A^{1/2}$, $\xi^{1/2}$, $\alpha^{1/2}$, and $\beta^{1/2}$ for the $5p^6$ subshell of Xe calculated in the RPAE[35] in a broad energy range, together with the experimental results available at present.[54,92,93] There are general properties of the parameters which can be demonstrated by this figure. First of all, of five values σ, β, A, α, ξ, only three are mutually independent in the nonrelativistic approximation. Therefore, there are two relations between the parameters ($l \neq 0$):

$$\alpha^j = A^j + \frac{2(-1)^{j-l-\frac{1}{2}}}{(2j+1)}(1 - \frac{1}{2}\beta^j) \quad (12)$$

$$(A^j + \frac{1}{2}\alpha^j)^2 + (2 \cdot \xi^j)^2 = \frac{2l(l+1)}{(2j+1)^2}(1+\beta^j)(1 - \frac{1}{2}\beta^j) \quad (13)$$

In the relativistic theory all five parameters are independent, but they can vary in the limits defined by the value of β^j as was demonstrated in Ref. 44. In particular, when β reaches its maximum value 2, all of the polarization parameters go to zero. This property is clearly seen in Figure 16.

According to (10), ξ^j is proportional to the sine of the phase shift difference, whereas β^j contains an analogous term but with the cosine. Therefore, there is the correlation in their behavior, namely, near each zero of ξ^j the parameter β^j has an extremum, and consecutive zeroes of ξ^j correspond in turn to a maximum and a minimum of β^j. This correlation is also demonstrated in Figure 16. The ξ^j parameter has zeroes at each point where the phase shift difference $(\delta_{l+1} - \delta_{l-1})$ is equal to the integer number of π, and also where either d^j_{l+1} or d^j_{l-1} goes through zero. In Figure 16 the first zero of $\xi^{1/2}$ at $\varepsilon = 0.84$ Ryd is due to the change of sign of the phase shift difference whereas the second zero at $\varepsilon = 2.88$ Ryd corresponds to the Cooper minimum of the cross section where d_{l+1} changes sign. As it follows from (9)–(11), three parameters ξ^j, α^j, and β^j simultaneously are equal to zero in the Cooper minimum, while A^j has the extremum.

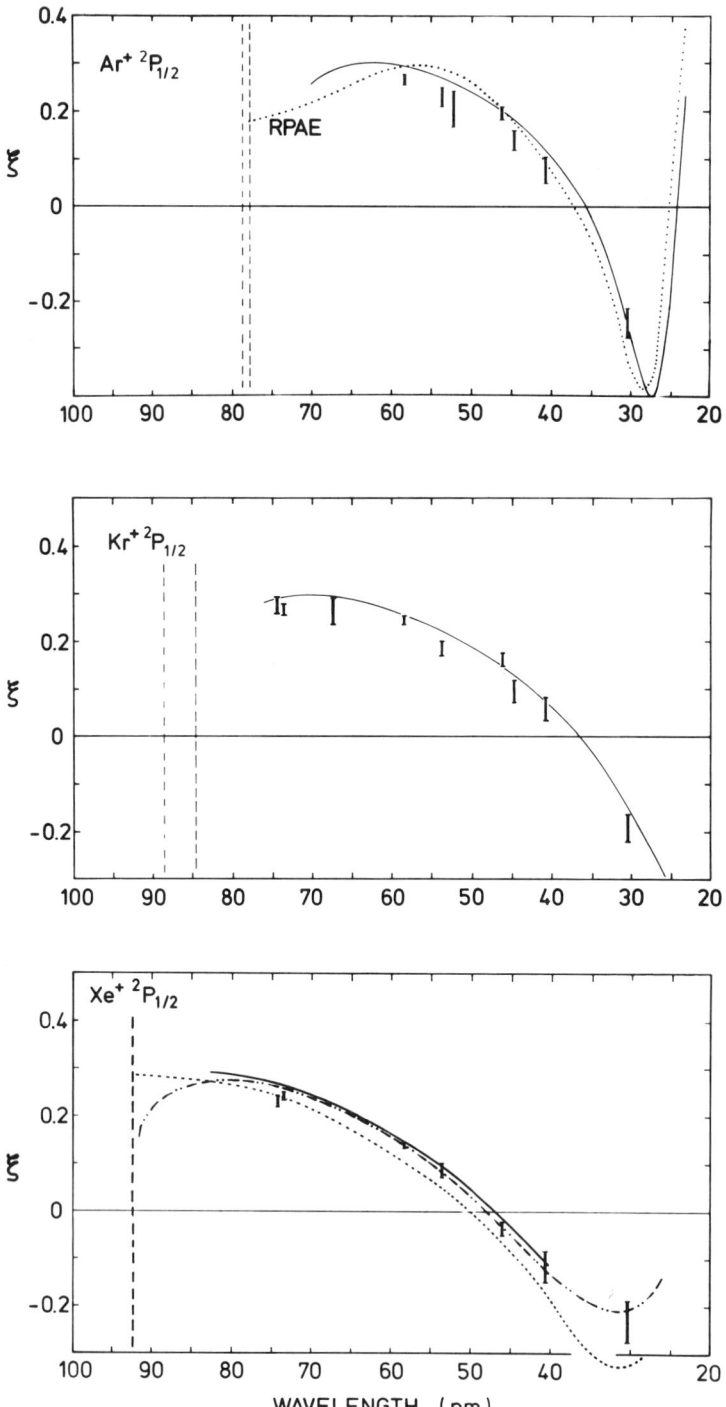

FIGURE 17. Experimental results (error bars) of the spin parameter ξ for photoelectrons corresponding to the ionic state $^2P_{1/2}$ of Ar[17,91] (upper panel), Kr[17,91] (middle panel), and Xe[17,27,91] (lower panel) in comparison with theoretical curves RRPA[98] (full), RPAE[26] (dashed), MQDT[27,64] (chained). The vertical dashed lines represent the ionization thresholds.

The polarization parameters have been calculated also for Group IIB elements using RRPA[94] as well as relativistic R-matrix methods.[95,96] Special investigations have been devoted to the autoionization resonances between the $^2P_{3/2}$ and $^2P_{1/2}$ thresholds in rare gases,[15,61] and in Tl atoms.[97]

3.3. Comparison between Theory and Experiment for Rare Gas Atoms

The comparison between theoretical RPAE calculations and experimental data in spin-resolved photoelectron spectroscopy has already been shown for xenon 5p in Figures 15 and 16. Figure 17 extends this cross comparison to the lighter rare gases Kr (4p) and Ar (3p). There is excellent agreement between the experimental results[17,91] of the ξ parameter and the corresponding RRPA[98] and RPAE[26] calculations. For xenon, the calculated results of an extension of the data in the discrete spectral range by means of the MQDT are also given in Figure 17 (chained)[27,64] showing excellent agreement with experiment, too. Figure 17 also demonstrates that the magnitude of the spin polarization values observed do not depend on the strength of the spin-orbit interaction. Argon has a weak spin-orbit interaction resulting in an ionic fine-structure splitting of 178 meV, while xenon is strongly influenced by the spin-orbit interaction with an ionic fine-structure splitting of 1.3 eV. There is no doubt, with a nonresolved spin-orbit interaction, one would not find a pronounced photoelectron spin polarization, since the opposite signs of the polarizations of the two peaks in the photoelectron spectrum referring to the $^2P_{1/2}$ and $^2P_{3/2}$ ionic states would cancel each other out. The overall shapes of polarization curves for the three rare gases in Figure 17 are very similar, they are only shifted and stretched when going from Ar to Xe. The maximum and minimum values are the same. The lighter the rare gas atom, the more difficult is the spin polarization spectroscopy experiment due to the higher spectroscopic resolution needed; when, however, the spin-orbit fine-structure splitting is resolved by experiment, the spin polarization values are no longer a measure for the strength of the spin-orbit interaction.

A very sensitive region for the cross comparison between experiment and theory is, however, the autoionization range between the first ($^2P_{3/2}$) and the second ($^2P_{1/2}$) photoionization threshold. Here there is a pronounced resonances region due to the interference of the open (Xe$^+$ $^2P_{3/2}$) continuum and the autoionizing Rydberg states converging to the (Xe$^+$ $^2P_{1/2}$) limit. The accuracies of theories and spin polarization experiment are especially demonstrated in the positions and heights of the resonances in the A spin parameter, which is the polarization measured for the angle-integrated total electron flux: in Figure 18 for krypton at about 88.25 and 87 nm or in Figure 19 for xenon at about 100 and 97 nm there is only a small energetic shift between *ab initio* theories[61,64] or [61,15,27,64] and experiment,[99,100] or [27] respectively, and nearly no differences in the heights of minima and maxima. The narrow s-resonances in Kr and Xe at 87.9 and 98.5 nm, respectively, are not completely resolved in the

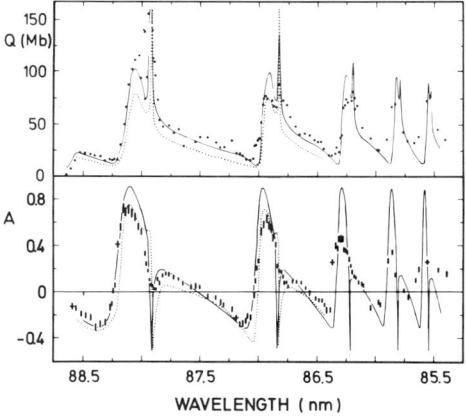

FIGURE 18. Kr $4p^6$ photoionization in the autoionization region. (Upper) Cross section measured, solid curve[101,102] and points.[99,100] (Lower) Spin polarization A[99,100] measured in comparison with theory RRPA[61] solid, MQDT[64] dotted.

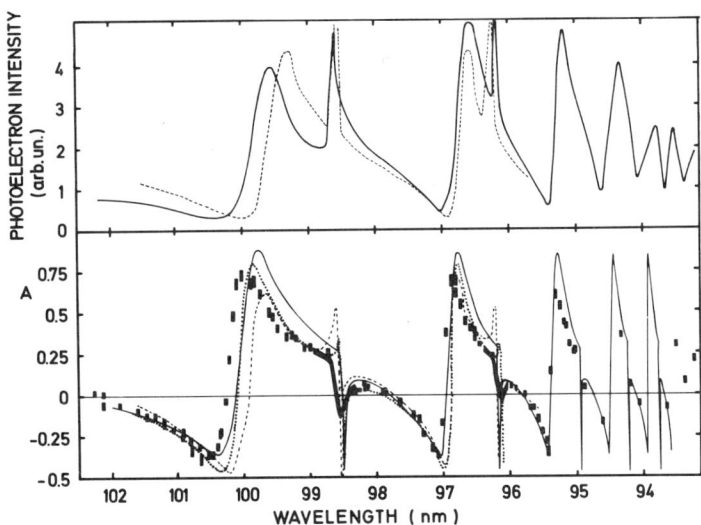

FIGURE 19. Photoionization of Xe atoms in the autoionization range: (top) cross section, photoelectron intensity; (bottom) spin polarization parameter A. Experimental results (error bar rectangles lower part and full curve upper part[27]; theoretical curves: dashed,[15] dotted,[27,64] full.[61]

experiment with the consequence of a weaker resonance oscillation observed than in theory. In summary, the data shown demonstrate that RPA theories and MQDT application excellently predict the spin polarization parameters of closed-shell atoms.

3.4. Experimental Results for Other Atoms and Comparison with Theory

Spin-polarized photoionization experiments have been performed over the last two decades with respect to different subshells: s subshell (Cs,[12] Ag,[103] and Hg[104–107]); p subshell (In,[49] Tl,[16,18,56] and Pb[88,51,108]; d subshell (Hg[48,107,109–111]); and f subshell (Yb[112]).

Photoionization of these atoms differs in several respects. Some do not show a fine-structure splitting in the ground state (Cs, Ag, Hg, Yb), or leave the remaining ion with closed shells or subshells, also without ionic fine-structure splitting (Cs, Ag, In, Tl) and in atoms with several outermost bound electrons (Pb, Yb), the selection rules $\Delta j = \pm 1, 0$ and $\Delta m_j = +1$ for the transitions of the electron to be ejected as a photoelectron have to be replaced by the selection rules of the overall transition $\Delta J = \pm 1, 0$ and $\Delta M_j = +1$, when circularly polarized σ^+ light is used. In the following, three particular cases of spin-resolved photoionization of Tl, Hg, Pb are discussed.

Figure 20 shows the five dynamical parameters for photoionization of thallium atoms in the autoionization region between 130 and 165 nm dominated by the broad $Tl(6s6p^2)^2D_{3/2}$ resonance and at 149 nm by the much narrower $Tl(6s6p^2)^2S_{1/2}$ resonance. The classification of these two resonances with respect to the J values was proved using spin polarization data.[16] There is an excellent agreement between the experimental results[56] and the theoretical RPAE calculation,[97] convoluted to the experimental radiation bandwidth for A, ξ, α, and β. In the resonance at 149 nm the $D_{1/2}$ matrix element describing the transition $5p_{1/2} \to \varepsilon s_{1/2}$ dominates with respect to $D_{3/2}(5p_{1/2} \to \varepsilon d_{3/2})$ due to autoionization amplification. Thus, A reaches up to +1 and β has a deep minimum. A sensitive test of the RPAE calculation are the phase-sensitive parameters ξ, β, and α; usually these quantities are strongly affected by autoionization resonances as shown in many studies.[109,106,54] The results given in Figure 20 are a data set which completely characterize the photoionization experiment; because of only two allowed transitions $5p_{1/2} \to \varepsilon s_{1/2}, \varepsilon d_{3/2}$ there are only three independent quantities—two are redundant. This additional experimental information served to check the consistency of the matrix elements and the phase shift difference, evaluated by experiment, given in Figure 21.[56] The relation of the quantum

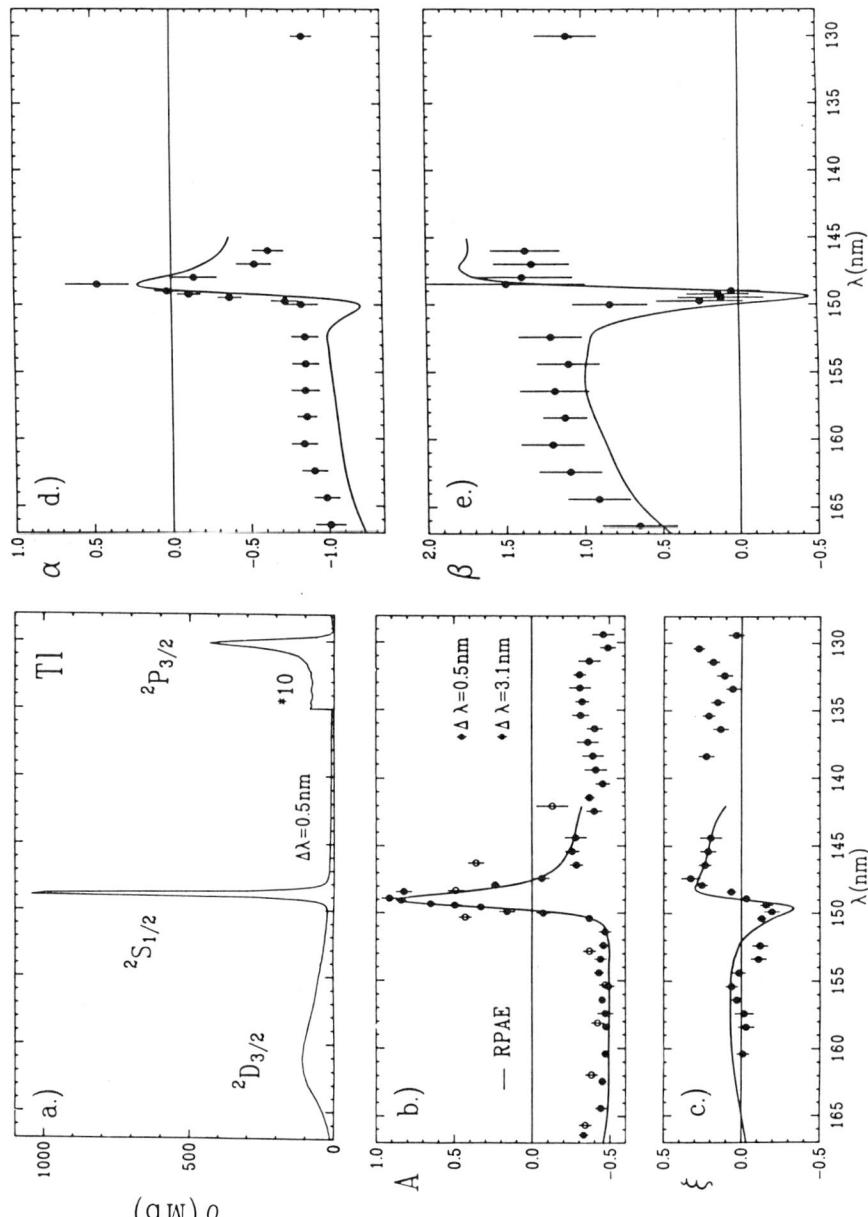

FIGURE 20. Photoionization of Tl$(6s^26p)^2P_{1/2}$ in the autoionization region $(6s6p^2)$. Experimental results for cross section, spin polarization parameters, and asymmetry parameter $\beta^{(56)}$ in comparison with RPAE calculation,[97] convoluted to the radiation bandwidth of the experiment $(\Delta\lambda = 0.5$ nm) as a function of wavelength.

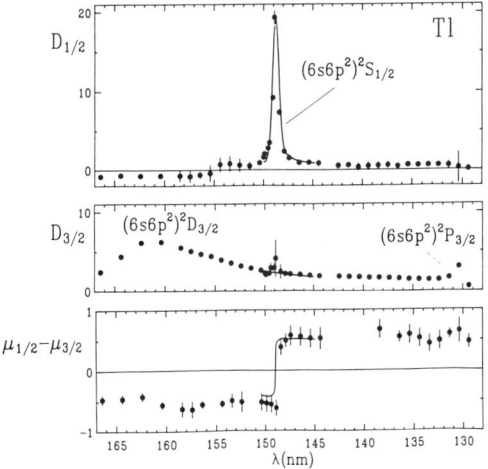

FIGURE 21. Dipole matrix elements $D_{1/2}$ and $D_{3/2}$ and quantum-defect difference $\mu_{1/2}-\mu_{3/2}$ for photoionization of $Tl(6s^2 6p)^2 P_{1/2}$ as a function of the wavelength in the $6s6p^2$ autoionization region. The full curve represents the RPAE[97] convoluted to the radiation bandwidth used in the experiment ($\Delta\lambda = 0.5$ nm).[56]

defect μ_i to the phase shift δ_i is given by[15] $\delta_i = \sigma_l + \pi\mu_i - \pi l/2$, σ_l being the Coulomb phase shift for an outgoing partial wave with angular momentum l in a pure Coulomb field. There is excellent agreement with the RPAE theory[97] again in all details; the matrix elements demonstrate clearly the channel, the autoionization interference falls in $J = 1/2$. There is no resonance contribution in the $D_{3/2}$ matrix element within the experimental uncertainty. The quantum-defect difference $\mu_{1/2}-\mu_{3/2}$ (lowest part of Figure 21) shows a change from negative to positive values by the order of 1 demonstrating a phase shift variation by π in accordance with Fano's theory for the interaction of an isolated state with a single continuum channel.[113] Below and above the resonance, the difference $|\mu_{1/2}-\mu_{3/2}|$ is nearly constant ($= 0.5$) in energy, indicating that the phase shift variation is only caused by the energy dependence of the Coulomb phase shift. Moreover, this example thus confirms the basic idea of the quantum defect theory (QDT).[114]

Figure 22 shows all experimental spin-resolved photoelectron spectroscopy data available for Hg 5d. The results are shown for both final ionic states (open symbols: $j = 3/2$; full symbols: $j = 5/2$); the vertical broken lines represent the D thresholds. The figure shows the results for the spin parameters

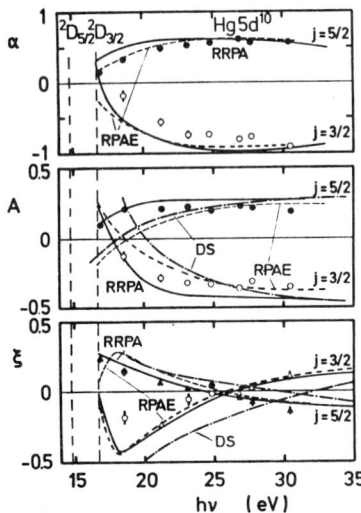

FIGURE 22. Photoionization of Hg 5d. Experimental results for the spin parameters α, A, and ξ for the final ionic state $^2D_{5/2}$ (full symbols) and $^2D_{3/2}$ (open symbols).[48] Triangles (for ξ)[109]; full curves, RRPA calculation[94]; broken curves, RPAE calculation[115]; chain curves, DS calculation.[116] The vertical broken lines indicate the ionization thresholds.

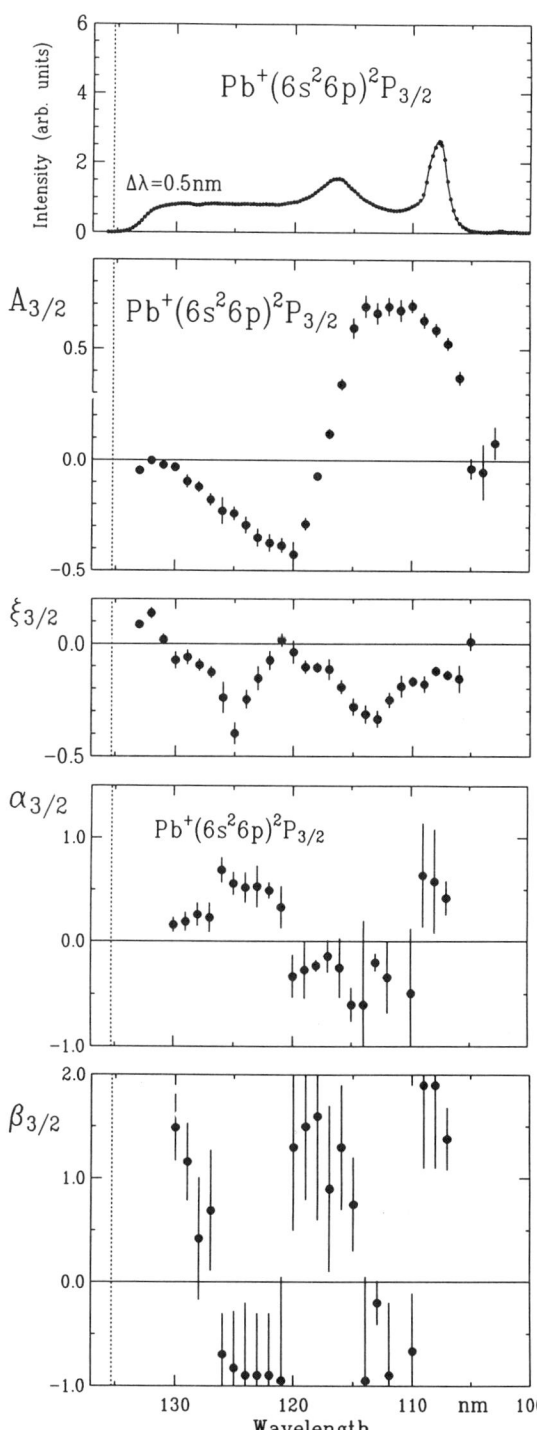

FIGURE 23. Experimental results of the spin polarization parameters $A_{3/2}$, $\xi_{3/2}$, $\alpha_{3/2}$ and the asymmetry parameter $\beta_{3/2}$ for photoionization of $Pb(6s^2 6p^2)_{J=0}$ with final ionic state $Pb^+(6s^2 6p)\ ^2P_{3/2}$ together with the corresponding photoelectron intensity and the photoionization threshold (vertical dotted).[51]

α, A, and ξ. The circles and triangles represent the set of measurements reported in Refs. 48 and 109, respectively.

The full curves represent the RRPA calculation[94] (experimental thresholds are used; correlations with the $6s$ and within the $5d$ shell are included); the broken curves reproduce the RPAE calculation.[115] The chain curves are a one-channel Dirac–Slater calculation (DS).[116] In the nonrelativistic approximation, neglecting the spin-orbit interaction in the continuum states, the ratio of the spin parameters is equal to the statistical ratio ($A_{3/2}/A_{5/2} = -1.5$, the same for α and ξ). In accordance with this the results for the two fine-structure components differ in sign.

The general energetic trend of the parameters can be explained by the following: all three parameters are complicated functions of the three transition matrix elements and the relative phases of the continuum wave functions. The parameter ξ is determined mainly by phase shift differences between p and f waves, which, due to the Coulomb part, vary rapidly at threshold and change sign at 25 eV.

The A parameter, on the other hand, is determined by the phase shifts between partial waves with the same l (spin-orbit interaction) which are known to be small and energetically nearly constant.[116] Therefore, A is influenced mainly by the ratio of the matrix elements, which is maximum in the shape resonance at approximately 20 eV above threshold.[117] This resonance, therefore, also causes the broad maximum of A starting approximately 10 eV above threshold.

All theoretical results reproduce the general energy dependence quite well, but in the threshold region systematic discrepancies remain. The overall agreement of the experimental data is best with the RRPA results, although considerable differences remain. The agreement with the DS results could be improved by introducing experimental thresholds.

In general, the agreement for the $j = 5/2$ level seems to be slightly better than with the $j = 3/2$ level. This seems to indicate indeed that the potential barrier effect which is dominant in the $j = 5/2$ level[116] is treated more adequately, while the spin-orbit effects between the outgoing p waves in the $j = 3/2$ level are underestimated.

The most complicated target atom for spin-resolved photoelectron spectroscopy from the understanding point of view has been found in experiment to be lead. There is also not yet a theory discussing the results observed in experiment.[51] Pb has a ground-state configuration $(6s^26p^2)_{J=0}$ with a configuration interaction in $j \cdot j$ coupling 93%$(6p_{1/2})^2$ and 7%$(6p_{3/2})^2$.[118,119] Thus, the $6p$ photoionization has two ionization thresholds corresponding to two final ionic states $Pb^+(6s^26p)^2P_{1/2}$ and $^2P_{3/2}$ at 167.2 and 135.3 nm, respectively.[120]

The most surprising result was the existence of "hidden" resonances: Figure 23 shows the photoelectron intensity and the four dynamical parameters $A_{3/2}$, $ξ_{3/2}$, $α_{3/2}$, and $β_{3/2}$ as functions of the radiation wavelength for the photoionization with the final ionic state $Pb^+(6s^26p)^2P_{3/2}$[51]: while the cross section is absolutely flat and without any structure between 120 and 130 nm, $ξ_{3/2}$ shows a pronounced resonance behavior, emanating very probably from a dramatic phase shift variation, since the other two phase shift-dependent parameters $α_{3/2}$ and $β_{3/2}$ also strongly vary in this energy range. The $A_{3/2}$ parameter varies between 130 and 100 nm in a very broad resonance between -0.5 and $+0.7$; the width of this resonance-variation is in no case connected with the intensity peak at about 117 nm. It is worth noting that at this resonance, the energetic position of the $^2P_{3/2}$ cross section, the spin parameter $ξ_{1/2}$ of the electrons corresponding to the other $^2P_{1/2}$ ionic channel, has "hidden" resonance too, although the corresponding cross section ($^2P_{1/2}$) does not show a resonance (not shown in Figure 23). To explain the physical reason for these hidden resonances and to treat the corresponding configuration interaction needs further theoretical work.

4. PHOTOIONIZATION OF MOLECULES

4.1. Differences between Atomic and Molecular Cases

Since the spin-orbit (or spin-axis) interaction is present also in molecules, it is natural to suppose that molecular photoelectrons are polarized too. The first experimental measurements were not very encouraging,[121] but subsequent theoretical[20] and experimental[19,21,122,123] investigations have cleared the situation. It was shown theoretically that for rotating molecules, if they are not chiral, the angular distribution of photoelectrons with defined spin polarization is described by the same equation (1) as in atoms, where the polarization parameters are given now by the equations presented in Refs. 20 and 25. As in atoms, the main effect of the order of 1 appears due to the fine-structure splitting of molecular levels. The ground state of molecules usually corresponds to zero values of all quantum numbers, therefore one had to separate photoelectrons corresponding to different fine-structure components of the final state. It is possible to do this in practice if a molecule contains at least one relatively heavy atom, as in hydrogen halides, used by both theorists[22,23] and experimentalists[25] as a test case. Figure 24 shows the photoelectron spectrum of I_2 ($h\nu$ = 21.22 eV) clearly resolved with respect to the ionic fine-structure splitting $^2\Pi_{1/2}$ and $^2\Pi_{3/2}$.[21]

In the nonrelativistic approximation the molecular polarization parameters satisfy the relations analogous to relations (7) in atoms, namely, the parameters have the opposite sign for different fine-structure components and are inversely proportional to the statistical weights of these states, which are equal in molecules. That this is the case only for lighter molecules in experiment and not for very heavy ones, is demonstrated in Figure 25, where the cross section, the asymmetry parameter β, and the spin parameter ξ are given for certain Br_2 and I_2 peaks in the photoelectron spectra.[21] While the ratios of cross sections and β parameters have been found to have the statistical value 1, this is not the case for the ξ parameter in I_2. While one fine-structure peak in the spectrum ($^2\Pi_{1/2g}$) is positively polarized, the other one ($^2\Pi_{3/2g}$) is unpolarized. This indicates a strong influence of the spin-orbit interaction on the continuum states for heavy molecules such as I_2, while the nonrelativistic approximation seems to be valid for Br_2 and lighter molecules. From the analysis of the explicit expression for the parameter A,[20] it follows that $|A| \leq 0.5$, that is, the greatest possible degree of spin polarization of the total photoelectron flux from molecules is equal to 50%, contrary to the atomic case where it can reach 100%. In general, the polarization of molecular photoelectrons is expected to be lower, due to the fact that even for absorption of polarized light, there is always the sum over all polarization states of a photon beam in the molecular frame, which leads to a greater number of allowed transitions as compared to the atomic case. Different transitions give contributions of different magnitude and even of different sign to the degree of spin polarization, so that the total polarization is not very high.

If molecules are chiral, that is, if they have neither plane of symmetry nor the center of inversion, the angular distribution of photoelectrons with defined spin polarization becomes more complicated and contains in addition to the terms given by equation (1) five new terms given below[35]

$$\Delta I_k^\lambda(\vec{q},\vec{s}) = \frac{\sigma_k(\omega)}{8\pi} \{\lambda D^k(\vec{q}\ \vec{k}) + \lambda C^k(\vec{s}[\vec{q}\ \vec{k}]) + B_1^k \cdot (\vec{q}\ \vec{s}) + B_2^k(\vec{q}\ \vec{k})(\vec{s}\ \vec{k}) + B_3^k(\vec{q}\ \vec{k})^2(\vec{q}\ \vec{s})\} \tag{14}$$

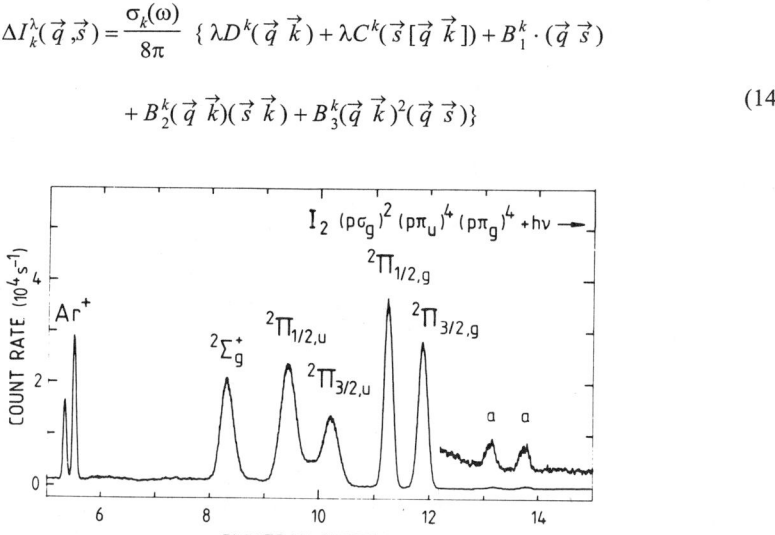

FIGURE 24. HeI(21.22 eV) photoelectron spectrum of I_2 at the magic angle; on the left are calibration peaks of argon, on the right two small peaks denoted (a) corresponding to the He Iα radiation (23.07 eV).[21]

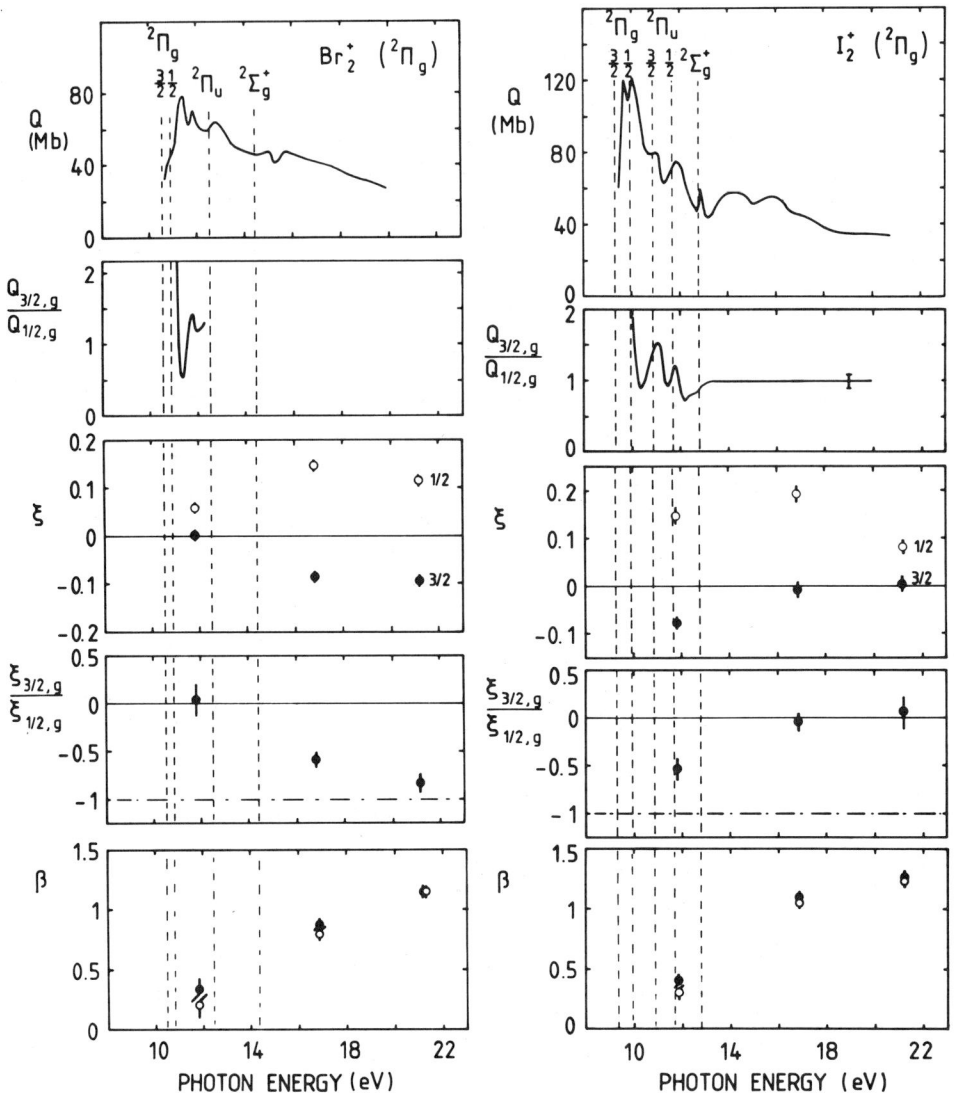

FIGURE 25. Partial photoionization cross section Q, spin parameter ξ, ratios of Q and ξ, and asymmetry parameter β for $^2\Pi_{g3/2,1/2}$ peaks of Br_2 and I_2.[21]

where k is a set of quantum numbers characterizing the final ionic state, and in the case of linearly polarized light the vector $\vec{k}$ should be replaced by the polarization vector $\vec{e}$. The first term in (14) gives the CDAD predicted for the first time in Ref. 124. The second term gives the component of the transverse polarization of photoelectrons which has an opposite sign for right and left circularly polarized light. And the last three terms in (14) characterize the longitudinal polarization of photoelectrons which appears for any polarized and unpolarized light. Thus, the angular distribution of photoelectrons with defined spin polarization ejected from chiral molecules is defined by ten independent parameters. Unfortunately, up to now there are no experimental results for chiral molecules.

It is worthwhile to note that all terms in (14) appear already in the electric dipole approximation, whereas the well-known phenomena of optical rotation and circular dichroism in the photoabsorption appear in the electric dipole–magnetic dipole interference terms which are α times smaller. CDAD is not connected with the spin-orbit interaction and to observe it one need not resolve the fine-structure splitting. Four other terms in (14) describe the spin polarization of photoelectrons and therefore the fine-structure splitting for their observation had to be resolved. All new parameters in (14) are proportional to the differences of pairs of dipole matrix elements having opposite signs of all projections of orbital momenta and spins. For achiral molecules these differences are identically equal to zero, whereas for chiral molecules they are different from zero and are proportional to the asymmetry parameter discussed in Ref. 125. The realistic estimation of this parameter without calculations is hardly possible.

The complete experiment in molecules is much more complicated than in atoms, since in molecules the orbital angular momentum is not a good quantum number. Therefore, all partial waves can contribute to the expansion of the ground-state wave function, and the number of allowed dipole transitions is, in principle, infinite. In practice, the number of partial waves contributing to this expansion can be restricted by few terms, which leads to the finite number of parameters required for the complete experiment in this approximation. Nevertheless, this number is usually more than five, so that the spin polarization measurements alone could not suffice for the complete experiment. The only way to increase the number of observable parameters is to investigate molecules fixed in space, or aligned molecules, which will be discussed in Section 5.

4.2. Influence of Molecular Vibration and Rotation onto Polarization of Photoelectrons

In addition to the electronic structures of isolated atoms including the influence of the spin-orbit interaction, molecules vibrate and rotate. Thus, the question arose whether molecular vibration and rotation influence the dynamical parameters describing the photoionization. Instinct would say one could expect an influence of the rotation onto the spin polarization of the photoelectrons (after absorption of circularly polarized light) since the overall molecular rotation is an angular momentum as the electron spin and orbital momentum; furthermore, one would not expect any influence of the vibration on the photoelectron spin polarization.

In this section two examples of molecules to be photoionized—NO and HI—are described; the first a light molecule with a spin-orbit splitting in the ground state and the second a heavy one with the spin-orbit splitting in the final ionic state. Spin- and angle-resolved photoelectron spectroscopy experiments have also been performed successfully with other halogen compounds CH_3Br, CH_3I, HBr, I_2, Br_2.[21,122,123,126]

NO is one of the few molecules with an open orbital showing a spin-orbit fine-structure splitting of the ground state $^2\Pi_{1/2}$, $^2\Pi_{3/2}$ similar to thallium atoms. However, the splitting of only 15 meV[127,128] for NO could not be resolved spectroscopically by means of an electron spectrometer; instead of this the population in only one of both states was thermally frozen out (≤ 40 K) by means of a helium-seeded supersonic NO beam so that the ratio of populations $^2\Pi_{3/2}/^2\Pi_{1/2} \approx 1\%$.

Using this spin-orbit state prepared NO and circularly polarized synchrotron radiation, the strongest autoionization resonance of NO occurs at a photon energy of 13.82 eV; by means of the electron spectrometer used, the vibrational structure of the final ionic state NO^+ X $^1\Sigma^+$ was resolved giving peaks for $v = 0$ till $v = 6$ with a maximum at $v = 3$ and a vibrational splitting of 300 meV.[128] The three spin polarization parameters A, α, and ξ have been determined for the $v = 2$ and $v = 3$ levels of NO^+ X $^1\Sigma^+$; the angular distribution obtained for the spin polarization component $A(\Theta)$ is given in Figure 26.[127,128]

Within the experimental uncertainties, the dynamical spin parameters have been found to be the same for both vibrational ionic states; the main interesting result, other than that of a nonvanishing spin polarization, is that only the α parameter seems to be different from zero. This means that the spin polarization vector lies in the reaction plane (i.e., $\xi = 0$) and the angle-integrated flux is unpolarized ($A = 0$). There is no influence of the vibration on the quantities of photoelectron spin polarization, as would be expected.

This is different for photoionization of the heavy HI molecule, which is isoelectronic to the xenon atom. The photoionization of HI has been studied extensively with respect to the spin polarization of photoelectrons by means of circularly polarized synchrotron radiation[25,129] as well as coherent VUV radiation.[86,130,131]

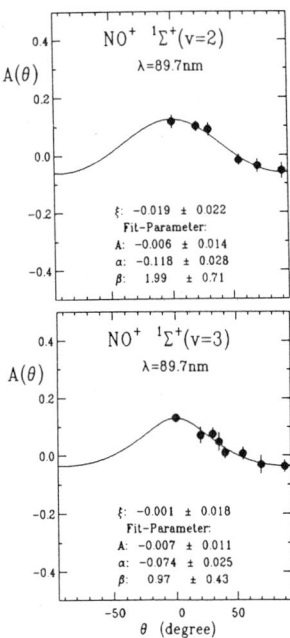

FIGURE 26. Spin polarization component $A(\Theta)$ as a function of the emission angle Θ for photoionization on NO molecules for vibrationally resolved final ionic states $^1\Sigma^+$ v = 2 and v = 3 in the autoionization resonance at 89.7 nm; least-squares fits yield the parameters given.[127,128]

Figure 27[129] shows the photoelectron spectrum of HI obtained at a photon energy of 12.08 eV; it reflects also levels v > 0 corresponding to the ionic fine-structure states $^2\Pi_{3/2,1/2}$. The "non-Franck–Condon" population of vibrationally excited levels arises due to the process of electronic autoionization converging to the $HI^+\ A\ ^2\Sigma^+_{1/2}$ state.[132] The upper panels of Figure 28 show the partial photoionization cross sections for the v = 0 and v = 1 vibrational ionic states.[129] We see differences between the cross sections v = 0 and v = 1 in the spin-orbit autoionization region between the corresponding $^2\Pi_{3/2}$ and $^2\Pi_{1/2}$ thresholds. But one also notices deviations above the $^2\Pi_{1/2}$ thresholds. There is a broad resonance structure near 12.4 eV whose broadened nature can be attributed to a strong predissociation of the Rydberg states with $A^2\Sigma^+$ core.[129,132] The cross sections indicate that the coupling via the resonances must be considerably different for v = 0 and v = 1. The lower panels of Figure 28 show the corresponding spin polarization A results. The nonrelativistic relationship of equal magnitude but opposite sign for A for the two spin-orbit states is fulfilled remarkably well, except for the energy region of 11.3–12 eV. When comparing the polarization data for $^2\Pi_{3/2}$ v = 0 and v = 1 channels, the most striking difference occurs at around 11.6 eV, where the polarization values for v = 0 show a dramatic drop down to almost zero; a similar effect has been observed for HBr $^2\Pi$ v = 0 in the region

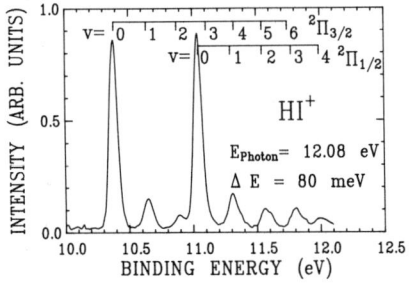

FIGURE 27. Vibrationally resolved photoelectron spectrum for $HI^+\ X\ ^2\Pi$ final ionic states.[129]

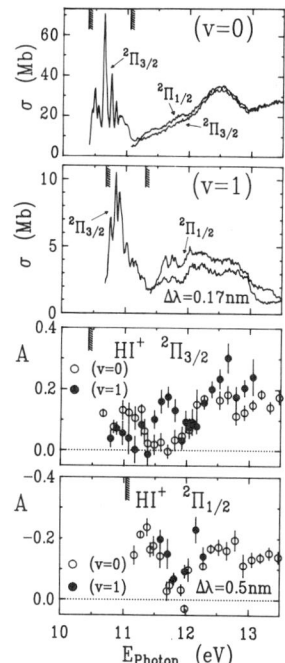

FIGURE 28. Partial photoionization cross sections and spin polarization parameters A for HI$^+$ $^2\Pi_{3/2,1/2}$ v = 0, 1 final ionic states.$^{(129)}$ Ionization limits are indicated by vertical lines.

between the $d\pi$ and $p\pi$ autoionization resonances.$^{(126)}$ Since the spin polarization A is directly proportional to the difference of partial cross sections $\sigma_\delta - \sigma_\sigma$ for transitions $\pi \to \varepsilon\delta$ and $\pi \to \varepsilon\sigma$, the results for v = 1 and v = 0 in the lower panels of Figure 28 can only be explained by an enlarged photoelectron $\varepsilon\delta$ contribution for v = 1 compared with the case of the ionic ground state (v = 0)$^{(129)}$ due to different autoionization coupling.

The autoionization region of HI (autoionizing Rydberg states converging to the $^2\Pi_{1/2}$ v = 0 ionic state which couple via spin-orbit interaction to the continua of the $^2\Pi_{3/2}$ v = 0 ionic state) shows pronounced resonance structures connected with HI Rydberg states $(5p\pi)^3 nl\lambda$. In the upper panel of Figure 29 the total photoelectron yield is shown for the Rydberg order 6 obtained by use of coherent VUV radiation.$^{(86)}$ The spectrum is dominated by a broad resonance which can be assigned to $d\pi$ and $d\delta$ states.$^{(133,134)}$ Sharp resonances due to $f\sigma$, π, $s\sigma$, $g\pi$, and $d\sigma$ states can be detected superimposed on the wing of the main d structure. It is worth noting that because of rotational structure the number of resonances observed greatly exceeds the number of electronic states assigned. For molecules, the orbital angular momentum of the electron is not a good quantum number and l mixing, i.e., coupling between states of the same λ and different l, has to be considered using this classification.$^{(133)}$

In the lower panel of Figure 29 the spin polarization measured$^{(86)}$ is plotted together with the results of the *ab initio* calculation.$^{(133)}$ The experimental data exhibit positive values of about +10%. Corresponding to the sharp features in the yield spectrum, many sharp resonances appear in the spin polarization data. The results of the *ab initio* calculation,$^{(133)}$ which neglects any rotational or vibrational effects, are shown as a dotted curve; they disagree with the experimental data in the absolute degree of the polarization and the position of the resonances which is not surprising in view of the effects neglected. They agree, however, in the occurrence of an overall positive background polarization and negative polarization values in the sharp resonances.

The most prominent structure of the experimental spin polarization data is a resonance at 86315 cm^{-1}. This can be assigned to the perturbed $d\sigma$ complex; hence, an easy rotational analysis is hampered. For this reason the further attention went to the resonances between 86000 and 86100 cm^{-1}. This spectral region is shown in more detail in Figure 30 on an enlarged scale. Curve a is a magnification of the central part of Figure 29(a), whereas curve c shows in detail the corresponding spin polarization

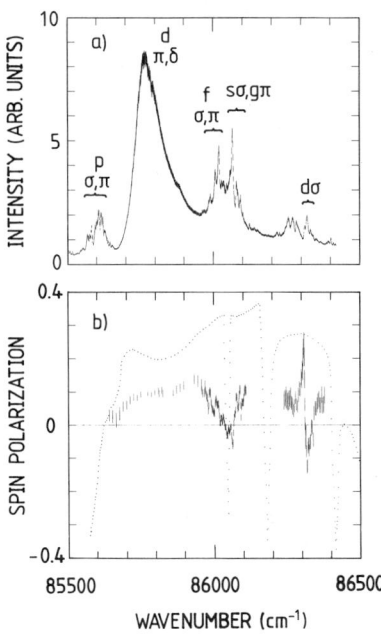

FIGURE 29. Photoelectron yield and corresponding spin polarization parameter A measured in photoionization of jet-cooled HI using coherent VUV radiation[86]; the dotted curve results from an *ab initio* calculation.[133]

of Figure 29(b). Both spectra exhibit many overlapping resonance structures. The bandwidth of the VUV radiation is at least three times smaller than the spacing between two data points in the plots shown and it is smaller than the autoionization linewidths. The largest obstacle for an analysis of the spin polarization data with respect to the influence of rotational effects is an overlap between rotational branches of different electronic states.[86]

In order to solve this problem, a simulation of the yield curve was performed (curve b), based on certain approximations[86]: the validity of the Born–Oppenheimer approximation was assumed. Starting from the quantum defects μ_a of the nine Rydberg series predicted by the calculation,[133] the positions of all lines were determined using the Rydberg formula and the J-dependent spacings[135] between rotational lines for Hund's case c.

The most intense peaks were reproduced quite well. However, deviations indicating the limitations of the approximations made are observed in three structures of the spectrum: the central part between the f and s peaks, where the measured intensity varies only slightly, and the structures connected with the sharpest resonances in the spin polarization data at 85985.8 and 86094.8 cm^{-1}. The assignments of the Q, R_0, and R_1 lines are indicated in Figure 30, curves b and c.[86]

The spin polarization (curve c) exhibits structure with energy intervals between the different peaks which are equal to the rotational spacings. Most of these peaks are strongly correlated with resonances in the yield curve. Compared to the spin polarization values in their direct neighborhood, some of these polarization peaks show an upward trend while others go downward. These observed changes of sign in the resonant contributions must be due to rotational effects, since the separation between the resonances is the rotational spacing as denoted in Figure 30, curve c. In particular, the data show that the two lines R_0 and R_1 of the same electronic Rydberg state ($s\sigma$) have spin polarization values of opposite signs.

Further detailed theoretical work is needed to explain this influence of the rotation to the photoelectron spin polarization quantitatively. A first and very important step in this direction has been made recently[136] in the rotationally resolved photoionization of HI.

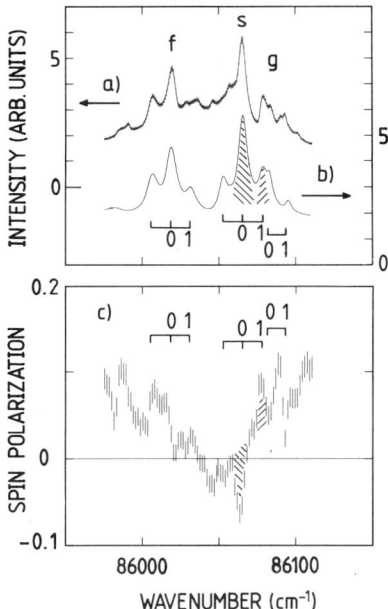

FIGURE 30. Curves of (a) and (c) show an expanded view of the central part of Figure 29(a) and (b), respectively[86]; curve (b) is the result of a simulation calculated (see text). In (b) and (c) the lines indicate the members Q, R_0, and R_1 of the rotational branches in photoionization of HI.

5. PHOTOIONIZATION AND AUGER DECAY OF ORIENTED SYSTEMS

5.1. Polarized Atoms

The most detailed information on the photoionization process can be obtained from ionization of polarized atoms by polarized radiation when the angular distribution and the spin polarization of photoelectrons are measured. Atoms with an open shell can be polarized by optical pumping, by magnetic deflection in the inhomogeneous field of a six-pole magnet.[30] Atoms adsorbed at a surface are usually aligned. Another way to create polarized (aligned or oriented) atoms is to excite them by linearly or circularly polarized light and then to ionize the excited atom. All of these methods are currently used in different experiments; therefore, it is worthwhile learning what new information can be obtained from experiments with polarized atoms.

The most general expression for the angular distribution of photoelectrons with defined spin polarization ejected from polarized atoms can be presented in the following form[68]:

$$I_j(\vec{q},\vec{s},\vec{n}) = \frac{\sigma_{nlj}(\omega)}{8\pi} \sum_{L,M} \sum_{x,\xi} \sum_{N,M_N} B^j(L,M,x,\xi,N,M_N)$$

$$\cdot \rho^{\vec{n}}_{NO} \cdot Y^*_{LM}(\vec{q}) Y^*_{x\xi}(\vec{s}) Y^*_{NM_N}(\vec{n}) \tag{15}$$

where B^j are the dynamical parameters expressed through the dipole matrix elements and phase shifts, $\rho^{\vec{n}}_{NO}$ are the state multipoles, and $\vec{n}$ is the unit vector in the direction of atomic polarization, that is, the direction along which the atomic density matrix is diagonal. The number of parameters here is rather high so that usually they are not mutually independent. Therefore, there is no reason to perform such complicated measurements. It is better instead to simplify the problem either by considering a particular geometry or by excluding the analysis of the spin polarization of photoelectrons. From the experimental point of view, the latter is much more preferable; therefore, we will restrict the consideration here to this case only.

There is an analogy between the angular distribution of photoelectrons with defined spin polarization ejected from unpolarized atoms with the angular distribution of photoelectrons averaged over spin projections, ejected from polarized atoms. But since the electron spin is equal to 1/2 while the total angular momentum of an atom can be larger than 1/2, the investigation of polarized atoms can give more than five parameters, as mentioned in Section 2.2. This is important for open-shell atoms, where more than five parameters can be needed. As was shown in Ref. 68, the angular distribution of photoelectrons ejected from polarized atoms by circularly polarized light is characterized by $(9J+1)$ or $(9J+1/2)$ parameters for integer or half-integer values of the total angular momentum of an atom J, respectively. For absorption of linearly polarized light the number of parameters is equal to $(6J+1)$ or $6J$ for integer or half-integer J, respectively.

The most concise expression for the angular distribution of photoelectrons ejected from polarized atoms is obtained in the density matrix formalism when the polarization state of both photons and atoms is described by state multipoles.[137] The result can be presented as follows[70]:

$$I_j(\vec{q},\vec{n}) = \sigma_{nlj}(\omega)\sqrt{3(2j+1)} \sum_{K,L,N} C^j_{KLN} \sum_{x,M,M_N} \rho^{\gamma}_{Kx}$$

$$\cdot \rho^{\vec{n}}_{N0} Y^*_{LM}(\vec{q}) Y^*_{NM_N}(\vec{n}) \begin{pmatrix} K & L & N \\ x & M & M_N \end{pmatrix} \quad (16)$$

where ρ^{γ}_{Kx} are the state multipoles of the photon beam. The parameters C^j_{KLN} are analogous to the spin polarization parameters (9)–(11). But while the parameters A^j, α^j, ξ^j have opposite signs for the fine-structure levels with $j = l + 1/2$ and $j = l - 1/2$, the parameters C^j_{KLM} have the same sign for the fine-structure levels and approximately the same value, like the angular asymmetry parameter β^j. Therefore, the fine-structure levels need not be resolved in order to observe the angular distribution described by (16), and these measurements can be performed also in light atoms where the photoelectron spin polarization measurements are hardly possible.

In many cases, and for one-electron subshells in particular, more than five independent parameters are not necessary. Therefore, it can be desirable to reduce further the number of parameters in (16). To this end, one can investigate the circular or linear dichroism in the angular distribution of photoelectrons.[70,71,138] CDAD and LDAD are described by different terms in (16), namely, CDAD by terms with $K=1$ while LDAD by terms with $K=2$. Integration of CDAD and LDAD over electron ejection angles gives the usual circular dichroism (CD) and linear dichroism (LD) of polarized atoms:

$$I^{CD}_j(\vec{n}) = I^{+1}_j(\vec{n}) - I^{-1}_j(\vec{n})$$
$$= -\sigma_{nlj}(\omega) \cdot \sqrt{6(2j+1)} C^j_{101} \rho^{\vec{n}}_{10} \cdot (\vec{n}\,\vec{k}) \quad (17)$$

$$I^{LD}_j(\vec{n}) = I^{\vec{e}_y}_j(\vec{n}) - I^{\vec{e}_x}_j(\vec{n})$$
$$= \sigma_{nlj}(\omega) \frac{3}{2}\sqrt{2(2j+1)} \cdot C^j_{202} \rho^{\vec{n}}_{20} [(\vec{n}\,\vec{e}_x)^2 - (\vec{n}\,\vec{e}_y)^2] \quad (18)$$

where $\vec{e}_x$ and $\vec{e}_y$ are the polarization vectors of linearly polarized light. So, polarized atoms are optically active and reveal either CD or LD depending on whether they have been oriented ($\rho^{\vec{n}}_{10} \neq 0$) or aligned ($\rho^{\vec{n}}_{20} \neq 0$).

Finally, if light is unpolarized, one can use the other simplification of (16) by considering the difference between two angular distributions corresponding to two orthogonal polarizations of atoms, say, along the X and Y axes of the laboratory frame[75]:

$$I_j^{XY}(\vec{q}) = I_j^{un}(\vec{q})|_{\vec{n}\|X} - I_j^{un}(\vec{q})|_{\vec{n}\|Y}$$

$$= \sigma_{nlj}(\omega) \frac{15(2j+1)}{8\pi} \left\{ -\frac{i}{5} \rho_{10}^{\vec{n}} C_{221}^j \cos\vartheta \sin\vartheta (\sin\varphi + \cos\varphi) \right.$$

$$+ \frac{1}{\sqrt{5}} \rho_{20}^{\vec{n}} [C_{022}^j + \frac{1}{\sqrt{7}} C_{222}^j] \sin^2\vartheta \cos 2\varphi$$

$$\left. + \frac{1}{4\sqrt{7}} \rho_{20}^{\vec{n}} C_{242}^j \sin^2\vartheta (7\cos^2\vartheta - 1) \cos 2\varphi \right\} \quad (19)$$

where ϑ and φ are the spherical angles of the vector $\vec{q}$ and terms with $N > 2$ have been omitted. Here we have three independent coefficients, which in many cases can be sufficient for the complete experiment.

5.2. Oriented Molecules

By analogy with polarized atoms, one can consider also oriented molecules. By oriented molecules we refer to both fixed-in-space and aligned or oriented molecules. Rotating molecules can be excited by polarized light and made aligned or oriented, like atoms, depending on light polarization. In this case there will be nonequal population of states with different projections of the total angular momentum including rotation. The other way to obtain oriented molecules is to stop their rotation. For example, molecules adsorbed at a surface or molecules in liquid crystals are not rotating but are fixed in space. Molecules can be oriented by an external field.[139,140] It is possible also to select the photoionization process of oriented-in-space molecules in a gas phase by coincidence measurements of fragment ions and photoelectrons provided the molecular ion decays rather quickly after ionization.[141]

The general expression for the angular distribution of photoelectrons with defined spin polarization ejected from oriented molecules is analogous to equation (15)[35,142] and contains a triple expansion over spherical functions. It is possible to consider this process either in the coordinate frame defined by the photon beam, or in the coordinate frame defined by the molecular orientation. The latter choice gives the simpler answer and therefore is preferable.[142]

Analysis of the general expression for linear molecules shows[142] that in Hund's cases a and b, which in the absence of rotation coincide, the photoelectrons are always polarized in the direction of the molecular axis irrespective of the direction of a photon beam, its polarization, or the direction of photoelectron ejection. In Hund's case c the photoelectron spin can have any direction and after integration over electron ejection angles, the total photoelectron flux has all three components of the spin polarization vector different from zero.

The general expression (15) is very complicated even for particular geometry and can hardly be investigated experimentally in the nearest future. To simplify the problem, one had to exclude the electron spin polarization measurement. The corresponding angular distributions have been derived in Ref. 143 and measured by several authors[144,145] for molecules adsorbed at a surface.

The angular distribution still contains rather many terms and it could be preferable to perform several measurements each giving only a few parameters. To this end one should measure, like in atoms, not only the angular distribution itself, but CDAD and LDAD as was proposed in Refs. 146–148. Measurements of CDAD for both aligned[149] and space-fixed[150] molecules fully support the theoretical predictions, and even the first "complete" experiment in molecules has been reported recently.[151] Oriented molecules are optically active and reveal also CD and LD, as well as the rotation

5.3. Polarized Auger Electrons

That the Auger electron emission of ferromagnetic materials is spin polarized and that the spin polarization of Auger electrons reflects at least qualitatively the magnetization of the sample, has been well known for some years.[155] Auger electron emission is part of a two-electron emission process or at least in treatment of a cascade the second step of it; thus, in principle it contains all information about the Auger process, the orientation and alignment of the hole state, and correlations of both electrons emitted.

The theoretical investigations into angular distribution and spin polarization of Auger electrons have increased rapidly during the past 15 years.[156–160] So far most publications have concentrated on the rare gas atoms. First experiments of spin-polarized Auger electron emission from atoms have also been performed after creation of the hole by means of unpolarized ion or electron bombardment[161,162]; the degree of polarization observed, however, was extremely small (1%), since the hole alignment was not very large and a hole orientation was not created.

It has been known for more than a decade[88] that photoionization of a closed shell using circularly polarized radiation produces not only polarized photoelectrons but also spin-oriented hole states. For example, ignoring any correlation effects, the ionic spin polarization of $Pb^+(6s^26p)$ has been predicted to be directly proportional to the photoelectron spin polarization A and to be $-A/3$.[88]

Very recently two experiments of Auger electron spin analysis after photoemission using circularly polarized synchrotron radiation have been performed with free Ba atoms[163] and (nonferromagnetic) adlayers [K, Rb, Cs on Pt(111)].[164,165] It is the purpose of this section to review briefly these activities.

First spin polarization measurements of Auger electrons produced by resonant photoionization of the $5p$ shell of free Ba atoms have been performed[163] using circularly polarized synchrotron radiation at the 6.5-m NI monochromator of BESSY.[80] The investigated two-step autoionization process leading to Auger emission occurs at photon energies above 21 eV:

absorption: $Ba(5p^66s^2)^1S_0 + h\nu \rightarrow Ba^*(5p^5)\,^2P n_0 l_0 n_1 l_2$
autoionization: $\rightarrow Ba^+(5p^5)\,^2P n_3 l_3 n_4 l_4 + e^-_{Ph}$
Auger process: $\rightarrow Ba^{2+}(5p^6)\,^1S_0 + e^-_{Ph} + e^-_{Au}$

The photoionization with circularly polarized light may produce hole states in inner shells with not only aligned but also oriented angular momentum.[88,166] This polarization can be partially passed to the outgoing Auger electron giving rise to nonvanishing angle-integrated polarization transfer.[158] For the case of a final $^1S_0$ state examined here, there is only one Auger amplitude and the polarization is entirely determined by the photoelectron matrix elements.

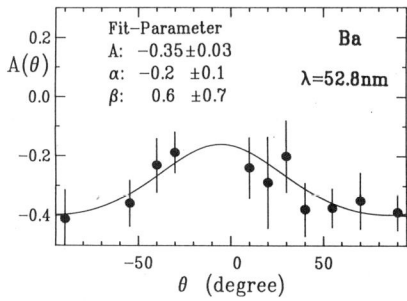

FIGURE 31. Angular dependence of the spin polarization component $A(\Theta)$ of the Auger electrons from free Ba atoms[163] including least-squares fit.

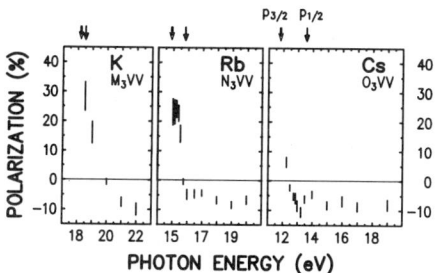

FIGURE 32. Spin polarization $A(\Theta = 180°)$ averaged across the Auger peaks in normal emission of thick K, Rb, and Cs layers on Pt(111) as a function of the energy of circularly polarized radiation creating the hole.[164,165] The arrows indicate the thresholds for excitation of the $p_{3/2}$ and $p_{1/2}$ hole states.

In angle- and spin-resolved measurements the angular distributions of the spin polarization components $A(\Theta)$ (Figure 31) and $P_\perp(\Theta)$ were obtained at a photon energy of 23.5 eV for the Auger line at $E_{kin} = 7.5$ eV [mainly $5p^56s^2\,^2P_{3/2} \to 5p^6\,^1S_0$, with admixture of $5p^55d(^1D)6s\,^2D_{3/2} \to 5p^6\,^1S_0$]. From least-squares fits of these distributions the spin polarization parameters A, ξ, α and the asymmetry parameter β for the Auger electrons were determined. Using these results and the formalism of Kabachnik and Lee,[158] the orientation parameter A_{10} and the alignment parameter A_{20} of the intermediate Ba$^+$ state and the ratio of the nonrelativistic reduced matrix elements for the autoionization process were determined approximately. From the data $A_{10} \approx -0.45 \pm 0.08$ and $A_{20} \approx 0.08 \pm 0.05$ and $D_s/D_d \approx 0.25 \pm 0.07$ was obtained,[163] where D_s and D_d are the reduced matrix elements for outgoing s and d electrons, respectively.

The spin-dependent CVV Auger decay of oriented $3p_{3/2}$-, $4p_{3/2}$-, and $5p_{3/2}$-hole states in thick layers of K, Rb, and Cs, respectively, on Pt(111) in a normal-emission setup have been studied.[164,165] The oriented hole states were excited by normal-incidence circularly polarized light from BESSY (6.5-m NIM) with energies from 12 to 22 eV. Figure 32 displays the dependence of the spin polarization averaged across the Auger peak versus photon energy for K, Rb, and Cs.[165]

Since the spin polarizations of the primary photoelectrons have not yet been measured in this first study of polarized Auger electrons, it is not clear from experiments whether the sign change of the Auger electron polarization observed as given in Figure 32 is correlated with a corresponding sign change of the spin polarizations of photoelectrons and thus of the holes, or whether it originated in the Auger decay alone by means of a change of the pure singlet coupling of both valence electrons participating in the decay to more triplet contribution at higher energies. Further experimental and theoretical activities are in progress and will shed light on this behavior.

6. CONCLUSION

This review should demonstrate that the existence of spin-polarized photoelectrons is rather the rule than the exception in atomic and molecular photoionization. In general, one has only to resolve the fine-structure splitting by use of an electron spectrometer in an angle-resolved photoemission study, or if it is too small, the fine-structure splitting of the atomic or molecular ground state can be resolved by means of cooling the beam to low temperatures for population of only the lowest fine-structure state. Theoretical studies have also shown, however, that it is not necessary to measure the photoelectron spin polarization when the target is spin polarized or geometrically oriented to get out the same information about the photoionization process. In many cases spin- and angle-resolved photoelectron spectroscopy studies give experimental data which form a complete set characterizing the photoionization process from the quantum mechanics point of view. These types of experiment are then also called "complete experiment."

The strong progress in the field of spin-polarized atomic and molecular photoionization has been enabled by the fast development of circularly polarized VUV radiation sources: synchrotron radiation as well as coherent laser radiation.

Further applications of circularly polarized undulator radiation to atomic and molecular photoionization will open this field for higher photon energies up to 250 eV in the future. There are many unresolved problems to be dealt with in theory: open-shell atoms, molecules including vibration and rotation, two-electron emissions and correlations as for example in Auger decay after photoionization.

There is, however, also the large field of spin-polarized photoemission from adsorbates (including submonolayer regimes and dilute adlayers) and solid surfaces which is not discussed in this chapter. There are a lot of atomic effects such as the spin polarization of photoelectrons created in autoionization processes[167,168] or obtained in photoemission on nonferromagnetic solids using unpolarized radiation,[169,170] an effect which should not exist in bulk photoemission from symmetry points of view. This field is reviewed elsewhere.[52,171–174] But also there the existence of spin-polarized photoelectrons is rather the rule than the exception, independent of using polarized or unpolarized radiation and of using a polarized or an unpolarized target.

ACKNOWLEDGMENTS. Our thanks go to N. Böwering, M. Drescher, R. Kuntze, N. Müller, M. Salzmann, and F. Schäfers for intensive discussions and to Kay Lofthouse for preparation and careful reading of the manuscript. Financial support of BMFT, DFG, and EEC is gratefully acknowledged.

REFERENCES

1. F. SAUTER, *Ann. Phys. (Leipzig)* **11**, 454 (1931).
2. B. C. H. NAGEL, *Ark. Fys.* **18**, 1 (1960).
3. B. C. H. NAGEL AND P. O. M. OLSSON, *Ark. Fys.* **18**, 29 (1960).
4. U. FANO, *Phys. Rev.* **178**, 131 (1969).
5. R. H. PRATT, A. RON, AND H. K. TSENG, *Rev. Mod. Phys.* **45**, 273 (1973).
6. Y. S. KIM, I. B. GOLDBERG, AND R. H. PRATT, *Phys. Rev. A* **45**, 4542 (1992).
7. P. R. S. GOMES AND J. BYRNE, *J. Phys. B* **13**, 3975 (1980).
8. J. KESSLER, *Rev. Mod. Phys.* **41**, 3 (1969).
9. M. J. SEATON, *Proc. R. Soc. London Ser. A* **208**, 408, 418 (1951).
10. J. W. COOPER, *Phys. Rev.* **128**, 681 (1962).
11. G. BAUM, M. S. LUBELL, AND W. RAITH, *Phys. Rev. Lett.* **25**, 267 (1970).
12. U. HEINZMANN, J. KESSLER, AND J. LORENZ, *Z. Phys.* **240**, 42 (1970); *Phys. Rev. Lett.* **25**, 1325 (1970).
13. N. A. CHEREPKOV, *Phys. Lett.* **40A**, 119 (1972).
14. N. A. CHEREPKOV, *J. Phys. B* **10**, L653 (1977).
15. C. M. LEE, *Phys. Rev. A* **10**, 1598 (1974).
16. U. HEINZMANN, H. HEUER, AND J. KESSLER, *Phys. Rev. Lett.* **34**, 441 (1975); **36**, 1444 (1976).
17. U. HEINZMANN, G. SCHÖNHENSE, AND J. KESSLER, *Phys. Rev. Lett.* **42**, 1603 (1979).
18. N. BÖWERING, M. SALZMANN, M. MÜLLER, H. W. KLAUSING, AND U. HEINZMANN, *Phys. Scr.* **41**, 429 (1990).
19. U. HEINZMANN, F. SCHÄFERS, AND B. A. HESS, *Chem. Phys. Lett.* **69**, 284 (1980).
20. N. A. CHEREPKOV, *J. Phys. B* **14**, 2165 (1981).
21. G. SCHÖNHENSE, V. DZIDZONOU, S. KAESDORF, AND U. HEINZMANN, *Phys. Rev. Lett.* **52**, 811 (1984).
22. H. LEFEBVRE-BRION, A. GIUSTI-SUZOR, AND G. RASEEV, *J. Chem. Phys.* **83**, 1557 (1985).
23. G. RASEEV, F. KELLER, AND H. LEFEBVRE-BRION, *Phys. Rev. A* **36**, 4759 (1987).
24. G. RASEEV AND N. A. CHEREPKOV, *Phys. Rev. A* **42**, 3948 (1990).
25. N. BÖWERING, M. MÜLLER, M. SALZMANN, AND U. HEINZMANN, *J. Phys. B* **24**, 4793 (1991).
26. N. A. CHEREPKOV, *J. Phys. B* **12**, 1279 (1979).
27. U. HEINZMANN, *J. Phys. B* **13**, 4353, 4367 (1980).
28. E. FUES AND H. HELLMANN, *Phys. Z.* **31**, 465 (1930).
29. G. BAUM AND U. KOCH, *Nucl. Instrum. Methods* **71**, 189 (1969).
30. V. W. HUGHES, R. L. LONG, JR., M. S. LUBELL, M. POSNER, AND W. RAITH, *Phys. Rev. A* **5**, 195 (1972).
31. W. VON DRACHENFELS, U. T. KOCH, T. M. MÜLLER, W. PAUL, AND H. R. SCHAEFER, *Nucl. Instrum. Methods* **140**, 47 (1977).
32. P. F. WAINWRIGHT, M. J. ALGUARD, G. BAUM, AND M. S. LUBELL, *Rev. Sci. Instrum.* **49**, 571 (1978).

33. M. J. Alguard, J. E. Clendelin, R. D. Ehrlich, V. W. Hughes, J. S. Ladish, M. S. Lubell, K. P. Schüler, G. Baum, W. Raith, R. H. Miller, and W. Lysenko, *Nucl. Instrum. Methods* **163**, 29 (1979).
34. R. Möllenkamp and U. Heinzmann, *J. Phys. E* **15**, 692 (1982).
35. N. A. Cherepkov, *Adv. At. Mol. Phys.* **19**, 395 (1983).
36. B. E. Kryłov and M. G. Kozlov, *Opt. Spectrosc.* **47**, 838 (1979).
37. D. T. Pierce, R. J. Celotta, G.-C. Wang, W. N. Unertl, A. Galejs, C. E. Kuyatt, and S. R. Mielczarek, *Rev. Sci. Instrum.* **51**, 478 (1980).
38. T. Maruyama, E. L. Garwin, R. Prepost, G. H. Zapalac, J. S. Smith, and J. D. Walker, *Phys. Rev. Lett.* **66**, 2376 (1991).
39. G. Laplanche, M. Jaouen, and A. Rachman, *J. Phys. B* **19**, 79, 101 (1986).
40. V. L. Jacobs, *J. Phys. B* **5**, 2257 (1972).
41. N. M. Kabachnik and I. P. Sazhina, *J. Phys. B* **9**, 1681 (1976).
42. N. A. Cherepkov, *Zh. Eksp. Teor. Fiz.* **65**, 933 (1973) [*Sov. Phys. JETP* **38**, 463 (1974)].
43. B. Brehm, *Z. Phys.* **242**, 195 (1971).
44. K.-N. Huang, *Phys. Rev. A* **22**, 223 (1980).
45. H. Klar, *J. Phys. B* **13**, 3117 (1980).
46. J. Kessler, *Polarized Electrons*, 2nd ed. (Springer, Berlin, 1985).
47. G. Schönhense, *Phys. Rev. Lett.* **44**, 640 (1980).
48. F. Schäfers, C. Heckenkamp, G. Schönhense, and U. Heinzmann, *J. Phys. B* **21**, 769 (1988).
49. M. Müller, N. Böwering, F. Schäfers, and U. Heinzmann, *Phys. Scr.* **41**, 42 (1990).
50. M. Müller, N. Böwering, M. Salzmann, H. W. Klausing, and U. Heinzmann, BESSY Jahresbericht 1988, pp. 122–123.
51. M. Müller, N. Böwering, F. Schäfers, and U. Heinzmann, *Phys. Scr.* **41**, 38 (1990).
52. U. Heinzmann, in *Photoemission and Absorption Spectroscopy of Solids and Interfaces with Synchrotron Radiation*, edited by M. Campagna and R. Rosei (North-Holland, Amsterdam, 1990), p. 469.
53. C. Heckenkamp, F. Schäfers, G. Schönhense, and U. Heinzmann, *Phys. Rev. Lett.* **52**, 421 (1984).
54. C. Heckenkamp, F. Schäfers, G. Schönhense, and U. Heinzmann, *Z. Phys. D* **2**, 257 (1986).
55. B. Kessler, N. Müller, B. Schmiedeskamp, B. Vogt, and U. Heinzmann, *Z. Phys. D* **17**, 11 (1990).
56. M. Müller, N. Böwering, A. Svensson, and U. Heinzmann, *J. Phys. B* **23**, 2267 (1990).
57. J. Kessler, *Comments At. Mol. Phys.* **10**, 47 (1981).
58. R. E. Huffman, Y. Tanaka, and J. C. Larrabee, *J. Chem. Phys.* **39**, 902 (1963).
59. J. A. R. Samson and J. L. Gardner, *Phys. Rev. Lett.* **31**, 1327 (1973).
60. C. Heckenkamp, F. Schäfers, G. Schönhense, and U. Heinzmann, *Phys. Rev. A* **32**, 1252 (1985).
61. W. R. Johnson, K. T. Cheng, K.-N. Huang, and M. le Dourneuf, *Phys. Rev. A* **22**, 989 (1980).
62. M. Y. Amusia and V. K. Ivanov, *Izv. Akad. Nauk SSSR Ser. Fiz.* **41**, 2509 (1977).
63. D. Dill, *Phys. Rev. A* **7**, 1976 (1973).
64. J. Geiger, *Z. Phys. A* **282**, 129 (1977).
65. J. Geiger, *Z. Phys. A* **276**, 219 (1976).
66. D. Dill and U. Fano, *Phys. Rev. Lett.* **29**, 1203 (1972).
67. U. Fano and D. Dill, *Phys. Rev. A* **6**, 185 (1972).
68. H. Klar and H. Kleinpoppen, *J. Phys. B* **15**, 933 (1982).
69. K.-N. Huang, *Phys. Rev. Lett.* **48**, 1811 (1982).
70. N. A. Cherepkov and V. V. Kuznetsov, *J. Phys. B* **22**, L405 (1989).
71. N. A. Cherepkov, V. V. Kuznetzsov, and V. A. Verbitskii, *J. Phys. B.* **28**, 1221 (1995).
72. H. Bethe, *Ann. Phys.* **5**, 325 (1930).
73. W. F. Chan, G. Cooper, and C. E. Brion, *Phys. Rev. A* **44**, 186 (1991).
74. A. Hamnett, W. Stoll, G. Branton, C. E. Brion, and M. J. van der Wiel, *J. Phys. B* **9**, 945 (1976).
75. N. A. Cherepkov and V. V. Kuznetsov, *J. Phys. B* **25**, L445 (1992).
76. W. Bußert and H. Klar, *Z. Phys. A* **312**, 315 (1983).
77. V. D. Ovsyannikov, *Zh. Eksp. Teor. Fiz.* **88**, 1168 (1985).
78. W. Mehlhorn, in *Today and Tomorrow in Photoionization*, edited by M. Y. Amusia and J. B. West (Daresbury, 1991), p. 15.
79. U. Heinzmann, B. Osterheld, and F. Schäfers, *Nucl. Instrum. Methods* **195**, 395 (1982).
80. F. Schäfers, W. Peatman, A. Eyers, C. Heckenkamp, G. Schönhense, and U. Heinzmann, *Rev. Sci. Instrum.* **57**, 1032 (1986).
81. C. Heckenkamp, A. Eyers, F. Schäfers, G. Schönhense, and U. Heinzmann, *Nucl. Instrum. Methods A* **246**, 500 (1986).

82. J. Bahrdt, A. Gaupp, W. Gudat, M. Mast, K. Wolter, W. B. Peatman, M. Scheer, T. Schroeter, and C. Wang, BESSY Jahresbericht 1991, pp. 446–452.
83. M. B. Moiseev, M. N. Nikitin, and N. I. Fedosov, *Sov. Phys.* **J21**, 332 (1978).
84. K. J. Kim, *Nucl. Instrum. Methods* **222**, 11 (1984).
85. R. Hilbig and R. Wallenstein, *IEEE J. Quantum Electron.* **19**, 194, 1759 (1982).
86. T. Huth-Fehre, A. Mank, M. Drescher, N. Böwering, and U. Heinzmann, *Phys. Rev. Lett.* **64**, 396 (1990).
87. A. Mank, M. Drescher, T. Huth-Fehre, G. Schönhense, N. Böwering, and U. Heinzmann, *J. Phys. B* **22**, L487 (1989).
88. U. Heinzmann, *J. Phys. B* **11**, 399 (1978).
89. M. Y. Amusia and N. A. Cherepkov, *Case Stud. At. Phys.* **5**, 47 (1975).
90. K.-N. Huang, W. R. Johnson, and K. R. Cheng, *At. Data Nucl. Data Tables* **26**, 33 (1981).
91. U. Heinzmann, G. Schönhense, and J. Kessler, *J. Phys. B* **13**, L153 (1980).
92. J. L. Dehmer, W. A. Chupka, J. Berkowitz, and W. T. Jivery, *Phys. Rev. A* **12**, 1966 (1975).
93. L. Torop, J. Morton, and J. B. West, *J. Phys. B* **9**, 2035 (1976).
94. W. R. Johnson, V. Radojević, P. Deshmukh, and K. T. Cheng, *Phys. Rev. A* **25**, 337 (1982).
95. K. Bartschat and P. Scott, *J. Phys. B* **18**, 3725 (1985).
96. K. Bartschat, *J. Phys. B* **20**, 5023 (1987).
97. N. A. Cherepkov, *Opt. Spectrosc.* **49**, 1067 (1980).
98. K.-N. Huang, W. R. Johnson, and K. T. Cheng, *Phys. Rev. Lett.* **43**, 1658 (1979).
99. U. Heinzmann and F. Schäfers, *J. Phys. B* **13**, L415 (1980).
100. F. Schäfers, G. Schönhense, and U. Heinzmann, *Phys. Rev. A* **28**, 802 (1983).
101. V. L. Carter and R. D. Hudson, *J. Opt. Soc. Am.* **63**, 733 (1973).
102. V. Saile, Ph.D. thesis, Universität München 1976 (unpublished).
103. U. Heinzmann, A. Wolcke, and J. Kessler, *J. Phys. B* **13**, 3149 (1980).
104. F. Schäfers, G. Schönhense, and U. Heinzmann, *Z. Phys. A* **304**, 41 (1982).
105. G. Schönhense, U. Heinzmann, J. Kessler, and N. A. Cherepkov, *Phys. Rev. Lett.* **48**, 603 (1982).
106. G. Schönhense, F. Schäfers, C. Heckenkamp, U. Heinzmann, and M. A. Baig, *J. Phys. B* **17**, L771 (1984).
107. F. Schäfers, C. Heckenkamp, M. Müller, V. Radojević, and U. Heinzmann, *Phys. Rev. A* **42**, 2603 (1990).
108. U. Heinzmann, G. Schönhense, and A. Wolcke, in *Coherence and Correlation in Atomic Collisions*, edited by H. Kleinpoppen and J. F. Williams (Plenum Press, New York, 1979), p. 607.
109. G. Schönhense, F. Schäfers, U. Heinzmann, and J. Kessler, *Z. Phys. A* **304**, 31 (1982).
110. G. Schönhense and U. Heinzmann, *Phys. Rev. A* **29**, 987 (1984).
111. M. Müller, F. Schäfers, N. Böwering, C. Heckenkamp, and U. Heinzmann, *Phys. Scr.* **35**, 459 (1987).
112. A. Svensson, M. Müller, N. Böwering, U. Heinzmann, V. Radojević, and W. Wijesundera, *J. Phys. B* **21**, L179 (1988).
113. U. Fano, *Phys. Rev.* **124**, 1866 (1961).
114. M. J. Seaton, *Proc. Phys. Soc.* **88**, 801, 815 (1966).
115. V. K. Ivanov, S. Y. Medvedev, and V. A. Sosnivker, A.F. Ioffe Physical Technical Institute, Leningrad, USSR, Internal Report No. 615 (1979).
116. F. Keller and F. Combet Farnoux, *J. Phys. B* **18**, 3581 (1985).
117. S. P. Shannon and K. Codling, *J. Phys. B* **11**, 1193 (1978).
118. S. Süzer, M. S. Banna, and D. A. Shirley, *J. Chem. Phys.* **63**, 3473 (1975).
119. M. O. Krause, P. Gerard, A. Fahlman, T. A. Carlson, and A. Svensson, *Phys. Rev. A* **33**, 3146 (1986).
120. W. R. S. Garton and M. Wilson, *Proc. Phys. Soc.* **87**, 841 (1966).
121. U. Heinzmann, J. Kessler, and E. Kuhlmann, *J. Chem. Phys.* **68**, 4753 (1978).
122. U. Heinzmann, B. Osterheld, F. Schäfers, and G. Schönhense, *J. Phys. B* **14**, L79 (1981).
123. F. Schäfers, M. A. Baig, and U. Heinzmann, *J. Phys. B* **16**, L1 (1983).
124. B. Ritchie, *Phys. Rev. A* **13**, 1411 (1976).
125. A. Rich, J. van House, and R. A. Hegstrom, *Phys. Rev. Lett.* **48**, 1341 (1982).
126. H. Lefebvre-Brion, M. Salzmann, H.-W. Klausing, M. Müller, N. Böwering, and U. Heinzmann, *J. Phys. B* **22**, 3891 (1989).
127. M. Salzmann, M. Müller, N. Böwering, and U. Heinzmann, 17 ICPEAC, Brisbane 1991, Abstracts of Contributed Papers, Book of Abstracts, edited by I. E. McCarthy, W. R. MacGillivray, and M. C. Standage, p. 34, ISBN0-7503-0139-2.
128. M. Salzmann, M. Müller, N. Böwering, and U. Heinzmann, BESSY Jahresbericht (1990), pp. 120–121, ISSN 0179-4159.
129. N. Böwering, M. Salzmann, M. Müller, H.-W. Klausing, and U. Heinzmann, *Phys. Rev. A* **45**, R11 (1992).
130. T. Huth-Fehre, A. Mank, M. Drescher, N. Böwering, and U. Heinzmann, *Phys. Scr.* **41**, 454 (1990).

131. A. Mank, M. Drescher, T. Huth-Fehre, N. Böwering, U. Heinzmann, and H. Lefebvre-Brion, *J. Chem. Phys.* **95**, 1676 (1991).
132. N. Böwering, H.-W. Klausing, M. Müller, M. Salzmann, and U. Heinzmann, *Chem. Phys. Lett.* **189**, 467 (1992).
133. H. Lefebvre-Brion, A. Giusti-Suzor, and G. Raseev, *J. Chem. Phys.* **83**, 1557 (1985).
134. J. H. D. Eland and J. Berkowitz, *J. Chem. Phys.* **67**, 5034 (1977).
135. G. Herzberg, *Molecular Spectra and Molecular Structure: I Spectra of Diatomic Molecules* (Van Nostrand, Princeton 1950).
136. M. Büchner, G. Raseev, and N. A. Cherepkov, *J. Chem. Phys.* **96**, 2691 (1992); and p. 51 in Ref. 127.
137. K. Blum, in *Density Matrix Theory and Applications* (Plenum Press, New York, 1981).
138. N. A. Cherepkov, in *XVII ICPEAC, Invited Papers* (Adam Hilger, Bristol, 1991), p. 153.
139. F. Harren, D. H. Parker, and S. Stolte, *Comments At. Mol. Phys.* **26**, 109 (1991).
140. S. Kaesdorf, G. Schönhense, and U. Heinzmann, *Phys. Rev. Lett.* **54**, 885 (1985).
141. A. V. Golovin, V. V. Kuznetsov, and N. A. Cherepkov, *Sov. Tech. Phys. Lett.* **16**, 363 (1990).
142. N. A. Cherepkov and V. V. Kuznetsov, *Z. Phys. D* **7**, 271 (1987).
143. D. Dill, *J. Chem. Phys.* **65**, 1130 (1976).
144. J. W. Davenport, *Phys. Rev. Lett.* **36**, 945 (1976).
145. R. J. Baird, *J. Electron Spectrosc.* **24**, 55 (1981).
146. N. A. Cherepkov, *Chem. Phys. Lett.* **87**, 344 (1982).
147. R. L. Dubs, S. N. Dixit, and V. McKoy, *J. Chem. Phys.* **85**, 656 (1986).
148. N. A. Cherepkov and G. Schönhense, *Europhys. Lett.* **24**, 79 (1993).
149. J. R. Appling, M. G. White, T. M. Orlando, and S. L. Anderson, *J. Chem. Phys.* **85**, 6803 (1986).
150. C. Westphal, J. Bansmann, M. Getzlaff, G. Schönhense, N. A. Cherepkov, M. Braunstein, V. McKoy, and R. L. Dubs, *Surf. Sci.* **253**, 205 (1991).
151. K. L. Reid, D. J. Leahy, and R. N. Zare, *Phys. Rev. Lett.* **68**, 3527 (1992).
152. N. A. Cherepkov and V. V. Kuznetsov, *J. Chem. Phys.* **95**, 3046 (1991).
153. G. Schönhense, *Phys. Scr.* **T31**, 255 (1990).
154. N. A. Cherepkov, *Adv. At. Mol. Opt. Phys.* **34**, 207 (1994).
155. R. Allenspach, D. Mauri, M. Taborelli, and M. Landolt, *Phys. Rev. B* **35**, 4801 (1987).
156. N. M. Kabachnik, *J. Phys. B* **14**, L337 (1981).
157. N. M. Kabachnik and I. P. Sazhina, *J. Phys. B* **21**, 267 (1988).
158. N. M. Kabachnik and O. V. Lee, *J. Phys. B* **22**, 2705 (1989).
159. B. Lohmann, *J. Phys. B* **23**, 3147 (1990).
160. N. M. Kabachnik, B. Lohmann, and W. Mehlhorn, *J. Phys. B* **24**, 2249 (1991).
161. U. Hahn, J. Semke, H. Merz, and J. Kessler, *J. Phys. B* **18**, L417 (1985).
162. H. Merz and J. Semke, in *X-Ray and Inner Shell Processes* (AIP Conf. Proc. 215), edited by T. A. Carlson, M. O. Krause, and S. T. Manson (AIP, New York, 1990), p. 719.
163. R. Kuntze, N. Böwering, M. Salzmann, and U. Heinzmann, 10th Int. Conf. VUV Radiat. Phys. Paris 1992, Book of Abstracts, p. Tu 75, and BESSY Jahresbericht 1991, pp. 112–113, ISSN 0179-4159.
164. P. Stoppmanns, B. Schmiedeskamp, B. Vogt, N. Müller, and U. Heinzmann, *Phys. Scr.* **T41**, 190 (1992).
165. P. Stoppmanns, B. Schmiedeskamp, B. Vogt, N. Müller, and U. Heinzmann, BESSY Jahresbericht 1991, pp. 245–246.
166. E. G. Berezhko and N. M. Kabachnik, *J. Phys. B* **13**, 959 (1980).
167. G. Schönhense, A. Eyers, U. Friess, F. Schäfers, and U. Heinzmann, *Phys. Rev. Lett.* **54**, 547 (1985).
168. G. Schönhense, B. Kessler, N. Müller, B. Schmiedeskamp, and U. Heinzmann, *Phys. Scr.* **35**, 541 (1987).
169. N. Irmer, R. David, B. Schmiedeskamp, and U. Heinzmann, *Phys. Rev. B* **45**, 3849 (1992).
170. B. Schmiedeskamp, N. Irmer, R. David, and U. Heinzmann, *Appl. Phys. A* **53**, 418 (1991).
171. U. Heinzmann and G. Schönhense, in *Polarized Electrons in Surface Physics*, edited by R. Feder (World Scientific, Singapore, 1985), p. 467.
172. J. Kirschner, *Polarized Electrons at Surfaces* (Springer, Berlin, 1985).
173. G. Schönhense, *Appl. Phys. A* **41**, 39 (1986).
174. U. Heinzmann, *Phys. Scr.* **T17**, 77 (1987).

CHAPTER 16

PHOTOIONIZATION OF EXCITED AND IONIZED SYSTEMS

F. J. WUILLEUMIER AND J. B. WEST

1. INTRODUCTION

The measurement of photoabsorption and photoionization cross sections has long been an area of fundamental interest to atomic physics. Its significance was realized at the end of the last century, when the first vacuum spectrograph was built by Schumann[1] to investigate the wavelength region below 2000 Å, but both optical and vacuum technology at that time were in their infancy and progress was slow. Rowland[2] provided a major advance with a means of ruling accurate diffraction gratings, providing reliable wavelength calibration in this spectral region, which came to be known as the vacuum ultraviolet. Beutler[3] used a helium discharge source for metal vapor absorption experiments below 1200 Å, and, together with advances in theoretical understanding of atomic structure through the development of the quantum theory and wave mechanics, it became possible to analyze and understand stellar and solar emission and absorption spectra. For example, Edlén[4] was able to identify and tabulate many of the transitions due to both neutral and highly ionized species in stellar spectra. In the x-ray region, i.e., at wavelengths below ~1 Å, crystal optics were used and experimental data were more readily obtainable, but there is a basic difference in the physics involved. At x-ray energies it is mainly the inner or core electrons which are ionized. The ejection of such electrons can, to a reasonable first approximation, be considered as a one-electron process or hydrogen-like, where the electron moves in the Coulomb field of the core and corrections for the outer electrons can be made by the use of a Coulomb-like screening potential. This simplification is not valid in the vacuum ultraviolet and soft x-ray regions, where the ionization or excitation process involves the more loosely bound outer electrons, which interact with each other strongly leading to major departures from the Coulomb field approximation. Two-electron excitations and autoionization are prominent in the outer-shell and low-lying inner-shell photoionization spectra of both atoms and molecules, and form major challenges to our understanding of the photoionization process. Classic experiments in this respect were made by Beutler[5] on the autoionizing lines between the $^2P_{1/2,3/2}$ limits of argon, and these measurements stimulated Fano's[6] analysis of autoionization in which interaction takes place between the bound state of the electron and its continuum channel.

Up to this time most measurements concentrated on the identification of structure, but the next major impetus to this area of research occurred during the late 1950s and 1960s, driven by new interest in space and upper atmosphere research. There was a need for intense laboratory sources of radiation in the region below 2000 Å, and this was met primarily by discharge sources which produced strong resonance lines, such as the helium lamp, or weak continua from the condensed discharge. Samson[7] has reviewed developments in both light sources and monochromators up to this point, as well as the

---

F. J. WUILLEUMIER • Laboratoire de Spectroscopie Atomique et Ionique, Unité de l'Université Paris Sud Associée au CNRS, 91405 Orsay Cedex, France.   J. B. WEST • Daresbury Laboratory, SERC, Daresbury, Warrington WA4 4AD, United Kingdom.

VUV and Soft X-Ray Photoionization. Edited by Uwe Becker and David A. Shirley. Plenum Press, New York, 1996.

experimental techniques in use at that time. Synchrotron radiation sources also emerged during this period, in particular at the National Bureau of Standards[8] in the United States where an extensive program in atomic and molecular physics began. Absolute cross section measurements for atoms in the VUV were still unavailable, apart from the rare gases,[9] and even with the advent of more intense continuum sources during the 1970s this remained the case. For the rare gases, reasonable agreement was achieved between current experiments[10] and theory[11] at that time; the ready availability of synchrotron radiation also permitted differential spectroscopies such as electron spectrometry to be used. This led to the positive identification of intershell electron correlation effects[12,13] and corresponding advances in theoretical analysis using the random phase approximation[14]; many more examples are given in the photoionization section of the invited papers at the sixth conference on Vacuum Ultraviolet Radiation Physics[15] and in a review presenting all results obtained using synchrotron radiation until 1980.[16]

Up to the 1980s, despite the abundance of such species in stellar atmospheres and the need for absolute data, there were very few measurements on the photoionization of excited atoms or ions. Those that were available came from the use of discharge sources[17,18] such as the Vodar or BRV source.[19] Since then two major developments have transformed this situation; the development of purpose-built synchrotron radiation sources and the use of laser-produced plasmas. In the first case, a gain of one order of magnitude in photon flux through the use of undulators[20] became available; in the second case, the potential of the dual laser-produced plasma, where the timing between the two pulses is varied, was first demonstrated by Carillon et al.[21] and fully exploited later by Carroll and co-workers.[22,23] Absolute measurements remained elusive, however, and, particularly for highly charged species, this persists to the present day. It is the aim of this chapter to review the present state of measurement and theory in this relatively new field, and to indicate the most likely developments which will lead to further progress in the future.

2. EXPERIMENTAL TECHNIQUES

In this section the emphasis is on the equipment used to make measurements on excited atoms and ions, leaving the experimental data to be covered in detail in the following section. It has already been pointed out that absolute cross section measurements are scarce; many of the measurements to be presented here are normalized, perhaps to a theoretical value or with the use of the oscillator strength sum rule. Results from the use of these procedures must be viewed with caution; interactions between electrons in different shells are difficult to treat quantitatively from a theoretical point of view, although there have been notable successes using, for example, the RPA method. It remains the case, however, that absolute measurements are urgently needed for a reliable interpretation of the electron correlation effects which dominate the photoionization spectrum of all but the lightest atoms, and are the ultimate test of the theoretical results. In addition, measurements concentrate on resonance regions for the obvious reason that cross sections are higher there, but there is also a need for data in the nonresonant continuum, particularly partial cross section measurements so that the behavior of individual channels can be clearly identified in regions where interchannel interactions will be weaker, and the difference between resonant and nonresonant behavior clearly understood.

2.1. Spark and Plasma Sources (DLPP)

From the earliest days of VUV spectroscopy arc discharges have been used as a means of studying the spectra of ionized species,[24] and we do not attempt to review this here, but restrict this chapter to work where the intent is to separate and clearly define the species of interest. The earliest measurements of photoionization spectra over an extended energy range were made using two vacuum spark sources synchronized in time with respect to each other. The first discharge was used to generate the vapor required, the second to provide the background continuum source. By altering the timing between the two sources, different stages of ionization can be observed, the most highly ionized stages being observed for the shortest delays. Figure 1 shows the experimental layout used by Mehlman-Balloffet and Esteva[25,26] for measurements on several ionic species. This technique can be much improved,

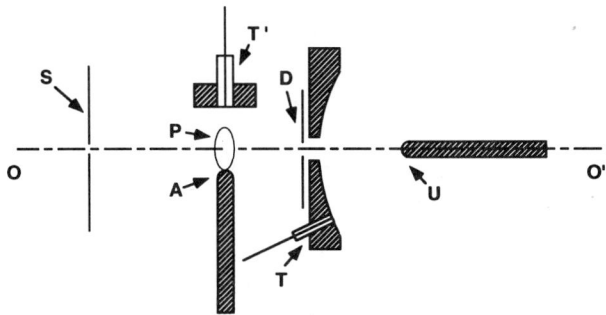

FIGURE 1. "Double Vodar source" experiment of Mehlman-Balloffet and Esteva[25]; OO′, optical axis; A, anode of the discharge generating the plasma P to be studied; T, ring-shaped triggering spark; T′, triggering spark; U, uranium anode; D, 3-mm-diameter diaphragm.

particularly the timing aspect, by using lasers rather than sparks to generate the plasmas. The system shown in Figure 2 was developed and later used by Carillon et al.[21] and Jaeglé et al.[27] to study population inversions in Al^{3+} ions, as one of the first sources of laser radiation in the soft x-ray region. In fact, these experiments used one laser, Nd glass, split into two beams with an optical delay between

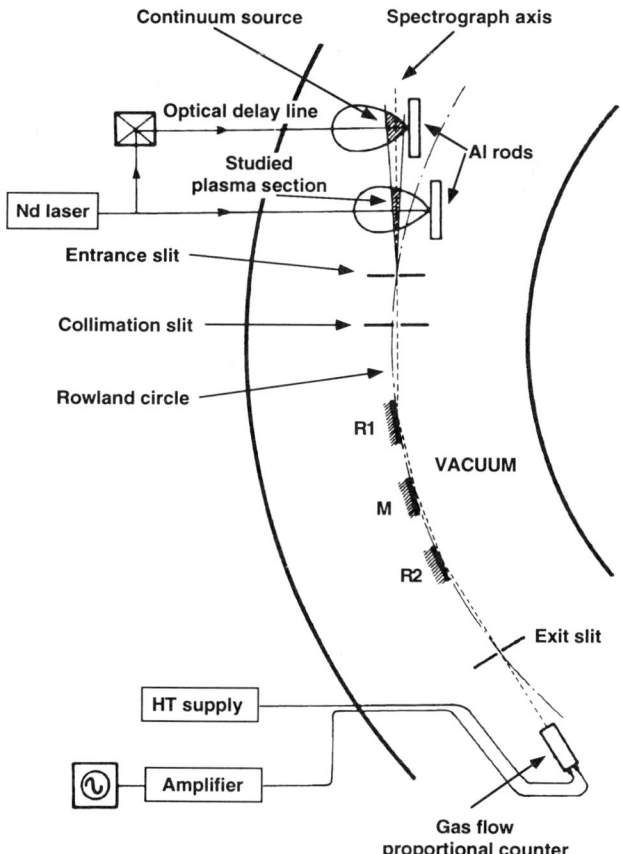

FIGURE 2. Laser-produced plasma experiment of Carillon et al.[21]; R1, R2, concave diffraction gratings, and M, concave mirror, all mounted on the Rowland circle.

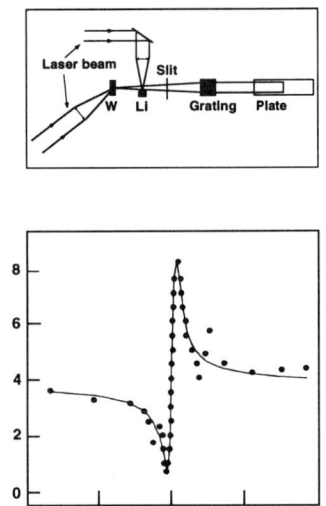

FIGURE 3. Laser-produced plasma experiment, using a tungsten target and a lithium sample, of Carroll and Kennedy[22] (upper panel), and their Li^+ spectrum (lower panel).

them as shown in the figure, and became known as the laser-produced plasma (LPP) technique. Carroll and Kennedy[22] introduced refinements by varying the focusing of the beams. In separate experiments, following the work of Ehler and Weissler,[28] they showed that, for this kind of source, highly focused beams on high-Z, targets gave the best continua in the soft x-ray region. To produce primarily ground-state Li^+, the beam producing the absorbing plasma had to be defocused, and the optimum distance from the Li target at which to observe the Li^+ spectrum through the spectrograph determined experimentally. Their success can be clearly seen in Figure 3 where their measurements of the

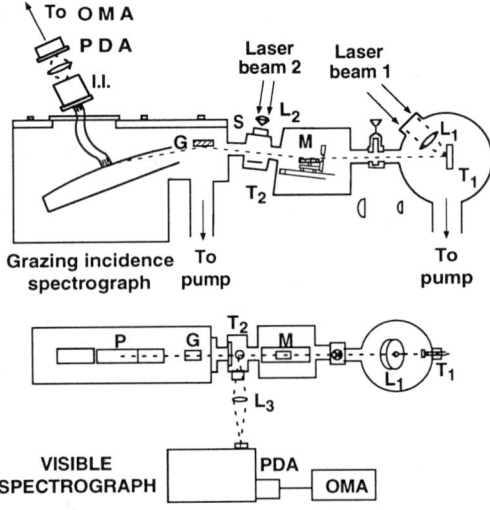

FIGURE 4. Laser-produced plasma experiment of Jannitti et al.[30]; upper and lower parts are vertical and horizontal views, respectively; L_1, aspheric lens; L_2, spherocylindrical lens; M, toroidal mirror; T_1, T_2, targets; S, entrance slit; G, grating; P, focal surface; I.I., image intensifier.

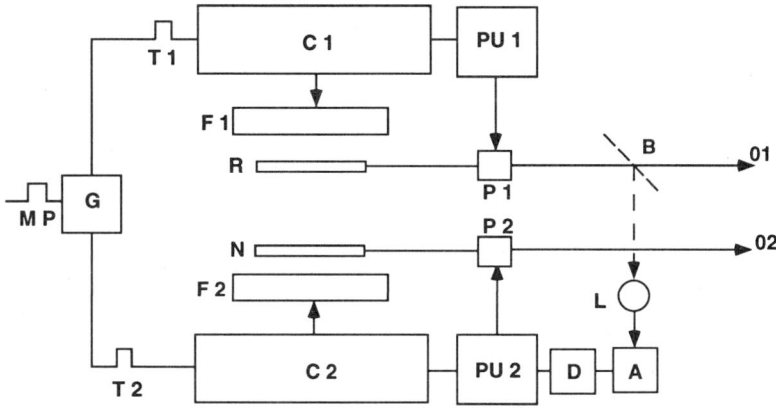

FIGURE 5. Dual laser-produced plasma experiment of Carroll and Costello[31]; MP, master pulse; G, delayed pulse generator; T1, trigger pulse to ruby laser control circuit; T2, trigger pulse to neodymium oscillator amplifier system; C1, C2, charging and control circuits; F1, F2, flashlamps; R, ruby rod; N, neodymium oscillator amplifier rods; P1, P2, Pockels cells; PU1, PU2, Pockels cells control units; B, beam splitter; L, light-sensitive diode; A, amplifier; D, delay circuit; O1, O2, optical output pulses.

absorption spectrum of the heliumlike ion Li^+ are shown below the experimental layout. Jannitti et al.[29] also used this technique, and introduced further technical developments[30] shown in the apparatus in Figure 4. The background source was a copper plasma, viewed end-on by the spectrometer to increase source brightness and focused onto the spectrograph entrance slit, through the absorbing plasma, by a toroidal mirror. The conventional photographic plate was replaced by a scintillator and optical multichannel analyzer (OMA) located on the Rowland circle, enabling quantitative measurements to be made. In this way absorption cross sections in the range 10–80 Å were measured for a series of light ions, e.g., Be^{2+}. Absolute values were obtained by normalization to the known oscillator strengths of the principal series members.

The next development in the use of laser plasmas was demonstrated for atomic thorium by Carroll and Costello,[31] in which dual lasers were used (DLPP), allowing more power onto both target and sample and more flexibility in the range of time intervals between the sample laser and the illuminating laser. This system is shown in Figure 5; a continuously variable delay between 250 ns and 10 ms is available. A ruby laser was used to generate the absorbing plasma, and a fraction of its output diverted to trigger the drive unit of the Nd glass laser's Pockels cell for the generation of the background continuum from a tungsten target. Such systems are ideal for the study of isoelectronic sequences, and are better suited for neutral species because of the availability of longer delays between sample generation and illumination. It has been used successfully by Costello et al.[32] for refractory elements, e.g., uranium.

The use of laser plasma sources is undergoing constant refinement, and careful timing can isolate the charged species of interest. Their use extends well into the soft x-ray region, and although they cannot in general be used for absolute cross section measurements, they are capable of providing good relative measurements, which can then be normalized to an absolute measurement, or to sum rules where intershell correlations are not significant. As will be seen later in the case of Mn^+, these techniques do leave some doubt over the purity of the absorbing species, but, particularly for highly charged ions, they provide data which are almost impossible to obtain in any other way. The review by Costello et al.[33] provides further and more detailed information on the development and future prospects for these sources.

2.2. Resonant Laser-Driven Ionization (RLDI)

An alternative method of generating ionized atoms is to use the high instantaneous intensity from a pulsed laser system to excite resonantly an atomic transition. The precise mechanism for generating ions from this point was not definitely identified at that time, but was soon attributed to ionization by

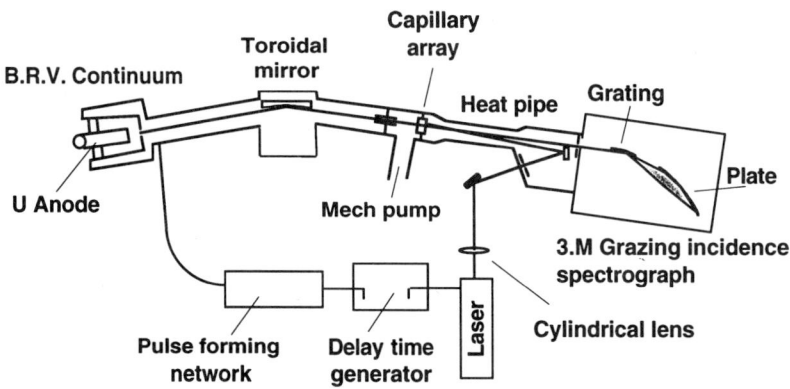

FIGURE 6. Resonant laser-driven ionization experiment of Lucatorto and McIlrath.[18]

fast electrons produced in the medium by superelastic collisions. Lucatorto and McIlrath[18] were the first to observe the ionization of sodium vapor by laser radiation. In their experiment, whose layout is shown in Figure 6, the $3s \rightarrow 3p$ transition was saturated by a pulsed dye laser; they then suggest a following stage in which the $4p$ level is populated by stimulated Raman scattering and a second photon ionizes the $4p$ or $4s$ levels. Alternatively, the collision of two excited sodium atoms occurring at the same time as a laser photon is absorbed can also yield ions, according to

$$\text{Na}(3p) + \text{Na}(3p) \rightarrow \text{Na}_2^+ + e^-$$

$$\text{Na}(nl) + \text{Na}(3p) \rightarrow \text{Na}(3s) + e^-(2.1 \text{ eV} - E_{nl})$$

The seeded electrons can then undergo what have been called superelastic collisions, gaining 2.1 eV in kinetic energy at each collision, according to

$$e^-(E_c) + \text{Na}(3p) \rightarrow e^-(E_c + 2.1 \text{ eV}) + \text{Na}(3s)$$

Then, the electrons have sufficient energy to ionize atoms in the excited state until full ionization of the medium occurs. In a following experiment on lithium, McIlrath and Lucatorto[34] concluded also that ions are most probably formed from electron impact ionization, where the electrons themselves are derived from free electrons in the vapor being superelastically heated by collisions with excited atoms, as described above in the sodium case. In this experiment they also varied the timing between the laser pulse and the vacuum spark BRV source used to provide the VUV continuum, and were in this way able to observe the evolution of the vapor from ground-state Li $\rightarrow$ excited-state Li* ($1s^2 2p$) $\rightarrow$ Li$^+$. The identification of excited Li atoms in their experiment tends to favor the collisional/electron impact mechanism as the source of the ionized atoms.

These experiments have been improved by Cromer et al.,[35] along similar lines to the DLPP experiments described above.[29] The technique has been used successfully for calcium[36] in the same way as above for lithium; it has also been used to examine the role played by configuration mixing involving the $3d$ electrons in Cr and the isoelectronic Mn$^+$, an experiment[37] which will be discussed in greater detail in a later section. In comparison with the DLPP technique this method is limited to absorbing species which can be produced as vapors of high density, and the resonant transitions must lie in regions accessible to the laser, but there are clearly specific problems for which it is ideally suited.

2.3. Ion Beam Experiments

The primary limitation in the use of the RLDI technique is the tuning range of the laser, and for the production of ions in a well-known state and in known densities, there is little alternative but to

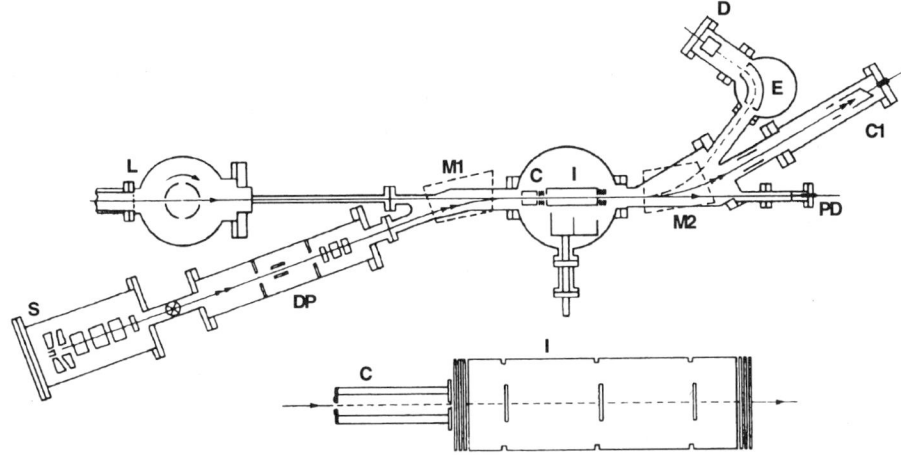

FIGURE 7. Merged beam experiment from Lyon et al.[38]; L, chopper vessel; S, surface ionization source; DP, differential pumping assembly; M1, M2, magnets; C, collimator; I, interaction region; PD, calibrated photodiode; C1, Faraday cup; E, electrostatic selector; D, Johnston multiplier.

use ion beams. This was the approach adopted by Lyon et al.,[38] whose primary interest was the measurement of absolute photoionization cross sections. These authors used the merged beam technique established previously[39] for electron impact ionization, in order to increase the effective size of the interaction region. The apparatus is shown in Figure 7; the ground-state ions, produced by a surface ionization source in the cases of Ca^+, Sr^+, and Ba^+, were mass analyzed in the 20° sector magnet M1 and merged with a parallel photon beam from a 5-m normal-incidence monochromator installed at the Daresbury Synchrotron Radiation Source. After passing through the interaction region I, shown enlarged in the lower part of the figure, the magnet M2 was used to separate the singly charged ions from the doubly charged ones. The photodiode PD measured the absolute value of the photon flux, the Faraday cup measured the intensity of the singly charged beam, and the doubly charged ions were detected after passing through the electrostatic selector E by a precalibrated Johnston multiplier at D. In order to measure an absolute cross section it is also necessary to know the overlap of the ion beam with the photon beam, and this was done by scanning slits across both the ion and photon beams at three points along the interaction region, shown in the enlarged section of Figure 7. This had to be done in both the x and y directions, z being the direction of the photon and ion beams. Since the variation in beam profile was very small along the length of the interaction region, an average overlap integral O for the two beams could be constructed from the three measurement positions shown. The absolute cross section σ is then given by

$$\sigma = \frac{Se^2 v \eta O}{IJL\Omega}$$

where S is the signal rate, e is the electronic charge, I and J are the ion and photodiode currents, respectively, L is the length of the interaction region, η and Ω are the photodiode and Johnston multiplier efficiencies, and v is the ion velocity. Even at the low background pressures (1×10^{-10} mbar) achieved in the experimental chamber, it was still necessary to discriminate against doubly charged ions generated from the residual gas and apertures outside the interaction region. To do this, cylinder I was biased so that doubly charged ions generated inside it would have a characteristic energy which could be selected by magnet M2 and the electrostatic selector E.

Using the above equipment, measurements were made on the singly charged ions of Ba,[38] Ca,[40] and Sr[41] with the surface ionization source described by Peart and Lyon.[42] Further measurements were made on Ga^+ and K^+ using a Kunsman[43,44] source, in which a mixture of $X_2CO_3:Al_2O_3:2SiO_2$

(X = Ga or K) proportioned by mass is adhered to a tungsten spiral, and on Zn^+ using an electron bombardment source.[43] In the last case there is a possibility of metastable states contaminating the beam, but because of the long flight times of the ions these were considered likely to have decayed before reaching the interaction region.

The ion beams used in the above experiments were limited by space charge to a maximum density of $\sim 10^6$ ions/cm^3, with the effect that cross sections $< 10^{-18}$ cm^2 were not measurable above the background noise. Because of autoionization, cross sections up to three orders of magnitude greater than this were observed in some spectral regions, and further measurements are clearly desirable for deeper core levels and for multiply charged ions. In the former case the cross sections are likely to be smaller as the photon energy increases, and furthermore the probability of multiple ionization increases with resultant further partition of the cross section. In the latter case, space charge effects will limit the ion beam density even more severely; raising the energy of the ion beam to overcome this is no solution, because the stripped background increases dramatically with energy, rapidly overwhelming the desired signal.

The first solution to these problems is a more intense photon source, and to some extent this is provided by synchrotron radiation from an undulator. This has enabled the first differential measurements to be made on an atomic ion, at the Orsay storage ring Super ACO, by Bizau et al.[45,46] The experimental details are shown in Figure 8; the radiation from an undulator is dispersed and focused by a toroidal grating monochromator into the source volume of an electron cylindrical mirror analyzer (CMA). The ion beam is introduced to the CMA along its axis and from the opposite direction to the VUV radiation. A plasma discharge ion source[47] was used to generate Ca^+ ions, selected by the magnet M1 and brought to a 2-mm-high by 4-mm-wide focus in the CMA by a toroidal electrostatic deflector and three focusing quadrupoles. After leaving the CMA the ions were refocused by a further quadrupole and deflected by a further magnet into two Faraday cups for Ca^+ and Ca^{2+}.

Another approach to preparing the ion species is to use ion traps. Church et al.[48] have used the white radiation from a beam line at the NSLS, Brookhaven, to generate highly charged ions and store them in a Penning trap. Rate coefficients were measured for charge-changing collisions, and a density of $\sim 10^4$ ions/cm^3 was measured. Later experiments[49] improved this to $\sim 5 \times 10^5$ ions/cm^3, but a further substantial increase in this density would be needed for photoelectron experiments. The small dimensions of the source and the fact that more than one species is present, exclude it from being used in its present form for total cross section measurements. It is nevertheless a good method of studying collisional processes in low-energy highly charged ions; other kinds of ion traps and ion sources are now being proposed for photoionization experiments and will be discussed in the future developments section at the end of this chapter.

2.4. Combined Laser–Synchrotron Experiments

Laser sources were used some time ago[50] to prepare the excited state of an atom, and subsequently ionize it, making use of the high power available from Nd glass lasers and frequency doubling/quadrupling to excite and ionize the atom. The majority of this work was concerned with the identification of Rydberg structure resulting from excited-state photoabsorption, but some cross section

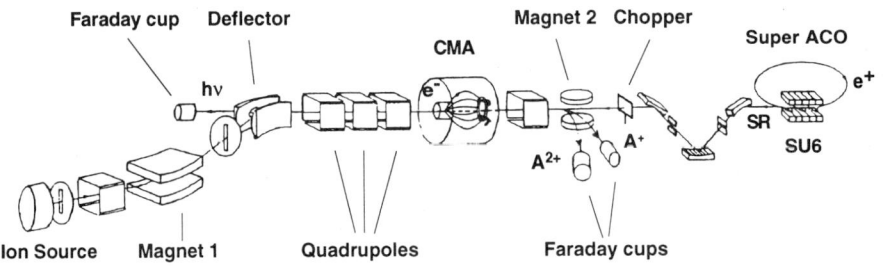

FIGURE 8. Atomic ion photoelectron spectroscopy experiment from Bizau et al.[45]

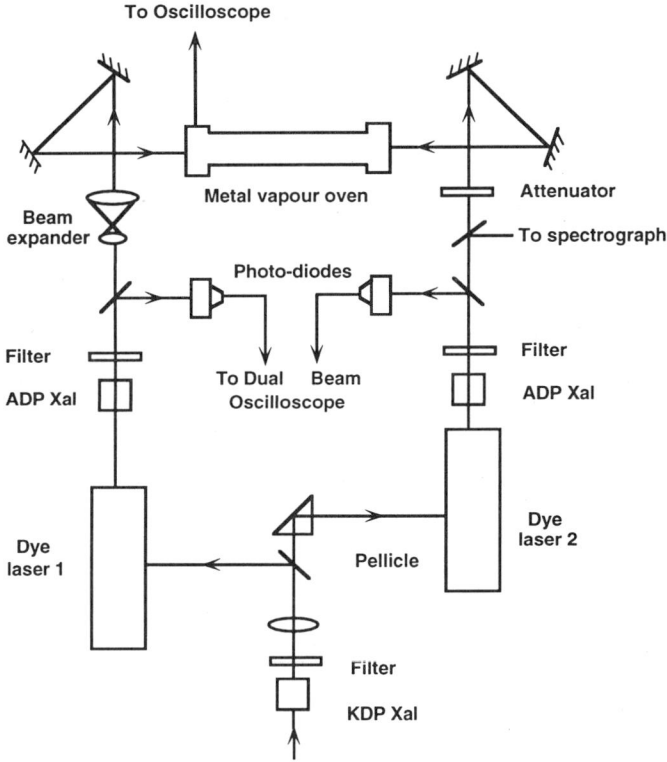

FIGURE 9. Laser excitation and ionization experiment of Bradley et al.[50]

measurements were also made, for the $1s2s$ $^{1,3}S$ levels of helium,[51] the ^3D levels of barium,[52] and the $6s\ ^2P$ levels of cesium.[53] Bradley et al.[54] measured the photoionization cross section of the $3s3p$ $^1P_1$ state of Mg in the region of the $3s3p \to 3p^2$ autoionizing transition at 3009 Å, using the equipment shown in Figures 9 and 10. More recently, photoelectron angular distributions have been measured, for example by Berry and his collaborators, using an excimer pumped dye laser to prepare excited states in the alkaline earth atoms and also to photoionize them (see e.g., Mullins et al.[55] and references therein).

Such experiments are limited by the narrow range of tunability of the laser sources, and a combination of these with the broad continuum provided by synchrotron radiation (SR) would seem the ideal solution for extended studies of this type.[56] Unfortunately, the time structure of the laser and that of an SR source are very poorly matched: SR is almost continuous in time, and has a low instantaneous intensity, whereas the pulsed laser has an extremely high intensity per pulse with a relatively low repetition frequency. As a result, measurements, particularly differential ones, combining the two types of source in the VUV were possible only with high-intensity CW dye lasers. The first measurements of this kind were made at the Orsay storage ring by Wuilleumier and co-workers on excited sodium atoms[57–59]; the experimental layout is shown in Figure 11 in its present configuration. An argon ion pumped ring dye laser excited the sodium atoms produced by the oven to the $2p^63p\ ^2P_{3/2}$ state, and the VUV photons from a toroidal grating monochromator are focused close to the oven nozzle. The oven itself is ingeniously placed inside the electron spectrometer, a cylindrical mirror analyzer (CMA), to maximize the efficiency of signal collection. By means of electron

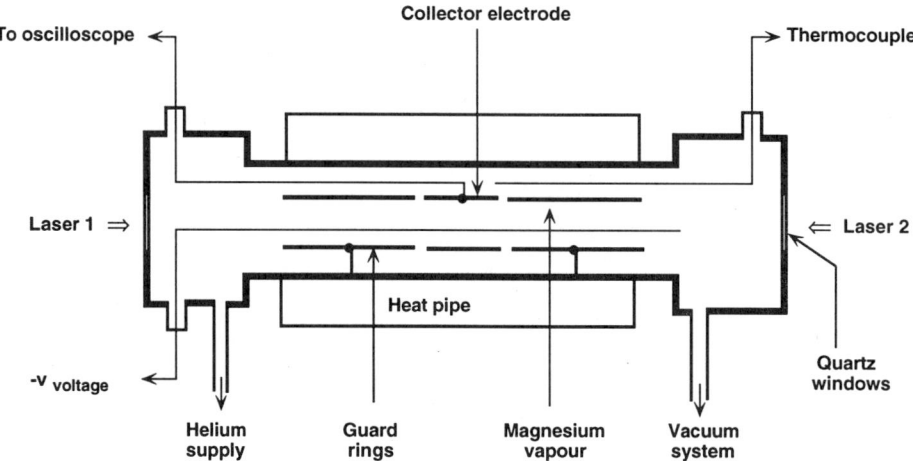

FIGURE 10. Heat pipe vapor cell used in the experiment of Bradley et al.[50]

spectroscopy these measurements were able to identify clearly transitions from the Na excited states.[60–63] Excited states of Ba* were also studied with this method.[60,62,64–66] The technique was refined further for Na by Ferry et al.,[67] and by Carré et al.,[68] who combined two laser beams with SR in order to produce autoionizing states in Na of the type $2p^5 3snl$, $n \geq 4$. The two-laser system is also represented in Figure 11. In a further development of this equipment, six channeltrons have been placed in a ring near the exit slits of the CMA, as shown in Figure 12, in order to measure photoelectron angular distributions.[69]

In the above experiments the density of the vapor beam was sufficiently high to ensure trapping of the resonant radiation from the decay of the excited states, thereby increasing the density of the excited states required for the experiment. This also had the effect of depolarizing the fluorescence emission, and thus any alignment of the excited species, derived from the fact that the laser radiation is strongly polarized, is lost. A different approach was followed by Meyer et al.,[70] with the advantage of access to undulator radiation. With approximately two orders of magnitude increase in the VUV light flux, they were able to offset the much lower vapor beam density required to maintain alignment

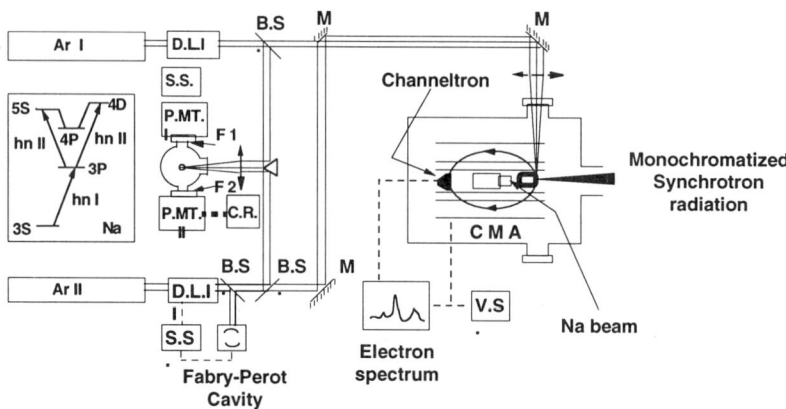

FIGURE 11. Combined laser and synchrotron experiment of Wuilleumier and co-workers, adapted from Carré et al.[68]

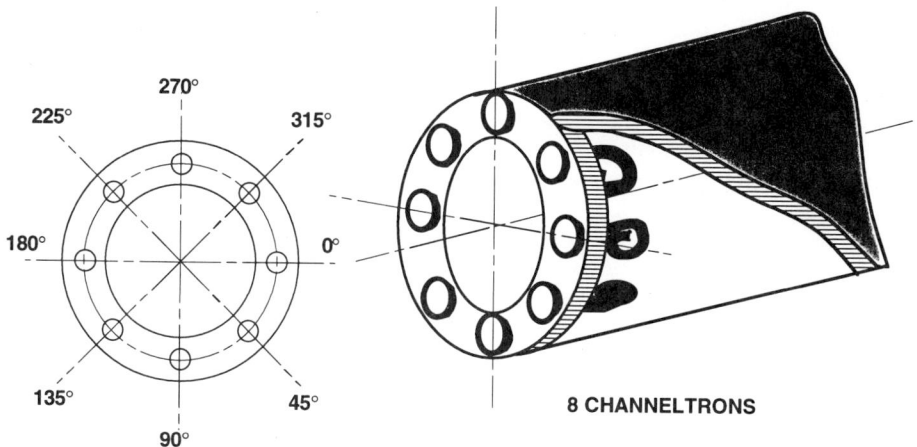

FIGURE 12. Modification of CMA to take six channeltron detectors from Bizau et al. [69]

in the sample. The most recent version of this experiment,[71] which allows the measurement of photoelectron angular distributions, is shown in Figure 13. VUV radiation from a toroidal grating monochromator on an undulator beam line fitted to the BESSY SR facility in Berlin illuminated the metal vapor beam from an effusive oven, and the laser beam was made collinear with the SR beam but propagating in the opposite direction. The electron spectrometer was able to rotate about the light beams as axis, and two angular parameters could be varied: the angle η between the polarization vector of the SR and that of the laser, and the angle θ between the emission direction of the electron and the SR polarization vector. These experiments give detailed information on the photoionization of atomic excited states, since these are prepared in a well-defined way, and Pahler et al.[72] have used it to determine the symmetries of the Li $1snln'l'$ autoionizing resonances.

The above experiments allow all of the parameters of the photoionization process to be determined, the laser being used to form the target species in a completely defined quantum state. This has been done in the past by deducing the initial state from measurements of the outgoing particles, e.g., the spin polarization, energies and angular distributions of the photoejected electrons,[73] angular distributions of Auger electrons,[74,75] fluorescence angular distribution[76] or polarization[77,78] in order to determine the alignment of the residual ion and specify fully the photoionization channel being measured. A review of this work is in general outside the scope of this chapter; although it does involve the decay of excited states of atoms, the main purpose is not the study of the spectroscopy of the excited

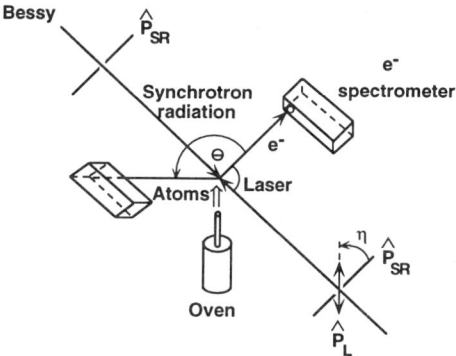

FIGURE 13. Combined laser and synchrotron experiment of Sonntag.[71]

3. EXPERIMENTAL RESULTS

3.1. Excited Atoms

The first absolute measurement of the cross section of a selected excited atomic state was made by Bradley et al.[54] for the $3s3p\ ^1P_1$ state of magnesium. A heat pipe was used to form a stable vapor column of known pressure, in which collecting electrodes were placed in order to measure the photoelectron current. Figure 14 shows the results for the $3s3p\ ^1P_1 \rightarrow 3p^2\ ^1S_0$ autoionizing transition; the error bars represent the relative error in the measurements, the absolute error being ~50% mainly because of the uncertainty in the calculation of the number density of the absorbing species in the heat pipe; the position of the resonance and its width were in excellent agreement with theory.[80] As pointed out earlier, lasers have also been used to make photoelectron angular distribution measurements from excited states; Mullins et al.[55] have done this for photoionization of the $(5d6p)\ ^3D_1$, $(5d6p)\ ^3P_1$, and $(6s6p)\ ^{1,3}P_1$ states of Ba to both the $^2S_{1/2}$ and $^2D_{3/2,5/2}$ states of Ba$^+$. Large differences were seen in the anisotropies for the different ionization channels, and the largest anisotropies resulted when one channel was dominant, e.g., ionization from the $(5d6p)$ states to the $5d$ ion core. This indicated that, in general, a single-channel theoretical analysis would be satisfactory, but intermediate state interactions had to be included to account for the significant effects of weaker channels, e.g., ionization of the $(6s6p)$ state to the $5d$ ion core where two electrons are involved.

The broad continuum of synchrotron radiation is the obvious source to continue with this type of work, but because of the time characteristics needed a CW laser to generate the excited states, rather than the pulsed laser used for the work described above for magnesium. As stated earlier, the first experiments of this kind were carried out on the LURE storage ring ACO and have already been extensively reviewed.[56] We show here an example[63] where the technique has been used for the measurement of oscillator strengths of autoionizing transitions in sodium,[62,63] using the equipment described earlier and shown in Figure 11. Photoelectron spectra were taken at photon energies of 32.73 and 32.50 eV, with laser on and off, as shown in Figure 15. Peaks 1, 2, and 3 are due to ionization by second-order radiation from the TGM, with peak 2 being due to ionization of a $2p$ electron from excited sodium. Peak 4, which appears only when on resonance at 32.73 eV, is due to autoionization of the $2p^5(^2P)3s3p(^1P)\ ^2D$ levels. The oscillator strength for peak 4 can then be calculated by normalization to peak 2 as follows: the area of peak 2 is given by

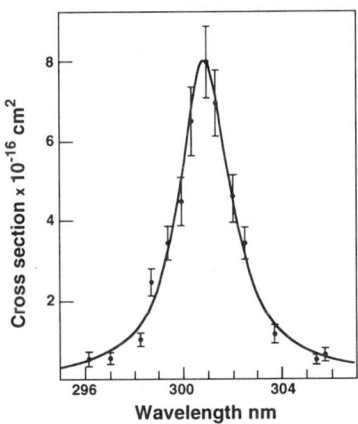

FIGURE 14. Cross section of $3s3p\ ^1P_1 \rightarrow 3p^2\ ^1S_0$ transition in magnesium, from Bradley et al.[50]

where K is a spectrometer constant, I_{hv} is the photon flux integrated over the spectrometer bandpass ΔE at energy hv, E_{hv} is the kinetic energy of the photoelectron, n_{3p} is the excited state density, and $\sigma(hv)$ is the photoionization cross section. Similarly for peak 4, its area N_R is given by

$$N_{hv} = KI_{hv}E_{hv}n_{3p}\,\sigma(hv)$$

$$N_R = 110K'\frac{I_R}{\Delta E_R}E_R n_{3p} \int_{\Delta E_R} \frac{df_R}{d\varepsilon}\,d\varepsilon$$

where the quantity K' is the product of the spectrometer constant and the branching ratio for autoionization and the factor 110 is the constant which relates oscillator strength to the product of cross section and photon energy. Since in this experiment the monochromator bandpass was much larger than the width of the resonance, the integral in the above equation can be replaced by the oscillator strength f_R, given from the above equations by

$$f_R = 0.0091\,\frac{K}{K'}\frac{I_{hv}}{I_R}\frac{E_{hv}}{E_R}[\sigma(hv)\,\Delta E_R]\frac{N_R}{N_{hv}}$$

All of the parameters in this equation are measured in the experiment, with the exception of the autoionization branching ratio K/K', assumed to be one since LS coupling is valid for sodium, and $\sigma(hv)$, which was taken from the literature.[81,82]

Both ^1P and ^3P terms can result from the exchange interaction between the $3s$ and $3p$ electrons, the spectra shown in Figure 15 being taken for the ^1P resonance. Each of these terms can couple to the $2p^5\,^2P_{1/2,3/2}$ core to form $^{2,4}D$, $^{2,4}P$, and $^{2,4}S$ terms, and each of these groups of terms was scanned, to give an excitation curve. Figure 16 shows the result of doing this for the ^1P configuration, where the data have been normalized to the peak counting rates. The continuous line is the sum of the fitted

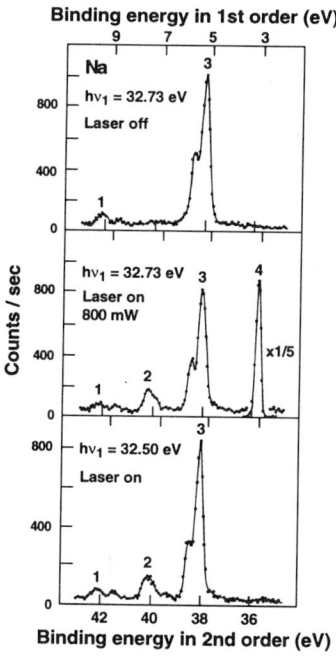

FIGURE 15. Electron spectra of excited sodium from Bizau et al.[63]

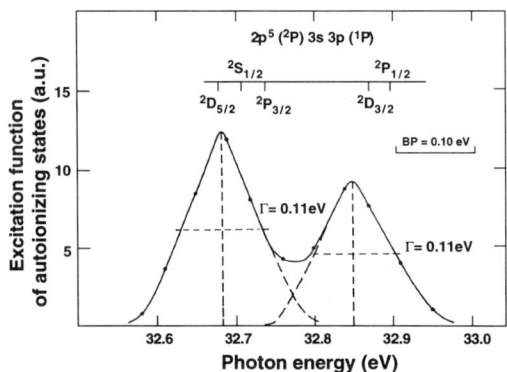

FIGURE 16. Excitation spectra for the $2p^5(^2P)3s3p$ (^1P) configuration of sodium, from Bizau et al.[63]

dashed lines, where the fit was made assuming that only the ^2D lines contribute to autoionization. Even parity ^2P terms could not autoionize because of parity considerations, and the ^2S terms were considered too weak. Similar measurements were made for the $2p^5(^2P)3s3p(^3P)$ configuration, where again most of the intensity was contained in the ^2D levels, with a small amount of intensity in the ^2S level in this case. The results for the oscillator strengths are shown in Table I, where it can be seen that there is also some intensity in the ^4P levels, which normally would not be expected to autoionize in strict LS coupling. This implies mixing with ^2D levels, for which autoionization is allowed.

The first measurements of a cross section for direct photoionization into the continuum of inner-shell electrons in an excited atom were made for $5p^66s5d$ $^{1,3}D_{3/2,5/2}$ excited barium atoms.[83] The 5d, 5p, and 5s cross sections were measured using synchrotron radiation from the Orsay storage ring. We show in Figure 17 the variation of these 5p and 5s photoionization cross sections. The strong maximum observed in both cross sections around 105 eV photon energy, i.e., just above the 4d ionization threshold in the excited barium atom, is due to strong intershell correlations between the 4d-, and the 5s- and 5p-subshell electrons. Such behavior is not reproduced by one-electron calculations[84] (dashed line in Figure 17). It was first observed in photoionization of xenon atoms in the ground state.[85]

TABLE I
Oscillator Strengths for Autoionizing Resonances in Laser-Excited Sodium $2p^63p\ ^2P_{3/2} \to 2p^53s3p^a$

Classification	$h\nu^b$	$h\nu^c$	f
$2p^5 (3s\ 3p\ ^3P)\ ^2S_{1/2}$	31.78	31.77	0.018 (3)^d
$2p^5 (3s\ 3p\ ^3P)\ ^2D_{5/2}$	31.40	31.40	
			0.070 (10)
$2p^5 (3s\ 3p\ ^3P)\ ^2D_{3/2}$	31.34	31.34	
$2p^5 (3s\ 3p\ ^3P)\ ^4P_{5/2}$	31.19	31.19	< 0.01
$2p^5 (3s\ 3p\ ^1P)\ ^2D_{5/2}$	32.68	32.69	0.051 (9)
$2p^5 (3s\ 3p\ ^1P)\ ^2D_{3/2}$	32.85	32.87	0.036 (2)
			$\Sigma = 0.18\ (3)$

^{a}From Bizau et al. (adapted from Ref. 63 using, for normalization, the more recent data on the 2p single photoionization cross section from Cubaynes et al.[86]).
^{b}Photoelectron spectroscopy measurement of resonance energy.
^{c}Resonance energy from Ref. 18.
^{d}The number in parentheses is the estimated probable error of the measurement.

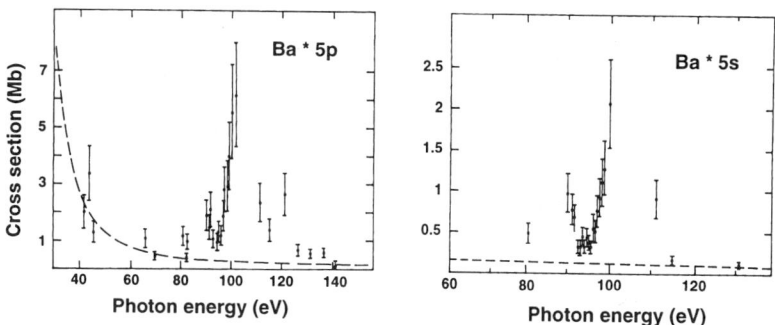

FIGURE 17. Photoionization cross sections for the 5p (left panel) and 5s (right panel) electrons of Ba excited to the $6s5d\ ^{1,3}D$ states, from Bizau et al.[83]

With the higher photon flux available from the super ACO storage ring at LURE, it has proved possible to measure more accurately inner-shell photoionization cross sections for excited alkali atoms.[86–89] Strong enhancements of the relative intensity of correlation satellites produced by shakeup transitions, such as:

$$2p^63p\ ^2P_{3/2} + h\nu \rightarrow 2p^54p\ ^2L_{2J+1} + e^-$$

were discovered first in excited sodium atoms.[86,87] Full photoelectron spectra of sodium atoms in the ground state and in the $2p^63p\ ^2P_{3/2}$ excited state are shown in Figure 18. It was established that the single 2p photoionization cross section, i.e., the cross section for the process:

$$2p^63s\ ^2S_{1/2} + h\nu \rightarrow 2p^53s\ ^2P_{1/2,3/2} + e^-$$

for sodium atoms in the ground state, and:

$$2p^63p\ ^2P_{3/2} + h\nu \rightarrow 2p^53p\ ^2L_{2J+1} + e^-$$

for sodium atoms in the $3p\ ^2P_{3/2}$ excited state are different, about 20% lower in the excited state, in apparent contradiction with the theoretical prediction[90,91] that the inner-shell cross section should be the same for atoms in both ground and excited states. However, by comparing the relative intensity of shakeup satellites in the ground state (photoelectron line at 46.2 eV binding energy in Figure 18, upper

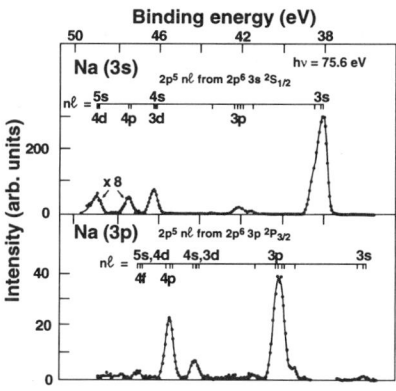

FIGURE 18. 2p photoelectron spectra of sodium showing enhanced satellite structure from excited sodium, from Cubaynes et al.[86]

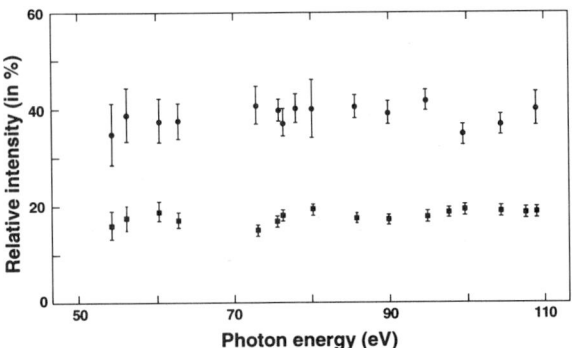

FIGURE 19. Ratio of satellite to 2p main line intensities for ground-state and excited Na, from Cubaynes et al.[86]

panel) and in the excited state (photoelectron line at 45.5 eV binding energy in Figure 18, lower panel), it was possible to conclude[86] that the partial cross section due to satellite structure was also different in both cases. The cross section for shakeup satellites, when compared to the cross section for single photoionization in the 2p subshell, was found to have twice the intensity in the excited state (40%) compared to the ground state (20%), as is evident in Figure 19. The cross section for producing other types of satellites was found to be about the same for both states. When all components of the photoionization cross section were added, the theoretical prediction was verified within 5%, although the theory did not account for the difference in the main components. This enhancement of the shakeup correlation satellites is not limited to 2p-subshell photoionization in excited sodium atoms. Other studies of 2s photoionization in 3p-excited atoms[92] revealed the same behavior. Further experiments on potassium atoms laser-excited into the $3p^64p\,^2P_{3/2}$ state and photoionized in the 3p subshell revealed a similar trend.[93]

Very recent investigations[89,94] on lithium atoms excited into the $1s^22p\,^2P_{3/2}$ state and photoionized into the 1s subshell show an even stronger enhancement of the relative intensity of the shakeup correlation satellites. Preliminary measurements[89] at 92.2 eV photon energy gave a relative intensity of about 70% for the processes:

$$1s^22p\,^2P_{3/2} + h\nu \rightarrow 1s\,(3s, 3p, 3d)\,^2L_{2J+1} + e^-$$

compared to about 25% for similar processes in the ground state[95]:

$$1s^22s^2S_{1/2} + h\nu \rightarrow 1s(3s, 3p, 3d)\,^2L_{2J+1} + e^-$$

Since the conjugate shakeup satellites, which are well resolved from the shakeup satellites in sodium (Figure 18), do not show any enhancement in the case of excited, sodium atoms, this increase in the intensity of the lithium satellites can be attributed entirely to a very strong enhancement (about a factor of 3) of the relative intensity of shakeup satellites.

We summarize in Figure 20 the Z-dependence of the relative intensity of shakeup satellites along the sequence of alkali atoms,[96] following 1s photoionization in $1s^22p\,^2P_{3/2}$ excited lithium (upper panel), 2p photoionization in $2p^63p\,^2P_{3/2}$ excited sodium (middle panel), and 3p photoionization in $3p^64p\,^2P_{3/2}$ excited potassium atoms (lower panel). As expected, the enhancement of shakeup satellites is strongest for excited lithium atoms. The change of one unit in the charge of the electron core $1s^2$ in lithium atoms decreases the shielding effect dramatically and produces a contraction of the orbital in the excited ion. In the Li$^+$ ion, the outer electron in the 3p-excited orbital sees a charge of two, instead of a charge of 1 when it is in the 2p orbital of lithium atoms in the ground state. Then the $3p^+$ wave function in Li$^+$ ion occupies roughly the same region of space as the 2p wave function in the initial state. Consequently, the $\langle 2p2p^+\rangle$ overlap, which is proportional to the probability for the outer electron not to be shaken up (or off) in the photoionization of the 1s inner-shell electron, is small while the

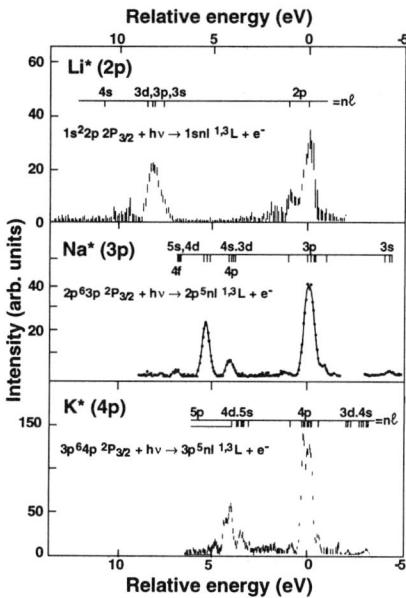

FIGURE 20. Comparison of Li, Na, and K excited-state photoelectron spectra showing Z dependence, from Journel et al.[96]

$\langle 2p3p^+\rangle$ overlap, which gives roughly the probability for the outer electron to be shaken up, is large. For higher-Z atoms, the relative variation in the shielding effect produced by the core electrons on the outer electron decreases with increasing Z values. Thus, the enhancement in the intensity of the shakeup correlation satellites decreases with increasing Z values, as it is observed experimentally in Figure 20.

Undulator sources have permitted a considerable refinement in combined laser/synchrotron experiments, the additional photon flux allowing measurements to be made on aligned atoms. Earlier in this chapter it was pointed out that, in order to preserve the alignment produced by the exciting laser, the vapor beam density must be low enough to avoid depolarization through collisions, with consequent greater demands on the intensity of the light source. Even so, using bending magnet radiation from the BESSY storage ring, Meyer et al.[70] were able to measure the partial autoionization cross sections for excited lithium in the $1s^2 2p\ ^2P_{1/2}$ and $^2P_{3/2}$ states, for the cases where the laser polarization vector was perpendicular and then parallel to the synchrotron radiation polarization vector. Since the $^2P_{1/2}$ state is not aligned, the partial cross section for this state resembled the absorption cross section seen in earlier measurement,[34] but the partial cross sections for the $^2P_{3/2}$ state depended strongly on the relative orientation of the two polarization vectors. This work has been reviewed by Sonntag and Zimmermann,[97] and we do not discuss it further here, but proceed to the development of this method made possible by access to undulator radiation.

In order to extend the ionizing wavelength range beyond that available to the earlier laser experiment,[55] Kerling et al.[98] used a rare gas discharge line source to ionize laser-aligned atoms, but the potential of the method was not fully realized until Pahler et al.[72] used the greater flexibility provided by undulator radiation at the BESSY storage ring. These authors focused on the autoionizing lines between 60 and 65 eV in excited lithium. Using the equipment shown in Figure 13 and a geometry identical to that used in the earlier laser experiments, the data shown in Figure 21 were taken at a photon energy of 61.06 eV, corresponding to the lithium $1s2p^2\ ^2D_{3/2,5/2}$ resonance. The intensity of the ejected electrons was measured as a function of spectrometer angle θ with respect to the polarization vector of the undulator radiation, for three different values of the angle η between the laser and undulator polarization vectors. The continuous lines are fits of the theoretical expression

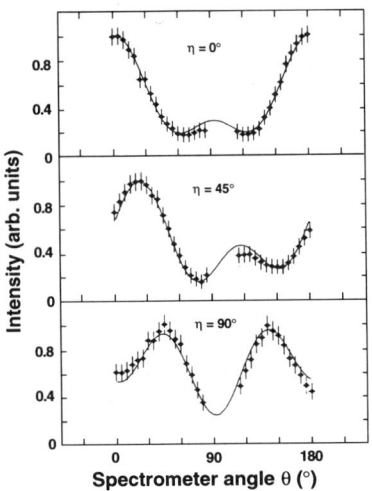

FIGURE 21. Photoelectron angular distributions of excited lithium, from Pahler et al. [72] (η fixed).

$$\frac{d\sigma}{d\Omega} = c[\alpha_{00}P_{00}(\theta) + \alpha_{20}P_{20}(\theta) + \alpha_{40}P_{40}(\theta) + \alpha_{21}P_{21}(\theta) + \alpha_{41}P_{41}(\theta) + \alpha_{22}P_{22}(\theta) + \alpha_{42}P_{42}(\theta)]$$

to the experimental data for the photoelectron angular distributions, LS coupling being assumed. In this expression the $P_{ij}(\theta)$ are normalized associated Legendre polynomials,[99] and it can be shown that the coefficients α_{ij} depend only on the angle η. The above expression is somewhat simpler than the general expression, in that it contains fewer terms, because for excited lithium the states which can decay to the ground state of the ion $^1S_0$ are limited to $^2S_{1/2}$ and $^2D_{3/2,5/2}$. For the transition $^2S_{1/2} \to {}^1S_0 + e^-$, all but the (00) coefficients vanish, and thus the photoelectron intensity does not depend on θ. It

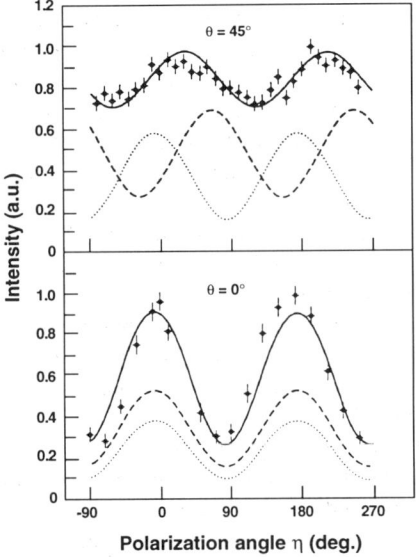

FIGURE 22. Photoelectron angular distributions of excited lithium, from Pahler et al. [72] (θ fixed).

is clear therefore from the angular distributions in Figure 21 that the resonance at 61.06 eV does not have simple S character, and in fact the data fit well to a D resonance. In contrast, the measurements taken on the resonance at 63.54 eV, shown in Figure 22, could not be fitted to a pure S or D resonance.[72] In this case the angle η was varied for two values of θ, and the solid lines were generated from the sum of the calculated angular distributions for an S and a D resonance. The good fit to the experimental data was achieved by assuming that the transitions $1s^2 2p\ ^2P_{3/2} \rightarrow 1s2p^2\ ^2S_{1/2}$ and $1s2s^2(^3S)4d\ ^2D_{3/2,5/2}$ were excited within the 0.3-eV bandpass of the monochromator, with equal probability.

The ability of these experiments to define precisely the transition being measured gives them great promise for the future, and by being able to make differential measurements they are transforming excited-state spectroscopy, much as was done during the 1970s for ground-state atoms. Although limited at present by the wavelength coverage of laser sources, future developments of this technology will give access to inner-shell excitations and provide fundamental data obtainable at the moment only from theoretical calculations.

3.2. Atomic Ions

Spectroscopic studies of ionized atoms have been made for a considerable period, using various kinds of discharge sources and techniques such as flash pyrolysis, and information on the energy levels of atoms and ions has been collected in the well-known Moore's tables,[100] compiled from the experimental data available at that time. As pointed out earlier in this chapter, emphasis is now moving toward cross section and differential measurements and representative examples of this kind of work are given here. All of the experimental techniques described above have made important contributions toward this, and in many respects are complementary: absolute cross section measurements can be used as a basis for normalization of data taken using "relative" methods, and to guide the more difficult differential experiments with regard to the signal level expected.

The RLDI technique has made a major contribution to the photoionization of atomic ions, and one of the earliest experiments by Lucatorto et al.,[101] discussed in the following section, demonstrated the potential of this method. A particularly interesting case which shows the power of the method is the absorption spectrum of Cr. Using a laser-produced plasma as the backlight, Cooper et al.[37] measured the absorption spectrum of Cr and that of Mn$^+$, where the Mn$^+$ was formed by RLDI. Their results are shown in Figure 23; in the region of the $3p \rightarrow 3d$ giant resonance in Cr a pronounced $3p \rightarrow nd$ ($n > 3$) Rydberg series is seen, which does not occur in the other transition metals. This series in Cr was at first thought to be due to the unpaired outer $4s$ electron; the lowest Rydberg levels will be those with the highest value for the total spin, which means that the spin of the outer electron will

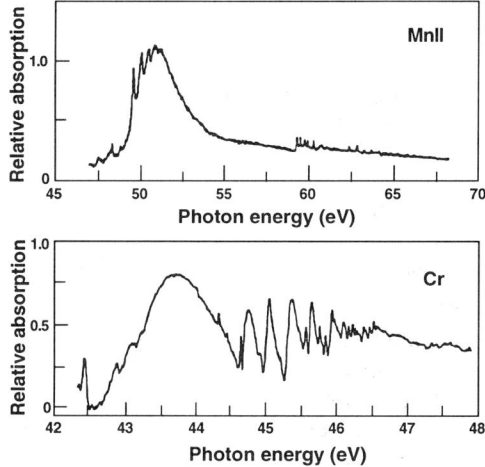

FIGURE 23. Mn$^+$ and Cr spectra from Cooper et al.[37]

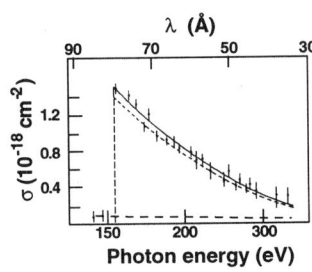

FIGURE 24. Be^{2+} cross section from Jannitti et al.[30]

be parallel to that of the 3p hole. Auger decay with the 4s electron filling the 3p hole is therefore forbidden, lengthening the lifetime of the Rydberg levels and resulting in the sharp structure seen. The absorption spectrum of the isoelectronic Mn$^+$ contains no similar prominent Rydberg structure and Cooper et al.[37] concluded that the explanation does not lie in the unpaired electron, but through coupling with the giant resonance. There are different interpretations of the structure seen in the Cr spectrum and these will be discussed in the theoretical section.

In addition to the RLDI method, the related dual laser plasma technique[33] has been used not only for spectroscopic work but also in favorable cases for the measurement of cross sections, although normalization was necessary to obtain absolute values. Jannitti et al.,[30] using the equipment shown in Figure 4, have measured the Be^{2+} absorption spectrum below the ionization threshold. From measurements of the absorption coefficients of the higher series members they modeled the absorption spectrum, using their known instrumental function. They also ensured that the instrumental width was not a significant part of the linewidth by observing the absorption spectrum close to the target where the lines would be Stark broadened. Using known values of the oscillator strengths[102] they obtained the cross section at the ionization threshold, and could extrapolate this into the continuum to give the results shown in Figure 24. Good agreement is seen between the results normalized to the threshold value and a theoretical Hartree–Slater calculation.[103] In the same experiment they also observed the He-like two-electron excitation $1s^2\ ^1S \to 2s2p\ ^1P$ and thereby were able to measure the Fano parameters q (line profile) and Γ (width) for the excited ion. Table II shows these values compared with theoretical values[104,105] for Γ and E, the energy of the resonance, and also a comparison between the q value for Be^{2+} and those from earlier measurements on helium[106] and singly ionized lithium.[22] The differences in q indicate varying intensities of the autoionizing transition compared to the direct ionization

TABLE II
Comparison of the Parameters q, Γ, and E of the Resonance $1s^2\ ^1S \to 2s2p\ ^1P^0$ for Be^{2+} Ions[a]

Experiment	
$\quad E$ (resonance)	281.25 ± 0.07 eV
$\quad \Gamma$ (resonance)	0.076 ± 0.01 eV
$\quad q$ (resonance)	-1.7 ± 0.3
Theory	
$\quad$ Drake and Dalgarno	
$\quad\quad E$ (resonance)	281.268 eV
$\quad\quad R$ (resonance)	0.074 eV
$\quad$ Chan and Stewart	
$\quad\quad E$ (resonance)	281.338 eV

[a]From Jannitti et al.[30]

TABLE III

Excitation Energies and Oscillator Strengths for Transitions from the $4d^{10}5s^25p^66s^2S_{1/2}$ Ground State of Ba^{+a}

Excited state	Excitation energy (eV)	Oscillator strength
$4d^{10}5p^56s^2\ ^2P$	19.5	0.26
$(4d^{10}5p^55d\ ^3P)6s\ ^2P$	16.7	0.001
$(4d^{10}5p^55d\ ^1P)6s\ ^2P$	25.0	10.0

^{a}From Lyon et al.[38]

transition, and it has been pointed out[30] that further measurements of q along this isoelectronic sequence, although difficult, would be very interesting and useful. Jannitti et al.[107] have made similar measurements on the carbon sequence and using a similar normalization technique measure a cross section of 0.47 ± 0.05 Mb at the 1s threshold of C^{4+}, also in good agreement with the same Hartree–Slater theoretical calculations.[103]

The merged beam technique shown in Figure 7 is to date the only method which has provided absolute photoionization cross sections for atomic ions without normalization to other measurements or theoretical values. It was used first for singly charged barium ions[108] at the Daresbury SRS, in the region of the strong autoionizing transition $5p \rightarrow 5d$. The cross section was expected to be large from this preliminary experiment with lower incident light flux, where an estimate of the cross section in excess of 10^{-16} cm^2 was obtained. Lyon et al.[108] also calculated oscillator strengths for transitions from the ground state of the ion to the $5p^56s^2\ ^2P$, $5p^55d(^3P)6s\ ^2P$, and $5p^55d(^1P)6s\ ^2P$ autoionizing levels, using a single-configuration Hartree–Fock wave function for the ground state. The excited-state wave function included configuration interaction among the three autoionizing levels. The results are shown in Table III; the value of 10 was later seen to be an overestimate for the total oscillator strength, but was close enough to confirm that strong autoionizing transitions were expected here.

The later experiment on Ba$^+$ revealed a wealth of structure, most of which remains unidentified, except where it could be correlated with series limits identified from absorption spectra of neutral barium. Table IV gives the limits of series which terminate in autoionizing levels of Ba$^+$, calculated by Connerade et al.[109] from their measurements on neutral barium. The data on Ba$^+$ are shown in Figure 25 where the series limits for neutral barium are indicated by the vertical arrows. In this figure the vertical bars below the data indicate the lines seen in an earlier flash pyrolysis experiment by Roig.[110] Although lines due to impurities and excited states were likely to be present in that experiment, the features in the two experiments correlate quite well. The identification of the lines seen in the singly ionized strontium[41] data was aided similarly by comparison with photoabsorption data,[111] and also with data from a photoelectron spectroscopy experiment.[112]

TABLE IV

Comparison with Series Limits in Neutral Barium Measured by Connerade et al.[109]

	Connerade et al.[109]			Lyon et al.[38]	
Possible assignment	Character	Energy (eV)		Energy (eV)	Cross section (cm^2)
$5p^56s^2\ ^2P_{1/2}$	Broad and intense	19.544		19.546 ± 0.008	3.3 × 10^{-16}
	Weak and sharp	18.305			
$(5p^55d\ ^3D)6s\ ^2D_{3/2}$	Strong and sharp	17.583		17.594 ± 0.008	2.7 × 10^{-16}
$5p^56s^2\ ^2P_{3/2}$	Intense and broad	17.477		17.485 ± 0.008	2.5 × 10^{-16}
	Weak and sharp	17.013		17.000 ± 0.03	0.4 × 10^{-16}
	Weak and sharp	16.777			
	Weak and sharp	16.431		16.430 ± 0.03	0.35 × 10^{-16}

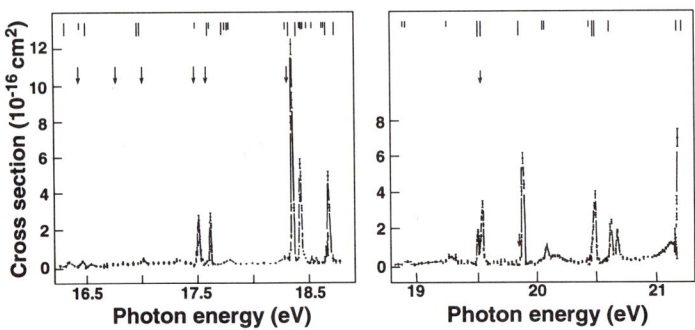

FIGURE 25. Absolute photoionization cross section of Ba^+, from Lyon et al.[38]

The $p \to d$ resonances in Ca^+, Sr^+, and Ba^+ are an interesting sequence, and are compared in Figure 26. Again, for Ba^+ the flash pyrolysis data[110] correlate well with the results from the ion beam experiment. The vertical bars below the Ca^+ data were taken from the experiment by Sonntag et al.[36] The total oscillator strength for these transitions is large, 3.74 for Ba^+, albeit less than the value of 10 obtained in the calculation.[108] Spin-orbit splitting is apparently evident for all of them, but these transitions are likely to consist of mixed configurations and thus cannot be described simply. Connerade et al.[113] have discussed the breakdown of the independent electron picture in the analysis of the inner-shell spectra of neutral calcium, strontium, and barium, finding evidence for many-body effects which lead to the existence of multiple series limits where normally two would be expected. This implies that additional states of the ion exist, states not normally allowed in a simple model. The complexity increases and becomes more visible as the size of the ion increases, an effect clearly seen in Figure 26, although a quantitative explanation for these data remains to be given. The case of Ca^+ is examined further below, where it will be evident that even for this relatively light ion many-body effects make it difficult to be sure of the identity of many of the features observed experimentally.

Measurements were also made on the singly charged ions of potassium,[43] gallium,[44] and zinc[44]; potassium and zinc are discussed below in relation to recent theoretical work. Data for gallium are shown in Figure 27, and this is an interesting case of interfering resonances, similar to those seen in the isoelectronic neutral zinc by Marr and Austin[114] many years ago. A theoretical analysis of overlapping resonances was given by Mies,[115] in an extension of Fano's earlier analysis[6]; when the widths of adjacent resonances are similar to their distance apart, there can be large changes in the observed line profiles. This is clearly the reason for the structure seen for Ga^+; the transitions involved are to the $^3P_1$ and $^3D_1$, configurations of the autoionizing level $3d^{10}4s^2 \to 3d^9 4s^2 4p$ ($^1P_1, ^3P_1, ^3D_1$). The

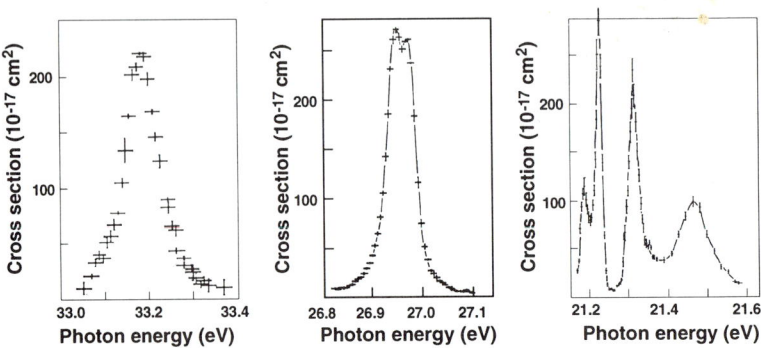

FIGURE 26. The $p \to d$ resonances in Ca^+, Sr^+, and Ba^+, from Lyon et al., adapted from Refs. 39, 40, and 38.

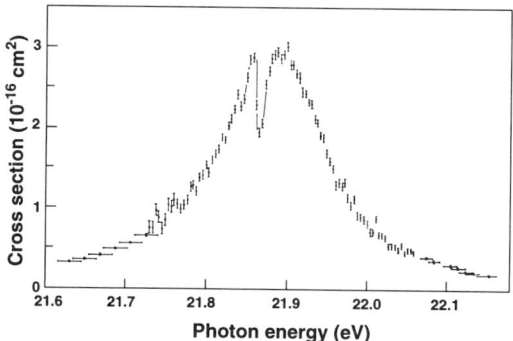

FIGURE 27. The ($^1P_1$, $^3D_1$) autoionizing lines in Ga$^+$, from Peart et a.[44]

$^1P_1$ configuration, in analogy with the work on zinc, was assumed to be the weaker and sharper transition seen at 20.975 eV.

It will be evident from Figures 25 to 27 that the minimum measurable cross section was 10^{-18} cm^2; the ions measured so far using the merged beam technique have been chosen because of the high cross sections expected in those spectral regions. Moving to short wavelengths will mean working with lower cross sections, and furthermore multiple ionization will also occur implying further partition of the cross section. The extension of this technique to other species and to higher photon energies requires intense radiation sources, and just recently Koizumi et al.,[116] working on an undulator beam line at the Photon Factory, have used the merged beam technique and apparatus almost identical to that shown in Figure 7 to measure the relative yield of Ba^{2+} and Ba^{3+} in the region of $4d$ ionization of Ba$^+$. These new results are shown in Figure 28. Further refinements of this experiment will lead to absolute cross section measurements; its extension to measurements on highly charged ions is also planned and will be discussed in the future developments section of this chapter.

The availability of undulator radiation at the Orsay storage ring has made it possible to make the first differential measurements on an atomic ion, using the apparatus shown in Figure 8 and described above. For their first experiment Bizau et al.[45] chose Ca$^+$, in the region of the $3p \to 3d$ giant resonance, and the results are shown in Figure 29. Despite the large background due to residual gas, the signal resulting from the autoionization of the excited Ca^{+*} ion in the region of the $3p \to 3d$ giant resonance was clearly seen. In Figure 30, the constant ionic state spectrum, taken by tracking the photoelectron energy with the photon energy, shows that good count rates were achieved. The detected energy of the

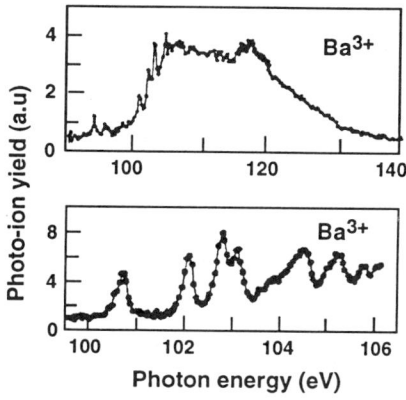

FIGURE 28. Relative Ba^{3+} photoion yield following photoionization of Ba$^+$ ions, from Koizumi et al.[116]

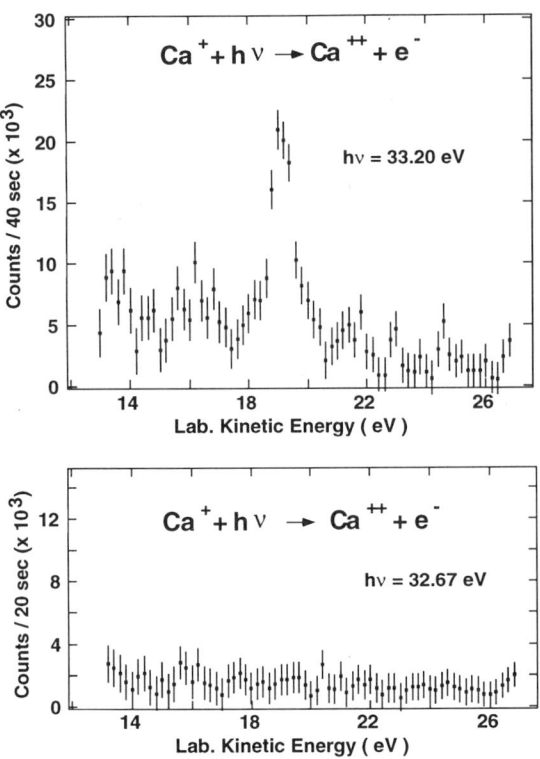

FIGURE 29. The photoelectron spectrum of Ca^+, from Bizau et al.[45]

photoelectrons differed from the simple value obtained by subtracting the binding energy of the 4s electron from the photon energy for two reasons: retardation by the positive charge of the ion beam itself, reducing the measured value, and a geometrical effect due to the component of the velocity of the ion beam in the detection direction of the CMA, which increased the measured value. In this instance these two corrections were similar and of opposite sign, resulting in a small final correction, but in general this could vary widely depending on the parameters of the experiment.

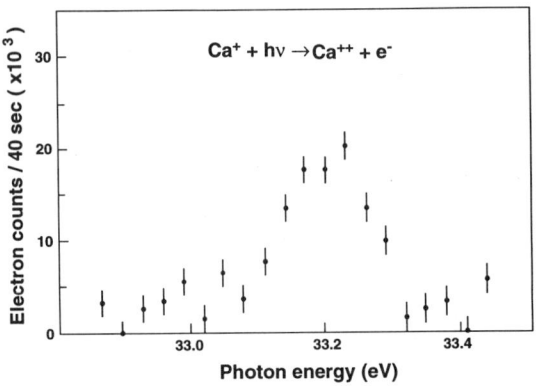

FIGURE 30. Constant ionic state spectrum measured in the region of the giant $3p \to 3d$ resonance in Ca^+, from Bizau et al.[45]

4. DEVELOPMENTS IN THEORETICAL INTERPRETATION

4.1. Inadequacy of the One-Electron Approximation

Because of the lack of experimental data on atomic ions, especially for inner-shell photoionization cross sections, most of our present knowledge comes from theoretical predictions mainly based on one-electron Hartree–Slater calculations.[103] Only in some special cases, especially when experimental data were available, have more sophisticated approximations been used, such as for the singly charged K^+, Ca^+, Mn^+, and Ba^+ ions.[118–122] However, it is well known that the Hartree–Slater approximation does not include any correlation effects. For atoms in the ground state, this approximation gives results in poor agreement with experimental data, in particular in the photon energy ranges where special features appear. In short, the inadequacy of the one-electron model has been demonstrated,[123,124] and quantitative data are needed, for the energy position of the ionization thresholds, of the Cooper minima and delayed maxima, as well as for the absolute values of the cross sections. Some of the experimentally observed features, which were qualitatively explained in the one-electron approximation, have been reproduced quantitatively only when electron correlations were properly introduced in the theoretical models.

One of the best examples is the so-called delayed maximum in photoionization cross sections for inner subshells with high-l orbital quantum numbers ($l \geq 2$). We will present, as an illustrative example, the first case that has been investigated in detail, i.e., the $4d$ photoionization cross section in xenon. In Figure 31, we show the variation of the total photoabsorption cross section in the energy range above the $4d$ ionization threshold as experimentally measured[125] (curve a), and the results of the first theoretical calculation[126] of the $4d$ photoionization cross section (curve b) using the one-electron Hartree–Slater approximation. Previous measurements of the total photoabsorption cross sections[127] for the closest Z elements, i.e., the lanthanides, in this energy range already showed that the maximum of the $4d$ cross section in these elements, expected to be at threshold in the hydrogenic approximation, is shifted to several tens of electron volts above threshold. These earlier observations were fully confirmed by the first continuous measurement of the $4d$ cross section in xenon,[125] using synchrotron radiation which showed (see Figure 31) that the cross section rises to a large maximum at about 30 eV above the threshold. This unusual behavior was attributed[126] to the fact that the potential

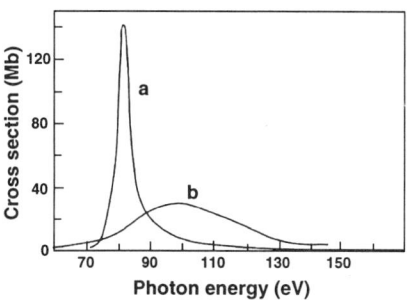

FIGURE 31. The photoabsorption cross section of Xe: (a) Hartree–Slater calculation by Cooper[126]; (b) measurements from Ederer.[125] Lukirskii et al.[127] also measured this cross section.

barrier existing at threshold, because of the high value of the centrifugal force for high-l electrons, prevents the εf wave function from penetrating the region of maximum charge density of the $4d$ electrons. Only as the photon energy increases can the f wave penetrate the centrifugal barrier so that the partial cross section for $4d \rightarrow \varepsilon f$ transitions rises rapidly and becomes the dominant contribution to the photoabsorption cross section, producing what was called the delayed onset of the $4d$ cross section. Quantitative agreement between experiment and theory was not good, either for the energy position of the maximum, or for the cross section value at the maximum, because the theoretical model used at that time was a rather crude, frozen core, one-electron model. It should be noted that the experiment measured at that time the total photoabsorption cross section, including the contribution of the $5p$, $5s$, and $4d$ subshells, whereas theory calculated only the total $4d$ photoionization cross section. In spite of that difference, the Hartree–Slater theory overestimated the cross section by at least a factor of five.

It took 25 years for theoreticians to improve the one-electron calculations in order to obtain an excellent agreement with experiment. In Figure 32 is shown the present status of experiment and theory for the single $4d$ photoionization cross section. The most reliable experimental determination of this cross section[128] using photoelectron spectroscopy has been chosen among the recent measurements for comparison. These results were obtained using a novel normalization procedure[128] that makes the results independent of any measured value of the total photoabsorption cross section. Only the latest theoretical results[129] are shown in Figure 32. They were obtained in the random phase approximation with exchange (RPAE) taking into account the proper self-energy part (SEP) of the photoelectron's Green's function and relaxation effects. The SEP describes the polarization potential acting on the photoelectron. Taking SEP into account means a reduced single $4d$ photoionization cross section, because of the inelastic scattering, by the outer electrons, of the electron ejected from the $4d$ subshell. This leads to a transfer of oscillator strength from the single photoionization channel to the two-electron photoionization channels.[129] These highly correlated calculations are in excellent agreement with the experimental values, and this fully demonstrates the importance of correlation and relaxation effects to describe what was originally considered as the manifestation of pure one-electron effects. In fact, these maxima are now called by some authors "correlated maxima." A similar conclusion was also drawn in comparing the most recent measurements[130] and calculations[129,131,132] of the $4d$ photoionization cross section in barium.

For atoms in excited states, while there are many calculations of cross sections for photoionization of the excited electron, few theoretical results are available for inner-shell photoionization. Evidently, the Hartree–Slater approximation cannot provide results in better agreement with experiment any more than it does for atoms in the ground state. In the case where comparison with experiment can be made, such as in the case of excited barium atoms,[66,83] the one-electron model[84,133] is unable to account for the experimentally observed behavior of the cross sections in the photon energy range where correlation effects have been shown to be important for atoms in the ground state.[130] In the case of alkali atoms in excited states, even the prediction made using many-body perturbation

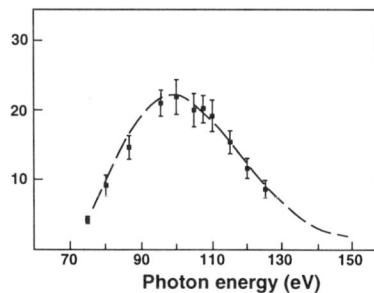

FIGURE 32. The $4d$ single photoionization cross section of Xe. Experiment: black squares from Kämmerling et al.[128]; theory: dashed line, RRPA calculations by Amusia et al.[129]

theory,[90,91] that the cross sections should be the same as for atoms in the ground state, is a first approximation, valid only for the total photoionization cross section and not for the partial subshell cross sections.[86–88,92,93] Only highly correlated calculations, such as the ones recently developed in the R-matrix approximation,[94,134,135] are able to reproduce the experimental data satisfactorily.

Indeed, for highly charged ions, such as heliumlike ions, the influence of correlation effects is expected to decrease in comparison with the attractive potential due to the nuclear charge. This explains why some agreement was observed between some measurements and the Hartree–Slater calculations, such as for $1s$ photoionization in the $1s^2\ ^1S$ state of Be^{2+} and C^{4+} ions.[30,107] However, for many-electron ions with low charge states (1 to 5 typically), there is an intermediate situation between the fully correlated description and the predominant nuclear attraction. In this case, Hartree–Slater calculations are not expected to give a good representation of the photoionization process. Correlated calculations and experimental determinations of energy and cross sections are definitely needed to provide an accurate representation of the situation.

4.2. Recent Theoretical Developments for Excited Atoms

Due to the scarcity of experimental data on photoionization of excited atoms until recently, there has been little incentive for advanced theoretical calculations, and those that do exist refer to a specific experimental problem. For example, Norcross and Stone[136] have calculated excited-state cesium photoionization cross sections, and reproduce the experimental finding, from recombination in a cesium plasma,[137] that the $6\ ^2P$ photoionization cross section is two orders of magnitude greater than the $6\ ^2S$ cross section. A more recent example is the $5d$ photoionization cross section of laser-excited barium atoms,[66] where a local density approximation (LDA) was used to calculate the cross sections. The LDA method has been described by Zangwill and Soven[138]: the external radiation field is replaced by an effective local field, which permits the response of the electronic charge density to the external field to be taken into account in a self-consistent way. In essence it is very similar to the RPAE approach (e.g., see Amusia et al.[139]), in which the external radiation field collectively excites the electrons in a particular shell, resonantly in the case of an absorption maximum. The main difference is the replacement of the Hartree–Fock orbitals by LDA orbitals in which the exchange operator is replaced by a local exchange correlation potential. The method has been successful in describing resonant photoionization cross sections in the rare earth metals and their compounds, and provides physical insight into the phenomenon. The experiment on laser-excited Ba atoms by Bizau et al.[66] is shown in Figure 33, where the experimental data are compared with LDA calculations of the $5d$ and $6s$ partial cross sections. Two types of LDA calculations are shown in the figure; the continuous line is derived from a Hartree–Fock–Slater calculation,[133] in which electron correlations within the $4d$ shell are not explicitly included. The dashed line is similar to that described by Zangwill and Soven,[138] where the external field excites collective oscillations of the $4d$ shell. It can be seen that the data taken in the lower-energy nonresonant region agree quite well with the simpler theoretical calculation. In the

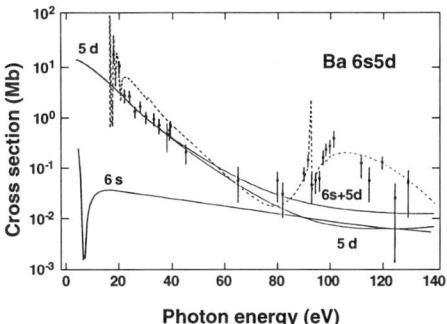

FIGURE 33. Photoionization cross section of the $5d$ electrons in excited barium; experimental data and calculations from Bizau et al.[66] Continuous lines HFS calculation, Ref. 133 dashed line, LDRPA calculation, Ref. 66.

higher-energy region where resonant excitation of the 4d shell takes place, only the LDRPA calculation reproduces the main features of the experiment, including the resonant enhancement of the 5d cross section around 92 eV due to interaction with discrete final states having a $4d^9 5s^2 5p^6 6s5dnl$ configuration.

The above example demonstrates that theoretical techniques applied to ground-state species can be successfully applied to excited atoms, particularly in those cases where the ground state has closed shells or subshells. As will be seen later, this can be applied to ions also, including cases where an s subshell is incomplete, but the applicability of these methods to open-shell atoms remains an open question in general through the lack of experimental data. For the identification of transition energies good calculations are now available, using relativistic wave functions and including configuration interaction; see, for example, Cowan.[140] A specific application of this theoretical method is described by Mansfield and Murnane,[141] for the case of autoionizing transitions in atomic cadmium, and in general good agreement is obtained for the transition energies, less good for oscillator strengths. The Cowan codes are now being applied as an aid to identification of transitions in the spectra of multiply charged ions taken using the dual laser plasma technique, and are proving very useful in this respect[142]; in addition, for some applications a more recent relativistic series of codes has been developed by Grant.[143]

For the purpose of identifying spectral lines in the solar corona and laboratory plasmas, there are many calculations of transition energies and linewidths for isoelectronic sequences, a complete review of which is beyond the scope of this chapter. Typical of this kind of work are the calculations of Fox and Dalgarno[144] using a Z-expansion variation of the Hartree–Fock method for the Li-like sequence, and their theoretical values give fair agreement with beam-foil experiments. McIlrath and Lucatorto[34] made use of multiconfigurational Hartree–Fock calculations to assign the lines observed in their laser-excited Li spectra by which they were able to identify transitions from the $1s^2 2p\ ^2P^0$ state of Li. Theoretical cross sections for excited states of atoms are still scarce, although methods are now available to calculate them, as will be seen below.

For inner-shell photoionization cross sections, the first calculations were made for excited sodium and potassium atoms, using many-body perturbation theory.[90,91] They did not predict large differences in the inner-shell photoionization cross sections between atoms in the ground state and in several optically excited states, nor strong variations in the angular distribution of the photoelectrons. The second theoretical finding seems to be confirmed by recent angular distribution measurements in excited sodium atoms,[145] but we have seen earlier (see Figure 19) that the experimentally measured single photoionization cross sections are quite different when the ground state is compared to the first excited state.[86,92] A large part of the oscillator strength is transferred from the single photoionization channel to the shakeup transition channel in the excited atoms, and the existing theory did not account for this. The effect is greatest in the lowest-Z excited atoms that have been studied, i.e., for lithium atoms excited to the $1s^2 2p\ ^2P_{3/2}$ state[89]: In this case, the removal of a 1s electron introduces the largest variation in the potential felt by the outer electron, in such a way that the orbital occupied by this outer electron, even after possible excitation to a more excited orbital, occupies a region of space closer to the nucleus than it does in the ground-state atom.

Two recent calculations give quite different results. The first calculation,[146] made 3 years after the discovery of the strong enhancement of the relative intensity of shakeup satellites in excited sodium atoms, uses the multiconfiguration Hartree–Fock approximation (MCHF) to calculate only the intensity of the shakeup satellites in excited lithium atoms (process: $1s^2 2p + h\nu \rightarrow 1s3p + e^-$). The energy dependence of the relative intensity of the shakeup satellites found from these results shows this intensity *decreasing* with increasing photon energy, starting from the satellite ionization threshold. This result is extremely surprising since, in all cases studied so far, it has always been found that this intensity has a finite value at threshold and increases with increasing photon energy, reaching a plateau value at high photon energy. Thus, the calculated behavior is *a priori* wrong and immediately throws some doubt on the apparent agreement of the theoretical value (70% at 92 eV photon energy) with the experimental value.[89] The second set of calculations[134,135] includes a more exhaustive theoretical analysis of all processes leading to correlation satellites in inner-shell photoionization of excited atoms. After having developed the R-matrix approximation for inner-shell photoionization,[147] Vo Ky *et al.*

used it for the calculations of partial cross sections in lithium and sodium atoms, in the ground state[148] as well as in the first excited state.[94,134,135] The Li$^+$ target was represented by 19 states using configuration interaction including up to 103 "basic" configurations, namely, 54 configurations exclusively with spectroscopic orbitals, and 49 configurations that contained nonspectroscopic orbitals. For the representation of the continuum, 38 basis functions were used for each l value of the orbital angular momentum of the scattered electron, up to l equal to 3. In sodium, the situation is more complicated because there are two inner-shell ionization thresholds in the photon energy range of interest (40 to 140 eV). The Na$^+$ target was represented by 39 states, including 30 basic configurations. The calculated values of the ionization thresholds and inner-shell excitation resonances are in extremely good agreement with the experimental data, within a few hundredths of an electron volt. The calculations provide partial cross sections for producing the various types of correlation satellites, i.e., for different values of the angular momentum of the ejected electron. For instance, in excited lithium atoms, partial cross sections have been calculated for each group of correlation satellites.[94,135] Compared to the intensity of single inner-shell photoionization, these relative intensities are 5% ($1s^22p + h\nu \rightarrow 1s3s + e^-$, $\Delta l = -1$), 16% ($1s^22p + h\nu \rightarrow 1s3d + e^-$, $\Delta l = +1$), and 54% ($1s^22p + h\nu \rightarrow 1s3p + e^-$, $\Delta l = 0$) at 92 eV photon energy. Since the corresponding correlation satellite lines are not resolved experimentally,[89] the total of these relative intensities, i.e., 75%, has to be compared to the experimental value of (70 ± 10)%. The agreement is excellent. The energy dependence calculated for these relative intensities do show the expected behavior and will be compared soon to the experimental measurements presently in progress.[94]

For excited sodium atoms, the experimentally observed enhancement (about a factor of 2) for the intensity of the shakeup correlation satellites following $2p$ photoionization is fully reproduced by the theoretical calculations,[94] as well as the energy dependence of the relative intensities of each group of correlation satellites. Calculations are in progress for excited potassium atoms. It should also be mentioned that a very recent calculation[149] of the absolute cross section for producing the shakeup correlation satellites in excited sodium atoms, using many-body perturbation theory to partition the total cross section into its different components, is also in agreement with the experimental data.

4.3. Calculations for Ions

4.3.1. R-Matrix and MQDT Calculations: The Stellar Opacity Project. A major impetus for such calculations has been the need for cross section data for astrophysical applications and the analysis of plasmas generated for the purposes of nuclear fusion. For this reason the stellar opacity project was set up by Seaton and co-workers, and the theoretical method and sample calculations were described in a series of papers.[150] These dealt mainly with photoionization from the ground and excited states of ions, singly and multiply charged, and are based on the use of the R-matrix method.[151] In the past, this method, which defines an inner region and an outer region around the target atom or molecule and enables the final-state wave functions to be matched across the boundary region, has been successful in calculating the cross sections for outer-shell photoionization in the rare gases.[11] Within the R-matrix framework, Berrington et al.[152] describe the application of the close coupling approximation, used extensively in electron–atom collision theory, where the target functions are constructed using configuration interaction theory and include correlation orbitals. The result has been a set of computer codes which can be used to calculate the outer-shell photoionization cross sections of the ground and excited states of the ions of light elements for which relativistic corrections are not needed. Although these have not as yet been published extensively in the literature, access to the data base is available.[153] Some examples of the calculations have been published; Fernley et al.[154] have published energy levels, oscillator strength values, and photoionization cross sections for the heliumlike ions and for the excited ^1S states of neutral helium.

4.3.2. Hartree–Slater and RPAE Calculations: Isoelectronic and Isonuclear Sequences. Of course, the vast bulk of these data remain untested by experiment, and this is a cause of some concern to both theorists and experimentalists alike. In the absence of experimental data, the tendency has been for calculations to concentrate on isoelectronic and isonuclear sequences to give better physical insight into the major factors which influence the structure and cross section. Attending first to work of a

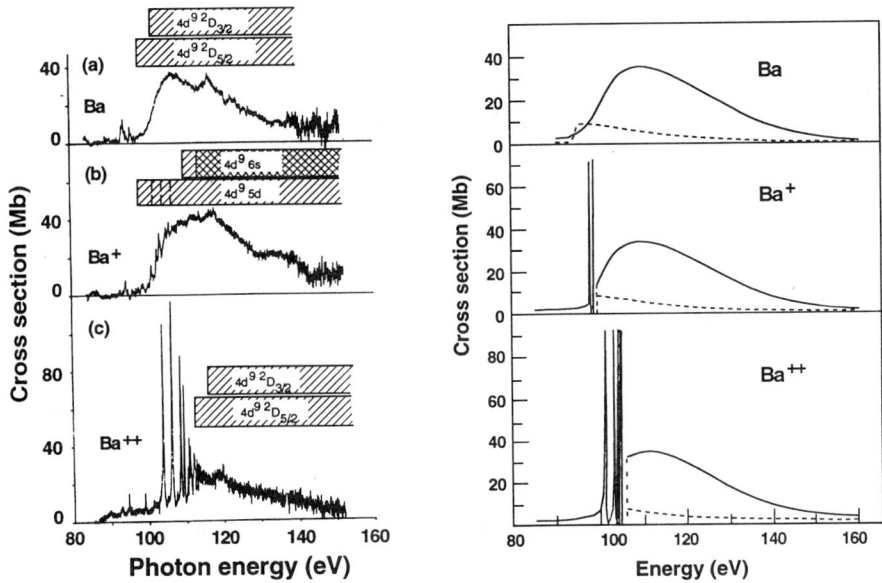

FIGURE 34. Photoabsorption data of Ba, Ba$^+$, and Ba^{2+}. (Left) Experimental data from Lucatorto et al.[101]; (right) theoretical calculations from Nuroh et al.[122] Solid line = TDLTDA; dashed line = HF calculations.

structural type, for which there are more experimental data, a comprehensive calculation of excitation energies and line strengths using a relativistic multiconfiguration Hartree–Fock technique has been published by Cheng and Johnson[155] for the Mg isoelectronic sequence, and good agreement obtained with other theoretical data for low-Z ions, with the expectation that for high Z their method would give accurate values. A striking example of the use of the theory for ions is seen in the left panel of Figure 34, where time-dependent LDA calculations[122] can be compared to the experimental data of Lucatorto et al.[101] for Ba, Ba$^+$, and Ba^{2+} in the right panel, showing clearly the appearance of $4d \to 4f$ excitations modified by collective interactions within the $4d$ shell. In effect, and as pointed out by Nuroh et al.,[122] the $4d$ "ionization threshold moves across the collective mode spectral distribution, unveiling the discretelike structure prominently displayed by Ba^{2+}." Calculations using a relativistic form of the RPA method have been carried out on the Be-like sequence of ions by Johnson and Huang[156]; the problem of including double electron excitations in this method was tackled by using a multiconfiguration wave function for the initial state. Good agreement with experiment was again achieved.

In the case of cross sections the situation is far less satisfactory. Early work was carried out by Combet Farnoux and Lamoureux[157,158] who calculated cross sections for the isoelectronic sequences of Ne, Ne$^+$, and Ne^{2+}, within a one-electron framework using Hartree–Fock functions and a central noncoulombic potential. Reilman and Manson[159] have calculated cross sections for the iron isonuclear sequence, using Hartree–Slater wave functions for the initial states and continuum wave functions based on the appropriate central HS potential. Their results for the iron $3p$ shell are shown in Figure 35 where it is clearly seen that until electrons begin to be removed from the $n = 3$ shell the cross section per $3p$ electron is unaffected; the Cooper minimum, and the peak in the cross section remain the same for the Fe, Fe$^+$, and Fe^{2+} species. Similar results were obtained for the Mg and Ar isonuclear sequences by Nasreen et al.,[160] who used the RRPA method. The conclusion drawn from the one-electron calculations, using the central field approximation, on the cross section of inner shells is that they remain virtually unaffected by the removal of outer-shell electrons.[161] Certainly, the studies on neutral potassium[162] and singly ionized potassium[43] verify experimentally that the absence of the outer electron does not affect the ionization cross section substantially. However, there is considerable difference between the cross section values calculated using Hartree–Slater orbitals and the RRPA

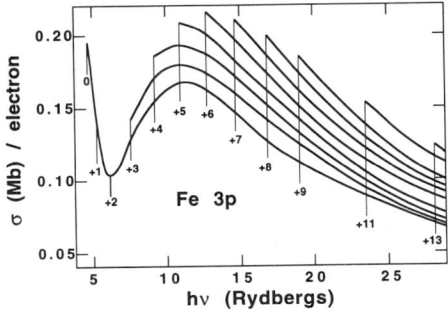

FIGURE 35. Theoretical data for the iron isonuclear sequence, from Reilman and Manson.[159]

method,[118] as can be seen in Figure 36, which shows the variation of the photoionization cross section of K^+. The experimental data favor quite clearly the RRPA results. In other cases, the results obtained using the one-electron[103] and RRPA[163] models are similar, as demonstrated in Figure 37 in the case of Al^+ ions. Note, however, that there are no experimental data available for this ion. In conclusion, it is clear that the simple central field model is inappropriate for determining absolute cross sections for many-electron ions, especially near ionization thresholds, in the vicinity of a Cooper minimum or a delayed maximum, and that more experimental data of this kind are needed.

4.3.3. Pseudorelativistic Calculations; Use of Cowan's Codes for K^+ and Zn^+ An interesting further point arises in the data for K^+ shown in Figure 36; Nasreen et al.[118] consider the structure seen in the experiment around 35.5 eV to be an experimental artifact. Although the error bars are large enough to make the energy positions of the two features uncertain, there is little doubt that the structures seen at ~36 and 39 eV are outside the limits of experimental error. By reexamining this region using structure calculations based on Cowan's codes, Cowan and Wilson[164] have been able to identify these features. Recognizing that there is strong interaction between the $3s3p^64p$ and $3s^23p^43d4p$ configurations, they found that the $3s3p^64p\ ^1P_1$ level should lie lower and close to 35.5 eV, rather than at 39.8 eV as suggested by Peart and Lyon[43] and based on an earlier electron impact experiment by Aizawa et al.[165] The identity of the feature at 39.8 eV could then be reconsidered; by calculating the energy separations they suggest that this is due to the $3s3p^65p\ ^1P_1$ level.

Cowan and Wilson[166] also used the same theoretical method to analyze the structure seen in Zn^+ by Peart et al.[44] Pseudorelativistic Hartree–Fock wave functions are used in this method and configuration interaction is taken into account, although some experience is needed in selecting the most important configurations to be included in the calculation. The results are shown in Figure 38,

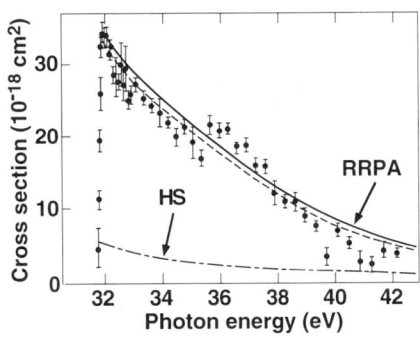

FIGURE 36. Photoionization cross sections and theory for K^+. Experimental data from Peart and Lyon[43]; theoretical data from a one-electron Hartree–Slater calculation (HS) and relativistic random phase approximation (RRPA) calculations by Nasreen et al.[118]

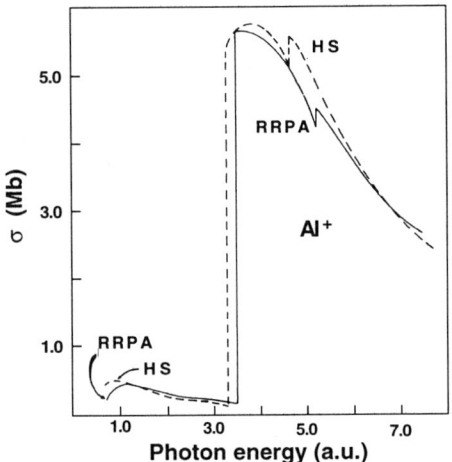

FIGURE 37. Theoretical photoionization cross sections for Al$^+$ ions, showing the Hartree–Slater[103] and RRPA[163] results. No experimental data are available.

for the transitions from the $3d^{10}4s\ ^2S_{1/2}$ ground state of Zn$^+$ to the $J=\frac{1}{2}$ and $\frac{3}{2}$ levels of the $3d^9 4s5p$ configuration; the vertical bars represent the calculated oscillator strengths. The energy scale has been adjusted so that the lowest $3d^9 4s4p$ calculated levels coincide with experiment, and in this way most of the structures could be identified. Transitions to the $3d^9 4s6p$, $7p$, $4f$, and $5f$ levels were also calculated, and there was clear correspondence between the calculated and experimental peak positions; in general, the agreement for the oscillator strengths was less satisfactory.

4.3.4. Comparison of Calculations with Experimental Data for Ca$^+$ and Mn.$^+$ As seen earlier, absolute cross section measurements have been made for the singly charged ions of calcium, strontium, and barium, but theoretical analysis has so far been undertaken only for calcium, and even for this

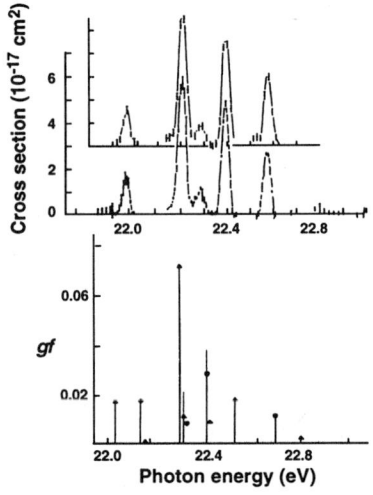

FIGURE 38. The $3d^{10}4s\ ^2S_{1/2} \to 3d^9 4s5p\ ^2P_{1/2,3/2}$ transitions of Zn$^+$. (Upper) Experimental data from Peart et al.[44]; (lower) theoretical oscillator strengths from Cowan and Wilson.[164]

atom much remains to be done. Miecznik et al.[167] used the R-matrix method to calculate the cross sections measured by Lyon et al.,[40] and achieved very good agreement for the $3p \to 3d$ resonance; some other structures were also tentatively identified. Very recently, a nonrelativistic, spin-polarized version of the RPAE method, developed by Amusia et al.,[168] has been used to carry this further. Full details of the method are given by Ivanov and West[120]; in summary, each filled subshell is divided into groups of electrons, one with spin "up" (spin quantum number $\mu = \frac{1}{2}$) and the other with spin "down" ($\mu = -\frac{1}{2}$). For the Ca$^+$ ground state, the electron structure is therefore

$$1s\uparrow\ 1s\downarrow\ 2s\uparrow\ 2s\downarrow\ 2p^3\uparrow\ 2p^3\downarrow\ 3s\uparrow\ 3s\downarrow\ 3p^3\uparrow\ 3p^3\downarrow\ 4s\uparrow \quad (^2S)$$

The spin-polarized Hartree–Fock (SPHF) equations are then solved to obtain the different energies $E_{nl\mu}$ and radial wave functions $R_{nl\mu}(r)$ for the two groups of electrons. The excited-state wave functions are obtained in the field of the SPHF frozen atomic state with a hole in the corresponding $nl\mu$ subshell. The SPRPAE dipole amplitude for the transition from the ground state $|i\rangle$ to the excited state $\langle f|$ is composed of two main terms, the SPHF dipole matrix element and a "correlation" element constructed by summing the interactions of the hole states with all of the discrete and continuum final states. The equation is ($h = m = e = 1$):

$$\langle f|\mathcal{D}|i\rangle = \langle f|d|i\rangle + \sum_{\substack{\eta > F \\ v \leq F}} \left[\frac{\langle \eta|\mathcal{D}|v\rangle\langle vf|\mathcal{U}|\eta i\rangle}{\omega + E_\eta - E_v - i\delta} - \frac{\langle v|\mathcal{D}|\eta\rangle\langle \eta f|\mathcal{U}|vi\rangle}{\omega + E_\eta - E_v - i\delta} \right], \delta \to 0$$

where i, f, v, η are the sets of quantum numbers (n,l,m,μ), $\langle f|d|i\rangle$ is the SPHF dipole matrix element with the d operator in "length" or "velocity" form, $\omega = E_f - E_i$ is the photon energy, and E_k is the SPHF electron energy of the state k. The combined Coulomb matrix element is

$$\langle vf|\mathcal{U}|\eta i\rangle = \langle vf|\mathcal{V}|\eta i\rangle - \langle fv|\mathcal{V}|\eta i\rangle\delta_{\mu_\mu\mu_v}$$

where $\mathcal{V} = 1/r$ and the factor $\delta_{\mu_\mu\mu_v}$ cancels the exchange interaction of electrons with different spin projections. The sum in the first equation above is performed over the discrete and continuous spectrum of η states (F is the Fermi level). If the integration is restricted to just one hole state ($n_v = n_i, l_v = i, \mu_v = \mu_i$), the electron correlations are taken into account within one spin-polarized subshell. The dipole amplitude $\langle f|\mathcal{D}|i\rangle$ obtained from the formula with $v \neq i$ describes intershell interaction effects.

The oscillator strength, partial and total photoionization cross section, and the angular distribution asymmetry parameter are obtained using the usual formulas

$$f_{nl\mu \to n'l\pm l\mu} = \frac{N_{nl\mu}}{(2l+1)\omega} |\langle n'l \pm l\mu|\mathcal{D}|nl\mu\rangle|^2$$

$$\sigma_{nl\mu \to \varepsilon l\pm l\mu}(\omega) = \frac{4\pi^2 \alpha N_{nl\mu}}{(2l+1)\omega} |\langle \varepsilon l \pm l\mu|\mathcal{D}|nl\mu\rangle|^2$$

$$\sigma_{tot}(\omega) = \sum_{n,l,\mu} \sigma_{nl\mu \to \varepsilon l\pm l\mu}(\omega)$$

$$\beta(\omega) = \frac{(l+2)|\mathcal{D}_{l+1}|^2 + (l-1)|\mathcal{D}_{l-1}|^2 + 6\sqrt{l(l+1)}\ \text{Re}\left[\mathcal{D}_{l+1}\mathcal{D}_{l-1}^*\ e^{i(\delta_{l+1} - \delta_{l-1})}\right]}{(2l+1)[|\mathcal{D}_{l+1}|^2 + |\mathcal{D}_{l-1}|^2]}$$

where $\alpha = 1/137$ is the fine-structure constant, $N_{nl\mu}$ is the number of electrons in the $nl\mu$ subshell, $\mathcal{D}_{l\pm1} = \langle \varepsilon l \pm l\mu|\mathcal{D}|nl\mu\rangle$, $\varepsilon = \omega - I_{nl\mu}$, and $\delta_{l\pm1}$ are the photoelectron scattering phase shifts.

TABLE V
Parameters of the $3p \uparrow \to nl \uparrow$ and $3p \downarrow \to nl \downarrow$ Autoionizing Resonances in Ca$^+$ Ions[a]

Transition	Transition energy (eV)				Width Γ (eV)	Oscillator strength	
	ΔE^{SPHF}	ΔE^{RPAE}	ΔE^{exp}	ΔE^R		f_{RPAE}	f_{exp}
$3p \to 4s \downarrow$	29.77	29.74	28.205	27.6	7.6×10^{-4}	0.265	0.13
			28.545				0.067
$3d \uparrow$	30.81	34.13	33.19	33.23	0.062	3.0	2.3
$3d \downarrow$	31.02			33.63; 33.2[b]			
$5s \uparrow$	36.85	36.82			5×10^{-5}	0.052	
$5s \downarrow$	37.30	37.30			2×10^{-5}	0.067	0.153
$4d \uparrow$	37.79	37.37		37.906	3×10^{-4}	0.002	
$4d \downarrow$	37.86	38.57			0.018	0.63	0.597
$6s \downarrow$	39.756	39.73			10^{-6}	0.028	
$6s \uparrow$	39.762	39.78			3.5×10^{-5}	0.034	0.067
$5d \downarrow$	39.99	39.87		37.607	4.9×10^{-5}	6×10^{-4}	
$5d \uparrow$	40.05	40.33			0.007	0.24	0.117
$7s \downarrow$	40.90	40.90			2×10^{-5}	5×10^{-4}	
$7s \uparrow$	40.97	40.97			7×10^{-5}	0.016	
$6d \downarrow$	41.02	40.99			6×10^{-4}	0.0012	0.084
$6d \uparrow$	41.11	41.24			3.6×10^{-3}	0.123	
$7d \uparrow$	41.70	41.79			3.6×10^{-3}	0.115[c]	0.083

[a] From Ivanov and West.[120]
[b] Theoretical energies calculated by Mansfield and Ottley[192] and Hansen.[193]
[c] Γ and f_{RPAE} likely to be an overestimate for this transition: see Ref. 120.

Where this summation is for just one hole state, the electron correlations are obtained for one spin-polarized subshell. This is fundamentally a one-electron calculation, and the prominent one-electron transitions in the Ca$^+$ spectrum were identified by comparison of the measured oscillator strengths with the calculated ones. The result is shown in Table V for transitions from the $3p$ shell, as are the calculated and experimental transition energies. There is a discrepancy of between 1 and 1.5 eV for the transition energies, to be expected from the one-electron calculation in this case where the

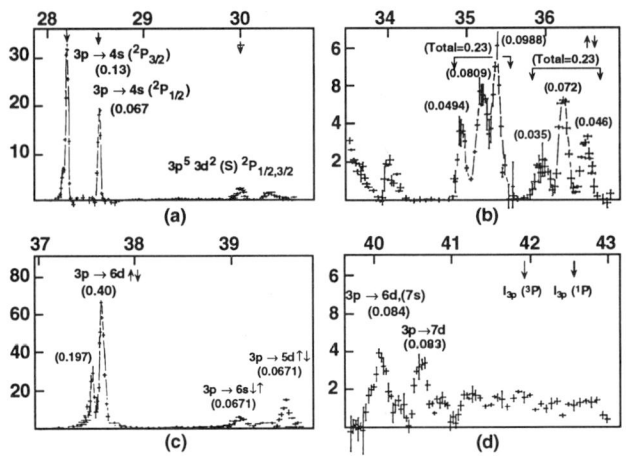

FIGURE 39. Identification of transitions in Ca$^+$, from Ivanov and West.[120]

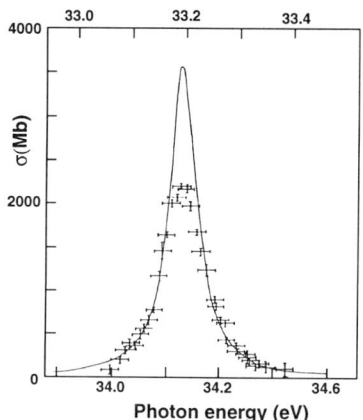

FIGURE 40. Theoretical cross section for the $3p \rightarrow 3d$ resonance in Ca$^+$, from Ivanov and West.[120]

core states are frozen. The RPAE oscillator strength values are in better agreement with experiment; in Table V, where spin-up (↑) and spin-down (↓) values are shown, these should be added to compare with the experimental value. Also, the theoretical values do not resolve spin orbit split components so the two values shown for the $3p \rightarrow 4s$ transition should be added together for comparison with theory. The identifications of the structure are also shown in Figure 39; the downward-pointing arrows in Figure 39(a) are derived from the series limits calculated by Mansfield and Newsom[169] from their absorption data on neutral Ca, and assuming that their identification is correct give reliable assignments for these lines in Ca$^+$. The unlabeled structures in Figure 39 have been tentatively assigned to two electron excitations[120]: the peak at 34 eV to the $3p^53d4p$ state and the group of peaks around 35.2 eV to the $3p^54p^2$ state. Figure 40 shows the cross section for the giant $3p \rightarrow 3d$ resonance, where the theoretical values have been convoluted with the experimental bandpass. The lower energy scale is the one for the theoretical values, and the one at the top of the figure is the experimental one; again, a shift of ~1 eV is apparent as expected. The agreement between the SPRPAE and experiment here is fair, and not as good as that obtained by the R-matrix calculation of Miecznik et al.[167] for this resonance.

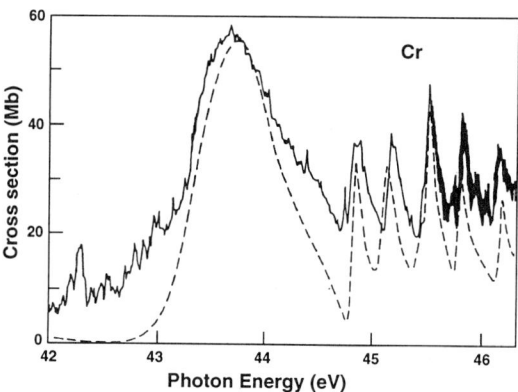

FIGURE 41. Comparison of the SPRPAE calculations of Dolmatov[170] (dashed line) with the ^3p absorption spectrum of Cr measured by Cooper et al. [37] (full jagged line).

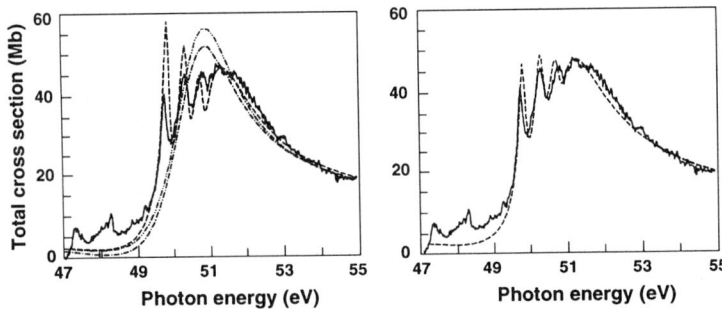

FIGURE 42. Theoretical curves for the $3p \to 3d$ giant resonance in Mn^+, from Dolmatov.[121] (Left) Total $3d + 4s$ cross section for Mn^+ (dashed line), Mn^{+*} (single chain dotted curve), and neutral Mn (double chain dotted curve). The full (jagged) curve is taken from the experimental data of Cooper et al.[37] and Costello et al.[174] (Right) The calculated "mixed" $3p$ absorption cross section with 10% Mn, 20% Mn^+, and 70% Mn^{+*} compared with the same experimental data

Two very interesting further examples of the application of this method to open-shell atoms have recently been published by Dolmatov.[121,170–173] The experimental spectra of Mn, Mn^+, and Cr, obtained from RLDI and DLPP experiments, show some very interesting structure and important differences, as seen earlier (Figure 23). Taking first the data shown in Figure 23, Cooper et al.[37] concluded, from calculations using Cowan's[140] code, that strong mixing could be expected between the $3d^54snd$ ($n > 3$) levels in Cr and the $3d^64s$ ^7P giant resonance, due to their close proximity. In contrast, in the isoelectronic Mn^+, this mixing is not expected to be strong because the two configurations are well separated in energy. The prominent $3p \to nd$ ($n > 3$) Rydberg series seen in Cr, but not in Mn^+, they suggest arises because the Rydberg series borrows intensity from the giant resonance. The SPRPAE calculations[121] give a rather different interpretation; the $3p \to 3d$ giant resonance actually absorbs oscillator strength from the Rydberg series in both Cr and Mn^+, so the explanation does not lie there. The reason that this series is seen in Cr is due to an "anticollapse" phenomenon,[173] because there is a critical change in relaxation effects for the nd orbitals in going from Cr ($Z = 24$) to Mn^+ ($Z = 25$). Now that the nd orbitals in Cr are no longer collapsed, their overlap with the $3p$ orbital is substantially greater than in Mn^+, and this accounts for the prominent Rydberg structure seen. The SPRPAE results are shown compared to the experimental data in Figure 41, where good agreement is evident.

In a further application of the SPRPAE method to the absorption spectra of ground-state Mn^+, Dolmatov[121] has shown that strong mixing is expected between the $3p \to 4s$ resonance and the $3p \to 3d$ giant resonance. His results are shown in the left panel of Figure 42, where the peaks in the Mn^+ spectrum around 50 eV are due to the $3p \to 4s$ resonance; without including this in the calculations, a broad peak of lower intensity is obtained in this region. Agreement with the experimental data,[37,174] which have been normalized to the theory at 55 eV, is quite good, apart from the relative intensities of the $3p \to 4s$ peaks. Much better agreement is obtained, as seen in the right panel of Figure 42 when it is assumed that the experimental vapor contained 10% Mn atoms and 20% Mn^{+*} excited ions, for which the $4s$ state is occupied and therefore there is no $3p \to 4s$ resonance. This example highlights the difficulties facing the experimental determination of cross sections for excited atoms and ions; in this case the cross sections could be clearly separated, because of the absence of the $3p \to 4s$ resonance in the unwanted species, but in those cases where structure from contaminants occurs in the same spectral region, separation of the true cross section may be impossible. The merged ion beam experiments described above do overcome this problem, but unfortunately they are rather limited in the range of atoms which can be studied and are restricted to low vapor densities, particularly for multiply charged ions.

5. FUTURE WORK

5.1. Extension of the Present Experiments

It is clear from the above that much remains to be done, both theoretically and experimentally, for singly charged atomic ions. The case outlined above for Ca shows that at present no one theoretical method is sufficient to give a satisfactory description of both cross sections (or oscillator strengths) and transition energies. Calcium is the largest atom which can be tackled with nonrelativistic theory, and even in this case *LS* coupling is not universally valid. Work is currency under way to extend the RPAE method so that two-electron excitations and spin-orbit coupling can be included, but for the identification of the structure, particularly for the heavier ions Sr^+ and Ba^+ for which some data are available, other methods where configuration interactions are included explicitly and more exactly may be required. Partial cross sections for the $4s$, $3p$, and $3s$ subshells of Ca^+, as well as angular distribution parameters for the $3p$ shell have been calculated,[119,120] but as yet there are no measurements of these parameters and early experiments in photoelectron spectroscopy of atomic ions[45] show just how difficult this will be. The priority will be for measurements to be made over a wide energy range in order to test and hopefully raise confidence in the theoretical data provided by calculations such as those, for example, in the stellar opacity project. Also, singly charged ions with electronic configurations such as $nd^{10}(n + 1)s^2(n + 1)p$ or $nd^{10}(n + 1)s^2(n + 1)p^2$ are of special interest for investigations by electron spectrometry in the near future. Several projects, continuing the former experiments by photoion spectrometry (at ASTRID),[175] or involving new designs to produce singly charged ions[176] and to study photoionization of heliumlike light ions[177] are in preparation.

To go further, i.e., to be able to study multiply charged ions using synchrotron radiation, there will need to be significant improvements in the performance of both photon and ion sources. However, the future development of photon–ion experiments will deal mainly with low- and medium-charge state ions, and not with highly charged ions, at least for the few next years. First, photoionization cross sections decrease dramatically with increasing charge state and photon energy. Photoionization cross sections are in general less than 0.1 Mb for neutral atoms in the hard x-ray region and will be significantly less for multiply charged ions, even though giant resonances are predicted to exist for heavy ions.[178] Thus, there will be a technical limitation in cross section measurements, due to signal-to-noise ratio, for photoion- and, even more, for photoelectron spectrometry. Photoabsorption experiments are the best candidates for the highest photon energies, provided it becomes feasible to probe a plasma of highly charged ions with synchrotron radiation, as has been recently proposed at LURE.[179] Fluorescence experiments may be the only differential experiments that can be performed in the hard x-ray region, taking advantage of the extremely low background which is a feature of this technique. Second, following the variation of correlation and relativistic effects along isonuclear or isoelectronic sequences requires a sufficient number of electrons to be left in the ion. As a result, multiply charged ions with charge states ranging from typically 2 to 5 will offer the most interesting cases to be studied. Third, many-electron multiply charged ions with high Z present the largest variety of interesting phenomena in the soft x-ray range.

5.2. New Storage Rings and New ECR/EBIT/EBIS Ion Sources

The design of the new third generation of electron (positron) storage rings presently in operation (Super ACO: 800 MeV, in Orsay; the Advanced Light Source: ALS, 1.5 GeV, in Berkeley; the European Synchrotron Radiation Facility: ESRF, 6 GeV, in Grenoble; ASTRID: 550 MeV in Aarhus) or under construction (ELETTRA: 2 GeV, in Trieste; the Advanced Photon Source: APS, 7 GeV, in Argonne; and the more recent Japanese project SPRING-8: 8 GeV, in Himeji) is based in the main on the production of synchrotron radiation from insertion devices (wigglers and undulators). After monochromatization, such undulators can provide photon fluxes in the 10^{13} photons/s range in a typical bandwidth lower than 0.1% of the photon energy, up to several hundredths of an electron volt. In the x-ray region, the photon beams emitted by multipole wigglers are also suitable candidates for photon–ion experiments in the high photon energy range.

From the ion point of view, two experimental designs can be adopted. Either photons and ions are extracted from their respective sources and are selectively prepared in two different beams, as in the experiments described in this review, or the photon beam is focused within the ion source itself. In the first solution, which one can call an *external beam* experiment, the two beams are quasi-continuously extracted and are crossed in the source volume of an x-ray or electron spectrometer, or they are merged over a long distance for total photoionization measurements by ion spectrometry. In the second solution, called an *internal beam* experiment, the photon beam is focused into a volume where the multiply charged ions are stored and the high repetition rate of the storage ring is matched, using adequate timing techniques, to the pulsed production of the ions to observe mostly fluorescence radiation emitted as the result of the interaction. In this geometry, the analysis of charged particles produced in the interaction will be most difficult. In this second geometry, the ions can be stored either in a storage ring or in an EBIS-type ion source.

For use in the external beam mode of operation, the electron cyclotron resonance ion source (ECRIS) is certainly the most reliable ion source, being able to deliver intense beams of ions in a large variety of charge states. In the ECR source, the source plasma is located in a magnetic "trap" allowing a long exposure time to electron bombardment, resulting in efficient ionization. The electrons are accelerated by a 6- to 16-GHz oscillating electromagnetic wave, at the frequency of the electron cyclotron resonance. The ionic species can be extracted easily and selected through magnetic optics to form a monoionic beam of low velocity. It can then be focused into an interaction zone or shaped to form a parallel beam being able to merge with a photon beam over a large distance. One of the first models[180] of this source, MINIMAFIOS, was used for several years in collision experiments. An improved model,[181] called CAPRICE, has been constructed and has been in operation for several years in Grenoble. Figure 43 shows a schematic view of this new source. An original coaxial microwave design allows a very compact two-stage source in an entirely removable vacuum chamber. With CAPRICE, electric ion currents of about 100 µA can be obtained for a large number of gaseous ions with charge states up to 8+. Metallic ions can also be produced with lower currents up to relatively high charge states. Several copies of this ion source are under construction or are in operation in many laboratories studying or using multiply charged ions. Combining the ECR ion source with the beam from an undulator fitted to a third-generation synchrotron radiation source will be the state of the art for the next generation of experiments. Such a project has already been suggested for the Super ACO storage ring. The schematic design of the experimental setup is shown in Figure 44. In order to make the best use of the ion source, the ICARIOS project[182] includes two ion beam lines: one for studying

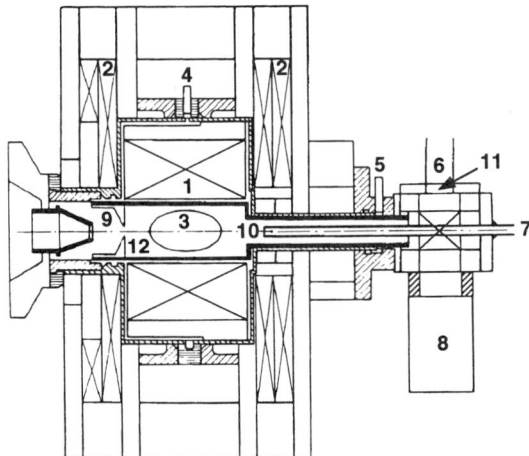

FIGURE 43. The CAPRICE ion source. 1, magnets; 2, solenoids; 3, ECR surface; 4, cooling water inlet; 5, cooling water outlet; 6, microwave inlet; 7, gas inlet; 8, turbomolecular pump; 9, extracting electrodes; 10, gas pipe; 11, microwave window; 12, removable vacuum chamber; from Bourg et al.[181]

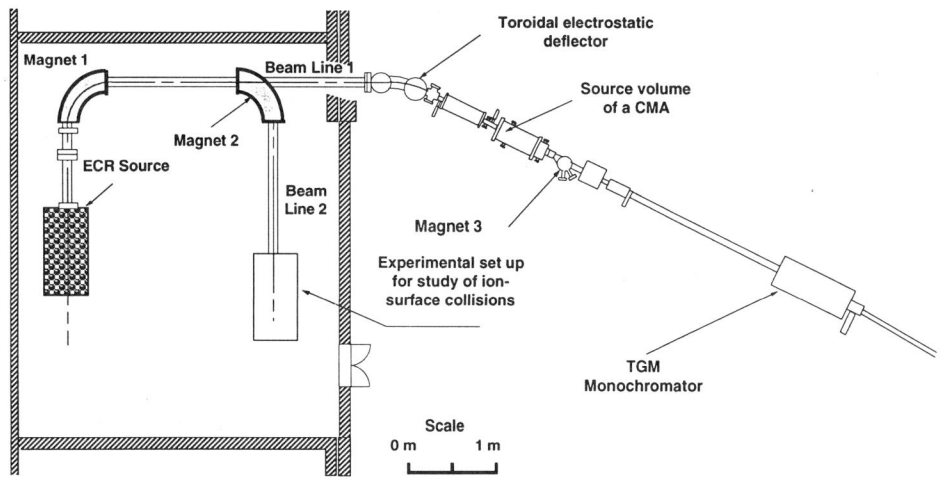

FIGURE 44. Schematic design of the ICARIOS experimental arrangement, combining an ECR ion source with the photon beam from an undulator, from Barat et al.[182]

the interaction of low-charge-state ions with photons in the source volume of an electron spectrometer or along a straight section of a photoion spectrometer, the other one for collision studies involving highly charged ions. Similar projects will undoubtedly be proposed in the near future at other advanced synchrotron radiation centers, such as the Advanced Light Source in Berkeley or the ESRF in Grenoble.

Ion sources for the internal mode of operation are of two types: the ions can be stored either in an EBIT-type source or in a heavy-ion storage ring. The EBIT (electron beam ion trap) apparatus[183,184] traps highly charged ions in the space charge of an electron beam, which also serves to ionize them. The ion trap consists of cylindrical drift tubes which contain the trapped ions. The electron beam, which follows the central magnetic field line of superconducting Helmholtz coils, is injected along the axis of the drift tubes. At the peak magnetic field of 3 tesla, the electron beam diameter is 70 μm. Ions

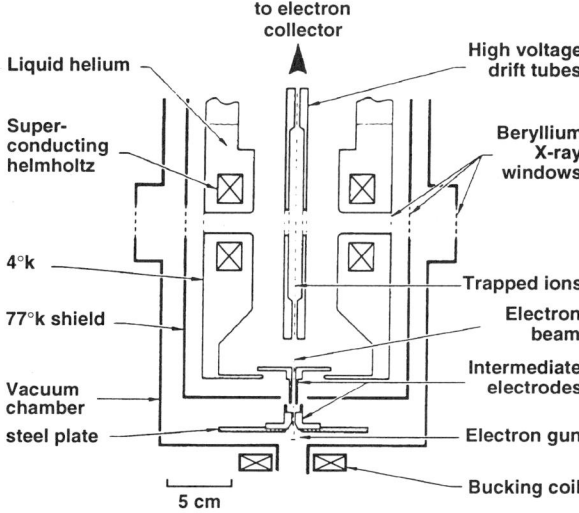

FIGURE 45. Scheme for an electron beam ion trap (EBIT) arrangement, from Marrs et al.[184]

in low charge states are loaded into the trap by injecting them from an auxiliary ion source which is fired periodically to refill the trap. Highly charged ions are then obtained by successive ionization in the electron beam. X rays can be observed through several different x-ray ports. Figure 45 shows an example of this source. EBIT has been used to study dielectronic recombination of highly charged ions.[185,186] Recently, a scheme for an EBIS (electron beam ion source) specially designed for synchrotron radiation experiments on multiply charged ions has been proposed[187] and constructed.[188] Its basic design employs the idea of cooling the solenoid coil with liquid nitrogen.[189,190] It is supposed to handle a maximum electron current of 100 mA with a maximum energy of 5 keV. While it seems feasible to use it in the internal beam mode for fluorescence experiments as proposed using photons from the 2.5-GeV Photon Factory in Tsukuba and focused into the source with a maximum interaction length of 30 cm, extracting the photoionized multiply charged ions and analyzing them by the time-of-flight technique appears to be questionable.

An alternative approach is to use an ion storage ring. The combination of a heavy-ion storage ring with a wiggler insertion device of the National Synchrotron Light Source at Brookhaven National Laboratory has been proposed by Jones et al.[191] Basically, ions produced with a tandem Van de Graaf accelerator would be injected into a cooled heavy-ion storage ring (CHISR). Storage rings for heavy ions are presently built or are in operation at several places, such as CERN, Heidelberg, Darmstadt, and Stockholm. A stored ion current of 10^{16}–10^{17} particles/s seems to be obtainable. Although the use of such a facility appears feasible as a means of measuring cross sections of multiply charged ions in the x-ray region, and the fact that it uses stored beams could overcome problems associated with metastable species, the enormous expense and scale of the technical problems involved may in the end rule it out.

The interest in carrying out such studies for highly stripped ions in the x-ray photon energy range, and their practicability need to be investigated further. Experience needs to be gained on low-charge-state ions with lower-energy storage rings in the soft x-ray range. The density of charged ions attainable in the source volume of a spectrometer diminishes rapidly with increasing values of the total charge. Photoelectron spectrometry experiments are the ones requiring the highest luminosity. Fluorescence and ion spectrometry studies require fewer photons and such experiments could go to higher charge states and higher photon energies. It is not completely clear, in the case where only fluorescence experiments are feasible, whether an ECRIS or an EBIS would be the most suitable ion source.

Finally, studies on atomic ions appear to have a more promising future than experiments on excited atoms. The lack of continuously tunable and sufficiently powerful CW dye lasers over extended photon energy ranges is a strong limitation on their development. Lasers mode-locked to the frequency of synchrotron radiation in storage rings are under active study, but the power of such lasers is still too low for experiments involving inner-shell excitations. The use of optical lasers in cascade to create highly excited atomic states is probably a better way to study outer-shell and inner-shell photoexcitation/photoionization processes in such atoms.

ACKNOWLEDGMENTS. The authors thank the editors for their patience in waiting for this manuscript. They are also very grateful to their collaborators, especially to Jean-Marc Bizau, Denis Cubaynes, and Jacques Obert, for their help in setting up and performing the experiments described in this review. They express their gratitude to Maurice Berland and Denise Huissier for their great help in preparing the figures presented in this chapter.

REFERENCES

1. V. Schumann, *Akad. Wiss. Wien* **102**, 625 (1893).
2. H. A. Rowland, *Philos. Mag.* **16**, 469 (1882).
3. H. Beutler, *Z. Phys.* **86**, 495 (1933).
4. B. Edlén, *Z. Astrophys.* **7**, 378 (1933).
5. H. Beutler, *Z. Phys.* **93**, 177 (1935).
6. U. Fano, *Phys. Rev.* **124**, 1866 (1961).
7. J. A. R. Samson, *Techniques of Vacuum Ultraviolet Spectroscopy* (Wiley, New York, 1967).

8. K. Codling and R. P. Madden, *Phys. Rev. Lett.* **10**, 516 (1963).
9. J. A. R. Samson, *Adv. At. Mol. Phys.* **2**, 177 (1966).
10. J. B. West and G. V. Marr, *Proc. Ry. Soc. London Ser. A* **349**, 397 (1976).
11. P. G. Burke and K. T. Taylor, *J. Phys. B.* **8**, 2620 (1975).
12. L. Torop, J. Morton, and J. B. West, *J. Phys. B* **9**, 2035 (1976).
13. M. Y. Adam, F. Wuilleumier, N. Sandner, S. Krummacher, V. Schmidt, and W. Mehlhorn, *Jpn. J. Appl. Phys.* **17**, 170 (1978).
14. G. Wendin, *J. Phys. B* **5**, 110 (1972) and references therein.
15. Proceedings of the VIth International Conference on Vacuum Ultraviolet Radiation Physics, *Appl. Opt.* **19**, 4042–91 (1980).
16. F. J. Wuilleumier, *At. Phys.* **7**, 481 (1981).
17. J.-M. Esteva and G. Mehlman-Balloffet, *Astrophys. J.* **193**, 747 (1974).
18. T. B. Lucatorto and T. J. McIlrath, *Phys. Rev. Lett.* **37**, 428 (1976).
19. G. Balloffet, J. Romand, and B. Vodar, *Compt. Rend. Ac. Sci.* **252**, 4139 (1961).
20. G. Brown, K. Halbach, J. Harris, and H. Winwick, *Nucl. Instrum. Methods* **208**, 65 (1983).
21. A. Carillon, P. Jaeglé, and P. Dhez, *Phys. Rev. Lett.* **25**, 140 (1970).
22. P. K. Carroll and E. T. Kennedy, *Phys. Rev. Lett.* **38**, 1068 (1977).
23. P. K. Carroll, E. T. Kennedy, and G. O'Sullivan, *Appl. Opt.* **19**, 1454 (1980).
24. Rydberg Centennial Conference on Atomic Spectroscopy, Lund University, edited by B. Edlén, *Årrskr. Avd.* 2 **50**, 1 (1955).
25. G. Mehlman-Balloffet and J. M. Esteva, *Astrophys. J.* **157**, 945 (1969).
26. J. M. Esteva, G. Mehlman-Balloffet, and J. Romand, *J. Quant. Spectrosc. Radiat. Transfer* **12**, 1291 (1972).
27. P. Jaeglé, G. Jamelot, A. Carillon, A. Sureau, and P. Dhez, *Phys. Rev. Lett.* **33**, 1070 (1974).
28. A. W. Ehler and G. L. Weissler, *Appl. Phys. Lett.* **8**, 89 (1966).
29. E. Jannitti, P. Nicolosi, and G. Tondello, *Physica* **124C**, 139 (1984).
30. E. Jannitti, P. Nicolosi, and G. Tondello, *Opt. Commun.* **50**, 225 (1984).
31. P. K. Carroll and J. T. Costello, *Phys. Rev. Lett.* **57**, 1581 (1986).
32. J. Costello, W. G. Lynam, and P. K. Carroll, *J. Phys. (Paris)* **49**, C1-243 (1988).
33. J. T. Costello, J.-P. Mosnier, E. T. Kennedy, P. K. Carroll, and G. O'Sullivan, *Phys. Scr.* **T34**, 77 (1991).
34. T. J. McIlrath and T. B. Lucatorto, *Phys. Rev. Lett.* **38**, 1390 (1977).
35. C. L. Cromer, J. M. Bridges, J. R. Roberts, and T. B. Lucatorto, *Appl. Opt.* **24**, 2996 (1985).
36. B. F. Sonntag, C. L. Cromer, J. M. Bridges, T. J. McIlrath, and T. B. Lucatorto, *AIP Conf. Proc. Ser.* **147** (American Institute of Physics, New York, 1986), p. 142.
37. J. W. Cooper, C. W. Clark, C. L. Cromer, T. B. Lucatorto, B. F. Sonntag, E. T. Kennedy, and J. T. Costello, *Phys. Rev. A* **39**, 6074 (1989).
38. I. C. Lyon, B. Peart, J. B. West, and K. Dolder, *J. Phys. B* **19**, 4137 (1986).
39. B. Peart, J. G. Stevenson, and K. Dolder, *J. Phys. B* **6**, 146 (1973).
40. I. C. Lyon, B. Peart, K. Dolder, and J. B. West, *J. Phys. B* **20**, 1471 (1987).
41. I. C. Lyon, B. Peart, and K. Dolder, *J. Phys. B* **20**, 1925 (1987).
42. B. Peart and I. C. Lyon, *J. Phys. E* **17**, 920 (1984).
43. B. Peart and I. C. Lyon, *J. Phys. B* **20**, L673 (1987).
44. B. Peart, I. C. Lyon, and K. Dolder, *J. Phys. B.* **20**, 5403 (1987).
45. J. M. Bizau, D. Cubaynes, M. Richter, F. J. Wuilleumier, J. Obert, J. C. Putaux, T. J. Morgan, E. Källne, S. Sorensen, and A. Damany, *Phys. Rev. Lett.* **67**, 576 (1991).
46. J. M. Bizau, D. Cubaynes, M. Richter, F. Wuilleumier, J. Obert, and J. C. Putaux, *Rev. Sci. Instrum.* **63**, 1389 (1992).
47. J. C. Putaux, J. Obert, G. Boissier, and P. Paris, *Nucl. Instrum. Methods Sect. B* **26**, 213 (1987).
48. D. A. Church, S. D. Kravis, I. A. Sellin, C. S. O, J. C. Levin, R. T. Short, M. Meron, B. M. Johnson, and K. W. Jones, *Phys. Rev. A* **36**, 2487 (1987).
49. S. D. Kravis, D. A. Church, B. M. Johnson, M. Meron, K. W. Jones, J. Levin, I. A. Sellin, Y. Azuma, N. Berrah-Mansour, H. G. Berry, and M. Druetta, *Phys. Rev. A* **45**, 6379 (1992).
50. D. J. Bradley, P. Ewart, J. Nicholas, J. R. D. Shaw, and D. Thompson, *Phys. Rev. Lett.* **31**, 263 (1973).
51. R. F. Stebbings, F. B. Dunning, F. K. Tittel, and R. D. Rundel, *Phys. Rev. Lett.* **30**, 815 (1973).
52. J. L. Carlsten, T. J. McIlrath, and W. H. Parkinson, *J. Phys. B* **7**, L244 (1974).
53. K. J. Nygaard, *IEEE J. Quantum Electron.* **9**, 1020 (1973).
54. D. J. Bradley, C. H. Dugan, P. Ewart, and A. F. Purdie, *Phys. Rev. A* **13**, 1416 (1976).
55. O. C. Mullins, R.-l. Chien, J. E. Hunter, D. K. Jordan, and R. S. Berry, *Phys. Rev. A* **31**, 3059 (1985).

56. F. J. WUILLEUMIER, D. L. EDERER, AND J. L. PICQUE, *Adv. At. Mol. Phys.* **23**, 197 (1988).
57. J.-M. BIZAU, J.-L. LE GOUËT, D. L. EDERER, P. KOCH, F. J. WUILLEUMIER, J.-L. PICQUÉ, AND P. DHEZ, in *Abstracts of Contributed Papers of the XII International Conference on the Physics of Electronic and Atomic Collisions*, edited by S. Datz, Post dead line papers, 1981, p. 1.
58. J.-M. BIZAU, F. J. WUILLEUMIER, P. DHEZ, D. L. EDERER, J. L. LE GOUËT, J. L. PICQUÉ, AND P. KOCH, in *Laser Techniques for Extreme Ultraviolet Spectroscopy*, edited by T. J. McIlrath and R. J. Freeman, AIP Conf. Proc. Ser. **90** (American Institute of Physics, New York, 1982), p. 331.
59. J.-L. LE GOUËT, J.-L. PICQUÉ, F. J. WUILLEUMIER, J.-M. BIZAU, P. DHEZ, P. KOCH, AND D. L. EDERER, *Phys. Rev. Lett.* **48**, 600 (1982).
60. F. J. WUILLEUMIER, in *X-Ray and Atomic Inner Shell Physics*, edited by B. Crasemann, AIP Conf. Proc. Ser. **94** (American Institute of Physics, New York, 1982), p. 615.
61. F. J. WUILLEUMIER, *J. Phys. (Paris)* **43**, C2-347 (1982).
62. F. J. WUILLEUMIER, in *Laser Techniques in the Extreme Ultraviolet*, edited by S. E. Harris and T. B. Lucatorto, AIP Conf. Proc. Ser. **119** (American Institute of Physic, New York, 1984), p. 220.
63. J.-M. BIZAU, F. J. WUILLEUMIER, D. L. EDERER, J.-C. KELLER, J.-L. LE GOUËT, J.-L. PICQUÉ, B. CARRÉ, AND P. M. KOCH, *Phys. Rev. Lett.* **55**, 1281 (1985).
64. J. M. BIZAU, B. CARRE, P. DHEZ, D. L. EDERER, P. GÉRARD, J. C. KELLER, P. KOCH, J. L. LE GOUËT, J. L. PICQUÉ, G. WENDIN, AND F. J. WUILLEUMIER, in *Abstracts of Contributed Papers of the XIII International Conference on the Physics of Electronic and Atomic Collisions*, edited by J. Eichler, W. Fritsch, I. V. Hertel, N. Stolterfoht, and U. Wille (North-Holland, Amsterdam, 1983), p. 27.
65. A. NUNNEMANN, T. PRESCHER, M. RICHTER, M. SCHMIDT, B. SONNTAG, S. BAIER, W. FIEDLER, B. R. MÜLLER, M. SCHULZE, AND P. ZIMMERMANN, *J. Phys. B* **18**, L337 (1985).
66. J.-M. BIZAU, D. CUBAYNES, P. GÉRARD, F. J. WUILLEUMIER, J.-L. PICQUÉ, D. L. EDERER, B. CARRÉ, AND G. WENDIN, *Phys. Rev. Lett.* **57**, 306 (1986).
67. M. FERRAY, F. GOUNAND, P. D'OLIVEIRA, P. R. FOURNIER, D. CUBAYNES, J.-M. BIZAU, T. J. MORGAN, AND F. J. WUILLEUMIER, *Phys. Rev. Lett.* **59**, 2040 (1987).
68. B. CARRÉ, P. D'OLIVEIRA, M. FERRAY, F. GOUNAND, D. CUBAYNES, J.-M. BIZAU, AND F. J. WUILLEUMIER, *Z. Phys. D* **15**, 177 (1990).
69. J.-M. BIZAU, D. CUBAYNES, B. ROUVELLOU, AND F. J. WUILLEUMIER, in *Abstracts of the 4th European Conference on Atomic and Molecular Physics*, 1992, Part I, p. 173.
70. M. MEYER, B. MÜLLER, A. NUNNEMANN, T. PRESCHER, E. VON RAVEN, M. RICHTER, M. SCHMIDT, AND P. ZIMMERMANN, *Phys. Rev. Lett.* **59**, 2963 (1987).
71. B. SONNTAG, in *Today and Tomorrow in Photoionisation*, Proceedings of the UK/USSR Seminar, Leningrad 1990, edited by M. Y. Amusia and J. B. West, Daresbury Laboratory Report Series DL/SCI/R29, p. 83.
72. M. PAHLER, C. LORENZ, E. VON RAVEN, J. RÜDER, B. SONNTAG, S. BAIER, B. R. MÜLLER, M. SCHULZE, H. STAIGER, P. ZIMMERMANN, AND N. M. KABACHNIK, *Phys. Rev. Lett.* **68**, 2285 (1992).
73. U. HEINZMANN, *J. Phys. B* **13**, 4367 (1980).
74. A. HAUSMANN, B. KÄMMERLING, H. KOSSMAN, AND V. SCHMIDT, *Phys. Rev. Lett.* **61**, 2669 (1988).
75. T. A. CARLSON, D. R. MULLINS, C. E. BEALL, B. W. YATES, J. W. TAYLOR, D. W. LINDLE, B. P. PULLEN, AND F. A. GRIMM, *Phys. Rev. Lett.* **60**, 1382 (1988).
76. J. JIMÉNEZ-MIER, C. D. CALDWELL, AND D. L. EDERER, *Phys. Rev. Lett.* **57**, 2260 (1986).
77. H. HAMDY, H.-J. BEYER, J. B. WEST, AND H. KLEINPOPPEN, *J. Phys. B* **24**, 4957 (1991).
78. H.-J. BEYER, H. HAMDY, K. UEDA, K. J. ROSS, N. KABACHNIK, J. B. WEST, AND H. KLEINPOPPEN, *J. Phys. B* **28**, L47 (1995).
79. V. SCHMIDT, *Rep. Prog. Phys.* **55**, 1483 (1992).
80. D. G. THOMPSON, A. HIBBERT, AND N. CHANDRA, *J. Phys. B* **7**, 1298 (1974).
81. K. CODLING, J. R. HAMLEY, AND J. B. WEST, *J. Phys. B* **10**, 2797 (1977).
82. S. KRUMMACHER, V. SCHMIDT, J. M. BIZAU, D. L. EDERER, P. DHEZ, AND F. WUILLEUMIER, *J. Phys. B* **15**, 4363 (1982).
83. J.-M. BIZAU, D. CUBAYNES, P. GÉRARD, F. J. WUILLEUMIER, AND J.-L. PICQUÉ, *J. Phys. (Paris)* **48**, C9-491 (1987).
84. D. SALZMANN AND R. PRATT, *Phys. Rev. A* **30**, 2767 (1984).
85. J. B. WEST, P. R. WOODRUFF, K. CODLING, AND R. G. HOULGATE, *J. Phys. B* **9**, 407 (1976).
86. D. CUBAYNES, J.-M. BIZAU, F. J. WUILLEUMIER, B. CARRÉ, AND F. GOUNAND, *Phys. Rev. Lett.* **63**, 2460 (1989).
87. D. CUBAYNES, J.-M. BIZAU, B. CARRÉ, AND F. J. WUILLEUMIER, in *The Physics of Electronic and Atomic Collisions*, AIP Conf. Proc. Ser. **205**, edited by A. Dalgarno, R. S. Freund, P. M. Koch, M. S. Lubell, and T. B. Lucatorto (American Institute of Physics, New York, 1990), p. 214.

88. F. J. WUILLEUMIER, D. CUBAYNES, AND J. M. BIZAU, in *Atomic and Molecular Physics*, edited by C. Cisneros, I. Alvarez, and T. L. Morgan (World Scientific, Singapore, 1991), p. 474.
89. L. JOURNEL, B. ROUVELLOU, D. CUBAYNES, J.-M. BIZAU, AND F. J. WUILLEUMIER, in *Abstracts of Contributed Papers of the XVIII International Conference on the Physics of Electronic and Atomic Collisions*, edited by T. Andersen, B. Fastrup, F. Folkmann, and H. Knudsen (Aarhus University, Denmark, 1993), Vol. I, p. 22.
90. T. N. CHANG AND Y. S. KIM, in *X-Ray and Atomic Inner-Shell Physics*, AIP Conf. Proc. Ser. **94**, edited by B. Crasemann (American Institute of Physics, New York, 1982), p. 633.
91. T. N. CHANG AND Y. S. KIM, *J. Phys. B* **15**, L835 (1982).
92. M. RICHTER, J.-M. BIZAU, D. CUBAYNES, T. MENZEL, F. J. WUILLEUMIER, AND B. CARRÉ, *Europhys. Lett.* **12**, 35 (1990).
93. J.-M. BIZAU, D. CUBAYNES, AND F. J. WUILLEUMIER, in *Vacuum Ultraviolet Radiation Physics*, edited by J. Wuilleumier, Y. Petroff, and I. Nenner (World Scientific, Singapore, 1993), p. 162.
94. F. J. WUILLEUMIER, L. JOURNEL, D. CUBAYNES, J.-M. BIZAU, B. ROUVELLOU, S. AL MOUSSALAMI, L. VOKY, AND Z. LIU, in *Proceedings of the Fourth US-Mexico Symposium on Atomic and Molecular Physics*, edited by I. Alvarez, C. Cisneros, and T. J. Morgan (World Scientific, Singapore, 1995), p. 385.
95. F. J. WUILLEUMIER, in *Review of Fundamental Processes and Applications of Atoms and Ions*, edited by C. D. Lin (World Scientific, Singapore, 1993), p. 283.
96. L. JOURNEL, B. ROUVELLOU, J.-M. BIZAU, F. J. WUILLEUMIER, AND M. Y. AMUSIA, Post-dead line abstract presented at the international conference *Inner-Shell Ionization Processes and X-Ray Phenomena*, X-93, Debrecen, Hungary, 1993.
97. B. SONNTAG AND P. ZIMMERMANN, *Rep. Prog. Phys.* **55**, 911 (1992).
98. C. KERLING, N. BÖWERING, AND U. HEINZMANN, *J. Phys. B* **23**, L629 (1990).
99. H. KLAR AND H. KLEINPOPPEN, *J. Phys. B* **15**, 933 (1982).
100. C. E. MOORE, *Atomic Energy Levels*, Vols. I–III, NSRDS-NBS (1971).
101. T. B. LUCATORTO, T. J. MCILRATH, J. SUGAR, AND S. M. YOUNGER, *Phys. Rev. Lett.* **47**, 1124 (1981).
102. W. L. WIESE, in *Progress in Atomic Spectroscopy*, Part B, edited by W. Hanle and H. Kleinpoppen (Plenum Press, New York, 1979).
103. R. F. REILMAN AND S. T. MANSON, *Astrophys. J. Suppl. Ser.* **40**, 815 (1979).
104. G. W. F. DRAKE AND A. DALGARNO, *Proc. R. Soc. London Ser. A* **320**, 549 (1971).
105. Y. M. C. SHAN AND A. L. STEWART, *Proc. Phys. Soc.* **90**, 619 (1967).
106. R. P. MADDEN AND K. CODLING, *Astrophys. J.* **141**, 364 (1965).
107. E. JANNITTI, P. NICOLOSI, AND G. TONDELLO, *Phys. Scr.* **41**, 458 (1990).
108. I. C. LYON, B. PEART, J. B. WEST, A. E. KINGSTON, AND K. DOLDER, *J. Phys. B* **17**, L345 (1984).
109. J.-P. CONNERADE, M. W. D. MANSFIELD, G. H. NEWSOM, D. H. TRACY, M. A. BAIG, AND K. THIMM, *Philos. Trans. R. Soc. London Ser. A* **290**, 30 (1979).
110. R. A. ROIG, *J. Opt. Soc. Am.* **66**, 1400 (1976).
111. M. W. D. MANSFIELD AND G. H. NEWSOM, *Proc. R. Soc. London Ser. A* **377**, 431 (1981).
112. M. D. WHITE, D. RASSI, AND K. J. ROSS, *J. Phys. B* **12**, 315 (1979).
113. J. P. CONNERADE, M. A. BAIG, AND M. SWEENEY, *J. Phys. B* **23**, 713 (1990).
114. G. V. MARR AND J. M. AUSTIN, *J. Phys. B* **2**, 107 (1969).
115. F. H. MIES, *Phys. Rev.* **175**, 164 (1968).
116. T. KOIZUMI, Y. ITOH, M. KIMURA, T. M. KOJIMA, S. KRAVIS, M. OURA, M. SANO, T. SEKIOKA, AND Y. AWAYA, *Atomic Collision Research in Japan—Progress Report* **19**, 98 (1993).
117. F. J. WUILLEUMIER, J.-M. BIZAU, B. ROUVELLOU, D. CUBAYNES, AND L. JOURNEL, *Nucl. Instrum. Methods*, **87**, 190 (1994).
118. G. NASREEN, P. C. DESHMUKH, AND S. T. MANSON, *J. Phys. B* **21**, L281 (1988).
119. V. K. IVANOV, in *Vacuum Ultraviolet Radiation Physics*, Proceedings of the 10th VUV Conference, edited by F. J. Wuilleumier, Y. Petroff, and I. Nenner (World Scientific, Singapore, 1993), p. 178.
120. V. K. IVANOV AND J. B. WEST, *J. Phys. B* **26**, 2099 (1993).
121. V. K. DOLMATOV, *J. Phys. B* **26**, L79 (1993).
122. K. NUROH, M. J. STOTT, AND E. ZAREMBA, *Phys. Rev. Lett.* **49**, 862 (1982).
123. F. J. WUILLEUMIER, in *International Research Facilities*, Proceedings of the 4th European Physical Society Seminar, edited by I. Slaus (European Physical Society, Zagreb, 1989), p. 259.
124. F. J. WUILLEUMIER, in *Correlations and Polarization in Electronic and Atomic Collisions and (e, 2e) Reactions*, edited by P. J. O. Teubner and E. Weigold, IOP Conference Series 122 (Institute of Physics, London, 1992), p. 203.
125. D. L. EDERER, *Phys. Rev. Lett.* **13**, 760 (1964).
126. J. W. COOPER, *Phys. Rev. Lett.* **13**, 762 (1964).

127. A. P. Lukirskii, I. A. Brytov, and T. M. Zimkina, *Opt. Spectrosc.* **17**, 234 (1964).
128. B. Kämmerling, H. Kossman, and V. Schmidt, *J. Phys. B* **22**, 841 (1989).
129. M. Y. Amusia, L. V. Chernysheva, V. K. Ivanov, and K. L. Tsemekhman, *J. Phys. B* **23**, 393 (1990).
130. J.-M. Bizau, D. Cubaynes, P. Gérard, and F. J. Wuilleumier, *Phys. Rev. A* **40**, 3002 (1989).
131. V. Radojevic, M. Kutzner, and H. P. Kelly, *Phys. Rev. A* **40**, 727 (1989).
132. M. Kutzner, Z. Altun, and H. P. Kelly, *Phys. Rev. A* **41**, 3612 (1990).
133. C. Theodosiou, *Phys. Rev. A* **33**, 2164 (1986).
134. L. Vo Ky, P. Faucher, F. Bely-Dubau, W. Eissner, and H. E. Saraph, in *Abstracts of Contributed Papers of the XVIII international Conference on the Physics of Electronic and Atomic Collisions*, edited by T. Andersen, B. Fastrup, F. Folkmann, and H. Knudsen (Aarhus University, Denmark, 1993), Vol. I, p. 27.
135. L. Vo Ky, A. Hibbert, L. Journel, B. Rouvellou, D. Cubaynes, J.-M. Bizau, and F. J. Wuilleumier, Post-dead line abstract presented at the international conference *Inner-Shell Ionization Processes and X-Ray Phenomena*, X-93, Debrecen, Hungary, 1993.
136. D. W. Norcross and P. M. Stone, *J. Quant. Spectrosc. Radiat. Transfer* **6**, 277 (1966).
137. F. L. Mohler, *J. Res. Natl. Bur. Stand.* **10**, 771 (1933).
138. A. Zangwill and P. Soven, *Phys. Rev. Lett.* **45**, 204 (1980).
139. M. Y. Amusia, N. A. Cherepkov, and L. V. Chernysheva, *Zh. Eksp. Teor. Fiz.* **60**, 160 (1971); for a review, see also M. Y. Amusia and N. A. Cherepkov, *Case Stud. At. Phys.* **5**, 47 (1975).
140. R. D. Cowan, *The Theory of Atomic Structure and Spectra* (University of California Press, Berkeley, 1981).
141. M. W. D. Mansfield and M. M. Murnane, *J. Phys. B* **18**, 4223 (1985).
142. J. T. Costello and G. O'Sullivan (private communication).
143. I. P. Grant, *Comput. Phys. Commun.* **55**, 425 (1989).
144. J. L. Fox and A. Dalgarno, *Phys. Rev. A* **16**, 283 (1977).
145. B. Rouvellou, J.-M. Bizau, D. Cubaynes, J. Novak, M. Pahler, L. Journel, F. J. Wuilleumier, L. Voky, P. Faucher, A. Hibbert, and N. Berrah, *Phys. Rev. Lett.* **75**, 33 (1995).
146. Z. Felfli and S. T. Manson, *Phys. Rev. Lett.* **68**, 1687 (1992).
147. L. Vo Ky, H. E. Saraph, W. Eissner, Z. W. Liu, and H. P. Kelly, *Phys. Rev. A* **46**, 3945 (1992).
148. L. Vo Ky, P. Faucher, F. Bely-Dubau, W. Eissner, and H. E. Saraph, in *Abstracts of Contributed Papers of the XVIII International Conference on the Physics of Electronic and Atomic Collisions*, edited by T. Andersen, B. Fastrup, F. Folkmann, and H. Knudsen (Aarhus University, Denmark, 1993), Vol. I, p. 26.
149. Z. Liu, private communication (1993).
150. M. J. Seaton, *J. Phys. B* **20**, 6409 (1987), and following papers.
151. D. C. S. Allison, P. G. Burke, and W. D. Robb, *J. Phys. B* **5**, 55 (1972).
152. K. A. Berrington, P. G. Burke, K. Butler, M. J. Seaton, P. J. Storey, K. T. Taylor, and Y. Yan, *J. Phys. B* **20**, 6379 (1987).
153. K. A. Berrington (private communication).
154. J. A. Fernley, K. T. Taylor, and M. J. Seaton, *J. Phys. B* **20**, 6457 (1987).
155. K. T. Cheng and W. R. Johnson, *Phys. Rev. A* **16**, 263 (1977).
156. W. R. Johnson and K.-N. Huang, *Phys. Rev. Lett.* **48**, 315 (1982), and references therein.
157. F. Combet Farnoux and M. Lamoureux, *Phys. Lett.* **43A**, 183 (1973).
158. M. Lamoureux and F. Combet Farnoux, *J. Phys. (Paris)* **35**, 205 (1974).
159. R. F. Reilman and S. T. Manson, *Phys. Rev. A* **18**, 2124 (1978).
160. G. Nasreen, S. T. Manson, and P. C. Deshmukh, *Phys. Rev. A* **40**, 6091 (1989).
161. S. T. Manson, *Phys. Rev. A* **38**, 506 (1988).
162. R. D. Driver, *J. Phys. B* **9**, 817 (1976).
163. G. Nasreen and S. T. Manson, *Phys. Rev. A* **38**, 504 (1988).
164. R. D. Cowan and M. Wilson, *J. Phys. B* **21**, L201 (1988).
165. H. Aizawa, K. Wakiya, H. Suzuki, F. Koike, and F. Sasaki, *J. Phys. B* **18**, 289 (1985).
166. R. D. Cowan and M. Wilson, *J. Phys. B* **21**, L275 (1988).
167. G. Miecznik, K. A. Berrington, P. G. Burke, and A. Hibbert, *J. Phys. B* **23**, 3305 (1990).
168. M. Y. Amusia, V. K. Dolmatov, and V. K. Ivanov, *Sov. Phys. JETP* **58**, 67 (1983).
169. M. W. D. Mansfield and G. H. Newsom, *Proc. R. Soc. London Ser. A* **357**, 77 (1977).
170. V. K. Dolmatov, *J. Phys. B* **25**, L629 (1992).
171. V. K. Dolmatov, in *Abstracts of the Sixteenth International Conference on X-ray and Inner Shell Processes*, Debrecen, Hungary, Book of Abstracts, edited by L. László and D. Berényi, p. 75 (1993).
172. V. K. Dolmatov, *Phys. Lett. A* **174**, 116 (1993).
173. V. K. Dolmatov, *J. Phys. B* **26**, L585 (1993).
174. J. T. Costello, E. T. Kennedy, B. F. Sonntag, and C. W. Clark, *Phys. Rev. A* **43**, 1441 (1991).

175. J. B. West, T. Andersen, F. Folkmann, and H. Knudsen (private communication).
176. A. Gottwald, T. Luhmann, and M. Richter, in *BESSY Jahresbericht*, 1993.
177. I. Alvarez, C. Brion, H. C. Bryant, C. Cisneros, P. Heimann, D. H. Jaecks, M. S. Lubbel, H. H. Michels, J. B. A. Mitchell, T. J. Morgan, A. S. Schlachter, and O. Yenen, *Bull. Am. Phys. Soc.* **39**, 1221 (1994).
178. T. Aberg and J. Tulkki, in *Proceedings of the ESRF–LURE Workshop on Photon–Ion Interactions*, edited by F. J. Wuilleumier and E. Källne (ESRF, Grenoble, 1988), p. 143.
179. M. Leduc, R. Marmoret, J. Lachkar, C. Nazet, F. Bugaut, F. Jequier, C. Blancard, J.-M. Constantini, B. Deslandes, L. Mancheron, and E. Mary, proposal submitted to LURE, Orsay, June 1995.
180. M. Delaunay, S. Dousson, R. Geller, B. Jacquot, D. Hitz, P. Ludwig, P. Sortais, and S. Bliman, *Nucl. Instrum. Methods B* **23**, 177 (1987).
181. F. Bourg, R. Geller, and B. Jacquot, *Nucl. Instrum. Methods A* **254**, 13 (1987).
182. M. Barat, J.-P. Briand, and F. J. Wuilleumier, Project ICARIOS, Report of Université Paris XI and Paris VI (1992).
183. M. A. Levine, R. E. Marrs, J. R. Henderson, D. A. Knapp, and M. B. Schneider, *Phys. Scr.* **T22**, 157 (1988).
184. R. E. Marrs, M. A. Levine, D. A. Knapp, and J. R. Henderson, *Phys. Rev. Lett.* **60**, 1715 (1988).
185. J. P. Briand, P. Charles, J. Arianer, H. Laurent, C. Goldstein, J. Dubau, M. Loulergue, and F. Bely-Dubau, *Phys. Rev. Lett.* **53**, 617 (1984).
186. D. A. Knapp, R. E. Marrs, M. A. Levine, C. L. Bennett, M. H. Chen, J. R Henderson, M. B. Schneider, and J. H. Scofield, *Phys. Rev. Lett.* **62**, 2104 (1989).
187. S. Kravis, Y. Awaya, T. Kambara, Y. Kanai, T. Kojima, K. Okuno, M. Kimura, and S. Ohtani, *AIP Conf. Proc. Ser.* **274** (American Institute of Physics, New York, 1992), p. 671.
188. M. Oura, S. Kravis, Y. Awaya, K. Okuno, and M. Kimura, *Atomic Collision Research in Japan—Progress Report* **19**, 95 (1993).
189. K. Okuno, *Jpn. J. Appl. Phys.* **28**, 1124 (1989).
190. K. Okuno, K. Soejima, and Y. Kaneko, *Nucl. Instrum. Methods* **B53**, 387 (1991).
191. K. W. Jones, B. M. Johnson, M. Meron, Y. Y. Lee, P. Thieburger, and W. C. Thomlinson, *Nucl. Instrum. Methods Phys. Res. Sect. B* **24**, 381 (1987).

CHAPTER 17

PHOTOIONIZATION OF ORIENTED SYSTEMS AND CIRCULAR DICHROISM

G. SCHÖNHENSE AND J. HORMES

1. INTRODUCTION

Circular dichroism (CD) spectroscopy is a modification of normal absorption spectroscopy using circularly polarized light instead of unpolarized light for determining the *difference* in the absorption coefficients for right and left circularly polarized light, respectively, in optically active samples. Thus, there is nothing mysterious about CD spectroscopy. However, relative to most other spectroscopic techniques there seems to be a large psychological barrier in its application and at least physicists often regard CD, and optical activity in general, as a rather obscure technique without any useful application. There are some understandable reasons for this attitude, manifesting themselves strikingly in the fact that CD is hardly ever treated in elementary physics textbooks:

- Though optical activity is one of the oldest known physical phenomena, it was only 30 years ago that instruments for measuring CD spectra became available.
- Samples which are optically active consist of so-called "chiral" crystals or "chiral" molecules; these are crystals or molecules which are not superimposable onto their mirror image. In the majority of cases, molecules with this property consist of so many atoms that physicists believe them not to be suitable for basic physical investigations.
- Up till now there exists no straightforward theory of optical activity; data are in most cases interpreted on the basis of more or less empirical rules with restricted validity and only in very few cases is it possible to extract from the data the structural information about the investigated molecules being included in the spectra. Furthermore there is still a lack of systematic and reliable data for a quantitative comparison with theory.

This chapter is not expected to remove all prejudices against CD but it is intended to give an easy-to-read and rather general introduction and to present some illustrative applications of the technique. If the reader is interested in finding out more about the field of optical activity, any of Refs. 1–7 will be useful. The main focus of this chapter is on electronic absorption phenomena. Emphasis is put on the UV–VUV region and on the use of synchrotron radiation at least when "typical" experiments and applications are discussed. This can be justified by the fact that the theoretically simple chiral molecules have their electronic absorption more or less exclusively in this region and that the investigation of most of the important molecules in biology is very much improved by the additional electronic transitions accessible in this energy region. However, the general field of "chiroptical" spectroscopy includes a number of other phenomena, e.g., vibrational CD,[8] circularly polarized

---

G. SCHÖNHENSE • Institut für Physik, Johannes-Gutenberg-Universität Mainz, 55099 Mainz, Germany.
J. HORMES • Physikalisches Institut der Universität Bonn, 53115 Bonn, Germany.

VUV and Soft X-Ray Photoionization. Edited by Uwe Becker and David A. Shirley. Plenum Press, New York, 1996.

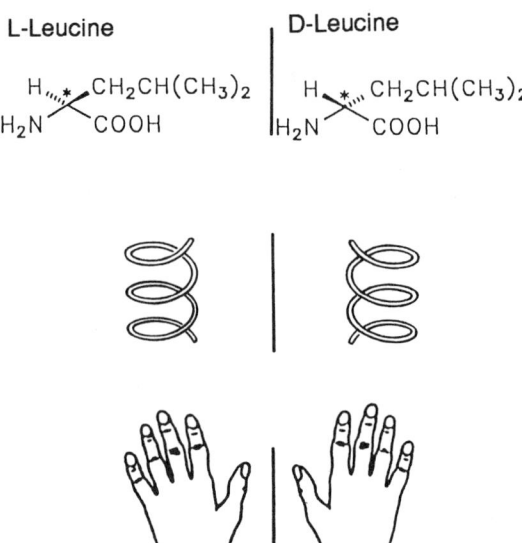

FIGURE 1. Typical objects with chirality, i.e., which are not superimposable onto their mirror images.

luminescence,[9] and circular intensity differential scattering.[10] All of these will not be discussed in this chapter.

Finally, it should be noted that the phenomenon of "chirality," or more precisely chiral excess, is not at all restricted to molecules. Chiral properties can be found in fundamental nuclear particles, in inanimate nature (crystals), in the macroscopic world of living nature (animals and plants), and also in the technical sphere, where screws are perhaps the best-known examples for chirality. These aspects are dealt with in a recently published book edited by Janoschek.[11] In Figure 1 three typical objects with "chirality," i.e., objects which are not superimposable onto their mirror images, are shown: an optically active molecule, a spiral, and the human hand.

2. WHAT IS CIRCULAR DICHROISM?

2.1. Empirical Description of CD

When linearly polarized monochromatic light of wavelength λ passes through a so-called "optically active" sample, e.g., solutions of sugar, the plane of polarization is rotated by an angle Φ which is given by the empirical equation

$$\Phi = cl\alpha \tag{1}$$

where c = concentration of the optical active sample in g cm^{-3}, l = path length of the absorption cell in decimeters for fluids, and α = specific rotation angle at the wavelength λ. This phenomenon was observed in 1817 by Biot who also concluded that optical rotation in solution is essentially a molecular property.[12]

A more generalized form of equation (1) is the following:

$$\hat{\Phi} = \Phi - i\Theta = \frac{\pi l}{\lambda}(\hat{n}_- - \hat{n}_+) \tag{2}$$

Here $\hat{\Phi}$ is now a complex rotation angle, the real part Φ of which is the rotation given by equation (1) whereas the imaginary part Θ is the ellipticity. The complex refractive indices are given by

$$\hat{n}_\mp = n_\mp - ik_\mp \tag{3}$$

with $n_\mp$ = real refractive index for right (−) and left (+) circularly polarized light, respectively, and $k_\mp$ = corresponding absorption coefficients.

From equation (2) it is obvious that two different types of experiments are possible depending on whether the real part Φ or the imaginary part Θ is to be determined. In the first type of experiment which is referred to as an optical rotatory dispersion (ORD) experiment, one measures according to

$$\Phi = \frac{\pi l}{\lambda}(n_- - n_+) \tag{4}$$

the frequency-dependent rotation of plane-polarized light which results from the difference in the refractive indices of the medium toward left and right circularly polarized light. The second type where the difference in the absorption of the two circularly polarized beams is determined

$$\Theta = \frac{\pi l}{\lambda}(k_- - k_+) \tag{5}$$

is called a CD measurement. In principle, ORD and CD provide the same information as they are related by a Kramers–Kronig transformation.[13] However, when measuring ORD in the transparent region of a sample, all electronic transitions contribute to the refractive indices. In general, a separation of these contributions is impossible and thus also calculations which have to include all of these transitions are complicated and difficult to verify. The interpretation of a CD spectrum is much simpler because CD is measured for each electronic transition separately by scanning the wavelength through the absorption band. As the absorption coefficient goes rather quickly to zero in the wings of the absorption line, it is in general possible to separate also close-lying transitions. These problems are demonstrated in a paper by Hirano et al.[14] where both types of data are compared for poly[(R)-oxypropylene].

2.2. Basis for a Quantum-Mechanical Description of CD

ORD and CD are both macroscopic properties of a sample. However, as was already mentioned, chirality of the molecules which make up the sample is the indispensable condition for both phenomena. Thus, ORD and CD are intimately related to the structure of a single molecule, a fact that explains the practical interest chemists take in these effects. It is highly desirable to find out whether structural information can be extracted from measured CD spectra. For this purpose, the macroscopic observables—Φ or Θ—are connected with microscopic properties of a single molecule—given by the wave functions and the position of the transition moments within the molecule—in three steps[15]:

1. The experimental observables are connected with the difference in the refractive indices for rcp and lcp light (as was already done in the previous section)

$$\Phi = \frac{\pi l}{\lambda}(n_- - n_+) \tag{6}$$

$$\Theta = \frac{\pi l}{\lambda}(k_- - k_+) \tag{7}$$

2. Making use of Maxwell's equations the refractive indices are connected with the electromagnetic moments induced in individual molecules by the light wave:

$$n = n(\vec{m}, \vec{\mu}) \tag{8}$$

$$k = k(\vec{m}, \vec{\mu}) \quad (9)$$

where $\vec{\mu}$ is the induced electrical moment and $\vec{m}$ is the induced magnetic moment. Contrary to the atomic case, for molecules either moment can be nonzero for a specific transition due to the lower symmetry.

3. These moments are then calculated quantum mechanically and the corresponding transition probabilities W_{if} are calculated:

$$W_{if} \sim \langle \Psi_f | \vec{\mu}, \vec{m} | \Psi_i \rangle \langle \Psi_f | \vec{\mu}, \vec{m} | \Psi_i \rangle \quad (10)$$

To first order in the electromagnetic fields and assuming an isotropic medium, i.e., solutions, the induced moments can be expressed as

$$\vec{\mu} = \alpha_0 \vec{E} + \beta_0 \vec{H} \quad (11)$$

$$\vec{m} = \gamma_0 \vec{E} + \chi_0 \vec{H} \quad (12)$$

It can be shown that[15]

$$\hat{\Phi} = \Phi - i\Theta \approx (\hat{n}_- - \hat{n}_+) \approx \beta_0 - \gamma_0 \quad (13)$$

To observe optical activity, i.e., rotation of the plane of polarization of the $\vec{E}$ vector, the fields of the electromagnetic wave have to induce moments perpendicular to themselves. This is also obvious from equations (11) and (12) showing that optical activity arises from those components of the induced moments perpendicular to the electric and magnetic field of the light wave and to the direction of propagation.

$\vec{m}$ and $\vec{E}$ are vectors, but $\vec{\mu}$ and $\vec{H}$ are pseudovectors. Thus, by going from the real system to its mirror image, $\vec{m}$ and components of the induced moments perpendicular to $\vec{E}$ change sign whereas $\vec{\mu}$ and $\vec{H}$ do not. Hence, β_0 and γ_0 are nonzero only if the system is different in the two frames. This is the aforementioned condition of chirality for optically active molecules which cannot have a center of inversion so that they are not superimposable onto their mirror image (see Figure 1).

The calculation of transition probabilities applying time-dependent perturbation theory on the basis of electric and magnetic dipole interaction with the radiation is worked out in a paper by Schellman.[16] With the interaction Hamiltonian

$$V = -\vec{\mu}\vec{E} - \vec{m}\vec{H} \quad (14)$$

the transition probability is obtained from

$$W_{if} \approx |\langle \Psi_f | V | \Psi_i \rangle|^2 = V_{if}^2 \quad (15)$$

The calculation for circularly polarized light gives

$$(V_{\pm if})^2 \approx D_{if} + G_{if} \mp R_{if} \quad (16)$$

The quantities given in this equation are defined as

$$D_{if} \approx |\langle \Psi_f | \vec{\mu} | \Psi_i \rangle|^2 \quad (17)$$

$$G_{if} \approx |\langle \Psi_f | \vec{m} | \Psi_i \rangle|^2 \tag{18}$$

$$R_{if} \approx \text{Im}(\langle \Psi_f | \vec{\mu} | \Psi_i \rangle \langle \Psi i_j | \vec{m} | \psi_i \rangle) \tag{19}$$

D_{if} is the electric dipole strength, G_{if} the magnetic dipole strength, and R_{if} the rotatory strength. R_{if} is the molecular measure of CD and is only nonzero when there is at the same time an electric and a magnetic dipole interaction of the light with the molecule. D_{if} is obtained from the experimental data by integrating an absorption band, R_{if}, by integrating the corresponding CD band.

A concrete calculation of R_{if} for a special molecule is not straightforward especially when the molecular orbitals of the initial and the final state of the electronic transition under consideration are localized on a limited part of the whole molecule or even on one atom. In these cases, hydrogenic wave functions with well-defined parity centered at the excited atom have to be used for the calculations. As $\vec{\mu}$ connects states of different parity and $\vec{m}$ connects only states of the same parity, $R_{if} \neq 0$ cannot be obtained with these wave functions. This is the reason why there is no unified theory of optical activity. To avoid the problems mentioned, for each group of optically active molecules a special model is made with the goal of separating the positions within the molecule where the electric and the magnetic dipole transition respectively take place. Some of these models will be discussed in Section 4.

2.3. Importance of Optically Active Molecules in Biology and Pharmacy

That there is a fundamental connection between molecular chirality and life was first established in a classic experiment by Louis Pasteur: a growing yeast consumed only the dextrorotatory acid when the ammonium salt of racemic tartaric acid was offered as a carbon source. Moreover, the present living matter on earth is "asymmetric" as it contains mainly dextrorotatory sugars and levorotatory amino acids. In almost all cases, nature uses these enantiomers to compound polysaccharides, proteins, and nucleic acids. Thus, also biopolymers are asymmetric. There are a few interesting examples where our organs of perception have the ability to distinguish the two enantiomers of a molecule. For example, L-leucine tastes bitter whereas D-leucine is sweet; R-limonene smells like oranges whereas S-limonene has the odor of lemons (the structure of the leucines is given in Figure 1). Neither the origins of optical activity in nature nor the basis for the ability of organisms to discriminate between the enantiomers of a molecule are known. Some of the models proposed are discussed in recent publications.[17,18]

Many drugs become active in the organism by interaction with chiral receptor proteins. If the drugs are also chiral, the interaction of the enantiomers with the receptor is in general different and so is the medical efficacy. More than half of all available synthetic drugs are chiral, but just 10–20% of these substances are being used as pure enantiomers. Thus, the majority of all pharmaca consists of about 50% of substances which are not really necessary and which might even be dangerous to the patient. The best-known example for this problem may be the so-called "Contergan catastrophe." In the 1950s this drug was used as a soporific. After a few years a connection between the drug and the deformity of newborns was recognized. In this case, it was the right-turning form of the active molecule thalidomide which was teratogenic whereas the left-turning form was not harmful.

These examples are intended to illustrate the importance of optically active molecules in biology and pharmacy. Thus, it is obvious that CD measurements can be very helpful in the investigation of biological and medical samples and are very often used in these fields. In Section 5.2 some applications in biology will be discussed in more detail.

3. HOW TO MEASURE CD SPECTRA?

3.1. Conventional Methods

The ORD method is applicable to all chiral substances, whereas CD is confined to enantiomers with absorption spectra in the accessible wavelength regions, as CD exists only in the regions of absorption. In general, CD constitutes only a small fraction of the absorption (typically about 1/10,000)

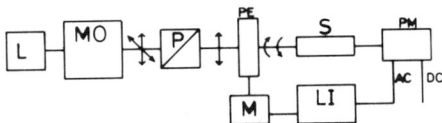

FIGURE 2. Block diagram of a CD instrument. L, light source; MO, monochromator; P, polarizer; PE, photoelastic modulator; S, sample; PM, photomultiplier; LI, lock-in detector; M, PEM power supply.

and as the light intensity is already diminished by the absorption itself, CD is measured favorably by a modulation technique with phase-sensitive detection and lock-in amplification. Such a setup, now called a Dichrograph, was first reported in 1960 by Grosjean and Legrand.[19] Historical and technical details of polarization modulation technique are summarized in a review article by Drake.[20] A schematic representation of the setup of a conventional modern CD instrument is given in Figure 2. Monochromatic light leaving the exit slit of a suitable monochromator passes through a polarizer, e.g., a Wollaston prism, separating plane-polarized beams, one of which is discarded. The remaining linearly polarized beam passes a modulator system where it is converted periodically to right and left circularly polarized light. The modulated beam then passes through the sample and is finally detected by a photomultiplier. If the sample is optically active, an AC signal with the frequency of the modulation superimposed on a DC signal is detected by the multiplier. To a good approximation[20] the desired CD signal is proportional to the intensity of the AC current divided by that of the DC current.

The crucial part of the setup described here is the light modulator producing circularly polarized light, a "dynamical" quarter-wave plate. In the first instrument constructed by Grosjean and Legrand[19] and its direct successors the modulator was an electro-optical device, a Pockels cell with a tetragonal crystal of $(NH_4)H_2PO_4$ (ADP) with glycerol electrodes. Depending on the applied voltage the ADP–Pockels cell could be used as a $\lambda/4$ plate between 600 nm (~ 5400 V) and 200 nm (~ 1100 V) whereas the short-wavelength limit of about 185 nm was given by the transmission of the ADP crystal and the electrodes.

The disadvantages of the electro-optical modulator, namely,

- A limited aperture (~ 1°) for the light beam because of the different refractive indices of the crystal in directions along and perpendicular to the tetragonal optical axis and
- The limited transmission range in the UV (~ 185 nm) and in the IR (~ 1.0 µm),

were overcome by an alternative means of modulating polarization which was described during the latter half of the 1960s, originally by Billardon and Badoz[21] and then by Kemp[22] and others.[23–25] This new device was based on stress-induced birefringence. The so-called photoelastic modulator (PEM) is made of a transparent optically isotropic element which is periodically stressed by piezoelectric ceramic disks[21] or a matched element resonator relying on the sympathetic oscillation in an optical element in contact with a piezoelectric driver excited into its natural oscillation frequency (between 10 and 50 kHz).[22,23] A standing wave is induced in the latter device with a nondestructive node at the junction of the two elements.

The use of calcium fluoride (CaF_2) as the optical element in a PEM in principle gave access to the VUV region down to about 125 nm. With LiF modulators[26] which recently became available, this limit is further shifted to about 105 nm. However, for successful experiments in this energy range, further technical developments were necessary:

- Vacuum monochromators had to be used.
- Light sources for the VUV had to be optimized. This was achieved by most groups through the use of a H_2 discharge in a modified Hinteregger design.[27,28]
- Linear polarizers suitable for the VUV had to be developed. Rochon and Wollaston prisms made from MgF_2 were used by Johnson[29] and by Drake and Mason.[30] To overcome the

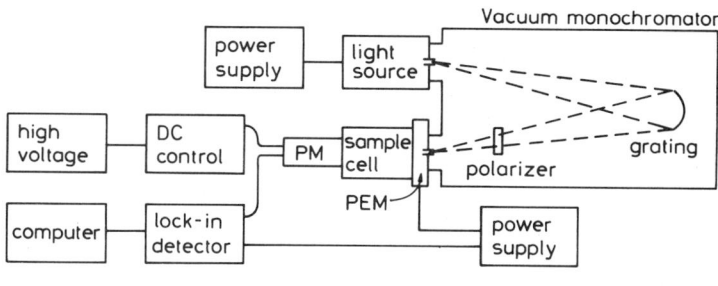

FIGURE 3. Schematic drawing of a VUV CD instrument.

wavelength limit of about 135 nm given by the use of MgF$_2$, also single reflection from an oriented biotite plate was used.[31–33]

Altogether, no more than eight instruments for CD (and also magnetic CD) measurements in this energy range have been built to date. The development of CD spectroscopy in the VUV range was reviewed by Johnson[34] and Snyder.[35] Here also the existing VUV instruments are characterized.

The typical design of a CD instrument for VUV experiments is shown in Figure 3. It employs the optimized Hinteregger discharge lamp and a vacuum monochromator, here a 1-m normal-incidence McPherson equipped with a grating of 600 or 1200 lines/mm. The prism is placed inside the monochromator before the exit slit, the height of which is adjusted to block one of the plane-polarized beams. The second linearly polarized beam passes through the photoelastic modulator and is converted alternately to right and left circularly polarized light. The modulated light passes through the sample and is detected by a photomultiplier. If the sample cell contains an optically active compound, the photomultiplier will detect an AC signal superimposed on a DC signal. In general, the DC signal is kept constant and the AC signal detected by a lock-in amplifier is directly proportional to the CD signal wanted.

The central problem of all conventional VUV CD instruments is the signal-to-noise ratio which is proportional to the square root of the intensity.[2,20,36] The limit of sensitivity with 2-mm slits, a grating of 1200 lines/mm, and a lock-in time constant of 30 s is about 10^{-5} in the wavelength range between 145 and 330 nm.[37] These figures illustrate that conventional CD measurements are time-consuming and limited in their energy range and in their resolving power. Thus, a more intense VUV light source, such as synchrotron radiation, is particularly attractive for this type of experiment.

3.2. Measurements with Synchrotron Radiation

Synchrotron radiation (SR) from electron storage rings has some special properties making it an ideal light source for CD measurements:

- In the plane of the electron orbit the radiation is highly linearly polarized.
- The spectrum is continuous from the IR to the x-ray region.
- The light is highly collimated.
- The intensity of the radiation in the VUV region is much higher than that of conventional VUV sources.

The properties of SR are discussed in more detail in several books and review articles.[38]

Because of the polarization properties of SR, no polarizer for producing linearly polarized light is necessary (see Figure 2). Thus, measurements are possible to the limit given by the λ/4 plate (1050

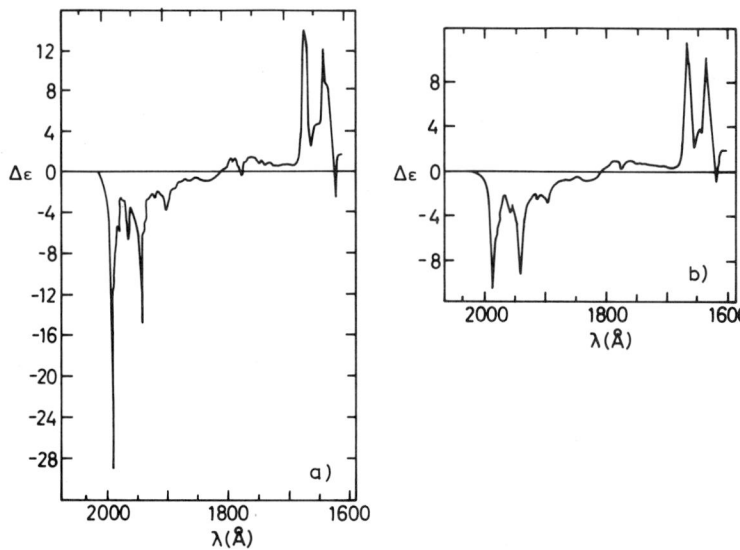

FIGURE 4. CD spectrum of (+)-3-methylcyclopentanone in the gas phase: (a) with synchrotron radiation; (b) with a conventional light source (redrawn from Ref. 39).

Å for LiF). As noise on CD spectra is inversely proportional to the square root of the number of photons, the use of SR makes it possible to reach this limit not only with improved signal-to-noise ratio but also with much better resolution. This was already clearly demonstrated by the first measurements of natural CD using SR which were carried out in 1980 by Snyder and Rowe at the Tantalus I storage ring in Wisconsin.[39] As an example of these measurements the CD spectrum of (+)-3-methylcyclopentanone in the vapor phase between 2050 and 1600 Å is shown in Figure 4 in comparison to a spectrum taken with a conventional light source. Here, the spectral bandwidth with SR was 2 Å as compared to typically 20 Å with conventional sources resulting in the observation of more peaks and dramatic peak height changes.

The design of a CD instrument for the use of SR is basically the same as for a conventional light source. Schematic drawings are given in the publications of the Wisconsin group[35,39] and the Bonn group,[42] respectively. To maintain the linear polarization of SR it is important that all reflections are s-type, i.e., out of the plane of the electron orbit. Thus, it is possible to obtain a degree of linear polarization better than 97%.[42] Besides the already mentioned pieces of apparatus, another instrument for CD measurements with SR was developed at the Surf ring of the National Bureau of Standards[43] and later moved to the NSL in Brookhaven.

Further details of all available SR CD instruments and an overview of the CD [and magnetic CD (MCD)] experiments carried out with SR are given in the review articles by Snyder[35,40] and Stevens.[41] An interesting application of the use of SR for CD measurements is described in a paper by Wickramaaratchi et al.[44] These authors have used CD as a detection technique in gas chromatography for specific monitoring of optically active samples absorbing in the VUV region, e.g., terpene compounds. Using for these experiments SR from the Brookhaven NSL source, a detection limit as low as some nanograms could be established for a selected group of hydrocarbons.

3.3. Present Experimental Limits

Because of the cutoff wavelength of the LiF modulator required for producing circularly polarized light, CD measurements using the linearly polarized central component of SR are limited to energies below ~11 eV. However, according to Schwinger's theory,[45] SR emitted into directions above and below the plane of the electron beam is elliptically polarized with a high fraction of circular

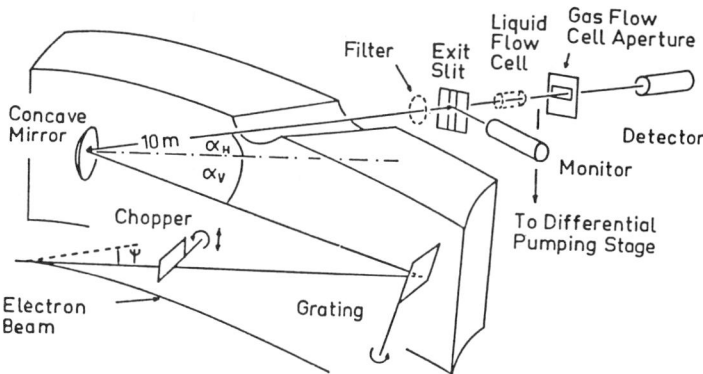

FIGURE 5. Schematic diagram of the experimental setup for measuring CD spectra with the circularly polarized components of synchrotron radiation.[50]

polarization. This was verified experimentally in the visible by Joos[46] and in the VUV by Heinzmann and co-workers.[47] Two monochromators, a 10-m normal-incidence instrument at the 2.5-GeV synchrotron in Bonn[47] and a 6.5-m normal-incidence device of the Gillieson type at BESSY I in Berlin,[48,49] have been built for the use of the circularly polarized radiation emitted out of the plane of the electron orbit. With the monochromator in Bonn, experiments for measuring CD spectra were also performed.[50] Figure 5 is a schematic view of the experimental setup used for these experiments. A spherical mirror projects the tangent point of the electron beam (= entrance slit) onto the exit slit of the monochromator. A reflecting holographic plane grating (4960 lines/mm) facing the tangent point in a Littrow mounting gives a resolution of 0.05 nm with an exit slit width of 1.3 mm. A flag in front of the grating is used to shut off either the upper (lcp) or the lower half (rcp) of the beam. As the modulation technique could not be given up for CD measurements, the beam flag in front of the grating which could be moved only very slowly was replaced by a chopper system. For measurements below 105 nm a windowless absorption cell was developed which in combination with an efficient differential pumping stage allowed a pressure times absorption length of 0.1 torr × 160 cm.

The experiments here described have demonstrated that it is feasible to use the circular polarization of SR for CD (and MCD) measurements. However, the results obtained have shown clearly that future monochromators have to be designed very carefully for their special task of obtaining CD spectra of a quality comparable to those taken in the region above 105 nm by conventional techniques and with SR.

4. WHAT CAN BE LEARNED FROM CD SPECTRA?

ORD and CD are macroscopic properties of macroscopic samples. The phenomenological equations (4) and (5), of course, give no clue as to why in an optically active medium n_- is different from n_+ or k_- is different from k_+. Regarding the quantum-mechanical definition of the rotatory strength R_{if} given in equation (19), it is obvious that the molecular source of optical activity is the mutual interference of induced electric dipole and magnetic dipole moments within the molecule. If it would be possible to construct accurate wave functions for chiral molecules, R_{if} could of course be calculated. An introduction to the theoretical basis and the computational problems of these calculations is given in a review by Hansen and Bouman.[51] Results of such calculations are discussed for methyloxirane in a recent paper by Carnell et al.[52] However, it should be mentioned that in ab initio calculations in which each molecule has to be treated as an individual case a great deal of generality, widely observed in the experiments, is lost.

In the following some of the semiempirical models for the origin of optical activity will be discussed. Though these models are in most cases restricted not only to a special group of molecules but also to a special electronic transition, they provide at least some physical insight. Furthermore, they are the basis for the semiempirical rules connecting sign (and sometimes magnitude) of CD signals with the molecular structure.

4.1. Models for the Origin of Optical Activity

For a theoretical description a chiral molecule is generally divided into two parts: the part where the light is absorbed, the so-called "chromophore," and one or more groups of substituents, the so-called "ligands." It is assumed that the chromophore and the ligands are quasi-independent systems within the molecule and that the overlap between their charge distributions can be neglected.

According to Moffitt and Moscowitz,[53] optically active molecules can be divided into two classes: those in which the chromophore is chiral itself (class A) and those for which it is achiral, while the rest of the molecule is chiral (class B). In the latter case, the achiral chromophore is chirally perturbed by its chiral environment. Class B is often subdivided into two groups[54]: the first group includes chiral rings (B I) and the second all of those molecules for which the nearest chiral perturbation is at least two bonds away from the chromophore (B II). For group B II molecules the so-called "sector rules" can be derived which will be discussed below whereas for all other cases "helicity rules"[55] have to be used to correlate CD spectra with molecular geometry.

In both classes of molecules there can be species with two or more chromophores (chiral or achiral) exhibiting strong electric transition moments interacting with each other in a chiral way ("exciton interaction"). In these cases two CD bands of opposite signs are obtained, a so-called "CD couplet," where in principle both bands should have the same magnitude. These molecules are summarized in a third class C.

For all classes and groups of molecules mentioned above, various models have been developed to rationalize the observed optical activity, i.e., to explain the simultaneous excitation of electric and magnetic moments within the molecule. An overview of these models is given in the books by Mason[6] and Barron.[56]

For class A molecules the inherently chiral chromophore model is used. Here it is assumed that the molecular orbitals are delocalized significantly over the chiral nuclear framework so that no selection rules restrict the corresponding electronic transitions induced by the light. Thus, electric dipole, magnetic dipole, and also higher-order transitions are allowed between all states so that $R_{if} \neq 0$.

For all other classes either the electric or the magnetic dipole moments within the chromophore are zero because of the high symmetry of the chromophore (or the moments are perpendicular to each other). Thus, the chromophore alone shows no optical activity. Optical activity arises from an interaction with the chiral molecular environment. The interaction between the two groups takes place through the Coulomb potential between the corresponding charge distributions. The models developed to describe this type of interaction are summarized as "coupling models." Basically three groups of these coupling models can be distinguished:

- The static coupling model (based on the one-electron theory of Condon et al.[57])
- The dynamic polarization model (based on the work of Kuhn,[58] Boys,[59] and Kirkwood[60])
- The two-group electric-dipole mechanism, which is a special case of the dynamic polarization model

In the following, two selected models will be discussed in more detail. They were chosen as in both cases the model correlates CD spectra to some molecular structural information in such a way that it is at least in principle possible to extract this information from the experimental spectra.

4.2. The Dynamic Polarization Mechanism and the Octant Rule

The starting point of the dynamic polarization model is the assumption that not only the chromophore but also the ligands in its environment play an active role in the absorption process. This model which is especially suited for group B II molecules was worked out in detail by Höhn and Weigang.[61] In its simplest form the model starts from a magnetic dipole transition ($l \to l$) of the chromophore resulting in a magnetic dipole moment m_{if}. This moment is accompanied by even 2^n electric multipole moments Θ_{if} ($n = 2, \ldots, 2l$). By these moments an electric dipole moment proportional to the corresponding dipole polarizability $\bar{\alpha}(L)$ is induced in each ligand group through the Coulomb interaction. The resulting total electric dipole moment of all ligands is given by

$$\mu_{if} = -\Theta_{if}\bar{\alpha}(L) G^L_{xyz} \qquad (20)$$

Here G^L_{xyz} is a geometric tensor describing the arrangement of the ligand groups in the molecule. μ_{if} is nonzero for non-centrosymmetric molecules. As the electric dipole moment is phase-locked to the multipole transition moments of the chromophore, it interacts with the incoming radiation and increases the isotropic absorption.

The model presented here has turned out to provide a generalized theoretical basis for several of the so-called "sector rules" which were found empirically for several groups of chromophores (e.g., see Ref. 62). These rules connect the position of a ligand group in the molecular environment of the chromophore with the sign and in some cases also with the magnitude of the CD signal for a special electronic transition. The perhaps most famous one is the octant rule for the $n \to \pi^*$ transition of optically active ketones near 300 nm. This rule was formulated by Moffitt et al.[63] on the basis of their experimental results. For applying this rule, the plane of the carbonyl group is regarded as lying in the yz coordinate plane with the carbon atom at the origin and the oxygen lying on the $+z$ axis as shown in Figure 6. If now the product of the coordinates (XYZ) of the various ligands is negative in this coordinate system, the CD signal of the $n \to \pi^*$ transition is positive and vice versa, provided that the average electric dipole polarizability of the ligand group is larger than that of the hydrogen atom being replaced.

This rule can be derived from the dynamic polarization model as was shown by Weigang and Höhn.[64] When m^z_{if} is the z component of the magnetic dipole moment of the transition in the coordinate system of Figure 6, the rotational strength R_{if} is given by

$$R_{if} \sim -\frac{XYZ}{R^7} \bar{\alpha}(L) \Theta^{xy}_{if} m^z_{if} \qquad (21)$$

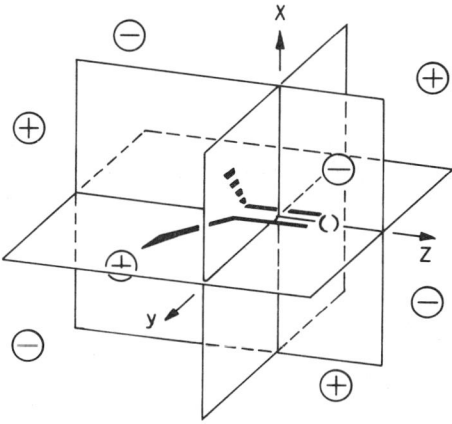

FIGURE 6. Coordinate system for the octant rule of the $n \to \pi^*$ transition of the carbonyl group in optically active ketones.

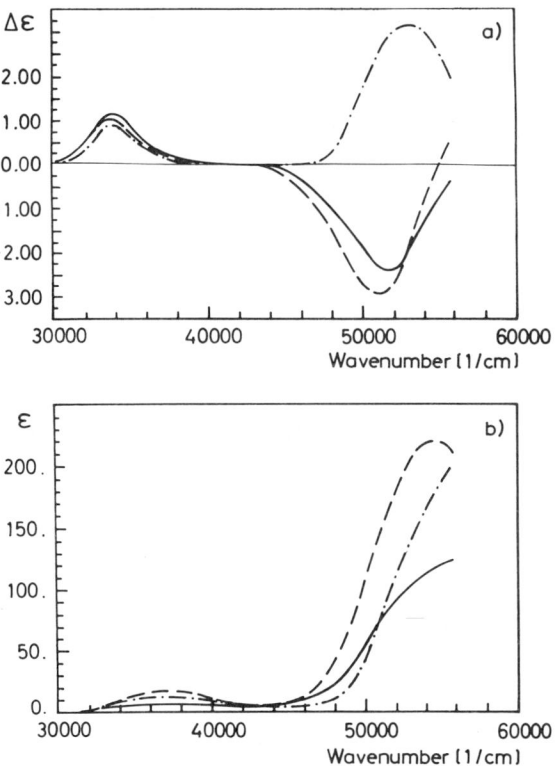

FIGURE 7. CD and absorption spectra of three camphor compounds in trifluorethanol: (+)-fenchone (dots and dashes), (+)-camphor (dashed line), and (+)-camphorsulfonic acid (solid line).

This equation where $(XYZ)/R^7$ has replaced the geometric tensor G^L_{xyz} from equation (20) shows the expected correlation between geometrical parameters and rotational strength. However, as neither $\overline{\alpha}(L)$ nor Θ^{xy}_{ij} are known exactly, it is hardly ever possible to extract detailed geometric informations from the measured rotational strength and the octant rule is just applied as an indication of the position of the ligands relative to the carbonyl group. Recently a more general theory for the $n \rightarrow \pi^*$ transition in carbonyl compounds was developed based on an independent system/perturbation approach by Rodger and Rodger.[65] These authors could show that all available experimental data including the observed "exceptions" can be understood in terms of the predictions made by their theory.

The octant rule is restricted to the magnetic dipole allowed $n \rightarrow \pi^*$ transition. For the electric dipole allowed $n \rightarrow \sigma^*$ transition a more complicated "octant sector" rule can be derived connecting again the position of the various ligands relative to the carbonyl chromophore to the sign of the rotational strength.[66] In Figure 7 the CD spectra and the absorption spectra of three camphor compounds in solution are shown.[67] The low-energy $n \rightarrow \pi^*$ transition shows the same (positive) sign in the CD spectra of all three molecules. However, the electric dipole allowed high-energy $n \rightarrow \sigma^*$ transition has a negative CD signal for two of these molecules whereas it is positive for the third one. This finding indicates again an "exception" from the relevant octant sector rule and further systematic investigations seem to be necessary for an improved understanding.

4.3. The Exciton Coupling Model

This theory treats a special case of the dynamic coupling model discussed above and it also originates from the coupled oscillator model studied by Kuhn[58] and the polarizability theory developed by Kirkwood.[60] Theoretical details and applications were worked out by Harada and

R-(+)-binaphthole   L-(−)-binaphthole

FIGURE 8. The structure of R-(+)-binaphthole and L-(−)-binaphthole.

Nakanishi.[7,68] The model is applicable when there are at least two identical chromophores in a molecule located in chiral position with respect to each other and exhibiting a strong electric dipole transition (group C molecules).

As a simple but typical molecule of this type, binaphthole is shown in Figure 8. Both naphthalene groups are chromophores for e.g. the electric dipole transition to the ^1B_b state of naphthalene at about 45,000 cm^{-1}. Here the electric dipole moment is localized over both chromophores.

As the chromophoric groups are identical, the wave function of the ground state Ψ_i has to be written as a direct product of the individual ground-state wave functions, and the excited state Ψ_f is given as the symmetric and antisymmetric combination of the individual ground and excited states:

$$\Psi_i = \Phi_i^a \Phi_i^b; \qquad \Psi_f = (1/\sqrt{2})(\Phi_f^a \Phi_i^b \pm \Phi_i^a \Phi_f^b) \tag{22}$$

This leads to two energy levels for the excited state:

$$E^\alpha = E_f - V_{ij}; \qquad E^\beta = E_f + V_{if} \tag{23}$$

The energy gap $2V_{ij}$ is called the Davydov splitting. It can be shown that the dipole strength for the two transitions is given by[7]

$$D^\alpha = (1/\sqrt{2})(\vec{\mu}_{if}^a - \vec{\mu}_{if}^b); \qquad D^\beta = (1/\sqrt{2})(\vec{\mu}_{if}^a + \vec{\mu}_{if}^b). \tag{24}$$

To rationalize the required magnetic moment, the following assumptions are made:

- The dipole group b located at $\vec{r}_b$ sees the charge transfer in group a with the electric dipole moment $\vec{\mu}_{if}^a$ as a "quasi rotation of charges."
- By this charge transfer a magnetic moment is induced at $\vec{r}_b$ proportional to $\vec{r}_b$ and the current density

$$\vec{m}_{if}^b \sim (v_{if}/c)(\vec{r}_b \times \vec{\mu}_{if}^a) \tag{25}$$

- The mutual influences between group a and b in the distance $\vec{r}_{ab}$ are nonzero for systems without symmetry inversion center.
- Thus, there is optical activity to be observed with the rotational strength[7]

$$R_{if} \sim \vec{r}_{ab}(\vec{\mu}_{if}^a \times \vec{\mu}_{if}^b) \tag{26}$$

It can be shown that the rotational strengths for the two transitions of equation (24) are equal with opposite signs:

$$R_{if}^{\alpha\beta} = \pm(\pi/2)v_0 \vec{r}_{ij}(\vec{\mu}_{if}^a \times \vec{\mu}_{if}^b) \tag{27}$$

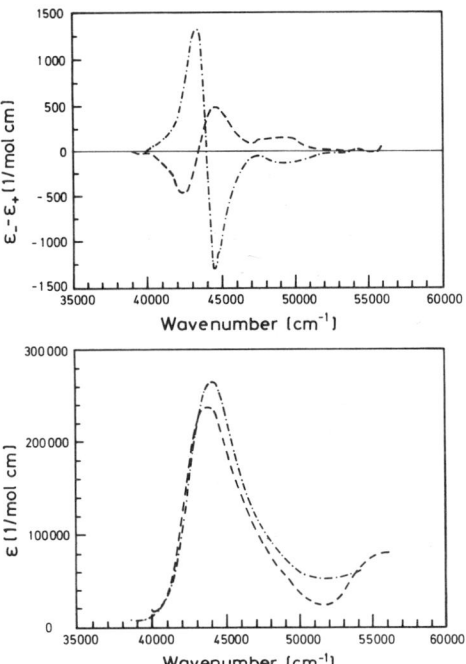

FIGURE 9. Absorption and CD spectra of R-(+)-binaphthole (dashed line) and L-(–)-binaphthole (dots and dashes) in the energy region between 35,000 and 60,000 cm^{-1}.

v_0 is the wavenumber of the electronic excitation in a single chromophore. As a typical example for a molecule with exciton coupling, the absorption and CD spectra of L-(–)-binaphthole and R-(+)-binaphthole are given in Figure 9. The CD spectra show quite clearly the predicted bisignate shape where both components of the transition have about the same rotational strength.

Using the structural information provided by equation (27) it is possible to determine the absolute configuration of the molecule provided, the direction of the transition moment in the chromophore is known. When the electric transition moments constitute a right-handed screwness the CD sign of the low-energy transition is positive and that of the high-energy one negative; for left-handed screwness it is the other way around. In some cases it is possible to numerically calculate CD spectra of various compounds from the known geometric parameters. This procedure is worked out in the book by Harada and Nakanishi.[7] They show that the Davydov splitting V_{ij}, the rotational strength, and the dipole strength depend on the dihedral angle between the planes of the chromophore. Thus, it is in principle also possible to extract the geometric parameters from the measured spectra or to calculate CD spectra from known geometry and dipole strength. Illustrative examples for these calculations are given in their book.

5. SOME TYPICAL APPLICATIONS OF CD MEASUREMENTS

Various workers measuring CD spectra in the VUV region have concentrated on different types of molecules and they have carried out their experiments with different fixings of their aims. In particular, the groups of Schnepp,[69] Mason,[6] Gedanken,[70,71] Snyder,[35] and the Bonn group[72] have investigated small molecules with simple chromophores in order to classify and assign electronic transitions of chemically important groups; most of the other groups have concentrated mainly on biological molecules. For example, nucleic acids have been investigated by Johnson[73] and Sutherland,[74] polysaccharides by Johnson[75] and Stevens[76] whereas CD spectra of proteins have been studied by the group of Johnson.[77]

5.1. Applications in Chemistry

In chemistry, CD spectroscopy is in most cases still used empirically to observe conformational changes in groups of related molecules. However, if the configuration and the unperturbed electronic properties of all groups within a molecule are known, at least in theory the conformation of the molecule can be extracted from CD spectra. On the other hand, for molecules where the exciton coupling model can be applied it is possible to determine the absolute configuration of the molecule (see Section 4.3).

If configuration and conformation are known, CD spectra can be used to assign electronic excited states. The electronic transitions of many molecules are lying in the VUV and they show very often broad and structureless continuous bands. This makes an unambiguous assignment of these transitions difficult and in many cases even impossible. Here, the additional information beyond normal absorption spectra as can be provided by CD spectra of optically active molecules can be very helpful.[37] As CD spectra contain both positive and negative signals, it is often possible to uncover transitions not being resolved in the absorption spectrum. As an example of this type of application, the gas-phase absorption and CD spectrum of S-(+3)-epoxy-3-methylbutane is shown in Figure 10: while in the CD spectrum at least three distinct electronic transitions are visible, the absorption spectrum is nearly structureless.[78]

From equations (17) and (19) it follows that the so-called Kuhn anisotropy factor $g = R_{if}/D_{if}$ is proportional to the quotient of the magnetic and electric dipole moment. Thus, CD bands with small g values represent electric dipole allowed and magnetic dipole forbidden transitions, whereas large values of g represent the opposite case. For the camphor-type molecules shown in Figure 7, g values of about 0.015 are found for the low-energy transition and about 0.005 for the high-energy one. These values support the assignment to a magnetic dipole and an electric dipole transition, respectively.[67]

In most cases the assignment of an electronic absorption spectrum is supported by a quantum-mechanical calculation. When CD spectra are available, the assignment can be further substantiated by calculating at least the sign of the rotational strength according to equation (19). On the other hand, CD spectra are also a sensitive test for the quality of a calculation. Both aspects are illustrated in publications by Rauk et al.[79,80] Here especially the problems are discussed in relation to the usual approach to the computation of chemical effects in large molecules by using small model systems which incorporate the effects under investigation. In general, such an approach is not suitable in the study of natural optical activity which is a second-order phenomenon and very sensitive to small changes in the charge distributions and to conformational effects. The importance of combining CD

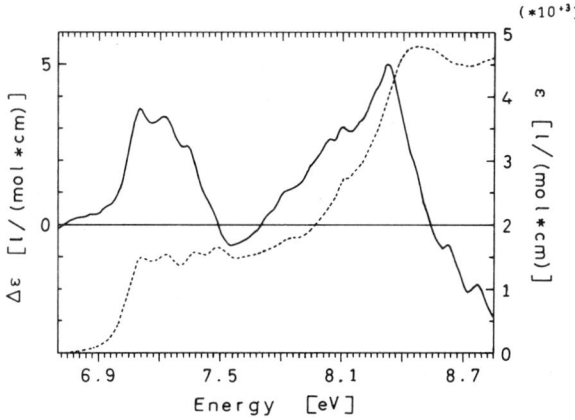

FIGURE 10. CD and absorption spectrum of S-(+3)-epoxy-3-methylbutane.

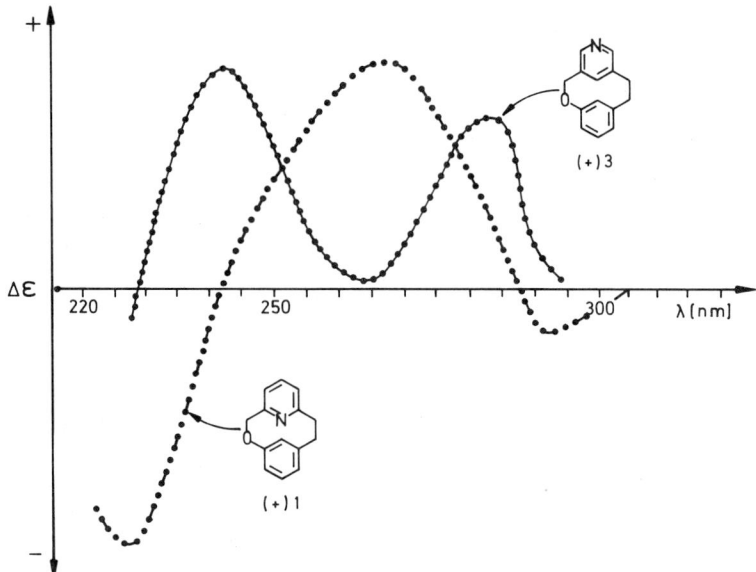

FIGURE 11. CD spectra of two strained helical oxa[2.2]phanes (redrawn from Ref. 87).

measurements with theoretical calculations for a conformational analysis is also demonstrated in the publications of Buss et al.[81,82] An instructive example is the investigation of chiral 11-*cis*-, 13-*cis*-, and all-*trans* retinal Schiff bases where CD measurements are used to gain structural information about these molecules playing an important role in basic biochemical processes.[83–85]

A great number of CD experiments especially in the VUV region are carried out with the aim of improving the understanding of the relationships between chiroptical properties and stereochemical features. An interesting new approach within these efforts is the concept of the "rotating" chromophore followed by Vögtle et al.[86,87] These researchers have synthesized groups of optically active molecules with high stereochemical rigidity. In these groups only the orientation of the chromophore is varied without changing the electronic and/or molecular structure. As an example of such investigations, the near-UV CD spectra of two strained helical oxa[2.2]phanes are given in Figure 11. The striking differences in the spectra are exclusively caused by the different orientation of the chromophore. A detailed explanation of these observations is still lacking although new and improved insights into the possibility of extracting structural information from CD spectra can be expected from these types of experiments in the future.

5.2. Applications in Biology

CD measurements have been widely used to study biological systems. Illustrative examples of these investigations are given in review articles on proteins, carbohydrates, and nucleic acids (see the references given above). Here the application of CD spectroscopy for studying the structure and conformation of proteins will be discussed in some detail.

The general approach in protein structural analysis by CD spectra is to find a correlation between measured spectra and basic secondary structure. That CD spectra are very sensitive to those structures is demonstrated in Figure 12 where the spectra of three basic structures are shown. Assuming that the spectra of suitable reference compounds are linearly independent functions, the CD spectrum of a protein $R_p(\lambda)$ can be calculated at each wavelength from

$$R_p(\lambda) = \sum_{i=1}^{N} a_i R_i(\lambda) \tag{28}$$

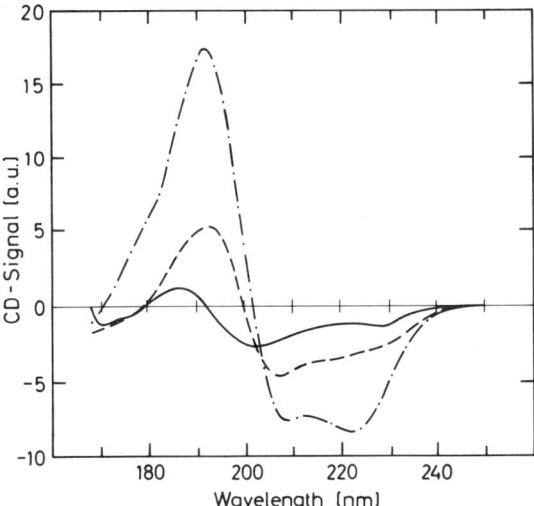

FIGURE 12. CD spectra of myoglobin (dots and dashes), lysozyme (dashed line), and α-chymotrypsin (solid line) (redrawn from Ref. 93).

where $R_i(\lambda)$ is the CD of the ith type of structure and a_i is the fraction of this structure in the protein under investigation. This type of analysis was carried out for the first time by Greenfield and Fasman.[88] However, there are a number of severe problems related to this approach as the influences of the length of segments of secondary structure, contributions from chromophoric side chains, and, e.g., the different known types of β turns are neglected.

All of these disadvantages can be avoided by using a set of CD spectra of proteins with secondary structure known from x-ray diffraction containing contributions from all factors affecting CD spectra. The problem with this alternative used for the first time by Saxena and Wetlaufer[89] and Chen et al.[90] is that there are quite often more proteins in the basis set than there is information content in the spectrum of the protein being analyzed. In general, five basis CD spectra, having a priori no direct physical significance, are needed to reconstruct a set of 16 reference protein spectra that extend into the VUV region to 178 nm. It has been shown that the CD spectrum of a protein truncated at 200 nm contains only the information content equivalent to two equations (most protein CD spectra have two bands in the region between 240 and 200 nm). A spectrum truncated at 190 nm, the wavelength limit of older CD apparatuses, contains three or at best four equations and only the spectrum truncated at about 178 nm contains five equations required to determine the secondary structure categories: α helix, antiparallel β sheet, parallel β sheet, β turns, and others.[91] A detailed discussion of this approach, its problems and limitations is given in papers by Hennessey and Johnson.[92] Applying this method to proteins of known structure gives results of such good agreement with x-ray data that it seems to be applicable with confidence to unknown systems. As an example of the fitting procedure, Figure 13 shows the reconstruction of the CD spectrum of papain (which is one protein from the reference set) with all five basis functions. The procedure described here is applied to monitor structural changes on denaturation, ligand binding, and with solvent variation. Examples of these analyses are given in a review by Johnson.[93]

Recently, Johnson's group has extended the wavelength range for their CD measurements down to about 168 nm, thus adding one more band to most of the spectra.[95] This additional band adds more information and the analysis carried out showed that the prediction of secondary structures of proteins was improved once again. These measurements were carried out with a conventional light source and due to intensity problems at short wavelengths a short path length of about 2 μm had to be used and several scans were added to improve statistics.

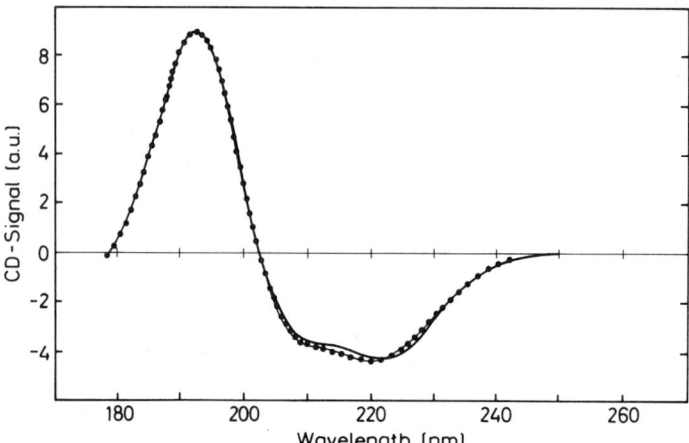

FIGURE 13. CD spectrum of papain: measured (solid line) and reconstructed (dotted line) using five basis spectra (redrawn from Ref. 92).

Using SR as a more intense VUV light source could perhaps improve the measurements discussed earlier and also a further extension into the VUV might be possible. That this is true for the investigation of nucleic acids was demonstrated by the CD measurements of double-stranded nucleic acids by Sutherland et al.[94] These authors could show that the CD signal below 200 nm of the investigated species were not only greater in amplitude than those above 200 nm but also from shape and amplitude of the signal a better indicator of the conformation.

6. CIRCULAR DICHROISM IN THE ANGULAR DISTRIBUTION OF PHOTOELECTRONS (CDAD)

For almost two decades it has been argued theoretically[96–99] that the phenomenon of CD, usually observed in photoabsorption, should be visible in photoemission as well. Moreover, CD in photoemission, or, more precisely, CD in the angular distribution of photoelectrons (CDAD), has been predicted to arise even for *nonchiral species such as diatomic molecules*, provided they are spatially oriented.[96] A substantial difference to the common CD is that CDAD occurs already in *the pure electric dipole approximation*,[97] whereas the common CD is due to an interference between electric and magnetic dipole transitions. Consequently, CDAD asymmetries were expected[98,99] to exceed CD asymmetries by several orders of magnitude.

In this section we will illustrate the physical origin of this new appearance of optical activity and explain the experimental technique used to detect CDAD. Case studies for oriented molecules will be given in the following section.

6.1. Physical Nature of CDAD in an Electric Dipole Transition

In its broadest sense the term *dichroism* refers to the selective absorption of one of the two orthogonal polarization states of an incoming photon beam. CD in photoabsorption of chiral (i.e., "handed") molecules having no plane or center of symmetry is a special case of natural optical activity.

An obvious conclusion is that CD should be visible in photoemission from chiral molecules as well. Ritchie[96] performed detailed theoretical studies of this problem and predicted a CDAD of photoelectrons also from nonchiral molecules provided they are spatially oriented. These calculations went beyond the electric dipole approximation and the predicted effects were relatively small. For chiral molecules, Ritchie derived a CDAD in the pure electric dipole approximation.

Subsequently, Cherepkov[97] obtained a CDAD in a simple electric dipole approximation without the inclusion of the spin-orbit interaction for oriented nonchiral molecules in a special geometry. Moreover, the magnitude of the expected photoelectron-intensity variation on reversal of photon helicity was estimated to exceed that of the "classical" CD by orders of magnitude. A first numerical model calculation for oriented CO by Dubs et al.[98] indeed predicted CDAD effects of up to almost 100% asymmetry that should be immediately visible in a photoemission experiment. These theoretical predictions of the Leningrad and Pasadena groups gave the impulse for experimental studies of the new phenomenon.

The criterion for the appearance of CD effects is that the combined system of photons and reaction partner or target exhibit a definite handedness. Thus, at first sight, it seems surprising that nonchiral species such as diatomics can have an asymmetric response to photons of different helicities. The point, however, is that, even if the target has no "natural" chirality, it is possible to "induce" it by means of the experimental geometry. In order to define a handedness, the three relevant axes, i.e., direction of photon impact **q**, photoelectron momentum **k**, and molecular axis **n**, must be noncoplanar.

This situation is illustrated in Figure 14: The molecule in (a) exhibits classical chirality, whereas the experimental arrangement in (b) is chiral because the two "enantiomorphic" geometries are not identical. Both (a) and (b) show mirror images, and thus a given circular polarization on the left has the opposite sense on the right. Hence, from basic symmetry considerations, the interaction of "handed" light with a target in a handed observation geometry may indeed give rise to dichroic effects. Chiral molecules show optical activity due to the lack of inversion symmetry in the structure of the molecules themselves, while the CDAD from linear or other nonchiral molecules appears due to the lack of inversion symmetry in the overall geometry of the experiment.

From these symmetry considerations we can immediately draw the following conclusions: (1) CDAD *will necessarily require a preferred molecular orientation in space*, because for a random mixture the axis **n** is missing. (2) CDAD *will not occur in an angle-integrated observation*, because

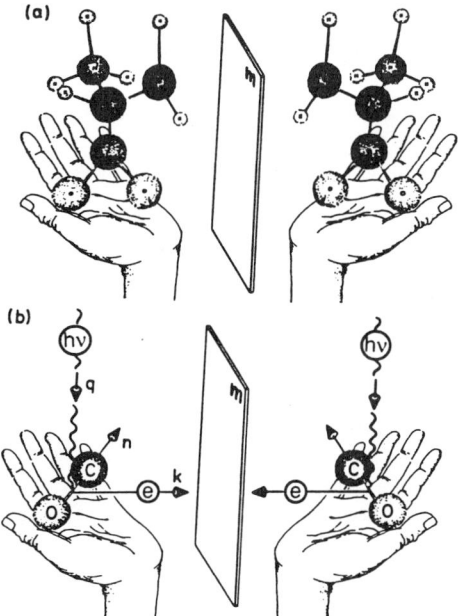

FIGURE 14. "Natural" handedness of a chiral molecule (a) and handedness "induced" by a disymmetric experimental arrangement (b). Photon propagation direction **q**, electron momentum **k**, and molecular axis **n** define a handed coordinate system. Both (a) and (b) exhibit a lack of inversion symmetry thus allowing for circular dichroic effects.

then a definite **k** is missing. (3) CDAD *will vanish in the plane through* **n** *and* **q** because then the three vectors are coplanar.

Of course, in an optical dipole transition at low photon energies, parity-violating processes are completely negligible. The result must therefore be invariant on application of the parity operation which changes both the photon helicity and the handedness of the molecules (Figure 14a) or the vector arrangement (Figure 14b). Hence, this invariance causes the interaction of right circularly polarized (RCP) light with a "right" enantiomer or a "right" vector frame to yield the identical result as left circularly polarized (LCP) light interacting with a "left" enantiomer or a "left" vector frame.

However, switching of the photon helicity and keeping the geometrical arrangement fixed obviously leads to *nonequivalent physical situations*, allowing for CD asymmetries in the photoelectron intensity along a definite direction **k**.

6.2. A Quantitative Example

Of course, such symmetry arguments alone cannot tell us anything about the size of a possible CDAD effect. Since the phenomenon arises in a pure electric dipole transition[97] and does not require inclusion of spin-orbit interaction or higher multipole contributions, it is rather simple to estimate its magnitude by means of a model calculation. In the following, we will treat photoemission from an aligned p_z-orbital as a quantitative example illustrating CDAD for an explicit case.

The differential photoemission cross section is proportional to the square of the total dipole matrix element

$$\frac{d\sigma}{d\Omega}(\theta, \phi) = \frac{4\pi}{3} \alpha a_0^2 h\nu \, |\langle \Psi_f | \mathrm{Op} | \Psi_i \rangle|^2 \tag{29}$$

where α is the fine-structure constant, a_0 is the Bohr radius, and Ψ_f and Ψ_i are the wave functions of the final and initial state, respectively. For aligned atomic orbitals, the differential cross section can be derived using the equations and tables given by Goldberg et al.[100] [note, however, that equation (22) of that paper is not valid for CP light]. The dipole operator corresponding to the beam of circularly polarized light in the experimental geometry used (see Section 6.3) is $\mathrm{Op} = z \sin 50° + x \cos 50° \pm iy$ for RCP or LCP light.

Employing symmetry selection rules (or by using the tables of Ref. 100) one can see that for this geometry the photoemission from the p_z-orbital leads to three final-state channels (the amplitude of the d_{xz} wave is zero in the plane of observation) with *three contributions to the dipole matrix element* (see Figure 15):

$$p_z(Y_{10}) \rightarrow \begin{cases} s & (Y_{00}) & [z] \\ d_{z^2} & (Y_{20}) & [z] \\ d_{yz} & \left(\frac{i}{\sqrt{2}}(Y_{21} + Y_{2-1})\right) & [\pm iy] \end{cases} \tag{30}$$

Here, the spherical harmonics on the right side denote the angular parts of the corresponding final-state partial waves. The components of the dipole operator which cause the excitation are given in square brackets.

Insertion of the actual wave functions into the matrix element yields

$$\langle \Psi_f | \mathrm{Op} | \Psi_i \rangle = 4\pi \left[\frac{\sin 50°}{\sqrt{3}} R_s \exp[i\delta_s] Y_{00}^* - \frac{2 \sin 50°}{\sqrt{15}} R_{d_{z^2}} \exp[i\delta_{d_{z^2}}] Y_{20}^* \pm \right.$$

$$\left. \pm \frac{1}{\sqrt{5}} R_{d_{yz}} \exp[i\delta_{d_{yz}}] \frac{1}{\sqrt{2}}(Y_{21} + Y_{2-1})^* \right] \tag{31}$$

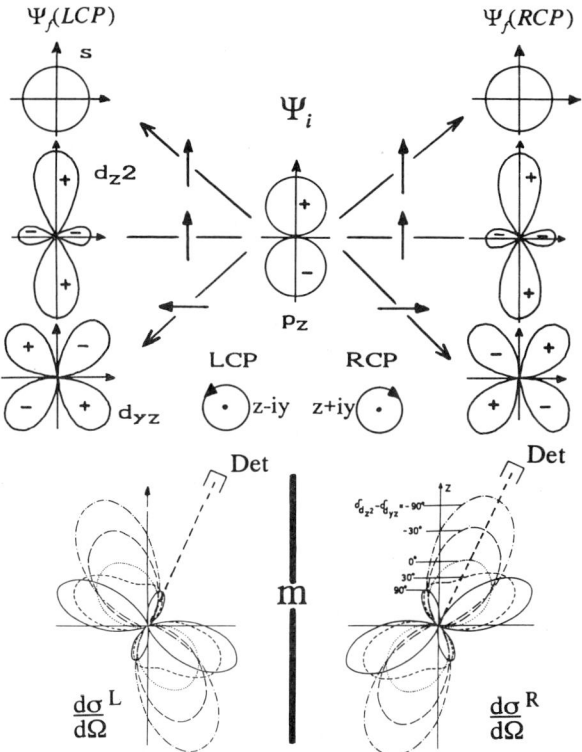

FIGURE 15. Circular dichroism in the angular distribution (CDAD) of photoelectrons derived for a p_z-orbital. Photon incidence is along x, perpendicular to the drawing plane. The upper panel shows the angular parts of the initial state (p_z) and the final-state partial waves (s, d_{z^2}, d_{yz}) excited by the electric dipole transition. The lower panel shows the photoelectron angular distribution pattern from a model calculation assuming equal radial matrix elements ($R_s/R_d = 1$) and certain values for the phase differences.

where the R_{lm} are the usual radial matrix elements[100] and δ_{lm} are the relative phases of the corresponding final-state partial waves, which we consider as both l and m dependent in order to account for anisotropic final-state interactions (e.g., due to the possible presence of the surface in an adsorbate system).

Equation (31) already reveals the origin of the CD in photoemission: *the ± sign of the dipole operator appears in the third term of the matrix element*, due to the d_{yz} partial wave excited by the $\pm iy$ component. The differential cross section is proportional to the square of the matrix element [see equation (29)]. The ± sign enters into $d\sigma/d\Omega$ in the form of *two interference terms which determine CDAD quantitatively*. Neglecting again the d_{xz} wave, which does not contribute in this geometry, the differential cross section is given by

$$\frac{d\sigma}{d\Omega}(\theta, \phi)^{R,L} \propto \sin^2 50°[R_s^2 + R_{d_{z^2}}^2(3\cos^2\theta - 1)^2] + R_{d_{yz}}^2(3\sin\theta\cos\theta\sin\phi)^2$$

$$- 2\sin^2 50° R_s R_{d_{z^2}} \cos(\delta_s - \delta_{d_{z^2}})(3\cos^2\theta - 1) \pm 3\sin 50° R_s R_{d_{yz}} \sin(\delta_s - \delta_{d_{yz}}) \sin 2\theta \sin\phi$$

$$\pm 3\sin 50° R_{d_{z^2}}^2 \sin(\delta_{d_{z^2}} - \delta_{d_{yz}})(3\cos^2\theta - 1)\sin 2\theta \sin\phi \quad (32)$$

The two CDAD terms vary in the azimuth like $\sin\Phi$ (Φ is counted from the x-axis); hence, $\sin\Phi = 1$ in the (y,z)-plane (the plane of photoelectron detection, see Section 6.3). The CDAD in photoemission from a spatially fixed p_z-orbital of an isolated atom (in isotropic space) has been theoretically predicted by Dubs et al.[98b]: its angular dependence should be proportional to $\sin 2\theta$. The first CDAD term agrees with their result, whereas the second term is absent in the free-atom-like treatment, which does not allow for a phase difference between partial waves with the same l values.

The resulting forms of $d\sigma/d\Omega$ are illustrated in Figure 15 as polar plots. The case of unpolarized light (not shown) would essentially reflect the structure of the initial-state orbital peaked along z. In contrast, the angular distributions for RCP and LCP light are considerably different. For the calculation of the angular distributions, numerical assumptions have been made. First, the matrix elements R_s and R_d were taken to be equal, which is likely to be true for some systems in the low-energy region, although with increasing energy the $l+1$ channel usually becomes dominant. Second, the phase-shift difference between s- and d-wave was chosen such that the CDAD effect is maximal, i.e., $\sin(\delta_s - \delta_{dz}) = 1$. This condition is also realistic for certain systems because this phase difference varies rapidly above threshold, with the main contribution resulting from the Coulomb phase shift which can vary by almost π within the first 100 eV. Third, the phase-shift difference between the two d partial waves was varied between $-90°$ and $+90°$ in order to demonstrate the strong phase dependence of CDAD. The free-atom case is represented by the $0°$ curve. Clearly, the circularly polarized light causes a *marked rotation of the angular distribution pattern, in the same sense as the electric vector of the ionizing radiation*, i.e., differently for RCP and LCP. The total photocurrents for RCP and LCP are identical because one pattern is the mirror image of the other one (mirror plane m). This means that here the total photocurrent does not show a CD effect and, as noted above, an angle-resolved measurement is required to observe it.

The at-first-sight minor effect of a sign change in a matrix element thus turned out to have drastic consequences for the photoelectron angular distribution. If the photoelectron detector is kept fixed ("Det" in Figure 15), the change of photon helicity will cause huge intensity changes, i.e., a CDAD. Only along z (coplanar geometry) and along y (since for p_z this is a nodal plane) no differences occur.

6.3. Experimental Technique

From the principle of CDAD illustrated in Figure 14 we see that the phenomenon can be observed by an angle-resolving photoelectron spectrometer with three special features: (1) the ionizing radiation must be *circularly polarized*, (2) besides photon and electron beam direction a third axis must be defined through an *oriented or aligned target*, and (3) the *experimental geometry must define a handedness*, e.g., by arranging the planes of photon incidence and electron detection perpendicular to each other [thus giving $\sin \Phi = 1$ in equation (32)].

Possible ways to orient a molecular target should be mentioned briefly. One is the application of an external hexapole field on a molecular beam.[101,102] This technique is rather involved and yields only relatively low target densities and an incomplete orientation. Another method for aligning a molecular target is by optical pumping with linearly polarized light. In fact, CDAD was first observed from NO aligned both in this way[103,104] and by photofragmentation of methyl nitrite, CH_3ONO.[105] The Bielefeld/Mainz group used the adsorption on a single-crystal surface as a technique of orientation, which can yield a perfectly oriented ensemble of molecules. This method leads to very high target densities, about two orders of magnitude larger than in the gas phase, resulting in huge photoelectron intensities.

Beyond the LiF cutoff at $h\nu = 11.5$ eV, conventional methods of producing circularly polarized (CP) light fail because of the lack of transparent materials. However, out-of-plane synchrotron radiation has proven to be an excellent source of CP light in the windowless VUV range. It combines all of the common advantages of synchrotron radiation such as high intensity and brightness, easy tunability and good ultrahigh vacuum conditions with a high degree of circular polarization. This method works not only in the wavelength region of normal-incidence VUV optics but also in the x-ray range (see Section 7.5).

The experimental arrangement for CDAD measurements in the low-energy range is shown schematically in Figure 16. It is operated at the 6.5-m normal-incidence monochromator[106] at the storage ring BESSY (Berlin). This beamline has been optimized for a high flux of CP radiation in the photon energy range from the visible up to 40 eV. It accepts angular ranges of ± 25 mrad horizontal and ± 5 mrad vertical; its virtual entrance slit is the electron beam at the tangential point in the bending magnet. The optical components are a spherical mirror in near-normal incidence (at $4°$) and inter-

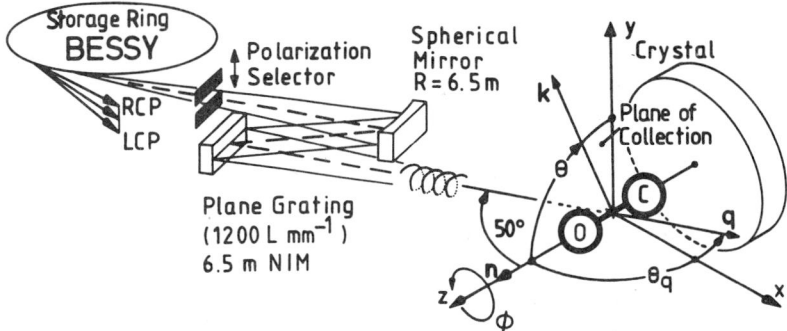

FIGURE 16. Schematic view of the experimental setup for the investigation of circular dichroism in photoemission using circularly polarized synchrotron radiation at BESSY. R and L denote right- and left-handed circular polarization. Photon propagation direction **q**, electron momentum **k**, and surface normal **n** define a handedness.

changeable plane holographic gratings of either 1200 lines mm^{-1} (Al + MgF$_2$ coated) or 3600 lines mm^{-1} (Os coated) placed in the convergent beam behind the mirror. Using a 2-mm exit slit the bandwidths are 0.4 or 0.2 nm, respectively. The degree of circular polarization can be selected by means of two movable beam stops defining the vertical angular range accepted. With the circular polarization being set to 92 ± 2%, a maximum intensity of 8 × 10^{11} photons s^{-1} is obtained at $h\nu = 30$ eV and 500 mA stored beam.

The rotatable photoelectron spectrometer has been attached to the beamline such that the plane of incidence is horizontal with the photon beam **q** and surface normal **n** forming an angle of $\theta_q = 130°$. The CDAD asymmetries are thus reduced by a factor of sin 130° = 0.77 as compared with the ideal case of photon incidence along x (which corresponds to grazing incidence). The plane of photoelectron detection is vertical and the spectrometer covers the whole possible range from $\theta = 90°$ to $-90°$. Its angular setting (via a stepper motor) and energy sweep, as well as the monochromator setting (including polarizer apertures) are computer controlled. Thus, the measuring procedure is fully automated and allows for a high data acquisition rate.

The photoelectron spectrometer[107] is a simulated hemispherical capacitor whose design parameters essentially follow those given by Jost.[108] Its angular acceptance is ± 3°, and the energy resolution was usually set to 250 meV FWHM (at 4 eV pass energy), a value which is sufficient for most adsorbate studies. Data are accumulated by a microcomputer in multichannel scaling mode at typical count rates of 10^5 counts s^{-1} in the peak maxima.

After the photoelectron spectrum for one light helicity was recorded, the light polarization was automatically switched by driving the polarizer slit to the opposite acceptance position for taking the second spectrum. Because of small variations of the spatial electron-beam position in the storage ring, intensity differences between RCP and LCP light could occur. These were checked and corrected by setting the intensities of the spectra for $\theta = 0°$ equal, since at this emission angle no CDAD is possible [see equation (32)].

6.4. What Can Be Learned from CDAD Measurements?

CDAD spectroscopy as an experimental technique is rather new, hence its usefulness is far from being fully explored. However, there are a few aspects which raise new possibilities and which will be discussed in this section.

CDAD is the result of a pure electric dipole transition and can be understood in terms of a nonrelativistic orbital picture neglecting all electron-spin effects. It is a consequence of the interference of final-state partial waves differing in their m quantum number by ± 1. Since its theoretical description is rather simple, it is straightforward to derive analytical expressions for the asymmetry. These contain a "geometrical" factor which depends on the angular parts of the wave functions involved and a "dynamical" factor which contains the basic quantities that govern the photoexcitation, i.e., the dipole matrix elements and relative phases of final-state partial waves.

This fact has three practical consequences:

1. It is relatively simple to *calculate CDAD numerically*. Provided that the relevant dipole matrix elements and relative phases of partial waves exist, they simply need to be inserted into the corresponding expressions. Hence, CDAD measurements may serve as a sensitive test for photoemission theories (see below).
2. It is possible to extract the (ratios of) matrix elements and phase shifts from the measured CDAD data by performing a fit using the analytical CDAD expressions. Similar approaches are often termed *perfect or complete photoemission experiments*, which means that all relevant matrix elements and phases are determined experimentally.[109] First results in this direction have been obtained for benzene,[110a] the π-band of graphite,[111] and atomic oxygen (see Section 7.4).
3. Owing to the geometrical factor in the CDAD expressions any *structural changes of adsorbate overlayers* show up in the CDAD[110a] (see also Section 7.2). In the gas phase, CDAD is ideally suited to probe the degree of alignment or orientation of molecules or fragments as originally proposed and discussed by Dubs et al.[98c,103–105]

A point of similar importance is that CDAD is not restricted to optically active ("chiral") molecules. Instead, it occurs *for all molecules, provided they are spatially fixed* (i.e., oriented), which means that the rotational degree of freedom is (at least partly) absent. It is well known that this is very often the case for molecules adsorbed on surfaces. Hence, molecular adsorbates are an ideal testing ground for this new observable quantity.

All molecular adsorbates studied so far (e.g., CO, NO, C_6H_6, CH_3I, COOH, C_{60}) and also clean surfaces of solids [e.g., Pd(111), graphite(0001), Ni(100), Ni(111), Fe(100), Pt(111)] clearly showed the effect, thus revealing CDAD to be a very common phenomenon.

Because of its extreme dependence on photoemission dynamics (see, e.g., the phase dependence of the differential cross section in Figure 15), CDAD in photoemission from solids constitutes a *sensitive probe for hybridization regions of electronic bands*. This was recently demonstrated for palladium: The A_{CDAD} parameter versus photon energy for a certain interband transition exhibits a sharp asymmetric ("autoionization-like") profile in the region of an avoided crossing between the free-electron-like and a flat f-like final-state band.[112]

Previously, the use of circularly polarized SR has proven to be a powerful tool for CD measurements (see Sections 1–5) or in connection with photoelectron spin-polarization spectroscopy (e.g., see Ref. 107) in order to fully exploit the relativistic selection rules. However, the latter method required experimental resolution of spin-orbit interaction thus ruling out the wide class of low-Z systems. The CDAD technique *does not require spin-orbit interaction*. Hence, it is applicable to all molecular species independent of atomic number, symmetry, or orbital structure.

Concluding this section, it should be mentioned that a similar effect occurs in photoemission with linearly polarized light.[113] Analogously, it has been termed *linear dichroism in the angular distribution of photoelectrons* (LDAD). Because no experiments on the LDAD have been performed, we will not discuss this technique further.

7. CASE STUDIES OF CDAD FOR ORIENTED MOLECULES AND COMPARISON WITH *AB INITIO* CALCULATIONS

7.1. Experiments on Gas-Phase Molecules with Laser-Induced Alignment

After the work of Ritchie[96] (see Section 6.1), Parzynski[114] made the first theoretical prediction of a CDAD for polarized alkali atoms in the ground state. Since there is only an orientation of the spin in the ns ground state, a CDAD in this case arises only as a consequence of the spin-orbit interaction. For aligned (and oriented) atoms, Dubs et al.[98c] and Cherepkov and Kuznetsov[97d] derived the general expressions of CDAD. For space-fixed or aligned molecules, the same authors[97,98] performed detailed theoretical treatments including numerical predictions.[98]

The first measurements of CDAD were performed by Appling et al.,[103,104] who used the new phenomenon to probe the alignment of NO molecules excited to the $A^2\Sigma^+$ $v = 0$ state by linearly polarized light [two-color (1+1) REMPI]. In these experiments two counterpropagating laser beams crossed the molecular beam and the photoelectron spectrometer was set up at an angle normal to the laser beams. The angular dependence of CDAD was measured by rotating the electric vector of the pump laser beam. These experiments demonstrated that CDAD is not a small effect and that it can be successfully utilized to detect molecular alignment.

The CDAD arising in the final ionization step by the circularly polarized laser beam directly probed the degree of alignment of the optically pumped A-state rotational levels. Resultant photoelectron angular distributions exhibited significant R–L asymmetries, the phase and magnitude of which are shown to be related to the curvature of the excited-state M_J distribution.

An example is shown in Figure 17. In these curves the CDAD intensity has been normalized to the intensity of the R or L spectra at 90°. The solid curves represent a least-squares fit of the measured

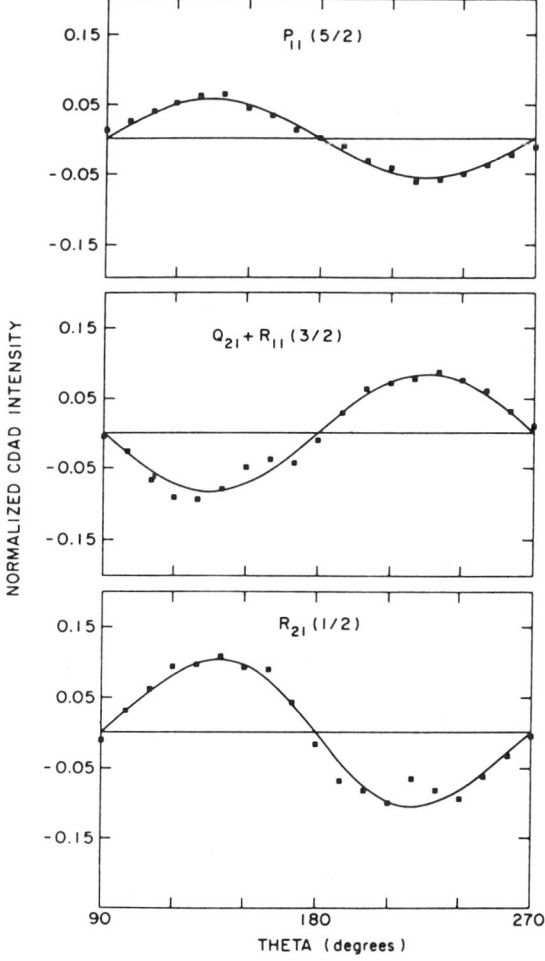

FIGURE 17. Experimental CDAD for three rotational transitions in REMPI of NO. All transitions lead to $J' = \frac{3}{2}$; curves are a least-squares fit. (From Ref. 104.)

points, thereby extracting the alignment parameters corresponding to the three rotational transitions shown (P, Q + R, and R branch).[104] The evaluation showed that the phases of the experimentally determined alignment parameters are clearly branch-dependent, in agreement with theory. The opposite phase for the Q branch can be directly seen in Figure 17 (central panel, dominating Q contribution) which reflects the opposite alignment of this state. Note that all spectra arise from the same $J' = \frac{3}{2}$ intermediate state. The influence of the hyperfine depolarization effect is also discussed in this paper.[104]

Cuéllar et al.[115] performed the first CDAD measurements for free atoms. In this case, CDAD spectra were employed, to examine the alignment of the $7P_{3/2}$ and $7P_{1/2}$ levels of cesium, prepared by absorption of linearly polarized laser radiation. In agreement with theory,[98c] the $7P_{1/2}$ level could not be aligned (spherically symmetric) and therefore showed no CDAD, whereas the $7P_{3/2}$ level exhibits a significant CDAD signal. Figure 18 shows the results for the photoelectron angular distributions for right and left circularly polarized light (a) and the CDAD asymmetry (b). The measured asymmetry is about four times lower than predicted by theory (solid curve). The authors attribute this difference between experiment and theory to the m_j-mixing due to hyperfine coupling during the finite ionization time of the laser pulse.

We conclude this section on results for gas-phase systems by recalling the general theoretical predictions of a very recent paper by Berakdar et al.[116] Although this paper deals with double photoionization of unoriented free atoms, it shows up strong analogies with CDAD, which may be important for future investigations. If an ensemble of randomly oriented free atoms is double-ionized by circularly polarized light, the photoelectron pair experiences the handedness of the coordinate frame spanned by the photon beam and the two electron momenta. This chirality, a new manifestation of electronic correlation, causes a *circular dichroism in multiply differential cross sections*. Numerical calculations[116] show that this dichroism is large under suitable kinematical conditions. The authors demonstrate that measurements of the triply differential cross section (TDCS) in double photoionization of unoriented atoms using circularly polarized light yield additional information which cannot be obtained from measurements with linear light polarization. The dichroism is defined as the difference between the TDCSs for left and right circularly polarized light: $\Delta = \text{TDCS}^L - \text{TDCS}^R$. Of course, in single photoionization of randomly oriented species this phenomenon cannot exist because a third axis besides photon beam and electron momentum is missing.

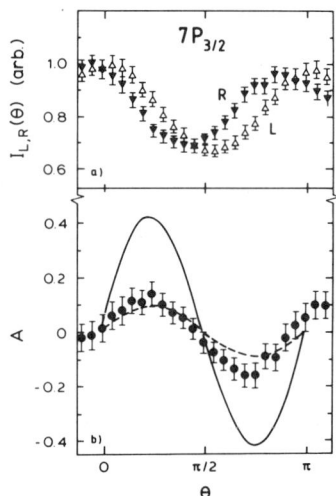

FIGURE 18. CDAD for the $7P_{3/2}$ level of Cs. (a) Photoelectron angular distributions for right and left circularly polarized light. (b) CDAD asymmetry in comparison with theoretical calculations without (full curve) and with (dashed curve) hyperfine interactions. (From Ref. 115.)

There is a striking similarity between this "chiral electron pair" effect in double photoionization and the CDAD for aligned or oriented atoms:

1. In both cases the chirality is experimentally induced by defining three noncoplanar axes.
2. Both CD effects disappear in the plane-wave limit for the final states.
3. Sign and size of both effects depend very sensitively on details of the initial- and final-state wave functions employed.
4. In both cases the dynamical factor of the crucial dichroism term can be expressed as the imaginary part of the product of two different complex dipole transition matrix elements (see Refs. 116 and 122).

The analogy could possibly be viewed as follows: One electron of the pair (in TDCS) plays the role of the axis of alignment or orientation (in CDAD), i.e., this electron defines the axis correlated with the target. The second electron of the pair plays the role of the photoelectron in CDAD.

7.2. Oriented CO and NO on Surfaces

Carbon monoxide adsorbed on various transition metal surfaces probably constitutes the best-characterized molecular chemisorption system. It is usually found in upright orientation with the carbon end pointing toward the surface, a geometry which has been verified by various methods (e.g., Refs. 117, 118). Its bonding mechanism via σ-donation/π-backdonation involving essentially the 5σ- and $2\pi^*$-orbitals[119] has been investigated in detail. *The first theoretical prediction of CDAD was for photoemission from the 4σ-orbital of isolated oriented CO*[98a] and later these calculations have been extended to NO and include the $5\sigma/1\pi$-orbitals and the corresponding orbitals for oriented NiCO,[120] thereby accounting in the simplest approximation for the chemisorption bond to a transition metal.

The Bielefeld Mainz group has thus started their studies with CO and NO adsorbed on a Pd(111) surface. Figure 19 shows a typical pair of photoemission spectra taken with RCP and LCP light (denoted by R and L, respectively). The spectra correspond to saturation coverage of NO on Pd(111) at 140 K. The adsorbate-induced features at binding energies of 14.5 and 9.5 eV correspond to the NO 4σ and overlapping $5\sigma/1\pi$-orbitals. Because the chemical shift is larger for 5σ, located at the C atom, than for the other orbitals, it overlaps with 1π in the adsorbate phase. The features close to the Fermi energy E_F result from the d-bands of the Pd substrate. The R and L spectra have been normalized to one another as described above.

Clearly, all direct transitions, i.e., the adsorbate-induced peaks and also the Pd $4d$ features, show huge CDAD asymmetries, thus confirming the general theoretical predictions of Cherepkov[97] and Dubs et al.[98] The smooth, structureless background of secondary electrons is identical for excitation with RCP and LCP light. This is a general behavior in all spectra. Owing to scattering processes, these electrons have no "memory" for the angular distribution of the primary electrons. Their intensity is essentially given by the total photocurrent, which cannot display a CD effect. This possibility of a distinction between direct transitions and secondary electrons is a special aspect of this new technique.

In order to extract the CDAD information quantitatively from a set of spectra like Figure 19, it is advantageous to introduce an asymmetry function A_{CDAD} as

$$A_{CDAD}(\theta,\phi) = \frac{\frac{d\sigma}{d\Omega}(\theta,\phi)^R - \frac{d\sigma}{d\Omega}(\theta,\phi)^L}{\frac{d\sigma}{d\Omega}(\theta,\phi)^R + \frac{d\sigma}{d\Omega}(\theta,\phi)^L} = \frac{I^R(\theta) - I^L(\theta)}{I^R(\theta) + I^L(\theta)} \quad (33)$$

where $I^R(\theta)$ and $I^L(\theta)$ represent the photoelectron intensity of a photoemission peak with the secondary electron background being subtracted. This relative quantity does not require absolute cross section measurements which are problematic, e.g., due to the unknown transmission functions of the electron optics and the spectrometer.

Analogously to the atomic p_z case discussed in Section 6.2, CDAD is also in the molecular case a direct manifestation of the interference of final-state partial waves differing by $\Delta m = \pm 1$ in their

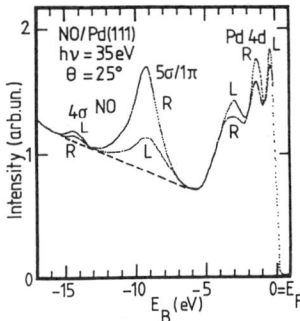

FIGURE 19. Typical photoelectron spectra of NO on Pd(111) illustrating CDAD. R and L denote the spectra for right and left circularly polarized light. The binding energy scale refers to the Fermi energy, E_F, of the substrate crystal. The dashed line denotes the Pd spectrum without NO. (From Ref. 129.)

magnetic quantum number.[98] For an initial σ-orbital two transition channels ($\sigma' \to \sigma$ and $\sigma' \to \pi$) are allowed, thus leading to a series of interference terms containing products of spherical harmonics $Y_{l0}Y_{l\pm1}$.

Figure 20 summarizes the experimental A_{CDAD} for the photoemission peaks of CO 4σ on Ni(100) as a function of the emission angle for several photon energies[123] in comparison with the theoretical calculations.[98a,120,122] The error bars contain both statistical and systematic (largely reproducibility) uncertainties. Each experimental point represents the mean value of several independent runs with different target preparations. For the determination of I^R and I^L a linear background as indicated in Figure 19 has been assumed and simple peak heights were taken; these gave the same results as integrated peak areas within the error limits.

A systematic effect on the emission angle arising in adsorbate and solid-state photoemission is the refraction of the outgoing photoelectron wave at the electrostatic surface barrier. When the photoelectron surmounts this barrier, its momentum parallel to the surface is conserved, whereas the perpendicular component is reduced. This causes an increase of the polar emission angle outside of the surface region, which has been corrected in Figure 20 (for details see Ref. 110a).

Since the calculation assumed 100% photon polarization and 45° angle of incidence (experimentally 92 ± 2% and 50 ± 1°), a further small correction factor entered into the figure. Comparison with the *ab initio* calculations shown as solid and dashed curves for CO and NiCO, respectively, reveals a reasonable agreement with experiment concerning the general energy and angular dependence. All CDAD curves go through zero at normal emission, as expected for a perfect upright orientation of the molecules. The 4σ-orbital is the lone pair being located at the oxygen atom and hence pointing away from the surface.

At photon energies above 30 eV, a marked minimum in the theoretical CDAD curves occurs at high angles. This is a consequence of the 4σ shape resonance[117,118] centered around 36 eV. This resonance is known to occur in the $l = 3$ partial wave of the outgoing photoelectrons in the σ-channel of the continuum. Within the resonance, CDAD is thus described by interference of $l = 3$, $m = 0$ and $l = 1, 2, 3, \ldots$, $m = 1$ continua. Indeed, at 36 eV the CDAD angular pattern is largely determined by the term $Y_{30} \cdot Y_{31}$, shown as the dotted curve.

The basic result of Figure 20 is that *photoemission dynamics and especially the continuum partial-wave phase-shift differences as calculated for free molecules are essentially retained on adsorption on a surface* even in a close-packed monolayer. The crude approximation of treating the chemisorption bond to a transition metal surface via the calculation for an isolated NiCO molecule (chain curves) seems to slightly improve the agreement with experiment as suggested by the results for 27.7 and 30 eV.

There are several possible reasons for the remaining discrepancies which will be examined in the future. On the theoretical side, the ionization potentials were simply reduced from their gas-phase values (19.7 eV for CO 4σ) to the experimental values (16.3 eV for CO 4σ on the surface) to account

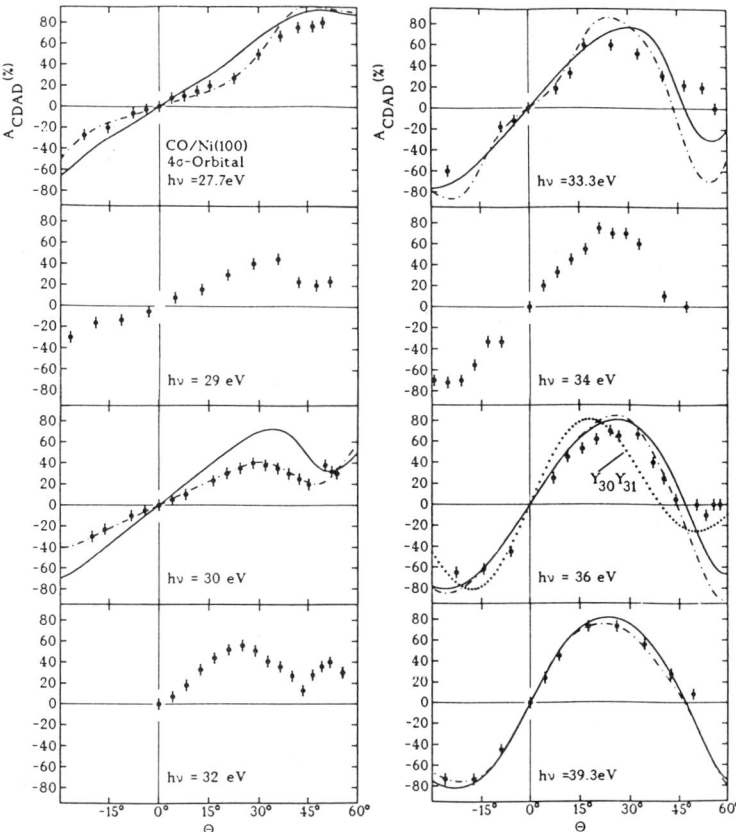

FIGURE 20. CDAD asymmetry for CO 4σ as a function of the photoelectron emission angle for various photon energies. Experimental points were taken for the saturated CO monolayer on a Ni(100) crystal surface at room temperature and include counting statistics and reproducibility. (From Ref. 123.) Solid and chain curves represent the theoretical *ab initio* calculations[120] for isolated spatially fixed CO and NiCO molecules, respectively.

for the relaxation of the ionic state which is essentially due to screening effects induced by the metal surface. Being solely based on energetic considerations, this approach may not be adequate for treating the dynamical excitation process itself. Of course, the "chemical state" of the molecules in terms of the Blyholder model,[119] as well as other many-electron correlation effects will probably quantitatively affect the sensitive CDAD parameter.

On the experimental side, the influence of the neighbor molecules (adsorbate-band formation) will affect the CDAD especially for high emission angles. However, in the asymmetry A_{CDAD} all influences which affect both sorts of electrons (i.e., those excited by RCP or by LCP light) identically, will exactly cancel when deriving the ratio of equation (33). This special property is an advantage for CDAD that does not exist in UPS.

Finally, a refraction of the photon beam could occur at the adsorbate overlayer. According to the Fresnel equations,[124] the effective angle of incidence θ_q (see Figure 16) would increase on entry of the optically thicker medium. Since this angle enters as $\sin \theta_q$ into the CDAD equation (32), the CDAD asymmetry would be diminished. Indeed, most experimental curves in Figure 20 display smaller asymmetries than predicted theoretically. In a systematic study with varying θ_q, CDAD may offer the fascinating possibility to investigate the refraction of VUV light as a function of $h\nu$ within the adsorbate overlayer. Such an investigation is in progress.

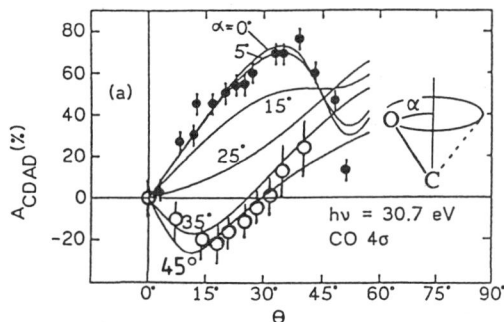

FIGURE 21. Dependence of the CDAD asymmetry on the tilt angle α between molecular axis and surface normal for CO 4σ. Full circles, experimental points for CO on Pd(111) (perpendicular adsorption)[121]; open circles, same for CO on Fe(100) (tilted adsorption)[125]; curves, *ab initio* calculation.[120]

For such adsorbate studies the knowledge of the adsorption geometry is an essential precondition. A very striking example of the dependence on geometry is shown in Figure 21. For the adsorption system CO/Pd(111) the molecules are standing upright on the surface, yielding high positive asymmetries of up to 75%. On the Fe(100) surface, however, CO adsorbs with a tilt angle of 45° with respect to the surface normal,[125] leading to negative asymmetries below θ = 30°. The *ab initio* calculation reveals the pronounced dependence of A_{CDAD} on the tilt angle α; already at α = 30° the asymmetry is changing its sign!

Figure 22 shows CDAD parameters versus collection angle at different photon energies for the NO 5σ/1π-orbitals. Systematic theoretical studies of the 5σ/1π CDAD for NO have not yet been carried out due to complex multiplet structure resulting from 1π ionization. This structure makes it necessary to use a potential averaged over six multiplets which would introduce additional uncertainties into the calculation. Furthermore, changes in electronic structure due to surface binding can be expected to make free NO a poorer model for photoemission from adsorbed NO than in the CO case. Calculated CDAD spectra for photoemission just from the 5σ-orbital of tilted NO for hν = 30.7 eV are shown along with the measured values. It is known from the CO studies[122] that 5σ photoemission dominates the CDAD 5σ/1π signal at angles below 40°. For the reasons discussed in Ref. 122, these calculated CDAD spectra for 5σ cannot be expected to be as useful a model of adsorbate photoemission as for 4σ, being located farther away from the surface. At best, these 5σ calculations should be viewed as suggesting a bent orientation.

7.3. Adsorbed Benzene and Formate as Model Systems of Totally Oriented Species

At low coverage, *benzene molecules* adsorb on most single-crystal surfaces with their molecular plane parallel to the surface.[126] This system thus constitutes a model case for a highly oriented polyatomic molecule. As substrate crystal served the Pd(111) face thus giving a significant chemisorption contribution via the π-orbitals so that the monolayer adsorbs already at room temperature; alternatively, C_6H_6 was physisorbed on the basal plane of graphite which results in a relatively small chemical effect on the molecule.

For polyatomic molecules the CDAD may generally become a quite complicated expression involving many partial waves, so that all details of photoemission dynamics will govern the CDAD behavior. Figure 23 shows a series of CDAD data for the $3a_{1g}$-orbital which is totally symmetric (see right panel). Clearly, the angular pattern of the asymmetry is strongly energy dependent, which reflects the spectral variation of the dipole matrix elements and the relative phases of continuum partial waves (see Section 6.1).

The analysis of photoemission spectra of adsorbed molecules in terms of selection rules and allowed transition channels is possible by means of the powerful group-theoretical approach (e.g., see Ref. 130).

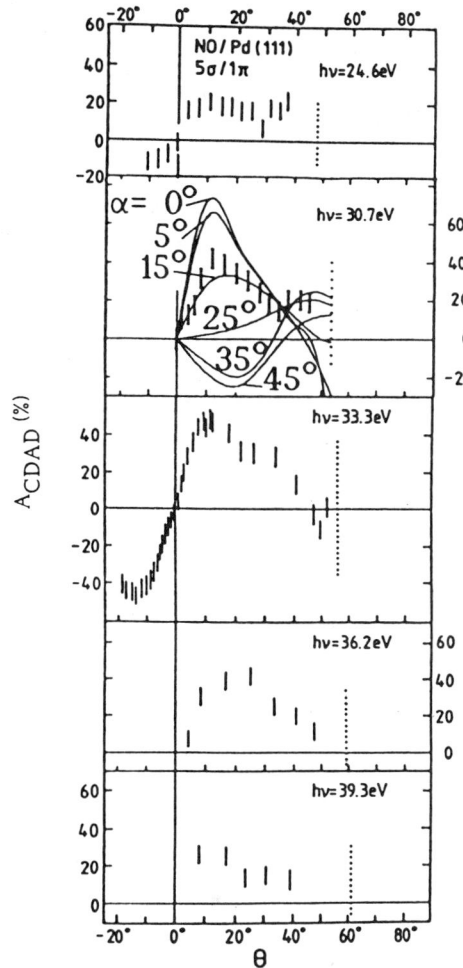

FIGURE 22. Same as Figure 20 but for the $5\sigma/1\pi$-orbitals of NO molecules on the Pd(111) surface at $T = 200$ K. The curves at $h\nu = 30.7$ eV represent an *ab initio* calculation for NO 5σ at various tilt angles α with respect to a given axis (here the surface normal).[122]

The point-group symmetry of monolayer benzene on surfaces like graphite is C_{6v} or C_{3v}.[126-128] Inspection of the corresponding character tables and multiplication tables reveals that, for the photon operator of circular polarization $z \pm iy$, there are usually several allowed transition channels, and these make the CDAD expressions rather involved. One exception is the *totally symmetric initial orbital* a_{1g} because the (total) initial and the final ionic hole states have the same symmetry. As a consequence, only two photoemission channels are allowed according to the scheme

$$a_{1g} \rightarrow \begin{cases} a_{2u} \text{ (excited by the } z\text{-component) } [Y_{10}, Y_{30},...] \\ e_{1u} \text{ (excited by the } iy\text{-component) } [Y_{1\pm1}, Y_{3\pm1},...] \end{cases} \quad (34)$$

Again, it is the well-defined phase relation between the z and $\pm iy$ components that causes the CDAD in the coherent excitation to the two continuum channels.

The final state in photoemission from molecules observed by the detector at infinity is readily described as a one-center wave function, since molecular dimensions are negligible. Hence, the asymptotic wave function corresponding to the a_{2u} and e_{1u} final states can be expanded in a series of spherical harmonics. Since the transformation properties of these states are known from group theory, one can extract the contributing Y_{lm}'s from the compatibility tables. It turns out that for $l < 5$ only few

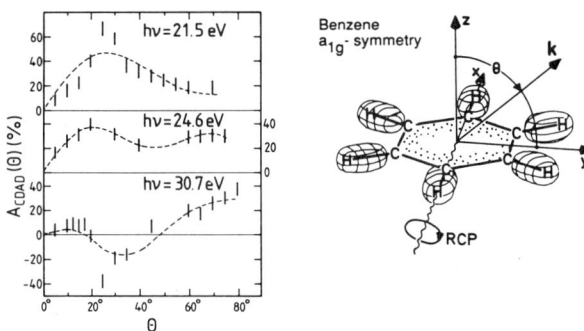

FIGURE 23. Energy dependence of the CDAD asymmetry pattern corresponding to the totally symmetric orbital $3a_{1g}$ of benzene physisorbed at low coverage on the graphite(0001) surface at 140 K. Error bars represent the experimental data; dashed curves show the result of a fit to the four lowest CDAD terms.[129] The right panel illustrates the a_{1g} symmetry.

p- and f-waves contribute, namely those given in equation (34). Thus, if the analysis is restricted to partial waves with l<5, only the following four CDAD interference terms are allowed:

$$\text{CDAD}(\theta) \propto aY_{10}Y_{11} + bY_{30}Y_{31} + cY_{30}Y_{11} + dY_{10}Y_{31} \tag{35}$$

where the coefficients a–d represent the dynamical factors consisting of products of two matrix elements times the sine of the phase-shift difference.

In order to find out whether these four partial waves are already capable of describing the basic behavior, and to see which term causes the marked variation around $\theta = 25°$, a four-parameter fit of equation (35) to the data of Figure 23 has been performed. The result is shown as dashed curves which reflect the systematic behavior reasonably. The basic contribution to the variation is due to the coefficients c and d, whereas the two other terms (equal l) are almost constant. This can only mean that the p-f partial-wave phase-shift difference varies rapidly. For free molecules, the Coulomb phase-shift difference would give a major contribution and an interesting question for further analysis will be: To what extent do dynamical properties like the partial-wave phases "survive" in the adsorbate overlayer?

There cannot be a perfect agreement between fit and experiment, because the fit represents only the numerator of A_{CDAD}, equation (33). The cross section denominator will tend to reduce the asymmetry values at small angles, thereby improving the agreement between experiment and fit curves. The fact that the angle of photon incidence is 50° (Figure 16) instead of 90° causes only a scaling factor.

Room-temperature adsorption of formic acid (HCOOH) on Cu(110) yields the reaction product *formate* (COOH), the structure of which has been studied extensively.[131] On this surface the formate species is standing upright (on its oxygen "legs") on the close-packed copper rows, as illustrated in Figure 24. The molecule is totally fixed in space and thus represents an ideal candidate for the investigation of a possible *azimuthal dependence of CDAD*, i.e., a dependence on the angle φ in Figure 16.

Figure 25 shows CDAD data for the $1b_2$-orbital taken in two different azimuthal observation directions. The structure model indicates that the [110] azimuth coincides with the molecular plane, whereas [001] is perpendicular to it. Obviously, the azimuth has a dramatic influence on the CDAD asymmetries: With the electron spectrometer placed in the [001] plane, a high CDAD of up to 60% is observed, whereas in the [110] plane CDAD is always zero within the error limits (lower right panel). It should be noted that in both cases the geometry is the one shown in Figure 16, i.e., it is always noncoplanar thus allowing for a CDAD.

In order to understand this behavior, we will again consider the allowed transition channels. For an initial orbital with b_2 symmetry, the following three channels are allowed:

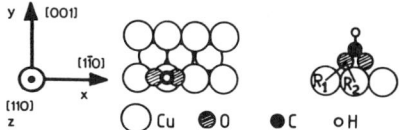

FIGURE 24. Adsorption geometry of formate chemisorbed on copper(110) at room temperature.[123]

$$1 b_2 \to \begin{cases} a_2 \text{ (electric vector along } [1\bar{1}0]) \\ a_1 \text{ (electric vector along } [1\bar{1}0]) \\ b_2 \text{ (electric vector along } [001]) \end{cases} \quad (36)$$

As usual, CDAD arises due to interference of degenerate wave functions of these three channels. Inspection of the character table of the corresponding point group C_{2v} reveals that both the a_2 and b_2 final states are odd with respect to the mirror operation in the molecular plane, whereas a_1 is even. Consequently, a_2 and b_2 have nodal planes along the [110] azimuth and when the electron spectrometer is placed in this plane, only the final-state channel with a_1 symmetry contributes to the photoelectron signal. In this case, however, also the CDAD interference terms must vanish because they represent product terms of two channels. This conclusion is clearly verified by the lower right panel of Figure 25.

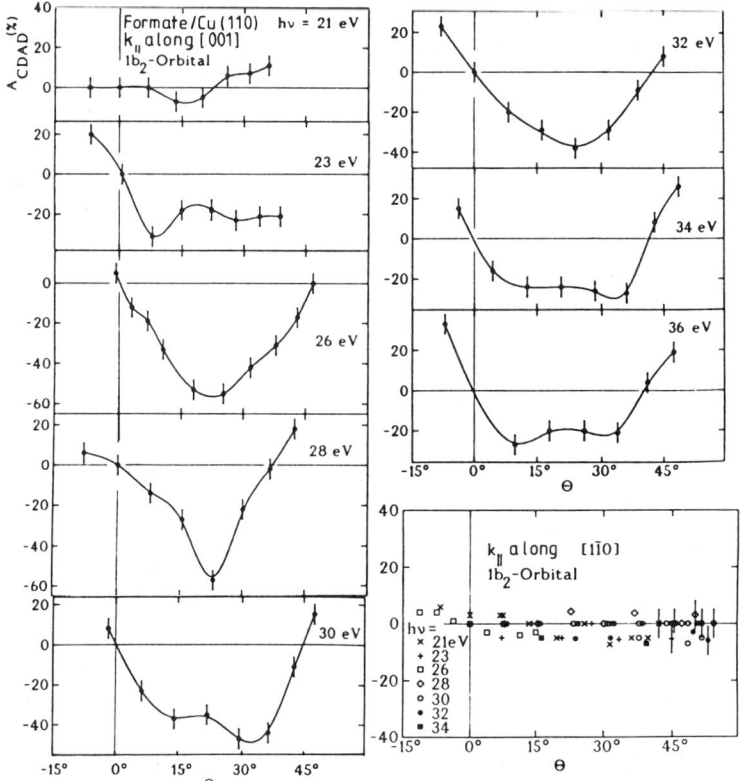

FIGURE 25. CDAD asymmetry for photoelectrons from the $1b_2$-orbital of the formate species. Perpendicular to the molecular plane (i.e., along [100]) large asymmetries occur, whereas in the molecular plane [1$\bar{1}$0] CDAD vanishes at all photon energies. (From Refs. 123, 133.)

In contrast, the [001] azimuth coincides with a nodal plane of a_2 only. Thus, both a_1 and b_2 contribute giving rise to interference due to coherent superposition of the degenerate continuum states. This interference leads to a CDAD as was illustrated in Section 6.2. High asymmetries up to –60% are the consequence of this effect.

These results demonstrate that CDAD is well suited *to probe the molecular orientation* with respect to both the polar angle (see Figure 21) and the azimuth. Of course, the same arguments hold for molecules in the gas phase oriented by optical, scattering, or coincidence techniques. Besides this practical aspect, the results for benzene and formate, which have been recorded for all valence orbitals,[123] bear detailed information on photoemission dynamics. The examples shown in Figures 23 and 25 give an impression of the spectral variation of partial-wave interference patterns. Numerical calculations like those for CO and NO would be highly desirable.

7.4. Atomic Oxygen on a Copper Surface

For the $2p$ signal of atomic oxygen a CDAD calculation does exist.[98b] Therefore, oxygen atoms produced by dissociative adsorption on the copper (110) surface have been investigated. On this surface the O atoms reside in the long-bridge positions [causing a p (2×1) missing-row (or "added-row") reconstruction]. On this surface with only twofold rotational symmetry the degeneracy of the O $2p$ level is completely lifted leading to well-separated $2p_x$, $2p_y$, and $2p_z$ states,[132] which have all been investigated by CDAD spectroscopy.[133]

In the following, we will consider only the p_z state, because this is exactly the model case treated in Section 6.2. The calculation by Dubs *et al.*[98b] assumes a spatially fixed p_z-orbital of an isolated atom in isotropic space. In this case, no phase difference between the two $l = 2$ final-state partial waves is allowed since for free atoms phases are not m-dependent. As a consequence, the second CDAD term (carrying a ± sign) in equation (32) must vanish. Then only the first term is different from zero which has the polar angular dependence $\sin 2\theta$ (in our geometry $\sin \phi = 1$). The result of the *ab initio* calculation[98b] in the framework of this approximation is shown in Figure 26 as the dashed curve. Since the CDAD asymmetry contains the sum of the R and L cross section in the denominator [equation (33)], the shape of the dashed curve is not exactly proportional to $\sin 2\theta$. This theoretical curve predicts the correct sign and magnitude of the CDAD; however, the shape of the curve obviously deviates systematically from the experimental points.

There are several general reasons for the disagreement between the calculation for a free oxygen atom and the measurement for the chemisorption system. First, the O atom experiences the strongly anisotropic environment due to the presence of the surface. In fact, this symmetry lowering gives rise to the lifting of the $2p$ degeneracy. In the theoretical description, we can take this effect into account

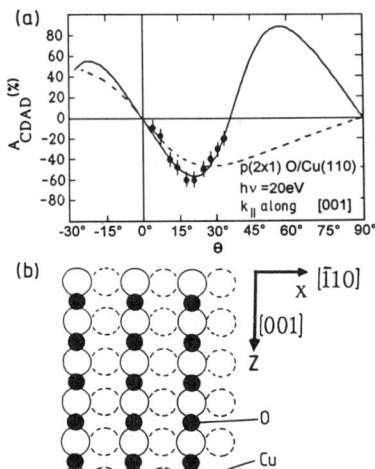

FIGURE 26. (a) CDAD asymmetry of the oxygen $2p_z$-orbital for atomic oxygen on copper(110) in comparison with an *ab initio* calculation for free oxygen[98b] (dashed curve) and a fit taking into account the symmetry reduction on the surface (full curve).[123,133] (b) Corresponding structure model of oxygen (full circles) on copper(110) (open circles); dashed circles denote the second copper layer owing to the missing row $p(2 \times 1)$ reconstruction.

by allowing for a phase shift difference between the two d partial waves with different symmetry, i.e., d_{z^2} ($m = 0$) and d_{yz} ($m = \pm 1$) (see Figure 15). According to Section 6.2 this leads to equation (32) *including* the last term.

In order to check whether this extended model is capable of describing the observed curve, a fit of equation (33) [including cross sections analogously to equation (32) slightly modified owing to the geometry shown in Figure 26b] to the experimental points has been performed. The full curve in Figure 26 represents the result of this fit. Numerically, the radial matrix elements R_s and R_d were taken from Goldberg et al.,[100] whereas the phase-shift differences $\delta_s - \delta_{dyz}$ and $\delta_{dz2} - \delta_{dyz}$ were taken as *free fit parameters*. Note that the third phase-shift difference is redundant. Obviously, there is perfect agreement between the experimental data and the result of the fit. This means that indeed the surface-induced symmetry lowering and its effect on partial-wave phases gives a major contribution, thus improving the agreement with theory.

At higher photon energies, above $hv = 30$eV, the measured curves[123,133] cannot be satisfactorily fitted by equations (33) and (32), which means that further contributions become quantitatively important. Besides the *final-state effect*, modifying the outgoing partial waves, there is also an effect on the *initial state* due to the chemical bond to the neighboring copper atoms. Actually, the linear oxygen–copper chain in the z-direction (see Figure 26b) gives rise to O$2p_z$ – Cu$3d_{z2}$ hybrid states.[132] Consequently, there is a d_{z2}-like admixture present in the initial p_z state. Taking into account this contribution improves the fit significantly.[123,133] The amount of d_{z2} admixture (hybridization) could thus be determined experimentally.

The fit revealed[123,133] that $\delta_{dz2} - \delta_{dyz}$ varied between 230 and 190° in the energy range between $hv = 20$ and 34 eV, respectively. In the same energy interval $\delta_s - \delta_{dyz}$ varied between 190 and 340° with some contribution due to the variation of the s-d Coulomb phase-shift difference.

One of the most fascinating aspects of this new type of spectroscopy is the possibility to determine such subtle features like partial-wave phases from the experimental CDAD data. This could even be carried out in solid-state photoemission from the π-orbital of graphite[111]. The evaluations showed that A_{CDAD} is very sensitive to the phase differences but much less to the dipole matrix elements [actually only ratios of matrix elements enter into equation (33)]. Especially the quantitative values of the partial-wave phase differences with equal l, which do not exist for free atoms, will shed new light into adsorbate photoemission dynamics via comparison with numerical ab initio calculations. To our knowledge such information is presently inaccessible by other spectroscopic methods.

7.5. CDAD in X-Ray Photoemission from Core Levels

The wave functions of deep-lying core levels in fixed molecules or adsorbates which do not experience crystal-field effects are statistically isotropic ($l \geq 1$) or spherically symmetric (s-levels). Consequently, no alignment (or orientation) is present in the ground-state wave function describing such a core level. Hence, a model like the one discussed in Section 6.2 is not applicable because this necessarily requires an alignment in the initial state. Also, early theoretical papers on CDAD predicted the new phenomenon only for $l > 0$ initial states. The question was, whether the *anisotropic final-state interactions* due to the fixed molecule or due to the presence of the surface may induce a CDAD effect.

Recently, Chandra[99] performed a theoretical treatment for a spatially oriented molecule of T_d symmetry, showing that also a pure s-like initial state may lead to a CDAD effect provided the final state is properly described by the symmetry-adapted wave functions. It thus appeared very challenging to check the general prediction of CDAD arising for initial states without significant alignment in a nonisotropic environment.

First CDAD measurements were restricted to photon energies up to $hv = 40$ eV, the cutoff of the normal-incidence monochromator used for previous studies. Later, the SX700/3 monochromator was installed at BESSY. It provides circularly polarized SR in the soft x-ray region, so that core levels can be investigated with respect to a possible CDAD effect. This monochromator also makes use of the circularly polarized out-of-plane SR. The photon energy range of the SX700/3 lies between $hv = 30$ and 2000 eV using the 1200 lines/mm grating.[134] New calculations for this monochromator, originally designed for linearly polarized light, showed that also circular polarization is available by employing a movable premirror.[135] The SX700/3 has a vertically movable plane mirror which accepts the

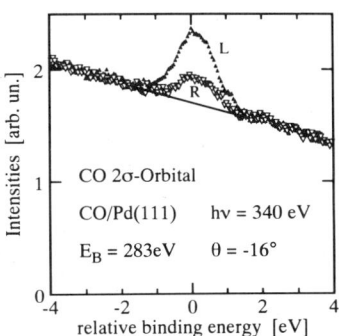

FIGURE 27. X-ray photoelectron spectra of CO/Pd(111) taken with right (R) and left circularly polarized (L) radiation from the SX 700/3 monochromator at BESSY. The observed peak corresponds to carbon 1s and shows a clear CDAD effect. The straight line indicates the background of secondary electrons. (From Ref. 137.)

elliptically polarized light above and below the plane of the storage ring and reflects it into the monochromator at a deviation angle of about 2°. The vertical acceptance of the system is 220 μrad, the horizontal acceptance 2 mrad.

The first experiments[137] were again performed for the adsorbate system CO on Pd(111) which has been well characterized by several groups (e.g., Ref. 136 and references therein) showing that CO is perpendicularly fixed on the surface with the carbon end pointing to the crystal. Figure 27 shows a typical pair of x-ray photoelectron spectra of the carbon 1s signal taken with right and left circularly polarized radiation. Clearly, the peak intensities for the two photon helicities are different, i.e., a CDAD is present. The background of secondary electrons, originating from inelastic scattering processes, does not show a CDAD asymmetry (straight line underlying the peaks).

Figure 28 displays the asymmetry A_{CDAD} versus emission angle θ for various photon energies. A_{CDAD} is an odd function of θ as one anticipates when changing the experimentally induced chirality from a "right-handed" to a "left-handed" system. The measured asymmetries have been normalized with respect to the degree of circular polarization P_{circ} = 70% in this photon energy range. The asymmetry reaches high values of up to 75% which suggests that CDAD in photoemission from core levels is an effect of first order in a pure electric dipole transition.

The existence of CDAD for s-like inner shells can be derived by a very simple orbital model[137] which we will outline in the following. In diatomic molecules the initial and final states are characterized by the projection m_l of the orbital momentum onto the molecular axis, denoted as σ, π, ... states. The carbon K-shell in CO (CO 2σ-orbital) is to a large extent s-like, the $l \geq 1$ admixture being less than 1%.[140] Hence, in photoemission the essential contribution to the dipole matrix element results from the following transition channels:

$$s\sigma(Y_{00}) \rightarrow \begin{cases} p\sigma(Y_{10}) & [z \text{ component}] \\ p\pi(Y_{1\pm 1}) & [\pm iy \text{ component}] \end{cases} \quad (37)$$

Of course, this picture is only an approximation because the σ- and π-continua may in principle consist of an infinite number of partial waves Y_{l0} (for σ) and $Y_{l\pm 1}$ (for π). However, it does already predict a CDAD in the correct order of magnitude.

In a molecule, partial-wave phase shifts δ_{lm} may depend on both l and m. Consequently, there will in general be a phase difference between the pσ and pπ continuum partial waves. *It is this m-dependence of partial-wave phases that gives rise to a CDAD for s-like states.* A derivation analogously to Section 6.2 reveals that the ± sign of the photon operator (describing the helicity) shows up in an interference term of pσ and pπ in the differential cross section. For the geometry of Figure 16 the differential cross section can be written as (with $R_{p\sigma}$ and $R_{p\pi}$ being the radial matrix elements):

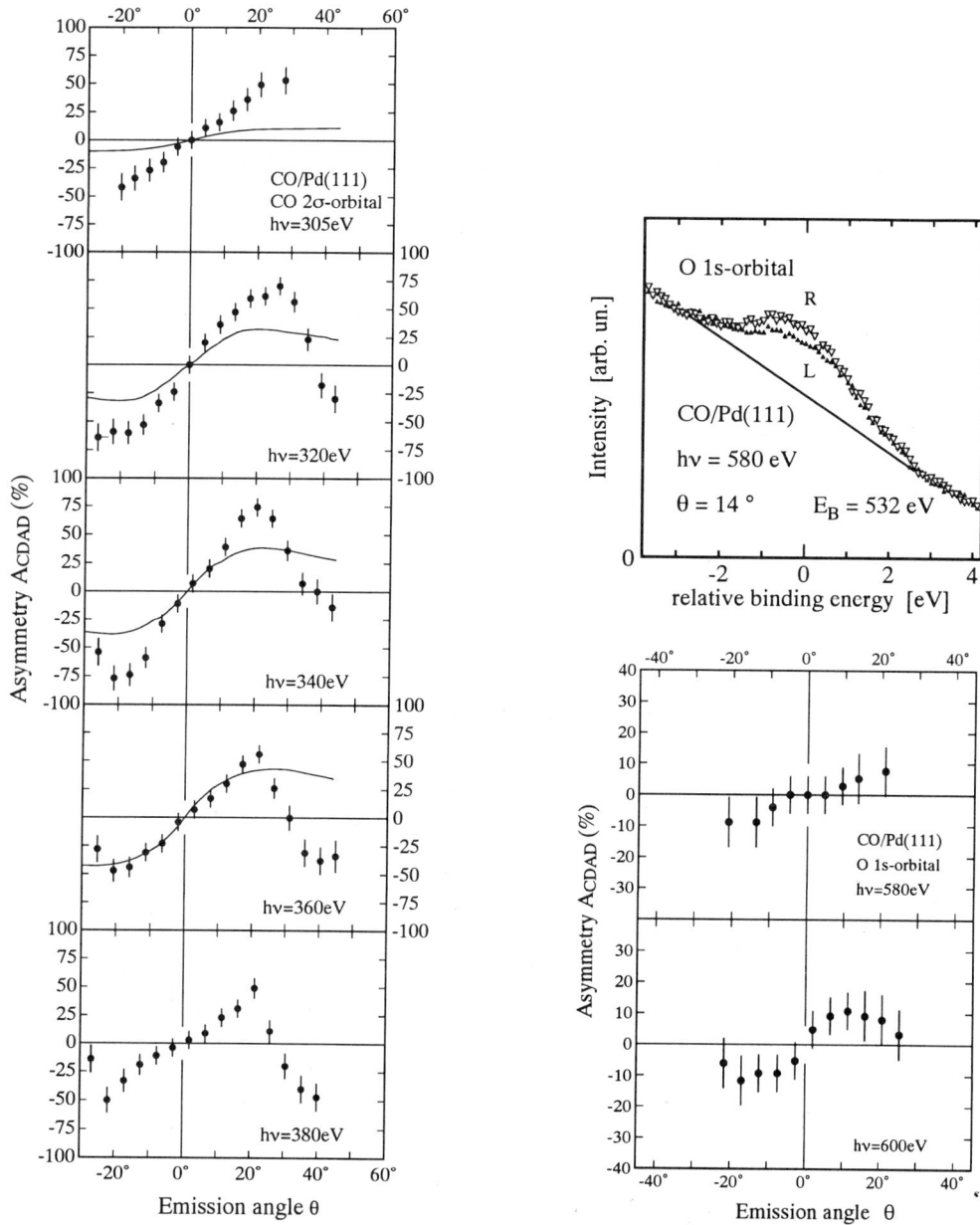

FIGURE 28. Circular dichroism in photoemission from the carbon 1s (left panel) and oxygen 1s level (right panel) for adsorbed CO in the soft x-ray region. All data have been corrected with respect to a degree of circular polarization of $P_{circ} = 70\%$. The error bars represent counting statistics, the solid lines display *ab initio* calculations from Pavlychev and Cherepkov.[139,141] Note that each data point corresponds to a pair of spectra like those in Figure 27.

$$\frac{d\sigma}{d\Omega}(\theta, \phi = 90°) \propto |\langle \psi_f | z \sin 50° + x \cos 50° \pm iy | \psi_i \rangle |^2$$

$$\propto 4\pi | \sin 50° R_{p\sigma} e^{i\delta_{p\sigma}} \cos\theta \pm iR_{p\pi} e^{i\delta_{p\pi}} \sin\theta |^2$$

$$\propto [\sin^2 50° R_{p\sigma}^2 \cos^2\theta + R_{p\pi}^2 \sin^2\theta \pm 2 \sin 50° R_{p\sigma} R_{p\pi} \sin\theta \cos\theta \sin(\delta_{p\sigma} - \delta_{p\pi})] \quad (38)$$

As a consequence of the last, helicity-dependent term, the angular distribution pattern of the differential cross section for circularly polarized radiation is rotated clockwise for RCP and counterclockwise for LCP radiation with respect to the z-axis. For unpolarized radiation the differential cross section has a spherically symmetric shape like the initial s-state. Similar to the treatment in Section 6.2, a change

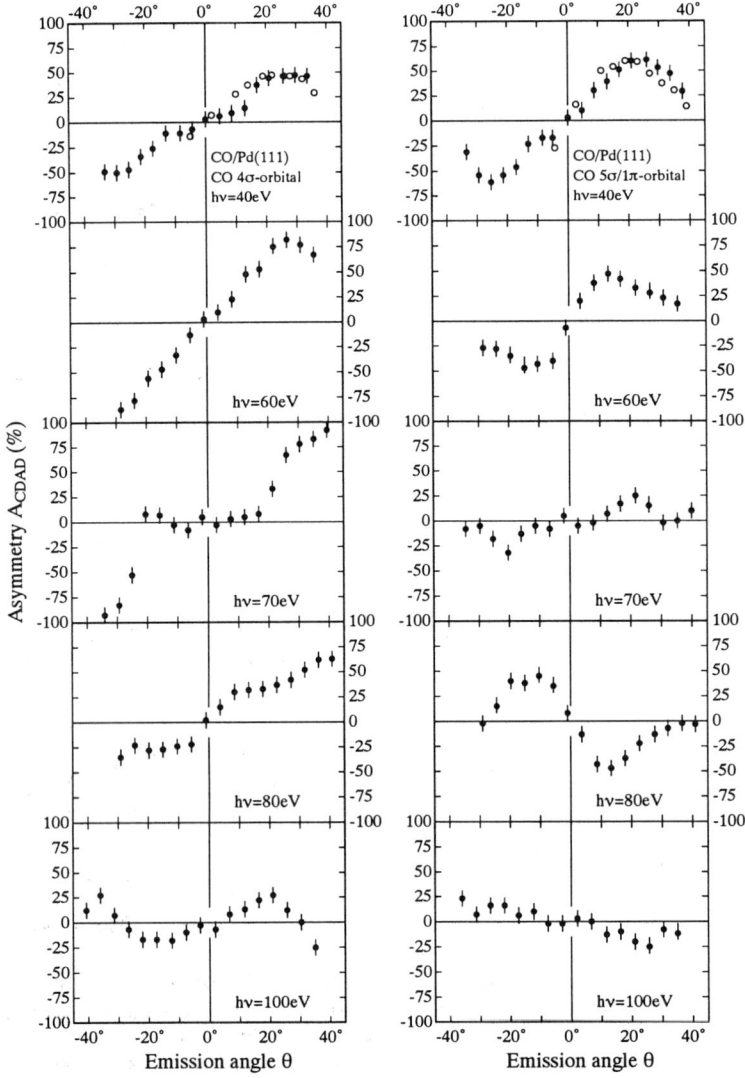

FIGURE 29. Same as Figure 20 but for the CO monolayer on Pd(111) taken at the higher photon energies of the SX700/3 monochromator at BESSY. Open circles at $h\nu = 40$ eV have been measured using the 6.5-m normal-incidence monochromator and are shown for comparison.

of photon helicity retains the phase of the $p\sigma$-wave (indicated by the + and − signs in the angular pattern of the wave function) but changes the phase of the $p\pi$-wave.

Since in equation (38) only one term is helicity-dependent, the CDAD, i.e., the difference between the R and L cross section, has a very simple form:

$$\left(\frac{d\sigma}{d\Omega}\right)^R - \left(\frac{d\sigma}{d\Omega}\right)^L \propto R_{p\sigma} R_{p\pi} \sin(\delta_{p\sigma} - \delta_{p\pi}) \sin 2\theta \qquad (39)$$

It is a single interference term containing the product of the dipole matrix elements and the sine of the corresponding phase-shift difference. Its angular dependence is given by $\sin 2\theta$. Note that this term represents the numerator of A_{CDAD}, equation (33). Obviously, the angular dependence of the experimental A_{CDAD} data in Figure 28 is more complicated. Our crude model of a local s-p excitation does, of course, not correctly describe photoemission dynamics from the CO molecule. In particular, the $l > 1$ final-state contributions as well as photoelectron diffraction effects are neglected. The latter effects should be significant for electron energies around $E_{kin} = 100$ eV because in this case the photoelectron wavelength is on the order of the interatomic distances.

The SX700/3 monochromator also allowed extending the CDAD investigations of the CO *valence orbitals*, discussed in Figure 20, to higher photon energies. The results between $hv = 40$ and 100 eV are shown in Figure 29 for the CO 4σ (left) and the overlapping CO $5\sigma/1\pi$-orbitals (right column).

One of the main questions for future investigations will be: To what extent do the presence of the surface and the chemical bond alter photoemission dynamics of an adsorbed molecule or atom? Since they experience the weakest bond (pure van der Waals forces) and owing to the wealth of data in the gas phase including so-called "complete" photoemission experiments,[109b,c] adsorbed rare gases will be the subject of new experiments which are just being prepared.

There is an interesting side aspect of the new measurements in the soft x-ray range: Due to its size reaching 75% asymmetry, CDAD is well suited to *calibrate the circular light polarization throughout a wide photon energy region*. This possibility has already been utilized to calibrate the polarization of the SX700/3 monochromator at $hv = 40$ eV by using CDAD data from the 6.5-m NIM (open circles in Figure 29).

8. FUTURE DEVELOPMENTS

As discussed in Section 3.3, CD measurements in the VUV region are still limited to the region above 105 nm by the transmission of the LiF modulator. Though the aforementioned test experiments with the CP out-of-plane components of synchrotron radiation[50] indicate that it should be possible to extend CD spectroscopy at least to the limit of the normal incidence range with a carefully designed monochromator, no significant progress is as yet evident. It turned out, for example, that the CD detection sensitivity of the Dragon monochromator which was used successfully for several soft x-ray MCD experiments (see Section 8.3) was only about 0.3% and thus far too low for CD measurements. Here, a new design has now been developed for a monochromator producing simultaneous RCP and LCP light which will be used with an ultrahigh-vacuum compatible chopper.[142] However, as there are several storage rings where insertion devices (wigglers and undulators) for the production of CP light are tested, these developments suitable to overcome the energy limit for CD measurements will be discussed in the following section.

Over the last decade, a variety of mainly theoretical treatments were published describing the different phenomena emerging from the interaction of polarized (especially CP) radiation with matter in a wide range of energies. Some of these new techniques, e.g., circular intensity differential scattering (CIDS) and circular differential microscopy, are discussed in Ref. 143. CIDS, which is closely related to CD and for which test experiments with synchrotron radiation have been carried out, will be discussed below. Finally, circular magnetic x-ray dichroism (CMXD) will also be presented. This technique using basically the same experimental setup as is required for CD measurements has turned out to be very successful and has provided valuable information, e.g., on the magnetism of transition metals.

Owing to its large asymmetries, CDAD can be used as a novel contrast mechanism in a photoelectron-emission microscope. The special features of such a "CDAD-PEEM" will be discussed in 8.2.

8.1. Circularly Polarized Radiation from Insertion Devices

Undulators and wigglers are arrangements providing spatially periodic magnetic fields, usually consisting of an appropriate sequence of, e.g., permanent magnet blocks or plates. These devices are put into the straight sections of a storage ring and make the electrons move transversely to their normal path. In this way, synchrotron radiation with improved properties is produced as compared to radiation from the bending magnets. At present, these insertion devices are used in many storage rings worldwide.

Radiation emitted from an ordinary planar wiggle or undulator is linearly polarized. However, as early as 1971 the first ideas regarding production of CP radiation with insertion devices were published.[144] An arrangement was proposed to have vertical and horizontal jaws of electromagnet arrays with the vertical jaw having a phase difference of one-quarter of the magnetic period relative to the horizontal. Thus, the magnetic field varies helically along the axis so that electrons may emit CP radiation during their helical motion. The first insertion device for CP radiation was built and tested by Onuki and co-workers at the storage ring TERAS in 1988.[145] Here, circular polarization in the visible part of the spectrum could be observed. Over the last few years, several different designs for undulators and wigglers for the production of CP radiation have been published. An overview of these activities is given in the review by Kitamura.[146] Most of these designs do not offer the possibility of rapid change between RCP and LCP light. Thus, these devices are not very suitable for CD experiments where a modulation technique for detection is indispensable. For MCD this is not a severe limitation as the magnetic field can be modulated. The so-called cross field undulator proposed independently by Moiseev et al.[147] and Kim[148,149] has the unique feature of offering a potentially rapid change between alternate polarizations.

The cross field undulator system produces two linearly polarized wave trains from two planar undulators arranged in series with perpendicular magnetic fields. The superposition of these two waves is achieved by coherent superposition on a monochromator grating. The relative phase of the two waves is controlled by the so-called modulator between the undulators. The modulator is retarding the electrons relative to the radiation by up to one optical period while maintaining the direction of the electron beam. The first cross field undulator based on this design has been built at BESSY I in Berlin.[150,151] Most of the properties predicted could be verified experimentally so that there is good hope that after completion of the required beam line CD measurements can be carried out over the energy range from 20 to 200 eV thus opening a new and exciting field for this type of experiments.

8.2. Related Techniques: Circular Intensity Differential Scattering (CIDS) and CDAD Microscopy

Chiral media interact differently with the light of opposite circular polarization. As previously discussed, ORD is caused by difference in the real part of the refractive index whereas CD is caused by the difference in the imaginary part. In classical electrodynamics, scattering of light is the result of the radiation emitted on changes in the charge distribution of a system, induced by an oscillating electromagnetic field. These changes are described by the polarizability of the medium which contains all of the symmetry properties necessary to account for the differential scattering of light of opposite circular polarization. An appropriate measure of this phenomenon is CIDS:

$$\text{CIDS}(\vartheta) = \frac{(I_L - I_R)}{(I_L + I_R)} \tag{29}$$

where I_L and I_R are the scattered intensities for the two circular polarizations of the incident radiation at a specified scattering angle ϑ.

Bustamante et al. have worked out the basic theory of CIDS for oriented helices as models for chiral structures.[152,153] They found the differential scattering patterns to be very sensitive to helical parameters showing lobes with alternating sign in the angular dependence. The number of lobes and the positions of zeros are a direct measure of the ratio of radius and pitch of the helices to the wavelength of the radiation. The signs depend on the sense of the helix. A major advantage of CIDS is that in principle measurements can be carried out at any wavelength. No absorption bands are required.

The first CIDS experiments were carried out on oriented liquid crystals[155] and biopolymers[156] in the visible region with a laser light source. Very recently it has been proposed that CIDS be used in the VUV region between 16 and 500 eV.[157] For estimating the feasibility of detecting CIDS in this energy range, a series of computations were carried out based on the theory of Bustamante et al.[152,153] It was found that the maximum CIDS depends strongly on the polarizability anisotropy according to

$$\text{CIDS} \approx (\text{polarizability anisotropy})^\mu$$

where μ is a scaling parameter. Because the magnitude of the atomic polarizability at very high frequencies as well as their intrinsic anisotropy is not known, suitable models had to be developed. The calculations indicate that successful CIDS measurements should be possible in the wanted energy region.

The first test measurements extending CIDS to the UV region were carried out at the SRC in Wisconsin for various samples. These experiments showed a strong dependence of the CIDS structure on the ratio of wavelength to diameter of the spheres which were investigated as model compounds. This indicates an extreme sensitivity to dimensional parameters not necessarily reflecting helical organization. Here also further theoretical work seems to be necessary.

Only a decade after the first theoretical predication of CDAD for isolated molecules with alignment or orientation, this new spectroscopic technique will find a promising application in a special CDAD photoelectron emission microscope (PEEM). This new instrument is just being tested by the Mainz group. In a PEEM[154] a solid sample is illuminated by an intense light source and the photoelectrons released are used to obtain an image on a fluorescent screen by using standard electron-optical techniques. A lateral resolution of 30 nm and better can be gained because the electrons (and not the photons) are used for imaging. In the CDAD-PEEM circularly polarized light is used for the excitation and a special contrast aperture defines the detected angular interval analogously to $\mathbf{k}$ in Figure 16. The CDAD contrast is generated by taking images for both photon helicities and deriving the difference image R–L. This contrast will be extremely sensitive upon the local geometrical and electronic structure. For molecular adsorbates domains with different orientation or structural phase transitions can thus be viewed, even if the angles involved vary only a few degrees (Figure 21). In material science, CDAD microscopy will be complementary to diffraction techniques because it yields structural information with high lateral resolution.

8.3. Measurements in the X-Ray Region

In 1986 for the first time the dependence of a spectrum on the polarization of the radiation was observed in the x-ray region. van der Laan et al. observed with linearly polarized radiation magnetic linear dichroism (MLD) at the terbium M_V edge at about 1200 eV in ferromagnetic Tb iron garnets.[158] MCD studies were started in 1987 by Schütz et al.[159] for Fe-K absorption in ferromagnetic iron. An overview of the various experiments performed with polarized radiation in the x-ray region for ferromagnets and ferrimagnets is given in Refs. 160 and 162. In these experiments a double slit in front of a normal double-crystal x-ray monochromator is used to separate RCP and LCP synchrotron radiation emitted above and below the plane of the electron orbit. Absorption spectra for both beams are recorded simultaneously but independently by special "double ionization chambers." Samples are placed inside a solenoid and the direction of the magnetic field is reversed every second. Thus, the absorption coefficients $\mu \uparrow\uparrow(E)$ for parallel and $\mu \downarrow\downarrow(E)$ for antiparallel magnetization with respect to the photon spin are determined. From $\mu_c(e) = \mu \uparrow\uparrow - \mu \downarrow\downarrow$ the spin-dependent part of the absorption coefficient $\mu^\pm = \mu \pm \mu_c$ can be determined which is proportional to the difference of the spin densities of unoccupied states near the Fermi level.

The CD detection sensitivity of all MXCD experiments carried out to date is rather low (around 0.1%). As CD signals of natural optically active samples are at least two orders of magnitude smaller, experiments to detect CD in the x-ray region have not yet been successful. However, some developments to improve this situation, e.g., by using a UHV compatible chopper making again lock-in detection possible, are under way.[142]

A different type of a CD experiment using x-ray radiation for excitation was carried out by Gauthier et al.[161] These authors measured CP x-ray excited optical luminescence (CP-XEOL) from chiral Eu^{3+} complexes. In these first experiments, the samples were excited with a beam of polychromatic x rays. The optical luminescence signal was analyzed with a combination of a PEM and a linear polarizer according to its circular polarization. From the obtained CP-XEOL spectra the Kuhn asymmetry factor g could be calculated and found to reasonably agree with results of ordinary CP luminescence.

ACKNOWLEDGMENTS. J.H. would like to thank W. C. Johnson, Jr., P. A. Snyder, A. Gedanken, A. Rauk, E. S. Stevens, V. Büs, and J. C. Sutherland for providing copies of their publications and also unpublished material on their CD and MCD experiments. Professor F. von Busch's critical reading of the manuscript is gratefully acknowledged. The early experiments of the Bonn group in the field of CD spectroscopy with synchrotron radiation were supported by the BMFT (Bundesminister für Forschung und Technologie) whereas our recent work is being supported by the Deutsche Forschungsgemeinschaft (DFG) as project B1 within the Sonderforschungsbereich 334.

G.S. would like to thank C. Westphal, M. Getzlaff and J. Bansmann, without whom the experiments never would have worked. Further thanks are due to N.A. Cherepkov, V. McKoy and G. Raseev for many fruitful discussions and to the BESSY staff for good cooperation. The experiments were supported by BMFT (Projects 05 5UMAAB and 05 431 AXB6) and in Mainz by the Materialwissenschaftliches Forschungszentrum (MWFZ).

REFERENCES

1. C. Djerassi, *Optical Rotatory Dispersion: Applications to Organic Chemistry* (McGraw–Hill, New York, 1960).
2. L. Velluz, M. Legrand, and M. Grosjean, *Optical Circular Dichroism: Principles, Measurements and Applications* (Verlag Chemie, Weinheim, 1965).
3. D. J. Caldwell and H. Eyring, *The Theory of Optical Activity* (Wiley–Interscience, New York, 1971).
4. P. Crabbé, *Optical Rotatory Dispersion and Circular Dichroism in Chemistry and Biochemistry* (Academic Press, New York, 1972).
5. E. Charney, *The Molecular Basis of Optical Activity, Optical Rotatory Dispersion and Circular Dichroism* (Wiley, New York, 1979).
6. S. F. Mason, *Molecular Optical Activity and the Chiral Discriminations* (Cambridge University Press, Cambridge, 1982).
7. N. Harada and K. Nakanishi, *Circular Dichroic Spectroscopy, Exciton Coupling in Organic Stereochemistry* (Oxford University Press, Oxford, 1983).
8. P. J. Stephens and M. A. Lowe, *Annu. Rev. Phys. Chem.* **36**, 213 (1985).
9. F. S. Richardson and J. P. Riehl, *Chem. Rev.* **77**, 773 (1977).
10. I. Tinoco, Jr., and A. L. Williams, Jr., *Annu. Rev. Phys. Chem.* **35**, 329 (1984).
11. R. Janoschek (Ed.), *Chirality, From Weak Bosons to the α-Helix* (Springer-Verlag, Berlin, 1991).
12. J. B. Biot, *Mem. Acad. Sci. Fr.* **2**, 41 (1817).
13. W. P. Healy, *J. Phys. B* **7**, 1633 (1974).
14. T. Hirano, A. Sato, T. Tsuruta, and W. C. Johnson, Jr., *J. Polym. Sci. Polym. Phys. Ed.* **17**, 1601 (1979).
15. A. D. Buckingham and P. J. Stephens, *Annu. Rev. Phys. Chem.* **17**, 399 (1966).
16. J. A. Schellman, *Chem. Rev.* **75**, 323 (1975).
17. D. C. Walker (Ed.), *Origins of Optical Activity in Nature* (Elsevier, Amsterdam, 1979).
18. A. J. MacDermott, *4th Int. Conf. on CD, Bochum 1991, Book of lectures and posters*, p. 75 (1991).
19. M. Grosjean and M. Legrand, *Compt. Rend. Paris* **251**, 2150 (1960).
20. A. F. Drake, *J. Phys. E* **19**, 170 (1986).
21. M. Billardon and J. Badoz, *Compt. Rend. Paris* **262**, 1672 (1966).

22. J. C. Kemp, *J. Opt. Soc. Am.* **59**, 950 (1969).
23. S. N. Jasperson and S. E. Schnatterly, *Rev. Sci. Instrum.* **40**, 761 (1969).
24. L. F. Mollenhauer, D. Downie, H. Engstrom, and W. B. Grant, *Appl. Opt.* **8**, 661 (1969).
25. J. C. Cheng, L. A. Nafie, S. D. Allen, and A. I. Braunstein, *Appl. Opt.* **15**, 1960 (1976).
26. Hinds International Inc., Photoelastic Modulator Systems, 1988.
27. D. E. Eastman and J. J. Donelon, *Rev. Sci. Instrum.* **41**, 1648 (1970).
28. W. C. Johnson, Jr., *Rev. Sci. Instrum.* **42**, 1283 (1971).
29. W. C. Johnson, Jr., *Rev. Sci. Instrum.* **35**, 1375 (1964).
30. A. F. Drake and S. F. Mason, *J. Phys. Colloq.* **39**, C4 212 (1978).
31. M. B. Robin, N. A. Keubler, and Y. H. Pao, *Rev. Sci. Instrum.* **37**, 922 (1966).
32. E. S. Pysh, *Annu. Rev. Biophys. Bioeng.* **5**, 63 (1976).
33. A. Gedanken and M. Levy, *Rev. Sci. Instrum.* **48**, 1661 (1977).
34. W. C. Johnson, Jr., *Annu. Rev. Phys. Chem.* **29**, 93 (1978).
35. P. A. Snyder, *Photochem. Photobiol.* **44**, 237 (1986).
36. E. Krausz and G. Cohen, *Rev. Sci. Instrum.* **48**, 1506 (1977).
37. A. Gedanken, in *Photophysics and Photochemistry in the Vacuum Ultraviolet*, edited by S. P. McGlynn et al. (Reidel, Dordrecht, 1985), p. 765.
38. D. E. Eastman and Y. Farge (Eds.), *Handbook of Synchrotron Radiation*, Vols. 1–4 (North-Holland, Amsterdam, 1983–1991); C. Kunz (Ed.), *Synchrotron Radiation, Techniques and Applications* (Springer, Berlin, 1979); C. R. A. Catlow and G. N. Greaves (Eds.), *Application of Synchrotron Radiation* (Blackie, Glasgow, 1990).
39. P. A. Snyder and E. M. Rowe, *Nucl. Instrum. Methods* **172**, 345 (1980).
40. P. A. Snyder, *Nucl. Instrum. Methods* **222**, 363 (1984).
41. E. S. Stevens, in *Handbook on Synchrotron Radiation*, Vol. 4, edited by S. Ebashi, M. Koch, and E. Rubenstein (Elsevier, Amsterdam, 1991).
42. J. Hormes, A. Klein, W. Krebs, W. Laaser, and J. Schiller, *Nucl. Instrum. Methods* **208**, 849 (1983).
43. J. C. Sutherland, E. J. Desmond, and P. Z. Takacs, *Nucl. Instrum. Methods* **172**, 195 (1980); J. C. Sutherland, P. C. Keck, K. P. Griffin, and P. Z. Takacs, *Nucl. Instrum. Methods* **195**, 375 (1982).
44. M. A. Wickramaaratchi, E. T. Premuzic, M. Lin, and P. A. Snyder, *J. Chromatogr.* **390**, 413 (1987).
45. J. Schwinger, *Phys. Rev.* **75**, 1912 (1949).
46. P. Joos, *Phys. Rev. Lett.* **4**, 558 (1960).
47. U. Heinzmann, *J. Phys. B* **13**, 4353 (1980); U. Heinzmann, B. Osterheld, and F. Schäfers, *Nucl. Instrum. Methods* **195**, 395 (1982).
48. A. Eyers, C. Heckenkamp, F. Schäfers, G. Schönhense, and U. Heinzmann, *Nucl. Instrum. Methods* **208**, 303 (1983).
49. C. Heckenkamp, F. Schäfers, G. Schönhense, and U. Heinzmann, *Phys. Rev. Lett.* **52**, 421 (1984).
50. J. Schiller and J. Hormes, *Nucl. Instrum. Methods* **A246**, 772 (1986).
51. A. E. Hansen and T. D. Bouman, *Adv. Chem. Phys.* **XLIV**, 545 (1980).
52. M. Carnell, S. D. Peyerimhoff, A. Breest, K. H. Gödderz, P. Ochmann, and J. Hormes, *Chem. Phys. Lett.* **180**, 477 (1991).
53. W. Moffitt and A. Moscowitz, *J. Chem. Phys.* **30**, 648 (1959).
54. G. Snatzke, *Chem. Unserer Zeit* **16**, 160 (1982).
55. G. Snatzke, in *Chirality, From Weak Bosons to the α-Helix*, edited by R. Janoschek (Springer-Verlag, Berlin, 1991), p. 59.
56. L. D. Barron, *Molecular Light Scattering and Optical Activity* (Cambridge University Press, Cambridge, 1982).
57. E. U. Condon, W. Altar, and H. Eyring, *J. Chem. Phys.* **5**, 753 (1937).
58. W. Kuhn, *Trans. Faraday Soc.* **26**, 293 (1930).
59. S. F. Boys, *Proc. R. Soc. London Ser. A* **144**, 655 (1934).
60. J. G. Kirkwood, *J. Chem. Phys.* **5**, 479 (1937).
61. E. G. Höhn and O. E. Weigang, Jr., *J. Chem. Phys.* **48**, 1127 (1968).
62. A. Rauk, *Origins of Optical Activity in Nature*, edited by D. C. Walker (Elsevier, Amsterdam, 1979), p. 193.
63. W. Moffitt, R. B. Woodward, A. Moscowitz, W. Klyne, and C. Djerassi, *J. Am. Chem. Soc.* **83**, 4013 (1961).
64. O. E. Weigang, Jr., and E. G. Höhn, *J. Am. Chem. Soc.* **88**, 3673 (1966).
65. A. Rodger and P. M. Rodger, *J. Am. Chem. Soc.* **110**, 2361 (1988).
66. O. E. Weigang, Jr., *J. Am. Chem. Soc.* **101**, 1965 (1979).
67. H. Lagier, Diplom-Thesis, Bonn University, BONN-IR-86-12, Bonn, 1986.

68. N. Harada and K. Nakanishi, *Acc. Chem. Res.* **5**, 257 (1972).
69. K. P. Gross and O. Schnepp, *Chem. Phys. Lett.* **36**, 531 (1975).
70. A. Gedanken, K. Hintzer, and V. Schurig, *J. Chem. Soc. Chem. Commun.* **1984**, 1695 (1984).
71. A. Gedanken and V. Schurig, *J. Phys. Chem.* **91**, 1324 (1987).
72. J. Schiller, H. Lagier, A. Klein, J. Hormes, F. Vögtle, A. Aigner, and R. Thomessen, *Phys. Scr.* **35**, 463 (1987).
73. W. C. Johnson, Jr., in *Landolt–Börnstein Series VII*, edited by W. Saenger (Springer, Berlin, 1989).
74. K. H. Johnson, D. M. Gray, and J. C. Sutherland, *Nucleic Acids Res.* **19**, 2275 (1991).
75. W. C. Johnson, Jr., *Adv. Carbohydr. Chem. Biochem.* **45**, 73 (1987).
76. E. S. Stevens, *Photochem. Photobiol.* **44**, 287 (1986).
77. W. C. Johnson, Jr., *Proteins, Structure, Function, and Genetics* **7**, 205 (1990).
78. K. Gödderz, Diplom-Thesis, Bonn University, BONN-IR-90-17, Bonn, 1990.
79. A. Rauk, J. O. Jarvie, H. Ichimura, and J. M. Barriel, *J. Am. Chem. Soc.* **97**, 5656 (1975).
80. G. V. Shustov, S. V. Varlamov, I. I. Chervin, A. E. Aliev, R. G. Kostyanovsky, D. Kim, and A. Rauk, *J. Am. Chem. Soc.* **111**, 4210 (1989).
81. P. Faupel and V. Buss, *Makromol. Chem. Theory Simul.* **1**, 311 (1992).
82. S. Wenzel and V. Buss, *J. Phys. Org. Chem.* **5**, 748 (1992).
83. V. Buss, V. Haas, and U. Wingen, *Z. Naturforsch.* **44b**, 333 (1989).
84. U. Wingen, L. Simon, M. Klein, and V. Buss, *Angew. Chem. Int. Ed. Engl.* **24**, 761 (1985).
85. V. Haas and V. Buss, *J. Chem. Soc. Chem. Commun.* **18**, 1320 (1991).
86. F. Vögtle, K.-J. Przybilla, A. Mannschreck, N. Pustet, P. Büllesbach, H. Reuter, and H. Puff, *Chem. Ber.* **121**, 823 (1988).
87. K.-J. Przybilla and F. Vögtle, *Chem. Ber.* **122**, 347 (1989).
88. N. Greenfield and G. D. Fasman, *Biochemistry* **8**, 4108 (1969).
89. V. P. Saxena and D. B. Wetlaufer, *Proc. Natl. Acad. Sci. USA* **68**, 969 (1971).
90. Y.-H. Chen, J. T. Yang, and H. M. Martinez, *Biochemistry* **11**, 4120 (1972).
91. P. Manavalan and W. C. Johnson, Jr., *Suppl. J. Biosci.* **8**, 141 (1985).
92. J. P. Hennessey, Jr., and W. C. Johnson, Jr., *Biochemistry* **20**, 1085 (1981); *Anal. Chem.* **125**, 177 (1982).
93. W. C. Johnson, Jr., *Annu. Rev. Biophys. Biophys. Chem.* **17**, 145 (1988).
94. J. C. Sutherland, B. Lin, J. Mugavero, J. Trunk, M. Tomasz, R. Santella, L. Marky, and K. J. Breslauer, *Photochem. Photobiol.* **44**, 295 (1986).
95. A. Toumadje, S. W. Alcorn, and W. C. Johnson, Jr., *Anal. Biochem.* in press.
96. B. Ritchie, *Phys. Rev. A* **12**, 567 (1975); **13**, 1411 (1976); **14**, 359 (1976); **14**, 1396 (1976).
97. (a) N. A. Cherepkov, *Chem. Phys. Lett.* **87**, 344 (1982); (b) N. A. Cherepkov and V. V. Kuznetsov, *J. Phys. B* **20**, L 159 (1987); (c) *Z. Phys. D* **7**, 271 (1987); (d) *J. Phys. B* **22**, L 405 (1989); (e) *J. Chem. Phys.* **95**, 3046 (1991).
98. (a) R. L. Dubs, S. N. Dixit, and V. McKoy, *Phys. Rev. Lett.* **54**, 1249 (1985); (b) *Phys. Rev. B* **32**, 8389 (1985); (c) *J. Chem. Phys.* **85**, 656 (1986); (d) *J. Chem. Phys.* **85**, 6267 (1986); (e) R. L. Dubs, V. McKoy, and S. N. Dixit, *J. Chem. Phys.* **88**, 968 (1988).
99. N. Chandra, *Phys. Rev. A* **39**, 2256 (1989).
100. S. M. Goldberg, C. S. Fadley, and S. Kono, *J. Electron Spectrosc. Relat. Phenom.* **21**, 285 (1981).
101. S. Stolte, K. K. Chakravorty, R. B. Bernstein, and D. H. Parker, *Chem. Phys.* **71**, 353 (1982); S. Stolte, *Ber. Bunsenges. Phys. Chem.* **86**, 413 (1982).
102. S. Kaesdorf, G. Schönhense, and U. Heinzmann, *Phys. Rev. Lett.* **54**, 885 (1985).
103. J. R. Appling, M. G. White, T. M. Orlando, and S. L. Anderson, *J. Chem. Phys.* **85**, 6803 (1986).
104. J. R. Appling, M. G. White, R. L. Dubs, S. N. Dixit, and V. McKoy, *J. Chem. Phys.* **87**, 6927 (1987).
105. J. W. Winniczek, R. L. Dubs, J. R. Appling, V. McKoy, and M. G. White, *J. Chem. Phys.* **90**, 949 (1989); R. L. Dubs, S. N. Dixit, and V. McKoy, *J. Chem. Phys.* **86**, 5886 (1987).
106. F. Schäfers, W. Peatman, A. Eyers, C. Heckenkamp, G. Schönhense, and U. Heinzmann, *Rev. Sci. Instrum.* **57**, 1032 (1986).
107. G. Schönhense, *Appl. Phys.* **A41**, 39 (1986).
108. K. Jost, *J. Phys. E* **12**, 1001, 1006 (1979).
109. (a) J. Kessler, *Comments At. Mol. Phys.* **10**, 47 (1981) and *Polarized Electrons*, 2nd ed. (Springer, Berlin, 1985); (b) F. Schäfers, G. Schönhense, and U. Heinzmann, *Phys. Rev. A* **28**, 802 (1983); (c) C. Heckenkamp, F. Schäfers, G. Schönhense, and U. Heinzmann, *Phys. Rev. A* **32**, 1252 (1985).
110. (a) G. Schönhense, *Phys. Scr.* **T31**, 255 (1990); (b) *Vacuum* **41**, 506 (1990).
111. G. Schönhense, C. Westphal, J. Bansmann, and M. Getzlaff, *Europhys. Lett.* **17**, 727 (1992).

112. G. Schönhense, C. Westphal, J. Bansmann, M. Getzlaff, J. Noffke, and L. Fritsche, *Surf. Sci.* **251/252**, 132 (1991).
113. N. A. Cherepkov and G. Schönhense, in *Synchrotron Radiation and Dynamic Phenomena*, edited by A. Beswick (American Institute of Physics, New York, 1991), p. 69 and *Europhys. Lett.* **24**, 79 (1993).
114. R. Parzynski, *Acta Phys. Pol. A* **57**, 49 (1980).
115. L. E. Cuéllar, C. S. Feigerle, H. S. Carman, Jr., and R. N. Compton, *Phys. Rev. A* **43**, 6437 (1991).
116. J. Berakdar and H. Klar, *Phys. Rev. Lett.* **69**, 1175 (1992); J. Berekdar, H. Klar, A. Huetz, and P. Selles, *J. Phys. B* **26**, 1463 (1993).
117. E. W. Plummer and W. Eberhard, *Adv. Chem. Phys.* **49**, 533 (1982).
118. H.-J. Freund and M. Neumann, *Appl. Phys. A* **47**, 3 (1988).
119. G. Blyholder, *J. Phys. Chem.* **68**, 2772 (1964); *J. Vac. Sci. Technol.* **11**, 865 (1974).
120. V. McKoy, R. L. Dubs, and M. Braunstein, private communication (1989) and Ref. 122.
121. C. Westphal, J. Bansmann, M. Getzlaff, and G. Schönhense, *Phys. Rev. Lett.* **63**, 151 (1989).
122. C. Westphal, J. Bansmann, M. Getzlaff, G. Schönhense, N. A. Cherepkov, M. Braunstein, V. McKoy, and R. L. Dubs, *Surf. Sci.* **253**, 205 (1991).
123. C. Westphal, Ph.D. Thesis, University of Bielefeld (1991).
124. M. Born and E. Wolf, *Principles of Optics* (Pergamon Press, Elmsford, NY, 1984).
125. C. Westphal, F. Fegel, J. Bausmann, M. Getzlaff, G. Schönhense, J. A. Stephens and V. McKoy, *Phys. Rev. B* **50**, 17534 (1994).
126. F. P. Netzer and J. U. Mack, *J. Chem. Phys.* **79**, 1017 (1983).
127. H.-P. Steinrück, C. Schneider, P. A. Heimann, T. Pache, E. Umbach, and D. Menzel, *Surf. Sci.* **208**, 136 (1989).
128. J. U. Mack, E. Bertel, and F. P. Netzer, *Surf. Sci.* **159**, 265 (1985).
129. C. Westphal, J. Bansmann, M. Getzlaff, and G. Schönhense, *J. Electron Spectrosc. Relat. Phenom.* **52**, 613 (1990).
130. M. Tinkham, *Group Theory and Quantum Mechanics* (McGraw–Hill, New York, 1964).
131. See, e.g., A. Puschmann, J. Haase, M. D. Crapper, C. E. Riley, and D. P. Woodruff, *Phys. Rev. Lett.* **54**, 2250 (1985).
132. R. Courths, B. Cord, H. Wern, H. Saalfeld, and S. Hüfner, *Solid State Commun.* **63**, 619 (1987).
133. C. Westphal, M. Getzlaff, J. Bansmann, and G. Schönhense, (to be published).
134. H. Petersen, *Opt. Commun.* **40**, 402 (1982) and *Nucl. Instrum. Methods A* **246**, 260 (1986); G. Kaindl, M. Domke, C. Laubschat, E. Weschke, and C. Xue, *Rev. Sci. Instrum.* **63**, 1234 (1992).
135. H. Petersen, F. Schäfers, and M. Willmann, BESSY annual report p. 425 (1990); H. Petersen, M. Willmann, F. Schäfers, and W. Gudat, *Nucl. Instr. and Methods A* **333**, 129 (1993).
136. H.-J. Freund and M. Neumann, *Appl. Phys. A* **47**, 3 (1988).
137. J. Bansmann, C. Ostertag, G. Schönhense, F. Fegel, C. Westphal, M. Getzlaff, F. Schäfers, and H. Petersen, *Phys. Rev. B* **46**, 13496 (1992).
138. V. McKoy, M. Braunstein, and R. L. Dubs, private communication (1992).
139. A. A. Pavlychev and N. A. Cherepkov, private communication (1992).
140. Theoretical calculation for the C 1s- and O 1s-orbitals of CO from T. Porwol and H. J. Freund, University of Bochum, Germany (unpublished).
141. See A. A. Pavlychev, A. S. Vinograd, V. N. Akimov, and S. V. Nekipelov, *Phys. Scr.* **42**, 160 (1990).
142. C. T. Chen, *Rev. Sci. Instrum.* **63**, 1229 (1992).
143. F. Allen and C. Bustamante (Eds.), *Applications of Circularly Polarized Radiation Using Synchrotron and Ordinary Sources* (Plenum Press, New York, 1985).
144. J. M. J. Madey, *J. Appl. Phys.* **42**, 1906 (1971).
145. H. Onuki, N. Saito, and T. Saito, *Appl. Phys. Lett.* **52**, 173 (1988).
146. H. Kitamura, *Synchrotron Radiation News* **5(1)**, 14 (1992).
147. M. B. Moiseev, M. N. Nikitin, and N. I. Fedosov, *Sov. Phys. J.* **21**, 332 (1978).
148. K.-J. Kim, *Nucl. Instrum. Methods* **219**, 425 (1984).
149. K.-J. Kim, in Ref. 143, p. 21.
150. J. Bahrdt, A. Gaupp, W. Gudat, M. Mast, K. Molter, W. B. Peatman, M. Scheer, T. Schroeter, and C. Wang, *Synchrotron Radiation News* **5(2)**, 12 (1992).
151. J. Bahrdt, A. Gaupp, W. Gudat, M. Mast, K. Molter, W. B. Peatman, M. Scheer, T. Schroeter, and C. Wang, *Rev. Sci. Instrum.* **63**, 339 (1992).
152. C. Bustamante, M. F. Maestre, and I. Tinoco, Jr., *J. Chem. Phys.* **73**, 4273 (1980).
153. C. Bustamante, M. F. Maestre, and I. Tinoco, Jr., *J. Chem. Phys.* **73**, 6046 (1980).
154. M. Mundschau, E. Bauer, and W. Swiech, *Surf. Sci.* **203**, 412 (1988).

155. K. Hall, K. S. Wells, D. Keller, B. Samori, M. F. Maestre, I. Tinoco, Jr., and C. Bustamante, in Ref. 143, p. 77.
156. C. Nicolini, in Ref. 143, p. 93.
157. M. F. Maestre, C. Bustamante, P. A. Snyder, E. Rowe, and R. Hansen, in *Production and Analysis of Polarized X-rays*, SPIE Proc. **1548**, in press, 1992.
158. G. van der Laan, B. T. Thole, G. A. Sawatzky, J. B. Goedkoop, J. C. Fuggle, J.-M. Esteva, R. C. Karnatak, J. P. Remeika, and H. A. Dabkowska, *Phys. Rev. B* **34**, 6529 (1986).
159. G. Schütz, *Phys. Rev. Lett.* **58**, 737 (1987).
160. A. Balerna, E. Bernieri, and S. Mobilio (Eds.), *2nd European Conference on Progress in X-ray Synchrotron Radiation Research*, Societa Italiana di Fisica, Conference Proceedings, Vol. 25, 1990, Chap. 3 and 5.
161. C. Gauthier, I. Ascone, J. Goulon, R. Cortes, J.-M. Barbe, and R. Guilard, in Ref. 160, p. 395.
162. T. Jo, *Synchrotron Radiation News* **5(1)**, 21 (1992).

AUTHOR INDEX

Abed, S., 513
Åberg, T., 67, 68, 202, 209, 311, 401, 406, 408, 416, 417, 424, 597
Adam, M. Y., 21, 72, 87–89, 148, 149, 154, 155, 181, 182, 196, 384, 385, 562
Afrosimov, V. V., 386
Ågren, H., 309, 409
Aizawa, H., 591
Ajello, J. M., 356
Akopyan, M. E., 265, 272, 273
Aksela, H., 209, 210, 324, 326, 401, 403, 405, 406, 408–414, 416, 418, 420, 421, 422, 423, 473, 477, 479
Aksela, S., 152, 209, 310, 311, 314, 401, 406, 411–413, 418, 421
Albiez, A., 208, 214, 215, 409, 410, 418
Alguard, M. J., 522
Allamendola, L., 251, 292
Allen, F., 645
Allenspach, R., 554
Allison, D. C. S., 589
Altick, P. L., 47, 54
Altun, Z., 67, 71, 72, 106, 154, 452, 453, 454
Alvarez, I., 597
Amusia, M. Ya., 2, 3, 5–8, 11, 12, 14–19, 23, 25, 27, 29, 31, 39, 40, 42, 47, 48, 51, 52, 54, 61–64, 68, 69, 72, 74, 82, 99, 106, 135, 142, 144, 146, 148, 149, 153, 162, 163, 240, 383, 386, 389, 390, 392, 506, 529, 530, 531, 535, 586, 587, 592
Andlauer, B., 251
Angonoa, G., 122, 125, 126, 302
Appell, J., 294
Appling, J. R., 553, 628, 630, 631, 632

Arkell, A., 222
Armen, G. B., 106, 150, 151, 153, 200, 202, 209, 213, 215, 315, 318, 407, 417, 420, 423
Asaad, W. N., 410
Asplund, L., 208, 320
Avaldi, L., 148, 362
Aymar, M., 72, 73, 455

Bachrach, R. Z., 138
Baer, T., 251, 254, 356, 496, 514
Bagus, P. S., 193
Bahrdt, J., 534, 646
Baig, M. A., 384–386, 388, 393–395
Baird, R. J., 553
Balasubramanian, K., 244
Baldwin, D. P., 286
Balerna, A., 647
Balloffet, G., 562
Baltzer, P., 167, 186, 188, 203, 402
Bancroft, G. M., 314
Bandarage, G., 170
Banichevich, A., 326
Bansmann, J., 642
Barat, M., 598, 599
Barron, L. D., 616
Barrus, D. M., 124
Barth, A., 127
Bartlett, R. J., 99
Bartschat, K., 539
Basco, N., 223
Bassett, P. J., 226
Bates, D. R., 81
Baum, G., 521, 522
Baumert, T., 318
Baumgärtel, H., 227, 231
Bauschlicher, C. W., Jr., 244
Bawagan, A. D. O., 145, 147
Bearden, J. A., 191
Beck, R. D., 256

Becker, J., 254
Becker, U., 21, 23, 25, 28, 29, 33, 62–64, 66–69, 72, 73, 145–149, 154, 155, 161, 164, 168, 171, 198, 209, 215, 298, 305, 307, 308, 311, 315, 326, 355, 359, 371, 372, 388, 389, 393, 394, 402, 404, 420, 423, 424, 428, 429, 466, 467, 470, 473, 474, 478, 479, 505, 515
Ben-Itzhak, I., 336
Benning, P., 172
Berakdar, J., 518, 632, 633
Berezhko, E. G., 424, 425, 428–431, 434, 506, 554
Berkowitz, J., 92, 136, 140, 142, 146, 148, 152, 155, 221, 222, 231, 237–240, 242–244, 247, 265, 269, 270–272, 274, 276, 280, 281, 285
Berman, M., 130
Berrah, N., 142, 151–153
Berrington, K. A., 589
Bertsch, G. F., 9, 171, 256
Besnard, M. J., 497
Beswick, J. A., 221
Bethe, H. A., 81, 82, 531
Beutler, H., 203, 561
Beyer, H.-J., 571
Beynon, J. H., 251
Bhatia, K. S., 460
Bieri, G., 284
Biester, H. W., 255
Billardon, M., 9, 612
Binkley, J. S., 245
Binning, R. C., 242
Biot, J. B., 608
Bishenden, E., 222
Bizau, J. M., 24, 68, 69, 75, 76, 143, 156, 371, 420, 568–576, 583, 584, 586, 587, 594, 597

Bjorklund, G. C., 442, 444
Blatt, J. M., 89
Blint, R. J., 284
Blum, K., 424, 426, 428, 433, 552
Blundell, S. A., 42
Blyholder, G., 633, 635
Bock, H., 227
Bodeur, S., 296, 297, 302, 303
Bogey, M., 246, 247
Born, M., 635
Borst, M., 213, 417
Bouisset, E., 297
Bouma, W. J., 248
Bourg, F., 598
Böwering, N., 522, 526, 528, 529, 540, 545, 547–549
Bowman, R. M., 318
Boyle, J., 54, 59, 63, 65, 66
Boys, S. F., 616
Bozek, J. D., 9, 298, 320, 321, 332, 405
Bracher, J., 294
Bradley, D. J., 568, 569, 570, 572
Braitbart, O., 251
Bransden, B. H., 112
Brauner, M., 98
Braunstein, M., 121
Bréchignac, C., 11, 291
Brehm, B., 496, 512, 513, 524
Briand, J. P., 600
Brion, C. E., 149, 150, 154, 156, 196, 198, 278
Brown, C. M., 455
Brown, E. R., 54, 57–62, 71
Brown, G., 562
Brown, G. S., 295, 315
Brueckner, K. A., 47, 51
Bruhn, R., 161
Bruneau, J., 410, 414
Brunt, R. J. van, 252, 254
Büchner, M., 550
Buckingham, A. D., 609
Bulgin, D. K., 225, 407
Burgers, P. C., 248
Burgt, P. J. M. van der, 383
Burke, P. G., 3, 47, 389, 562, 589
Burkov, S. M., 71, 72
Bustamante, C., 646, 647
Büttgenbach, S., 469
Buß, V., 622
Bußert, W., 432, 533

Cafolla, A. A., 324, 377
Cairns, R. B., 157
Caldwell, C. D., 89–91, 188, 202, 204, 209, 210, 404, 420, 424–426, 607
Cameron, A. E., 465
Carillon, A., 562, 563
Carlson, R. W., 278
Carlson, T. A., 75, 77, 98, 152, 182, 196, 208, 210, 291, 294, 304, 307, 314, 315, 317, 318, 326, 334, 403, 420, 424– 429, 571
Carlsson-Göthe, M., 188
Carlsten, J. L., 569
Carnell, M., 615
Carravetta, V., 409
Carroll, P. K., 562, 564, 565, 580
Carroll, T. X., 308, 314, 321, 423
Carré, B., 570
Carter, S. L., 66, 67, 74–77, 158
Carter, V. L., 539
Casanova, S., 292
Cederbaum, L. S., 109, 117, 165, 224, 308, 309, 311, 321, 326
Cermák, V., 324
Cerrina, R., 327
Cesar, A., 417
Chamberlain, R. N., 480
Chan, W. F., 137, 148, 531
Chan, Y.-M., 580
Chandra, N., 305, 318, 345, 624, 641
Chang, E. S., 97
Chang, J., 62
Chang, T. N., 76, 575, 587, 588
Charney, E., 607
Chase, M. W., Jr., 242
Chaudhry, M. A., 511
Chen, C. T., 166, 188, 296, 297, 298, 302, 319, 405, 645, 647
Chen, M. H., 408, 409, 416, 426
Chen, Y.-H., 623
Cheng, J. C., 612
Cheng, K. T., 167, 590
Cherepkov, N. A., 5, 6, 521–524, 529, 531, 533, 535–542, 544, 545, 552–554, 624–626, 630, 631, 633
Chupka, W. A., 265
Church, D. A., 568
Chutjian, A., 356
Cini, M., 308
Cleff, B., 425, 426
Codling, K., 146, 369, 377, 387, 388, 390, 392, 393, 396, 573
Colbourne, D., 229
Cole, B. E., 166, 169
Colegrove, B. T., 245, 247
Colle, R., 409
Combet-Farnoux, F., 590
Condon, E. U., 616
Connerade, J. P., 302, 449, 450–452, 454, 459, 480, 489, 581, 582
Continetti, R. E., 242
Cook, J. P. D., 72
Coon, J. B., 224
Cooper, J. W., 61, 62, 82–84, 135, 138, 144, 426, 487, 521, 566, 579, 580, 585, 595, 596
Cornaggia, C., 295, 337, 338
Cornford, A. B., 223
Correia, N., 308, 309, 311, 321
Costello, J. T., 63, 65, 66, 565, 580, 588, 596
Coulman, D., 324
Courths, R., 640, 641
Coville, M., 304
Cowan, R. D., 588, 591, 592, 596
Cox, P. A., 92
Cox, R. A., 222
Crabbé, P., 607
Craig, B. I., 95
Crasemann, B., 304, 328, 417
Cromer, C. L., 566
Cubaynes, D., 94, 96, 97, 99, 574–576, 587, 588
Cubric, D., 189, 190, 191, 210, 213
Cuéllar, L. E., 632
Curtis, D. M., 497
Curtiss, L. A., 244, 245, 247
Cutler, J. N., 304, 318, 320, 321, 405
Cvejanovic, S., 356

Daasch, W. R., 105
Dadouch, A., 274–276, 287
Dalgarno, A., 99
Davidovits, P., 267
Davis, H. F., 222
Davis, L. C., 164
de Brito, A. N., 423
de Miranda, M. P., 320, 334
de Souza, G. G. B., 307, 311, 314, 324
Decman, D. J., 208, 209
Dehmer, J. L., 86–89, 105, 122, 123, 146, 148, 152, 155, 165, 167, 265, 299, 301, 304, 305, 308, 512, 516, 537
Dehmer, P. M., 206, 265
Delaunay, M., 598
Delwiche, J., 278
Denis, A., 117
Derby, R. I., 222
Derenbach, H., 87, 88, 146, 152, 155, 404
Deshmukh, P. C., 591, 592
Desjardins, S. J., 171
Deslattes, R. D., 30, 302, 380, 382
Deutsch, C., 295
Deutsch, M., 151
Devons, S., 424

Dewar, M. J. S., 254
Dibeler, V. H., 226, 230, 265, 272, 273
Dill, D., 82, 123, 125, 138, 308, 529, 530, 531
Dill, J., 553
Ding, A., 172
Ditchburn, R. W., 72, 73
Djerassi, C., 607
Dolmatov, V. K., 585, 595, 596
Domke, M., 83, 85, 121, 166, 188, 204, 296, 302, 319, 370, 387, 407
Doppelfeld, J., 27, 29, 30, 150
Drachenfels, V. W., 522
Drake, A. F., 612, 613
Drake, G. W. F., 580
Dreizler, R. M., 42
Driver, R. D., 590
Dubs, R. L., 553, 624, 625, 627, 630, 631, 632, 633, 634, 640
Dudde, R., 291, 314
Dujardin, G., 271, 286, 316, 328, 338, 497
Duley, W. W., 252
Dunlavey, S. J., 240
Dunning Jr., T. H., 121
Dupré, C., 503
Durup, J., 266, 268, 269, 291
Dyall, K. G., 410
Dyke, J. M., 188, 189, 192, 193, 194, 243, 244, 248, 282, 284
Dzionk, Ch., 487, 488, 491
D'Angelo, P., 302
Dézarnaud, C., 297

Eastman, D. E., 612, 613
Eberhardt, W., 304, 311, 312, 314, 317, 321, 326, 328, 329, 332, 339, 417, 420, 423, 470, 471, 496, 497, 515
Ederer, D. L., 585
Edlén, B., 447, 561, 562
Edqvist, O., 276
Ehler, A. W., 564
Ehresmann, A., 35, 152, 390, 391, 392
Ehrhardt, H., 294
Eichler, J., 426
El-Sherbini, Th. M., 469, 470, 474
Eland, J. H. D., 221, 294, 308, 328, 329, 332, 335, 343, 404, 470, 478, 496, 497, 498, 502, 512, 513, 515, 517, 518, 549
Engelking, P. C., 92
Erickson, J., 280
Ervin, K. M., 285
Esteva, J.-M., 562

Ewing, J. J., 266, 267
Eyers, A., 615

Fadley, C. S., 304
Faegri, K., Jr., 409
Fahlman, A., 87, 88, 155, 156
Fano, U., 48, 49, 52, 82, 138, 203, 239, 364, 408, 424, 449, 469, 474, 521, 523, 530, 542, 561, 574, 582
Farmer, J. B., 465
Farooqi, S. M., 454, 456, 457
Farren, R. E., 129
Faupel, P., 622
Feagin, J. M., 85, 516
Feldhaus, J., 404
Felfli, Z., 96, 588
Ferguson, E. E., 248
Fernley, J. A., 589
Ferrand-Tanaka, L., 330
Ferray, M., 570
Ferrer, J. L., 302
Ferrett, T. A., 301, 307, 311, 314, 315
Field, T., 155, 497
Fielder, W. R., 59, 60
Filipponi, A., 303
Flemming, M. G., 184, 188, 205, 206
Flesch, R., 6, 223
Flores-Riveros, A., 321, 323
Flügge, S., 423
Fox, J. L., 588
Frankel, S., 502
Franklin, J. L., 236
Frasinski, L. J., 328, 337, 497, 503
Frenking, G., 18, 234
Freund, H.-J., 312, 313, 633, 634, 641
Frey, R., 497, 513
Friedrich, H., 287, 319
Froese-Fischer, Ch., 20
Frye, D., 71
Fuchs, R., 294
Fues, E., 522
Furusawa, Y., 292

Gadea, F. X., 320
Ganz, J., 188, 203
Garnier, V. von, 157, 158
Garton, W. R. S., 450, 544
Garvin, L. J., 61, 62, 65, 161, 162
Gauthier, Ch., 647
Gay, J. C., 461
Gedanken, A., 613, 620, 621
Geiger, J., 529, 531, 538–540
Gelius, U., 105, 402
Gérard, P., 202
Gibson, S. T., 240, 242

Gil, T. J., 167
Ginter, M. L., 458
Glushko, V. P., 281
Gmelin, L., 5, 222
Goddard III, W. A., 240
Gödderz, K. H., 621
Goeppert Mayer, M., 490
Goldberg, M., 39, 626, 627
Goldstone, J., 47, 51
Gole, J. L., 226
Golovin, A. V., 553
Gomes, P. R. S., 521
Goodman, G. L., 92
Goscinski, O., 116
Gottwald, A., 597
Grant, I. P., 406, 408, 409, 588
Greene, C. H., 47, 142, 363, 367, 379, 452–454
Greenfield, N., 623
Grev, R. S., 247
Gribakin, G. F., 14, 15, 20, 21
Grice, R. G., 267
Griesmann, U., 446, 452
Grimm, F. A., 123, 125, 307
Grosjean, M., 612
Gross, K. P., 620
Guet, C., 39
Gunn, S. R., 247
Gustafsson, T., 88, 89, 278, 280
Guyon, P. M., 140, 265

Haas, V., 622
Habenicht, W., 307, 315, 329, 344
Haber, K. S., 265, 276
Haensel, R., 63, 65, 66
Hagan, D. A., 497
Hahn, U., 433, 554
Halim, H., 230
Hall, K., 647
Hall, R. I., 143, 144, 147, 308, 356, 357, 359, 360, 362–367, 371, 372, 376, 377, 384, 385, 388–390, 393, 396, 496, 507, 508, 514, 515
Hallmeier, K.-H., 300
Hamdy, H., 571
Hamnett, A., 167, 531, 553
Hansen, A. E., 55, 615
Hansen, J. E., 87, 480
Hanson, D. M., 304, 328, 330, 339, 344, 497
Harada, N., 607, 619, 620
Harren, F., 553
Harris, P. G., 83, 85
Hartmann, E., 304, 318
Hartree, D., 2, 3
Hatfield, J. V., 189
Hausmann, A., 424–426, 434, 571
Hawley, M., 236

Haworth, A., 369
Hayaishi, T., 138, 156, 210, 423, 466, 470, 472, 475–478, 510
Hayasaka, N., 291
Hayes, M. A., 370, 371
Hayes, R. G., 339
Hayes, W., 319
Healy, W. P., 609
Heaven, M. C., 274
Heckenkamp, Ch., 527–535, 537, 540, 542, 615
Hedin, L., 116
Heimann, P. A., 142, 147, 188, 191, 198, 305, 319, 356, 388, 389, 393–395, 405, 473, 477–479
Heinzmann, U., 136, 203, 434, 521, 522, 528–531, 533, 535–540, 544, 547, 554, 556, 571, 615
Heiser, F., 145, 149–151, 198, 356, 359, 360, 393, 395
Hemmers, O., 167–170, 308, 316
Hennessey Jr., J. P., 623, 624
Hergenhahn, U., 428, 429
Herrick, D .R., 370
Hertel, I. V., 11, 171, 173, 256
Hertz, H., 135
Herzberg, G., 243, 271, 274, 282, 284, 295, 550
Hicks, P. J., 370
Hieda, K., 291, 327
Hilber, G., 443
Hilbig, R., 442, 443, 535
Hillier, I. H., 309
Hills, A. J., 222, 241
Hinds International Inc., 612
Hirano, T., 609
Hitchcock, A. P., 121, 125–129, 167, 296, 301, 304, 316, 317, 319, 324, 326, 328, 339, 340
Höhn, E. G., 617
Holland, D. M. P., 75, 77, 466, 469, 470, 479, 487
Hollander, J. M., 297
Hormes, J., 614
Hotokka, M., 311
Hotop, H., 91, 281
Houlgate, R. G., 72, 74, 148, 394
Howat, G., 90, 414
Huang, K.-N., 16, 87, 88, 148, 149, 152, 407, 433, 524, 530, 531, 535, 537, 538
Huber, K. P., 225, 274, 319
Hudson, E., 296, 297, 319
Huetz, A., 144, 356, 366, 367, 393, 506, 507
Huffmann, R. E., 529
Hughes, V. W., 522, 551

Humphries, C. M., 222
Huster, R., 213
Hutchinson, M. H. R., 443
Huth-Fehre, T., 535, 536, 547, 549–551
Huzinaga, S., 50

Illing, G., 314
Imamura, T., 340
Imhof, R. E., 501
Irmer, N., 556
Ishihara, T., 75
Itoh, Y., 480
Ivanov, V. K., 14, 542, 544, 585, 593, 594, 595, 597

Jacobs, V. L., 142, 143, 523
Jaegle, P., 563
Jannitti, E., 564–566, 580, 581, 587, 857
Janoschek, R., 608
Jans, W., 383
Jasperson, S. N., 612
Jean, Y., 293, 294
Jennison, D. R., 309, 409
Ji, D., 314
Jiménez-Mier, J., 160–162, 184, 193, 202, 210, 211, 379, 380, 571
Jo, T., 647
Jochims, H. W., 234, 252, 254, 255
Johnson III, R. D., 244
Johnson Jr., W. C., 612, 613, 620, 621, 623
Johnson, K. H., 621
Johnson, W. R., 47, 87, 88, 146, 157–159, 203, 529, 530, 539, 540, 542, 544, 590
Jolly, W. L., 128
Jones, K. W., 600
Joos, P., 615
Joshi, R. M., 236
Jost, K., 629
Journel, L., 575–577, 587–589

Kabachnik, N. M., 426–434, 506, 523, 554, 555
Kaesdorf, S., 553, 628
Kagan, R. L., 263
Kakar, S., 166
Kämmerling, B., 21, 23, 66, 68, 154, 385, 424–431, 434, 496, 503, 506, 510, 511, 586
Kanamori, H., 167, 307, 311
Karlsson, L., 318, 416, 417
Kaspar, F., 321
Katayama, D. H., 461
Kaufel, R., 232, 233, 236
Kauschka, G., 236

Kay, R. B., 121–123, 317
Kazansky, A. K., 31
Kelber, J. A., 308
Keller, F., 158, 542, 544
Keller, P. R., 307
Kellokumpu, M., 418
Kelly, H. P., 3, 36, 48, 50, 55–57, 68, 73, 74, 106, 112, 414
Kemp, J. C., 612
Kennedy, D. J., 146, 148, 152, 386
Kerkhoff, H. G., 163, 164
Kerling, C., 577
Kessler, B., 527, 528
Kessler, J., 434, 521, 525, 529, 534, 630, 645
Kilcoyne, A. L. D., 126, 127, 129, 130, 307
Kilin, V. A., 386
Killgoar Jr., P. C., 276
Kim, D. Y., 298
Kim, K. J., 534, 646
Kim, Y. S., 521
Kimura, K., 224, 281, 282
King, G. C., 136, 138, 296, 319, 356, 364
Kirkwood, J. G., 616, 618
Kirschner, J., 556
Kitamura, H., 646
Klapstein, D., 513
Klar, D., 15, 30, 185, 188, 203
Klar, H., 379, 432, 433, 506, 524, 530, 551, 552, 578
Klein, R., 285, 286
Klonover, A., 130
Knapp, D. A., 600
Kobayashi, K., 292, 327
Kobrin, P. H., 150, 158, 159, 161
Koch, A., 297
Koenders, B. G., 195
Koizumi, T., 481, 583
Komiha, N., 513
Körber, H., 404
Kossmann, H., 72, 149, 196, 198, 359, 503, 516
Kosugi, N., 298
Kowalczyk, S. P., 157
Krailler, R. E., 251
Krässig, B., 496, 507, 508
Krätschmer, W., 171
Kratzat, M., 265, 276–280
Kraus, B., 382
Kraus, A., 265
Krause, M. O., 92, 93, 135, 136, 147, 149, 150, 152–156, 161, 181, 182, 185, 186, 188, 191, 194, 196–199, 204, 208, 214, 215, 315, 318, 404, 405, 417, 496, 544
Krausz, E., 613

Kravis, S., 600
Kravis, S. D., 568
Kronast, W., 379–381, 425
Kronig, R. de L., 274
Kroto, H. W., 171, 255
Krummacher, S., 168, 171, 573
Krylov, B. E., 522
Kuchiev, M. Yu., 32
Kudo, T., 243
Kuetgens, U., 302
Kühlewind, H., 251, 252
Kuhn, W., 616, 618
Kuiper, P., 324
Kuntze, R., 434, 554, 555
Kunz, C., 135
Kutina, R. E., 249, 250
Kutzner, M., 36, 68, 69, 153–155, 157, 158, 586
Kuyragi, H., 291
Kvaran, Á., 265, 272, 273
Ky, L. Vo., 587–589

Laan, G. van der, 647
Lablanquie, P., 16, 23, 168, 317, 326, 328, 336, 359, 420, 466, 467, 470, 475, 477, 478, 496, 515
Lagier, H., 618, 621
Lammertsma, K., 294
Lamoureux, M., 59, 60, 590
Landau, L. D., 40
Landolt, M., 432
Lane, A. M., 451
Langer, B., 98, 142, 148, 150, 201, 202, 393, 395
Langhoff, P. W., 168, 170, 299, 301, 305
Lapiano-Smith, D. A., 318, 326, 340
Laplanche, G., 522
Laramore, G. E., 309
Larkins, F. P., 106, 308, 311, 314, 406, 409
Larsson, M., 294, 308, 317, 417
Lavollée, M., 329, 343, 517
Lavrentjev, S. V., 12
Lawrence, W. G., 222
Lawson, J. S. Jr., 502
Lawton, E. A., 227
Le Gouët, J. L., 569
Leach, S., 255
Learner, R. C. M., 447
LeBrun, T., 324, 329, 332, 505
Leduc, M., 597
Lee, C. M., 522–524, 529, 530, 539, 540, 542
Lee, K., 298, 319
Lefebvre-Brion, H., 522, 545, 547, 549, 550

Léger, A., 251, 252
Lermé, J., 513
Leung, K. T., 72
Leväsalmi, M., 409, 410, 415, 418
Levasseur, N., 317, 332
Levin, J. C., 75, 99, 141, 150, 511
Levine, M. A., 599
Lewandowski, B., 480
Lias, S. G., 154, 225, 236, 251, 286
Libermann, D. A., 491, 492
Lichtenberger, D. L., 171, 255
Liebsch, A., 553
Liebsch, T., 171–173
Liegener, C.-M., 309, 409
Lifshitz, C., 251, 256
Lindau, I., 137
Lindle, D. W., 71, 72, 142, 143, 148, 149, 152, 153, 155, 168, 205, 298, 307, 311, 318, 344, 370, 371, 380, 473, 474, 477
Lineberger, W. C., 91
Lipei, M. M., 265
Lippmann, B. A., 49
Liu, X.-H., 85
Liu, Z. F., 295, 315, 320, 589
Lohmann, B., 426–428, 433, 439, 554
Lucartorto, T. B., 562, 566, 579, 590
Lucchese, R. R., 105, 106, 112–114, 278
Luke, T. M., 147, 393, 394
Lukirskii, A. P., 585
Lynch, D. L., 105, 106, 123, 308
Lyon, I. C., 567, 581, 582, 593

Ma, C. I., 330, 344
Ma, Y., 122, 296, 319, 320
MacDermott, A. J., 611
Macek, J., 83, 84
Mack, J. U., 637
Madden, R. P., 56, 83, 369, 371, 384–386, 388, 393–395, 469, 562, 580
Madey, J. M. J., 646
Maeda, K., 185
Maestre, M. F., 647
Magnusson, E., 299
Mahalingam, M., 318, 340
Maier, J. P., 497, 512, 513
Malutzki, R., 162, 413
Manaa, M. R., 248
Manavalan, P., 623
Mank, A., 535, 547
Mansell, P. I., 226, 230
Mansfield, M. W. D., 167, 210, 211, 581, 588, 595

Manson, S. T., 82, 86, 88, 97, 590
Mäntykenttä, A., 414, 418, 419
Marangos, J. P., 443, 445
Marr, G. V., 135, 140, 146, 149, 157, 167, 369, 471, 582
Marrs, R. E., 599
Mårtensson, A.-M., 71
Mårtensson, N., 416
Maruyama, T., 522
Mason, S. F., 607, 616, 620
Masuoka, T., 497
Mathur, D., 336, 337
Matsuoka, T., 278, 280
Matthew, J. A.D., 308
Maulbetsch, F., 142, 143, 367
Mayer, R., 317
Mazeau, J., 496, 506, 507
McCarthy, I. E., 72
McCulloh, K. E., 294, 518
McElvany, S. W., 256
McFarlane, S. C., 429
McIlrath, T. J., 566, 577, 588
McKoy, V., 633–635, 644
McLaren, R., 128, 129
Medhurst, L. J., 149, 150, 166, 302, 393, 395
Mehlhorn, W., 200, 201, 209, 213, 215, 401, 404, 418, 423–426, 433, 533
Mehlman-Balloffet, G., 562, 563
Menzel, A., 324
Menzel, D., 291, 314, 324
Merz, H., 427, 433, 554
Meulen, P. van der, 184, 205, 206, 207
Meyer, H.-D., 130
Meyer, M., 33, 62, 63, 64, 97, 161, 193, 209, 332, 340, 403, 420, 421, 473, 483, 570, 577
Miecznik, G., 592, 595
Mies, F. H., 582
Migdal, A. B., 42
Miller, D. L., 152
Minelli, D., 311
Minkwitz, R., 227
Minnhagen, L., 359
Mitsuke, K., 265, 276, 278, 281, 283–286
Möbus, B., 148, 379, 381, 384, 385, 387
Moddeman, W. E., 304, 305, 308, 309
Moffitt, W., 616, 617
Mohler, F. L., 294, 587
Moiseev, M. B., 534, 646
Mollenhauer, L. F., 612
Möllenkamp, R., 522
Molodtsov, S. L., 173

Moore, C. E., 191, 274, 275, 359, 372, 376, 409, 579
Morin, P., 314, 316, 324–326, 328, 332–335, 337, 340, 368, 423, 498
Morioka, Y., 276
Morisson, J. D., 265
Mott, N. F., 266
Mowrey, R. C., 256
Müller, N., 526, 528, 529, 540, 541, 543, 544
Müller-Dethlefs, K., 329, 340, 496, 498
Mullins, O. C., 569, 572, 577
Munakata, T., 282
Mundschau, M., 647
Murakami, E., 466, 470, 474
Murakami, J., 332, 340
Murphy, R., 321, 339, 496, 497
Murray, A. J., 501

Nagaoka, S., 318, 328, 340
Nagata, T., 27, 480, 481, 491
Nagel, B. H. C., 521
Nahon, L., 209, 420
Nasreen, G., 585, 590, 591
Naves de Brito, A., 324
Neeb, M., 321, 322, 323
Nelson, M. C., 340
Nenner, I., 165, 221, 304, 315, 324, 326, 328, 339, 340, 344, 359, 423
Netzer, F. P., 636, 637
Neumark, D. M., 274
Ng, C. Y., 221
Nicolini, C., 647
Nicovich, J. M., 222, 248
Nishimura, T., 514
Norcross, D. W., 587
Nordgren, J., 304, 323, 409, 410, 421
Norwood, K., 343
Nourbakhsh, S., 237, 249, 250
Nunnemann, A., 570
Nuroh, N., 585, 590
Nygaard, K. J., 569
Niessen, W. von, 284

Oertel, H., 265, 274, 275
Ohashi, H., 291
Ohmichi, N., 251
Ohrendorf, E. M.-L., 302, 303
Okuno, K., 600
Okuyama, K., 479, 509
Ong, W., 87, 88
Onuki, H., 646
Orel, A. E., 301, 305
Oura, M., 600
Ovsyannikov, V. D., 533

Padial, N., 105, 122–124, 167, 278
Pahler, M., 215, 571, 577–579
Pan, C., 62–64, 76, 77, 92, 93, 148, 163–165, 194
Pankratov, A. V., 227
Parks, E. K., 263, 267
Parzynski, R., 630
Pavlychev, A. A., 300, 643
Pearson, K., 184
Peart, B., 12, 567, 568, 582, 583, 590–592
Peatman, W. B., 140, 357
Perry, W. B., 304
Petersen, H., 188, 641
Peterson, K. A., 222, 226
Petrie, S., 256
Petrini, D., 91
Pettersson, L. G. M., 417
Peyerimhoff, S. D., 269, 270
Piancastelli, M. N., 129, 301, 307
Pickup, B. T., 115
Pierce, D. T., 522
Pines, D., 2
Plummer, E. W., 167, 169, 368, 633, 634
Poliakoff, E. D., 221, 307, 318, 323, 340
Pople, J. A., 242, 243
Porwol, T., 642
Potts, A. W., 243, 280, 284
Pratt, R. H., 521
Pratt, S. T., 195, 196
Prescher, T., 406, 468, 469
Preses, J. M., 72, 73
Price, S. D., 496, 505, 507, 514, 515
Proceedings of Conf., 182, 203
Przybilla, K.-J., 622
Puschmann, A., 638
Putaux, J. C., 568
Pysh, E. S., 613

Qian, Z.-D., 59, 61

Rabe, R., 66
Rabus, H., 319
Radeloff, J., 499
Radi, P. P., 256
Radojevic, V., 23, 204, 586
Randall, K. J., 166, 169–171, 320
Rao, P. T., 268
Raseev, G., 522, 545
Rau, A. R. P., 91, 92
Rauk, A., 617, 621
Raven, E. von, 210, 403, 420, 421, 479, 509
Read, F. H., 356, 363–365
Rehr, J. J., 300, 301
Reich, T., 169, 170

Reid, K. L., 554
Reilman, R. F., 580, 581, 585, 590–592
Reimer, A., 169
Reinke, D., 232
Reintjer, J. F., 443
Remmers, G., 296, 297, 319
Ren, S., 172
Rescigno, T. N., 105, 123
Reynaud, C., 297–300, 302, 303
Rich, A., 547
Richard, E. C., 222
Richard-Viard, M., 252, 512, 513
Richardson, F. S., 608
Richter, M., 62, 63, 64, 97, 164, 209, 405, 420, 489, 492, 576, 587, 588, 594
Ritchie, B., 546, 624, 630
Robicheaux, F., 58, 59, 60, 92, 93, 206
Robin, M. B., 227, 232, 233, 301, 613
Rodger, A., 618
Roig, R. A., 581, 582
Rosenberg, R. A., 338, 340
Rosenstock, H. M., 236, 251
Rosenzweig, N., 462
Rost, J. M., 85
Rottke, H., 221
Rouvellou, B., 588
Rowland, H. A., 561
Rühl, E., 18, 222, 225, 233, 252, 255, 291, 340, 342, 345
Ruščić, B., 58, 243–249, 265, 285
Russek, A., 213, 417
Rye, R. R., 311

Sadeghpour, H. R., 84–86
Saha, H. P., 47, 106, 386
Saile, V., 539
Saito, N., 146, 153, 298, 308
Salman, I., 343, 344
Salomonson, S., 71, 72, 142, 143, 205, 370, 371
Salzmann, D., 574, 586
Salzmann, M., 547
Sampoll, G., 336, 337
Samson, J. A. R., 59–61, 72–75, 77, 88, 89, 94, 95, 99, 135, 138, 140, 149, 152, 153, 155, 167, 182, 198, 372, 380, 381, 383, 385, 389, 390, 392, 396, 465, 529, 530, 561, 562
Sandler, P., 256
Sandner, W., 213
Sato, Y., 328, 330, 344, 479
Sauter, F., 521
Sax, A. F., 247
Saxena, V. P., 623

Schaefer, S. H., 267
Schäfers, F., 526, 528, 533, 534, 539, 540, 542, 544, 547, 554, 628
Schaphorst, S. J., 19, 206, 301, 324
Schartner, K.-H., 93, 94, 146, 371, 372, 377, 380, 381, 383, 385, 387–389, 391–395
Schellman, J. A., 610
Schenk, H., 265, 284, 286
Schiller, J., 615, 620, 645
Schimmelpfennig, B., 297
Schirmer, J., 92, 106, 118, 125, 170, 194, 297, 308, 312, 319
Schmelz, H. C., 340, 341
Schmidbauer, M., 121–124, 170, 307
Schmidt, E., 161, 162, 193
Schmidt, V., 72, 75, 77, 142–144, 149, 182, 196, 198, 203, 213, 304, 305, 308, 355, 363, 417, 429, 511, 572
Schmiedekamp, A., 229
Schmiedeskamp, B., 556
Schmitt, A., 131
Schmoranzer, H., 153, 381, 382, 390, 392, 393, 396
Schneider, B. I., 130
Schnetz, M., 501
Schnopper, H. W., 302
Schönhense, G., 157, 158, 522, 525, 527, 540, 542, 544–547, 554, 556, 629, 630, 634
Schumann, V., 561
Schütz, G., 647
Schwarz, W. H. E., 297, 324
Schwarzkopf, O., 144, 506
Schwinger, J., 614
Scott, L. T., 254
Scott, P., 452
Seaton, M. J., 521, 542
Seetula, J. A., 247
Selles, P., 506
Sette, F., 167, 301
Sevier, K. D., 182, 208
Sevin, A., 299
Shaginyan, V. R., 42
Shahabi, S., 59, 60
Sham, T. K., 121, 122
Shannon, S. P., 544
Shanti, N., 152
Shaw, D. A., 319, 324, 325
Shaw, R. W., 311
Sheehy, J. A., 278, 301
Sheen, S. H., 263, 267
Shigemasa, E., 298, 299
Shustov, G. V., 621
Shyu, J. S., 158, 160
Siegbahn, H., 105, 409

Siegbahn, K., 135, 138, 182, 196, 208, 291, 304, 305
Siegbahn, P. E. M., 409
Siggel, M. R. F., 166
Silfvast, W. T., 72
Silverstone, H. J., 50
Simon, M., 298, 328, 329, 335, 340–343
Slater, J. C., 116
Smid, H., 198, 355
Smith, A. L., 514
Smith, D., 502
Snatzke, G., 616
Snyder, L. C., 105
Snyder, P. A., 613, 614, 620
Sonntag, B., 135, 182, 209, 420, 566, 571, 577, 582
Sorensen, S. L., 416
Southworth, S. H., 146, 148, 152, 155, 156, 404, 471, 474
Spears, D. P., 72, 75
Spielberger, L., 141
Spohr, R., 356
Stanton, R. E., 256
Stapelfeldt, H., 97
Starace, A. F., 9, 10, 47, 82, 106, 182
Starke, K., 214
Stebbings, R. F., 569
Steger, H., 255, 256
Steinrück, H.-P., 637
Stenhagen, E., 255
Stephens, J. A., 105
Stephens, P. J., 607
Stevens, E. S., 615, 621
Stevenson, D. P., 232
Stockbauer, R., 166
Stöhr, J., 293, 296, 300, 301
Stolte, S., 628
Stoppmanns, P., 554, 555
Stranges, S., 314
Strauss, C. E. M., 5, 516
Sugiura, M. Y., 452
Sukhorukov, V. L., 25, 149, 150, 385–388, 390, 392, 393, 394, 395, 396, 397
Sutherland, D. G. J., 320, 321
Sutherland, J. C., 614, 624
Süzer, S., 544
Suzuki, S., 265, 280, 281, 283–285
Svensson, A., 327, 540
Svensson, S., 72, 145, 149, 150, 154, 196–199, 214, 215, 324, 326, 359, 416–418
Svensson, W. A., 156, 158
Swalen, J. D., 229
Swanson, J. R., 47
Szepes, L., 343

Tambe, B. R., 158, 159
Tarantelli, F., 308–311, 409
Terenin, A., 263
Terminello, L. J., 173
Theodosiou, C. E., 158, 586
Thissen, R., 324, 328, 340
Thomas, T. D., 147, 296, 304, 308, 319, 321
Thompson, D. G., 572
Thorson, W., 269
Thouless, D. J., 3, 39
Tinkham, M., 636
Tinoco Jr., I., 608
Tonkyn, R. G., 188, 203
Torop, L., 155, 537, 562
Toumadje, A., 623
Tracy, D. H., 487
Trainham, R., 281
Tronc, M., 297, 319
Truesdale, C. M., 121, 124, 125, 168, 170, 298, 307, 308, 311
Tsai, B. P., 138, 255
Tsuchiya, T., 294
Tulkki, J., 12, 13, 87–89, 106, 144, 148, 149, 152, 154–156, 377, 390, 408–410, 414, 415, 417, 418, 426, 470
Tully, F. P., 266
Tyson, T. A., 300, 301

Ueda, K., 150, 314, 318, 328, 329, 339, 340
Ungier, L., 169, 311
Urisu, T., 291, 327

Vager, Z., 295, 334
Vaida, V., 222
Varga, D., 404
Väyrynen, J., 412, 413
Végh, L., 506
Velluz, L., 607, 613
Vidal, C. R., 443
Vögtle, F., 622
Völkel, M., 501

Wachter, A., 155
Wagman, D. D., 222, 225, 236
Wahl, A. C., 294
Wahner, A., 222
Wainwright, P. F., 522
Walker, D. C., 611
Walker, T. E. H., 158, 265
Waller, I., 343
Walter, C. W., 21
Walters, D. L., 304
Wannberg, B., 417
Wannier, G. H., 31, 355, 363, 473, 506

Wapstra, A. H., 501
Wark, D. L., 198–200
Weaver, J. H., 171
Weber, W., 418, 421
Wehlitz, R., 140–144
Weigang Jr., O. E., 617, 618
Weiss, M. J., 226
Wen, C.-R., 327
Wendin, G., 68, 87, 157, 172, 562
Wentzel, G., 407
Wenzel, S., 622
Werme, L. O., 409, 414
Werner, H., 172
West, J. B., 156, 380, 385, 471, 562, 574, 597
Westphal, C., 553, 633–641
Wetmore, R. W., 317, 339
Wexler, S., 291, 294
White, M. D., 581
White, M. G., 87, 88, 138
Whitfield, S. B., 154–156, 162, 184, 202, 209–212, 403, 417, 420, 421, 426, 427

Wickramaaratchi, M. A., 614
Wiel, M. J. van der, 466, 469, 470
Wiese, W. L., 580
Wight, G. R., 75, 77, 121
Wijesundera, W., 25, 59, 72, 74, 75, 148, 149, 198, 385–387, 393, 396
Wiley, W. C., 328, 466, 501
Williams, A. R., 3
Wills, A. A., 94, 95, 147, 149, 150, 198, 360, 362, 371–374, 376, 377, 384, 388, 393, 395, 396
Wingen, U., 622
Winkler, P., 76
Winniczek, J. W., 628, 630
Wittel, K., 229, 233
Woodruff, P. R., 71, 93, 142, 380
Wu, C. Y. R., 223
Wu, J. Z., 209
Wuilleumier, F. J., 97, 136, 140, 142, 145, 146, 203, 302, 304, 389, 562, 569, 570, 572, 575, 576, 585, 587

Wurz, P., 256
Wu, J.-Z., 91, 172

Yagishita, A., 188, 211, 298, 308, 420, 480
Yan, Yu., 589
Yang, C. N., 82
Yates, B. W., 404
Yeh, J.-J., 146, 152, 162
Yencha, A. J., 265, 272
Yi, J.-Y., 256
Yoo, R. K., 237, 243, 255, 256
Yoshino, M., 480
Yudin, N. P., 506

Zähringer, K., 311, 345, 409
Zangwill, A., 3, 47, 62–64, 68, 587
Zewail, A. H., 319
Zimkina, T. M., 489
Zimmermann, J. A., 255
Zimmermann, P., 27, 491
Zubek, M., 364, 368–370

SUBJECT INDEX

Absolute cross sections, 562, 565, 567, 580, 581
Absorption spectra, 384, 385
Adsorbates, 630, 633
Adsorbed layers of alkaline metals, 554, 555
Adsorbed monolayers of Xenon, 528
Aligned atoms, 570, 571, 577
Aligned molecules, 551, 552, 553, 630
Alignment in,
  double photoionization, 511
  photoexcitation, 424, 425, 428, 433
  photoionization, 424, 425, 432, 433, 434
Alignment parameter, 424, 425, 426, 427, 428, 431, 434, 435
Alkali halides, 266, 267, 268
Alkali metals, 479, 480, 481, 482
Alkaline earth, 71, 479, 480, 481, 482
Aluminium, 591, 592
Angle resolved photoelectron spectroscopy (ARPES), 136
Angle- and spin-resolved photoelectron spectroscopy, 525, 526, 527, 528, 529
Angular distribution, 82, 85–89, 137, 138, 366, 505, 506, 507, 516, 530, 531, 553
  asymmetry parameter, 6, 54, 82–89, 138, 366, 370, 371, 529, 535, 537, 540–546
Angular momentum transfer, 82, 138
Argon, 12, 19, 55, 56, 72, 75, 94, 146, 148, 181, 182, 198, 209, 213, 357–360, 371–375, 379, 380, 381, 383–386, 393, 395, 402, 410, 414, 416, 421, 428, 429, 525, 538, 539
Arsenic, 238
AsH, 237, 240
AsH$_2$, 237, 240
Atomic beam technique, 467, 468
Auger electron
  angular distribution, 403, 404, 423–429, 431, 433, 434, 435
  anisotropy of, 423, 424
    normal, 426, 428, 429
    resonant, 428, 429
  anisotropy parameter, 424–431, 433, 434, 435
  cascade, 418, 419, 420, 421, 423
  coincidences with, 496, 505, 508, 511, 515

Auger electron (*cont.*)
  spectrometry (AES), 208
  spin polarization, 432–435, 554, 555
Auger line, intensity of, 401, 402
  position of, 401
  width of, 315
Auger process, 203, 208, 209, 286, 287, 469, 473
  Participator, 471–474
  Spectator, 472–475, 477, 478; *see also* Spectator electron
  vibrational effects in, 312
Auger shake off, 315, 316, 317, 473
Auger spectrum, 308–311, 409–413, 416, 417
Auger transitions
  energies of, 406, 407, 409
  following electron beam excitation, 404, 411
  photon excitation, 402
  synchrotron radiation excitation, 404, 405
  probability (rate) of, 407, 408, 414, 421
  resonant, 311–315, 403, 420–423, 428, 429, 433
Autoionization, 47, 52, 56, 184, 185, 203, 215, 368, 370, 384, 386, 469, 471, 473, 508, 509, 514
  in double ionization, 508, 514
  in valence ionization, 145, 153
Azulene, 251, 255

Bandpass, 184
Barium, 23, 66, 68, 100, 101, 420, 434, 554, 555, 567, 569, 570, 572, 574, 581–583, 585–588, 590
Benzene, 636–640
Beryllium, 48, 76, 89, 90, 91, 565, 580, 587, 590
BF$_3$, 339
BiF, 237, 243
BiF$_2$, 237, 243
Biorthogonalization, 108
Born–Oppenheimer approximation, 131, 295, 319, 550,
Born–Oppenheimer states, 268
Br$_2$, 271–274, 302, 545
Branching ratio, 89–93, 375–379

Breit interaction, 407
Broadening of photolines, 183, 186
Bromine, 195
Brueckner orbitals, 54
BRV (Ballofet–Romand–Vodar) source, 562, 566
Buckminsterfullerene, 9, 38, 171, 222, 255, 256; *see also* Fullerenes, C60

$C\equiv C$ triple bond, 244
$C=C$ double bond, 244
C-C single bond, 244
$C_{10}H_8$ isomers, 251–255
C_2H_2, 285, 286
C_2H_4, 319, 320
C_2H_4BrCl, 341
C_2H_6, 297
C_2H_6BrCl, 341
C_3H_8, 297
C_4H_4S, 339
C_{60}, 171, 172, 222, 255, 256; *see also* Buckminsterfullerene, Fullerenes
C_6H_6 isomers, 251
Cadmium, 66, 157, 158, 412, 413, 588
Calcium, 20, 421, 567, 568, 582, 583, 584, 592–595
Carbon, 195, 581, 587
CDAD microscopy, 646, 647
Centrifugal barrier, 144, 491, 459, 586
Cesium, 15, 569, 587
CF_2Cl_2, 284, 285
CF_3Cl, 284, 285, 339
CF_4, 284, 285
$CFCl_3$, 284, 285
CH, 243, 244
CH_2, 243, 244
CH_2BrCl, 341
CH_2OH, 247–251
CH_2SH, 247–251
CH_3, 243, 244
CH_3Br, 280–284, 324
CH_3CF_3, 329, 344
CH_3Cl, 280–284, 324, 339
CH_3F, 280–284
CH_3I, 317, 324, 326, 334
CH_3O, 247–251
CH_3S, 247–251
CH_3SH, 299
CH_4, 280–284, 296, 297
Chalcogen hydrides, 240–243
Chalcogens, 187, 194, 202, 205, 206, 209
Channelplate detector, 369, 370, 382
Chaos, 461, 462
Chiral electron pair, 632
  geometry, 625
  molecules, 546, 624
Chirality, 607, 608, 609, 624
  natural, 624, 625
Chlorine, 12, 57–59, 92, 93, 194, 205
Chromium, 12, 579, 580, 595, 596

Circular dichroism (CD), 552, 553, 607
  application in biology, 622–624
  application in chemistry, 621–622
  empirical description, 608, 609
  quantum mechanical description, 609–611
  in the angular distribution (CDAD), 531, 533, 546, 547, 552, 553
Circular intensity differential scattering (CIDS), 645, 646, 647
Circularly polarized light, 612, 626, 645
  from synchrotron radiation, 613, 614, 629
  from undulators and wigglers, 645, 646
ClO_2, 222–226
Cl_2, 269–274, 297, 298, 324
Closed-Shell Atoms, 194, 197, 209
CO, 165, 168–171, 275, 276, 311, 312, 313, 316, 317, 319, 336, 337, 339
CO_2, 278, 280, 334, 337, 338
Coherent VUV radiation, 535
Coincidence measurements, 361, 467, 475–479
Coincidence (technique), 328–332, 421, 429–432, 435, 495, 496
  photoelectron–Auger electron, 403, 429–432, 435
  accidental, 498–504
  electron–electron (PEPECO), 496, 505–510, 514, 515
  electron–ion (PEPICO), 251, 257, 360, 361, 362, 496, 497, 502–508, 510, 512–515
  electron–ion–ion (PEPIPICO), 330–332, 497, 503, 504, 516–518
  electron–photon (PEFCO), 340, 497, 512, 513
  ion–ion (PIPICO), 316, 317, 497, 504, 514–516
  ion–photon (PIFCO), 497, 505, 512, 513, 515
  methods, 495, 496, 498, 518
  multiple, 498, 503, 504, 505
  notations, 496
  optimization, 501, 502, 503
  pulsed ionization (fields), 502, 503
  resolving time, 498–502
  three-fold, 495, 497, 498, 503, 504, 505
  true, 498–502
  two-fold, 495, 503, 518
  threshold-electron ion (T-PEPICO), 221, 252–254, 358, 360–362, 496, 505–510, 515
Collection efficiency, 359, 466, 467
Collision-induced ion pair formation, 266
Collisions, 460, 461
Complete Auger recombination experiments, 434, 435
Complete photoionization experiments, 425, 434, 630
  with atoms, 522, 529–533, 630, 645
  with molecules, 547
Complete set of states, 50
Configuration Interaction (CI), 47, 72, 73, 408, 409, 411, 418–421
  in the final continuum state (FCSCI), 408, 409, 411, 414, 415, 421

Subject Index

Configuration Interaction (CI) (*cont.*)
  in the final ionic state (FISCI), 87, 359, 408–411, 414, 418
  in the initial state (ISCI), 419
  in the initial ionic state (IISCI), 408, 418
Constant final state spectroscopy (CFS), 140, 190, 359, 367
Constant ionic (or initial) state spectroscopy (CIS), 185, 208, 216
Constant kinetic energy (CKE), 189, 359
Continuity theorem, 452
Convolution, 183
Cooper minimum, 12, 29, 59, 144, 425, 434, 452, 521, 529, 537, 585, 590
Cooper–Zare model, 138
Core equivalent approximation, 293, 297, 303, 319
Core excited resonances, 420, 423
Core levels, 641
Core valence separation (CVS) approximation, 108, 109, 110, 114–121, 131
Core-hole spectrum, 166
Correlation
  effects, 462, 562, 565, 574
  Hamiltonian, 50, 51
  satellites, 359, 360, 575, 576, 577, 588, 589
  in the final state, 51, 56, 57, 75
  in the ground state, 47, 51, 56, 58, 75
Correlational decay, 34
COS, 276, 278, 280
Coster–Kronig transitions, 401, 416
Coulomb Green's function, 113
Coupled equation method, 54, 57, 58, 61, 62, 65, 71
Cowan code, 588, 591, 592, 596
Cross section, 137–139, 368–377, 370, 377, 506, 507, 511

Dead time, 499
Decay channel, 373–379
Degeneracy, 107, 110
Delayed maxima, 585, 591
Density matrix, 523, 551
Detectors, 498, 495, 502, 503
Diagram line, 200, 208, 214; *see also* Auger diagram line
Dichroism, circular, 518; *see also* Circular dicroism
Differential cross section, 627
Dipole matrix element, 525, 530, 532, 540, 542
Dirac–Slater approximation, 544
Direct double ionization, 480
Direct knock out, 21
Discharge sources, 561, 562
Dissociative double ionization, 316, 317
Double Auger process, 469, 470, 473–475; *see also* Auger process
Double photoexcitation, 70–72, 302–304, 450
Double photoionization, 22, 54, 73–76, 98, 99, 141, 168, 506–518, 359, 360, 362, 363, 495, 499, 502, 504–511, 514, 515, 516
  direct, 505, 506, 514, 517
  indirect, 505–507, 510, 514, 515, 517

Double-well potential, 490
Doubly excited states, 83, 84, 85, 370, 373, 384, 386–397
  hyperspherical coordinate representation, 83, 84, 85
  molecular orbital representation, 85
Delta self-consistent field (SCF) approximation, 109, 116, 117
Dual laser produced plasma (DLPP), 562–566, 580
Dynamic polarization model, 617
Dynamical effects in photoionization, 130, 131
Dyson equation, 20

Effective interelectron photoionization, 3, 28
Effective potential, 59, 60
Effusive source, 467
Electric dipole transition, 624
Electric-dipole approximation, 522
Electron
  affinity, 20
  beam ion source (EBIS), 598, 600
  beam ion trap (EBIT), 599, 600
  bombardment, 208–213
  correlation, 196, 356, 568; *see also* Correlation effects
  cyclotron resonance ion source (ECR), 598, 599
  energy analyzer (spectrometer), 183, 195, 200, 208, 356, 357, 404
  spectrometry, 181
  with Synchrotron Radiation (ESSR), 135, 181, 216
  spectroscopy for chemical analysis (ESCA), 135
  angular distribution, 317, 511
  satellites, 507, 508; *see also* Satellites
  zero-kinetic energy, 356–368, 496, 497, 507, 510, 512, 513, 514
Emission cross section, fluorescence, 378
Enantiomer, 626
Energy degeneracy, 458, 459
Equivalent core holes, 110, 111, 112, 130, 293, 297, 303, 319
Europium, 162–165
Exchange integral, 240
Excitation cross section, 378
Excited electrons, 14
Exciton coupling model, 618–620
Extended X-ray absorption fine structure (EXAFS), 299, 300, 301

F_2, 272, 274
Fano
  effect, 521, 522, 529
  formula, 239
  profile, Beutler–Fano, 184, 188, 203, 215, 483, 484
Faraday rotation, 447, 448, 449, 452
$Fe(CO)_2(NO)_2$, 341, 342, 343

Fermis golden rule, 48
Fine Structure, 186, 192, 194, 203, 206, 211
  splitting, 522, 523, 539, 540, 545, 547
  transitions, 91–93
Fluorescence spectroscopy, 93–94, 377, 378, 379
Fluorides, Group V, 243, 340
Fluoroamines, 226–232
Formate species, 638
Fourier transform spectrometer, 447
Four-wave mixing, 442, 443
Franck–Condon approximation, 131, 295, 319
Frozen orbital approximation, 407
Frozen Core Hartree–Fock (FCHF) approximation, 106, 108, 121–125
Fullerenes, 3, 38, 171, 222, 255, 256

GeH_2, 244
GeH_3, 244
Generalized RPAE, 16; *see also* RPAE
Gerade and ungerade states, 268, 272, 274, 288
Giant resonances, 9, 60, 62, 66, 161, 489, 490, 583, 584
Goddard–Harding model, 240–243
Gold, 521
Gradual orbital collapse, 163
Grice–Herschbach formula, 267
Gallium, 567, 568, 582, 583, 596
Grant code, 588

H^-, 83, 84, 85
H_2, 268, 269
$H_2CO(D_2CO)$, 296, 311
$H_2S(D_2S)$, 296, 302, 303, 304, 317, 319, 339
Haloethylenes, 232–236
Halogens, 193, 194, 206, 209, 320
Halogencompounds, 340
Handedness, 625, 626
Hartree–Fock approximation, 6, 50–52, 535
Hartree–Slater wave functions, 426
HBr, 302, 320, 324, 325, 326
HCl, 297, 320, 324, 326, 339
HCN, 285
Helium, 71, 74, 83–85, 93, 98, 99, 140, 142, 144, 187, 204, 363–367, 369–371, 569, 580
Heterogeneous coupling, 271, 274
2,4-hexadyne, 251
HF, 311, 324
HI, 320, 324, 326
Higher-order processes, 212
Hollow cathode, 447
Homogeneous coupling, 274
Hund's case, 274
Hydrides, Groups V and VI, 240–243
Hydrogen chlorine, 416, 417
Hydrogen iodide, 547–550

I_2, 271, 272, 274, 545, 546
Indium, 526, 528
Induced chirality, 625

Inner-shell
  excitation, 459, 460
  pair formation, 271, 286, 287
  photoionization, 94–97, 504, 506, 508
  resonance decays, 470–479
Interchannel
  coupling, 47, 52, 144
  interaction, 3, 87, 99–101
Interference minimum, 11
Intermediate coupling, 408, 409, 411, 412, 413, 414, 420, 421, 426
Iodine, 209, 238
Ion spectrometry, 583
Ion storage ring, 600
Ion traps, 568, 599, 600
Ions, 562, 565–568, 579–585, 596, 597
Iron, 590, 591
Iron group elements, 482–487
Isomerization, 226, 254

Jahn–Teller distortion, 229, 243, 284
Jet beam, 467

K-matrix evaluation, 54, 58
K-shell satellites, 145, 147, 150, 151; *see also* Satellites
Kramers–Kronig transformation, 449
Krypton, 150–153, 190, 200, 208, 209, 210, 213, 390–392, 396, 397, 410, 411, 420, 506, 538, 539

Lanthanum, 11, 17
Lead, 526, 528, 540, 543, 544
Length, velocity and acceleration forms, 5, 6
Level width, 186, 187, 208, 318
LiF, 311
Line profile, 183, 184, 203; *see also* Photoline
Linear dichroism, (LD), 552, 553
Linear dichroism in the angular distribution of photoelectrons (LDAD), 532, 533, 552, 553, 630
Lippmann–Schwinger equation, 113
Lithium, 96, 97, 202, 564–566, 571, 576–579, 588, 589
Local density approximation (LDA), 587, 590
LS-coupling, 6
LS-multiplets, 192, 194, 196, 198, 209

Magnesium, 71, 72, 209, 421, 569, 572, 590
Magnetic X-ray circular dichroism (MXCD), 645, 647
Magneto-optical vernier, 449
Manganese, 12, 61, 160–162, 193, 202, 211, 212, 565, 566, 579, 580, 592, 593, 595, 596
Many body perturbation theory (MBPT), 3, 40, 47, 48, 586–590
Mass spectrometer, 466
Massey adiabatic criterion, 266
Mean free path, 467
Mean ion charge, 30

Subject Index

Mercury, 158, 159, 526, 528, 542, 544
Merged beams, 567, 583, 598
Metastable, 568
Mo(CO)$_6$, 297
Molecular flow condition, 467
Molecular ionization
   angular distribution, 512, 516
   branching ratios (pathways), 511–514
   Coulomb explosion, 294, 327, 333, 334, 337, 517
   deferred charge separation, 343, 517
   dissociation, 294, 323–326, 511–517
   Franck–Condon principle, 295, 514
   initial momentum distributions, 517
   kinetic energy release, 511, 513, 514
   predissociation, 512, 513
   quasi-equilibrium theory, 513, 514
   reaction mechanisms, 517
   secondary dissociation, 517
   statistical theory of dissociation, 343, 344, 513, 514
Molecular orientation, 625
Molecules, photoionization of aligned, 553
   chiral, 546, 547
   oriented (fixed-in-space), 553
Monochromator, 183, 185, 188, 200, 533, 534
Mott detector, 534, 535
Multi-channel detector, 188, 189, 210
Multiatomic formation, 1
Multichannel quantum defect theory (MQDT), 58, 72, 529, 530, 539, 540
Multi configuration Hartree/Dirac–Fock method (MCH/DF), 47, 406, 408, 409, 412, 426, 428
   Multichannel (MMCDF), 408, 409, 414
Multidetection Techniques, 372
Multielectron ejection, 469, 474, 475
Multiple ionization, 31, 47, 470, 475–479
Multiple scattering method (MSM), 105, 123, 125, 300, 301, 307, 308

N$_2$, 296–299, 302, 323, 337, 339, 340
N$_2$O, 276, 307, 330–334, 338–340
Naphthalene, 251–255
Negative ions, 12
Neon, 76, 94, 144–146, 186, 188, 197, 202, 203, 208, 214, 215, 360–363, 367, 368, 375–377, 387–394, 410, 414, 415, 416, 590
NF, 243
NF$_2$, 237, 243
NF$_3$, 226–232, 516–518
NH$_2$, 240
NH$_2$F, 226–232
NH$_3$, 226–232, 324
NHF$_2$, 226–232
Nickel, 193
Nickel-carbonyl, 633–635
Nitric oxide, 633, 634, 636
Nitrogen, 206
NO, 275, 297, 321, 547, 548
NO$_2$, 276, 277
Nonlocality, 6

σ-ω coupling, 272
O$_2$, 274, 275, 321, 322, 324
Octant rule, 617, 618
One-electron model, 5, 361, 585, 586, 590
One-step decay, 474–479
Open-shell atoms, 110, 191, 192, 194, 201, 238, 239, 240
Optical activity, 608
Optical potential, 130, 132
Optical rotatory dispersion (ORD), 609
Orientation parameter, 434
Orthogonality problem, 106, 108
Oscillator strength, 565, 572–574, 580–582, 588, 592–596
Out-of-plane synchrotron radiation, 628
Outer-shell electrons, 370, 383–392; *see also* Electrons
Overlap integrals, 67
Oxygen, 187, 194, 202, 205, 206, 209, 640, 641

P$_2$, 237
Palladium, 412, 413
Parity-unfavored transition, 149, 530
Partial cross section, 137, 138; *see also* Cross section
Partial ion yield, 316, 317, 469–471, 477, 478
Partial wave expansion, 113
Partial wave phases, 634, 638
Post collision interaction (PCI), 184, 200, 213
PCl$_3$, 297
Pearson function, 184
Perturbation theory, 48, 50, 51, 73, 99–101, 114–121
Perturbed series, 452
Photoelectron spectrometry (PES), 135, 183, 189; *see also* Electron spectrometry
PF, 243
PF$_2$, 237, 243
PH, 237
PH$_2$, 237
Phase matching, 444
Phase shift, 54
Phosphorus, 237
Photoelastic modulator (PEM), 612
Photoelectron
   spectra;, 139–141, 145, 147, 151, 154, 157, 159, 160, 163, 167, 171, 183, 190, 223, 224, 226; *see also* Photoline
   spectroscopy, 135, 183, 189; *see also* Electron spectroscopy
   with defined spin polarization, 524, 525, 545–547
   Auger electron angular correlation, 429–432,
   Auger electron coincidence, 403, 429–432, 435; *see also* Coincidences
   emission microscope, 646, 647
   See also electron
Photoemission, 528, 624, 633; *see also* Photoelectron

Photoionization cross section, 4, 11–15, 17, 19, 20, 23, 137, 138, 142, 146, 148, 152, 155, 158, 159, 161, 162, 164, 165, 168–170, 172, 524, 525; *see also* Cross section
Photoionization with excitation; 47, 69, 70; *see also* Satellites
Photoline, shape of, 183, 193
Photolysis, laser induced, 237
π^*-resonance, 166, 167, 168
Pnicogen fluorides, 243
Pnicogen hydrides, 240–243
Polar stratosphere, 222
Polarization, 2, 54, 67, 69, 71, 104, 113, 120–122, 125–128, 130, 379
Polarized
  atoms (photoionization of), 530
  electrons, 522
  residual ions, 533
Position sensitive detector, 369, 382
Positive ions, photoionization of, 12
Post collision interaction (PCI), 22, 31, 362, 407, 417, 510, 511
  multi step, 32
  radiative, 34
Potassium, 13, 421, 567, 576, 577, 582, 585, 590, 591
Potential barrier, 585, 586; *see also* Centrifugal barrier
Predissociation rate, 270, 271
Propensity rules, 85
Pulsed acceleration field technique, 467
Photon-induced fluorescence spectroscopy (PIFS), 377, 381

Q-reversal, 454
Quantum yield, 497, 499, 512, 513
Quasi-Auger process, 24
Quasi equilibrum theory (QET), 271, 513, 514
Quantum defect difference, 529, 542

R-matrix, 47, 58, 72, 454, 457, 539, 587–589, 593, 595
Radial (rotational) coupling, 269, 272, 274, 288
Radiation trapping, 568
Radiationless Process, 203
Random phase approximaton (RPA), 131, 153, 540, 590
  with exchange (RPAE), 7, 8, 16, 47, 48, 52–54, 61, 63, 66, 68, 69, 72, 161, 529, 530, 535, 539–541, 590, 591
  Relativistic (RRPA), 47, 48, 529, 530, 535, 539–341
Rare earths, 62, 487–492
Rare gases, 140–156, 381, 383
Reaction plane, 525, 527, 529
Rearrangement, 2, 14
Relativistic interaction, 87
Relaxation, 2, 94–96, 105–107, 114–125, 130, 408, 418
  energy, 111, 114–125
Relaxed core Hartree–Fock (RCHF) approximation, 106, 108–110, 121–125
Resonance-enhanced multiphoton ionization (REMPI), 188, 195

Residual internal energy, 282
Resolution (resolving power), 182, 185, 191, 192
Resonant Auger process, 203, 209; *see also* Auger process
Resonances, 368, 371, 380ff, 384–387, 521, 522, 539, 540, 542, 544
  in atoms, 529, 530, 539, 540, 541, 544
  in molecules, rotationally resolved, 547–551
  in molecules, vibrationally resolved, 547
Resonant enhancement, 443
Resonant laser driven ionization (RLDI), 565, 566, 579, 580
Resonant photoelectron, 471–475, 477–479
Resonant photoionization, 89–91
Resonant Raman effect, 315, 407, 417
Resonant shake-off, 472–474, 477–479
RRKM calculations, 251, 254
Rydberg states, 222–224, 231, 233

Satellites, 3, 24, 66, 67, 72, 87–91, 93–98, 141, 149–150, 156, 160, 162, 197, 208, 507, 508
Satellite production, 368, 371, 374, 402, 414, 418–420
  due to Auger cascades, 418–420
  due to initial state effects, 418
  due to multiple ionization, 402, 418
  due to configuration interaction, 418, 419
  due to shake up, 418, 575, 576
Scattering theory, 451
Schwinger variational principle, 106, 113
SCl_2, 302
Screening energy, 113, 115, 116, 118, 119, 128
  operator, 115, 117
  induced by relaxation, 111, 119, 124, 126, 130
Secondary electrons emission, 21
Secondary photons emission, 22
Select-detect-type experiments, 495, 499
Selenium, 238
Self energy, 36, 586
Self-consistent field, 2
SF_5Cl, 298, 300, 301
SF_6, 285, 286, 296, 298, 300, 301, 315, 316, 339
Shadow lines, 4, 24
Shake off, 67, 510
Shake transitions, 419, 421; *see also* Satellites
  shake down, 420, 421, 472, 474, 479
  shake off, 67, 200, 208, 212, 418, 421, 576
  shake up, 24, 72, 197, 200, 202, 208, 212, 215, 418–421, 472, 474, 477–479, 575–577, 621, 627, 636, 642, 644
Shape resonance, 66, 144, 168, 276, 286, 287, 308, 510, 634
Si-Si, 244
Si_2H_2, 243, 244
Si_2H_3, 244
Si_2H_4, 243, 244
Si_2H_5, 243
$SiCl_4$, 340

SUBJECT INDEX 667

SiF_4, 287, 310, 311, 314, 315, 340
SiH_2, 237, 244
SiH_3, 237, 244
SiH_4, 324
Silver, 413, 414
Silicon, 14
Silicon molecules, 302, 320
Simulation of photoionization by (e, 2e) scattering, 531, 533
Singlet-triplet splitting, 224, 244
SO_2, 223, 224, 279, 280, 286, 296, 338, 339
Sodium, 94–97, 566, 569, 567, 572–577, 588, 589
Solid angle, 502
Source strength, 500, 501, 503
Spectator electron, 402, 403, 420, 428, 429, 472, 477
  model, 420, 428, 429, 433
Spectroscopic factor, 109, 125, 126, 129, 131
Spin polarization
  by polarization transfer, 432–434
  coefficients (parameters) of, 432–435
  dynamic, 432, 433, 435
  of Auger electrons, 432, 433, 434, 554, 555
  of photoelectrons, 521–551
Spin-orbit interaction, 58, 61, 65, 522, 523, 528, 535, 537, 539, 544, 545
Spin-orbit splitting, 523, 539, 544, 545
State multipoles, 533
State selectivity, 498, 507, 511
Static-exchange potential, 112
Stereoisomers, 232–236
Stieltjes–Tchebycheff moment theory (STMT), 105, 125, 129
Stokes parameters, 506
Strontium, 71, 72, 210, 567, 582
Subthreshold excitation, 34
Sulfur, 238
Super Coster–Kronig transition, 157, 161, 401
Synchrotron radiation, 135, 533, 597
$Si_2(CH_3)_3$, 343, 344
Sudden approximation, 67, 73
Surface ionization ion source, 567

Tamm–Dancoff approximation (TDA), 54, 131
Tellur, 238
Temperature dependence, 271, 272
Thallium, 529, 540, 541, 542
Thallium halides, 266–268
Thermionic diode, 445
Thioethers, 297
Three-particle dissociation, 271, 285

Threshold coincidence techniques, 358
Threshold electron, 473, 475
Threshold laws, 361
Threshold photoelectron spectroscopy (TPES), 356–360
Threshold region, 506, 507
Time-dependent local density approximation, 3, 42, 47, 68, 491
Time-of-flight (technique), 138, 466, 496, 502, 503, 511, 517
Total ion yield, 469, 470
Transient species, 236–251
Transition elements, 156, 193, 202, 203, 209, 211
Transition metals, 61, 482–487
Transition operator method, 116
Triple Auger process, 473, 475; see also Auger process
Two electron–two vacancy excitation, 19, 26, 28
Two-step double photoionization, 362, 429, 430
  decay of resonances, 472–479
  model of Auger decay, 203, 213, 407, 409, 424, 428
Tungsten, 63

Undulator, 568, 570, 583, 597, 645, 646
Upconversion (frequency), 442

Vacancy, 2
Vacuum, 2
Vanadium, 189, 192
Vanishing spectral fluctuations, 457
  width effect, 452
Vaporization, 405, 406
VUV absorption, 222, 223, 233
VUV laser, 442

Wannier law, 31
Wavelength calibration, 447
Width, 183
Window resonance, 56, 206

X-ray photoemission, 641–645
Xenon, 12, 23, 66, 67, 85–89, 153–156, 185, 203, 205, 209, 210, 377, 390–392, 410, 411, 414, 420, 426, 427, 430, 528, 529, 530, 531, 537–540

Zero electron kinetic energy (ZEKE), 140, 181, 188, 315, 356
Zinc, 410, 412, 582, 591, 592

Series Publications

Below is a chronological listing of all the published volumes in the *Physics of Atoms and Molecules* series.

ELECTRON AND PHOTON INTERACTIONS WITH ATOMS
Edited by H. Kleinpoppen and M. R. C. McDowell

ATOM–MOLECULE COLLISION THEORY: A Guide for the Experimentalist
Edited by Richard B. Bernstein

COHERENCE AND CORRELATION IN ATOMIC COLLISIONS
Edited by H. Kleinpoppen and J. F. Williams

VARIATIONAL METHODS IN ELECTRON–ATOM SCATTERING THEORY
R. K. Nesbet

DENSITY MATRIX THEORY AND APPLICATIONS
Karl Blum

INNER-SHELL AND X-RAY PHYSICS OF ATOMS AND SOLIDS
Edited by Derek J. Fabian, Hans Kleinpoppen, and Lewis M. Watson

INTRODUCTION TO THE THEORY OF LASER–ATOM INTERACTIONS
Marvin H. Mittleman

ATOMS IN ASTROPHYSICS
Edited by P. G. Burke, W. B. Eissner, D. G. Hummer, and I. C. Percival

ELECTRON–ATOM AND ELECTRON–MOLECULE COLLISIONS
Edited by Juergen Hinze

ELECTRON–MOLECULE COLLISIONS
Edited by Isao Shimamura and Kazuo Takayanagi

ISOTOPE SHIFTS IN ATOMIC SPECTRA
W. H. King

AUTOIONIZATION: Recent Developments and Applications
Edited by Aaron Temkin

ATOMIC INNER-SHELL PHYSICS
Edited by Bernd Crasemann

COLLISIONS OF ELECTRONS WITH ATOMS AND MOLECULES
G. F. Drukarev

THEORY OF MULTIPHOTON PROCESSES
Farhad H. M. Faisal

PROGRESS IN ATOMIC SPECTROSCOPY, Parts A, B, C, and D
Edited by W. Hanle, H. Kleinpoppen, and H. J. Beyer

RECENT STUDIES IN ATOMIC AND MOLECULAR PROCESSES
Edited by Arthur E. Kingston

QUANTUM MECHANICS VERSUS LOCAL REALISM: The Einstein-Podolsky-Rosen Paradox
Edited by Franco Selleri

ZERO-RANGE POTENTIALS AND THEIR APPLICATIONS IN ATOMIC PHYSICS
Yu. N. Demkov and V. N. Ostrovskii

COHERENCE IN ATOMIC COLLISION PHYSICS
Edited by H. J. Beyer, K. Blum, and R. Hippler

ELECTRON–MOLECULE SCATTERING AND PHOTOIONIZATION
Edited by P. G. Burke and J. B. West

ATOMIC SPECTRA AND COLLISIONS IN EXTERNAL FIELDS
Edited by K. T. Taylor, M. H. Nayfeh, and C. W. Clark

ATOMIC PHOTOEFFECT
M. Ya. Amusia

MOLECULAR PROCESSES IN SPACE
Edited by Tsutomu Watanabe, Isao Shimamura, Mikio Shimizu, and Yukikazu Itikawa

THE HANLE EFFECT AND LEVEL CROSSING SPECTROSCOPY
Edited by Giovanni Moruzzi and Franco Strumia

ATOMS AND LIGHT: INTERACTIONS
John N. Dodd

POLARIZATION BREMSSTRAHLUNG
Edited by V. N. Tsytovich and I. M. Ojringel

INTRODUCTION TO THE THEORY OF LASER–ATOM INTERACTIONS
(Second Edition)
Marvin H. Mittleman

ELECTRON COLLISIONS WITH MOLECULES, CLUSTERS, AND SURFACES
Edited by H. Ehrhardt and L. A. Morgan

THEORY OF ELECTRON–ATOM COLLISIONS, Part 1: Potential Scattering
Philip G. Burke and Charles J. Joachain

POLARIZED ELECTRON/POLARIZED PHOTON PHYSICS
Edited by H. Kleinpoppen and W. R. Newell

INTRODUCTION TO THE THEORY OF X-RAY AND ELECTRONIC SPECTRA OF FREE ATOMS
Romas Karazija

VUV AND SOFT X-RAY PHOTOIONIZATION
Edited by Uwe Becker and David A. Shirley